Peter M.A. Sloot   C.J. Kenneth Tan
Jack J. Dongarra   Alfons G. Hoekstra (Eds.)

Computational Science – ICCS 2002

International Conference
Amsterdam, The Netherlands, April 21-24, 2002
Proceedings, Part II

Springer

Volume Editors

Peter M.A. Sloot
Alfons G. Hoekstra
University of Amsterdam, Faculty of Science, Section Computational Science
Kruislaan 403, 1098 SJ Amsterdam, The Netherlands
E-mail: {sloot,alfons}@science.uva.nl

C.J. Kenneth Tan
University of Western Ontario, Western Science Center, SHARCNET
London, Ontario, Canada N6A 5B7
E-mail: cjtan@acm.org

Jack J. Dongarra
University of Tennessee, Computer Science Department
Innovative Computing Laboratory
1122 Volunteer Blvd, Knoxville, TN 37996-3450, USA
E-mail: dongarra@cs.utk.edu

Cataloging-in-Publication Data applied for

Die Deutsche Bibliothek - CIP-Einheitsaufnahme

Computational science : international conference ; proceedings / ICCS 2002,
Amsterdam, The Netherlands, April 21 - 24, 2002. Peter M. A. Sloot (ed.). -
Berlin ; Heidelberg ; New York ; Barcelona ; Hong Kong ; London ; Milan ;
Paris ; Tokyo : Springer
Pt. 2 . - (2002)
  (Lecture notes in computer science ; Vol. 2330)
  ISBN 3-540-43593-X

CR Subject Classification (1998): D, E, G, H, I, J, C.2-3
ISSN 0302-9743
ISBN 3-540-43593-X Springer-Verlag Berlin Heidelberg New York

Springer-Verlag Berlin Heidelberg New York
a member of BertelsmannSpringer Science+Business Media GmbH

http://www.springer.de

© Springer-Verlag Berlin Heidelberg 2002
Printed in Germany

Typesetting: Camera-ready by author, data conversion by PTP-Berlin, Stefan Sossna e.K.
Printed on acid-free paper   SPIN: 10869723   06/3142   5 4 3 2 1 0

Preface

Computational Science is the scientific discipline that aims at the development and understanding of new computational methods and techniques to model and simulate complex systems.

The area of application includes natural systems – such as biology, environmental and geo-sciences, physics, and chemistry – and synthetic systems such as electronics and financial and economic systems. The discipline is a bridge between 'classical' computer science – logic, complexity, architecture, algorithms – mathematics, and the use of computers in the aforementioned areas.

The relevance for society stems from the numerous challenges that exist in the various science and engineering disciplines, which can be tackled by advances made in this field. For instance new models and methods to study environmental issues like the quality of air, water, and soil, and weather and climate predictions through simulations, as well as the simulation-supported development of cars, airplanes, and medical and transport systems etc.

Paraphrasing R. Kenway (R.D. Kenway, Contemporary Physics. 1994): 'There is an important message to scientists, politicians, and industrialists: in the future science, the best industrial design and manufacture, the greatest medical progress, and the most accurate environmental monitoring and forecasting will be done by countries that most rapidly exploit the full potential of *computational science*'.

Nowadays we have access to high-end computer architectures and a large range of computing environments, mainly as a consequence of the enormous stimulus from the various international programs on advanced computing, e.g. HPCC (USA), HPCN (Europe), Real-World Computing (Japan), and ASCI (USA: Advanced Strategie Computing Initiative). The sequel to this, known as 'grid-systems' and 'grid-computing', will boost the computer, processing, and storage power even further. Today's supercomputing application may be tomorrow's desktop computing application.

The societal and industrial pulls have given a significant impulse to the rewriting of existing models and software. This has resulted among other things in a big 'clean-up' of often outdated software and new programming paradigms and verification techniques. With this make-up of arrears the road is paved for the study of real complex systems through computer simulations, and large scale problems that have long been intractable can now be tackled. However, the development of complexity reducing algorithms, numerical algorithms for large data sets, formal methods and associated modeling, as well as representation (i.e. visualization) techniques are still in their infancy. Deep understanding of the approaches required to model and simulate problems with increasing complexity and to efficiently exploit high performance computational techniques is still a big scientific challenge.

The International Conference on Computational Science (ICCS) series of conferences was started in May 2001 in San Francisco. The success of that meeting motivated the organization of the meeting held in Amsterdam from April 21–24, 2002.

These three volumes (Lecture Notes in Computer Science volumes 2329, 2330, and 2321) contain the proceedings of the ICCS 2002 meeting. The volumes consist of over 350 – peer reviewed – contributed and invited papers presented at the conference in the Science and Technology Center Watergraafsmeer (WTCW), in Amsterdam. The papers presented reflect the aims of the program committee to bring together major role players in the emerging field of computational science.

The conference was organized by The University of Amsterdam, Section Computational Science (http://www.science.uva.nl/research/scs/), SHARCNET, Canada (http://www.sharcnet.com), and the Innovative Computing Laboratory at The University of Tennessee.

The conference included 22 workshops, 7 keynote addresses, and over 350 contributed papers selected for oral presentation. Each paper was refereed by at least two referees.

We are deeply indebted to the members of the program committee, the workshop organizers, and all those in the community who helped us to organize a successful conference. Special thanks go to Alexander Bogdanov, Jerzy Wasniewski, and Marian Bubak for their help in the final phases of the review process. The invaluable administrative support of Manfred Stienstra, Alain Dankers, and Erik Hitipeuw is also acknowledged. Lodewijk Bos and his team were responsible for the local logistics and as always did a great job.

ICCS 2002 would not have been possible without the support of our sponsors: The University of Amsterdam, The Netherlands; Power Computing and Communication BV, The Netherlands; Elsevier Science Publishers, The Netherlands; Springer-Verlag, Germany; HPCN Foundation, The Netherlands; National Supercomputer Facilities (NCF), The Netherlands; Sun Microsystems, Inc., USA; SHARCNET, Canada; The Department of Computer Science, University of Calgary, Canada; and The School of Computer Science, The Queens University, Belfast, UK.

Amsterdam, April 2002   Peter M.A. Sloot,
                                                              Scientific Chair 2002,

                                                           on behalf of the co-editors:
                                                                  C.J. Kenneth Tan
                                                                  Jack J. Dongarra
                                                                  Alfons G. Hoekstra

Organization

The 2002 International Conference on Computational Science was organized jointly by The University of Amsterdam, Section Computational Science, SHARCNET, Canada, and the University of Tennessee, Department of Computer Science.

Conference Chairs

Peter M.A. Sloot, Scientific and Overall Chair ICCS 2002 (University of Amsterdam, The Netherlands)
C.J. Kenneth Tan (SHARCNET, Canada)
Jack J. Dongarra (University of Tennessee, Knoxville, USA)

Workshops Organizing Chair

Alfons G. Hoekstra (University of Amsterdam, The Netherlands)

International Steering Committee

Vassil N. Alexandrov (University of Reading, UK)
J. A. Rod Blais (University of Calgary, Canada)
Alexander V. Bogdanov (Institute for High Performance Computing and Data Bases, Russia)
Marian Bubak (AGH, Poland)
Geoffrey Fox (Florida State University, USA)
Marina L. Gavrilova (University of Calgary, Canada)
Bob Hertzberger (University of Amsterdam, The Netherlands)
Anthony Hey (University of Southampton, UK)
Benjoe A. Juliano (California State University at Chico, USA)
James S. Pascoe (University of Reading, UK)
Rene S. Renner (California State University at Chico, USA)
Kokichi Sugihara (University of Tokyo, Japan)
Jerzy Wasniewski (Danish Computing Center for Research and Education, Denmark)
Albert Zomaya (University of Western Australia, Australia)

Local Organizing Committee

Alfons Hoekstra (University of Amsterdam, The Netherlands)
Alexander V. Bogdanov (Institute for High Performance Computing and Data Bases, Russia)
Marian Bubak (AGH, Poland)
Jerzy Wasniewski (Danish Computing Center for Research and Education, Denmark)

Local Advisory Committee

Patrick Aerts (National Computing Facilities (NCF), The Netherlands Organization for Scientific Research (NWO), The Netherlands
Jos Engelen (NIKHEF, The Netherlands)
Daan Frenkel (Amolf, The Netherlands)
Walter Hoogland (University of Amsterdam, The Netherlands)
Anwar Osseyran (SARA, The Netherlands)
Rik Maes (Faculty of Economics, University of Amsterdam, The Netherlands)
Gerard van Oortmerssen (CWI, The Netherlands)

Program Committee

Vassil N. Alexandrov (University of Reading, UK)
Hamid Arabnia (University of Georgia, USA)
J. A. Rod Blais (University of Calgary, Canada)
Alexander V. Bogdanov (Institute for High Performance Computing and Data Bases, Russia)
Marian Bubak (AGH, Poland)
Toni Cortes (University of Catalonia, Barcelona, Spain)
Brian J. d'Auriol (University of Texas at El Paso, USA)
Clint Dawson (University of Texas at Austin, USA)
Geoffrey Fox (Florida State University, USA)
Marina L. Gavrilova (University of Calgary, Canada)
James Glimm (SUNY Stony Brook, USA)
Paul Gray (University of Northern Iowa, USA)
Piet Hemker (CWI, The Netherlands)
Bob Hertzberger (University of Amsterdam, The Netherlands)
Chris Johnson (University of Utah, USA)
Dieter Kranzlmüller (Johannes Kepler University of Linz, Austria)
Antonio Lagana (University of Perugia, Italy)
Michael Mascagni (Florida State University, USA)
Jiri Nedoma (Academy of Sciences of the Czech Republic, Czech Republic)
Roman Neruda (Academy of Sciences of the Czech Republic, Czech Republic)
Jose M. Laginha M. Palma (University of Porto, Portugal)

James Pascoe (University of Reading, UK)
Ron Perrott (The Queen's University of Belfast, UK)
Andy Pimentel (The University of Amsterdam, The Netherlands)
William R. Pulleyblank (IBM T. J. Watson Research Center, USA)
Rene S. Renner (California State University at Chico, USA)
Laura A. Salter (University of New Mexico, USA)
Dale Shires (Army Research Laboratory, USA)
Vaidy Sunderam (Emory University, USA)
Jesus Vigo-Aguiar (University of Salamanca, Spain)
Koichi Wada (University of Tsukuba, Japan)
Jerzy Wasniewski (Danish Computing Center for Research and Education, Denmark)
Roy Williams (California Institute of Technology, USA)
Elena Zudilova (Corning Scientific, Russia)

Workshop Organizers

Computer Graphics and Geometric Modeling
Andres Iglesias (University of Cantabria, Spain)
Modern Numerical Algorithms
Jerzy Wasniewski (Danish Computing Center for Research and Education, Denmark)
Network Support and Services for Computational Grids
C. Pham (University of Lyon, France)
N. Rao (Oak Ridge National Labs, USA)
Stochastic Computation: From Parallel Random Number Generators to Monte Carlo Simulation and Applications
Vasil Alexandrov (University of Reading, UK)
Michael Mascagni (Florida State University, USA)
Global and Collaborative Computing
James Pascoe (The University of Reading, UK)
Peter Kacsuk (MTA SZTAKI, Hungary)
Vassil Alexandrov (The Unviversity of Reading, UK)
Vaidy Sunderam (Emory University, USA)
Roger Loader (The University of Reading, UK)
Climate Systems Modeling
J. Taylor (Argonne National Laboratory, USA)
Parallel Computational Mechanics for Complex Systems
Mark Cross (University of Greenwich, UK)
Tools for Program Development and Analysis
Dieter Kranzlmüller (Joh. Kepler University of Linz, Austria)
Jens Volkert (Joh. Kepler University of Linz, Austria)
3G Medicine
Andy Marsh (VMW Solutions Ltd, UK)
Andreas Lymberis (European Commission, Belgium)
Ad Emmen (Genias Benelux bv, The Netherlands)

Automatic Differentiation and Applications
H. Martin Buecker (Aachen University of Technology, Germany)
Christian H. Bischof (Aachen University of Technology, Germany)
Computational Geometry and Applications
Marina Gavrilova (University of Calgary, Canada)
Computing in Medicine
Hans Reiber (Leiden University Medical Center, The Netherlands)
Rosemary Renaut (Arizona State University, USA)
High Performance Computing in Particle Accelerator Science and Technology
Andreas Adelmann (Paul Scherrer Institute, Switzerland)
Robert D. Ryne (Lawrence Berkeley National Laboratory, USA)
Geometric Numerical Algorithms: Theoretical Aspects and Applications
Nicoletta Del Buono (University of Bari, Italy)
Tiziano Politi (Politecnico-Bari, Italy)
Soft Computing: Systems and Applications
Renee Renner (California State University, USA)
PDE Software
Hans Petter Langtangen (University of Oslo, Norway)
Christoph Pflaum (University of Würzburg, Germany)
Ulrich Ruede (University of Erlangen-Nürnberg, Germany)
Stefan Turek (University of Dortmund, Germany)
Numerical Models in Geomechanics
R. Blaheta (Academy of Science, Czech Republic)
J. Nedoma (Academy of Science, Czech Republic)
Education in Computational Sciences
Rosie Renaut (Arizona State University, USA)
Computational Chemistry and Molecular Dynamics
Antonio Lagana (University of Perugia, Italy)
Geocomputation and Evolutionary Computation
Yong Xue (CAS, UK)
Narayana Jayaram (University of North London, UK)
Modeling and Simulation in Supercomputing and Telecommunications
Youngsong Mun (Korea)
Determinism, Randomness, Irreversibility, and Predictability
Guenri E. Norman (Russian Academy of Sciences, Russia)
Alexander V. Bogdanov (Institute of High Performance
Computing and Information Systems, Russia)
Harald A. Pasch (University of Vienna, Austria)
Konstantin Korotenko (Shirshov Institute of Oceanology, Russia)

Sponsoring Organizations

The University of Amsterdam, The Netherlands
Power Computing and Communication BV, The Netherlands
Elsevier Science Publishers, The Netherlands
Springer-Verlag, Germany
HPCN Foundation, The Netherlands
National Supercomputer Facilities (NCF), The Netherlands
Sun Microsystems, Inc., USA
SHARCNET, Canada
Department of Computer Science, University of Calgary, Canada
School of Computer Science, The Queens University, Belfast, UK.

Local Organization and Logistics

Lodewijk Bos, MC-Consultancy
Jeanine Mulders, Registration Office, LGCE
Alain Dankers, University of Amsterdam
Manfred Stienstra, University of Amsterdam

Table of Contents, Part II

Modern Numerical Algorithms

Network Support and Services for Computational Grids

Stochastic Computation: From Parallel Random Number Generators to Monte Carlo Simulation and Applications

Global and Collaborative Computing

Climate Systems Modelling

Parallel Computational Mechanics for Complex Systems

Tools for Program Development and Analysis

Automatic Differentiation and Applications

Table of Contents, Part I

Computer Science – Computer Systems Models

Scientific Computing – Stochastic Algorithms

Complex Systems Applications 2

Computer Science – Networks

Scientific Computing – Domain Decomposition

Complex Systems Applications 3

Computer Science – Code Optimization

Methods for Complex Systems Simulation

Grid and Applications

Problem Solving Environment 1

Data Mining

Computer Science – Scheduling and Load Balancing

Problem Solving Environment 2

Problem Solving Environments 3

Computational Fluid Dynamics 2

Complex Systems Applications 4

Scientific Computing – Computational Methods 2

Scientific Computing – Computational Methods 3

Table of Contents, Part III

Computing in Medicine

High Performance Computing
in Particle Accelerator Science and Technology

Geometric Numerical Algorithms: Theoretical Aspects and Applications

Soft Computing: Systems and Applications

PDE Software

Numerical Models in Geomechanics

Education in Computational Sciences

Computational Chemistry and Molecular Dynamics

Geocomputation and Evolutionary Computation

Modeling and Simulation in Supercomputing and Telecommunications

Determinism, Randomness, Irreversibility, and Predictability

Workshop Papers I

Inverse Direct Lighting with a Monte Carlo Method and Declarative Modelling

Vincent Jolivet[1], Dimitri Plemenos[2], Patrick Poulingeas[3]

Laboratoire MSI.
83, rue d'Isle. 87000 Limoges. France.
[1] jolivet@unilim.fr
[2] plemenos@unilim.fr
[3] poulinge@msi.unilim.fr

Abstract. In inverse lighting problems, a lot of optimization techniques have been used. A new method in the framework of radiosity is presented here, using a simple Monte-Carlo method to find the positions of the lights in a direct lighting. Declarative modelling will also be used to allow the designer to describe in a more intuitive way his lighting wishes. Declarative modelling will propose to the user several solutions, and probably some interesting scenes not previously imagined by the designer.

1 Introduction

Usually, in a rendering process, the user gives the physical characteristics of the elements of the scene. After this, he chooses a position for the lights and describes their properties. Then, using an illumination model, an algorithm is able to render the scene.

The aim of inverse lighting is to help the designer to set the lights in a more intuitive manner. An inverse lighting software should automatically compute the characteristics of the lights from the designer's wishes. This approach would be very interesting if the designer could specify high-level properties (like the ambience in a room, or the brightness of an object). These possibilities are given by declarative modelling [11]. This could drastically shorten the classical design cycle where the designer have to adjust empirically the lighting parameters.

In section 2 of this paper, we will specify the inverse lighting problem and will present the state of the art. The section 3 will introduce declarative modelling, its motivations, concepts and methods. We will then expose in section 4 our work combining radiosity, a basic Monte-Carlo method and declarative modelling, and finally we will conclude in section 5.

P.M.A. Sloot et al. (Eds.): ICCS 2002, LNCS 2330, pp. 3–12, 2002.
© Springer-Verlag Berlin Heidelberg 2002

2   The Inverse Lighting Problem

In computer graphics, the problems of inverse lighting consist in determining, from a given scene, the characteristics of one or several lights to obtain certain type of lighting. An algorithm finds the various characteristics of lights which, integrated into the scene, will allow to create a specific ambience wished by the designer. The characteristics are for example :

- The light position.
- The light orientation (for example if it is a spotlight).
- The physical parameters of the light, as the intensity.
- The geometrical nature of the light (its shape).
- Etc.

Mostly, for a lighting asked by the user, there is not a single possible solution. A set of scenes corresponding more or less to the designer's wishes is generally generated by an algorithm of inverse lighting.

To clarify the studied problem, one is generally brought to specify three fundamental characteristics which are going to influence the proposed algorithms :

1. The nature of the lighting (direct or global).
2. The physical model of illumination (Phong's model, radiosity, computation of the radiance at a point, etc.).
3. The a priori constraints on used lights (The position of lights can be fixed, and then only the physical parameters of these lights have to be found. One can also have a set of authorized positions, or simply no constraint on the positions of the lights).

In [13], [14] and [12] are developed taxonomies from the criteria given in the previous paragraph. By following [12], one can take as classification criterion the physical model of illumination chosen in the researches made until now.

- For Phong's model: [10] and [15].
- For radiosity: [16], [5], [10], [1] and [9].
- For radiance computation at a point: [2] and [3].

In these papers, a lot of optimization techniques are used. The underlying idea is to carry out at best the designer's wishes, supposed very precise. With declarative modelling, we shall see how this presupposition can be an obstacle to the discovery of interesting solutions of lighting.

3   Declarative Modelling

Declarative modelling [11] is an investigation method of a set of solutions corresponding to a linguistic description of a problem. The studied problem is generally in scene modelling, but the paradigm of declarative modelling could apply to other areas than computer graphics. Declarative modelling tries to facilitate the work of design by finding scenes corresponding to the general idea of the designer's project.

It's frequently admitted that declarative modeling requires 3 phases:
1. The description phase where the designer specifies his desires, not by means of precise geometrical or numeric data, but by means of linguistic terms. This allows to give a vague description of certain properties of the scene.
2. The generation phase where an algorithm generates scenes corresponding to the description done in the previous stage.
3. The exploration phase where the software displays solutions, and the user retains the best ones for him (This choice can be taken into account for the continuation of this phase: One will present to the user only scenes "looking like" the previously selected ones).

In the usual inverse lighting methods, the user specifies in a very precise way the objectives (for example by giving to certain elements of the scene the color which they should have after lights will have been found by the algorithm), what justifies the use of optimization techniques to satisfy as well as possible the user's requirements.

With declarative modelling, the designer does not need to strongly detail his desires in lighting. He can give a rough description of the ambience for the scene, or for some parts of it. The resolution engine (occurring in the generation phase) of the declarative modeller is then going to propose him several solutions, among which some that he can consider interesting while he wouldn't have think about them a priori (It is moreover likely that it would not have obtained them if he had had to supply an extremely precise description of his wishes by means of numerical values, as it is the case with usual methods).

Thus, declarative modelling seems an excellent designing process because it does not enclose the user into the a priori ideas (more or less conscious) that he can have of the scene to be produced.

4.   Inverse Lighting and Declarative Modelling

The inverse lighting algorithm that we propose takes place within the framework of radiosity and limits temporarily itself to direct lighting. Using a Monte-Carlo ray-casting technique [17], it extends the work of [9] made in the field of radiative exchanges and tested only for environments in two dimensions.

Moreover, we would like to use declarative description of the objectives, that is, description where only expected properties have to be given to the system. In the case of inverse lighting, the property "This part of object O has to be lighted" is an imprecise declarative description and can produce several solutions. It's possible to reduce the imprecision of a property by using fuzzy sets based modifiers like "very", "enough", etc. [4].

4.1   A New Method of Inverse Lighting

From a set O of patches to be lit, one tries to position a light, which will be a new patch added to the scene, whose shape is a quadrilateral and which should satisfy some conditions of lighting expressed in the form of a numerical interval for the luminous flow of each patch of O.

Unlike other methods of inverse lighting radiosity, the light does not belong to the patches of the initial meshing of the scene.

To search the patches which could be lights for a scene, a simple Monte-Carlo method is used, as described in [17]. Stochastic rays are shot from every patch to be lighted, and the patches of the scene reached at first by the rays are marked.

Patches able to light directly the patches to be illuminated (belonging to the list O) are collected in a list L. When no more new patches are added to the list L after maxRays sent, no more rays are shot. It's then admitted that L contains all the directly visible patches from the patch to be lit.

The form factors between the patch to be illuminated and the patches in the list L can then be estimated. This will help us for the calculation of the emittance of the patch which will act as a light.

The proposed algorithm is the following one :

```
Procedure findLightSourcesForPatch
  // P_i is the patch to be lighted
  L ← Ø    // L is the list of the directly visible
           // patches from P_i
  numRays ← 0
  repeat
    newPatch ← false
    Shoot a ray stochastically from patch P_i
    Let P_j be the first patch hit by the ray
    if P_j ∉ L then
      L ← L ∪ P_j
      newPatch ← true
    endif
    numRaysGot(P_j) ← numRaysGot(P_j)+1
    numRays ← numRays+1
    if newPatch = false then
      alreadyHit ← alreadyHit+1
    else
      alreadyHit ← 0
    endif
  until alreadyHit = maxRays
  for all patches P_j ∈ L do
    formFactor(i,j) ← numRaysGot(P_j) / numRays
  endforall
endProcedure
```

Remark on the results:
From a certain threshold value for maxRays, the list L increases little. The patches that are added then in L have not enough influence on the patch to be lit. They present not enough importance:

- because they can lead to a contradiction by lighting a zone which the user would like to leave in shadow.

- because if they have a real importance for the objectives to achieve, they will be taken into account by the patches to be lit near the one that has just been studied (It is indeed a rather common situation: the designer is generally interested in the lighting of a set of patches forming an object or a piece of an object of the scene, and not in the lighting of a single patch).

To limit the calculation time, it is necessary to find a good value for maxRays. However determining a good quality heuristic for maxRays is difficult, because this value depends strongly on the geometry of the scene.

That is why we developed a new version of the previous algorithm in which maxRays does not appear any more.

In every stage, we shoot only a limited number of rays (of the order of 100). If at the end of a stage, the number of new patches hit by the rays is greater than 20 % of the total number of rays sent, the process is repeated. In practice, we notice that we get, in the same calculation time, almost all of the patches obtained by adjusting empirically maxRays' value for the previous algorithm.

Generally, the surface to be lighted is made of several patches. So, the problem to resolve is the lighting of a set of surface patches. In this case, the obtained lights must lie in the intersection of all the lists obtained for all the patches to be lighted. Let us call this list Λ.

Remark:

If we want to light a surface made of several patches, it is not interesting to look for the possible lights for all these patches. With a small number of patches selected on the surface, almost the whole set of the patches able to be lights for direct lighting can be found. With this process, a lot of calculation time can be saved.

To achieve this, it is necessary to choose some "good" (i.e. representative) patches of a surface (or of an object) to be lighted.

The following ideas could be applied, although they have not yet been tested in our prototype:

For every vertex of the surface, the patch (or one of the patches) including this vertex is selected. The patch containing the centre of the surface is added in the list of "good" patches.

If a whole object (or a part of an object) has to be lighted, the previous process has to be applied to all the surfaces of the object. The value of the angle formed by two surfaces will allow new optimizations. If this angle is small, it will be possible to take into account only one time some neighbor patches in the list of the patches of the scene to be lit.

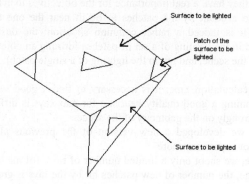

Fig. 1. Choice of the representative patches of the surface of the object to be lit.

When one deals with scenes containing few objects, it is not unusual that for 5 patches to be lit, a list Λ of much more than 300 patches able to be lights is obtained. A lot of patches belonging to the surfaces of the bounding box of the scene are indeed added to Λ if there is no object put between these surfaces and the patch to be lit, even by fixing a priori constraints to the position of the potential lights.

We thus decide to delete a patch j from the list Λ if the form factor F_{ij} is less than a threshold value (minThreshold(i)) for at least one of the patches i to light. The proposed heuristic for minThreshold(i) is :

$$\text{minThreshold}(i) = \frac{1}{\text{Number of visible patches from } i}. \tag{1}$$

With this heuristic, the number of elements of Λ decreases drastically.

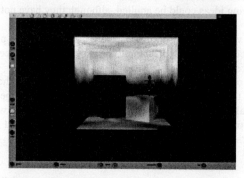

Fig. 2. All the potential lights are painted in white. The surface to be lighted lies on the top of the left side box.

Fig. 3. Remaining lights when removing patches with low value form factors and limiting lights to be to the ceiling.

The obtained area for lights in Fig. 3 is now limited enough to start the research of a position of a light.

After having determined the list Λ, the problem is to fully determine a light source on the chosen area.

Currently, we limited ourselves to a single emitting patch (that we will note P_l), whose shape is a quadrilateral. To put this patch, a convex polygon is built from the patches of the list Λ. For this purpose, the Graham's algorithm [6] is used and applied to the vertices of the patches of the list Λ. We sweep then the convex polygon to find the suitable positions for the emitting patch.

When a suitable position has been obtained, an emittance L^e is given to the lighting patch P_l, taking into account values allowed for the luminous flow Φ_i of patches P_i to be lit.

For every patch P_i, we have the inequalities:

$$\min_i \leq \Phi_i \leq \max_i . \tag{2}$$

Which become:

$$\frac{\min_i}{\pi \rho_i F_{li} A_l} \leq L^e \leq \frac{\max_i}{\pi \rho_i F_{li} A_l} . \tag{3}$$

Where A_l indicates the area of the patch P_l.

It is obvious that the form factors F_{li} between the patch P_l and the patches to be lit have to be calculated. For that, we use again the previous Monte-Carlo method.

By taking into account the set O of patches to be illuminated, the acceptable values for L^e can be found, or conclude that we have no possible solution for this position of the illuminating patch. With a discretization of the interval of the suitable values for L^e, we will be able to give the user several solutions, for the same position of P_l.

Figures 4 and 5 show examples of lighting solutions found by our algorithm.

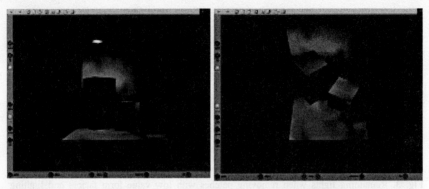

Fig. 4. First solution found. The patches to be lighted are on the top of the left side box.

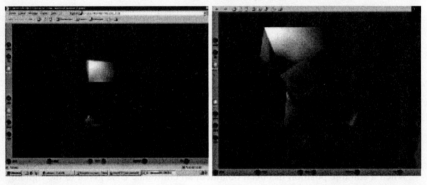

Fig. 5. Last position found for the lighting patch. The patches to be lighted are still on the top of the left side box.

Remark on the computational time of these scene: The previous scenes are composed of 5938 patches. The determination of the light source has taken about 10 seconds.

4.2   Within the Framework of Declarative Modelling

4.2.1   The Description and Generation Phases

With declarative modelling, the user is not obliged to specify details such as "patches 19 and 70 should have an energy of 1889 watts". The model of description supplied by the declarative modeller allows the designer to manipulate high-level concepts (the "table", a "weak lighting", etc.). He can so express his desires with a linguistic form more intuitive than raw numeric data. The properties which the user is going to require for the scene correspond to his still vague idea at the beginning of the modelling work.

The resolution engine will generate, from these properties, several solutions. This way of proceeding can supply to the designer a lighting that he could find interesting although he did not imagine it at the beginning of the description. The property asked by the user can be realized in several ways, especially for the number and the size of the lights:
- by a single big light put in the scene,
- by a single light, but of smaller size and put in another position,
- with several lights,
- etc.

4.2.2   The Representation Model of the Declarative Modeller

To handle the lighting properties which will be expressed with qualitative form (for example: "the top of the table is very lit"), we intend to use the works of E. Desmontils[4] where the fuzzy subsets theory is used to manage the properties of a domain (A domain is defined as the set of values that a scene, or a part of a scene, can take for a certain characteristic).

In the case of the illumination of an object, we have a "simple" property that can be modified (The object can be "very", "little", etc. lit). This property is represented by a fuzzy interval. The possible modifier can be itself a fuzzy operator (for example: "the table is more or less lit"). It is also possible to propose a "comparison" property between two parts of a scene. A modifier can then be applied to such a property (for example: "the chair is slightly less lit than the table").

To obtain a membership function (which is here a measure of $\Re$ in $[0,1]$) for the illumination property, it will probably be necessary to use psycho-physical tests. The interval of values for L^e can thus be an α-cutting, where α will correspond to an acceptance threshold value ($\alpha \in [0,1]$).

5   Conclusion

We have proposed in this paper an efficient algorithm for inverse direct lighting within the framework of radiosity. The computational times are good enough for reasonable sizes of scenes and the method can also propose several solutions to the problem.

With declarative modelling, we will be able to qualitatively improve the design cycle in the lighting process. Declarative modelling, with its linguistic description of the goals to achieve, allows us to deal with many types of design constraints (as aligned luminaires) that conventional tools ignore.

In the future, we will also try to take into account global illumination supplied by radiosity.

References :

[1]   M. Contensin M., J.-L. Maltret. Computer Aided Lighting for Architects and Designers. IKM'97, Weimar, Février 1997.

[2]   A.C. Costa, A.A. Sousa, F.N. Ferreira. Optimization and Lighting Design. WSCG'99.

[3]   A.C. Costa, A.A. Sousa, F.N. Ferreira. Lighting Design: a Goal based Approach using Optimization. 10th EG Workshop on Rendering, 1999.

[4]   E. Desmontils. Formulation des Propriétés en Modélisation Déclarative à l'Aide des Ensembles Flous. Rapport de Recherche IRIN – 106. Décembre 1995.

[5]   J. Elorza, I. Rudomin. An Interactive System for Solving Inverse Illumination Problems using Genetic Algorithms. Computacion Visual 1997.

[6]   R.L. Graham. An Efficient Algorithm for Determining the Convex Hull of a Finite Set of Points in the Plane. Information Processing Letters, 1. 1972.

[7]   V. Harutunian, J.C. Morales, J.R. Howell. Radiation Exchange within an Enclosure of Diffuse-Gray Surfaces: The Inverse Problem. ASME/AIAA International Heat Transfer Conference, 1995.

[8]   J.K. Kawai, J.S. Painter, M.F. Cohen. Radioptimization – Goal Based Rendering. SIGGRAPH'93.

[9]   M. Oguma, J.R. Howell. Solution of the Two-Dimensional Blackbody Inverse Radiation Problem by Inverse Monte-Carlo Method. ASME/JSME Thermal Engineering Conference, 1995.

[10]   P. Poulin, A. Fournier. Lights from Highlights and Shadows. March 1992, Symposium on Interactive 3D Graphics.

[11]   D. Plemenos. Declarative Modelling by Hierarchical Decomposition.The Actual State of the MultiFormes Project. GraphiCon'95, St Petersbourg, 1-5 juillet 1995.

[12]   P. Poulingeas. L'Eclairage Inverse – Etat de l'Art. Rapport de Recherche MSI 01-01. Octobre 2001.

[13]   G. Patow, X. Pueyo. A Survey on Inverse Reflector Design and Light Source Distribution Problems. Institut d'Informàtica i Aplicacions, Universitat de Girona. Private Communication.

[14]   G. Patow, X. Pueyo. A Survey on Inverse Emittance and Inverse Reflectometry Computation Problems. Institut d'Informàtica i Aplicacions, Universitat de Girona. Private Communication.

[15]   P. Poulin, K. Ratib, M. Jacques. Sketching Shadows and Highlights to position Lights. Proceedings of Computer Graphics International 1997.

[16]   C. Schoeneman, J. Dorsey, B. Smits, J. Arvo, D. Greenberg. Painting with Light. SIGGRAPH'93.

[17]   P. Shirley. Radiosity via Ray Tracing. Graphics Gems II, p. 306-310, James Arvo Ed., Academic Press, San Diego. 1991

Light Meshes—Original Approach to Produce Soft Shadows in Ray Tracing

Victor A. Debelov and Igor M. Sevastyanov

Institute of Computational Mathematics and Mathematical Geophysics SB RAS,
Russia, Novosibirsk
(debelov,sevastyanov)@oapmg.sscc.ru

Abstract. The given paper is devoted to the application of the original approach of Light Meshes to the recursive ray tracing algorithm. It
is a sort of a simulation of a solution for the global illumination problem. It consumes not so much extra computational time and, nevertheless, allows: a) to speedup a process of shadow calculations, b) to build
soft shadows, c) to simulate main effects of diffuse interreflections, d)
to omit the ambient term. In order to obtain the goals listed above we
introduce the notion of space light meshes. A light mesh is a point set
(mesh points) in 3D space of a scene. At each mesh point we compute
illuminance, which characterizes some kind of "light field intensity" in
the scene. Then important part of resulting illumination of a scene object point (object point) is obtained via interpolation of stored values of
nearby mesh points. Our experiments show that the suggested approach
reduces computational time with respect to ray tracing algorithm if a
scene contains a lot of light sources and/or an image has a large resolution.

1 Introduction

In the given work we consider the problem of photorealistic images synthesis of
3D scenes using a light-backwards recursive ray tracing algorithm (e.g. [1]) in
the following context. Scenes are characterized as follows.

- Without participating media.
- Contain any number of point light sources.
- Consist of opaque one-sided surfaces. Each surface has diffuse emissive, diffuse and specular reflectance properties. We assume a quite simple model,
  i.e. each object point y is assigned: the diffuse reflection coefficient $k_d(y)$;
  the specular reflection coefficient $k_s(y)$; the diffuse emission $Emiss(y)$.
- A position of a camera and a final image resolution are given too.

Here we suggest an approach that should improve known drawbacks of ray tracing:

- Absence of soft shadows.
- Necessity to assign some value as the ambient term while calculations of an
  object point illumination.

P.M.A. Sloot et al. (Eds.): ICCS 2002, LNCS 2330, pp. 13–21, 2002.
© Springer-Verlag Berlin Heidelberg 2002

– Poor accuracy of a diffuse interreflection calculation, even an absence of color bleeding (e.g. see [2] about the scene "Constructive Wood").

It is known that all these drawbacks are solved (at least theoretically) by heavy global illumination techniques. Since the ray tracing method is still the most popular and widespread used, we developed the modification of it in order to give a certain solution to posed problems.

1.1   Scalar Light Field

The irradiance in a scene is a function of five variables, three for the position and two for the direction. In the Stanford project "Light Field Rendering" light field is represented as 4D functions. In order to improve the accuracy of a diffuse interreflection calculation Ward and Heckbert [3] introduced the notion of the irradiance gradients; which is direction-dependent. They keep two vectors and a diffuse irradiance value for points on scene surfaces. In our approach we decided to reduce the amount of information up to a single value in each space point. Let's assume that light induces intensity in each point of a space of a scene similar to electrical or pressure (potential) fields. We suggest considering "a light scalar field" which is characterized by a single value of light field intensity $I(x)$, $x \in R^3$. This value does not depend on directions. Probably by comparing of values in two close points we can obtain directional information. Basing on ideas of the bidirectional Monte Carlo ray tracing algorithm, especially on illumination maps [4], photon maps [5] that are fulfilled in 2 steps: *forward)* rays (photons) are cast from light sources and stored in a certain structure in a scene surfaces; *backward)* path tracing from a camera to gather incoming intensities. As a rule a proper data structures are created taking into account a consideration of scene geometry. In our case we decide to accumulate light energy in every point of scene space. We do 2 steps too but while first step we issue backward rays from the space point to the point light sources in order to accumulate the direct illumination. In the paper [6] a scene space is divided into small volumes. Photons are traced from light sources and stored into volumes together with information on directions. Stored information is used while backward phase. In a paper [7] authors suggest to compute irradiance in space points with directional information also. In our algorithm we save computer memory because we do not keep information on directions at all, nevertheless we obtain quite realistic simulation of scene images.

1.2   Direct Light Field

Let's define the scalar field DirectLight(x) that accounts for direct illumination of arbitrary point $x \in R^3$:

$$\text{DirectLight}(x) = \sum_{i=1}^{Nl} V_i(x) \cdot E_i \cdot \frac{1}{\varphi_1(r_i(x))} \qquad (1)$$

where:

Nl is a number of point light sources;

$V_i(x)$ is the visibility. It takes 1, if ith point light source sees point x, otherwise 0.

E_i is the intensity (color) of ith point light source;

$r_i(x)$ is the distance from the point to ith point light source;

$\varphi_1(t)$ in the given work has the classical view $Const + t$.

Thus we accumulate direct illumination that reaches the point x from all directions. Then a direct illumination of an object point y – a point of a scene surface S – is defined via formula:

$$I(y) = \bar{I}(y) \cdot \sum_{i=1}^{Nl} \varphi_2(n_y, lp_i - y) \cdot E_i \cdot \frac{1}{\varphi_1(r_i(y))} \qquad (2)$$

where:

n_y is the normal vector in y,

lp_i is the position of ith light source,

$$\varphi_2(n, v) = \begin{bmatrix} 0 & \text{if } \cos \angle(n, v) < 0, \\ \cos \angle(n, v) & \text{otherwise,} \end{bmatrix}$$

$$\bar{I}(y) = \frac{1}{\mu(D_y)} \int_{x \in D_y} \text{DirectLight}(x)\, dV \qquad (3)$$

$D_y \subset R^3$ is a certain neighborhood of the point y, an interpolation set. Really the shape of the neighborhood should be selected accordingly to the reflectance properties at the point y and geometric properties of a surface. But in the limits of the given work we do not consider it.

$\mu(D)$ is the measure (volume) of the set D. In other words we simply integrate and average out intensities of all points belonging to selected neighborhood.

1.3   Indirect Light Field

The diffuse illumination of a space point $x \in R^3$ is defined as:

$$\text{IndirectLight}(x) = \int_\Omega L(x, \omega)\, d\omega \qquad (4)$$

where: $L(x, \omega)$ is the incoming indirect radiance to the point x in the direction ω from an object point y. $L(x, \omega) = \bar{I}(y)$ if y exists on a ray; otherwise 0. Ω is the unit sphere of directions. Instead of integration we calculate a sum using a fixed number of samples in a uniformly weighted, stratified Monte Carlo sampling analogously to [3]. As a rule a scene space point does not belong to a scene surface; so we integrate along a whole sphere of directions. If a ray hits a certain scene surface in the point y then the incoming diffuse intensity is calculated using formula:

$$\bar{I}_d(y) = \frac{1}{\mu(D_y)} \int_{x \in D_y} \text{IndirectLight}(x)\varphi_5(x, y)\, dV \qquad (5)$$

$\varphi_5(x, y) = 1$ in the given work.

1.4 Full Intensity of Object Point

Let's suppose that we have computed both direct and indirect fields in each point of a scene space. We suggests the following simulation of a full local intensity of the object point y:

$$I_{\text{full}}(y) = I(y) + k_d(y)\tilde{I}_d(y) \tag{6}$$

where $k_d(y)$ is the coefficient of the diffuse reflection. It is a sum of the direct part and the indirect part. The former represents the irradiance accumulated from all "almost" visible point light sources and the latter represents the term obtained as a diffuse interreflection accumulated in surrounding points of a scene space.

Remark 1. If a scene surface emits the light in the point y then the expression (3) is slightly modified:

$$\tilde{I}(y) = \text{Emiss}(y) + \frac{1}{\mu(D_y)} \int_{x \in D_y} \text{DirectLight}(x)\, dV \tag{7}$$

1.5 Light Meshes

Obviously that continuous functions $\text{DirectLight}(x)$ and $\text{IndirectLight}(x)$ are represented as samples given in finite point sets. The corresponding point sets DLM and ILM we call light meshes although they may be arbitrary point sets without any regular structure. In a work [3] the meshless caching scheme was used when points are selected from consideration of geometric properties – the location of the computed indirect irradiance values is determined by the proximity and curvature of the surfaces, and does not fall on a regular grid. In the presented work we use almost a similar fashion but essentially we select the light mesh arbitrary, i.e. without geometric considerations. We wish the light mesh lives, i.e. senses scene geometry and adapts itself to it. It is our main goal (for future), and in the given paper we will try to show the feasibility of the approach that it allows reaching goals declared in the beginning of the paper.

Remark 2. Light meshes DLM and ILM may differ as a set of points.

Remark 3. In the tests given in the paper we used regular point sets.

Remark 4. It should be stressed that here we describe rather the simulation of illumination but the approximation of the global illumination solution. Although our approach was found during the study of global illumination its main formulations were derived basing on intuition and experiments too.

1.6 Recursive Gathering of Intensity

In Fig. 1 we show a standard path which used in a backward recursive ray tracing algorithm: the initial ray R1, the reflected ray R2 in the point P, the next reflected ray R3 in the point Q. Although light meshes DLM and ILM are quite dense in the space, only a few of their points are used for calculations – those points that belong to neighborhoods D_P and D_Q. We have the following algorithm that is similar to one described in [1].

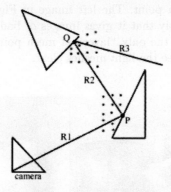

Fig. 1. The path traced from a camera

1. Initially all points of DLM and ILM are assigned the unknown value.
2. Analogously to [1] we build the path from a camera: camera $\to P \to Q \ldots$
3. Gather light energy from the end of the path.
4. At the point P we have intensity $I(Q)$ incoming along the ray R2.
5. Define the set DLM(P) – points of DLM belonging to D_P.
6. Calculate values for all points of DLM(P) with the unknown value using formula (1).
7. Calculate the direct illumination of P using formulas (3) and (2).
8. Define the set ILM(P) – points of ILM belonging to D_P.
9. Calculate values for all points of ILM(P) with the unknown value using formulas (4) and (5).
10. Calculate the full local intensity of P using formula (6).
11. Calculate the intensity outcoming from P to a camera using the expression:

$$I_{\text{out}}(P) = I(Q) + k_s(P)I(Q) \tag{8}$$

12. Repeat the procedure for all pixels of the image.

1.7 The Ambient Term

A user assigns the ambient term arbitrary and it is an artificial trick. The original method [8] allows defining it via the radiosity algorithm that estimates the diffuse interreflections between scene faces. Our light meshes carry out the similar work. Thus the formulas of the light meshes algorithm have no ambient terms.

2 Interpolation

During experiments with different interpolation schemes we selected two of them: Nearest and Spherical which gives acceptable images. The former scheme means that discrete neighborhoods DLM(P) and ILM(P) of the object point P include a single nearest mesh point. The left image in Fig. 2 illustrates this interpolation mode. Obviously that it gives images of bad quality but renders scenes very quickly because the only "layer" of mesh points nearby scene surfaces is computed. One can see the light meshes clear.

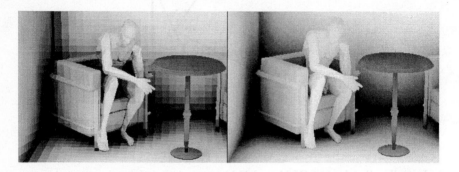

Fig. 2. Nearest interpolation mode (left). Spherical interpolation mode (right)

Spherical interpolation. The neighborhood D_P of the object point P is a full sphere of the definite radius. The right image in Fig. 2 is obtained using spherical interpolation over the same coarse meshes as the left image. The images in Fig. 3 illustrates that the method works quite well if a scene contains: specular surfaces (left image), diffuse surfaces (right image).

Fig. 3. Scene with specular surfaces (left). Diffuse surfaces (right)

3   Soft Shadows

It is known that soft shadows increase realism of synthesized images. Ray tracing is the method working with scenes illuminated by point light sources, and thus from the theoretical point of view it can produce only sharp shadows. Many specialists tried to enhance realism by developing techniques for simulations of soft shadows, it could be found in the Internet discussion. Main efforts concern two goals: a) to reduce a computational cost of shadows [9], b) to simulate soft shadows from point light sources. As one can see from formulas above we never use shadow rays from object points. It is implicitly done while calculating values of light meshes and does not depend on the depth of the recursion but on the number of actually used mesh points.

The cost of ray tracing depends on the number of light sources and consequently on the number of shadow rays. For example using of photon maps reduces this cost [10] but nevertheless shadow tests are done. In the paper [11] an inexpensive algorithm for soft shadows was introduced. It supposed the simulation of point light sources by spherical bulbs. Note that it works improper if an object casting shadows is too small. In order to create the effect of a soft shadow we calculate illuminations in DLM mesh points and then interpolate the values of nearby mesh points onto rendered surfaces of a scene. And it works! The soft shadows obtained looks quite realistic, namely: a shadow diffuses not only outside of the theoretical contour of a sharp shadow but inside it too, as well. Thus one could see the effect of an enlarged area of a point light source. It is important that the presented algorithm provides plausible results when too small scene objects (leaves) practically do not cast noticeable shadows, see example in Fig. 4. By the proper control of the shape and size of the interpolation neighborhood $D(y)$ and the density of light meshes a user can vary shadows from soft to sharp. A generation of soft shadows does not make the tracer slower, it allows computing faster. The difference becomes noticeable while increasing a number of light sources and/or scene objects and image resolution. Let's consider an example of a simple scene, consisting of 23542 triangles and 100 light sources. A conventional tracer calculates an image of 600×600 pixels for 277 seconds. A light mesh tracer takes 40 seconds. Experiments with scenes consisting of 100 light sources and 20000–50000 triangles show that a light mesh tracer is from 6 to 10 times faster than a conventional tracer.

4   Conclusion

We hope that we has shown the feasibility of the light meshes approach, and that the technique of light meshes extends the available arsenal of techniques enhancing the ray tracing algorithm. The indirect light mesh provides color interaction only between those surfaces that "see" each other. So the ray tracing problem of diffuse color interreflections is not solved yet by our algorithm, for example the mentioned scene "Constructive wood" is rendered wrong. Probably the reason of this relict drawback results from the independence of the direct and indirect meshes. We are going to continue investigations of this question. The second direction of our interests is an adaptive meshing and interactive control. The usage of light meshes allows step-by-step improvements of the image similar to the progressive radiosity algorithm [2]. After adding of new points to a mesh the values assigned to existing mesh points are not recalculated. As the adaptation

Fig. 4. Soft shadows of the forest leaves. Three point light sources are placed just above trees

Fig. 5. Conventional ray tracing, sharp shadows

of a mesh we mean the process of increasing its density in certain regions of a scene. For example, if the interpolation sphere contains the neighboring points that differ in values more than some threshold then new points are added to the mesh. A user can control the process of image refinement interactively. Beginning from the coarse mesh and the "nearest" mode of interpolation he could: a) switch the interpolation mode; b) change the radius of an interpolation sphere; c) use more dense meshes in the next step; d) select regions of a scene where the meshes must be more dense; etc. And the final remark concerns possibilities to a specific parallelization of the presented algorithm. Calculations of all unknown values (steps 6 and 9 of the algorithm) can be done in parallel. Obviously we do not consider all questions concerning suggested space light meshes, we left beyond the paper such problems as fighting with artifacts on images, adaptive interpolation schemes, mathematics of light meshes.

5   Acknowledgements

The presented work was supported by Russian Foundation for Basic Research grants 99-0100577, 0101-06197.

References

1. Whitted, T.: An Improved Illumination Model for Shaded Display. Commun. ACM. Vol. 23, No. 6, (1980) 343–349.
2. Cohen, M.F., Wallace, J.R.: Radiosity and Realistic Image Synthesis. – Academic Press, New York (1993).
3. Ward, G.J., Heckbert P.S.: Irradiance Gradients. Proc. of the Third Eurographics Workshop on Rendering, Bristol, UK (1992) 85–98.
4. Arvo, J.: Backward Ray Tracing. Developments in Ray Tracing, SIGGRAPH'86 Course Notes Developments in Ray Tracing, 12 (1986).
5. Jensen, H. Wann, Christensen, N.J.: Photon Maps in Bidirectional Monte Carlo Ray Tracing of Complex Objects. Computers and Graphics. 19, 2(1995) 215–224.
6. Chiu, K., Zimmerman, K., Shirley, P.: The Light Volume: an Aid to Rendering Complex Environments. Seventh Eurographics Workshop on Rendering, Porto, Portugal, 1996.
7. Greger, G., Shirley, P., Hubbard, P.M., Greenberg, D.P.: The irradiance volume. IEEE Computer Graphics and Applications, 18(2), 1998, 32–43.
8. Castro, L., Neumann, L., Sbert, M.: Extended Ambient Term. J. of Graphics Tools. 5, 4(2000), 1–7.
9. Ghali, S., Fiume, E., Seidel, H.P.: Shadow Computation: a Unified Perspective. Proc. EUROGRAPHICS'2000 CD. shadows.pdf.
10. Jensen, H. Wann, Christensen, N.J.: Efficiently Rendering Shadows using the Photon Map. Proc. of Compugraphics'95. Alvor (1995) 285–291.
11. Parker, S., Shirley, P., Smits, B.: Single Sample Soft Shadows. Tech. Rep. UUCS-98019, Computer Science Department, University of Utah (1998).

Adding Synthetic Detail to Natural Terrain Using a Wavelet Approach

Mariano Perez, Marcos Fernandez and Miguel Lozano

Department of Computer Science, University of Valencia,
Valencia, Spain
{Mariano.Perez, Marcos.Fernandez, Miguel.Lozano}@uv.es

Abstract. Terrain representation is a basic topic in the field of interactive graphics. The amount of data required for good quality terrain representation offers an important challenge to developers of such systems. For users of these applications the accuracy of geographical data is less important than their natural visual appearance. This makes it possible to mantain a limited geographical data base for the system and to extend it generating synthetic data.

In this paper we combine fractal and wavelet theories to provide extra data which keeps the natural essence of actual information available. The new levels of detail(LOD) for the terrain are obtained applying an inverse Wavelet Transform (WT) to a set of values randomly generated, maintaining statistical properties coherence with original geographical data.

1 Introduction

Terrain representation is a basic topic in the field of training simulators, in both military and civil applications. It is obvious that, in these representations, the amount of geographic data provided to the subject is clearly related to the feeling of visual immersion achieved by the system. However it, is not so obvious that accuracy of the geographical data is less important than their natural visual appearance. Usually the terrain representation is based on a digital elevation model (DEM) plus a set of textures mapped on the mesh. The viewer integrates this geographic information without realizing about the real source of the data presented but about the natural apparency of the final representation.

Other important topic in the field of simulation and interactive graphics is the freedom of movement within the synthetic environment, that should be as similar as possible to the real world. Additionally, is expected to have an homogeneous visual quality from every possible point of view. This introduces an important drawback in most of the current terrain representation models, which have a limited resolution to shown to the viewer. For instance, a flight simulator offers a good quality appearance of the terrain representation when observed from high altitude. This also allows the coverage of large scale terrains. However, for low height fly mission, the visual quality of the representation suffers an important

P.M.A. Sloot et al. (Eds.): ICCS 2002, LNCS 2330, pp. 22–31, 2002.

degradation, because the original terrain model is not accurate enough to provide the required detail of the representation.

A possible solution to the previously indicated problem could be to resize the geographical data base by adding extra data for those special purposes. Nevertheless, there will be always some limitations which come mainly from: budget restrictions, availability (it is not always possible to have access to the data with the required accuracy), resources consumption (storage space, computing, etc.). An approach to solve the resize problem could be the addition of extra resolution, to real terrain meshes, by means of the natural generation of new synthetic data. In order to maintain the natural appearance and the fidelity to the original terrain, the extra data should statistically follow the properties extracted from the original data. To accomplish this goal we propose a combination of the fractal (in concrete $1/f$ processes) and wavelet theory.

Considering the limits in storage capacities, our solution will not store the generated data; on the contrary it will be dynamically generated when needed. This means that generated data has to be compatible with the temporal contrainsts of the simulation system.

2   Fractal and Multifractal Functions

The family of $1/f$ fractals has been successfully used to model a wide variety of natural phenomena and even sociological aspects of the human behavior. Well known examples are: natural landscapes, distribution of a river turbulent flow, evolution of the stocks in the markets, etc. [12]

An important characteristic of the $1/f$ processes is that its spectral density presents an exponential behavior. The spectral density decades on frequency according to the following equation:

$$S(f) \sim \frac{1}{f^\gamma} \tag{1}$$

where f is the frequency and γ is a constant.

This means that a log-log representation of this density with respect the frequency is an straight line with a slope of $-\gamma$.

Fractional Brownian motion (fBm) is probably the best known mathematical characterization of the $1/f$ processes [7]. This theory has been frequently studied due to its simplicity and the wide range of natural phenomena that is able to model.

Based on the works from Mandelbrot, is possible to represent fractals objects using "fractal dimension". This parameter has an integer value for non-fractals objects (1 for curves, 2 surfaces,etc.), whereas for fractal objects the fractal dimension has a non integer value bigger than its topological dimension. The fractal dimension can be used as a roughness index. For instance, in case of curves, values close to one mean low roughness in the function, and values close to two correspond with extremely rough functions. FBm fractal functions are characterized by having only one fractal dimension with an homogeneous distribution over the whole domain. This type of fractal objects are known as monofractals.

There are other kinds of fractals which present a variation of the fractal dimension along its domain; this family of fractals is usually referred as multi-fractals. The Multifractals were introduced firstly to model energy dissipation turbulence's [8][5]. They have been proved to be adequate to model a wide range of non-homogeneous processes [11] [10]. One important consequence of the lack of homegenity in the fractal dimension is that the increments are not stationary, so there is no accurate value for the fractal dimension parameter. It will change in an erratic way along the domain.

3   Wavelets on the Generation of fBm

Two of the most interesting properties of the fBm noise are its non stationary behavior, with stationary increments, and its self-similarity at different scales. The stationarity property implies a time-dependent analysis, while self-similarity requires some scale-dependent analysis. These two characteristics are intrinsic to in the wavelet transform [6] [2], which makes them a powerful tool for the analysis and synthesis of fBm's.

An orthogonal wavelet decomposition of a function $X(t)$ generates detail and coarse coefficients recursively by using the equations:

$$c_{j,i} = \sum_k c_{j+1,k} h[-i+2k]$$

$$d_{j,i} = \sum_k c_{j+1,k} g[-i+2k]$$

being $c_{j,i}$ and $d_{j,i}$ the coarse and the detail coefficients respectively. Being h and g the low and high filter coefficients associated to the wavelet base.

The statistical behavior of these coefficients was previously analyzed by Faldrin [4]. Based on the results of this studies it is possible to assume in practical applications that the correlation amongst the detail coefficients can be ignorated as Wornell point out in [13]. Therefore, it is possible to consider these coefficients as a set of independent Gaussian variables. As a summary, what Wornell discovered is a quite simple procedure to synthesize nearly-fBm functions. The similarity between the results using this method and a pure fBm is directly related to the number of "vanish moments" of the wavelet base selected. Even though, it is not necessary to use basis with high number of vanish moments. In fact, the only important restriction is to select wavelet bases with the regularity property. [12].

4   Terrain Representation Based on fBm

Natural landscapes are examples of self-similar fractal phenomena where the geometrical structure is invariant when examined at different scales. This qualitative characterization is the base for most approaches to synthetic terrain generation [3], [1]. The first attempts to the generation of fractal landscapes were made by Mandelbrot several years ago [9]. .

If we consider $V_H(x, y)$ a Brownian surface with a Holder parameter H $(0 < H < 1)$, every section of this surface obtained by a vertical plane, generates a fBm curve with H parameter. Based on this property, if we move a distance $\Delta r = \Delta x^2 + \Delta y^2$ over the surface, the expected value of the function variation will be:

$$\text{Var}(V_H) \propto \Delta r^{2H}$$

The fractal dimension of this surface is bigger than the topological dimension of the corresponding non-fractal surface.

5 Approaches to Increase the Resolution of a Terrain Mesh

After the introduction of some basic concepts about fractals functions and wavelets, we will explain our approach, based on these principles, to increase the resolution of a natural terrain mesh by adding new levels of detail (LOD). The new synthetic LODs preserve the statistical behavior intrinsic to the real data.

5.1 Global Scope Approach (GSA)

The proposed approach is based on the generation of nearly-fBm functions using the results of the statistical behavior analysis of the detail coefficients in pure fBm [13].

As previously indicated, the fBm has self-similarity at every scale. If we assume that the terrain meshes are fBm, or at least they are close enough to it, the new LODs generated using this technique maintain the statistical properties of the original natural mesh. The assumption includes that the variance of detail coefficients follows a decreasing power-law.

Taking into account the bidimensional nature of the meshes for the terrain representation of the WT, three different types of coefficients will be generated: horizontal $(d_{(h)j, \vec{k}})$, vertical $(d_{(v)j, \vec{k}})$ and diagonal $(d_{(d)j, \vec{k}})$ ones (equation 3). The variances of these coefficients $(\sigma_{(h)}, \sigma_{(v)}$ and $\sigma_{(d)})$ -equation 2- are independent among them. This implies the definition of three different γ values: $(\gamma_{(h)}, \gamma_{(v)}, \gamma_{(d)})$ (equation 3).

GSA Algorithm . The algorithm consists of the following steps:

1. Evaluate the associated variances to the wavelet coefficients in each of the levels and for each of the bands (horizontal, vertical and diagonal). It is assumed that these values follow a Gaussian distribution centred at cero ($\mu = 0$). The dispersion σ at level j is obtained by using the following equations:

$$\left(\sigma_{(\cdot)j}\right)^2 = \text{Var}\left[d_{(\cdot)j, \vec{k}}\right] \tag{2}$$

2. Adjust the logarithm of the variances, obtained in the previous step, to a straight line[12]:

$$\log_2 \left(\text{Var} \left[d_{(\cdot)j,\, \vec{k}} \right] \right) = -j\gamma_{(\cdot)} + a_{(\cdot)} \qquad (3)$$

We have three kinds of coefficients, so the process has to be repeated for each of them, producing three values of slope: $(\gamma_{(h)}, \gamma_{(v)}, \gamma_{(d)})$, and three ordinate values $((a_{(h)}, a_{(v)}, a_{(d)})$. As low levels have not sufficient number of coefficient values to consider them statistically significant, we can reject them. In this way, we consider only variances from level 3 and up.

3. Randomly generate the values of the coefficients for the new levels. The generated values follow a Gaussian distribution centred at cero and with dispersion obtained using equation 2. This guarantees the same statistical behavior as for the original levels.

4. Generate the new level of the terrain mesh by calculating the inverse wavelet transform of the coefficients obtained in step 3.

Figure 1 shows a visual example where we have applied this approach has been applied to generate a finer resolution of a natural terrain mesh.

This approach offers good results when the fractal dimension of the original mesh is more or less homogenous across the whole surface (figure 1, top). However, most of the practical cases do not follow the previous conditions. For instance, landscapes including rough mountains and smooth valleys (figure 1, bottom), have no homogeneous fractal dimension. Figure 1 shows that the results of GSA in this case of heterogeneous terrains is not so good as those obtained where the the homogenty in the fractal dimension is accomplished.

5.2   Local Scope Approach (LSA)

As indicated at the end of the next section, the GSA method presents serious deficiencies to its application to general meshes. To solve these problems we have introduced the local scope approach that can be used in the common DEMs that are better represented as multifractal objects.

Even though multifractals have been extensively studied, there are not too many real applications using this theory. A possible cause is the mathematical complexity involved in its use. The last remark is important because our goal is to develop an algorithm efficient enough to dynamically generate new LODs. To accomplish both objectives: good characterization of the fractal nature of DEMs, and good computational performance, we assume the following constraints:

1. The mesh is divided into regions, each of them has a more or less homogeneous fractal dimension.
2. The fractal dimension transition across neighbour regions is smooth.

The first assumption does not impose important constraints because it has been well established in the literature that DEMs present a locally monofractal behavior. In fact, most of the analysis and synthesis algorithms related to terrain meshes are based on fBm. The second condition might be more difficult to accomplish in some cases, but its influence in the final result is not so critical; partial violation of it produces still good enough final meshes.

Local Fractal Dimension Estimation. One of the key points in LSA, as it was in GSA, is the estimation of the γ parameter. In LSA we have a different γ parameter at each location. The γ estimation is based on the use of the variance of the detail coefficients of the WT. However, we only consider the coefficients which have a spatial domain close to the location where we are estimating the γ value.

Being $X(x, y)$ a function (accomplishing the previously stated conditions)to which we have previously applied the WT, we can conclude that around each point $\vec{r} = (x, y)$ there is a region that satisfies equation 3 with a single value of the γ parameter.

As in the GSA, our meshes are discrete functions parametrized by two parameters, then WT produces three different types of detail coefficients: horizontal $(d_{(h)j, \vec{k}})$, vertical $(d_{(v)j, \vec{k}})$ and diagonal $(d_{(d)j, \vec{k}})$ ones. The variances associated to each type of coefficients are independent among themselves , and can be derived from equation:

$$\left(\sigma_{(\cdot)j}(\vec{r})\right)^2 = \mathrm{Var}\left[d_{(\cdot)j, \vec{k}}\right](\vec{r}) = A_{(\cdot)}(\vec{r})\, 2^{-j\,\gamma_{(\cdot)}(\vec{r})}$$

being $A_{(\cdot)}(\vec{r})$ a constant through the different levels j, and $\gamma_{(\cdot)}(\vec{r})$ the local γ parameter at point $\vec{r}$.

The next step will be able the definition of the spatial regions with homogeneous fractal dimension. This will determine the particular sets of coefficients to be used. To solve it in a accurate way is not trivial. It implies an additional computational cost that will be incompatible with our temporal restrictions. To overcome this limitation we make an important simplification: actual regions are not determined. Instead we suppose that the fractal dimesion is more or less homogeneous in a square window centred at the current point, so only coefficients inside the window are used to estimate the $\gamma_{(\cdot)}(\vec{r})$ parameters.

The previous simplification is not free of risk, because the final result depends on the proper selection of the window size. The selection of the appropiate windows size may require some kind of trial and error test.

Once the window size has been selected, we always consider the same number of coefficients, no matter the level we are working on, to obtain the local variance.

Values ($\gamma_{(h)}$, $\gamma_{(v)}$, $\gamma_{(d)}$) are calculated as the slope of the straight line represented by the equation:

$$\log_2\left(\mathrm{Var}\left[d_{(\cdot)j, \vec{k}}\right]\right)(\vec{r}) = -j\gamma_{(\cdot)}(\vec{r}) + a_{(\cdot)}(\vec{r}) \tag{4}$$

Due to the exponential increase of spatial domain covered by detail coefficients when the level decreases, only variance of the coefficients at higher levels is considered.

New LODs Generation. Once the set of $\gamma_{(.)}(\overrightarrow{r})$ and $a_{(.)}(\overrightarrow{r})$ values have been evaluated, we are ready to generate the additional levels of detail. To generate the vertex that conforms with the new LOD mesh at level n we use, as in the GSA, the inverse wavelet transform applied to the synthetic coefficients generated randomly. The random generation is based on a Gaussian distribution centred at cero and having its dispersion expressed by the following equation:

$$\sigma^2_{(.)n}(\overrightarrow{r}) = 2^{a_{(.)}(\overrightarrow{r})}2^{-m_{(.)}(\overrightarrow{r})} \tag{5}$$

The process can be repeated as many times as needed until we achieve the desired resolution in the final terrain mesh. However, a degradation in the quality of the final result has been observed when the number of extra detail levels is too high.

LSA Algortim. As a summary of LSA we will present the specific steps involved in this solution. If we have m initial levels generated using actual data, and if n is the number of extra levels of detail (the final number of levels will be $m + n$) the algorithm associated to LSA will repeat the following steps for each point at level m:

1. Evaluate the local variance of the detail coefficients at levels $j < m$ (this process has to be done for the three types of coefficients).
2. Calculate the values of the slopes ($\gamma_{(h)}(\overrightarrow{r})$, $\gamma_{(v)}(\overrightarrow{r})$, $\gamma_{(d)}(\overrightarrow{r})$) and the ordinate values ($a_{(h)}(\overrightarrow{r})$, $a_{(v)}(\overrightarrow{r})$, $a_{(d)}(\overrightarrow{r})$), using equation 3.
3. Generate randomly, using a Gaussian distribution centred at cero and with the dispersion obtained from equations 5, the new details coefficients for levels from m to $m + n - 1$ related to the current point $\overrightarrow{r}$.

Once we have applied the previous algorithm to every point at level m we calculate the inverse wavelet transform for the new added coefficients obtaining the extra terrain levels of detail.

5.3   Results

To test the presented approaches we have used two different meshes. The first mesh contains only rough mountain areas (figure 1,top), and the second one includes rough mountains and smooth valleys (figure 1,bottom).

The quality of the results will depend on three main parameters: the approach selected (local or global), the selected wavelet base, and, in the case of local scope approach, the window size. To test the last parameter we have selected square windows of four different window sizes: 1×1, 3×3, 5×5 and 7×7.

The estimation of the goodness of the results will be based on two main criteria, related to our original objectives: visual quality and computing performance. The first aspect is somewhat subjective while the second will be based on measuring of the time spent for the new level generation.

The two original terrain meshes have a size of 16 × 16 with 4 actual level of detail (figures 1, left). We have increased the mesh resolution up to 64 × 64 points, which implies the generation of two extra levels of detail.

Figure 2 shows the results of applying the LSA, with a linear-spline wavelet base, to the hetereogeneous test mesh. Four different window sizes have being used for the test. Table 1 presents the time to generate the extra levels of detail for both approaches. Time measures have been performed using a Pentium II 233 MHz processor.

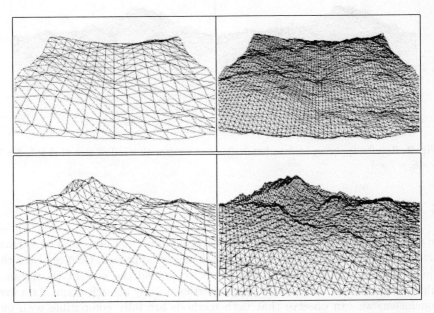

Fig. 1. Top-Left: Original homogeneous mesh, (16×16) points. Top-Right: Results after applying GSA to homogeneous mesh . Synthetic mesh has (64 × 64) points. Bottom-Left: Original hetereogeneous mesh, (16×16) points. Bottom-Right:Results of applying GSA to hetereogeneous mesh. Synthetic mesh has (64 × 64) points.

Table 1. Spent Time (ms.) for the different approaches (linear-spline wavelet base).

	GSA	LSA 1x1	LSA 3x3	LSA 5x5	LSA 7x7
Homogeneous M.	78	95	100	122	132
Heterogeneous M.	77	95	100	120	130

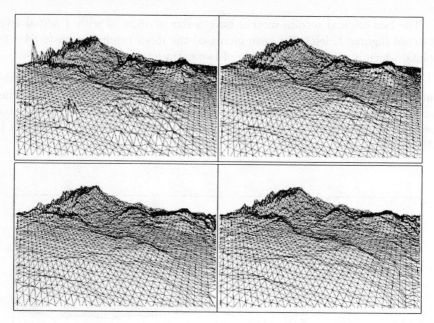

Fig. 2. Results applying LSA to heterogeneous mesh with window size of :1 × 1, 3 × 3, 5 × 5, y 7 × 7.

6   Conclusions and future work

In this paper, two new methods (GSA and LSA) have been presented to increase actual terrain resolution based on a fractal-wavelet approach. GSA analyses the global statistical properties to decide how to extend the new LODs terrains, while LSA analyses these statistical properties locally at each point. As a first conclusion, we can observe that both methods are fully compatible with our temporal restrictions imposed for real-time terrain representation, allowing a dynamical generation of extra levels of detail on demand.

GSA has been proved to be a good approach when used on terrains with uniform geographical properties(figure 1, top), but it fails in not so much homogeneous terrains (figure 1, bottom). This behavior is not strange because global statistical analysis tends to minimize specific local properties having as a result a uniform distribution for the whole mesh. To overcome this problem, the special analysis used in LSA preserves local properties of terrain using only coefficients close to the particular location (figure 2). The 25% extra time of this method (table1),considering a 3 × 3window size, is justified taking into account the achieved improvement in the final visual appearance.

As indicated in section three, the vanish moments of the wavelet bases have influence on the achieved degree of proximity to pure fBm. Tests performed using others symetric different wavelet bases have proved that this influence has no very important consequences in the final visual appearance. For cost reasons the recomended base for interactive applications is the linear-spline.

The window size in LSA is another important aspect for the quality results. Tests indicate that window sizes inferior to 3 × 3 have got no enough values to perform reliable statical analysis, this is clearly reflected in the visual appearance. On the contrary, too large window size (superior to 5 × 5) are not capable reflecting local essence at each point of the surface, becoming in a behavior similar to GSA (figure 2).

As indicated in the introduction of this paper, the other important component in terrain representation is the realistic textures mapped to the terrain mesh. The fractal nature of these terrain textures suggest a possible extension of the presented works to generate finer resolution images from original textures.

References

1. M. F. Barnsley, R. L. Devaney, B. B. Mandelbrot, H. O. Peitgen, D. Saupe, and R. F. Voss. *The Science of fractal images*. Springer-Verlag, 1988. ISBN 0 387 96608 0.
2. I. Daubechies. Orthogonal bases of compactly supported wavelets. *Commun. Pure Appl. Math*, 41:909–996, 1988.
3. C. E. Bolb F. K. Musgrave and R. S. Mace. The synthesis and renderin of eroded fractal terrains. In *SIGGRAPH'89, Computer Graphics Proceedings*, pages 41–50, 1989.
4. Patrick Flandrin. Wavelet analysis and synthesis of fractional brownian motion. *IEEE Transactions on Information Theory*, 38(2):910–917, 1992.
5. U. Frisch and G. Parisi. Fully developed turbulence and intermittency. In *Int. Summer School on Turbulence and Predictability in GeoPysical Fluid Dynamics and Climate Dynamics*, pages 84–88, 1985.
6. S. G. Mallat. A theory for multiresolution signal decomposition: the wavelet representation. *IEEE Trans. Patt. Anal. Machine Intell.*, 11:674–693, 1989.
7. B. B. Mandelbrot. Fractional brownian motion, fractional noises and applications. In *SIAM Review 10*, pages 422–437, 1968.
8. B. B. Mandelbrot. Intermittent turbulence in self similar cascades: Divergence of high moments and dimensionof the carrier. *Fluid Mech.*, 62(3):331, 1974.
9. B. B. Mandelbrot. *The Fractal Geometry of Nature*. W. H. Freeman and Co., New York, 1982.
10. B. B. Mandelbrot. *Fractals and scaling in Finance*. Springer, New York, 1997.
11. B. B. Mandelbrot and C. J. Evertsz. Multifactility of the armonic measure on fractak aggregates and extended self-similarity. *Physica*, A 177:386–393, 1991.
12. Gregory W. Wornell. Wavelet-based representations for the 1/f family of fractal procesess. *Proceedings of the IEEE*, 81(10):1428–1450, 1993.
13. Gregory W. Wornell and Alan V. Oppenheim. Estimation of fractal signals from noisy measurements using wavelets. *IEEE Transactions on Signal Prcessing*, 40(3):785–800, 1992.

The New Area Subdivision Methods for Producing Shapes of Colored Paper Mosaic

Sang Hyun Seo[1], Dae Wook Kang[1], Young Sub Park[2], and Kyung Hyun Yoon[2]

[1] Department of Image Engineering Graduate School of Advanced Imaging Science, Multimedia, and Film, University of Chung-Ang, DongJak-Gu, Seoul, Korea
{ddipdduk, gigadeath}@cglab.cse.cau.ac.kr
http://cglab.cse.cau.ac.kr
[2] Department of CS&E University of Chung-Ang, DongJak-Gu, Seoul, Korea
{cookie, khyoon}@cglab.cse.cau.ac.kr

Abstract. This paper proposes a colored paper mosaic rendering technique based on image segmentation that can automatically generate a torn and tagged colored paper mosaic effect. The previous method[12] did not produce satisfactory results due to the ineffectiveness of having to use pieces of the same size. The proposed two methods for determination of paper shape and location that are based on segmentation can subdivide an image area by considering characteristics of image. The first method is to generate a Voronoi polygon after subdividing the segmented image again using a quad tree. And the second method is to apply the Voronoi diagram on each segmentation layer. Through these methods, the characteristic of the image is expressed in more detail than the previous colored paper mosaic rendering method.

1 Introduction

The trend of current computer graphic techniques, such as radiosity and ray tracing, puts its emphasis on generating a photo realistic image by considering the relationship between light and 3D geometry data in the rendering process. Images obtained through these classical rendering methods do not show any individual character and tend to give a super-realistic expression. Recently, there has been a great deal of interest in NPR (Non-Photorealistic Rendering) techniques. Several computer graphics researchers are studying various ways to simulate conventional artistic expressions such as pen-and-ink illustrations, watercolor, and impressionism. Therefore, the main focus of the NPR methods is to create a special rendering style similar to those created freehand by people. Images created using these methods show an enhanced artistic beauty, more than the images created using traditional computer graphic techniques[4][7].

The colored paper mosaic rendering method, a non-photorealistic rendering method[12], is introduced in order to simulate man-made mosaic work. Various features that produce mosaic effects in mosaic works must be considered. For example, how the paper is attached by hand, the irregular lines of torn paper, and white paper effects that show a cutting plane. These rendering techniques are

P.M.A. Sloot et al. (Eds.): ICCS 2002, LNCS 2330, pp. 32–41, 2002.

introduced in Seo's paper and will be briefly explained later in this paper. The previous method did not produce satisfactory results due to the ineffectiveness of having to use pieces of the same size. The input image was divided into grids of equal size before the Voronoi sites could be placed, which caused a loss of the characteristics of the image[12].

This paper introduces a method of acquiring a Voronoi diagram by segmenting the input image using the segmentation technique, and then placing the Voronoi site into the area created by splitting the segmented input. The first proposed algorithm locates the Voronoi sites on the location of each tree node by subdividing the segmented area with the quad tree technique. In the second algorithm the Voronoi polygon is generated by the method of applying the Voronoi diagram on each segmented layer. The advantages of this proposed method are that pieces of paper can be aligned along the borderline of the object, and the size of attached paper can be adjusted according to the size of a similar colored image area in order to create a more man-made effect on the produced image.

2   Previous Works

The colored paper mosaic rendering method proposed in this paper is the method used to create an image by using pieces of paper torn by hand. Application software that is currently in use, such as Photoshop, provides methods that are similar to the mosaic rendering technique (ex, Mosaic filter, Crystallize filter, Stained Glass filter, Mosaic Tile filter and so on). However, it is difficult to express the colored paper mosaic effect with the filtering methods provided by such graphic tools. The resulting images created using current technology produce fairly artificial and mechanical images, and therefore are not suitable for being applied to create natural images, which is the main focus of NPR.

Hausner introduced the technique that simulates decorative mosaics using CVD(Centroidal Voronoi Diagram), which normally arranges points in a regular hexagonal grid. The direction of the tiles is determined by creating a gradient map after the edges have been determined by placing the tiles[8].

Seo has introduced a new algorithm of automatically simulating the torn colored paper effect with a computer. The shape of the colored-paper is determined by using the Voronoi diagram and the torn colored-paper effect is expressed through the random fractal method[12]. However, in this method, the quality of the resulting image depends largely on the size of the attached paper and the complexity of the input image. Also, the method is limited in its ability to express a natural colored paper mosaic rendering effect due to its limitation of having to use pieces of regular size to express the image.

In this paper, we introduce a new algorithm that determines the shape of the attached paper by taking the edge and area of the object into account by analyzing the input image.

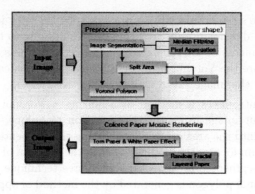

Fig. 1. Mosaic rendering flowchart

3   The New Area Subdivision Methods for Producing Shapes

Fig. 1 is the flowchart of our system, which is introduced in this paper. The median filtering and pixel aggregation for the segmentation are processed in the preprocessing stage before the quad tree or the Voronoi diagram is applied. The colored paper mosaic rendering process is applied to each Voronoi polygon that is generated in the preprocessing stage. The details of each process will be illustrated in the following section.

3.1   Extraction of Features for Determination of Paper Shape

The object in the input image should be extracted for the colored paper mosaic rendering that expresses the main feature of the image. The edge of the object on the input image can be obtained through extracting the image feature. Edge detection, a method for connecting the detected edge pixels, and the image segmentation method are used to divide the image into objects. In this paper, we use the image segmentation technique. The problem caused by a piece of paper crossing the boundary (edge) of the object that consists of input image can be prevented through image segmenting. The pixel aggregation algorithm[6] has been used as the segmentation method in this paper. The pixel aggregation algorithm is a simple method for segmentation using gray scale images that have been preprocessed through smoothing. This algorithm compares the intensity of the pixel with its neighbor pixels based on the initial coordinate for segmentation. The area is included in the same area if the value for the intensity is smaller than the threshold, and segmented if the value is above the threshold. The edges are used to extract the image feature in determining the polygon to be used to express the mosaic rendering. The effect of putting pieces of the mosaic together manually along the edge of the image can be expressed through this method. Fig. 2 is an example of the segmented image. Fig. 2(a) is input image. Fig. 2(b)

(a) Input image   (b) Smoothed Gray Scale Image   (c) Segmented Image

Fig. 2. Input and segmented image

is smoothed gray scales image. Quarterly segmented images are shown in Fig. 2(c).

3.2   Determination of Paper Shape

In the work of Seo et al, the Voronoi diagram[9] has been used to generate the shape of an attached paper. The input image is divided into even grid sizes to generate the Voronoi polygon. Random points are generated in each divided area. After creating a Voronoi diagram using these input points, the Voronoi polygons have passed the mosaic rendering stage. In this paper, the Voronoi diagram has also been used to determine the shape of the attached paper. But two new algorithms have been applied to locate the sites of the Voronoi diagram onto the segmented images.

The first method locates the Voronoi sites on each leaf node by subdividing the segmented image using quad tree. In the second method the Voronoi polygon is generated by randomly locating the Voronoi sites into each segmented layer. The Voronoi polygons created through the first method will be smaller in size due to the level of the tree nodes obtained around the edge of the object. This method is effective in expressing the detailed parts of an image by having the effect of illustrating finely torn paper. The second method applies the Voronoi diagram on each layer of the segmented image. Voronoi polygons will be created along the edge (boundary) through the segmentation. Therefore, the feature of the image can be maintained and the borderline formed by curved lines can be expressed precisely.

Voronoi Diagram Based On Quad Tree Subdivision The Voronoi diagram based on Quad Tree[2] further subdivides the segmented image by using the quad tree algorithm. The size of each paper can be controlled for the simulation by increasing or decreasing the subdivision (tree level) rate around the border through the quad tree. The size of each Voronoi polygon can be determined by the size of the node of the quad tree through randomly locating the Voronoi sites for each node. The dense shape of a paper can be expressed more effectively along the border of the image by using this method.

The root node that contains the whole image is created to acquire the nodes of the quad tree in the combined segmented image. Each node further subdivides itself into four child nodes if the number of segmented layers that it occupies is more than two. The same algorithm is applied repeatedly on the child node obtained through this method until it occupies only a single segmented layer. However, segmentation is disrupted if the size of the node reaches a minimum threshold in order to prevent creating pieces that are too small(Table 1: MINI-MUM_SIZE). Table 1 explains a simple pseudo code of the algorithm. Voronoi

Table 1. Pseudo code of quad tree algorithm

```
Quadtree_split( cell, size){
    if(size>MINIMUM_SIZE){
        if(existing two or more segmented area in cell){
                Quadtree_split (left_top of cell, size/4);
                Quadtree_split (right_top of cell, size/4);
                Quadtree_split (left_bottom of cell, size/4);
                Quadtree_split (right_bottom of cell, size/4);
                }
        else{ this cell is leaf node return; }
    else {
        this cell is leaf node
        return;
    }}}
```

polygons of various sizes are created through random location of Voronoi sites on each node. The shape of each Voronoi polygon determines the shape of the paper. Fig. 3(a) shows the input image. Fig. 3(b) is the subdivided image by quad tree. Finally, Fig. 3(c) is the resulting image of the Voronoi polygon obtained by randomly locating the Voronoi site on the leaf nodes of the quad tree.

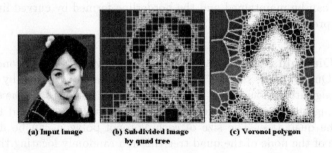

(a) Input Image   (b) Subdivided Image   (c) Voronoi polygon
                        by quad tree

Fig. 3. Voronoi diagram based on quad tree

Constraint Voronoi Diagram on Each Segmented Layer The algorithm, which is explained in this section, is a method of applying the Voronoi diagram on each segmented layer. In general, when the actual mosaic is created by hand, the boundary of objects in the image and edge of the attached paper are fit to clearly express the objects. However, it is difficult to express the boundary of the objects in the previous colored paper mosaic rendering algorithm because the Voronoi sites are randomly located around the boundary, and generated Voronoi polygons cross the boundary. The polygons created by the Voronoi diagram should not overlap with the boundary of the object so as to express the features of the image. To prevent overlapping, each and every segmented area should be separated and then the Voronoi diagram applied.

This algorithm makes it possible to attach a piece of paper along the boundary of objects in the input image. The Voronoi sites should first be placed randomly and then the Voronoi diagram should be applied on each segmented layer. The Voronoi diagram is applied to each layer of the segmented image to express the boundary of the object. The process of determining the shape of the paper along the borderline is simulated when each Voronoi edge is fitted to the border of the object. The shape of the entire area that the Voronoi diagram should be applied to is either rectangular or regular. Whereas, the shape of the segmented layer that the Voronoi diagram should be applied to is irregular. Now, the boundary of the segmented layer has a curved shape. In this paper, the edge

Table 2. Pseudo code of edge map generating algorithm

```
VoronoiCell&Edge( ){
    width = image width;
    height = image height;
    Site = Current site list;
    Map = array of index[width*height];
    Repeat
        Map[each position of image] = index of near site;
    Until width*height
    Repeat
        If(existing another site around from current position in the Map)
                This position is Edge
    Until width*height;
    }
```

map is used to express the edge of Voronoi polygons generated along the curved boundary. The edge map includes information about the location of pixels and the index of the Voronoi site. The location of pixels is stored in the form of lists, and these lists are approximated as having one or several edges. To construct an edge map, an array of the same size as that of the input image is allocated. Each location of an array stores the index of its closest Voronoi site. The edge

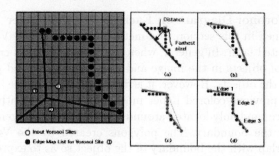

Fig. 4. Edge map for one Voronoi site and approximation of Edge map

map is computed again to determine edge pixel by comparing the number of Voronoi sites stored in the neighbor index. If the number of Voronoi sites stored in the current index is different from that of the neighbor index, the current index is detected as edge pixels. The detected edge pixels are stored as a linked list, which is used to determine the final shape of the paper. By doing so, the edge map for each Voronoi site is completed. Table 2 shows pseudo code of the proposed algorithm.

Fig. 4 shows an edge map for one Voronoi site and the approximation process. The black points in Fig. 4 are the edges map for Voronoi Site 1. The approximation process below is briefly explained.

- **Step 1:** Initial edge map is approximated as one edge by linking both ends of the list (Fig. 4(a)).
- **Step 2:** The pixel, which is farthest from the corresponding edge, should be located. If the maximum value of distance is smaller than the defined threshold, the process should cease (Fig. 4(a)).
- **Step 3:** Based on the located pixel, which has the maximum distance, two sub-edges are constructed (Fig. 4 (b)).
- **Step 4:** Step 2 is repeated on each newly approximated sub-edge (Fig. 4(c) (d)).

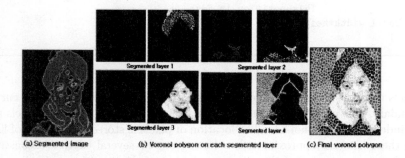

Fig. 5. Constraint Voronoi diagram on each segmented layer

Fig. 5(a) shows the segmented input image and initial Voronoi sites. In Fig. 5(b), each segmented layer that the Voronoi diagram is applied to is classified (four layers). Finally, Fig. 5 (c) is the Voronoi diagram where four segmented layers are combined.

3.3   Mosaic Rendering: Torn and White Paper Effect

Torn Paper Effect.   The colored paper mosaic expression is the method of implementing a torn paper effect. The shape of the edge of a torn piece of paper is irregular. A method for adding an irregular height to the straight line connecting the two points on the edges that form the Voronoi polygon is required to express this effect. The random midpoint displacement algorithm, a random fractal[5][10] method, is used to express the irregular shape more naturally in this paper (Fig. 6(a)). Random fractal methods are used for modelling irregular curved lines or curved surfaces that appear in nature. The torn paper effect is implemented by controlling the iteration number and the random amount by the length of the edge line[12].

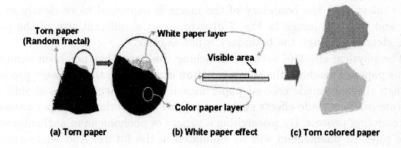

(a) Torn paper   (b) White paper effect   (c) Torn colored paper

Fig. 6. Irregular torn paper and white paper effect

White Paper Effect.   A piece of colored paper consists of many different layers of fiber. Therefore, when it is torn apart, it doesn't show an even surface but a rough white surface. In this paper, we assume that a colored paper consists of two different layers. The top layer is the colored layer and the bottom layer is the white paper. Paper colors should represent the image region they cover. Each paper color may be either a point sample form the pixel at the paper's center, or an average of the pixels covered by the paper. The size and the shape of each layer of paper using the random midpoint displacement algorithm are different due to its randomness. With the continued process of attaching these random piece of paper, there comes a case when the white layer is larger than the colored layer, which is the method used to express the white paper effect of the torn paper (Fig. 6(b) Fig. 6(c)) shows the torn colored paper effect created using the proposed method [12].

4   Conclusion and Future Works

A new algorithm to more naturally determinating the shape of colored paper used for rendering has been introduced in this paper. The image has been classified through segmentation algorithm according to objects. A method of applying the Voronoi diagram on the quad tree, and a method of applying the Voronoi diagram directly onto each segmented layer have been implemented to determine the shape of the colored paper. In the case of the method using the quad tree, the rendered image is similar to actual hand-made mosaic work because smaller pieces of paper are used around the edge of the object while larger pieces of paper are used in the rest of the area. With respect to the Voronoi diagram method based on segmented layer, the boundaries of the object are expressed more clearly. However, The difference in the size of each paper cannot be thoroughly expressed.

These two methods propose that the pieces of paper can be aligned along the border line of the object, and that the size of attached paper can be adjusted according to the size of a similar colored image area to create a more man-made effect on the produced image. Fig. 7 compares the resulting mosaic images created using the basic Voronoi diagram and the methods proposed in this paper. It is evident that the boundary of the image is expressed more clearly in Fig. 7(b) and that the image in Fig. 7(c) expresses the different size of the pieces more clearly, although the boundary is not as clear.

The physical effect of actually attaching the paper has not been simulated in this paper. Therefore, the result is not an exact match to the result produced through the man-made colored paper mosaic effect. Future works should concentrate on man-made effects such as wrapped and overlapped paper caused in the rendering process. By researching a variety of phenomena in authentic work, these physical phenomena will be simulated in the future. And while approximating edge map to geometrical edge, approximated edges sometimes do not fit completely around the borderline between two adjacent segmented layers. These inconsistencies don't affect the rendered image a lot, but will be considered. Also, the shapes of the attached paper used for the simulation should also be diverse for various mosaic effects.

Acknowledgement

This work was partially supported by the Chung-Ang University Special Research Grants No.961173 in 1996

References

1. Aaron Hertzmann: Paintery Rendering with Curtved Brush Strokes of Multiple Sizes, In SIGGRAPH 98 Proceedings, (1998) 453-460
2. Alan Watt: 3D Computer Graphics, Addison Wesley (2000)

3. Cassidy J. Curtis: Loose and Sketchy Animation, In SIGGRAPH'98 Abstracts and Applications (1998) 317
4. Daniel Teece: Three Dimensional Interactive Non PhotoRealistic Rendering, PhD. thesis, University of Sheffield, England (1998)
5. A. Fournier, D. Fussell, and L. Carpenter: Computer Rendering ofStochastic Models, Communication of ACM, Vol.25, No.6 (1982)371-384
6. Gonzalez, Woods: Digital Image Processing, Addison Wesley (1993)
7. Stuart Green: Non-Photorealistic Rendering, SIGGRAPH'99 Course Note#17 (1999)
8. Alejo Hausner: Simulating Decorative Mosaics, In SIGGRAPH'2001 Proceedings (2001) 573-580
9. Mark de Berg, M. V. Kerveld, M. Overmars and O. Schwarzkopf: Computational Geometry Algorithms and Applications, Springer-Verlag (1997) 145-161
10. Michael F. Barnsley: Fractals Everywhere. 2nd Edition, AP Professional (1993)
11. Miyata, K.: A method of generating stone wall patterns, In SIGGRAPH'90 Proceedings (1990) 387-394
12. Sang Hyun Seo, Young Sub Park, Sung Ye Kim and Kyung Hyun Yoon: Colored Paper Mosaic Rendering, In SIGGRAPH'2001 Sketchy and Applications (2001) 156
13. Young Sub Park, Sung Ye Kim, Chung Woon Cho and Kyung Hyun Yoon: Manual Color Paper Mosaic Technique , Journal of Computer Graphics Vol 4 (2002) 123-130, In Korean

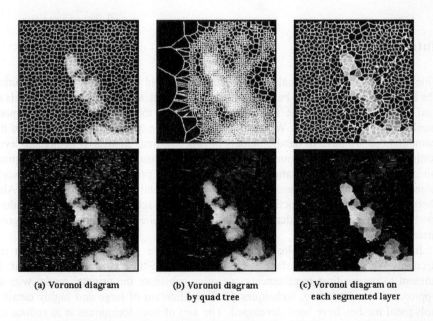

(a) Voronoi diagram   (b) Voronoi diagram   (c) Voronoi diagram on
                          by quad tree   each segmented layer

Fig. 7. Result images 1: Voronoi diagram and images of colored paper mosaic rendering

Fast Algorithm for Triangular Mesh Simplification Based on Vertex Decimation

Martin Franc, Vaclav Skala

Department of Computer Science and Engineering
University of West Bohemia
Univerzitni 8, 306 14 Pilsen, Czech Republic
{marty|skala}@kiv.zcu.cz   http://herakles.zcu.cz

Abstract. A common task in computer graphics is the visualisation of models of real world objects. These models are very large and complex and their surfaces are usually represented by triangular meshes. The surface of complex model can contain thousands or even million of triangles. Because we want fast and interactive manipulation with these models, we need either to improve our graphics hardware or to find a method how to reduce the number of triangles in the mesh, e.g. mesh simplification. In this paper we will present a fast algorithm for triangular mesh reduction based on the principle of vertex decimation.

Introduction

Due to the wide technological advancement in the field of computer graphics during the last few years, there has been an expansion of applications dealing with models of real world objects. For the representation of such models polygonal (triangular) meshes are commonly used. With growing demands on quality, the complexity of the computations we have to handle models having hundreds thousands or perhaps even millions of triangles. The source of such models are usually 3D scanners, computer vision and medical visualisation systems, which can produce models of real world objects. CAD systems commonly produce complex and high detailed models. Also there are surface reconstruction or iso-surface extraction methods, that produce models with a very high density of polygonal meshes displaying almost regular arrangement of vertices.

In all areas which employ complex models there is a trade off between the accuracy with which the surface is modelled and the time needed to process it. In attempt to reduce time requirements, we often substitute the original model with an approximation. Therefore, techniques for simplification of large and highly detailed polygonal meshes have been developed. The aim of such techniques is to reduce the complexity of the model whilst preserving its important details.

We shall present a new fast and simple algorithm for the simplification of very large and complex triangular meshes (hundreds of thousands of triangles). The algorithm is based on the combination of commonly used decimation techniques.

P.M.A. Sloot et al. (Eds.): ICCS 2002, LNCS 2330, pp. 42–51, 2002.
© Springer-Verlag Berlin Heidelberg 2002

This paper is structured as follows: In section 2 we discuss our previous work and fundamental techniques for the mesh simplification especially decimation. We shall also mention the aspect of parallelization. This section includes an overview of the simplification techniques we recently used. The bucketing (hash) function is described in section 3 and it is used to avoid vertex sorting in the decimation process in order to decrease algorithm complexity. In section 4 we present our new approach in detail – main hypothesis, data structures and the algorithm. Section 5 presents results obtained and section 6 provides a conclusion.

Our Previous Work

As already mentioned, our approach is based on the vertex decimation. We chose the vertex decimation because of its simplicity, stability and fast processing. This method is also easy to parallelize.

Decimation

In general, decimation techniques can be divided into three main categories according to the mesh primitives, which the method is dealing with. Therefore we recognize the following cases:
- Vertex decimation,
- Edge decimation (contraction or collapse),
- Triangle (patch) decimation.

The principle of all the aforementioned methods is similar. Initially we have to evaluate the importance of the primitive (vertices, edges, patches) in the mesh. Then, the least important one is removed and after removal the resulting hole is triangulated. We therefore have a new mesh with a smaller number of triangles.

In our method we started with the vertex decimation proposed by Schroeder[1]. Each vertex is evaluated according to its distance from an average plane (or line) given by its neighbouring vertices, Fig. 1(left).

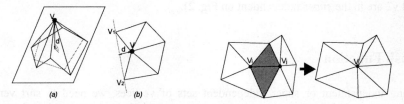

Fig. 1. Vertex importance computation (*left*), edge collapse (*right*)

The second step is to remove the least importance vertex and make a new triangulation. Since we are in 3D, the triangulation is quite complex and time consuming. Considering this, we decided to continue with the edge decimation instead. We evaluated all of the edges going out of the vertex and the least important edge is contracted to zero length. This means that the vertex is moved to the endpoint of chosen edge, which is to be removed from the mesh together with two adjacent triangles, Fig. 1(right).

This method preserves the mesh topology as well as a subset of original vertices.

Parallel Processing

The idea of progressive vertex removal leads to parallel processing. If we imagine one step of iteration with more than one vertex removal processed by several processors, we get simple and efficient parallelization.

The only condition, which must be maintained, is to obtain only one hole per vertex removal; otherwise the triangulation could increase the program complexity and run time significantly. There is a mechanism called independent set of vertices[4, 12] to carry out such a condition. Two vertices can be in the independent set of vertices if they don't share an edge, as shown in Fig. 2.

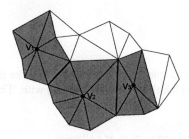

Fig. 2. Independent set of vertices

We assign an importance value to each vertex and then select an independent set to remove vertices with the lowest importance[6]. To construct an independent set from the assigned importance values we process all vertices in order of their importance and select a vertex if none of its neighbours have been taken.

Due to the special data structure, which we used and in order to avoid critical sections in parallel code, we have had to use a stricter rule. We have used a *super* independent set of vertices, where none of two triangles share an edge (vertices v1 and v2 are in the *super* independent on Fig. 2).

Hash Function

Using independent or super independent sets of vertices, we need to sort vertices according to their importance and also to create the independent set of vertices. This appeared to be a critical part of the previous approaches. Since we did not want to use sorting algorithms[7] because of their time complexity, we used a special function to threshold vertices and let only the least few important vertices be considered as candidates for the reduction.

The initial idea was to divide data set (vertices) according to the number of free processors and run decimation as several independent parts. As we already mentioned, most of data was produced by 3D scanners or iso-surface extraction methods such as Marching Cubes[8]. Considering the principle of both techniques, we can suppose that such triangular mesh is a sequence of strips, where neighbouring vertices are also very close to each other in data file or memory. In other words, if we divide a data set into, let's say, five groups, according to vertices index, there is a good probability that vertices in each group will be close to each other, so the program can run without critical sections except vertices on the *boundary* of each group. Those vertices will not be processed.

This approach brought surprisingly good results for the real object models in the sense of processing time and acceptable quality of approximation. On the other hand there were some problems with the control of simplification degree (instead of 90% reduction you can get 99% as well). Vertices on the border of groups had to be handled in a special way. Another problem arose with artificially generated data or data changed on purpose. Such models don't fit to the assumption about strips and the algorithm is quite ineffective in this case. This experience led us to use a hash function that provides vertex bucketing. This approach enabled us to avoid sorting.

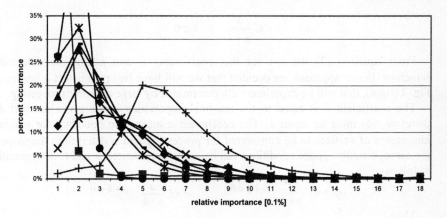

Fig. 3. Vertex importance histogram

New Approach

Instead of vertex sorting or thresholding we can use a hash function and with enough memory we can "sort" vertices in O(N) time complexity, where N is a number of vertices. It is necessary to notice that we do not use modulo operation in the hash function for the address computation to the hash table.

A histogram of vertex importance for tested data sets is shown in Fig. 3. It is obvious that over 80-90% of all vertices have the importance below 1% of maximum importance value.

We used a simple hash function $\quad y = \dfrac{x}{|x| + k}\quad$ (1)

shown in Fig. 4 (left).

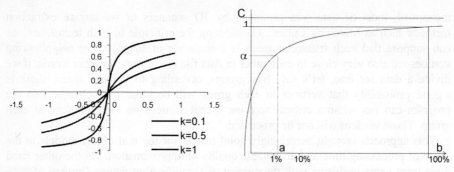

Fig. 4. Graph of the function used (*left*), definition of the hash function (*right*)

This hash function enables us to map the interval $<0,\infty)$ to $<0,1)$ non-linearly. Because we need to map only 1% of important vertices the hash function (1) must be modified accordingly, see Figure 4 (right), and scaling coefficient C must be introduced.

$$y = C\frac{x}{x+k} \qquad x \geq 0 \tag{2}$$

From equation (2) we can see that coefficients C and k must be determined somehow. In our approach we decided that we will have two parameters a and α, see Fig. 4 (right), that will be experimentally determined by large data sets processing.

The coefficient b is equal to the maximal importance in the given data set and therefore $f(b)$ must be equal 1. The coefficient a means the boundary for maximal importance of vertices to be considered for processing and α determines the slope of the curve, actually. Those conditions can be used for parameter k and C determination as follows:

$$1 = C\frac{b}{b+k}, \quad b+k \neq 0 \qquad \alpha = C\frac{a}{a+k}, \quad a+k \neq 0$$

Then

$$C = \frac{b+k}{b} \qquad \text{and} \qquad \alpha = \frac{b+k}{b}\frac{a}{a+k}$$

Solving that we get:

$$k = \frac{ab(1-\alpha)}{\alpha b - a} \qquad\qquad C = \frac{b+k}{b} = 1 + \frac{k}{b}$$

for $b = 1$

$$k = \frac{a(1-\alpha)}{\alpha - a}, \qquad C = 1 + k$$

According to our experiments on large data sets, we have found that the optimal values of coefficients are following: $a = 2$, $\alpha = 80$. This means that 2 percents of the least important vertices will be mapped onto 80% of the whole hash table. The user has to set values of both parameters. The parameter α has a direct influence to the cluster length in the hash data structure, where the cluster length is equal to number of vertices in the same bucket.

Algorithm and Data structure

We described above a framework upon which our algorithm is based. Having described the specific details of the method, we can present our new algorithm now:
1. Evaluate importance of all vertices
2. Make clusters according to the vertex importance
3. Remove vertex from the first cluster, if it is empty continue with the next one
4. Evaluate changed importance of neighbouring vertices and insert vertices in the proper bucket
5. Repeat steps 3 and 4 until desired reduction reached

To make the algorithm more efficient we proposed a special data structure. We use a table of triangles, where we store indexes of all vertices for each triangle. In table of vertices we store each vertex coordinates, a list of triangles sharing this vertex and the address of the cluster where the vertex currently belongs. We also have a vector of clusters where indexes of vertices are stored.

Results

In this section we present results of our experiments that compare achieved time, quality of approximation and also show some examples of reduced models.

We have used several large data sets but we mention experimental results only with 7 different data sets, see Table 1.

Table 1. Data sets used

Model name	No. of triangles	No. of vertices
Teeth	58,328	29,166
Bunny	69,451	35,947
Horse	96,966	48,485
Bone	137,072	60,537
Hand	654,666	327,323
Dragon	871,414	437,645
Happy Buddha	1,087,716	543,652
Turbine blade	1,765,388	882,954

Methods Comparison

Fig. 5 shows running time for 96% reduction for three different approaches of vertex ordering. We can see that using the hash function we obtained the best running time in comparison to the other methods. It is necessary to point out that the methods using thresholding or sorting algorithm were implemented in parallel. The run time of hash function is equal to the performance of 2-3 processors running thresholding and it is faster than 8 processors running the method with sort algorithm. It is because both the sort algorithm and the creation of the independent set of vertices were implemented sequentially. With the thresholding we removed the sort algorithm completely, but independent set of vertices remained.

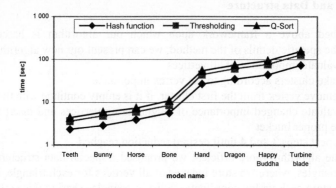

Fig. 5. Achieved time comparison for three mentioned approaches

Unfortunately our results are not directly comparable with other known algorithms, due to the different platforms. To make the results roughly comparable at least, we use the official benchmarks presented by SPEC as shows Table 2, where η presents the superiority of DELL computer against the SGI. Table 3 presents our results according to results obtained recently taking the ratio η into the consideration.

Table 2: Benchmark test presented by Standard Performance Evaluation Corporation.

Benchmark test / machine	SGI R10000	DELL 410 Precision	η (DELL/SGI)
SPECfp95	8.77	13.1	1.49
SPECint95	10.1	17.6	1.74

Table 3. Rough comparison of running times of reduction of the Bunny model.

Algorithm	Time for a reduction from 69.451 to 1.000 triangles [sec]
Proposed algorithm	4,1 * η = 4,1* 1,49 = **6,11**
Garland[3]	10,4
Lindstrom & Turk[11]	2585
Hoppe[2]	500
JADE[10]	325

It is obvious that our algorithm is quite fast, however, we do not know the approximation quality reached by the other algorithms. The quality of an approximation can be measured by several approaches. Probably the most popular way is to compute a geometric error using E_{avg} metric[9] derived from Hausdorff distance. As our method keeps the subset of original vertices, we use more simple formula (3):

$$E_{avg}(M_1, M_2) = \frac{1}{k_1} \sum_{v \in X_1} d_v^2(M_2), \quad X_1 \subset P(M_1) \tag{3}$$

where M_1 and M_2 are original and reduced model, k_1 is original number of vertices, $P(M_1)$ is a set of original vertices and $d_v(M_2)$ is the distance between original set of vertices and reduced model.

If we compare all three above described approaches, we will find that the error values are almost the same. It is also hard to say which method gives the best results, because for each model we get different behaviour of error. Examples of reduced models are presented in Fig. 6,7,8.

Fig. 6. A bunny model (courtesy GaTech) at different resolutions; the original model with 69,451 triangles on the left, reduced to approx. 1,000 triangles on the right

Fig. 7. A teeth model (courtesy Cyberware) at different resolutions; the original model with 58,328 triangles on the left, reduced to approx. 29,000 triangles in the middle and 6,000 triangles approximation on the right

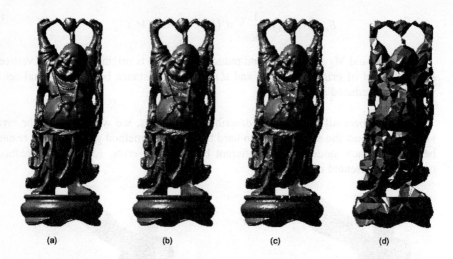

Fig. 8. The Happyb model (courtesy GaTech) at different resolutions; the original model with 1,087,716 triangles (a), reduced to 105,588 triangles (b), 52,586 triangles (c), 10,974 triangles (d)

Conclusions

We have described our superior algorithm for simplification of triangular meshes, which is capable of producing good approximations of polygonal models.

The algorithm combines Schroeder's decimation (its vertex importance evaluation) with edge contraction to simplify object models in a short time. We have introduced a hash function, which we use instead of expensive vertex sorting. Our algorithm has proved its high speed and simplicity.

Acknowledgements

The authors would like to thank all who contributed to this work, especially to colleagues, MSc. and PhD. students at the university of West Bohemia in Plzen who have stimulated this work. This paper benefits from several discussions with them.

We also benefited from Cyberware.com model gallery and also the large model repository located at Georgia Institute of Technology;
URL: http://www.cc.gatech.edu/projects/large_models.

This work was supported by The Ministry of Education of the Czech Republic: project MSM 235200002

References

1. W. Schroeder, J. Zarge, W. Lorensen. *Decimation of Triangle Meshes.* In SIGGRAPH 92 Conference Proceedings, pages 65-70, July 1992.
2. H. Hoppe, T. DeRose, T. Duchamp, J. McDonald, W. Stuetzle. *Mesh optimization.* In SIGGRAPH 93 Conference Proceedings, pages 19-26, 1993.
3. M. Garland, P. Heckbert. *Surface Simplification Using Quadric Error Metrics.* In SIGGRAPH 97 Conference Proceedings, 1997.
4. M. Garland. Multiresolution Modeling: *Survey & Future Opportunities.* In the SIGGRAPH 97 course notes, 1997.
5. D. Kirkpatrick. *Optimal Search in Planar Subdivisions.* SIAM J. Comp., pages 12:28-35, 1993.
6. B. Junger, J. Snoeyink. *Selecting Independent Vertices for Terrain Simplification.* In WSCG 98 Proceedings, Pilsen University of West Bohemia, pages 157-164, February 1998.
7. M. Franc, V. Skala. *Triangular Mesh Decimation in Parallel Environment.* In 3rd Eurographics Workshop on Parallel Graphics & Visualization Conference Proceedings, Universitad de Girona, September 2000.
8. W. Lorensen, H. Cline. *Marching Cubes: A High Resolution 3D Surface Construction Algorithm.* Computer Graphics (SIGGRAPH 87 Proceedings), 21(4):163-169, July 1987.
9. R. Klein, G. Liebich, W. Straser. *Mesh reduction with error control.* Proceedings of Visualization '96, 1996.
10. A. Ciampalini, P. Cignoni, C. Montani, R. Scopigno. *Multiresolution decimation based on global error.* Technical Report CNUCE: C96021, Istituto per l'Elaborazione dell'Informazione - Condsiglio Nazionale delle Richere, Pisa, ITALY, July 1996.
11. P. Lindstrom, G. Turk. *Fast and memory efficient polygonal simplification.* IEEE Visualization 98 Conference Proceedings, 1998.
12. A.W.F. Lee, W. Sweldens, P. Schroder, L. Cowsar. *MAPS: Multiresolution Adaptive Parameterization of Surfaces.* In SIGGRAPH '98 Proceedings. 1998.

Geometric Determination of the Spheres which Are Tangent to Four Given Ones

E. Roanes-Macías[1], E. Roanes-Lozano[1]

Universidad Complutense de Madrid, Dept. Algebra,
Edificio "La Almudena", c/ Rector Royo Villanova s/n, 28040-Madrid, Spain
roanes@mat.ucm.es ; eroanes@fi.upm.es

Abstract. Apollonius' problem (find the tangent circles to three given ones) has attracted many mathematicians and has been solved using different methods along more than 22 centuries. Nowadays computers allow to mechanize the solving process and to treat its generalization to higher dimension using algebraic methods. Starting from the classical Vieta-Steiner solution for dimension 2, we have developed a method valid for dimension n, that, thanks to the use of an original coding, allows to choose in advance the relative position of the solution sphere w.r.t. the given ones (i.e., if each tangency is exterior or interior). Moreover, the possible degeneracy of some of the solution (hyper-)spheres in (hyper-)planes and the existence of configurations with an infinity number of solutions are considered.

1 Introduction

More than 22 centuries ago Apollonius of Perga proposed in his book *Contacts* (Tangencies) the problem of constructing the tangent circles to three given ones.

The problem has interested many great mathematicians from antiquity up to date. The first known solution was found by F. Vieta [12] in 1600. Later, other solutions were found by Descartes, Newton, Euler, Steiner, Poncelet, Gergonne, Mannheim, Fouché... In 1898 Lemoine compared the simplicity and exactness of the different solutions found till then [4]. A discussion about the number of solutions of the problem was described by Hadamard [3].

The interest of the problem is not only aesthetic; it has also had practical applications. In 1936, Nobel award winner Soddy [11] particularized the problem to the study of the relation between the radius of a molecule and the radii of its atoms, finding a curious relation between them. This result was extended to dimension n by Pedoe [6] and Coxeter [2] in 1968.

Also with a chemical motivation in the background, the extension of the problem to 3D (substituting circles by spheres and ellipsoids) has been recently solved by performing an algebraic reduction that makes use of Dixon resultant and Groebner bases [5].

Similarly, Maple's *Geometry* package uses a pure algebraic method to solve the problem. But these approaches are neither constructive nor allow the user

P.M.A. Sloot et al. (Eds.): ICCS 2002, LNCS 2330, pp. 52–61, 2002.

to choose in advance the relative position of the solution circle w.r.t. the given ones (i.e., if it is externally or internally tangent to each given one).

A constructive solution of the 3D problem is given in this article. It uses a method, derived from Vieta-Steiner's, that allows to choose in advance the relative position of the solution sphere w.r.t. the four given ones. Also following this line, a constructive solution to the 2D case was developed by the authors some years ago [9, 10]

2   Solving the Problem in a Way that Allows to Choose the Solution in Advance

Algebraic methods solve the problem by stating a polynomial system whose unknowns are the center and radius of the solution sphere and which coefficients depend on the coordinates of the centers and the radii of the four given spheres. The system is solved using a method that can deal with non-linear polynomial systems, like those based on Groebner bases.

As said above, these methods do not allow to choose in advance the relative position of the solution sphere w.r.t. the four given ones (i.e., if it is externally or internally tangent to each given one).

To achieve this goal we have decided to apply a geometric method, before coding it properly, in order to directly obtain the chosen solution when using the implementation.

2.1   3D Extension of Vieta-Steiner's Method

Among the many existing geometric resolution methods, we have chosen the one that we consider fits best, according to its computational adaptability. It combines F. Vieta's reduction to a simpler problem, with the application of the inversion-based method due to J. Steiner.

So, Vieta-Steiner's solution of Apollonius' problem will be extended to 3D. The description of the 2D case can be found in any classic Euclidean Geometry book [1, 8, 7]. We shall introduce afterwards those concepts related to inversion that will make this computational approach possible.

In an Euclidean real space, the inversion of pole O and power k is the involutive transformation where two points, P and P', different from O, do correspond iff O, P, P' are collinear points such that $\overrightarrow{OP} \cdot \overrightarrow{OP'} = k$.

In this inversion, the sphere such that P and O are endpoints of a diameter, is the inverse of the perpendicular plane to OP through P' (where P' is the inverse of P).

In the same inversion, the inverse of a sphere that doesn't pass through the pole of inversion, O, and such that P and Q are endpoints of a diameter and are collinear with O, is the sphere such that points P' and Q' (inverse of P and Q, respectively) are endpoints of a diameter.

Let us state the problem precisely. Given four spheres S_0, S_1, S_2, S_3 of different centers C_0, C_1, C_2, C_3 and radii r_0, r_1, r_2, r_3 (respectively), another sphere,

S, tangent to the four given ones and which center and radius will be denoted C and r (respectively), is to be determined.

Let us suppose $r_0 = min\{r_0, r_1, r_2, r_3\}$. When subtracting or adding r_0 from the radii of the four spheres S_0, S_1, S_2, S_3, another four spheres are obtained: S_0', S_1', S_2', S_3', of centers C_0, C_1, C_2, C_3 and radii $r_0' = r_0 - r_0 = 0, r_1' = r_1 \pm r_0, r_2' = r_2 \pm r_0, r_3' = r_3 \pm r_0$ (whether adding or subtracting r_0 will depend on the inclusions between S and S_1, S_2, S_3, as will be detailed below). They will be tangent to sphere S', of center C and radius $r' = r \pm r_0$ (depending on whether S includes S_0 or not). This way the problem is reduced to determine the sphere S' through C_0 tangent to S_1', S_2', S_3'. We shall refer to this first part of the process as *Vieta's reduction*.

Applying now an inversion, I, of pole C_0, the spheres $S_1'' = I(S_1'), S_2'' = I(S_2')$ and $S_3'' = I(S_3')$ are obtained. If precisely the geometric power of C_0 w.r.t. S_1' is chosen as the power of I, then $S_1'' = S_1'$.

As S' passes through the pole of inversion C_0 and I is an angle-preserving transformation, figure $S'' = I(S')$ must be a plane, tangent to spheres S_1'', S_2'' and S_3'' (curiously, the same configuration of a plane and three spheres is used in the 3D proof of a 2D theorem due to Gaspard Monge).

Consequently, the image by I of a plane, S'', tangent to S_1'', S_2'' and S_3'', is a sphere or a plane (depending on whether $C_0 \notin S''$ or $C_0 \in S''$), S', that passes through C_0 and is tangent to S_1', S_2' and S_3'. Unapplying Vieta's reduction to S', a solution sphere or plane, tangent to S_0, S_1, S_2, S_3, is obtained.

Clearly, from each plane S'', tangent to S_1'', S_2'' and S_3'', a solution sphere or plane would be obtained. As, in the usual case, there are at most 8 of such planes, and S can include S_1 or not, the maximum number of solutions is 16 (S_0, S_1, S_2, S_3 are supposed to be exterior two by two).

There are special cases where an infinite number of solutions exist. This is the case, for instance, when there is a cylinder or a cone that is tangent to the four spheres simultaneously or when C_0, C_1, C_2, C_3 are vertices of a square and $r_0 = r_1 = r_2 = r_3$.

Up to here we have described a straightforward extension to 3D of Vieta-Steiner's method for solving the problem in the Euclidean plane. From here onwards our original contribution to solve the proposed problem begins. The main difficulty of the process lies on the appropriate selection of the tangent plane S'', in order to obtain precisely the desired solution for S, among the 16 possible ones. This will be achieved through an adequate coding.

2.2   Coding the Geometric Elements

A point P of coordinates (p_1, p_2, p_3) will be codified as the list of its coordinates: $[p_1, p_2, p_3]$.

The plane through point P and perpendicular to the unitary vector v, of coordinates (v_1, v_2, v_3), will be codified by the list $[P, v]$, where P and v are the sublists $[p_1, p_2, p_3]$ and $[v_1, v_2, v_3]$, i.e. by the list of lists $[[p_1, p_2, p_3], [v_1, v_2, v_3]]$.

The spherical surface of center P and radius r will be codified by list $[P, r]$, where P is the sublist $[p_1, p_2, p_3]$ and r is a positive real number. Therefore, the four spheres given in the problem will be introduced as a list $[S_0, S_1, S_2, S_3]$, where each S_i is the sublist $[C_i, r_i]$, of its center and radius.

In order to determine the relative position of each of the given spheres, S_i ; $i = 0, 1, 2, 3$, w.r.t the solution sphere, S, i.e., if they are included or not in S, an *inclusion code*, g_i, valued in $\{-1, 1\}$ will be used. We shall assign $g_i = -1$ or $g_i = 1$, depending on whether S_i is external or internal tangent to the solution sphere S (respectively). The relative position of the spheres in list $[S_0, S_1, S_2, S_3]$ w.r.t. the sphere solution, S, will be determined by a list of codes $[g_0, g_1, g_2, g_3]$.

2.3 Selecting the Tangent Plane According to the Chosen Solution

Once the four spheres S_0, S_1, S_2, S_3 are given, a right selection of the tangent plane S'' must be performed, in order to obtain precisely the desired solution for S, among all possible ones.

Let us denote by v a unitary vector, ortogonal to this plane (director vector) and by T_1, T_2, T_3 the intersection points of this plane with the spheres S_1'', S_2'', S_3'' (respectively).

As T_1, T_2, T_3 belong to the tangent plane S'', vectors $\overrightarrow{T_1 T_i}; i = 2, 3$ are perpendicular to vector v, and therefore their dot products by vector v are zero, so the following equalities will have to be verified.

$$v \cdot v = 1 \ , \ \overrightarrow{T_1 T_i} \cdot v = 0 \ ; \ i = 2, 3 \tag{1}$$

On the other hand, according to the properties of inversion, sphere S' includes $S_i' = I(S_i'')$ iff S_i'' is in a different half-space than the pole of inversion C_0 w.r.t. plane $S'' = I(S')$. Therefore, for each pair of spheres among C_1'', C_2'', C_3'', they must be in the same or in different half-spaces of border S_i'', depending on whether their inclusion codes have the same or different signs. Consequently, the centers, C_1'', C_2'', C_3'' of the respective spheres S_1'', S_2'', S_3'', will be in one or the other half-space of border the tangent plane S'', depending on the values of g_i ; $i = 1, 2, 3$. So, if r_1'', r_2'', r_3'' are the radii of the spheres S_1'', S_2'', S_3'' (respectively), their tangent points, T_1, T_2, T_3, can be expressed, initially, as

$$T_i = C_i'' + g_i r_i'' v \ ; \ i = 1, 2, 3 \tag{2}$$

Substituting now in (1) the values of T_1, T_2, T_3 given by (2), a polynomial system of degree 2 whose unknowns are the coordinates of vector v is obtained. It is straightforward that this system is equivalent to another one where all equations are linear except one of degree 2, yielding two possible solutions for vector v.

This is logical. Let us observe that, once the values of the g_i have been fixed, there are still two possible planes S'', tangent to C_1'', C_2'', C_3'', that keep these three spheres at the corresponding half-space of border S'' (in accordance with those values of g_i). Therefore there are two possible directions for vector v.

One of this two possible vectors v has to be chosen. This selection must be done according to whether sphere S_1' is included or not in S (what depends on whether g_1 is 1 or -1). Consequently, sphere $S_1'' = I(S_1')$ and the pole of inversion C_0 must be in different or the same half-space w.r.t. to the tangent plane S'', depending on whether g_1 is 1 or -1.

But S_1'' and C_0 are in the same or in different half-space w.r.t. the tangent plane S'', depending on whether the inner products $\overrightarrow{T_1 C_1''} \cdot v$ and $\overrightarrow{T_1 C_0} \cdot v$ are of different or of the same sign.

Therefore, among the two different vectors v, the one such that the inner products $\overrightarrow{T_1 C_1''} \cdot v$ and $\overrightarrow{T_1 C_0} \cdot v$ are of different or the same sign will be chosen, depending on whether g_1 is 1 or -1.

2.4   Extension to Dimension n

The whole process described above for dimension 3 is also valid for any dimension $n \geq 2$.

3   Algorithm

Input: $[S_0, S_1, S_2, S_3]$, $[g_0, g_1, g_2, g_3]$ (list of spheres and list of inclusion codes)

Output: S (solution sphere(s), solution plane(s) or *Without solution* string)

(1) $S_i' := [C_i, r_i'],\ r_i' = r_i - r_0$; $i = 1, 2, 3$ (Vieta's reduction)

(2) $S_i'' := I(S_i') = [C_i'', r_i'']$; $i = 1, 2, 3$ (I = inversion of center C_0)

(3) $T_i := C_i'' + g_i r_i'' v$; $i = 1, 2, 3$, where $v := [v_1, v_2, v_3]$

(4) $[v^*, v^{**}]$:=solutions of system $\{v \cdot v = 1\ ,\ \overrightarrow{T_1 T_i} \cdot v = 0\ ;\ i = 2, 3\}$

(5) IF sign$(\overrightarrow{T_1 C_1''} \cdot v^*) \neq$ sign$(g_1(\overrightarrow{T_1 C_0} \cdot v^*))$ THEN $w := v^*$, ELSE $w := v^{**}$

(6) IF w imaginary THEN RETURN *Without solution* string

(7) $T_1 := C_1'' + g_1 r_1'' w$ (tangent point of S_1'')

(8) $S'' := [T_1, w]$ (tangent plane)

(9) IF $C_0 \in S''$ THEN RETURN $S := [C_0, w]$ (solution plane)

(10) $S' := I(S'') = [C, r']$ (inverse of the tangent plane)

(11) $S := [C, r' + g_0 r_0]$ (solution sphere; similarly solution spheres or plane(s))

4   Implementation

The previous algorithm does not require symbolic calculations but exact arithmetic (to be sure that the right decision is taken in the conditionals and for checking solutions). Anyway, as a system that contains a non-linear equation is to be solved, it is advisable to use a Computer Algebra System (CAS). Among

them, we have chosen Maple, taking into account its comfort, diffusion, portability and calculation power.

The package we have developed works in any dimensión (for circles, spheres
or hyper-spheres, respectively). It has 26 procedures, of which we shall briefly
describe the main ones (those mentioned in the examples), omitting all subpro-
cedures where the steps of the algorithm above are implemented.

From here onwards and for the sake of brevity, we shall talk of *spheres* really
meaning circles or 3D-spheres or hyper-spheres, and we shall talk of *planes* really
meaning lines or planes or hyperplanes (according to the dimension of the space).

ApoSol(SphereList,CodeList) is the main procedure. Its first argument is
the list of spheres $[S_0, S_1, S_2...]$ and its second argument is the list $[g_0, g_1, g_2...]$
of inclusion codes of these spheres. It returns the solution sphere, S, as mentioned
above:

- list of center and radius, $[C, r]$, if the solution is a sphere

- list of point and vector, $[P, v]$, if the solution is a plane

- string *Without solution*, if there is no real solution

- list of center and radius, expressed as a function of parameters (to which
particular values should be given in order to obtain particular solutions), if there
were infinite solutions.

ApoComp(Solution,SphereList,CodeList) is the procedure that allows to
check if the solution obtained is correct and the right one. Its first argument is
the solution, S, previously obtained, its second argument is the list $[S_0, S_1, S_2...]$
of the given spheres and the third one is the list $[g_0, g_1, g_2...]$ of inclusion codes.
This procedure calculates the distances from the centers of the given spheres to
the center of the solution sphere or to the solution plane, and subtracts from it
the sum/difference of radii or distances, depending on the values of the g_i. If the
solution is correct, such results must be zero all of them. So, a list of zeros must
be obtained.

Equ(figure,vars) is a procedure that allows to obtain the equation of a figure
of those considered in section 2.2 (if they are introduced in the way explained
there). Using the terminology of 2.2, its first argument is of the form $[P, r]$ (if
the figure is a sphere) or of the form $[P, v]$ (if the figure is a plane). Its second
argument is the list of names of the coordinate axes $[x, y, z...]$ w.r.t. which the
equation of S is to be expressed (the output).

ApoDib(Solution,SphereList) allows to plot in Maple both the 2D and 3D
cases. Its first argument is the solution (sphere or plane) and the second one is
the list of given spheres. It returns the plot of the solution figure, S, together
with all the given spheres.

5   Gallery of Examples

In the following examples the code is written in Maple 7. The solutions are
obtained using *ApoSol*, and are allocated in variable *Sol*. They are checked af-

terwards using procedure *ApoComp*. They are represented in Maple 7 using procedure *ApoDib* or using DPGraph2000 (a package specialized in graphing implicit 3D functions; see http://www.davidparker.com/index.html for details). When an equation is needed it is obtained using *Equ.*

Example 1. Given three circles, each one exterior to each other, determine the circle that is externally tangent to the first two ones and internally tangent to the third one.

```
> S0:=[[2,-4],3]: S1:=[[-5,3],1]: S2:=[[2,5],2]:
> S:=[S0,S1,S2]; G:=[-1,-1,1]:
```

$$S := [[[2, -4], 3], [[-5, 3], 1], [[2, 5], 2]]$$

```
> Sol:=ApoSol(S,G);
```

$$Sol := [\,[\frac{-3126}{2455} + \frac{68}{2455}\sqrt{3619},\ \frac{-103}{491} + \frac{28}{491}\sqrt{3619}],\ \frac{-4364}{2455} + \frac{252}{2455}\sqrt{3619}]$$

```
> ApoComp(Sol,S,G);
```

$$[0,0,0]$$

```
> ApoDib(S,Sol);
```

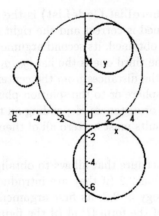

Fig. 1. Figure of Example 1

Example 2. Given four spheres, each one exterior to each other, determine the sphere externally tangent to them all.

```
> S0:=[[0,5,0],2]: S1:=[[0,-5,0],3]: S2:=[[5,0,0],2]:
> S3:=[[0,0,5],1]: S:=[S0,S1,S2,S3]; G:=[-1,-1,-1,-1]:
```

$$[[[0,5,0],2],[[0,-5,0],3],[[5,0,0],2],\ [[0,0,5],1]]$$

> Sol:=ApoSol(S,G);

$$Sol := [\ [\ \frac{1}{8048270}(4940 + 3\sqrt{220667})(-279 + 2\sqrt{220667}),\ \frac{1}{8048270}(4940+$$

$$3\sqrt{220667})(-279 + 2\sqrt{220667}),\ \frac{-98}{445} + \frac{3}{890}\sqrt{220667}\ ],\ \frac{-457}{178} + \frac{1}{89}\sqrt{220667}\ ]$$

> ApoComp(Sol,S,G);

$$[0, 0, 0, 0]$$

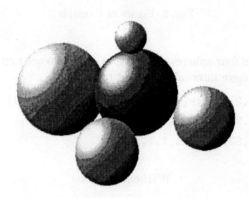

Fig. 2. Figure of Example 2

Example 3. Case where the solution sphere degenerates into a plane.

> S0:=[[0,0,4],3]: S1:=[[2,-7,-3],4]: S2:=[[2,11,-3],4]:
> S3:=[[-7,2,-3],4]: S:=[S0,S1,S2,S3]: G:=[-1,1,1,1]:
> Sol:=ApoSol(S,G);

$$Sol:=[\ [2,-7,1],\ [0,0,-1]]$$

> ApoComp(Sol,S,G);

$$[0, 0, 0, 0]$$

> Equ(S0,[x,y,z]), Equ(S1,[x,y,z]), Equ(S2,[x,y,z]),
Equ(S3,[x,y,z]);

$$x^2 + y^2 + z^2 - 8z + 7, x^2 + y^2 + z^2 - 4x + 14y + 6z + 46,$$

$$x^2 + y^2 + z^2 - 4x - 22y + 6z + 118, x^2 + y^2 + z^2 + 14x - 4y + 6z + 46$$

> Equ(Sol,[x,y,z]);

$$-z + 1$$

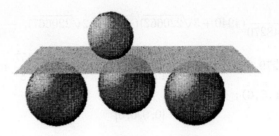

Fig. 3. Figure of Example 3

Example 4. Given four spheres, of which at least two are mutually external, try to determine a sphere internally tangent to them all.

```
> S0:=[[5,0,0],2]: S1:=[[-5,0,0],10]: S2:=[[0,5,0],2]:
> S2:=[[0,5,0],2]: S3:=[[0,0,0],2]:
> S:=[S0,S1,S2,S3]: G:=[1,1,1,1]:
> ApoSol(S,G);
```
Without solution

Example 5. Given four non-intersecting spheres of the same radii and which centers are the vertices of a square, determine the spheres externally tangent to them all.

```
> S0:=[[2,0,0],1]: S1:=[[-2,0,0],1]: S2:=[[0,2,0],1]:
> S2:=[[0,2,0],1]: S3:=[[0,-2,0],1]:
> S:=[S0,S1,S2,S3]: G:=[1,1,1,1]:
> Sol:=ApoSol(S,G);
```

$$Sol := [\ [\ 0,0,2\frac{t\sqrt{1-t^2}}{-1+t^2}\ ], 2\sqrt{-\frac{1}{-1+t^2}-1}\ ]$$

(They are spheres which center lies on line $x = 0 = y$ and which radii are ≥ 1)
```
> ApoComp(Sol,S,C);
```
$$[0,0,0,0]$$

Example 6. Given five hyper-spheres in 4D, determine a hyper-sphere tangent to them all.

```
> S0:=[[0,5,0,0],2]: S1:=[[0,0,0,0],1]: S2:=[[0,0,5,0],2]:
> S3:=[[-5,0,0,0],1]: S4:=[[0,0,0,5],1]:
> S:=[S0,S1,S2,S3,S4]: G:=[1,-1,1,-1,-1]:
```

```
> Sol:=ApoSol(S,G);
```

$$Sol := [\ [\ \frac{-5}{2}, \frac{1}{12334}(375 + 8\sqrt{862})(96 + 5\sqrt{862}), \frac{1}{12334}(375 + 8\sqrt{862})$$

$$(96 + 5\sqrt{862}), 5/2\ ], \frac{41}{7} + \frac{5}{14}\sqrt{862}\ ]$$

```
> ApoComp(Sol,S,G);
```

$$[0, 0, 0, 0, 0]$$

6   Conclusions

Mixing new and old techniques has made possible to generalize Apollonius' problem to higher dimension. The approach maintains Vieta-Steiner's solution's elegance. The original coding of each solution (hyper-)sphere is another advantage of this approach. Both the resolution method and the implemented program are valid for dimension n (hyper-spheres in dimension n tangent to n+1 given ones). Moreover, the possible degeneracy of some of the solution (hyper-)spheres and the possibility of an infinity of solutions are considered. Therefore we think this is one more step forward in the study of Apollonius' problem.

Acknowledgements

This work is partially supported by project TIC2000-1368-C03-01 (Ministry of Science and Technology, Spain).

References

1. Berger M.: *Geometry I*, Springer-Verlag, Berlin-Heidelberg, 1987.
2. Coxeter H.S.M.: The Problem of Apollonius, *Am. Math. Monthly*, Vol. **75** (1968) 5-15.
3. Hadamard J.: *Lecons de Géométrie Elementaire*, A. Colin, Paris, 1947-49.
4. Lemoine E.: Application de d'une méthode d'évaluation de la simplicité des constructions a la comparaison de quelques solutions du probléme d'Apollonius, *Nouvelles Ann. Math.* (1892) 453-474.
5. Lewis R. H.: Apollonius Meets Computer Algebra. In: *Proceedings of ACA'2001*, http://math.unm.edu/ACA/2001/Proceedings/NonStd/
6. Pedoe D.: On a theorem in Geometry, *Am. Math. Monthly*, Vol. **74** (1967) 627-640.
7. Pedoe D.: *Geometry*, Dover Pub., New York, 1988.
8. Ogilvy C.S.: *Excursions in Geometry*, Dover Pub., New York, 1990.
9. Roanes Lozano E.: El Problema de Apolonio, *Bol. Soc. Puig Adam*, Vol. **14** (1987) 13-41.
10. Roanes Macías E., Roanes Lozano E.: *Nuevas tecnologías en Geometría*, Editorial Complutense, Madrid, 1994.
11. Soddy F.: The Kiss Precise, *Nature*, **137** (1936) 1021.
12. Vieta F.: *Varia Responsa. IX: Apollonius Gallus*, Real Academia de Ciencias, Madrid, not dated edition (Reprint of the original dated 1600).

Metamorphosis of Non-homeomorphic Objects

Mehdi Elkouhen and Dominique Bechmann

Université Louis Pasteur – LSIIT, Pôle API –
Boulevard Sébastien Brant, 67400 Strasbourg France
elkouhen,bechmann@dpt-info.u-strasbg.fr

Abstract. In this article, we present our axial four dimensional deformation tool. This tool is defined in the context of geometrical modeling of animations, where animations are represented by four dimension polyhedrons, and it permits to control the shape/topology of animations.

We illustrate this tool by deforming animations representing a motionless object into the merging of two similar objects. And we show how controlling the shape of the tool enables to control the path the objects follow during the merging but also the smoothness of the merging.

1 Introduction

A metamorphosis is the process of transforming continuously a source object into a target object, while keeping the main features of both objects. The solutions are generally classified in volume-based and boundary-representation based approaches [17]. In a simplified way, the volume-based approach of objects metamorphosis can be stated as a "blend" of objects. In the boundary-based approach, the process is harder : it requires to construct a supermesh that collapses on the source and target object. So in this approach, the metamorphosis is the continuous deformation of the supermesh from the first collapsed shape to the second.

The algorithms of construction of object metamorphosis are various. They are guided by the user who can set correspondence points [18], skeletons[22, 8] or features [9]. The algorithms take in account the geometry of the objects : they deal differently for star-shaped objects [13, 14], cylinder like objects [16] or objects Homeomorphic to a disk [12]. The shape of the inbetween objects can be controlled by influence shapes[19, 21]. Except in [7, 23], the boundary-representation based techniques deal only with homeomorphic objects or at least homeomorphic to a sphere while the volume-based approach techniques deal with arbitrary objects. But none of these techniques permit to control precisely the topology of the inbetween objects. For example, how can a designer make new features (e.g. holes) appear only on the inbetween objects? In techniques that use influence shapes, one can use a toroidal influence shape but this strongly modifies the shape of the inbetween objects by giving them the shape of a torus. Geometrical modeling of animation (GMA) permits to overcome this. The main

P.M.A. Sloot et al. (Eds.): ICCS 2002, LNCS 2330, pp. 62–71, 2002.

reason is that the animations are modeled as four dimensional polyhedrons and many operations of control of shape/topology can be adapted to the fourth dimension.

In GMA, animations are visualized as polyhedrons [11]. The movies are extracted by displaying cuts of the object by a set of planes. Different algorithms of construction of animations have been studied. In [4], Brandel constructs the animations by extruding surfaces following an arbitrary polyline. When the extrusion is linear and the set of cut planes are chosen normal to the line, the animations represent the initial surface translating along the extrusion line. In [5], the animations are constructed by thickening 3D graphs so that the animations represent a sequence of merging and scission of spheres. In [20], Skapin discusses of the creation of animations by computing the Cartesian product of two objects. This technique involves that users must have a strong intuition on the way to choose the objects. As the topology of a cut depends not only on the topology of the 4D polyhedron but also on its shape, deformation tools have been defined. Aubert and Brandel [1, 4] defined animations with topological changes using Dogme [3] the N-dimensional deformation tool . To increase the control on the shape of the animations, we defined a set of tools [2] exhibiting different kinds of control.

The main goal of our work is the creation of metamorphosis of non homeomorphic objects; we deal with this problem by deforming 4D polyhedrons. In this article, we explain how to deform an animation defined by a motionless object in the merging of two similar surfaces. For this, we use our axial deformation tool, and we show how a simple control on the shape of the tool permits to control the path of the merging and its smoothness. We illustrate this technique by explaining the different steps of construction of the merging of two tori. In section 2, we describe our axial deformation tool. In section 3, we give the different steps of construction of the animation. In section 4, we explain how to create sharp mergings.

2   Deformation Tool

This tool is an adaptation, of Chang and Rockwoods [6] axial deformation tool, to a four dimensional tool. It is defined by a control polygon P_i and a set of handles. The polygon defines a Bézier curve, which is the axis of deformation. The handles are introduced for a four dimensional control of the deformation. As shown in Fig. 1(a), there are three handles S_i, T_i and L_i (represented by yellow, red and blue vectors) associated to each control point P_i except the last one.

In section 2.1, we describe the techniques to create and manipulate such a tool and in Sect. 2.2 we give the algorithm of deformation.

2.1   Creation and Manipulation

A designer can create a tool in two ways. The first one consists in specifying a linear control polygon and three directions A, B and C : the tool created is linear (Fig. 1(a)) and all the handles S (respectively T and L) share the same direction A (resp. B and C). The user can also create a tool by simply inputting a Bézier curve (Fig. 1(b)); the handles are computed automatically by the algorithm described in Sect. 3.4.

When manipulating an axial deformation tool, one manipulates three kinds of parameters : Position of the control points, Length of the handles, Orientation of the handles. To show the link between the modification of each of these parameters and the shape of the deformations, three examples are presented. In order to stay as clear as possible, the objects considered are cubes (embedded in the xyz space) and only the handles S and T are displayed (as the time component of the vertices of the object are null, the handles L are not taken into account by the deformation algorithm 2.2).

The Position of the control points : In Figure 2(d), the designer moved the control points defining the initial linear axis in order to define a bent axis : the cube is thus warped along the bent axis (Fig. 2(a)).

The Length of the handles : In Figure 2(e), the designer created a linear tool and then pulled the handles : the cube is tapered (Fig. 2(b)).

The Orientation of the handles : In Figure 2(f), the designer created a linear tool and then turned the handles around the axis of deformation : the cube is twisted (Fig. 2(c)).

2.2   Algorithm

The algorithm is a simple extension of the one given by Chang and Rockwood[6], obtained by taking into account the new handle L and the time component of the input vertices. We have modified the original algorithm in order to make the axis of deformation represent the evolution of time instead of representing the x axis and this by interpolating in the time direction instead of interpolating in the x direction.

The algorithm takes in input a point M, of coordinates (x, y, z, t), and interpolates the result of affine transformations applied to the point; the result $P_0^n(M)$ is the deformed point.

$$
\begin{cases}
P_i^0(M) = P_i & 0 \leq i \leq n \\
P_i^1(M) = \Theta[P_{i-1}^0(M), P_i^0(M), S_i, T_i, L_i](M) & 0 \leq i \leq n-1 \\
P_i^j(M) = \Theta[P_{i-1}^{j-1}(M), P_i^{j-1}(M), \mathbf{0}, \mathbf{0}, \mathbf{0}](M) & \begin{cases} 1 \leq j \leq n \\ 0 \leq i \leq n-j \end{cases}
\end{cases}
$$

With Θ defined as following :

$$\Theta[O, T, \boldsymbol{i}, \boldsymbol{j}, \boldsymbol{k}] \begin{pmatrix} x \\ y \\ z \\ t \end{pmatrix} = O + tOT + x\boldsymbol{i} + y\boldsymbol{j} + z\boldsymbol{k}$$

In this section we adapted Chang and Rockwoods mathematical formalism. We will now show how to instantiate this tool and to define merging of polyhedrons in a smooth way.

3   Construction of the Smooth Merging

Intuitively, the steps to construct the merging of two similar shapes consists in extruding an initial shape, bending the resulting object and then extracting the animation. The bending step enables "symmetric" sections of the initial object to be defined in the same plane.

For example, if we wish to merge two circles, we extrude a circle (the result is a cylinder); and then we bend the cylinder 3(a). Figures 3(b)–3(d), represents a set of cuts of the cylinder.

In the following, we show how to merge two tori. The reason we define the merging of two similar objects comes from the way we construct our four-dimensional polyhedrons by extruding 3D shapes : the extrusion operator naturally associates to each vertex of one object a vertex of the other object. The techniques we present in next section are not dependant on the way we construct our objects, so they could be applied on more general objects [5, 20, 21].

3.1   Construction of the Four Dimensional Polyhedron

We construct our four dimensional polyhedron by extruding a torus, embedded in the xyz space, following the time direction. The points having their time component null are called "bottom" points, and symmetrically the points having their time component equal to 1 are called "top" points.

3.2   Sketching the Curve

The designer sketches a Bézier curve, which represents the path the objects follow during the merging. For the current animation, the designer sketched a curve having the shape of an "S". This curve (Fig. 4(a)) defines the axis of our deformation tool. In the two next subsections, we will explain how to compute the final parameters of the tool.

3.3   Send the Inner Control Points to the Future

The two endpoints of the tool represent the positions of the two objects at the beginning of the animation (time=0). The inner part of the curve represents the motion path of the two objects, so we have to modify the position of the control points by sending them to the future.

The strategy we followed in this example, for a curve defined by $n+1$ control points P_i, consists in sending the points of index $i < n/2$ to the time i and sending the points of index $i > n/2$ to the time $n - i$.

The strategy can be chosen freely; each strategy generates a different kind of merging. It is by choosing another strategy that we show, in Sect. 4, how to control the smoothness of the merging.

3.4   Compute the Handles

The most important stage in the definition of the animation is to compute the handles correctly. Otherwise the 4D polyhedron self-intersects or is flattened; in these cases the animations are not appealing. The solution consists in computing the handles using the rotation minimizing frame [15]. This algorithm ensures the change of direction between two successive handles of same kind is minimal. In this way, we eliminate sudden changes of direction of handles and so we avoid self intersecting deformations. In this subsection, we present a four dimensional version of the way we compute our handles.

The first control point and the last one represent the beginning of the animation (time=0). In order to make the "top" and "bottom" points of the constructed object share the same time, the time component of the handles, associated to the point P_0 and P_{n-1}, have to be null. This is a consequence of the fact that the result of the deformation of a point of coordinates $(x, y, z, 0)$ is $P_0 + xS_0 + yT_0 + zT_0$ and the result for a point of coordinates $(x, y, z, 1)$ is $P_n + xS_{n-1} + yT_{n-1} + zT_{n-1}$.

The first three handles S_0, T_0 and L_0 are set by the user in order to form an orthonormal set of vectors and that the time component of these vectors is null. In this way, the four vectors $P_0 P_1$, S_0, T_0 and L_0 form a orthogonal basis of $\mathbb{R}^4$.

The other handles are computed in an iterative way. The handle H_{i+1} is computed by turning the handle H_i. It is the rotation defined in the plane spanned by the two vectors $P_{i-1} P_i$ and $P_i P_{i+1}$; the angle of rotation is the angle between the vectors $P_{i-1} P_i$ and $P_i P_{i+1}$. For a precise presentation of four dimensional rotations, one can refer to Hanson's article on N-dimensional rotations [10].

As we compute iteratively the new set of vectors $P_i P_{i+1}$, S_i, T_i and L_i by turning the previous set, these sets always form an orthonormal basis of $\mathbb{R}^4$.

As the three vectors S_{n-1}, T_{n-1} and L_{n-1} are normal to the time vector $P_{n-1}P_n$, they are space vectors. This is the condition we needed, so the algorithm is valid.

3.5   Apply the Deformation and Extract the Animation

Figures 4(b)–4(e) represent different cuts of this object, by a set of hyperplanes normal to time.

An advantage of controlling the shape of the metamorphosis by a deformation tool, is that we can control the continuity of the mergings with visual parameters. This will be the subject of next section where we will discuss of another strategy for the step 3.3, and of the continuity between axial tools.

4   Controlling the Smoothness of the Merging

The animations extracted from the curves constructed using the strategy described in 3.3, represent two points that merge softly. In order to make the points merge sharply, one can, for example, use another strategy that flattens the curve and extract the animation using a set of planes parallel to the flat area.

For example, we defined a tool defined by two continuously connected Bézier curves. In order to flatten the axis around the connection point, we set the control points, neighboring the connection point, to be on a same line.

We deform our extruded torus, using the two tools defined by the two previous curves. Figures 4(f)–4(i) represent different steps of the sharp merging of the tori.

As the definition of the continuity between two axial 3D De Casteljau Generalized deformation tools was discussed in [6], we only precise, in next paragraph, the way we deal with two continuously linked tools and the computing of the handles over the global curve.

Let the first tool be defined by $n + 1$ control points P_i^0 and the handles S_i^0, T_i^0 and L_i^0, and the second tool be defined by $m + 1$ control points P_i^1 and the handles S_i^1, T_i^1 and L_i^1. The points P_n^0 and P_0^1 are merged in order to make the three control points P_{n-1}^0, P_n^0 and P_0^1 collinear. The user specifies the handles for the control point P_0^0; the rotation minimizing frame algorithm computes the other handles of the tool. For continuity reasons, the first handles associated to the second tool have to be set to be equal to their "equivalent" handles of index $n - 1$ of the first tool. Finally, the other handles of the tool are computed automatically.

5   Conclusion

In this article, we presented our four dimensional extension of Chang and Rockwood's axial deformation tool and discussed the way it can be used to control sharp/smooth merging of shapes. This work will be continued by a study of the merging of N different objects and the control of the areas of contact during the mergings. The first point should require to create an algorithm, of construction of 4D polyhedrons, that brings together the strength of the Cartesian product [20] and the interpolation of objects as proposed by Turk and O'Brien[21]. The second point should require to use deformation tools of higher topological dimension such as tools defined by surfaces [2] etc.

References

[1] F. Aubert and D. Bechmann. Animation by deformation of space-time objects. *Eurographics'97*, 16(3):57–66, 1997.

[2] D. Bechmann and M. Elkouhen. Animating with the "multidimensional deformation tool". *Eurographics Workshop on Computer Animation and Simulation'01*, pages 29–35, 2001.

[3] P. Borrel and D. Bechmann. Deformation of n-dimensional objects. *International Journal of Computational Geometry & Applications*, 1(4):427–453, 1991.

[4] S. Brandel, D. Bechmann, and Y. Bertrand. STIGMA: A 4-dimensional modeller for animation. *Eurographics Workshop on Computer animation and simulation 1998*, 1998.

[5] S. Brandel, D. Bechmann, and Y. Bertrand. Thickening: an operation for animation. *The Journal of Visualization and Computer Animation*, 11(5):261–277, 2000.

[6] Y. Chang and A. P. Rockwood. A generalized de casteljau approach to 3d free-form deformation. *Proceedings of SIGGRAPH 94*, pages 257–260, 1994.

[7] D. DeCarlo and J. Gallier. Topological evolution of surfaces. In *Graphics Interface '96*, pages 194–203, 1996.

[8] E. Galin and S. Akkouche. Blob metamorphosis based on minkowski sums. *Eurographics'96*, 15(3):C143–C154, 1996.

[9] A. Greogory, A. State, M. Lin, D. Manocha, and M. Livingston. Feature-based surface decomposition for correspondence and morphing between polyhedra. *Proceedings of computer animation'98*, pages 64–71, 1998.

[10] A J. Hanson. Rotations for n-dimensional graphics. *Graphics Gems V*, pages 55–64, 1995.

[11] S. R. Hollasch. Four-space visualization of 4d objects. Master's thesis, Arizona state university, 1991.

[12] T. Kanai, H. Suzuki, and F. Kimura. 3d geometric metamorphosis based on harmonic maps. *Proceedings of Pacific Graphics'97*, pages 97–104, 1997.

[13] James Kent, Richard Parent, and Wayne E. Carlson. Establishing correspondences by topological merging: A new approach to 3-D shape transformation. *Proceedings of Graphics Interface '91*, pages 271–278, 1991.

[14] James R. Kent, Wayne E. Carlson, and Richard E. Parent. Shape transformation for polyhedral objects. *Computer Graphics*, 26(2):47–54, 1992.

[15] F. Klok. Two moving coordinate frames for sweeping along a 3d trajectory. *Computer Aided Geometry Design*, 3:217–229, 1986.

[16] F. Lazarus and A. Verroust. Metamorphosis of cylinder-like objects. *The Journal of Visualization and Computer Animation*, 8(3):131–146, 1997.

[17] F. Lazarus and A. Verroust. Three-dimensional metamorphosis: A survey. *The Visual Computer*, 14:373–389, 1998.

[18] R. E. Parent. Shape transformation by boundary representation interpolation: a recursive approach to establishing face correspondences. *The Journal of Visualization and Computer Animation*, 3(4):219–239, 1992.

[19] J. Rossignac and A. Kaul. AGRELs and BIPs: Metamorphosis as a bezier curve in the space of polyhedra. *Eurographics'94*, pages C179–C184, 1995.

[20] X. Skapin and P. Lienhardt. Using cartesian product for animation. *Eurographics Workshop on Computer Animation and Simulation 2000*, pages 187–201, August 2000.

[21] G. Turk and J. O'Brien. Shape transformation using variational implicit functions. *Proceedings of SIGGRAPH 99*, pages 335–342, 1999.

[22] B. Wyvill, J. Bloomenthal, T. Beier, J. Blinn, A. Rockwood, and G. Wyvill. Modeling and animating with implicit surfaces. *Siggraph course notes*, 23, 1990.

[23] M. Zöckler, D. Stalling, and H-C. Hege. Fast and intuitive generation of geometric shape transitions. *The Visual Computer*, 16(5):241–253, 2000.

(a) Linear tool   (b) Bent tool

Fig. 1. Axial tools

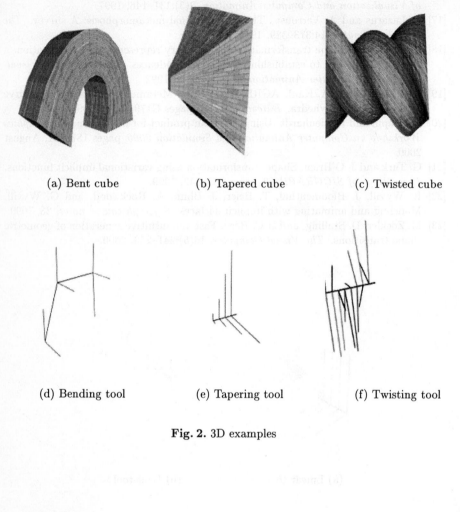

(a) Bent cube   (b) Tapered cube   (c) Twisted cube

(d) Bending tool   (e) Tapering tool   (f) Twisting tool

Fig. 2. 3D examples

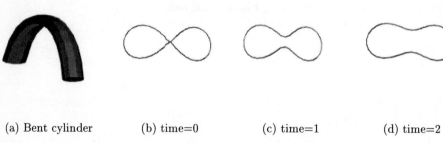

(a) Bent cylinder   (b) time=0   (c) time=1   (d) time=2

Fig. 3. Merging of 2 circles

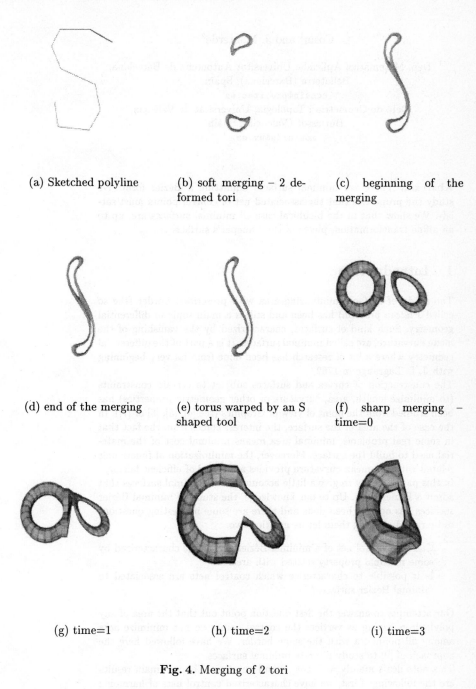

(a) Sketched polyline

(b) soft merging – 2 deformed tori

(c) beginning of the merging

(d) end of the merging

(e) torus warped by an S shaped tool

(f) sharp merging – time=0

(g) time=1

(h) time=2

(i) time=3

Fig. 4. Merging of 2 tori

Bézier Surfaces of Minimal Area

C. Cosín[1] and J. Monterde[2]

[1] Dep. Matemàtica Aplicada. Universitat Autònoma de Barcelona,
Bellaterra (Barcelona), Spain,
ccosin@pie.xtec.es
[2] Dep. de Geometria i Topologia, Universitat de València,
Burjassot (València), Spain
monterde@uv.es

Abstract. There are minimal surfaces admitting a Bézier form. We study the properties that the associated net of control points must satisfy. We show that in the bicubical case all minimal surfaces are, up to an affine transformation, pieces of the Enneper's surface.

1 Introduction

The study of surfaces minimizing area with prescribed border (the so called *Plateau problem*) has been and still is a main topic in differential geometry. Such kind of surfaces, characterized by the vanishing of the mean curvature, are called minimal surfaces. It is a part of the differential geometry where a lot of research has been done from its very beginning with J. L. Lagrange in 1762.

The construction of curves and surfaces subject to certain constraints (to minimize length, area, curvature or other geometric properties) has been studied from the point of view of Graphics (see [4], [5], [6] or [7]). In the case of the area of the surface, the interest comes from the fact that in some real problems, minimal area means minimal cost of the material used to build the surface. Moreover, the minimization of functionals related with the mean curvature provides a method of efficient fairing.

In this paper we try to give a little account of the minimal surfaces that admit a Bézier form. Up to our knowledge, the study of minimal Bézier surfaces has not yet been done and there are some interesting questions to be raised. Among them let us mention two:

- Can the control net of a minimal Bézier surface be characterized by some minimal property related with areas?
- Is it possible to characterize which control nets are associated to minimal Bézier surfaces?

Our attempts to answer the first question point out that the area of any polyhedron having as vertices the control points do not minimize area among all polihedra with the same border. We have followed here the approach of [9] to study discrete minimal surfaces.

This note deals mainly with the second question. The two main results are the following: First, we have characterized control nets of harmonic

P.M.A. Sloot et al. (Eds.): ICCS 2002, LNCS 2330, pp. 72–81, 2002.

Bézier surfaces, and second, we have proved that any bicubical polynomial minimal surface is, up to an affine transformation, a piece of a well known minimal surface: the Enneper's surface.

The consequence of our results is that minimal surfaces are too rigid to be useful as candidates for blendings between arbitrary surfaces. Only for some configurations of the border control points we can assure that a Bézier surface exists with minimal area.

The connection between the two topics, Bézier and minimal surfaces, is not new. Let us recall some of them. First, Sergei Bernstein, who defined the now called Bernstein polynomials, was a prolific researcher in the realm of minimal surfaces at the beginning of the twentieth century. See for instance [1] and [2]. One of its most celebrated results was to prove that if a minimal surface is the graph of a differentiable function defined on the whole $\mathbb{R}^2$, i.e, $\vec{x}(u,v) = (u, v, f(u,v))$, then it is a plane.

Second, the solutions to some Plateau problems, for example, the Gergonne surface, resemble Bézier surfaces. (Look at Figure I)

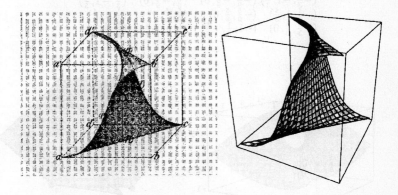

Figure I. Left. The Schwarz's solution (1865) to the Gergonne's problem (1816): find the minimal surface dividing the cube into two equal parts and joining the inverse diagonals of two opposed faces. Right. A Bézier surface with a similar shape.

And third, both kind of surfaces share some crucial properties: A Bézier surface is always included in the convex hull of its net of control points. Analogously, a minimal surface is always included in the convex hull of its border.

2   Definitions

Definition 1. *Given a net of control points in* $\mathbb{R}^3$, $\{P_{ij}\}_{i,j=0}^{n,m}$, *the associated Bézier surface,* $\vec{x} : [0,1] \times [0,1] \to \mathbb{R}^3$, *is given by*

$$\vec{x}(u,v) = \sum_{i=0}^{n} \sum_{j=0}^{m} B_i^n(u) B_j^m(v) P_{ij}. \tag{1}$$

Definition 2. *A surface S is minimal if its mean curvature vanishes.*

Equivalently, S is a minimal surface iff for each point $p \in S$ one can choose a small neighbourhood, U_p which has minimal area among other patches V having the same boundary as U.

Example 1. The first non trivial example of minimal surface with polynomial coordinate functions is the Enneper's surface (Figure II): $\vec{x}$: $\mathbb{R}^2 \to \mathbb{R}^3$ defined by

$$\vec{x}(u,v) := (u - \frac{u^3}{3} + uv^2, v - \frac{v^3}{3} + vu^2, u^2 - v^2).$$

The control net for the portion of the Enneper's surface defined by $u, v \in [-1, 1]$, is given by

$$
\begin{array}{llll}
(-\frac{5}{3}, -\frac{5}{3}, 0) & (-1, -\frac{1}{3}, -\frac{4}{3}) & (1, -\frac{1}{3}, -\frac{4}{3}) & (\frac{5}{3}, -\frac{5}{3}, 0) \\
(-\frac{1}{3}, -1, \frac{4}{3}) & (-\frac{5}{9}, -\frac{5}{9}, 0) & (\frac{5}{9}, -\frac{5}{9}, 0) & (\frac{1}{3}, -1, \frac{4}{3}) \\
(-\frac{1}{3}, 1, \frac{4}{3}) & (-\frac{5}{9}, \frac{5}{9}, 0) & (\frac{5}{9}, \frac{5}{9}, 0) & (\frac{1}{3}, 1, \frac{4}{3}) \\
(-\frac{5}{3}, \frac{5}{3}, 0) & (-1, \frac{1}{3}, -\frac{4}{3}) & (1, \frac{1}{3}, -\frac{4}{3}) & (\frac{5}{3}, \frac{5}{3}, 0)
\end{array}
$$

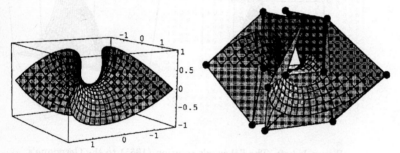

Figure II. Left: a piece of the Enneper's surface ($u, v \in [-1, 1]$). Right: Its control net as a Bézier surface.

3   Minimal surfaces with isothermal coordinates

Let us recall that a chart $\vec{x} : U \to S$ on a surface, S, is said to be isothermal the map $\vec{x}$ is a conformal map, i.e, if angles between curves in the surface are equal to the angles between the corresponding curves in the coordinate open subset U. It is easy to check that for an isothermal chart the coefficients, E, F, G, of the first fundamental form satisfy $E = G$ and $F = 0$.

Note that this implies that the two families of coordinate curves of the chat $\vec{x}$ are orthogonal because $F = 0$, and that the length of the coordinate curve from $\vec{x}(u_0, v_0)$ to $\vec{x}(u_0, v_0 + h)$ is equal to the length of the coordinate curve from $\vec{x}(u_0, v_0)$ to $\vec{x}(u_0 + h, v_0)$.

A well known result of the theory of minimal surfaces is the following (see [10]): if $\vec{x}$ is an isothermal map then $\vec{x}$ is minimal iff $\Delta \vec{x} = 0$, where Δ is the usual Laplacian operator. The relation between the mean curvature and the chart is due to the fact that for an isothermal map

$$\vec{x}_{uu} + \vec{x}_{vv} = 2\lambda^2 HN,$$

where $\lambda = E = G$ and N is the unitary normal vector of the surface associated to the chart.

The conditions that a net of control points must satisfy in order to have an isothermal associated Bézier surface are more difficult to handle (they can be expressed as a system of quadratic equations) than the conditions in order to be harmonic (in this case, the equations are linear). So, let us study first that second condition.

We will compute the Laplacian of a Bézier surface (1).

$$\Delta \vec{x}(u,v) = \left(\tfrac{\partial^2}{\partial u^2} + \tfrac{\partial^2}{\partial v^2} \right) \vec{x}(u,v)$$

$$= n(n-1) \sum_{i=0}^{n-2} \sum_{j=0}^{m} B_i^{n-2}(u) B_j^m(v) \Delta^{2,0} P_{ij}$$

$$+ m(m-1) \sum_{i=0}^{n} \sum_{j=0}^{m-2} B_i^n(u) B_j^{m-2}(v) \Delta^{0,2} P_{ij},$$

where $\Delta^{2,0} P_{ij} = P_{i+2,j} - 2P_{i+1,j} + P_{ij}$, $\Delta^{0,2} P_{ij} = P_{i,j+2} - 2P_{i,j+1} + P_{ij}$.

We shall rewrite the last expression again as a Bézier surface of degrees n and m. In order to do this, we will need the following relation

$$B_i^{n-2}(t) = \tfrac{1}{n(n-1)} \left((n-i)(n-i-1) B_i^n(t) \right.$$

$$\left. + 2(i+1)(n-i-1) B_{i+1}^n(t) + (i+1)(i+2) B_{i+2}^n(t) \right).$$

Let us define, for $i \in \{0, \ldots, n-2\}$

$$a_{in} = (n-i)(n-i-1), \quad b_{in} = 2(i+1)(n-i-1), \quad c_{in} = (i+1)(i+2),$$

and $a_{in} = b_{in} = c_{in} = 0$ otherwise.

Therefore,

$$\Delta \vec{x}(u,v) = \sum_{i=0}^{n} \sum_{j=0}^{m} B_i^n(u) B_j^m(v)$$

$$\left(n(n-1)(a_{in} \Delta^{2,0} P_{ij} + b_{i-1,n} \Delta^{2,0} P_{i-1,j} + c_{i-2,n} \Delta^{2,0} P_{i-2,j}) \right.$$

$$\left. + m(m-1)(a_{jm} \Delta^{0,2} P_{ij} + b_{j-1,m} \Delta^{0,2} P_{i,j-1} + c_{j-2,m} \Delta^{0,2} P_{i,j-2}) \right).$$

This expression can be seen as the Bézier surface associated to a net of control points $\{Q_{ij}\}_{i,j=0}^{n,m}$. Thus, due to the fact that $\{B_i^n(u) B_j^m(v)\}_{i,j=0}^{nm}$ is a basis of polynomials, we get that $\vec{x}$ is harmonic iff $Q_{ij} = 0$ for all i,j.

Substituting the discrete operators $\Delta^{2,0}$ and $\Delta^{0,2}$ by its definitions and sorting terms we get that for any i,j the following expression vanish:

$$n(n-1)(P_{i+2,j} a_{in} + P_{i+1,j}(b_{i-1,n} - 2a_{in})$$

$$+ P_{i-1,j}(b_{i-1,n} - 2c_{i-2,n}) + P_{i-2,j} c_{i-2,n})$$

$$+ m(m-1)(P_{i,j+2} a_{jm} + P_{i,j+1}(b_{j-1,m} - 2a_{jm})$$

$$+ P_{i,j-1}(b_{j-1,m} - 2c_{j-2,m}) + P_{i,j-2} c_{j-2,m})$$

$$+ P_{ij}((a_{in} - 2b_{i-1,n} + c_{i-2,n}) n(n-1)$$

$$+ (a_{jm} - 2b_{j-1,m} + c_{j-2,m}) m(m-1)).$$

In the case of a quadratic net ($n = m$) we can state the following theorem

Theorem 1. *Given a quadratic net of points in* $\mathbb{R}^3$, $\{P_{ij}\}_{i,j=0}^{n}$, *the associated Bézier surface,* $\vec{x} : [0,1] \times [0,1] \to \mathbb{R}^3$, *is harmonic, i.e,* $\Delta \vec{x} = 0$ *iff*

$$
\begin{aligned}
0 = & P_{i+2,j} a_{in} + P_{i+1,j}(b_{i-1,n} - 2a_{in}) + P_{i-1,j}(b_{i-1,n} - 2c_{i-2,n}) \\
& + P_{i-2,j} c_{i-2,n} + P_{i,j+2} a_{jm} + P_{i,j+1}(b_{j-1,m} - 2a_{jm}) \\
& + P_{i,j-1}(b_{j-1,m} - 2c_{j-2,m}) + P_{i,j-2} c_{j-2,m} \\
& + P_{ij}(a_{in} - 2b_{i-1,n} + c_{i-2,n} + a_{jm} - 2b_{j-1,m} + c_{j-2,m}).
\end{aligned} \tag{2}
$$

Let us study Equation (2) in the simplest cases: biquadratic and bicubical Bézier patches.

3.1 Biquadratic harmonic Bézier patches

In the case $n = m = 2$ from the equations in (2) it is possible to find an expression of four of the control points in terms of the other five. In fact, using Mathematica, we have obtained that the null space of the coefficient matrix of (2) is of dimension four. Moreover, it is possible to choose as free variables points in the first and last column of the control net.

Corollary 1. *A biquadratic Bézier surface is harmonic iff*

$$
\begin{aligned}
P_{01} &= \tfrac{1}{2}(2P_{00} + P_{02} - 2P_{10} + P_{20}), \\
P_{11} &= \tfrac{1}{4}(P_{00} + P_{02} + P_{20} + P_{22}), \\
P_{21} &= \tfrac{1}{2}(P_{00} + -2P_{10} + 2P_{20} + P_{22}), \\
P_{12} &= \tfrac{1}{2}(-P_{00} + P_{02} + 2P_{10} - P_{20} + P_{22}).
\end{aligned} \tag{3}
$$

A way of writing for example the equation involving the inner control point, P_{11}, is using a mask

$$
P_{11} = \frac{1}{4} \times \begin{matrix} 1 & 0 & 1 \\ 0 & \bullet & 0 \\ 1 & 0 & 1 \end{matrix} \tag{4}
$$

Remark 1. In [6], the author presents a method to improve an initial blending, F_0, through a sequence of blending surfaces minimizing some fairing functionals. In section 3.3, the author suggests the following modification: instead of using the initial blending surface, to use a modified surface obtained by averaging the inner control points. The averaging method suggested there, after an analysis of its implementation, is given precisely by the mask (4). Therefore, the use of this mask can be now justified from Equations (3): the inner point of a quadratic harmonic Bézier surface must verify such a mask.

Remark 2. Note that mask (4) is a kind of dual of the mask associated to the Laplace operator. It can be found in [4] that the mask

$$
\frac{1}{4} \times \begin{matrix} 0 & 1 & 0 \\ 1 & \bullet & 1 \\ 0 & 1 & 0 \end{matrix} \tag{5}
$$

is the discrete form of the Laplacian operator. Such a mask is used in the cited reference to obtain control nets resembling minimal surfaces that fit between given boundary polygons.

In general, the authors define in [4] the notion of permanence patches to be those generated by masks of the form

$$\frac{1}{4} \times \begin{matrix} \alpha & \beta & \alpha \\ \beta & \bullet & \beta \\ \alpha & \beta & \alpha \end{matrix} \tag{6}$$

with $4\alpha + 4\beta = 1$. Therefore, mask (4) is a particular case with $\alpha = 0.25$, whereas mask (5) corresponds to $\alpha = 0$. Anyway, as it is said there, any of such masks do not produce control nets of minimal surfaces.

In fact, let us recall that we are not trying to produce Coons nets. We try to characterize control nets of minimal surfaces. We have found that in the biquadratic case Eqs. (3) must be satisfied. But in order to obtain a minimal patch, we have to impose also the isothermal conditions. It is just a matter of computation to show that any control net verifying Eqs. (3) and the isothermal conditions is a piece of a plane.

3.2 Bicubical harmonic Bézier patches

In the case $n = m = 3$ from the equations in (2) it is possible to put half of the control points in terms of the other eight. In fact, using Mathematica, we have obtained that the null space of the coefficient matrix of (2) is of dimension eight. Moreover, it is possible to choose as free variables exactly the eight points in the first and last column of the control net.

Corollary 2. *A bicubic Bézier surface is harmonic iff*

$$P_{11} = \tfrac{1}{9}(4P_{00} + 2P_{03} + 2P_{30} + P_{33}),$$
$$P_{21} = \tfrac{1}{9}(2P_{00} + P_{03} + 4P_{30} + 2P_{33}),$$
$$P_{12} = \tfrac{1}{9}(2P_{00} + 4P_{03} + P_{30} + 2P_{33}),$$
$$P_{22} = \tfrac{1}{9}(P_{00} + 2P_{03} + 2P_{30} + 4P_{33}),$$
$$P_{10} = \tfrac{1}{3}(4P_{00} - 4P_{01} + 2P_{02} + 2P_{30} - 2P_{31} + P_{32}), \tag{7}$$
$$P_{20} = \tfrac{1}{3}(2P_{00} - 2P_{01} + P_{02} + 4P_{30} - 4P_{31} + 2P_{32}),$$
$$P_{13} = \tfrac{1}{3}(2P_{01} - 4P_{02} + 4P_{03} + P_{31} - 2P_{32} + 2P_{33}),$$
$$P_{23} = \tfrac{1}{3}(P_{01} - 2P_{02} + 2P_{03} + 2P_{31} - 4P_{32} + 4P_{33}).$$

Remark 3. This means that given the first and last columns of the control net (eight control points in total), the other eight control points are fully determined by the harmonic condition. In other words, any pair of two opposed borders of a harmonic Bézier surface determines the rest of control points.

Remark 4. This fact is analogous to what happens in the Gergonne surface: given two border lines, the inverse diagonals of two opposed faces of a cube, the Gergonne surface is fully determined (See Figure I). In our case, given two opposed lines of border control points, the harmonic Bézier surface is fully determined.

Remark 5. The first four equations in (7) imply that the four inner control points are fully determined by the four corner points. So, there are two different kind of masks depending if the point is an inner control point or not:

$$P_{11} = \frac{1}{9} \times \begin{matrix} 4 & \bullet & \bullet & 2 \\ 0 & \bullet & \bullet & 0 \\ 0 & \bullet & \bullet & 0 \\ 2 & \bullet & \bullet & 1 \end{matrix}, \qquad P_{10} = \frac{1}{3} \times \begin{matrix} 4 & \bullet & \bullet & 2 \\ -4 & \bullet & \bullet & -2 \\ 2 & \bullet & \bullet & 1 \\ 0 & \bullet & \bullet & 0 \end{matrix}. \qquad (8)$$

The other points have similar masks.

Remark 6. Let us insist in the fact that harmonic chart need not to be minimal. We have obtained the conditions to be harmonic. In order to be minimal, more conditions are needed. Let us split the control points of a Bézier surface into two subsets: the inner points $\{P_{ij}\}_{i,j=1}^{n-1,m-1}$ and the border points. It is not true that given an arbitrary set of border points, there is a unique set of inner control points such that the Bézier surface associated to the whole control net is of minimal area. What we can say is that given just a few border control points, the rest of control points are determined. In the next section we will find which bicubical Bézier surfaces are minimal.

4   Bicubical minimal Bézier patches

We have seen before that the unique biquadratic minimal Bézier patch is the plane. In the cubical case we know that at least there is another minimal Bézier surface, the Enneper's surface. What we want to determine is if this is the only possibility. In order to do that we need to change the methods to attack the problem. The second great moment in the theory of minimal surfaces was the introduction of the methods of complex variable. Let us recall here the main results.

Let $\vec{x}(u, v)$ be an isothermal minimal chart and let us define

$$(\phi_1, \phi_2, \phi_3) = \frac{\partial \vec{x}(u, v)}{\partial u} - i \frac{\partial \vec{x}(u, v)}{\partial v}.$$

The functions (ϕ_1, ϕ_2, ϕ_3) verify

$$\phi_1^2 + \phi_2^2 + \phi_3^2 = 0. \qquad (9)$$

Lemma 1. *([10] Lemma 8.1) Let D be a domain in the complex z-plane, g(z) an arbitrary meromorphic function in D and f(z) an analytic function in D having the property that at each point where g(z) has a pole of order m, f(z) has a zero of order at least 2m. Then the functions*

$$\phi_1 = \frac{1}{2} f(1 - g^2), \quad \phi_2 = -\frac{i}{2} f(1 + g^2), \quad \phi_3 = fg \qquad (10)$$

will be analitic in D and satisfy (9). Conversely, every triple of analytic functions in D satisfying (9) may be represented in the form (10), except for $\phi_1 = i\phi_2, \phi_3 = 0$.

Note that both functions can be computed by

$$f = \phi_1 + i\phi_2, \qquad g = \frac{\phi_3}{\phi_1 + i\phi_2}. \qquad (11)$$

Note also that Equations (10) are not exactly those of Lemma 8.1 in [10]. There is an slight difference in the sign of ϕ_2. Anyway the statement is equivalent.

Lemma 2. *([10] Lemma 8.2) Every simple-connected minimal surface in $\mathbb{R}^3$ can be represented in the form*

$$\vec{x}(u,v) = (Re \int_0^z \phi_1(z)dz, Re \int_0^z \phi_2(z)dz, Re \int_0^z \phi_3(z)dz) + P_0, \quad (12)$$

where the ϕ_k are defined by (10), the functions f and g having the properties stated in Lemma 1, the domain D being either the unit disk or the entire plane, and the integral being taken along an arbitrary path from the origin to the point $z = u + iv$.

So, a minimal surface is determined by the pair of complex functions f and g. For example, the most obvious choice: $f(z) = 1, g(z) = z$, leads to the Enneper's surface.
We are going to consider now the following problem: to determine all bicubical polynomial minimal surfaces.
The number of possible choices of the two functions f and g in such a way that the chart given by (12) is a polynomial of degree 3 is not reduced just to $f(z) =$ constant and g a degree 1 polynomial in z. Another possibility is $f(z) = (p(z))^2$ and, $g(z) = \frac{q(z)}{p(z)}$, where $p(z), q(z)$ are polynomials of degree 1. Therefore, the problem we are interested in is not so obvious.

Theorem 2. *Any bicubical polynomial minimal chart, $\vec{x} : U \to \mathbb{R}^3$, is, up to an affine transformation of $\mathbb{R}^3$, an affine reparametrization of the general Enneper's surface of degree 3, i.e, there are $H_3 \in Aff(\mathbb{R}^3)$ and $H_2 \in Aff(\mathbb{R}^2)$ such that*

$$\vec{x}(u,v) = H_3(\vec{x}^{Enneper}(H_2(u,v))),$$

for any $(u,v) \in U$.

Proof. We can suppose that the chart is isothermal. On the contrary, a well known result of the theory of minimal surfaces states that the chart is a reparametrization of an isothermal chart.
All along the proof we will consider polynomial patches not in the Bernstein polynomial basis, but in the usual polynomial basis.
Let us consider a bicubical, polynomial, isothermal and harmonic chart

$$\vec{x}(u,v) = (\sum_{i,j=0}^{3} a_{ij}u^i v^j, \sum_{i,j=0}^{3} b_{ij}u^i v^j, \sum_{i,j=0}^{3} c_{ij}u^i v^j),$$

where $a_{ij}, b_{ij}, c_{ij} \in \mathbb{R}$.
Let us denote by $\vec{v_{ij}}$ the vector (a_{ij}, b_{ij}, c_{ij}).

Note first that thanks to a translation, we can suppose that $\overrightarrow{v_{00}} = 0$.
As the chart is orthogonal ($F = 0$), then, by inspection on the higher
degree terms, it is possible to deduce the following relations

$$\overrightarrow{v_{33}} = \overrightarrow{v_{32}} = \overrightarrow{v_{23}} = \overrightarrow{v_{31}} = \overrightarrow{v_{13}} = \overrightarrow{v_{22}} = 0.$$

For example, the coefficient of $u^5 v^5$ in F is the norm of $3\overrightarrow{v_{33}}$, therefore if
$F = 0$ then $\overrightarrow{v_{33}} = 0$. To deduce the other relations proceed analogously
with the coefficients of $u^5 v^3$, $u^3 v^5$, $u^5 v^1$, $u^1 v^5$ and $u^3 v^3$ in that order.
Now, $\Delta \overrightarrow{x} = 0$ iff

$$\overrightarrow{v_{20}} = -\overrightarrow{v_{02}}, \qquad \overrightarrow{v_{21}} = -3\overrightarrow{v_{03}}, \qquad \overrightarrow{v_{12}} = -3\overrightarrow{v_{30}}.$$

As the chart is isothermal, from the coefficients of v^4 in $E = G$ and
$F = 0$, respectively, we obtain

$$\|\overrightarrow{v_{30}}\| = \|\overrightarrow{v_{03}}\|, \qquad < \overrightarrow{v_{30}}, \overrightarrow{v_{03}} > = 0.$$

At this point the deduction splits into two cases:

Case A. $\overrightarrow{v_{30}} = 0$

In this case, and after some computations, the chart is a piece of the plain
$z = 0$. But the plane can be parametrized using a polynomial chart of
degree 1, so, it cannot be considered as a proper solution of the problem.

Case B. $\overrightarrow{v_{30}} \neq 0$

In this case, thanks to a rotation and an uniform scaling, we can suppose
that

$$\overrightarrow{v_{30}} = (1, 0, 0), \qquad \overrightarrow{v_{03}} = (0, 1, 0).$$

Therefore, from the coefficient of v in $F = 0$ and $E = G$, we obtain

$$a_{11} = -2b_{02}, \qquad b_{11} = 2a_{02}.$$

It can be proved that the isothermal conditions can be now reduced to
just four equations involving the coordinates of the vectors $\overrightarrow{v_{ij}}$. Nevertheless, it is easier at this point of the proof to introduce the use of complex
numbers.
Let us consider

$$(\phi_1, \phi_2, \phi_3) = \frac{\partial \overrightarrow{x}}{\partial u} - i \frac{\partial \overrightarrow{x}}{\partial v},$$

where

$$\phi_1 = a_{10} - 2a_{02}u + 3u^2 - 2b_{02}v - 3v^2 - i(a_{01} - 2b_{02}u + 2(a_{02} - 3u)v),$$
$$\phi_2 = b_{10} - 2b_{02}u + (2a_{02} - 6u)v - i(b_{01} + 2a_{02}u - 3u^2 + 2b_{02}v + 3v^2),$$
$$\phi_3 = c_{10} - 2c_{02}u + c_{11}v - i(c_{01} + c_{11}u + 2c_{02}v).$$

$$(13)$$

The chart $\overrightarrow{x}$ is isothermal iff Equation (9) is verified.
Now, we can compute the pair of complex functions, $f^{\overrightarrow{x}}, g^{\overrightarrow{x}}$, according
to (11).
In our case, we obtain that $f^{\overrightarrow{x}}$ is a constant function and that $g^{\overrightarrow{x}}$ is a
polynomial in z of degree 1. Indeed,

$$f^{\overrightarrow{x}}(z) = a_{10} + b_{01} + i(-a_{01} + b_{10}),$$

$$g^{\overrightarrow{x}}(z) = \frac{c_{10} - ic_{01} + -(2c_{02} + ic_{11})z}{a_{10} + b_{01} + i(-a_{01} + b_{10})} = mz + n$$

Let us denote the coefficient m by $\rho e^{it} \in \mathbb{C}$, where $\rho > 0$ and $t \in [0, 2\pi[$. Now, let us consider the chart

$$\vec{y}(u,v) = \vec{x}\left(\frac{1}{\rho}(\cos(t)u + \sin(t)v), \frac{1}{\rho}(-\sin(t)u + \cos(t)v)\right).$$

Note that $\vec{y}$ is also an isothermal chart. It is easy to check that the pair of complex functions, $f^{\vec{y}}, g^{\vec{y}}$ associated to $\vec{y}$ are now

$$f^{\vec{y}}(z) = a \in \mathbb{C}, \qquad g^{\vec{y}}(z) = z.$$

Let us also denote a in the form $\mu e^{is} \in \mathbb{C}$, where $\mu > 0$ and $s \in [0, 2\pi[$, and let T be the linear transformation of $\mathbb{R}^3$ defined as the composition of the uniform scaling with factor $\frac{1}{\mu}$ and the spatial rotation with respect to the z-axis and angle $-s$. A well known property of the Enneper surface says that the minimal surface defined by $f(z) = a, g(z) = z$ is the image by T of the Enneper surface. $\qquad\square$

5 Acknowledgements

We would thank J. Nuo Ballesteros for a careful reading of the manuscript and to the referees for their valuable comments and suggestions.

References

1. S. Bernstein, *Sur les surfaces définies au moyen de leur courboure moyenne ou totale*, Ann. Sci. l'Ecole Norm. Sup., 27, 233-256 (1910).

2. S. Bernstein, *Sur les equations du calcul des variations*, Ann. Sci. l'Ecole Norm. Sup., 29, 431-482 (1912).

3. G. Farin, *Curves and surfaces for computer aided geometric design. A practical guide*, Morgan Kaufmann, 5th. ed. (2001).

4. G. Farin, D. Hansford *Discrete Coons patches*, CAGD 16, 691–700 (1999).

5. G. Greiner, *Blending Surfaces with Minimal Curvature*, in W. Straßer and F. Wahl (eds.), Proc. Dagstuhl Workshop Graphics and Robotics, pp. 163-174, Springer, 1994

6. G. Greiner, *Variational Design and Fairing of Spline Surfaces*, in Computer Graphics Forum (Proc. Eurographics '94), 143-154, 1994.

7. H. P. Moreton, C. H. Séquin, *Functional Minimization for Fair Surface Design*, preprint, (2001).

8. J. C. C. Nitsche, *Lectures on minimal surfaces*, vol. 1, Cambridge Univ. Press, Cambridge (1989).

9. K. Polthier, W. Rossman, *Discrete constant mean curvature and their index*, preprint (2001).

10. R. Osserman, *A survey of minimal surfaces*, Dover Publ., (1986).

Transformation of a Dynamic B-spline Curve Into Piecewise Power Basis Representation

Joonghyun Ryu[1], Youngsong Cho[1] and Deok-Soo Kim[1]

[1] Department of Industrial Engineering, Hanyang University
17 Haengdang-Dong, Seongdong-Ku, Seoul, 133-791, South Korea
dskim@hanyang.ac.kr

Abstract. In the fields of computer aided geometric design and computer graphics, B-spline curves and surfaces are often adopted as a geometric modelling tool and their evaluation is frequently required for a various geometric processing. In this paper, we present a new algorithm to convert B-spline curves into piecewise polynomials in power form. The proposed algorithm considers recursive B-spline basis function to be a pile of linear functions and collects all the necessary linear functions in each knot span. Then, algorithm computes the power form representation of B-spline basis functions and the required transformation of B-spline curve is obtained through the linear combination of B-splines in power form and the corresponding control points in each knot span.

1. Introduction

In the applications of computer graphics and computer aided geometric design, shapes are often modeled in terms of freeform curves and surfaces represented in B-spline form and their evaluation is frequently required for a various geometric processing. A B-spline curve of degree p with (m + 1) knots is defined by

$$\mathbf{C}(t) = \sum_{i=0}^{m-p-1} \mathbf{P}_i N_{i,p}(t) \text{ for } 0 \le t \le 1 \tag{1}$$

P.M.A. Sloot et al. (Eds.): ICCS 2002, LNCS 2330, pp. 82–91, 2002.

where $\mathbf{P}_i$ and $N_{i,p}(t)$ are control points and B-spline basis functions of degree p on a knot vector $\mathbf{U} = \{0,...,0, \mathrm{t}_{p+1},..., \mathrm{t}_{m-p-1}, 1,...1\}$, $\mathrm{t}_i \leq \mathrm{t}_{i+1}$, respectively [1]. In general, $N_{i,p}(t)$ is defined as the following recurrence formula in Equation (1)

$$N_{i,p}(t) = \frac{t - \mathrm{t}_i}{\mathrm{t}_{i+p} - \mathrm{t}_i} N_{i,p-1}(t) + \frac{\mathrm{t}_{i+p+1} - t}{\mathrm{t}_{i+p+1} - \mathrm{t}_{i+1}} N_{i+1,p-1}(t) \qquad (2)$$

$$N_{i,0}(t) = \begin{cases} 1 & if \quad \mathrm{t}_i \leq t < \mathrm{t}_{i+1} \\ 0 & otherwise \end{cases}$$

where $i = 0, 1, ..., m\text{-}p\text{-}1$. Equation (2) shows that the evaluation of $C(t)$ requires inevitably a recursive function evaluation which makes the evaluation slower. Even though a faster implementation in a non-recursive form may exist for periodic curve [2][3], a further reduction of the computation time is desirable especially when the B-spline curve changes its shape continuously by moving some of the control points.

Once a curve or surface is represented in power form, a point evaluation can be made faster due to Horner's rule even though the issue of numerical stability remains [4]. In this paper, we propose a faster algorithm for the evaluation of a B-spline curve based on the conversion of curve into a piecewise polynomial in a power form. It is also known that faster computation of the characteristic points on a curve, such as inflection points and cusps, can be facilitated by the conversion of a B-spline curve into a set of piecewise polynomial curves in power form. Note that the subdivision of a parametric curve at these characteristic points facilitates the fast computation of intersection points between curves [5]. In addition, IGES supports a free-form curve as a piecewise polynomial in power form with an entity type 112 [6]. Due to the relative advantages of implicit representation of curves or surfaces over parametric one in some geometric calculations such as a point inclusion problem, it is sometimes necessary that a parametric form be converted to an implicit form [7]. The implicitization process, which uses a resultant, usually requires the curve to be represented in power form [8]. Since this operation is computationally demanding, the reduction of computation should not be ignored.

Discussed in this paper is the transformation of B-spline curve into a set of piecewise polynomials in power form, which is known to be a tedious task [9]. Especially, the focus of the paper is made on dynamic curves in the sense that one or more of the control points of the curves are moving. On the other hand, a curve with fixed control points is called static.

Since a static B-spline curve can be converted into a set of piecewise Bezier curves by a knot refinement [1][10][11][12][13], applying a basis conversion operation to each piece will produce a set of piecewise polynomials in power form. This approach is called the KR-approach in this paper. The required transformation can also be obtained through applying Taylor expansion of the B-spline at each knot span with the appropriate number of terms depending on the degree of the curve [14], which is denoted by the TE-approach in this paper.

The main idea of the proposed algorithm, called a direct expansion (DE) algorithm, is as follows: after collecting all linear terms that make up the basis functions in a knot span, the algorithm directly obtains the power form representation of basis functions in the knot span by expanding the summation of products of appropriate linear terms. Then, the polynomial curves in power form in the knot span can easily be obtained by multiplying the basis functions in power form with corresponding control points. Repeating this operation for each knot span, a B-spline curve are transformed into a set of piecewise polynomials in power form. Experiments show that the proposed DE algorithm significantly outperforms the existing approaches for the case of dynamic curves. Hence, the proposed algorithm can be very useful for the curve implicitization as well as the computation of intersections when the curves are dynamically changing.

2. Direct Expansion of A Static Curve

Definition 1. A truncated basis function, $\tau_{i,w}(t)$, $i=w-p,w-p+1,..,w$, $w=p,p+1,..,m-p-1$, be an active polynomial segment among $N_{i,p}$'s, in $\left[t_w, t_{w+1}\right)$.

Figure 1a shows cubic B-splines with $m = 11$, and Figure 1b illustrates the corresponding truncated basis functions. Note that there are $p+1$ truncated basis functions in $\left[t_i, t_{i+1}\right)$ for a B-spline curve of degree p. If we collect all the truncated basis functions with same i value, then we can obtain a well-known B-spline basis function. For example, there exist four truncated basis functions $\tau_{0,3}(t)$, $\tau_{1,3}(t)$, $\tau_{2,3}(t)$ and $\tau_{3,3}(t)$ in $\left[t_3, t_4\right)$ and $N_{3,3}$ consists of four truncated basis functions, $\tau_{3,3}(t)$, $\tau_{3,4}(t)$, $\tau_{3,5}(t)$ and $\tau_{3,6}(t)$. Thus, B-spline basis function $N_{i,p}$ can be represented by the following equation.

$$N_{i,p}(t) = \sum_{w=i}^{i+p} \tau_{i,w}(t) N_{w,0}(t) \tag{3}$$

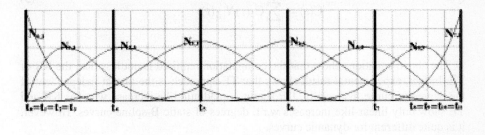

Fig. 1. Cubic basis splines and truncated basis functions with five knot spans of nonzero length

The truncated basis function in power form can be computed summation of products among appropriate linear terms, which can be obtained by Equation (2) and the enumeration of 0-1 sequences [15]. Once all of the truncated basis functions are computed, the B-spline curve in the knot span can easily be transformed into a polynomial curve in power form by the summation of the multiplications between appropriate control points and truncated basis functions. Thus, the power form polynomial curve for $[t_i, t_{i+1})$ is given as

$$C_i(t) = \sum_{j=i-p}^{i} \tau_{j,i}(t)P_j \qquad (4)$$

where $\tau_{j,i}(t)$ is a truncated basis function and P_j is the corresponding control point and each truncated basis function is already of power form. Thus, if the previous operation is performed for each knot span, then a static B-spline curve can be transformed into a set of piecewise polynomial curve in power form which can be formulated as follows.

$$C(t) = \sum_{i=p}^{m-p-1} C_i(t)N_{i,0}(t) \tag{5}$$

The above equations show that the computation required is not less than the KR or TE. In fact, experimental result shows that TE-approach is the fastest independent of the degree of curve. Especially, DE shows a quadratic-like increase whereas KR and TE show only linear-like increases w.r.t. degrees of static B-spline curves. However, it is quite different for dynamic curves.

3. Direct Expansion of A Dynamic Curve

Definition 2. A dynamic B-spline curve, $C_d(t)$, is a B-spline curve with more than one control point moving. Thus, $C_d(t)$ can be represented by the following equation.

$$C_d(t) = \sum_{i \in I} N_{i,p}(t)P_i + \sum_{j \in J} N_{j,p}(t)\tilde{P}_j \tag{6}$$

where I and J are the index sets of fixed control points, P_i and moving control points, $\tilde{P}_j$, respectively.

A naive approach to transform $C_d(t)$ to piecewise polynomials in power form is to recalculate the curve segment in every knot span whenever some control points are moving. This method is obviously unsatisfactory since it wastes computing time for the knot spans with unchanged curve shape. Let $C(t)$ be a B-spline curve before any control point moves and $C_d(t)$ be a dynamic curve counterpart of $C(t)$. Then, $C_d(t)$ can be now rewritten as the following equation using difference vectors, starting at old control points and ending at new control points.

$$C_d(t) = \sum_{k \in K} N_{k,p}(t)P_k + \sum_{j \in J} N_{j,p}(t)D_j \tag{7}$$

Where $K \equiv I \bigcup J$. That is, $P_k, k \in K$, is all control points of $C(t)$, and $D_j = (\tilde{P}_j - P_j)$ corresponds to the displacement of the moving control point. Thus, Equation (7) means that $C_d(t)$ can be obtained by the summation of original curve $C(t)$ and difference vectors multiplied by the corresponding basis functions. It is required to detect knot spans that are affected by the second term of Equation (7) so that the transformation can be done more efficiently. In the case of KR-approach, a knot refinement and a basis conversion are performed for all knot spans of curve segments whose shapes are changed by moving control points. Similarly, TE-approach can recalculate the coefficients of polynomial curves for the knot spans of curve segments whose shapes are changed. The derivative information and factorial evaluation are needed for each coefficient of the polynomial.

However, the computational behavior of DE-algorithm is quite different. Regardless of whether control points are moving or not, the truncated basis functions are fixed. It turns out that the computational gain of DE algorithm for a dynamic curve is more significant than that of others if that a static curve, $C(t)$, is provided as Equation (5) through DE algorithm, as a pre-processing tool for a dynamic curve.

On the other hand, $C_d(t)$ can also be divided into two groups: the first group is the set of curve segments whose shapes are fixed, and the second is the set of curve segments whose shapes are changed by moving control points. Hence the following equation holds.

$$C_d(t) = \sum_{m \in M} C_m(t) N_{m,0}(t) + \sum_{n \in N} \tilde{C}_n(t) N_{n,0}(t) \tag{8}$$

where M and N are index sets for knot spans of curve segments whose shapes are fixed and changed by moving control points, respectively. In addition, $\tilde{C}_n(t)$ can be again rewritten by using truncated basis functions as follows since $\tilde{C}_n(t)$ may also have both fixed and moving control points.

$$\tilde{C}_n(t) = \sum_{q \in Q} \tau_{q,n}(t) P_q + \sum_{r \in R} \tau_{r,n}(t) \tilde{P}_r \tag{9}$$

where Q and R are index sets for fixed and moving control points for, $\tilde{C}_n(t)$ respectively. Therefore,

$$\tilde{C}_n(t) = \sum_{s \in S} \tau_{s,n}(t) P_s + \sum_{r \in R} \tau_{r,n}(t)(\tilde{P}_r - P_r) \tag{10}$$

where $S \equiv Q \cup R$ and $|S| = p+1$. $P_s, s \in S$, is all the control points of $C_n(t)$ before they move. Thus, Equation (10) can be rewritten as Equation (11) using difference vector and truncated basis function.

$$\tilde{\mathbf{C}}_n(t) = \mathbf{C}_n(t) + \sum_{r \in R} \tau_{r,n}(t)\mathbf{D}_r \qquad (11)$$

where $\mathbf{D}_r = \tilde{\mathbf{P}}_r - \mathbf{P}_r$ is a difference vector whose value is the displacement of the moving control point. Thus, for a particular knot span, a changed curve segment in power form, $\tilde{\mathbf{C}}_n(t)$, can be obtained by summing the original polynomial $\mathbf{C}_n(t)$ in the form of Equation (4) and difference vector multiplied by the corresponding truncated basis function. Performing the operation in Equation (11) for the knot spans that are influenced by the moving control points completes the desired transformation for $\mathbf{C}_d(t)$.

4. Experiments for Dynamic Curves

Transformation of a dynamic curve into a set of piecewise polynomial curves in power form through DE algorithm consists of two steps: i) pre-processing, and ii) the operation of Equation (11) for all the knot spans with changed curve segments. While TE and KR-approach gets much more computational burdens, DE takes only $(p + 1)$ multiplications and $(p + 1)$ additions for a knot span. The computation time for each approach is provided in Figure 2 where pre-processing time is not considered. DE algorithm outperforms the other approaches and the computational gain of DE algorithm gets significant as the degree of curve and the number of control points increase. Since the relative time portion of DE algorithm is negligible, the trend of computation time of DE algorithm is provided separately in Figure 3. In each degree of curve, it seems that the computation time increases in linear pattern.

Although the implementation details may affect to the experimental results, we believe that our implementation considers the possible minimum operations for the KR and TE-approach.

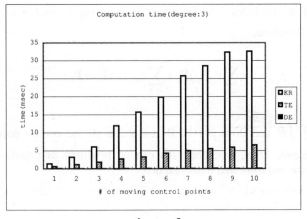

a degree: 3

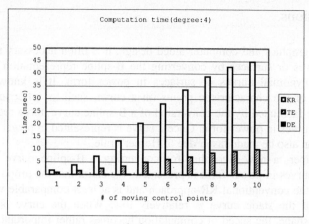

b degree: 4

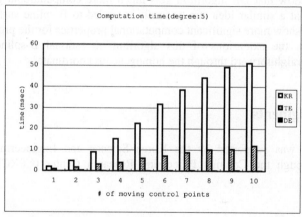

c degree: 5

Fig. 2. Computation time vs. the number of moved control points

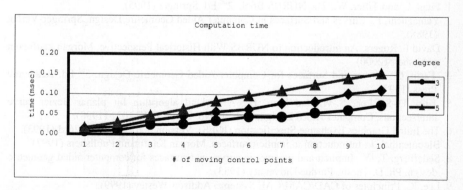

Fig. 3. Computation time vs. the number of moving control points for degree 3,4 and 5

5. Conclusions

In computer graphics and computer aided design, it is often necessary to manipulate B-spline curves or surfaces by converting the B-spline representation into a set of piecewise polynomial curves or surfaces in power form. It is known that faster computation of the characteristic points on a curve, such as inflection points and cusps, can be facilitated by the conversion of a B-spline curve into a set of piecewise polynomial curves in power form. Once a curve is represented in power form, a point evaluation can also be made faster due to Horner's rule.

In this paper, a new algorithm for converting a B-spline curve to piecewise polynomial curves in power form is presented. We claim that the proposed algorithm outperforms the conventional KR-approach and is at least comparable with TE when the degree of the static curve is relatively low. When the curve is dynamically changing its shape, the speed of computation becomes rather important. In this case, experiments show that DE algorithm gets much more computational gain. It is our expectation that a similar idea can be easily extended to B-spline surfaces, and our approach will show more significant computational properties for the problem.

In addition, the extensions of this algorithm to rational B-spline curves and surfaces are straightforward through the homogeneous coordinate.

Acknowledgments

This work was supported by the Korea Science and Engineering Foundation (KOSEF) through the Ceramic Processing Research Center (CPRC) at Hanyang University.

References

1. Piegl, L. and Tiller, W.: The NURBS Book. 2nd Ed. Springer (1995).
2. Yamaguchi, F.: Curves and Surfaces in Computer Aided Geometirc Design. Springer-Verlag (1988).
3. David F. Rogers: An introduction to NURBS:With Historical Perspective. Morgan kaufmann publishers (2000).
4. Farin, G.: Curves and Surfaces for Computer-Aided Geometric Design. 3rd Ed. Academic Press (1997).
5. Kim, D.-S., Lee, S.-W. and Shin, H.: A cocktail algorithm for planar Bezier curve intersections, Computer Aided Design, Vol. 30, No. 13, pp.1047-1051 (1998).
6. The Initial Graphics Exchange Specification (IGES). Version 5.2, ANSI Y14.26M (1993).
7. Bloomenthal, J.: Introduction to implicit surfaces. Morgan Kaufmann Publishers (1997).
8. Sederberg, T. W.: Implicit and parametric curves and surfaces for computer aided geometric design, Ph. D. Thesis. Purdue University (1983).
9. Lee, K.: Principles of CAD/CAM/CAE Systems. Addison-Wesley (1999).
10. Boehm, W. and Prautzsch, H.: The insertion algirthm, Computer-Aided Design, Vol. 12, No. 4, July, pp.58-59 (1985).

11. Boehm, W.: On the efficiency of knot insertion algorithms, Computer Aided Geometric Design, Vol. 2, Nos. 1-3, July, pp.141-143 (1985).
12. Cohen, E., Lyche, T. and Riesenfeld, R.: Discrete B-splines and subdivision techniques in computer-aided geometric design and computer graphics, Computer Graphics and Image Processing Vol. 14, No. 2, pp.87-111 (1980).
13. Goldman, R.N.: Blossomming and knot insertion algorithm for B-spline curves, Computer Aided Geometric Design, Vol.7, pp.69-81 (1990).
14. Lasser, D. and Hoschek, J.: Fundamentals of Computer Aided Geometric Design. A. K. Peters (1993).
15. Kim, D.-S., Ryu, J., Jang, T., Lee, H., and Shin, H.: Conversion of a B-spline curve into a set of piecewise polynomial curves in a power form, Korea Israel Bi-National Conference on Geometric Modeling and Computer Graphics in the World Wide Web Era, pp.195-201, Seoul, Korea, 1999.

Rapid Generation of C^2 Continuous Blending Surfaces

Jian J Zhang and Lihua You

National Centre for Computer Animation, Bournemouth University,
Poole BH12 5BB, United Kingdom

Abstract. Most surface-blending methods are able to blend surfaces with tangent continuity. However, curvature continuity has become increasingly important in geometric modelling and its applications, such as computer animation, computer-aided design and virtual reality. In this paper, we present a method which is able to achieve C^2 continuity based on the use of partial differential equations (PDE). A sixth order partial differential equation with one vector-valued parameter is introduced to produce such blending surfaces. Since computational efficiency is crucial for interactive computer graphics applications, we have developed a unified closed form (analytical) method for the resolution of this sixth order PDE. Therefore blending surfaces of up to C^2 smoothness can be generated in real time.

Key Words. surface blending, C^2 continuity, sixth order partial differential equation, closed form solution

1 Introduction

Surface blending is an important topic of surface modelling. There has been a great deal of effort devoted to this topic leading to the development of a number of surface blending methods. A comprehensive survey on existing blending methods for parametric surfaces was made by Vida et al. [20].

The rolling-ball blends, probably the most popular and classic surface blending method, generate blending surfaces with a moving ball. Depending on whether the radius of the ball changes or not, such blending method is classified as constant-radius blending [18], [10], [4] and variable-radius blending [11], [12], [16]. The main advantage of rolling-ball blending is that, since it is defined by a simple physical motion, the shape is generated in an intuitive manner. It is also attractive from the modelling point of view that the spine, the trimlines, the assignment and the profile are automatically generated. However, the surface swept by the moving ball is of a high algebraic degree, even in relatively simple cases.

Blends with cyclides were also investigated by some researchers [1], [2], [3], [19]. Cyclides can be regarded as generalisations of the torus and can be described by implicit quartic equations or in parametric form using trigonometrical parameterisation or rational biquadratic Bézier equations. When used for blending, the required cyclide pieces can be easily constructed by identifying circles on them, especially if these lie on the planes of symmetry. Cyclides are normally used in simple blends, such as where a cylinder obliquely meets a plane.

P.M.A. Sloot et al. (Eds.): ICCS 2002, LNCS 2330, pp. 92–101, 2002.
© Springer-Verlag Berlin Heidelberg 2002

Surface blending with the solution to a partial differential equation is also an effective method. This method was first proposed by Bloor and Wilson [5]. In the reported literature, however, the PDE based approach has only been used to construct a C^1 continuous blending surface. Another challenge is how to solve a PDE efficiently without compromising the geometric accuracy. Due to the difficulty of obtaining an analytical solution, various numerical methods are developed and used as mainstream solutions, which are inevitably very slow. Cheng [9] proposed a finite difference method and carried out the blending between two cylinders, and between a cylinder and a cone. Li et al. developed a boundary penalty finite element method for surface blending [13], [14], [15]. In addition to these, collocation method [6] and finite element method [8] were developed for free form surface generation. Finite difference method was also applied to solve dynamic PDEs for cloth simulation [21], [24]. Although they are effective in finding the solutions to the PDEs, solving a large set of linear algebra equations is generally very time-consuming. As a result, these numerical methods are unsuitable for the graphics applications requiring high computational efficiency. In order to generate surfaces faster, Bloor and Wilson developed a pseudospectral method [7]. However, this method has been found not accurate enough for certain applications [27].

The vector-valued parameter in a PDE has a strong influence on the shape of the generated blending surfaces. In order to provide designers with more shape control options, You and Zhang proposed a more general fourth order partial differential equation. It has three vector-valued parameters and covers all existing fourth order partial differential equations used for surface generation [22], [26]. Zhang and You also discussed the impact of the orders of the PDEs used for free form surface creation [25].

All the above-mentioned partial differential equations are of the fourth order which can only cope with positional and tangential boundary conditions. Therefore, they can only guarantee tangential continuity. In practical engineering design, however, blending surfaces with curvature continuity are required in many situations. For example, high-speed cams without curvature continuity can cause harmful impact. Due to the importance of this issue, the curvature continuity of blending surfaces has also attracted substantial attention. Pegna and Wolter [17] presented the Linkage Curve theorem for the design of curvature continuous blending surfaces. However, since the blending with curvature continuity is much more difficult to achieve than that with tangential continuity, rapid generation of curvature continuous blending surfaces remains an open research issue.

Blending based on PDEs has unique advantages over the conventional approaches, because curvature continuity can be readily incorporated into the boundary conditions of the PDEs. In this paper, we will introduce a sixth order partial differential equation to accommodate the requirement of C^2 continuity. Unlike other (numerical) resolution methods reported in the literature, we will develop a unified closed form solution (analytical solution) to the PDE, which is able to produce a blending surface interactively.

2 Partial Differential Equation and Closed Form Solution

It is known that a fourth order partial differential equation can meet both positional and tangential boundary conditions. To satisfy the curvature boundary conditions, a sixth order PDE has to be used. Using the operator defined in (3), the following sixth order partial differential equation in a vector form can be applied for this purpose

$$\frac{\partial^6 \mathbf{x}}{\partial u^6} + 3\mathbf{a}^2 \frac{\partial^6 \mathbf{x}}{\partial u^4 \partial v^2} + 3\mathbf{a}^4 \frac{\partial^6 \mathbf{x}}{\partial u^2 \partial v^4} + \mathbf{a}^6 \frac{\partial^6 \mathbf{x}}{\partial v^6} = \mathbf{0} \tag{1}$$

where $\mathbf{a} = \begin{bmatrix} a_x & a_y & a_z \end{bmatrix}^T$ is a vector-valued shape parameter, and $\mathbf{x} = \begin{bmatrix} x & y & z \end{bmatrix}^T$ is a vector-form positional function.

An arbitrary 3D surface can be represented with the solution to PDE (1) subject to position, tangent and curvature boundary conditions:

$$u = 0 \quad \mathbf{x} = \mathbf{G}_1(v) \quad \frac{\partial \mathbf{x}}{\partial u} = \mathbf{G}_2(v) \quad \frac{\partial^2 \mathbf{x}}{\partial u^2} = \mathbf{G}_3(v)$$
$$u = 1 \quad \mathbf{x} = \mathbf{G}_4(v) \quad \frac{\partial \mathbf{x}}{\partial u} = \mathbf{G}_5(v) \quad \frac{\partial^2 \mathbf{x}}{\partial u^2} = \mathbf{G}_6(v) \tag{2}$$

where $\mathbf{G}_i(v)$ $(i = 1, 2, \cdots, 6)$ represent the vector-valued functions which describes the position, tangent and curvature on the boundaries.

The blending surface generated with the solution to PDE (1) subject to boundary conditions (2) will guarantee curvature continuity. In fact, boundary conditions (2) are stronger than those of curvature continuity. Therefore, we here call the blending surfaces defined with boundary conditions (2) C^2 continuous blending surfaces.

PDE (1) under boundary conditions (2) can be solved with various numerical methods, such as the above mentioned finite element method, finite difference method, and collocation method as well as the weighted residual method [23]. Considering the importance of computational efficiency in interactive computer graphics, we will in this paper develop a unified closed form solution of PDE (1).

To facilitate the description, let us define a vector operator whose two operands are two column vectors, which produces a new column vector whose each element is the product of the corresponding elements of the two column vectors, i.e.

$$\mathbf{st} = \begin{bmatrix} s_x & s_y & s_z \end{bmatrix}^T \begin{bmatrix} t_x & t_y & t_z \end{bmatrix}^T = \begin{bmatrix} s_x t_x & s_y t_y & s_z t_z \end{bmatrix}^T \tag{3}$$

Then decomposing the functions (2) into basic functions which are not in a polynomial form, boundary conditions (2) can be rewritten as follows

$$u = 0 \quad \mathbf{x} = \sum_{i=1}^{l} \mathbf{a}_{1i} \mathbf{g}_i(v) \quad \frac{\partial \mathbf{x}}{\partial u} = \sum_{i=1}^{l} \mathbf{a}_{2i} \mathbf{g}_i(v) \quad \frac{\partial^2 \mathbf{x}}{\partial u^2} = \sum_{i=1}^{l} \mathbf{a}_{3i} \mathbf{g}_i(v)$$
$$u = 1 \quad \mathbf{x} = \sum_{i=1}^{l} \mathbf{a}_{4i} \mathbf{g}_i(v) \quad \frac{\partial \mathbf{x}}{\partial u} = \sum_{i=1}^{l} \mathbf{a}_{5i} \mathbf{g}_i(v) \quad \frac{\partial^2 \mathbf{x}}{\partial u^2} = \sum_{i=1}^{l} \mathbf{a}_{6i} \mathbf{g}_i(v) \tag{4}$$

where $\mathbf{a}_{ji}$ $(j = 1, 2, \cdots, 6;\ i = 1, 2, \cdots, I)$ are known vector-valued coefficients, and $\mathbf{g}_i(v)$ $(i = 1, 2, \cdots, I)$ are basic functions in a vector form.

Assuming that all $\mathbf{g}_i(v)$ $(i = 1, 2, \cdots, I)$ can be expressed in the following forms (if some $\mathbf{g}_i(v)$ cannot be expressed in this way, they can be transformed into Fourier series to satisfy the requirements), then we have

$$\mathbf{g}_i^{(2)}(v) = \mathbf{b}_{2i}\mathbf{g}_i(v)$$
$$\mathbf{g}_i^{(4)}(v) = \mathbf{b}_{4i}\mathbf{g}_i(v)$$
$$\mathbf{g}_i^{(6)}(v) = \mathbf{b}_{6i}\mathbf{g}_i(v)$$
$$(i = 1, 2, \cdots, I)$$

(5)

where $\mathbf{g}_i^{(k)}(v) = \dfrac{d^{(k)}\mathbf{g}_i(v)}{dv}$ $(k = 2, 4, 6)$ and $\mathbf{b}_{ki}$ $(k = 2, 4, 6)$ are known vector-valued coefficients.

Corresponding to the basic functions, we can construct in the following a unified closed form solution of PDE (1) subject to boundary conditions (4) with the method of variable separation

$$\mathbf{x}(u, v) = \sum_{i=1}^{I} \mathbf{f}_i(u)\mathbf{g}_i(v)$$

(6)

Substituting (6) into PDE (1), the form of function $\mathbf{f}_i(u)$ can be determined. Depending on the values of $\mathbf{b}_{ki}$ $(k = 2, 4, 6)$, $\mathbf{f}_i(u)$ has three different equations. Here we take the x component as an example to present its equation.

If $b_{x2i} = b_{x4i} = b_{x6i} = 0$,

$$f_{xi}(u) = c_{xi0} + c_{xi1}u + c_{xi2}u^2 + c_{xi3}u^3 + c_{xi4}u^4 + c_{xi5}u^5$$

(7)

If $b_{x2i} = b_{xi}^2,\ b_{x4i} = b_{xi}^4,\ b_{x6i} = b_{xi}^6$,

$$f_{xi}(u) = (c_{xi0} + c_{xi1}u + c_{xi2}u^2)\cos a_x b_{xi}u + (c_{xi3} + c_{xi4}u + c_{xi5}u^2)\sin a_x b_{xi}u$$

(8)

And if $b_{x2i} = -b_{xi}^2,\ b_{x4i} = b_{xi}^4,\ b_{x6i} = -b_{xi}^6$,

$$f_{xi}(u) = (c_{xi0} + c_{xi1}u + c_{xi2}u^2)e^{a_x b_{xi}u} + (c_{xi3} + c_{xi4}u + c_{xi5}u^2)e^{-a_x b_{xi}u}$$

(9)

where b_{xi} is the coefficient before the parametric variable v in the basic functions $\mathbf{g}_i(v)$.

The unknown constants in equations. (7), (8) and (9) can be determined by substituting them into (6), then substituting (6) into boundary conditions (4).

3 Application Examples

In order to demonstrate the application of the above closed form solution in surface blending, in this section, we will create blending surfaces of three blending examples.

The first example is to blend an elliptic cylinder and a sphere which is used to explain the application of closed form solution (9). Suppose that the boundary conditions for this blending task have the forms of

$$u = 0 \quad x = a\cos v \qquad\qquad \frac{\partial x}{\partial u} = 0 \qquad\qquad \frac{\partial^2 x}{\partial u^2} = 0$$

$$y = b\sin v \qquad\qquad \frac{\partial y}{\partial u} = 0 \qquad\qquad \frac{\partial^2 y}{\partial u^2} = 0$$

$$z = h_0 - h_1 \qquad\qquad \frac{\partial z}{\partial u} = -h_1 \qquad\qquad \frac{\partial^2 z}{\partial u^2} = 0$$

$$u = 1 \quad x = r\sin u_0 \cos v \qquad \frac{\partial x}{\partial u} = r\cos u_0 \cos v \qquad \frac{\partial^2 x}{\partial u^2} = -r\sin u_0 \cos v$$

$$y = r\sin u_0 \sin v \qquad \frac{\partial y}{\partial u} = r\cos u_0 \sin v \qquad \frac{\partial^2 y}{\partial u^2} = -r\sin u_0 \sin v$$

$$z = r\cos u_0 \qquad\qquad \frac{\partial z}{\partial u} = -r\sin u_0 \qquad\qquad \frac{\partial^2 z}{\partial u^2} = -r\cos u_0$$

(10)

Taking the x component as an example and comparing (10) with (4) and (5), we have $g_{x1}(v) = \cos v$ and $b_{x21} = b_{x61} = -1,\ b_{x41} = 1$. Therefore, the closed form solution of PDE (1) for the x component is

$$x = \left(\sum_{j=0}^{2} c_{x1j} u^j e^{a_x u} + \sum_{j=3}^{5} c_{x1j} u^{j-3} e^{-a_x u} \right) \cos v \qquad (11)$$

 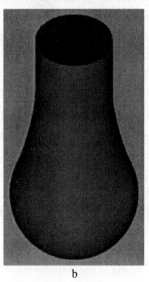

a   b

Fig. 1. Blending between an elliptic cylinder and a sphere

Substituting (11) into boundary conditions (10), all the unknown constants in (11) can be determined. With this closed form solution, we obtain the blending surfaces given in Fig. 1a and 1b. Figure 1a is the same as 1b except that three different colours are used to distinguish the three surface patches. As the position, tangent and curvature continuities are all guaranteed at the boundary curves, as expected it looks very smooth (Figure 1b).

The second example is to blend a circular cylinder and a plane at a specified straight line. Its main aim is to demonstrate the application of the closed form solution (7) in surface blending. The boundary conditions for this blending problem are

$$
\begin{aligned}
u = 0 \quad x = a_0 - a_1 v \qquad & \frac{\partial x}{\partial u} = 2(a_0 - a_1 v) \qquad & \frac{\partial^2 x}{\partial u^2} = 2(a_0 - a_1 v) \\
y = b_0 + b_1 v \qquad & \frac{\partial y}{\partial u} = 2(b_0 + b_1 v) \qquad & \frac{\partial^2 y}{\partial u^2} = 2(b_0 + b_1 v) \\
z = 0 \qquad & \frac{\partial z}{\partial u} = 0 \qquad & \frac{\partial^2 z}{\partial u^2} = 0 \\
u = 1 \quad x = r \cos v \qquad & \frac{\partial x}{\partial u} = 0 \qquad & \frac{\partial^2 x}{\partial u^2} = 0 \\
y = r \sin v \qquad & \frac{\partial y}{\partial u} = 0 \qquad & \frac{\partial^2 y}{\partial u^2} = 0 \\
z = h_0 + h_1 \qquad & \frac{\partial z}{\partial u} = h_1 \qquad & \frac{\partial^2 z}{\partial u^2} = 0
\end{aligned}
\tag{12}
$$

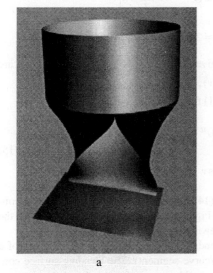 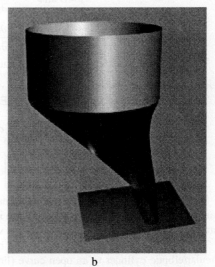

a   b

Fig. 2. Blending between a circular cylinder and a plane at a specified straight line

For the x component, we have $g_1(v) = 1$, $g_2(v) = v$ and $g_3(v) = \cos v$. Therefore, the closed form solution of PDE (1) for the x component takes the following form

$$x(u,v) = \sum_{j=0}^{5} c_{x1j}u^j + \sum_{j=0}^{5} c_{x2j}u^j v + \left(\sum_{j=0}^{2} c_{x3j}\, u^j e^{a_x u} + \sum_{j=3}^{5} c_{x3j}u^{j-3} e^{-a_x u} \right) \cos v \quad (13)$$

In the same way as above, all the unknown constants in closed form solution (13) can be determined by substituting it into boundary conditions (12). The blending surface generated with (13) is depicted in Fig. 2a and 2b, which are from different viewing angles of the same blending surface.

The final example is to blend an elliptic cylinder and a plane at a specified curve. The boundary conditions for this blending task can be written as follows

$$
\begin{array}{llll}
u = 0 & x = c\cosh dv & \dfrac{\partial x}{\partial u} = 0 & \dfrac{\partial^2 x}{\partial u^2} = 2c\cosh dv \\[2mm]
& y = c\sinh dv & \dfrac{\partial y}{\partial u} = 0 & \dfrac{\partial^2 y}{\partial u^2} = 2c\sinh dv \\[2mm]
& z = 0 & \dfrac{\partial z}{\partial u} = 0 & \dfrac{\partial^2 z}{\partial u^2} = 0 \\[2mm]
u = 1 & x = a\cos v & \dfrac{\partial x}{\partial u} = 0 & \dfrac{\partial^2 x}{\partial u^2} = 0 \\[2mm]
& y = b\sin v & \dfrac{\partial y}{\partial u} = 0 & \dfrac{\partial^2 y}{\partial u^2} = 0 \\[2mm]
& z = h_0 & \dfrac{\partial z}{\partial u} = 2h_1 & \dfrac{\partial^2 z}{\partial u^2} = 2h_1
\end{array}
\quad (14)
$$

The basic functions for the x component given by the above boundary conditions are $g_{x1}(v) = \cosh dv$ and $g_{x2}(v) = \cos v$. For $g_{x1}(v)$, $b_{x21} = d^2$, $b_{x41} = d^4$, $b_{x61} = d^6$.

Therefore, the closed form solution of PDE (1) for the x component is

$$x(u,v) = \left(\sum_{j=0}^{2} c_{x1j}u^j \cos a_x du + \sum_{j=3}^{5} c_{x1j}u^{j-3}\sin a_x du \right)\cosh dv + \left(\sum_{j=0}^{2} c_{x2j}u^j \right.$$

$$\left. e^{a_x u} + \sum_{j=3}^{5} c_{x2j}u^{j-3}e^{-a_x u} \right)\cos v \quad (15)$$

Substituting (15) into boundary conditions (14) and determining all its unknown constants, we obtain the blending surface shown in Figure 3a and 3b. These two images are the same blending surface viewed from different viewing angles.

The last two examples are complex. The blending surface connects a closed curve of a circular/elliptic cylinder to an open curve (line/curve segment). The blending surface constructed using the partial differential equation smoothly merges from a closed shape at one end to an open shape at the other. We have not found similar blending surfaces in existing literature. They suggest that our method based on the solution of a PDE have ability to generate complex blending surfaces.

Because the solution is analytical, the developed closed form method is computationally very efficient. We have timed the process of determining the unknown constants in the closed form solutions and found the process took less than 10^{-6} of a second on a 800 MHz PC for all three blending tasks, no problem for interactive computer graphics applications.

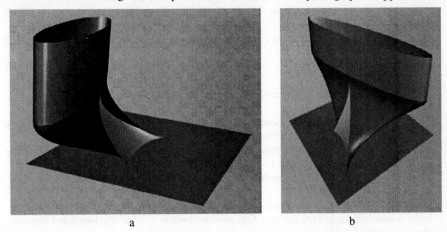

a   b

Fig. 3. Blending between an elliptic cylinder and a plane at a specified curve

4   Conclusions

Surface blending with C^2 continuity has been investigated in this paper. Different from the existing blending methods considering curvature continuity, we introduced a sixth order partial differential equation which provides enough degrees of freedom to accommodate curvature boundary conditions.

Since the computational speed is an important factor in interactive computer graphics applications, we developed a unified closed form resolution method. Depending on the differential properties of the boundary functions, the closed form solution of the PDE has three different forms.

By applying the developed closed form solution, we have given three examples to illustrate the use of our method. They are the C^2 continuous blending between an elliptic cylinder and a sphere, between a circular cylinder and a plane at a specified straight line, and between an elliptic cylinder and a plane at a given curve. These examples demonstrate the application of the three forms of the PDE solutions in surface blending (7), (8), (9), respectively. It is found that the developed blending method can tackle complex blending problems, some of which have not been seen in the literature. In addition, this method is computationally very fast, good enough for interactive applications.

References

1. Allen, S., Dutta, D.: Cyclides in pure blending I, Computer Aided Geometric Design, 14 (1997) 51-75
2. Allen, G., Dutta, D.: Cyclides in pure blending II, Computer Aided Geometric Design, 14 (1997) 77-102
3. Allen, G., Dutta, D.: Supercyclides and blending, Computer Aided Geometric Design, 14 (1997) 637-651
4. Barnhill, R. E., Farin, G. E., Chen, Q.: Constant-radius blending of parametric surfaces, Computing Supple., 8 (1993) 1-20
5. Bloor, M. I. G., Wilson, M. J.: Generating blend surfaces using partial differential equations, Computer Aided Design, 21(3) (1989) 165-171
6. Bloor, M. I. G., Wilson, M. J.: Representing PDE surfaces in terms of B-splines, Computer Aided Design, 22(6) (1990) 324-331
7. Bloor, M. I. G., Wilson, M. J.: Spectral approximations to PDE surfaces, Computer Aided Design, 28(2) (1996) 145-152
8. Brown, J. M., Bloor, M. I. G., Bloor, M. S., Wilson, M. J.: The accuracy of B-spline finite element approximations to PDE surfaces, Computer methods in Applied Mechanics and Engineering, 158 (1998) 221-234
9. Cheng, S. Y., Bloor, M. I. G., Saia A., Wilson, M. J.: Blending between quadric surfaces using partial differential equations, in Ravani, B. (Ed.), Advances in Design Automation, Vol. 1, Computer Aided and Computational Design, ASME, (1990) 257-263
10. Choi, B. K., Ju, S. Y.: Constant-radius blending in surface modeling, Computer Aided Design, 21(4) (1989) 213-220
11. Chuang, J.-H., Lin, C.-H., Hwang, W.-C.: Variable-radius blending of parametric surfaces, The Visual Computer, 11 (1995) 513-525
12. Chuang, J.-H., Hwang, W.-C.: Variable-radius blending by constrained spine generation, The Visual Computer, 13 (1997) 316-329
13. Li, Z. C.: Boundary penalty finite element methods for blending surfaces, I. Basic theory, Journal of Computational Mathematics, 16 (1998) 457-480
14. Li, Z. C.: Boundary penalty finite element methods for blending surfaces, II. Biharmonic equations, Journal of Computational and Applied Mathematics, 110 (1999) 155-176
15. Li, Z. C., Chang, C.-S.: Boundary penalty finite element methods for blending surfaces, III, Superconvergence and stability and examples, Journal of Computational and Applied Mathematics, 110 (1999) 241-270
16. Lukács, G.: Differential geometry of G^1 variable radius rolling ball blend surfaces, Computer Aided Geometric Design, 15 (1998) 585-613
17. Pegna, J., Wolter, F.-E.: Geometrical criteria to guarantee curvature continuity of blend surfaces, Journal of Mechanical Design, Transactions of the ASME, 114 (1992) 201-210
18. Rossignac, J. R., Requicha, A. A. G.: Constant-radius blending in solid modeling, Computers in Mechanical Engineering, (1984) 65-73
19. Shene, C.-K.: Blending two cones with Dupin cyclides, Computer Aided Geometric Design, 15 (1998) 643-673
20. Vida, J., Martin, R. R., Varady, T.: A survey of blending methods that use parametric surfaces, Computer-Aided design, 26(5) (1994) 341-365
21. You, L. H., Zhang, J. J., Comninos P.: Cloth deformation modelling using a plate bending model, The 7th International Conference in Central Europe on Computer Graphics, Visualisation and Interactive Digital Media, (1999) 485-491

22. You, L., Zhang, J. J.: Blending surface generation with a fourth order partial differential equation, The Sixth International Conference on Computer-Aided Design and Computer Graphics (CAD/Graphics'99), Shanghai, China, 1-3 December, (1999) 1035-1039
23. You, L. H., Zhang, J. J., Comninos, P.: A solid volumetric model of muscle deformation for computer animation using the weighted residual method, Computer Methods in Applied Mechanics and Engineering, 190 (2000) 853-863
24. Zhang, J. J, You, L. H., Comninos P.: Computer simulation of flexible fabrics, The 17th Eurographics UK Conference, (1999) 27-35
25. Zhang, J. J., You, L. H.: Surface representation using second, fourth and mixed order partial differential equations, International Conference on Shape Modelling and Applications, Genova, Italy, 7-11, May, (2001) 250-256
26. Zhang, J. J., You, L. H.: PDE based surface representation, Computers & Graphics, 25(6) (2001), in print.
27. Zhang, J. J., You, L. H.: Surface blending using a power series solution to fourth order partial differential equations, Computer-Aided Design, under review, 2001

Interactive Multi-volume Visualization

Brett Wilson, Eric B. Lum, and Kwan-Liu Ma

Department of Computer Science
University of California at Davis
One Shields Avenue, Davis, CA 95616, U.S.A.
{wilson,lume,ma}@cs.ucdavis.edu,
WWW home page: http://www.cs.ucdavis.edu/~ma

Abstract. This paper is concerned with simultaneous visualization of two or more volumes, which may be from different imaging modalities or numerical simulations for the same subject of study. The main visualization challenge is to establish visual correspondences while maintaining distinctions among multiple volumes. One solution is to use different rendering styles for different volumes. Interactive rendering is required so the user can choose with ease an appropriate rendering style and its associated parameters for each volume. Rendering efficiency is maximized by utilizing commodity graphics cards. We demonstrate our preliminary results with two case studies.

1 Introduction

Volume rendering has been accepted as an effective method for visualizing physical phenomena or structures defined in 3-space. The advent of texture hardware algorithms [2,5] and special hardware like the VolumePro [14] make possible interactive volume rendering which is very attractive to many disciplines. While previous volume visualization algorithms were mostly designed for looking at one volume at a time or for animating time-varying volume data [11]. the ability to simultaneously visualize multiple volumes becomes increasingly desirable for many application areas.

In medical research and practice, many imaging modalities including X-ray imaging, Doppler ultrasound, CT scans, MRI scans, etc. are available to physicians for making better diagnoses and surgical plans. While generally the images from different modalities are looked at individually or side by side, it would often beneficial to see two types of data sets simultaneously by spatially superimposing them. For example, it is helpful to visualize functional data overlaid on anatomical data.

In scientific computing, to gain a better understanding of the intrinsic properties of certain physical or chemical process, scientists often try to simulate and study different aspects of the process. The capability to visualize different variables simultaneously describing the same spatial domain and to determine their correlations is thus desirable. For example, in a multi-disciplinary computing environment, several engineering analysis programs, such as structural and flow solvers, run concurrently and cooperatively to perform a multi-disciplinary design. The goal may be to identify the relevant design variables that can explain the causes of a particular phenomenon, like vortices in a flow field.

P.M.A. Sloot et al. (Eds.): ICCS 2002, LNCS 2330, pp. 102–110, 2002.

Multimodality volume visualization in medical imaging generally must proceed with a registration step because volumes from different modalities almost always have either different resolutions or various degree of distortions. Registration can be as simple as performing a set of linear transformations but can also be a very complex process involving extensive human intervention [1, 15].

Similarly, in many scientific or engineering studies, simulation data and experimental data of different resolutions or on different kinds of computational mesh structures need to be looked at together. Resampling is often done to match one resolution to the other. Registration/resampling by itself can present many challenges. The work reported in this paper focuses on the visualization problems after registration/resampling has been done.

We aim to develop techniques that can generate effective visualizations which reveal both correspondence and distinction between multimodality volume data sets. Our approach to the multi-volume visualization problem is to use:

- highly interactive rendering,
- mixed rendering styles, and
- user-controlled data fusion.

Interactive rendering allows the user to freely change rendering and visualization parameters, as well as data fusion schemes. Rendering different volumes with different styles, if done appropriately, may enhance perception of shape, structure, and spatial relationship. The data fusion problem here is to determine how to display multiple data values defined at the same spatial location. For example, a CT image and MRI image from the same patient can be combined to show both bones and fat clearly in a single image. We feel that the user should be allowed to select a particular method to combine images according to the properties under study. The rest of the paper discusses these topics. Finally, two different data sets are used to discuss the techniques and processes we have developed to achieve more efficient multi-volume visualization.

2   Interactive Volume Rendering

Highly interactive volume rendering can be achieved by using either graphics hardware-assisted methods or a parallel computer. We are particularly interested in utilizing consumer PC graphics cards like the Nvidia GeForce 3. For large data sets, a cluster of PCs is used to distribute both the data and the the rendering calculations.

2.1   Hardware-accelerated volume rendering

Specialized volume rendering hardware such as the VolumePro [14] has been available for realtime volume rendering for several years. Recent lower-cost consumer graphics cards such as the Nvidia GeForce 3 also support volume rendering through the use of volumetric textures [5]. To render a volume using volumetric textures, a series of planes parallel to the screen plane are drawn. These planes intersect the volumetric texture, which is transformed using OpenGL's texture transformation matrix, and the graphics hardware interpolates the texture in three dimensions. The closer together the planes are drawn, the more accurate (but slower) the resulting volume rendering becomes. This method is limited by the low amount of memory typically available on consumer graphics cards. For example, the GeForce 3 has 64 MB of memory, limiting the volume texture size to $256 \times 256 \times 256$. For higher-resolution data, parallel approaches must be used [12].

A more widely-supported method for hardware-accelerated volume rendering is to map large numbers of 2-d textures onto axis-aligned parallel planes [2]. Generally, 8-bit indexed-color textures are used to minimize memory usage and to allow realtime modification of the colormap and transfer function by palette modification. This method produces images with more visual artifacts than the 3-d texture method, and sets of slices for each axis must be kept to prevent the slices from appearing edge-on. However, 2-d textures are somewhat faster than 3-d textures on current hardware, and can be straightforwardly paged in and out of video memory, so data sets can be displayed which exceed the memory of the video card (although with a significant performance penalty).

Realtime lighting can enhance geometry and structure of data. We currently implement two methods for lighting in hardware, both of which encode normal information in a separate set of textures. The first method encodes the normal at each data point as an 8-bit index, indicating one of 256 direction vectors uniformly distributed in 3-space. These textures are drawn in a separate pass using a palette which maps each normal index to a specular light value for the current view vector. The second method uses the texture-shader features of the Nvidia GeForce 3. The normal vector at each point is encoded in the three color values of an RGB texture. The GeForce 3 interpolates these values for each pixel and computes the dot product with the current view vector. The result is used as a lookup into a texture that encodes a reflection map. This method provides better precision than the first method at a cost of memory usage.

2.2   Parallel volume rendering

Many software algorithms for parallel volume rendering have been developed [9, 10, 13]. In this work, we intend to use hardware-accelerated rendering. One significant limitation of volume rendering using consumer PC graphics hardware is the relatively small amount of video memory. For example, the Nvidia Geforce 3's 64 megabytes of video memory is shared between the frame buffer and texture memory. It is very desirable to fit the volume being rendered entirely in texture memory in order to avoid swapping data into the graphics card from main memory over the relatively slow graphics bus. By subdividing the volume spatially and distributing it across a cluster of PCs equipped with graphics cards, it is possible to render significantly larger volumes into the aggregated video memory of the entire cluster. In addition to the larger amounts of texture memory provided by a PC cluster, performance improvements also result from the combined fill-rate of multiple graphics cards.

A PC cluster we have built recently consists of 17 PCs. The 16 node PCs are connected with a 100-base-T fast Ethernet while the connection to the host PC, which is used for final display and user interface control, is through the cluster's switch with a gigabit Ethernet. Each PC has an AMD Athlon 1.3 Ghz processor, one gigabyte of PC133 SDRAM and a GeForce 3.

Our current implementation of the renderer subdivides and distributes the volume using k-d tree subdivision. Each node PC resamples it's subvolume to a size of $256 \times 256 \times 256$ regardless of the dimensions of the actual volume. For smaller volumes this permits higher order resampling to be done in software prior to the tri-linear interpolation done in hardware. During rendering, for each frame, every node on the cluster renders its subvolume and composites the resulting subimage using binary-swap [13] with the final image being sent to the host for display. Presently, using 8 of the node PCs, we are able to render $512 \times 512 \times 512$ voxels to a 512×512 window at about 1-2 frames per second. The current implementation can switch between high and lower resolution modes based user's need for interactivity. For example, rendering a scaled down version of the data instead like $256 \times 256 \times 256$ voxels, over 10 frames per second can be achieved.

3   Mixing Rendering Styles

When rendering multiple volumes, it is important that visual cues be present to help differentiate the volumes. This can be achieved by varying the color, lighting, as well as rendering style used for each of the different volumes. For example, Hauser et al. [7] show how to efficiently perform both maximum intensity projection and direct volume rendering to generate effective visualizations. In addition, the mixing of photorealistic and non-photorealistic rendering styles can be particularly effective in not only differentiating the two volumes, but also in drawing focus to features of interest.

3.1   Nonphotorealistic rendering

Non-photorealistic rendering involves the application of techniques used by artists for the creation of computer generated imagery. These techniques have been applied to scientific visualization for the creation of imagery that can be more meaningful than that generated with more traditional photorealistic techniques. Non-photorealistic rendering is usually associated with the representation of surfaces, but has also been applied to direct volume rendering [3, 12]. Through the application of several non-photorealistic rendering techniques it is possible to accentuate key features in a volume, while deemphasizing structures that might obstruct or detract from those features.

An artist rarely depicts shading by simply making colors darker, but instead relies on variations in both light intensity as well as color temperature or tone to indicate illumination. For example, shadows are often shown in cooler blues, while directly illuminated regions are shown with warmer yellows and reds. Gooch et al. [6], describe how tone shading can be applied to the rendering of surfaces for the creation of illustrations. We implement tone shading by modulating the rendered volume with a volumetric tone texture. This texture consists of a paletted texture of gradient directions with a palette that varies from cool to warm depending on the lighting of each gradient direction as calculated using a standard Phong shading model.

Silhouette rendering consists of adding dark lines around an object and can be very effective in enhancing fine structures. They can also aid in depth perception when viewing overlapping structures of similar color. Silhouette rendering is accomplished in a second rendering pass that modulates the transfer function with a texture that is dark and highly opaque in those regions with a normal perpendicular to the view, and highly transparent in those regions parallel to the viewer. This texture is obtained by using a paletted texture with gradient directions and specifying a palette such that each normal has an opacity that increases with degree silhouette should be shown.

Through the variation of color based on distance, depth perception can be improved [4]. In particular, color can be varied in temperature as well as intensity depending on the distance from the viewer. We implement this non-photorealistic rendering style by modulating each textured polygon drawn in the rendering processes with a color that is selected based on the distance from the viewpoint. The user is able to interactively specify how color is manipulated based on distance using a transfer function styled interface that maps distance to a color and opacity map. Thus, closer objects can be mapped to warmer colors compared to cooler distant objects, improving perception of the depth relationship between these objects.

4   Data Fusion Schemes

When rendering multiple volumes simultaneously, we must decide how to treat multiple values defined at the same spatial location. There are basically three approaches to this data fusion problem:

1. using one of the values based on some criterion
2. using one value which is weighted by a function of some or all of other values
3. using one value for each color channel

An example for the first approach is the alternating sampling used in [8] for rendering two volumes. The second approach gives us more freedom. For example, we could use the opacity transfer function for one volume to enhance or de-enhance some aspects of the other volume. This is similar to the common practice of volume visualization in which gradient magnitude is used to enhance boundary surface. Furthermore, a weighted sum might be used with scalings that reflect a desired property, such as distance from the viewer [12]. The third approach is probably the simplest for one to implement and to verify its results. While it is limited to visualization of three or fewer volumes, in practice we hardly need to see more than three volume simultaneously.

We have implemented a suite of combining operations based on these three approaches. Most importantly, our system allows the user to interactively select a particular way to present multimodality data. We found this capability aids to the multi-volume data exploration process.

5   Results

We have been studying multi-volume visualization using data sets generated from medical imaging, biological data imaging, and computational fluid dynamics simulations. In the following sections, case studies on a mouse data set and a plant microbiology data set are presented to summarize our initial experience. Alternating sampling was used for data fusion for both data sets.

5.1   Case study I: a mouse data set

We use two volumes from small animal imaging: a Positron Emission Tomography (PET) data and an MRI scan of a mouse head. Each of these data sets consists of 47 slices at 256×256 pixels per slice. Prior to visualization, the data was cropped to remove a significant amount of empty space, and resampled to 100 slices at 256×256. On a single PC, we can render to 500×500 pixels window without lighting at 25 frames per second, with one volume lit at 7 frames per second, and with both volumes lit at 1 frame per second.

The left image in Figure 1 shows high density regions of the PET data set in green, indicating the region of higher brain activity. The blue areas show a narrow range of values present in the MRI data set. This allows some anatomical structure to be shown while keeping most of the PET scan data visible. White specular lighting is applied to the MRI data to help define its shape.

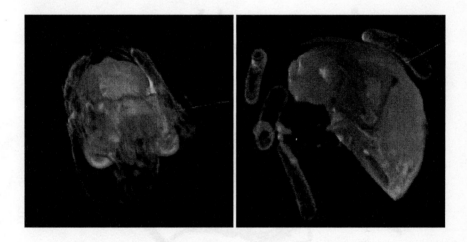

Fig. 1. Simultaneous visualization of PET and MRI mouse data.

The right image in Figure 1 shows high density PET data in red with specular highlighting. The large tube structures around the data are alignment markers to aid in image registration. The MRI data is shown in gray with a clipping plane to allow a portion of the PET data to be visible. This MRI data is more difficult to visualize because it consists of large regions of relatively constant value separated by small regions of other values. This makes it hard to show structure while keeping the image transparent enough to reveal deeper structure. In addition, noise causes specular lighting to be grainy and distracting.

In each case, the PET data was much easier to visualize than the MRI data. The PET data is smooth and present in large, mostly convex areas. This makes it easy to see desired structure by selecting a cutoff for displayed densities, and the smooth contours produce smooth, consistent specular highlights. By contrast, the MRI data has a lot of fine detail at many different densities. A partially transparent MRI image reveals too many levels of detail to be useful, while a mostly opaque MRI image obscures other data we may want to see (such as the PET data).

5.2   Case study II: roots data set

In our second case study we examine a root data set generated using confocal microscopy. Two volumes were obtained based on the reflected light emission of two different fluorescences. The first volume consists of a root from the species Melilotus alb, which has the common name white clover. The other volume contains mycorrhizal fungi attached to the root. Each volume has $512 \times 512 \times 256$ voxels. The scientists we worked with were interesting in studying the symbiotic association that exist between the plant and fungus.

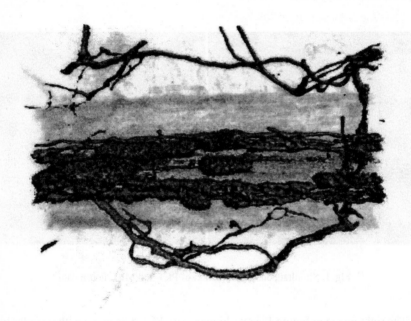

Fig. 2. Mixed-style rendering of plant root data from confocal microscopy. The red part displays the cover of the fungi and the green part shows the extent of root.

In Figure 2 we see a hardware rendered image of the root and fungus generated with our technique. The fungus, shown in red, is rendered using a transfer function that makes it appear opaque, allowing it's fine tubular structures to be readily visible. Lighting parameters have also been specified to make the shape more obvious. The root, rendered in green, is shown to give the fungus structure context. It is of less importance to the scientist and is therefore shown with relatively transparent opacity map. Lighting is less visible for the root since it would draw attention away from the fungus being studied.

Using eight nodes of our PC cluster to render this data, we are able to achieve over three frames per second. Using a single PC, we would have to scale down the data to maintain the same interactivity.

6   Conclusions

Simultaneous visualization of volumetric data from multiple modalities is challenging because of the disparity in data resolutions, and the higher storage and processing requirements. In our preliminary study using data from medical imaging and plant microbiology, we address some of the most relevant issues including rendering models, data fusion, and high-performance rendering.

We have shown that multi-modality volume visualization can aid in the understanding of volumetric medical data. The additional information provided by simultaneous visualization allows areas such as tumors or active regions of the brain to be more precisely located relative to each other and to the anatomy of the animal.

We have also shown that appropriate shading models and use of transparency in multi-volume visualization provide information not available from the use of a single, traditional rendering model. The non-photorealistic rendering we used helps bring out information about the geometry and boundary of a structure, and the customizable transparency provides information about the relative positions and size of two or more structures, and allows the researcher to highlight the most important details of an image.

We intend to create a system framework for further study. Other future work directions include refining our mixed-style rendering model to derive more effective illustrations, conducting a comprehensive study of the data fusion problem, and studying other types of multi-volume data,

Acknowledgments

This work has been sponsored in part by NSF PECASE, NSF LSSDSV, and DOE SciDAC. The authors are grateful to Dr. Juan Jose Vaquero, Dr. Michael Green, the National Institutes of Health, and UCLA Ann Hirsch Laboratory for providing the test data sets.

References

1. Maintz, A. and Viergever, M.: A survey of medical image registration. Medical Image Analysis, vol. 2, no. 1. (1998) 1–36.
2. Cabral, B., Cam, N., and Foran, J.: Accelerated volume rendering and tomographic reconstruction using texture mapping hardware. 1994 Workshop on Volume Visualization (October 1994) 91–98.
3. Ebert, D. and Rheingans, P.: Volume illustration: non-photorealistic rendering of volume models, IEEE Visualization 2000 Conference. (October 2000) 195–202.
4. Foley, D. J., Van Dam, A., Feiner, S. K., and Hughes, J. F.: Computer Graphics: Principles and Practice. Addison Wesley, 1996.

5. Van Gelder, A. and Hoffman, U.: Direct Volume Rendering with Shading via Three-Dimension Textures, 1996 Symposium on Volume Visualization. (1996) 23–30.
6. Gooch, A., Gooch, B., Shirley, P., and Cohen, E.: A non-photorealistic lighting model for automatic technical illustration, SIGGRAPH '98 Conference. (July 1998) 447–452.
7. Hauser, H., Mroz, L., Bischi, G.-I., and Groller, E.: Two-level volume rendering - fusing MIP and DVR. IEEE Visualization 2000 Conference. (2000) 211–218.
8. Hastreiter, P., and Ertl, T.: Integrated registration and visualization of medical image data. Computer Graphics International (1998) 78–85.
9. Lacroute, P.: Real-time volume rendering on shared memory multiprocessors using the shear-warp factorization. 1995 Parallel Rendering Symposium (October 1995) 15–22.
10. Li, P. P., Whitman, S., Mendoza, R., and Tsiao, J.: ParVox – a parallel splatting volume rendering system for distributed visualization. 1997 Symposium on Parallel Rendering (October 1997) 7–14.
11. Lum, E. B., Ma, K.-L., and Clyne, J.: Texture hardware assisted rendering of time-varying volume data. IEEE Visualization 2001 Conference. (October 2001) 263–270.
12. Lum, E. B., and Ma, K.-L.: Hardware-accelerated parallel non-photorealistic volume rendering. International Symposium on Non-Photorealistic Animation and Rendering, Annecy, France (June 2002).
13. Ma, K.-L., Painter, J., Krogh, M., and Hansen, C.: Parallel volume rendering using binary-swap compositing. IEEE Computer Graphics & Applications. (July 1994) 59–68.
14. Pfister, H., Hardenbergh, J., Knittel, J., Lauer, H., and Seiler, L.: The VolumePro real-time ray-casting system. ACM SIGGRAPH '99 Conference. (1999) 251–260.
15. Woods, R. , Grafton, S., Holmes, C., Cherry, S., and Mazziotta, J.: Automated image registration: II. intersubject validation of linear and nonlinear models. Journal of Computer Assisted Tomography. (1998) 22:155–165.

Efficient Implementation of Multiresolution Triangle Strips *

Óscar Belmonte[1], Inmaculada Remolar[1], José Ribelles[1],
Miguel Chover[1], and Marcos Fernández[2]

[1] Departamento de Lenguajes y Sistemas Informáticos, Universitat Jaume I,
12071 Castellón, Spain
{belfern,remolar,ribelles,chover}@lsi.uji.es
[2] Departamento de Informática, Universitat de València,
46100 València, Spain
{marcos}@robotica.uv.es

Abstract. Triangle meshes are currently the most popular standard model to represent polygonal surfaces. Drawing these meshes as a set of independent triangles involves sending a vast amount of information to the graphic engine. It has been shown that using drawing primitives, such as triangle fans or strips, dramatically reduces the amount of information. *Multiresolution Triangle Strips* (MTS) uses the connectivity information to represent a mesh as a set of *multiresolution triangles strips*. These strips are the basis of both the storage and rendering stages. They allow the efficient management of a wide range of levels of detail. In this paper, we have taken advantage of the coherence property between two levels of detail to decrease the visualisation time. MTS has been compared against *Progressive Meshes and Multiresolution Ordered Meshes with Fans*, the only model that uses the triangle fan as an alternative to the triangle primitive. In all cases, Multiresolution Triangle Strips obtains a better frame rate.

1 Introduction

Triangle meshes have become the standard model to represent polygonal surfaces in Computer Graphics. The main reasons for this are the simplicity of the algorithms for drawing triangles, which are easily implemented in hardware, and the fact that any polygon with a number of sides greater than three can be broken down into a set of triangles.

Nowadays, highly detailed geometric models are necessary in many Computer Graphics applications. In these cases, the objects are represented by large triangle meshes, which are expensive to visualise. Some objects in the scene could be replaced by an approximation, adapting the number of triangles of a mesh to the needs of each application. This approximation is said to have a lower *Level of Detail* (LOD).

---

* Supported by grant P1.1B2000-21 (Fundació Caixa Castelló - Bancaixa)

P.M.A. Sloot et al. (Eds.): ICCS 2002, LNCS 2330, pp. 111–120, 2002.

Multiresolution models support the representation and processing of geometric entities at different levels of detail, depending on specific application needs. The common criteria to determine the most suitable LOD are the distance of the object from the viewer, the projected area of the object on the screen, the eccentricity of this object on the screen and the intrinsic importance of the object. Current graphics systems can render more triangles than they receive. The bottleneck at the rendering stage is the throughput of the graphics systems in receiving the information to visualise. This information decreases considerably if the connectivity information between triangles is used in the mesh representation. Triangle fans and strips, which appear in the majority of graphics libraries such as *OpenGL*, make use of this property. Modelling a mesh as a set of triangle fans or strips saves having to send a large amount of redundant information to the graphics system.

All the multiresolution models in the literature, except *Multiresolution Ordered Meshes with Fan* (MOM-FAN), base the storage and rendering process on the triangle primitive. *Multiresolution Triangle Strips* (MTS) is the first model that represents a multiresolution object using the triangle strips primitive in these two stages. An MTS model is made up of a set of multiresolution strips. A multiresolution strip represents the original strip and all its LODs.

In section 2, the concept of triangle strips and the problem of searching strips in a triangle mesh are reviewed. The simplification method by Garland and Heckbert, and multiresolution models that make use of triangle fans or strips at the rendering stage are commented in this section. In section 3, MTS is presented, together with its data structure and the LOD recovery algorithm. It is shown, through the use of an example, how a multiresolution strip is constructed. In section 4, it is explained how the coherence property is incorporated into the model. In section 5, time results of MTS are shown and are compared against the results obtained for PM and for MOM-FAN. Finally, in section 6, conclusions and future work are presented.

2   Previous Work

In this section, some works relative to the multiresolution model presented in this paper are reviewed. The concept of triangle strips and an algorithm for searching strips over a polygonal surface is discussed. After that, we review the simplification algorithm of Garland and Heckbert [4], which has been used to obtain the coarse meshes from the original model. Finally, other multiresolution models that use triangle fans or strips, in either the storage or the rendering stage, are discussed.

2.1   Triangle strips

A strip encodes a sequence of triangles where every two sequential triangles share a common inner edge. Making use of this connectivity information decreases the amount of information sent to the graphic engine at the rendering stage. A

Fig. 1. Left, sequential strip. Right, generalized strip

triangle strip is encoded by a sequence of vertices. The vertex sequence for the strip in Figure 1.a is 0,1,2,3,4,5, where triangle i is made up of the vertices i, $i+1$ and $i+2$. In this way, it is only necessary to send $T+2$ vertices to the graphic engine in order to render T triangles, as opposed to the $3T$ vertices required if the triangles are rendered independently. The sequence of vertices for the strip in Figure 1.b is 0,1,2,3,4,3,5,6. In this case, it is necessary to send vertex 3 twice. This operation is called *swap*. A triangle strip that can be represented without any swap operation is called a *sequential strip*. A triangle strip that needs this operation is called a *generalized strip*.

2.2 Searching Triangle Strips

A triangle mesh can be drawn as a set of triangle strips. The best set of triangle strips that describes a mesh sends the lowest number of vertices to the graphics system. Searching this best set of strips in a mesh is an NP-complete problem [1]. Thus, it becomes necessary to use heuristic strategies in the search for strips in a rational time. The STRIPE algorithm [3] has been used for extraction of the triangle strips. This algorithm is publicly available at http://www.cs.sunysb.edu/~stripe/. The STRIPE algorithm allows control over some parameters in the searching process, such as the use of swap operations.

2.3 Simplification Using Vertex Pair Contraction

A simplification method provides a version of the original object with fewer triangles while maintaining the visual appearance. Important surveys about simplification methods [4] [5] [9], have been published. The simplification method used in Multiresolution Triangle Strips is the method proposed by Garland and Heckbert [4] based on vertex pair contraction. This method, called *Qslim*, has been used for several reasons: i) it returns the sequence of contractions applied to the original object, ii) the simplified mesh error is sufficiently small , and iii) it is public domain and can be found at http://www.cs.cmu.edu/~garland/quadrics/.

2.4 Multiresolution Modelling

Garland [6] defines a multiresolution model as a model representation that captures a wide range of approximations of an object and which can be used to reconstruct any one of them on demand. Multiresolution models can be classified in two broad groups: *discrete multiresolution models*, which contain a discrete number of LODs and a control mechanism to determine which LOD is the most adequate in each moment; and *continuous multiresolution models*, which capture a vast range of, virtually continuous, approximations of an object. These can be subdivided into two main classes according to their structure: tree-like models, and historical models [9]. There is no relationship between the LODs of the object in a discrete multiresolution model. Thus, the size of these models increases quickly when new levels of detail are included. They usually store between five and ten levels of detail. Graphics standards such as *VRML* or *OpenInventor* use discrete multiresolution models. These models are easily implemented and can be edited by users and optimised for rendering. The main disadvantage is the *visual artefact* that occurs during the change between two levels of detail. One solution to decrease this visual artefact is to draw both levels of detail of the object using transparency methods.

Two consecutive levels of detail differ by a lower number of triangles in a continuous multiresolution model. These little changes introduce a minimal visual artefact. The size of the model decreases as compared to the discrete models because no duplicate information is stored. The Progressive Meshes of Hugues Hoppe [7] is the most well known continuous multiresolution model available nowadays. It is included in Microsoft's graphics library DirectX 8.0.

Multiresolution Models Using Triangle Fans or Strips Primitives. Hoppe [8] has used strips in the rendering stage of a viewdependent multiresolution model. After selecting which triangles to draw, strips of triangles are searched. Through experimentation, Hoppe concludes that the fastest triangle strip search algorithm is a greedy one. In order to reduce the strip fragmentation, strips are grown in a clockwise spiral manner.

El-Sana et al. [3] have developed a view-dependent multiresolution model based on an edge-collapsing criterion. The first step in constructing the model is to search triangle strips on it. These triangle strips are stored in a data structure called a skip list. Once the multiresolution model has determined which triangles to visualise, the skip list is processed. If none of its triangles has been collapsed, the strip is drawn and if not, the skip list is processed in order to update the strips. The triangle strips are not the basic primitive of the model, they are only used to speed up the rendering process. The work presented in [10] uses triangle fans as its basic representation primitive. Using this primitive, the storage cost is reduced, but the behaviour of this new model as regards its visualisation time is similar to its predecessor. A short average fan length, the high percentage of degenerate triangles, and the need to adjust the fans to the required LOD in real time all contribute to produce overall results which do not bring about a global improvement in visualisation time.

Neither of the previous models uses the triangle strips primitive in both the storage and the rendering stage. Hoppe searches the strips over the simplified model prior to rendering it. In El-Sana's work, the triangles Skip-Strips are updated for the collapsed edges before rendering the model.

3   MTS Model

An MTS model represents a mesh as a set of multiresolution triangle strips. Each multiresolution triangle strip is made up of the original strip and all its LODs, and a list containing the vertices at the beginning of the strip for each LOD. The sequence of vertices in a strip induces an order relationship between them. It can be concluded that these strips may be represented by a directed graph. Due to this fundamental structure, the LOD recovery algorithm is based on graph traversal.

3.1   Data Structure

A multiresolution strip is represented by a directed graph and a list with the vertices at the beginning of the strip.

Conceptually, each vertex of the strip is identified by a node on the graph and each inner edge with an arc on the graph. Two nodes are said to be *adjacent* if they are joined by an arc. If it is a directed graph, the arc direction is determined by the order of the vertices in the sequence that represents the strip. In practice, a graph is represented by an adjacency list [2], (see Fig. 2.c). In this representation, each node v has a list with all of its adjacent nodes w_i for any LOD. Each node w_i on the adjacency list of node v represents an arc from v to w_i. Each strip in the model is an instance of the *class MultiresolutionStrip*.

Each node on the graph has three fields. The first field (*vIndex*) is an index to the memory address containing the geometric data of the vertex. The second field (*neighbours*) is a pointer to the adjacency list. The third field (*currentNeighbour*) is an index to the next neighbour that will be visited by the LOD extraction algorithm. Each node on the graph is an instance of *class ColumnNode*, (see Listing 1).

```
class ColummNode
  unsigned long vIndex;
  RowNode * neighbours;
  unsigned int currentNeighbour;
end class
```

```
class MultiresolutionStrip
  RowNode * sBegin;
  ColummNode * colVertices;
end class

class RowNode
  unsigned int colIndex;
  unsigned long res;
end class
```

Listing 1: Data structures of the MTS model.

Each arc on the graph has two fields. The first field (*colIndex*) is an index to the next node in the vertex sequence that represents the strip. The second field (*res*) is an integer that indicates the maximum LOD at which the arc may be traversed by the LOD extraction algorithm. Each arc of the graph is an instance of *class RowNode*.

The vertex at the beginning of the strip could change as the strip is simplified. The list of strip beginnings indicates, for each LOD, which is the vertex at the beginning of the strip. Each element on the list of strip beginnings has two fields. The first field (*colIndex*) is an index to the node that begins the strip. The second field (*res*) is an integer indicating the maximum LOD at which the vertex is valid as the beginning of the strip. These fields are the same that those in the *class RowNode*, so each element on the list of strip beginnings is an instance of *class RowNode*.

Construction Example. The construction of a multiresolution strip starts from the original strip. As the strip is simplified, the sequence of vertices changes. These changes in the vertex sequence should be introduced into the graph and the list of strip beginnings that represents the strip.

Let's take the strip in Fig. 2.a) as an example. In this figure, the initial graph and the strip beginning list are shown. Let's label the maximum level of detail at which a strip can be represented as level of detail 0 (LOD 0). The sequence of vertices at LOD 0 is 0,1,2,3,4,5,6,-3. The special label -3 specifies that the end of the strip has been reached. After the first vertex pair contraction ($2 \rightarrow 3$), LOD 1, the resulting vertex sequence is 0,1,3, -1,3,4,5,6,-3. The special node labelled -1 indicates that vertex 3 is the end of one strip and the beginning of the next one. There are two new edges (1,3) and (3,-1,3,4). The new edge (1,3) is stored in the multiresolution strip by adding a new arc to the list of adjacencies of node 1. The field colIndex of this new arc is initialised to 3 and the res field is initialised to 1. The special arc (3,-1,3,4) is stored as follows: first, a new arc is added to the list of adjacencies of node 3. This arc has its colIndex field initialised to -1 and its res field initialised to 1. Another new arc is also added to node 3, in this case with its colIndex field initialised to 4 and its res field initialised to 1. The arcs that remain, (4,5) and (5,6), update their res field to the new LOD 1.

The second vertex pair contraction ($0 \rightarrow 3$), LOD 2, produces the vertex sequence 3,4,5,6,-3. The beginning of the strip has become vertex 3. This change is stored in the multiresolution triangle strip and a new object of *class RowNode* with its colIndex field initialised to 3 and its res field initialised to 2 is added to the list of strip beginnings. Three edges disappear while the others remain. This fact is stored in the graph by updating the res field of the remaining arcs to the new LOD 2.

A new strip having only one triangle is produced after the last contraction ($6 \rightarrow 4$), LOD 3. The vertex sequence is 3,4,5,-3. The arcs (3,4) and (4,5) remain at this LOD so its res field is updated to 3. The final vertex of the strip at this LOD is vertex 5, and a new node with the special label -3 has to be added to the list of adjacencies of this vertex.

3.2   Uniform Level of Detail Recovery Algorithm

The LOD recovery algorithm traverses the graph in order to extract the demanded LOD. The algorithm proceeds in two steps. First, the algorithm finds

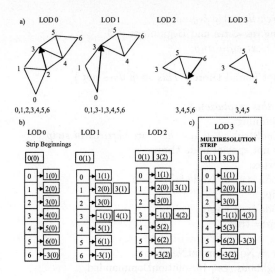

Fig. 2. Construction example of a multiresolution strip. a) Original strip and the sequence of vertex pair contractions. b) The detailed process of a multiresolution strip construction. c) The final multiresolution strip.

the vertex at the beginning of the strip that is compatible with the demanded LOD. Second, the graph is traversed from the vertex at the beginning through those arcs that are compatible with the demanded LOD until a special -3 node is reached. In this case, being compatible means that the arc in the adjacency list has a res field that is bigger or equal to the demanded LOD. The pseudo code of the algorithm is shown in Listing 2.

4 Coherence

Coherence in a multiresolution model means taking advantage of the last information extracted in the extraction of the new LOD demanded. The use of coherence decreases the time consumed by the recovery algorithm, thus saving the need to repeat calculations already done. Coherence in an MTS model means only processing those multiresolution strips that have changed with respect to the last LOD extraction. The sequences for those multiresolution strips that remain will be the same. Each sequence of a multiresolution strip has an LOD interval of validity. The minimum of the interval (minRes) is the maximum value of those arcs that are not compatible with the demanded LOD. The maximum of the interval (maxRes) is the minimum value of those arcs that are compatible with the demanded LOD. In this way, if the new LOD demanded is between these two values the vertex sequence remains valid. The minimum and maximum of the interval are retrieved when the graph is traversed. This interval is recalculated every time it becomes invalid.

```
{ First we search the strip beginning }
while beginning.res < res and Beginnings ≠ ∅
    beginning = next₆beginning;
end while
if beginning Not Found then { This strip does not }
    exit
else { exist at this resolution }
    Node = beginnning;
    while not EndStrip   { While there are vertices in strip }
        if Node is not special node then
            DrawVertex;
        else if Node is -1 or Node is -2 then
            NewStrip;
        else if Node is -3 then
            EndStrip = true;
        end if
        Neighbour = Node.currentNeigbour;
        while Neighbour.res ¡ ResolutionDemmanded
            Neighbour = nextNeighbour;
        end while
        Node = Neighbour;
    end while
end if
```

Listing 2: Pseudo-code of the level of detail recovery algorithm.

5   Results

The MTS model has been subjected to several tests. These tests are addressed
at evaluating the visualisation time in a real time visualisation application. The
results are compared with those of the PM and MOM-FAN. PM uses the triangle
primitive both in the data structures and at the rendering stage. MOM-FAN uses
the triangle fans primitive both in the data structures and at the rendering stage.
A proper implementation of PM is used. Previously, this implementation has
been verified. The time obtained in the extraction is the same as that published
by the author. The polygonal objects used in the test come from the *Stanford
University Computer Graphics Laboratory* (http://www-graphics.stanford.edu/
data/ 3Dscanrep/) and *Cyberware* (http://www.ciberware.com/ models/). The
tests were performed on an HP Kayak XU with two Pentium III processors at
550 MHz and 1 Gb. of main memory. A GALAXY by Evans & Sutherland with
15 Mb. of video memory was used.

5.1   Time Extraction Reduction Using Coherence

Table 2 shows the extraction time with and without coherence for the MTS models. This time remains almost constant independently of the number of LODs
extracted for models without coherence. This is due to the fact that every mul-

Table 1. Characteristics of the objects and model sizes in Mb.

Model	#Strips	#Triangles	#Vertices	MTS	PM	MOM-FAN
Cow	136	5804	2904	0.263	0.256	0.202
Sphere	173	30624	15314	1.301	1.318	1.104
Bunny	1229	69451	34834	3.095	2.993	2.245
Phone	1747	165963	83044	16.995	16.770	13.309

Table 2. Time extraction reduction using coherence in the MTS model. Time in milliseconds.

Model	Linear		Exponential		Random	
	Coherence	Without	Coherence	Without	Coherence	Without
Cow	0.469	1.519	0.389	1.713	1.560	1.640
Sphere	7.364	10.209	6.234	10.830	10.700	10.080
Bunny	15.197	27.120	13.052	29.067	27.500	27.965
Phone	44.264	65.539	38.613	70.254	65.470	66.410

tiresolution strip is processed for each new LOD demanded. If coherence is used, only multiresolution strips that have changed are processed.

5.2   Visualisation time

The MTS model has been compared against PM and MOM-FAN models. These models have been subjected to two kinds of experiments. The first experiment is addressed at evaluating the performance of each model in terms of number of frames per second when a series of LODs are demanded and the object is immediately visualised. The second experiment is more realistic. The user of a real time application needs to feel smooth, not jerky, motion. In this experiment a new LOD is demanded each second. After the extraction, the model is visualised until a new LOD is demanded. Results are shown in Figure 3.

6   Conclusions and Future Work

The main contribution of the multiresolution model presented in this paper is the use of the triangle strips primitive as the basis of the data structure and in the rendering stage. The main benefit of using the triangle strip primitive is the decrease in the amount of data sent to the graphic engine and the resulting acceleration of the rendering stage.

The MTS model has been compared with PM and with MOM-FAN. The model sizes created with MTS are comparable with those of PM. As regards visualisation time, MTS provides significantly higher frame rates than those offered by PM. MTS model sizes are bigger than those created by MOM-FAN. Therefore, MTS frame rates are also significantly higher than the results obtained

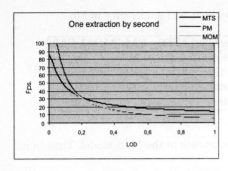 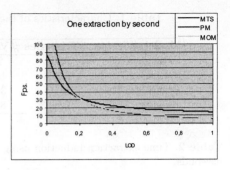

Fig. 3. The plot on the left shows the frame rate for each model as a new LOD is demanded and visualised. The plot on the right shows the frame rate for each model when an LOD is demanded each second. One hundred LODs were extracted.

by MOM-FAN. This is mainly due to the fact that triangle fans in an MOM-FAN model have between three and four triangles per triangle fan. The number of triangles per strip in MTS is about forty. The main conclusion is that the use of triangle strips in a multiresolution model provides frame rates far beyond those offered by models based on the triangle or triangle fans primitive. The drawback of the MTS model is the high extraction time. Introducing coherence inside each strip is something to be dealt with in future work in order to reduce the extraction time. View-dependent visualisation and progressive transmission of the models through a computer network are also lines of future work.

References

1. Arkin E.M., Helod, M., Mitchell J. B. S., Skiena, S. S.: Hamiltonian Triangulation for Fast Rendering, Visual Computer 12(9), 429-444, 1996.
2. Brassard, G., Bratley P.: Fundamentals of Algorithmics, Prentice Hall, 1996.
3. El-Sana, J. Evans, F., Varshney, A., Skiena S., Azanli, E.: Efficiently Computing and Updating Triangle Strips for Real-Time Rendering. The Journal Computer-Aided Design, Vol(32), IS(13), 753-772.
4. Garland, M., Heckbert P.: Surface Simplification Using Quadratic Error Metrics. Proc. of SIGGRAPH'97 (1997) 209-216
5. Garland, M., Heckbert, P.: Survey of polygonal surface simplification algorithms, Multiresolution Surface Modeling Course Notes of SIGGRAPH'97, 1997.
6. Garland, M.: Multiresolution Modeling: Survey & Future Opportunities. State of the Art Reports of EUROGRAPHICS '99 (1999) 111-131
7. Hoppe, H.: Progressive Meshes, Proceedings of SIGGRAPH '96, 99-108, 1996.
8. Hoppe, H.: View-Dependent Refinement of Progresive Meshes. Proc. of SIG-GRAPH'97 (1997) 189-198
9. Puppo, E., Scopigno, R.: Simplification, LOD and Multiresolution - Principles and Applications, Tutorial Notes of EUROGRAPHICS'99, 1999.
10. Ribelles, J., López, A., Remolar, I., Belmonte, Ó., Chover M.: Multiresolution Modelling of Polygonal Surface Meshes Using Triangle Fans, Proceedings of 9th Discrete Geometry for Computer Imagery Conference, 431-442, 2000.

The Hybrid Octree:
Towards the Definition of a Multiresolution Hybrid Framework

Imma Boada[1] and Isabel Navazo[2]

[1] Institut Informàtica i Aplicacions, Universitat de Girona, Spain
imma@ima.udg.es
[2] Dep. LSI , Universitat Politècnica de Catalunya, Spain
isabel@lsi.upc.es

Abstract. The Hybrid Octree (HO) is an octree-based representation scheme for coding in a single model an exact representation of a surface and volume data. The HO is able to efficiently manipulate surface and volume data independently. Moreover, it facilitates the visualization and composition of surface and volume data using graphic hardware. The HO definition and its construction algorithm are provided. Some examples are presented and the goodness of the model is discussed.

1 Introduction

The visualization of scenes that integrate surface and volume data in a single image plays an important role in a large number of scientific visualization applications. In surgical planning, for example, the volume information captured from input devices needs to be visualized, manipulated and analyzed along with objects such as osteotomy surfaces, prosthetic devices or scalpels; in meteorology clouds have to be rendered over terrain data, etc,.

We define a *hybrid scene* as an scene composed of surface and volume data. The different techniques that have been proposed to deal with hybrid scenes are based on one of the two following approaches. The *data conversion* approach that reduces surface and volume data to a common codification scheme, by applying a voxelization process or a polygonalization strategy. Then volume and polygonal data can be rendered using a *classical* rendering pipeline [5, 6, 15, 18]. The *independent representation* approach that preserves original surface and volume data in their independent representation schemes. Data is rendered independently, by the application of two separate rendering processes, and data integration is part of the visualization process which requires the definition of specialized hybrid render algorithms able to composite surface and volume data in depth sorted order. Such an approach usually requires a costly process to properly composite the data [9, 17, 7, 19, 8].

P.M.A. Sloot et al. (Eds.): ICCS 2002, LNCS 2330, pp. 121–130, 2002.
© Springer-Verlag Berlin Heidelberg 2002

In this paper, we introduce a new approach to deal with hybrid scenes: the Hybrid Octree (HO). The HO is an octree-based data structure able to maintain surface and volume data simultaneously and implicitly ordered. The HO avoids conversion artifacts and loss of information of data conversion methods. The capabilities of the HO to represent and efficiently manipulate surface and volume data, either *independently*, taking advantage of classical surface and volume visualization approaches, or in an *integrated* manner, combining surface and volume data in a single image have been presented in detail in [1].

The paper has been structured as follows. In Section 2 we review our previous work related to surface and volume data octree based representations. In Section 3 a detailed description of the HO and its construction process are provided. The capabilities of the HO to manipulate surface and volume data are analyzed in Section 4. Finally, conclusions and future work are given in Section 5.

2   Octree based Codifications. Previous Work.

An octree, originally introduced for solid representations [11, 16, 14], is a tree that codes the recursive subdivision of a finite cubic universe. The root of the tree represents the universe, a cube with 2^n edge. This cube is divided into eight identical cubes, called octants, with an edge length of 2^{n-1}. Each octant is represented by one of the eight descendants of the root. If the information of an octant can not be represented in an exact way, it is labeled as a *Grey node*. Each Grey node is divided into another eight identical cubes which are represented as descendants of the octant in question. This process is repeated recursively until octants contain data that can be represented exactly (named Terminal Octree Nodes) or octants have a minimum edge length called resolution (Minimal Resolution Nodes). According to the data that we would like to represent by means of an octree, it is needed the definition of criteria to determine which information have to be stored in each node and when a node of the octree is a terminal node.

Our study has been restricted to volume data sampled in 3D regular grids and to fitted surfaces reconstructed from the volume data by the Discretized Marching Cubes (DiscMC) [12, 13]. Therefore, the HO has to be designed to support these data in an integrated manner in its terminal nodes.

Previously to the description of the HO we are going to review the octree based structures that we have proposed to represent regular distributed volume data (the Volume Octree, (VO))[2] and a decimated DiscMC fitted surface (the Surface Octree (SO)) [3].

2.1   The Volume Octree

The VO [2] is an octree based codification used to code homogeneous regions of 3D regular sampled data in a compressed way. In the VO, a node is considered *homogeneous* if each sample data inside its associated octant can be approximated by a trilinear interpolation of the eight values on the corners of the octant. The maximum error which is introduced by this trilinear interpolation is the *nodal error*, denoted as ε_0. The nodal error is related to the degree of homogeneity in the area covered by the node.

The construction of the VO starts with the construction of an octree initially complete. Then, during a bottom-up octree traversal nodal errors are computed and all sibling nodes which can be merged in the parent node without introducing an approximation error are purged. Therefore, the VO terminal nodes are those for which their nodal error $\varepsilon_0 \leq \epsilon_u$ with ϵ_u the user required degree of accuracy. The VO allows to obtain multiresolution volume representations by applying error-driven adaptive traversals. These multiresolution representations are the basis of a texture memory assignation policy that reduces the space of texture memory required to obtain 3D texture-based volume data visualizations.

2.2   The Surface Octree

The SO [3] is an octree based codification used to maintain a decimated codification of a fitted surface without introducing error. In addition to the classical White, Black and Grey nodes (octants that are external, internal and crossed by the surface), the SO incorporates five new terminal surface nodes (TS nodes) (see Fig. 1).

To codify the surface in a TS node n_i we maintain the configuration of the node (i.e. F, E, DE, B^n or B^*) and the set of plane equations defining the surface. All the patterns of the Marching Cubes algorithm [10] can be codified as a TS node, therefore the SO is able to maintain any fitted surface obtained with the Marching Cubes. In case of using the DiscMC the plane codification is reduced to a set of integers.

The SO construction process starts with an octree initially complete in which maximal and minimal values covered by each node are stored [20]. The octree is traversed and the DiscMC is applied to the terminal nodes intersected by the surface to provide the surface codification. Afterwards, on a bottom-up octree traversal a merging process is applied to merge coplanar planes and codify them when is possible in upper SO nodes without introducing any simplification error. All details of this merging process can be found in [3]. The capabilities of the SO to support multiresolution surface reconstructions are described in [4].

3   The Hybrid Octree Construction

The HO can be interpreted as the integration of a VO and a SO. Its construction process is composed of two phases: the *Volume Data Integration* and the *Surface Data Integration* (we are going to consider DiscMC fitted surfaces).

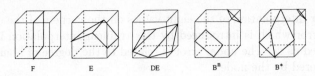

F   E   DE B^n B^*

Fig. 1. Terminal Surface nodes: (F) Face node, a node crossed by a single planar face. (E) Edge node, a node crossed by two planar faces that converge into the node generating an edge. (DE) Double Edge node, a node crossed by three planar faces that converge into the node generating two edges. (B^n) Band^n node, a node crossed by n planar faces that never intersect into the node, with $2 \leq n \leq 4$. (B^*) Special Band node, a node obtained from the union of a F and an E node.

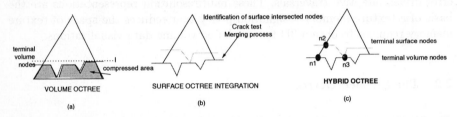

Fig. 2. (a) The Volume Octree is constructed. (b) The construction of the Surface Octree starts at terminal volume nodes. (c) The hybrid octree maintains volume and surface terminal nodes.

3.1   Volume Data Integration

The first phase of the HO construction process is concerned with volume data. This phase starts with the construction of an octree initially complete. Voxel values are not represented explicitly by the octree nodes, but implicitly (the extents of the corresponding region in the voxel space are directly computed from the node index).

For each node n_i of the HO we maintain:

- the maximum (max) and minimum (min) values of the region covered by node (to improve the surface extraction process required in the next phase [20]);

- a list of *nodal errors* $\varepsilon_0, .., \varepsilon_{l_{max}-k}$ where l_{max} is the depth of the octree and k the level of n_i (see Sect. 2.1). These *nodal errors* are computed during a bottom-up octree traversal in which all sibling nodes which can be merged in the parent node without introducing an approximation error are purged (i.e. nodes where $\epsilon_0 \leq \epsilon_u$ with ϵ_u the user allowed accuracy). The list of nodal errors is used to perform multiresolution volume visualizations.

This phase ends when terminal HO nodes with respect to the volume criteria have been detected. (see Fig. 2(a)). Note that these nodes are distributed at different levels of the HO.

3.2   Surface Data Integration

The second phase of the HO construction process codifies the surface in a set of HO nodes. The surface codification strategy is similar to the one used for the SO construction and is composed of the following steps:

(i) *Detect the volume HO terminal nodes intersected by the surface.* During a top-down HO traversal terminal nodes such that $max \leq isovalue \leq min$, where *isovalue* identifies the surface, are detected.

(ii) *Use the TS node configurations (i.e. F, E, DE, B^n or B^*) to codify the surface in the volume HO terminal nodes.* There is a correspondence between the DiscMC patterns and the TS configurations [1]. Therefore applying the DiscMC to the node the TS configuration is directly obtained.

Note that as a consequence of the pruning process that has been applied in the volume integration phase, the terminal HO nodes can be distributed at different levels of the octree. To codify the surface in a node we evaluate its neighbor nodes intersected by the surface: (a) if all evaluated nodes are in the same level no cracks can appear (see Fig. 3(a)); (b) conversely, if the nodes are from different levels cracks can appear (see Fig. 3(b)). In this case the solution we have adopted, that experimentally generates nice results, adapts the intersection points of the processed node to the point position of the deeper neighbor node. This is an straightforward process as the DiscMC allows to determine in advance all possible intersection point positions.

(iii) *Surface Compression.* Once the surface has been codified, a bottom-up octree traversal is performed and the merging process is applied to detect and represent coplanar facets in upper HO levels *without* introducing error. The resultant surface is codified in a set of TS nodes distributed at different upper levels of the HO (see Fig. 2(b)).

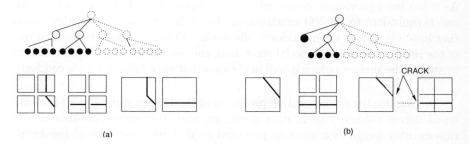

(a)   (b)

Fig. 3. Black circles identify intersected terminal volume nodes. (a) When all the nodes are at the same level the continuity of the surface is preserved. (b) When terminal volume nodes correspond to different octree levels, the application of the DiscMC leads holes on the final surface.

3.3   Terminal HO nodes

If we compare the VO, SO and HO, it can be seen that unlike VO and SO, the HO construction has applied two different criteria to identify terminal HO nodes. Consequently, we can identify terminal HO nodes with respect to the volume or the surface octree construction criterion (see Fig. 2(c)). In particular:

- HO nodes such that $\varepsilon_0(n_i) \leq \epsilon_u$ are terminal nodes with respect to the volume criterion (see node n1 of figure 2(c)).
- HO nodes intersected by the surface with a TS configuration are terminal nodes with respect to the surface criterion (see node n2 of Fig. 2(c)).
- HO nodes such that $\varepsilon_0(n_i) \leq \epsilon_u$, intersected by the surface and with a TS configuration are terminal nodes with respect to the volume and also with respect to the surface construction criteria (see node n3 of Fig. 2(c)).

4   Hybrid Octree Evaluation

The goal of the testing sessions has been proving the capabilities of the HO to manipulate surface and volume data either independently or in an integrated way. In the case that only surface or only volume data has to be explored the use of the HO has not much sense. In these situations is more appropriate the use of the VO or the SO. However, although we are conscious of this fact, we have considered opportune, in the first tests, to evaluate the suitableness of the HO to perform only volume or only surface data explorations.

4.1   HO volume data exploration

As it has been previously described the first phase of the HO construction process is equivalent to the VO construction. The only difference is the information that has to be stored in the nodes: while in the VO we maintain the nodal error, in the HO we maintain the nodal error, max and min values and the information related to the surface (which is null in the case that only volume data is codified).

To visualize the volume data represented in the HO we apply the 3D texture-based octree volume visualization algorithm with the importance-driven texture memory assignation strategy proposed in [2]. Given some user-defined constraints, the HO octree is traversed at rendering time and a set of nodes is selected. This set of nodes determines the current brick decomposition to be used to render the volume data and how the volume data has to be represented in texture memory. The experimental results achieved with the HO volume visualization are equivalent to the ones achieved with the VO [2] in quality and efficiency.

4.2   HO surface data exploration

In the case of surface data explorations the comparison of the SO and the HO is based on the evaluation of the surface codification and the surface reconstruction processes. The models used for the tests are: a CT-vertebra of 128x128x80 represented as (1) in Table 1; a CT-head of 96^3 represented as (2) and a CT-scanned jaw of $128x128x40$ represented as (3). The tests have been performed on a Pentium III at 450MHz with 512Kb of cache memory and a NVidia graphics card.

The results reported in Table 1 compares the distribution of TS nodes before and after the application of the merging process in case of using the HO or the SO presented in [3]. The small difference in the number of nodes of each configuration is due to that in the HO the surface codification process is done in terminal volume HO nodes that not necessarily are of minimal resolution (homogeneous regions have been compacted in the Volume Integration phase described in Sect. 3.1). Note that the HO and the SO achieve the same degree of decimation after the application of the merging process. The HO requires more time to codify the surface. This increase is consequence of the crack test that has to be applied to codify the surface with no error. Memory requirements are comparable with the SO ones.

Table 1. Columns (2) to (6) represent the number of TS nodes according its configuration before the application of the merging process; columns (7) to (11) TS nodes after the application of the merging process; column (12) simplification rate; column (13) times (in seconds) including the codification of the surface in the terminal intersected nodes and the application of the merging process.

Model	F	E	DE	B^n	B^*	F	E	DE	B^n	B^*	Dec. rate	time
(1)(HO)	60.124	5.215	59	57	0	30.095	5.094	64	70	0	44%	7,9
(1)(SO)	62.250	5.191	59	57	0	30.890	5.079	64	70	0	44%	7,1
(2)(HO)	47.654	13.075	556	2.306	200	34.127	13.013	570	2.554	214	15%	4,7
(2)(SO)	48.927	12.975	556	2.326	200	34.336	13.413	570	2.554	214	15%	3,8
(3)(HO)	95.892	22.367	1.122	2.432	196	37.641	25.832	1.142	2.462	210	30%	7,4
(3)(SO)	99.902	25.522	1.122	2.432	196	38.824	25.888	1.142	2.462	210	30%	6,9

To evaluate the capabilities of the HO to support multiresolution surface reconstructions we have applied the importance driven multiresolution algorithm presented in [4]. This algorithm starts with the definition of the region of interest (ROI) by a simple 3D subregion of the dataset domain. Then the HO is traversed and terminal surface nodes contained in the ROI are detected. The surface of these nodes is reconstructed with maximal accuracy. Then, it is performed a process to reconstruct, in a simplified manner, the outer ROI intersected nodes.

Figure 4 presents visual results related to multiresolution surface reconstructions of a CT-head model codified in the HO. When changing the ROI previous computations can be used, i.e. not all the HO nodes have to be recalculated.

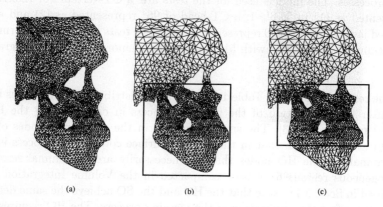

(a)   (b)   (c)

Fig. 4. (a) Lateral view of the CT-skull at full resolution composed by 69.899 planes. (b) Simplified representation composed of 48.108 polygons. The simplification rate if compared with (a) is 30% with 35% of TS nodes into the ROI. (c) Simplified representation composed of 32.213 polygons. The simplification rate if compared with (a) is 53% with 20% of TS nodes into the ROI.

4.3 HO volume and surface data exploration

The most attractive feature of the HO is its capability to maintain surface and volume data simultaneously codified in its nodes. The main advantage of such a capability is that all the sorting phase required by any hybrid renderer is implicitly made when the HO is built, thus during the hybrid visualization process the sorting information can be obtained directly from the HO.

To perform multiresolution hybrid visualizations we have defined a 3D texture-based hybrid visualization function [1]. This function takes advantage of the multiresolution texture memory assignation policy presented in [2], using a compressed texture representation of homogeneous and no importance areas of the volume. The surface is located between the textured polygons according to the position of the TS node inside the HO nodes.

The renderings of figure 5 are obtained from the CT-head dataset. In these images the rendered surface polygons correspond to the dental region. The volume data has been rendered by adopting different voxel-to-texel ratio. At each step the required texture memory is reduced (8Mb in (a), 4 Mb in (b) and 2Mb in (c)). This reduction improves rendering speed.

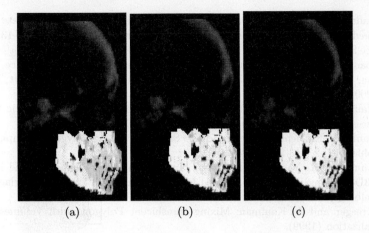

(a)   (b)   (c)

Fig. 5. Different renderings obtained on the CT-head. Surface polygons represent the teeth region.

5   Conclusions and Future Work

We have presented the Hybrid Octree, an extension of the classical octree able to support volume and surface data codified in its nodes and therefore, able to maintain all the data defining an hybrid scene. The main properties of the HO are that: (i) it maintains original surface and volume information. Thus, it avoids conversion artifacts and loss of information of data conversion methods; (ii) it maintains order information implicitly represented in its nodes. Therefore, it reduces the complexity of the depth sorting process required by hybrid renderers [1]; (iii) it preserves all the capabilities of the SO and the VO, therefore it is a well-suited framework for supporting multiresolution surface and volume data explorations.

Our future work is centered on the codification of any fitted surface (not only DiscMC surfaces) by modifying the number of discrete planes used to represent the surface. This modification will produce more compact and smooth surfaces and probably with less error.

Acknowledgements: This work was partially financed by the project 2FD97-1511 of the Spanish Government.

References

1. I. Boada, Towards Multiresolution Integrated Surface and Volume Data Representations. PhD Thesis, Universitat Politècnica de Catalunya, September (2001).
2. I. Boada, I. Navazo, and R. Scopigno, Multiresolution Volume Visualization with a Texture-based Octree. The Visual Computer, Springer International, 17 (3), 185-197, (2001).

3. I. Boada and I. Navazo, An Octree Isosurface Codification based on Discrete Planes. Proceedings IEEE Spring Conference on Computer Graphics (2001) 130-137, Budmerice,Slovakia.
4. I. Boada and I. Navazo, Multiresolution Isosurface Fitting on a Surface Octree. Vision, Modeling and Visualization 2001 proceedings, 318-324, Stuttgart, November(2001).
5. A. Kaufman, Efficient Algorithms for 3D Scan-Conversion of parametric Curves, Surfaces and Volumes, Computer Graphics, 21, 4, 171-79, July (1987).
6. A. Kaufman, Efficient Algorithms for 3D Scan-Converting Polygons, Computer and Graphics, 12, 2, 213-219,(1988).
7. A. Kaufman, R. Yagel and D. Cohen, Intermixing Surface and Volume Rendering. 3D Imaging in Medicine, Edited by K.H. Hohne et al. Springer Verlag Berlin Heidelberg (1990).
8. K. Kreeger and A. Kaufman, Mixing translucent Polygons with Volumes. IEEE Visualization (1999).
9. M. Levoy, A Hybrid Ray Tracer for Rendering Polygon and Volume Data. IEEE Computer Graphics and Applications, 10, 33-40, March (1990).
10. W.Lorensen and H.Cline, Marching cubes a high resolution 3D surface construction algorithm,. ACM Computer Graphics (Proceedings of SIGGRAPH '87), vol.21, n 4, 163-170, (1987).
11. D.Meagher, Geometric modeling using octree encoding. Computer Graphics and Image Processing, 19(2):129-147,(1982).
12. C.Montani, R.Scateni and R.Scopigno, Discretized Marching Cubes, in Visualization '94 Proceedings, R.D. Bergeron and A.E.Kaufman, Eds. (1994), 281-287, IEEE Computer Society Press.
13. C.Montani, R.Scateni and R.Scopigno, Decreasing Isosurface Complexity via Discrete Fitting, Computer Aided Geometric Design, 17 (2000) 207-232.
14. I.Navazo, Extended Octree Representation of General Solids with Plane Faces: Model Structure and Algorithms. Computer and Graphics, vol 13, 1, (1989), 5-16.
15. B.A. Payne and A.W. Toga, Distance Field manipulation of surface models. IEEE Computer Graphics and Applications, 12(1), 65-71. January (1992).
16. H.Samet, Applications of Spatial Data Structures. Addison Wesley, Reading, MA, (1990).
17. L.M.Sobierajski and A. Kaufman, Volumetric Ray-tracing. In Proceedings of 1994 Symposium on Volume Visualization, 11-18. ACM Press, October (1994).
18. M. Sramek, Non-binary Voxelization for Volume Graphics. In Proceedings of Spring Conference on Computer Graphics, (2001) 35-51.
19. D.Tost, A.Puig and I.Navazo, Visualization of mixed scenes based on volume and surface. In Proceedings of the Fourth Eurographics Workshop on Rendering, 281-294, (1993).
20. J. Wilhems, A. Van Gelder. Octrees for Faster Isosurface generation. ACM Transactions on Graphics, 11(3):201-297, July (1992).

Interactive Hairstyle Modeling
Using a Sketching Interface

Xiaoyang Mao[1], Kouichi Kashio[2], Hiroyuki Kato[1] and Atsumi Imamiya[1]

[1] Department of Computer and Media Engineering, Yamanashi University, 4-3-11 Takeda,
Kofu, Yamanashi 400-0016, Japan
{Mao,kato,imamiya}@hci.media.yamanashi.ac.jp
[2] Sony Corporation, Mobile Network Company, Japan.
Hirokazu.Kashio@dvpj.sony.co.jp

Abstract. Modeling and rendering human hair remains to be one of the most challenging computer graphics problems. This paper presents a new interactive hair modeling system featured with a user-friendly sketching interface. With the sketching interface, any user, even a first time user, can create a hair model of his/her desired style just in a few minutes simply by interactively drawing several strokes illustrating the global and local features of the hairstyle. The result of the system is a generic geometry-based representation of hair which can be post manipulated and rendered with most existing systems.

1 Introduction

Needless to say, hair is one of the most important visual elements featuring a human character. A person can look having a completely different expression just by changing his/her hairstyle. To synthesis realistic hair images, researchers have been challenged by many difficult problems inherently caused by the special properties of human hair, such as the huge number of hair strands, the extremely small diameter of individual strand compared with an image pixel, and the complex interaction of light and shadow among strands. In the past decade, a number of novel techniques have been developed for successfully shading hair strands and eliminating alising artifacts[1~6]. In this paper, we presents a new system for allowing users to interactively create 3D hair models of their desired styles. Modeling different 3D hairstyles is even a more difficult problem as the huge variation of hairstyles is determined by extremely complex interactions of many factors. These factors include gravity, static charges, strand-to-strand and strand-to-head interaction, styling tools such as hair pins, hair bands and mousse. Recently, several papers have been published addressing this hairstyle modeling issue[7~11]. A common feature of those systems is the using of cluster model. Because of the cosmetics and static electricity, nearby hair strands tend to form clusters. Use a cluster instead of a single strand as the smallest unit of shape control makes it possible for allowing users to interactively design their desired hairstyles. The trigonal prism wisp model presented by Chen et al.[7] represents each cluster(wisp) with a trigonal prism and the hair strands within a cluster is modeled by the distribution map defined on the cross-sections of the prism.

P.M.A. Sloot et al. (Eds.): ICCS 2002, LNCS 2330, pp. 131–140, 2002.

An editing tool is provided for allowing users to interactively position clusters on the scalp and change the shapes of clusters. The hierarchical cluster model and V-hairStudio developed by Xu et al.[8,10] uses a generalized cylinder to represent a hair cluster. Each cross-section of a cylinder is associated with a density distribution of the hair strands inside the cluster and the output of the system is a volume density representation of hair. Hairstyles are created by interactively changing the shape of the center lines of cylinders. Other researchers explored the possibility of using vector fields for hairstyle modeling[9,11]. Yu succeeded in creating very realistic and complicated hairstyles by modeling the flow and curliness of hair using procedurally defined vector fields[11].

Although most existing hair modeling systems do provide users with some kinds of interactive design tools, the interfaces of those tools, however, are not user friendly. It can be very difficult and tedious for users to construct a hairstyle by adjusting the shape of each hair cluster. For example, it takes about one to two hours for a user familiar with the system to create a typical hairstyle with the hierarchical cluster model[8]. The largest advantage of our new hairstyle modeling system is its user-friendly sketching interface. With this interface, any user, even a first time user, can create a hair model of his/her desired style just in a few minutes. Our technique is based on the consideration that when we look at someone's hairstyle, we pay more attention to its overall design instead of the detailed shape of individual hair strand or cluster. Therefore, instead of asking a user to put strands or clusters together to build up the whole hairstyle, we ask a user to start with specifying the silhouette of the target hairstyle. Based on the silhouette, the system automatically generates a hairstyle and then the user is allowed to locally modify it to obtain the ideal one. In our system, the hairstyle silhouette and all other operations are specified interactively with 2D mouse or tablet as freeform strokes.

The remaining part of the paper is organized as follows: Section 2 gives the overview of the system. Section 3 describes how to specify the silhouette and partition lines of hair and also the area for growing hair. Section 4 describes the algorithm for defining the surface boundary of a hairstyle and Section 5 introduces the rule for growing hair. Section 6 demonstrates some hairstyles generated with our system. Finally, in Section 7, we conclude the paper by showing the future research directions.

2   Overview

A typical hairstyle modeling process in our system consists of the following 6 steps:
1. Loading 3D head model to the view window
2. Specifying the area to grow hair
3. Specifying the partition line
4. Specifying the silhouette of the hairstyle
5. Growing hair
6. Locally modifying the hairstyle

After the 3D head model is displayed on the view window, a user is asked to interactively draw the boundary of the region to grow hair and also the partition line of hair. Fig. 1(a) is an example of such a user-drawn boundary and a partition line. Next, the user is asked to draw a stroke representing the silhouette of the target hairstyle(Fig. 1(b)). Based on the silhouette line, the system automatically generates a surface representing the boundary of the hairstyle(Fig. 2(b)). We call this surface *silhouette surface* hereafter. 3D hair model is then obtained by growing hair strands to fill in the space between the silhouette surface and the surface pf the scalp(Fig. 1(d)). Finally, the user is allowed to locally modify the hairstyle using operations such as cutting, combing and perm. Currently, only the cutting operation is supported and other local modification operations will be available in the near future. We will discuss the detail of each step in the following sections.

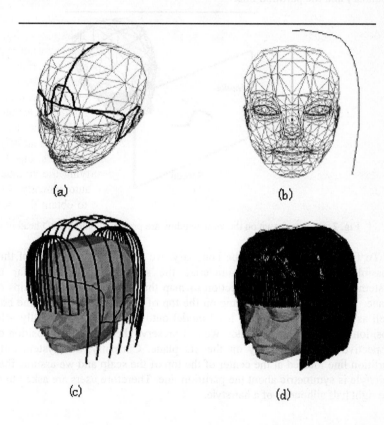

Fig. 1. A typical process of hairstyle modeling. (a) A user draws the strokes specifying the boundary of the region to grow hair and the partition line of hair. (b) The user also draw the silhouette line of the target hairstyle. (c) The system automatically creates the silhouette surface of the hairstyle. (d) The system grows hair strands to fill in the space between the silhouette surface and the surface of the scalp.

3   Specifying the Region to Grow Hair, the Partition Line and the Silhouette Line

The boundary of the region to grow hair, the partition line of hair and the silhouette of a hairstyle are all specified through drawing freeform strokes either using a 2D mouse or a tablet. The strokes representing the boundary of the region to grow hair and the partition line are drawn on the displayed 3D head model directly. A user can rotate the head model to any angle suitable for the drawing task. As shown in Fig. 2, the drawn strokes are projected onto the head model to get the 3D representation of the boundary and the partition line.

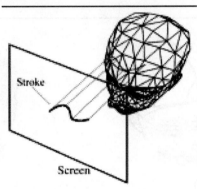

Fig. 2. Strokes drawn on the view window are projected onto the 3D head model.

To find the region inside the boundary, we need a flattened map of the scalp. A possible approach is to parameterize the scalp using a 2D polar coordinate system[11]. Here we use projection to map the boundary and the scalps onto a 2D plane. We virtually place a plane on the top of the head and project the boundary as well as all the vertices of the head model onto the plane. By carefully selecting the position of the projection center, we can preserve the relative 3D position of vertices respective to the boundary on the 2D plane. Currently, our system only accepts partition line located at the center of the top of the scalp and we assume the resulting hairstyle is symmetric about the partition line. Therefore users are asked to draw only the right half silhouette of a hairstyle.

4   Silhouette Surface

Based on the silhouette line, the system automatically creates the silhouette surface of the hairstyle. The shape of the silhouette surface should be decided by taking into consideration both the shape of silhouette line and the shape of scalp. To do this, as shown in Fig. 3, we first obtain a set of evenly spaced cross-sections of the head model along the Z-axis and create a silhouette line for each cross-section by deforming the input silhouette line to match the shape of the contour line of the cross-section. The cross-section starts at the front end point of the partition line and ends at the back end point. Interpolating the silhouette lines for all cross-sections gives the silhouette surface of the side area of the hairstyle. The cross-section with the largest width in X-direction is called *basic cross-section*. The user drawn silhouette line is first projected onto the plane of the basic cross-section to obtain the silhouette line for the basic cross-section. We call this silhouette line *basic silhouette line*. To create the silhouette line for other cross-sections, we calculate the height and width difference of the cross-section and the basic cross-section. Then the silhouette line is obtained by first translating the basic silhouette line in Y-direction the amount of the height difference followed by a shrinking in X-direction with the amount of the width difference.

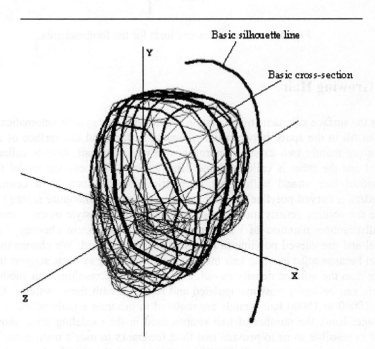

Fig. 3. Cross-sections and silhouette lines.

To generate the silhouette surface for the forehead and the back area, we approximately assume hair strands grow radially from the front and back end points of the partition line. As shown in Fig. 4, to obtain the silhouette lines for the front area, we sweep the silhouette line for the cross-section at the front end point of the partition line through the forehead while keeping the distance from the silhouette lines to the forehead to be constant. The silhouette lines for the back area is created in the same way by sweeping the silhouette line for the cross-section at the back end point of partition line. Fig. 1(c) is the silhouette surface generated with the described method based on the user drawn silhouette line shown in Fig. 1(b).

Fig. 4. Creating silhouette lines for the forehead area.

5 Growing Hair

After the surface silhouette of a hairstyle is generated, the system automatically grows hair to fill in the space between the silhouette surface and the surface of the scalp. There are mainly two existing approaches to represent hair. One is called explicit model and the other is called volume density model. The explicit model represents individual hair strand with some geometrical object, such as a connection of cylinders, a curved polyline or a connected segments of triangular prism[1,2,4,5,11] while the volume density model represents the entire hair style macroscopically as a 3D distribution function of hair density[3,8,10]. Our system chooses the explicit model and use curved polyline to represent each hair strand. We choose the explicit model because most existing hair modeling and rendering systems support this model rather than the volume density model, and therefore the resulting hair models of our system can be easily post manipulated and rendered with those systems. Usually at least 10000 to 15000 hair strands are required to generate a realistic hair image. On the other hand, the number of hair strands used in the modeling stage should be as small as possible so as to provide real time feedbacks to user's operations. To fulfill these two requirements simultaneously, we allow users to specify the number of hair strands by themselves. As we will demonstrate in the next section, 1000 strands seems to be enough for the purpose of previewing and with this number users can get a real time feedback of their operations on a standard PC. We also support a trigonal prism representation for the final hair model.

After the number of hair strands is specified, the positions for growing these hair strands are decided by uniformly distributing this number of points over the scalp within the region to grow hair. Each hair strand is generated following the rule that it passes through the space in between the silhouette surface and the surface of the scalp. The higher its growing position is, the closer it is from the silhouette surface. This rule is depicted in Fig. 5. We first generate the vertical line segment between the top of the scalp and the lowest position of hair growing region. Then we bisect the segment with the position to grow the strands to obtain two segments of length a and b, respectively. The polyline of the hair strand is then constituted by the vertices bisecting the distance between the silhouette surface and scalp surface by the ration of a/b.

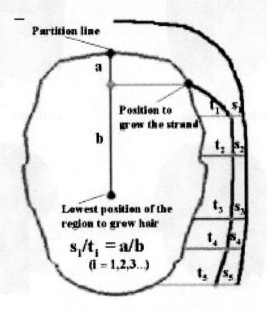

Fig. 5. Rule for growing a hair strand.

6   Results

We have implemented our interactive hairstyle modeling system on Windows environment with Visual C++ and OpenGL graphics library. Fig. 6 shows several images of the hairstyle generated with the stroke inputs shown in Fig. 1. Fig. 6(a) and (b) are the images when 1000 and 10000 hair strands are used, respectively. Fig. 6(c) is the image generated by locally applying a cutting operation to the hair model shown in Fig. 6(b). Cutting operation is specified by drawing the freeform stroke of the cutting line directly on the 3D rendered hair model. The Fig. 6(d) is the image rendered with 3D StudioMax[1]. Each hair strand is represented as a trigonal prism in this case. Times required for rendering images of 1000 and 10000 hair strands on a Pentium machine with two Xeon 933 MHz CPU are about 0.4 and 9 seconds, respectively. Fig. 7 and 8 demonstrate that how curly and wavy hairstyle can also be easily created simply by drawing a curly or wavy silhouette line.

---

[1] 3D StudioMax is the trade mark of Autodesk, Inc.

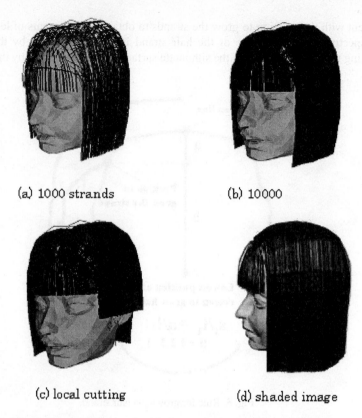

(a) 1000 strands   (b) 10000

(c) local cutting   (d) shaded image

Fig. 6. Images generated with the stroke inputs shown in Fig. 1.

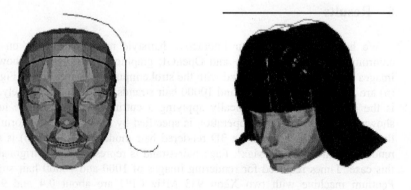

Fig. 7. A curly hairstyle generated with our system. Left: user specified silhouette. Right: resulting hairstyle.

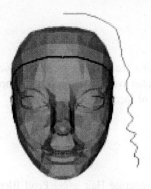

Fig.8. A wavy hairstyle generated with our system. Left: user specified silhouette. Right: resulting hairstyle.

7   Concluding Remarks

We have presented an interactive hairstyle modeling system featured with a sketch interface. The idea of using sketching interface is also inspired by Teddy, a successful sketching interface for 3D freeform design[12]. As we mentioned in Section 1, the variation of hairstyles is determined by extremely complex interactions of many factors, such as gravity, strand-to-strand and strand-to-head interaction and styling tools. A faithful physical simulation of all those factors can be very difficult and time consuming. In many applications, however, an expressive presentation of personalized features of hairstyles is of most important. Our sketching interface provides an easy to use expressive design tool to those applications. For example, as the potential applications, our system can be used to smooth the communication between a customer and a stylist at hair salons and also used for in house CG creation for visually present the a designer's requirements to a professional CG modeling specialist.

The results presented in this paper are just the very initial results of our prototype system and the variation of hairstyles is limited currently. We are now implementing other local modification operations such as combing and perm. To make the hair image look more realistic, we need to add some randomness to the positions and shapes of hair strands generated following the rule introduced in Section 5. We are now implementing a method which groups hair strands near by into clusters and adds randomness to the shape of each hair cluster. Also we need some special consideration for generating short cut hairstyles.

Acknowledgements

The authors would like to thank Issei Fujishiro from Ochanomizu University for his helpful comments and continuous support. This work was partially supported by Telecommunications Advancement Organization of Japan.

References

1. LeBlanc A. M., Turner R. and Thalmann D.: Rendering Hair using Pixel Blending and Shadow Buffers, *The Journal of Visualization and Computer Animation*, Vol. 2, (1991) 92-97
2. Rosenblum R.. R., Carlson W. E. and Tripp III E.: Simulating the Structure and Dynamics of Human Hair: Modeling, Rendering and Animation, The Journal of Visualization and Computer Animation, Vol. 2, (1991)141-148
3. Kajiya T. and Kay T.L.: Rendering Fur with Three Dimensional Textures, Computer Graphics, Vol.23, No.3, (1989) 271-280
4. Watanabe Y. and Suenaga Y.: A Trigonal Prism-Based Method for Hair Image Generation", IEEE Computer Graphics & Applications, (1992) 47-53
5. Anjyo K., Usami Y. and Kurihara T.:A Simple Method for Extracting the Natural Beauty of Hair, Computer Graphics, Vol.26, No.2, (1992) 111-120
6. Magnenat Thalmann N., Kurihara T., Thalmann D.: An Integrated System for Modeling, Animating, and Rendering Hair., Proceedings of EUROGRAPHICS '93, 211-221
7. Chen L.H., Saeyor S., Dohi H. and Ishizuka M.: A System of 3D Hair Style Synthesis based on the Wisp Model, Visual Computer, Vol. 15, (1999) 159-170
8. Yang, Xue Dong, Tao Wang Zhan Xu, and Hu Qiang: Hierarchical Cluster Hair Model, Proceedings of the IASTED International Conference on Computer Graphics and Imaging, (2000) 75-81
9. Hadap S. and Magnenat-Thalmann N.: Interactive Hair Styler Based on Fluid Flow, Proceedings of Eurographics Workshop on Computer Animation and Simulation '2000
10. Xu Z. and Yang X. D: V-HairStudio: An Interactive Tool for Hair Design, IEEE Computer Graphics & Applications, Vol. 21, No. 3 (2001)36-43
11. Yu Yizhou: Modeling Realistic Virtual Hairstyles,Proceedings of Pacific Graphics'01, (2001) 295-304
12. Igarashi T., Matsuoka S., Tanaka H.: Teddy: A Sketching Interface for 3D Freefrom Design, ACM SIGGRAPH 99, 409-416

Orthogonal Cross Cylinder Using Segmentation Based Environment Modeling

Seung Taek Ryoo, Kyung Hyun Yoon

Department of Image Engineering
Graduate School of Advanced Imaging Science, Multimedia and Film
ChungAng University, Seoul, Korea
{bluelancer, khyoon}@cglab.cse.cau.ac.kr
http://cglab.cse.cau.ac.kr

Abstract. Orthogonal Cross Cylinder (OCC) mapping and segmentation based modeling methods have been implemented for constructing the image-based navigation system in this paper. The OCC mapping method eliminates the singularity effect caused in the environment maps and shows an almost even amount of area for the environment occupied by a single texel. A full-view image from a fixed point-of-view can be obtained with OCC mapping although it becomes difficult to express another image when the point-of-view has been changed. The OCC map is segmented according to the objects that form the environment and the depth value is set by the characteristics of the classified objects for the segmentation-based modeling. This method can easily be implemented on an environment map and makes the environment modeling easier through extracting the depth value by the image segmentation.

1 Introduction

With the development of image-based representation technology, it has become possible to navigate through the surrounding environment without actually modeling and constructing the virtual environment just by acquiring the image of the area to navigate. Photogrammetric modeling, image based modeling using corresponding points, interpolation from dense samples, cylinder based rendering, and the vanishing point based modeling method are generally used to construct an image-based navigation system [1]. With previous methods, it is difficult to acquire an image with a full-view when navigating through the environment. Also, the earlier methods are difficult to apply on environment maps. A new environment mapping method, and an environment modeling method through image segmentation are introduced in this paper to resolve such a problem.

The Orthogonal Cross Cylinder (OCC) is the object expressed by the intersection area that occurs when a cylinder is orthogonal with another (Fig. 1). The OCC mapping method eliminates the singularity effect caused in the environment maps and displays an almost even amount of area for the environment occupied by a single texel. The full-view image can be obtained through such a mapping method but it is difficult to express the image when the point of view

P.M.A. Sloot et al. (Eds.): ICCS 2002, LNCS 2330, pp. 141–150, 2002.

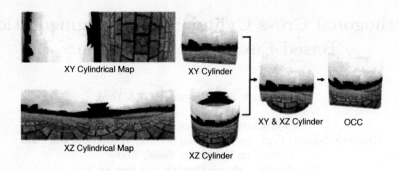

Fig. 1. Orthogonal Cross Cylinder

has been moved. The environment model must be composed using the depth value from the environment map to acquire the image from the changed point of view. The OCC map is divided by the object that forms the environment, and the depth value is set by the characteristics of the object in the segmentation based modeling method, which is used for this purpose. This method is easily applied on the environment map and can easily be used for the environment modeling by extracting the depth value through image segmentation. It becomes possible to develop an environment navigation system with a full-view through these methods.

2   Related Work

Environment mapping is a method of adding realism to a scene by using special texture indexing methods. This method refers to the process of reflecting a surrounding environment within a shiny object and it was originally introduced as a cheaper alternative to ray tracing [2]. In image-based rendering, the virtual viewer replaces the shiny object, and that part of the map intercepted by the viewer's field of vision is presented to him as a two-dimensional projection. Environment mapping is essentially the process of pre-computing a texture map and then sampling texels from this texture during the rendering of a model. The texture map is a projection of 3D space to 2D space. The methods used for the projection are sphere mapping, latitude mapping, cube mapping, cylindrical mapping, and paraboloid mapping [3].

Blinn/Newell Latitude Mapping used a latitude-longitude map indexed by the reflected ray [2]. The texture map's U coordinate represents longitude(from 0 to 360 degrees) and the V coordinate represents latitude(from -90 to 90 degrees). A latitude-longitude map's distortion will become apparent if the environment map is applied to planar objects with consistent neighborhood relationships such as a mirror. An object with curved surfaces will alleviate the distortion inherent in the environment map image. Furthermore, there is a singularity in the mapping at the pole. Sphere mapping is a type of environment mapping in which the irradiance image is equivalent to that which would be seen in a perfectly reflective hemisphere when viewed using an orthographic projection. Sphere mapping uses a single texture image. This means that look-ups are fast and the environment is represented compactly. But, there is a singularity in the mapping at the outer edge and it does not use all the texture in the texture map. There is a waste of texture memory. Cube mapping was invented by Ned Greene in 1986 [4]. The cube map stores the environment as six textures. Each texture represents the view out of one face of the cube as seen from the center of the cube. A cube map is easier to generate than a sphere map and results in less distortion than the other mappings. But, cube mapping requires indexing through several textures in order to compose the final image. Also, discontinuity is introduced near the edges of the map and this results in seams occurring along the object. For cylindrical mapping, a cylinder can be easily unrolled into a simple planar map. The surface is without boundaries in the azimuth direction. One shortcoming of a projection on a finite cylindrical surface is the boundary conditions introduced at the top and bottom [5]. Dual paraboloid environment mapping was invented in 1998 by Heidrich and Seidel [6]. Dual-paraboloid mapping requires two textures to store the environment, one texture for the front environment and another texture for the back environment. The underlying geometry of the mapping is based on a parabola, which has the desirable property of containing a single point of focus for incoming rays. A dual-paraboloid map compared to the sphere map approach is better sampling of the texture environment with the elimination of sphere mapping's sparkle artifacts. But, constructing the dual-paraboloid map requires warping two textures instead of just one and also requires two rendering passes or the use of multi-texturing. Also, almost 25% of the pixels are not used, and edge pixels between maps need to perform special blending in order to reduce artifacts from using two maps. In this paper, we propose the use of OCC mapping, which is a new type of mapping that eliminate the singularity problem occurring at the pole of the environment map and the problem of the size of the area (solid angle) subtended by a texel environment map of various forms. The properties of OCC mapping are presented in chapter 3.

Environment modeling using environment information is required to freely navigate through the environment. Photogrammetric modeling, a modeling using the disparity acquired through a matching algorithm, and modeling using the vanishing point method are a couple of modeling methods [1]. A 3D model can be constructed through the method using the disparity, but it requires a high cost for searching the corresponding point from the image. The vanishing point based modeling method only constructs a rough model, which makes it difficult to express the details of the environment. With the photogrammetric modeling method, it is difficult to match the object and the primitive due to the deformation of the image when applied on the environment map. The segmentation based modeling method is used in this study to solve such a problem. This method can easily be applied on environment maps and can be used to model the environment easily. The segmentation based modeling method is discussed in detail in chapter 4.

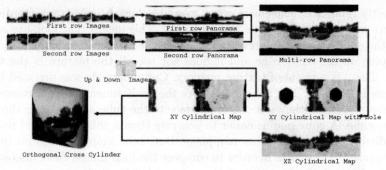

First row Images

First row Panorama

Second row Images

Second row Panorama

Multi-row Panorama

Up & Down Images

XY Cylindrical Map

XY Cylindrical Map with Hole

Orthogonal Cross Cylinder

XZ Cylindrical Map

Fig. 2. Creation of OCC map from real scene

3   Orthogonal Cross Cylinder

OCC has been developed for evenly sampling the environment map by the pixels on the screen and its effective storage while maintaining the characters of the cylindrical map. It is an object created by intersecting the XZ cylinder of Y-axis and the XY cylinder of the Z-axis (Fig. 1).

The process of expressing the surrounding environment using OCC mapping consists of creating an XZ and XY cylindrical map, reproducing the OCC map through eliminating redundancy, and rendering through OCC map sampling. The environment map is created by using a virtual line camera in the environment which creates three dimensions when created from computer-generated images and by using the stitching algorithm when created from real-world images. The XZ and XY cylindrical maps created through these methods contain ineffective samples that overlap. Therefore, a step for eliminating such redundant samples is required in the process of constructing a more effective OCC map. The OCC map is composed in a way to be compatible with other environment maps. An image can be created in any direction through rendering based on texture mapping and ray casting using the OCC map created.

3.1   Creation of OCC Map

Figure 2 is a flow diagram of the process to construct an OCC from a real scene. An image is acquired by pitching up (using a tripod) the camera with a semi fisheye lens attached, by 30 degrees off the horizontal plane and then taking the six images by rotating anti-clockwise around the Y-axis at 30 degree intervals. The camera is then pitched down by 30 degrees off the horizontal plane and rotated around the Y-axis to acquire the six images in 30 degree intervals. Finally, the bottom and the top of the scene, after the tripod has been removed, are captured as an image. The images acquired from the first rotation are stitched together as a cylindrical map using the panorama tool. The images acquired from the next rotation are put together to form the second cylindrical map. A panoramic image with a wide angle can be obtained by stitching two different cylindrical

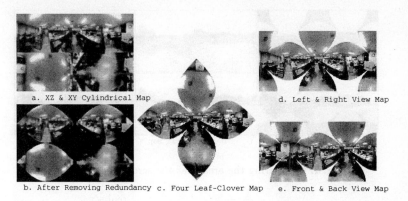

a. XZ & XY Cylindrical Map

d. Left & Right View Map

b. After Removing Redundancy   c. Four Leaf-Clover Map   e. Front & Back View Map

Fig. 3. Various environment maps using OCC

maps together in the same way. However, the panoramic image obtained through this method cannot include its entire surrounding environment due to the lack of information on its surrounding environment around each pole. The panoramic images are transformed into an XZ and an XY cylindrical map to resolve this problem. The transformed XZ cylindrical map produces a perfect solution by acquiring all the necessary information on its surrounding environment, whereas the solution from the XY cylindrical map would have holes due to the lack of information on the top and bottom of the image. These holes can be filled by using the image of the top and bottom acquired in the first step of the process and an image-editing program.

An OCC can be constructed using the XZ and XY cylindrical maps created through this process. However, the XZ and XY cylindrical maps used to create the OCC are not suitable to be used as OCC maps because they contain redundant environment data. This problem is discussed in the following chapter.

3.2   The Shape of OCC Map

The easiest way of mapping the OCC onto a 2D screen coordinate is to use the XZ and XY cylindrical map (Fig. 3-a). However, since the two cylindrical maps contain redundant environment data, it is necessary to discard such redundancy beforehand. The following equation can be obtained by finding the intersection area of the two cylinders using the radius (r) of the cylinder and the coordinate of the cylinder (θ, v).

$$v = r \pm r sin\frac{\theta}{r} \tag{1}$$

The redundant data of the two cylinders can be discarded using equation 1 (Fig. 3-b). But, this cylindrical map is ineffective by having 35% of its storage space left as empty space. The XZ and XY cylindrical map is similar in its shape to a four-leaf clover (Fig. 3-c). Each end of the leaf of the clover meets at a single point. A map that maps the right side and the left side of the OCC (Fig. 3-d)

a. OCC Map   b. Octahedral Map   c. Spherical Map

Fig. 4. Comparing the area that the screen pixels occupy

and a map that maps the front and the backside of the OCC (Fig. 3-e) can be constructed through this characteristic of the OCC map. The space for storing the samples needed to construct an OCC can be reduced by 25% and wasted space can be reduced by 19% through this method. The front and the backside of the map can be used through warping for even more effective use of space (Fig. 4). The octahedral map is formed through warping the pixels towards the straight line to fill the space left empty in the front and the back-view map.

3.3   Comparing Sampling Pattern

In this study, the accuracy of each map has been evaluated by comparing the texture area corresponding to a certain solid angle and the solid angle corresponding to each texel. The size of the area that the screen pixel occupies is more even at the pole and near the equator for the OCC map and the octahedral map, than the spherical map, as shown in Figure 4.

Also, the solid angle for each texel has been calculated to measure the effectiveness of the sampling on each map. The solid angle of the texel for the cube map, cylindrical map, spherical map, OCC map, and the octahedral map have each been measured in this study. The solid angle is expressed as the area that the ray has been projected onto the sphere for each texel

Table 1. Comparing solid angle

	Sample #	Total	Min.	Max.	Avg.	Dev.	Area
Cube	239756	12.573205	0.000019	0.000100	0.000052	0.000020	100 %
Cylinder	125600	8.888219	0.000036	0.000100	0.000071	0.000021	71 %
Sphere	197192	12.566737	0.000001	0.000100	0.000064	0.000031	100 %
OCC	159764	12.559790	0.000036	0.000100	0.000079	0.000019	100 %
Octahedron	197192	12.566234	0.000023	0.000100	0.000064	0.000020	100 %

Table 1 shows the minimum and maximum value, average value, and the deviation of the solid angle on each map. If the average value is small, the area

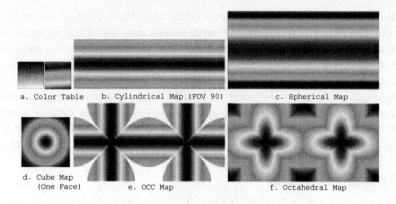

a. Color Table   b. Cylindrical Map (FOV 90)   c. Spherical Map

d. Cube Map
(One Face)   e. OCC Map   f. Octahedral Map

Fig. 5. Comparing solid angle per single texel

of the environment that the texel represents is of a small size and vice versa. The deviation value represents the change in the value of the solid angle. The change in the value of the angle becomes more even as the deviation value becomes smaller. Therefore, an effectively sampled map has a large average value and a small distribution value. The cube map has the smallest average value of all the maps, but has a small deviation value, which causes little change in the value of the solid angle of a texel. The cylindrical map shows a good sampling pattern with a large average value and a small distribution value, but has a problem of not being able to cover all the area. The spherical map has a sampling pattern that shows a rapid decrease of the value for the solid angle at the pole, and a large difference in the value between the minimum and maximum value of the solid angle. The OCC map has the best sampling pattern of them all, with the largest average value and the smallest distribution value. A spectrum color table has been used to visually express the difference between the solid angle of each texel for the various maps, instead of using the grey shade color table, in Figure 5-a. Figure 5 shows that the OCC map provides the best sampling pattern, which expresses the surrounding environment most effectively.

4   Segmentation Based Modeling

The OCC method, explained above, is used to navigate its surrounding environment from a fixed point of view. However, a model that uses the environment map is required to freely navigate through the virtual environment. The depth value must be extracted from the OCC map to construct such a model. The segmentation based modeling method is used for this purpose in this study.

4.1   Image Segmentation

An indoor virtual environment consists of a floor and ceiling and the outdoor of ground and sky. From such features, the environment map is divided into a floor (ground), ceiling (sky), and the surrounding objects. A method for image segmentation is required to divide the environment map into these characteristics. The image segmentation method can be divided into a region-based method and an edge-based method [7][8]. Region based methods take the basic approach of dividing the image into regions and classifying pixels as inside, outside, or on the boundary of a structure based its location and the surrounding 2D regions. On the other hand, the edge-based approach classifies pixels using a numerical test for a property such as image gradient or curvature. The environment map can be segmented by the characteristics of the object through these methods.

4.2 Depth Calculation

The depth can easily be acquired from the segmented environment map using the following conditions:

- The floor (ground) and the ceiling consist of a plane that is parallel to the XZ plane in the real-world coordinate.
- The sky consists of a semi-sphere with an infinite radius.
- The surrounding objects are perpendicular to the ground.

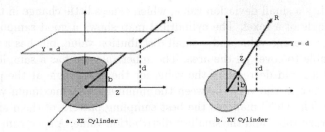

a. XZ Cylinder   b. XY Cylinder

Fig. 6. Depth calculation using the relationship between ceiling (floor) and OCC

Figure 6 shows the process of extracting the depth value of the ceiling. The depth can be obtained by testing for the intersection point between the ceiling that is parallel to the XZ plane, and the ray that crosses the actual 3D coordinate mapped onto the pixel on the environment map. The relationship between the XZ cylinder and the ray (Fig.6-a), and between the XY cylinder and the ray (Fig. 6-b) must be considered to get the depth value from the OCC map. The following equation can be obtained by calculating the radius (r) of the cylinder, height (d) of the XZ plane, and the y coordinate value (b) of a point on the cylinder with the triangle equation.

$$Z(Depth) = r \times \frac{d}{b} \qquad (2)$$

The depth of the floor and the ceiling of the segmented environment map can be extracted using Equation 2. Also, because the surrounding objects are perpendicular to the ground, the depth image can be obtained by setting the depth value of the floor on the objects surrounding the floor. Figure 7 shows the process of environment mapping from OCC map and environment modeling from depth image. Figure 8 compares three frames that have been rendered using the OCC model and the Environment model using depth image. The image above is the OCC model image, and the image below is the environment model expressed by adding a depth image to the OCC model. It should be evident that the OCC model with depth, expresses the environment well by observing the occlusion in the wall in the middle.

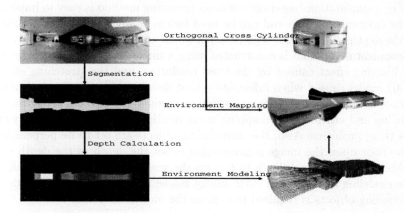

Fig. 7. Environment mapping from OCC map and Environment modeling from depth image

Fig. 8. Comparing of OCC and segmentation based model

5 Conclusion & Future Works

The OCC mapping method, a new method used for effective environment mapping and a segmentation-based environment modeling method for freely navigating through the environments are introduced in this study. The OCC mapping method eliminates the singularity effect caused in the environment maps and shows an almost even amount of area for the environment occupied by a single texel. The OCC mapping method is suitable for application on environment navigation systems due to its effective storage of the environment and faster sampling time.

The segmentation-based environment modeling method is easy to implement on the environment map and can be used for environment modeling by extracting the depth value by the segmentation of the environment map. However, an environment model that is constructed using a single OCC map has the problem of a blurring effect caused by the fixed resolution, and the stretching effect of the 3D model caused when information that does not exist on the environment map occurs due to the occlusion. A study on voronoi-diagram based environment modeling and environment mapping using multiple OCC maps is underway to solve these problems. Also, the surrounding objects are set to be perpendicular to the floor upon the image segmentation to set the depth as the depth of the neighboring floor. It is difficult to express the surrounding objects in detail with these modeling methods. An environment modeling method that subdivides the surrounding objects is required to express the objects in detail.

Acknowledgement

This work was partially supported by the National Research Laboratory program of the Korean Ministry of Science and Technology (No. 2000-N-NL-01-C-285)

References

1. Bregler, C., Cohen, M. F., Debevec, P., McMillan L., Sillion, F. X., Szeliski, R.: Image-based Modeling, Rendering, and Lighting. Siggraph 2000 Course 35, (2000)
2. Blinn J., Newell M.: Texture and reflection in computer generated images. Communications of the ACM, (1976) 19:456–547
3. Blythe, D., Grantham, B., McReynolds, T., Nelson, S. R.: Advanced Graphics Programming Techniques Using OpenGL. Siggraph '99 Course 29, (1999)
4. Greene, N.: Environment Mapping and Other Applications of World Projections. Computer Graphics and Applications, (1986) 6(11):21–29
5. McMillan, L., Bishop, G.: Plenoptic modeling : An image-based rendering system. Siggraph '95, (1995) 39–46
6. Heidrich, W., Seidel, H. -P.: View independent Environment Maps. Eurographics/ACM Siggraph Workshop on Graphics Hardware '98, (1998) 39–46
7. Imielinska, C., Laino-Pepper, L.: Technical Challenges of 3D Visualization of Large Color Data Sets. The Second Visible Human Project Conference Proceedings, (1998)
8. Mortensen, E. N., Reese, L. J., Barrett, W. A.: Intelligent Selection Tools. IEEE Conference on Computer Vision and Pattern Recognition '00, (2000) 776–777

Helping the Designer in Solution Selection: Applications in CAD

Caroline Essert-Villard

Laboratoire des Sciences de l'Image, de l'Informatique et de la Télédétection
UMR CNRS 7005 – Université Louis Pasteur
Pôle API, Boulevard Sébastien Brant – F-67400 Illkirch, France
essert@dpt-info.u-strasbg.fr

Abstract. In CAD, symbolic geometric solvers allow to solve constraint systems, given under the form of a sketch and a set of constraints, by computing a symbolic construction plan describing how to build the required figure. But a construction plan does not usually define a unique figure, and the selection of the expected figure remains an important topic. This paper expose three methods, automatic or interactive, to help the designer in the exploration of the solution space. These methods guide him towards the expected solution, by basing the construction on the observation of the sketch. A set of examples from a large range of application domains illustrate the different methods.

1 Introduction

Over the past 20 years, many studies have been done about solving geometric problems. Geometric solvers offer the possibility to solve constraint systems graphically and interactively given by a designer under the form of a dimensioned sketch. Several kinds of methods, numeric, symbolic, or even combined, are used to solve such geometric constraint systems (see R. Anderl and R. Mendgen for a survey [1]). In the case of symbolic solvers, and if the dimensioned sketch is well constrained, the result is a symbolic *construction plan*. Its numerical interpretation enables to generate all the solutions of the system.

This approach has been chosen in our team because of its exactitude, and the ability to propose not only one solution satisfying the constraints, but the whole solution space. The solution space provided by the symbolic geometric solver *YAMS* [2], that was developed from this approach, can be represented as a tree. Indeed, as the construction plan is a list of definitions, in wich the function may generate several results (such as the intersection between a line and a circle), we can build a branching in the *solutions tree* at each time the function actually generates several results, becoming a *multifunction*.

Even in the well constrained case, a constraint system generally defines a large number of figures, whereas the designer usually wants the unique figure corresponding to his will. This sets the problem of how to select the "expected" solution ? We think that a designer does not draw a dimensioned sketch haphazardly, but has already in mind the kind of solution he wants. We take advantage of this idea to help him in the selection of the pertinent solution.

P.M.A. Sloot et al. (Eds.): ICCS 2002, LNCS 2330, pp. 151–160, 2002.
© Springer-Verlag Berlin Heidelberg 2002

In this paper, we will illustrate our methods by presenting some results of experimentations in several applicative domains. These examples show at once the advantages of the methods, how they are limitated, and which answers to these limitations would be helpful. Note that all figures presented in this paper are snapshots from our prototype *SAMY*, extending *YAMS*.

2   Solutions Selection and Solution Space Browsing

Whatever the solving method, the standard response given by most of the solvers is to browse the whole solution space, and/or to use some simple pruning heuristics. But in the case of large systems, such an exploration may be both tedious and very long. Let us see what propositions have been made so far by various authors to the problem of the selection of a solution amongst many.

B. Brüderlin briefly observed in [3] the existence of ambiguous solutions, and proposed to introduce additional constraints to help clearing the ambiguity. But doing this may lead to an over-constrained system.

J. Owen exposed in [4] an idea, yet oulined by B. Aldefeld ([5]), to solve the ambiguities. He proposed to compare the relative placements of the elements of the solution figure with those of the given sketch. However, for this kind of method, the number of criteria to give is too important, and furthermore these criteria are, to our mind, very insufficient to find the expected solution (see [6]).

C. Hoffmann *et al.* were some of the firsts to be really interested in the multiple solutions problem. In [7], they proposed 3 ways to find the expected solution. The first one, quickly abandoned, was to move the misplaced elements of a figure. The second one was to over-constrain the system. The third one was to propose an "incremental" mode, allowing the user to modify an element at a "level" of the construction if it is not suitable. But this method didn't seem to be very intuitive yet. Moreover, after a first change, the figure had to be "regenerated" before another change could be done. From that point of view, the use of a symbolic solving method would be costless in time and processes.

In the end, the problem of selecting the most relevant solution amongst the whole solution space, instead of simply one solution complying with the constraints, has not been studied yet in a satisfactory way. That is why we propose to improve in some way the outset of an answer brought by C. Hoffmann *et al.*, with a deeper study of some automatic and interactive methods. For this work, we will take advantage of the properties of the solving method followed in the existing symbolic geometric solver of our team, *YAMS* [2], and stay in that framework.

3   Automatic and Interactive Tools

We propose two categories of methods and associated tools: the first one is automatic (selection with no intervention of the user) and the second and third ones are interactive (letting the user choose).

3.1   Branch Freezing

As we explained before, we think that the designer has in mind the kind of figure he wants to obtain when he draws his dimensioned sketch, and that in most cases he takes into account his idea to draw a sketch having likeness to his will. That is why we intend to use the sketch in order to find the solution that has the best likeness with it. Let us precise that we work on plans yet computed by *YAMS*.

Likeness Let us first define our notion of likeness. Authors often use comparisons of relative placements of the elements of the figure to determine if two figures are similar or not. We only notice that most of these properties comparisons can be held in check by some simple examples (see[6]). So, we proposed a better criterion based on homotopy, called *S-Homotopy*. Homotopy as in [8] is too general for our case, whereas S-homotopy takes exactly into account geometric figures solutions of constraint systems, by including continuous deformation of constraint systems themselves to ensure homogeneity. It also includes a very important point that imposes the continuous deformation not to go through a degenerate case. If that occured, we could jump to another solution.

This led us to propose a numbering for multifunctions values, which is compatible with the continuous deformation of the geometric constraint systems, and which we defined as *continuous numbering*. This means that whatever the continuous deformation of the considered constraint system, the result numbered n will always keep the same geometric properties. The link between the notions of likeness and continuous numbering is that two figures e and f are S-homotopic if and only if they have the same number in their solutions tree. In addition, given a sketch e, there is at most a unique solution f such that e and f are S-homotopic, *i.e.* that have, in our sense, the best likeness.

Freezing Technique We first point out the fact that the sketch can be seen as a particular solution of the constraint system instantiated by the values read on the sketch. So, it corresponds to a particular branch of the solutions tree, that we call the *sketch branch*. Then, we suppose that the user gave a sketch and some constraints that look like his expectations, that is, in the sense we defined earlier, there is a S-homotopy between the sketch and the solution. Thus, if we can find a solution having the same number than the sketch, then we advance the idea that it is the intended solution. The operation consisting in storing the number of the sketch branch is called *freezing* of a branch. When this number is known, we can launch an interpretation with the given dimensions, guided by the number of the *frozen* branch.

One of the advantages of this method is its speed. Indeed, instead of potentially comparing some geometric properties of all the objects of the figure with each other, we only compare, at each junction of the tree, the objects that are brought into play in the concerned multifunction. Since the treatment is made as the interpretation goes along, this method reduces significantly the processing time in comparison with a systematic method. Another advantage is that this method is very general, thanks to its continuity justification by S-homotopy.

Limitations The method we exposed in previous section works fine when all constraints are metric. But another type of constraints is also used in *YAMS*: Boolean constraints. As examples, we can cite tangency, or equality of objects. In the case of Boolean constraints, some information is missing to find the intended solution. In that case, the freezing of a branch produces, instead of a unique branch, a small sub-tree of the initial solutions tree.

Fig. 1. Neveu *et al.* examples: sketches and solutions

Applications Geometric constructions under constraints can be used to represent objects in various domains. We present here several kinds of examples, illustrating the use of the branch freezing to find the expected solution in varied situations.

First, let us take an example from literature. In [9], B. Neveu *et al.* presented geometric figures made of triangles, that we recall on Fig.1. The constraint systems, containing only distance constraints, provide respectively 128 and 64 solutions. If we use our branch freezing technique, we obtain an instantaneous good answer for both (second and fourth parts of Fig.1), after an also instantaneous symbolic resolution. This technique avoids to compute all the solutions and review them. To have an idea of what reviewing them one by one would cost at worst, the authors cite the following approximative process times (not including viewing): 9 sec. and 1.4 sec. on a Pentium III 500, respectively.

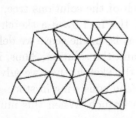 

Fig. 2. Triangulation example: the sketch, the right solution, and one of the rejected solutions

The next example is a representative triangles problem, as we can meet in triangulation studies. The sketch of Fig.2 is composed of 34 adjacent triangles, with constrained sides. This kind of problem is often cited, for instance by Owen in [4], and well known to have 2^{p-2} solutions if p is the number of points, that is in our case 33554432 potential solutions. On the middle of Fig.2, we can see

the result found instantaneously with the branch freezing method. On the right
of Fig.2 is presented one of the solutions rejected by our method.

Let us take now an example describing a more concrete object, and that will
illustrate the limitation of this method. The sketch on top of Fig.3 represents
a lever, as it could be drawn for manufacturing, given with the appropriate
constraints on distances, angles, and two tangency constraints involving two of
the arcs on the right. As all the constraints are not metric, the branch freezing
can't reduce the solution space to one single branch, but reduces it to a small
sub-tree with only 4 branches representing the 4 remaining solutions (see Fig.3).

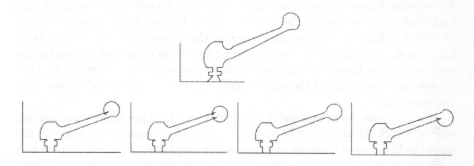

Fig. 3. The lever: the sketch and 4 remaining solutions

If the remaining sub-tree is, like this one, very small, it is quite easy to find
the good solution by viewing all of them one by one. But if the sub-tree is too
large to be reviewed this way, or if none of them correspond to the user's intend,
either because the sketch was too close to a limit case, or because he wants to
see other solutions, it becomes useful to provide other tools to help him. That
is why we propose next two interactive tools that can constitute a complement
for this first method, or be used alone as well.

3.2 Step by Step Interpretation

First, let us recall that the whole solution space may be very wide, and that
exporing it may be quite tedious. In order to help the user exploring efficiently
the solution space, a first proposal is to let the user take the choices in hand.
This is possible because our approach is formal and we have a construction plan,
generating a solutions space structured as a tree labeled with the numerical results of multifunctions. Thanks to that, it is easy to do a step by step evaluation,
allowing the user to choose, at each fork of the tree, a value among the available
results. This technique is quite similar to the one explained in [7], but our symbolic approach allow us a more efficient use because the figure does not have to
be regenerated between each step.

Preconditions The solver may not provide automatically the construction plan
in the appropriate form, and it may be necessary to make a *topological sort*

before performing the step by step interpretation. The topological sort is made by placing first the definitions corresponding to visible elements, called *sketch definitions*, according to the current order, and then interleaving the definitions of auxiliary construction elements just before the first sketch definition that needs it (i.e. that contains it as an argument). That way, if a backtracking has to be done, it remains located within a layer delimited by two sketch definitions. So, if none of the results satisfy the user, then we are sure that a previous sketch definition has to be thrown back into question.

Technique When a construction plan is sorted as we explained above, the only backtracking to be done in the solutions tree is located in the subtree between to sketch definitions $d1$ and $d2$, excluding the highest one (if the root is on top), say $d1$. If the user is not satisfied with any of the numerical interpretations proposed for the lowest one ($d2$), and wishes to see other possible solutions, then we are sure that some of the other sketch definitions have to be thrown back into question.

In such a case, we browse the sketch definitions that have been defined earlier, and on which $d2$ depends. We suggest to the user to reconsider some of the values he had chosen for these previous sketch definitions. First, we propose him to review only a few of them, those that are placed closer in the tree. Then, progressively we put into consideration more definitions, including those that were defined a longer time ago.

Limitations This method is the most precise and allows a total control over the construction. However, if the example is complex, the construction plan is long and contains a lot of multifunctions, and the solutions tree is very large, then it may be quite long to examine each step of the construction one by one to guide it from the beginning to the end. In addition, if the construction plan is the result of the assembling process of several sub-figures solved in various auxiliary coordinate systems, then it may be difficult to have an idea of the final result in the course of the process.

The technique can be enhanced with several kinds of breakpoint tools, for instance to offer the opportunity to freeze a part of a solution tree between two breakpoints, and then to skip this part that has become a big step. But this implies knowledge on the construction plan configuration and on the definitions on the user's part. That led us to introduce a more intuitive tool, that we will present in next Section.

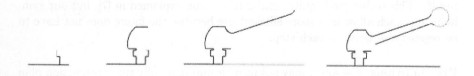

Fig. 4. The lever: 4 of the 29 steps

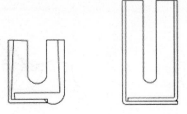

Fig. 5. The support: sketch and expected solution

Applications In the case of the lever, no decomposition was necessary to compute the construction plan, so there isn't any discontinuity during the step by step construction, making it quite easy. On Fig.4, we can see some of the 29 steps leading to the final lever.

On the opposite, a decomposition in two sub-figures was necessary to compute the construction plan of the example of Fig.5, taken from a blueprint from building industry. As this rail support is the result of the assembling of two sub-figures built in two different coordinate systems, there are jumps of some parts of the figure during the step by step process. As we can see on Fig.6, the first part of the figure is easily built with this method. But when we start to build the second part, some elements of the first part automatically move to fit the second part. This is due to the displacements applied to assemble the subfigures in the same coordinate system. This implies a temporary incoherence in the figure.

This phenomenon is increased with the number of subfigures. In the triangulation example, the construction plan is made of 4 subfigures, so there are 4 different temporary coordinate systems. Then, the step by step construction is quite difficult to control, as we can see on steps n.20 and n.21 shown on right hand side of Fig.7. Moreover, the coordinate systems may be more or less far from the coordinate system used in the sketch, so the user may be mislead when appreciating the directions, see on left hand side of Fig.7 where the beginning of subfigure 2 is in grey.

Therefore, we tried to find another way to explore the solution space, still involving the user's interaction, but more intuitive, that would free the user from the subfigures problems.

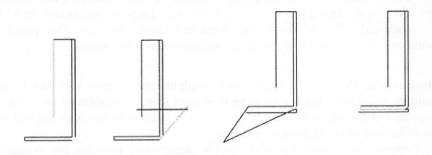

Fig. 6. The support: 4 of the 22 steps

Fig. 7. Triangulation. Left: starting the second subfigure. Right: jump in step by step process (steps 20 and 21)

3.3   Interactive Graphical Manipulation

This approach is based on a completely computed solution. When this is done, we propose the user to review it by moving the misplaced elements with his mouse towards other available places.

Preconditions This method needs the same topological sort as the step by step interpretation, for identical reasons. Auxiliary definitions are placed in the construction program only when they are needed by a sketch definition, and not before. That way, if we change of branching in a layer, we can go on following the same numbering in the layers below if they are not dependent on this layer.

Principle The process starts with a computed solution figure. If the user thinks that an element of the given figure, or a part of it, is not at the right place, then he can point his mouse on the misplaced element. Then, other available values for this element are shown on the figure, according to the initial constraints. The user can drag-and-drop the element towards the new place, and the rest of the figure is updated, by following the same scheme in the tree. It may happen that the element the user moved is linked with some others because of a dependence between them. If so, the whole set of linked elements will be moved altogether.

  To propose other solutions for one element, we simply take advantage of the structure of the solution space, by looking in the solutions tree which result can be found by doing a backtracking in the smallest sub-tree containing the element and other values. Doing this ensures that the other proposed values comply with the constraints. The solutions tree allows us to easily find the other possible solutions for one element by jumping from one branch to another.

Limitations However, in some cases, manipulating a figure may not be as intuitive as we wish. For instance, we may want to move an element on which a large part of the figure depends. If so, it is very difficult to predict what will be the behaviour of the dependent elements.

  Moreover, we remain dependent on the construction program. For example, if the user wants to move the element used as the origin of the system, he will not be able to do it unless he makes an equivalent inverse displacement on all other elements of the figure, or unless the system is solved again starting from another origin. We plan to search a way to rearrange dynamically the construction plan during the process, according to the needs.

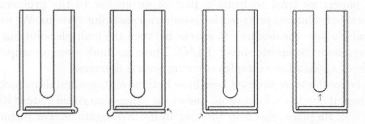

Fig. 8. The support: 4 steps of the interactive manipulation

Applications Let us take again the significative, and quite simple to understand, example of the rail support. The arcs that make up the left and right bends and the central arc are built using a multifunction that can provide up to 2 solutions, according to the parameters. Thus, suppose that we did not use the branch freezing method, and that the computed solution is the one presented on top left part of Fig.8. In that case, the user can successively move the 3 misplaced arcs to their respective alternative solution (see steps on Fig.8), in order to obtain the final result shown on right of Fig.8. In the same way, points can also be moved with this method. Most of the time, these manipulations of elements are intuitive enough to allow a quick convergence to the expected solution.

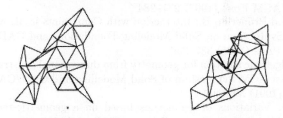

Fig. 9. Triangulation: jump in interactive manipulation of one point

Unfortunately, sometimes interactive manipulation may not be intuitive enough. As we said in Section 3.2, the triangulation of the space is computed using 4 subfigures. This is also a drawback for this manipulation method, because the displacement of one point may have consequences over several subfigures and may imply an update of a large part of the final figure. On Fig.9, we show a jump caused by the displacement of one single point (part of the bold triangle) that made a large number of other points move with the update: a part of the figure is like "bended" after the change. However, this example was only reused here as a case study, since it was perfectly handled by branch freezing (recall Section 3.1).

4   Conclusion

In this paper, we tried to bring a part of an answer to the problem of the multiplicity of solutions provided for geometric constraint systems. We presented 3 methods to help the designer to choose between the multiple solutions yielded by the symbolic geometric solver *YAMS*. These methods were accompanied by a sample of application examples covering several domains.

All the methods we exposed here have both advantages and drawbacks. Most of the time, if one has a limitation, one of the others can either reduce its effects or even take its place, and help the user in his investigations. For instance, the interactive methods handle the cases of Boolean constraints and help in browsing a remaingin sub-tree, result of a branch freezing. These multiple combinations of tools are currently being studied, as well as other new heuristics.

This quite applicative presentation points out at once that symbolic solvers, and most particularly *YAMS*, allow to handle a large range of problems in many fields, and that they lead in most cases quickly to the expected solution without redrawing the sketch.

References

1. Anderl, R. and Mendgen, R.: Parametric design and its impact on solid modeling applications. In proceedings of 3rd ACM/IEEE Symposium on Solid Modeling and Applications, ACM Press (1995) 1–12
2. Dufourd, J.-F. and Mathis, P. and Schreck, P.: Formal resolution of geometric constraint systems by assembling. In proceedings of the ACM-Siggraph Solid Modelling Conference, ACM Press (1997) 271–284
3. Sohrt, W. and Brüderlin, B.: Interaction with Constraints in 3D Modeling. In proceedings of Symposium on Solid Modeling Foundations and CAD/CAM Applications, ACM Press (1991) 387–396
4. Owen, J.: Algebraic solution for geometry from dimensional constraints. In proceedings of the 1st ACM Symposium of Solid Modelling and CAD/CAM Applications, ACM Press (1991) 397–407
5. Aldefeld, B.: Variations of geometries based on a geometric-reasoning method. Computer-Aided Design, Elsevier Science (1988) 20(3):117–126
6. Essert-Villard, C. and Schreck, P. and Dufourd, J.-F.: Sketch-based pruning of a solution space within a formal geometric constraint solver. Artificial Intelligence, Elsevier Science (2000) 124:139–159
7. Bouma, W. and Fudos, I. and Hoffmann, C. and Cai, J. and Paige, R.: Geometric constraint solver. Computer-Aided Design, Elsevier Science (1995) 27(6):487–501
8. Lamure, H. and Michelucci, D.: Solving Constraints Systems by Homotopy. In proceedings of 3rd ACM/IEEE Symposium on Solid Modeling and Applications, ACM Press (1995) 263–269
9. Jermann, C. and Trombettoni, G. and Neveu, B. and Rueher, M.: A Constraint Programming Approach for Solving Rigid Geometric Systems. In proceedings of CP2000, Springer LNCS 1894 (2000) 233–248

Polar Isodistance Curves on Parametric Surfaces

J. Puig-Pey, A. Gálvez, A. Iglesias*

Department of Applied Mathematics and Computational Sciences, University of
Cantabria, Avda. de los Castros, s/n, E-39005, Santander, Spain
{puigpeyj,galveza,iglesias}@unican.es

Abstract. In this paper, a new method for interrogation of parametric
surfaces is introduced. The basic idea is to consider the distance measured
on certain curves on a surface as an interrogation tool. To this aim, two
different sets of characteristic curves are considered: the *normal section
curves* and the *geodesic curves*. The differential equations of these sets of
curves starting radially from a given point of the surface are stated. Then,
they are solved numerically, introducing the arc-length on the surface as
the integration variable. Associated with those curves we construct the
polar isodistance curves which are obtained by joining the points at the
same distance from a given point of the surface along the section or
geodesic curves. Finally, some illustrative examples for NURBS surfaces,
by far the most common surfaces in industry, are also described.

1 Introduction

An important and widely researched issue in CAGD (Computer-Aided Geometric Design) is that of *surface interrogation*. Roughly speaking, it consists of
questioning geometrically understandable characteristics of already constructed
surfaces. The surface interrogation techniques attempt to illuminate those characteristics that are not easily discernible by conventional rendering. Of course,
many different methods can be applied to this purpose [1, 2, 4–6, 8, 19]. Among
them, a very popular and interesting body of research is the so-called *characteristic curves* on a surface. These are curves reflecting either the visual or the
geometric properties of the surface. For the visual properties we can use the
reflection lines [7] and the *isophotes* [16], which help to evaluate the behavior
and aesthetics of the surface under illumination models. On the other hand, the
geometric properties can be accurately analyzed through the *contour lines* [1,
14, 17], the *lines of curvature* [1], *geodesic paths* [1, 12], *asymptotic lines* [19], etc.

A major advantage of the characteristic curves on a surface is that they
do not depend on the surface parameterization. Unfortunately, for complicated
surfaces (such as the NURBS surfaces) these curves can only be described as the
solutions of ordinary differential equations (ODEs). Hence, working with these
curves implies that those equations must be numerically solved.

* This research was supported by the CICYT of the Spanish Ministry of Education
(projects TAP1998-0640 and DPI2001-1288) and the European Fund FEDER (Contract 1FD1997-0409).

On the other hand, the notion of distance on a surface has been applied to several machining problems. For example, it has been shown that there is no simple way to describe the guiding surface for the cutter of a five-axis machining tool [6]. In addition to some other standard techniques, curves of equal distance have been proposed for the tool paths. While avoiding the troublesome question of computing intersections [3, 9] or offset curves [13], a procedure for easily calculating distances on a surface is absolutely necessary. Futhermore, the notion of distance plays a fundamental role in outstanding problems of NC (numerical controlled) tool-path generation, namely the determination of the step-forward distance, the orientation of the tool or the interference checking between the tool and the surface.

In this paper, a combination of some characteristic curves and the distance on a surface is proposed as a new surface interrogation tool. Firstly, we introduce two new families of characteristic curves on a surface: the *normal section curves* and the *geodesic curves*. In our approach, both families of curves are assumed to start at a given point on the surface, which will usually be chosen by its relevance for interrogation. In both cases, the curves are obtained by solving systems of ordinary differential equations with their corresponding initial conditions and using parametric representations $(u(s), v(s))$, where s is the arc-length parameter of the curve on the surface. Associated with the section and geodesic curves we define what we call *polar isodistance curves* which are obtained by joining the points at the same distance from a given point of the surface along the normal section or geodesic curves. These new curves will allow us to introduce the distance as a significant element for surface interrogation.

The structure of this paper is the following: in Section 2 some basic mathematical concepts to be used throughout the paper are introduced. Section 3 describes both the normal section curves and the geodesic curves. Section 4 discusses the numerical procedure to solve the systems of ODEs obtained in the previous section. Finally, Section 5 presents some examples of section and geodesic curves and their associated polar isodistance curves for the case of NURBS surfaces.

2   Mathematical Preliminaries

In this paper we restrict ourselves to the case of differentiable parametric surfaces. Therefore, they are described by a vector-valued function of two variables:

$$\mathbf{x}(u, v) = (x(u, v), y(u, v), z(u, v)), \qquad u, v \in \Omega \subset \mathbb{R}^2 \qquad (1)$$

where u and v are the surface parameters. Expression (1) is called a parameterization of the surface $\mathbf{x}$. At regular points, the partial derivatives $\mathbf{x}_u(u, v)$ and $\mathbf{x}_v(u, v)$ do not vanish simultaneously. For $\{u = u_0, v = v_0\}$, $\mathbf{x}_u$ and $\mathbf{x}_v$ are vectors on the tangent plane to the surface at the point $\mathbf{x}(u_0, v_0)$, each being tangent to the parametric or coordinate curve $v = v_0$ and $u = u_0$, respectively. These vectors define the unit normal vector $\mathbf{N}$ to the surface at $\mathbf{x}(u_0, v_0)$ as:

$$\mathbf{N} = \frac{\mathbf{x}_u \times \mathbf{x}_v}{|\mathbf{x}_u \times \mathbf{x}_v|} \qquad (2)$$

where $\times$ denotes the cross product.

A curve in the domain Ω can be described by means of its parametric representation $\{u = u(t), v = v(t)\}$. This expression defines a three-dimensional curve on the surface $\mathbf{x}$ given by $\mathbf{x}(t) = \mathbf{x}(u(t), v(t))$. Applying the chain rule, the tangent vector of the curve $\mathbf{x}$ at a point $\mathbf{x}(t)$ becomes:

$$\mathbf{x}'(t) = \mathbf{x}_u \, u'(t) + \mathbf{x}_v \, v'(t) \qquad (3)$$

In this work the curve $\mathbf{x}$ will usually be parameterized by the arc-length s on the surface. Its geometric interpretation is that a constant step s traces a constant distance along an arc-length parameterized curve. Since some industrial operations require an uniform parameterization, this property has several practical applications. For example, in computer controlled milling operations, the curve path followed by the milling machine must be parameterized such that the cutter neither speeds up nor slows down along the path. Consequently, the optimal path is the one parameterized by the arc-length. In this case, the differential relation between s, t, u and v is given by the First Fundamental Form of the surface:

$$s'(t)^2 = E \, u'(t)^2 + 2F \, u'(t) \, v'(t) + G \, v'(t)^2 \qquad (4)$$

where

$$E = \mathbf{x}_u.\mathbf{x}_u \quad , \quad F = \mathbf{x}_u.\mathbf{x}_v \quad , \quad G = \mathbf{x}_v.\mathbf{x}_v \qquad (5)$$

and "." indicates the dot product. For the sake of clarity, in this paper the parameter s will be strictly used to refer to a curve parameterized by the arc-length on the surface.

3 New Characteristic Curves for Surface Interrogation

3.1 Section Curves

Many authors have dealt with the problem of sectioning parametric surfaces [3, 6, 9, 10, 14]. This problem has several practical applications. For instance, the main goal of the contouring algorithms in NC tool-path generation is to generate ordered sequences of points on planar sections of the given surface, from which tool path approximations can be computed. In this section, a description of a planar section of a parametric surface in terms of ODEs is discussed. From here, the concept of polar isodistance section curve is introduced.

The differential equation of the section $\mathbf{x}(t)$ at a point $\mathbf{x}(u, v)$ of a parametric surface by a plane which is normal to a given vector $\mathbf{D}$ can be written as:

$$\mathbf{x}'(t).\mathbf{D} = 0$$

where, from (3) we obtain:

$$\mathbf{x}_u.\mathbf{D} \, u'(t) + \mathbf{x}_v.\mathbf{D} \, v'(t) = 0 \qquad (6)$$

In this paper we do not consider singular cases; this implies that all the points on the section are regular points and that vector $\mathbf{D}$ is never normal to the surface along the section $\mathbf{x}(t)$.

Since a primary goal is to measure distances on curves on a surface, it is more interesting to consider the arc-length surface parameterization for Eq. (6). In this case, the corresponding differential equations for $\dfrac{du}{ds}$ and $\dfrac{dv}{ds}$ are obtained from (4) and (6) as a system of first-order explicit ordinary differential equations:

$$\frac{du}{ds} = \pm \frac{\mathbf{x}_v.\mathbf{D}}{\sqrt{E\,(\mathbf{x}_v.\mathbf{D})^2 - 2\,F\,(\mathbf{x}_u.\mathbf{D})(\mathbf{x}_v.\mathbf{D}) + G(\mathbf{x}_u.\mathbf{D})^2}}$$

$$\frac{dv}{ds} = \mp \frac{\mathbf{x}_u.\mathbf{D}}{\sqrt{E\,(\mathbf{x}_v.\mathbf{D})^2 - 2\,F\,(\mathbf{x}_u.\mathbf{D})(\mathbf{x}_v.\mathbf{D}) + G(\mathbf{x}_u.\mathbf{D})^2}}$$

(7)

with the initial conditions:

$$u(0) = u_0 \quad , \quad v(0) = v_0 \tag{8}$$

associated with the point $\mathbf{P} = \mathbf{x}(u_0, v_0)$. Each solution of (7)-(8) defines a section curve $(u(s), v(s))$ in the parametric domain (u, v) of the surface. The three-dimensional section curve on the surface is given by $\mathbf{x}(u(s), v(s))$, which obviously passes through the point $\mathbf{P}$. In addition, it should be remarked that because s is essentially a non-negative parameter, only two possible combinations for the signs in (7), namely (+) with (-) or alternatively (-) with (+), are allowed. Each of these feasible choices lead to a different solution for (7)-(8), associated with each of the pieces of the section curve starting from $\mathbf{P}$, joined with continuity at $\mathbf{P}$. Further, solving the system (7) for the same initial condition (that is, for the same point $\mathbf{P}$) and different directions on the tangent plane at $\mathbf{P}$ provides an interesting procedure for interrogation on the local behavior of the surface around this point. Such directions can be defined in terms of only one variable representing the angle between a partial derivative of the surface at $\mathbf{P}$, let us say $\mathbf{x}_v$, and the corresponding vector of direction. In the following, let $\mathbf{x}_v^{\perp}$ denote a vector of the tangent plane at a point $\mathbf{P}$ orthogonal to $\mathbf{x}_v$, which is given by:

$$\mathbf{x}_v^{\perp} = -G\,\mathbf{x}_u + F\,\mathbf{x}_v \tag{9}$$

These vectors $\mathbf{x}_v$ and $\mathbf{x}_v^{\perp}$ define an orthogonal reference system on the tangent plane at the point $\mathbf{P}$. Therefore, any unit vector $\mathbf{n}$ lying on such a tangent plane can be written as:

$$\mathbf{n} = \cos(\alpha)\,\frac{\mathbf{x}_v}{|\mathbf{x}_v|} + \sin(\alpha)\,\frac{\mathbf{x}_v^{\perp}}{|\mathbf{x}_v^{\perp}|} \tag{10}$$

where α represents the angle between $\mathbf{x}_v$ and $\mathbf{n}$. With this simple procedure, we can obtain the section of the surface by a plane containing the normal vector to the surface, $\mathbf{N}$ (see Eq. (2)), and whose intersection with the tangent plane at $\mathbf{P}$ is a line forming an angle α with the direction of $\mathbf{x}_v$ at $\mathbf{P}$. This section will be obtained by numerical integration of (7) and (8), where $\mathbf{n}$ is defined as in (10) and $\mathbf{D}$ is given by:

$$\mathbf{D} = \mathbf{N} \times \mathbf{n}$$

Varying this angle α, a set of normal sections $\mathbf{x}(u(s), v(s))$ passing through $\mathbf{P}$ are calculated, all of them resulting from the rotation of the normal planes around the normal vector $\mathbf{N}$. Fixing now a value $s = s_L$ for the arc-length parameter s and assuming that the trajectories have not reached the limits of the parametric domain of the surface, one might obtain the set of points at distance s_L from $\mathbf{P}$ along all the section curves, defining a *polar isodistance normal section curve* associated with this value s_L. This curve can be understood as the polar representation of a pseudo circumference lying on the surface with center at $\mathbf{P}$ and radius s_L, which is measured from $\mathbf{P}$ to the points of the pseudo circumfererence along the normal section curves.

3.2   Geodesic curves

Given a curve $\mathbf{C}$ on a surface $\mathbf{x}$, its *geodesic curvature* at a point $\mathbf{P}$ is the curvature at this point of the projection curve of $\mathbf{C}$ on the tangent plane to $\mathbf{x}$ at $\mathbf{P}$. *Geodesics* are defined as the curves on a surface with zero geodesic curvature. It is well known that if a curve connects two points on a surface following a minimum length trajectory, then it must be a geodesic [18]. That makes the geodesics particularly attractive for surface interrogation, especially when distances measured along curves on the surface are considered. On the other hand, the geodesic curvature has been widely applied to surface interrogation [19]. Thus, the geodesics are an excellent tool to identify the boundaries between regions with positive and negative geodesic curvatures.

The geodesics of a parametric surface $\mathbf{x}(u, v)$ can be expressed through a system of two second-order ODEs [18] or alternatively as a system of four explicit first-order ODEs [1] given by:

$$\begin{cases} \dfrac{du}{ds} = u' \\[2mm] \dfrac{dv}{ds} = v' \\[2mm] \dfrac{du'}{ds} = -\Gamma_{11}^1 \, u'^2 - 2\Gamma_{12}^1 \, u'v' - \Gamma_{22}^1 \, v'^2 \\[2mm] \dfrac{dv'}{ds} = -\Gamma_{11}^2 \, u'^2 - 2\Gamma_{12}^2 \, u'v' - \Gamma_{22}^2 \, v'^2 \end{cases} \qquad (11)$$

where Γ_{ij}^k, $i, j, k = 1, 2$ are the first kind Christoffel symbols which can be calculated from the E, F, G coefficients of the First Fundamental Form of the surface (see Eq. (5)) and their first derivatives [18]. Once again, s represents the arc-length parameter on the surface such that the functions u, v, u' and v' depend solely on it. Observe again that the equations are formulated in the u-v parameters and hence the integration of the system (11) will give a curve expressed as

$(u(s), v(s))$ in the parametric domain Ω for u,v. Such a curve corresponds to a three-dimensional curve on the surface, given by $\mathbf{x}(u(s), v(s))$. The domain for the variable s is an interval $[0, L]$, L being the length of the geodesic. The initial conditions for u and v can be easily established as:

$$u(0) = u_{in} \quad , \quad v(0) = v_{in} \tag{12}$$

where $\mathbf{x}(u_{in}, v_{in})$ is the point $\mathbf{P}$ where the geodesic starts. On the contrary, the initial conditions for u' and v' require further discussion. Firstly, note that the relation $u'(t) = c\, v'(t)$ defines a direction on the tangent plane of the surface at a point $\mathbf{P}$. Indeed, inserting this relation in (3) we obtain

$$\mathbf{x}'(t) = c\, \mathbf{x}_u\, v'(t) + \mathbf{x}_v\, v'(t) = (c\, \mathbf{x}_u + \mathbf{x}_v)\, v'(t)$$

leading to the direction:

$$\mathbf{d} = c\, \mathbf{x}_u + \mathbf{x}_v \tag{13}$$

This direction $\mathbf{d}$ can also be characterized in terms of the arc-length by giving $u'(s)$ and $v'(s)$ in $\mathbf{P}$ for the curve $\mathbf{x}(s) = \mathbf{x}(u(s), v(s))$. Combining the previous relation $u'(t) = c\, v'(t)$ with (4) we obtain:

$$u'(s) = \frac{\pm c}{\sqrt{E\, c^2 + 2\, F\, c + G}} \quad , \quad v'(s) = \frac{\pm 1}{\sqrt{E\, c^2 + 2\, F\, c + G}} \tag{14}$$

Note also that, in order for the relation $u'(t) = c\, v'(t)$ to hold, the same sign must be taken in (14) for both $u'(s)$ and $v'(s)$. On the other hand, different values of the parameter c are associated through (13) with different tangent directions $\mathbf{d}$ at $\mathbf{P}$. It should be remarked, however, that there is not a direct geometric interpretation of the values of this parameter c. Fortunately, a description of c in terms of the angle α can be easily derived. From (9) and (10), the vector $\mathbf{n}$ can also be written as:

$$\mathbf{n} = -sin(\alpha)\frac{G}{|\mathbf{x}_v^{\perp}|}\mathbf{x}_u + \left(\frac{cos(\alpha)}{|\mathbf{x}_v|} + \frac{F\, sin(\alpha)}{|\mathbf{x}_v^{\perp}|}\right)\mathbf{x}_v \tag{15}$$

For the vectors $\mathbf{d}$ and $\mathbf{n}$ to be parallel, it is enough to establish the proportionality of the coefficients of $\mathbf{x}_u$ y $\mathbf{x}_v$ in (13) and (15). From the corresponding equation, we obtain

$$c = \frac{-sin(\alpha)\, G|\mathbf{x}_v|}{cos(\alpha)\, |\mathbf{x}_v^{\perp}| + sin(\alpha)\, F|\mathbf{x}_v|} \tag{16}$$

This equation allows us to rewrite a direction $\mathbf{d}$ on the tangent plane in terms of the angle α, thus gaining a geometrical insight. Now, the following initial conditions for u' and v' are considered:

$$u'(0) = \left(\frac{du}{ds}\right)_{s=0} = u'_{in} \quad , \quad v'(0) = \left(\frac{dv}{ds}\right)_{s=0} = v'_{in} \tag{17}$$

where u'_{in}, v'_{in} are obtained from (14) and (16) by giving the starting angle α between the geodesic trajectory to be calculated and vector $\mathbf{x}_v$ at the initial point $\mathbf{P} = \mathbf{x}(u_{in}, v_{in})$.

Analogously to the comments given for the normal sections, one can choose a set of initial angles α from the initial point $\mathbf{P}$. The integration of the system (11)-(12)-(17) allows us to construct the corresponding set of geodesics starting from $\mathbf{P}$ and forming such initial angles with vector $\mathbf{x}_v(u_{in}, v_{in})$ at $\mathbf{P}$. The concept of polar isodistance curve also extends to geodesics: a fixed value $s = s_L$ for the arc-length parameter s yields a set of points at distance s_L from $\mathbf{P}$ along all the geodesic curves, defining a *polar isodistance geodesic curve* associated with this value s_L. The meaning of this curve is very similar to that of the section curve: due to the minimal distance property of geodesics, a polar isodistance geodesic curve for a given value $s = s_L$ is the geometric place of points on the surface which are at a minimum distance s_L from $\mathbf{P}$, measured on the surface.

4   Numerical Process

Although (7)-(8) and (11)-(12)-(17) are both systems of first-order ODEs, the calculation of the analytical expression for u and v (and subsequently for $\mathbf{x}(u, v)$) as functions of the parameter s can be, in general, a very hard task, if not impossible, which is beyond the scope of this work. Hence, to deal with the most typical surfaces in design, such as the NURBS surfaces (see [15] for a description), we are forced to use numerical techniques.

Fortunately, both systems of equations can be numerically integrated by applying some standard techniques. In particular, all the numerical work has been performed by using Runge-Kutta methods with the ODE solvers of Matlab [11]. It should be noticed, however, that when the surface consists of several patches (for example, the piecewise bicubic splines or the NURBS surfaces) some kind of continuity conditions must be imposed to assure that the differential model is still valid in the neighbourhood of the patch boundaries. On the contrary, the model must be strictly applied inside each patch treating it separately and reconstructing the initial conditions when the considered trajectory crosses from one patch to the next.

On the other hand, the domain Ω of the parametric surface should be taken into account during the integration process. The parametric domain of the most typical surfaces in design exhibits a rectangular structure [2, 6], whose boundaries impose additional restrictions to the integration method. In particular, we must check whether the trajectories reach the limits of the parametric domain. For example, in the case of nonperiodic NURBS surfaces [15], these intervals are limited by the first and last values of the corresponding knot vectors for u and v. On the contrary, in the case of working with surfaces which are partially or completely closed, additional tests for self-intersection must be performed.

The procedure for building a polar isodistance normal section or geodesic curve associated with a given distance s_L with respect to a reference point $\mathbf{P}$ on a surface consists of making a loop giving regularly-spaced α values for the initial directions from **P**. Then, a step-by-step integration is performed for each normal section or geodesic trajectory, until the value $s = s_L$ is attained. The transversal isodistance curve is obtained by joining the end points, one for each normal section or geodesic, in the (u, v) domain and obtaining its three-dimensional image on the surface.

5   Some Illustrative Results

In this section the performance of the proposed method is discussed by means of some illustrative examples from the NURBS surface family [15]. In general, the method deals with any parametric surface. However, because of their advantages in industrial environments, their flexibility and the fact that they can represent well a wide variety of shapes (such as conics and quadrics), we will focus here on NURBS surfaces.

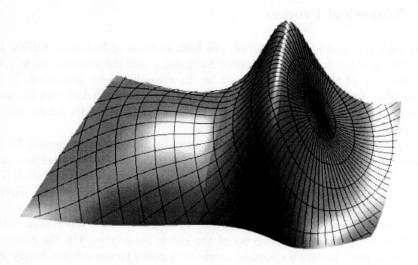

Fig. 1. Section curves and associated polar isodistance curves for a NURBS surface

As a first illustration, Fig. 1 shows a set of 80 normal section curves (obtained following Sec. 3.1) starting from a given point on a $(4, 4)$-order NURBS surface defined by a grid of 4×6 control points and nonperiodic knot vectors for both u and v. The surface has a single patch in the direction u and three patches in the direction v. The figure also shows the polar isodistance normal section curves associated with different values for s_L from 0.3 to 4.5 with step 0.3. Of course, if the chosen distance s_L always leads to points inside the domain of the surface, the polar isodistance curve will be closed. In this case, we can generate a continuous representation by simply interpolating the calculated points with a closed cyclic cubic spline, thus obtaining its image on the surface $\mathbf{x}(u, v)$. In

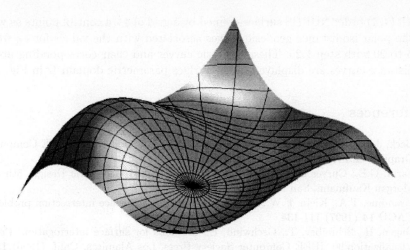

Fig. 2. Geodesic curves and associated polar isodistance curves for a NURBS surface

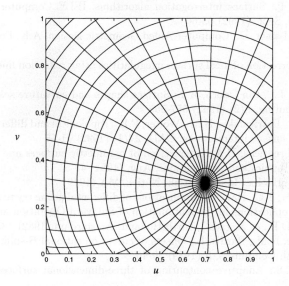

Fig. 3. Geodesic curves and associated polar isodistance curves in the $u - v$ parametric domain for the NURBS surface in Fig. 2

the present example, this only happens for $s_L = 0.3$ and $s_L = 0.6$. On the contrary, the curves will be made of unconnected pieces and only a continuous approximation of each piece can be obtained (for example, by using open cubic splines with the points calculated in the integration process).

As the second example, Fig. 2 shows a set of 40 geodesic curves (obtained following Sec. 3.2) referring to the point given by $\mathbf{x}(u = 0.7, v = 0.3)$ of a single-patch $(4, 4)$-order NURBS surface defined by a grid of 4×4 control points as well as the polar isodistance geodesic curves associated with the values for s_L from 1.25 to 20 with step 1.25. These geodesic curves and their corresponding polar isodistance curves are displayed in the surface parametric domain Ω in Fig. 3.

References

1. Beck, J.M., Farouki, R.T., Hinds, J.K.: Surface analysis methods. IEEE Computer Graphics and Applications, Dec. (1986) 18-36
2. Farin, G.E.: Curves and Surfaces for Computer-Aided Geometric Design, 5th ed. Morgan Kaufmann, San Francisco (2001)
3. Grandine, T.A., Klein, F.W.: A new approach to the surface intersection problem. CAGD **14** (1997) 111-134
4. Hagen, H., Schreiber, T., Gschwind, E.: Methods for surface interrogation. Proc. Visualization'90, IEEE Computer Society Press, Los Alamitos, Calif. (1990) 187-193
5. Hagen, H., Hahman, S., Schreiber, T., Nakajima, Y., Wördenweber, B., Hollemann-Grundstedt, P.: Surface interrogation algorithms. IEEE Computer Graphics and Applications, Sept. (1992) 53-60
6. Hoschek, J., Lasser, D.: Computer-Aided Geometric Design, A.K. Peters, Wellesley, MA (1993)
7. Klass, R.: Correction of local surface irregularities using reflection lines. CAD, **12**(2) (1980) 73-77
8. Koenderink, J.J., van Doorn, A.J.: Surface shapes and curvature scales. Image and Vision Computing, **8**(2) (1992) 557-565
9. Kriezis, G.A., Patrikalakis, N.M., Wolter, F.E.: Topological and differential equation methods for surface intersections. CAD, **24**(1) (1992) 41-55
10. Lee, R.B., Fredricks, D.A.: Intersection of parametric surfaces and a plane. IEEE Computer Graphics and Applications, Aug. (1984) 48-51
11. The Mathworks Inc: Using Matlab. Natick, MA (1999)
12. Munchmeyer, F.C., Haw, R.: Applications of differental geometry to ship design. In: Computer Applications in the Automation of Shipyard Operation and Ship Design IV, Rogers, D.F., et al. (eds.) North Holland, Amsterdam (1982) 183-188
13. Patrikalakis, N.M., Bardis, L.: Offsets of curves on rational B-spline surfaces. Engineering with Computers, **5** (1989) 39-46
14. Petersen, C.S.: Adaptive contouring of three-dimensional surfaces. CAGD **1**(1) (1984) 61-74
15. Piegl, L., Tiller, W.: The NURBS Book, Springer Verlag, Berlin Heidelberg (1997)
16. Poeschl, T.: Detecting surface irregularities using isophotes. CAGD **1**(2) (1984) 163-168
17. Satterfield, S.G., Rogers, D.F.: A procedure for generating contour lines from a B-spline surface. IEEE Computer Graphics and Applications, Apr. (1985) 71-75
18. Struik, D.J.: Lectures on Classical Differential Geometry, 2nd ed., Dover Publications, New York (1988)
19. Theisel, H., Farin, G.E.: The curvature of characteristic curves on surfaces. IEEE Computer Graphics and Applications, Nov./Dec. (1997) 88-96

Total Variation Regularization for Edge Preserving 3D SPECT Imaging in High Performance Computing Environments *

L. Antonelli[1], L. Carracciuolo[1], M. Ceccarelli[2], L. D'Amore[1], and A. Murli[1]

[1] University of Naples, Federico II
and Center for Reserach on Parallel Computing and Supercomputers(CPS)- CNR,
Complesso M.S.Angelo Via Cintia, 80126 Naples, Italy
(laura.antonelli,luisa.damore,almerico.murli)@dma.unina.it,
carracci@pixel.dma.unina.it
[2] University of Sannio, Via Port'Arsa 11, Benevento, Italy
ceccarelli@unisannio.it

Abstract. Clinical diagnosis environments often require the availability of processed data in real-time, unfortunately, reconstruction times are prohibitive on conventional computers, neither the adoption of expensive parallel computers seems to be a viable solution.
Here, we focus on development of mathematical software on high performance architectures for Total Variation based regularization reconstruction of 3D SPECT images. The software exploits the low-cost of Beowulf parallel architectures.

1 Introduction

The problem of reconstructing a 3D image from a set of its 2D projections from different angles of view arises in medical imaging for 3D tomography. 3D tomography requires huge amount of data and long computational times. Indeed, the reconstruction times are often prohibitive (of order of hours) on conventional machines. Hence, parallel computing seems to be an effective way to manage this problem, especially in a clinical environment.

However, there are still many barriers to an effective use of High Performance Computing (HPC) technology in medical imaging. First it can be employed in medical devices only to an extent which is determined by the overall cost of the system. On the other hand, the development and set up of efficient methods is a complex task which requires the joint effort of researchers in the fields of computing, industries, and medical communities.

This work is developed within a collaborative national project among computational scientists. Within this context, our focus is development of mathematical software on high performance architectures for reconstruction of 3D SPECT images by using edge preserving regularization. The main contribution of the paper

---

* The activity has been developed within the Project: "Inverse Problems in Medical Imaging" partially supported by MURST, grant number MM01111258

P.M.A. Sloot et al. (Eds.): ICCS 2002, LNCS 2330, pp. 171–180, 2002.

is, from one point, the introduction of the Total Variation-based edge preserving regularization method in an accurate 3D SPECT imaging model, and from the other point, due to the enormous computing requirements of the corresponding deconvolution problem, the development of a scalable parallel algorithm for the solution of the corresponding non-linear Euler Lagrange equation. Our algorithm exploits the parallelism in terms of the number of projections (rows of the system matrix). The proposed parallel algorithm and the experiments are based on low cost architectures made of Pentium Processors connected by a Fast Switch which could be really employed in a clinical environment

2   Regularized Image Reconstruction

Single photon emission computed tomography (SPECT) is a tomographic technique which gives the spatial distribution of radioactive tracers injected into the patient's body for a variety of diagnostic purposes. Data are collected by means of a small number of large detectors which provide a sequence of M 2D projections evenly spaced over over 2π around the patient.

The basic mathematical description of SPECT imaging relates the observed data u_0 with the original image u through a linear model:

$$u_0(s, \phi, z) = \int \int \int u(\mathbf{x}, z')\mathcal{K}(s - \mathbf{x} \cdot \theta, \mathbf{x} \cdot \theta^{\perp}, z - z')\mathrm{d}\mathbf{x}\mathrm{d}z' \qquad (1)$$

where $\mathbf{x} = \{x, y\}$ are plane coordinates where the body of the patient is represented by a disk centered at the origin, z-axis is the axis of rotation of the detectors; $\theta = \{\cos\phi, \sin\phi\}$, ϕ being the projection angle of the detector plane with respect to a fixed position, s is the radial distance and $\theta^{\perp} = \{-\sin\phi, \cos\phi\}$. The kernel κ models the physical structure of the measurement device which introduce an unavoidable blurring effect due to the fact that the rays pass through the holes of a collimator. This model is called the Blurred Radon Transform [2], in which the integration kernel κ, which is usually called *Point Spread Function (PSF)*, differs from Radon's one for two reasons: the PSF is no longer an impulse function, it is defined on the projection plane as a function of two variables, s and z, and also depends on the source-detector distance $t = \mathbf{x} \cdot \theta^{\perp}$.

In compact form the above mathematical model of 3D SPECT imaging can be stated as follows:

$$u_0 = K\,u + \eta \qquad (2)$$

where η is assumed to be white noise, and K is a linear operator, the unknown vector u represents the 3D image and u_0 is the so-called *3D sinogram*. By this way it belongs to the class of linear inverse and ill posed problems. Indeed, due to the compactness of the operator $\mathcal{K}$, the computation of $\mathcal{K}^{-1}$ has the effect of noise amplification with the risk of obtaining unuseful solutions. These problem have been widely investigated, and the regularization approach produces an estimate, u^{λ} of u as the solution of

$$u^{\lambda} = \mathrm{argmin}_u\{||\mathcal{K}u - u_0|| + \lambda\mathcal{R}(u)\} \qquad (3)$$

where $\mathcal{R}(u)$ serves as regularization functional and λ is a parameter controlling the weight given to the regularization term relative to the residual norm.

The Total Variation (TV) norm, was first used by Rudin *et al.* [8], it is defined as

$$TV(u) = \int \int \sqrt{u_x^2 + u_y^2} dx dy. \tag{4}$$

The main advantage of the TV norm as regularization functional is that it does not require the solution to be continuous, thus allowing the presence of "jumps" in the solution, with a preservation of sharp edges. This is due to the fact that the TV functional is defined over the space of Bounded Variation functions [5], which can eventually contain step discontinuities. The solution of (3) with $\mathcal{R}$ given by (4) corresponds to the solution of the Euler-Lagrange equation

$$\mathcal{K}^*(\mathcal{K}u - u_0) - \lambda \nabla \cdot \left(\frac{\nabla u}{|\nabla u|} \right) = 0, \quad \frac{du}{dn} = 0 \tag{5}$$

The main difficulty in solving (5) is related to the highly non-linear second order elliptic term $\left(\frac{\nabla u}{|\nabla u|} \right)$ representing a diffusion operator with coefficient given by $\frac{1}{|\nabla u|}$. A number of approaches has been proposed for solving (5), which can be classified in three main categories

- Time marching [8]
- Fixed Point [9, 4]
- Newton method [3].

In this paper we focus on the Fixed Point method as it provides the best compromise between robustness (dependence on the initial solution, parameters, ...) and speed of convergence (it can be shown to be a quasi-newton method). In particular, the solution is obtained by solving, until convergence, a sequence of linear systems arising from the discretization of the following equation

$$\left(\mathcal{K}^*\mathcal{K} - \lambda \nabla \cdot \left(\frac{\nabla}{|\nabla u^n|} \right) \right) u^{n+1} = \mathcal{K}^* u_0 \tag{6}$$

starting with $u^0 = u_0$.

It can be shown, [4], that the sequence of solutions $\{u^n\}_{n \in \mathcal{N}}$ converges as $n \to \infty$, to a global minimum of (3). The discretization of (6) leads to a linear system

$$(K^*K - \lambda L(u^n))u^{n+1} = K^* u_0 \tag{7}$$

where $L(u)$ represents the discretization of the operator $\nabla \cdot (\frac{\nabla}{|\nabla u|})$. In order to solve (7) we adopt the iterative conjugate gradient (CG) method whose main computational kernel is the multiplication of the system matrix $(K^*K - \lambda L(u^n))$ with the approximate solutions. The subject of the next section is therefore devoted to description of how parallelism has been introduced into solution of such system.

3 Parallel Reconstruction Algorithm

Our work has been essentially devoted to obtain a parallel version of the existing code by exploiting parallelism inside those modules representing the main
computational bottlenecks. Actually, the most intensive calculation is due to the
products which involve operators K^* and K respectively. Then in the following
we mainly focus on these operations.

The matrix K is a $L \times L$ block Toeplitz matrix, where L is the number of
object slices. Let us introduce the vector $\{u_0^i\}_{i=1,...,L}$, $u_0^i \in \Re^{J \cdot M}$, M being the
number of angular projections, and J the number of beans. Moreover, let each
block $u^i \in \Re^{N \cdot N}$ represent the 2D slice, finally $[E_i] \in \Re^{(M \cdot J) \times (N \cdot N)}$ represents
the projection matrix.

By this way, product Ku can be rewritten[1] as:

$$
\begin{pmatrix} u_0^1 \\ \vdots \\ u_0^L \end{pmatrix} = \begin{pmatrix} [E_0] & \cdots & [E_{L-1}] \\ \vdots & \vdots & \vdots \\ [E_{L-1}] & \cdots & [E_0] \end{pmatrix} \begin{pmatrix} u^1 \\ \vdots \\ u^L \end{pmatrix}
\tag{8}
$$

Standard acquisition parameters are $M = 120$, $J = 128$ and $L = 64$.

The way in which concurrency has been introduced was strongly induced
by the physical meaning of the SPECT reconstruction process. Specifically, we
observe that the $E_j \cdot u^i$ products, i.e. the *projection* process of SPECT imaging
on each 2D slice, can be described as a set of distinct projections each one
corresponding to a projection angle. In this way the whole projection process
synthesized by equation (2) is obtained by repeatedly performing the projection
operation for each projection angle. Starting from this consideration, we can
distribute projection angles among processors. In other words, each processors
is assigned a set of projection angles ϕ over which the projection is performed. As
a consequence, the whole set of sinograms and image slices is shared among the
processors. The first phase of the reconstruction process performs a distribution
of the sinograms among all the processor. Then, each iteration of the algorithm
requires a global sum in order to keep updated the current solution of the system
(7).

Let us now state the parallel algorithm using a linear algebra setting. We
note that the most natural way to accomplish the parallelism as inspired by the
projection/retroprojection processes is to perform parallelization of products

$$
f^i = \sum_{j=0}^{L-1} E_{(j+i-1) \bmod L} \cdot u^{j+1} \quad i = 1, ..., L
$$

in a row-wise fashion. In other words, concurrency has been introduced by using
a row-block cyclic distribution of the matrices $E_{(j+i-1) \bmod L}$. By this way each

---

[1] It is worth noting that the block representation of K as described in (10) is not
actually available. In practice, the matrix K is rather sparse with a sparsity between
5% and 10%. The coefficients of each row of the matrix K are given under form of
look-up tables [2].

processor performs the product between a row-block of $E_{(j+i-1) \bmod L}$ and vector $\mathbf{u}^{j+1}$. In the same way, parallelization of the products

$$\mathbf{z}^i = \sum_{j=0}^{L-1} E^*_{(j+i-1) \bmod L} \cdot \mathbf{f}^{j+1} \quad i = 1, ..., L$$

is exploited in a column-wise fashion. This means that each process acts a partial update over the whole data, requiring a global sum at the end of each step.

As already underlined, the above choice was motivated by several reasons, the first is due to the structure of the look-up tables for accessing the elements of the projection matrix. Indeed, the look-up table containing the coefficients of K is given in a closed form in such a way that we can access independently to each row of K corresponding to a given angular projection over the whole data set. The second was due to the fact that a further colomnwise distribution of the matrix K, from one hand could allow a distribution of the data, but, on the other hand, it would require a further global communication step at each computation of $K\mathbf{u}$. Considering the actual dimentions of the matrices, and the obtained results, with the processors arranged as a ring, we believed this last option the most suited for our application. In addition, this choice required minor adjustment of existing projection code.

For what concerns the parallelization of the matrix product $L(u^n)u^{n+1}$, it is important to point out that it requires about the 0.1% of the whole processing time for each iteration, therefore since we adopted a global input data distribution, $i.\ e.$ the input sinogram and the current solution are shared among all the processors, we simultaneously perform this operation locally as it does not influence the whole computing time. Other choices are of course possible, however the parallelization of this operation would introduce a further communication overhead without improving the processing time.

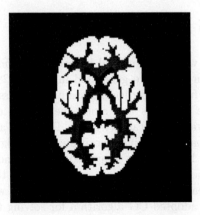

Fig. 1. The Hoffman phantom used in the experiments

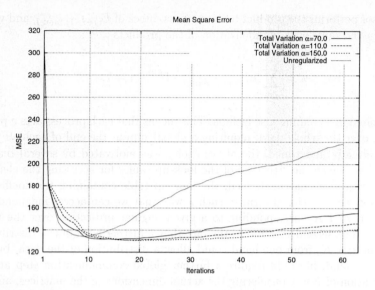

Fig. 2. The MSE between the computed and ideal solution for the unregularized problem and for the regularized problem with various regularization parameters. Note the *semiconvergence* property of the unregularized iterative algorithm.

4   Experiments and Results

The above algorithm was experimented by using a Hoffman phantom depicted in figure 1. Projection data were generated in a 128x128 image array for 120 views over $[0, \pi)$, the fully 3D projector assumes a Gaussian model of collimator blurr both in the 2D slices and along the staking direction of the slices [1]. Figure 2 demonstrates the stability of the TV-based algorithm with respect to the standard CG-based iterative algorithm [1], the figure reports the Mean Squared Error (MSE) between the computed solution and the ideal solution as function of the iterations. Note that, in order to make the comparison homogeneous, the MSE measure for the TV-based algorithm was computed at each inner iteration of the adopted CG algorithm for solving the linear system (7). The behaviour of the MSE clearly shows the classical *semiconvergence* property of iterative reconstruction algorithms, in particular the unregularized algorithm has an optimal number of iterations after which the quality of the reconstruction

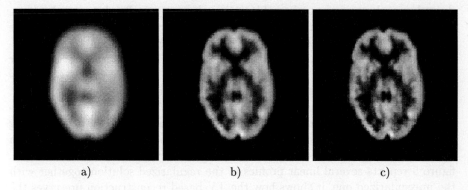

a)   b)   c)

Fig. 3. The TV-regularized solutions. Figure a) is obtained with 1 TV iteration, b) refers to 4 TV iteration and c) to 7 iterations. Note that each TV iteration requires 7 inner CG iterations.

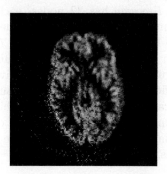

Fig. 4. The unregularized solution after 60 CG iterations.

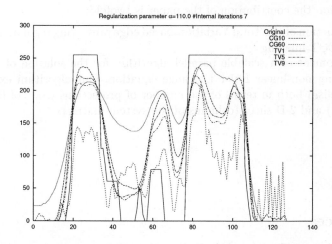

Fig. 5. Cross-section of the solution for various iteration number of the regularize and unregularized problem.

degrades, such optimal number of iterations being typically chosen with experimentation, depending on the amount of noise and the kind of images. On the contrary, the TV-regularized algorithm offers the warranty to converge to an accurate and stable solution for a wide range of regularization parameters. Neither such regularization tends to destroy image features. We plot in figure 3 the regularized solutions for different regularization parameters and iteration number. The unregularized solution is reported in figure 4. What it is evident here is the property of the TV regularization to produce a stable solution while mantaining the features of the images, *i.e.* the regularization is *discontinuity adaptive* [6] in the sense that the sharp changes (edges) are preserved. As a further example, figure 5 reports several linear profiles of the regularized solution together with the unregularized one, it shows how the TV-based reconstruction preserves the stepwise behavior of the original image.

For what concerns the computing environment, we implemented the algorithm on two Beowulf systems, available at CPS-CNR (Naples). The first, referred in the figures as *Beocomp* has 32 nodes, each node being a 450 Mhz Pentium II processor running Linux Red Hat 6.0. The second, referred as *Vega*, consisting of 18 Pentium IV running Red Hat Linux 7.2. The program uses the MPI and the Basic Linear Algebra Communication Subprograms (BLACS) communication libraries. Figures 7 and 8 report respectively the achieved speedup and the corresponding efficiency on both systems. Results show that the problem of 3D SPECT imaging can be efficiently solved on high performance architectures. The time reduction gained with the adoption of such parallel algorithm is in accordance with diagnostic times, allowing the use of accurate iterative reconstruction algorithms.

5   Conclusions

In conclusion the contribution of the paper is twofold:

- introduction of the Total Variation-based edge preserving regularization method in SPECT imaging
- development of a scalable parallel algorithm for the solution of the corresponding non-linear Euler Lagrange equation. Our algorithm exploits the parallelism both in terms of the number of projections (rows of the system matrix) and 2-D slices (columns of the system matrix).

References

1. D. Baldini, P. Clavini and A. R. Formiconi, 'Image reconstruction with conjugate gradient algorithm and compensation of the variable system response for an annular SPECT system', *Phys. Medica*, vol 14, pp. 159-173, 1998.

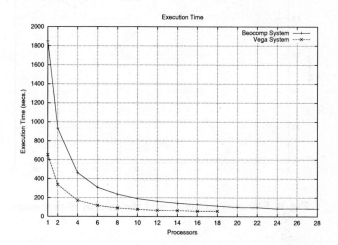

Fig. 6. Execution times as function of the number of processors

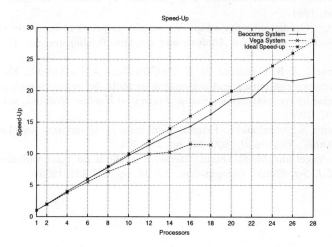

Fig. 7. Speedup as function of the number of processors

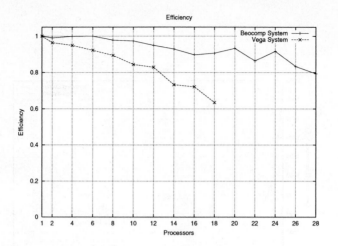

Fig. 8. Efficiency as function of the number of processors

2. P. Boccacci, P. Bonetto, P. Calvini and A. R. Formiconi, 'A simple model for the efficient correction of collimator blur in 3D SPECT imaging', *Inverse Problems*, vol. 15, pp. 907-930, 1999.
3. T. Chan, G. Golub and P. Mulet, 'A primal-dual method for total variation-based image reconstruction', UCLA CAM Report n. CAM-95-43, 1995.
4. P. Charbonnier, L. Blanc-Ferlaud, G. Aubert, M. Barlaud, 'Deterministic edge-preserving regularization in computed imaging', *IEEE Trans. on Image Processing*, vol. 6, pp. 298-311.
5. E. Giusti, *Minimal surfaces and functions of bounded variation*, Birkhauser, Boston, 1984.
6. S. Z. Li, 'On discontinuity-adaptive smoothness priors in computer vision", *IEEE Transactions on Pattern Analysis and machine Intelligence*, vol. 17, pp. 576-586, 1995.
7. A. Passeri, A. R. Formiconi, M. T. De Cristofaro, A. Pupi and U. Meldolesi, 'High performance computing and networking as tools, for accurate emission computed tomography reconstruction', *Europ. Journal of Nuclear Medicine*, vol. 24, n. 4, pp. 390-397, 1997.
8. L. Rudin, S. Osher, and E. Fatemi, 'NonLinear total variation based noise removal algorithms', *Physica D*, 1992, **60**, pp. 259–268.
9. C. R. Vogel and M. E. Oman, 'Iterative methods for total variation denoising', *SIAM J. Sci. Statist. Computation*, vol. 17, pp. 227-238, 1996.

Computer Graphics Techniques for Realistic Modeling, Rendering and Animation of Water. Part I: 1980-88

A. Iglesias

Department of Applied Mathematics and Computational Sciences, University of Cantabria, Avda. de los Castros, s/n, E-39005, Santander, Spain
iglesias@unican.es
http://personales.unican.es/iglesias

To Jack Bresenham, with admiration

Abstract. The realistic representation of natural phenomena has a long tradition in computer graphics. Among the natural objects, one of the most interesting (and most difficult to deal with) is water. During the last two decades a number of papers on computer graphics techniques for modeling and rendering of water have been published. However, the computer graphics community still lacks a survey classifying the vast literature on this topic, which is certainly unorganized and dispersed and hence, difficult to follow. This is the first of a series of two papers (both included in this volume) intended to fill this gap. Our aim is to offer a unifying survey of the most relevant computer graphics techniques for realistic modeling, rendering and animation of water. In this paper we focus on the methods developed during the period 1980-88.

1 Introduction

Modeling natural phenomena has always been among the most challenging problems in computer graphics. Because natural objects are frequently quite asymmetric and nonrigid, they have an inherent complexity far beyond that of most artificial objects. Of all the natural objects, one of the most interesting (but most difficult to deal with) is water. In fact, the modeling and rendering of water has been a traditional problem in computer graphics that has been addressed by many authors during the last two decades. However, the computer graphics community still lacks a survey classifying the vast literature on this topic. This is the first in a series of two papers (both included in this volume) whose aim is to offer a unifying survey of the most relevant computer graphics techniques for realistic modeling, rendering and animation of water. The whole survey has a chronological character; the two papers are devoted to the methods developed during the periods 1980-88 and 1989-97, respectively. This (somewhat artificial) classification has some obvious pros and cons. A major advantage is the opportunity to gain a clear perspective on how computer graphics advances have improved the rendering and animation of water over time. On the contrary, a

P.M.A. Sloot et al. (Eds.): ICCS 2002, LNCS 2330, pp. 181–190, 2002.

major shortcoming is that models from different years might be organized into different sections, even though they are based on the same principles and/or methods. Finally, we should remark that both papers have been strongly influenced by the limitations of space. Some references have had to be omitted and many explanations have been reduced to the minimum. However, we hope that we have included enough comments and references to make the papers useful to our readers.

2   Earlier works: 1980-85

The early 80s marked the starting point of research on the computer modeling and rendering of natural phenomena. However, relatively little time was spent on modeling the appearance and behavior of water. At that time, research on this topic was basically concentrated on representing a large mass of water without boundaries, such as the ocean. On the other hand, water was seen as a compact fluid rather than a mixture of individual droplets.

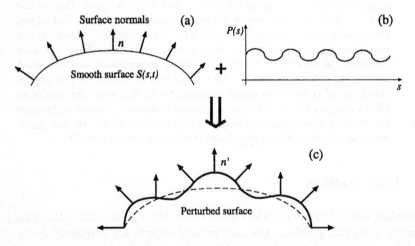

Fig. 1. 2D-scheme of the bump mapping technique: a smooth surface (a) is perturbed by adding a *bump mapping* function $P(s)$ (b) to the surface in the direction of the normal vector. This process allows a realistic rough-textured surface to be obtained (c)

The first attempts to render waves in water were based on the *bump mapping* technique developed by J. Blinn in 1978 [4]. This method allows realistic rough-textured surfaces to be obtained by perturbing the surface normal. Blinn realized that true rough-textured surfaces have a small random component in the surface normal and hence in the light reflection direction, which can be well reproduced by adding a perturbation function to the surface in the direction of the normal vector, as shown in Fig. 1. This solution is better than adding texture patterns to smooth surfaces because, in this last case, the resulting surfaces also appear smooth.

Fig. 2. Random waves perturbation function to simulate a choppy ocean through the bump mapping tecnique

Of course, different perturbation functions can be used to produce different effects. For instance, Schachter [32] proposed a model for fields of random waves (similar to those shown in Fig. 2) that involved a table look-up of precomputed narrow band noise waveforms (a technique intended for real-time applications, and which was implemented in hardware). Other early references were two sets of Siggraph slides not described elsewhere. The first one is the "Pyramid" slide by Gary Demos *et al.* in 1981 [27], where the waves in the sunset are obtained by bump mapping the flat surface with cycloidal waveforms. The second one is the "Night Castles" slide, by Ned Green, in the 1982 collection which used sine waves for the bump mapping technique.

These beautiful images lack some important realistic effects, such as the reflection of objects in the water. To overcome these limitations, Witthed's film "The Compleat Angler" combined bump mapping with *ray tracing*, a well-established technique to render transparencies taking into account reflection and refraction of light (see [14] for a nice introduction to the field). In that film, ray tracing[1] was applied to animate realistic reflections from ripples in a small pool, where the ripples were created by applying a single sinusoidal function to perturb the surface normal [37].

Although based on ray tracing too, Max used a different approach to render the wave surfaces for his famous film "Carla's Island" [20]. His scheme [19] was a ray tracing model in which ocean waves and islands were rendered by different but related algorithms. His work was based on the Fishman and Schachter algorithm [8] for rendering the raster images of the *height field*, i.e., single valued functions of two variables. Max modified the algorithm (one vertical scan line at a time) to reproduce the detail of small ripples near the eye without wasting time near the horizon [19]. The same height field was used for the islands, which were represented as elliptical paraboloids with superimposed cosine terms to give the rolling hills. The wave model was represented by an approximate solution (valid

---

[1] It should be remarked here that as early as in the 70s there were some references to procedural models for ray tracing, such as [17, 23]. See also [31].

only for waves of small amplitude) in which the wave velocity v is proportional to the square root of the wavelength λ; that is, $v \propto \sqrt{\lambda}$. He also assumed a first linear approximation for the wave surface, meaning that the wave trains pass through each other without modification (clearly, a non-realistic situation). Under this assumption, the model for the wave surface was given by a sum of cosine functions corresponding to individual trains of waves of low amplitude. In the model, additional terms from Stokes' approximation [34] of the wave equation might be added to large-amplitude waves. Unfortunately, the author considered the second-order term should be added to the largest wave only, so these (basically) linear waves exhibited a notable deficiency: they formed a self-replicant pattern when viewed over any reasonably large area. On the other hand, his linear small-amplitude theory was restricted to deep water, but it is in shallow water where waves break and this theory fails to predict this phenomenon. Finally, as pointed out by the author, this scheme did not address the problem of rendering ocean scenes with clouds, later analyzed in [11, 12, 21].

To solve these problems, Perlin applied bump mapping through a set of 20 cycloidal waveforms, each radiating in a circular fashion from a randomly placed center point [26]. Then, the appearance of the ocean surface was improved by combining bump mapping with a rich texture map based on nonlinear functions, the so-called *solid texture*. This technique, which is independent of the surface geometry or the coordinate system, defines a texture throughout a three-dimensional volume, where the object to be textured is embedded. The textured surface is obtained as the intersection of the object and the three-dimensional texture volume [7]. The solid texture was applied by Perlin to generate realistic images of clouds, which could be hereby incorporated into the ocean scenes. Another approach used clamped analytical functions to antialias textured surfaces [24]. These schemes work well when the viewer is some distance away from the water so that the ocean surface appears flat.

Although these initial developments were able to produce very beautiful images, they still needed further research to be suitable for realistic rendering and animation. This is due to the techniques the previous methods were based on, mostly bump mapping (such as [15, 26, 32, 37]) and ray tracing (such as [19, 20, 37]). Bump mapping is inexpensive, but it is not sufficient to simulate and animate waves in general. Since the actual surface is flat, bump mapped waves do not exhibit realistic silhouette edges or intersections with other surfaces. In addition, these waves cannot shadow one another or cast shadows on other surfaces. Thus, these methods could not produce realistic scenes containing a seashore; they are forced to restrict themselves to simple images of the ocean surface as one might see it from an aircraft well out the sea. However, when the viewer is near the surface, the three-dimensional nature of the surface becomes more important. On the other hand, ray tracing also exhibits some remarkable disadvantages, such as its tendency to alias arising from point sampling distant waves near the horizon, the difficult of rendering complex procedural models and the most important one, its immense requirements of calculation time, making animation unapproachable. In fact, although many improved methods have been proposed since then [1], ray tracing is still not adequate for computer animation.

3   Mid Period: 1986-88

During this period the realism of water animations was improved in several directions. On one hand, considerable effort was devoted to simulating the interaction between the fluid and a solid. For example, in [10, 25] the authors proposed a combination of particle systems and hydrodynamics. The *particle systems* were introduced by Reeves [29] (see also [30]) to describe natural phenomena which cannot be well described by means of polygons or curved surfaces (trees, grass, wind, etc.). Their ability to model complex phenomena is given by the fact that these particles do not have smooth, well-defined surfaces but rather irregular, complex surfaces with variable shapes. In addition, the characteristics and shape of the particles change with time under the action of physical or stochastic models. For example, in [30] stochastic wind functions and wind maps were employed for shading and rendering particle systems representing both gusts of wind and the motion of wind-blown blades of grass. Finally, the particles are generated, move within the system and die or leave it over time.

These particle systems were generally applied to model the foam and the spray generated by wave breaking and collisions with obstacles. In [25] a particle system for each breaking wave was generated, where the initial position of each particle is at the crest of the wave and its initial velocity is in the same direction as the wave motion. Then, to avoid excessively uniform particle behavior, stochastic perturbations [9] with a Gaussian distribution were added to the velocity of the particles. Finally, another particle system model was generated to simulate the spray from the collision of waves with partially sumerged obstacles (rocks, piers, etc.). A similar approach can be found in [10]. There, the authors derived a rule for the generation of spray and foam: spray is generated when the difference between the particle speed and the surface speed projected in the direction of the normal to the surface exceeds a threshold; otherwise, foam is generated. When generated, spray is sent in the direction of the normal, whereas the foam is sent sliding along the wave surface.

On the other hand, the theory of hydrodynamics was applied to model the ocean waves and many of their associated effects. It should be remarked, however, that there is no hydrodynamic model able to fully and realistically describe the behavior of any real ocean waves. In addition, a model that might be completely useless to a physicist or an oceanographer might be very good for computer graphics and vice versa. Therefore, these approaches must be understood as simple (although often dramatic) approximations carried out for rendering and animation purposes only, with no physical meaning in general.

The general model of wave motion is highly nonlinear, and has no convenient solutions, so many simplified and idealized models have been usually applied instead. For example, Peachey [25] considered the Airy (linear) model of sinusoidal waves of small amplitudes. In his model, the velocity and the wavelength of a wave depend on the depth of the water. On the other hand, the ocean surface was modeled with a height field, thus preventing waves whose crests curl forward. From this point of view, Peachey's model seems to be closely related to Max's model (see Sect. 2). However, Peachey also dealt with wave refraction due to wave velocity changing with depth. *Wave refraction* implies that when the waves approach the shore from deep water, their crests tend to become parallel to the shoreline regardless of their initial orientation. This effect has important consequences in water rendering and animation: the models ignoring wave refraction may produce implausible situations, such as the crests running perpendicular to the beach in [19]. As remarked by several authors, in Max's film "Carla's Island" the ocean waves appear to cut right through the islands as if the islands were made of air.

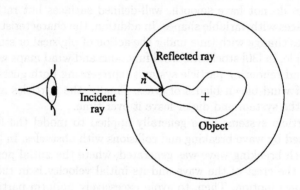

Fig. 3. Scheme of the environment mapping technique

Wave refraction was also considered by Fournier and Reeves [10], whose scheme was based on the Gerstner-Rankine model, proposed long ago in oceanography [13, 28]. Roughly speaking, this model establishes that particles of water describe circular or elliptical stationary orbits. Realistic wave shapes and other needed effects, such as those due to the depth (like wave refraction and surf) and the wind can be easily reproduced by simply varying some parameters of the orbit equations. In [10] the ocean surface was modeled as a parametric surface which obviously allows traditional rendering techniques, including ray tracing [16, 35] and adaptive subdivision [6]. The color and shading of the ocean surface were simulated through an *environment map*, a technique introduced by Blinn and Newell in 1976 [5]. Their basic assumption is that the scene environment is composed of objects and lights distant from the object to be rendered (the wave's surface, in this case). Then, each image of the surrounding environment is mapped onto the inside of a gigantic sphere with the rendered object at its center. The color of a point on the wave is determined by tracing a ray from the camera to the object, reflecting it over the surface normal and tracing the reflected ray outward to the sphere, thus giving us the index of the image and hence the color to be reflected (see Fig. 3).

To control the shape of the ocean the authors introduced several *trains of waves*, i.e. groups of waves sharing the same basic characteristics (heights, periods and wavelengths) and the same phase origin. Of course, each train of waves has its own set of parameters and optional stochastic elements. They allow the "variability" and "randomness" characteristics of the sea to be introduced via a combination of small variations within a train and large variations between trains. Since these variations were basically applied to the height and the wavelength, which can be made a function of time, animation could be easily incorporated in this way. Finally, bump mapping was applied to obtain some additional effects, such as water shimmering and water chop. For animation purposes, the bump map was translated over time just like trains of waves were translated over time.

Almost at the same time, other methods for modeling water surfaces based on Fourier synthesis [22], generalized stochastic subdivision [18], etc., were suggested. These methods worked well for producing still images, but were unsuitable for animation. The reason is that they did not include realistic models for the evolution of the surface over time, so actual phenomena such as wave refraction and depth effects were totally passed over. For example, in [22] the authors used an empirical wind-driven sea spectrum model to filter white-noise images. The ocean images were generated from these white-noise images by using the forward FFT (Fast Fourier Transform) and the inverse FFT, thus representing a fully developed sea in nature. The images were then rendered with a ray tracing algorithm to produce realistic ocean scenes. This work focused on synthesizing fully developed seas in deep water rather than seas in shallow water (where effects such as wave refraction and wave breaking exist). The result has a good appearance although it is valid only for deep water waves and difficult to animate correctly.

Another earlier[2] remarkable reference dealing with wave refraction is [36]. In this approach, Snell's law was applied to control the changes in direction of the ocean waves due to wave refraction. The wave orthogonals (the normals to the wave front) were wave-traced similarly to ray tracing. A ray representing a wave's front direction of propagation was followed and tested for the nearest intersection with the contour lines of the ocean floor. The angle of refraction was obtained from Snell's law and the resulting ray was tested for intersection, and so on. Then this process was simplified taking into account that wave refraction is independent of the viewer's viewpoint (hence, it must be computed only once) and it is essentially a two-dimensional effect. To improve realism, the ocean surface was treated as a true three-dimensional surface. It was represented by Beta-splines [2], the tension shape parameter $\beta2$ (see [3] for a description) being used to add more complexity to the surface: higher values correspond to sharper edges within the surface. This parameter was also changed locally to create smooth troughs and sharper crests. Instead of using ray tracing, two texture

---

[2] Although finally published in 1987, most of the work in this paper was essentially completed before the appearance of the papers [10, 22, 25] (see Guest's Editor Introduction, ACM Transactions on Graphics 6(3) (1987) 165-166).

mapping schemes (both including viewer orientation, surface normal orientation and some laws of optics) were employed to render the ocean waves. Both texture maps were created by hand and contained the reflected and the refracted color information, respectively. These schemes were not as physically accurate as ray tracing but were computationally quicker.

4   Conclusions

Table 1 summarizes the main conclusions of this paper. The left column includes

Table 1. Main water models (left column) and computer graphics techniques (right column) developed during the 80s for water modeling, rendering and animation

Water Models	Computer Graphics Techniques
EARLY WORKS: 1980-85 (*Simple models, Good still images, Oceans without boundaries*)	**Bump mapping, Ray tracing Height fields, Solid texture**
1980: Schachter [32]	Bump mapping (with random waves)
1980: Witthed [37]	Bump mapping (with sine waves) Ray tracing
1981: Pyramid [27]	Bump mapping (with cycloidal waves)
1981: Max [19],[20]	Height fields Ray tracing
1982: Night Castles	Bump mapping (with sine waves)
1985: Perlin [26]	Bump mapping (with cycloidal waves) *Solid texture*
MID PERIOD: 1985-88 (*Hydrodynamics, Wave refraction, Solid-fluid interaction, Animation*)	**Particle systems, Wave tracing Environment map, Fourier filter**
1986: Peachey [25]	Particle systems Height fields Ray tracing, A-buffer
1986: Fournier & Reeves [10]	Particle systems Environment maps, Bump mapping Ray tracing, Adaptive subdivision, etc.
1987: Mastin et al. [22]	*Fourier-domain filter* Ray tracing *Wind-driven spectrum model*
1987: T'so & Barsky [36]	Beta-splines *Fresnel texture mapping Wave-tracing*

the references of the most relevant computer graphics techniques for modeling and rendering of water developed during the 80s. They have been organized for clarity into two different periods, 1980-85 and 1986-88, each associated to a set of techniques listed in bold in the right-hand column. Italics in this column are used to indicate that the corresponding technique was originally described in the reference on the left. We should finally remark that, in spite of these significant advances, there were still a number of unsolved problems. Computer graphics techniques to solve them will be treated in the second part of this survey.

Acknowledgements

The author would like to acknowledge the CICYT of the Spanish Ministry of Education (project TAP98-0640) for partial support of this work. Many thanks are also given to the three referees for their careful reading of the initial version of the manuscript and their helpful suggestions which led to a substantial improvement of the paper.

References

1. Arvo, J., Kirk, D.: A survey of ray tracing acceleration techniques. In: A. Glassner (ed.): An Introduction to Ray Tracing, Academic Press, London, San Diego (1989)
2. Barsky, B.A.: Computer Graphics and Geometric Modelling Using Beta-Splines. Springer-Verlag, Heidelberg (1987)
3. Barsky, B.A., Beatty, J.C.: Local control of bias and tension in Beta-splines. Proc. of SIGGRAPH'83. Computer Graphics 17(4) (1983) 193-218. Also in: ACM Transactions on Graphics 2(2) (1983) 109-134
4. Blinn, J.F.: Simulation of wrinkled surfaces. Proc. of SIGGRAPH'78. Computer Graphics 12(3) (1978) 286-292
5. Blinn, J.F., Newell, M.E.: Texture and reflection in computer generated images. Communications of the ACM 19 (1976) 542-547
6. Dippé, M., Swensen, J.: An adaptive subdivision algorithm and parallel architecture for realistic image synthesis. Proc. of SIGGRAPH'84. Computer Graphics 18(3) (1984) 149-158
7. Eber, D.S., Musgrave, F.K., Peachey, D., Perlin, K., Worley, S.: Texturing and Modeling. A Procedural Approach, Academic Press, Boston (1994)
8. Fishman, B., Schachter, B.: Computer display of height fields. Computers and Graphics 5 (1980) 53-60
9. Fournier, A., Fussell, D., Carpenter, L.: Computer rendering of stochastic models. Communications of the ACM 25(6) (1982) 371-384
10. Fournier, A., Reeves, W.T.: A simple model of ocean waves. Proc. of SIGGRAPH'86. Computer Graphics 20(4) (1986) 75-84
11. Gardner, G.Y.: Simulation of natural scenes using textured quadric surfaces. Proc. of SIGGRAPH'84. Computer Graphics 18(3) (1984) 11-20
12. Gardner, G.Y.: Visual simulation of clouds. Proc. of SIGGRAPH'85. Computer Graphics 19(3) (1985) 297-303
13. Gerstner, F.J.: Theorie der wellen. Ann. der Physik 32 (1809) 412-440
14. Glassner, A. (ed.): An Introduction to Ray Tracing, Academic Press, London, San Diego (1989)

15. Haruyama, S., Barsky, B.A.: Using stochastic modeling for texture generation. IEEE Computer Graphics and Applications 4(3) (1984) 7-19. Errata: IEEE Computer Graphics and Applications 5(2) (1985) 87
16. Kajiya, J.T.: New techniques for ray-tracing procedurally defined objects. ACM Transactions on Graphics 2(3) (1983) 161-181
17. Kay, D.S.: Transparency, Refraction and Ray Tracing for Computer Synthesized Images. Master's Thesis, Cornell University, Ithaca, New York (1979)
18. Lewis, J.: Generalized stochastic subdivision. ACM Transactions on Graphics 6(3) (1987) 167-190
19. Max, N.L.: Vectorized procedural models for natural terrain: waves and islands in the sunset. Proc. of SIGGRAPH'81. Computer Graphics, 15(3) (1981) 317-324
20. Max, N.L.: Carla's island. Issue 5 of the Siggraph Video Review (1981)
21. Max, N.L.: The simulation of natural phenomena. Proc. of SIGGRAPH'83 (panel). Computer Graphics, 17(3) (1983) 137-139
22. Mastin, G.A., Watterberg, P.A., Mareda, J.F.: Fourier synthesis of ocean scenes. IEEE Computer Graphics and Applications 7(3) (1987) 16-23
23. Newell, M.E.: The utilization of procedure models in Digital Image Synthesis. Ph. D. Thesis, University of Utah, Salt Lake City, Utah (1975)
24. Norton, A., Rockwood, A.P., Skolmoski, P.T.: Clamping: a method for antialiasing textured surfaces by bandwidth limiting in object space. Proc. of SIGGRAPH'82. Computer Graphics 16(3) (1982) 1-8
25. Peachey, D.R.: Modeling waves and surf. Proc. of SIGGRAPH'86. Computer Graphics 20(4) (1986) 65-74
26. Perlin, K.: An image synthesizer. Proc. of SIGGRAPH'85. Computer Graphics 19(3) (1985) 287-296
27. Pyramid Catalogue: Pyramid, Box 1048, Santa Monica (1981)
28. Rankine, W.J.W.: On the exact form of waves near the surfaces of deep water. Phil. Tran. Roy. Soc. A 153(4), (1863) 127-138
29. Reeves, W.T.: Particle systems - a technique for modeling a class of fuzzy objects. Proc. of SIGGRAPH'83. Computer Graphics 17(3), (1983) 359-376; also in ACM Transactions on Graphics 2(2), (1983) 91-108
30. Reeves, W.T., Blau, R.: Approximate and probabilistic algorithms for shading and rendering structured particle systems. Proc. of SIGGRAPH'85. Computer Graphics 19(3), (1985) 313-322
31. Rubin, S., Whitted, T.: A 3-dimensional representation for fast rendering of complex scenes. Proc. of SIGGRAPH'80. Computer Graphics 14(3) (1980) 110-116
32. Schachter, B.: Long crested wave models. Computer Graphics and Image Processing 12 (1980) 187- 201
33. Sims, K.: Particle dreams (Video). Segment 42, Siggraph Video Review 38/39 (1988)
34. Stokes, G.G.: Mathematical and Physical Papers, Vol. 1. Cambridge University Press, Cambridge (1880)
35. Toth, D.L.: On ray tracing parametric surfaces. Proc. of SIGGRAPH'85, Computer Graphics 19(3) (1985) 171-180
36. Ts'o, P., Barsky, B.A.: Modeling and rendering waves: wave-tracing using beta-spline and reflective and refractive texture mapping. ACM Transactions on Graphics 6(3) (1987) 191-214
37. Witthed, T.: An improved illumination model for shaded display. Communications of the ACM 23(6) (1980) 343-349

Computer Graphics Techniques for Realistic Modeling, Rendering and Animation of Water. Part II: 1989-1997

A. Iglesias*

Department of Applied Mathematics and Computational Sciences, University of
Cantabria, Avda. de los Castros, s/n, E-39005, Santander, Spain
iglesias@unican.es
http://personales.unican.es/iglesias

Abstract. This work concerns the realistic modeling, rendering and
animation of water. Specifically, the paper surveys many of the most
outstanding references on computer graphics techniques regarding this
problem developed during the period 1989-97. The paper, which has been
organized for clarity into two different periods, follows on from a previous
one dedicated to the same topic during the period 1980-88 [11].

1 Early nineties: 1989-92

Although the previous works [11] provided adequate models for waves hitting
the beach, to quote just one example, a significant range of physical phenomena
still remained unexplored. A major question was the accurate description of
fluid dynamics. Great interest was also placed upon several physical phenomena
related to water such as:

- state of matter changes such as melting and freezing,
- complex natural phenomena such as wetting and drying,
- mass transport or flow including meandering,
- the behavior of individual droplets and their streams,
- interaction with static and dynamic buoyant obstacles, etc.

On the other hand, a number of interesting rendering effects, such as the
simulation of reflected waves, the interaction between light and water, the analysis of caustics, etc. had not yet been considered by the previous computer
graphics techniques. Finally, water realistic animation was still in its infancy and
it had to be greatly improved. To overcome these limitations, several new models
were proposed. Roughly, they can be grouped into two different alternatives
which are analyzed in the following sections.

---

* This research was supported by the CICYT of the Spanish Ministry of Education
(projects TAP1998-0640 and DPI2001-1288).

P.M.A. Sloot et al. (Eds.): ICCS 2002, LNCS 2330, pp. 191–200, 2002.

1.1   Interaction of a Large Number of Particles

The first alternative tried to simulate the fluid dynamics by the interaction of a large number of particles [7, 17, 26, 27]. Thus, in [17] and [27] the authors studied the attraction and repulsion forces between particles to simulate various degrees of fluid viscosity and state of matter change such as melting. For example, the approach in [17] presented a particle based model for fluids, powders and gelatinous solids. This model was based on what the authors called *globules*, a term intended to avoid connotations with words such as particle or blob, and used to designate the elements of the connected particle system. Globules can be applied for the detection of soft collisions between the particles and obstacles, involving forces which vary gradually with distance, thus allowing globules to flow over one another. The forces acting on the particles depend on two scaling factors, one for attraction/repulsion, s_r, and another one, s_d, (drag) to attenuate the inter-globule force based on distance. For powders and fluids the globules interact with a short range repulsive force and drag. In the solid state the attraction term is also considered, so globules interact with a short range repulsion, a medium range attraction and long range indifference. The temperature values for a given pair of globules can be used to simulate changes in their interaction behavior, thus allowing the melting of solids and the freezing of liquids. This temperature effect was later analyzed by Tonnesen in [27], through a model describing the changes in geometry and movement of elemental volumes as a consequence of thermal energy and external forces. This model simulates both the liquid and solid states by varying the shape of the potential energy curve as a function of temperature. Objects at hot temperatures behave like fluids, with rapidly varying geometry. At cold temperatures they resemble solids, with a stable shape which can be modified, however, under the influence of appropriate external forces. Changes in topology are modelled by using particles to represent elemental volumes with potential energies between pairs of particles. Finally, another remarkable reference using this particle interaction scheme is due to Sims [26]. In his work, a waterfall was simulated by applying gravity to thousands of non-interacting water droplet particles and bouncing them off obstacles made of planes and spheres. The water droplets were created at the top of the waterfall, flowed over the last edge at the bottom and then recycled back to the top of the waterfall. These particles of different shapes, sizes, colors and transparencies were rendered by using a parallel rendering particle system incorporating some techniques for increasing the image quality, such as anti-aliasing, hidden surfaces and motion-blur. The combination of some *ad hoc* tricks (such as the combination of white and blue particles to avoid lighting calculations and exaggerated motion blur to give the flow a smoother look) allow realistic pictures of the waterfall to be obtained. Additional effects such as the deformations produced on a falling droplet as it enters in contact with a surface are also analyzed in [20].

1.2   Partial Differential Equations for the Fluid Dynamics

The second approach is to directly solve a partial differential equation (PDE) system describing the fluid dynamics [14]. Of course, this alternative is much more reliable and realistic in terms of physical simulation. The main problem is that a truly accurate simulation of fluid dynamics requires the computation of the fluid motion throughout a volume. This implies that the computation time for each iteration is at least proportional to the cube of the resolution, making the computation prohibitively expensive. Fortunately, for rendering and animation purposes, to obtain very accurate results is less important than other factors, such as the speed of the simulation and the stability of the numerical methods involved within. With this idea in mind, Kass and Miller [14] considered a very simplified subset of water flow where:

- the water surface can be represented by a height field,
- the vertical component of the velocity of the water particles can be ignored (so motion is uniform through a vertical column) and
- the horizontal component of the velocity of the water in a vertical column is approximately constant.

With these simplifications, the motion equations of the water were approximated in terms of a grid of points on the height field, obtaining a wave equation in which the wave velocity is proportional to the square root of the depth of the water. Thus, this method can generate wave refraction with depth and take into account other features, such as wave reflections, net transport of water and situations in which the boundary conditions change through time altering the topology of the water. Due the height-field representation of the water surface, the two-dimensional motion equations were numerically integrated by using a finite-difference technique. This technique converts the PDE into ODEs, which can be integrated through a first-order implicit method. The algorithm is very easy to implement and gives rise to tridiagonal linear systems. For the three-dimensional case, one alternating-direction method was applied, allowing the 3D iteration to be divided into two 2D sub-iterations, so complexity does not radically increase. Additional simplifications were added for rendering, making high-resolution simulations possible. The water was rendered using *caustic shading* that simulates the refraction of illuminating rays at the water surface. The work also incorporated realistic appearance for water in sand through a *wetness map* that computes the wetting and drying of sand as the water passes over it. However, this wetness map had to be filtered to avoid aliasing artifacts in the boundary between wet and dry areas before using it for shading.

Another interesting reference using caustics for water rendering is that in [29]. This work, cited here for its applications to realistic water rendering, presented a new method (the so-called *backward beam tracing*) for computing specular to diffuse transfer of light. This problem appears when light reflecting from, or refracting through, one surface (the specular surface) hits a diffuse surface where, by definition, it is emitted equally in all directions. Early 90s' traditional global illumination techniques [9], such as those based on ray tracing and radiosity, did not handle it. This fact was not so surprising when using ray tracing, best suited to solving the diffuse and specular to specular mechanisms of light transport. After all, the specular to diffuse transfer is view independent and hence, more suitable for radiosity. But even radiosity techniques generally ignored this problem. Remarkable exceptions were given by the *hybrid techniques* [25, 28], where a combination of ray tracing and radiosity is considered. The hybrid method uses two passes for illumination calculation: a view independent pass, similar to diffuse radiosity, computing the diffuse illumination of a scene, and a view dependent pass, similar to distributed ray tracing, that uses the previous one to provide global illumination information and subsequently the specular components. Unfortunately, these methods only captured the softest of specular and diffuse effects varying slowly over pixels, and they became impractical for capturing the (more interesting) higher frequency effects.

The approach in [29] is also a two-pass algorithm, the first pass being a variation of the backward ray tracing [1] called backward beam tracing, then combined with a rendering phase (the second pass) incorporating a test for caustic polygons associated with diffuse polygons. The method was applied to simulate the light-water interaction where the water surface acts as the specular surface. Its performance was illustrated through some frames from an animated underwater caustic sequence, showing the familiar sinuous shifting patterns of light on objects underwater as they intersect the caustic. As these patterns are driven directly by the water surface (described by a polygonal mesh of triangles displaced by a height field), animating them is merely a question of animating the water surface. Therefore, this animation process consists of deciding upon an appropriate time interval and frequency/amplitude rates. On the other hand, if the triangles in the polygonal mesh are small enough, the beams can be represented as polygonal illumination volumes, rendered using a modified version of a light volume rendering technique proposed by Nishita et al. [21]. The results show that this technique is particularly effective to describe the light-water interaction.

2   1993-97

During these years, the two approaches introduced in the previous years (namely, the interaction of a large number of particles and the system of PDEs to describe the fluid dynamics) were extended and improved. In addition, several interesting effects for substantially improving both rendering and animation were described.

2.1   Improving Particle System Methods

The early water models developed for computer graphics and based on particle systems (see [11], Sec. 3) assumed that particles of water moved in circular and elliptical orbits around their initial positions. Such an assumption was intended to render large bodies of water such as the ocean. A very different situation is that of animating the flow of very small amounts of water, in which this particle motion model is no longer able to realistically reproduce the fluid dynamics.

Some interesting references to overcome this problem appeared in this period. For instance, in [16] the use of string textures for rendering large waterfalls was proposed. Particle systems usually require a large number of particles, thus leading to unreasonably large amounts of memory usage. This is because the position and velocity are stored explicitly for each particle. On the contrary, implicit methods are used to store parameters in solid texture, by means of addressing small lookup tables. Mallinder suggested a new modelling method which he called *string texture*, which is basically a development from particle systems and solid texture, using a method of implicitly storing particles. From this point of view, string texture is a technique to save memory for storing information on particles.

A strictly geometrical approach to calculate the rolling of a rigid convex object on a smooth biparametric surface is reported by Hégron in [10]. This author assumed that there is a single contact point (determined through a simple iterative algorithm) between the object and the surface which is maintained during the animation. Then, the motion was computed in terms of discrete displacements on the tangent plane at the contact point and projections onto the surface. Since this projection is time-consuming, an alternative prediction-correction scheme (based on projecting the motion onto the tangential plane and then making corrections to ensure that the points are on the surface) was proposed instead. Although the proposal does not take into account the physics of the object and it was not intended specifically for water animation purposes, it has a potential application to the problem of modeling the water droplets on a curved surface.

Other relevant references for rendering the flow of small amounts of water are described in [4, 15]. They included a physically based model for simulating icicle growth realistically [15] and the rendering of water currents using particle sytems [4]. Unfortunately, since these authors did not consider the interfacial dynamics, it is difficult to animate the streams of water droplets.

Another interesting alternative was given by Kaneda et al. in [12]. These authors proposed a method for animating water droplets and their streams on a glass plate taking into account the dynamics between fluid and solid. In this scheme, the shape and the motion of the water droplets depend on the gravity force, the surface tensions of the glass plate and the water and the interfacial tension between them.

To simulate the stream the surface model of the glass plate was discretized so that the water droplets travel from one mesh point to the next according to a set of rules describing their dynamics as a function of their mass, the angle of the plate's inclination, φ, and affinity. A droplet placed on a lattice point (i, j) moves to one of the three lower lattice points $(i - 1, j - 1)$, $(i - 1, j)$ or $(i - 1, j + 1)$. Because of the nature of the wetting phenomena, some amount of water remains behind the route of the stream so the mass of the droplet decreases with time and finally the flow stops. The model also includes equations for the speed of the running water droplet as a function of the wetness and the angle φ and for the speed and mass of the new droplet resulting from merging of two original droplets. The final algorithm returns the positions and masses of all water droplets for every frame of the animation.

To render the water droplets and the streams the authors proposed a high-speed rendering method which takes into account the reflection and the refraction of the light. Basically, it is an extension of the environment map by Greene [8] in which objects in the scene are projected onto the planes of a cuboid, whose center is on the glass plate. The method consists of calculating background textures by projecting objects in the scene onto the faces of a cuboid whose center is on the glass plate, then calculating the directions of rays reflected or refracted by water droplets on a glass plate, and finally determining pixel colors by using the background textures and the intersection of the ray and the cuboid. The method finally generates an image with these pixel colors at these intersection points, thus avoiding calculating the intersections between the ray and the objects (undoubtely, the most time-comsuming process in ray tracing).

This model was firstly applied to animate water droplets meandering on an inclined glass plate [12] and later extended to water droplets on curved surfaces [13]. In this last case, the surface was described by Bézier patches and then converted to a discrete surface model for which each quadrilateral mesh was approximated by a plane. In [13] the applications of water droplet animation were classified into two categories: those mainly pursuing rendering speed (such as in drive simulators in which interactivity is the goal) and those pursuing photo-reality. According to this classification, two different rendering algorithms were proposed. The first one (for which the stream is modeled as a group of spheres) is essentially an extension of the simple and fast rendering method described in [12] to handle Bézier patches. The second one is a more sophisticated high-quality rendering method intended for photo-reality and based on metaballs.

The metaball technique was first introduced by Blinn in [2] and he called it *blobs*. Subsequently, this technique was improved by several authors [19, 30–34] who coined the terms *metaballs* and *soft objects*. This technique is widely used to represent "soft objects" like liquids [32]. Its popularity comes from the fact that the model is defined by just a few simple constraints providing free-form deformations. In addition, it is used for water modeling because the water droplets merge smoothly together in this model.

In [13] the metaball is defined by its center, its density distribution (the authors applied that of [34]) and a threshold value defining the surface. To overcome the conflict of working with both implicit surfaces (given by the metaball technique) and parametric surfaces (the Bézier surfaces of the problem), a combination of two rendering methods, namely the one in [22] for Bézier surfaces and the one in [23] for metaballs, is proposed. In this new scheme the density distribution along a ray was converted into a Bézier function and the intersection between the ray and the metaball is calculated using a Bézier clipping [22].

2.2   Improving PDE Fluid Dynamics Methods

One of the first attempts to simulate faithfully the behavior of the fluid dynamics was the work of Kass and Miller [14] already analyzed in Sec. 1.2. This model is limited in the sense that it does not address the full range of three-dimensional motion (including rotational and pressure effects) found in a liquid. In addition, since the velocity of the fluid is known only on the surface and internal pressure is not calculated at all, the model cannot easily incorporate dynamic or buoyant objects.

A more realistic model can be obtained by considering the Navier-Stokes (NS) equations, notably the most comprehensive of all fluid models. Basically, these equations describe completely the motion of a fluid at any point within a flow at any instant of time. Because of their ability to capture the turbulent or stable behavior of the fluid with arbitrary viscosity in three dimensions, they are often used to simulate accurately fluid phenomena.

The Navier-Stokes equations have been applied to create models of water motion for computer graphics [3, 5]. In [3] a simplified version of the Navier-Stokes equations in two dimensions was considered. Such a simplification was obtained by removing the vertical dependence and solving the resulting two-dimensional system. In other words, they assumed that the fluid has zero depth, thus treating it as being completely flat during the computation. Although still able to model some kind of interactions between moving objects and the flow, such an assumption considerably restricts the range of phenomena of the model. For example, since the obstacles must be two-dimensional, the model does not deal with submerged objects. In addition, neither convective wave effects nor mass transport can be incorporated into this model.

Another comprehensive methodology based on the NS equations for animating liquid phenomena was introduced in [5] and later extended in [6]. These authors noticed that the real fluid dynamics can only be captured by solving the three-dimensional Navier-Stokes equations. To this end, they considered a two-stage calculation over an environment of static obstacles surrounded by fluid. Firstly, a finite difference approximation to the NS equations was applied to a low-resolution discretized representation of the scene. This scheme was coupled with an iterative step to refine the velocities and determine the pressure field, which was combined with the Lagrange equations of motion to simulate dynamic buoyant objects. At this stage, the fluid position in 2D was tracked by convecting massless marker particles with local fluid velocity. These marker particles are very useful to highlight many internal fluid motion effects, such as rotation and splashing, as well as for animating violent phenomena such as overtunning waves. Additional effects such as vorticity were also taken into account in this step. In the original work, a secondary calculation was performed to accurately determine the liquid surface position through a height field equation. Since this equation describes surfaces as single-valued functions, it can be effectively applied to situations such as puddles, rivers or oceans. For dramatic effects such as crashing waves or splashing, a combination of height field and marker particles was applied. The performance of the method was illustrated through some realistic 2D and 3D examples, such as the animation titled *Moonlight Cove*, a $50 \times 15 \times 40$ mesh used to show the effect of two large ocean waves crashing into a shallow cove. Scene complexity was increasing by including submerged rocks and irregular sea bottom, thus leading to interesting features on the water surface.

The dynamics behavior of splashing fluids was also analyzed in [24]. Essentially, the authors introduced a model that simulates the behavior of a fluid when objects impact or float on its surface. To this purpose, they used a three-part system where each subsystem corresponds to a physical area of the fluid body: the volume, the surface and the spray (see Figure 1).

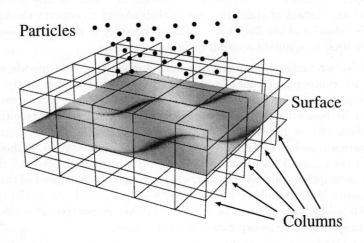

Fig. 1. Scheme of the fluid model in reference [24]. The model is a three-part system containing subsystems for volume (columns), surface and spray (particles)

The water volume that makes up the main body of the fluid was simulated by a collection of vertical tubes or columns connected by virtual pipes, thus allowing here to be a flow between these columns according to the hydrostatics laws for pressure. In addition, flow conditions must be specified to model boundary conditions, such as barriers (flow set to zero) or fluid sources or sinks (flow set to a positive or negative constant). The surface subsystem allows external objects to interact with the fluid system and consists of mesh of control points whose vertical positions are found by averaging the heights of the surrounding columns. Finally, the spray subsystem is a particle system to model water droplets disconnected from the main body of the fluid. These droplets fall under the influence of the gravity and their dynamics is quite simple since the particles are noninteracting. To preserve the total volume of the system, the volume of each particle was substracted from the column from which it was created. Although the model cannot deal with submerged objects, the impact problem can be treated in a natural way. The impact of the object on the surface is transferred to the pressure of the affected columns. When thresholding the vertical velocity of the column, spray is generated yielding a particle system until it is reabsorbed by the fluid, encounters a ground plane or strikes some other object. The model is able to describe a wide range of phenomena such as waves, impacts, splashes or floating objects. On the contrary, those phenomena related to vertical effects such as turbulence or underwater effects, are not adequately described in this model.

A remarkable extension of this proposal was described in [18], including an explicit underwater terrain and the removal of the vertical isotropy, thus making the interaction between particles and fluid possible. The model consists of a volume system (containing the information about the main body of fluid) coupled with a particle system. Although some effects such as friction do not follow the actual physical laws and additional parameters with no physical meaning have been introduced to simulate spreading, the model is overall based on the physics of fluids. The model simulates accurately many different situations: the propagation of the surface waves, the variation of the propagation of the speed with terrain, the variation of wave patterns with the depth of water, the dynamics of the water droplets, the floating and submerged objects, splashing, etc. Some shortcomings of the model are that a height field for the water surface was considered (clearly an unrealistic situation in the real world) and that the force from an object striking the surface is strictly applied in the vertical direction, even although the direction of the object has some angle with the vertical. Extensions of these models to overcome these limitations were subsequently described. They will be analyzed in a future work and reported elsewhere.

References

1. Arvo, J.: Backward ray tracing. Developments in ray tracing. SIGGRAPH'86 course notes, Vol. 12 (1986)
2. Blinn, J.F.: Generalization of algebraic surfaces drawing. ACM Transactions on Graphics, Vol. 2(3) (1980) 235-256
3. Chen, J.X., Lobo, N.V.: Toward interactive-rate simulation of fluids with moving obstacles using Navier-Stokes equations. Graphical Models and Image Processing 57(2) (1995) 107-116
4. Chiba, N., Sanakanishi, S., Yokoyama, K., Ootawara, I., Maruoka, K., Saito, N.: Visual simulation of water currents using a particle-based behavioural model. Journal of Visualization and Computer Animation, Vol. 6(3) (1995) 155-171
5. Foster, N., Metaxas, D.: Realistic animation of liquids. Proc. of Graphics Interface'96, Calgary, Canada (1996) 204-212; also in: Graphical Models and Image Processing 58(5) (1996) 471-483
6. Foster, N., Metaxas, D.: Controlling fluid animation. Proc. of Computer Graphics International CGI'97, IEEE Computer Society Press, Menlo Park, CA (1997) 178-188
7. Goss, M.E.: A real-time particle system for display of ship wakes. IEEE Computer Graphics and Applications 10(3) (1990) 30-35
8. Greene, N.: Environment mapping and other applications of world projections. IEEE Computer Graphics and Applications 6(11) (1986) 21-29
9. Hall, R.: Illumination and Color in Computer Generated Imagery. Springer-Verlag (Series: Monographs in Visual Communication) New York (1989)
10. Hégron, G.: Rolling on a smooth biparametric surface. The Journal of Visualization and Computer Animation 4 (1993) 25-32
11. Iglesias, A.: Computer graphics techniques for realistic modeling, rendering and animation of Water. Part I: 1980-88. (this volume)

12. Kaneda, K., Kagawa, T., Yamashita, H.: Animation of water droplets on a glass plate. Proc. of Computer Animation'93 (1993) 177-189
13. Kaneda, K., Zuyama, Y., Yamashita, H., Nishita, T.: Animation of water droplets on curved surfaces. Proc. of Pacific Graphics'96, IEEE Computer Society Press, Los Alamitos, Calif. (1996) 50-65
14. Kass, M., Miller, G.: Rapid, stable fluid dynamics for computer graphics. Proc. of SIGGRAPH'90. Computer Graphics 24(4) (1990) 49-57
15. Kharitonsky, D., Gonczarowski, J.: A physical based model for Icicle growth. The Visual Computer, Vol. 10(2) (1993) 88-100
16. Mallinder, H.: The modeling of large waterfalls using string texture. Journal of Visualization and Computer Animation, Vol. 6(1) (1995) 3-10
17. Miller, G., Pearce, A.: Globular dynamics: a connected particle system for animating viscous fluids. Computers and Graphics 13(3) (1989) 305-309
18. Mould, D., Yang, Y.-H.: Modeling water for computer graphics. Computers and Graphics 21(6) (1997) 801-814
19. Murakami, S., Ichihara, H.: On a 3d display method by metaball technique. Journal of the Electronic Communications 70(8) (1987) 1607-1615
20. Nishikawa, N., Abe, T.: Artificial nature in splash of droplets. Compugraphics'91 1 (1991) 457-466
21. Nishita, T., Miyawaki, Y., Nakamae, E.: A shading model for atmosferic scattering considering luminous intensity distribution of light sources. Proc. of SIGGRAPH'87. Computer Graphics 21(4) (1987) 303-310
22. Nishita, T., Sederberg, T.W., Kakimoto, M.: Ray tracing rational trimmed surface patches. Proc. of SIGGRAPH'90. Computer Graphics 24(4) (1990) 337-345
23. Nishita, T., Nakamae, E.: A method for displaying metaballs by using Bézier clipping. Proc. of EUROGRAPHICS'94. Computer Graphics Forum 13(3) (1994) 271-280
24. O'Brien, J.F., Hodgins, J.K.: Dynamic simulation of splashing fluids. Proc. of Computer Animation'95 (1995) 198-205
25. Sillion, F., Puech, C.: A general two-pass method integrating specular and diffuse reflection. Proc. of SIGGRAPH'87. Computer Graphics 23(3) (1989) 335-344
26. Sims, K.: Particle animation and rendering using data parallel computation. Proc. of SIGGRAPH'90. Computer Graphics 24(4) (1990) 405-413
27. Tonnesen, D.: Modeling liquids and solids using thermal particles. Proc. of Graphics Interface'91 (1991) 255-262
28. Wallace, J.R., Cohen, M.F., Greenberg, D.P.: A two-pass solution to the rendering equation: a synthesis of ray tracing and radiosity methods . Proc. of SIGGRAPH'87. Computer Graphics 21(4) (1987) 311-320
29. Watt, M.: Light-water interaction using backward beam tracing. Proc. of SIGGRAPH'90. Computer Graphics 24(4) (1990) 377-385
30. Wyvill, B., McPheeters, C., Wyvill, G.: Data structure for soft objects. The Visual Computer 2(4) (1986) 227-234
31. Wyvill, B., McPheeters, C., Wyvill, G.: Animating soft objects. The Visual Computer 2(4) (1986) 235-242
32. Wyvill, B.: Soft. Proc. of SIGGRAPH '86, Electronic Theater and Video Review 24 (1986)
33. Wyvill, G., Wyvill, B., McPheeters, C.: Solid texturing of soft objects. IEEE Computer Graphics and Applications, December (1987) 20-26
34. Wyvill, G., Trotman, A.: Ray-tracing soft objects. Proc. of Computer Graphics International CGI'90, Springer Verlag (1990) 469-476

A Case Study in Geometric Constructions

Étienne Schramm and Pascal Schreck*

Laboratoire des Sciences de l'Image, de l'Informatique et de la Télédétection
UMR CNRS - Université Louis Pasteur 7005
Pôle API, Boulevard Sébastien Brant
F-67400 Illkirch - France
etienne@axis.u-strasbg.fr, schreck@dpt-info.u-strasbg.fr

Abstract. The geometric constructions problem is often studied from a combinatorial point of view: a pair data structure + algorithm is proposed, and then one tries to determine the variety of geometric problems which can be solved. Conversely, we present here a different approach starting with the definition of a simple class of geometric construction problems and resulting in an algorithm and data structures. We show that our algorithm is correct, complete with respect to the class of simply constrained polygons, and has a linear complexity. The presented framework is very simple, but in spite of its simplicity, this algorithm can solve non-trivial problems.

1 Introduction

Handling geometric objects declaratively described by a system of geometric constraints remains an important issue in CAD, even in the 2D case. To make feasible this manipulation, a geometric solver is always needed to find one, some, or all the solutions of a constraint system. Geometric constructions in CAD have been studied by various authors (see e.g.[2, 6, 3, 4, 9, 11, 10, 12, 7]). To this date, the most famous methods are based on a combinatorial analysis of the problem to be solved: the geometric side is quickly forgotten by giving a graph structure which is used to do constraint propagation [6, 1], maximal flow search [7] and/or problem decomposition [6, 10, 7]. In these approaches, different notions of correctness are defined and studied (e.g. in [6, 7]). We consider here "semantic" correctness, that is: the discovered figures are solutions of the constraint system.

We present a complete study in a very simple framework: we consider *simply constrained polygons* where the constraints can only concern the lengths of the edges and the angles between two edges of the polygon. We give an algorithm allowing to solve all the constraint schemes in this framework. We prove that our algorithm is correct and complete (*all* the solutions of the constraint system are found, in the degenerate cases zero solutions are found and explanations can be given). Moreover, we show that the class of complexity is $O(n)$ where n is the number of vertices. Our method is very simple and this paper can be

---

* To whom correspondence should be addressed

P.M.A. Sloot et al. (Eds.): ICCS 2002, LNCS 2330, pp. 201–210, 2002.
© Springer-Verlag Berlin Heidelberg 2002

regarded as a simple case study introducing the domain of geometric construc-
tions.Nevertheless, the class of the solved problems is not trivial (e.g. see the
example given on Figure 4) Furthermore, it seems possible to use this proper
method in a multi-agent approach like the one described in [4]. Please, note that
simply constrained polygons constitute a subclass of the problems solved by the
Sunde method (see [13]), but, as far we know, there is no efficient implementation
of this method.

In Sections 2 and 3 of this paper, we expose the geometric framework of
the simply constrained polygons. In Section 4, we present and justify our resolution algorithm. In Section 5, we show how that algorithm can be efficiently
implemented. And, finally, we give some concluding remarks in section 6.

2   Definition of Polygons by Lengths and Angles

A polygon is classically defined as a finite sequence of points in the Euclidean
space with a fixed reference, that is , a polygon P is a function $P : [1, C] \rightarrow \mathbb{R}^2$
where $[1, C]$ is a finite interval of integers, $C > 2$, and C is the number of vertices
of the polygon. Let us recall some usual notations: $|P| = C$ is the number of
vertices, which is also the number of edges of P, $P(n)$ or simpler P_n denotes the
n^{th} point of polygon P, $E(n)$ denotes the nth oriented edge of P (i.e the oriented
segment $[P(n), P(n+1)]$ if $n < |P|$ and $[P(n), P(1)]$ if $n = |P|$) and $\|E(n)\|$ its
length. At last, we note $\mathbb{P}$ the set of the polygons of the plane.

For the purposes of this paper, we consider two functions families over the
polygons namely the *polygon constructive functions* and the *permutation functions* which are defined as follows:

Definition 1. *The set of the polygon constructive functions is the set $CF = \{f_{n,a,l} | n \in \mathbb{N}^*, a \in \mathbb{R}, l \in \mathbb{R}_+^*\}$ where each function $f_{n,a,l}$ is defined by:*

$$f_{n,a,l} : \mathbb{P} \rightarrow \mathbb{P}$$

$$\forall m \in [1, |P| + 1] \qquad f_{n,a,l}(P)(m) = \begin{cases} P_m & \text{if } m < n \\ Q & \text{if } m = n \\ P_{m-1} & \text{if } m > n \end{cases}$$

with point Q defined by:

$$\begin{pmatrix} x_Q \\ y_Q \end{pmatrix} = \frac{l}{P_{n-1}P_n} \begin{pmatrix} \cos(a) & -\sin(a) \\ \sin(a) & \cos(a) \end{pmatrix} \begin{pmatrix} x_{P_n} - x_{P_{n-1}} \\ y_{P_n} - y_{P_{n-1}} \end{pmatrix} + \begin{pmatrix} x_{P_{n-1}} \\ y_{P_{n-1}} \end{pmatrix}$$

Figure 1 a) illustrates such a construction. It is easy to verify that point Q
is the unique point such that $P_{n-1}Q = l$ and $\angle(\overrightarrow{P_{n-1}P_n}, \overrightarrow{P_{n-1}Q}) = a$.

Let us define the other family of functions over the polygons:

Definition 2. *The set of permutation functions is the set $PF = \{g_n | n \in \mathbb{N}^*\}$
where each function g_n is defined by:*

$$g_n : \mathbb{P} \rightarrow \mathbb{P}$$

$$\forall m \in [1, |P|] \qquad g_n(P)(m) = \begin{cases} P_m & \text{if } m \neq n + 1 (mod|P|) \\ S & \text{otherwise} \end{cases}$$

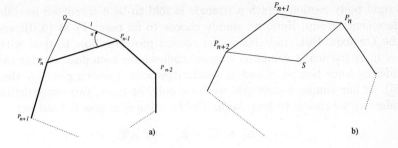

Fig. 1. Action of a polygon constructive function (a) and a permutation function (b)

with point S defined by:

$$\begin{pmatrix} x_S \\ y_S \end{pmatrix} = \begin{pmatrix} x_{P_{n+2}} + x_{P_n} - x_{P_{n+1}} \\ y_{P_{n+2}} + y_{P_n} - y_{P_{n+1}} \end{pmatrix}$$

Figure 1 b) shows the effect of the permutation function g_n over a polygon. Note that S is the unique point verifying $\overrightarrow{P_n S} = \overrightarrow{P_{n+1} P_{n+2}}$.

It is easy to see that any polygon P can be generated by the polygon constructive functions applied successively to the triangle (P_1, P_2, P_3) (see Figure 2 for a "graphic idea" of the proof). This assertion can be transformed into: let $C = |P|$, knowing triangle (P_1, P_2, P_3), the distances $P_3 P_4$, ..., $P_{C-1} P_C$ and the oriented angles $\angle(\overrightarrow{P_3 P_2}, \overrightarrow{P_3, P_4})$, ... $\angle(\overrightarrow{P_{C-1} P_{C-2}}, \overrightarrow{P_{C-1}, P_C})$, one can re-construct the polygon P using the functions $f_{a,n,l}$

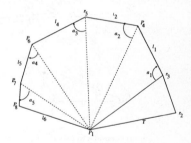

Fig. 2. Construction of a polygon using the polygon constructive functions

We will see below that, thanks to the permutation functions, this constraint scheme can be enlarged to the simply constrained polygons. But beforehand, we have to consider the way to get the triangle (P_1, P_2, P_3) from lengths and angles.

The construction of a triangle knowing the length of its three edges (resp. two lengths and one angle, or one length and two angles) is easy and well known, but this simplicity conceals two important problems of the geometric constructions in CAD. For instance, let us consider the function $t_1 : \mathbb{R}_+^* \times \mathbb{R}_+^* \times \mathbb{R}_+^* \rightarrow \mathbb{P}$ computing a triangle having the three given lengths. First problem, there is an infinity of triangles having these lengths: each of them can be deduced from one of two particular solutions using a direct isometry (also called a displacement or a rigid body motion). Such a triangle is told to be constructed modulo the displacements group. Here, we simply choose to fix point P_1 at $(0,0)^t$ and P_2 on the Ox axis. But, and this is the second problem, how to deal with the two or more particular solutions ? Some studies have been done on this subject considering some heuristics and/or similarities with a sketch given by the user ([6, 5]). In our simple framework, we have only, at most, two constructions to consider, so, we choose to keep both. The function t_1 is now defined by:

$$t_1 : \mathbb{R}_+^* \times \mathbb{R}_+^* \times \mathbb{R}_+^* \to \mathbb{P} \times \mathbb{P}$$

with $(P', P'') = t_1(l_1, l_2, l_3)$, $P_1' = P_1'' = (0,0)^t$, $P_2' = P_2'' = (l_1, 0)^t$ and

$$P_3' = \begin{pmatrix} \frac{l_1^2 + l_3^2 - l_2^2}{2l_1} \\ \sqrt{l_3^2 - \left[\frac{l_1^2 + l_3^2 - l_2^2}{2l_1}\right]^2} \end{pmatrix} \text{ and } P_3'' = \begin{pmatrix} \frac{l_1^2 + l_3^2 - l_2^2}{2l_1} \\ -\sqrt{l_3^2 - \left[\frac{l_1^2 + l_3^2 - l_2^2}{2l_1}\right]^2} \end{pmatrix}$$

Figure 3 shows the four cases of triangle construction (with, eventually some permutations of edges) and the construction of the two triangles in the cases of functions t_1 and t_3.

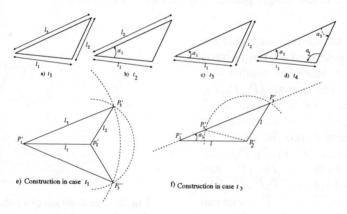

a) t_1 b) l_2 c) l_3 d) l_4

e) Construction in case t_1 f) Construction in case t_3

Fig. 3. Four triangle cases (a) to (d) and construction for t_1 (e) and t_3 (f)

The following definition makes the things clearer. In this definition, all the triangles (P_1, P_2, P_3) verify $P_1 = (0,0)^t$ and $P_2 = (l_1, 0)^t$.

Definition 3. *A triangle constructive function is one of the four partial functions t_1, t_2, t_3 and t_4 which are defined by:*

$$t_1 : \mathbb{R}_+^* \times \mathbb{R}_+^* \times \mathbb{R}_+^* \to \mathbb{P} \times \mathbb{P}$$
$$t_2 : \mathbb{R}_+^* \times \mathbb{R}_+^* \times \mathbb{R} \to \mathbb{P} \times \mathbb{P}$$
$$t_3 : \mathbb{R}_+^* \times \mathbb{R}_+^* \times \mathbb{R} \to \mathbb{P} \times \mathbb{P}$$
$$t_4 : \mathbb{R}_+^* \times \mathbb{R} \times \mathbb{R} \to \mathbb{P} \times \mathbb{P}$$

- t_1 *is defined as before,*
- $t_2(l_1, l_2, a_1) = (P, P)$ *with P such that $P_1P_3 = l_2$ and $\angle(\overrightarrow{P_1P_2}, \overrightarrow{P_1P_3}) = a_1$*
- $t_3(l_1, l_2, a_1) = (P', P'')$ *where P' and P'' are the two polygons such that $P_2'P_3' = P_2''P_3'' = l_2$ and $\angle(\overrightarrow{P_1''P_2''}, \overrightarrow{P_1''P_3''}) = \angle(\overrightarrow{P_1'P_2'}, \overrightarrow{P_1'P_3'}) = a_1$ (see Figure 3.f)*
- $t_4(l_1, a_1, a_2) = (P, P)$ *with polygon P such that $\angle(\overrightarrow{P_1P_2}, \overrightarrow{P_1P_3}) = a_1$ and $\angle(\overrightarrow{P_2P_3}, \overrightarrow{P_2P_1}) = a_2$*

Thus, we extend classically our families of functions CF and PF to the ordered pairs of polygons in order to have the following result: *every polygon can be constructed thanks to the functions of CF, PF and t_1, t_2, t_3 and t_4 modulo the displacements group.* We can now examine the notion of constraint scheme.

3   Constraint Schemes and Simply Constraint Polygons

A length constraint (over a polygon) is an ordered pair (n, l) where n is a natural integer and l a non negative real. We say that a polygon P satisfies the constraint (n, l) if $n \leq |P|$ and $\|E(n)\| = l$.

Definition 4. *A set of length constraints L is* coherent *if $(n, l) \in L$ and $(n, k) \in L$ leads to $k = l$.*

We note $C_L(n)$ the predicate indicating if the constraint (n, l) is in L or not. When $C_L(n)$ is true, we note $\texttt{length}(n, L)$ the imposed value in L for the edge $E(n)$.

An angle constraint is a triple (n, m, a) where n and m are natural integers, with $n \neq m$ and a a real number. We say that a polygon P satisfies the constraint (n, m, a) if $n \leq |P|$, $m \leq |P|$ and the oriented angle between $E(n)$ and $E(m)$ is equal to a, i.e. $\angle(E(n), E(m)) = a$. Given a set of angle constraints A, we can define the *associated unoriented angle graph* $G(A)$ whose nodes are the segments $E(i)$ and whose unoriented edges correspond to the angle constraints between the segments *i.e.* $\{E(i), E(j)\} \in G(A)$ if there is a such that $(i, j, a) \in A$ or $(j, i, a) \in A$. The transitive closure of the angle graph corresponds to the Chasles relation $(\angle(\overrightarrow{OA}, \overrightarrow{OB}) + \angle(\overrightarrow{OB}, \overrightarrow{OC}) = \angle(\overrightarrow{OA}, \overrightarrow{OC}))$. This fact justifies the definition:

Definition 5. *A set of angle constraints A is* coherent *if there is no circuit in the associated angle graph.*

As before, we note $C_A(n_1, n_2)$ the predicate indicating if the edges $E(n_1)$ and $E(n_2)$ are angle constrained or not. That is, $C_A(n_1, n_2)$ is true if and only if there is a path between $E(n_1)$ and $E(n_2)$ in the unoriented graph $G(A)$. Then, we note $\texttt{angle}(n_1, n_2, A)$ the value imposed by the angle constraints in A and computed using the Chasles relation.

A scheme of constraints is an ordered triple $S = (L, A, n)$ containing a set of length constraints, a set of angle constraints and the number n of edges considered (which is noted $N(S)$). We impose that n is greater than the maximal edge number appearing in the sets of constraints. A constraint scheme S is well-constrained if there is a finite number of polygons P such that P satisfies all the constraints in S, $|P| = N(S)$, $P_0 = (0,0)^t$ and P_1 is on Ox. This definition leads classically (e.g. see [6]) to the definition of a *structurally well-constrained* scheme of constraints which is *purely combinatorial* and which meets the previous definition when the constraints values are algebraically independent and the complex solutions are considered. Surprisingly, due to the simplicity of the framework, simply constrained polygons have always at most two solutions (possibly zero). So, if a constraint scheme is not a simply constrained polygon: either there are redundant constraints, or the system is not well-constrained. Then, we have the following theorem:

Theorem 1. *A constraint scheme* (L, A, n) *is structurally well-constrained if and only if* L *and* A *are coherent and either* $|L| = n$ *and* $|A| = n - 3$, *either* $|L| = n - 1$ *and* $|A| = n - 2$ *or either* $|L| = n - 2$ *and* $|A| = n - 1$. *We call such a constraint scheme a* simply constrained polygon.

Sketch of the proof: The necessary condition comes immediately from the facts $|A| + |L| = 2n - 3$, $|A| \le n$ and $1 \le |L| \le n$. It is relatively easy, but too long for this paper, to prove by recurrence that this cardinality condition is sufficient: the idea is to use on a constraint scheme the equivalent operations than the ones used on the polygons constructive functions and the permutation functions). This idea is also used in the analysis of a constraint scheme given below. □

Figure 4 shows, graphically, an example of a simply constrained polygon $S = (L, A, n)$ where $L = \{(1, k_1), (2, k_2), \ldots, (6, k_6)\}$ and $A = \{(1, 4, a_1), (2, 5, a_2), (3, 6, a_3)\}$. It becomes a well constrained scheme when the formal parameters k_1, k_2 $\ldots k_6$ and a_1, $\ldots a_3$ are instantiated with appropriate values.

The objective is then to take a well constrained schema S in order to yield one (or two) polygons P which satisfies all the constraints in S.

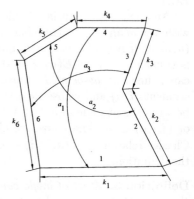

Fig. 4. A structurally well-constrained polygon

4   Analysis and Construction

From the Theorem 1 of the previous section, the following result can be deduced:

Theorem 2. *Given $S = (L, A, n)$ a simply constrained polygon with $n > 3$, there are two edges $n_1 \in \{1, 2, 3, 4\}$ and $n_2 \in \{1, 2, 3, 4\}$, with $n_1 \neq n_2$, such that $C_L(n_1)$, $C_L(n_2)$ and $C_A(n_1, n_2)$.*

Proof. Since $G(A)$ has no circuit, each addition of an edge decreases the number of connected components by 1. So, in case $|A| = n - 3$, there are exactly three connected components in $G(A)$. The pigeon hole principle shows that $\exists n_1 \in \{1, 2, 3, 4\}$ and $n_2 \in \{1, 2, 3, 4\}$ such that $C_A(n_1, n_2)$ and $n_1 \neq n_2$. Since $|L| = n$ in this case, the result is obvious. The other two cases ($|A| = n-2$ and $|A| = n-1$) can be treated by the same use of the pigeon hole principle. $\square$

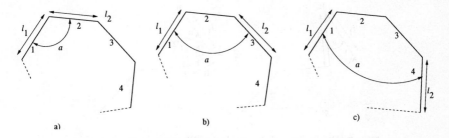

Fig. 5. The three cases

Thanks to Theorem 2, we can see that there are fundamentally three cases examining the constraints over the first four edges: the angle and length constrained edges are consecutive, or separated by one edge, or separated by two edges. These cases are graphically represented on Figure 5 .Let us examine the c) case and explain how to simplify it. The c) case corresponds to the constraint scheme $L = \{(1, l_1), (4, l_2), \ldots\}$ and $A = \{(1, 4, a), \ldots\}$. This constraint scheme can be transformed in a way that is similar to the permutation function: $S = (L, A, n)$ becomes $S' = (L', A', n)$ where L' and A' are obtained from L and A replacing 3 by 4 and vice-versa. Next, we apply the same operation exchanging 3 with 2 (See Figure 6) giving the scheme S''. It is easy to demonstrate that P is a solution of S'' if and only if $g_2(g_3(P))$ is a solution of S. Then, we are in the situation described on Figure 7 left: edges 1 and 2 are constrained in length and in angle. The construction of triangle (ABC) is quite usual and requires only standard formulae: so the distance AC and the angle b can easily be computed. Then, we obtain a new constraint scheme S''' computed from S'' by forgetting edges 1 and 2, adding a new edge (say $1'$) and transferring all the angle constraints between edge 1 or 2 and the other edges to the new edge $1'$ using computed angle b and, finally, by adding the length constraint $(1, l_1')$ to L''. It is still easy to demonstrate that polygon P satisfies the scheme constraint S''' if and only if $f_{1,b,l_1}(P)$ satisfies S''. Note that the resulting scheme is still structurally well constrained and has one edge fewer than S.

The other cases are very similar. This leads us to propose the following algorithm to solve a simply constrained polygon:

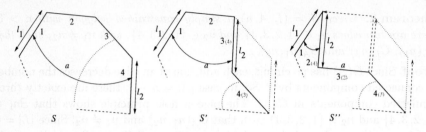

Fig. 6. Beginning of the analysis (1): permutations

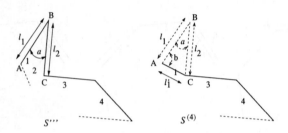

Fig. 7. Beginning of the analysis (2): triangle construction

```
input: simply constrained polygon
output: polygon
solve(S) = if triangle(S) then solve_triangle(S)
   elif case_a(S) then build_case_a(coefs_a(S),solve(simplify_case_a(S)))
   elif case_b(S) then build_case_b(coefs_b(S),solve(simplify_case_b(S)))
   elif case_c(S) then build_case_c(coefs_c(S),solve(simplify_case_c(3)))
fi
```

where functions triangle() and solve_triangle() are in charge of the triangle
cases resolution, functions case_X() test if the constraint scheme is in case a), b)
or c). Then, functions simplify_case_X() make the simplification following the
description above given, while functions coefs_X() compute the corresponding
dimensions in order to apply the correct construction and permutation functions
thanks to *meta-functions* build_case_X(). We have then the following theorem:

Theorem 3. *The previous algorithm is correct and complete with respect to the
simply constrained polygons.*

That means the algorithm finds all the solutions whatever the simply constrained
polygon to be solved. In the degenerate cases, no solutions are found (this is correct) and explanations can be given to the user. Since, in this paper, the auxiliary
functions, and hence the algorithm, are not formally described, it is impossible
to prove rigorously here this theorem. Let us say that the correctness comes
from the facts that, first, the functions triangle() and solve_triangle() are
correct and, second, that if S is in the case X (with $X \in \{a, b, c\}$) then this is correctly discovered by the case_X() function and, build_case_X(coef_X(S), P)

is a solution of S if and only if P is a solution of `simplify_case_X(S)`. The description of the simplification above given should convince the reader of these facts. The completeness comes from Theorem 2 and the fact that all the cases are handled by the algorithm.

5   Implementation

Although we have functionally and recursively described our algorithm, we have carried out an imperative procedural implementation. We have shown that, with adequate data structures, all the auxiliary functions given in previous section have a $O(1)$ complexity in space and time.

Note first, that the algorithm can easily be put under a procedural form using a stack in order to keep in memory the construction functions to apply and which were discovered during the analysis of the constraint scheme. By this way, the transformed algorithm consists in two phases: the analysis of the constraint scheme and the re-construction of the resulting polygon.

In the first phase, the algorithm has to detect if the simply constrained polygon is a triangle or not. In the first case, the resolution is well known and is done in constant time. If it is not a triangle, the system searches, with the first four edges, what is the simplification case. This can be done easily by examining the data structures containing the constraints. We chose to implement the set of angle constraints with a forest where each tree has a depth 1 and each node contains a pointer to the root. The construction of such a structure is achieved in linear time and the search for an angle constraint between two edges is achieved in constant time. So, it is easy to see that the functions `case_X` are executed in constant time. The analysis and the simplification are implemented according to the description given above. It is obvious that each step can be performed in constant time. Since there are $n - 3$ steps, the complexity of the analysis/simplification phase is $O(n)$.

In the second phase, the system uses all the informations collected during the first phase and which are kept in the stack. These informations indicate which functions, with which parameters and in which order to call, in order to build the polygon from the "initial" triangle yet constructed. Each step of the construction consists in the application of one, two or three functions. So, it is achieved in constant time. Thus, the entire construction is done in linear time.

It is easy to see that with our data structures the space complexity is $O(n)$ as well.

This algorithm has been implemented under the form of a C++ prototype. It was experimented on small examples, $i.e.$: simply constrained polygons with about ten points. For each example, the resolution was immediate.

6   Conclusion

We presented in this paper a simple framework in the geometric constraint solving domain. Examining the family of the simply constrained polygons, we found a general algorithm to solve the structurally well-constraint scheme. This algorithm is correct, complete and has a linear complexity.

Of course, this framework is only a case study and the resulting algorithm is not very powerful. But it can solve difficult problems. Furthermore it can be used by other solvers which perform decomposition as the one described in [4]. One can also imagine to generalize this approach to "constrained complex cellular" consisting in an assembling of simply constrained polygons.

References

1. S. Ait-Aoudia, R. Jegou and D. Michelucci: Reduction of constraint systems. In: *Proceedings of the Compugraphics Conference*. Compugraphics Alvor, Portugal(1993) 83-92.

2. B. Aldefeld, H. Malberg, H. Richter and K. Voss: Rule-based variational geometry in computer-Aided Design. In: D.T. Pham (ed):*Artificial Intelligence in Design*. Springer-Verlag (1992) 27-46.

3. B. Brüderlin: Automatizing geometric proofs and constructions. In:*Proceedings of Computational Geometry '88*. Lecture Notes in Computer Science, Vol. 333. Springer-Verlag (1988) 232-252.

4. J.-F. Dufourd, P. Mathis, and P. Schreck: Formal resolution of geometric constraint systems by assembling. In:*Proceedings of the ACM-Siggraph Solid Modelling Conference, Atlanta*. ACM Press (1997) 271-284.

5. C. Essert-Villard, P. Schreck, and J.-F. Dufourd: Sketch-based pruning of a solution space within a formal geometric constraint solver. In: *Journal of Artificial Intelligence*, Num. 124. Elsevier (2000) 139-159.

6. I. Fudos and C. M. Hoffmann: Correctness proof of a geometric constraint solver. In: *International Journal of Computational Geometry and Applications* Num. 6. World Scientific Publishing Company(1996) 405-420.

7. C. M. Hoffmann, A. Lomonosov, M. Sitharam: Decomposition Plans for Geometric Constraint Systems, Part I: Performance Measures for CAD. In: *Journal of Symbolic Computation* Num. 31. Academic Press(2001) 367-408.

8. G.A. Kramer: A geometric constraint engine. In: *Artificial Intelligence* Num. 58. Elsevier(1992) 327-360.

9. H. Lamure and D. Michelucci: Solving constraints by homotopy. In: *Proceedings of the ACM-Siggraph Solid Modelling Conference*. ACM Press(1995) 134-145.

10. J. Owen: Algebraic solution for geometry from dimensional constraints. In: *Proceedings of the 1st ACM Symposium of Solid Modelling and CAD/CAM Applications*. ACM Press(1991) 134-145.

11. G. Sunde: Specification of shape by dimensions and other geometric constraints. In: *Proceedings of the Eurographics Workshop on Intelligent CAD systems*. Noordwisjkerout(1987).

12. I.E. Sutherland. Sketchpad: A man-machine graphical communication system. In: *Proceedings of the IFIP Spring Joint Computer Conference*.Detroit, Michigan(1963) 329-36.

13. A. Verroust, F. Schoneck and D. Roller: Rule-oriented method for parameterized computer-aided design. In: *Computer-Aided Design* Vol. 10 Num. 24. Elsevier(1992) 329-36.

Interactive versus Symbolic Approaches to Plane Loci Generation in Dynamic Geometry Environments

Francisco Botana

Departamento de Matemática Aplicada
Universidad de Vigo
Campus A Xunqueira, 36005 Pontevedra, Spain
fbotana@uvigo.es

Abstract. This paper reviews current approaches to plane loci generation within dynamic geometry environments. Such approaches are classified as interactive, when just a plot of the locus is shown, and symbolic,
if, in addition to plotting the locus, its equation is also given. It is shown
how symbolic approaches outperform the interactive ones when dealing
with loci which are algebraic curves. Additionally, two experimental improvements are reported: i) an efficient computer algebra system allows
symbolically generated loci to behave as dynamic objects, and ii) a general purpose computer algebra system is used to remove spurious parts
of some loci.

1 Introduction

Dynamic geometry software refers to computer programs where accurate construction of geometric configurations can be done. The key characteristic of this
software is that unconstrained parts of the construction can be moved and, as
they do, all other elements automatically self–adjust, preserving all dependent
relationships and constraints [13]. An inmediate consequence of this behavior
allows us to keep track on the path of an object that depends on another object
while this one is dragged. If the dependent object is a point, its trace gives a
locus, whereas if it has higher dimensionality, the path can be used to suggest
related geometric elements, such as envelopes. Most dynamic geometry software
implements loci generation just from a graphic point of view, returning them
as states of the screen. The loci obtained in this way are restricted by the allowed transformations in the system, no algebraic information about them is
known, and, sometimes, they become aberrant for particular positions of the
construction. On the other side, the application of symbolic methods for loci
generation, although restricted to algebraic curves, generalizes the class of ob-
tainable loci, returns their algebraic expression, and behaves in a uniform way
for all construction instances.

The structure of the paper is as follows. Section 2 describes current approaches to loci generation in classic dynamic geometry environments and in

P.M.A. Sloot et al. (Eds.): ICCS 2002, LNCS 2330, pp. 211–218, 2002.

two more recent programs. Symbolic methods for the problem are briefly recalled in Section 3, and a successful implementation of one of such methods into a dynamic geometry program is reported. Finally, some ways for further development in this field are proposed.

2   Interactive Approaches

The introduction, in the late eighties, of The Geometer's Sketchpad [9] and CABRI [16] marked the birth of dynamic geometry software. The Geometer's Sketchpad, GSP, was a project developed by Jackiw, whereas CABRI, called today Cabri Geometry, was designed by an interdisciplinary team led by Laborde. Both programs share the strategy for tracing loci: selecting an object, the *driver object*, with a predefined path, the locus of another object depending on the former is drawn by sampling the path and plotting the locus object for each sample. The driver object, in both GSP and Cabri, must be a point, whereas the path can be any linear object, arc, or even a locus, containing the driver object.

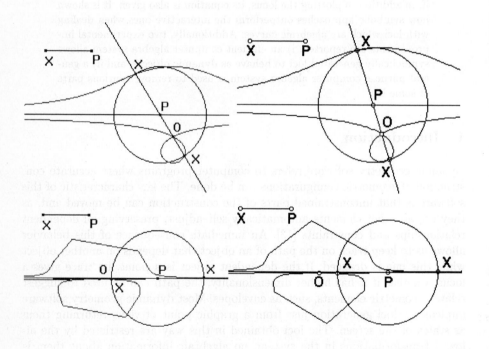

Fig. 1. Conchoid of Nicomedes in Cabri (left) and GSP (right)

If the locus object is a point (other options are linear objects and arcs), an option allows the user to fit a curve, by linear interpolation, to the locus points, thus returning the locus as a usually continuous curve. Nevertheless, the uniform division of the path can produce anomalous loci. For the sake of illustration, let us consider the conchoid of Nicomedes as a locus. Given a point O and another point P lying on a line, the conchoid of Nicomedes is the locus of points X such that X, O and P are collinear, and the distance between X and P is constant. Both GSP and Cabri easily find the curve (Fig. 1). But note that if O is *close* to the line, the linear interpolation fails to return a correct locus, even setting the number of samples to its maximum value. Some contiguous positions of P on its path, the horizontal line, produce successive not–near positions of X.

Geometry Expert [7] tries to avoid this problem returning loci only as sequences of points and widening the range of sampling (Fig. 2). But this behavior cannot be corrected within this graphic approach, since small changes of some elements in a construction can sometimes produce sudden jumps of dependent objects.

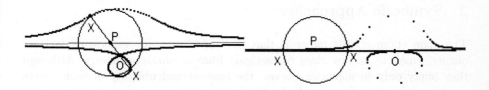

Fig. 2. Conchoid of Nicomedes in Geometry Expert

Using complex analysis, Cinderella [14, 18] has solved the problem of continuity, but, with regard to loci generation, it shares the problems of the interpolating approach of GSP and Cabri, as illustrated with the conchoid in Fig. 3. Furthermore, its strategy of returning loci as the positions only accessible by real continuous moves [14, p. 137] sometimes avoids their correct generation. Fig. 4 shows, in Cinderella, the locus of a point whose product of distances to F_1 and F_2 is constant (the similar triangles ABC and DBE in the left are used to multiply). The program finds just one of the pair of Cassinian ovals for this configuration.

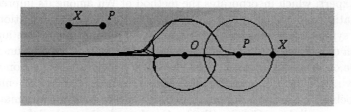

Fig. 3. Conchoid of Nicomedes in Cinderella

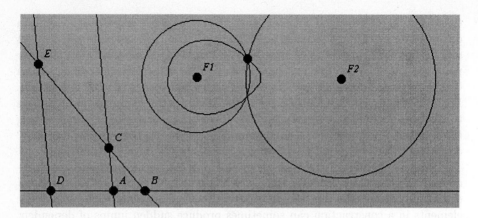

Fig. 4. Cinderella finds just a Cassinian oval

3   Symbolic Approaches

The main claim in this paper is that the above ways for generating loci are outperformed by a new class of methods from algebraic geometry. Although they apply only to algebraic curves, the amount and ubiquity of such curves would justify using these methods. Furthermore, there exists another reason to support their use. Up to now, no dynamic geometry software could obtain the locus of a non–dependent object, thus excluding a huge class of them. Symbolic approaches deal with locus points in a uniform way: it does not matter whether the point is dependent or there are other objects that depend on it.

Symbolic generation of loci can be seen as a subfield of geometric discovery, which is also related to automatic theorem proving. Constructive methods such as Groebner bases [4] or Wu's method [23] have been much more successful in automatic proving than earlier attempts based on logical approaches.

On the one hand, the method of Wu has been used for loci discovery by Chou [6], where Steiner theorem is rediscovered, and by Roanes and Roanes in the plane [19] and the space [20]. Both uses are purely algebraic in the sense that no graphical environment for diagram construction is provided, and they require human intervention, mainly when setting the order of algebraic operations. Geometry Expert, which incorporates the method of Wu among its impressive set of automatic provers, has not been designed for automatic generation of loci using its symbolic capabilities. The use of Wu's method for a true automatic generation of loci within a dynamic geometry environment remains unexplored.

On the other side, Groebner bases have been widely used for automatic theorem proving [10–12,15]. A recent work due to Recio and Vélez [17] emphasizes using Groebner bases for automatic discovery, rather than for automatic proving. It proposes linking Cabri with the Groebner basis method for automatic discovery in an intelligent program for learning Euclidean geometry. Preliminary steps in this direction, with a new dynamic geometry environment, have been reported in [1, 2]. Narrowing the goal of automatic geometry discovery to loci discovery, Groebner bases allow the easy obtaining of equations and plots of algebraic curves that are the locus of a point. Lugares [21] links a dynamic geometry software with Mathematica [22] and CoCoA [5]. Basically, it allows the user to draw a geometric construction and it returns the locus of a point on which some extra conditions are stated. In this way, the locus point can constrain some other elements of the construction. The construction and the locus point conditions are translated into polynomial equations, and an elimination process, using Groebner bases, is carried out on them. The geometric–dependent variables are eliminated, leaving a set of polynomials in the independent variables. This set, seeing the locus point variables as indeterminates and all of the rest as parameters, is the locus searched for (see [3] for a technical description). Let us consider, for example, the locus of a point such that its perpendicular projections on the sides of a triangle determine another triangle whose oriented area is k. There is no way to find this locus with the interactive approach taken in most dynamic geometry systems, whereas Lugares finds, in no appreciable time, that the locus is a circle (the circumcircle for $k = 0$ –Simson theorem–, and a concentric circle for $k \neq 0$ –Steiner theorem), as shown in Fig. 5 for Simson theorem.

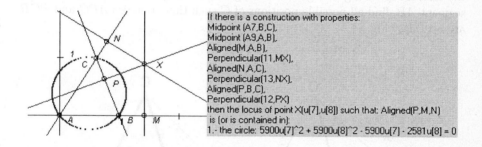

Fig. 5. The locus of X such that M, N, P are collinear is the circumcircle

Despite the doubly exponential computational cost of Buchberger's algorithm, most loci in elementary geometry are quickly computed by CoCoA, thus allowing their dynamic behavior when dragging any other element. Fig. 6 illustrates this assertion with a particular case of a recent extension of Simson theorem [8]: Given a triangle ABC and three directions, the locus of points X such that its projections on the sides are collinear is a conic. For a certain instance of the construction, the conic is an ellipse, while dragging C it becomes a hyperbola. Although the elapsed time between both drawings cannot be generally stated (it depends on the actual coordinates of the construction points), empirical findings show that the time between successive drawings moves between 0 and 2 seconds on a conventional PC.

In turn, Mathematica is considerably slower than CoCoA when dealing with Groebner bases, hence it is not used to show dynamic loci. Nevertheless, being

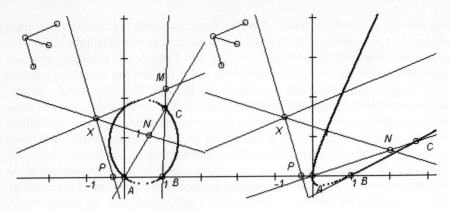

Fig. 6. Dynamic behavior of a locus

a general–purpose symbolic system, it allows to move a step forward in loci obtaining. As is well known, geometric relations such as *between* can not be expressed through polynomial equations. Let us take into account, for example, the task of separating triangles, that is, given a triangle ABC and a point P on segment AB, find all possible positions of C such that triangles APC and PCB are isosceles (Fig. 7).

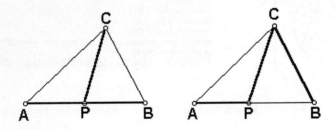

Fig. 7. Two configurations for separable triangles: $AP = PC = PB$, $AP = PC = CB$

While discovering that C lies on a circle centered in P with radius PA for the first configuration (where P is AB midpoint) is trivial, the answer to the second one is not so easy. Note that the locus we are searching for is a case where the locus point imposes conditions on other construction elements, so interactive approaches will generally fail when trying to looking for it. Furthermore, no polynomial–based symbolic approach can deal with the restriction on P lying on segment AB. Using CoCoA as the symbolic engine, Lugares returns an hyperbola and a circle as locus of C (Fig. 8, left). Nevertheless, it is clear that not all points on the hyperbola satisfy the locus requirements. No polynomial equation can express that P is on segment AB, but on line AB. Using the numerical capabillities of Mathematica, the locus lines are sampled and a member of each sample is tested in order to see if it satisfies all construction properties: the points lying on the hyperbola and exterior to the circle derive from positions of P not in segment AB, hence these samples are rejected from the locus, which is returned as shown in the right of Figure 8.

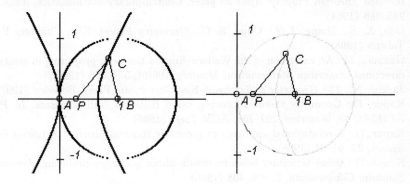

Fig. 8. The locus found with CoCoA (left) and with Mathematica (right)

4 Conclusion

In this paper, interactive and symbolic approaches to the problem of plane loci generation within dynamic geometry environments are compared. It is shown that Groebner bases method can be efficiently used to discover loci, allowing to automate this task in a graphical interface. The ability for finding the equations of loci balances the restriction of the method to algebraic curves. Furthermore, the symbolic approach is more general than the interactive one, in the sense that it allows to search for loci of points constraining other elements. It is also shown how efficient implementations of Groebner bases method compete with interactive approaches when simulating dynamic behavior of loci. First steps in generalizing the symbolic approach to deal with inequations are illustrated by means of a simple problem.

References

1. Botana, F., Valcarce, J. L.: A dynamic-symbolic interface for geometric theorem discovery. *Computers and Education,* to appear (2002)
2. Botana, F., Valcarce, J. L.: Cooperation between a dynamic geometry environment and a computer algebra system for geometric discovery. In V. G. Ganzha, E. W. Mayr, E. V. Vorozhtsov, *Computer Algebra and Scientific Computing CASC 2001,* Springer, Berlin, 63–74 (2001)
3. Botana, F., Valcarce, J. L.: A software tool for investigation of plane loci. Technical Report, University of Vigo, Pontevedra (2001)

4. Buchberger, B.: Groebner Bases: an Algorithmic Method in Polynomial Ideal Theory. In N.K. Bose, *Multidimensional systems theory*, Reidel, Dordrecht, 184-232 (1985)

5. Capani, A., Niesi, G.,L. Robbiano: CoCoA, a system for doing Computations in Commutative Algebra. Available via anonymous ftp from: cocoa.dima.unige.it

6. Chou, S. C.: Proving Elementary Geometry Theorems Using Wu's Algorithm. *Automated Theorem Proving: After 25 years*, Contemporary Mathematics, AMS, 29, 243-286 (1984)

7. Gao, X. S., Zhang, J. Z., Chou, S. C.: *Geometry Expert*. Nine Chapters Publ. Taiwan (1998)

8. Guzmán, M.: An extension of the Wallace–Simson theorem: projecting in arbitrary directions. *American Mathematical Monthly*, 106(6), 574–580 (1999)

9. Jackiw, N.: *The Geometer's Sketchpad*. Key Curriculum Press, Berkeley (1997)

10. Kapur, D.: Geometry theorem proving using Hilbert's Nullstellensatz. In *Proc. SYMSAC'86*, Waterloo, 202–208, ACM Press (1986)

11. Kapur, D.: A refutational approach to geometry theorem proving. *Artificial Intelligence*, 37, 61–94 (1988)

12. Kapur, D.: Using Groebner bases to reason about geometry problems. *Journal of Symbolic Computation*, 2, 399–408 (1986)

13. King, J., Schattschneider, D.:*Geometry Turned On*. MAA, Washington (1997)

14. Kortenkamp, U.: Foundations of dynamic geometry. Ph. D. Thesis, ETH, Zurich (1999)

15. Kutzler, B., Stifter, S.: On the application of Buchberger's algorithm to automated geometry theorem proving. *Journal of Symbolic Computation*, 2, 389–397 (1986)

16. Laborde, J. M., Bellemain, F.: *Cabri Geometry II*. Texas Instruments, Dallas (1998)

17. Recio, T., Vélez, M. P.: Automatic discovery of theorems in elementary geometry. *Journal of Automated Reasoning*, 23, 63–82 (1999)

18. Richter–Gebert, J., Kortenkamp, U.: *The Interactive Geometry Software Cinderella*. Springer, Berlin (1999)

19. Roanes–Macías, E., Roanes–Lozano, E.: Búsqueda automática de lugares geométricos. *Proceedings Spanish Educ. Session IMACS–ACA'99, Boletín Sociedad Puig Adam de Prof. de Matemáticas*, 53, 67–77 (1999)

20. Roanes–Macías, E., Roanes–Lozano, E.: Automatic determination of geometric loci. 3D–extension of Simson–Steiner theorem. *Proceedings of Artificial Intelligence and Symbolic Computation 2000, Lecture Notes in Artificial Intelligence*, 1930, 157–173 (2000)

21. Valcarce, J. L., Botana, F.: Lugares. Manual de Referencia. Technical Report, University of Vigo, Pontevedra (2001) http://rosalia.uvigo.es/sdge/lugares

22. Wolfram, S.: *The Mathematica book*. Cambridge University Press, Cambridge (1996)

23. Wu, W. T.: *Mechanical Theorem Proving in Geometries*. Springer, Vienna (1994)

Deformations Expressed as Displacement Maps

An Easy Way to Combine Deformations

Hubert Peyré and Dominique Bechmann

Université Louis Pasteur — LSIIT
Pôle API — Boulevard Sébastien Brant
67400 Strasbourg — France
peyre,bechmann@dpt-info.u-strasbg.fr

Abstract. Up to now, assuming a geometric object as being a set of points, a deformation was defined as a mathematical function which mapped the original object set of points into a new one. In short, a new point was linked with each object point in the zone of influence of the deformation. With such a statement, it was not possible to apply several deformations simultaneously as soon as their zones of influence overlapped each other. A new formal definition of a deformation is brought out in this paper. Instead of a new point, this definition links a displacement with each object point in the zone of influence of the deformation. Thus, applying several deformations simultaneously is made feasible by blending, for each point of the original object, the displacement due to each deformation.

1 Introduction

In geometric modeling, deformations are both used to refine or to change the shape of a geometric object. Deformations have been introduced a long time ago into the field of geometric modeling. They have quickly and widely been accepted as a powerful technique and many deformation models have been worked out. Formally, in an Euclidean n-dimensional space, if Ω is a geometric object set of points and if Γ is the subset of points of Ω with regard to a given deformation, then this deformation is represented as the mathematical function denoted f, which maps each point from Γ into a new point in the Euclidean n-dimensional space.

$$f : \Gamma \subseteq \Omega \to \mathbb{R}^n$$
$$P \quad \mapsto P' = f(P) \ . \tag{1}$$

In the remainder of this paper, for convenience reasons, whereas any point P in the domain Γ of the deformation f will be labeled *point to be deformed*, any point P' in the range $f(\Gamma)$ of this deformation will be denoted *deformed point*. If all the points of the geometric object Ω are in the domain Γ of the deformation, the deformation is said global (i.e. the whole geometric object is modified), else the deformation is said local (i.e. only a part of the geometric object is modified).

Afterwards, deformations have been divided into two classes depending on whether they resorted to a deformation tool or not. Barr's deformations, put

P.M.A. Sloot et al. (Eds.): ICCS 2002, LNCS 2330, pp. 219–228, 2002.

forward in [1], belong to the deformations class which require no deformation tool since they are matrix functions of the points location. As for the deformation models which have recourse to a deformation tool, most of them are laid-out in [2].

In its most basic form, a couple of geometric entities are enough to define a deformation tool : one will be labeled the *initial geometric entity* (IGE), the other being denoted the *final* or *deformed geometric entity* (DGE). The geometric entities may be of various types. For instance, the initial geometric entity of Sederberg and Parry's free-form deformations [8] is a parametric volume, the control points of which belong to a parallelepipedical lattice.

The process to perform a deformation which requires a deformation tool always follows the same broad outlines.

- First and foremost, the object is embedded into the initial geometric entity of the deformation tool. In other words, the location of the points to be deformed, so far expressed in the Euclidean coordinate system, is now expressed in a suitable coordinate system for the initial geometric entity of the deformation tool. This way, although the location of the points to be deformed remains unchanged, these points are possessed of new coordinates with regard to a new space coordinate system. These new coordinates are labeled *local coordinates with regard to the deformation tool.*
- Next, the local coordinates with regard to the deformation tool of each object point to be deformed are used in an appropriate coordinate system for the final geometric entity of the deformation tool in order to determine its new location in space. Therefore, to make a deformation intuitive, the coordinate system linked to the initial geometric entity of the deformation tool which characterizes this deformation must be consistent with the one associated to the final geometric entity.

From the user's point of view, the only tasks to carry out are first to specify the initial geometric entity and next to modify it into the final one. Both of these tasks may be achieved in an efficient interactive manner.

It has previously been pointed out that many deformation tools have been worked out and that the scope of their potential is indisputably wide. Nonetheless, combining several *auxiliary deformation models*, each based on an *auxiliary deformation tool*, in order to build up a new deformation model based on what might be called a *deformation multi-tool*, would be susceptible to grow up the possibilities of the existing deformation models. This paper aims to describe a deformation model based on such a deformation multi-tool.

The main problem is to specify the way the auxiliary deformation models have to be combined together. In fact, this problem is indistinguishable of the one which consists in defining the expected result of the achievement of a deformation, resorting to a deformation multi-tool, on a geometric object. Applying the auxiliary deformations successively would be an easy answer. This answer is not satisfactory at all however. Effectively, the fact that a deformation multi-tool combines several auxiliary deformation tools involves both their initial and final geometric entities have to be defined. Assuming that the deformation multi-tool combines two auxiliary deformation tools and that one object point to be deformed is concerned with both the two auxiliary deformation tools, nothing guarantees that, once one of the two auxiliary deformations have been applied, the corresponding deformed point is still concerned with the other auxiliary deformation. On the other hand, even though the deformed point is still concerned with the other auxiliary deformation, whatever the auxiliary deformation applied first, the result of the successive auxiliary deformations depends on the order in which these auxiliary deformations are applied. Stating deformations as in (1), this remark can be proved easily since functional composition is not commutative with most of the existing deformation functions.

So a deformation model based on a deformation multi-tool have to meet both the two following properties.

- Any point to be deformed, concerned with several auxiliary deformations, must really undergo the effect of each auxiliary deformation.
- The result of a deformation based on a deformation multi-tool must not depend on the order in which the auxiliary deformations are achieved.

Stating a deformation as in (1) does not allow the first property to be met. Thus, another definition of a deformation is proposed in section 2. This definition describes a deformation through the expedient of the point displacement it generates. This way, when an object point to be deformed is concerned with several auxiliary deformations, the corresponding displacements can be blended so that the result of the deformation does not depend on the order in which the auxiliary deformations are applied. Section 3 depicts a deformation model based on a deformation multi-tool expanding the displacements blend concept. Implementation results of the deformation model based on a deformation multi-tool are finally set out in section 4.

2   Another Approach to the Definition of a Deformation

According to its own definition, a deformation maps each object point with regard to the deformation into a new point of the Euclidean n-dimensional space. In other words, any object point to be deformed undergoes a displacement from its initial location within the original object, to its final location within the deformed object. Consequently, any deformation can be defined through the expedient of the displacements it generates within the set of points of the geometric object undergoing this deformation. Formally stated, in an Euclidean n-dimensional space, if Ω is the set of points of a geometric object and if Γ is the subset of points of Ω with regard to a given deformation f, then the displacement of the points of Ω, generated by f, can be defined with the following mathematical function denoted g.

$$g : \Omega \to \mathbb{R}^n$$

$$P \mapsto g\left(P\right) = \begin{cases} \overrightarrow{PP'} = f\left(P\right) - P & \forall P \in \Gamma \\ \overrightarrow{0} = P - P & \forall P \in \Omega \setminus \Gamma \ . \end{cases} \tag{2}$$

where f is the function stated in (1). From now on g will be called the *displacement function linked with the deformation f*. The fact that f is representative of a single deformation (i.e. of a single deformation tool) involves that g is also representative of a single deformation (i.e. of a single deformation tool).

Several works deal with displacement mapping [3, 6, 7]. Lewis [6] is the only one to use this concept to define a deformation model however.

It is significant to notice the displacement function puts on an intermediary act in the deformation definition. As soon as a displacement function is linked with a given deformation, the mathematical function which really describes this deformation is:

$$f : \Omega \to \mathbb{R}^n$$
$$P \mapsto P' = f(P) = P + g(P) \ . \tag{3}$$

Thus, the location of a deformed point is the vectorial sum between the location of the corresponding point to be deformed and the displacement generated by the deformation for this point to be deformed.

As a matter of fact, any deformation can be formulated through the expedient of a displacement function. Effectively, the traditional expression (1) of the definition of a deformation is inferred straight from the new one (3), substituting $g(P)$ for the expression (2) of its definition.

3   Definition of a Multi-Tool Deformation Model

This section aims to describe a multi-tool deformation model, that is a deformation model likely to combine several auxiliary deformation models, each based on an auxiliary deformation tool. As mentioned in the introduction, the accepted issue to reach that goal consists in blending, for each object point, its displacements resulting from each auxiliary deformation. The operator used to blend the displacements will be called *blend operator* henceforth and will be denoted $\oplus$. This way, assuming the multi-tool combines m auxiliary deformation tools, each described with its representative displacement function g_i, $(i \in \{1, \cdots, m\})$, the deformation based on this multi-tool and applied to a geometric object Ω is defined with the following mathematical function denoted $\tilde{f}$.

$$\tilde{f} : \Omega \to \mathbb{R}^n$$
$$P \mapsto P' = \tilde{f}(P) = P + \bigoplus_{i=1}^{m} g_i(P) \ . \tag{4}$$

The blend operator needs necessarily to be commutative so that the result of the deformation does not depend on the order in which the auxiliary deformations are applied. This property is the only one this operator really needs to meet. Formally stated, assuming that Ω have to be modified by m auxiliary deformations, each resorting to an auxiliary deformation tool described by a displacement function g_i $(i \in \{1, \cdots, m\})$:

$$\forall k, l \in \{1, \cdots, m\}, \ \forall P \in \Omega : g_k(P) \oplus g_l(P) = g_l(P) \oplus g_k(P) \ . \tag{5}$$

The vectorial sum is an attractive blend operator for displacements since, besides the fact it is a commutative operator, each point of the original object is moved to the accurate location expected by the user.

$$\forall P \in \Omega, \bigoplus_{i=1}^{m} g_i(P) = \sum_{i=1}^{m} g_i(P) \ . \tag{6}$$

Another example of blend operator for displacements might be the weighted average of the displacements.

$$\forall P \in \Omega, \bigoplus_{i=1}^{m} g_i(P) = \frac{\sum_{i=1}^{m} \alpha_i g_i(P)}{\sum_{i=1}^{m} \alpha_i} \ . \tag{7}$$

where α_i is a weight assigned to the auxiliary deformation tool the displacement function of which is g_i.

In fact, any blend operator can be used as long as it meets the commutative property (5).

As regards the computation process of a deformation based on a deformation multi-tool, its pseudocode is the one of Fig. 1. This pseudocode shows that achieving a deformation based on a deformation multi-tool is hardly more time-consuming than if all the auxiliary deformations was achieved sequentially. Effectively, in addition to the computation of each auxiliary deformation, the computation process only requires, for each point of the object:

- to compute, for each auxiliary deformation, the displacement (which is, actually, only a vectorial subtraction),
- to blend in the computed displacements,
- to translate the original point along the computed blend displacement (which is, actually, only a vectorial sum).

As regards blending in the computed displacements, it might seem absolutely necessary to compute, in a first stage and for each point of the object, the displacement generated by each auxiliary deformation before to blend them. This would be particularly inefficient. In fact, if the blend operator is of the type of those defined above (Equations (6) and (7)), the various sums can be updated each time a new displacement is computed, adding themselves in specific variables initially set to zero. This way, once all the sums are processed it only remains, if the case arises, to compute the quotients.

4   Results

The deformations based on deformation multi-tools utmost interest is to broaden the deformation scope. The modeling of a teapot, proposed in Fig. 2 and reached by deforming a sphere, a cylinder and a tube, shows the power of that new kind of deformations. The multi-tool used to arrive at this teapot combines nine deformation tools: three 1D-deformation tools for the lid, another 1D-deformation tool

```
For each object point Pi{
  For each auxiliary deformation{
    - Compute the deformation mapping from Pi
    - Compute the displacement of Pi
  }
  - Blend the computed displacements
  - Translate Pi along the blend displacement
}
```

Fig. 1. Pseudocode of the computation process of a deformation based on a deformation multi-tool.

for the base, a 1D-deformation tool, a 2D-deformation tool and a 3D-deformation tool for the spout, and both a 1D-deformation tool and a 3D-deformation tool for the handle. Besides, the vectorial sum is used as the blend operator.

Resorting to a deformation multi-tool certainly forces another way of handling deformations on the user. Most of the existing deformation tools provide the user with a strong intuition about the result they produce. As soon as several deformation tools are combined into a deformation multi-tool, the user must take into account the result locally produced by each auxiliary deformation in order to have an intuition of the one given by the multi-tool based deformation. Fig. 3, which displays the modeling of the teapot spout by deforming a tube, illustrates this remark. On the one hand, the tube is bent in a complex way with an axial deformation such as the one put forward in [5] and on the other hand it is tapered along two perpendicular directions with a 2D-deformation such as the one proposed in [4] ; both of the two deformation tools being combined into one deformation multi-tool. This new "philosophy" may seem constraining for the user. Experimentation has proved the careful thought required from the user is rather slight however and that it quickly turns into a habit, nay a reflex, as soon as the user knows the behavior of the auxiliary deformations and of the blend operator.

Despite this fact, resorting to a deformation multi-tool affords a decisive advantage. Racked shapes which could only be reached handling a very complex deformation tool, can easily be produced splitting up this complex deformation tool into several auxiliary ones (easily handled) and combining them into one deformation multi-tool. Fig. 5 shows that there are three different ways for modeling the racked shaped object of Fig. 5(g), 5(h) and 5(i). The first way resorts to a rather complex 3D-deformation tool which is much too hard to define. The second and the third ways both resort to two easier to build 3D-deformation tools. Whereas the second way forces the user to apply the deformations in a given order, the third way combines both the deformation tools in one deformation multi-tool. The obvious advantage of the third way, compared with the second, is that the user need not to bother to determine the order in which the deformations have to be applied.

Each deformation model presents particular interests so that it is generally used to carry out a specific task. Deformation multi-tools concept obviously dissociates itself from that limitation. Cases where the expected shape can not be reached easily resorting to a deformation multi-tool are likely to happen however. Fig. 4 illustrates such a case.

5   Conclusion

This paper aimed at combining several deformation tools, the geometric entities of which might be of various topologies or space dimensions. To reach this target, the developed technique consists, at first, in defining a deformation as a mathematical function which maps each point of the object to modify into a displacement instead of describing it the traditional way, that is as a mathematical function which maps each point of that object into a new one. Thus, while several deformations are combined, blending the displacements generated by these deformations, for each point of the object to modify, is enough.

This technique has led to the concept of deformations based on deformation multi-tools. A deformation multi-tool is a deformation tool abstraction susceptible of combining several traditional deformation tools. One of the points of great interest regarding deformation tools lies in the fact they can combine any type of deformation tools. Therefore, deformation multi-tools can be extended endlessly since any deformation tool which would be worked out in the future could be combined with existing deformation tools into a deformation multi-tool.

References

1. Barr, A. H.: Global and Local Deformations of Solid Primitives. ACM Computer Graphics (SIGGRAPH'84). **18** (3) 21–34, 1984.
2. Bechmann D.: Multidimensional Free-Form Deformation Tools. State of the Art Report (EUROGRAPHICS'98), Lisbon (Portugal), 1998.
3. Cook R. L.: Shade Trees. ACM Computer Graphics (SIGGRAPH'84). **18** (3) 223–231, 1984.
4. Feng J., Ma L. and Peng Q.: A new Free-Form Deformation Through the Control of Parametric Surfaces. Computer and Graphics, Pergamon Press. **20** (4) 531–539, 1996.
5. Lazarus F., Coquillart S. and Jancène P.: Axial Deformations: An intuitive Technique. Computer-Aided Design. **26** (8) 607–613, 1994.
6. Lewis J. P.: Algorithms for Solid Noise Synthesis. ACM Computer Graphics (SIGGRAPH'89). **23** (3) 263–270, 1989.
7. Pedersen H. K.: Displacement Mapping Using Flow Fields. ACM Computer Graphics Proceedings (SIGGRAPH'94), Annual Conference Series. 279–286, 1994.
8. Sederberg T. W. and Parry S. R.: Free-Form Deformation of Solid Geometric Models. ACM Computer Graphics (SIGGRAPH'86). **20** (4) 151–160, 1986.

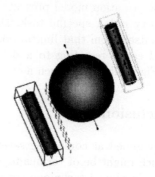

(a) Solid primitives: a sphere, a cylinder and a tube.

(b) Nine deformation tools IGEs — 6 1D-deformation tools (3 of them are superposed at the top of the sphere), 1 2D-deformation tool and 2 3D-deformation tools — combined into one deformation multi-tool.

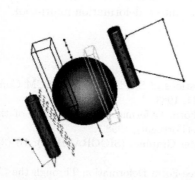

(c) Nine deformation tools DGEs combined into the deformation multi-tool.

(d) Result of the deformation, based on the deformation multi-tool, applied onto the solid primitives.

Fig. 2. Modeling of a teapot by deforming three solid primitives (a sphere, a cylinder and a tube) with a deformation multi-tool combining nine deformation tools.

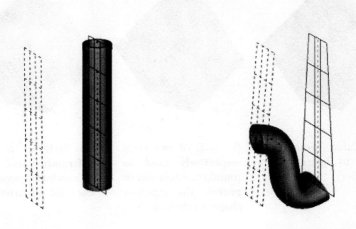

(a) Tube and IGEs of the multi-tool.   (b) DGEs of the multi-tool and result
of the deformation based on it.

Fig. 3. Modeling of the spout of the teapot displayed in Fig. 2 by deforming a tube with a deformation multi-tool combining both a 2D-deformation tool and 1D-deformation tool. The user must take into account the result locally produced by each auxiliary deformation tool to have an intuition of the one given by the deformation multi-tool.

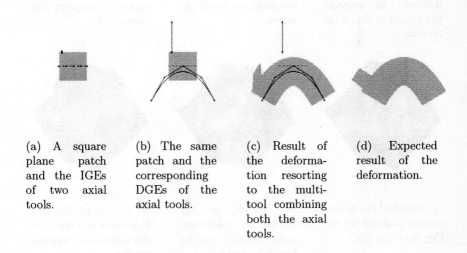

(a) A square plane patch and the IGEs of two axial tools.

(b) The same patch and the corresponding DGEs of the axial tools.

(c) Result of the deformation resorting to the multi-tool combining both the axial tools.

(d) Expected result of the deformation.

Fig. 4. A typical case where the expected result can not be reached easily resorting to a deformation multi-tool. In this case the expected result is reached more easily applying the "bumping" deformation then the "bending" deformation.

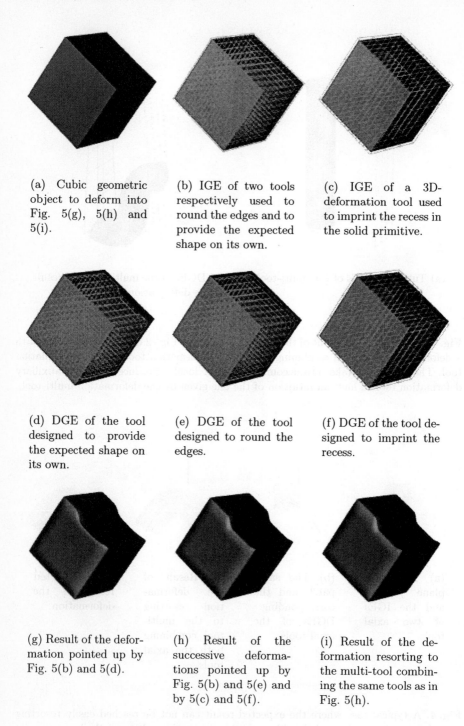

(a) Cubic geometric object to deform into Fig. 5(g), 5(h) and 5(i).

(b) IGE of two tools respectively used to round the edges and to provide the expected shape on its own.

(c) IGE of a 3D-deformation tool used to imprint the recess in the solid primitive.

(d) DGE of the tool designed to provide the expected shape on its own.

(e) DGE of the tool designed to round the edges.

(f) DGE of the tool designed to imprint the recess.

(g) Result of the deformation pointed up by Fig. 5(b) and 5(d).

(h) Result of the successive deformations pointed up by Fig. 5(b) and 5(e) and by 5(c) and 5(f).

(i) Result of the deformation resorting to the multi-tool combining the same tools as in Fig. 5(h).

Fig. 5. Three different ways to reach nearly the same shape.

A Property on Singularities of NURBS Curves

A. Arnal[1], A. Lluch[1], and J. Monterde[2]

[1] Dep. de Matemàtiques,Universitat Jaume I
Castelló, Spain parnal@mat.uji.es,lluch@mat.uji.es
[2] Dep. de Geometria i Topologia, Universitat de València,
Burjassot (València), Spain monterde@uv.es

Abstract. We prove that if an open Non Uniform Rational B-Spline curve of order k has a singular point, then it belongs to both curves of order $k-1$ defined in the $k-2$ step of the de Boor algorithm. Moreover, both curves are tangent at the singular point.

1 Introduction

There are some references in the literature (see [1],[7] or [2]) dealing with methods and algorithms to detect singularities on Bézier curves and its generalizations. Another approach to the study of singularities in polynomial curves is to deduce some properties of the curve when a singularity is present. This is what is done in [5] where an analysis of the behavior of rational Bézier curves under the change of one of its control points was realized. One of the properties of singular Rational Bézier curves shown in [5] is that given a nth degree singular Rational Bézier curve the singular point belongs to the $(n-1)$th degree rational Bézier curve associated with the first n control points and the corresponding weights and also to the $(n-1)$th degree rational Bézier curve associated to the last n control points and corresponding weights. Moreover, both curves, which are both $(n-1)$th degree rational Bézier curves defined in the $(n-1)$th step of the de Casteljau algorithm, intersect tangentially.

This note presents a generalization of some of the properties on [5] for the case of B-Spline curves and Nurbs curves. In particular we obtain that if an open Nurbs curve of order k has a singular point, then it belongs, analogously to the rational Bézier case, to both curves of order $k-1$ defined in the $k-2$ step of the de Boor algorithm. Moreover, we found that both curves are also tangent at the singular point.

2 Definitions and properties of Nurbs curves

2.1 Definition of Nurbs and B-Spline curves

According to [4] a **B-Spline of order k** is made up of pieces of polynomials of degree $k-1$, joined together with C^{k-2} continuity at the break points, so,

P.M.A. Sloot et al. (Eds.): ICCS 2002, LNCS 2330, pp. 229–238, 2002.

corresponding to the set of break points $t_0 < t_1 < t_2 < ... < t_{m-1} < t_m$ of a B-Spline function, it's defined the associated **knot vector**

$$T = (t_0, t_1, t_2, ..., t_{m-1}, t_m).$$

We refer to the individual points t_k of T as **knots**.

Thus, given a knot vector $T = (t_0, t_1, ..., t_{n-1}, t_n, t_{n+1}, ..., t_{n+k})$ the associated **normalized B-Spline**, $N_{i,k}$, of order k (degree $k-1$) is defined to be the following function:

$$N_{i,1}(t) = \begin{cases} 1 & \text{for } t_i \leq t < t_{i+1} \\ 0 & \text{otherwise} \end{cases},$$

for $k = 1$, and

$$N_{i,k}(t) = \frac{(t - t_i)}{(t_{i+k-1} - t_i)} N_{i,k-1}(t) + \frac{(t_{i+k} - t)}{(t_{i+k} - t_{i+1})} N_{i+1,k-1}(t), \qquad (1)$$

$\forall k > 1$ and $\forall i = 1, ..., n$.

The main properties of the basis functions we will use in this note are:

1. $N_{i,k}(t) > 0$ for $t \in [t_i, t_{i+k}]$ and $N_{i,k}(t) = 0$ otherwise.
2. $\sum_{i=0}^{n} N_{i,k}(t) = 1 \quad \forall t \in [t_{k-1}, t_{n+1}]$.

Now, let us introduce the definition of a Nurbs curve. According to [6], a **Non Uniform Rational B-Spline curve of order k (Nurbs)**, is defined by

$$X(t) = \frac{\sum_{i=0}^{n} N_{i,k}(t) w_i D_i}{\sum_{i=0}^{n} N_{i,k}(t) w_i} \qquad \forall t \in [t_{k-1}, t_{n+1}] \text{ and } n \geq k - 1 \qquad (2)$$

where the $\{D_i\}_{i=0}^{n}$ are the control points, the $\{w_i\}_{i=0}^{n}$ are the weights and the $N_{i,k}$ are the normalized B-Spline functions of order k defined on the knot vector

$$T = (t_0, t_1, ..., t_{n-1}, t_n, t_{n+1}, ..., t_{n+k}).$$

Along this note we have chosen the knot vector

$$T = (t_0 = t_1 = ... = t_{k-1} < t_k < ... < t_n < t_{n+1} = t_{n+2} = ... = t_{n+k}) \qquad (3)$$

to get an open and clamped Nurbs curve of order k with endpoints D_0 and D_n.

Remark 1. If we choose weights $w_i = 1 \; \forall i$, then, from the second property of the basis functions we get that Nurbs curves are a generalization of B-Spline curves.

Therefore a B-Spline curve is defined as follows:

Given the control points $\{D_i\}_{i=0}^{n}$ a **B-Spline curve of order k**, $k \leq n - 1$, associated to the knot vector, T, as above, is defined as

$$X(t) = \sum_{i=0}^{n} N_{i,k}(t) D_i, \qquad \forall t \in [t_{k-1}, t_{n+1}].$$

2.2   The de Boor algorithm for Nurbs curves

For the practical evaluation of Nurbs curves with given weights, we can use the **de Boor algorithm**, (sce [4]), which allows the computation of points on a Nurbs curve without explicit knowledge of the B-Spline basis functions $\{N_{i,k}\}$. The algorithm construction is based on the recursive definition of the normalized B-Spline.

Taking the initial values

$$D_j^0(t) = D_j, \qquad\qquad w_j^0(t) = w_j,$$

and defining

$$\alpha_i^j(t) = \frac{t - t_i}{t_{i+k-j} - t_i},$$

the scheme of computation is the following

$$w_i^j(t)\, D_i^j(t) = \left(1 - \alpha_i^j(t)\right) w_{i-1}^{j-1}(t)\, D_{i-1}^{j-1}(t) + \alpha_i^j(t)\, w_i^{j-1}(t)\, D_i^{j-1}(t)$$

$$w_i^j(t) = \left(1 - \alpha_i^j(t)\right) w_{i-1}^{j-1}(t) + \alpha_i^j(t)\, w_i^{j-1}(t) \qquad\qquad j > 0$$

where, in addition to the weighted de Boor points, we also apply the algorithm to the weights.

Then, after $k - 1$ steps, it finally leads to

$$X(t) = \frac{w_r^{k-1}(t)\, D_r^{k-1}(t)}{w_r^{k-1}(t)} = D_r^{k-1}(t), \qquad t \in [t_r, t_{r+1}]. \qquad (5)$$

On the other hand, according to [6], we will express the first derivative of a Nurbs curve

$$X(t) = \frac{\sum_{i=0}^n N_{i,k}(t)\, w_i D_i}{\sum_{i=0}^n N_{i,k}(t)\, w_i} = \frac{A(t)}{w(t)},$$

by the following formula

$$X'(t) = \frac{A'(t) - w'(t)\, X(t)}{w(t)} \qquad\qquad (6)$$

where

$$A'(t) = (k-1) \sum_{i=1}^n \frac{w_i D_i - w_{i-1} D_{i-1}}{t_{i+k-1} - t_i} N_{i,k-1}(t) \qquad\qquad (7)$$

$$w'(t) = (k-1) \sum_{i=1}^n \frac{w_i - w_{i-1}}{t_{i+k-1} - t_i} N_{i,k-1}(t).$$

2.3   The blossoming principle for Nurbs curves

According to the blossoming principle for B-Spline curves at [4], we can state a generalization for Nurbs curves. This blossoming principle is based on the construction of the polar form associated to the B-Spline basis functions, and it uses its recursive formula (1).

If we define

$$w_i^j [u_0, u_1, ..., u_{j-1}] :=$$

$$\frac{(t_{i+k-j} - u_{j-1}) \, w_{i-1}^{j-1} [u_0, u_1, ..., u_{j-2}] + (u_{j-1} - t_i) \, w_i^{j-1} [u_0, u_1, ..., u_{j-2}]}{t_{i+k-j} - t_i},$$

and $D_i^j [u_0, u_1, ..., u_{j-1}]$ such that

$$w_i^j [u_0, u_1, ..., u_{j-1}] \, D_i^j [u_0, u_1, ..., u_{j-1}] =$$

$$\frac{(t_{i+k-j} - u_{j-1}) \, w_{i-1}^{j-1} [u_0, u_1, ..., u_{j-1}] \, D_{i-1}^{j-1} [u_0, u_1, ..., u_{j-2}]}{t_{i+k-j} - t_i}$$

$$+ \frac{(u_{j-1} - t_i) \, w_i^{j-1} [u_0, u_1, ..., u_{j-1}] \, D_i^{j-1} [u_0, u_1, ..., u_{j-2}]}{t_{i+k-j} - t_i}$$

then, the blossoming principle asserts that the Nurbs curve is given by

$$X(t) = \frac{w_r^{k-1} \left[t, .^{(k-1)}., t \right] D_r^{k-1} \left[t, .^{(k-1)}., t \right]}{w_r^{k-1} \left[t, .^{(k-1)}., t \right]} = D_r^{k-1} \left[t, .^{(k-1)}., t \right]. \qquad (8)$$

3   Singularities of Nurbs curves

A differentiable curve $\alpha : I \longrightarrow \mathrm{IR}^m$ is said to be **regular** if $\alpha'(t) \neq 0$ for all $t \in I$. Otherwise, the curve is said to be **singular**, and then, if $\alpha'(\bar{t}) = 0$ for $\bar{t} \in I$, the image for this parameter value, $\alpha(\bar{t})$, is said to be a **singular point** or **singularity**.

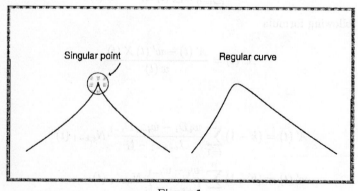

Figure 1

Given the control points $\{D_i\}_{i=0}^n$ and weights $\{w_i\}_{i=0}^n$, let $X(t)$ be the open and clamped Nurbs curve of order k associated to a knot vector

$$T = (t_0 = t_1 = ... = t_{k-1} < t_k < ... < t_n < t_{n+1} = t_{n+2} = ... = t_{n+k}).$$

In this section we will found that if a Nurbs curve has a singular point $X(\bar{t})$, at the parameter value $\bar{t} \in [t_r, t_{r+1}]$, then it belongs to both previous curves of order $k-1$ at the de Boor algorithm, i.e., $D_{r-1}^{k-2}(t)$ and $D_r^{k-2}(t)$.

Theorem 1. *Let $X(t)$ be an open and clamped Nurbs curve of order k with control points $\{D_i\}_{i=0}^n$, weights $\{w_i\}_{i=0}^n$ and associated to a knot vector T. If $X'(\bar{t}) = 0$ for $\bar{t} \in [t_r, t_{r+1}]$ where $r \in \{k-1, ..., n\}$, then*

$$X(\bar{t}) = D_r^{k-1}(\bar{t}) = D_r^{k-2}(\bar{t}) = D_{r-1}^{k-2}(\bar{t}).$$

Proof. First of all, we apply the recursive definition of the normalized B-Spline functions $N_{i,k}$ to the definition (2) of a Nurbs curve, then, shifting the indices in the second term by $i = i-1$ and taking $D_{-1} = D_{n+1} = 0$ we get

$$X(t) = \frac{\sum_{i=0}^n w_i D_i N_{i,k}(t)}{\sum_{i=0}^n w_i N_{i,k}(t)}$$

$$= \frac{t\left(\sum_{i=0}^{n+1} \frac{w_i D_i - w_{i-1} D_{i-1}}{t_{i+k-1}-t_i} N_{i,k-1}(t)\right) + \sum_{i=0}^{n+1} \frac{w_{i-1}D_{i-1}t_{i+k-1}-w_i D_i t_i}{t_{i+k-1}-t_i} N_{i,k-1}(t)}{t\left(\sum_{i=0}^{n+1} \frac{w_i - w_{i-1}}{t_{i+k-1}-t_i} N_{i,k-1}(t)\right) + \sum_{i=0}^{n+1} \frac{w_{i-1}t_{i+k-1}-w_i t_i}{t_{i+k-1}-t_i} N_{i,k-1}(t)}.$$

$$(9)$$

Having in mind (7), the de Boor algorithm and property 1 of the normalized B-Spline functions, we evaluate (9) at $\bar{t} \in [t_r, t_{r+1}]$ and we obtain that:

$$X(\bar{t}) = \frac{\frac{\bar{t} A'(\bar{t})}{k-1} + \sum_{i=1}^n w_i^1(0) D_i^1(0) N_{i,k-1}(\bar{t})}{\frac{\bar{t} w'(\bar{t})}{k-1} + \sum_{i=1}^n w_i^1(0) N_{i,k-1}(\bar{t})}.$$

Moreover, formula (6) let us assert that if $X(\bar{t})$ is a singular point at the parameter value $\bar{t} \in [t_{k-1}, t_{n+1}]$, then

$$X(\bar{t}) = \frac{A'(\bar{t})}{w'(\bar{t})},$$

so we can establish the following relation

$$\frac{\frac{\bar{t} A'(\bar{t})}{k-1} + \sum_{i=1}^n w_i^1(0) D_i^1(0) N_{i,k-1}(\bar{t})}{\frac{\bar{t} w'(\bar{t})}{k-1} + \sum_{i=1}^n w_i^1(0) N_{i,k-1}(\bar{t})} = \frac{A'(\bar{t})}{w'(\bar{t})}$$

hence

$$X(\bar{t}) = \frac{\sum_{i=1}^n w_i^1(0) D_i^1(0) N_{i,k-1}(\bar{t})}{\sum_{i=1}^n w_i^1(0) N_{i,k-1}(\bar{t})}.$$

Now if we apply again the recurrence formula for the normalized B-Spline functions and we use the blossoming principle we get

$$X\left(\bar{t}\right) = \frac{\sum_{i=1}^{n} w_i^1\left(0\right) D_i^1\left(0\right) N_{i,k-1}\left(\bar{t}\right)}{\sum_{i=1}^{n} w_i^1\left(0\right) N_{i,k-1}\left(\bar{t}\right)}$$

$$= \frac{\sum_{i=1}^{n} \frac{(t_{i+k-2}-\bar{t})w_{i-1}^1(0)D_{i-1}^1(0)+(\bar{t}-t_i)w_i^1(0)D_i^1(0)}{t_{i+k-2}-t_i} N_{i,k-2}\left(\bar{t}\right)}{\sum_{i=1}^{n} \frac{(t_{i+k-2}-\bar{t})w_{i-1}^1(0)+(\bar{t}-t_i)w_i^1(0)}{t_{i+k-2}-t_i} N_{i,k-2}\left(\bar{t}\right)}$$

$$= \frac{\sum_{i=1}^{n} w_i^2\left[0,\bar{t}\right] D_i^2\left[0,\bar{t}\right] N_{i,k-2}\left(\bar{t}\right)}{\sum_{i=1}^{n} w_i^2\left[0,\bar{t}\right] N_{i,k-2}\left(\bar{t}\right)}.$$

Therefore, the re-indexing can be continued by repeated insertion of the recurrence formula for the basis functions, and also, we can apply the blossoming principle for the $k-1$ steps, so we obtain

$$X\left(\bar{t}\right) = \frac{\sum_{i=1}^{n} w_i^{k-1}\left[0,\bar{t},..^{(k-2)}..,\bar{t}\right] D_i^{k-1}\left[0,\bar{t},..^{(k-2)}..,\bar{t}\right] N_{i,1}\left(\bar{t}\right)}{\sum_{i=1}^{n} w_i^{k-1}\left[0,\bar{t},..^{(k-2)}..,\bar{t}\right] N_{i,1}\left(\bar{t}\right)}$$

$$= D_r^{k-1}\left[0,\bar{t},..^{(k-2)}..,\bar{t}\right].$$

Hence, the above result let us to establish the following relation

$$X\left(\bar{t}\right) = D_r^{k-1}\left[\bar{t},..^{(k-1)}..,\bar{t}\right] = D_r^{k-1}\left[0,\bar{t},..^{(k-2)}..,\bar{t}\right]. \tag{10}$$

Taking into account the symmetry of the blossom functions

$$D_r^{k-1}\left[0,\bar{t},..^{(k-2)}..,\bar{t}\right] = D_r^{k-1}\left[\bar{t},..^{(k-2)}..,\bar{t},0\right],$$

and otherwise its recursive definition, then, from the equality

$$D_r^{k-1}\left[\bar{t},..^{(k-1)}..,\bar{t}\right] = D_r^{k-1}\left[\bar{t},..^{(k-2)}..,\bar{t},0\right],$$

we can obtain the following relation

$$D_{r-1}^{k-2}\left[\bar{t},..^{(k-2)}..,\bar{t}\right] = D_r^{k-2}\left[\bar{t},..^{(k-2)}..,\bar{t}\right].$$

On the other hand, applying again the symmetry at (10), we get

$$X\left(\bar{t}\right) = D_r^{k-1}\left[\bar{t},..^{(k-1)}..,\bar{t}\right] = D_r^{k-1}\left[\bar{t},..^{(k-2)}..,\bar{t},0\right] =$$

$$= \frac{\left(t_{r+1}w_{r-1}^{k-2}\left[t,..^{(k-2)}..,t\right] - t_r w_r^{k-2}\left[t,..^{(k-2)}..,t\right]\right) D_r^{k-2}\left[t,..^{(k-2)}..,t\right]}{t_{r+1}w_{r-1}^{k-2}\left[t,..^{(k-2)}..,t\right] - t_r w_r^{k-2}\left[t,..^{(k-2)}..,t\right]}$$

$$= D_r^{k-2}\left[\bar{t},..^{(k-2)}..,\bar{t}\right].$$

Therefore

$$X\left(\bar{t}\right) = D_r^{k-1}\left[\bar{t}, ..^{(k-1)}.., \bar{t}\right] = D_r^{k-2}\left[\bar{t}, ..^{(k-2)}.., \bar{t}\right] = D_{r-1}^{k-2}\left[\bar{t}, ..^{(k-2)}.., \bar{t}\right],$$

that is

$$X\left(\bar{t}\right) = D_r^{k-1}\left(\bar{t}\right) = D_r^{k-2}\left(\bar{t}\right) = D_{r-1}^{k-2}\left(\bar{t}\right).$$

Figure 2 illustrates an open B-Spline curve $X(t)$ of order 4 with control points $\{D_i\}_{i=0}^{5}$ and uniform knot vector $T = (0,0,0,0,1,2,3,3,3,3)$ which has a singular point for the parameter value $\bar{t} \in [t_4, t_5] = [1,2]$.

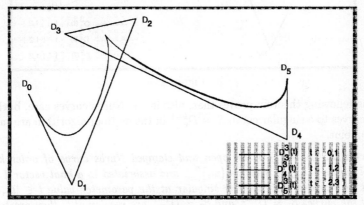

Figure 2

Figure 3 illustrates both previous curves to $X(t) = D_4^3(t)$ in the de Boor algorithm. These curves, $D_4^2(t)$ and $D_3^2(t)$, are represented in this picture only for parameter values on the interval $[t_4, t_5] = [1,2]$, and we can observe that they are tangent at the singular point $X\left(\bar{t}\right)$.

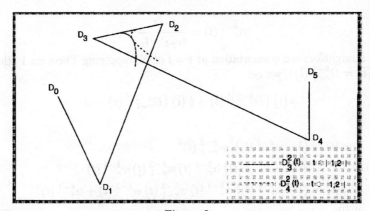

Figure 3

Figure 4 illustrates the three curves described above, $X(t) = D_4^3(t)$, the B-Spline curve and also, $D_4^2(t)$ and $D_3^2(t)$ for the parameter values at $[t_4, t_5] = [1, 2]$.

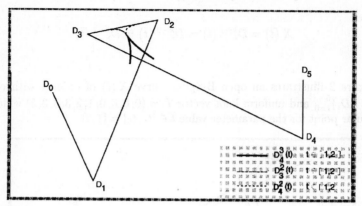

Figure 4

The following theorem proves that, also in the Nurbs curves case, both previous curves to a singular curve $X = D_r^{k-1}$ in the de Boor algorithm are tangent at this point.

Theorem 2. *Let $X(t)$ be an open and clamped Nurbs curve of order k with control points $\{D_i\}_{i=0}^n$, weights $\{w_i\}_{i=0}^n$ and associated to a knot vector T.*

Let us suppose that $X(t)$ is singular at the parameter value $\bar{t} \in [t_r, t_{r+1}]$, then if the curves $\beta_1 = D_r^{k-2}$ and $\beta_2 = D_{r-1}^{k-2}$ are regular at $\bar{t}$, they are tangent at that point.

Proof. The last step of the de Boor algorithm for constructing Nurbs curves involves

$$X(t) = \frac{\left(1 - \alpha_r^{k-1}(t)\right) w_{r-1}^{k-2}(t) D_{r-1}^{k-2}(t) + \alpha_r^{k-1}(t) w_r^{k-2}(t) D_r^{k-2}(t)}{\left(1 - \alpha_r^{k-1}(t)\right) w_{r-1}^{k-2}(t) + \alpha_r^{k-1}(t) w_r^{k-2}(t)}$$

where

$$\alpha_r^{k-1}(t) = \frac{t - t_r}{t_{r+1} - t_r}.$$

An straightforward computation at $t = \bar{t}$ of X', applying Theorem 1 (that is $D_r^{k-2}(\bar{t}) = D_{r-1}^{k-2}(\bar{t})$), we get

$$a(\bar{t}) \left(D_r^{k-2}\right)'(\bar{t}) + b(\bar{t}) \left(D_{r-1}^{k-2}\right)'(\bar{t}) = 0$$

where

$$a(t) = \left(1 - \alpha_r^{k-1}(t)\right)^2 w_{r-1}^{k-2}(t)^2$$
$$+ \left(1 - \alpha_r^{k-1}(t)\right) \alpha_r^{k-1}(t) w_{r-1}^{k-2}(t) w_r^{k-2}(t)$$
$$b(t) = \left(1 - \alpha_r^{k-1}(t)\right) \alpha_r^{k-1}(t) w_{r-1}^{k-2}(t) w_r^{k-2}(t) + \alpha_r^{k-1}(t)^2$$

that is $\beta_1'(\bar{t})$ and $\beta_2'(\bar{t})$ are proportional, so β_1 and β_2 are tangent at $X(\bar{t})$.

Taking into account that Nurbs curves are a generalization of B-Spline curves choosing weights $w_i = 1$ $\forall i$, we can assert the following corollaries for B-Spline curves and so for Bézier curves.

Corollary 1. *Let $X(t)$ be an open and clamped B-Spline curve of order k with control points $\{D_i\}_{i=0}^n$ and associated to a knot vector T.*
If $X'(\bar{t}) = 0$ for $\bar{t} \in [t_r, t_{r+1}]$ where $r \in \{k-1, ..., n\}$, then

$$X(\bar{t}) = D_r^{k-1}(\bar{t}) = D_r^{k-2}(\bar{t}) = D_{r-1}^{k-2}(\bar{t}).$$

i.e, the singular point belongs to both previous curves to $X(t)$ at the de Boor algorithm, and moreover if both curves are regular at $t = \bar{t}$, then they are also tangent at that point.

Remark 2. Along this note we have considered a non uniform and clamped knot vector to get an open Nurbs curve, nevertheless, we could also consider open curves with unclamped knot vectors. In fact, all the results obtained above can be easily generalized for non uniform and unclamped Nurbs curves.

The following corollary, proved in [5], particularizes the Nurbs case theorem to Bézier curves.

Corollary 2. *Let us suppose that the Bézier curve $\alpha = B[P_0, ..., P_n]$ is not regular and let $\alpha(\bar{t})$ be a singular point, then*

$$\alpha(\bar{t}) = B[P_0, ..., P_{n-1}](\bar{t}) = B[P_1, ..., P_n](\bar{t}),$$

i.e., the singular point belongs to both previous curves to $B[P_0, ..., P_n]$ in the de Casteljau algorithm, and moreover if they are regular at the parameter value $\bar{t}$, then they are tangent at that point.

4   Conclusions

The main results on this note let us assert that if an open Nurbs curve of order k has a singular point, then it belongs, analogously to the rational Bézier case, to both curves of order $k-1$ defined in the $k-2$ step of the de Boor algorithm, and also that these curves are tangent at the singular point. These results are generalizations of some of the properties obtained in [5] for rational Bézier curves.

Differently from previous works which aim is to detect singularities, this note characterizes them in terms of the control points involved in the de Boor algorithm. This new approach results in the possibility of constructing curves with such singular points.

Then, one of the applications of these results, would be the possible construction of a singular shape using just one Nurbs curve (or B-Spline curve), instead of obtaining the desired singularity by joining regular Nurbs curves with the singular point at the juncture, and this would lead us to a computational saving. The following figure shows how using one Bézier curve, instead of two regular Bézier curves to obtain a singular shape, would let to reduce the number of points required to define the curve.

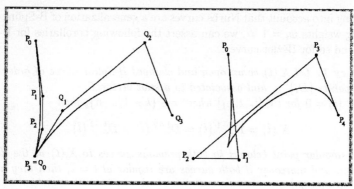

Figure 5

5 Acknowledgements

We would thank the referees for their valuable comments and suggestions.

References

1. S. Bu-Qing and L. Ding-Yuan, *Computational Geometry. Curve and Surface Modeling*, Academic Press, Inc., San Diego, 1989.
2. D. Manocha and J. F. Canny, *Detecting cusps and inflection points in curves*, Computer Aided Geometric Design 9, 1–24 (1992).
3. G. Farin, *Curves and surfaces for computer aided geometric design. A practical guide*, Morgan Kaufmann, 5th. ed. (2001).
4. J. Hoscheck and D. Lasser, *Fundamentals of Computer Aided Geometric Design*, A.K.Peters, Wellesley,1993.
5. J. Monterde, (2001), *Singularities of rational Bézier curves*, Computer Aided Geometric Design, 18, 805-816.
6. L. Piegl and W. Tiller, *The Nurbs Book*, 2a Ed., Springer Verlag, Berlin, 1997.
7. M. Sakai, (1999), *Inflection Points and Singularities on Planar Rational Cubic Curve Segments*, Computer Aided Geometric Design, 16, 149-156.

Interactive Deformation of Irregular Surface Models

Jin Jin Zheng and Jian J. Zhang

National Centre for Computer Animation
Bournemouth University
Poole, Dorset BH12 5BB, UK
jj_zheng@lycos.com   jzhang@bournemouth.ac.uk
http://ncca.bournemouth.ac.uk/main/staff/jian/

Abstract. Interactive deformation of surface models, which consist of quadrilateral (regular) and non-quadrilateral (irregular) surface patches, arises in many applications of computer animation and computer aided product design. Usually a model is mostly covered by regular patches such as Bézier or B-spline patches and the remaining areas are blended by irregular patches. However, the presence of irregular surface patches has posed a difficulty in surface deformation. Although regular patches can be easily deformed, the deformation of an irregular patch, however, has proven much trickier. This is made worse by having to maintain the smoothness conditions between regular and irregular patches during the process of deformation. By inserting extra control points, we have proposed a technique for the deformation of irregular surface patches. By setting continuity conditions as constraints, we also allow a surface model of an arbitrary topology, consisting of both regular and irregular surface patches, to be deformed smoothly.

1 Introduction

Interactive deformation of surface models is an important research topic in surface modelling, with numerous applications in computer animation, virtual reality and computer aided product design. Traditionally geometric models are represented by parametric surfaces, such as Bézier and B-spline surfaces. NURBS is so popular that it has become a de-facto standard. However, these surface models suffer from one significant setback — an inability of coping with surfaces of irregular topology, such as holes and branches. To remedy this weakness, two main alternatives have been proposed and also have taken increasing popularity in geometric modelling: subdivision surfaces [2], [4] and surfaces with combined regular (quadrilateral) and irregular (non-quadrilateral) patches. The latter is to blend regular patches with irregular patches (known as the blending surface patches) to form an overall smooth surface model. In this paper, our interest is in the interactive deformation of an irregular surface model (possibly with holes and branches) represented using the second approach.

Blending surface has long been an important research subject in geometric modelling. It is one of the most often used surface types for the representation of aesthetic features of computer generated models. A large number of methods have been developed which include the rolling ball method [23], the cyclide based method [24] and the partial differential equation method [22]. Another large family of techniques which also attract enormous attention are the control point based irregular surfaces. With this approach, a model is mostly covered by regular patches such as Bézier or NURBS patches. The remaining areas are blended by irregular n-sided patches. One technique is to use several regular control point patches to generate an irregular blending patch [6], [11], [12], [15], [18]. Another scheme is to produce a complete n-sided patch for the blending task. Loop and DeRose [14] proposed a method using S-patches

P.M.A. Sloot et al. (Eds.): ICCS 2002, LNCS 2330, pp. 239–248, 2002.
© Springer-Verlag Berlin Heidelberg 2002

for an *n*-sided hole. Sabin [17] tackled the same problem with a B-spline like control point patch. Zheng and Ball [20] generalised Sabin's patches to an arbitrary *m* degree.

A necessary feature of a modern geometric modelling system is the facility for interactive deformation of both regular and irregular surface models. A high degree of user control and interactivity is a practical requirement on nowadays geometric modelling systems, especially in computer animation and product design where people are expecting more and more visual realism from the CG models and characters.

Given a surface model consisting of both regular and irregular surface patches with at least G^1 continuity, interactive deformation involves the shape change of both types of patches, with possibly the following user-controlled deformation operations:

- moving control points of a patch;
- specifying geometric constraints for a patch, such as positional interpolation, i.e. letting the surface interpolate a given point;
- deforming a patch by exerting virtual forces, which can act as sculpting tools.

All three operations have been studied by various researchers for a regular patch. Things will however get more complicated when those operations are applied to irregular patches. But by far the most difficult task is to allow all these operations for both types of surface patches without violating their connection smoothness, such as the continuity conditions between a regular patch and an irregular patch. In this paper we propose a technique which will enable all these operations.

2 Background

Surface deformation is a desirable facility in both computer animation and product design. To date the majority of research has focused on the deformation of regular surface patches.

Deforming regular surfaces

One useful sculpting operation is to deform a surface patch by specifying positional constraints, i.e. spatial points that the surface has to interpolate. By moving these interpolated points interactively, one can deform the surface with a greater degree of direct user control, compared with the ordinary control-point based deformation approach. With the original configuration of control points, however, one often finds that there are not enough degrees of freedom (DOF) to satisfy the sufficient number of constraints. This difficulty can be overcome by producing extra control points for a surface patch. Here let us briefly review two types of most often used regular surfaces in this context: Bézier and B-spline surfaces.

A well-known property of Bézier surfaces is that the same surface can be described with a higher order representation. The process of obtaining a higher order representation is called degree elevation [25]. For instance, a cubic Bézier patch can be equivalently represented by a quartic patch without changing the geometry of the original surface. As a result of degree elevation, more control points are produced, which in effect provide more DOF for the surface. These extra degrees of freedom can then satisfy extra constraints.

B-spline surfaces, especially NURBS, again can be treated to satisfy extra constraints in a similar fashion. A NURBS surface patch can be refined by inserting node points in its node vector. The inserted node points are used to calculate the corresponding control points. The number of node points to be inserted depends on the number of constraints to be satisfied.

At a slightly higher level, a designer might find the ability to perform virtual sculpting operations attractive, whereby the designer deforms a surface by applying virtual forces [3], [21], [26].

Deforming irregular surfaces

To our knowledge, the deformation of irregular surfaces is a neglected research topic. One possible reason is that irregular surfaces are more difficult to deal with. It is also true that irregular surfaces, such as blending surfaces, are not as often encountered as regular surfaces, and in the past people were easily satisfied by a crude look of a CG model. With the rapid improvement of hardware performance, development of computer graphics and perfection of rendering techniques, however, the pursuit of realism of CG models has become a common requirement. Irregular patches whose roles are often to smoothly connect regular patches are no longer inferior to the mainstream regular surfaces.

In our previous attempt [21], we have tried to deform an irregular patch by applying virtual forces on it. Although the result is promising, we only considered a simple scenario that there is only one irregular patch and it is assumed there are enough DOF to start with. To make this technique practically useful, this model is clearly too simple and restrictive.

Outline of the proposed research

In this paper, we propose a technique for the deformation of a surface model that consists of both regular and irregular surface patches. This technique will have the following special contributions:

- no assumption is made for the degrees of freedom of an irregular patch. If extra DOF is needed, they will be produced without altering the original geometry;
- both regular and irregular surface patches can be deformed in the unified form by both geometric constraints and virtual forces;
- during deformation process, the smoothness conditions between patches (regular and irregular) will be maintained.

Since the deformation of a regular patch is already in the public knowledge, in this paper we will concentrate on the issues of irregular surface patches and the connection between different patches. For the modelling of irregular patches, we employ the model proposed by Zheng and Ball [20], which represents a generic n-sided surface patch. This surface model is control-point based and to a large extent similar to Bézier surfaces. To ensure enough degrees of freedom are present, we will formulate an explicit formula to elevate the degree of the surface and to insert a necessary number of extra control points. To guarantee that the smooth conditions between surface patches are not violated, they will be incorporated into our deformation formula as constraints which have to be satisfied.

3. Introduction to Zheng-Ball Patches

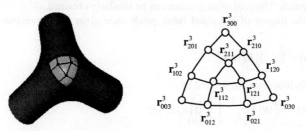

Fig. 1. 3-sided cubic Zheng-Ball patch with its control points.

As it is the basic model we employ for the proposed technique of surface deformation, in this section the basics of the n-sided Zheng-Ball patches are presented. For details, the reader is referred to [20]. An n-sided patch of degree m is defined by the following equation:

$$\mathbf{r}^m(u) = \sum_{j=0}^{[m/2]} \sum_{\min\lambda=j} B_\lambda^m(u)\mathbf{r}_\lambda^m \tag{1}$$

where $\lambda = (\lambda_1, \lambda_2, \cdots, \lambda_n)$ represents the n-ple subscripts, $\lambda_i \leq n$. $u = (u_1, u_2, \cdots, u_n)$ are the n parameters. $\mathbf{r}_\lambda^m$ denotes the control points in 3D space $\mathbf{R}^3$, as shown in Figure 1, and $B_\lambda^m(u)$ are the associated basis functions. The parametrisation of $u = (u_1, u_2, \cdots, u_n)$ can be in [20].

For example, for a cubic triangular patch, i.e. $n=3$, $m=3$, we have

$$B_\lambda^3(u) = F_\lambda^3(u) + \binom{3}{\lambda_{i-1}}\binom{3}{\lambda_i}\prod_{j=1}^3 u_j^2 P_\lambda^3(u) \tag{2}$$

in which

$$F_{3\delta_i}^3(u) = u_i^3(1 - u_{i+1}u_{i+2}), \quad F_{\delta_i+2\delta_{i+2}}^3(u) = 3u_{i+2}^3 u_i(1 - (1+u_i)u_{i+1}), \tag{3a}$$

$$F_{\delta_i+\delta_i+2\delta_{i+2}}^3(u) = 9u_i u_{i+1}u_{i+2}^2, \quad F_{\delta_i+\delta_i+2\delta_{i+2}}^3(u) = 9u_i u_{i+1}u_{i+2}^2 \tag{3b}$$

$$P_{3\delta_i}^3(u) = 2(u_i - 1), \quad P_{\delta_i+2\delta_{i+2}}^3(u) = \tfrac{2}{3}, \quad F_{\delta_i+\delta_i+2\delta_{i+2}}^3(u) = -\tfrac{2}{3} \quad i = 1,2,3. \tag{4}$$

Here the 3-ple indices $\delta_i = (\delta_{i1}, \delta_{i2}, \delta_{i3})$, $\delta_{ij} = \begin{cases} 1 & i = j \\ 0 & i \neq j \end{cases}$ and the subscript i is circularly taken modulo 3.

This patch model can have any number of sides and is able to smoothly blend the surrounding regular patches. In Figure 1, the 3-sided cubic patch, with 12 control points, is connected with the surrounding patches. However, once such an n-sided patch is constructed, there is no room for further deformation, i.e. there are no spare DOF are present. This fact will pose an even greater difficulty if one wants to maintain the smoothness at the patch boundaries during deformation.

4. Explicit Formula of Degree Elevation

A useful feature of this n-sided patch is its similarity to a Bézier patch. Control points can be inserted by elevating the degree of the surface. Therefore it allows more control points to be generated without changing the geometry of the surface. But before this feature is made of use, we have to solve one problem. That is to obtain a degree elevation formula such that the inserted control points are explicitly generated, as [2] only provides an implicit, recursive formula. In the interests of space, in the following we will derive the formulas of degree elevation for a 3-sided patch. Those of other patches can be similarly obtained.

Elevating the degree of a 3-sided cubic patch once gives the expression of the resulting quartic patch,

$$\mathbf{r}^4(u) = \sum_{j=0}^2 \sum_{\min\lambda=j} B_\lambda^4(u)\mathbf{r}_\lambda^4 \tag{5}$$

where the basis functions are defined by

$$B_\lambda^4(u) = F_\lambda^4(u) + \binom{4}{\lambda_{i-1}}\binom{4}{\lambda_i}\prod_{j=1}^3 u_j^2 P_\lambda^4(u) \tag{6}$$

and the functions $F_\lambda^4(u)$ are defined by

$$F_{4\delta_i}^4(u) = u_i^4(1 - 4u_{i+1}u_{i+2}), \quad F_{3\delta_i+\delta_{i+2}}^4(u) = 4u_i^3 u_{i+2}(1 - (1+2u_{i+2})u_{i+1}), \tag{7a}$$

$$F_{2\delta_i+2\delta_{i+2}}^4(u) = 6u_i^2 u_{i+2}^2(1 - 2u_{i+1}), \quad F_{3\delta_i+\delta_{i+1}+\delta_{i+2}}^4(u) = 16u_i^3 u_{i+1}u_{i+2} \tag{7b}$$

$$F_{2\delta_i+\delta_{i+1}+2\delta_{i+2}}^4(u) = 24u_i^2u_{i+1}u_{i+2}^2, \quad i=1,2,3, \quad F_{222}^4(u) = 36u_1^2u_2^2u_3^2,$$ (7c)

The remainder functions $P_\lambda^4(u)$ are unknown. To derive them, we first calculate the following auxiliary functions $R_\lambda^3(u)$:

$$R_{121}^3(u) = \frac{1}{9\prod\limits_{t=1}^3 u_t^2}\left(F_{121}^3(u) - \frac{1}{16}\left(9F_{131}^4(u) + 6F_{122}^4(u) + 6F_{221}^4(u) + \frac{4}{3}F_{222}^4(u)\right)\right)$$

$$= \frac{1}{9\prod\limits_{t=1}^3 u_t^2}\left(9u_1u_2^2u_3 - \frac{1}{16}\left(9\cdot16u_1u_2^3u_3 + 6\cdot24u_1u_2^2u_3^2 + 6\cdot24u_1^2u_2^2u_3 + \frac{4}{3}36u_1^2u_2^2u_3^2\right)\right) = -\frac{1}{3}(1+6u_2)$$ (8a)

$$R_{021}^3(u) = \frac{1}{3\prod\limits_{t=1}^3 u_t^2}\left(F_{021}^3(u) - \frac{1}{16}\left(12F_{031}^4(u) + 3F_{131}^4(u) + 8F_{022}^4(u) + 2F_{122}^4(u)\right)\right)$$

$$= \frac{1}{3\prod\limits_{t=1}^3 u_t^2}\left(3u_2^2u_3(1-(1+u_3)u_1) - \frac{1}{16}\left(12\cdot4u_2^3u_3(1-(1+2u_3)u_1) + 3\cdot16u_1u_2^3u_3\right.\right.$$

$$\left.\left. +8\cdot6u_2^2u_3^2(1-2u_1) + 2\cdot24u_1u_2^2u_3^2\right)\right) = 0$$ (8b)

$$R_{030}^3(u) = \frac{1}{\prod\limits_{t=1}^3 u_t^2}\left(F_{030}^3(u) - \frac{1}{16}\left(16F_{040}^4(u) + 4F_{130}^4(u) + 4F_{031}^4(u) + F_{131}^4(u)\right)\right)$$

$$= \frac{1}{\prod\limits_{t=1}^3 u_t^2}\left(u_2^3(1-3u_1u_3) - \frac{1}{16}\left(16u_2^4(1-4u_1u_3) + 4\cdot4u_1u_2^3(1-u_3(1+2u_1))\right.\right.$$

$$\left.\left. +4\cdot4u_2^3u_3(1-u_1(1+2u_3)) + 16u_1u_2^3u_3\right)\right) = 4u_2^2$$ (8c)

We are now in a position to formulate functions $P_\lambda^4(u)$.

Considering the symmetry of the functions, we can choose

$$P_{222}^4(u) = 0, \qquad P_{2j2}^4(u) = 0 \qquad j = 0,1$$ (9)

Following the recursive degree elevation algorithm [2], we have

$$P_{131}^4(u) = P_{121}^3(u) + R_{121}^3(u) - P_{221}^4(u) - P_{122}^4(u) - \frac{4}{3}P_{222}^4(u)$$ (10a)

$$P_{031}^4(u) = P_{021}^3(u) + R_{021}^3(u) - P_{131}^4(u) - P_{122}^4(u) - P_{022}^4(u)$$ (10b)

$$P_{040}^4(u) = P_{030}^3(u) + R_{030}^3(u) - P_{031}^4(u) - P_{130}^4(u) - P_{131}^4(u)$$ (10c)

Inserting (4c), (8a) and (9) into (10a), we get

$$P_{131}^4(u) = -(1+2u_2)$$ (11a)

Inserting (4b), (8b), (9) and (11a) into (10b), we obtain the following

$$P_{031}^4(u) = \frac{5}{3} + 2u_2$$ (11b)

By symmetry, we have

$$P_{130}^4(u) = \frac{5}{3} + 2u_2$$ (11c)

Substituting (4a), (8c), (11b), (11c) and (10a) into (10c), we get

$$P_{040}^4(u) = 4u_2^2 - \frac{13}{3}$$ (11d)

By symmetry, we have the following complete expression for the functions $P_\lambda^4(u)$:

$$P_{4\delta_i}^4(u) = 4u_i^3 - \frac{13}{3}, \quad P_{3\delta_i+\delta_{i+1}}^4(u) = \frac{5}{3} + 2u_i, \quad P_{3\delta_i+\delta_{i+1}+\delta_{i+2}}^4(u) = -(1+2u_i),$$

$$P_\lambda^4(u) = 0, \text{ for other } \lambda \text{ 's} \tag{12}$$

So far, we have obtained the explicit expression of the remainder functions $P_\lambda^4(u)$. Substituting formulae (7) and (12) into (6) results in the complete expression of the basis functions $B_\lambda^4(u)$ of degree 4.

The control points r_λ^4 of the quartic surface (5) are shown in Figure 2 and are expressed in terms of the control points r_λ^3 for cubic surface (1), as following:

$$r_{\lambda_i^{k,j}}^4 = \frac{1}{16}\left(kj r_{\lambda_i^{k-1,j-1}}^3 + k(4-j) r_{\lambda_i^{k-1,j}}^3 + (4-k)j r_{\lambda_i^{k,j-1}}^3 + (4-k)(4-j) r_{\lambda_i^{k,j}}^3 \right) \tag{13a}$$

$$k \le 2, j < 2, \qquad\qquad i = 1, \cdots n,$$

$$r_{\lambda_i^{2,2}}^4 = \frac{1}{n}\sum_{i=1}^n r_{\lambda_i^{1,1}}^3 \tag{13b}$$

where $\lambda_i^{k,j} = k\delta_i + j\delta_{i+1} + \lambda_{i+2}\delta_{i+2}$.

The generated extra control points will be used to modify the shape of the 3-sided patch as shown in Figure 2. With G^1 continuity being satisfied by the control points near the boundaries, the extra central control point can be moved freely to modify the shape of the blending surface. And the process of degree elevation can continue to generate further extra control points until enough degrees of freedom are created.

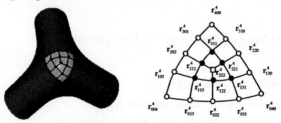

Fig. 2. Quartic patches with control points after degree elevation. The circles represent the control points contributing to the C^0 condition, the black dots represent the control points contributing to the G^1 condition, and the square in the middle represents the free central control point.

Comparing Figure 1 with Figure 2 it is clear that, after degree elevation, one central control point is obtained, which has provided an extra degree of freedom. Since this control point does not contribute to the current continuity conditions, moving this control point will deform the shape of the blending patch intuitively without violating the continuity conditions with the surrounding patches.

5. Surface Deformation

To deform an irregular patch, [21] proposed a technique based on minimising the following energy functional:

$$E = V^T K V - 2V^T F, \tag{14}$$

where V is the control vector whose entries are 3D control point vectors for an n-sided patch. K is the stiffness matrix whose entries are functions of Zheng-Ball base functions. F is the force vector whose entries are functions of both Zheng-Ball base functions and the physical forces applied.

To deform all regular and irregular surface patches, the same idea is applied, with the boundary conditions being coded as constraints.

For an arbitrary patch Π_i — regular or irregular — we may define a similar energy functional to (14):

$$E_i = V_i^T K_i V_i - 2V_i^T F_i \tag{15}$$

where V_i, K_i and F_i are the control point vector, stiffness matrix and force vector, respectively, with respect to each patch Π_i. For the whole surface model, we have

$$V = (V_1, V_2, \cdots)^T, \quad K = \begin{pmatrix} K_1 & 0 & \\ 0 & K_2 & \\ & & \ddots \end{pmatrix}, \quad F = (F_1, F_2, \cdots)^T \tag{16}$$

Thus the new global energy functional is given by

$$E = V^T K V - 2V^T F \tag{17}$$

The continuity constraints are defined by the following linear matrix equation

$$AV = b \tag{18}$$

Minimising the quadratic form (17) subject to constraint (18) leads to the production of a deformed model consisting of both regular and irregular patches.

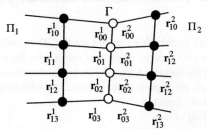

Fig. 3. Two cubic patches Π_1 and Π_2 share a common boundary Γ.

Remarks. Suppose two cubic patches Π_1 and Π_2 (either regular or irregular) share a common boundary Γ as show in Fig. 3. Typical G^1 continuity constraints for the two patches Π_1 and Π_2 can be expressed by the following linear equation which can be embedded into (18):

$$A_1 \tilde{V} = b_1 \tag{19}$$

where $\tilde{V} = (r_{00}^1, r_{01}^1, r_{02}^1, r_{03}^1, r_{10}^1, r_{11}^1, r_{12}^1, r_{13}^1, r_{00}^2, r_{01}^2, r_{02}^2, r_{03}^2, r_{10}^2, r_{11}^2, r_{12}^2, r_{13}^2)^T$,

$$A_1 = \begin{pmatrix} 1 & & & & -1 & & & & & & & & & & & \\ & 1 & & & & -1 & & & & & & & & & & \\ & & 1 & & & & -1 & & & & & & & & & \\ & & & 1 & & & & -1 & & & & & & & & \\ 1 & & & & -1 & & & & -1 & & & & 1 & & & \\ & 1 & & & & -1 & & & & -1 & & & & 1 & & \\ & & 1 & & & & -1 & & & & -1 & & & & 1 & \\ & & & 1 & & & & -1 & & & & -1 & & & & 1 \end{pmatrix}, \quad b_1 = 0 \tag{20}$$

A surface model with irregular patches is given in Figure 4. There are eight triangular patches on the outer corners of the model, and eight pentagonal patches on the inner corners of the model. All the remaining parts are covered by regular bi-cubic Bézier patches (the blue ones). The original model is on the left of Figure 4. The model is deformed by the algorithm presented above. The resulting models are shown on the middle and right of Figure 4. We can see that both regular patches and irregular blending patches are deformed.

Fig. 4. Model with 3- and 5-sided patches (green patches). (Middle and Right) Deformed models.

6. Algorithm for Interactive Deforming

In surface deformation, linear constraints are also useful deformation tools, such as to let the surface interpolate specified points, curves and norms [3], [21]. Linear constraints can be expressed as:

$$AV = b \qquad (21)$$

where V is the column vector whose entries are the 3D co-ordinates of l control points. A is a $k \times l$ matrix of coefficients. k is the number of constraints. Each row of the matrix A represents a linear constraint on the surface. Assuming that redundant constraints in (21) are eliminated.

If physical forces are applied to the surface, the following linear system is generated by minimising the quadratic form (18):

$$KV = F_0 + \iint Z \cdot \Delta f(u) du dv \qquad (22)$$

where K is an $l \times l$ positive semi-definite stiffness matrix. Z is the vector whose entries are the base functions. F_0 is generated by the initial forces. $\Delta f(x)$ is the density of the distributed forces used to deform an initial n-sided surface to a new shape with new vector V. As mentioned in [21], if the matrix K is singular, then least square solution will be used.

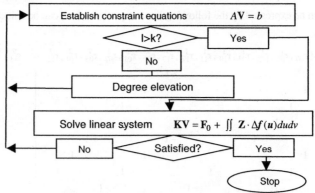

Fig. 5. Algorithm of interactive deformation.

The process of deforming the surface is equivalent to solving linear system (22) with respect to vector V subject to constraint (21).

In the literature, it is always assumed that there are enough degrees of freedom with

respect to **V** in constraint (21). In fact it is not always so. There are two different cases:
- $l>k$. There are free variables left in (21). So linear system (22) can be solved.
- $l\leq k$. There is no free variable left in (21). So linear system (22) is not solvable.

In the latter case, extra degrees of freedom are needed to solve linear system (22). This is achieved by degree elevation. Since it is not known beforehand how many extra degrees of freedom are needed, the process will be iterative, as illustrated by the above diagram in Figure 5:

Fig. 6. A smooth model with 3- and 5-sided cubic surface patches (left). Deformed model after twice degree elevation (right). Arrows indicate the forces applied on the surface points.

Figure 6 shows a smooth model with 3- and 5-sided cubic surface patches. Without degree elevation, the 5-sided patch cannot be deformed (left). With degree elevation, the model is deformed. The boundary conditions are maintained during deformation.

7. Conclusions

The deformation of a quadrilateral surface model is a routine method for surface modelling. But in the literature, no reported methods have been able to deform a surface model interactively if n-sided surface patches are involved without compromising the smoothness conditions with the surrounding surfaces.

There are two main contributions in this paper. Firstly, we have proposed a surface deformation technique, which is able to deform a surface model consisting of connected regular and irregular surface patches without violating the smoothness conditions at the patch boundaries. Secondly, we have derived an explicit formula for degree elevation of irregular patches, and this derivation is applied to the deformation of irregular patches. As most CAD systems produce quadrilateral cubic patches, an irregular patch can be initially set as an n-sided cubic patch, which satisfies G^1 continuity condition when connecting to other regular patches. This irregular patch is then degree elevated to produce a required number of extra control points for it to be deformed together with the others while still maintaining the smoothness constraints between the patches. The algorithm of degree elevation has been given as an iterative procedure to satisfy the user requirements in the applications of interactive model deformation.

References

1. Ball A A and Zheng, J J (2001). Degree elevation for n-sided surfaces. CAGD 18(2). 135-147.
2. Catmull E and Clark, J (1978). Recursively generated B-spline surfaces on arbitrary topological meshes, CAD 10, No 6, 350-355.
3. Celniker G and W Welch (1992). Linear constraints for deformable B-spline surfaces. ACM Proceedings of the 1992 Symposium on Interactive 3D Graphics. 165-170.

4. Doo D W H and Sabin, M A (1978). Behaviour of recursive division surfaces near extraordinary points, CAD 10, 356-360.
5. Gregory J A and Yuen, P K (1992). An arbitrary mesh network scheme using rational splines, in Lyche, T. and Schumaker, L. L. eds., *Mathematical Methods in Computer Aided Geometric Design II*, Academic Press.
6. Gregory J A and Zhou, J (1994). Filling polygonal holes with bi-cubic patches, CAGD 11, 391-410.
7. Hosaka M and Kimura, F (1984). Non-four-sided patch expressions with control points, CAGD 1, 75-86.
8. Juhász I (1999). Weight-based shape modification of NURBS curves, Computer Aided Geometric Design 16 (1999), 377–383
9. Kocic L M (1991). Modification of Bézier curves and surfaces by degree-elevation technique, Computer-Aided Design 23, 692-699.
10. Lin J, Ball A A and Zheng J J (1998). Surface modelling and mesh generation for simulating superplastic forming, Journal of Materials Processing Tech., Vol. 80-81, 613-619.
11. Liu D (1986). A geometric condition for smoothness between adjacent rectangular Bézier patches, Acta. Math. Appl. Sinica 9, No. 4, 432-442.
12. Liu D and Hoschek J (1989), GC^1 continuity conditions between adjacent rectangular and triangular Bézier surface patches, CAD 21, 194-200.
13. Liu W (1997). A simple, efficient degree raising algorithm for B-spline curves, CAGD 14, 693-698.
14. Loop C T and DeRose, T D (1989). A multisided generalisation of Bézier surfaces, ACM Transaction on Graphics 8, No 3, 204-234.
15. Peters J (1990). Smooth mesh interpolation with cubic patch, CAD 22(2). 109-120.
16. Piegl L and Tiller W (1994). Software-engineering approach to degree elevation of B-spline curves, Computer-Aided Design 26, 17-28.
17. Sabin M A (1983). Non-rectangular surfaces suitable for inclusion in a B-spline surface, in T. Hagen, ed., *Eurographics'83*, 57-69.
18. Van Wijk J J (1986). Bi-cubic patches or approximating non-rectangular control-point meshes, CAGD 3, 1-13.
19. Varady T (1991). Overlap patches: a new scheme for interpolating curve networks with n-sided regions, CAGD 8, No 1, 7-27.
20. Zheng J J and Ball, A A (1997). Control point surfaces over non-four-sided areas, CAGD 14, No. 9, 807-821.
21. Zheng, J J and J J Zhang (2001). Physically Based Sculpting of n-sided Surfaces. CAD-Graphics 2001. Kunming, International Academic Publishers. 63-69.
22. Zhang, J J and You, L H (2001). PDE based surface representation, Computers and Graphics, 25(6), to appear.
23. Barnhill, R. E., Farin, G. E. and Chen, Q(1993). Constant-radius blending of parametric surfaces, Computing Supple., 8:1-20.
24. Shene, C.-K., 1998, Blending two cones with Dupin cyclides, Computer Aided Geometric Design, 15:643-673.
25. G Farin (1997). *Curves and Surfaces for Computer Aided Geometric Design*. Academic Press, fourth edition.
26. Terzopoulos D and H Qin (1994). Dynamic NURBS with geometric constraints for interactive sculpting. ACM Transaction on Graphics, Vol 13, No 2, 103-136.

Bandwidth Reduction Techniques for Remote Navigation Systems

Pere-Pau Vázquez and Mateu Sbert

Institut d'Informàtica i Aplicacions, University of Girona,
Campus Montilivi, EPS, E-17071 Girona, Spain
pvazquez@ima.udg.es

Abstract. In this paper we explore a set of techniques to reduce the bandwidth in remote navigation systems. These systems, such as exploration of virtual 3D worlds or remote surgery, usually require higher bandwidth than the common Internet connection available at home. Our system consists in a client PC equipped with a graphics card, and a remote high-end server, which hosts the remote environment and serves information for several clients. Each time the client needs a frame, the new image is predicted by both the client and the server and the difference with the exact one is sent to the client. To reduce bandwidth we improve the prediction method by exploiting spatial coherence and wiping out correct pixels from the difference image. This way we achieve up to 9:1 reduction ratios without loss of quality. These methods can be applied to head-mounted displays or any remote navigation software.

1 Introduction

Recent advances in computer science have allowed dramatic breakthroughs at the beginning of the new century, such as remote computer-assisted surgery, 3D gaming, or the creation of compelling special effects in films. However, there are still some problems to be solved to obtain attractive applications such as 3D remote exploration or gaming. One of the most important problems comes from the limited bandwidth of common internet connections. Many real time rendering systems require the transmission of images at 15-30 frames per second to guarantee a continuous sensation of movement. Moreover, these frame rates can only be obtained with high-end systems, despite the latest chipset releases of graphics hardware.

In this paper we present some techniques that can help to reduce the necessary bandwidth without any noticeable cost in the final quality. We have taken advantage of Image-Based Rendering methods and used a collaborative scheme between client and server. Reduction ratios of 5:1 in rotations and up to 9:1 in translations over previous systems that also use image compression and view prediction can be achieved. The ratios refer to the amount of information sent through the network. The increase in computing cost is perfectly assumably by the client processor, as the only required operation is a forward warping plus an image update.

P.M.A. Sloot et al. (Eds.): ICCS 2002, LNCS 2330, pp. 249–257, 2002.
© Springer-Verlag Berlin Heidelberg 2002

The remainder of the paper is organized as follows: in Section 2 we review the previous work, in Section 3 we present our techniques for bandwidth reduction, the results obtained are discussed in Section 4, and finally in Section 5 we conclude and analyze possible future work.

2   Previous Work

During the second half of the last decade, a set of new rendering methods, called Image-Based Rendering, have been developed. All of them have in common the use of images in some stage to partly or completely substitute the geometry of the scene [1–3]. The use of precomputed images simplifies the rendering algorithms thus obtaining realistic effects at low cost. Although most of these methods have huge memory requirements and are therefore not suitable for dealing with very complex scenes in a low-end PC, some of these techniques can be used to accelerate rendering or to compensate for network latencies in remote navigation tasks. Mark *et al* [4] use two images generated in a server and warp them together to generate new views at interactive frame rates. Some commercial products such as QuickTime VR [5] offer panoramas over the network which are divided into pieces, so that the client can view them before they are totally downloaded, however, a large portion of the panorama has to be received before it may be viewed. Moreover, these systems only allow for camera rotation and zooming, and it is difficult to extend them to handle translations.

Our approach is a technique to reduce bandwidth which involves image compression and image-based rendering. A cooperative scheme is described in [6], where the server generates a high-quality rendering and a low-quality one and subtracts them, sending the result to the client. The client generates a low-quality image and adds the one sent by the server. This way the author obtains better compression and avoids the low quality of JPEG scheme in edges and smooth shading. On the other hand, this method requires that the geometry is also sent to the client in order to allow it to generate the low-quality image. Biermann *et al* [7] use a similar collaborative scheme. To reduce bandwidth, the client performs a forward warping to predict the new view, the server does the same and also computes the correct view. Then, the predicted view is subtracted from the exact one and the difference view is compressed. This way they achieve reduction ratios of 55% on small movements of the camera. Cohen-Or *et al* [8] use a compression scheme which reduces up to one order of magnitude the bandwidth, compared to a MPEG postrendered sequence with equivalent quality. Their compression scheme requires also that the client has access to the geometric data and exploit spatial coherence in order to improve compression ratios.

3   Bandwidth Reduction Techniques

In this section we present two bandwidth reduction techniques, a two-level forward warping and a compressing method which avoids sending information corresponding to correctly predicted pixels to the client. We use a similar architecture to the one presented in [6], but improving the tasks on the client and on the server side. Figure 1 depicts this architecture and the tasks. The client predicting system is enhanced by adding a simple and low cost two-level of detail forward warping. The server side reduces bandwidth by only transmitting the pixels to be updated, not the whole image (with the aid of a bitmap that encodes the pixels that have to be changed). This way we achieve a reduction of bandwidth of 5 to 1 over the previous method for rotations of up to 30 degrees and a reduction of up to 9 to 1 for translations.

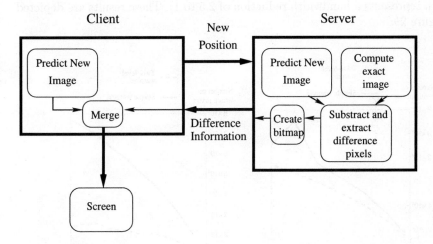

Fig. 1. The client-server remote navigation system. Instead of sending every frame through the network the server only sends the difference pixels, together with a bitmap which indicates the pixels to be modified. The client predicts the new view using a two-level forward warping.

3.1   Two-Level Forward Warping Known Pixels

In a similar way to Biermann *et al* [7] we use a forward warping to predict the next image. However, we will exploit spatial coherence. Some possible techniques are: to create a mip-map of the resulting image (after warping) and fill in the gaps, or to compute correct splat sizes for each point. On the other hand, these techniques can be quite costly and we want a very cheap method. Our system performs a two-level forward warping of the valid pixels in previous image. The first one consists into a reprojection of each point using the OpenGL ([9]) primitive *GL_POINT* assigning to the points a size (*GL_SIZE*) of four. Then, the depth buffer is cleared and the points reprojected again with a size of one. As some of the present gaps should have the same colour as closer pixels, this method reduces the visually unpleasant holes, and correct pixels of the predicted image increase by a factor of up to a 40% percent, and an average of a 20% percent (bandwidth reduction of 1.25:1 ratio) in rotations. This method is not new in the sense that it can be considered a particular case of splatting, but it is cheap and only uses graphics hardware commonly available in most personal computers. Figure 2a depicts the difference in number of wrong pixels for simple and two-level warping strategies for different rotations. Note that the number of wrong pixels in the second case grows slower and the percentage of reduction is almost constant in rotations of up to 30 degrees. We have also tested our method for translation transformations, in this case, as the pixels are reusable longer, the amount of correct pixels grows up to the 68% and an average of 60% (which represents a bandwidth reduction of 2.5 to 1). These results are depicted in Figure 2b.

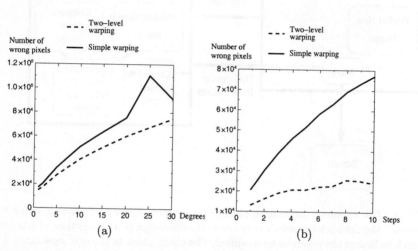

(a)   (b)

Fig. 2. Comparison of the number of wrong pixels for simple warping and two-level warping in rotations (a) and translations (b). Note that for translations the improvement is higher as the projected colours are correct longer.

In Figure 3 we can see an example of rotation. In this case the camera rotates ten degrees. Figure 3a has been created by only warping the pixels of the previous image, while 3b has been created with our two-level forward warping, the amount of difference pixels has been reduced in a 20 percent. Moreover, image 3b can be used if latency is high as it does not present as many noticeable artifacts as 3a does. Figures 3c and 3d show the difference images which have to be sent through the network.

Figure 4 shows the images resulting from a translation of five steps in world coordinates. Note that the amount of reused pixels is higher than in rotations and thus our two-level warping performs better than simply warping known points. Moreover, the predicted image could be used in case of high network latency. Figure 4a shows the image obtained with simple warping and 4b with

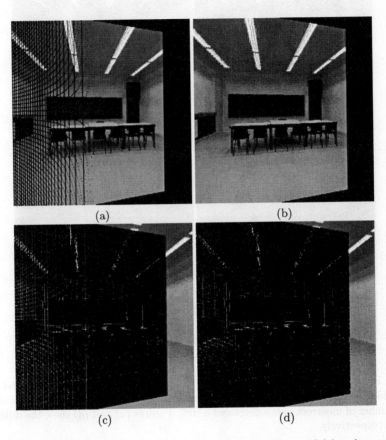

(a) (b)

(c) (d)

Fig. 3. Predicted images for a rotation of ten degrees. Figure *(a)* has been created using a simple forward warping and Figure *(b)* was created using a 2-level forward warping. The number of invalid pixels decreases a 20%. Figures *(c)* and *(d)* show the difference images respectively.

our two-level warping. Figures4c and 4d show the respective difference images. We obtain an improvement of up to a 68% and the average is about the 60%.

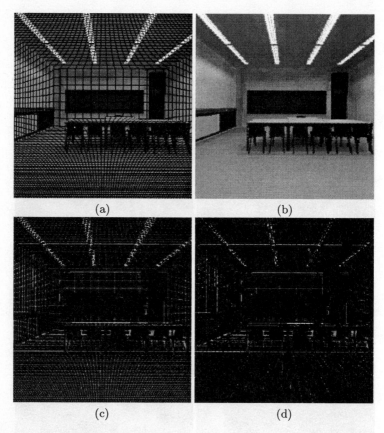

(a)   (b)

(c)   (d)

Fig. 4. Predicted images for a translation of 5 steps. Figure *(a)* has been created using a simple forward warping and Figure *(b)* was created using a 2-level forward warping. The number of incorrect pixels decreases a 60%. Figures *(c)* and *(d)* show the difference images respectively.

3.2   View Compression

Up to now we have seen how to reduce the amount of information to be sent through the network thanks to the use of spatial coherence. The previous method in [7] sent the total difference image, although compressed, and achieved reductions of up to 55% with small movements of the camera. However, with this method, some pixels that do not contribute to the final image are sent through the network. Our purpose in this paper is not to study new compression techniques, but we are going to take advantage of the fact that, as some of the information will not be needed, such as the pixels which are not different, we can wipe them out, and only send through the network the different ones. These pixels can also be compressed. With this method we can obtain up to 8:1 reduction ratios for small movements and an average of 5-6:1 reduction over the JPEG compressed difference image.

In order to do it we have to inform the client which pixels have to be updated. This can be achieved by creating a bitmap containing values of 1 for the pixels which have to be changed and zeroes elsewhere. The size of the bitmap will be 1/24th of the total image, as each pixel colour is usually encoded with three bytes and a single bit is enough to determine if it has to be changed. Thus, if the difference pixels are less than 23/24 of the total (about 96%) pixels in the image, we will save memory. In our tests we have found that the number of *valid pixels* is far higher than this 4 per cent for rotations of more than 30 degrees. In order to perform warping at every frame, the depth values of the updated pixels should also be transmitted. With our system this poses no problem, as the same bitmap may also be used to encode the Z values to be sent and thus we will have equivalent savings respect to the previous methods, which send these depth values. Our method achieves good results with no loss of quality because the number of pixels that can be reused is in general higher than a 4% and thus, our masking method can save memory. A lossy scheme could be competitive with ours, but our method has the advantage that we are not losing quality. Moreover, we could further reduce the amount of information to be sent by also using a lossy method [10] to compress the resulting data, which is smaller than the original one if the number of pixels to reuse is higher than a 4%.

4 Results

We have tested our method with different scenes and different movements of the camera, rotations and translations. The comparison between the size of information with our method and JPEG compressed difference images is depicted in Figure 5a. We can see that for the classroom scene, we achieve bandwidth reductions of up to 8:1 for small rotations, and the average reduction stays high for wide movements: 5:1 for rotations of 30 degrees.

We have also tested translations. As we have already mentioned, translations show a better behavior due to the fact that the pixels can be reused longer. With our method we obtain improvements of up to 89.5% (ratio of 9.58:1) and an average of 85% of savings. In Figure 5b we see the reduction ratio for movements of 1 to 10 steps in the viewing direction.

5 Conclusions and Future Work

We have presented two techniques for bandwidth reduction in remote navigation systems. These techniques can be applied to any system which has a limited bandwidth. Our method consists in two parts, one concerning the client side, and another one which uses the information available on the server side. In the first

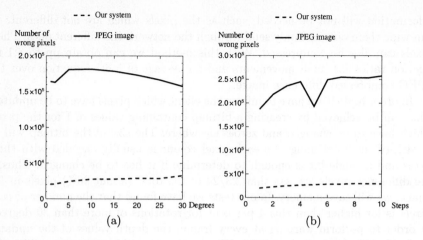

Fig. 5. Comparison of results. With our method we achieve an improvement of up to a 80% on rotations, that is, theoretically we can send 5 times more information with the same bandwidth. For translations improvements are better, up to a 85% of savings.

one a two-level forward warping is performed in order to exploit spatial coherence. The second task consists in reducing the information to be transmitted by only sending the pixels to be updated, with the aid of a bitmap that encodes their positions. This way we achieve up to a 9.5 to 1 reduction ratios. The average ratio is about 5:1 for wide movements of even 30 degrees. Our contribution is the combination of forward warping (adding a previous step changing pixel sizes to exploit spatial coherence) and the use of masking tailored to reduce the amount of information sent through the network. In spite of their simplicity, these methods lead to very good bandwidth reduction ratios.

In the future we want to apply our method to dynamically varying environments. In this case the server should send information about the variation in the visible pixels (such as a vector field, combined with the bitmap that indicates the pixels to be moved) to obtain temporally valid views in the client side.

Acknowledgments

This work has been partially supported by BR98/1003 grant of Universitat de Girona, SIMULGEN ESPRIT project #35772, and TIC 2001-2416-C03-01 of the Spanish Government.

References

1. Marc Levoy and Pat Hanrahan. Light field rendering. In *Computer Graphics Proceedings (Proc. SIGGRAPH '96)*, pages 31–42, August 1996.
2. Steven J. Gortler, Radek Grzeszczuk, Richard Szeliski, and Michael F. Cohen. The lumigraph. In *Computer Graphics Proceedings (Proc. SIGGRAPH '96)*, pages 43–54, 1996.

3. L. McMillan and G. Bishop. Plenoptic modeling: An image-based rendering system. *Proc. of SIGGRAPH 95*, pages 39–46, August 1995.
4. L. McMillan W. R. Mark and G. Bishop. Post-rendering 3d warping. In *Proc. of 1997 Symposium on Interactive 3D Graphics*, pages 7–16, New York, April 1997. ACM Press.
5. Shenchang Eric Chen. Quicktime vr - an image-based approach to virtual environment navigation. In *Computer Graphics Proceedings (Proc. SIGGRAPH '95)*, pages 29–38, 1995.
6. Marc Levoy. Polygon-assisted jpeg and mpeg compression of synthetic scenes. In *Computer Graphics Proceedings (Proc. SIGGRAPH '95)*, pages 21–28, August 1995.
7. H. Biermann, A. Hertzmann, J. Meyer, and K. Perlin. Stateless remote environment navigation with view compression. Technical Report TR1999-784, Media Research Laboratory, New York University, New York, NY, 1999.
8. Daniel Cohen-Or, Yair Mann, and Shachar Fleishman. Deep compression for streaming texture intensive animations. In Alyn Rockwood, editor, *Siggraph 1999, Computer Graphics Proceedings*, Annual Conference Series, pages 261–268, Los Angeles, 1999. ACM Siggraph, Addison Wesley Longman.
9. Silicon Graphics Inc. Open gl web page, 2001. Specification document, available from http://www.opengl.org.
10. Hee Cheol Yun, Brian K. Guenter, and Russell M. Mersereau. Lossless compression of computer-generated animation frames. *ACM Transactions on Graphics*, 16(4):359–396, October 1997. ISSN 0730-0301.

OSCONVR: An Interactive Virtual Reality Interface to an Object-Oriented Database System for Construction Architectural Design

Farhi Marir, Karim Ouazzane, and Kamel Zerzour

Knowledge Management Group (KMG), School of Informatics and Multimedia Technology, University of North London, Holloway Road, London N7 8DB, UK
{f.marir, k.ouazzane, k.zerzour}@unl.ac.uk

Abstract. The paper presents the OSCONVR system, which is an interactive interface to an Object database for construction architecture design. The OSCONVR system is part of the OSCON (Open Systems for Construction) funded project. The aim of the OSCONVR system is to explore the potentials of using virtual reality as an interface for an integrated project database using the World Wide Web. It describes the steps taken to link the OSCON integrated database to a VRML environment.

Virtual Reality has often been looked at as a visualisation tool. This paper puts the argument that VR should be used as the user interface. For instance, the user should interact with a 3D column in VR rather than a column in traditional database environment. This will allow the construction practitioners better access to information, which will motivate them to use integrated databases. In addition, VR should be used as a vehicle for classifying information. The user should be able to look at the design, costing, time planning, and facilities management views according to his/her requirements. This may prove helpful in projecting information from a project integrated database.

1. Introduction

The need for remote accessing of information within an integrated environment has become a necessity in this changing world. Information needs to be queried and interrogated in order to assist the various professions in the construction sector. For instance, site engineers should be able to access and query the remote database from their sites if they have access to a modem and the Internet. Technologies are now available which can be used to this end. The Internet and its facilities should be exploited for the benefits of better management and retrieval of construction information. The VRML (Virtual Reality Modelling Language) which is a web-based standard will be explored as a means of remotely interrogating information stored within an integrated database. The paper highlights the main findings of the OSCON integrated database project, which has managed to link the database to the VRML environment [3]. A VRML application, which reads information about design produced in AutoCAD, has been developed within the OSCON project. The VRML over the Internet will allow practitioners within the construction industry better access to the OSCON integrated database. The main benefits associated with the use of

P.M.A. Sloot et al. (Eds.): ICCS 2002, LNCS 2330, pp. 258–267, 2002.

VRML are the integration of project databases and VRML technologies, which will ultimately result in better productivity through the effective retrieval of information.

This paper aims at demonstrating the benefits of using the Internet to access construction information. The VRML facility will be used to demonstrate how information can be remotely accessed from a database. In order to achieve this, a VRML-based prototype has been developed. Common Gateway Interface programs can then be embedded in web pages. When the page is accessed, the program is automatically executed on the server and results are sent over the net to the user. These facilities are freely available on the web. If exploited properly, the construction industry will benefit tremendously in terms of information exchanges and management. It is expected that the user will be able to query information about specifications, design information, cost estimating and time planning information. In addition, he/she will be able to access information about suppliers and materials, which are freely available on the web.

2. Virtual Reality and Architecture Design

One of the major problems associated with construction project integrated databases is the shear information involved and the complexity of storing such information. Traditionally, the user interface provided within databases is used to query information stored in these databases. This may prove difficult in browsing through the many records of instances of entities developed within the scope of a certain construction application. One way of browsing and querying is through the use of VR interfaces. It is a more natural way of interfacing with information as the user can visually identify the objects of interests and retrieve information about them using the VR interface. This approach was adopted by OSCON in order to respond to the user needs for a more user friendly environment. Within OSCON, the user can navigate through the VR model and identify elements by clicking on them. Information is then obtained about the specific objects depending on the view or requirements of a particular participant of the construction process. For instance, the designer can retrieve information about the specifications of a cavity wall, the QS can obtain cost information about the cavity wall, and the time planner queries the model about duration of building the cavity wall. This process will be described in detail in one of the forthcoming sections. To conclude, this paper suggests that VR technologies should be used as means of providing better information interface, in addition to their visualisation capabilities.

An awareness workshop on VR and rapid prototyping for engineering was organised in the UK by the EPSRC (Engineering and Physical Sciences Research Council) has highlighted the benefits of using VR as the technology for visualisation and interactions with 3D models [2, 10, 13]. The proceedings of this workshop include some of the most up to date literature within VR in the UK. Most of the papers discuss how VR can be used as design tools [4]. Animation and simulation was another theme [14, 12]. Training was also mentioned where VR can play a major role. Alshawi & Faraj [1] was amongst the few researchers who suggested that VR should be used as the interface for 3D models and databases. This paper builds on the work done by the previous researchers in order to develop a VR tool, which can be used as a combination of visualisation, animation and simulation and as an interface for an object oriented integrated database. The user should be able to interact with objects rather than using DXF files as suggested by Alshawi and Faraj [1] and Griffin [4]. Some useful web sites are included in the references [5, 6, 7, 8, 9].

3. The OSCONVR System

The OSCONVR work is part of the OSCON funded project, which aims at the development of a framework for integration of construction information. The integrated database within OSCON supports the functions of design, estimating and planning by allowing these phases to effectively share information dynamically and intelligently. The system revolves around a central object-oriented information model. This model consists of domain models, which support integration of information within a specific domain, e.g. estimating. The information model also contains a core model, which captures knowledge about the means of transferring information across domains. All the models in the system are fully independent of specific applications, and each domain model provides support for general classes of a given application. In order to demonstrate how an integrated approach can benefit construction projects, the OSCON team has developed a suite of software applications, e.g. CAD application, cost estimating and planning, wrapper software for CA-Superproject®, a VRML interface which will actively share construction information via a central object-oriented project database as shown in Figure 1. The CAD application allows a user to create and manipulate architectural components of a building.   The components are stored as instances of classes in the object oriented database. These instances are read by the VRML interface in order to create a 3D view of the building which gives the user a better environment for navigation and walkthrough. Time and cost estimates are also generated automatically based on the design information stored in the database. The applications are being developed on PCs running under Microsoft Windows NT and are implemented in Microsoft Visual C++. The database is implemented using ObjectStore® OODBMS in conjunction with Object Engineering Workbench (OEW®) modelling software.   OEW is useful for generating OSCON models and the associated code in C++.

The integration approach developed by OSCON is generic and thus adaptable to any specific requirements of the industry. The system can, for example, be easily tailored to solving problems within the civil engineering or any other industry. This system can enhance the efficiency of the industry as a whole, improve productivity and consequently speed up the process of design and construction through the rapid prototyping facilities provided by the OSCON database.

The overall architecture of the OSCON system is shown below.

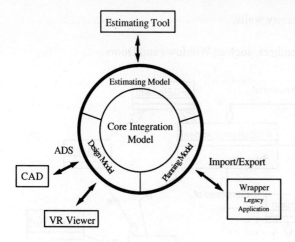

Fig. 1. The OSCON Overall architecture

3.1 OSCONVR Architectural Design Model

The root cause of the different problems in construction projects is the information or data upon which all participants depend. Because of fragmentation within the industry, there are many different interpretations of the semantics of the data in use. The formal ways in which to describe information semantics are referred to as models, which represent a formal description of a view of a domain of interest. Several researchers have suggested models for space and space enclosure and have also developed object-oriented product models that could be used to store richer kinds of data and knowledge about a product, including knowledge about its design, manufacturing and operational parameters [15]. None of these models has a universal acceptance. This is due to the fact that a data model must be readily accessible to all construction-related applications and users. It must also be compatible with systems across AEC and other industry [16]. Recently a group of software companies and users have come together to form the Industry Alliance for Interoperability and define specifications for a set of Industry Foundation Classes [17]. The aim of the OSCONVR system is not to recommend another design model but to derive classes from the IFC Generic classes and enhance them with knowledge about the detailed design stage that can be shared with other applications, such as cost and time planning. The detailed design model is a description of the design components and their specifications with very limited topological properties incorporated as shown in Figure 2. The model offers the representation of several structural members:

- Foundation: Raft, pad and strip.
- Slab: Ground slab, floor slab
- Columns,
- Beams,

- Walls: solid and cavity walls,
- Roof, and
- Non-structural members, such as Windows and Doors.

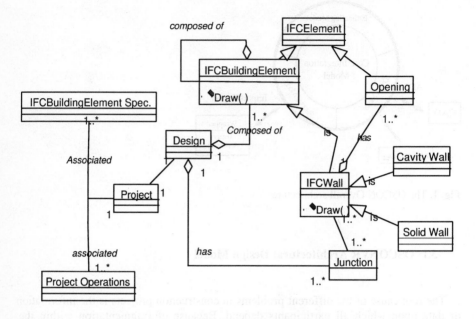

Fig. 2. A subset of the OSCONVR architectural design model.

3.2 The OSCONVR Abstract Design Classes

The OSCONVR model is designed to be generic and abstract enough to allow separation between the design model and the specific implementations of commercial CAD tools. For this it uses the Abstract Factory Design Pattern which provides an interface for creating families of related objects without specifying their concrete classes (Gamma E et al., 1994). As shown in Figure 3, the OSCONVR system defines:

- an Abstract ShapeFactory Class that declares the interface for creating each basic kind of design element,
- an Abstract Design Class (IfcBuildingElement) that represents an abstraction for each kind of architectural design element and provides abstractions that the OSCONVR design model classes can use to draw and render themselves in any graphical display environment
- a set of Concrete Subclasses of the ShapeFactory Class which implement the design elements for a specific CAD or graphical display environment. These subclasses can be used by the model classes to

'draw' themselves without knowing which application is actually doing the drawing. In OSCONVR application the VRML_Drawer are the concrete classes designed specifically for the VRML applications respectively, and

- a set of client classes that encapsulate the commands the users will want to issue. Instances of these classes can then be used to implement command functionality in an environment independent manner.

The ShapeFactory's interface has an operation that returns a new design object for each Abstract Design Class of the design component. The client class (from the user application) uses the IfcBuildingElement Abstract Design Class, which calls operations to obtain instances of a design element and draw themselves without being aware of the Concrete Classes they are using. In other words, the client class has to commit to an interface defined by the Abstract Design Class, not to a particular Concrete Class.

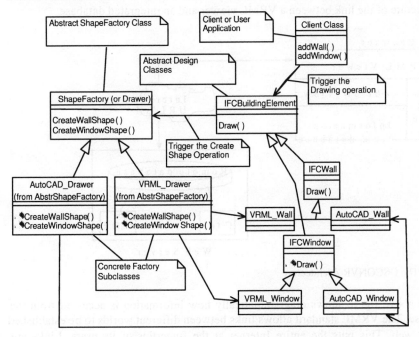

Fig. 3. The OSCONVR Abstract ShapeFactory and Abstract Design Mode

In practice, all classes in the OSCONVR design model inherit from the base class *IFCBuildingElement* that defines a common interface. This interface includes a *draw()* operation which is redefined in subclasses to allow instances of the design classes to draw themselves in a given display environment. The *draw()* operation passes a pointer to the ShapeFactory (or Drawer) object which encapsulates a set of simple 3D drawing operations as virtual member functions. These functions are implemented in the ShapeFactory concrete subclasses *i.e. VRML_Drawer*, to display the drawing in specific environments. As a result of this modelling approach the design of building components stay independent of the prevailing VR display environment and the VRML applications discussed below become graphical front ends for the instances of the OSCONVR architectural design model.

4. The OSCONVR Prototype

The Virtual Reality Modelling Language (VRML) is a developing standard for describing interactive three-dimensional scenes developed across the internet. A VRML browser is needed to load VRML files, which allow users to navigate through VRML worlds. The VRML file is a textual description of the VRML World. It contains nodes that describe shapes and their properties. VRML's four primitive shapes include cube, sphere, cone and cylinder. Figure 4 illustrates the overall architecture of the link between a VRML viewer and an integrated database

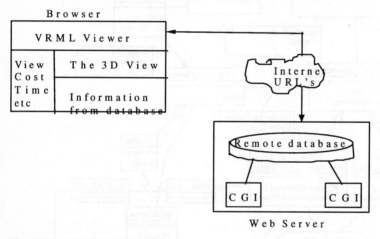

Fig. 4. The OSCONVR Architecture

The above figure shows diagrammatically how information is accessed from the database. The VRML standard allows links between different worlds to be established on the web. This puts the entire Internet at the fingertips of its users. Links are anchored to specific objects such as walls, beams, etc. By clicking on an anchored object you request information from other worlds using a URL(Universal Resource Locator) that specifies the address of a file on the web. World loaded from the Internet are delivered by the web server running on the remote host at the remote Internet site. In our case, the URL specifies a CGI script, which is a c++ program to run on the remote host under the control of the web server. The CGI program returns information about objects being queried in the VRML browser.

As previously mentioned, Virtual Reality has mostly been looked at as a visualisation tool. This paper puts the argument that VR should be used as the user interface. For instance, the user should interact with a 3D column in VR rather than a column in traditional database environment. This will allow the construction practitioners better access to information, which will motivate them to use integrated databases. In Figure 5 shown below, a screen shot of a house shown in VRML. The information is read from the OSCON object oriented database and displayed in this VR environment. The user can navigate inside the building clicking on design objects and retrieving information about their properties which include geometrical, cost and time data

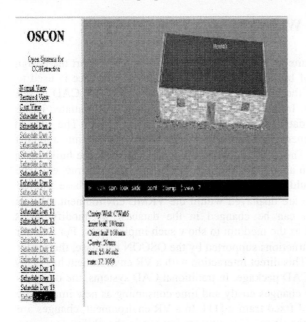

Fig. 5. An OSCONVR house model

The VRML environment to retrieve information is split into three windows using frames. The top window displays the VR representation of the building. This where the user interacts with the VRML image. The bottom windows display the information about objects queried in the VRML environment. The left-hand side windows include the views, which are supported by the system. These are described as follows:

- The normal view displays the basic representation of the building without any textures. It is mainly used for visualisation purposes.
- The textured view shows the textures associated with building elements such as walls, roofs, doors, etc..
- The cost view displays a coloured representation of the building according a colour criteria. For instance, an not estimated item or an expensive unit will appear in yellow. This would help to chase mistakes in the estimating model through the VRML environment.
- The planning view shows the various phases of the project plan in the 3D environment. For instance, in schedule day 1 only the slab is shown whereas in

schedule day 5 the slab and cavity walls are shown according to information stored in the project plan.

It has to be said that there is not limit on the views, which the system can support. For instance, facilities management, structural design, or any other view can be added easily.

5. How the System Works

The user is provided with a suite of integrated applications, which support the design and time and cost planning of buildings. The AutoCAD design interface is used to generate the design layout of the building. The user interacts with AutoCAD as the graphical display environment. In reality, the design information is instantiated in the integrated object oriented database and displayed in AutoCAD. The design information is then used in generating quantities which are used by the time and cost planning prototype software. The VRML application is used to show the building in 3D and to retrieve information about specific objects in terms of cost, time, etc. The textures of design objects could be changed in the object-oriented database and the implications on cost and time are displayed within the VRML environment. This is one example of how design can be changed in the database by modifying its specifications and VR is used as the medium to show such implications. For a better description of the numerous functions supported by the OSCON database, the reader is referred to Aouad et al [3]. This direct interaction with a VR environment has many advantages over the use of a CAD package. In traditional CAD systems, the design is relatively static, which makes changes costly and time consuming as new images are re-generated from sequences of fixed frames [11]. In a VR environment, changes are handled efficiently by the technology. The ultimate benefit is the ability to create walkthroughs, which can facilitate collaboration between clients, designers, contractors and suppliers. This is the main objective of the OSCON database and it is strongly believed that VR is the medium for communication and convergence.

6. Conclusions

This paper discussed the development of a VRML prototype, which will be used as an interface for a project integrated database (OSCON). This prototype is a web-based application, which can be run from any web site. This will allow for construction information to be readily communicated between head offices and construction sites and any other locations. This will ensure that information is communicated in a much better format with a lot more visualisation capabilities. This paper has demonstrated that a VRML can be used as an interface to a complex object oriented database. This interface has more navigation capabilities. The user will ultimately find it simpler to navigate in a VR environment rather than browsing through thousands of records in a crude database environment.

1. References Alshawi, M & Faraj, I. Integrating CAD and VR in construction. Proceedings of the Information Technology Awareness Workshop. January 1995, University of Salford.
2. Ames, A et al. The VRML source book, John Wiley and sons, 1996.
3. Aouad G., Marir F., Child T., Brandon P. and Kawooya A. (1997). A Construction integrated databases- Linking Design, Planning and Estimating. International Conference on Rehabilitation and Development of civil engineering infrastructure systems, June 9-11, 1997, American University of Beirut, Lebanon.
4. Griffin, M. Applications of VR in architecture and design. Proceedings of the Information Technology Awareness Workshop. January 1995, University of Salford.
5. http://www.construct.rdg.ac.uk/ITProjects/proje123.htm(Ashworth: linking Kappa, AutoCAd and WTK)
6. http://www.strath.ac.uk/Departments/CivEng/vcsrg.html (Retik visual scheduling)
7. http://www.construct.rdg.ac.uk/ITsearch/Projects/project12.html(James Powell: visualisation)
8. http://wquoll.maneng.nott.ac.uk/Research/virart/industry/maintran.html(Nottingham, maintenance training: VIRART)
9. http://wquoll.maneng.nott.ac.uk/Research/virart/industry/rp.html(Nottingham,Rapid prototyping: VIRART)
10. Hubbold, R and Stone, R. Virtual reality as a design tool in Rolls Royce. Proceedings of the Information Technology Awareness Workshop. January 1995, University of Salford.
11. Larijani, L.C. The virtual reality primer. McGraw-Hill, USA, 1994.
12. Lorch, R. Animation in communication. Proceedings of the Information Technology Awareness Workshop. January 1995, University of Salford.
13. Penn, A et al. Intelligent architecture: rapid prototyping for architecture and planning. Proceedings of the Information Technology Awareness Workshop. January 1995, University of Salford.
14. Retik, A & Hay, R. Visual simulation using VR. Arcom 10th conference, Vol 12, 1994. pp 537-546.
15. Svensson K. and Aouad G. (1997), Developing standardised architectural building product models for spaces and space enclosures. Submitted to Automation in Construction.
16. Froese T. and Paulson B. (1994). OPIS: An object model-based project information system. *Microcomputers in Civil Engineering*, 9, pp 13-28.
17. IFC (Industry Foundation Classes) (1997), Release 1.0, Vol 1-4.

Internet Client Graphics Generation Using XML Formats

Javier Rodeiro and Gabriel Pérez

Depto. Informática ,
Escuela Superior Ingeniería Informática,
Edificio Politécnico, Campus As Lagoas s/n
Universidad de Vigo,
32004 OURENSE, Spain
jrodeiro@uvigo.es
http://www.ei.uvigo.es/~jrodeiro/

Abstract. For a long time abstractions have been commonly used for representing graphical elements. The use of network systems (in particular INTERNET) for this type of elements has the difficulty of trying to submit the enormous volume of information contained. One possible solution to this could be the transference of an abstracted definition of the charts and to perform the drawing in the client which is requesting the graphical form. Previous approaches have been based either on SVG [1] or XML [2] but they have the inconvenient of being unsuitable for future incremental developments. In the work presented here a new approach based on XML and Java is proposed. It does provide not only a working tool but it is also an open and incremental system regarding the possible techniques which can be applied during the drawing process in the client and during the transference of information as a XML document.

1   Introduction

When data are registered in any kind of source (electronic, paper, etc.) we can foresee that any type of consult of the stored information is intended. In order to make those data more accessible and to facilitate operations with them there are a wide range of informatics' applications, from the simplest worksheet to the most complex database management systems.

Once this friendly and efficient system is created, problems arise in relation to enable data presentations in such ways which are complete and easy to understand and that they do not lose their reliability. To achieve this there are graphics and data diagrams.

One of the common difficulties when performing queries to a remote data base or worksheet to obtain a graphic result (i.e. as a reliable image which represents the consulted data) occurs when trying to send the resulting chart through the network. The main problem is that they are big in size. Reductions in the size for a better handling inside the network can be sorted out with file compressors (generally based in losses of information) or by using vector graphics [1].

P.M.A. Sloot et al. (Eds.): ICCS 2002, LNCS 2330, pp. 268–274, 2002.

In this project we intend to solve this problem by reducing the data fluxes during the communication between the client and the server when representing the results. To achieve this, instead of submitting the final document as JPEG, BMP, etc., the server will send a graphic representation as XML (eXtensible Marked Language), through an applet Java, to the client. This has been defined previously to generate the graphic from the consulted database.

2   Proposed project

2.1   Description

Graphic description was performed in XML, by means of the following DTD (Data Type Definition) [3].

```
<!ELEMENT grafico (angular | circular)>
 <!ELEMENT angular(nombre,ymax,nelementos,distancia,ancho,barra*)>
  <!ELEMENT nombre (#PCDATA)>
  <!ELEMENT ymax (#PCDATA)>
  <!ELEMENT nelementos (#PCDATA)>
  <!ELEMENT distancia (#PCDATA)>
  <!ELEMENT ancho (#PCDATA)>
  <!ELEMENT barra EMPTY>
   <!ATTLIST barra
     valor CDATA #REQUIRED
     indice CDATA #REQUIRED>
```

Implementation was carried out by using a Java platform, for both XML documents generation and rendering, originated from the XML definition. In this implementation it is remarkable the independence of this platform, between the server and the clients. For this reason the use of applets was chosen due to the fact that they can be executed within a web navigating environment and only the address (URL) of the XML document is needed. Thus the described method provides to the client clear information about the server which has processed and sent the data.

The communication between the client and the server is as follows:

1. The client (for example from a website) logs into the server and requests one chart based on certain data and specifies the graphic type and the format desired.
2. Based on this request the server originates a XML file which contains the components definitions of the image. At the same time a web page is created in which the correspondent applet to the required graph is included.
3. This applet creates a communication channel with the initial document, then extracts the information contained and represents this in the navigator.

2.2   Interpreting XML

When handling XML documents it is of great importance to follow the definition indicated in the DTD which corresponds to the type of documents under use. Within the client-server structure, both elements directly operate in XML. In addition it should be taken into account that they are completely independent and therefore because of the lack of direct communication between them it is essential to ensure that they both work on the same XML base.

Regarding the server component, we are currently implementing reference servers in order to establish the foundations of later versions which will allow to work with different information sources (databases, keyboard directly, from text files...). It is also noticeable that this system has been built by using an open methodology, so that this project can be easily improved by third persons in relation to either rendering methods or to the XML structure of the representation.

The API Java for XML and JAXP, provides great advantages in the development of these servers and, indeed the necessary methods to originate the result contained in XML documents are common to all servers.

However, in relation to the client, this is not the case mainly due to the restrictions in size imposed to the applets (they must be around 5KB, including the subclass gdXML, which has been employed here and it is described below). In order to reduce the applets size we have focused the development in following two premises: firstly, the subclasses gdXML should implement their own XML interpreter for their correspondent DTD; and secondly, the structure of the file containing the XML document must be as much simple as possible (i.e. it should not allow line breaks in the label space, for example).

When performing the implementation the task of building a XML interpreter in every subclass created increases slightly the effort but allows to reduce the resulted coding substantially.

Additionally, preparing the client to work with a simple structure in the archive also influences in the final size. This is achieved by increasing the complexity at the server level, since that it is in charge of creating the XML document in the most simple way.

2.3   Client

Initially one single applet with the function of interpreting and representing all chart types was used (gdXMLapp). In order to do this we employed the class renderer which creates an URL object for the XML file, opens the flux of information to it and represents the final output in the navigator.

The main problem with this procedure was the size of the class renderer which increased enormously. This increment in size was due to the implementation of the necessary methods to represent the new graphic solutions. In addition to this, more procedures should be added for the applet to call the class renderer and to indicate the type of graphic to be used. In view of this we decided to change the classes hierarchy so that from the mother class, gdXML (which implements all the necessary methods to interpret the XML document) the classes representing the graphics are subsequently developed.

The class gdXML includes abstract methods [1] for drawing procedures. This allows the generation of new subclasses following a pre-established format, which facilitates the programming of both the applets and these new subclasses included in the formers (Fig. 1).

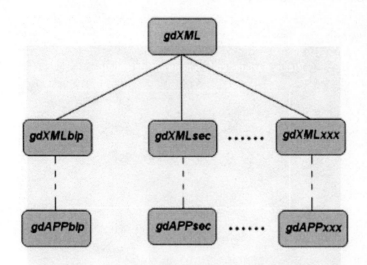

Fig. 1. Structure of the client classes hierarchy

The abstracts methods listed below are proposed:

rname(Graphics g) Chart name.
rLegend(Graphics g) It places the legend within the applet environment.
rElements(Graphics g) It draws the elements which conform the graphic (columns, sectors,...).

To date two classes have been implemented(including the corresponding applets), which were gdXMLblp (creation of columns, lines and scatter charts) and gdXMLsec (pie charts). They can be described as an specialisation of the class renderer because each one is able to interpret a specific chart.

The main advantage provided here is to reduce the final size of the file as a consequence of including only the minimum number of drawing functions to represent the requested chart. In addition to this, the implementation effort is also reduced by inheriting from the class gdXML the necessary methods for interpretation purposes.

These applets work as follow:

---

[1] These methods were not implemented in gdXML, instead each subclass created from it is in charge of its own implementation

1. The applet receives the name of the file to be interpreted from the web page code.
2. The applet creates a flux from the URL of the file.
3. The applet runs the class builder procedure and uses this flux as parameter.
4. The builder procedure obtains the necessary values to build the chart.
5. The applet uses the necessary visualization methods to create the resulting image in the navigator (Fig. 2).

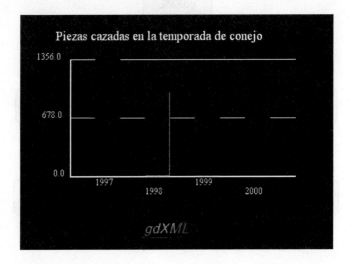

Fig. 2. Example of the resulting image in the navigator

2.4   The applets gdAPPblp and gdAPPsec

The applet gdAPPsec (through the class gdXMLsec) represents the pie charts and separates the different sectors which constitute the total figure. To do this the centre of the circumference including the sector desired has to be moved. The implemented solution in the circular class is to divide the trigonometric circumference into eight zones, each one of 45° and drag the centre of each sector according to its location.

In order to know the sector location the 'cumulative angle' is the reference adopted [2]. Depending on its location it is dragged to the point which is considered to be the centre. This movement must be accumulative, i.e. if two sectors are in the same area we cannot apply a fix step moving because this will result in superimposed sectors in the final figure appearing in the navigator.

This technique of cumulative movements presents the problem of knowing where the centre should be moved to. For instance, in the first zone (from 0° to

---

[2] The starting angle in a sector is the sum of the length of the preceding arches

45°)if we cumulatively move the centre along the axis X all the sectors starting at that point will be superimposed, and therefore they should be moved along the axis Y.

The applet gdAPPblp has the function of representing the scatter, line and column charts. This task is executed in conjunction with the class gdXMLblp. Despite its reduced size (less than 5KB between the applet and the class) has the following characteristics:

- It allows to change the font size. For example, the chart title can have a bigger font size than the text in the legend (this is a common characteristic to all applets)
- It adjust the scale of the axis Y. The method realises a direct interpretation of the height of the column, i.e., if the value to be represented is equal to 20 the column will be 20 pixels high. This results in some of the charts having a deviated height-width ratio. To avoid this the input values are corrected so that the length of axis Y is half of that of axis X.

3   Conclusions

One of the most important aspects of this project is the size of the applets interpreting the XML files. For this reason, much effort in elaborating the coding has been derived to their optimisation. In general terms, reducing the size of the client involves more work in the development of the server component.

Another important aspect in the client element is the fact of not using Java classes to handle XML documents. Instead we have created the classes gdXMLblp and gdXMLsec, both being able to play the roles of XML interpreters, DOM (Document Object Model) handlers and including representation methods. Also it is noticeable that the use of a DOM neither of a XML interpreter is never necessary.

With a modem connection of 56600bits/s (this is a minimal requirement nowadays) these applets can be downloaded in less than a second. To this we should add the time to open the website which contains them, and obviously, the execution time of this applet in the client. Generally the total waiting time, since the graphic is requested to when appears on screen is slightly over one second.

This work was partially funded by the C.I.C.Y.T. under the projects TEL-1999-0958 (Plan Nacional de I+D) and TEL99-0335-C04-03 (Plan Nacional de I+D).

References

1. Jan Christian Herlitz, *Drawml and svg*, Proceedings of the First XML Europe conference (XML Europe 99), 1999, Granada, Spain, 27-30 April, pp. 61–70.
2. Mark Roberts, *Graphic Element Markup*, Proceedings of the First XML Europe conference (XML Europe 99), 1999, Granada, Spain, 27-30 April, pp. 547–557.

3. Javier Rodeiro and Gabriel Pérez, *Generación de gráficos en arquitecturas cliente-servidor mediante xml*, Proceedings of XI Congreso Español de Informática Gráfica, 2001, Girona, Spain, 4-6 July.

The Compression of the Normal Vectors of 3D Mesh Models Using Clustering

Deok-Soo Kim, Youngsong Cho, and Donguk Kim

Department of Industrial Engineering, Hanyang University, 17 Haengdang-Dong,
Sungdong-Ku, Seoul, 133-791, South Korea
dskim@hanyang.ac.kr, {ycho1971, donguk}@ihanyang.ac.kr

Abstract. As the transmission of 3D shape models through Internet becomes more important, the compression issue of shape models gets more critical. The issues for normal vectors have not yet been explored as much as it deserves, even though the size of the data for normal vectors can be significantly larger than its counterparts of topology and geometry. Presented in this paper is an approach to compress the normal vectors of a shape model represented in a mesh using the concept of clustering. It turns out that the proposed approach has a significant compression ratio without a serious sacrifice of the visual quality of the model.

1. Introduction

As the use of Internet is an every day practice, the rapid transmission of data becomes one of the most critical factors to make the business successful. To transmit data faster through network, the compression of the data is the fundamental technology to obtain. Since the advent of its concept, the compressions of text, sound and image have been investigated in depth and we enjoy the result of research and development of such technologies in everyday life. The research on the compression of shape model, however, has been started very recently. In 1995, Deering published the first and noble paper discussing the issue [3].

The issues related to the topology and/or geometry of shape models have been extensively investigated. However, its counterpart for normal vectors has not been explored as much as it deserves, even though the size of the data for normal vectors can be significantly larger than its counterparts of topology and geometry.

Normal vectors may or may not exist in a model even though they are necessary for the visualization of the model. If normals do not exist in a model, usually rendering software creates the normals from the face definitions. However, there are frequently cases that normals are necessary. Once normal vectors exist in a shape model, the file size of the normals is quite big as shown in Fig. 1 compared to the size of topology and/or geometry data. In the examples shown in the figure, normals take almost half of whole data of shape models. It should be noted that the ratio may vary depending on the model and the implementation of the authoring tools. The models used in this paper are represented in VRML and were created using a CAD system called ProEngineer.

P.M.A. Sloot et al. (Eds.): ICCS 2002, LNCS 2330, pp. 275–284, 2002.
© Springer-Verlag Berlin Heidelberg 2002

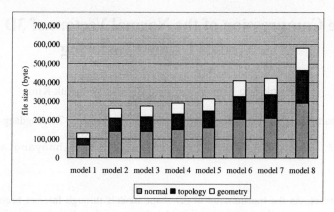

Fig. 1. The sizes of geometry, topology and normal vectors for various mesh models represented in VRML

A normal vector usually consists of three real numbers represented by float numbers, and a float number uses 32-bits in IEEE standard. It has been known that two normals with a discrepancy of 2^{-46} radian can be distinguished with normals represented in the float numbers, and it is generally agreed that this is too much detail information than necessary for the general graphics objects [3]. Hence, the compression of the normal vectors is an important issue for the exchange of a shape model through Internet or graphics pipeline.

The presented algorithm takes advantage of the well-known technique of clustering. The normals of a model are automatically grouped in a set of clusters where each cluster has a representative normal which is computed as a mean value of normals in the cluster. Each representative normal vector has a unique identification associated with it and the normals in a cluster are represented by the cluster identification for its use.

It turns out that the proposed algorithm compresses the normal vectors of a model in less than 10 % of the size of the original model without a serious sacrifice of the visual quality of the appearance of the models.

2. Related Works

Ever since Deering's noble work on the compression of shape model, there have been several researches on various aspects of shape model compression. One line of researches is the compression of topology [2][3][5][8][10][11][13], geometry [3][9][11][13], colors [12], normal vectors [3][12], and so on.

Among several algorithms regarding on the topology compression, Edgebreaker [10] reserves a special attention since our algorithm uses the information produced by Edgebreaker. Edgebreaker transforms a mesh structure into a string of C, L, E, R, and S symbols and redefines the sequence of faces in the original mesh model without creating any change in the model.

The research for normal vector compression has not yet been explored as much as it deserves. Deering's approach is the table-based approach to allow any useful normal to be represented by an index. To convert a normal vector into an index of normal on the unit sphere, the unit sphere is split up into 48 segments with identical surface area. In the each segment, the normal vectors are represented by two n-bits integers which two angles using spherical coordinates are converted into [3].

In Taubin's proposal for VRML compressed binary format, normal vectors are quantified with a subdivision scheme. An octant of the base octahedron is divided by recursively subdividing the base triangle n times. A normal vector is then encoded into a sequence of $3+2n$ bits, where n is the subdivision level [12].

In Hoppe's Progressive mesh, he introduced a wedge to represent discontinuities of normals on mesh. It is a set of vertex-adjacent corners whose attributes are the same. In vertex split recodes, a field of 10 bit encodes the wedges to which new corners created by vertex split are assigned and the deltas of normals is encoded [6].

3. Representations of normal vectors in VRML model

Among the several representations of shape models, a mesh representation, a triangular mesh model in particular, is the main representation in this paper, and the mesh model is assumed to be orientable manifold.

A mesh model is usually obtained from a surface model or the boundary of a solid model, and normal vectors are assigned at the vertices or faces of the model depending on the requirements of the visual appearance of the model. The principle use of normals in a shape model lies in the rendering process so that a model can be visualized more realistically. Generally, the shading algorithms, such as Gouraud shading and Phong shading [4], require the normal be known for each vertex of a shape model.

Usually there are four different ways to assign normal vectors to a mesh model reflecting two factors. First, normal vectors may be assigned at the vertices or faces. Second, the coordinate values of the normal vectors may be explicitly represented where they should appear or the indices of the vectors may be used instead of the coordinates themselves. In fact, this categorization lies under the design concept of VRML itself.

The node to define a triangular mesh in VRML 97 is *IndexedFaceSet*, and the node has two fields named *normalPerVertex* and *normalIndex* [1]. If the value *normalPerVertex* field is TRUE, the normal vectors in the mesh model are assigned at the vertices of the model. Otherwise, the normal vectors are defined at the faces. If the *normalIndex* field exists in the mesh definition, the coordinate values of a vertex may be defined only once and they can be referenced as many as needed via the indices. Hence, the possible ways that normals are related to a mesh model in VRML 97 can be illustrated as shown in Fig. 2. Even though Fig. 2(a) and (c) look similar, they are in fact different method in the sense that (a) does not use indices while (c) does.

Among these four possible configurations of assignments, we will be presenting a compression algorithm and file format for Case 4 that the normal vectors are assigned

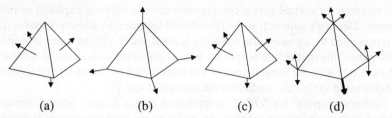

Fig. 2. Categorization of normal vector representation in VRML (a) Case 1 : *normalPerVertex* (FALSE) *normalIndex*(non-existing) (b) Case 2 : *normalPerVertex*(TRUE) *normalIndex*(non-existing) (c) Case 3 : *normalPerVertex*(FALSE) *normalIndex*(existing) (d) Case 4 : *normalPerVertex*(TRUE) *normalIndex*(existing)

at the vertices and the normal indices are used. Note that this case is the most general configuration among the four cases.

4. Clustering of the normal vectors

The approaches to compress normal vectors should be investigated in three aspects: bit-efficiency, the amount of information losses, and the decompression time.

Note that the approaches by Deering and Taubin yield a number of surface (this surface is a vector surface on which the end points of normals lie) segments with identical area. Provided that the normal vectors are usually unevenly distributed in the space (in its angular distribution), some of the surface segments are less frequently used than the others. Both approaches also use a fixed length code to reference a surface segment and therefore the bit-efficiency can decrease significantly if the distribution of normals is seriously skewed.

Based on this observation, the bit-efficiency can be improved in one of two ways: With surface segments with identical area, the entropy encoding using Huffman code may be used so that the frequency of the normal occurrences in the segment may be accounted. The other way could be to make a number of regions so that each region may contain normal vectors with identical or closer to identical number of normals. In this case, hence, the areas of related surface regions may differ. In this way, the bit-utilization of fixed length code may be maximized. Theoretically speaking, both approaches will reach similar compression ratio, except that the first approach needs Huffman tree itself to be stored. In addition, the first approach needs to navigate Huffman tree in the decompression process.

Considering these factors, we propose an approach that reflects the concept to assign identical number of normals to a segment. In our algorithm, clustering technique is used to compress the normal vectors. We use the standard and simplest K-means algorithm for the clustering. In the future, it may be necessary to analyze the pros and cons of different clustering algorithms for the compression of normal vectors. K-means algorithm is an iterative clustering algorithm. Suppose that the number of clusters and the initial mean of each cluster are given. K-means algorithm then assigns each data to the closest cluster based on a dissimilarity measure and

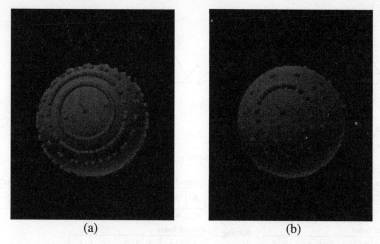

(a)   (b)

Fig. 3. (a) Normal vectors of original model (b) Cluster means after clustering

updates cluster mean values by incorporating the new data until the cluster means are stabilized [7]. The dissimilarity measure is the angle between normal vectors. The cluster mean is the axis direction of the smallest enclosing cone.

Shown in Fig. 3 (a) is 1,411 normal vectors from a bolt model. The small balls on a larger sphere represent the end points of unit normal vectors. Note that several normals may coincide at a ball on the sphere. Fig. 3 (b) illustrates the mean normal vectors of each cluster, where all normal vectors are grouped into 64 clusters.

5. Normal Vector Compression

As was discussed, we will be presenting a compression algorithm for Case 4 that the normal vectors are assigned at the vertices and the normal indices are used.

5.1 Encoding of normal vector

Consider an example that has three faces and six normals as illustrated in Fig. 4. Suppose that the clustering process yielded three clusters A, B, and C, where n_1 and n_4 are in the cluster A, n_2 is in the cluster B, and the others, n_3, n_5, and n_6, are in the cluster C. In Fig. 4, f_1 is initially related with three normals n_1, n_2, and n_3, and this fact is reflected in the first three integers, 1, 2, and 3, respectively, in *Normal Index* array. These integers denote that f_1 is related with the first, the second, and the third vectors in *Cluster Pointer* array which again point to the cluster mean values, $\overline{n}_A$, $\overline{n}_B$, and $\overline{n}_C$, of clusters A, B, and C, respectively.

Suppose that *Cluster Pointer* array is rearranged via a sorting operation so that the key values of the elements in the array are in an ascending order. Then, the representation of the model is identical to the one before the sorting was applied, if the index values in *Normal Index* array are appropriately adjusted.

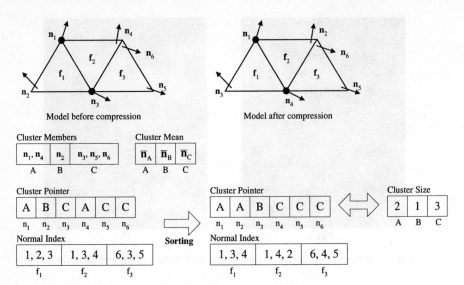

Fig. 4. Encoding of normal vector

This observation is well-illustrated in the figure. Suppose that *Cluster Pointer* array *before* sorting is rearranged to *Cluster Pointer* array *after* sorting. Then, the information related to a face f_1 remains identical if the normals n_2 and n_3 (represented by integers 2 and 3) in *Normal Index* array *before* sorting is changed to n_3 and n_4 in the rearranged *Normal Index* array *after* sorting, respectively. Hence, f_1 is related with n_1, n_3, and n_4 after the rearrangement of *Cluster Pointer* array without any topological change in the model.

Then, we further encode *Cluster Pointer* array by devising another array named *Cluster Size* which contains the number of occurrence of each cluster in a sequential order as shown in the figure. In the given example, *Cluster Size* array contains three elements where the first, the second, and the third elements indicate the occurrences of normals associated with the clusters A, B, and C, respectively.

The first integer 2 in *Cluster Size* array, for example, indicates that the first two elements in *Cluster Pointer* array are pointing to the cluster A. Note that this also means that the cluster A consists of two normal vectors that are n_1 and n_2 from the clustering operation. Similarly, the second integer 1 means that the cluster B has one normal that is n_3. Then, it is easy to verify that model before compression and model after compression altogether defines identical model without any ambiguity as shown in Fig. 4.

5.2 Encoding of normal index

Our algorithm assumes that topology of the mesh model is compressed by Edgebreaker. Note that Edgebreaker redefines the sequence of faces of the original mesh model and the changes should be reflected in the indices of the normals. In addition, it should be reminded that the normal indices were changed once more when the sorting operation was performed.

It is interesting to observe that a normal index is more likely to appear within a few *topological distances* after its first appearance. The topological distance is defined here as the integer that represents the relative distance between the currently recurring normal index value and the latest occurrence of the same normal index value in *Normal Index* array.

Suppose that a triangle in a model has an associated unique normal vector. Then, normal indices at three vertices will refer to an identical normal, and the index values will be identical. The normal indices of a face may show up in a consecutive order in *Normal Index* array. Also, suppose that three triangles share one vertex, and there is a unique normal at the vertex (which is frequently the case in many models). Then, *Normal Index* array will have an identical index value three times with two topological intervals in-between. Since Edgebreaker creates a triangle chain by producing a new neighboring triangle, a normal index at the vertex of a face is likely to appear after one or two (or more) triangles are created. We separate the normal indices into two distinct groups: *absolute* and *relative* indices.

Suppose that an index to a particular normal vector appears at a particular position in *Normal Index* array. When the second occurrence of the index happens within a prescribed maximum topological distance, denoted as r, the second occurrence of the index is represented by relative index that has the topological distance as the value. On the other hand, the second occurrence of the index may happen beyond r or a new index, not existing beforehand, the index is represented by absolute index and has the index itself as the value.

Fig. 5 (a) is *Normal Index* array after sorting in Fig. 4. Suppose that $r = 4$. Normal indices of 1 and 4 in f_2 have topological distance 3 and 2, respectively, and normal index of 4 in f_3 has topological distance 3. These three indices are represented by relative indices as shown at the gray cells in Fig. 5 (b).

On the other hand, the normal indices of 1, 3, and 4 in f_1, 2 in f_2, and 6 in f_3 are completely new indices in the array. In other words, they do not exist in the array beforehand. Hence, they have to be represented by absolute indices. In the case of the normal index 3 (the very last integer in the array) in f_3, the topological distance is 7 since n_3 was shown up as the second integer in the array which is far beyond the predefined r (=4) and therefore it is also represented by an absolute index.

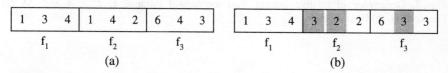

Fig. 5. (a) usage of normal indices (b) mixed usage of absolute and relative index ($r = 4$)

To effectively use the mixed indices of absolute and relative, it is necessary to consider two factors: the bit-size of relative indices and the coverage of indices with relative indices. Depending on these two factors, the mixed use may compress the model or may not.

Shown in Fig. 6 is the frequency distribution of normals that can be covered by the absolute and the relative indices. Note that $r = \infty$ for the relative indices in this experiment meaning that all recurring indices are counted. As shown in the figure, approximately 80 % of the normal indices are recurring. Hence, the normal indices can be better compressed if the recurring normals can be more efficiently encoded, if the appropriate *r* is prescribed.

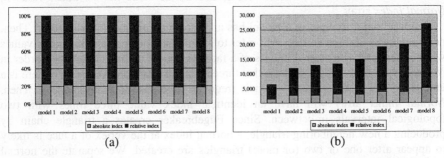

(a)   (b)

Fig. 6. Distribution of absolute index vs. relative index (a) Frequencies, (b) Percentiles

To decide appropriate number for r, we performed another experiment that is summarized in Fig. 7. The figure shows the coverage of relative indices which can be covered by the topological distances 1, 2, 3, 4, and higher than 5. Note that the relative indices with topological distances 1, 2, and 3 altogether cover roughly 80% of total relative indices. This observation suggests that most of the relative indices can refer to the latest occurrence of the same normal index using 2-bits of memory.

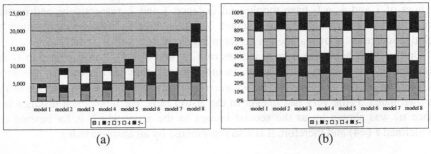

(a)   (b)

Fig. 7. Occurrence of relative indices with topological distance 1, 2, 3, 4 and, 5+ (a) Frequencies, (b) Percentiles

6. Experiment

The models used in this paper are created using a commercial CAD system called ProEngineer. Table 1 shows the comparison between file sizes compressed by different approaches. Column A shows the size of VRML file in ASCII which contains the normal vectors only, and the column B is the size of compressed files using an application called WinZip. Min Required Memory, denoted as MRM, means the theoretical minimum memory size to store normal vectors in a binary form.

Table 1. Comparison of file size

	VRML File (A)	Zip of A (B)	MRM (C) (calculated)	Proposed Algorithm (D)	D/A	D/B	D/C
model 1	68,392	18,730	41,892	5,934	8.68%	31.68%	14.16%
model 2	132,491	36,888	77,952	10,706	8.08%	29.02%	13.73%
model 3	139,411	38,637	82,524	11,450	8.21%	29.63%	13.87%
model 4	150,678	40,990	89,736	11,781	7.82%	28.74%	13.13%
model 5	158,900	43,664	92,796	13,001	8.18%	29.78%	14.01%
model 6	205,110	54,436	122,172	16,218	7.91%	29.79%	13.27%
model 7	210,446	55,578	124,500	17,190	8.17%	30.93%	13.81%
model 8	291,132	77,532	168,348	24,789	8.51%	31.97%	14.72%

The compression ratio of the proposed algorithm is approximately 10 % and 30 % of VRML files in ASCII and WinZip compressed files of corresponding VRML files, respectively.

Fig. 8 shows the visual difference between the original model and the model with normal vectors compressed by the proposed algorithm. Note that original model and compressed model have the same geometry information and the same topology information. Fig. 8 (a) shows the visualization of original model 5 with 2,861 normal vectors and Fig. 8 (b) shows the visualization of model 5 compressed by the proposed algorithm and the normal vectors in model 5 are represented by 64 cluster means. It is not easy to find the remarkable distinction between (a) and (b).

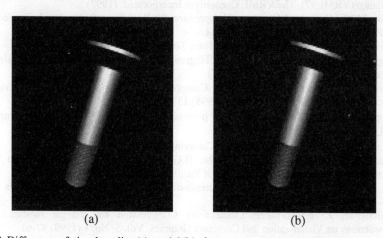

(a)   (b)

Fig. 8. Difference of visual quality (a) model 5 before compression (158,900 bytes) (b) model 5 after compression (13,001 bytes)

7. Conclusion

Normal vectors are necessary to create more realistic visualization in a shape model. When the normals exist in the definition of model, the file size could be significantly large compared to the size of geometry and topology data. Hence, the file size can be an obstacle to transmit the model data seamlessly through a network.

The proposed compression technique obtains a significant compression ratio of roughly 10% for original normal vectors without a serious sacrifice of the visual quality of the model. In particular, we have presented techniques related to the absolute and relative indices for compressing normal vectors.

However, there are a few issues to be further studied. Using a better algorithm than K-means algorithm for the clustering may be one of the important topics. In addition, we believe that the mean values of the representative normal vectors in the clusters could be also compressed.

Acknowledgements

This work was supported by the Korea Science and Engineering Foundation (KOSEF) through the Ceramic Processing Research Center(CPRC) at Hanyang University.

Reference

1. Carey, R., Bell, G., Marrin, C.: ISO/IEC DIS 1477-1: 1997 Virtual Reality Modeling Language(VRML97). The VRML Consortium Incorporated (1997)
2. Chow, M.M.: Optimized Geometry Compression for Real-Time Rendering. In Proceedings of IEEE Visualization '97 (1997) 347-354
3. Deering, M.: Geometry Compression. In Proc. SIGGRAPH' 95 (1995) 13-20
4. Foley, J.D., van Dam, A., Feiner, S.K., Hughes, J.F.: Computer Graphics : Principles and practice 2nd edn. Addison Wesley (1987)
5. Gumhold, S., Strasser, W.: Real-Time Compression of Triangle Mesh Connectivity. In Proceedings of ACM SIGGRAPH '98 (1998) 133-140
6. Hoppe, H: Efficient implementation of progressive meshes: Computers and Graphics, Vol. 22, No. 1 (1998) 27-36
7. Jain, A.K., Dubes, R.C.: Algorithms for Clustering Data. Prentice-Hall (1988)
8. Kim, Y.S., Park, D.G., Jung, H.Y., Cho, H.G.: An Improved TIN Compression Using Delaunay Triangulation. In Proceedings of Pacific Graphics '99 (1999) 118-125
9. Lee, E.S., Ko, H.S.: Vertex Data Compression For Triangle Meshes. Eurographics 2000, Vol. 19, No. 3 (2000) 1-10
10. Rossignac, J.: Edgebreaker: Connectivity Compression for triangle meshes. IEEE Transactions on Visualization and Computer Graphics, Vol. 5, No. 1 (1999) 47-61
11. Taubin, G., Rossignac, J.: Geometric Compression Through Topological Surgery. ACM Transactions on Graphics, Vol. 17, No. 2 (1998) 84-115
12. Taubin, G., Horn, W.P., Lazarus, F., Rossignac, J.: Geometric Coding and VRML. Proceedings of the IEEE, Vol. 86, Issue 6 (1998) 1228 -1243
13. Touma, C., Gotsman, C.: Triangle Mesh Compression. In Proceedings of Graphics Interface '98 (1998) 26-34

Semi-metric Formal 3D Reconstruction from Perspective Sketches

Alex Sosnov[1], Pierre Macé[2], and Gérard Hégron[3]

[1] Ecole des Mines de Nantes, Département d'Informatique,
44307 Nantes Cedex 3, France
[2] Tornado Technologies Inc., 7–11 av. Raymond Feraud, 06200 Nice, France
[3] CERMA UMR CNRS 1563, Ecole d'Architecture de Nantes,
44319 Nantes Cedex 3, France

Abstract. We present a new approach for accurate and fast reconstruction of 3D models from hand-drawn *perspective* sketches and imposed geometric constraints. A distinctive feature of the approach is the decomposition of the reconstruction process into the stages of correction of the 2D sketch and elevation of the 3D model. All 3D *constraints* that describe the spatial structure of the model are *strictly* satisfied, while *preferences* that describe the model projection are treated in *relaxed* manner. The *constraints* are subdivided into the projective, affine and metric ones and expressed in algebraic form by using the Grassmann-Cayley algebra. The constraints are resolved one by another following the order of their types by using the *local* propagation methods. The *preferences* allow to apply linear approximations and to systematically use formal methods.

1 Introduction

Problems of using traditional conception tools in conjunction with new computer-aided design, modeling and engineering instruments arise nowadays in many domains close to architectural design. Thus, the task of analyzing and understanding 3D shapes from freehand drawings is receiving increased attention.

During the last years, several methods for reconstruction of 3D scenes either from hand-drawn sketches or photographs have been proposed. To reconstruct a 3D scene from a sketch, one seeks, by using numerical methods, the position of the center of projection and the 3D model so that its projection is maximally fits to the sketch ([3], [8]). Such approaches include the analysis of the sketch and even intentions of the user [3] to determine correct forms that the user means to do and spatial constraints that the reconstructing 3D model have to satisfy. However, most of these methods involve axonometric projections, while perspective ones provide more 3D information. On the other hand, the methods [2], [5] allow to reconstruct 3D models from perspective photographs. Moreover, the system [2] provide the user with a set of 3D primitives and allows to set constraints among them. However, such approaches can not be applied to imprecise hand-drawn sketches. All the above methods rely on heavy numerical computations that are time consuming and subject to numerical instabilities.

P.M.A. Sloot et al. (Eds.): ICCS 2002, LNCS 2330, pp. 285–294, 2002.

In this paper, we demonstrate that the problem of reconstruction of 3D models from perspective sketches can be significantly simplified by serializing the resolution of constraints and using geometric algebra. Our approach allows to perform the reconstruction faster and more accurately than before even for imprecise drawings and provides new possibilities for organization of user interfaces.

2   The Approach

Our method of reconstruction of 3D models from perspective hand-drawn sketches is based on separation of **constraints** and **preferences** [4], **serialization** of constraints and their **local** resolution, and application of **formal** methods.

In contrast to other approaches, we do not attempt to find the 3D model that maximally fits to the given sketch. Instead, we demand the user to set constraints that describe spatial structure of the given scene. We find the *correct sketch* that is *merely* close to the original one while *strictly* satisfies all the projective consequences of the imposed constraints. Then we elevate the 3D model from the obtained true projection by using the projective geometric Grassmann-Cayley algebra (Sect. 3). On the other hand, during such an elevation, any 3D element is determined by using its projection only if there is no way to determine it from incident 3D elements. Thus, we rigorously respect all the 3D *constraints* while consider the sketch only as the source of *preferences*.

Each of the 3D constraints represents a projective (*incidence, collinearity*), affine (*parallelism* of lines and planes) or metric (*orthogonality* of lines) relation. The constraints are resolved following the order of their types. Namely, we estimate vanishing points and lines to satisfy projective consequences of all the parallelism constraints with the minimum of distortion of the sketch. Then we correct vanishing points and evaluate the position of the center of projection so that all the 3D orthogonality constraints would be satisfied during the elevation of the 3D model. Next, the sketch is redrawn from the corrected vanishing points and certain free points to respect all the projective constraints. Finally, the 3D model is elevated from the obtained true projection and 3D coordinates of certain scene points. This stepwise resolution process is called *serialization*. x Unlike to other approaches, we do not try to solve arising systems of constraints globally. Instead, constraints are presented as a graph and resolved by the *local methods*, mainly by propagation. On the other hand, we use some global information to control propagation processes. Namely, we establish *operation priorities* to choose the most stable and computationally effective solution. Moreover, degenerate local solutions are rejected by using global relation tables. We decompose resolution processes into the *formal* and *numerical* stages and postpone calculations as late as possible. Namely, for most of the serialized stages we firstly construct a formal solution by using only constraint graphs. Next, we evaluate the concrete solution from the obtained formal one and the corresponding numerical parameters. Having the formal solution, we can distinguish between inaccuracies and inconsistencies and rapidly re-evaluate the reconstruction once the user changes numerical parameters or enriches the scene with new details.

3   The Grassman-Cayley Algebra

To build a formal solution of a reconstruction problem in stepwise manner, we need a constructive "coordinate-free" formalism that allows to express projective geometric statements by invariant algebraic ones. Consider a projective space P furnished with a multilinear alternating form called a *bracket*. Using this form, it is possible to define a double algebra of affine subspaces of P (i.e., points, lines and planes) [6]. This algebra is called the *Grassmann-Cayley algebra* (GC algebra). It provides binary operators *join* $\vee$ and *meet* $\wedge$ and the unary *duality* operator *. A *join* represents a sum of disjoint subspaces, while a *meet* represents intersection of subspaces and is dual to the join of the their duals. For example,

$$\text{point } a \vee \text{ point } b = \text{line } ab$$
$$(\text{plane } \pi_1 \wedge \text{ plane } \pi_2)^* = (\text{plane } \pi_1)^* \vee (\text{plane } \pi_2)^*$$
$$(a \vee b \vee c \vee d)^* = [a\,b\,c\,d] = s \ .$$

Since the scalar s identifies the volume, it is impossible to determine a distance in P in terms of this algebra. On the other hand, under certain assumptions it is possible to solve orthogonality problems using the notion of duality. Once we have chosen any plane π to represent the infinite plane π_∞, it is possible to formally represent all the constructions that imply parallelism and certain orthogonality constraints. Therefore our approach is called *semi-metric*.

To perform computations from the GC algebra formal expressions, 3D points are represented by 4D vectors of homogenous coordinates, while 3D lines and planes are represented by antisymmetric matrices containing their Plücker coordinates. All the calculations are reduced to evaluations of exterior products for joins and simple matrix products for meets. These products represent coordinate expressions for the corresponding GC algebra operations. If a result of join or meet is null (i.e., a null tensor), the arguments of the operator are linearly dependent. This fact used is to detect contradictions and imprecisions (Sect. 9).

4   Creation of Geometry

4.1   Elementary Objects and Constraints

We represent any 3D scene with the following elementary objects: *points, lines* and *planes*, and constraints: *collinearity* and *coplanarity* of points and lines, *parallelism* and *orthogonality* of lines and planes. The user creates such objects and constraints via high-level primitives or directly on a sketch.

Since we use the Grassman-Cayley algebra, we express any n-ary projective or affine constraint as a set of *incidences*. Incidence (signed below by the symbol ε) is the binary projective relation defined as:

$$\forall A, B \quad A \,\varepsilon\, B \Leftrightarrow A \in B \text{ or } B \in A \text{ or } A \subset B \text{ or } B \subset A \ .$$

Any coplanarity or collinearity constraint just should be binarized, e.g.:

$$\text{points } a, b, c \text{ are collinear } \textbf{iff } \exists \text{ line } L \ \{a \,\varepsilon\, L, b \,\varepsilon\, L, c \,\varepsilon\, L\} \ .$$

Any parallelism can be represented as a triple of incidences:

$$\text{lines } L_1 \parallel L_2 \text{ iff } \exists \text{ point } i_\infty \{L_1 \varepsilon i_\infty, L_2 \varepsilon i_\infty, i_\infty \varepsilon \pi_\infty\} \ ,$$

where π_∞ is the infinite plane, while the infinite point i_∞ unambiguously determines the whole pencil of the lines. Therefore, we can present all the projective and affine constraints in the form of a *constraint graph* $CG(V, E)$. Its vertices represent elementary 3D objects while its edges represent established constraints:

$$\forall u, v \in V(CG) \quad (u, v) \in E(CG) \Leftrightarrow u \varepsilon v \ .$$

On the other hand, we do not represent metric constraints by incidences. Instead, we create an *orthogonality graph* $OG(V, E)$ so that its vertices correspond to pencils of parallel lines while its edges represent orthogonalities.

4.2   Consistency of Constraints

We define a *geometry law* as a function from a constraint to its *consequences*, $G : u \varepsilon v \Rightarrow \{x \varepsilon y\}$. For geometry of incidences, we have only 3 basic laws:

1. $L \varepsilon \pi \Rightarrow \forall a \varepsilon L, \ a \varepsilon \pi$
2. $a \varepsilon L \Rightarrow (a \varepsilon \pi, \ b \varepsilon L, \ b \varepsilon \pi \Rightarrow L \varepsilon \pi) \text{ and } \forall \pi \varepsilon L, \ a \varepsilon \pi$
3. $a \varepsilon \pi \Rightarrow (a \varepsilon L, \ b \varepsilon L, \ b \varepsilon \pi \Rightarrow L \varepsilon \pi) \text{ and } (a \varepsilon \tilde{\pi}, L \varepsilon \pi, L \varepsilon \tilde{\pi} \Rightarrow a \varepsilon L)$

where a, L and π are any point, line and plane.

A constraint graph is *saturated* if it contains all the consequences of each of the imposed constraints. Use of only saturated graphs allows to improve efficiency of the reconstruction and to analytically reject degenerate solutions. Moreover, during creation of such a graph it is possible to detect logical contradictions. Thus, a constraint graph is *consistent* if it is saturated and not contradictory.

The consistency is maintained by update procedures that represent the basic geometry laws and are called by the system every time it creates the corresponding incidences. An update procedure infers new constraints that are consequences of the created one and the applied law. Since the geometry laws recursively depend on each other, the update procedures are called recursively. To prevent infinite cycles, one member of the created incidence is blocked as source, other as target so that it is possible to establish new constraints on the target member but not on the source one. Performance of update procedures depends on efficiency of searching geometry law elements and testing whether constraints are already established. Since it is estimated as $O(E/V)$ and for saturated graphs $|E| \gg |V|$, we use global *relation maps* to find constraints in constant time.

4.3   Projection of Constraints

Since projections of 3D points are required for the elevation, they are created directly in the constraint graph. The graph contains the center of projection o and the sketch plane π_{sk}. For any 3D point, its projection $\text{proj}(p)$ and the *line of sight op* are constructed so that $op \varepsilon p$, $op \varepsilon \text{proj}(p)$, $op \varepsilon o$, and $\text{proj}(p) \varepsilon \pi_{sk}$.

Projections of 3D lines do not provide any information for the reconstruction, while they are required for the resolution of orthogonalities and the sketch correction. Thus, they are created just in the *sketch graph* $SG(V, E)$ that represents the projection of the scene $CG(V, E)$ on the sketch plane:

$$V(SG) = \{\text{proj}(v) : v \in V(CG)\} \text{ and } E(SG) = \{(\text{proj}(u), \text{proj}(v)) : u \,\varepsilon\, v\} \;.$$

All the projections and their incidences are constructed by the system every time it creates an usual or infinite 3D point or line, or sets their incidence by following a declared or inferred constraint. Thus, projections of any parallel lines have a common vanishing point. This point is aligned with the center of projection and the infinite point that determines the pencil. Vanishing points of projections of coplanar pencils of parallel lines are aligned.

5   Formal Reconstruction

Once we have built a scene constraint graph, we can construct a formal solution of the reconstruction problem for any scene described by the graph. The solution is represented as the *formal reconstruction plan*. For each of the elementary 3D objects of the scene, the plan contains a coordinate-free expression of the GC algebra determining the object so that it would satisfy the constraints imposed on it. For instance, if a point a has constraints $a \,\varepsilon\, L_1$, $a \,\varepsilon\, L_2$, the plan contains expression $a = L_1 \wedge L_2$, since a meet of two lines defines their intersection.

The algorithm to build a reconstruction plan is based on *propagation of known data*. Indeed, using the GC algebra we can *determine* any 3D object once we have *known* a sufficient number of 3D objects incidents to it. If an object has a redundant number of known neighbors, it can be determined in different ways. To choose the best solution, we set the *operation priorities*. Since the sketch is used just as the source of preferences, any 3D point is determined by using its projection only if it is impossible to determine it via other incident 3D elements. Furthermore, operations of GC algebra differ in accuracy and numerical computation cost. For example, a meet of a line and a plane requires 12 multiplications and is always valid while a meet of two lines requires 36 multiplications and may become degenerate due to precision errors. We obtain the following operations table, where priorities are decrementing from top to bottom, left to right:

line	$a \vee b$	$\pi_1 \wedge \pi_2$		
plane	$L \vee a$	$a \vee b \vee c$		
point	$L \wedge \pi$	$\pi_1 \wedge \pi_2 \wedge \pi_3$	$L_1 \wedge L_2$	$L \wedge op$ $\pi \wedge op$.

To start the algorithm, we should have some known points. The center of projection, all the sketch points (i.e., visible projections of 3D points), and all the vanishing points are set as known and propagated to determine the lines of sight. We also have to know certain 3D objects. The number of these objects is the number of degrees of freedom of the scene. This number and required objects are determined by the algorithm. All the current determined objects are stored in the priority queue according to priorities of their determining operations. If the queue is empty, the scene is underconstrained. Therefore, we increment the number of degrees of freedom and demand the user for new constraints (on the minimally undetermined objects). An object becomes known once it is extracted from the queue. Its determining operation is added to the reconstruction plan. All the already known objects incidents to the current known one set the alternatives to determine it. These *alternative evaluators* are added to the *alternatives graph* that is used later on the numerical evaluation stage. The current known object is propagated to its unknown (but possibly determined!) neighbors. Each of these neighbors is tested to be (re-)determined. To avoid linear dependency of arguments of determining operations, we use the fact that the constraint graph contains all the consequences of all the constraints. Namely, the constraint graph is analyzed to test whether the arguments of the determining operation are inferred as dependent. For example, if a plane can be determined by join of three points, the constraint graph is searched for a line incident to all these points. If we have a correct operation, we obtain its priority. If the object is determined first time, we just put it in the queue. Otherwise, the new priority is compared with the previous one and the object is moved close to the start of the queue if the new priority is higher. Thus, each of the scene objects becomes known only once it is determined with the maximum possible priority. Therefore, the order of reconstruction does not depend on the order of establishing constraints.

6   Formal Sketch Correction

To allow the elevation of a 3D model, a sketch should be a true perspective projection. Namely, projective consequences of all the imposed constraints should be respected. For example, projections of 3D parallel lines should have exactly one common vanishing point. Certainly, it is not a case for perspective hand drawings due to user errors and imprecisions.

The sketch is corrected by redrawing its points and lines in the order that all the incidence constraints imposed in the sketch graph would be *exactly* satisfied. The idea is to redraw the sketch from the most constrained to the less constrained points. The order is determined by the propagation of degrees of freedom (DOF). Namely, the points with the minimal DOF and their incident lines are erased from the sketch one by another. The DOF of a point is the number of its incident lines not erased yet. All the already erased lines incidents to the current erasing point depend on it, while the not erased incident lines set the constraints that the point has to satisfy. Thus, if the DOF of the point is 0, the point is considered to be *free*. On the other hand, if the DOF is more than 2, there is an overconstrained system (Fig. 1), since for any point it is impossible to exactly satisfy more than 2 incidence constraints. In this case, the reconstruction plan (Sect. 5) is searched for a 3D point that does not implicitly determine any 3D point while itself is determined without using the projection. This projection does not affect the reconstruction, thus, once it is found, it is erased (with its incident lines) from the sketch and the new point with the minimal DOF is considered.

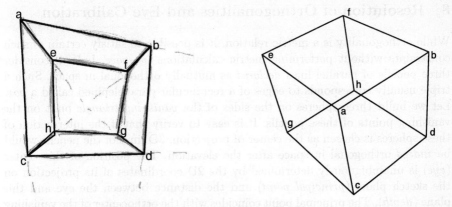

Fig. 1. The correction (left) and the reconstruction (right) of an overconstrained system. The projection of the point g is redundant, thus, it does not participate in the correction (black lines)

Once all the sketch elements are erased, they are redrawn in the *inverse* order and the formal correction plan is constructed. For each of the sketch elements (except free points), the plan contains coordinate-free expression determining the element so that it would exactly satisfy all the imposed incidence constraints. Firstly, the free points are determined, since they were most constrained. Usual free points remain unchanged, while the vanishing ones are evaluated later by weighted least squares (Sect. 7). Any sketch line is determined as a join of two firstly determined points that the line depends on. Since a line may be incident to many points, a determined line may depend on a point that is not determined yet. Such a line is added to the set of lines that constrain this point. Any point is determined as the projection or the intersection of its constraining lines once all these lines are determined. If the point has more than 3 constraining lines, there is again an overconstrained system. In this case, any point with the redundant projection is erased and the correction is restarted from scratch.

7   Estimation of Vanishing Points and Lines

In a true perspective sketch, projections of 3D parallel lines should intersect in exactly one vanishing point. On the other hand, if pencils of parallel lines contain coplanar lines, their vanishing points are aligned.

For each of the pencils of 3D parallel lines, its vanishing point is estimated by finding the closest point to the projections (i.e. sketch lines) of all the pencil lines. This point minimizes the sum of weighted squared distances to these projections where each of the weights is the maximum squared distance between all the sketch points declared as incident to the corresponding sketch line. Each of the vanishing lines that has more than 2 incident vanishing points is estimated by linear regression on these points. Finally, vanishing points that are incident to the obtained vanishing lines are projected on these lines.

8   Resolution of Orthogonalities and Eye Calibration

While orthogonality is a metric relation, it is possible to satisfy certain of such constraints without performing metric calculations *in space*. Indeed, consider three pencils of parallel lines *declared* as mutually orthogonal in space. Such a triple usually corresponds to edges of a rectangular parallelepiped called a *box*. Let we build three spheres on the sides of the *vanishing triangle* built on the vanishing points of these pencils. It is easy to verify that if the intersection of these spheres is chosen as the center of projection, 3D lines of the pencils would be *indeed* orthogonal in space after the elevation. The position of this center (*eye*) is unambiguously determined by the 2D coordinates of its projection on the sketch plane (*principal point*) and the distance between the eye and this plane (*depth*). The principal point coincides with the orthocenter of the vanishing triangle [1]. Furthermore, it is easy to verify that the depth is also determined by the coordinates of vanishing points. Therefore, we can treat scenes with any number of mutually orthogonal pencils of 3D parallel lines. Namely, to satisfy simultaneously all these orthogonalities it is sufficient to correct the sketch so that all the triples of mutually orthogonal vanishing points would define the vanishing triangles with the same orthocenters and the same depths of the eye.

8.1   Oriented Boxes

Two boxes having two parallel faces are called *oriented*. Their vanishing triangles have a common *vertical* vertex while all the other *horizontal* vertices are aligned. Such a configuration appears in urban scenes, where all the "like a box" buildings have vertical walls. Once we have aligned vanishing triangles, we set the principal point in the barycenter of their orthocenters. We have proved that it is possible to *exactly* find the minimal correction of the horizontal vertices of each of the triangles so that its orthocenter would coincide with the principal point [7]. On the other hand, it is easy to verify that aligned triangles with coincident orthocenters define the same depths of the eye.

8.2   Ordinary Boxes

Two boxes are called *ordinary* if there are no any relations between their vanishing lines. Once we have ordinary vanishing triangles, we set the principal point in the barycenter of their orthocenters. Then, for each of the triangles we find a minimal correction so that its orthocenter would coincide with the principal point. Such a correction implies a system of non-linear equations. Since its numerical solution is not exact, we reject the most corrected vertex and construct a vanishing triangle on the remaining ones with the principal point as the orthocenter. Thus, we construct the unambiguous *exact* solution [7]. Then we set the common eye depth d as the medium of the depths d_k defined by the corrected triangles. To correct each of the triangles so that it would define the depth d, we scale it with the factor $s = d/d_k$ by using the orthocenter as the center of transformation.

9   Numerical Evaluation

Once we have constructed a formal solution for a sketch correction and evaluated all the vanishing points to satisfy parallelism and orthogonality constraints, we evaluate corrections of other sketch elements by using coordinate expressions for their determining operations. Free points remain unchanged.

Once we have constructed a formal reconstruction plan, calculated the position of the center of projection and corrected all the sketch points, we evaluate the Plücker coordinates of all the elementary 3D objects composing the scene. We firstly set the eye to the obtained position and immerse it and the sketch into 3D projective space. Then the user is demanded to set in space all the points that were required to be known during the formal reconstruction (Sect. 5). The user may also set additional points. To set a point in space it is sufficient to set, for instance, its depth. Then, each of the scene objects is evaluated by applying a coordinate expression for its determining GC algebra operation. The result of evaluation of any object is invalid if it is a null tensor or represents an object at infinity while should represent the usual one. For example, a point is evaluated as infinite if its 4th coordinate is close to zero. Thus, if a result of evaluation of any object is invalid, we attempt to re-evaluate the object via all the combinations of its alternative evaluators presented in the alternative graph (Sect. 5). If the number of the alternative results close to each other is more than a certain threshold, there is just an error of precision and one of the results is returned as correct. Otherwise, there is a contradiction and the reconstruction is canceled.

It is possible to rapidly obtain reconstructions of a scene when the user incrementally enriches it with new details. Indeed, if the new elements have no any relations with the old ones, the correction and reconstruction algorithms are applied only to the new isolated subgraph of the constraint graph. Otherwise, the formal reconstruction is applied to the whole graph, while the correction algorithm is applied only to the new part of the sketch and orthogonal vanishing points are corrected so that the old part of the sketch remains unchanged. Thus, it is possible to evaluate only the changed part of the reconstruction plan.

10   Implementation and Results

Our system has been implemented on PC workstations using C++ for the kernel and Java for the UI. To reconstruct a 3D model, the user sets 3D constraints by drawing directly on a sketch. To simplify this process, the system provides the set of primitives. Then, the kernel rapidly corrects the sketch and elevates the model. The figure 2 represents the example of the reconstruction. Once all the constraints were explicitly or implicitly imposed by the user and inferred by the system, the model was obtained in 2 seconds on the PIII 733 MHz workstation.

11   Conclusion

We have presented a method that allows to accurately reconstruct 3D models even from so "unreliable sources" as perspective hand-drawn sketches once 3D

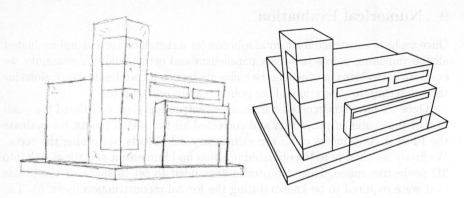

Fig. 2. The perspective hand-drawn sketch (left) and its reconstruction (right)

constraints that describe spatial structure of the models are known. The reconstruction is greatly simplified by firstly correcting the sketch and then elevating the 3D model from the obtained true perspective projection. The separation of 3D constraints and 2D preferences, the serialization of constraints and the usage of local methods to their resolution allow to avoid expensive numerical computations and thus improves the accuracy, the stability and the efficiency. The usage of projective geometric algebra allows to construct formal solutions. Having the formal solution, it is possible to compute the reconstruction in any available precision, distinguish between errors of precision and contradictions in data and rapidly reconstruct 3D scenes in the incremental manner. Future work includes the generalization of the approach to be suitable for any metric relations.

References

1. Caprile, B., Torre, V.: Using vanishing points for camera calibration. International Journal of Computer Vision, **4(2)** (1990) 127–140
2. Debevec, P.E., Taylor, C.J., and Malik, J.: Modeling and rendering architecture from photographs: a hybrid geometry- and image-based approach. In: Proceedings of ACM SIGGRAPH (1996) 11–20
3. Eggli, L., Hsu, Ch., Brüderlin, B., Elbert, G.: Inferring 3D models from freehand sketches and constraints. Computer-Aided Design, **29(2)** (1997) 101–112
4. Lhomme, O., Kuzo, P., and Macé, P.: A constraint-based system for 3D projective geometry. In: Brüderlin, B., Roller, D. (eds.): Geometric Constraint Solving and Applications. Springer Verlag, Berlin Heidelberg New York (1998)
5. Liebowitz, D., Criminisi, A., Zisserman, A.: Creating architectural models from images. In: Proceedings of EuroGraphics **18(3)** (1999) 39–50
6. Macé, P.: Tensorial calculus of line and plane in homogenous coordinates. Computer Networks and ISDN Systems **29** (1997) 1695–1704
7. Sosnov, A., Macé, P.: Correction d'un dessin en perspective suivant des contraintes d'orthogonalité. In: Actes de la conférence AFIG'01 (2001) 125–132
8. Ulupinar, F., Nevatia, R.: Constraints for interpretation of line drawings under perspective projection. CVGIP: Image Understanding **53(1)** (1991) 88–96

Reconstruction of Surfaces from Scan Paths

Claus-Peter Alberts

Department of Computer Science VII,
University of Dortmund
D-44227 Dortmund, Germany
alberts@ls7.cs.uni-dortmund.de

Abstract. Given a finite set of points sampled from a surface, the task of surface reconstruction is to find a triangular mesh connecting the sample points and approximating the surface. The solution presented in this paper takes into consideration additional information often available in sample data sets in form of scan paths. This additional information allows to cope with edges or ridges more reliably than algorithms developed for unstructured point clouds.

1 Introduction

Let us given a set of scan paths covering a surface. The task is to create a triangular mesh which includes the points of the scan paths and approximates the surface.

Figure 1 shows a cloud of points sampled from a surface in which scan paths can be recognized. A scan path is a sequence of neighboring points which are for instance obtained by moving the scanning sensor along a smooth curve over the surface and sampling its location at a certain rate. We assume that the sampling points are linearly arranged in the input data set according to their occurrence on the scan paths. However, we do not assume to have the information about locations at which the scan process is interrupted in order to move the sensor to a new location where scanning continues with a new scan path. Data sets of this type are typically delivered by tactile scanners like the Cyclon [13].

Figure 2 shows a part of a triangular mesh reconstructed from the point set of Fig. 1. Most of the vertices of the mesh are points of the input sampling set, but some additional points, often called Steiner points in literature, may occur, too. In a general setting, the construction of a triangular mesh representing the surface topologically and geometrically correctly is a non-trivial task. A considerable number of approaches have been suggested, many of which are compiled in the survey by Mencl and Müller [10]. We assume that the surface can be described by a function over a plane, in the following chosen as the x-y-plane. Under this condition a triangulation can be obtained by triangulating the flat set of points obtained by orthogonally projecting the sampling points onto this plane. However, although the problem of finding the correct topology is solved by this assumption, it turns out that such triangulations are not always geometrically

P.M.A. Sloot et al. (Eds.): ICCS 2002, LNCS 2330, pp. 295–304, 2002.

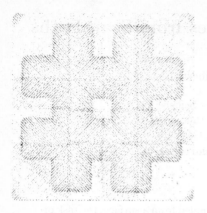

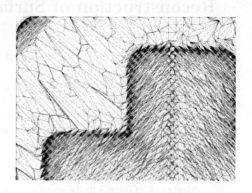

Fig. 1. Sample points of an object.

Fig. 2. Part of a triangulation.

Fig. 3. The Delaunay triangulation.

Fig. 4. The path adaptive triangulation considering feature trees.

satisfactory if the height information (z-coordinate) is not taken into consideration. Figures 3 and 4 illustrate the difference between a non-favorable and a favorable triangulation. Several data-dependent triangulation schemes have been proposed in the past which take into account spatial information [3, 4] (Fig. 5 and 6). A major criterion is curvature. In particular, a surface might have sharp edges or ridges which need special treatment like demonstrated for a different kind of sampling by [9]. In this case a typical approach is to decompose the surface into segments which are free of those artifacts, find a smooth triangulation of each of those segments, and compose them into a complete solution [7].

The solution to the surface reconstruction problem presented in this paper takes into consideration the additional information present in the sampling data set in form of scan paths. This additional information allows to cope with edges or ridges more reliably than algorithms developed for unstructured point clouds. For instance, consecutive points of a scan path should usually be connected by an edge because it can be expected that an edge of this type is close to the

Fig. 5. The ABN triangulation.   **Fig. 6.** The TAC triangulation.

surface. Furthermore, the relative dense points on a scan path are favorable to find points of high curvature which are characteristic for edges and ridges. Those points can be connected by edges to a polygonal path which represents a ridge, and thus should also be part of the triangulation.

The rest of this paper is organized along the steps of our approach. The first step, scan path extraction from the data set, is outlined in Sect. 2. The second step is filtering in order to reduce noise typical for scanned data (Sect. 3). The third step is to detect shape features along a scan path (Sect.4). In the fourth step an initial triangulation is determined which takes the scan paths into consideration (Sect. 5). In the fifth step, feature paths are calculated from the feature points of step 4 and are worked into the triangulation (Sect. 6).

2   Path Extraction

First we have to discover the start and end points of the scan paths. We use a heuristics that assumes some kind of fuzzy collinearity of the scan paths. This is a reasonable assumption for two reasons. Firstly, scan paths are often indeed linear in their projection on the x-y-plane. Secondly, for samplings of high density, even on a curved path two consecutive line segments are almost collinear.

Two consecutive points p_i and p_{i+1} of the input sequence are considered to belong to the same path if either

(1) the projected distance of p_i and p_{i+1} in the x-y-plane is sufficiently small, or
(2) the distance is large but the line segment $\overline{p_i p_{i+1}}$ is a fuzzy collinear continuation of $\overline{p_{i-1} p_i}$.

The angle between the two consecutive line segments under which they are still accepted as "collinear" depends on the length of $\overline{p_i p_{i+1}}$. It is specified by a function like that one of Fig. 7 which assigns the largest acceptable angle to an edge length.

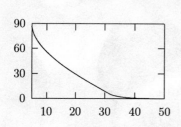

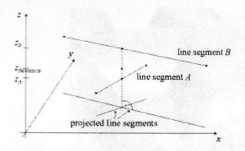

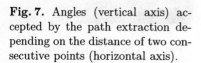

Fig. 7. Angles (vertical axis) accepted by the path extraction depending on the distance of two consecutive points (horizontal axis).

Fig. 8. Two crossing line segments.

The set of edges of the selected scan paths are checked for intersection in the projection on the x-y-plane (Fig. 8). For every pair of intersecting edges the z-coordinates z_A and z_B of the corresponding points of the intersection point on both edges are calculated. They are compared with a z-value $z_{theoretical}$ which is calculated as an average with weights reversely proportional to the edge lengths. If the difference exceeds a threshold the correlated edge is removed, that is a scan path is split into two paths. The background of this heuristics is that in this case longer edges cannot be a good approximation of the surface. They are removed preferably because it is more likely that short edges belong to a scan path.

3   Filtering

Sampled data usually are noisy because of mechanical or electrical properties of the sampling device. Typically the noise is high-frequent. High-frequent noise can be removed by low pass filters. The danger, however, is that sharp bends are smoothed out, too. We try to diminish this effect by using two different filters, one for smoothing the points according to their x- and y-coordinates, and one for smoothing the z-coordinate.

For filtering the x-y-coordinates, we assume the scan paths to be approximately straight-line except at sharp bends, that is at points where a path changes its direction significantly.

In order to maintain the sharp bends we use an adaptive Laplacian filter:

$$filter_{al}(p_i) = \mu_i \cdot \frac{1}{2}(p_{i-1} + p_{i+1}) + (1 - \mu_i) \cdot p_i.$$

An example of a non-adaptive Laplacian filter which sets μ_i, $0 \leq \mu_i \leq 1$, to constant value can be found in [14]. "Adaptive" means that the original point p_i is considered in a way that the larger a bend at p_i is the less it will be moved. μ_i depends on the angle the line segments incident to p_i include, on the property whether there is a sharp bend or not (within the x-y-plane) at p_i, and on the distance between the examined points [1].

Although we limit the possible movement of a point p_i, there is still the possibility that p_i is moved too far. In order to avoid this we introduce a second filter, the adaptive limited gravitation filter:

$$filter_{alg}(p_i) = \vartheta_i \cdot \frac{1}{2}(p_{i+1} + p_i + \nu_i(p_{i-1} - p_{i+1})) + (1 - \vartheta_i) \cdot p_i.$$

$\vartheta_i, 0 \le \vartheta_i \le 1$, models "adaptivity" and is calculated similar to μ_i. $\nu_i, 0 \le \nu_i \le 1$, is a measure for the attraction of the points p_{i-1} and p_{i+1} to p_i. It is reverse proportional to the length of the incident line segments to p_i. A constant value would smooth out flanks, especially if the filter is applied iteratively. The attraction avoids spreading out an area with many points which indicate significant changes.

The complete xy-filter is the average of these two filters.

The filter for the z-values should handle two situations. On the one hand it should smooth noisy parts and on the other hand it should keep sharp bends and slopes. For the first purpose a low pass filter seems to be appropriate while for the second purpose the median filter seems to be suitable [5]. So we use an adaptive combination of them:

$$filter_z(z_i) = \zeta_i filter_{lowpass}(z_i) + (1 - \zeta_i) filter_{median}(z_i),$$

with $\zeta_i, 0 \le \zeta_i \le 1$, measuring the difference between $filter_{lowpass}(z_i)$ and $filter_{median}(z_i)$.

4   Shape Features

The shape features of the scan paths we are interested in are sharp bends. In order to find sharp bends it is sufficient to investigate the z-profile and ignore the behavior of a path with respect to the x- and y-coordinates. We call the points representing sharp bends within a path "(sharp) feature points".

We use three methods to detect sharp bends properly which result in different valuations of every point p_i, the "angle valuation", the "compensation valuation", and the "curvature valuation". These valuations are aggregated and then representative points are selected. The three methods have different weaknesses and strengths. The combination just described yields a better overall result.

The first way of detecting a sharp bend at a point p_i is to use the angle α (Fig. 9) between the two line segments incident to p_i. A real number $0 \le r_{angle} \le 1$ is calculated which is based on α and indicates the possibility of a sharp bend at p_i. Because of improper scanning it is possible that the actual sharp bend lies between p_i and p_{i+1}. Therefore a second number $0 \le r_{2^{nd}angle} \le 1$ is calculated based on the angle γ (Fig. 9). Then $r_{2^{nd}angle}$ is shared between p_i and p_{i+1}. The "angle valuation" of p_i is chosen as the maximum of r_{angle} and $r_{2^{nd}angle}$.

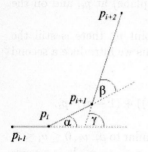
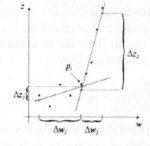
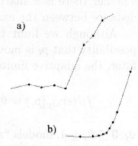

Fig. 9. Angles for determination of sharp bends within scan paths.

Fig. 10. Example of lines fitted into the profile of a path.

Fig. 11. Two different types of sharp bends.

In order to level out slight unevenness we calculate a further valuation of p_i. Therefore we take two regression lines through a number of points, one line compensating for unevenness before and one line compensating for it after p_i (Fig. 10). Then the "compensation valuation" is chosen as a function that maps the angle between these lines onto the real interval $[0, 1]$.

The last method detecting a sharp bend is to take the curvature of a path profile into account. We just calculate the curvature on the lines fitted into the profile of a path (see compensation valuation) because the curvature on the points without compensation is too sensitive to slight unevenness. The curvature k of a point p_i is measured by

$$k = \frac{\frac{\Delta z_1}{\Delta w_1} - \frac{\Delta z_2}{\Delta w_2}}{\sqrt{(\Delta w_1 + \Delta w_2)^2 + (\Delta z_1 + \Delta z_2)^2}}$$

with $w_i = \sum_{k=r_j+1}^{i} \sqrt{(x_k - x_{k-1})^2 + (y_k - y_{k-1})^2}$ if p_{r_j} is the start point of the containing path and z_i is the z-coordinate of p_i in the profile of this path. For the definition of the variables see Fig. 10. The definition of k is a modified version of that in [8] which favors special kinds of profiles unsuitably [1]. Then the "curvature valuation" is chosen as a function which maps k onto $[0, 1]$.

In order to calculate a single value for each point p_i the three valuations must be combined. We do this in fuzzy logic manner [6] by evaluating the phrase "a sharp bend is at p_i $\Longleftrightarrow$ the angle valuation *or* the compensation valuation *or* the curvature valuation gives a hint". In order to get a value of the left side of the arrow we have to aggregate the values on the right side. For our purposes we have chosen the "algebraic or" $(vel_a(r, s) = r + s - r \cdot s)$.

In this manner sharp bend a) in Fig. 11 will be detected properly, but sharp bend b) will get just a small valuation. The reason is that many points get a medium valuation instead of one point getting a big one. Because of our observation, we give every point the chance to increase its valuation taking the points in its neighborhood into account.

This procedure results in several points with more or less high valuations. Because a sharp bend can be represented by only one point, we have to select a representative. This is done by choosing the point with the highest valuation in a selection interval. Not only the valuation of a point but also its z-coordinate is considered by the selection since the sharp bend b) in Fig. 11 leads to a feature point with a relative high z-coordinate otherwise. If there is more than one point with maximal valuation and there is sufficient space between the left- and rightmost point both are selected as representatives.

5   Triangulation

Because we assume that scan paths describe the surface of a scanned object properly we are interested in triangular meshes which comprehend the scan paths. We call a triangulation with this property "path-adaptive triangulation". Using our special configuration of functional surfaces we basically can construct an approximating triangular surface mesh by triangulating the projected sampling points in the x-y-plane and then connecting two original points if they are adjacent in the triangulation in the x-y-plane. However, path-adaptation makes the procedure somewhat more complicated.

The first step of construction of a path-adaptive triangulation is to split path edges intersecting in their projection on the x-y-plane, by insertion of additional points. The reason is that intersecting edges are forbidden in a triangulation. The additional points are the intersection points of the projected edges. The pairs of intersecting edges are efficiently found by first testing the line segments of every path for intersection. This test is implemented by processing the projected edges in lexicographically sorted order of their vertices. The edges of different paths are tested by processing the axes-parallel bounding boxes of the projected paths in lexicographical order of their vertices with the goal of finding the pairs of intersecting bounding boxes. For the pairs of intersecting boxes the pairs of intersecting edges are determined by processing the edges in lexicographical order. In this manner practical computation times are achieved for data sets with hundred thousands of points. Furthermore, the procedure is easier to implement than well-known worst-case efficient algorithms of computational geometry [11].

The second step is to calculate a first triangulation of the projected sampling points and the projected additional points of the first step. We use the classical Delaunay triangulation for this purpose [11]. We calculate the Delaunay triangulation by sorting the points lexicographically according to their x- and y-coordinates, combining the first three points into a triangle, and then successively connecting the other points to the convex hull of the already inserted points, in sorted order. If necessary, edges are flipped until they satisfy the empty circle condition [2]. Edge flipping means to replace an edge e with the diagonal different from e in the quadrilateral defined by the two triangles incident to e (if it exists), if it is inside the quadrilateral.

The third step is to convert the Delaunay triangulation into a path-adaptive triangulation by edge flipping. Let the edges of the triangulation be colored black and those on scan paths not being part of the triangulation be colored red. The goal of edge flipping is to make the red edges to black edges, too. For this purpose we process the still existing red edges iteratively according to algorithm 1.

Algorithm 1. *Let $\mathfrak{L}$ be the set of edges crossing the red edge $\{p_s, p_e\}$. Furthermore, let $\mathfrak{L}$ be sorted according to the distance of the intersection points of its edges and $\{p_s, p_e\}$ to p_s.*
while $\mathfrak{L}$ *is not empty* **do**
    if *there is a "flipable edge" in $\mathfrak{L}$* **then** *flip it and remove it from $\mathfrak{L}$* **else**
        1. Search $\mathfrak{L}$ in increasing order for a "forceable edge" e_l.
        2. Flip e_l and successively all edges after e_l until a "flipable edge" or an "unforceable edge" is reached.
        3. If an "unforceable edge" is reached then flip back the edges in reverse order except e_l. Flip successively all edges before e_l until a "flipable edge" or an "unforceable edge" is reached. If an "unforceable edge" is reached again then flip all edges back to their initial position (including e_l), look for a "forceable edge" after e_l and continue with step 2.
        4. Flip the "flipable edge" and remove it from $\mathfrak{L}$.
    fi
od

Algorithm 1 uses three types of edges which are defined as follows: *flipable edges* do not intersect the red edge after flipping; *forceable edges* cross the red edge after flipping as well as before; *unforceable edges* have incident triangles forming a concave quadrilateral. After flipping unforceable edges would lie outside the quadrilateral, and therefore, they must not be flipped. This distinction is necessary because there are cases in which no flipable edges exist [1]. The existence of a forceable edge (if there is no flipable edge in $\mathfrak{L}$) and the correctness of algorithm 1, that means that the red edge is colored black in the end, have been proved in [1].

The average time of algorithm 1 seems to be quite difficult to estimate. Due to algorithm 1 every edge can be flipped at most three times within the steps 1 to 4. So the worst case is limited by $O(n^2)$ where n is the initial number of black edges in $\mathfrak{L}$.

6   Sharp Feature Trees

The next step is to construct curve-like sharp ridges of the scanned surface. For this purpose we use trees which connect feature points determined in previous steps of the algorithm. We call those trees "(sharp) feature trees". Up to now we do not consider closed paths because we have not yet developed criterions for closing a path to a loop.

A feature tree is obtained by building a minimum spanning tree (MST) in the triangulation, starting with an arbitrary feature point and using an algorithm similar to the MST-algorithm of Prim [12]. The algorithm of Prim takes one vertex as an MST and adds successively the point with the smallest distance to all points of the MST to the present tree. Since feature points need not be adjacent in the triangulation (there might be gaps) and we calculate MSTs only on feature points we had to modify the algorithm of Prim using intermediate points. We also stop adding points to the MST if the distances of all vertices not yet in the tree to the vertices of the MST are too large. The details are given by algorithm 2.

Algorithm 2. *Let Q be a priority queue containing vertices and let Q be sorted according to the distance between the vertices and the corresponding feature point from which they have been found (possibly using intermediate points, see below). Furthermore let p_s be the initial feature point.*
Initialize Q with all points adjacent to p_s and close enough to p_s.
while *Q is not empty* **do**
   remove the first entry p_r of Q;
   if *p_r is not marked* **then**
      mark p_r
      if *p_r is a feature point* **then**
         add $\overline{p_r p_u}$ to the feature tree where p_u is the feature point from which p_r has been found (possibly using intermediate points);
      fi
      add all points to Q which are adjacent to p_r, not marked, and close enough to p_r or p_u, respectively;
   fi
od

If feature points not belonging to an MST are left, a further MST is established with one of those points as start point.

Finally the path-adaptive triangulation is adapted so that the feature trees become part of the triangulation. This is achieved by applying the edge flipping algorithm of Sect. 5.

7   Empirical Evaluation and Conclusions

The presented algorithm is a new approach to detect and consider sharp features for triangular meshes. Neither the Delaunay triangulation (Fig. 3) nor the ABN or TAC triangulation (Fig. 5 and 6) model sharp features as adequate as the presented triangulation (Fig. 4). This is not surprising since between 33.5 and 86.7% of the edges of the investigated feature trees were not part of the Delaunay triangulation.

Although the worst case of the run time of algorithm 1 might be quadratic, often linear run time was observed. Furthermore, in the special case we yield at it is not clear if an example can be constructed leading to the upper bound.

The computation time was up to 18 seconds for a data set with 33K points (Fig. 1) and up to 5 minutes for a data set with 220K points. We used a Pentium III with 500MHz and needed 64MB and 250MB RAM, respectively. Our algorithm has been implemented in JAVA2. Most of the computation time was spent on calculating the Delaunay triangulation. The Delaunay, the ABN, and the TAC triangulation (Fig. 3, 5 and 6) were computed by another program which has been implemented in C++ in 3.5, 8.3, and 54 seconds, respectively. One reason for smoothing paths is to get a better starting point for bend detection. However, it has a smoothing effect on the overall surface, too. Surface-oriented smoothing techniques might improve the result if necessary.

The algorithm has several control parameters. Experience shows that there are reasonable default values. Sometimes an improvement can be achieved by a data-adaptive choice. Currently the values are chosen interactively based on a visualization of the mesh. Automatic approaches to parameter optimization can be an interesting topic of future research.

More detailed discussions can be found in [1].

Acknowledgements.
I like to thank Prof. Heinrich Müller for scientific discussions.

References

1. Alberts, Cl.-P.: Triangulierung von Punktmengen aus Abtastdaten unter Berücksichtigung von Formmerkmalen. Masters Thesis, Department of Computer Science VII, University of Dortmund, Germany, 2001.
2. Bern, M.: Triangulations. In Goodman, J.E., O'Rourke, J. (editors): Handbook of Discrete and Computational Geometry. CRC Press, New York, 1997, 413–428.
3. Van Damme, R., Alboul, L.: Tight Triangulations. In Dæhlen, M., Lyche, T., Schumaker, L.L. (editors): Mathematical Methods for Curves and Surfaces. Vanderbilt University Press, Oxford, 1995, 185–192.
4. Dyn, N., Levin, D., Rippa, S.: Data Dependent Triangulations for Piecewise Linear Interpolation. IMA Journal of Numerical Analysis 10 (1990) 137–154.
5. Gonzalez, R.C., Woods, R.E.: Digital Image Processing. Addison-Wesley, 1992.
6. Gottwald, S.: Fuzzy Sets and Fuzzy Logic. Foundations of Application – from a Mathematical Point of View. Artificial Intelligence Series, Vieweg, 1993.
7. Hoppe, H., DeRose, T., Duchamp, T., Halstead, M., Jin, H., McDonald, J., Schweitzer, J., Stuetzle, W.: Piecewise Smooth Surface Reconstruction. Proceedings of SIGGRAPH 1994, ACM Press, New York, 1994, 295–302.
8. Hoschek, J., Lasser, D.: Fundamentals of Computer Aided Geometric Design. A.K. Peters, Wellesley, MA, 1993.
9. Kobbelt, L.P., Botsch, M., Schwanecke, U., Seidel, H.-P.: Feature Sensitive Surface Extraction from Volume Data. Proceedings of SIGGRAPH 2001, ACM Press, New York, 2001, 57–66.
10. Mencl, R., Müller, H.: Interpolation and Approximation of Surfaces from Three-Dimensional Scattered Data Points. Proceedings of the Scientific Visualization Conference, Dagstuhl'97, IEEE Computer Society Press, 2000.
11. Preparata, F.P., Shamos, M.I.: Computational Geometry. An Introduction. Springer, New York, 1985.
12. Prim, R.C.: Shortest connection networks and some generalizations. Bell System Tech. Journal 36 (1957) 1389–1401.
13. http://www.renishaw.com/client/ProdCat/prodcat_levelthree.asp?ID=1229
14. Taubin, G.: A signal processing approach to fair surface design. Computer Graphics, Proceedings of SIGGRAPH'95, ACM Press, New York, 1995, 351–358.

Extending Neural Networks for B-Spline Surface Reconstruction

G. Echevarría, A. Iglesias, and A. Gálvez*

Department of Applied Mathematics and Computational Sciences, University of
Cantabria, Avda. de los Castros, s/n, E-39005, Santander, Spain
iglesias@unican.es

Abstract. Recently, a new extension of the standard neural networks,
the so-called *functional networks*, has been described [5]. This approach
has been successfully applied to the reconstruction of a surface from a
given set of 3D data points assumed to lie on unknown Bézier [17] and
B-spline tensor-product surfaces [18]. In both cases the sets of data were
fitted using Bézier surfaces. However, in general, the Bézier scheme is no
longer used for practical applications. In this paper, the use of B-spline
surfaces (by far, the most common family of surfaces in surface modeling and industry) for the surface reconstruction problem is proposed
instead. The performance of this method is discussed by means of several
illustrative examples. A careful analysis of the errors makes it possible
to determine the number of B-spline surface fitting control points that
best fit the data points. This analysis also includes the use of two sets of
data (the training and the testing data) to check for overfitting, which
does not occur here.

1 Introduction

The problem of recovering the 3D shape of a surface, also known as *surface reconstruction*, has received much attention in the last few years. For instance, in [9,
20, 21, 23, 24] the authors address the problem of obtaining a surface model from
a set of given cross-sections. This is a typical problem in many research and application areas such as medical science, biomedical engineering and CAD/CAM,
in which an object is often known by a sequence of 2D cross-sections (acquired
from computer tomography, magnetic resonance imaging, ultrasound imaging,
3D laser scanning, etc.).

Another different approach consists of reconstructing surfaces from a given
set of data points (see, for example, [10, 14, 15, 19, 22, 27]). In this approach, the
goal of the surface reconstruction methods can be stated as follows: *given a set of
sample points X assumed to lie on an unknown surface U, construct a surface
model S that approximates U.* This problem has been analyzed from several
points of view, such as parametric methods [3, 4, 11, 27], function reconstruction
[7, 28], implicit surfaces [19, 26], B-spline patches [22], etc.

* This research was supported by the CICYT of the Spanish Ministry of Education
(projects TAP1998-0640 and DPI2001-1288).

P.M.A. Sloot et al. (Eds.): ICCS 2002, LNCS 2330, pp. 305–314, 2002.

One of the most striking and promising approaches to this problem is that based on neural networks. After all, artificial neural networks have been recognized as a powerful tool for learning and simulating systems in a great variety of fields (see [8] and [12] for a survey). Since the behavior of the brain is the inspiration behind the neural networks, these are able to reproduce some of its most typical features, such as the ability to learn from data. This feature makes them specially valuable for solving problems in which one is interested in fitting a given set of data. For instance, the authors in [10] propose to fit surfaces through a standard neural network. Their approach is based on training the neural network to learn the relationship between the parametric variables and the data points. A more recent approach can be found in [13], in which a Kohonen neural network [16] has been applied to obtain free-form surfaces from scattered data. However, in this approach the network is used exclusively to order the data and create a grid of control vertices with quadrilateral topology. After this pre-processing step, any standard surface reconstruction method (such as those referenced above) has to be applied. Finally, a very recent work using a combination of neural networks and PDE techniques for the parameterization and reconstruction of surfaces from 3D scattered points can be found in [2].

It should be remarked, however, that the neural network scheme is not the "panacea" for the surface reconstruction problem. On the contrary, as shown in [17], some situations might require more sophisticated techniques. Among them, an extension of the "neural" approach based on the so-called *functional networks* has been recently proposed [5, 17]. These functional networks are a generalization of the standard neural networks in the sense that the weights are now replaced by neural functions, which can exhibit, in general, a multivariate character. In addition, when working with functional networks we are able to connect different neuron outputs at convenience. Furthermore, different neurons can be associated with neural functions from different families of functions. As a consequence, the functional networks exhibit more flexibility than the standard neural networks [5]. The performance of this new approach has been illustrated by its application to fit given sets of data from Bézier [17] and B-spline tensor-product surfaces [18].

In spite of these good results, the previous scheme is very limited in practice because the sets of data were fitted by means of Bézier surfaces in both cases. This is a drastic limitation because, in general, the Bézier scheme is not longer used for practical applications. The (more flexible) piecewise polynomial scheme (based on B-spline and NURBS surfaces) is usually applied in surface modeling and industry instead. The present paper applies this recently introduced functional network methodology to fit sets of given 3D data points through B-spline surfaces. In Sect. 2 we briefly describe the B-spline surfaces. Then, in Sect. 3 the problem to be solved is introduced. Application of the functional network methodology to this problem is described in Sect. 4. Sect. 5 reports the results obtained from the learning process for different examples of surfaces as well as a careful analysis of the errors. It includes the use of two sets of data (the training and the testing data) to check for overfitting. As we will show, this analysis makes it possible to determine the number of B-spline surface fitting control points that best fit the data points. Finally, Sect. 6 closes with the main conclusions of this work.

2   Some Basic Definitions

In this section we give some basic definitions required throughout the paper. A more detailed discussion about B-spline surfaces can be found in [25].

Let $\mathcal{S} = \{s_0, s_1, s_2, \ldots, s_{r-1}, s_r\}$ be a nondecreasing sequence of real numbers called *knots*. $\mathcal{S}$ is called the *knot vector*. The *ith B-spline basis function $N_{ik}(s)$ of order k* (or degree $k-1$) is defined by the recurrence relations

$$N_{i1}(s) = \begin{cases} 1 & \text{if } s_i \le s < s_{i+1} \\ 0 & \text{otherwise} \end{cases} \tag{1}$$

and

$$N_{ik}(s) = \frac{s - s_i}{s_{i+k-1} - s_i} N_{i,k-1}(s) + \frac{s_{i+k} - s}{s_{i+k} - s_{i+1}} N_{i+1,k-1}(s) \tag{2}$$

for $k > 1$. With the same notation, given a set of three-dimensional *control points* $\{\mathbf{P}_{ij}; i = 0, \ldots, m; j = 0, \ldots, n\}$ in a bidirectional net and two knot vectors $\mathcal{S} = \{s_0, s_1, \ldots, s_r\}$ and $\mathcal{T} = \{t_0, t_1, \ldots, t_h\}$ with $r = m + k$ and $h = n + l$, a *B-spline surface* $\mathbf{S}(s, t)$ *of order* (k, l) is defined by

$$\mathbf{S}(s,t) = \sum_{i=0}^{m} \sum_{j=0}^{n} \mathbf{P}_{ij} N_{ik}(s) N_{jl}(t), \tag{3}$$

where the $\{N_{ik}(s)\}_i$ and $\{N_{jl}(t)\}_j$ are the B-spline basis functions of order k and l respectively, defined following (1) and (2).

3   Description of the Problem

In this section we describe the problem we want to solve. It can be stated as follows: we look for the most general family of parametric surfaces $\mathbf{P}(s,t)$ such that their isoparametric curves (see [6] for a description) $s = \tilde{s}_0$ and $t = \tilde{t}_0$ are linear combinations of the sets of functions: $\mathbf{f}(s) = \{f_0(s), f_1(s), \ldots, f_m(s)\}$ and $\mathbf{f}^*(t) = \{f_0^*(t), f_1^*(t) \ldots, f_n^*(t)\}$ respectively. In other words, we look for surfaces $\mathbf{P}(s,t)$ such that they satisfy the system of functional equations

$$\mathbf{P}(s,t) \equiv \sum_{j=0}^{n} \boldsymbol{\alpha}_j(s) f_j^*(t) = \sum_{i=0}^{m} \boldsymbol{\beta}_i(t) f_i(s) \tag{4}$$

where the sets of coefficients $\{\boldsymbol{\alpha}_j(s); j = 0, 1, \ldots, n\}$ and $\{\boldsymbol{\beta}_i(t); i = 0, 1, \ldots, m\}$ can be assumed, without loss of generality, as sets of linearly independent functions.

This problem cannot be solved with simple standard neural networks: to represent it in terms of a neural network, we should allow some neural functions to be different, while the neural functions in neural networks are always identical. Moreover, the neuron outputs of neural networks are different; however, in our scheme, some neuron outputs in the example are coincident. This implies that the neural networks paradigm should be generalized to include all these new features, which are incorporated into the functional networks (see [5]). To be more precise, our problem is described by the functional network in Fig. 1(left) which can be simplified (see [17] for details) to the expression:

$$\mathbf{P}(s,t) = \sum_{i=0}^{m} \sum_{j=0}^{n} \mathbf{P}_{ij} f_i(s) f_j^*(t) \tag{5}$$

where the $\mathbf{P}_{ij}$ are elements of an arbitrary matrix $\mathbf{P}$; therefore, $\mathbf{P}(s,t)$ is a tensor product surface. Eq. (5) shows that the functional network in Fig. 1(left) can be simplified to the equivalent functional network in Fig. 1(right).

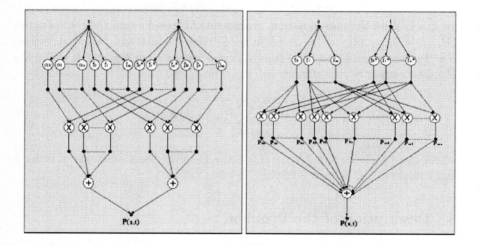

Fig. 1. (left) Graphical representation of a functional network for the parametric surface of Eq. (4); (right) Functional network associated with Eq. (5). It is equivalent to the functional network on the left

This functional network is then applied to solve the surface reconstruction problem described in Sect. 1. In order to check the flexibility of our proposal, we have considered sets of 256 3D data points $\{T_{uv}; u, v = 1, \ldots, 16\}$ (from here on, the *training points*) in a regular 16×16 grid from four different surfaces. The first one (*Surface I*) is a B-spline surface given by (3) with the control points listed in Table 1, $m = n = 5$, $k = l = 3$ and nonperiodic knot vectors (according to the classification used in [1]) for both directions s and t. The other three surfaces

Table 1. Control points used to define *Surface I*

(x,y,z)	(x,y,z)	(x,y,z)	(x,y,z)	(x,y,z)	(x,y,z)
$(0,0,1)$	$(0,1,2)$	$(0,2,3)$	$(0,3,3)$	$(0,4,2)$	$(0,5,1)$
$(1,0,2)$	$(1,1,3)$	$(1,2,4)$	$(1,3,2)$	$(1,4,3)$	$(1,5,2)$
$(2,0,1)$	$(2,1,4)$	$(2,2,5)$	$(2,3,5)$	$(2,4,4)$	$(2,5,3)$
$(3,0,3)$	$(3,1,4)$	$(3,2,5)$	$(3,3,1)$	$(3,4,2)$	$(3,5,3)$
$(4,0,2)$	$(4,1,2)$	$(4,2,4)$	$(4,3,4)$	$(4,4,3)$	$(4,5,2)$
$(5,0,1)$	$(5,1,2)$	$(5,2,3)$	$(5,3,3)$	$(5,4,2)$	$(5,5,1)$

(labelled as *Surface II, III* and *IV*) are explicit surfaces defined by the equations $z = y^3 - x^3 - y^2 + x^2 + xy$, $z = 2(x^4 - y^4)$ and $z = \dfrac{0.8y^2 - 0.5x^2}{x^2 + y^2 + 0.1}$, respectively. In order to check the robustness of the proposed method, the third coordinate of the 256 three-dimensional points (x_p, y_p, z_p) was slightly modified by adding a real uniform random variable ϵ_p of mean 0 and variance 0.05. Therefore, in the following, we consider points given by (x_p, y_p, z_p^*), where

$$z_p^* = z_p + \epsilon_p \quad , \quad \epsilon_p \in (-0.05, 0.05). \tag{6}$$

Such a random variable plays the role of a measure error to be used in the estimation step to learn the functional form of $\mathbf{P}(s,t)$.

4   Applying the functional network

To solve the problem described in the previous section, the neural functions of the network must be estimated (learned) by using some minimization method. In functional networks, this learning process consists of obtaining the neural functions based on a set of data $D = \{(I_i, O_i) | i = 1, \ldots, n\}$, where I_i and O_i are the i-th input and output, respectively, and n is the sample size. To this end, each neural function f_i is approximated by a linear combination of functions in a given family $\{\phi_{i1}, \ldots, \phi_{im_i}\}$. Thus, the approximated neural function $\hat{f}_i(\mathbf{x})$ becomes

$$\hat{f}_i(\mathbf{x}) = \sum_{j=1}^{m_i} a_{ij} \phi_{ij}(\mathbf{x}), \tag{7}$$

where $\mathbf{x}$ are the inputs associated with the i-th neuron. In the case of our example, the problem of learning the above functional network merely requires the neuron functions $x(s,t)$, $y(s,t)$ and $z(s,t)$ to be estimated from a given sequence of triplets $\{(x_p, y_p, z_p), \ p = 1, \ldots, 256\}$ which depend on s and t so that $x(s_p, t_p) = x_p$ and so on. For this purpose we build the sum of squared errors function:

$$Q_\alpha = \sum_{p=1}^{256} \left(\alpha_p - \sum_{i=0}^{M-1} \sum_{j=0}^{N-1} a_{ij} \phi_i(s_p) \psi_j(t_p) \right)^2 \tag{8}$$

where, in the present example, we must consider an error function for each variable x, y and z. This is assumed by α in the previous expression, so (8) must be interpreted as three different equations, for $\alpha = x$, y and z respectively. Applying the Lagrange multipliers to (8), the optimum value is obtained for

$$
\frac{\partial Q_\alpha}{2\partial a_{\gamma\mu}} = \sum_{p=1}^{256} \left(\alpha_p - \sum_{i=0}^{M-1} \sum_{j=0}^{N-1} a_{ij}\phi_i(s_p)\psi_j(t_p) \right) \phi_\gamma(s_p)\psi_\mu(t_p) = 0
$$

(9)

$$
\gamma = 0, 1, \ldots, M-1 \quad ; \quad \mu = 0, 1, \ldots, N-1.
$$

On the other hand, a B-spline function is basically a piecewise polynomial function whose number of spans r is given by $r = m + k - 1$, where m and k are the number of the control points and the order, respectively. Hence, we need to make a decision between the following two possibilities:

— to fix the number of control points and to change the order of the B-spline or
— to fix the order and then change the number of the control points.

In this paper we have considered the second option: to fit the 256 data points of our examples we have used nonperiodic third-order B-spline basis functions $\{N_{i3}(s)\}_i$ and $\{N_{j3}(t)\}_j$, that is, we have chosen $\{\phi_i(s) = N_{i3}(s) | i = 0, 1, \ldots, M-1\}$ and $\{\psi_j(t) = N_{j3}(t) | j = 0, 1, \ldots, N-1\}$ in (8). We remark that this choice is very natural: the B-spline functions are frequently used in the framework of both surface reconstruction and approximation theory. In particular, the third-order B-spline functions are the most common curves and surfaces in research and industry. Finally, nonperiodic knot vectors mean that we force the B-spline surfaces to pass through the corner points of the control net, a very reasonable constraint in surface reconstruction.

Therefore, we allow the parameters M and N in (9) to change. Of course, every different choice for M and N yields to the corresponding system (9), which must be solved. Note that, since third-order B-spline functions are used, the minimum value for M and N is 3. However, this value implies that the B-spline surface is actually a Bézier surface [1], so we have taken values for M and N from 4 to 8. Solving the system (9) for all these cases, we obtain the control points associated with the B-spline surfaces fitting the data. The results will be discussed in the next section.

5 Results

To test the quality of the model we have calculated the mean and the root mean squared (RMS) errors for M and N from 4 to 8 and for the 256 training data points from the four surfaces described in Sect. 3.

Table 2 refers to *Surface I*. As the reader can appreciate, the errors (which, of course, depend on the values of M and N) are very small, indicating that the approach is reasonable. The best choice (indicated in bold in Table 2) corresponds to $M = N = 6$, as expected because data points come from a B-spline surface defined through a net of 6×6 control points. In this case, the mean and the RMS errors are 0.0085 and 0.00071 respectively.

Table 2. Mean and root mean squared errors of the z-coordinate of the 256 training points from the *Surface I* for different values of M and N

	$N = 4$	$N = 5$	$N = 6$	$N = 7$	$N = 8$
$M = 4$	0.1975	0.1000	0.0941	0.0945	0.0943
	0.00919	0.00798	0.00762	0.00764	0.00763
$M = 5$	0.1229	0.0939	0.0885	0.0888	0.0886
	0.00873	0.00743	0.00700	0.00703	0.00702
$M = 6$	0.0676	0.0354	**0.0085**	0.0115	0.0093
	0.00528	0.00265	**0.00071**	0.00090	0.00082
$M = 7$	0.0691	0.0387	0.0208	0.0221	0.0217
	0.00547	0.00301	0.00163	0.00172	0.00168
$M = 8$	0.0678	0.0356	0.0117	0.0139	0.0131
	0.00531	0.00270	0.00093	0.00109	0.00103

Table 3. Control points of the reconstructed *Surface I*

(x, y, z)	(x, y, z)	(x, y, z)	(x, y, z)	(x, y, z)	(x, y, z)
$(0, 0, 1.0382)$	$(0, 1, 1.9897)$	$(0, 2, 3.047)$	$(0, 3, 2.9435)$	$(0, 4, 2.0411)$	$(0, 5, 0.9777)$
$(1, 0, 2.0048)$	$(1, 1, 2.9945)$	$(1, 2, 4.0228)$	$(1, 3, 1.9602)$	$(1, 4, 2.9981)$	$(1, 5, 2.0028)$
$(2, 0, 1.007)$	$(2, 1, 3.9777)$	$(2, 2, 4.9951)$	$(2, 3, 5.0357)$	$(2, 4, 3.9554)$	$(2, 5, 3.0221)$
$(3, 0, 2.9866)$	$(3, 1, 4.004)$	$(3, 2, 5.0283)$	$(3, 3, 0.9122)$	$(3, 4, 2.0926)$	$(3, 5, 2.968)$
$(4, 0, 2.0302)$	$(4, 1, 1.9729)$	$(4, 2, 4.0344)$	$(4, 3, 4.004)$	$(4, 4, 2.9637)$	$(4, 5, 2.0087)$
$(5, 0, 0.9757)$	$(5, 1, 2.0232)$	$(5, 2, 3.0047)$	$(5, 3, 3.009)$	$(5, 4, 1.9567)$	$(5, 5, 1.0423)$

Table 3 shows the control points for the reconstructed *Surface I* corresponding to the best case $M = N = 6$. They were obtained by solving the system (9) with a floating-point precision and removing the zeroes when redundant. A simple comparison with Table 1 shows that the corresponding x and y coordinates are exactly the same, which is as expected because they were not affected by the noise. On the contrary, since noise was applied to the z coordinate, the corresponding values are not obviously the same but very similar, indicating that we have obtained a very good approximation. The approximating surface is shown in Fig. 2(top-left) and it is virtually indistinguishable from the original surface.

To cross validate the model we have also used the fitted model to predict a new set of 1024 testing data points, and calculated the mean and the root mean squared (RMS) errors, obtaining the results shown in Table 4. The new results confirm our previous choice for M and N. A comparison between mean and

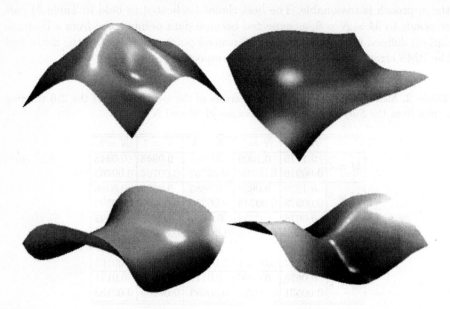

Fig. 2. (top-bottom,left-right) B-spline approximating surfaces of the surfaces labelled *Surface I, II, III* and *IV*, respectively. Their corresponding equations are described in Sect. 3

RMS error values for the training and testing data shows that, for our choice, they are comparable. Thus, we can conclude that no overfitting occurs. Note that a variance for the training data significantly smaller than the variance for the testing data is a clear indication of overfitting. This does not occur here.

Table 4. Mean and root mean squared errors of the z-coordinate of the 1024 testing points from the *Surface I* for different values of M and N

	$N = 4$	$N = 5$	$N = 6$	$N = 7$	$N = 8$
$M = 4$	0.1118	0.0943	0.0887	0.0889	0.0889
	0.00441	0.00384	0.00366	0.00367	0.00366
$M = 5$	0.10599	0.0888	0.0830	0.0833	0.0832
	0.00422	0.00363	0.00342	0.00343	0.00342
$M = 6$	0.0649	0.0341	**0.0078**	0.0109	0.0093
	0.00252	0.00130	**0.00032**	0.00042	0.00038
$M = 7$	0.0668	0.0381	0.0203	0.0216	0.0213
	0.00263	0.00149	0.00081	0.00085	0.00084
$M = 8$	0.0651	0.0345	0.0111	0.0133	0.0125
	0.00253	0.00133	0.00043	0.00051	0.00049

A similar analysis was carried out for the other surfaces described in Sect. 3. The corresponding tables of results are not included here because of limitations of space. It is enough to say that the best choice of M and N for the *Surface II* is $M = N = 4$. The mean and RMS errors are 0.0052 and 0.00040 respectively for the training points and 0.0049 and 0.00019 for the testing points. The *Surface III* has a more complex shape so larger values of M and N, namely $M = 6$ and $N = 7$, are required for the best fitting. For these values the mean and RMS errors are 0.0095 and 0.00073 respectively for the training points and 0.0090 and 0.00035 for the testing points. Finally, *Surface IV* is best fitted for $M = N = 7$. In this last case, the mean and RMS errors are 0.0143 and 0.00114 respectively for the training points and 0.0139 and 0.00055 for the testing points. The approximating B-spline surfaces are displayed in Fig. 2.

6 Conclusions

In this paper a powerful extension of neural networks, the so-called functional networks, has been applied to the surface reconstruction problem. Given a set of 3D data points, the functional network returns the control points and the degree of the B-spline surface that best fits these data points. We remark that the data points do not necessarily have to belong to a parametric surface. In fact, some examples of the performance of this method for both parametric and explicit surfaces have been given. A careful analysis of the error as a function of the number of the control points has also been carried out. The obtained results show that all these new functional networks features allow the surface reconstruction problem to be solved in several cases. Nevertheless, in order to assess the limitations of our proposal further research is required. This future work will be reported elsewhere.

References

1. Anand, V.: Computer Graphics and Geometric Modeling for Engineers. John Wiley and Sons, New York (1993)
2. Barhak, J., Fischer, A.: Parameterization and reconstruction from 3D scattered points based on neural network and PDE techniques. IEEE Trans. on Visualization and Computer Graphics **7**(1) (2001) 1-16
3. Bolle, R.M., Vemuri, B.C.: On three-dimensional surface reconstruction methods. IEEE Trans. on Pattern Analysis and Machine Intelligence **13**(1) (1991) 1-13
4. Brinkley, J.F.: Knowledge-driven ultrasonic three-dimensional organ modeling. IEEE Trans. on Pattern Analysis and Machine Intelligence **7**(4) (1985) 431-441
5. Castillo, E.: Functional Networks. Neural Processing Letters **7** (1998) 151-159
6. Farin, G.E.: Curves and Surfaces for Computer-Aided Geometric Design (Fifth Edition). Morgan Kaufmann, San Francisco (2001)
7. Foley, T.A.: Interpolation to scattered data on a spherical domain. In: Cox, M., Mason, J. (eds.), Algorithms for Approximation II, Chapman and Hall, London (1990) 303-310

8. Freeman, J.A.: Simulating Neural Networks with Mathematica. Addison Wesley, Reading, MA, (1994)
9. Fuchs, H., Kedem, Z.M., Uselton, S.P.: Optimal surface reconstruction form planar contours. Communications of the ACM, **20**(10) (1977) 693-702
10. Gu, P., Yan, X.: Neural network approach to the reconstruction of free-form surfaces for reverse engineering. CAD, **27**(1) (1995) 59-64
11. Hastie, T., Stuetzle, W.: Principal curves. JASA, **84** (1989) 502-516
12. Hertz, J., Krogh, A., Palmer, R.G.: Introduction to the Theory of Neural Computation. Addison Wesley, Reading, MA (1991)
13. Hoffmann, M., Varady, L.: Free-form surfaces for scattered data by neural networks. Journal for Geometry and Graphics, **2** (1998) 1-6
14. Hoppe, H., DeRose, T., Duchamp, T., McDonald, J., Stuetzle, W.: Surface reconstruction from unorganized points. Proc. of SIGGRAPH'92, Computer Graphics, **26**(2) (1992) 71-78
15. Hoppe, H.: Surface reconstruction from unorganized points. Ph. D. Thesis, Department of Computer Science and Engineering, University of Washington (1994)
16. Kohonen, T.: Self-Organization and Associative Memory (3rd. Edition). Springer-Verlag, Berlin (1989)
17. Iglesias, A., Gálvez, A.: A new Artificial Intelligence paradigm for Computer-Aided Geometric Design. In: Artificial Intelligence and Symbolic Computation. Campbell, J. A., Roanes-Lozano, E. (eds.), Springer-Verlag, Lectures Notes in Artificial Intelligence, Berlin Heidelberg **1930** (2001) 200-213.
18. Iglesias, A., Gálvez, A.: Applying functional networks to fit data points from B-spline surfaces. In: Proceedings of the Computer Graphics International, CGI'2001, Ip, H.H.S., Magnenat-Thalmann, N., Lau, R.W.H., Chua, T.S. (eds.) IEEE Computer Society Press, Los Alamitos, California (2001) 329-332
19. Lim, C., Turkiyyah, G., Ganter, M., Storti, D.: Implicit reconstruction of solids from cloud point sets. Proc. of 1995 ACM Symposium on Solid Modeling, Salt Lake City, Utah, (1995) 393-402
20. Meyers, D., Skinnwer, S., Sloan, K.: Surfaces from contours. ACM Transactions on Graphics, **11**(3) (1992) 228-258
21. Meyers, D.: Reconstruction of Surfaces from Planar Sections. Ph. D. Thesis, Department of Computer Science and Engineering, University of Washington (1994)
22. Nilroy, M., Bradley, C., Vickers, G., Weir, D.: G^1 continuity of B-spline surface patches in reverse engineering. CAD, **27**(6) (1995) 471-478
23. Park, H., Kim, K.: 3-D shape reconstruction from 2-D cross-sections. J. Des. Mng., **5** (1997) 171-185
24. Park, H., Kim, K.: Smooth surface approximation to serial cross-sections. CAD, **28** (12) (1997) 995-1005
25. Piegl, L., Tiller, W.: The NURBS Book (Second Edition). Springer Verlag, Berlin Heidelberg (1997)
26. Pratt, V.: Direct least-squares fitting of algebraic surfaces. Proc. of SIGGRAPH'87, Computer Graphics, **21**(4) (1987) 145-152
27. Schmitt, F., Barsky, B.A., Du, W.: An adaptive subdivision method for surface fitting from sampled data. Proc. of SIGGRAPH'86, Computer Graphics, **20**(4) (1986) 179-188
28. Sclaroff, S., Pentland, A.: Generalized implicit functions for computer graphics. Proc. of SIGGRAPH'91, Computer Graphics, **25**(4) (1991) 247-250

Computational Geometry and Spatial Meshes

César Otero, Reinaldo Togores

Dept. of Geographical Engineering and Graphical Expression Techniques
Civil Engineering School, University of Cantabria, Avda. de los Castros, s/n,
E-39005, Santander, Spain
oteroc@unican.es

Abstract. This paper establishes a relation between the field of study of *Geometric Design of the Structures* known as "Spatial Meshes" and *Computational Geometry*. According to some properties of Voronoi Diagrams and Convex Hulls, and by means of two transformations: one inversive, the other projective, a way for generating polyhedra approximating spheres and other second order surfaces (ellipsoids, paraboloids and hyperboloids), starting from a simple Voronoi Diagram, is shown. The paper concludes with the geometric design of C-TANGENT domes: a new way for generating Spatial Meshes that can become a possible field of application for Computational Geometry.

1   Introduction: What is a Spatial Mesh?

Fig. 1. Spatial structures: (left) spatial mesh; (right) plate structure

Structural design and calculus procedures can be made in a bi- or three-dimensional fashion. Structures like those in Fig. 1 do not have a dominant bi-dimensional stress-resisting element; stresses are transmitted along the whole structure in such a way that it is impossible to study the isolated parts. Its unitary character prevails during the calculus and design process in this kind of structures. What marks a difference between them is their macroscopic character,

P.M.A. Sloot et al. (Eds.): ICCS 2002, LNCS 2330, pp. 315–324, 2002.

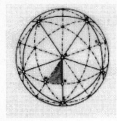
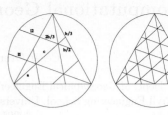

Fig. 2. (left) A geodesic mesh derived from a regular isosahedron; (right) two tesselations of the $\{3,5+\}_{b,c}$ type, proposed into a triangle

discrete in the case of the first illustration (Fig. 1-left) and continuous in the second one (Fig. 1-right). Both are Spatial Structures [3, 4]; the first one of the Spatial Mesh type, the second one of Plate type. By definition [7] *a Spatial Mesh is a structural system composed by linear elements, joined in such a way that stresses are transmitted in a tridimensional manner. In some cases, part of the component elements can be bi-dimensional. Macroscopically considered, a spatial structure can adopt planar or curved surface forms.*

2   Notions About the Geometry of Dome Shaped Spatial Meshes

Among the different possibilities for generating by means of nodes, bars and panels a spherical form, what we refer to as the Polyhedrical System has received a vast amount of contributions. Essentially it deals with generating a triangular mesh over the face of a regular or semi-regular polyhedron or over the portion of the sphere that is circumscribed around this polyhedron. The great amount of symmetry planes and rotation symmetry axes present in a geometric body of these characteristics means that the number of different elements in the triangulation (vertices, edges or faces) will be a minimum, as what happens on one of the faces - like face A-D-E in Fig. 2 left - of the regular polyhedron can be reproduced identically on every other face. All of the symmetry planes in a regular polyhedron produce great circles over its circumscribed sphere; the strong influence that these great circles have on the design of the triangular mesh account for the name it receives: Geodesic Dome.

The truth is that the symmetry planes and axis in a regular polyhedron affect every one of its faces, so the proposition stated in the preceding paragraph can be improved: only one part of a regular polyhedron's face requires different elements for its triangulation; the rest propagates around the entire polyhedron or its circumscribed sphere according to its symmetry conditions. In Fig. 2-left the elementary fragment of a regular icosahedron is shaded: 120 of such triangles are reproduced to generate the circumscribed sphere.

The basic model for the generation of a triangulation of the sphere is established through very diverse ways according to the approach taken by the different designers. We are proposing in this paper a general approach derived from Coxeter's mosaic [1] of type $\{3, 5+\}_{b,c}$ in which the edges of the mosaic do not have to be parallel to the edges of the polyhedron neither the vertices have to belong to the symmetry planes. Figure 2-right shows the development pattern for the $\{3, 5+\}_{2,1}$ and $\{3, 5+\}_{4,2}$ models. In this way, the pattern of a triangle of $1/120$ of the sphere's surface is reproduced all around it.

Figure 1-left shows an example of a spherical dome of the geodesic type. In it the procedure for generating the triangular mesh has followed these rules: (i) Selection of the initial polyhedron (usually a regular icosahedron); (ii) Determination of the triangular network on the polyhedron's face (or to be more precise, on a triangle like A-M-N in Fig. 2-left). The triangular network can be formulated in many diverse ways but Coxeter's notation is valid for all of them, that is, any procedure can be formulated by means of the $\{3, 5+\}_{b,c}$ notation; (iii) Projection of this network on the sphere circumscribing the polyhedron. In any of these cases, the design process is aided by numerical calculation methods for obtaining the positions and dimensions of the bars. References [1] and [2] are recommended for a more detailed view of the aforementioned questions.

3   An Introduction to the Geometry of Plate Meshes. Duality

Types of spatial meshes like the one shown in Fig. 1-left are named lattice structures (the basic structural element is a bar). As contrasted with the lattice structures, Wester [9] has proposed an interesting alternative suggesting that a *plate structure* (see Fig. 1-right), where plates (plane elements which are able to resist forces in their own plane only) are stabilized by shear forces, constitutes a new type and concept of structure with an applicability that is, perhaps, deeper than the former.

From the point of view of Statics, lattice and plate designs are equivalent: the equilibrium of force vectors on a node of a lattice structure is formulated in the same way as the equilibrium of moments (created by forces around a reference point) on a face of a plate structure. This equivalence is known as *Duality*.

From the point of view of Morphology, there can be found a relationship between a polyhedrical approach obtained via lattice and its corresponding one via plate (and vice versa). Wester proposes this correspondence in terms of Duality; three types of dual transformation, DuT, DuM and DiT [8] are, basically, able to supply the mentioned equivalence lattice-plate. It is remarkable that some of these dual transformations give, from a spherical lattice, dual solutions such as double curved shapes different from the sphere (the reciprocal one is possible too).

Besides Wester, some other researchers have studied structural dualism. Concerning the procedure to obtain the shape of the structure, it can be assumed that its geometrical support is the transformation known as polarity.

4   Voronoi Diagram and Delaunay Triangulation on the Plane

The topological similarity between *Plate Structures* and the *Voronoi Diagram* cannot pass unnoticed. The same happens when someone considers a *lattice mesh* (a geodesic dome as Fuller, i.e.) and contemplates the tessellation arising from a *Delaunay Triangulation*. Computational Geometry can supply some new ideas to the structural design. We will make use of them in the creation of a new way of designing of structural shapes. The most important feature is that the procedure works in 2D space, since a flat Voronoi Diagram will become a polyhedron that approximates a double curved structure (duality guarantees that an inversive transformation applied to a Delaunay Triangulation becomes a lattice grid inscribed into a Sphere).

5   Some useful properties of the Voronoi diagram

Assisted by Fig. 3-left, consider [6] the stereographic projection which describes the restriction of the inversion on E^3 to the plane $z = 1$ (or equivalently, to the sphere E, inversive image of this plane). Suppose then a plane P holding C (the center of E). Whatever the angle that this plane forms with the OXY, it is well known that the inversive shape of the circumference F_1 (arising from the intersection of P with E) is another circumference F'_1, the center of which is C', resulting from the intersection between the plane $z = 1$ and the line perpendicular to P from O (pole of the inversive transformation).

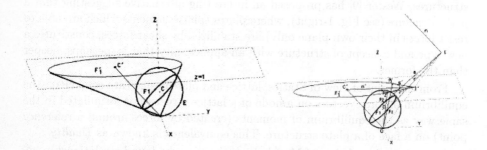

Fig. 3. The projection of n on the plane $z = 1$ from O is the perpendicular bisector of the pair of points P'_1 and P'_2 (transformed from P_1 and P_2 in the inversion)

Suppose, then, two points P_1 and P_2 on the sphere (Fig. 3-right) and consider also the planes π_1 and π_2, tangent to E at the points P_1 and P_2. Consider, at last, the line n of intersection of π_1 and π_2.

PROP. 1. The projection of n on the plane $z = 1$ from O is the perpendicular bisector of the pair of points P'_1 and P'_2 (transformed from P_1 and P_2 in the inversion)

Let then $S' = \{P'_1, P'_2, ..., P'_n\}$ be a set of points in the plane $z = 1$. It results that:

PROP. 2. The inversive image of the Voronoi Diagram of S' is a polyhedron that approximates the sphere $E[x^2+y^2+(z-1/2)^2 = 1/4]$ in such a way that each one of its faces is tangential to the sphere. There is a symmetric correspondence between each Voronoi polygon and each face of the polyhedron.

These properties are proven in [5].

6   A 2D Procedure for Creating Meshes Made Up by Non-Triangular Faces

This last property supposes a more intuitive way for choosing the final shape of a Spatial Mesh (lattice or plate) because its topology can be proposed by means of a 2D Voronoi Diagram at the plane $z = 1$. We can handle different hypotheses easily, as we illustrate in this sample case:

Fig. 4-left: a set of points belonging to the plane $z = 1$, (viewed from the point of the infinite of the axis OZ) is defined. Next, the Voronoi diagram associated with this set is constructed. Fig. 4-right: once property 2 is applied, the polyhedron derived from this diagram is obtained. Fig. 5-left shows the relation between the edges of Voronoi and the edges of the polyhedron: the south pole of the sphere is the center of the inversive transformation.

Fig. 4. (left) The inversive image of the Voronoi diagram of S' is (right) a polyhedron that approxiamtes the sphere $E[x^2 + y^2 + (z - 1/2)^2 = 1/4]$

Even though the set of points (Fig. 4-left) does not need to follow any rule, we have proposed a distribution by means of concentric rings in order to have different horizontal ones in the sphere: each of them is made up by identical polygons (1 or 2 per ring). A realistic image of the obtained body has been displayed in Fig. 5-right.

7   A polyhedron approximating a rotation paraboloid

Suppose [5, 6] that we now apply a suitable projective transformation (1), to E^3 and let (X, Y, Z, T) be the coordinates in the transformed space to the point (x, y, z, t). Specifically, this relation is proposed:

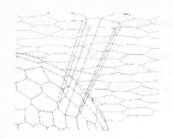

Fig. 5. (left) The relation between the edges of Voronoi and the edges of the spherical polyhedron; (right) a realistic image of the latter

$$\begin{pmatrix} X \\ Y \\ Z \\ T \end{pmatrix} = \begin{pmatrix} 1 & 0 & 0 & 0 \\ 0 & 1 & 0 & 0 \\ 0 & 0 & 0 & 1 \\ 0 & 0 & 1 & 0 \end{pmatrix} \cdot \begin{pmatrix} x \\ y \\ z \\ t \end{pmatrix} \qquad (1)$$

where t and T are, respectively, the homogeneous coordinates of the related points. We denote M_{HOM} to this matrix. Such a transformation:

(i) Is a rotation in E^4; (ii) Maps the plane $z = 0$ to the plane at infinity; (iii) Converts the sphere $E[x^2 + y^2 + (z - 1/2)^2 = 1/4]$ into the rotation paraboloid $P[X^2 + Y^2 + T^2 = Z.T]$ that, written in inhomogeneous coordinates, results: $Z = X^2 + Y^2 + 1$; (iv) Maps the plane $z = 1$ to the plane $Z = 1$; (v) Converts the sphere $E_2[x^2 + y^2 + z^2 = 1]$ into a rotation hyperboloid H_2 that, written in inhomogeneous coordinates, results: $X^2 + Y^2 - Z^2 + 1 = 0$.

Then, if we consider the set of points $S' = \{P'_1, P'_2, ..., P'_n\}$ (see point 5), on the plane $z = 1$, it follows that:

PROP 3. The projection from the point $(0, 0, -\infty)$ of the Voronoi Polygons of S' to the paraboloid $P[Z = X^2 + Y^2 + 1]$ makes up a polyhedron the faces of which are tangent to P. Each point of the set S' is transformed to the point of contact between the face of the polyhedron and the surface. The edges of the polyhedron are those transformed from the edges of the Voronoi Diagram of S'.

Indeed, the transformation keeps all the points of tangency and intersection. So, the same relation proposed in PROP. 2 between the sphere and the plane $z = 1$ remains but, now, the pole of projection has gone to the infinite $(0, 0, -\infty)$ and the sphere has became a paraboloid.

8   CR-Tangent Meshes

This last property relating the edges of the Voronoi Diagram and the edges of the Convex Hull (3D) of the set of generators of a Voronoi Diagram projected on the paraboloid $Z = X^2 + Y^2 + 1$ is a well known property in Computational Geometry, as proposed in [6] and it has been, in short, the clue needed to reach the proposals shown in point 7. However, PROP.3 is a particular case of this one:

PROP 4. The projection from the point $(0, 0, 1/2z_c - 1)$ of the Voronoi Polygons of S' to the second order surface $SF[X^2 + Y^2 - Z^2(1 - 2z_c) - 2z_cZ + 1 = 0]$ makes up an approximating polyhedron the faces of which are tangent to SF. Each point of set S' is transformed to the point of contact between the face of the polyhedron and the surface. The edges of the polyhedron are those transformed from the edges of the Voronoi Diagram of S'.

- When $z_c = 1/2$, SF is a rotated paraboloid.
- When $z_c < 1/2$, SF is a rotated hyperboloid.
- When $z_c > 1/2$, SF is a rotated ellipsoid.

Fig. 6. CR-Tangent models

For the polyhedra approximating the sphere, the paraboloid and the hyperboloid of two sheets represented in Fig. 6, all of them surfaces of revolution, we have proposed the name of CR-TANGENT meshes, CR being the abbreviation of the Spanish term "Cuádricas de Revolución". The term "tangent" is added because the approximating polyhedron is related in this way to a quadric surface (as opposed to lattice meshes in which the approximating polyhedron is inscribed in a sphere).

9   Matricial Treatment in the Generation of CR-Tangent Meshes: C-Tangent Meshes

9.1   Generation of the Revolution Paraboloid

It is widely known that the equations for an inversion whose center of inversion is the coordinate system's origin and whose inversion Ratio is 1 are the following:

$$
\begin{pmatrix} X' \\ Y' \\ Z' \\ T' \end{pmatrix} = \begin{pmatrix} \frac{1}{D^2} & 0 & 0 & 0 \\ 0 & \frac{1}{D^2} & 0 & 0 \\ 0 & 0 & \frac{1}{D^2} & 0 \\ 0 & 0 & 0 & 1 \end{pmatrix} \cdot \begin{pmatrix} X \\ Y \\ Z \\ T \end{pmatrix} \tag{2}
$$

where

$$D^2 = X^2 + Y^2 + Z^2 \tag{3}$$

We shall name this matrix M_{INV}. Given a set of points $S' = \{P'_1, P'_2, ..., P'_n\}$ lying on the $z = 1$ plane, and being $V(S)$ the Voronoi Diagram for such a set, The CR-Tangent mesh of the revolution paraboloid type arises from applying to the vertices of the Diagram the following transformation sequence:

$$P' = M_{HOM}.M_{INV}.P \tag{4}$$

Expression whose development is shown below:

$$
\begin{pmatrix} X' \\ Y' \\ Z' \\ T' \end{pmatrix} =
\begin{pmatrix} 1 & 0 & 0 & 0 \\ 0 & 1 & 0 & 0 \\ 0 & 0 & 0 & 1 \\ 0 & 0 & 1 & 0 \end{pmatrix}
\begin{pmatrix} \frac{1}{D^2} & 0 & 0 & 0 \\ 0 & \frac{1}{D^2} & 0 & 0 \\ 0 & 0 & \frac{1}{D^2} & 0 \\ 0 & 0 & 0 & 1 \end{pmatrix} \cdot
\begin{pmatrix} X \\ Y \\ Z \\ T \end{pmatrix} \tag{5}
$$

9.2 Generation of Other Quadric Surfaces of Revolution

The possibility for the generation of quadric surfaces of revolution other than the paraboloid arises from the fact that a projective transformation (1), being its matrix M_{HOM}, can be applied to the polyhedron approximating a sphere which is tangent to the $z = 1$ plane and whose center is on the Z axis but having a radius r different of $1/2$. Let's suppose that, as it is shown there, the center of the sphere tangent to plane $z = 1$ is at the point $(0, 0, Z_c)$.

We can reduce these cases to the preceding one if we produce two axis-wise transformations:

I. A translation to the south pole of the circumference $(0, 0, 2Z_c - 1)$. We shall name this translation matrix M_{TRA}.

II. A homothecy being its center the coordinate system's origin (which by now is on the sphere's south pole) and with a ratio of $1/2\,r$, that will transform the sphere of radius $r = 1 - Z_c$ into one of radius $1/2$ and return the generating points for the Voronoi Diagram to the $z = 1$ plane. We shall name this transformation matrix M_{ESC}. Once applied this two transformations, the problem is reduced to the one stated in the previous point; then the expression (5) applied to the Voronoi Diagram vertices will yield as a result the polyhedron which approximates a sphere of radius $1/2$ and then it will only be necessary to undo transformations I and II, which we attain by applying the operative matrices which we shall identify as $M_{ESC(-)}$ y $M_{TRA(-)}$. The result obtained can be summarized as:

$$P' = [M_{TRA(-)}.M_{ESC(-)}.M_{INV}.M_{ESC}.M_{TRA}].P \tag{6}$$

that can be developed as:

$$
\begin{pmatrix} X' \\ Y' \\ Z' \\ T' \end{pmatrix} =
\begin{pmatrix} 1 & 0 & 0 & 0 \\ 0 & 1 & 0 & 0 \\ 0 & 0 & 1 & S \\ 0 & 0 & 0 & 1 \end{pmatrix}
\begin{pmatrix} 2r & 0 & 0 & 0 \\ 0 & 2r & 0 & 0 \\ 0 & 0 & 2r & 0 \\ 0 & 0 & 0 & 0 \end{pmatrix} . M_{INV} .
\begin{pmatrix} \frac{1}{2r} & 0 & 0 & 0 \\ 0 & \frac{1}{2r} & 0 & 0 \\ 0 & 0 & \frac{1}{2r} & 0 \\ 0 & 0 & 0 & 1 \end{pmatrix}
\begin{pmatrix} 1 & 0 & 0 & 0 \\ 0 & 1 & 0 & 0 \\ 0 & 0 & 1 & -S \\ 0 & 0 & 0 & 1 \end{pmatrix}
\begin{pmatrix} X \\ Y \\ Z \\ T \end{pmatrix}
$$

(7)

where $S = 2Z_c - 1$; thus, now the homology transformation is applied:

$$
\begin{pmatrix} X'' \\ Y'' \\ Z'' \\ T'' \end{pmatrix} =
\begin{pmatrix} 1 & 0 & 0 & 0 \\ 0 & 1 & 0 & 0 \\ 0 & 0 & 0 & 1 \\ 0 & 0 & 1 & 0 \end{pmatrix} .
\begin{pmatrix} X' \\ Y' \\ Z' \\ T' \end{pmatrix}
$$

(8)

Points P'' are the vertices of the polyhedron approximating a quadric surface of revolution different from the paraboloid.

9.3   Attainment of C-Tangent Meshes

We shall name as C-TANGENT meshes (the name in this case referring to polyhedra which are tangent to quadric surfaces in general, including but not limited to those obtained by revolution) those obtained by applying the homology transformation to a sphere tangent to plane z=1 but placed in any position in space. This added degree of freedom means, as regarding the case studied in the previous point, simply that the center of the sphere can be placed in point $(X_c \ Y_c \ Z_c)$, always maintaining the condition that $r = 1 - z_c$. Summarising, the following transformation expressions result:

$$
\begin{pmatrix} X' \\ Y' \\ Z' \\ T' \end{pmatrix} =
\begin{pmatrix} 1 & 0 & 0 & X_c \\ 0 & 1 & 0 & Y_c \\ 0 & 0 & 1 & S \\ 0 & 0 & 0 & 1 \end{pmatrix}
\begin{pmatrix} 2r & 0 & 0 & 0 \\ 0 & 2r & 0 & 0 \\ 0 & 0 & 2r & 0 \\ 0 & 0 & 0 & 0 \end{pmatrix} . M_{INV} .
\begin{pmatrix} \frac{1}{2r} & 0 & 0 & 0 \\ 0 & \frac{1}{2r} & 0 & 0 \\ 0 & 0 & \frac{1}{2r} & 0 \\ 0 & 0 & 0 & 1 \end{pmatrix}
\begin{pmatrix} 1 & 0 & 0 & -X_c \\ 0 & 1 & 0 & -Y_c \\ 0 & 0 & 1 & -S \\ 0 & 0 & 0 & 1 \end{pmatrix}
\begin{pmatrix} X \\ Y \\ Z \\ T \end{pmatrix}
$$

(9)

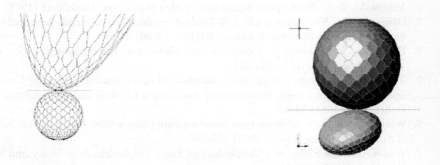

Fig. 7. C-Tangent models: (left) paraboloid; (right) ellipsoid

Points P'', obtained by (8), are the vertices for the polyhedron approximating a quadric surface. It can be easily demonstrated that the two preceding points present only special cases of this expression. Two of the surfaces obtained this way are shown in Fig. 7.

10   Summarising

Properties and methods of Computational Geometry lead to a simple 2D procedure, with a very simple computer implementation, that can be used for generating new Spatial Structure models (of the Plate type, see point 3) shaped as rotated ellipsoid, paraboloid or hyperboloid. The approximating polyhedron holds non-triangular polygons tangent to the surface it approaches. It is not difficult to notice that, following the procedure described, local changes in this resulting polyhedron are really easy to carry out. On the other hand, the properties of the Voronoi Diagram guarantee the structural stability of the body obtained.

Following a similar way as the one we have shown in this paper it is very easy to obtain the lattice type of structure transforming a Delaunay Triangulation instead of the Voronoi Diagram. We remark again that: our design effort is to select the most adequate simple set of points on the plane, whose associated tessellation (Voronoi Diagram o Delaunay triangulation) provides the basis for the generation of a Spatial Dome, making it possible to select a lattice type or a plate one. Wherever these points are placed, we find a polyhedron as the solution; as designers, we need only to study the position of the generators (or locally move some of them) in order to obtain a "good design".

References

1. Coxeter, H.S.M.: Regular Complex Polytopes Cambridge University Press, London (1974) 9-11
2. Critchlow, K.: Order in Space, Thames and Hudson, Norwich (1969) 76
3. Makowski, Z. S.: Analysis, Design and Construction of Braced Domes. Cambridge University Press, Cambridge (1984)
4. Makowski, Z. S.: Steel Space Structures. Verlag Stahleisen, Dusseldorf (1968)
5. Otero, C. Gil, V. , Alvaro, J.I.: CR-Tangent meshes. International Association for Shell and Spatial Structures Journal **21**(132) (2000) 41-48
6. Preparata, F., Shamos, I.: Computational Geometry: An Introduction. Springer-Verlag, New York (1985) 244-247
7. Tsuboi, Y.: Analysis, design and realization of space frames. (Working Group of Spatial Steel Structures), International Association for Shell and Spatial Structures Bulletin **84** (1984) 11-30
8. Wester, T.: A geodesic dome-type based on pure plate action. International Journal of Space Structures **5**(3-4) (1990) 155-167
9. Wester, T.: The Structural Morphology of Basic Polyhedra. John Wiley and Sons, New York (1997) 301-342

A Combinatorial Scheme for
Developing Efficient Composite Solvers*

Sanjukta Bhowmick, Padma Raghavan, and Keita Teranishi

Department of Computer Science and Engineering
The Pennsylvania State University
220 Pond Lab, University Park, PA 16802-6106
{bhowmick,raghavan,teranish}@cse.psu.edu

Abstract. Many fundamental problems in scientific computing have
more than one solution method. It is not uncommon for alternative solution methods to represent different tradeoffs between solution cost and
reliability. Furthermore, the performance of a solution method often depends on the numerical properties of the problem instance and thus
can vary dramatically across application domains. In such situations,
it is natural to consider the construction of a multi-method composite
solver to potentially improve both the average performance and reliability. In this paper, we provide a combinatorial framework for developing
such composite solvers. We provide analytical results for obtaining an
optimal composite from a set of methods with normalized measures of
performance and reliability. Our empirical results demonstrate the effectiveness of such optimal composites for solving large, sparse linear
systems of equations.

1 Introduction

It is not uncommon for fundamental problems in scientific computing to have
several competing solution methods. Consider linear system solution and eigenvalue computations for sparse matrices. In both cases several algorithms are
available and the performance of a specific algorithm often depends on the numerical properties of the problem instance. The choice of a particular algorithm
could depend on two factors: (i) the cost of the algorithm and, (ii) the probability
that it computes a solution without failure. Thus, we can view each algorithm as
reflecting a certain tradeoff between a suitable metric of cost (or performance)
and reliability. It is often neither possible nor practical to predict a priori which
algorithm will perform best for a given suite of problems. Furthermore, each
algorithm may fail on some problems. Consequently it is natural to ask the following question: Is it possible to develop a robust and efficient composite of

---

* This work has been funded in part by the National Science Foundation through
  grants NSF CCR-981334, NSF ACI-0196125, and NSF ACI-0102537.

P.M.A. Sloot et al. (Eds.): ICCS 2002, LNCS 2330, pp. 325–334, 2002.

multiple algorithms? We attempt to formalize and answer this question in this paper.

An illustrative example is the problem of solving sparse linear systems. We have a variety of algorithms for this problem, encompassing both direct and iterative methods [3]. Direct methods are highly reliable, but the memory required grows as a nonlinear function of the matrix size. Iterative methods do not require any additional memory but they are not robust; convergence can be slow or fail altogether. Convergence can be accelerated with preconditioning, but that leads to a larger set of preconditioning methods in addition to the basic iterative algorithms. In such cases there is often no single algorithm that is consistently superior even for linear systems from a specific application domain. This situation leads us to believe that rather than relying on a single algorithm, we should try to develop a multi-algorithmic composite solver. The idea of multi-algorithms has been explored earlier in conjunction with a multiprocessor implementation [1]; the multi-algorithm comprises several algorithms that are simultaneously applied to the problem by exploiting parallelism. We provide a new combinatorial formulation that can be used on uniprocessors and potentially generalized to multiprocessor implementations.

In our model, the composite solver comprises a sequence of different algorithms, thus endowing the composite with the higher cumulative reliability over all member algorithms. Algorithms in the sequence are executed on a given problem until it is solved successfully; partial results from an unsuccessful algorithm are not reused. We provide a combinatorial formulation of the problem in Section 2. Section 3 contains our main contribution in the form of analytical results for obtaining the optimal composite. We provide empirical results on the performance of composite solvers for large sparse linear systems in Section 4 and concluding remarks in Section 5.

2   A Combinatorial Model

We now formalize our problem using a combinatorial framework. The composite solver comprises several algorithms and each algorithm can be evaluated on the basis of two metrics: (i) performance or cost, and (ii) reliability. The former can be represented by a normalized value of the performance using either execution time or the number of operations. The reliability is a number in the range $[0, 1]$ reflecting the probability of successfully solving the problem. For example, if an iterative linear solver fails to converge on average on one fourth of the problems, its failure rate is 0.25 and its reliability is 0.75. In some situations, it may be possible to derive analytic expressions for both metrics. In other situations, these metrics can be computed by empirical means, i.e., by observing the performance of each algorithm on a representative set of sample problems.

Consider generating a composite solver using n distinct underlying methods (or algorithms) $M_1, M_2, \ldots, M_n$. Each method M_i, is associated with its normalized execution time t_i (performance metric) and reliability r_i; r_i is the success rate of the method and its failure rate is given by $f_i = 1 - r_i$. We define the *utility*

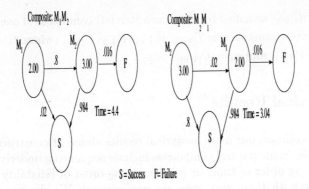

Fig. 1. Composites of two methods $M_1, t_1 = 2.0, r_1 = .02$ and $M_2, t_2 = 3.0, r_2 = .80$; both composites have reliability .984 but the composite M_2M_1 has lower execution time.

ratio of method M_i as $u_i = t_i/ri$. Let **P** represent the set of all permutations (of length n) of $\{1, 2, \ldots, n\}$. For a specific $\hat{P} \in \mathbf{P}$, we denote the associated composite by $\hat{C}$. $\hat{C}$ comprises all the n underlying methods $M_1, M_2, \ldots, M_n$ in the sequence specified by $\hat{P}$. If $\hat{P}_k$ denotes the $k-th$ element of $\hat{P}$, the composite $\hat{C}$ consists of methods $M_{\hat{P}_1}, M_{\hat{P}_2}, \cdots, M_{\hat{P}_n}$. Now for any $\hat{P} \in \mathbf{P}$, the total reliability (success percentage) of the composite $\hat{C}$ is independent of the permutation and invariant at $1 - \Pi_{i=1}^{i=n}(1 - r_i)$, a value higher than that of any component algorithm. Next, observe that $\hat{T}$, the worst case execution time of $\hat{C}$, is:

$$\hat{T} = t_{\hat{P}_1} + f_{\hat{P}_1} t_{\hat{P}_2} + \cdots + f_{\hat{P}_1} f_{\hat{P}_2} \cdots f_{\hat{P}_{n-1}} t_{\hat{P}_n}.$$

Thus, the execution times of different composites can indeed vary depending on the actual permutation. A two-method example in Figure 1 shows how the permutation used for the composite affects the execution time. Our goal is to determine the optimal composite, i.e., the composite with minimal worst-case execution time.

We now introduce some additional notation required for the presentation of our analytical results. Consider the subsequence $\hat{P}_k, \hat{P}_{k+1}, \cdots, \hat{P}_l$ ($\hat{P} \in \mathbf{P}$) denoted by $\hat{P}_{(k:l)}$. Now $\hat{P}_{(k:l)}$ can be associated with a composite comprising some $l - k$ methods using the notation $\hat{C}_{(k:l)}$. The total reliability of $\hat{C}_{(k:l)}$ is denoted by $\hat{R}_{(k:l)} = 1 - \prod_{i=k}^{l} f_{\hat{P}_i}$. Similarly the percentage of failure, $\hat{F}_{(k:l)} = \prod_{i=k}^{l} f_{\hat{P}_i}$. Observe that both these quantities depend only on the underlying set of methods specified by $\hat{P}_k, \hat{P}_{k+1}, \cdots, \hat{P}_l$ and are invariant under all permutations of these methods. We next define $\hat{T}_{(k:l)}$ as the worst-case time of $\hat{C}_{(k:l)}$; we can see that $\hat{T}_{(k:l)} = \sum_{i=k}^{l} [t_{\hat{P}_i} \prod_{m=k}^{l-1} f_{\hat{P}_m}]$. A final term we introduce is the total utility ratio of $\hat{C}_{(k:l)}$ denoted by $\hat{U}_{(k:l)} = \hat{T}_{(k:l)}/\hat{R}_{(k:l)}$.

For ease of notation, we will drop explicit reference to $\hat{P}_i$ in expressions for $\hat{R}, \hat{F}, \hat{T},$ and $\hat{U}$ for a specific $\hat{C}$ and $\hat{P}$. Now the expression for $\hat{T}_{(k:l)}$ simplifies to $\sum_{i=k}^{l} [t_i \prod_{m=k}^{l-1} f_m]$. Additionally, in an attempt to make the notation consistent, we will treat $\hat{C}_{(k:k)}$ specified by $\hat{P}_{(k:k)}$ as a (trivial) composite of one method and use related expressions such as $\hat{T}_{(k:k)}$, $\hat{R}_{(k:k)}$, $\hat{F}_{(k:k)}$ $\hat{U}_{(k:k)}$ (where $t_k = \hat{T}_{(k:k)}$,$r_k = \hat{R}_{(k:k)}$, $f_k = \hat{F}_{(k:k)}$, and $u_k = \hat{U}_{(k:k)}$.

3   Analytical Results

This section contains our main analytical results aimed at constructing an optimal composite. Some natural composites include sequencing underlying methods in (i) increasing order of time, or (ii) decreasing order of reliability. Our results indicate that both these strategies are non-optimal. We show in Theorems 1 and 2 that a composite is optimal if and only if its underlying methods are in increasing order of the utility ratio.

We begin by observing that for any $\hat{P} \in \mathbf{P}$ the composite $\hat{C}$, can be viewed as being formed by the sequential execution of two composites, $\hat{C}_{(1:r)}$ and $\hat{C}_{(r+1:n)}$. We can also easily verify that $\hat{T}_{(1:n)} = \hat{T}_{(1:r)} + \hat{F}_{(1:r)}\hat{T}_{(r+1:n)}$. We use this observation to show that for any composite, the overall utility ratio is bounded above by the largest utility ratio over all underlying methods.

Lemma 1. *For any $\hat{P} \in \mathbf{P}$, the utility ratio of the composite $\hat{C}$ satisfies $\hat{U} \leq \max\{\hat{U}_{(i:i)} : 1 \leq i \leq n\}$.*

Proof. We can verify (with some algebraic manipulation) that the statement is true for the base case with two methods ($n = 2$). For the inductive hypothesis, assume that the statement is true for any composite of $n - 1$ methods, that is, $\hat{U}_{(1:n-1)} \leq \max\{\hat{U}_{(i:i)} : 1 \leq i \leq n - 1\}$. Now consider $\hat{C}$, a composite of n methods with $\hat{P}$ as the associated sequence. By our earlier observation, we can view it as a composite of *two methods* with execution times $\hat{T}_{(1:n-1)}$ and $\hat{T}_{(n:n)}$, reliabilities $\hat{R}_{(1:n-1)}$ and $\hat{R}_{(n:n)}$, and utility ratios $\hat{U}_{(1:n-1)}$ and $\hat{U}_{(n:n)}$. If $\hat{U}_{(1:n-1)} \leq \hat{U}_{(n:n)}$, then by the base case, $\hat{U}_{(1:n)} \leq \hat{U}_{(n:n)}$ and by the induction hypothesis, $\hat{U}_{(1:n)} \leq \max\{\hat{U}_{(i:i)} : 1 \leq i \leq n\}$. It is also easy to verify that the statement is true if $\hat{U}_{(n:n)} \leq \hat{U}_{(1:n-1)}$.   □

Theorem 1. *Let $\tilde{C}$ be the composite given by the sequence $\tilde{P} \in \mathbf{P}$. If $\tilde{U}_{(1:1)} \leq \tilde{U}_{(2:2)} \leq \ldots \leq \tilde{U}_{(n:n)}$, then $\tilde{C}$ is the optimal composite, i.e., $\tilde{T} = \min\{\hat{T} : \hat{P} \in \mathbf{P}\}$.*

Proof. It is easy to verify that the statement is indeed true for the base case for composites of two methods ($n = 2$). We next assume that the statement is true for composites of $n - 1$ methods. Now we extend the optimal composite of $n - 1$ methods to include the last method; let this sequence be given by $\tilde{P}$ and the composite by $\tilde{C}$. For the sake of contradiction, let there be a permutation $\acute{P} \in \mathbf{P}$, such that $\acute{T} \leq \tilde{T}$ and the utility ratios $\{\acute{U}_{(i:i)} : 1 \leq i \leq n\}$ are not in increasing order of magnitude.
Let the k-th method in $\acute{C}$ be the n-th method in $\tilde{C}$. Therefore $\acute{T}_{(k:k)} = \tilde{T}_{(n:n)}$

and $\acute{F}_{(k:k)} = \tilde{F}_{(n:n)}$. Using the earlier observations:

$$\tilde{T} = \tilde{T}_{(1:k)} + \tilde{F}_{(1:k)}\tilde{T}_{(k+1:n-1)} + \tilde{F}_{(1:k)}\tilde{F}_{(k+1:n-1)}\tilde{T}_{(n:n)} \tag{1}$$

$$\acute{T} = \acute{T}_{(1:k-1)} + \acute{F}_{(1:k-1)}\acute{T}_{(k:k)} + \acute{F}_{(1:k-1)}\acute{F}_{(k:k)}\acute{T}_{(k+1:n)}$$

$$= \acute{T}_{(1:k-1)} + \acute{F}_{(1:k-1)}\tilde{T}_{(n:n)} + \acute{F}_{(1:k-1)}\tilde{F}_{(n:n)}\acute{T}_{(k+1:n)} \tag{2}$$

We know that $\tilde{T}_{(1:n-1)}$ is the optimal time over all composites of $n-1$ methods and thus lower than the time for composite obtained by excluding the k-th method in $\acute{C}$ and the n-th method in $\tilde{C}$. Thus $\acute{T}_{(1:k-1)} + \acute{F}_{(1:k-1)}\acute{T}_{(k+1:n)} \geq \tilde{T}_{(1:k)} + \tilde{F}_{(1:k)}\tilde{T}_{(k+1:n-1)}$, to yield:

$$\acute{T}_{(1:k-1)} + \acute{F}_{(1:k-1)}\acute{T}_{(k+1:n)} - \tilde{T}_{(1:k)} - \tilde{F}_{(1:k)}\tilde{T}_{(k+1:n-1)} \geq 0 \tag{3}$$

According to our assumption $\acute{T} \leq \tilde{T}$; we expand this relation using Equations 1 and 2 to show that $\acute{T}_{(1:k-1)} + \acute{F}_{(1:k-1)}\tilde{T}_{(n:n)} + \acute{F}_{(1:k-1)}(1 - \tilde{R}_{(n:n)})\acute{T}_{(k+1:n)}$ is less than or equal to $\tilde{T}_{(1:k)} + \tilde{F}_{(1:k)}\tilde{T}_{(k+1:n-1)} + \tilde{F}_{(1:k)}\tilde{F}_{(k+1:n-1)}\tilde{T}_{(n:n)}$. We can then rearrange the terms on either side to show that the left-hand side of Equation 3 is less than or equal to $\tilde{F}_{(1:k)}\tilde{F}_{(k+1:n-1)}\tilde{T}_{(n:n)} - \acute{F}_{(1:k-1)}\tilde{T}_{(n:n)} + \acute{F}_{(1:k-1)}\tilde{R}_{(n:n)}\acute{T}_{(k+1:n)}$. Thus,

$$0 < \tilde{F}_{(1:k)}\tilde{F}_{(k+1:n-1)}\tilde{T}_{(n:n)} - \acute{F}_{(1:k-1)}\tilde{T}_{(n:n)} + \acute{F}_{(1:k-1)}\tilde{R}_{(n:n)}\acute{T}_{(k+1:n)}.$$

By rearranging terms and using the equation $\tilde{F}_{(1:k)}\tilde{F}_{(k+1:n-1)} = \acute{F}_{(1:k-1)}\acute{F}_{(k+1:n)}$ to simplify, we obtain:

$$\acute{F}_{(1:k-1)}\tilde{T}_{(n:n)} - \tilde{F}_{(1:k)}\tilde{F}_{(k+1:n-1)}\tilde{T}_{(n:n)} \leq \acute{F}_{(1:k-1)}\tilde{R}_{(n:n)}\acute{T}_{(k+1:n)}.$$

$$\acute{F}_{(1:k-1)}\tilde{T}_{(n:n)} - \acute{F}_{(1:k-1)}\acute{F}_{(k+1:n)}\tilde{T}_{(n:n)} \leq \acute{F}_{(1:k-1)}\tilde{R}_{(n:n)}\acute{T}_{(k+1:n)}.$$

Cancelling the common terms on either side yields $\tilde{T}_{(n:n)}(1 - \acute{F}_{(k+1:n)}) \leq \tilde{R}_{(n:n)}\acute{T}_{(k+1:n)}$. Observe that this is equivalent to $\tilde{U}_{(n:n)} \leq \acute{U}_{(k+1:n)}$. By the definition of $\tilde{C}$, $\tilde{U}_{(n:n)}$ is the largest utility ratio among all the n methods. But if $\tilde{U}_{(n:n)} \leq \acute{U}_{(k+1:n)}$, there is a composite whose overall utility is higher than the maximum utility ratio of its component methods, thus contradicting Lemma 1. This contradiction occurred because our assumption that $\acute{T} \leq \tilde{T}$ is not true; hence the proof. □

We next show that if a composite is optimal, then its component methods are in increasing order of the utility ratio. The proof uses shortest paths in an appropriately weighted graph.

Theorem 2. *If $\tilde{C}_{(1:n)}$ is the optimal composite then the utility ratios are arranged in increasing order, i.e., $\tilde{U}_{(1:1)} \leq \tilde{U}_{(2:2)} \leq \ldots \leq \tilde{U}_{(n-1:n-1)} \leq \tilde{U}_{(n:n)}$.*

Proof. Consider a graph constructed with unit vertex weights and positive edge weights as follows. The vertices are arranged in levels with edges connecting vertices from one level to the next. There are a total of $n+1$ levels numbered 0 through n. Each vertex at level l ($0 \le l \le n$) denotes a subset of l methods out of n methods. Assume that the vertex is labeled by the set it represents. Directed edges connect a vertex V_S at level l to a vertex $V_{\bar{S}}$ only if $|\bar{S} \setminus S| = 1$ and $\bar{S} \cap S = S$, i.e., the set $\bar{S}$ has exactly one more element than S. Let F_S denote the total failure rate over all methods in the set S. If $\bar{S} \setminus S = \{i\}$, the edge $V_S \to V_{\bar{S}}$ is weighted by $F_S T_{(i:i)}$, the time to execute method i after failing at all previous methods. It is easy to verify that any path from V_0 (representing the empty set) to $V_{\{1,2,\cdots n\}}$ represents a particular composite, one in which methods are selected in the order in which they were added to sets at subsequent levels. Now the shortest path represents the optimal composite.

Assume we have constructed the shortest path in the graph. Consider a fragment of the graph, as shown in Figure 2. We assume that V_S is a node on the shortest path, and $V_{\hat{S}}$ is also a node on the shortest path, such that $\hat{S} - S = \{i, j\}$. There will be only 2 paths from V_S to $V_{\hat{S}}$, one including the node $V_{\bar{S}}$ ($\bar{S} - S = \{i\}$) and the other including the node V_{S^*} ($S^* - S = \{j\}$). Without loss of generality, assume V_{S^*} is the node on the shortest path; thus method j was selected before method i in the sequence. Let the time from V_0 to V_S be denoted by T_S and the failure rate by F_S. Using the optimality property of the shortest path:

$$T_S + F_S T_{(j:j)} + F_S F_{(j:j)} T_{(i:i)} \le T_S + F_S T_{(i:i)} + F_S F_{(i:i)} T_{(j:j)}$$

After canceling common terms we get $T_{(j:j)} + F_{(j:j)} T_{(i:i)} \le T_{(i:i)} + F_{(i:i)} T_{(j:j)}$. This can be simplified further using the relation $F_{(j:j)} = 1 - R_{(j:j)}$ to yield: $R_{(j:j)} T_{(i:i)} \ge R_{(i:i)} T_{(j:j)}$ and thus $U_{(j:j)} \le U_{(i:i)}$. This relationship between utility ratios holds for any two consecutive vertices on the shortest path. Hence, the optimal composite given by the shortest path is one in which methods are selected in increasing order of the utility ratio.   □

4   Empirical Results

Our experiments concern composite solvers for large sparse linear systems of equations. We use a suite of nine preconditioned Conjugate Gradient methods labeled $M_1, \ldots, M_9$. M_1 denotes applying Conjugate Gradients without any preconditioner. M_2 and M_3 use Jacobi and SOR preconditioning schemes respectively. Methods M_4 through M_7 use incomplete Cholesky preconditioners with 0,1,2 and 3 levels of fill. Methods M_8 and M_9 use incomplete Cholesky preconditioners with numerical drop threshold factors of .0001 and .01.

For our first experiment we used a set of six bcsstk sparse matrices from finite element methods in structural mechanics. We normalized the running time of each method by dividing it by the time required for a sparse direct solver. The geometric mean of the normalized running time was used as our estimate of t_i for each M_i. We assumed that the method was unsuccessful if it failed to converge in 200 iterations. We used the success rate as the reliability metric r_i for method

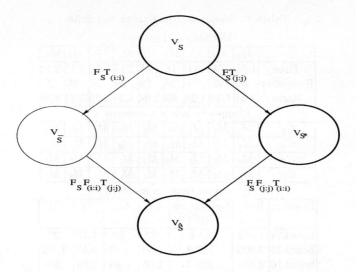

Fig. 2. Segment of the graph used in the proof of Theorem 2.

M_i. These two measures were used to compute the utility ratio $u_i = t_i/r_i$ for each method M_i. We created four different composite solvers C_T, C_R, C_X, C_O. In C_T underlying methods are arranged in increasing order of execution time. In C_R the underlying methods are in decreasing order of reliability. The composite C_X is based on a randomly generated sequence of the underlying methods. The composite C_O is based on the analytical results of the last section; underlying methods are in increasing order of the utility ratio. The overall reliability of each composite is .9989, a value significantly higher than the average reliability of the underlying methods. We applied these four composite solvers to the complete set of matrices and calculated the total time for each composite over all the test problems. The results are shown in Table 1; our optimal composite C_O has the least total time.

In our second experiment, we considered a larger suite of test problems consisting of matrices from five different applications. To obtain values of the performance metrics we used a sample set of 10 matrices consisting of two matrices from each application type. We constructed four composites solvers C_T, C_R, C_X, C_O as in our first experiment. Results in Table 2 indicate that our composite solver still has the least total execution time over all problems in the test suite. The total execution time of C_O is less than half the execution time for C_T, the composite obtained by selecting underlying methods in increasing order of time.

These preliminary results are indeed encouraging. However, we would like to observe that to obtain a statistically meaningful result it is important to use much larger sets of matrices. Another issue concerns normalization; we normalized by the time for a sparse direct solver but other measures such as the mean or median of observed times could also be used. These statistical aspects merit further study.

Table 1. Results for the `bcsstk` test suite.

Methods and metrics

	M_1	M_2	M_3	M_4	M_5	M_6	M_7	M_8	M_9
Time	1.01	.74	.94	.16	1.47	2.15	3.59	5.11	2.14
Reliability	.25	.50	.75	.25	. 50	. 50	.75	1.00	.25
Ratio	4.04	1.48	1.25	.63	2.94	4.30	4.79	5.11	8.56

Composite solver sequences

C_T	M_4	M_2	M_3	M_1	M_5	M_9	M_6	M_7	M_8
C_R	M_8	M_3	M_7	M_2	M_5	M_6	M_1	M_4	M_9
C_X	M_9	M_8	M_1	M_5	M_3	M_2	M_7	M_6	M_4
C_O	M_4	M_3	M_2	M_5	M_1	M_6	M_7	M_8	M_9

Execution time (in seconds)

Problem	Rank	Non-zeroes (10^3)	C_T	C_R	C_X	C_O
bcsstk14	1,806	63.4	**.25**	.98	1.19	.27
bcsstk15	3,908	117.8	1.88	5.38	9.45	**1.22**
bcsstk16	4,884	290.3	1.05	6.60	2.09	**.98**
bcsstk17	10,974	428.6	57.40	**12.84**	16.66	37.40
bcsstk18	11,948	149.1	4.81	5.70	12.40	**2.80**
bcsstk25	15,439	252.2	1.60	21.93	36.85	1.59
Total execution time			66.99	53.43	78.64	**44.26**

5   Conclusion

We formulated a combinatorial framework for developing multi-method composite solvers for basic problems in scientific computing. We show that an optimal composite solver can be obtained by ordering underlying methods in increasing order of the utility ratio (the ratio of the execution time and the success rate). This framework is especially relevant with the emerging trend towards using component software to generate multi-method solutions for computational science and engineering applications [2]. Our results can be extended to develop interesting variants; for example, an optimal composite with reliability greater than a user specified value, using only a small subset of a larger set of algorithms. Such "subset composites" can effectively reflect application specific tradeoffs between performance and robustness. Another potential extension could include a model where partial results from an unsuccessful method can be reused in a later method in the sequence.

Table 2. Results for the test suite with matrices from five applications.

Methods and metrics									
	M_1	M_2	M_3	M_4	M_5	M_6	M_7	M_8	M_9
Time	.77	.73	.81	.20	1.07	1.48	2.10	.98	.76
Reliability	.50	.60	.90	.50	.70	.60	.60	1.00	.40
Ratio	1.54	1.23	.90	.40	1.53	2.47	3.50	.98	1.91

Composite solver sequences									
C_T	M_4	M_2	M_9	M_1	M_3	M_8	M_5	M_6	M_7
C_R	M_8	M_3	M_5	M_2	M_6	M_7	M_1	M_4	M_9
C_X	M_9	M_8	M_7	M_6	M_5	M_3	M_2	M_1	M_4
C_O	M_4	M_3	M_8	M_2	M_5	M_1	M_9	M_6	M_7

Execution time (in seconds)						
Problem	Rank	Non-zeroes (10^3)	C_T	C_R	C_X	C_O
bcsstk14	1,806	63.4	**.31**	1.06	1.18	.37
bcsstk16	4,884	290.3	**.97**	6.35	2.07	.99
bcsstk17	10,974	428.6	35.7	**13.3**	16.3	23.4
bcsstk25	15,439	252.2	1.61	22.8	36.8	**1.60**
bcsstk38	8032	355.5	35.8	33.5	51.5	**2.39**
crystk01	4875	315.9	**.44**	4.03	.84	.47
crystk03	246,96	1751.1	**2.55**	35.8	5.45	2.56
crystm02	139,65	322.90	.32	.40	5.38	**.32**
crystm03	246,96	583.77	.60	.72	.73	**.60**
msc00726	726	34.52	.13	1.39	.23	**.13**
msc01050	1050	29.15	.80	**.10**	.23	.27
msc01440	1440	46.27	2.91	.79	2.39	**.5**
msc04515	4515	97.70	10.5	**1.95**	6.10	4.45
msc10848	10848	1229.77	75.6	101	163	**26.3**
nasa1824	1824	39.21	2.46	**1.15**	1.3	1.80
nasa2146	2146	72.25	**.09**	.64	2.29	.09
nasa2910	2910	174.29	10.9	**2.34**	6.69	2.80
nasa4704	4704	104.756	13.4	13.4	13.40	**4.61**
xerox2c1	6000	148.05	.27	1.92	**.23**	18.1
xerox2c2	6000	148.30	**.24**	.41	.48	.25
xerox2c3	6000	147.98	.27	.41	**.21**	.24
xerox2c4	6000	148.10	.23	.40	**.22**	.23
xerox2c5	6000	148.62	.25	.42	.24	**.23**
xerox2c6	6000	148.75	.29	.62	.90	**.23**
Total execution time			196.64	244.9	318.16	**90.6**

References

1. Barrett, R., Berry, M., Dongarra, J., Eijkhout, V., Romine, C.: Algorithmic Bombardment for the Iterative Solution of Linear Systems: A PolyIterative Approach. Journal of Computational and applied Mathematics, **74**, (1996) 91-110
2. Bramley, R., Gannon, D., Stuckey, T., Villacis, J., Balasubramanian, J., Akman, E., Berg, F., Diwan, S., Govindaraju, M.:Component Architectures for Distributed Scientific Problem Solving. To appear in a special issue of IEEE Computational Science and Eng., 2001
3. Golub, G.H., Van Loan, C.F.: Matrix Computations (3rd Edition). The John Hopkins University Press, Baltimore Maryland (1996)

Parallel and Fully Recursive
Multifrontal Supernodal Sparse Cholesky*

Dror Irony, Gil Shklarski, and Sivan Toledo

School of Computer Science, Tel-Aviv Univsity
http://www.tau.ac.il/~stoledo

Abstract We describe the design, implementation, and performance of a new parallel sparse Cholesky factorization code. The code uses a supernodal multifrontal factorization strategy. Operations on small dense submatrices are performed using new dense-matrix subroutines that are part of the code, although the code can also use the BLAS and LAPACK. The new code is recursive at both the sparse and the dense levels, it uses a novel recursive data layout for dense submatrices, and it is parallelized using Cilk, an extension of C specifically designed to parallelize recursive codes. We demonstrate that the new code performs well and scales well on SMP's.

1 Introduction

This paper describes the design and implementation of a new parallel direct sparse linear solver. The solver is based on a multifrontal supernodal sparse Cholesky factorization (see, e.g., [20]). The multifrontal supernodal method factors the matrix using recursion on a combinatorial structure called the elimination tree (etree). Each vertex in the tree is associated with a set of columns of the Cholesky factor L (unknowns in the linear system). The method works by factoring the columns associated with all the columns associated with proper decsendants of a vertex v, then updating the coefficients of the unknowns associated with v, and factoring the columns of v. The updates and the factorization of the columns of v are performed using calls to the *dense* level-3 BLAS [7, 6]. The ability to exploit the dense BLAS and the low symbolic overhead allow the method to effectively utilize modern computer architectures with caches and multiple processors. Our solver includes a newly designed and implemented subset of the BLAS/LAPACK, although it can use existing implementations, such as ATLAS [26] and BLAS produced by computer vendors [1, 2, 5, 15–17, 23].

While the multifrontal supernodal method itself is certainly not new, the design of our solver is novel. The novelty stems from aggressive use of recursion in all levels of the algorithm, which allows the solver to effectively utilize complex advanced memory systems and multiple processors. We use recursion in three ways, one conventional and two new:

---

* This research was supported in part by an IBM Faculty Partnership Award, by grants 572/00 and 9060/99 from the Israel Science Foundation (founded by the Israel Academy of Sciences and Humanities), and by a VATAT graduate fellowship.

P.M.A. Sloot et al. (Eds.): ICCS 2002, LNCS 2330, pp. 335–344, 2002.

- The solver uses a recursive formulation for both the multifrontal sparse factorization and for the new implementation of the BLAS. This approach is standard in multifrontal sparse factorization, and is now fairly common in new implementations of the dense linear algebra codes [3,9,10,13,12,14,25,26]. A similar approach was recently proposed by Dongarra and Raghavan for a non-multifrontal sparse Cholesky method [8]. This use of recursive formulations enables us to exploit recursion in two new ways.
- The solver exploits parallelism by declaring, in the code, that certain function calls can run concurrently with the caller. That is, the parallel implementation is based entirely on recursive calls that can be performed in parallel, and not on loop partitioning, explicit multithreading, or message passing. The parallel implementation uses Cilk [24, 11], a programming environment that supports a fairly minimal parallel extension of the C programming language and a specialized run-time system. One of the most important aspects of using Cilk is the fact that it performs dynamic scheduling that leads to both load balancing and locality of reference.
- The solver lays out dense submatrices recursively. More specifically, matrices are laid out in blocks, and the blocks are laid out in memory using recursive partitioning of the matrices. This data layout, originally proposed by Gustavson et al. [12] ensures automatic effective utilization of all the levels of the memory hierarchy and can prevent false sharing and other memory-system problems. The use of a novel indirection matrix enables low-overhead indexing and sophisticated memory management for block-packed formats.

The rest of the paper is organized as follows. Section 2 describe the design of the new dense subroutines. Section 3 describes the design of the parallel sparse Cholesky factorization code. Section 4 describes the performance of the new solver, and Section 5 presents our conclusions.

2   Parallel Recursive Dense Subroutines

Our solver uses a novel set of BLAS (basic linear algebra subroutines; routines that perform basic operations on dense blocks, such as matrix multiplication; we informally include in this term dense Cholesky factorizations). The novelty lies in the fusion of three powerful ideas: recursive data structures, automatic kernel generation, and parallel recursive algorithms.

2.1   Indirect Block Layouts

Our code stores matrices by block, not by column. Every block is stored contiguously in memory, either by row or by column. The ordering of blocks in memory is based on a recursive partitioning of the matrix, as proposed in [12]. The algorithms use a recursive schedule, so the schedule and the data layout match each other. The recursive layout allows us to automatically exploit level 2 and 3 caches and the TLB. The recursive data layout also prevents situations in which a single cache line contains data from two blocks, situations that lead to false sharing of cache lines on cache-coherent multiprocessors.

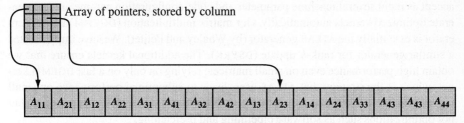

Figure 1. The use of indirection in the layout of matrices by block. The matrix is represented by an array of pointers to blocks (actually by an array of structures that contain pointers to blocks).

Our data format uses a level of indirection that allows us to efficiently access elements of a matrix by index, to exploit multilevel memory hierarchies, and to transparently pack triangular and symmetric matrices. While direct access by index is not used often in most dense linear algebra algorithms, it is used extensively in the extend-add operation in multifrontal factorizations.

In our matrix representation, shown in Figure 1, a matrix is represented by a two-dimensional array of structures that represent submatrices. The submatrices are of uniform size, except for the submatrices in the last row and column, which may be smaller. This array is stored in a column major order in memory. A structure that represents a submatrix contains a pointer to a block of memory that stores the elements of the submatrix, as well as several meta-data members that describe the size and layout of the submatrix. The elements of a submatrix are stored in either column-major order or row-major order. The elements of all the submatrices are normally stored submatrix-by-submatrix in a large array that is allocated in one memory-allocation call, but the order of submatrices within that array is arbitrary. It is precisely this freedom to arbitrarily order submatrices that allows us to effectively exploit multilevel caches and non-uniform-access-time memory systems.

2.2 Efficient Kernels, Automatically-Generated and Otherwise

Operations on individual blocks are performed by optimized kernels that are usually produced by automatic kernel generators. In essence, this approach bridges the gap between the level of performance that can be achieved by the translation of a naive kernel implementation by an optimizing compiler, and the level of performance that can be achieved by careful hand coding. The utility of this approach has been demonstrated by ATLAS, as well as by earlier projects, such as PHiPAC [4]. We have found that on some machines with advanced compilers, such as SGI Origin's, we can obtain better performance by writing naive kernels and letting the native optimizing compiler produce the kernel. On SGI Origin's, a compiler feature called the loop-nest optimizer delivers better performance at smaller code size than our automatically-generated kernels.

We currently have kernel generators for two BLAS routines: DGEMM and DSYRK, and we plan to produce two more, for DTRSM, and DPOTRF. The kernel generators accept as input several machine parameter and code-configuration parameters and generate optimized kernels automatically. Our matrix multiplication (DGEMM) kernel generator is essentially the ATLAS generator (by Whaley and Petitet). We have implemented a similar generator for rank-k update (DSYRK). The additional kernels ensure that we obtain high performance even on small matrices; relying on only on a fast DGEMM kernel, which is the strategy that ATLAS uses, leads to suboptimal performance on small inputs. Our DSYRK kernel is simpler than ATLAS's DGEMM kernel: it uses unrolling but not optimizations such as software pipelining and prefetching.

The flexibility of our data structures allows us one optimization that is not possible in ATLAS and other existing kernels. Our data structure can store a submatrix either by row or by column; a bit in the submatrix structure signals whether the layout is by row or by column. Each kernel handles one layout, but if an input submatrix is laid out incorrectly, the kernel simply calls a conversion subroutine that transposes the block and flips the layout bit. In the context of the BLAS and LAPACK calls made by the sparse factorization code, it is never necessary to transpose a block more than once. In other BLAS implementations that are not allowed to change the layout of the input, a single block may be transposed many times, in order to utilize the most efficient loop ordering and stride in each kernel invocation (usually in order to perform the innermost loop as a stride-1 inner product).

2.3 Parallel Recursive Dense Subroutines

The fact that the code is recursive allows us to easily parallelize it using Cilk. The syntax of Cilk, illustrated in Figure 2 (and explained fully in [24]) allows the programmer to specify that a function call may execute the caller concurrently with the callee. A special command specifies that a function may block until all its subcomputations terminate. Parallelizing the recursive BLAS in Cilk essentially meant that we added the spawn keyword to function calls that can proceed in parallel and the sync keyword to wait for termination of subcomputations. The full paper will contain a full example. We stress that we use recursion not just in order to expose parallelism, but because recursion improves locality of reference in the sequential case as well.

3   Multifrontal Supernodal Sparse Cholesky Factorization

Our multifrontal supernodal sparse Cholesky implementation is fairly conventional except for the use of Cilk. The code is explicitly recursive, which allowed us to easily parallelize it using Cilk. In essence, the code factors the matrix using a postorder traversal of the elimination tree. At a vertex v, the code spawns Cilk subroutines that recursively factor the columns associated with the children of v and their descendants. When such a subroutine returns, it triggers the activation of an extend-add operation that updates the frontal matrix of v. These extend-add operations that apply updates from the children of v are performed sequentially using a special Cilk synchronization mechanism called *inlets*.

```
cilk matrix* snmf_factor(vertex v) {
  matrix* front = NULL;
  inlet void extend_add_helper(matrix* Fc) {
    if (!front) front = allocate_front(v);
    extend_add(Fc, front);
    free_front(Fc);
  }

  for (c = first_child[v]; c != -1; c = next_child[c]) {
    extend_add_helper( spawn snmf_factor(c) );
  }
  sync; // wait for the children & their extend-adds
  if (!front) front = allocate_front(v); // leaf

  // now add columns of original coefficient matrix to
  // frontal matrix, factor the front, apply updates,
  // copy columns to L, and free columns from front

  return front;
}
```

Figure 2. Simplified Cilk code for the supernodal multifrontal Cholesky factorization with inlets to manage memory and synchronize extend-add operations.

3.1   Memory Management and Synchronization using Inlets

Inlets are subroutines that are defined within regular Cilk subroutines (similar to inner functions in Java or to nested procedures in Pascal). An inlet is always called with a first argument that is the return value of a spawned subroutine, as illustrated in Figure 2. The runtime system creates an instance of an inlet only after the spawned subroutine returns. Furthermore, the runtime system ensures that all the inlets of a subroutine instance are performed atomically with respect to one another, and only when the main procedure instance is either at a spawn or sync operation. This allows us to use inlets as a synchronization mechanism, which ensures that extend-add operations, which all modify a dense matrix associated with the columns of v, are performed sequentially, so the dense matrix is not corrupted. This is all done without using any explicit locks.

The use of inlets also allows our parallel factorization code to exploit a memory-management technique due to Liu [18–20]. Liu observed that we can actually delay the allocation of the dense frontal matrix associated with vertex v until after the first child of v returns. By cleverly ordering the children of vertices, it is possible to save significant amounts of memory and to improve the locality of reference. Our sequential code exploits this memory management technique and delays the allocation of a frontal matrix until after the first child returns. In a parallel factorization, we do not know in advance which child will be the first to return. Instead, we check in the inlet that the termination of a child activates whether the frontal matrix of the parent has already been allocated. If not, then this child is the first to return, so the matrix is allocated and initialized. Otherwise, the extend-add simply updates the previously-allocated frontal matrix.

Since Cilk's scheduler uses on each processor the normal depth-first C scheduling rule, when only one processor works on v and its descendants, the memory allocation pattern matches the sequential one exactly, and in particular, the frontal matrix of v is allocated after the first-ordered child returns but before any of the other children begin their factorization process. When multiple processors work on the subtree rooted at v, the frontal matrix is allocated after the first child returns, even if it is not the first-ordered child.

3.2   Interfaces to the Dense Subroutines

The sparse Cholesky code can use both traditional BLAS and our new recursive BLAS. Our new BLAS provide two advantages over traditional BLAS: they exploit deep memory hierarchies better and they are parallelized using Cilk. The first advantage is not significant in the sparse factorization code, because it affects only large matrices, in particular matrices that do not fit within the level-2 cache of the processor. Since many of the dense matrices that the sparse factorization code handles are relatively small, this issue is not significant. The second advantage is significant, as we show below, since it allows a single scheduler, the Cilk scheduler, to manage the parallelism in both the sparse factorization level and the dense BLAS/LAPACK level.

On the other hand, the recursive layout that our BLAS use increases the cost of extend-add operations, since computing the address of the (i, j) element of a frontal matrix becomes more expensive. By carefully implementing data-access macros (and functions that the compiler can inline on the SGI platform), we have been able to reduce the total cost of these operations, but they are nonetheless significant.

4   Performance

Figure 3 shows that the uniprocessor performance of our new dense matrix subroutines is competitive and often better than the performance of state-of-the-art vendor libraries (SGI's SCSL version 1.3 in this case). The graphs show the performance of the vendor's routine declines when the matrix grows above the size of the processor's level-2 cache, but that the performance of our routine does not. The graphs in the figure also show that even though the cost of copying to and from column-major order is significant, on large matrices our routine outperforms the vendor's routine even when this cost is included. Measurements on Pentium-III machines running Linux, now shown here due to lack of space, indicate that our new routines are similar in performance and sometimes faster than ATLAS. They are faster especially on small matrices, where the benefit of using multiple automatically-generated kernels is greatest. The data in the figure shows the performance of dense Cholesky factorization routines, but the performance characteristics of other routines are similar.

In experiments not reported here, we have found that on the Origin 3000, the uniprocessor performance of our new dense codes does not depend on the ordering of blocks. That is, as long as we lay out matrices by block, performance is independent of the ordering of blocks (recursive vs. block-column-major). It appears that the spatial locality that laying out matrices by block provides is sufficient, and that the additional coarser-grained spatial locality that we achieve by recursive layout of blocks does not contribute significantly to performance.

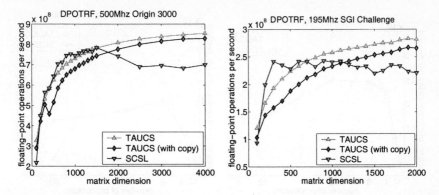

Figure 3. The uniprocessor performance of the new dense Cholesky factorization (denoted TAUCS) compared to the performance of SCSL, SGI's native BLAS/LAPACK. Each plot shows the performance of our new subroutine with recursive layout, the performance of the new subroutine when the input and output is in column-major order (in which case we copy the input and output to/from recursive format), and the performance of SCSL.

Figure 4 shows that our new dense matrix routines scale well unless memory access times vary too widely. The graphs show the performance of the dense Cholesky factorization routines on a 32-processor SGI Origin 3000 machine. The entire 32-processor machine was dedicated to these experiments. On this machine, up to 16 processors can communicate through a single router. When more than 16 processors participate in a computation, some of the memory accesses must go through a link between two routers, which slows down the accesses. The graphs show that when 16 or fewer processors are used, our new code performs similarly or better than SCSL. The performance difference is especially significant on 16 processors. But when 32 processors are used, the slower memory accesses slow our code down much more than it slows SCSL (but even SCSL slows down relative to its 16-processors performance). We suspect that the slowdown is mostly due to the fact that we allocate the entire matrix in one memory-allocation call (so all the data resides on a single 4-processor node) and do not use any memory placement or migration primitives, which would render the code less portable and more machine specific.

Figure 5 (left) shows that our overall sparse Cholesky code scales well with up to 8 processors. The code does not speed up further when it uses 16 processors, but it does not slow down either. The graph also shows the benefit of parallelizing the sparse and dense layers of the solver using the same parallelization mechanism. The code speeds up best (green circles) when both the sparse multifrontal code and the dense routines are parallelized using Cilk. When we limit parallelism to either the sparse layer or to the dense routines (red/blue triangles), performance drops significantly.

We acknowledge that the absolute performance of the sparse factorization code, as shown in Figure 5 (right), appears to be lower than that of state-of-the-art sparse Cholesky codes. We have not measured the performance of the code relative to the performance of other codes, but from indirect comparisons it appears that the code is slower than PARADISO [22], for example. We have received reports that the code is

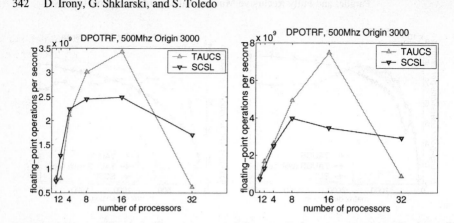

Figure 4. The parallel performance of the new dense Cholesky factorization (without data copying) on matrices of dimension 2000 (left) and 4000 (right). The lack of speedup on 2 processors on the 2000-by-2000 matrix appears to be a Cilk problem, which we have not been able to track down yet.

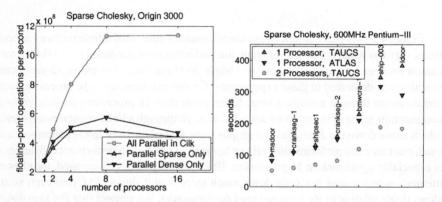

Figure 5. The performance of the parallel supernodal multifrontal sparse Cholesky in Cilk. The graph on the left shows the performance of the numerical factorization phase of the factorization on a 500 MHz Origin 3000 machine, on a matrix that represents a regular 40-by-40-by-40 mesh. The symbolic factorization phase degrades performance by less than 5% relative to the performance shown here. The three plots show the performance of our Cilk-parallel sparse solver with our Cilk-parallel dense routines, the performance of our Cilk-parallel sparse solver with sequential recursive dense routines, and of our sequential sparse solver with SCSL's parallel dense routines.

The plot on the right shows the performance on a dual-processor Pentium-III machine for some of the PARASOL test matrices. Performance is given in seconds to allow comparisons with other solvers. The plot shows the performance of our code with recursive dense routines on 1 and 2 processors, and also the performance of our sparse code when it calls ATLAS.

We acknowledge that the absolute performance of the source factorization code, as shown in Figure 5 (right), appears to be lower than that of state-of-the-art sparse

significantly faster on a uniprocessor than the sequential code of Ng and Peyton [21]. The sequential code is 5-6 times faster than Matlab 6's sparse chol, which is written in C but is not supernodal. We believe that using relaxed supernodes will bring our code to a performance level similar to that of the fastest codes (currently the code only exploits so-called fundamental supernodes). We plan to implement this enhancement in the near future.

5 Conclusions

Our research addresses several fundamental issues: Can blocked and possibly recursive data layouts be used effectively in a large software project? In other words, is the overhead of indexing the (i, j) element of a matrix acceptable in a code that needs to do that frequently (e.g., a multifrontal code that performs extend-add operations on frontal matrices)? How much can we benefit from writing additional automatic kernel generators? We clearly benefit from the DSYRK generator, but will additional ones help, e.g. for extend-add? Can Cilk manage parallelism effectively in a multilevel library that exposes parallelism at both the upper sparse layer and the lower dense layer?

So far, it seems clear that the Cilk can help manage parallelism and simplify code in complex parallel codes. However, the slowdowns on 32 processors suggest that Cilk codes should manage and place memory carefully on ccNUMA machines. It also appears that additional optimized kernels are beneficial, particularly since many of the dense matrices that sparse solvers handle are quite small. It is not clear yet whether the improved spatial locality of blocked layouts justifies the additional overhead of random accesses to matrix elements.

Our research not only addresses fundamental questions, but it also aims to provide users of mathematical software with state-of-the-art high-performance implementations of widely-used algorithms. A stable version of our sequential code is freely available at www.tau.ac.il/~stoledo/taucs. This version includes the sequential multifrontal supernodal solver, but not the recursive BLAS or the parallelized Cilk codes. We plan to freely distribute all the codes once we reach a stable version that includes them.

References

1. R. C. Agarwal, F. G. Gustavson, and M. Zubair. Exploiting functional parallelism of POWER2 to design high-performance numerical algorithms. *IBM Journal of Research and Development*, 38(5):563–576, 1994.
2. R. C. Agarwal, F. G. Gustavson, and M. Zubair. Improving performance of linear algebra algorithms for dense matrices using algorithmic prefetch. *IBM Journal of Research and Development*, 38(3):265–275, 1994.
3. B. S. Andersen, J. Waśniewski, and F. G. Gustavson. A recursive formulation of cholesky factorization of a matrix in packed storage. *ACM Transactions on Mathematical Software*, 27:214–244, June 2001.
4. J. Bilmes, K. Asanovic, C. W. Chin, and J. Demmel. Optimizing matrix multiply using PHIPAC: a portable, high-performance, ANSI C coding methodology. In *Proceedings of the International Conference on Supercomputing*, Vienna, Austria, 1997.
5. Compaq. Compaq extended math library (CXML). Software and doc-umuntation available online from http://www.compaq.com/math/, 2001.
6. J. J. Dongarra, J. D. Cruz, S. Hammarling, and I. Duff. Algorithm 679: A set of level 3 basic linear algebra subprograms. *ACM Transactions on Mathematical Software*, 16(1):18–28, 1990.

7. J. J. Dongarra, J. D. Cruz, S. Hammarling, and I. Duff. A set of level 3 basic linear algebra subprograms. *ACM Transactions on Mathematical Software*, 16(1):1–17, 1990.

8. J. J. Dongarra and P. Raghavan. A new recursive implementation of sparse Cholesky factorization. In *Proceedings of the 16th IMACS World Congress 2000 on Scientific Computing, Applications, Mathematics, and Simulation*, Lausanne, Switzerland, Aug. 2000.

9. E. Elmroth and F. Gustavson. Applying recursion to serial and parallel QR factorization leads to better performance. *IBM Journal of Research and Development*, 44(4):605–624, 2000.

10. E. Elmroth and F. G. Gustavson. A faster and simpler recursive algorithm for the LAPACK routine DGELS. *BIT*, 41:936–949, 2001.

11. M. Frigo, C. E. Leiserson, and K. H. Randall. The implementation of the Cilk-5 multi-threaded language. *ACM SIGPLAN Notices*, 33(5):212–223, 1998.

12. F. Gustavson, A. Henriksson, I. Jonsson, B. Kågström, and P. Ling. Recursive blocked data formats and BLAS's for dense linear algebra algorithms. In B. Kågström, J. Dongarra, E. Elmroth, and J. Waśniewski, editors, *Proceedings of the 4th International Workshop on Applied Parallel Computing and Large Scale Scientific and Industrial Problems (PARA '98)*, number 1541 in Lecture Notes in Computer Science Number, pages 574–578, Ume, Sweden, June 1998. Springer.

13. F. G. Gustavson. Recursion leads to automatic variable blocking for dense linear-algebra algorithms. *IBM Journal of Research and Development*, 41:737–755, Nov. 1997.

14. F. G. Gustavson and I. Jonsson. Minimal-storage high-performance Cholesky factorization via blocking and recursion. *IBM Journal of Research and Development*, 44:823–850, Nov. 2000.

15. IBM. Engineering and scientific subroutine library (SCSL). Software and documuntation available online from http://www-1.ibm.com/servers/eservers/pseries/software/sp/essl.html, 2001.

16. Intel. Math kernel library (MKL). Software and documuntation available online from http://www.intel.com/software/products/mkl/, 2001.

17. C. Kamath, R. Ho, and D. P. Manley. DXML: a high-performance scientific subroutine library. *Digital Technical Journal*, 6(3):44–56, 1994.

18. J. W. H. Liu. On the storage requirement in the out-of-core multifrontal method for sparse factorization. *ACM Transactions on Mathematical Software*, 12(3):249–264, 1986.

19. J. W. H. Liu. The multifrontal method and paging in sparse Cholesky factorization. *ACM Transactions on Mathematical Software*, 15(4):310–325, 1989.

20. J. W. H. Liu. The multifrontal method for sparse matrix solution: Theory and practice. *SIAM Review*, 34(1):82–109, 1992.

21. E. G. Ng and B. W. Peyton. Block sparse Cholesky algorithms on advanced uniprocessor computers. *SIAM Journal on Scientific Computing*, 14(5):1034–1056, 1993.

22. O. Schenk and K. Gärtner. Sparse factorization with two-level scheduling in PARADISO. In *Proceedings of the 10th SIAM Conference on Parallel Processing for Scientific Computing*, page 10 pages on CDROM, Portsmouth, Virginia, Mar. 2001.

23. SGI. Scientific computing software library (SCSL). Software and documuntation available online from from http://www.sgi.com/software/scsl.html, 1993–2001.

24. Supercomputing Technologies Group, MIT Laboratory for Computer Science, Cambridge, MA. *Cilk-5.3 Reference Manual*, June 2000. Available online at http://supertech.lcs.mit.edu/cilk.

25. S. Toledo. Locality of reference in LU decomposition with partial pivoting. *SIAM Journal on Matrix Analysis and Applications*, 18(4):1065–1081, 1997.

26. R. C. Whaley and J. J. Dongarra. Automatically tuned linear algebra software. Technical report, Computer Science Department, University Of Tennessee, 1998. available online at www.netlib.org/atlas.

Parallel Iterative Methods in Modern Physical Applications *

X. Cai[1], Y. Saad[2], and M. Sosonkina[3]

[1] Simula Research Laboratory and University of Oslo, P.O. Box 1080, Blindern,
N-0316 Oslo, Norway
xingca@ifi.uio.no
[2] University of Minnesota, Minneapolis, MN 55455, USA
saad@cs.umn.edu
[3] University of Minnesota, Duluth, MN 55812, USA
masha@d.umn.edu

Abstract. Solving large sparse linear systems is a computationally-intensive component of many important large-scale applications. We present a few experiments stemming from a number of realistic applications including magneto-hydrodynamics structural mechanics, and ultrasound modeling, which have become possible due to the advances in parallel iterative solution techniques. Among such techniques is a recently developed Parallel Algebraic Recursive Multilevel Solver (pARMS). This is a distributed-memory iterative method that adopts the general framework of distributed sparse matrices and relies on solving the resulting distributed Schur complement systems. We discuss some issues related to parallel performance for various linear systems which arise in realistic applications. In particular, we consider the effect of different parameters and algorithms on the overall performance.

1 Distributed Sparse Linear Systems

The viewpoint of a distributed linear system generalizes the Domain Decomposition methods to irregularly structured sparse linear systems. A typical distributed system arises, e.g., from a finite element discretization of a partial differential equation on a certain domain. To solve such systems on a distributed memory computer, it is common to partition the finite element mesh by a graph partitioner and assign a cluster of elements representing a physical sub-domain to a processor. The general assumption is that each processor holds a set of equations (rows of the global linear system) and the associated unknown variables. The rows of the matrix assigned to a certain processor have been split into two parts: a *local* matrix A_i which acts on the local variables and an *interface* matrix X_i which acts on the external interface variables. These external interface variables must be first received from neighboring processor(s) before a distributed

---

* This work was supported in part by NSF under grants NSF/ACI-0000443 and NSF/INT-0003274, and in part by the Minnesota Supercomputing Institute.

P.M.A. Sloot et al. (Eds.): ICCS 2002, LNCS 2330, pp. 345–354, 2002.

matrix-vector product can be completed. Thus, each local vector of unknowns x_i $(i = 1, \ldots, p)$ is also split into two parts: the sub-vector u_i of interior variables followed by the sub-vector y_i of inter-domain interface variables. The right-hand side b_i is conformally split into the sub-vectors f_i and g_i. The local matrix A_i residing in processor i is block-partitioned according to this splitting. So the local equations can be written as follows:

$$\begin{pmatrix} B_i & F_i \\ E_i & C_i \end{pmatrix} \begin{pmatrix} u_i \\ y_i \end{pmatrix} + \begin{pmatrix} 0 \\ \sum_{j \in N_i} E_{ij} y_j \end{pmatrix} = \begin{pmatrix} f_i \\ g_i \end{pmatrix}. \tag{1}$$

The term $E_{ij} y_j$ is the contribution to the local equations from the neighboring sub-domain number j and N_i is the set of sub-domains that are neighbors to sub-domain i. The sum of these contributions, seen on the left side of (1), is the result of multiplying the interface matrix X_i by the external interface variables. It is clear that the result of this product will affect only the inter-domain interface variables as is indicated by the zero in the upper part of the second term on the left-hand side of (1). For practical implementations, the sub-vectors of external interface variables are grouped into one vector called $y_{i,ext}$ and the notation

$$\sum_{j \in N_i} E_{ij} y_j \equiv X_i y_{i,ext}$$

will be used to denote the contributions from external variables to the local system (1). In effect, this represents a local ordering of external variables to write these contributions in a compact matrix form. With this notation, the left-hand side of (1) becomes

$$w_i = A_i x_i + X_{i,ext} y_{i,ext}. \tag{2}$$

Note that w_i is also the local part of a global matrix-vector product Ax in which x is a distributed vector which has the local vector components x_i.

Preconditioners for distributed sparse linear systems are best designed from the local data structure described above. Additive and (variants of) multiplicative Schwarz procedures are the simplest preconditioners available. Additive Schwarz procedures update the local solution by the vector obtained from solving a linear system formed by the local matrix and the local residual. The exchange of data is done through the computation of the residual. The local systems can be solved in three ways: (1) by a (sparse) direct solver, (2) by using a standard preconditioned Krylov solver, or (3) by performing a backward-forward solution associated with an accurate ILU (e.g., ILUT) preconditioner.

Schur complement techniques refer to methods which iterate on the inter-domain interface unknowns only, implicitly using interior unknowns as intermediate variables. These techniques are at the basis of what will be described in the next sections. Schur complement systems are derived by eliminating the variables u_i from (1). Extracting from the first equation $u_i = B_i^{-1}(f_i - F_i y_i)$ yields, upon substitution in the second equation,

$$S_i y_i + \sum_{j \in N_i} E_{ij} y_j = g_i - E_i B_i^{-1} f_i \equiv g'_i, \tag{3}$$

where S_i is the "local" Schur complement $S_i = C_i - E_i B_i^{-1} F_i$. The equations (3) for all sub-domains i $(i = 1, \ldots, p)$ constitute a global system of equations involving only the inter-domain interface unknown vectors y_i. This global reduced system has a natural block structure related to the inter-domain interface points in each sub-domain:

$$
\begin{pmatrix}
S_1 & E_{12} & \cdots & E_{1p} \\
E_{21} & S_2 & \cdots & E_{2p} \\
\vdots & & \ddots & \vdots \\
E_{p1} & E_{p-1,2} & \cdots & S_p
\end{pmatrix}
\begin{pmatrix}
y_1 \\ y_2 \\ \vdots \\ y_p
\end{pmatrix}
=
\begin{pmatrix}
g_1' \\ g_2' \\ \vdots \\ g_p'
\end{pmatrix}.
\tag{4}
$$

The diagonal blocks in this system, the matrices S_i, are dense in general. The off-diagonal blocks E_{ij}, which are identical with those involved in (1), are sparse.

The system (4) can be written as $Sy = g'$, where y consists of all inter-domain interface variables $y_1, y_2, \ldots, y_p$ stacked into a long vector. The matrix S is the "global" Schur complement matrix. An idea proposed in [13] is to exploit methods that *approximately solve the reduced system* (4) to develop preconditioners for the original (global) distributed system. Once the global Schur complement system (3) is (approximately) solved, each processor will compute the u-part of the solution vector by solving the system $B_i u_i = f_i - E_i y_i$ obtained by substitution from (1).

For convenience, (3) is rewritten as a preconditioned system with the diagonal blocks:

$$
y_i + S_i^{-1} \sum_{j \in N_i} E_{ij} y_j = S_i^{-1} \left[g_i - E_i B_i^{-1} f_i \right].
\tag{5}
$$

This can be viewed as a block-Jacobi preconditioned version of the Schur complement system (4). This global system can be solved by a GMRES-like accelerator, requiring a solve with S_i at each step.

2 Parallel Implementation of ARMS (pARMS)

Multi-level Schur complement techniques available in pARMS [9] are based on techniques which exploit block independent sets, such as those described in [15]. The idea is to create another level of partitioning of each sub-domain. An illustration is shown in Figure 1, which distinguishes one more type of interface variables: *local interface variables*, where we refer to local interface points as interface points between the sub-sub-domains. Their couplings are all local to the processor and so these points do not require communication. These sub-sub-domains are not obtained by a standard partitioner but rather by the *block independent set* reordering strategy utilized by ARMS [15].

In order to explain the multilevel techniques used in pARMS, it is necessary to discuss the sequential multilevel ARMS technique. In the sequential ARMS, the matrix coefficient of the at the l-th level is reordered in the from

$$
P_l A_l P_l^T = \begin{pmatrix} B_l & F_l \\ E_l & C_l \end{pmatrix},
\tag{6}
$$

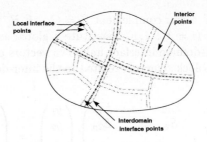

Fig. 1. A two-level partitioning of a domain

where P_l is a "block-independent-set permutation", which can be obtained in a number of ways. At the level $l = 0$, the matrix A_l is the original coefficient matrix of the linear system under consideration. The above permuted matrix is then approximately factored as

$$P_l A_l P_l^T \approx \begin{pmatrix} L_l & 0 \\ E_l U_l^{-1} & I \end{pmatrix} \times \begin{pmatrix} U_l & L_l^{-1} F_l \\ 0 & A_{l+1} \end{pmatrix}, \tag{7}$$

where I is the identity matrix, L_l and U_l form the LU (or ILU) factors of B_l, and A_{l+1} is an approximation to the Schur complement with respect to C_l,

$$A_{l+1} \approx C_l - (E_l U_l^{-1})(L_l^{-1} F_l). \tag{8}$$

During the factorization process, approximations to the matrices for obtaining the Schur complement (8) are computed. The system with A_{l+1} is partitioned again in the form (6) in which l is replaced by $l + 1$. At the last level, the reduced system is solved using GMRES preconditioned with ILUT [12]. In the parallel version of ARMS, the same overall strategy is used except that now the global block-independent sets are across domains. Consider a one-level pARMS for simplicity. In the first level reduction, the matrix A_1 that is produced, will act on all the interface variables, whether local or inter-domain. Thus, a one-level pARMS would solve for these variables and then obtain the interior variables in each processor without communication. We denote by *expanded Schur complement* the system involving the matrix A_1 that acts on inter-domain and local interface unknowns. For a more detailed description of pARMS see [9].

2.1   Diagonal Shifting in pARMS

Extremely ill-conditioned linear systems are difficult to solve by iterative methods. A possible source of difficulty is due to the ineffective preconditioning of such systems. The preconditioner may become unstable (i.e., has large norm of its inverse). To stabilize the preconditioner, a common technique is to shift the matrix A by a scalar and use this shifted matrix $A + \alpha I$ during preconditioning, see, e.g., [10]. Because the matrix is shifted, its preconditioner might be a rather accurate approximation of $A + \alpha I$. It is also more likely to be stable. However, for large shift values, the preconditioner might not represent accurately the original matrix A. So the choice of the shift value is important and leads to a trade-off between accuracy and stability of the preconditioner. We have considered this trade-off in [6, 14]. In [3], a strong correlation between stability of the preconditioner and the size of $\mathcal{E} = \log\left(\|(LU)^{-1}\|_{\text{inf}}\right)$ is shown and is suggested as a practical means of evaluating the quality of a preconditioner. We can inexpensively compute $\mathcal{E}_\alpha = \log\left(\|(LU)^{-1}e\|_1\right)$, where e is a vector of all ones and LU are incomplete LU factors of $A + \alpha I$. The estimate $\mathcal{E}_\alpha$ can be used in choosing a shift value: if this estimate is large, then we increase shift value and recompute (adjust) the preconditioner. Note that efficient techniques for updating a preconditioner when a new shift value is provided are beyond the scope of this paper. One such technique has been outlined in [4].

In the pARMS implementation, we have adapted a shifting technique for a distributed representation of linear system. Specifically, we perform the shifting and norm $\mathcal{E}_\alpha$ calculation in each processor independently. Thus, each processor i can have a different shift value depending on the magnitude of its $\mathcal{E}_{\alpha_i}$. Such an implementation is motivated by the observation that shifting is especially important for diagonally non-dominant rows, which can be distinguished among other rows by a local procedure. In each processor, the choice of shift value is described by the following pseudo-code:

ALGORITHM 21 *Matrix shifting*
1.   *Select initial shift $\alpha \geq 0$: $B = A + \alpha I$.*
2.   *Compute parallel preconditioner M for B.*
3.   *Calculate local $\mathcal{E}_\alpha$.*
4.   *If $\mathcal{E}_\alpha$ is large,*
5.   *Choose $\alpha' > \alpha$;*
6.   *Adjust preconditioner.*

Note that in Line 6 of Algorithm 21, depending on the type of preconditioner, the adjustment operation may be either local or global. For example, Additive Schwarz type preconditioners may perform adjustments independently per processor, whereas all the processors may need to participate in the adjustment of a Schur complement preconditioner. In addition, Lines 3 – 6 may be repeated several times.

3   Numerical Experiments

In this section we describe a few realistic applications, which give rise to large irregularly structured linear systems that are challenging to solve by iterative methods. The linear systems arising in ultrasound simulation were generated using Diffpack, which is an object-oriented environment for scientific computing, see [5, 8]. The magnetohydrodynamics application has been provided by A. Soulaimani and R. Touihri from the "Ecole de Technologie Superieure, Université du Québec", and the linear systems arising in tire design have been supplied by J. T. Melson of Michelin Americas Research and Development Corporation. For the sake of convenience, let us introduce some notation.

add_ilut. Additive Schwarz procedure without overlapping in which ILUT is used as a preconditioner for solving the local systems. These systems can be solved with a given number of GMRES inner iterations or by just applying the preconditioner.

add_iluk. Similar to **add_ilut** but uses ILU(k) as a preconditioner instead of ILUT.

add_arms. Similar to **add_ilut** but uses ARMS as a preconditioner for local systems.

sch_gilu0. This method is based on approximately solving the expanded Schur complement system with a global ILU(0)-preconditioned GMRES. The ILU(0) preconditioning requires a global order (referred to as a schedule in [7]) in which to process the nodes. A global multicoloring of the domains is used for this purpose as is often done with global ILU(0).

The suffixes **no_its** or **sh** are added to the above methods when no local (inner) iterations are used or when the shifted original matrix is used for the preconditioner construction, respectively.

3.1   Simulation of 3D Nonlinear Acoustic Fields

The propagation of 3D ultrasonic waves in a nonlinear medium can be modeled by a system of nonlinear PDEs.

The numerical scheme consists of using finite elements in the spatial discretization and finite differences for the temporal derivatives. At each time level, the discretization of gives rise to a system of nonlinear algebraic equations involving φ from three consecutive time levels. We apply Newton-Raphson iterations for the nonlinear system. We refer to [2] and the references therein for more information on the mathematical model and the numerical solution method. As a particular numerical test case, we use a 3D domain: $(x, y, z) \in [-0.004, 0.004] \times [-0.004, 0.004] \times [0, 0.008]$. On the face of $z = 0$, there is a circular transducer with radius $r = 0.002$, i.e., the pressure p is given within the circle. On the rest of the boundary we use a non-reflective boundary condition.

We consider solving the linear system during the first Newton-Raphson iteration at the first time level. The linear system has $185, 193$ unknowns and $11, 390, 625$ nonzero entries. Figure 2 presents the iteration numbers (left) and solution times (right) of this linear system on the IBM SP at the Minnesota Supercomputing Institute. Four 222 MHz Power3 processors share 4GB of memory per (Nighthawk) node and are connected by a high performance switch with other nodes. Two preconditioning techniques, **sch_gilu0_no_its** and **add_arms_no_its** have been tested on various processor numbers. Original right hand side and random initial guess have been taken.

It is observed that **sch_gilu0_no_its** preconditioning consistently leads to a faster convergence than **add_arms_no_its**. Both methods, however, show almost no increase in iterations with increase in processor numbers. The timing results are slightly better for **sch_gilu0_no_its** preconditioner except for the 16-processor case.

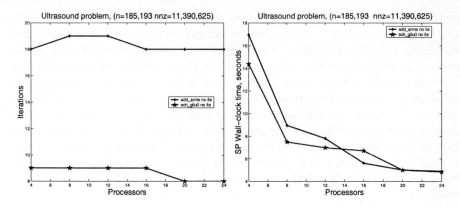

Fig. 2. Iterations (left) and timings results (right) for the ultrasound problem

3.2   A Problem Issued From Magnetohydrodynamic Flow

In [9], we have described the solution of a rather hard problem which arises from Magnetohydrodynamic (MHD) flows. The flow equations are represented as coupled Maxwell's and the Navier-Stokes equations. Here, we provide only a brief outline of a sample problem along with its solution and note the solution process when shifting techniques are used.

We solve linear systems which arise from the Maxwell equations only. In order to do this, a pre-set periodic induction field is used in Maxwell's equation. The physical region is the three-dimensional unit cube $[-1, 1]^3$ and the discretization uses a Galerkin-Least-Squares discretization. The magnetic diffusivity coefficient is $\eta = 1$. The linear system (denoted by MHD1) has $n = 485, 597$ unknowns and $24, 233, 141$ nonzero entries. The gradient of the function corresponding to Lagrange multipliers should be zero at steady-state. Though the actual right-hand side was supplied, we preferred to use an artificially generated one in order to check the accuracy of the process. A random initial guess was taken. Little difference in performance was seen when the actual right-hand and a zero vector initial guess were used instead. For the details on the values of the input parameters see [9].

We observed that all the methods without inner iterations experienced stagnation for the MHD1 problem. Additive Schwarz (add_arms_no_its) with or without overlap does not converge for any number of processors while the Schur global ILU(0) (sch_gilu0_no_its) stagnates when executed on more than nine processors. On four and nine processors, sch_gilu0 no its converges in 188 and 177 iterations, respectively. On an IBM SP, this amounts to 2,223.43 and 1,076.27 seconds, respectively. This is faster than 2,372.44 and 1,240.23 seconds when five inner iterations are applied and the number of outer iterations decreases to 119 and 109 on four and nine processors, respectively. The benefits of iterating on the global Schur complement system are clear since the Schur complement-based preconditioners converge for all the processor numbers tested as indicated in Figure 3, which presents the timing results (left) and outer iteration numbers (right). This positive effect can be explained by the fact that the Schur complement system is computed with good accuracy. Figure 3 also shows the usage of the shift value $\alpha = 0.1$ in the sch_gilu0_sh preconditioner construction. For this problem, shifting does not help convergence and results in larger numbers of outer iterations. Since a good convergence rate is achieved without shifting of the original matrix, the shift value applied in sch_gilu0_sh may be too large and the resulting preconditioner may not be a good approximation of the original matrix. The number of nonzeros in sch_gilu0_sh, however, is smaller than in sch_gilu0. Therefore, the construction of sch_gilu0_sh is always cheaper, and sch_gilu0_sh appears to be competitive for small processor numbers.

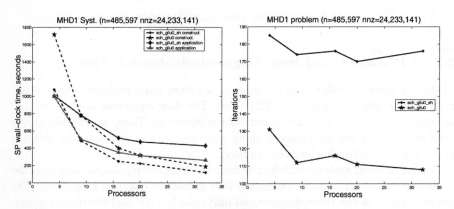

Fig. 3. Solution times (left) and outer iterations (right) for the (fixed-size) MHD1 problem with and without diagonal shifting

3.3   Linear Systems Arising in Tire Design

Tire static equilibrium computation is based on a 3D finite element model with distributed loads. Computation of static equilibrium involves minimizing the potential energy $\Pi(u)$ with respect to finite element nodal displacements $u^i (i = 1, 2, 3)$ subject to nonlinear boundary conditions, which change the symmetry of a tire. The equilibrium equations of the model are obtained by setting the variation $\delta\Pi(u)$ to zero. equivalently The Jacobian matrix of the equilibrium equations is obtained by finite difference approximations. The distributed load is scaled by a (loading) parameter λ, and as λ varies the static equilibrium solutions trace out a curve. The difficulty of the finite element problems and concomitant linear systems varies considerably along this equilibrium curve, as well as within the nonlinear iterations to compute a particular point on this curve. In [17], the problems of varying matrix characteristics are considered. All of the problems pose a challenge for iterative methods since the treatment of stationary solutions of rotation makes the systems extremely ill-conditioned during the nonlinear convergence process. It has been observed that an acceptable convergence was achieved *only* when a rather large shift was applied to the matrix diagonal to stabilize preconditioner. The size of the shift is very important: while making the preconditioner more stable, large shift values cause the preconditioner to be a poor approximation of the original matrix.

In this paper, we show (Table 1) the results of a few experiments with using pARMS on an example of a linear system, medium tire model $\mathcal{M}$, in which $n = 49,800$ and the number of nonzeros is approximately $84n$. In pARMS, a shift α is chosen *automatically*: starting with the zero shift, the preconditioner is reconstructed with a new shift (augmented by 0.1) if the estimate $\mathcal{E}_\alpha$ of the preconditioner inverse is large (greater than seven). In Table 1, we state the final value of α, the maximum $\mathcal{E}_\alpha$ among all the processors when $\alpha = 0$, the number Iter of iterations to converge, and the preconditioner application time Time spent when running on four processors. Due to the difficulty of this problem,

Table 1. Solution of tire model $\mathcal{M}$ on four processors

Method	α_{fin}	$\max E_{\alpha=0}$	Iter	Time
add_ilu(2)	0.1	46	543	287.13
add_ilut	0.1	116	537	211.45
sch_gilu0	0.2	146	575	369.00

which is also unpredictably affected by partitioning, the convergence was not observed consistently on any processor numbers. For example, no convergence has been achieved on eight processors for moderate shift values.

4   Conclusion

In this paper, we have illustrated the performance of the recently developed parallel ARMS (pARMS) code on several realistic applications. For all the problems considered, it is beneficial to use preconditioners based on Schur complement techniques, enhanced by a local multi-level procedure. In addition, a few inner (local to a sub-domain) preconditioning iterations enhance convergence for a problem arising from a magneto-hydrodynamics application.

We have also proposed an implementation of matrix shifting in the framework of distributed linear systems which allows a shift value to be assigned independently in each sub-domain. An automatic procedure for the shift value selection has also been implemented, which stabilizes the distributed preconditioner and often overcomes stagnation. We would like to underline the flexibility of the pARMS framework, which, with a proper selection of input parameters, allows to choose among many available options for solving real-world problems.

References

1. E.F.F. Botta, A. van der Ploeg, and F.W. Wubs. Nested grids ILU-decomposition (NGILU). *J. Comp. Appl. Math.*, 66:515–526, 1996.
2. X. Cai and Å. Ødegård. Parallel simulation of 3D nonlinear acoustic fields on a Linux-cluster. *Proceedings of the* Cluster 2000 *conference*.
3. E. Chow and Y. Saad. Experimental study of ILU preconditioners for indefinite matrices. *Journal of Computational and Applied Mathematics*, 87:387–414, 1997.
4. E. Chow and Y. Saad. Approximate inverse preconditioners via sparse-sparse iterations. *SIAM Journal on Scientific Computing*, 19:995–1023, 1998.
5. Diffpack World Wide Web home page. *http://www.nobjects.com*.
6. P. Guillaume, Y. Saad, and M. Sosonkina. Rational approximation preconditioners for general sparse linear systems. Technical Report umsi-99-209, Minnesota Supercomputer Institute, University of Minnesota, Minneapolis, MN, 1999.
7. D. Hysom and A. Pothen. A scalable parallel algorithm for incomplete factor preconditioning. Technical Report (preprint), Old-Dominion University, Norfolk, VA, 2000.
8. H. P. Langtangen. *Computational Partial Differential Equations – Numerical Methods and Diffpack Programming*. Springer-Verlag, 1999.
9. Z. Li, Y. Saad, and M. Sosonkina. pARMS: A parallel version of the algebraic recursive multilevel solver. Technical Report UMSI-2001-100, Minnesota Supercomputer Institute, University of Minnesota, Minneapolis, MN, 2001.
10. T.A Manteuffel. An incomplete factorization technique for positive definite linear systems. *Mathematics of computation*, 32:473–497, 1980.
11. Y. Saad. ILUM: a multi-elimination ILU preconditioner for general sparse matrices. *SIAM Journal on Scientific Computing*, 17(4):830–847, 1996.
12. Y. Saad. *Iterative Methods for Sparse Linear Systems*. PWS publishing, New York, 1996.
13. Y. Saad and M. Sosonkina. Distributed Schur Complement techniques for general sparse linear systems. *SIAM J. Scientific Computing*, 21(4):1337–1356, 1999.
14. Y. Saad and M. Sosonkina. Enhanced preconditioners for large sparse least squares problems. Technical Report umsi-2001-1, Minnesota Supercomputer Institute, University of Minnesota, Minneapolis, MN, 2001.
15. Y. Saad and B. Suchomel. ARMS: An algebraic recursive multilevel solver for general sparse linear systems. Technical Report umsi-99-107-REVIS, Minnesota Supercomputer Institute, University of Minnesota, Minneapolis, MN, 2001. Revised version of umsi-99-107.
16. Y. Saad and J. Zhang. BILUTM: A domain-based multi-level block ILUT preconditioner for general sparse matrices. Technical Report umsi-98-118, Minnesota Supercomputer Institute, University of Minnesota, Minneapolis, MN, 1998. appeared in SIMAX, vol. 21, pp. 279-299 (2000).
17. M. Sosonkina, J. T. Melson, Y. Saad, and L. T. Watson. Preconditioning strategies for linear systems arising in tire design. *Numer. Linear Alg. with Appl.*, 7:743–757, 2000.

Solving Unsymmetric Sparse Systems of Linear Equations with PARDISO

Olaf Schenk[1]* and Klaus Gärtner[2]

[1] Department of Computer Science, University of Basel, Klingelbergstrasse 50,
CH-4056 Basel, Switzerland
oschenk@ifi.unibas.ch
http://www.ifi.unibas.ch
[2] Weierstrass Institute for Applied Analysis and Stochastics, Mohrenstr. 39,
D-10117 Berlin, Germany
gaertner@wias-berlin.de
http://www.wias-berlin.de

Abstract. Supernode pivoting for unsymmetric matrices coupled with supernode partitioning and asynchronous computation can achieve high gigaflop rates for parallel sparse LU factorization on shared memory parallel computers. The progress in weighted graph matching algorithms helps to extend these concepts further and prepermutation of rows is used to place large matrix entries on the diagonal. Supernode pivoting allows dynamical interchanges of columns and rows during the factorization process. The BLAS-3 level efficiency is retained. An enhanced left–right looking scheduling scheme is uneffected and results in good speedup on SMP machines without increasing the operation count. These algorithms have been integrated into the recent unsymmetric version of the PARDISO solver. Experiments demonstrate that a wide set of unsymmetric linear systems can be solved and high performance is consistently achieved for large sparse unsymmetric matrices from real world applications.

1 Introduction

The solution of large sparse linear systems is a computational bottleneck in many scientific computing problems. When partial pivoting is required to maintain numerical stability in direct methods for solving nonsymmetric linear systems, it is challenging to develop high performance parallel software because partial pivoting causes the computational task-dependency graph to change during runtime. It has been proposed recently that permuting the rows of the matrix prior to factorization to maximize the magnitude of its diagonal entries can be very effective in reducing the amount of pivoting during factorization [1,9,10,15]. The proposed technique, static pivoting, is an efficient alternative to partial pivoting for parallel sparse Gaussian elimination.

* This work was supported by the Swiss Commission of Technology and Innovation KTI under contract number 5648.1.

P.M.A. Sloot et al. (Eds.): ICCS 2002, LNCS 2330, pp. 355–363, 2002.
© Springer-Verlag Berlin Heidelberg 2002

This paper addresses the issues of improved scalability and robustness of sparse direct factorization within a supernode pivoting approach used in the PARDISO solver[1]. The original aim of the PARDISO project [17, 19, 20] was to develop a scalable parallel direct solver for sparse matrices arising in semiconductor device and process simulations [4]. These matrices are in general unsymmetric with a symmetric structure, and partial pivoting was not the primary issue during the project. The underlying data structure of the solver is highly optimized and scalability has been achieved with a left-right looking algorithm on shared memory parallel computers [20]. However, after completing the first version of the solver, the authors realized the potential of static pivoting [15] and the use of prepermutations of rows to place large entries on the diagonal. Therefore, the current version is extended towards the efficient solution of large unsymmetric sparse matrices in a shared-memory computing environment.

The suite of unsymmetric test matrices that are used in the experiments througout this paper is shown in Table 1. All matrices are due to real world applications and are publicly available. The table also contains the dimension, the number of nonzeros, and the related application area. It is impossible to solve these linear systems without any pivoting or preordering in many cases.

2   Algorithmic features

In this section the algorithms and strategies that are used in the analysis and numerical phase of the computation of the LU factors are described.

Supernode pivoting

Figure 1 outlines the approach to solve an unsymmetric sparse linear system of equations. According to [15] it is very beneficial to precede the ordering by performing an unsymmetric permutation to place large entries on the diagonal and then to scale the matrix so that the diagonals entries are equal to one. Therefore, in step (1) the diagonal matrices D_r and D_c are chosen in such a way that each row and each column of $D_r A D_c$ have a largest entry equal to 1 in magnitude. The row permutation matrix P_r is chosen to maximize the product of the diagonal entries in $P_r D_r A D_c$ with the MC64 code [10]. In step (2) any symmetric fill-reducing ordering can be computed based on the structure of $A + A^T$, e.g. minimum degree or nested dissection. All experiments reported in this paper with PARDISO were conducted with a nested dissection algorithm [14]. Like other modern sparse factorization packages [2, 5, 7, 8, 13, 16], PARDISO takes advantage of the supernode technology — adjacent groups of rows and columns with the same structure in the factors L and U are treated as one supernode. An interchange among these rows of a supernode has no effect on the overall fill-in and this is the mechanism for finding a suitable pivot

---

[1] A prebuilt library of the unsymmetric solver PARDISO will be available for several architectures for research purposes at www.ifi.unibas.ch/~PARDISO in spring 2002.

Table 1. Unsymmetric test matrices with their order (N), number of nonzeros (NNZ), and the application area of origin.

Number	Matrix	N	NNZ	Application
1	af23560	23560	484256	Fluid dynamics
2	av41092	41092	1683902	Finite element analysis
3	bayer01	57735	277774	Chemistry
4	bbmat	38744	1771722	Fluid dynamics
5	comp2c	16783	578665	Linear programing
6	e40r5000	17281	553956	Fluid dynamics
7	ecl32	51993	380415	Circuit Simulation
8	epb3	84617	463625	Thermodynamics
9	fidap011	16614	1091362	Fluid dynamics
10	fidapm11	22294	623554	Fluid dynamics
11	invextr1	30412	1793881	Fluid dynamcis
12	lhr34c	35152	764014	Chemical engineering
13	mil053	530238	3715330	Structural engineering
14	mixtank	29957	1995041	Fluid dynamics
15	nasarb	54870	2677324	Structural engineering
16	onetone1	36057	341088	Circuit Simulation
17	onetone2	36057	227628	Circuit Simulation
18	pre2	659033	5959282	Circuit Simulation
19	raefsky3	21200	1488768	Fluid dynamics
20	raefsky4	19779	1316789	Fluid dynamics
21	rma10	46835	2374001	Fluid dynamics
22	tib	18510	1451491	Circuit simulation
23	twotone	120750	1224224	Circuit simulation
24	wang3	26064	177168	Semicond. dev. simulation
25	wang4	26068	177196	Semicond. dev. simulation

in PARDISO. However, there is no guarantee that the numerical factorization algorithm would always succeed in finding a suitable pivot within the supernode blocks. When the algorithm reaches a point where it can not factor the supernode based on the predescribed inner supernode pivoting, it uses a static pivoting strategy. The strategy suggests to keep the pivotal sequence chosen in the analysis phase and the magnitude of the potential pivot is tested against a threshold of $\sqrt{\epsilon} \cdot ||A||$, where ϵ is the machine precision and $||A||$ is the norm of A. Therefore, in step (3), any tiny pivots encountered during elimination is set to $\sqrt{\epsilon} \cdot ||A||$ — this trades off some numerical stability for the ability to keep pivots from getting to small. The result is that the factorization is in general not exact and iterative refinement may be needed in step (4). If iterative refinement does not converge, an iterative CGS algorithm [21] with the perturbed factors L and U as a preconditioner is used.

The numerical behavior of this approach is illustrated in Table 2, where the number of steps of iterative refinement required to reduce the componentwise relative backward error Berr $= \max_i \frac{|Ax-b|_i}{(|A|\cdot|x|+|b|)_i}$ [3] to machine precision is

---

(1) Row/column equilibration $A \leftarrow P_r \cdot D_r \cdot A \cdot D_c$, where D_r and D_c are diagonal matrices and P_r is a row permutation to maximize the magnitude of the diagonal entries.

(2) Find a permutation P_c to preserve sparsity: $A \leftarrow P_c \cdot A \cdot P_c^T$ based on $\hat{A} = A + A^T$.

(3) Level-3 Block BLAS factorization $A = P_s L U Q_s$, with supernode pivoting permutations P_s and Q_s. The growth of diagonal elements is controlled with:

    if $(|l_{ii}| < \sqrt{\epsilon} \cdot ||A||)$ **then**
       set $l_{ii} = \sqrt{\epsilon} \cdot ||A||$
    endif

(4) Solve $Ax = b$ using the block L and U factors, the permutation matrices P_c, P_s, P_r, P_s, and Q_s and iterative refinement.

(5) If iterative refinement fails, try CGS with the perturbed factors L and U as a preconditioner.

Fig. 1. Pseudo-code of the supernode pivoting algorithm for general unsymmetric sparse matrices.

shown and the true error is reported as Err $= \frac{||x_{true}-x||}{||x_{true}||}$ (computed from a constant solution $x_{true} = 1$.) It can be seen from the table that it is possible to solve nearly all unsymmetric matrices with the predescribed algorithm. A '*' behind the matrix name indicates that supernode pivoting is necessary to obtain convergence. For matrix pre2 three iteration of the CGS algorithm are necessary to reduce the error by a factor of 10^4, hence the missed subspace is really small.

Parallel LU algorithm with a two-level scheduling

The details of the dynamic scheduling algorithm are described in [18]. The left–right looking approach is writing synchronization data to supernodes that will be factored in the future (right-looking phase) but is reading all numerical data from supernodes processed in the past. To reduce the number of synchronization events and to introduce a smooth transition from tree level to pipelining parallelism the often used central queue of tasks is split into two: the first queue is used to schedule complete subtrees which have a local root node sufficiently far from the root node of the elimination tree. The supernodes inside the subtrees need not any synchronization and hence update the synchronization data of the supernodes of the second kind only (those close to the root node — which are in general large). These supernodes are scheduled by a second queue — contrary to the first task queue the second one keeps track of individual outer supernode updates. Whenever a process can not continue with processing outer updates due missing factorization results it puts the partially processed supernode back into the second queue of tasks and fetches any other supernode la-

Table 2. The numerical behavior of the PARDISO supernode pivoting approach. Err indicates the error, Berr the backward error, Res the norm of the residual, and Nb the number of steps of iterative refinement. A '*' after the matrix name indicates that supernode pivoting is necessary to obtain convergence. 'CGS' indicates that a CGS iterations is used to improve the solution.

Matrices	Err	Berr	Res	Nb	Err	Berr	Res
af23560	1.4e-12	2.8e-13	5.4e-11	2	2.5b-13	3.2e-16	1.6e-11
av41092	fail	fail	fail	fail	fail	fail	fail
bayer01*	1.0e-05	1.0e-08	2.6e-10	2	1.4e-06	3.6e-16	6.4e-14
bbmat	1.1e-08	3.6e-16	1.0e-11	1	1.2e-08	3.5e-16	8.6e-12
comp2c	2.7e+01	2.7e-04	5.1e-03	4	6.9e-07	2.0e-16	1.0e-12
e40r5000	1.8e+03	1.5e-02	1.6e+00	4	1.9e-10	1.1e-16	3.2e-11
ecl32	1.6e-03	2.5e-08	4.6e-06	3	6.2e-11	2.6e-16	3.8e-12
epb3	6.0e-12	8.9e-14	1.2e-14	2	9.6e-13	2.5e-16	5.1e-15
fidap011*	2.6e-04	2.8e-15	2.3e-06	2	2.6e-06	4.1e-16	1.2e-06
fidapm11*	4.6e-01	1.4e-04	6.0e-06	11	3.4e-12	2.8e-16	4.0e-15
invextr1*	1.4e+04	2.0e-05	5.4e+0	5	1.8e-06	1.1e-15	1.9e-06
mil053	3.7e-10	2.5e-12	8.7e-10	2	9.2e-10	2.3e-16	2.5e-11
mixtank*	8.8e-04	1.7e-14	2.5e-09	4	5.7e-11	3.5e-16	1.9e-14
nasarb*	8.5e-08	8.8e-14	6.5e-06	2	8.5e-10	4.8e-16	6.5e-06
onetone1	4.2e-10	6.6e-13	4.0e-10	2	6.9e-11	3.2e-16	1.3e-11
onetone2	5.1e-10	3.3e-12	2.5e-10	2	6.9e-11	2.3e-16	1.3e-11
pre2	2.3e-01	1.9e-01	2.6e-01	3 CGS	4.3e-05	4.0e-15	3.4e-05
raefsky3	2.9e-09	1.3e-10	5.5e-14	2	3.1e-16	4.6e-16	2.6e-16
raefsky4	8.3e-08	4.4e-09	1.2e-02	2	1.4e-12	4.0e-16	1.6e-04
rma10	1.3e-10	4.4e-16	1.4e-07	2	1.3e-10	4.3e-16	1.43-07
tib	2.2e-09	1.6e-12	5.5e-12	2	6.9e-12	5.5e-16	5.6e-13
twotone	2.2e-08	7.8e-13	1.4e-10	2	1.0e-10	7.8e-16	5.8e-11
wang3	6.4e-10	1.1e-14	3.4e-15	2	1.1e-11	2.5e-16	4.3e-16
wang4	1.0e-11	5.6e-15	4.2e-15	2	1.7e-12	2.6e-16	4.2e-16

beled as executable. This scheme exploits the larger granularity of outer updates (down to single dense matrix-matrix multiply operations), does not force small outer updates to be handled as individual tasks and is open to priority control strategies.

Unfortunately the complexity is already large. The introduction of factorization time dependence is possible, but the additional updates due to out of supernode pivoting have to be handled without a serious degradation of the parallel performance reached by the scheme up to now.

3   Experimental results

Table 3 list the performance numbers of some state-of-the art packages for solving large sparse systems of linear equations on a single IBM Power 3 processor.

Table 3. LU factorization times (in seconds) on a single 375 Mhz IBM power 3 processor for UMFPACK 3, MUMPS, WSMP, and PARDISO (with prereordering MC64 and METIS) respectively. The best time is shown in boldface, the second best time is underlined, and the best operation count is indicated by . The last row shows the approximate smallest relative pivot threshold that yielded a residual norm close to machine precision after iterative refinement for each package ([12, 13]).

Matrices	MUMPS time (in sec)	MUMPS ops ×10⁹	UMFPACK 3 time (in sec)	UMFPACK 3 ops ×10⁹	WSMP time (in sec)	WSMP ops ×10⁹	PARDISO time (in sec)	PARDISO ops ×10⁹
af23560	**3.89**	2.56	9.07	3.46	3.96	3.27	4.52	3.33
av41092	21.0	10.9	128.	37.4	4.56	2.14	fail	fail
bayer01	2.54	.697	1.11	.024	0.95	.040	**0.57**	.126
bbmat	54.3	41.6	88.3	39.1	**22.9**	20.1	52.6	27.9
comp2c	10.5	4.84	597.	113.	**1.64**	0.78	9.20	2.08
e40r5000	14.5	5.43	6.76	2.09	1.08	.521	**0.85**	.456
ecl32	64.2	64.6	191.	112.	**23.1**	21.0	29.9	22.4
epb3	2.84	1.17	5.77	1.34	1.66	.452	**1.54**	.441
fidap011	8.58	7.01	18.5	8.51	**3.93**	3.20	4.51	3.65
fidapm11	11.9	10.0	40.9	20.0	**6.50**	5.21	11.9	8.91
invextr1	80.7	71.5	178.	89.4	**9.93**	6.90	16.7	12.3
mil053	43.5	31.8	107.	46.2	23.0	14.4	**22.9**	13.7
mixtank	151.	141.	398.	243.	**21.9**	19.5	88.5	80.1
nasarb	12.8	9.45	55.9	28.2	**6.98**	5.41	9.42	7.02
onetone1	17.1	8.19	5.58	2.33	2.25	1.25	**4.14**	2.25
onetone2	1.67	.605	.760	.080	**.720**	.191	.901	.438
pre2	fail	fail	fail	fail	**127.**	96.3	fail	fail
raefsky3	4.44	2.90	16.0	7.87	**3.16**	2.57	3.22	2.63
raefsky4	107.	74.4	26.6	12.9	**4.91**	4.11	5.83	4.73
rma10	4.00	1.39	8.83	3.44	**2.47**	1.48	3.20	1.92
tib	.560	.122	28.1	.203	**.350**	.064	.201	.032
twotone	56.5	38.3	31.6	10.8	**13.5**	9.46	59.3	37.4
wang3	72.9	57.8	40.6	24.2	6.65	5.91	**5.88**	4.58
wang4	11.8	10.5	53.4	30.7	6.84	6.09	**3.94**	3.02
Thresh	0.01		0.20		0.01		–	

This table is shown to locate the performance of PARDISO against other well-kown software packages. A detailed comparison can be found in [12, 13]. A "fail" indicates that the solver ran out of memory, e.g. MUMPS [2], UMFPACK [6], or the iterative refinement did not converge, e.g PARDISO. The default option of the PARDISO was a nested dissection ordering and a prepermutation with MC64 for all matrices.

Table 4. Operation count (Ops), LU factorization time in seconds, and speedup (S) of WSMP and PARDISO on one (T_1) and four (T_4) 375 MHz IBM Power 3 processors with default options. The best time with four processors is shown in boldface, the best time with one processor is underlined.

Matrices	WSMP				PARDISO			
	ops $\times 10^9$	T_1 (s)	T_4 (s)	S $\frac{T_1}{T_4}$	ops $\times 10^9$	T_1 (s)	T_4 (s)	S $\frac{T_1}{T_4}$
af23560	3.27	<u>3.96</u>	2.27	1.8	3.33	4.52	**1.27**	3.6
bayer01	1.57	0.95	0.95	1.0	.126	<u>.571</u>	**0.31**	1.8
bbmat	20.1	<u>22.9</u>	**8.26**	2.8	27.9	52.6	13.6	3.9
comp2c	0.78	<u>1.64</u>	**0.67**	2.4	2.08	9.20	2.81	3.3
ecl32	21.0	<u>23.1</u>	**7.41**	3.1	22.4	29.9	8.81	3.4
epb3	.452	1.66	1.23	1.4	.441	<u>1.54</u>	**.811**	1.9
fidap011	3.20	<u>3.93</u>	1.78	2.2	3.65	4.51	**1.54**	2.9
invextr1	6.90	<u>9.93</u>	**4.67**	2.1	12.3	16.7	5.23	3.2
lhr34c	.163	<u>0.92</u>	0.93	1.0	.552	2.60	**0.90**	2.9
mil053	14.4	23.0	10.6	2.2	13.7	<u>22.9</u>	**9.91**	2.3
nasarb	5.41	<u>6.98</u>	3.37	2.1	7.02	9.42	**2.49**	3.8
onetone1	1.25	<u>2.25</u>	**1.52**	1.5	2.25	4.14	1.85	2.2
onetone2	.191	<u>0.72</u>	0.72	1.0	.438	.901	**0.31**	2.9
raefsky3	2.57	<u>3.16</u>	1.40	2.3	2.63	3.22	**0.92**	3.5
raefsky4	4.11	<u>4.91</u>	2.34	2.1	4.73	5.83	**1.69**	3.4
rma10	1.48	<u>2.47</u>	**0.99**	2.5	1.92	3.20	1.04	3.1
twotone	9.46	<u>13.5</u>	**9.05**	1.5	37.4	59.3	17.8	3.3
venkat50	1.75	<u>2.83</u>	1.13	2.5	1.83	3.21	**0.93**	3.4
wang3	5.91	6.65	3.50	1.9	4.58	<u>5.88</u>	**1.82**	3.2
wang4	6.09	6.84	3.08	2.2	3.02	<u>3.94</u>	**1.27**	3.1

For the parallel performance and scalability, the LU factorization of PAR-DISO is compared with that of WSMP in Table 4. The experiments were conducted with one and four IBM 375 Mhz Power 3 processors. The four processors all have a 64 KB level-1 cache and a four MB level-2 cache. WSMP uses the Pthread library and PARDISO uses the OpenMP parallel directives. Both solver always permute the original matrix to maximize the product of diagonal elements and nested-dissection based fill-orderings has been used [11,14]. Two important observation can be drawn from the table. The first is that WSMP needs in most of the examples less operations than PARDISO. It seems that the algorithm based on [11] produces orderings with a smaller fill-in compared with [14], which is used in PARDISO. The second observation is that the factorization times are affected by the preprocessing and WSMP is in most cases faster on a single Power 3 processor. However, the two-lewel scheduling in PARDISO provides better scalability and hence better performance with four Power 3 processors.

4   Concluding remarks

The focus of the comparison is mainly on the WSMP and the PARDISO packages[2] and their different approaches:

— stability with a general pivoting method and the dynamic directed acyclic task dependency graphs in WSMP and

— supernode pivoting with a preordering, and undirected graph partioning in PARDISO, where the efficient dynamic scheduling on precomputed communication graphs results in better speedup.

For different application areas the PARDISO approach results in the needed stability. It has reached a development status that can be improved mainly in the following directions:

1. Better reordering schemes to reduce the operation count.
2. Adding dynamic pivoting on the basis of introducing exceptions (what seems to be a justified assumption for the test matrices used – this set produces only a few necessary pivots outside the supernodes at the very beginning of the factorization process), and
3. Excluding some of the systematic operations with zeros in a postprocessing step without loosing the advantages of the left–right looking supernodal approach.

The different techniques used in both approaches may stimulate further improvements but it may be hard to reach the robustness and the operation counts of WSMP for unsymmetric matricces and the better scalability of PARDISO on SMPs.

Acknowledgments

The authors wish to thank Anshul Gupta, IBM T.J. Watson Research Center, for providing his large benchmark set of unsymmetric matrices, Iain Duff for the possibility to use the MC64 graph matching code, and the Computing Center at the University of Karlsruhe for supporting access to the IBM NightHawk-II parallel computers.

References

1. P. R. Amestoy, I. S. Duff, J.-Y. L'Excellent, and X. S. Li. Analysis and comparison of two general sparse solvers for distributed memory computers. Technical Report TR/PA/00/90, CERFACS, Toulouse, France, December 2000. Submitted to *ACM Trans. Math. Softw.*

2. Patrick R. Amestoy, Iain S. Duff, Jean-Yves L'Excellent, and Jacko Koster. A fully asynchronous multifrontal solver using distributed dynamic scheduling. *SIAM J. Matrix Analysis and Applications*, 23(1):15–41, 2001.

[2] The data presented go directly back to the authors at sufficiently close points in time.

3. Mario Arioli, James W. Demmel, and Iain S. Duff. Solving sparse linear systems with sparse backward error. *SIAM J. Matrix Analysis and Applications*, 10:165–190, 1989.
4. R.E. Bank, D.J. Rose, and W. Fichtner. Numerical methods for semiconductor device simulation. *SIAM Journal on Scientific and Statistical Computing*, 4(3):416–435, 1983.
5. T. A. Davis and I. S. Duff. An unsymmetric-pattern multifrontal method for sparse LU factorization. *SIAM J. Matrix Analysis and Applications*, 18(1):140–158, 1997.
6. Timothy A. Davis. UMFPACK. Software for unsymmetric multifrontal method. In *NA Digest, 01(11), March 18, 2001.*, http://www.cise.ufl.edu/research/sparse/umfpack.
7. J. Demmel, J. Gilbert, and X. Li. An asynchronous parallel supernodal algorithm to sparse partial pivoting. *SIAM Journal on Matrix Analysis and Applications*, 20(4):915–952, 1999.
8. J. W. Demmel, S. C. Eisenstat, J. R. Gilbert, X. S. Li, and J. W.-H. Liu. A supernodal approach to sparse partial pivoting. *SIAM J. Matrix Analysis and Applications*, 20(3):720–755, 1999.
9. I. S. Duff and J. Koster. The design and use of algorithms for permuting large entries to the diagonal of sparse matrices. Technical Report TR/PA/97/45, CERFACS, Toulouse, France, 1997. Also appeared as Report RAL-TR-97-059, Rutherford Appleton Laboratories, Oxfordshire.
10. I. S. Duff and J. Koster. The design and use of algorithms for permuting large entries to the diagonal of sparse matrices. *SIAM J. Matrix Analysis and Applications*, 20(4):889–901, 1999.
11. A. Gupta. Fast and effective algorithms for solving graph partitioning and sparse matrix ordering. *IBM Journal of Research and Development*, 41(1/2):171–183, January/March 1997.
12. A. Gupta. Improved symbolic and numerical factorization algorithms for unsymmetric sparse matrices. Technical Report RC 22137 (99131), IBM T. J. Watson Research Center, Yorktown Heights, NY, August 1, 2001.
13. A. Gupta. Recent advances in direct methods for solving unsymmetric sparse systems of linear equations. Technical Report RC 22039 (98933), IBM T. J. Watson Research Center, Yorktown Heights, NY, April 20, 2001.
14. G. Karypis and V. Kumar. A fast and high quality multilevel scheme for partitioning irregular graphs. *SIAM Journal on Scientific Computing*, 20(1):359–392, 1998.
15. X.S. Li and J.W. Demmel. A scalable sparse direct solver using static pivoting. In *Proceeding of the 9th SIAM conference on Parallel Processing for Scientic Computing*, San Antonio, Texas, March 22-34,1999.
16. E.G. Ng and B.W. Peyton. Block sparse Cholesky algorithms on advanced uniprocessor computers. *SIAM Journal on Scientific Computing*, 14:1034–1056, 1993.
17. O. Schenk. *Scalable Parallel Sparse LU Factorization Methods on Shared Memory Multiprocessors.* PhD thesis, ETH Zürich, 2000.
18. O. Schenk and K. Gärtner. Two-level scheduling in PARDISO: Improved scalability on shared memory multiprocessing systems. Accepted for publication in *Parallel Computing.*
19. O. Schenk and K. Gärtner. PARDISO: a high performance serial and parallel sparse linear solver in semiconductor device simulation. *Future Generation Computer Systems*, 789(1):1–9, 2001.
20. O. Schenk, K. Gärtner, and W. Fichtner. Efficient sparse LU factorization with left-right looking strategy on shared memory multiprocessors. *BIT*, 40(1):158–176, 2000.
21. P. Sonneveld. CGS, a fast Lanczos-type solver for nonsymmetric linear systems. *SIAM Journal on Scientific and Statistical Computing*, 10:36–52, 1989.

A Multipole Approach for Preconditioners

Ph. Guillaume, A. Huard and C. Le Calvez

MIP, UMR 5640
INSA, Département de Mathématiques
135 Avenue de Rangueil, 31077 Toulouse Cedex 4, France
guillaum,huard,lecalvez@gmm.insa-tlse.fr

Abstract. A new class of approximate inverse preconditioners is presented for solving large linear systems with an iterative method. It is at the intersection of multipole, multigrid and SPAI methods. The method consists in approximating the inverse of a matrix by a block constant matrix, instead of approximating it by a sparse matrix as in SPAI methods. It does not require more storage, or even less, and it is well suited for parallelization, both for the construction of the preconditioner and for the matrix-vector product at each iteration.

1 Introduction

Recently, multipole methods [9, 4, 2, 19] have dramatically improved the solution of scattering problems in Electromagnetism. The basic idea behind multipole methods consists in a low rank approximation of far field interactions. The matrix A obtained when using integral equations can at the first glance be considered as an approximation of the Green function $G(x, y)$ of the problem, that is $A_{ij} \simeq G(x_i, x_j)$ if the x_i's are the discretization points. The matrix A is dense, and for large problems it becomes impossible to store the whole matrix. When two points x^* and y^* are distant from each other, the starting point of the multipole approach is a separate variables approximation

$$G(x, y) \simeq \sum_{n=1}^{r} u_n(x)v_n(y)$$

which is valid for x close to x^* and y close to y^*. It leads to a low-rank approximation of the associated block of the matrix A :

$$A_{IJ} \simeq UV^T.$$

Here I and J are sets of indices of points respectively close to x^* and y^*. The sizes of A_{IJ}, U and V are respectively $|I| \times |J|$, $|I| \times r$ and $|J| \times r$. Hence both memory and computational time are saved if $r << |I|$ and $r << |J|$. When x^* and y^* are close to each other, the above approximation is not valid anymore, and $G(x^*, y^*)$ must be computed more carefully. This approach is general, and relies on the fact that the Green function associated to a pseudo-differential operator is singular on the diagonal, but regular outside.

P.M.A. Sloot et al. (Eds.): ICCS 2002, LNCS 2330, pp. 364–373, 2002.

The context looks quite different when considering finite element methods: the matrix A issued from the discretization of a partial differential equation is usually sparse, and there is a priori no interest for using a low-rank approximation of almost zero blocks (see however [3]). But its inverse A^{-1} is a dense matrix and, like the matrix issued from an integral equation, it is associated to a Green function $G(x, y)$ which is singular on the diagonal $x = y$ but smooth outside the diagonal. In the context of an approximate inverse used as a preconditioner in an iterative method, one can think of approximating off-diagonal blocks of A^{-1} by low-rank matrices. Since it leads to a number of unknowns significantly larger than in the original problem and because we do not need such a good approximation as in the case of integral equations, we can go even further in the approximation: off-diagonal blocks $(A^{-1})_{IJ}$ can be simply approximated by constant blocks. The size of the constant blocks can vary: smaller when they are close to the diagonal, and getting larger away from it. This approach makes sense in the case of non oscillatory Green functions associated to an elliptic equation like Poisson's equation or elasticity equations. This relies on the fact that piecewise constant functions can well approximate the Green function of the problem. It would not be well suited for Green functions arising for example from Helmholtz equations.

Like in the multipole method (see *e.g.* [7] [15]), a crucial point is the ordering of the unknowns. They need to be sorted by proximity, that is, in such a way that unknowns associated to neighboring points must be grouped together, and vice-versa. When the nodal table which has been used for assembling the matrix is available, a simple way to achieve this is to use a recursive coordinates bisection, but more sophisticated methods are available like recursive graph bisections or recursive spectral bisection [16] [18] [17] which do not require the nodal table.

In this paper, we focus our attention on the solution of systems issued from the discretization of elliptic partial differential equations, leading to a sparse symmetric and positive definite (SPD) matrix A. Our goal is to obtain a block constant approximate inverse C of the matrix A, used for preconditioning an iterative method.

Section 2 describes how to determine such a Block Constant Preconditioner (BCP) together with it's relation to multipole, multigrid and sparse approximate inverse (SPAI) methods. Section 3 reports some numerical experiments: the BCP is compared to a basic SPAI using the sparsity pattern of the original matrix A [6] and to the incomplete Cholesky factorization with no fill-in IC(0).

2   Description of the Block Constant Preconditioner

Consider a linear system

$$Ax = b, \quad x, b \in \mathbb{R}^n$$

where $A \in \mathcal{M}_n(\mathbb{R})$ is an SPD matrix. As mentioned in the introduction, the unknowns must be ordered by proximity, like in multipole methods. In the sequel, we suppose that this reordering has been done.

When the dimension n is large, an iterative method like the preconditioned conjugate gradient (PCG) is often used for solving such a system. It consists in applying the conjugate gradient algorithm to a system of the form $MAx = Mb$. Here M is an explicit left preconditioner. It should also be SPD. Right preconditioners can be used in a similar way, and will not be discussed here. The BCP preconditioner M is of the form

$$M = C + \omega I, \quad \omega > 0 \tag{1}$$

where I is the identity matrix and C is a block constant matrix (BCM), which consists in rectangular blocks of variable size whose elements are constant. The steps for computing the BCP of the matrix A are the following:

- determine the pattern of the BCM, i.e., location and size of the different constant blocks,
- compute the constants associated to this pattern by minimizing some Frobenius norm of $CA - I$ over the set of matrices having the same pattern,
- choose the parameter ω.

The way of computing C resembles SPAI methods [13] [11] [5]. The computation of C as well as the matrix-vector preconditioning operation is highly parallelizable. The difference lies in the fact that the approximation of the Green function $G(x, y)$ by a sum of discrete Dirac functions (SPAI methods) is replaced by a piecewise constant function, which is likely to offer a better approximation outside the diagonal $x = y$. For some particular cases of pattern of the BCM, the method becomes very close to a two-level multigrid method [10] [14]. In the general case, the difference lies in an attempt to simulate a cycle over several grids in a single operation. Hence, the BCP method is at the intersection of multipole, multigrid and SPAI methods.

2.1   Determination of the pattern of the BCM

A BCM pattern is obtained by a recursive splitting of the initial matrix A. The depth of recursiveness (the level of refinement) is determined by three parameters l_c, l_d and l_o:

- l_c is the coarsest level of refinement, and fixes the size of the largest blocks of the BCM,
- l_o is an intermediate level of refinement and determines the size of the smallest off-diagonal blocks of the BCM,
- l_d is the finest level of refinement and determines the size of the blocks of the BCM containing the diagonal elements.

Let d be the integer defined by $2^{d-2} < n \leq 2^{d-1}$. It is supposed here that $l_c \leq l_o \leq l_d \leq d$. The two extreme situations correspond to $l_d = 1$, in which case the whole matrix is constant, and to $l_c = d$, in which case each block of the BCM is of size 1×1.

Different values of these parameters lead to different sequences of refinement, and consequently to different patterns. Each refinement consists in splitting a block K into four blocks of equal size if possible. The algorithm is the following:

REFINEMENT ALGORITHM FOR DETERMINING THE PATTERN

$lev = 1$; K is a $n \times n$ block of level 1;

for $lev = 2 : l_c$

    split each $k \times l$ block K of level $lev - 1$ into 4 blocks of size $k_i \times l_j$

    where $k_1 \geq k_2$, $k_1 + k_2 = k$, $k_1 - k_2 \leq 1$, and $l_1 \geq l_2$, $l_1 + l_2 = l$, $l_1 - l_2 \leq 1$;

endfor

for $lev = l_c + 1 : \max(l_o, l_d)$

    for each block K of level $lev - 1$

        if

            K is a diagonal block and $lev \leq l_d$

            or

            K is not a diagonal block and $lev \leq l_o$

        and

            the corresponding block of A has nonzero elements

        then

            split the block K into 4 blocks following the previous rule as long as possible: if $\min(k, l) = 1$, just split it into 2 blocks when $k \neq l$ and of course do not split it if $k = l = 1$;

        endif

    endfor

endfor

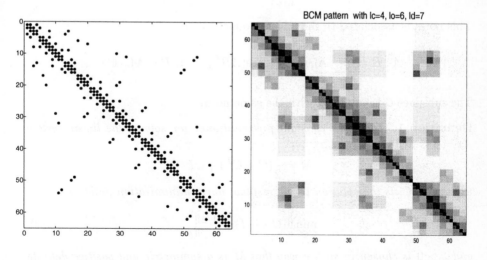

Fig. 1. Reordered A and example of BCM pattern

Note that the obtained pattern is also symmetric if A is symmetric. Following this algorithm, the BCM in Fig. 1 (right) has been obtained from the classical Poisson's matrix of size 64×64 reordered by a recursive coordinates bisection method (Fig. 1, left). With $l_c = 4$, $l_o = 6$ and $l_d = 7$, the largest blocks are 8×8, the smallest off-diagonal blocks are 2×2 and the smallest diagonal blocks are 1×1 (as all A_{ii} are nonzero, all diagonal blocks are in fact of size 1×1). The parameter l_d makes it possible to have a finer refinement on the diagonal where the Green function is singular. Each color in Fig. 1 (right) corresponds to a constant block. One can see that the blocks are smaller at the locations where the matrix A has nonzero elements. As in SPAI methods, many other algorithms can be proposed. For example, instead of using the pattern of A, one could use the pattern of A^s for a certain integer s.

2.2   Definition of the BCP

Once the pattern of the BCM has been determined, it remains to compute the values c_i of the constants for each block. Let $\mathcal{C} \subset \mathcal{M}_n(\mathbb{R})$ denote the linear space of matrices which satisfy a given pattern. A basis of this space consists in matrices E_i which have a single nonzero block (corresponding to the given pattern) with value 1. In this basis, a matrix $C \in \mathcal{C}$ can be written

$$C = \sum_{i=1}^{n_b} c_i E_i$$

where n_b is the dimension of $\mathcal{C}$ (the number of blocks). The space $\mathcal{M}_n(\mathbb{R})$ is equipped with the Frobenius norm defined by its square

$$\|A\|_F^2 = \sum_{i,j=1}^{n} A_{ij}^2, \quad A \in \mathcal{M}_n(\mathbb{R}).$$

This norm is associated to the scalar product

$$A : B = \sum_{i,j=1}^{n} A_{ij} B_{ij} = \text{trace}(AB^T), \quad A, B \in \mathcal{M}_n(\mathbb{R}).$$

The subspace $\mathcal{C}$ is equipped with the same norm.

Definition 1. *The block constant preconditioner for solving the linear system* $Ax = b$ *is defined by*

$$M = \frac{1}{2}(C + C^T) + \omega I$$

where $C \in \mathcal{C}$ *is the solution to the residual norm minimization problem*

$$\min_{C \in \mathcal{C}} \left\| (CA - I)A^{-1/2} \right\|_F^2 \tag{2}$$

and $\omega > 0$ *is chosen in such a way that* M *is a symmetric and positive definite matrix.*

If $l_c = l_d$ and if A is symmetric, then it can be shown that C is also symmetric, thus $M = C + \omega I$. When $l_d = d$, it may not be necessary to add a diagonal term ωI, because in that case C contains already the diagonal matrices. The larger is l_d, the larger is the number n_b of unknowns involved in the BCP, however the numerical results of Section 3 show that taking $l_d < d$ performs well for a reasonable number $n_b < n$ of constant blocks. When $l_d < d$, adding some diagonal terms becomes necessary because, as in multigrid methods, CA is not invertible and some smoother must complete the preconditioning operation CA. More general definitions are possible, as for example $M = C + D$ with D diagonal, and M minimizing $\|(MA - I)A^s\|_F$. The value $s = -1/2$ has a precise meaning only for SPD matrices, although the optimality condition (3) can be used for any kind of matrices, without the guaranty of a minimum. A simple choice for non SPD matrices is to take $s = 0$.

2.3 Computation of the BCM

The computation of the constants c_i follows from the next proposition.

Proposition 1. *If the matrix A is symmetric and positive definite, Problem (2) has a unique solution, which is also the solution to the linear system of equations*

$$(CA - I) : H = 0, \quad \forall H \in \mathcal{C}. \tag{3}$$

This linear system has a unique solution, and for $C = \sum_{i=1}^{n_b} c_i E_i$, it reads

$$Nc = g, \quad N \in \mathcal{M}_{n_b}(\mathbb{R}), \, c, g \in \mathbb{R}^{n_b} \tag{4}$$

where for $i, j = 1, \ldots, n_b$

$$N_{ij} = A : E_j^T E_i, \quad g_i = \text{trace}(E_i).$$

Proof. Equ. (3) is the optimality condition associated to Problem (2). Furthermore, the associated homogeneous system $CA : H = 0$ for all $H \in \mathcal{C}$ has the unique solution $C = 0$ (taking $H = C$ yields $\left\|CA^{1/2}\right\|_F^2 = 0$). Hence (3) has a unique solution, which is given by (4) when using the basis $(E_i)_{i=1}^{n_b}$ defined in Section 2.2.

The matrix N is block diagonal with $p = 2^{l_c - 1}$ sparse blocks on the diagonal. Hence its computation as well as solving Equ. (4) are parallelizable. Depending on the distribution of the matrix A among the processors, its parallelization may be complicated but still remains possible. Similarly, the matrix-vector preconditioning operation is parallelizable.

3 Numerical Results

We present some numerical experiments with the Poisson equation on a square with an homogeneous Dirichlet boundary condition, solved by using the five-point finite difference stencil reordered by the recursive coordinate bisection algorithm [16].

The BCP is compared to the SPAI(1) (same pattern as A for the sparse approximate inverse) and IC(0) preconditioners. The incomplete Cholesky factorization is of the form $A = U^T U + R$. The three preconditioners are used with the conjugate gradient algorithm. The experiments presented here were performed

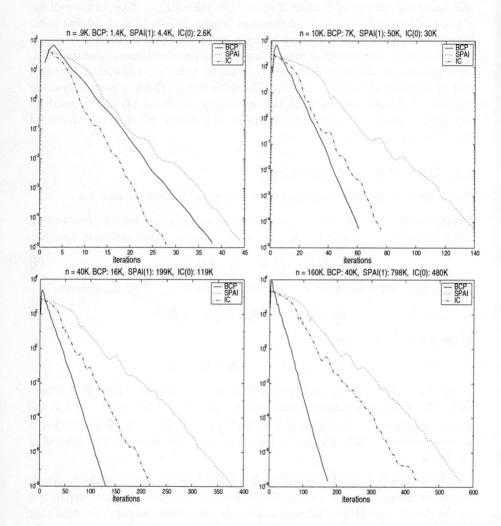

Fig. 2. Comparison BCP - SPAI - IC(0), $n = 900, 10000, 40000, 160000$.

with MATLAB and they all used the following parameters for the BCP:

```
omega = 1.5/normest(A);
d = ceil(log2(n))+1;
lc = ceil(d/2)-2;
lo = lc+3; ld = lo+2;
```

The function normest(A) computes an approximation of $||A||_2$ and ceil(x) is the smallest integer $i \geq x$. The letters omega, d, lc, ld and lo denote respectively ω, d, l_c, l_d and l_o (see Section 2 for the notation).

A first series of experiments is reported in Fig. 2 which shows the residual 2-norm history for four $n \times n$ matrices with increasing size $n = 900$, 10000, 40000 and 160000. At the top of each graphic is indicated the amount of storage (number of blocks or of nonzero elements) which was used by each preconditioner. One can observe that the BCP takes a clear advantage when n becomes large, and moreover it uses ten times less storage than IC(0) in the last case $n = 160000$. For an adequate coding of the BCP matrix-vector multiplication which will not be discussed here, the number of elementary operations per iteration is comparable.

A second series of experiments is reported in Fig. 3, which plots the number of iterations needed for obtaining a given precision (relative residual norm $\leq 10^{-6}$) versus the size of the system. One can see that the number of iterations with BCP is of order

$$O(h^{-1/2}) = O(n^{1/4}).$$

The results for BCP match the line $y = 6\,n^{1/4}$, and the results with IC(0) match the line $y = 2.5\,n^{0.35}$. For these experiments, the chosen values for n_b (number

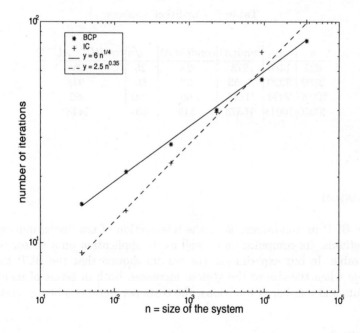

Fig. 3. Number of iterations with respect to the size n.

of blocks) and nnz(U) (number of nonzero elements in the matrix U) are given in Table 1. They indicate the amount of storage used by the preconditioners and they satisfy asymptotically

$$n_b \simeq \frac{n}{4}, \quad \text{nnz}(U) \simeq 3\,n.$$

Table 1. Storage used by the preconditioners.

n	36	144	576	2300	9200	36900
n_b	208	616	1300	2900	6300	15000
nnz(U)	90	408	1700	6900	27600	111000

A third series of experiments is reported in Table 2, which shows the spectral condition numbers $\lambda_{\max}/\lambda_{\min}$ of the matrices A, MA and $U^{-T}AU^{-1}$, with $M = C+\omega I$ and U defined above. We used the MATLAB function eigs fore computing them. One can observe that the condition number of MA is of order $\sqrt{n}$.

Table 2. Condition numbers.

n	n_b	cond(A)	cond(MA)	$\sqrt{n}$	cond($U^{-T}AU^{-1}$)
625	1480	273	23	25	25
2500	3220	1053	38	50	93
10000	7024	4133	65	100	366
40000	16048	16373	119	200	1448

Conclusion

The new BCP preconditioner is at the intersection of multipole, multigrid and SPAI methods. Its computation as well as its application on a vector is highly parallelizable. In our experiments the results showed that the BCP takes the advantage when the size of the system increases, both in terms of memory requirements and number of iterations, the costs per iterations being comparable.

References

1. O. Axelsson. Iterative Solution Methods. Cambridge University Press, 1994.
2. S. S. Bindiganavale and J. L. Volakis. Comparison of three fmm techniques for solving hybrid fe-bi systems. *IEEE Trans. Antennas Propagat. Magazine*, vol. 4 No. 4 (1997), 47-60.

3. R. Bramley and V. Meñkov. Low rand off-diagonal block preconditioners for solving sparse linear systems on parallel computers. Tech. Rep. 446, Department of Computer Science, Indiana University, Bloomington,1996.

4. R. Coifman, V. Rokhlin, and S.W Wandzura. The fast multipole method for the wave equation: A pedestrian description. *IEEE Trans. Antennas Propagat. Mag.*, vol. AP-35 No 3 (1993), 7-12.

5. E. Chow and Y. Saad. Approximate inverse preconditioners via sparse-sparse iterations, SIAM Journal on Scientific Computing, 19 (1998), 995-1023.

6. J.D.F. Cosgrove, J.C. Dias. Fully parallel preconditionings for sparse systems of equations, Proceedings of the Second Workshop on Applied Computing, ed. S. Uselton, Tulsa, Oklahoma, The University of Tulsa, 1988, 29-34.

7. E. Darve. Méthodes multipôles rapides : résolution des équations de Maxwell par formulations intégrales. Thèse de doctorat de l'Université Paris 6 (France), 1999.

8. T.F. Dupont, R.P. Kendall, H.H. Rachford. An approximate factorization procedure for solving self-adjoint elliptic difference equations, SIAM J. Numer. Anal., vol 5 (1968), 559-573.

9. N. ENGHETA, W. D. MURPHY, V. ROKHLIN AND M. S. VASSILIOU. The fast multipole methode (fmm) for electromagnetic scattering problems. IEEE Trans. Antennas Propagat., 40 No. 6 (1992) 634–641

10. A. Greenbaum. Iterative methods for solving linear systems. SIAM, 1997.

11. M. Grote, H.S. Simon. Parallel preconditioning and approximate inverse on the Connection Machine, Proceedings of the Sixth SIAM conference on parallel processing for scientific computing, SINCOVEC, R. Ed. SIAM, 519-523.

12. I. Gustafsson. A class of 1st order factorization methods, BIT, 18 (1978), 142-156.

13. L. Yu. Kolotina, A. Yu. Yeremin. On a family of two-level preconditionings of the incomplete block factorization type. Sov. J. Numer. Anal. Math. Modelling 1 (1986), 292-320.

14. G. Meurant. Computer solution of large linear systems. North Holland, 1999.

15. J.R. Poirier. Modélisation électromagnétique des effets de rugosité surfacique. Thèse de doctorat de l'INSA Toulouse (France), 2000.

16. A. Pothen, Alex, Horst D. Simon, Kang-Pu Liou. Partitioning sparse matrices with eigenvectors of graphs. Sparse matrices (Gleneden Beach, OR, 1989). SIAM J. Matrix Anal. Appl. 11 (1990), no. 3, 430-452.

17. Y. Saad. Iterative Methods for Sparse Linear Systems, PWS publishing, New York, 1996.

18. Horst D. Simon, Shang-Hua Teng. How good is recursive bisection ? SIAM J. Sci. Comput. 18 (1997), no. 5, 1436-1445.

19. J. Song, C.-C. Lu, and W. C. Chew. Multilevel fast multipole algorithm for electromagnetic scattering by large complex objects. *IEEE Trans. Antennas Propagat.*, vol. 45 no. 10 (1997), 1488-149.

Orthogonal Method for Linear Systems. Preconditioning

Henar Herrero[1], Enrique Castillo[1,2] and Rosa Eva Pruneda[1]

[1] Departamento de Matemáticas, Universidad de Castilla-La Mancha,
13071 Ciudad Real, Spain.
[2] Departamento de Matemática Aplicada y Ciencias de la Computacin,
Universidad de Cantabria, Santander, Spain.

Abstract. The complexity of an orthogonal method for solving linear systems of equations is discussed. One of the advantages of the orthogonal method is that if some equations of the initial linear system are modified, the solution of the resulting system can be easily updated with a few extra operations, if the solution of the initial linear system is used. The advantages of this procedure for this updating problem are compared with other alternative methods. Finally, a technique for reducing the condition number, based on this method, is introduced.

1 Introduction

Castillo *et al* [1] have introduced a pivoting transformation for obtaining the orthogonal of a given linear subspace. This method is applied to solve a long list of problems in linear algebra including systems of linear equations. Nowadays interest is focussed on iterative methods [2], [3], [4], [5] because they are more suitable for large systems. However, they present some difficulties such as conditioning, and for some problems a direct method can be more satisfactory. The direct methods arising from this transformation have complexity identical to that associated with the Gaussian elimination method (see Castillo *et al* [1]). However, they are specially suitable for updating solutions when changes in rows, columns, or variables are done. In fact, when changing a row, column or variable, a single step of the process allows obtaining (updating) the new solution, whithout the need of starting from scratch again. Therefore a drastic reduction in computational effort is obtained.

The paper is structured as follows. In Section 2 the pivoting transformation is introduced. In Section 3 this transformation is applied to find the general solution of a linear system of equations. In Section 4 the main advantage of the method, i.e., the updating of solutions, is analyzed. In Section 5 the complexity of the method related to its updating facilities is studied. Finally, in Section 6 an strategy to reduce the condition number based on this orthogonal procedure is given.

P.M.A. Sloot et al. (Eds.): ICCS 2002, LNCS 2330, pp. 374–382, 2002.

2 Pivoting Transformation

The main tool to be used in this paper consists of the so called *pivoting transformation*, which transforms a set of vectors $\mathbf{V}^j = \{\mathbf{v}_1^j, \ldots, \mathbf{v}_n^j\}$ into another set of vectors $\mathbf{V}^{j+1} = \{\mathbf{v}_1^{j+1}, \ldots, \mathbf{v}_n^{j+1}\}$ by

$$\mathbf{v}_k^{j+1} = \begin{cases} \mathbf{v}_k^j / t_j^j; & if \ k = j, \\ \mathbf{v}_k^j - \dfrac{t_k^j}{t_j^j} \mathbf{v}_j^j; & if \ k \neq j, \end{cases} \tag{1}$$

where $t_j^j \neq 0$ and $t_k^j; k \neq j$ are arbitrary real numbers. Note that the set $\{t_1^j, t_2^j, \cdots, t_n^j\}$ defines this transformation. In what follows the vectors above are considered as the columns of a matrix $\mathbf{V}^j$.

This transformation can be formulated in matrix form as follows. Given a matrix $\mathbf{V}^j = \left[\mathbf{v}_1^j, \ldots, \mathbf{v}_n^j\right]$, where $\mathbf{v}_i^j, i = 1, \ldots, n$, are column vectors, a new matrix $\mathbf{V}^{j+1}$ is defined via

$$\mathbf{V}^{j+1} = \mathbf{V}^j \mathbf{M}_j^{-1}, \tag{2}$$

where $\mathbf{M}_j^{-1}$ is the inverse of the matrix

$$\mathbf{M}_j = (\mathbf{e}_1, \ldots, \mathbf{e}_{j-1}, \mathbf{t}_j, \mathbf{e}_{j+1}, \ldots, \mathbf{e}_n)^T, \tag{3}$$

and $\mathbf{e}_i$ is the ith column of the identity matrix. In this paper this transformation is asociated with a vector $\mathbf{a}_j$, so that the transpose of $\mathbf{t}_j$ is defined by

$$\mathbf{t}_j^T = \mathbf{a}_j^T \mathbf{V}^j. \tag{4}$$

Since $t_j^j \neq 0$, the matrix $\mathbf{M}_j$ is invertible. It can be proved that $\mathbf{M}_j^{-1}$ is the identity matrix with its jth row replaced by

$$\mathbf{t}_j^* = \frac{1}{t_j^j}\left(-t_1^j, \ldots, -t_{j-1}^j, 1, -t_{j+1}^j, \ldots, -t_n^j\right).$$

This transformation is used in well-known methods, such as the Gaussian elimination method. However, different selections of the t-values lead to completely different results. In this paper this selection is based on the concept of orthogonality, and a sequence of m transformations associated with a set of vectors $\{\mathbf{a}_1, \ldots, \mathbf{a}_m\}$ is assumed.

The main properties of this pivoting transformation can be summarized as follows (see [6]):

1. Given a matrix $\mathbf{V}$, the pivoting transformation transforms its columns without changing the linear subspace they generate, i.e., $\mathcal{L}(\mathbf{V}^j) \equiv \mathcal{L}(\mathbf{V}^{j+1})$.

2. The pivoting process (2) with the pivoting strategy (4) leads to the orthogonal decomposition of the linear subspace generated by the columns of $\mathbf{V}^j$ with respect to vector $\mathbf{a}_j$. Let $\mathbf{a}_j \neq \mathbf{0}$ be a vector and let $t_k^j = \mathbf{a}_j^T \mathbf{v}_k^j; k = 1, \ldots, n$. If $t_j^j \neq 0$, then

$$\mathbf{a}_j^T \mathbf{V}^{j+1} = \mathbf{e}_j^T. \tag{5}$$

In addition, the linear subspace orthogonal to $\mathbf{a}_j$ in $\mathcal{L}(\mathbf{V}^j)$ is

$$\{\mathbf{v} \in \mathcal{L}(\mathbf{V}^j) | \mathbf{a}_j^T \mathbf{v} = 0\} = \mathcal{L}\left(\mathbf{v}_1^{j+1}, \ldots, \mathbf{v}_{j-1}^{j+1}, \mathbf{v}_{j+1}^{j+1}, \ldots, \mathbf{v}_n^{j+1}\right),$$

and its complement is $\mathcal{L}(\mathbf{v}_j^{j+1})$. In other words, the transformation (2) gives the generators of the linear subspace orthogonal to $\mathbf{a}_j$ and the generators of its complement.

3. Let $\mathcal{L}\{\mathbf{a}_1, \ldots, \mathbf{a}_n\}$ be a linear subspace. Then, the pivoting transformation (2) can be sequentially used to obtain the orthogonal set to $\mathcal{L}\{\mathbf{a}_1, \ldots, \mathbf{a}_n\}$ in a given subspace $\mathcal{L}(\mathbf{V}^1)$. Let t_i^j be the dot product of $\mathbf{a}_j$ and $\mathbf{v}_i^j$. Then assuming, without loss of generality, that $t_j^j \neq 0$, the following is obtained

$$\mathcal{L}(\mathbf{V}^j) = \mathcal{L}\left(\mathbf{v}_1^j - \frac{t_1^j}{t_q^j}\mathbf{v}_q^j, \ldots, \mathbf{v}_q^j, \ldots, \mathbf{v}_n^j - \frac{t_n^j}{t_q^j}\mathbf{v}_q^j\right)$$
$$= \mathcal{L}\left(\mathbf{v}_1^{j+1}, \ldots, \mathbf{v}_n^{j+1}\right) = \mathcal{L}(\mathbf{V}^{j+1}),$$

and

$$\mathcal{L}(\mathbf{a}_1, \ldots, \mathbf{a}_j)^\perp \equiv \{\mathbf{v} \in \mathcal{L}(\mathbf{V}^1) | \mathbf{a}_1^T \mathbf{v} = 0, \ldots, \mathbf{a}_j^T \mathbf{v} = 0\} = \mathcal{L}\left(\mathbf{v}_{j+1}^{j+1}, \ldots, \mathbf{v}_n^{j+1}\right).$$

In addition, we have

$$\mathbf{a}_r^T \mathbf{v}_k^{j+1} = \delta_{rk}; \ \forall r \leq j; \ \forall j, \tag{6}$$

where δ_{rk} are the Kronecker deltas.

4. The linear subspace orthogonal to the linear subspace generated by vector $\mathbf{a}_j$ is the linear space generated by the columns of $\mathbf{V}^k$; for any $k \geq j + 1$ with the exception of its pivot column, and its complement is the linear space generated by this pivot column of $\mathbf{V}^k$.

5. The linear subspace, in the linear subspace generated by the columns of $\mathbf{V}^1$, orthogonal to the linear subspace generated by any subset $W = \{\mathbf{a}_k | k \in K\}$ is the linear subspace generated by the columns of $\mathbf{V}^\ell, \ell > \max_{k \in K}$ with the exception of all pivot columns associated with the vectors in W, and its complement is the linear subspace generated by the columns of $\mathbf{V}^\ell, \ell > \max_{k \in K}$ which are their pivot columns.

The following theorem shows how the orthogonal transformation can be used, not only to detect when a vector is a linear combination of previous vectors used in the pivoting process, but to obtain the coefficients of such combination.

Theorem 1 (Linear combination). *Let* $\mathcal{L}\{a_1, \ldots, a_n\}$ *be a linear subspace. Applying sequentially the pivoting transformation (2), if* $t_k^j = 0$ *for all* $k = j, \ldots, n$, *then the* a_j *vector is a linear combination of previous vectors used in the process,*

$$a_j = \rho_1 a_1 + \ldots + \rho_{q-1} a_{j-1}, \tag{7}$$

where $\rho_i = t_i^j = a_j \bullet v_i^j$, *and* v_i^j *is the corresponding pivot column associated with the vector* a_i, *for all* $i = 1, \ldots, j - 1$.

Proof. If $t_k^j = 0$ for all $k = j, \ldots, n$,

$$a_j \in \mathcal{L}^{\perp}\{a_1, \ldots, a_{j-1}\}^{\perp} \equiv \mathcal{L}\{a_1, \ldots, a_{j-1}\}, \tag{8}$$

then $\exists \ \rho_1, \ldots, \rho_{j-1}$, such that $a_j = \rho_1 a_1 + \ldots + \rho_{j-1} a_{j-1}$. For $i = 1, \ldots, n$, we calculate the dot product

$$a_j \bullet v_i^j = (\rho_1 a_1 + \ldots + \rho_{q-1} a_{q-1}) \bullet v_i^j = \rho_1 (a_1 \bullet v_i^j) + \ldots + \rho_{j-1}(a_{j-1} \bullet v_i^j) \tag{9}$$

and using property 4 of Section 2, $a_i \bullet v_i^j = 1$ and $a_k \bullet v_i^j = 0$ for all $k \neq i$, $k = 1, \ldots, j - 1$, we obtain $a_j \bullet v_i^j = \rho_i$.

3 Solving a linear system of equations

Consider now the complete system of linear equations $Ax = b$:

$$
\begin{array}{llll}
a_{11}x_1 & +a_{12}x_2 & +\cdots & +a_{1n}x_n & = b_1 \\
a_{21}x_1 & +a_{22}x_2 & +\cdots & +a_{2n}x_n & = b_2 \\
\cdots & \cdots & \cdots & \cdots & \cdots \\
a_{m1}x_1 & +a_{m2}x_2 & +\cdots & +a_{mn}x_n & = b_m
\end{array}
\tag{10}
$$

Adding the artificial variable x_{n+1}, it can be written as:

$$
\begin{array}{lllll}
a_{11}x_1 & +a_{12}x_2 & +\cdots & +a_{1n}x_n & -b_1 x_{n+1} = 0 \\
a_{21}x_1 & +a_{22}x_2 & +\cdots & +a_{2n}x_n & -b_2 x_{n+1} = 0 \\
\cdots & \cdots & \cdots & \cdots & \cdots \\
a_{m1}x_1 & +a_{m2}x_2 & +\cdots & +a_{mn}x_n & -b_m x_{n+1} = 0 \\
a_{m1}x_1 & +a_{m2}x_2 & +\cdots & +a_{mn}x_n & -b_m x_{n+1} = 0 \\
& & & & x_{n+1} = 1
\end{array}
\tag{11}
$$

System (11) can be written as:

$$
\begin{array}{ll}
(a_{11}, \cdots, a_{1n}, -b_1)(x_1, \cdots, x_n, x_{n+1})^T & = 0 \\
(a_{21}, \cdots, a_{2n}, -b_2)(x_1, \cdots, x_n, x_{n+1})^T & = 0 \\
\cdots\cdots\cdots\cdots\cdots\cdots\cdots\cdots\cdots\cdots\cdots & \cdots \\
(a_{m1}, \cdots, a_{mn}, -b_m)(x_1, \cdots, x_n, x_{n+1})^T & = 0
\end{array}
\tag{12}
$$

Expression (12) shows that $(x_1, \ldots, x_n, x_{n+1})^T$ is orthogonal to the set of vectors:

$$\{(a_{11}, \ldots, a_{1n}, -b_1)^T, (a_{21}, \ldots, a_{2n}, -b_2)^T, \ldots, (a_{m1}, \ldots, a_{mn}, -b_m)^T\}.$$

Then, it is clear that the solution of (11), is the orthogonal complement of the linear subspace generated by the rows of matrix

$$\bar{A} = (A| - b), \tag{13}$$

i.e., the column **-b** is added to the matrix **A** :

$$\mathcal{L}\{(a_{11},\ldots,a_{1n},-b_1)^T,(a_{21},\ldots,a_{2n},-b_2)^T\ldots,(a_{m1},\ldots,a_{mn},-b_m)^T\}^\perp$$

Thus, the solution of (10) is the projection on $\boldsymbol{R}^n$ of the intersection of the orthogonal complement of the linear subspace generated by

$$\{(a_{11},\ldots,a_{1n},-b_1)^T,(a_{21},\ldots,a_{2n},-b_2)^T,\ldots,(a_{m1},\ldots,a_{mn},-b_m)^T\}.$$

and the set $\{\mathbf{x}|x_{n+1} = 1\}$.

To solve system (12) we apply the orthogonal algorithm to the **A** rows with $\mathbf{V}^1 \equiv \mathcal{L}\{\mathbf{v}_1^1,\ldots,\mathbf{v}_{n+1}^1\}$ where $\mathbf{v}_i^1 = \mathbf{e}_i$; $i = 1,\ldots,n+1$, and $\mathbf{e}_i$ is the vector with all zeroes except for the ith component, which is one.

If we consider the j−th equation of system (11), i.e., $\mathbf{a}_j\mathbf{x} = \mathbf{b}_j$, after pivoting with the corresponding associated vector, we can obtain the solution of this equation, $\mathbf{X}^j$,

$$\mathbf{X}^j \equiv \{\mathbf{x} \in \mathbf{V}^k|x_{n+1} = 1 \wedge \mathbf{a}_j^T \bullet \mathbf{x} = 0, k = j+1,\ldots m\}. \tag{14}$$

In fact, we consider the solution generated by the columns of the corresponding table except the column used as pivot (see Section 2, property 5), and we impose the condition $x_{n+1} = 1$.

After this sequential process we have $\mathbf{X}^m \equiv \mathcal{L}\{\mathbf{v}_1^m,\ldots,\mathbf{v}_{n_m}^m\}$, and then the solution of the initial system becomes

$$\hat{\mathbf{v}}_{n_m}^m + \mathcal{L}\{\hat{\mathbf{v}}_1^m,\ldots,\hat{\mathbf{v}}_{n_m-1}^m\}, \tag{15}$$

where $\hat{\mathbf{v}}$ is the vector obtained from $\mathbf{v}$ by removing its last component (see table 2 for an example).

When $\mathcal{L}\left(\hat{\mathbf{v}}_1^m,\ldots,\hat{\mathbf{v}}_{n_m-1}^m\right)$ degenerates to the empty set, we have uniqueness of solution. For this to occur, we must have $m = n$ and $|A| \neq 0$; that is, the coefficients matrix must be a square nonsingular matrix.

If we are interested in obtaining the solution of any subsystem, we will take the intersections of the corresponding solutions of each equation. Note that we have to keep the orthogonal and the complement set in each step for this process to be applied.

4   Modifying equations

Once the redundant equations have been detected in a system of linear equations, which is easy using the orthogonal method for solving the system (see Theorem 1), we can suppose without lost of generality, that the linear system is not redundant. In this section we show how to update the solution of this system after modifying one equation, with only one extra iteration of the orthogonal method.

Consider a not redundant initial linear system of equations $\mathbf{Ax} = \mathbf{b}$, and use the proposed method to solve it, but keeping the complement spaces. If the jth equation is modified, to obtain the new solution, the orthogonal subspace corresponding to the new equation is needed, instead of the orthogonal to the old equation. Since the orthogonal method can be started with any space $\mathbf{V}$, the last table of the initial process can be taken as the initial table for solving the modified system and the modified equation can be introduced to make the extra pivoting transformation. Using as pivot the pivot column associated with the modified row, and taking into account properties 1 and 4 from Section 2, the solution of the modified system is obtained.

5   Complexity

Castillo *et al* [1] have studied the number of operations required by the orthogonal method for solving a linear system of equations. The conclusion of this study is that when this method is used for obtaining the final solution of a linear system of equations and in each step the complement space is removed, the number of operations is the same as the number of operations required for the Gaussian elimination method, that is, around $2n^3/3$ operations (see [7]). The aim in this section is to study the number of operations of the complete process, keeping the complement space, in order to allow the updating of solutions after modifying an equation, and to compare this results with de Gaussian process.

The number of operations required to solve a linear system of equations with the Gaussian elimination method (see [8]) is $2n^3/3 + (9n^2 - 7n)/6$, and with the orthogonal method it is $(2n^3 - 5n + 6)/3$ (see Ref.[1]). However, the number of operations of the orthogonal method keeping the complement space is $2n^3 - n$.

When an equation is modified, an extra iteration is required to update the solution, which implies $4n^2 - 2n$ extra operations.

If several modifications in the original system are needed with the Gaussian elimination method we have to repeat the complete process each time to obtain the new solution. However, with the orthogonal method we need to make the complete process only once ($2n^3 - n$ operations) and one update for each modification ($4n^2 - 2n$ operations). Then, the number of total required operations when the size of the system is n and the number of updates is k, is

$$2n^3 - n + [k(4n^2 - 2n)]$$

for the orthogonal method, and

$$k[(2/3)n^3 + (9n^2 - 7n)/6]$$

for the Gaussian elimination method.

¿From these formulas we can conclude that for $k > 3$ the number of operations required by the orthogonal method is smaller and then, it outperforms

Table 1. Required number of products and sums for making an extra iteration of the orthogonal method.

	Extra iteration	
	Products or divisions	**Sums**
Dot Products	n^2	$n(n-1)$
Normalization	n	–
Pivoting	$n(n-1)$	$n(n-1)$
Total	$2n^2$	$2(n^2-n)$

the Gaussian elimination method. This fact can be useful for nonlinear problems solved by means of iterations on linear approaches, i.e., for Newton methods.

6   Reducing the condition number

In this section we propose a method for obtaining an equivalent system of linear equations with a very small condition number for a full matrix. This kind of matrices appears very often when using spectral methods for solving partial differential equations. The idea uses the procedure explained in section 3 for solving a linear system. It is based on detecting that the matrix is ill conditioned when several of the rows of the matrix $\mathbf{A}$ are in very close hyperplanes, i.e., when $\mathrm{tg}(\widehat{\mathbf{a}_i, \mathbf{a}_j}) \ll 1$. The solution consists of rotating one of the vectors, but keeping it into the same hyperplane $\mathcal{L}(\bar{\mathbf{a}}_i, \bar{\mathbf{a}}_j)$.

As an example, let $\bar{\mathbf{A}}$ be the matrix corresponding to the coefficient matrix whit the independent vector added of the following system:

$$0.832x_1 + 0.448x_2 = 1,$$
$$0.784x_1 + 0.421x_2 = 0. \tag{16}$$

Using the orthogonal method (pivoting transformation (1)), we can see in the second iteration of Table 2, that $\mathcal{L}(\bar{\mathbf{a}}_1^T)^\perp = \mathcal{L}(\mathbf{v}_2^1, \mathbf{v}_3^1)$. When the vector $\bar{\mathbf{a}}_2^T$ in introduced into the pivoting process, it can be observed that it is almost orthogonal to $\mathbf{v}_2^1$, one of the generators of $\mathcal{L}(\bar{\mathbf{a}}_1)^\perp$, and this is the source of ill conditioning character ($K(\mathbf{A}) = 1755$) of the given system.

To solve this problem the vector $\mathbf{a}_2$ can be replaced by a $\pi/2$ rotation, $\mathbf{a}_2^g = g_{\pi/2}(\mathbf{a}_2)$, of $\mathbf{a}_2$ such that $\bar{\mathbf{a}}_2^g = (g_{\pi/2}(\mathbf{a}_2), x_{23}) \in \mathcal{L}(\bar{\mathbf{a}}_1, \bar{\mathbf{a}}_2)$. In this way the linear space $\mathcal{L}(\bar{\mathbf{a}}_1, \bar{\mathbf{a}}_2)^\perp$ is the same as the linear space $\mathcal{L}(\bar{\mathbf{a}}_1, \bar{\mathbf{a}}_2^g)^\perp$.

To calculate $\bar{\mathbf{a}}_2^g$ the linear system $\mathbf{a}_2^g = \alpha \mathbf{a}_1 + \beta \mathbf{a}_2$ is solved, and then x_{23} is calculated as $x_{23} = -\alpha b_1 - \beta b_2$.

Table 2. Iterations for solving Eq. (16). Pivot columns are boldfaced.

	Iteration 1				Iteration 2		
$\bar{\mathbf{a}}_1^T$	$\mathbf{v}_1$	$\mathbf{v}_2$	$\mathbf{v}_3$	$\bar{\mathbf{a}}_2^T$	$\mathbf{v}_1^1$	$\mathbf{v}_2^1$	$\mathbf{v}_3^1$
0.832	**1**	0	0	0.784	*1*	**-0.538**	-1.202
0.448	**0**	1	0	0.421	*0*	1	0
−1	**0**	0	1	0	*0*	0	1
	0.832	0.448	−1		*0.784*	**−7.92e-004**	−0.942

Output		
$\mathbf{v}_1^2$	$\mathbf{v}_2^2$	$\mathbf{v}_3^2$
1	*-0.538*	-438.542
0	*1*	816.667
0	*0*	1

Table 3. Order of the condition number of the matrices equivalent to the Hilbert matrix of size $n = 10$ rotated in the rows from m to n.

m	9	8	7	6	5	4	3	2
K	10^{10}	10^8	10^7	10^6	10^4	10^3	10^2	10^2

If this technique is applied to the system (16), the new coefficient matrix is obtained

$$\mathbf{A}' = \begin{pmatrix} 0.8320 & 0.4480 \\ -0.2261 & 0.4210 \end{pmatrix}$$

and $K(\mathbf{A}') = 1.9775$, i.e., the condition number has been drastically reduced. The new independent term becomes $\mathbf{b}' = (1 \ 442.959)^T$. It is straightforward to prove that the system $\mathbf{A}'\mathbf{x} = \mathbf{b}'$ is equivalent to the system (16).

In the case of a $n \times n$ system, to be sure that all the vectors are in different hyperplanes the following method can be applied:

Starting by $i = m \, (1 < m < n)$ till $i = n$, the pairs (a_{ii}, a_{ii+1}) are compared with the pairs (a_{ji}, a_{ji+1}), $j = i + 1, ..., n$. If $\text{tg}(\widehat{\mathbf{a}_i, \mathbf{a}_j}) \ll 1$ the corresponding vector (a_{ji}, a_{ji+1}) is rotated to $(-a_{ji+1}^2/a_{ji}, a_{ji+1})$ in such a way that the subspaces in $\mathbf{R}^{n+1}$ are kept, i.e.,

$$\bar{\mathbf{a}}_j^g = (a_{j1}, ..., a_{ji-1}, -a_{ji+1}^2/a_{ji}, a_{ji+1}, x_{ji+2}, ..., x_{jn+1}) \in \mathcal{L}\{\bar{\mathbf{a}}_1, \bar{\mathbf{a}}_2, ..., \bar{\mathbf{a}}_j\}$$

To calculate $\bar{\mathbf{a}}_j^g$ the following linear system, with unknowns α_i, $i = 1, ..., j$, is solved:

$$\bar{\mathbf{a}}_j^{gtj} = \alpha_1 \bar{\mathbf{a}}_1^{tj} + ... + \alpha_j \bar{\mathbf{a}}_j^{tj},$$

where the superscript ^{tj} means truncated up to the index j, and then

$$x_{jk} = \alpha_1 a_{1k} + ... + \alpha_j a_{jk}.$$

In this way one can be sure that any two row vectors of the new matrix are in very different hyperplanes and the condition number is reduced. This fact can be observed in table 3 : $K(\mathbf{A}) = O(10^{13})$ for the Hilbert matrix, where it can be seen that if this technique is applied with increasing number of rows, the condition number reduces to $K = O(10^2)$ when all the rows are involved.

References

1. Castillo, E., Cobo, A., Jubete, F., Pruneda, R.E.: *Orthogonal Sets and Polar Methods in Linear Algebra: Applications to Matrix Calculations, Systems of Equations and Inequalities, and Linear Programming*. John Wiley and Sons, New York (1999)
2. Golub, G.H., van Loan, C.F.: *Matrix Computations*. Johns Hopkins University Press, London (1996)
3. Axelsson, O.: *Iterative Solution Methods*. Cambridge University Press, Cambridge (1996)
4. Kelley, C.T.: *Iterative Methods for Linear and Nonlinear Equations*. SIAM, Phyladelphia (1995)
5. Duff, I.S., Watson, G.A. (eds): *The state of the art in numerical analysis*. Oxford University Press, Oxford (1997)
6. Castillo, E., Cobo, A., Pruneda, R.E. and Castillo, C.: An Orthogonally-Based Pivoting Transformation of Matrices and Some Applications. *SIAM Journal on Matrix Analysis and Applications* **22** (2000) 666-681
7. Atkinson, K.E.: *An Introduction to Numerical Analysis*. Jonhn Wiley and Sons, New York (1978)
8. Infante del Río, J.A., Rey Cabezas, J.M.: Métodos numéricos. *Teoría, problemas y prácticas con Matlab*. Pirámide, Madrid (1999)

Antithetic Monte Carlo Linear Solver

Chih Jeng Kenneth Tan

School of Computer Science
The Queen's University of Belfast
Belfast BT7 1NN
Northern Ireland
United Kingdom**
cjtan@acm.org

Abstract. The problem of solving systems of linear algebraic equations by parallel Monte Carlo numerical methods is considered. A parallel Monte Carlo method with relaxation parameter and dispersion reduction using antithetic variates is presented. This is a report of a research in progress, showing the effectiveness of this algorithm. Theoretical justification of this algorithm and numerical experiments are presented. The algorithms were implemented on a cluster of workstations using MPI.
Keyword: Monte Carlo method, Linear solver, Systems of linear algebraic equations, Parallel algorithms.

1   Introduction

One of the more common numerical computation task is that of solving large systems of linear algebraic equations

$$Ax = b \tag{1}$$

where $A \in \mathbb{R}^{n \times n}$ and $x, b \in \mathbb{R}^n$. A great multitude of algorithms exist for solving Equation 1. They typically fall under one of the following classes: direct methods, iterative methods, and Monte Carlo methods. Direct methods are particularly favorable for dense A with relatively small n. When A is sparse, iterative methods are preferred when the desired precision is high and n is relatively small. When n is large and the required precision is relatively low, Monte Carlo methods have been proven to be very useful [5, 4, 16, 1].

As a rule, Monte Carlo methods are not competitive with classical numerical methods for solving systems of linear algebraic equations, if the required precision is high [13].

In Monte Carlo methods, statistical estimates for the components of the solution vector x are obtained by performing random sampling of a certain random variable whose mathematical expectation is the desired solution [14, 18]. These

---

** Now at **SHARCNET**, Western Science Center, The University of Western Ontario, London, Ontario, Canada N6A 5B7.

P.M.A. Sloot et al. (Eds.): ICCS 2002, LNCS 2330, pp. 383–392, 2002.

techniques are based on that proposed by von Neumann and Ulam, extended by Forsythe and Liebler [13, 6].

Classical methods such as non-pivoting Gaussian Elimination or Gauss-Jordan methods require $\mathcal{O}(n^3)$ steps for a $n \times n$ square matrix [2]. In contrast, to compute the full solution vector using Monte Carlo the total number of steps required is $\mathcal{O}(nNT)$, where N is the number of chains and T is the chain length, both quantities independent of n and bounded [1]. Also, if only a few components of x are required, they can be computed without having to compute the full solution vector. This is a clear advantage of Monte Carlo methods, compared to their direct or iterative counterpart.

In addition, even though Monte Carlo methods do not yield better solutions than direct or iterative numerical methods for solving systems of linear algebraic equations as in Equation 1, they are more efficient for large n. Also, Monte Carlo methods have been known for their embarrassingly parallel nature. Parallelizing Monte Carlo methods in a coarse grained manner is very often straightforward. This characteristic of Monte Carlo methods has been noted even in 1949, by Metropolis and Ulam [12].

This paper presents an Antithetic Monte Carlo algorithm for the solution of systems of linear algebraic equations. Antithetic variates have been used in Monte Carlo algorithms for integral problems and solution of differential equations. But there are no known previous attempts to use antithetic variates in Monte Carlo algorithm for the solution of systems of linear algebraic equations.

2   Stochastic Methods for Solving Systems of Linear Algebraic Equations

Consider a matrix $A \in \mathbb{R}^{n \times n}$ and a vector $x \in \mathbb{R}^{n \times 1}$. Further, A can be considered as a linear operator $A\left[\mathbb{R}^{n \times 1} \to \mathbb{R}^{n \times 1}\right]$, so that the linear transformation

$$Ax \in \mathbb{R}^{n \times 1} \tag{2}$$

defines a new vector in $\mathbb{R}^{n \times 1}$.

The linear transformation in Equation 2 is used in iterative Monte Carlo algorithms, and the linear transformation in Equation 2 is also known as the iteration. This algebraic transform plays a fundamental role in iterative Monte Carlo algorithms.

In the problem of solving systems of linear algebraic equations, the linear transformation in Equation 2 defines a new vector $b \in \mathbb{R}^{n \times 1}$:

$$Ax = b, \tag{3}$$

where A and b are known, and the unknown solution vector x is to be solved for. This is a problem often encountered as subproblems on in various applications such as solution of differential equations, least squares solutions, amongst others.

It is known that system of linear algebraic equation given by Equation 3, can be rewritten in the following iterative form [2, 18, 4]:

$$x = Lx + b, \tag{4}$$

where

$$(I - L) = A. \tag{5}$$

Assuming that $\|L\| < 1$, and $x^0 \equiv 0$, the von Neumann series converges and the equation

$$\lim_{k \to \infty} x^{(k)} = \lim_{k \to \infty} \sum L^m b = (I - L)^{-1} b = A^{-1} b = x \tag{6}$$

holds.

Suppose now $\{s_1, s_2, \ldots, s_n\}$ is a finite discrete Markov chains with n states. At each discrete time $t = 0, 1, \ldots, N$, a chain S of length T is generated:

$$k_0 \to k_1 \to \ldots \to k_j \to \ldots \to k_T$$

with $k_j \in \{s_1, s_2, \ldots, s_n\}$ for $j = 1, \ldots, T$.

Define the probability that the chain starts in state s_α,

$$\mathbf{P}\,[k_0 = s_\alpha] = p_\alpha \tag{7}$$

and the transition probability to state s_β from state s_α

$$\mathbf{P}\,[k_j = s_\beta | k_{j-1} = s_\alpha] = p_{\alpha\beta} \tag{8}$$

for $\alpha = 1, \ldots, n$ and $\beta = 1, \ldots, n$.

The probabilities $p_{\alpha\beta}$ thus define the transition matrix P. The distribution $(p_1, \ldots, p_n)^T$ is said to be acceptable to vector h, and similarly that the distribution $p_{\alpha\beta}$ is acceptable to L, if [14]

$$\begin{cases} p_\alpha > 0 \text{ when } h_\alpha \neq 0 \\ p_\alpha \geq 0 \text{ when } h_\alpha = 0 \end{cases} \text{ and } \begin{cases} p_{\alpha\beta} > 0 \text{ when } l_{\alpha\beta} \neq 0 \\ p_{\alpha\beta} \geq 0 \text{ when } l_{\alpha\beta} = 0 \end{cases} \tag{9}$$

Define the random variables W_j according to the recursion

$$W_j = W_{j-1} \frac{l_{k_{j-1}k_j}}{p_{k_{j-1}k_j}}, \; W_0 \equiv 1 \tag{10}$$

The random variables W_j can also be considered as weights on the Markov chain. Also, define the random variable

$$\eta_T(h) = \frac{h_{k_0}}{p_{k_0}} \sum_{j=0}^{T-1} W_j b_{k_j}. \tag{11}$$

From Equation 6, the limit of $\mathbf{M}\,[\eta_T(h)]$, the mathematical expectation of $\eta_T(h)$ is

$$\mathbf{M}\,[\eta_T(h)] = \left\langle h, \sum_{m=0}^{T-1} L^m b \right\rangle = \left\langle h, x^{(T)} \right\rangle \Rightarrow \lim_{T \to \infty} \mathbf{M}\,[\eta_T(h)] = \langle h, x \rangle \tag{12}$$

Knowing this, one can find an unbiased estimator of $\mathbf{M}\left[\eta_\infty\left(h\right)\right]$ in the form

$$\theta_N = \frac{1}{N}\sum_{m=0}^{N-1}\eta_\infty\left(h\right) \qquad (13)$$

Consider functions $h \equiv h^j = (0,0,\dots,1,\dots,0)$, where $h_i^j = \delta_i^j$ is the Kronecker delta. Then

$$\langle h,x\rangle = \sum_{i=0}^{n-1} h_i^j x_i = x_j. \qquad (14)$$

It follows that an approximation to x can be obtained by calculating the average for each component of every Markov chain

$$x_j \approx \frac{1}{N}\sum_{m=0}^{N-1}\theta_T^m\left[h^j\right]. \qquad (15)$$

In summary, N independent Markov chains of length T is generated and $\eta_T\left(h\right)$ is calculated for each path. Finally, the j-th component of x is estimated as the average of every j-th component of each chain.

3   Minimal Probable Error

Let I be any functional to be estimated by Monte Carlo method, θ be the estimator, and n be the number of trials. The probable error for the usual Monte Carlo method is defined as [14]:

$$\mathbf{P}\left[|I-\theta|\ge r\right] = \frac{1}{2} = \mathbf{P}\left[|I-\theta|\le r\right]. \qquad (16)$$

Equation (16) does not take into consideration any additional a priori information regarding the regularity of the solution.

If the standard deviation $(\mathbf{D}\left[\theta\right])^{\frac{1}{2}}$ is bounded, then the Central Limit Theorem holds, and

$$\mathbf{P}\left[|I-\theta|\le x\left(\frac{\mathbf{D}\left[\theta\right]}{n}\right)^{\frac{1}{2}}\right] \approx \Phi\left(x\right). \qquad (17)$$

Since $\Phi\left(0.6745\right)\approx\frac{1}{2}$, it is obvious that the probable error is

$$r \approx 0.6745\left(\mathbf{D}\left[\theta\right]\right)^{\frac{1}{2}}. \qquad (18)$$

Therefore, if the number of Markov chains N increases the error bound decreases. Also the error bound decreases if the variance of the random variable θ decreases.

This leads to the definition of almost optimal transition frequency for Monte Carlo methods. The idea is to find a transition matrix P that minimize the second moment of the estimator. This is achieved by choosing the probability proportional to the $|l_{\alpha\beta}|$ [5]. The corresponding almost optimal initial density vector, and similarly the transition density matrix $P = \{p_{\alpha\beta}\}_{\alpha,\beta=1}^n$ is then called the almost optimal density matrix.

4   Parameter Estimation

The transition matrix P is chosen with elements $p_{\alpha\beta} = \frac{|l_{\alpha\beta}|}{\sum_\beta |l_{\alpha\beta}|}$ for $\alpha, \beta = 1, 2, ..., n$. In practice the length of the Markov chain must be finite, and is terminated when $|W_j b_{k_j}| < \delta$, for some small value δ [14]. Since

$$|W_j b_{k_j}| = \left| \frac{l_{\alpha_0\alpha_1} \cdots l_{\alpha_{j-1}\alpha_j}}{\frac{|l_{\alpha_0\alpha_1}|}{\|L\|} \cdots \frac{|l_{\alpha_{j-1}\alpha_j}|}{\|L\|}} \right| |b_{k_j}| = \|L\|^i \|b\| < \delta, \tag{19}$$

it follows that

$$T = j \leq \frac{\log\left(\frac{\delta}{\|b\|}\right)}{\log \|L\|} \tag{20}$$

and

$$\mathbf{D}\left[\eta_T\left(h\right)\right] \leq \mathbf{M}\left[\eta_T^2\right] = \frac{\|b\|^2}{\left(1 - \|L\|\right)^2} \leq \frac{1}{\left(1 - \|L\|\right)^2}. \tag{21}$$

According to the Central Limit Theorem,

$$N \geq \left(\frac{0.6745}{\epsilon}\right)^2 \frac{1}{\left(1 - \|L\|\right)^2} \tag{22}$$

is a lower bound on N.

5   Antithetic Monte Carlo Method

It has been discussed in Section 3 that, from Equation 18, the error bound may be decreased by increasing the number of samples, or by decreasing the dispersion in the random variable θ. As previously mentioned, that decreasing the error bound by decreasing the dispersion in the random variable θ is a more feasible solution. The emphasis on efficient Monte Carlo sampling dates back to the early days of digital computing [8], but are still as important today, since the problems solved are considerably large. A major part of Monte Carlo numerical methods research has, thus been dedicated to dispersion reduction. Note that dispersion reduction is also commonly known as variance reduction.

One of the techniques of dispersion reduction is by using antithetic variables. Consider the problem of estimating the integral

$$\theta = \int_a^b f\left(x\right) dx. \tag{23}$$

Define a random variable Y within the range (a, b), with a corresponding density $p\left(y\right)$, and a function g such that

$$\mathbf{M}\left[g\left(Y\right)\right] = \int_a^b g\left(y\right) p\left(y\right) dy = \int_a^b f\left(y\right) dy = \theta. \tag{24}$$

If the function g is taken to be f, and Y is defined to have a uniform density over the range $[a, b]$, therefore

$$\theta = (b - a) \, \mathbf{M} \, [f \, (Y)] \, . \tag{25}$$

By drawing N samples of $f \, (y_i)$, an estimate of θ, $\hat{\theta}$,

$$\hat{\theta} = \frac{1}{N} \left((b - a) \sum_{i=0}^{N-1} f \, (Y_i) \right), \tag{26}$$

may be obtained. The random variable θ is called the crude Monte Carlo estimate of θ.

It follows that the dispersion of the estimate $\hat{\theta}$, $\mathbf{D} \left[\hat{\theta} \right]$ is

$$
\begin{aligned}
\mathbf{D} \left[\hat{\theta} \right] &= \frac{1}{N^2} \, (b - a)^2 \sum_{i=0}^{N-1} \mathbf{D} \, [f \, (Y_i)] \\
&= \frac{(b - a)^2}{N} \, \mathbf{D} \, [f \, (Y)] \\
&= \frac{(b - a)}{N} \int_a^b (f \, (x) - \theta)^2 \, dx.
\end{aligned} \tag{27}
$$

If $\hat{\theta}$ is an estimator of θ, it is possible to find another estimator, $\hat{\theta}'$, having the same unknown expectation as θ and a strong negative correlation with θ [7]. The resulting estimate,

$$\tilde{\theta} = \frac{1}{2} \left(\hat{\theta} + \hat{\theta}' \right), \tag{28}$$

will be an unbiased estimator of θ. The dispersion of $\tilde{\theta}$ is

$$\mathbf{D} \left[\tilde{\theta} \right] = \frac{1}{4} \mathbf{D} \left[\hat{\theta} \right] + \frac{1}{4} \mathbf{D} \left[\hat{\theta}' \right] + \frac{1}{2} \mathbf{C} \left[\hat{\theta}, \hat{\theta}' \right], \tag{29}$$

where $\mathbf{C} \left[\hat{\theta}, \hat{\theta}' \right]$ is the covariance between $\hat{\theta}$ and $\hat{\theta}'$. Since $\hat{\theta}'$ is negatively correlated with $\hat{\theta}$, then the value of $\mathbf{C} \left[\hat{\theta}, \hat{\theta}' \right]$ will be negative, leading to an overall smaller $\mathbf{D} \left[\tilde{\theta} \right]$ by choosing $\hat{\theta}'$ suitably.

The random variable $\hat{\theta}'$ is the antithetic variate, and this technique of dispersion reduction is classified as antithetic dispersion reduction. In general, any set of estimators which mutually compensate each other's dispersions are termed antithetic variates [7].

In Monte Carlo methods for solving linear systems of algebraic equations, antithetic dispersion reduction may be used by generating a Markov chain which is negatively correlated with another.

As usual, the random variables W_j are defined according to the recursion in Equation 10 which is given here again:

$$W_j = W_{j-1} \frac{l_{k_{j-1} k_j}}{p_{k_{j-1} k_j}}, \ W_0 \equiv 1.$$

In addition, define another set of random variables, W'_j as

$$W'_j = W'_{j-1} \frac{l_{k'_{j-1}k'_j}}{p_{k'_{j-1}k'_j}}, \ W'_0 \equiv 1. \tag{30}$$

Denote the estimate of the solution, x, computed by using the first Markov chain as $\tilde{x}$, and the estimate computed by using the antithetic Markov chain as $\hat{x}$.

Also, define the random variable

$$\eta'_T(h) = \frac{h_{k'_0}}{p_{k'_0}} \sum_{j=0}^{T-1} W'_j b_{k'_j}, \tag{31}$$

in line with to Equation 11.

The limit of mathematical expectation of $\eta'_T(h)$, $\mathbf{M}[\eta'_T(h)]$ is the same as that of the mathematical expectation of $\eta_T(h)$:

$$\mathbf{M}[\eta'_T(h)] = \left\langle h, \sum_{m=0}^{T-1} L^m b \right\rangle = \left\langle h, \hat{x}^{(T)} \right\rangle \Rightarrow \lim_{T \to \infty} \mathbf{M}[\eta'_T(h)] = \langle h, \hat{x} \rangle \tag{32}$$

Then it follows that the unbiased estimator required will be will be one in the form

$$\theta'_N = \frac{1}{N} \sum_{m=0}^{N-1} \eta'_\infty(h). \tag{33}$$

With the functions $h \equiv j^i = (0, 0, \ldots, 1, \ldots, 0)$, where $h_i^j = \delta_i^j$ is the Kronecka delta, then the dot product of vectors h and x is

$$\langle h, \hat{x} \rangle = \sum_{i=0}^n h_i^j \hat{x}_i = \hat{x}_j, \tag{34}$$

as in Equation 14.

It the follows that the approximation to $\hat{x}$, using the antithetic Markov chain, can be obtained by calculating the average for each component of every Markov chain

$$\hat{x}_j \approx \frac{1}{N} \sum_{m=0}^{N-1} \theta'^m_T[h^j]. \tag{35}$$

The approximation to the solution vector x can be calculated by taking the average of the estimate $\tilde{x}$, computed with the first Markov chain, and $\hat{x}$, computed with the second, antithetic, Markov chain:

$$x = \frac{1}{2}(\tilde{x} + \hat{x}). \tag{36}$$

It does not take much effort to see that the Antithetic Monte Carlo method can be used on its own, or it can be used in combination with some, or all, of the previously proposed methods, namely, the almost optimal Monte Carlo method proposed by Megson, Alexandrov and Dimov in [9], the Monte Carlo method with chain reduction and optimization [3], and the Relaxed Monte Carlo method [15].

6   Numerical Experiments

A parallel version of the Antithetic Monte Carlo algorithm was developed using Message Passing Interface (MPI) [11, 10]. Version 1.2.0 of the MPICH implementation of the Message Passing Interface was used. As the programs were written in C, the C interface of MPI was the natural choice interface.

Table 1 show the results for experiments with the Monte Carlo method and Table 2 show the results for experiments with the Antithetic Monte Carlo method. The matrices used in these experiments were dense (general) randomly populated matrices, with a specified norm. The stochastic error, ϵ, and the deterministic error parameters were both set to 0.01. The PLFG parallel pseudo-random number generator [17] was used as the source of randomness for the experiments conducted.

Data set	Norm	Solution time (sec.)	RMS error	No. chains
100-A1	0.5	0.139	4.76872e-02	454900
100-A2	0.6	0.122	4.77279e-02	454900
100-A3	0.7	0.124	4.78072e-02	454900
100-A4	0.8	0.127	4.77361e-02	454900
100-A5	0.5	0.137	3.17641e-02	454900
100-A6	0.6	0.124	3.17909e-02	454900
100-A7	0.7	0.124	3.17811e-02	454900
100-A8	0.8	0.119	3.17819e-02	454900
100-B1	0.5	0.123	3.87367e-02	454900
100-B2	0.6	0.126	3.87241e-02	454900
100-B3	0.7	0.134	3.88647e-02	454900
100-B4	0.8	0.125	3.88836e-02	454900
100-B5	0.5	0.121	2.57130e-02	454900
100-B6	0.6	0.119	2.57748e-02	454900
100-B7	0.7	0.120	2.57847e-02	454900
100-B8	0.8	0.126	2.57323e-02	454900

Table 1. Monte Carlo method with PLFG, using 10 processors, on a DEC Alpha XP1000 cluster.

7   Acknowledgment

I would like to thank M. Isabel Casas Villalba from Norkom Technologies, Ireland for the fruitful discussions and the MACI project at the University of Calgary, Canada, for their support, providing part of the computational resources used.

References

[1] ALEXANDROV, V. N. Efficient Parallel Monte Carlo Methods for Matrix Computations. *Mathematics and Computers in Simulation 47* (1998).

Data set	Norm	Solution time (sec.)	RMS error	No. chains
100-A1	0.5	0.123	4.77496e-02	454900
100-A2	0.6	0.124	4.75949e-02	454900
100-A3	0.7	0.122	4.77031e-02	454900
100-A4	0.8	0.119	4.77248e-02	454900
100-A5	0.5	0.125	3.19065e-02	454900
100-A6	0.6	0.118	3.17608e-02	454900
100-A7	0.7	0.120	3.17462e-02	454900
100-A8	0.8	0.127	3.17776e-02	454900
100-B1	0.5	0.116	3.88795e-02	454900
100-B2	0.6	0.124	3.88260e-02	454900
100-B3	0.7	0.121	3.87219e-02	454900
100-B4	0.8	0.114	3.88401e-02	454900
100-B5	0.5	0.114	2.57260e-02	454900
100-B6	0.6	0.097	2.58110e-02	454900
100-B7	0.7	0.130	2.57297e-02	454900
100-B8	0.8	0.123	2.57341e-02	454900

Table 2. Antithetic Monte Carlo method with PLFG, using 10 processors, on a DEC Alpha XP1000 cluster.

[2] BERTSEKAS, D. P., AND TSITSIKLIS, J. N. *Parallel and Distributed Computation: Numerical Methods.* Athena Scientific, 1997.

[3] CASAS VILLALBA, M. I., AND TAN, C. J. K. Efficient Monte Carlo Linear Solver with Chain Reduction and Optimization Using PLFG. In *High-Performance Computing and Networking, Proceedings of the 9th. International Conference on High Performance Computing and Networking Europe* (2001), B. Hertzberger, A. G. Hoekstra, and R. Williams, Eds., vol. 2110 of *Lecture Notes in Computer Science*, Springer-Verlag.

[4] DIMOV, I. Monte Carlo Algorithms for Linear Problems. In *Lecture Notes of the 9th. International Summer School on Probability Theory and Mathematical Statistics* (1998), N. M. Yanev, Ed., SCT Publishing, pp. 51 – 71.

[5] DIMOV, I. T. Minimization of the Probable Error for some Monte Carlo Methods. In *Mathematical Modelling and Scientific Computations* (1991), I. T. Dimov, A. S. Andreev, S. M. Markov, and S. Ullrich, Eds., Publication House of the Bulgarian Academy of Science, pp. 159 – 170.

[6] FORSYTHE, S. E., AND LIEBLER, R. A. Matrix Inversion by a Monte Carlo Method. *Mathematical Tables and Other Aids to Computation 4* (1950), 127 – 129.

[7] HAMMERSLEY, J. M., AND HANDSCOMB, D. C. *Monte Carlo Methods.* Methuen's Monographs on Applied Probability and Statistics. Methuen and Company, 1964.

[8] KAHN, H., AND MARSHALL, A. W. Methods of Reducing Sample Size in Monte Carlo Computations. *Journal of Operations Research Society of America 1* (1953), 263 – 278.

[9] MEGSON, G., ALEXANDROV, V., AND DIMOV, I. Systolic Matrix Inversion Using a Monte Carlo Method. *Journal of Parallel Algorithms and Applications 3, 3 – 4* (1994), 311 – 330.

[10] MESSAGE PASSING INTERFACE FORUM. *MPI: A Message-Passing Interface Standard*, 1.1 ed., June 1995.

[11] MESSAGE PASSING INTERFACE FORUM. *MPI-2: Extensions to the Message-Passing Interface*, 2.0 ed., 1997.

[12] METROPOLIS, N., AND ULAM, S. The Monte Carlo Method. *Journal of the American Statistical Association*, 44 (1949), 335 – 341.

[13] RUBINSTEIN, R. Y. *Simulation and the Monte Carlo Method*. John Wiley and Sons, 1981.

[14] SOBOL', I. M. *Monte Carlo Numerical Methods*. Moscow, Nauka, 1973. (In Russian.).

[15] TAN, C. J. K., AND ALEXANDROV, V. Relaxed Monte Carlo Linear Solver. In *Computational Science (Part II)* (2001), V. N. Alexandrov, J. J. Dongarra, B. A. Juliano, R. S. Renner, and C. J. K. Tan, Eds., vol. 2074 of *Lecture Notes in Computer Science*, Springer-Verlag, pp. 1289 – 1298.

[16] TAN, C. J. K., AND BLAIS, J. A. R. PLFG: A Highly Scalable Parallel Pseudo-random Number Generator for Monte Carlo Simulations. In *High Performance Computing and Networking, Proceedings of the 8th. International Conference on High Performance Computing and Networking Europe* (2000), M. Bubak, H. Afsarmanesh, R. Williams, and B. Hertzberger, Eds., vol. 1823 of *Lecture Notes in Computer Science*, Springer-Verlag, pp. 127 – 135.

[17] TAN, C. J. K., CASAS VILLALBA, M. I., AND ALEXANDROV, V. N. Accuracy of Monte Carlo Method for Solution of Linear Algebraic Equations Using PLFG and rand(). In *High Performance Computing Systems and Application* (2000), N. Dimopoulos and K. F. Li, Eds., Kluwer International Series in Engineering and Computer Science, Kluwer Academic Publishers. (To appear.).

[18] WESTLAKE, J. R. *A Handbook of Numerical Matrix Inversion and Solution of Linear Equations*. John Wiley and Sons, 1968.

Restarted Simpler GMRES Augmented with Harmonic Ritz Vectors

Ravindra Boojhawon and Muddun Bhuruth

Department of Mathematics, University of Mauritius
Reduit, Mauritius
{r.boojhawon, mbhuruth}@uom.ac.mu

Abstract. We describe a method for improving the convergence of the Simpler GMRES method for problems with small eigenvalues. We augment the Krylov subspace with harmonic Ritz vectors corresponding to the smallest harmonic Ritz values. The advantage over augmented GMRES is that the problem of finding the minimal residual solution reduces to an upper triangular least-squares problem instead of an upper-Hessenberg least-squares problem. A second advantage is that harmonic Ritz pairs can be cheaply computed. Numerical tests indicate that augmented Simpler GMRES(m) is superior to Simpler GMRES(m) and requires a lesser amount of work than augmented GMRES(m).

1   Introduction

Restarting the GMRES method after each cycle of m iterations slows down the convergence of the iterative method as each restart discards the information on the smallest eigenvalues which compose the solution. There are basically two approaches to reduce the negative effects of a restart: a preconditioning approach [1] and an augmented subspace approach [3]. These two approaches have been shown to improve the convergence of GMRES for problems with small eigenvalues. A cheaper implementation of GMRES can be achieved via the Simpler GMRES algorithm of Walker and Zhou [5]. The advantage is that the problem for finding the minimal residual solution reduces to an upper triangular least-squares problem instead of an upper-Hessenberg least-squares problem for GMRES. We show that for problems with small eigenvalues, it is possible to improve the convergence of restarted Simpler GMRES(m) by augmenting the Krylov subspace with harmonic Ritz vectors corresponding to the smallest harmonic Ritz values which are the roots of the GMRES polynomial.

2   Augmented Simpler GMRES

Consider the nonsymmetric linear system

$$Ax = b\,, \tag{1}$$

P.M.A. Sloot et al. (Eds.): ICCS 2002, LNCS 2330, pp. 393–402, 2002.

where $A \in \mathbb{R}^{n \times n}$ is nonsingular and $x, b \in \mathbb{R}^n$. Let x_0 be an initial guess such that $r_0 = b - Ax_0 \neq 0$. Suppose that we have applied the Arnoldi process with $w_1 = Ar_0/\|Ar_0\|_2$ to produce the orthonormal set of vectors $\{w_1, w_2, \ldots, w_{p-1}\}$ for some $p > 1$. Let $\mathcal{K}_p(w)$ denote the Krylov subspace

$$\mathcal{K}_p(w) \equiv \text{span}\left\{w, Aw, \ldots, A^{p-1}w\right\} . \tag{2}$$

If the GMRES iterate x_{p-1} is not a solution of (1), then $\mathcal{K}_p(w_1) = A\left(\mathcal{K}_p(r_0)\right)$ has dimension p [5]. Suppose we want to augment the Krylov subspace with k harmonic Ritz vectors $u_1, \ldots, u_k$.

Denote $m = p+k$ and let Y_m be the $n \times m$ matrix whose last k vectors are the harmonic Ritz vectors u_i for $i = 1, \ldots, k$. Also let W_m to be the $n \times m$ matrix whose first p columns are the vectors $w_1, \ldots, w_p$ and whose last k columns are formed by orthogonalizing the vectors $\tilde{u}_i = Au_i$, for $i = 1, \ldots, k$ against the previous columns of W_m by the following algorithm:

AugmentedSArnoldi

1. *Start:* Input k approximate eigenvectors U_k, $\tilde{U}_k = AU_k$, m and k. Let $p = m - k$; $w_1 = Ar_0$; $R_{1,1} = \|w_1\|_2$; $w_1 = w_1/R_{1,1}$;
2. *Iterate:* For $i = 2 : m$
   If $i \le p$, $w_i = Aw_{i-1}$; else, $w_i = \tilde{u}_{i-p}$;
   For $j = 1, \ldots, i-1$, $R_{j,i} = w_i^{\mathrm{T}}w_j$; $w_i = w_i - R_{j,i}w_j$;
   $R_{i,i} = \|w_i\|_2$; $w_i = w_i/R_{i,i}$;
3. *Output:* If $k = 0$, $Y_m = [r_0 \ \ W_{m-1}]$; else, $Y_m = [r_0 \ \ W_{p-1} \ \ U_k]$.

The result of the above algorithm is the decomposition

$$AY_m = W_m R_m , \tag{3}$$

where R_m is an $m \times m$ upper triangular matrix. If $k = 0$, $Y_m = [r_0 \ \ W_{m-1}]$ else $Y_m = [r_0 \ \ W_{p-1} \ \ U_k]$ where U_k is a matrix whose k columns are the k harmonic Ritz vectors.

Writing the restarted problem in the form $A(\tilde{x} - x_0) = r_0$, we find that the residual vector is given by

$$r_m = r_0 - W_m \tilde{w} , \tag{4}$$

where $\tilde{w} = R_m \hat{y} = (\xi_1, \xi_2, \ldots, \xi_m)^{\mathrm{T}}$. From (4) and by the minimal residual criterion [2], we have $W_m^{\mathrm{T}} r_m = 0$ and thus $r_m = r_{m-1} - \xi_m w_m$. The residual norm can be easily updated using

$$\|r_m\|_2 = \sqrt{(\|r_{m-1}\|_2^2 - \xi_m^2)} = \|r_{m-1}\|_2 \sin\left(\cos^{-1}(\xi_m/\|r_{m-1}\|_2)\right) . \tag{5}$$

The approximate solution $\tilde{x} - x_0$ is a combination of the columns of Y_m such that $\tilde{x} - x_0 = Y_m \hat{y} = z_m$ and $\hat{y} = R_m^{-1}\tilde{w} = (\eta_1, \eta_2, \ldots, \eta_m)^{\mathrm{T}}$. It then follows that the augmented Simpler GMRES(m) correction is given by $z_m = Y_m \widehat{y}$ where

$$
z_i = \begin{cases}
r_0 \widehat{y}_1, & \text{if } i = 1, \\
(r_0, w_1, w_2, \ldots, w_{i-1}) \widehat{y}_i, & \text{if } 1 < i \le p, \\
(r_0, w_1, w_2, \ldots, w_{p-1}, u_1, \ldots, u_{i-p}) \widehat{y}_i, & \text{if } p < i \le m .
\end{cases} \tag{6}
$$

In the Gram-Schmidt implementation of the augmented Simpler GMRES(m) algorithm, the correction z_i can be rewritten as

$$
z_i = \begin{cases}
\eta_1 r_0, & \text{if } i = 1, \\
\eta_1 r_{i-1} + \sum_{j=1}^{i-1}(\eta_{j+1} + \eta_1 \xi_j) w_j, & \text{if } 1 < i \le p, \\
\eta r_{i-1} + \sum_{j=1}^{p-1}(\eta_{j+1} + \eta_1 \xi_j) w_j \\
\quad + \sum_{j=p}^{i-1}(\eta_1 \xi_j w_j + \eta_{j+1} u_{j-p+1}), & \text{if } i > p .
\end{cases} \tag{7}
$$

Harmonic Ritz vectors are computed by applying the minimal residual criterion. This criterion for finding an approximate eigenvector $u = Y_m \widetilde{g}_m$ of A in $A\mathcal{K}_m(r_0)$ can be written as [2]

$$
(Au - \lambda u) \perp A\mathcal{K}_m(r_0) \Leftrightarrow W_m^{\mathrm{T}}(AY_m \widetilde{g}_m - \lambda Y_m \widetilde{g}_m) = 0 . \tag{8}
$$

Using (3), we have

$$
R_m \widetilde{g}_m = \lambda W_m^{\mathrm{T}} Y_m \widetilde{g}_m . \tag{9}
$$

Denoting $(W_m^{\mathrm{T}} Y_m)^{-1}$ by T_m, (9) becomes

$$
T_m R_m \widetilde{g}_m = \lambda \widetilde{g}_m . \tag{10}
$$

Noting that $T_m(R_m T_m) T_m^{-1} = T_m R_m$, the projected eigenvalue problem (9) becomes

$$
R_m T_m g_m = \widetilde{R}_m g_m = \lambda g_m , \tag{11}
$$

and the approximate eigenvector becomes $u = Y_m \widetilde{g}_m = Y_m(T_m g_m)$.

If $k = 0$, $Y_m = [r_0 \ W_{m-1}]$, and thus T_m and $\widetilde{R}_m$ can be easily obtained as shown in [2]. Now if $k > 0$, then $Y_m = [r_0 \ W_{p-1} \ U_k]$ and

$$
T_m^{-1} = \begin{pmatrix}
\xi_1 & 1 & & & & w_1^{\mathrm{T}} u_1 & \ldots & w_1^{\mathrm{T}} u_k \\
\xi_2 & 0 & 1 & & & w_2^{\mathrm{T}} u_1 & \ldots & w_2^{\mathrm{T}} u_k \\
\vdots & & \ddots & \ddots & & \vdots & & \\
\xi_{p-1} & & & 0 & 1 & w_{p-1}^{\mathrm{T}} u_1 & \ldots & w_{p-1}^{\mathrm{T}} u_k \\
\xi_p & & & & 0 & w_p^{\mathrm{T}} u_1 & \ldots & w_p^{\mathrm{T}} u_k \\
\vdots & & & & & \vdots & & \vdots \\
\xi_m & & & & & w_m^{\mathrm{T}} u_1 & \ldots & w_m^{\mathrm{T}} u_k
\end{pmatrix} . \tag{12}
$$

We then find that the matrix T_m has the following structure:

$$T_m = \begin{pmatrix} 0 & & & & \tau_{1,1} & \cdots & \tau_{1,k+1} \\ 1 & 0 & & & \tau_{2,1} & \cdots & \tau_{2,k+1} \\ & 1 & 0 & & \tau_{3,1} & \cdots & \tau_{3,k+1} \\ & & \ddots & \ddots & \vdots & & \vdots \\ & & & 1 & 0 & \tau_{m-1,1} & \cdots & \tau_{m-1,k+1} \\ & & & & 1 & \tau_{m,1} & \cdots & \tau_{m,k+1} \end{pmatrix}. \tag{13}$$

It is thus seen that the matrix $\widetilde{R}_m = R_m T_m$ is an upper Hessenberg matrix whose entries can be cheaply computed as follows: $\widetilde{R}_{i,j} = \sum_{q=i}^{q=m} R_{i,q} \tau_{q,j-p+1}$, for $i = 1, \ldots, m$, $j = p, \ldots, m$ and $\widetilde{R}_{i,j} = R_{i,j+1}$, for $j = 1, \ldots, p-1$, $i = 1, \ldots, j+1$.

In the following, we describe the Modified Gram-Schmidt implementation of the augmented Simpler GMRES algorithm:

SGMRESE

1. *Start:* Choose m, maximum size of the subspace, and k, the desired number of approximate eigenvectors. Let ϵ be the convergence stopping criteria. Let $x_0 = 0$ such that $r_0 = b$, $\rho_0 = \|r_0\|_2$, $\rho = 1$, $p = m-k$ and $w_1 = Ar_0/\|Ar_0\|_2$, $r_0 = r_0/\rho_0$.

2. *First cycle:* Apply the standard Simpler GMRES to produce Y_m, W_m, R_m, r_m, x_m, $\widetilde{w}$. Let $r_0 = r_m$ and $x_0 = x_m$. If $\|r_m\|_2$ is satisfied then stop, else $r_0 = r_0/\|r_0\|$.

3. *Form approximate eigenvectors:* Compute $T_m = (W_m^T Y_m)^{-1}$ and $\widetilde{R}_m = R_m T_m$ in order to solve $\widetilde{R}_m g_j = \lambda_j g_j$, for appropriate g_j. Form $u_j = Y_m(T_m g_j)$ and $\widetilde{u}_j = W_m(\widetilde{R}_m g_j)$ for $j = 1:k$.

4. *Other cycles:* Apply $AugmentedSArnoldi$ with $\widetilde{U}_k$, U_k, r_0, A, k and m as inputs to generate Y_m, W_m, R_m. Find $\widetilde{w} = [\xi_1, \xi_2, \ldots, \xi_m]^T$ and $r_m = r_0 - W_m \widetilde{w}$ during the Simpler Arnoldi orthogonalization, that is, at each ith orthogonalization step we find $\xi_i = w_i^T r_{i-1}$ and $r_i = r_{i-1} - \xi_i w_i$. We also update ρ using $\rho = \rho \sin(\cos^{-1}(\xi_i)/\rho)$ and if $\rho\rho_0 \leq \epsilon$ we go to **5**.

5. *Form the approximate solution:* $\widehat{y} = R_m^{-1}\widetilde{w}$ and $x_m = x_0 + \rho_0 Y_m \widehat{y} = x_0 + \rho_0 z_m$ where z_m is given by (7).

6. *Restart:* If $\|r_m\|_2 \leq \epsilon$ is satisfied then stop, else $x_0 = x_m$, $\rho_0 = \|r_m\|_2$ and $r_0 = r_m/\rho_0$. Set $\rho = 1$ and go to **3**.

2.1   Implementation Using Householder Reflectors

In order to maintain orthogonality of the vectors to a high accuracy, it is preferable to use Householder orthogonalizations. Let Q_j denote the Householder reflector that transforms only the jth through n^{th} entries of the vectors on which it acts. Suppose $Q_1, Q_2, \ldots, Q_i$ are such that

$$
\binom{R_i}{0} = \begin{cases} Q_i A r_0, & \text{if } i = 1, \\ Q_i \cdots Q_1 A(r_0, w_1, \ldots, w_{i-1}), & \text{if } 1 < i \le p, \quad (14) \\ Q_i \cdots Q_1 A(r_0, w_1, \ldots, w_{p-1}, u_1, \ldots, u_{i-p}), & \text{if } p < i \le m. \end{cases}
$$

Since R_i is invertible and

$$
\begin{cases} A r_0 = w_1 R_1, & \text{if } i = 1, \\ A(r_0, w_1, \ldots, w_{i-1}) = (w_1, \ldots, w_i) R_i, & \text{if } 1 < i \le p, \quad (15) \\ A(r_0, w_1, \ldots, w_{p-1}, u_1, \ldots, u_{i-p}) = (w_1, \ldots, w_i) R_i, & \text{if } p < i \le m. \end{cases}
$$

Then,

$$
Q_i \cdots Q_1(w_1, \ldots, w_i) = \binom{I_i}{0}, \qquad I_i \in \mathbb{R}^{i \times i}. \tag{16}
$$

and $w_j = Q_1 \cdots Q_i e_j = Q_1 \cdots Q_j e_j$ for $j = 1, \ldots, i$ where e_j is the jth column of identity matrix of order n. Letting $\widehat{y}_i = (\eta_1, \ldots, \eta_i)^{\text{T}}$, and using (6) and (16), we have

$$
z_i = \begin{cases} Q_1(\eta_1 Q_1 r_0), & \text{if } i = 1, \\ Q_1 \cdots Q_i[\eta_1 r_i + (\eta_2, \ldots, \eta_i, 0, \ldots)^{\text{T}}], & \text{if } 1 < i \le p, \\ Q_1 \cdots Q_i \left[\eta_1 r_i + (\eta_2, \ldots, \eta_p, 0, \ldots)^{\text{T}}\right] + \sum_{j=1}^{i-p} \eta_{p+j} u_j, & \text{if } p < i \le m. \end{cases}
\tag{17}
$$

Since $(Q_i \cdots Q_1 r_i)^T e_j = (Q_i \cdots Q_1 r_i)^T (Q_i \cdots Q_1 w_j) = r_i^T w_j = 0$ for $j = 1, \ldots, i$, it follows from (4) and (16) that $\xi_i = e_i^T Q_i \cdots Q_1 r_0$, and

$$
r_i = Q_1 \cdots Q_i \binom{0}{I_{n-i}} r_{i-1}, \qquad I_{n-i} \in \mathbb{R}^{(n-i) \times (n-i)}. \tag{18}
$$

The following algorithm describes orthogonalization using Householder transformations.

AugmentedSHArnoldi

1. *Start:* Input U_k, k approximate eigenvectors, $\widetilde{U}_k = A U_k$, m and k. Let
   $p = m - k$; $w_1 = A r_0$;
   Find Q_1 such that $Q_1 w_1 = [R_{1,1}, 0, \ldots, 0]^T$;
   $w_1 = Q_1 e_1$;
2. *Iterate:* For $i = 2 : m$
   If $i \le p$, $w_i = Q_{i-1} \cdots Q_1(A w_{i-1})$;
   else, $w_i = Q_{i-1} \cdots Q_1 \tilde{u}_{i-p}$;
   Find Q_i such that $Q_i w_i = [R_{1,i}, \ldots, R_{i,i}, 0, \ldots, 0]^T$;
   $w_i = Q_1 \cdots Q_i e_i$;
3. *Output:* If $k = 0$, $Y_m = [r_0 \ W_{m-1}]$; else, $Y_m = [r_0 \ W_{p-1} \ U]$.

For the augmented Simpler GMRES algorithm using Householder reflections, we give below the steps that are different from the Gram-Schmidt version of the algorithm.

SHGMRESE

2. *First cycle:* Apply standard Simpler Householder GMRES.

4. *Other cycles:* Apply *AugmentedSHArnoldi* with $\tilde{U}_k$, U_k, r_0, A, k and m as inputs to generate Y_m, W_m, R_m. Find $r_i = Q_i r_i$ and $\xi_i = e_i^T r_i$ at each ith orthogonalization step. We also update ρ using $\rho = \rho \sin(\cos^{-1}(\xi_i)/\rho)$ and if $\rho\rho_0 \leq \epsilon$, we go to **5.**

5. *Form the approximate solution:* $\hat{y} = R_m^{-1}\tilde{w}$ and $\tilde{x}_m = x_0 + \rho_0 Y_m \hat{y} = x_0 + \rho_0 z_m$ where z_m is given by (17).

3   Numerical Experiments

We describe the results of numerical experiments using Matlab. We compare restarted Simpler GMRES, SGMRES(m) with the two augmented versions of the algorithm, namely, SGMRESE(p, k), the implementation using Gram-Schmidt orthogonalization and SHGMRESE(p, k), the Householder version. We also give the results for GMRES augmented with eigenvectors, GMRESE(p, k) [3]. For each method, we compare the relative residual norm against the number of matrix-vector products (matvecs) and amount of work required (flops). We use $p = 21$ Krylov vectors and $k = 4$ harmonic Ritz vectors, and therefore $m = 25$. In all experiments, the initial guess is the zero vector.

Experiment 1. We consider the first four test problems considered by Morgan [3]. The matrices are bidiagonal with 0.1 in each superdiagonal position. In Problem 1, the matrix has entries 1, 2, 3, ... , 999, 1000 on the main diagonal. For Problem 2, the diagonal entries are 0.01, 0.02, 0.03, 0.04, 10, 11, ... , 1004, 1005, for Problem 3, we have $-2, -1, 1, 2, \ldots, 997, 998$ on the main diagonal and for Problem 4, the diagonal entries are 1, 1.01, 1.02, 1.03, 1.04, 2, 3, 4, ... , 995, 996. The right-hand side vector for each problem is a vector of ones.

We show in Fig. 1 the convergence history for Problem 2. This problem is difficult because of the small eigenvalues 0.01, 0.02, 0.03, 0.04. In this case, SGMRES(25) stagnates because the Krylov subspace is not large enough to develop approximations to the small eigenvalues. For roughly the same amount of work, SGMRES reaches a relative residual norm of 2.02E-2 whereas SGMRESE reaches a norm of 8.95E-11. On the other hand, we observe that the three augmented methods are convergent, each reaching a relative residual norm of 2.8E-5 after 207 matvecs. However, the work required by SGMRESE, the Gram-Schmidt implementation, is 2.29E7 which is less than that required by GMRESE, approximately 3.48E7 flops. Also, SHGMRESE requires 6.12E7 flops to reach the same relative residual norm.

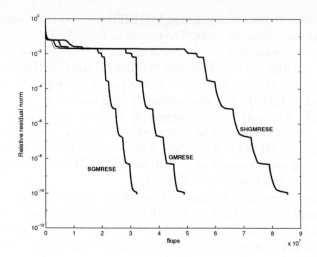

Fig. 1. Convergence history for example 2 of [3]

Table 1. Results for problems 1 to 4 of [3]

Problem	Method	matvecs	$\|r_i\|_2/\|r_0\|_2$	flops
	SGMRES	523	9.73E-11	3.30E7
1	GMRESE	239	9.38E-11	4.28E7
	SGMRESE	239	9.38E-11	2.73E7
	SHGMRESE	239	9.38E-11	7.07E7
	SGMRES	500	2.02E-2	3.17E7
2	GMRESE	280	8.95E-11	4.89E7
	SGMRESE	280	8.95E-11	3.22E7
	SHGMRESE	280	8.95E-11	8.54E7
	SGMRES	500	8.89E-3	3.17E7
3	GMRESE	399	9.64E-11	6.92E7
	SGMRESE	399	9.64E-11	4.55E7
	SHGMRESE	399	9.64E-11	1.18E8
	SGMRES	511	9.86E-11	3.40E7
4	GMRESE	425	9.78E-11	7.66E7
	SGMRESE	425	9.78E-11	4.95E7
	SHGMRESE	425	9.78E-11	1.32E8

Table 1 shows the results for the four test problems. For Problem 1, we note that SGMRES requires 523 matrix-vector products and a work of 3.30E7 to reach a relative residual norm of 9.73E-11 whereas the augmented algorithm SGMRESE needs 239 matrix-vector products and only 2.73E7 flops to reach a norm of 9.38E-11. Also observe that SGMRESE performs better than GMRESE.

The Householder version of augmented SGMRES requires more work because of the higher orthogonalization costs. The matrix in Problem 3 is indefinite and the results show that after 500 matrix-vector products, the relative residual norm reached by SGMRES is 8.89E-3 whereas SGMRESE and GMRESE reach a residual norm of 9.64E-11 after 399 matrix-vector products. However, note that the work required by GMRESE is approximately 50% more than the work required by SGMRESE. As mentioned by Morgan [3], Problem 4 is designed to be difficult for augmented methods. The results seem to confirm this as the augmented version of Simpler GMRES is not better than SGMRES. We note that SGMRESE takes a greater amount of work than SGMRES to reach approximately the same relative residual norm of 9.8E-11.

For Problem 3, we show in Table 2 the residual norm reached and the amount of flops required by SGMRESE and GMRESE after 5, 10, 15 and 20 runs. Note that a run is the iteration between two successive restarts. Both GMRESE and SGMRESE, require the same number of matrix-vector products for the different runs. However we remark that in each case, the amount of work required by SGMRESE is less than the work required by GMRESE to reach the same residual norm.

Table 2. Residual norm and flops after different runs for Problem 3

	After 5 runs		After 10 runs		After 15 runs		After 20 runs	
Method	$\|r_i\|_2$	flops	$\|r_i\|_2$	flops	$\|r_i\|_2$	flops	$\|r_i\|_2$	flops
SGMRESE	0.22	1.14E7	8.3E-5	2.38E7	2.1E-8	3.62E7	5.4E-12	4.86E7
GMRESE	0.22	1.74E7	8.3E-5	3.65E7	2.1E-8	5.55E7	5.4E-12	7.46E7

Similar results for Problem 2 show that, after 12 runs, GMRESE requires approximately 4.32E7 flops to reach a residual norm of 1.69E-7, whereas SGMRESE requires 2.88E7 flops to reach the same norm.

Experiment 2. We consider matrices, SHERMAN4, SHERMAN1 and SAYLR3 from the Harwell-Boeing sparse matrix collection. SHERMAN1 is nonsymmetric of dimension 1000 and has 3750 nonzero entries, SHERMAN4 is nonsymmetric, has dimension 1104 and 3786 nonzero entries and SAYLR3 is a nonsymmetric matrix, has dimension 1000 and 375 nonzero entries. The convergence history for SHERMAN4 in Fig. 3 shows that SGMRESE performs better than the other methods. In Table 3, we show the number of matrix-vector products, the relative residual norm and the amount of work required for the three matrices. For SHERMAN1, we observe that Simpler GMRES requires 5032 matrix-vector products to reach a relative residual norm of 9.99E-11 whereas the augmented version SGMRESE requires only 1304 matrix-vector products to reach approximately the same norm. Comparing the amount of work for the two methods, we observe that SGMRESE requires a lesser amount. For all the three test problems, we observe that SGMRESE performs better than GMRESE.

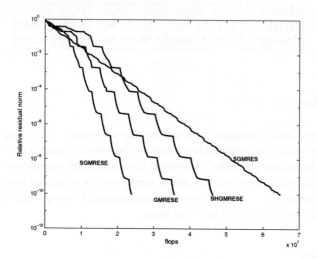

Fig. 2. Convergence for the SHERMAN4

Table 3. Results for Harwell-Boeing matrices

Matrix	Method	matvecs	$\|r_i\|_2/\|r_0\|_2$	flops
	SGMRES	891	9.96E-11	6.48E7
SHERMAN4	GMRESE	208	9.11E-11	3.57E7
	SGMRESE	208	9.11E-11	2.39E7
	SHGMRESE	208	9.11e-11	4.63E7
	SGMRES	5032	9.99E-11	3.36E8
SHERMAN1	GMRESE	1304	9.90E-11	2.26E8
	SGMRESE	1304	9.90E-11	1.48E8
	SHGMRESE	1304	9.90e-11	3.18E8
	SGMRES	3951	9.99E-11	2.64E8
SAYLR3	GMRESE	1142	9.98E-11	2.06E8
	SGMRESE	1142	9.98E-11	1.37E8
	SHGMRESE	1142	9.98E-11	3.48E8

We show in Table 4 the residual norm reached and the amount of work required after different runs for SHERMAN4. Observe that the amount of work required by augmented GMRES is roughly 40% more than the work required by augmented Simpler GMRES.

Table 4. Residual norm and flops after different runs for SHERMAN4

Method	After 5 runs		After 8 runs		After 10 runs	
	$\|r_i\|_2$	flops	$\|r_i\|_2$	flops	$\|r_i\|_2$	flops
SGMRESE	4.2E-3	1.21E7	7.2E-7	1.98E7	4.8E-9	2.39E7
GMRESE	4.2E-3	1.69E7	7.2E-7	2.88E7	4.8E-9	3.57E7

4   Conclusion

We have described a method to improve the convergence of Simpler GMRES for problems with small eigenvalues. Numerical tests indicate that using Gram-Schmidt orthogonalizations to implement the algorithm, significant reduction in the amount of work required for convergence can be achieved over augmented GMRES. More numerical experiments using preconditioning techniques for other problems will be reported in another work.

References

1. Erhel, J., Burrage, K., Pohl, B.: Restarted GMRES preconditioned by deflation. J. Comput. Appl. Math. **69** (1996) 303–318
2. Goossens, S., Roose, D.: Ritz and harmonic Ritz values and the convergence of FOM and GMRES. Numer. Lin. Alg. Appl. **6** (1999) 281–293
3. Morgan, R.B.: A restarted GMRES method augmented with eigenvectors. SIAM J. Matrix Anal. Appl. **16** (1995) 1154–1171
4. Saad, Y., Schultz M.H.: GMRES: A generalized minimal residual algorithm for solving nonsymmetric linear systems. SIAM J. Sci. Statist. Comput. **7** (1986) 865–869
5. Walker, H.F., Zhou L.: A Simpler GMRES. Numer. Lin. Alg. with Appl. **1** (1994) 571–581

A Projection Method for a Rational Eigenvalue Problem in Fluid-Structure Interaction

Heinrich Voss

Section of Mathematics, TU Hamburg-Harburg, D – 21071 Hamburg, Germany
voss @ tu-harburg.de   http://www.tu-harburg.de/mat/hp/voss

Abstract. In this paper we consider a rational eigenvalue problem governing the vibrations of a tube bundle immersed in an inviscid compressible fluid. Taking advantage of eigensolutions of appropriate sparse linear eigenproblems the large nonlinear eigenvalue problem is projected to a much smaller one which is solved by inverse iteration.

1   Introduction

Vibrations of a tube bundle immersed in an inviscid compressible fluid are governed under some simplifying assumptions by an elliptic eigenvalue problem with non-local boundary conditions which can be transformed to a rational eigenvalue problem. In [13] we proved that this problem has a countable set of eigenvalues which can be characterized as minmax values of a Rayleigh functional.

To determine eigenvalues and eigenfunctions numerically this approach suggests to apply the Rayleigh-Ritz method yielding a rational matrix eigenvalue problem where the matrices typically are large and sparse. Since for nonlinear eigenvalue problems the eigenvectors do not fulfill an orthogonality condition Lanczos type methods do not apply but each eigenpair has to be determined individually by an iterative process where each iteration step requires the solution of a linear system of large dimension. In this paper we propose a projection method where the ansatz vectors are constructed from the solutions of suitable linear eigenvalue problems. The resulting small dimensional nonlinear eigenproblem then is solved by inverse iteration.

The paper is organized as follows: Section 2 describes the model for the fluid-structure interaction problem under consideration, and Section 3 summarizes the minmax characterization for nonlinear eigenvalue problems and its application to the fluid-structure interaction problem of section 2. In Section 4 we consider numerical methods for nonlinear eigenvalue problems, and Section 5 contains a numerical example.

2   A Spectral Problem in Fluid-Solid Structures

In this paper we consider a model which governs the free vibrations of a tube bundle immersed in a compressible fluid under the following simplifying assumptions: The tubes are assumed to be rigid, assembled in parallel inside the fluid,

P.M.A. Sloot et al. (Eds.): ICCS 2002, LNCS 2330, pp. 403–411, 2002.

and elastically mounted in such a way that they can vibrate transversally, but they can not move in the direction perpendicular to their sections. The fluid is assumed to be contained in a cavity which is infinitely long, and each tube is supported by an independent system of springs (which simulates the specific elasticity of each tube). Due to these assumptions, three-dimensional effects are neglected, and so the problem can be studied in any transversal section of the cavity. Considering small vibrations of the fluid (and the tubes) around the state of rest, it can also be assumed that the fluid is irrotational.

Mathematically this problem can be described in the following way (cf. [2]). Let $\Omega \subset \mathbb{R}^2$ (the section of the cavity) be an open bounded set with locally Lipschitz continuous boundary Γ. We assume that there exists a family $\Omega_j \neq \emptyset$, $j = 1, \ldots, K$, (the sections of the tubes) of simply connected open sets such that $\bar{\Omega}_j \subset \Omega$ for every j, $\bar{\Omega}_j \cap \bar{\Omega}_i = \emptyset$ for $j \neq i$, and each Ω_j has a locally Lipschitz continuous boundary Γ_j. With these notations we set

$$\Omega_0 := \Omega \setminus \bigcup_{j=1}^{K} \Omega_j.$$

Then the boundary of Ω_0 consists of $K + 1$ connected components which are Γ and Γ_j, $j = 1, \ldots, K$.

We consider the rational eigenvalue problem

Find $\lambda \in \mathbb{R}$ and $u \in H^1(\Omega_0)$ such that for every $v \in H^1(\Omega_0)$

$$c^2 \int_{\Omega_0} \nabla u \cdot \nabla v \, dx = \lambda \int_{\Omega_0} uv \, dx + \sum_{j=1}^{K} \frac{\lambda \rho_0}{k_j - \lambda m_j} \int_{\Gamma_j} un \, ds \cdot \int_{\Gamma_j} vn \, ds. \quad (1)$$

Here

$$H^1(\Omega) = \{u \in L^2(\Omega_0) : \nabla u \in L^2(\Omega_0)^2\}$$

denotes the standard Sobolev space equipped with the usual scalar product

$$(u, v) := \int_{\Omega_0} (u(x)v(x) + \nabla u(x) \cdot \nabla v(x)) \, dx,$$

and n denotes the outward unit normal on the boundary of Ω_0.

Obviously $\lambda = 0$ is an eigenvalue of (1) with eigenfunction $u = \text{const}$. We reduce the eigenproblem (1) to the space

$$H := \{u \in H^1(\Omega_0) : \int_{\Omega_0} u(x) \, dx = 0\}$$

and consider the scalar product

$$\langle u, v \rangle := \int_{\Omega_0} \nabla u(x) \cdot \nabla v(x) \, dx.$$

on H which is known to define a norm on H which is equivalent to the norm induced by $(\cdot, \cdot)$.

By the Lax–Milgram lemma the variational eigenvalue problem (1) is equivalent to the nonlinear eigenvalue problem

Determine λ and $u \in H$ such that

$$T(\lambda)u := (-I + \lambda A + \sum_{j=1}^{K} \frac{\rho_0 \lambda}{k_j - \lambda m_j} B_j)u = 0 \tag{2}$$

where the linear symmetric operators A and B_j are defined by

$$\langle Au, v \rangle := \int_{\Omega_0} uv \, dx \quad \text{for every } u, v \in H \text{ and}$$

$$\langle B_j u, v \rangle := \int_{\Gamma_j} un \, ds \cdot \int_{\Gamma_j} vn \, ds \quad \text{for every } u, v \in H.$$

A is completely continuous by Rellich's embedding theorem and $w := B_j u$, $j = 1, \dots, K$, is the weak solution in H of the elliptic problem

$$\Delta w = 0 \text{ in } \Omega_0, \quad \frac{\partial}{\partial n} w = 0 \text{ on } \partial \Omega_0 \setminus \Gamma_j, \quad \frac{\partial}{\partial n} w = n \cdot \int_{\Gamma_j} un \, ds \text{ on } \Gamma_j.$$

By the continuity of the trace operator B_j is continuous, and since the range of B_j is twodimensional spanned by the solutions $w_i \in H$ of

$$\Delta w_i = 0 \text{ in } \Omega_0, \quad \frac{\partial}{\partial n} w = 0 \text{ on } \partial \Omega_0 \setminus \Gamma_j, \quad \frac{\partial}{\partial n} w = n_i \text{ on } \Gamma_j, \ i = 1, 2,$$

it is even completely continuous.

In Conca et al. [2] it is shown that the eigenvalues are the characteristic values of a linear compact operator acting on a Hilbert space. The operator associated with this eigenvalue problem is not selfadjoint, but it can be symmetrized in the sense that one can prove the existence of a selfadjoint operator which has the same spectrum as the original operator. The following section describes a framework to prove the existence of countably many eigenvalues taking advantage of a minmax characterization for nonlinear eigenvalue problems.

3   Characterization of eigenvalues

We consider the nonlinear eigenvalue problem

$$T(\lambda)x = 0 \tag{3}$$

where $T(\lambda)$ is a selfadjoint and bounded operator on a real Hilbert space H for every λ in an open real interval J. As in the linear case $\lambda \in J$ is called an eigenvalue of problem (3) if equation (3) has a nontrivial solution $x \neq 0$.

We assume that

$$f : \begin{cases} J \times H \to \mathbb{R} \\ (\lambda, x) \mapsto \langle T(\lambda)x, x \rangle \end{cases}$$

is continuously differentiable, and that for every fixed $x \in H^0$, $H^0 := H \setminus \{0\}$, the real equation

$$f(\lambda, x) = 0 \tag{4}$$

has at most one solution in J. Then equation (4) implicitly defines a functional p on some subset D of H^0 which we call the Rayleigh functional.

We assume that

$$\frac{\partial}{\partial \lambda} f(\lambda, x)\Big|_{\lambda = p(x)} > 0 \quad \text{for every } x \in D.$$

Then it follows from the implicit function theorem that D is an open set and that p is continuously differentiable on D.

For the linear eigenvalue value problem $T(\lambda) := \lambda I - A$ where $A : H \to H$ is selfadjoint and continuous the assumptions above are fulfilled, p is the Rayleigh quotient and $D = H^0$. If A additionally is completely continuous then A has a countable set of eigenvalues which can be characterized as minmax and maxmin values of the Rayleigh quotient.

For nonlinear eigenvalue problems variational properties using the Rayleigh functional were proved by Duffin [3] and Rogers [9] for the finite dimensional case and by Hadeler [4], [5], Rogers [10], and Werner [15] for the infinite dimensional case if the problem is overdamped, i.e. if the Rayleigh functional p is defined in the whole space H^0. Nonoverdamped problems were studied by Werner and the author [14]. In this case the natural numbering for which the smallest eigenvalue is the first one, the second smallest is the second one, etc. is not appropriate, but the number of an eigenvalue λ of the nonlinear problem (3) is obtained from the location of the eigenvalue 0 in the spectrum of the linear operator $T(\lambda)$.

We assume that for every fixed $\lambda \in J$ there exists $\nu(\lambda) > 0$ such that the linear operator $T(\lambda) + \nu(\lambda)I$ is completely continuous. If $\lambda \in J$ is an eigenvalue of $T(\cdot)$ then $\mu = 0$ is an eigenvalue of the linear problem $T(\lambda)y = \mu y$, and therefore there exists $n \in \mathbb{N}$ such that

$$0 = \max_{V \in H_n} \min_{v \in V^1} \langle T(\lambda)v, v \rangle$$

where H_n denotes the set of all n–dimensional subspaces of H and $V^1 := \{v \in V : \|v\| = 1\}$ is the unit ball in V. In this case we call λ an n-th eigenvalue of (3).

With this numbering the following minmax characterization of the eigenvalues of the nonlinear eigenproblem (3) was proved in [14]:

Theorem 1. *Under the conditions given above the following assertions hold:*

(i) For every $n \in \mathbb{N}$ there is at most one n-th eigenvalue of problem (3) which can be characterized by

$$\lambda_n = \min_{\substack{V \in H_n \\ V \cap D \neq \emptyset}} \sup_{v \in V \cap D} p(v). \tag{5}$$

The set of eigenvalues of (3) in J is at most countable.

(ii) *If*

$$\lambda_n = \inf_{\substack{V \in H_n \\ V \cap D \neq \emptyset}} \sup_{v \in V \cap D} p(v) \in J$$

for some $n \in \mathbb{N}$ then λ_n is the n-th eigenvalue of (3) and (5) holds.

(iii) *If there exists the m-th and the n-th eigenvalue λ_m and λ_n in J and $m < n$ then J contains a k-th eigenvalue λ_k for $m < k < n$ and*

$$\inf J < \lambda_m \leq \lambda_{m+1} \leq \ldots \leq \lambda_n < \sup J.$$

(iv) *If $\lambda_1 \in J$ and $\lambda_n \in J$ for some $n \in \mathbb{N}$ then for every $j \in \{1, \ldots, n\}$ the space $V \in H_j$ with $V \cap D \neq \emptyset$ and $\lambda_j = \sup_{u \in V \cap D} p(u)$ is contained in D, and the characterization (5) can be replaced by*

$$\lambda_j = \min_{\substack{V \in H_j \\ V_1 \subset D}}, \max_{v \in V_1} p(v) \quad j = 1, \ldots, n.$$

For the nonlinear eigenproblem (2) the general conditions obviously are satisfied for every open interval $J \subset \mathbb{R}_+$ which does not contain k_j/m_j for $j = 1, \ldots, K$. Namely, $T(\lambda)$ is selfadjoint and bounded on H and $T(\lambda) + I$ is completely continuous for every $\lambda \in J$. Moreover for fixed $u \in H^0$

$$f(\lambda, u) = -c^2 \int_{\Omega_0} |\nabla u|^2 \, dx + \lambda \int_{\Omega_0} u^2 \, dx + \sum_{j=1}^{K} \frac{\lambda \rho_0}{k_j - \lambda m_j} \left| \int_{\Gamma_j} un \, ds \right|^2 \quad (6)$$

is monotonely increasing with respect to λ. Hence, every open interval J such that $k_j/m_j \notin J$ for $j = 1, \ldots, K$ contains at most countably many eigenvalues which can be characterized as minmax value of the Rayleigh functional p defined by $f(\lambda, u) = 0$ where f is defined in (6).

Comparing p with the Rayleigh quotient of the linear eigenvalue problem

Determine $\lambda \in \mathbb{R}$ and $u \in H^0$ such that for every $v \in H$

$$c^2 \int_{\Omega_0} \nabla u \cdot \nabla v \, dx = \lambda \Big(\int_{\Omega_0} uv \, dx + \sum_{j=1}^{K} \frac{\rho_0}{k_j - \kappa m_j} \int_{\Gamma_j} un \, ds \cdot \int_{\Gamma_j} vn \, ds \Big). \quad (7)$$

we obtained the following inclusion result for the eigenvalues in $J_1 := (0, \min_j \frac{k_j}{m_j})$.

Theorem 2. *Let $\kappa \in J_1$. If the linear eigenvalue problem (7) has m eigenvalues $0 \leq \mu_1 \leq \mu_2 \leq \ldots \leq \mu_m$ in J_1 then the nonlinear eigenvalue problem (1) has m eigenvalues $\lambda_1 \leq \lambda_2 \leq \ldots \leq \lambda_m$ in J_1, and the following inclusion holds*

$$\min(\mu_j, \kappa) \leq \lambda_j \leq \max(\mu_j, \kappa), \quad j = 1, \ldots, m.$$

λ_j is a j-th eigenvalue of (1).

Similarly, for eigenvalues greater than $\max_j \frac{k_j}{m_j}$ we obtained the inclusion in Theorem 3. Here we compared the Rayleigh functional of (1) in the interval $(\max_j \frac{k_j}{m_j}, \infty)$ with the Rayleigh quotient of the linear eigenproblem:

Find $\lambda \in \mathbb{R}$ and $u \in H^0$ such that for every $v \in H$

$$c^2 \int_{\Omega_0} \nabla u \cdot \nabla v \, dx + \sum_{j=1}^{K} \frac{\kappa \rho_0}{\kappa m_j - k_j} \int_{\Gamma_j} un \, ds \cdot \int_{\Gamma_j} vn \, ds = \lambda \int_{\Omega_0} uv \, dx. \qquad (8)$$

Theorem 3. *Let $\kappa > \max_j \frac{k_j}{m_j}$. If the m-smallest eigenvalue μ_m of the linear eigenvalue problem (8) satisfies $\mu_m > \max_j \frac{k_j}{m_j}$ then the nonlinear eigenvalue problem (1) has an m-th eigenvalue λ_m, and*

$$\min(\mu_m, \kappa) \leq \lambda_m \leq \max(\mu_m, \kappa).$$

Theorems 2 and 3 are proved in [13]. Inclusions of eigenvalues λ of problem (1) which lie between $\min_j \frac{k_j}{m_j}$ and $\max_j \frac{k_j}{m_j}$ are under investigation.

4   Algorithms for nonlinear eigenvalue problems

In this section we consider the finite dimensional nonlinear eigenvalue problem

$$T(\lambda)x = 0 \qquad (9)$$

(for instance a finite element approximation of problem (1)) where $T(\lambda)$ is a family of real symmetric $n \times n$-matrices satisfying the conditions of the last section, and we assume that the dimension n of problem (9) is large.

For sparse linear eigenvalue problems the most efficient methods are iterative projection methods, where approximations of the wanted eigenvalues and corresponding eigenvectors are obtained from projections of the eigenproblem to subspaces which are expanded in the course of the algorithm. Methods of this type for symmetric problems are the Lanczos method, rational Krylov subspace methods and the Jacobi-Davidson method, e.g. (cf. [1]).

Generalizations of these methods to nonlinear eigenvalue problems do not exist. The numerical methods for nonlinear problems studied in [6], [7], [8], [11], [12] are all variants of inverse iteration

$$x^{k+1} = \alpha_k T(\lambda_k)^{-1} T'(\lambda_k) x^k \qquad (10)$$

where α_k is a suitable normalization factor and λ_k is updated in some way. Similarly as in the linear case inverse iteration is quadratically convergent for simple eigenvalues, and the convergence is even cubic if $\lambda_k = p(x_k)$ where p denotes the Rayleigh functional of problem (9).

An essential disadvantage of inverse iteration is the fact that each eigenvalue has to be determined individually by an iterative process, and that each step of this iteration requires the solution of a linear system. Moreover, the coefficient matrix $T(\lambda_k)$ of system (10) changes in each step, and in contrast to the linear case replacing (10) by

$$x^{k+1} = \alpha_k T(\sigma)^{-1} T'(\lambda_k) x^k$$

with a fixed shift σ results in convergence to an eigenpair of the linear system $T(\sigma)x = \gamma T'(\tilde{\lambda})x$ ($\gamma \neq 0$ depending on the normalization condition) from which we can not recover an eigenpair of the nonlinear problem (9).

A remedy against this wrong convergence was proposed by Neumaier [8] who introduced the so called residual inverse iteration which converges linearly with a fixed shift, and quadratically or cubically if the coefficient matrix changes in every iteration step according to reasonable updates of λ_k.

For large and sparse nonlinear eigenvalue problems inverse iteration is much too expensive. For the rational eigenvalue problem (1) the proof of the Inclusion Theorems 2 and 3 demonstrates that eigenvectors of the linear systems (7) and (8), respectively, are good approximations to eigenvectors of the nonlinear problem, at least if the shift κ is close to the corresponding eigenvalue. This suggests the following projection method if we are interested in eigenvalues of the nonlinear problem (1) in an interval $J \subset (0, \min_j \frac{k_j}{m_j})$ or $J \subset (\max_j \frac{k_j}{m_j}, \infty)$.

Projection method
1. Choose a small number of shifts $\kappa_1, \ldots, \kappa_r \in J$.
2. For $j = 1, \ldots, r$ determine the eigenvectors u_{jk}, $k = 1, \ldots, s_j$, of the linear problem (7) with shift κ_j corresponding to eigenvalues in J.
3. Let U be the matrix with columns u_{jk}, $j = 1, \ldots, r$, $k = 1, \ldots, s_j$. Determine the QR factorization with column pivoting which produces the QR factorization of UE where E denotes a permutation matrix such that the absolute values of the diagonal elements of R are monotonely decreasing.
4. For every j with $|r_{jj}| < \tau \cdot |r_{11}|$ drop the j-th column of Q where $\tau \in [0, 1)$ is a given tolerance, and denote by V the space that is spanned by the remaining columns of Q.
5. Project the nonlinear eigenvalue problem (1) to V and solve the projected problem by inverse iteration with variable shifts.

5   A numerical example

Consider the rational eigenvalue problem (1) where Ω is the L-shaped region $\Omega := (-8, 8) \times (-8, 8) \setminus ([0, 8) \times (-8, 0])$, Ω_j, $j = 1, 2, 3$ are circles with radius 1 and centers $(-4, -4)$, $(-4, 4)$ and $(4, 4)$, and $k_j = m_j = 1$, $j = 1, 2, 3$.

We discretized this eigenvalue problem with linear elements obtaining a matrix eigenvalue problem of dimension $n = 10820$ which has 25 eigenvalues $\lambda_1 \leq \ldots \leq \lambda_{25}$ in the interval $J_1 = (0, 1)$ and 16 eigenvalues $\tilde{\lambda}_{20} \leq \ldots \leq \tilde{\lambda}_{35}$ in $J_2 := (1, 2)$.

With 2 shift parameters $\kappa_1 := 0$ and $\kappa_2 = 0.999$ and the drop tolerances $\tau = 1e - 1$, $1e - 2$, $1e - 3$ we obtained eigenvalue approximations the relative errors of which are displayed in Figure 1. The dimensions of the projected eigenvalue problems were 25, 32 and 37 respectively.

With Gaussian knots $\kappa_1 = \frac{1}{2}(1 - \frac{1}{\sqrt{3}})$ and $\kappa_2 = \frac{1}{2}(1 + \frac{1}{\sqrt{3}})$ we obtained smaller relative errors, however in this case the projected problem found only 24 approximate eigenvalues missing $\lambda_{25} = 0.9945$.

Figure 2 shows the relative errors which were obtained with shift parameters $\kappa_1 = 1.001$ and $\kappa_2 = 2$ in problem (8) and dimensions 16, 24 and 29, respectively, of the nonlinear projected problem.

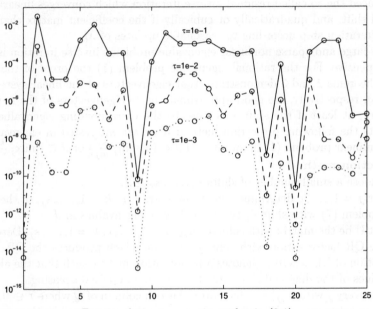

Fig. 1: relative errors; eigenvalue in (0,1)

References

1. Bai, Z., Demmel, J., Dongarra, J., Ruhe, A., van der Vorst, H., editors: *Templates for the Solution of Algebraic Eigenvalue Problems: A Practical Guide.* SIAM, Philadelphia, 2000
2. Conca, C., Planchard, J., Vanninathan, M.: Existence and location of eigenvalues for fluid-solid structures. Comput. Meth. Appl. Mech. Engrg. **77** (1989) 253–291
3. Duffin, R.J.: A minmax theory for overdamped networks. J. Rat. Mech. Anal. **4** (1955) 221–233
4. Hadeler, K.P.: Variationsprinzipien bei nichtlinearen Eigenwertaufgaben. Arch. Rat. Mech. Anal. **30** (1968) 297–307
5. Hadeler, K.P.: Nonlinear eigenvalue problems. Internat. Series Numer. Math. **27** 111–129

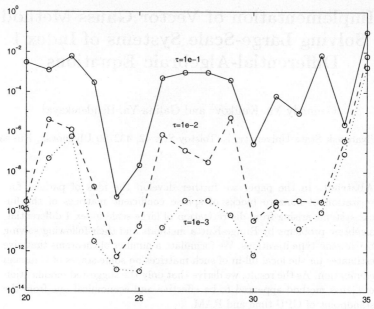

Fig. 2: relative errors; eigenvalues in (1,2)

6. Kublanovskaya, V.N.: On an application of Newton's method to the determination of eigenvalues of λ-matrices. Dokl. Akad. Nauk. SSR **188** (1969) 1240–1241

7. Kublanovskaya, V.N.: On an approach to the solution of the generalized latent value problem for λ-matrices. SIAM. J. Numer. Anal. **7** (1970) 532–537

8. Neumaier, A.: Residual inverse iteration for the nonlinear eigenvalue problem. SIAM J. Numer. Anal. **22** (1985) 914–923

9. Rogers, E.H.: A minmax theory for overdamped systems. Arch. Rat. Mech. Anal. **16** (1964) 89–96

10. Rogers, E.H.: Variational properties of nonlinear spectra. J. Math. Mech. **18** (1968) 479–490

11. Ruhe, A.: Algorithms for the nonlinear eigenvalue problem. SIAM J. Numer. Anal. **10** (1973) 674–689

12. Voss, H.: Computation of eigenvalues of nonlinear eigenvalue problems. pp. 147–157 in G.R. Joubert, editor: *Proceedings of the Seventh South African Symposium on Numerical Mathematics.* University of Natal, Durban 1981

13. Voss, H.: A rational spectral problem in fluid–solid vibration. In preparation

14. Voss, H., Werner,B.: A minimax principle for nonlinear eigenvalue problems with application to nonoverdamped systems. Math. Meth. Appl. Sci. **4** (1982) 415–424

15. Werner, B.: Das Spektrum von Operatorenscharen mit verallgemeinerten Rayleighquotienten. Arch. Rat. Mech. Anal. **42** (1971) 223–238

On Implementation of Vector Gauss Method for Solving Large-Scale Systems of Index 1 Differential-Algebraic Equations

Gennady Yu. Kulikov[1] and Galina Ya. Benderskaya[1]

Ulyanovsk State University, L. Tolstoy Str. 42, 432700 Ulyanovsk, Russia

Abstract. In the paper we further develop the idea of parallel factorization of nonzero blocks of sparse coefficient matrices of the linear systems arising from discretization of large-scale index 1 differential-algebraic problems by Runge-Kutta methods and their following solving by Newton-type iterations. We formulate a number of theorems that give estimates for the local fill-in of such matrices on some stages of Gaussian elimination. As the result, we derive that only the suggested modification of Gauss method appeared to be effective and economical one from the standpoint of CPU time and RAM.

1 Introduction

One of the main trends of modern computational mathematics is the development of effective numerical methods and software packages for the solution of index 1 differential-algebraic systems of the form (see, for example, [1], [2], [4]):

$$x'(t) = g\big(x(t), y(t)\big), \tag{1a}$$

$$y(t) = f\big(x(t), y(t)\big), \tag{1b}$$

$$x(0) = x^0, \ y(0) = y^0, \tag{1c}$$

where $t \in [0, T]$, $x(t) \in \mathbf{R}^m$, $y(t) \in \mathbf{R}^n$, $g : D \subset \mathbf{R}^{m+n} \to \mathbf{R}^m$, $f : D \subset \mathbf{R}^{m+n} \to \mathbf{R}^n$, and the initial conditions (1c) are consistent; i. e., $y^0 = f(x^0, y^0)$. Note that we consider only autonomous systems because any nonautonomous system may be converted to an autonomous one by introducing a new independent variable.

Let problem (1) satisfy the smoothness, nonsingularity and inclusion conditions introduced in [6]. Then, it has a unique solution $z(t) \overset{\text{def}}{=} \big(x(t)^T, y(t)^T\big)^T \in \mathbf{R}^{m+n}$ (see [6]). Besides, according to the Gear's definition in [3] problem (1) is of index 1.

To solve problem (1) numerically, we apply an l-stage implicit Runge-Kutta (RK) method given by the Butcher's table

$$\begin{array}{c|c} c & A \\ \hline & b^T \end{array},$$

* This work was supported in part by the Russian Foundation of the Basic Research (grants No. 01-01-00066 and No. 00-01-00197).

P.M.A. Sloot et al. (Eds.): ICCS 2002, LNCS 2330, pp. 412–421, 2002.

where A is a real matrix of dimension $l \times l$, b and c are real vectors of dimension l, to problem (1) and obtain the following discrete analogue:

$$x_{ki} = x_k + \tau \sum_{j=1}^{l} a_{ij} g(x_{kj}, y_{kj}), \tag{2a}$$

$$y_{ki} = f(x_{ki}, y_{ki}), \quad i = 1, 2, ..., l, \tag{2b}$$

$$x_{k+1} = x_k + \tau \sum_{i=1}^{l} b_i g(x_{ki}, y_{ki}), \tag{2c}$$

$$y_{k+1} = f(x_{k+1}, y_{k+1}), \quad k = 0, 1, ..., K - 1. \tag{2d}$$

Here $x_0 = x^0$, $y_0 = y^0$, and τ is a stepsize which may be variable. Algebraic system (2) is solved then by an iterative method at each time point t_k. Usually, the iterative process is taken in the form of simple or Newton-type iterations with trivial (or nontrivial) predictor [1], [2], [4], [7], [8],[10], [14].

In [11] it has been substantiated why Newton (or modified Newton) iteration is more preferable for solving differential-algebraic equations (1) than simple one. By this reason, we further consider only Newton-type iterations in the paper.

In order to solve problem (2) it is not necessary to apply an iterative process to all the equations of the discrete problem simultaneously. It is evident that equations (2a,b) do not depend on equations (2c,d). From this standpoint, it makes sense first to solve problem (2a,b) that contains $(m+n)l$ equations. Then, we have to find the solution of problem (2c) which will be a known function of the solution of (2a,b). Finally, we apply an iterations to system (2d) of dimension n. The advantage of this approach for the full and modified Newton methods was investigated in [7]. Moreover, when solving the larger problem (2a,b) we can limit ourselves by less quantity of iterations at the grid points on the segment $[0, T]$ (for details, see [12]).

Thus, from the above approach it follows necessity to solve discrete systems (2a,b) of dimension $(m+n)l$ many times during the integration. The latter allows us to conclude that Newton-type iterations put severe requirements on RAM and CPU time caused by the increasing dimension of discrete problem (2) in l times when an l-stage implicit RK method has been used. Therefore the basic problem is how to simplify and speed up the numerical solving of the linear systems with sparse coefficient matrices of the special form arising from the application of Newton method (full or modified) to problem (2a,b).

In [11], [13] we gave rise to the idea of modification of Gaussian elimination for parallel factorization of nonzero blocks of the matrices. In the present paper we first introduce a new term to denote this modification and it will be referred to further as *vector Gauss method*. This modification allows RAM and CPU time to be significantly reduced in the numerical integration of problem (1) by an implicit RK method. Now we discuss full details associated with pivoting in the course of implementation of the vector Gauss method. In addition, we give a simple example which illustrate theoretical results of the paper.

2   Preliminary Notes for Vector Gauss Method

As we have derived above, the basic part of Newton iteration applied to problem
(2) consists of solving linear systems of the form

$$\partial \bar{F}_k^\tau (Z_{k+1}^{i-1})(Z_{k+1}^{i-1} - Z_{k+1}^i) = \bar{F}_k^\tau Z_{k+1}^{i-1} \tag{3}$$

where lower indices mean time points, and upper ones denote iterations. Here
$Z_{k+1} \overset{\text{def}}{=} \left((z_{k1})^T, ..., (z_{k,l-1})^T, (z_{kl})^T \right)^T \in \mathbf{R}^{(m+n)l}$, where the vector $z_{kj} \overset{\text{def}}{=}$
$\left((x_{kj})^T, (y_{kj})^T \right)^T \in \mathbf{R}^{m+n}$, $j = 1, 2, \ldots, l$, unites components of the j-th stage
value of l-stage RK formula (2). The mapping $\bar{F}_k^\tau$ is the nontrivial part of dis-
crete problem (2) for computing the stage values Z_{k+1}, and $\partial \bar{F}_k^\tau (Z_{k+1}^{i-1})$ denotes
the Jacobian of the mapping $\bar{F}_k^\tau$ evaluated at the point Z_{k+1}^{i-1}.

From system (2a,b) it follows that the matrix $\partial \bar{F}_k^\tau (Z_{k+1}^{i-1})$ has the following
block structure:

$$\partial \bar{F}_k^\tau (Z) \overset{\text{def}}{=} \begin{pmatrix} \partial \bar{F}_k^\tau (Z)_1 \\ \partial \bar{F}_k^\tau (Z)_2 \\ \vdots \\ \partial \bar{F}_k^\tau (Z)_l \end{pmatrix} \tag{4}$$

where each block $\partial \bar{F}_k^\tau (Z)_j, j = 1, 2, \ldots, l$, is an $(m+n) \times (m+n)l$-matrix of the
form

$$\left(\begin{array}{ccccccccccc} O(\tau) & \cdots & O(\tau) & 1+O(\tau) & \cdots & O(\tau) & O(\tau) & \cdots & O(\tau) & O(\tau) & \cdots & O(\tau) \\ \vdots & \ddots & \vdots & \vdots & \ddots & \vdots & \vdots & \ddots & \vdots & \vdots & \ddots & \vdots \\ O(\tau) & \cdots & O(\tau) & O(\tau) & \cdots & 1+O(\tau) & O(\tau) & \cdots & O(\tau) & O(\tau) & \cdots & O(\tau) \\ 0 & \cdots & 0 & z & \cdots & z & z & \cdots & z & 0 & \cdots & 0 \\ \vdots & \ddots & \vdots & \vdots & \ddots & \vdots & \vdots & \ddots & \vdots & \vdots & \ddots & \vdots \\ 0 & \cdots & 0 & z & \cdots & z & z & \cdots & z & 0 & \cdots & 0 \end{array} \right) \begin{array}{l} \Big\} m \\ \\ \Big\} n \end{array}$$

$$\underbrace{\qquad\qquad}_{(m+n)(j-1)} \quad \underbrace{\qquad\qquad}_{m} \quad \underbrace{\qquad\qquad}_{n} \quad \underbrace{\qquad\qquad}_{(m+n)(l-j)}$$

Here z means in general a nontrivial element.

Having used the structure of matrix (4), Kulikov and Thomsen suggested
in [7] to exclude the zero blocks situated below the main diagonal from LU-
factorization and forward substitution. That reduces the number of arithmetical
operations and, hence, CPU time significantly when linear system (3) is solved.
This approach was called *Modification I* in [9]. Moreover, the advantage of the
new version of Gaussian elimination becomes greater if we have interchanged
the x-components of the vector Z_{k+1} with the corresponding y-components. It
means that we have interchanged the first m rows with the last n ones in each
submatrix $\partial \bar{F}_k^\tau (Z)_j$. In this case we can exclude all the zero blocks of matrix
(4) from the LU-factorization, forward and backward substitutions and solve
problem (3) very effectively (see *Modification II* of Gauss method in [9]).

However, the discussed modifications may appear useless when we solve large-
scale semi-explicit index 1 differential-algebraic systems. As an example, we may
take the model of overall regulation of body fluids [5]. This model is a problem

of the form (1) containing about two hundred variables. Having applied any implicit 3- or 4-stage RK method to the model we encounter the situation when the dimension of discrete problem (2) is too high to solve it by Newton-type iterations (see the numerical example in [11]). On the other hand, the Jacobian of a large-scale differential-algebraic system is often a sparse matrix. For instance, the model mentioned above is the case. Thus, the main problem must be how to implement Modification II of Gaussian elimination for matrix (4) effectively.

As in [11], we rearrange the variables and define the vectors:

$$X_{k+1} \stackrel{\text{def}}{=} \left((x_{k1})^T, ..., (x_{k,l-1})^T, (x_{kl})^T \right)^T \in \mathbf{R}^{ml},$$

$$Y_{k+1} \stackrel{\text{def}}{=} \left((y_{k1})^T, ..., (y_{k,l-1})^T, (y_{kl})^T \right)^T \in \mathbf{R}^{nl}.$$

Now $Z_{k+1} \stackrel{\text{def}}{=} \left((Y_{k+1})^T, (X_{k+1})^T \right)^T$ and matrix (4) has the form

$$\partial \bar{F}_k^\tau(Z) \stackrel{\text{def}}{=} \begin{pmatrix} \partial \bar{F}_k^\tau(Z)^Y \\ \partial \bar{F}_k^\tau(Z)^X \end{pmatrix}. \tag{5}$$

Here each submatrix has also the block structure:

$$\partial \bar{F}_k^\tau(Z)^Y \stackrel{\text{def}}{=} \begin{pmatrix} \partial \bar{F}_k^\tau(Z)_1^Y \\ \partial \bar{F}_k^\tau(Z)_2^Y \\ \vdots \\ \partial \bar{F}_k^\tau(Z)_l^Y \end{pmatrix} \quad \text{and} \quad \partial \bar{F}_k^\tau(Z)^X \stackrel{\text{def}}{=} \begin{pmatrix} \partial \bar{F}_k^\tau(Z)_1^X \\ \partial \bar{F}_k^\tau(Z)_2^X \\ \vdots \\ \partial \bar{F}_k^\tau(Z)_l^X \end{pmatrix}$$

where

$$\partial \bar{F}_k^\tau(Z)_i^Y \stackrel{\text{def}}{=} \left. \begin{pmatrix} 0 & \cdots & 0 & z & \cdots & z & 0 & \cdots & 0 & z & \cdots & z & 0 & \cdots & 0 \\ \vdots & \ddots & \vdots & \vdots & \ddots & \vdots & \vdots & \ddots & \vdots & \vdots & \ddots & \vdots & \vdots & \ddots & \vdots \\ 0 & \cdots & 0 & z & \cdots & z & 0 & \cdots & 0 & z & \cdots & z & 0 & \cdots & 0 \end{pmatrix} \right\} n$$

$$\underbrace{\qquad}_{n(i-1)} \underbrace{\qquad}_{n} \underbrace{\qquad}_{n(l-i)+m(i-1)} \underbrace{\qquad}_{m} \underbrace{\qquad}_{m(l-i)}$$

$$\partial \bar{F}_k^\tau(Z)_i^X \stackrel{\text{def}}{=} \left. \begin{pmatrix} O(\tau) & \cdots & O(\tau) & 1+O(\tau) & \cdots & O(\tau) & O(\tau) & \cdots & O(\tau) \\ \vdots & \ddots & \vdots & \vdots & \ddots & \vdots & \vdots & \ddots & \vdots \\ O(\tau) & \cdots & O(\tau) & O(\tau) & \cdots & 1+O(\tau) & O(\tau) & \cdots & O(\tau) \end{pmatrix} \right\} m$$

$$\underbrace{\qquad}_{nl+m(i-1)} \underbrace{\qquad}_{m} \underbrace{\qquad}_{m(l-i)}$$

Now we note that, when solving linear system (3) with matrix (5) by Modification II of Gaussian elimination, LU-factorization of any submatrix $\partial \bar{F}_k^\tau(Z)_i^Y$ does not influence the submatrices $\partial \bar{F}_k^\tau(Z)_j^Y$ for $j \neq i$. This means that the factorization of the matrix $\partial \bar{F}_k^\tau(Z)^Y$ falls into l independent LU-factorizations of the submatrices $\partial \bar{F}_k^\tau(Z)_i^Y$, $i = 1, 2, ..., l$. Moreover, structures of the similar nonzero blocks of all the submatrices $\partial \bar{F}_k^\tau(Z)_i^Y$ (i.e., the number and the places of nonzero elements) coincide, if the stepsize τ is sufficiently small.

It is evident that packing suggested in [11], [13] allows the parallel factorization of the matrix $\partial \bar{F}_k^\tau(Z)^Y$ to be implemented in practice. To store the matrix $\partial \bar{F}_k^\tau(Z)^X$, we can use any packing appropriate for sparse matrices in a general case because the matrix A of coefficients of the RK method may have zero elements and, hence, $\partial \bar{F}_k^\tau(Z)^X$ also contains zero blocks.

3   Implementation of Vector Gauss Method

As it is well-known, each step of Gauss method is split up into two independent stages. The first stage is pivoting in the active submatrix, and at the second one we have to eliminate the next variable from the remaining equations of linear system. We start with the discussion of different ways of pivoting for linear system (3) with the sparse coefficient matrix (5).

When solving a linear system with sparse coefficient matrix by Gauss method we have to determine a set of *admissible elements* of the active submatrix; i.e., the elements which are greater or equal to some $\epsilon > 0$. As the pivot, we then take the admissible element giving the minimum (or small) *local fill-in* (i.e., quantity of zero elements of the coefficient matrix which become nonzero ones) for the next step of Gaussian elimination. There are several ways to implement this idea in practice. In the paper we consider two of them and show how this process should be optimized for linear system (3) with the packed coefficient matrix (5).

The first strategy of pivoting is based on theorem 2.5.5 in [15]. According to this theorem, if we take the $(\mu+p, \mu+q)$-th element of matrix $\partial \bar{F}_k^\tau(Z)^{(\mu)}$ derived for μ steps of the Gauss method as a pivot $(p, q = 1, 2, \ldots, (m + n)l - \mu)$ then the local fill-in at the $\mu + 1$-th step is given by the (p, q)-th element of matrix $G^{(\mu)}$, where

$$G^{(\mu)} \stackrel{\text{def}}{=} B^{(\mu)} \left(\bar{B}^{(\mu)} \right)^T B^{(\mu)} \tag{6}$$

and the square matrix $B^{(\mu)}$ of dimension $(m + n)l - \mu$ is built from the elements of active submatrix of $\partial \bar{F}_k^\tau(Z)^{(\mu)}$ by replacing nonzero elements with the unity ones, and $\left(\bar{B}^{(\mu)} \right)^T$ is a transposed matrix to $\bar{B}^{(\mu)} \stackrel{\text{def}}{=} M_{(m+n)l-\mu} - B^{(\mu)}$ ($M_{(m+n)l-\mu}$ is a square matrix of dimension $(m+n)l-\mu$ with the unity elements). Then, as the pivot for the $\mu+1$-th step we choose that admissible element which gives the minimum local fill-in. In practice, implementation of the first strategy means calculation of the matrix $G^{(\mu)}$ before each step of Gauss method; i.e., for $\mu = 0, 1, \ldots, (m + n)l - 1$. As the result, the cost of this strategy (in terms of multiplication operations) can be determined by the following formula:

$$2 \sum_{\mu=0}^{(m+n)l-1} \left((m + n)l - \mu \right)^3 = \frac{(m + n)^2 l^2 ((m + n)l + 1)^2}{2}. \tag{7}$$

According to the second strategy, if we take the $(\mu + p, \mu + q)$-th element of matrix $\partial \bar{F}_k^\tau(Z)^{(\mu)}$ as a pivot then the maximum possible local fill-in at the $\mu + 1$-th step of Gauss method is given by the (p, q)-th element of matrix $\hat{G}^{(\mu)}$, where

$$\hat{G}^{(\mu)} \stackrel{\text{def}}{=} (B^{(\mu)} - I_{(m+n)l-\mu}) M_{(m+n)l-\mu} (B^{(\mu)} - I_{(m+n)l-\mu}) \tag{8}$$

with the matrices $B^{(\mu)}$ and $M_{(m+n)l-\mu}$ having the above sense (see theorem 2.5.14 in [15]). Then, as the pivot for the $\mu + 1$-th step of Gauss method we choose that admissible element which gives the minimum from the maximum possible local fill-ins. Thus, the computational cost of this strategy (in terms of multiplication operations) can be expressed as follows:

$$\sum_{\mu=0}^{(m+n)l-1} ((m+n)l - \mu)^2 = \frac{(m+n)l((m+n)l+1)(2(m+n)l+1)}{6}. \tag{9}$$

Obviously, the second strategy is not as exact as the first one since it does not allow the real local fill-in to be computed, but just gives some upper estimate. However, its practical implementation is much cheaper than that of the first strategy (compare (7) and (9)).

Let us now refer to the special structure of matrix $\partial \bar{F}_k^\tau(Z)$. It allows the factorization process to be split up into two stages. At the first stage we will eliminate y-components of the vector Z_{k+1}, and the x-components will be eliminated at the second one. Consider pivoting for the first stage of LU-factorization.

As we remember, the LU-factorization of matrix $\partial \bar{F}_k^\tau(Z)^Y$ falls into l independent LU-factorizations of the submatrices $\partial \bar{F}_k^\tau(Z)_i^Y$, $i = 1, 2, \ldots, l$, if when factorizing each submatrix $\partial \bar{F}_k^\tau(Z)_i^Y$ at the $\mu+1$-th step of Gauss method we take the pivot from the active submatrix which is the right-hand side minor located on intersection of the last $n - \mu$ rows and columns of the block $\partial_{y_{ki}} \bar{F}_k^\tau(z_{ki})_i^{Y(\mu)}$. By virtue of the nonsingularity condition (see [6]) we can always find at least one admissible element among the elements of the above minor if $\partial_{y_{ki}} \bar{F}_k^\tau(z_{ki})_i^Y$ is not an improperly stipulated matrix. So, if we choose the pivot by this way then, taking into account the structure of matrix $\partial \bar{F}_k^\tau(Z)$, the local fill-in may influence only the matrix of dimension $(n - \mu + ml) \times (n - \mu + m)$

$$\partial_{z_{ki}} \bar{F}_k^\tau(z_{ki})^{(\mu)} \stackrel{\text{def}}{=} \begin{pmatrix} \partial_{z_{ki}} \bar{F}_k^\tau(z_{ki})_i^{Y(\mu)} \\ \partial_{z_{ki}} \bar{F}_k^\tau(z_{ki})_1^{X(\mu)} \\ \partial_{z_{ki}} \bar{F}_k^\tau(z_{ki})_2^{X(\mu)} \\ \vdots \\ \partial_{z_{ki}} \bar{F}_k^\tau(z_{ki})_l^{X(\mu)} \end{pmatrix}. \tag{10}$$

To speed up the computation process, we suggested the simultaneous factorization of all the submatrices $\partial \bar{F}_k^\tau(Z)_i^Y$, $i = 1, 2, \ldots, l$. Thus, every step of the Gauss method now implies elimination not one but l variables; i. e., all the arithmetical operations are implemented in vector form with vectors of dimension l. Moreover, in this case we have to choose *the vector of pivots* (or shortly *the pivot vector*). That is why we call this approach *the vector Gauss method*. The packing suggested in [11], [13] makes it possible to realize the vector Gauss method in practice if we additionally require all the components of pivot vectors to be on the same places in active submatrices of the blocks $\partial_{y_{ki}} \bar{F}_k^\tau(z_{ki})_i^{Y(\mu)}$, $i = 1, 2, \ldots, l$, $\mu = 0, 1, \ldots, n - 1$.

As we have discussed above, in the process of numerical solving of linear systems with sparse coefficient matrices by Gauss method it is necessary first to determine some set of admissible elements and then to choose the pivot. In the context of vector Gauss method, this approach requires determination of the set of *admissible vectors*; i.e., the vectors with all their components being admissible elements. In this case we have to guarantee that this set will not prove to be empty for the nonsingular matrix $\partial \bar{F}_k^\tau(Z)$ (i.e., the sets of admissible elements of active submatrices of the blocks $\partial_{y_{ki}} \bar{F}_k^\tau(z_{ki})_i^{Y(\mu)}$ will produce nonempty intersection for any $\mu = 0, 1, ..., n-1$). By virtue of the smoothness, nonsingularity and inclusion conditions the above requirement can be easily met in the process of solving system (3) with the coefficient matrix (5), if the stepsize τ is sufficiently small. The detailed explanation of this fact will appear in [13].

The next step in the procedure of pivoting for the vector Gauss method is determination of the admissible vector which gives the minimum (or sufficiently small) local fill-in. In general, this problem tends to be very hard to solve. However, the situation is simplified significantly if RK methods with *dense coefficient matrix A* (i.e., $a_{ij} \neq 0$, $i, j = 1, 2, ..., l$) have been applied to problem (1). Then the blocks $\partial_{z_{ki}} \bar{F}_k^\tau(z_{ki})_i^{Y(\mu)}$ and $\partial_{z_{ki}} \bar{F}_k^\tau(z_{ki})_j^{X(\mu)}$, $j = 1, 2, ..., l$, are of the same structure. Therefore matrices $\partial_{z_{ki}} \bar{F}_k^\tau(z_{ki})^{(\mu)}$, $i = 1, 2, ..., l$, also have the same structure if we put the block $\partial_{z_{ki}} \bar{F}_k^\tau(z_{ki})_i^{X(\mu)}$ on the last position in all the matrices (see (10)). Thus, all the components of any admissible vector lead to one and the same fill-in and, hence, we actually can choose the pivot vector with all its components giving the minimal estimate for the local fill-in. In such a way, to determine the pivot vector it is enough just to find any of its components. Moreover, from the two theorems below we conclude that to compute such a component we may use not the whole matrix $\partial_{z_{ki}} \bar{F}_k^\tau(z_{ki})^{(\mu)}$ but only the matrix

$$\begin{pmatrix} \partial_{z_{ki}} \bar{F}_k^\tau(z_{ki})_i^{Y(\mu)} \\ \partial_{z_{ki}} g(z_{ki})^{(\mu)} \end{pmatrix}$$

of dimension $n - \mu + m \times n - \mu + m$, the last m rows of which are the corresponding block of the matrix

$$\partial g^l(Z) = \left(\partial_{y_{k1}} g(z_{k1}) \quad \cdots \quad \partial_{y_{kl}} g(z_{kl}) \quad \partial_{x_{k1}} g(z_{k1}) \quad \cdots \quad \partial_{x_{kl}} g(z_{kl}) \right).$$

This reduces execution time essentially.

The proofs of both theorems will appear in [13].

Theorem 1 *Let an l-stage RK formula with dense coefficient matrix A be used for constructing matrix (5). Suppose that the $\mu+1$-th step of vector Gauss method is being implemented and the $(\mu+p, \mu+q)$-th element of the block $\partial_{y_{ki}} \bar{F}_k^\tau(z_{ki})_i^{Y(\mu)}$, $\mu = 0, 1, ..., n-1$, $i = 1, 2, ..., l$, $p, q = 1, 2, ..., n-\mu$, is chosen as the i-th component of pivot vector. Then the local fill-in of the matrix $\partial_{z_{ki}} \bar{F}_k^\tau(z_{ki})^{(\mu)}$ from (10) is given by the (p, q)-th element of the matrix*

$$G^{(\mu)} = \left(B_{yi}^{Y(\mu)} \left(\bar{B}_{yi}^{Y(\mu)} \right)^T + B_{xi}^{Y(\mu)} \left(\bar{B}_{xi}^{Y(\mu)} \right)^T \right) B_{yi}^{Y(\mu)}$$
$$+ l \cdot B_{yi}^{Y(\mu)} \left(\bar{B}_y^{(\mu)} \right)^T B_{yi}^{(\mu)} + B_{xi}^{Y(\mu)} \left(\bar{B}_x^{l(\mu)} \right)^T B_y^{(\mu)},$$

(11)

where the matrices $B_{yi}^{Y\,(\mu)}$, $B_{xi}^{Y\,(\mu)}$, $B_{y}^{(\mu)}$, $B_{x}^{(\mu)}$ are obtained from the active submatrices of $\partial_{y_{ki}}\bar{F}_{k}^{\tau}(z_{ki})_{i}^{Y\,(\mu)}$, $\partial_{x_{ki}}\bar{F}_{k}^{\tau}(z_{ki})_{i}^{Y\,(\mu)}$, $\partial_{y_{ki}}g(z_{ki})^{(\mu)}$, $\partial_{x_{ki}}g(z_{ki})^{(\mu)}$ by replacing nonzero elements with the unity ones, $\bar{B}_{yi}^{Y\,(\mu)} = M_{(n-\mu)\times(n-\mu)} - B_{yi}^{Y\,(\mu)}$, $\bar{B}_{xi}^{Y\,(\mu)} = M_{(n-\mu)\times m} - B_{xi}^{Y\,(\mu)}$, $\bar{B}_{y}^{(\mu)} = M_{m\times(n-\mu)} - B_{y}^{(\mu)}$, $\bar{B}_{x}^{(\mu)} = M_{m\times(n-\mu)} - B_{x}^{(\mu)}$, and the matrix $\bar{B}_{x}^{l\,(\mu)}$ is obtained by multiplication of the diagonal elements of matrix $\bar{B}_{x}^{(\mu)}$ by $l-1$ and the nondiagonal elements by l.

Theorem 2 Let all the conditions of Theorem 1 hold. Then the maximum possible local fill-in of the matrix $\partial_{z_{ki}}\bar{F}_{k}^{\tau}(z_{ki})^{(\mu)}$ from (10) (not necessarily coinciding with the real one) is given by the (p,q)-th element of the matrix

$$\hat{G}^{(\mu)} = \left(\left(B_{yi}^{Y\,(\mu)} - I_{n-\mu}\right) M_{(n-\mu)\times(n-\mu)} + B_{xi}^{Y\,(\mu)} M_{m\times(n-\mu)}\right)\left(B_{yi}^{Y\,(\mu)}\right.$$
$$\left. -I_{n-\mu}\right) + l\left(\left(B_{yi}^{Y\,(\mu)} - I_{n-\mu}\right) M_{(n-\mu)\times m} + B_{xi}^{Y\,(\mu)} M_{m\times m}\right) B_{y}^{(\mu)}. \tag{12}$$

Theorems 1 and 2 not only allow the pivot vector to be determined but also make it possible to estimate the cost of the first and the second strategies (in terms of multiplication operations). Thus, taking into account that only n steps of the vector Gauss method are required for the full factorization of the matrix $\partial \bar{F}_{k}^{\tau}(Z)^{Y}$, the whole number of multiplications for the first strategy is

$$\sum_{\mu=0}^{n-1}\left(2(n-\mu)^{3} + (4m+1)(n-\mu)^{2} + m^{2}(n-\mu)\right) = \frac{n^{2}(n+1)^{2}}{2}$$
$$+(4m+1)\frac{n(n+1)(2n+1)}{6} + m^{2}\frac{n(n+1)}{2}. \tag{13}$$

Similarly, from formula (12) it follows that the quantity of multiplication operations for the second strategy is estimated by formula:

$$3\sum_{\mu=0}^{n-1}(n-\mu)^{2} = \frac{n(n+1)(2n+1)}{2}. \tag{14}$$

To illustrate the theoretical result, we consider a simple example. Let $m = n = 2$ and $l = 1, 2, 3, 4$. Now using formulas (7), (9), (13), (14) of the paper and (3.11), (3.12) from [13] we calculate the number of multiplications required to implement the first and the second strategies in the frames of standard Gauss method, Modification II and the vector Gauss method for RK methods with dense coefficient matrices. Table 1 gives the data for the first strategy, and Table 2 shows the result for the second one. Even from this small example, we can see that only in the vector Gauss method computational cost of pivoting does not depend on the number of stages of RK formulas and remains rather minor.

At the end of this section we stress the following aspect. It was noted earlier that the elimination of variables from system (3) is split up into two stages. We eliminated the y-components by using parallel factorization of the matrix

Table 1. The number of multiplication operations in the first strategy of pivoting for the factorization of matrix $\partial \bar{F}_k^\tau (Z)^Y$ $(m = n = 2)$

number of stages	Gauss method	Modification II	vector Gauss method
1	182	60	75
2	2392	184	75
3	11286	372	75
4	34400	624	75

Table 2. The number of multiplication operations in the second strategy of pivoting for the factorization of matrix $\partial \bar{F}_k^\tau (Z)^Y$ $(m = n = 2)$

number of stages	Gauss method	Modification II	vector Gauss method
1	25	10	15
2	174	30	15
3	559	60	15
4	1292	100	15

$\partial \bar{F}_k^\tau (Z)^Y$ and by applying the vector Gauss method. Then we eliminated the x-components. The first stage is more important for optimization because the most part of arithmetical operations falls on it. Therefore in [11], [13] we gave a way to further reduction of the number of operations at this stage. The whole version was called *Modification III*. It incorporates the approach of vector Gauss method with the idea to use the reduced matrix

$$\begin{pmatrix} \partial \bar{F}_k^\tau (Z)^Y \\ \partial g^l (Z) \end{pmatrix}$$

of dimension $(nl + m) \times (m + n)l$ instead of the full matrix (5) while eliminating the y-components. We refer to [11] or [13] for the numerical example.

4   Conclusion

In the paper we further developed the idea suggested in [11], [13] for parallel factorization of the submatrices $\partial \bar{F}_k^\tau (Z)_i^Y$, $i = 1, 2, \ldots, l$. We showed that the most effective algorithm will be obtained only if we use the vector Gauss method. The main advantage of the new method is that we can significantly reduce the number of arithmetical operations required to solve problem (1). Moreover, the computation cost of pivoting in the vector Gauss method does not depend on the number of stages of dense RK formulas applied to (1). Finally, we note that vector computers are the most useful to implement the vector Gauss method.

References

1. Ascher, U.M., Petzold, L.P.: Computer methods for ordinary differential equations and differential-algebraic equations. SIAM, Philadelphia, 1998
2. Brenan K.E., Campbell S.L., Petzold L.R.: Numerical solution of initial-value problems in differential-algebraic equations. North-Holland, N.Y., Amsterdam, L., 1989
3. Gear, C.W.: Differential-algebraic equations index transformations. SIAM J. Sci. Stat. Comput., **9** (1988) 39–47
4. Hairer, E., Wanner, G.: Solving ordinary differential equations II: Stiff and differential-algebraic problems. Springer-Verlag, Berlin, 1996
5. Ikeda, N., Marumo, F., Shiratare, M., Sato, T.: A model of overall regulation of body fluids. Ann. Biomed. Eng. **7** (1979) 135–166
6. Kulikov, G.Yu.: The numerical solution of the autonomous Cauchy problem with an algebraic relation between the phase variables (non-degenerate case). (*in Russian*) Vestnik Moskov. Univ. Ser. 1 Mat. Mekh. (1993) No. 3, 6–10; *translation in* Moscow Univ. Math. Bull. **48** (1993) No. 3, 8–12
7. Kulikov, G.Yu., Thomsen, P.G.: Convergence and implementation of implicit Runge-Kutta methods for DAEs. Technical report 7/1996, IMM, Technical University of Denmark, Lyngby, 1996
8. Kulikov, G.Yu.: Convergence theorems for iterative Runge-Kutta methods with a constant integration step. (*in Russian*) Zh. Vychisl. Mat. Mat. Fiz. **36** (1996) No. 8, 73–89; *translation in* Comp. Maths Math. Phys. **36** (1996) No. 8, 1041–1054
9. Kulikov, G.Yu., Korneva, A.A.: On effective implementation of iterative Runge-Kutta methods for differential-algebraic equations of index 1. (*in Russian*) In: Melnikov, B.F. (ed.): Basic problems of mathematics and mechanics. **3** (1997). Ulyanovsk State University, Ulyanovsk, 103–112
10. Kulikov, G.Yu.: Numerical solution of the Cauchy problem for a system of differential-algebraic equations with the use of implicit Runge-Kutta methods with nontrivial predictor. (*in Russian*) Zh. Vychisl. Mat. Mat. Fiz. **38** (1998) No. 1, 68–84; *translation in* Comp. Maths Math. Phys. **38** (1998) No. 1, 64–80
11. Kulikov, G.Yu., Korneva, A.A.: On efficient application of implicit Runge-Kutta methods to large-scale systems of index 1 differential-algebraic equations. In: Alexandrov, V.N. et all (eds.): Computational Science — ICCS 2001. International Conference, San Francisco, CA, USA, May 28-30, 2001. Proceedings, Part I. Lecture Notes in Computer Science, **2073** (2001), 832–841
12. Kulikov, G.Yu.: On effective implementation of implicit Runge-Kutta methods for systems of differential and differential-algebraic equations of index 1. (*in Russian*) In: Andreev, A.C. (ed.): Basic problems of mathematics and mechanics. Ulyanovsk State University, Ulyanovsk. (to appear)
13. Kulikov, G.Yu., Korneva, A.A., Benderskaya G.Ya.: On numerical solution of large-scale systems of index 1 differential-algebraic equations. (*in Russian*) Fundam. Prikl. Mat. (to appear)
14. Kværnø, A.: The order of Runge-Kutta methods applied to semi-explicit DAEs of index 1, using Newton-type iterations to compute the internal stage values. Technical report 2/1992, Mathematical Sciences Div., Norwegian Institute of Technology, Trondheim, 1992
15. Tewarson R.P.: Sparse matrices. Academic Press, New York and London, 1973

One Class of Splitting Iterative Schemes

Čiegis and V. Pakalnytė

Vilnius Gediminas Technical University
Saulėtekio al. 11, 2054, Vilnius, Lithuania
rc@fm.vtu.lt

Abstract. This paper deals with the stability analysis of a new class of iterative methods for elliptic problems. These schemes are based on a general splitting method, which decomposes a multidimensional parabolic problem into a system of one dimensional implicit problems. We use a spectral stability analysis and investigate the convergence order of two iterative schemes. Finally, some results of numerical experiments are presented.

1 Introduction

Iterative methods for solving discrete elliptic problems can be viewed as finite difference schemes for non stationary parabolic problems. The most important difference is that for elliptic problems we select the time step parameter according the requirements of the convergence to the stationary solution and can ignore the approximation error. For splitting methods for elliptic problems we refer to [4, 6]. In particular, [5, 6] involve an alternating direction method, [4] presents factorization schemes. The convergence rate can be increased if optimal non-stationary parameters are used for the definition of each iteration, see e.g. [6].

Recently the multicomponent versions of alternating direction method were proposed in [1]. These schemes are also used for solving multidimensional elliptic problems [2]. While in the previous papers the stability of these splitting schemes was investigated by the energy method, we will use the spectral stability analysis. For symmetric problems it gives necessary and sufficient convergence conditions and enables us to find optimal values of iterative parameters. Such analysis was used also in [3]

The content of this paper is organized as follows. In Section 2 we formulate the multicomponent iterative scheme. The convergence of 2D scheme is investigated in Section 3. In section 4 we investigate the stability of 3D iterative scheme, the analysis is done using numerical experiments. Finally, in Section 5 we study the convergence of the p-dimensional multicomponent iterative scheme and prove the energy stability estimates. In Section 6, the Seidel type scheme is formulated and investigated.

P.M.A. Sloot et al. (Eds.): ICCS 2002, LNCS 2330, pp. 422–431, 2002.

2   Multicomponent iterative scheme

Let the strictly elliptic problem be given by

$$-\sum_{i=1}^{p} \frac{\partial^2 u}{\partial x_i^2} = f(x), \quad x \in Q,$$

$$u(x) = 0 \quad x \in \partial Q,$$

where $Q = (0,1)^p$. We introduce a uniform grid in Q and approximate the elliptic problem by the following finite difference scheme

$$\sum_{\alpha=1}^{p} A_\alpha y = f, \tag{1}$$

here A_α denotes the approximation of a differential operator using standard central difference formula.

Let $(,)$ and $\| \cdot \|$ denote the inner product and the L_2 norm of discrete functions, respectively. Generally, we introduce the following assumptions:

$$A_\alpha^* = A_\alpha,$$

$$0 < m_\alpha \|y\| \le (A_\alpha y, y) \le M_\alpha \|y\|.$$

It is well known that for the Laplace operator these assumptions are satisfied with

$$8 \le m_\alpha, \quad M_\alpha \le \frac{4}{h^2}, \tag{2}$$

where h is the spatial step size.

In order to solve the system of linear equations (1) we define p unknown functions y_α, $\alpha = 1, 2, \ldots, p$. Then the *multicomponent alternating direction* (MAD) scheme is given as (see, [1]):

$$\frac{\overset{s+1}{y_\alpha} - \overset{s}{\tilde{y}}}{\tau} + p A_\alpha \left(\overset{s+1}{y_\alpha} - \overset{s}{y_\alpha} \right) + \sum_{\beta=1}^{p} A_\beta \overset{s}{y_\beta} = f, \quad \alpha = 1, 2, \ldots, p, \tag{3}$$

$$\overset{s}{\tilde{y}} = \frac{1}{p} \sum_{\alpha=1}^{p} \overset{s}{y_\alpha},$$

where $\overset{s}{y_\alpha}$ is the s-th iteration of y_α. We note that all equations (3) can be solved in parallel.

We will investigate the convergence of the MAD scheme by using the following error norms:

$$\| \overset{s}{e} \| = \| \overset{s}{\tilde{y}} - y \|, \quad \| \overset{s}{r} \| = \| \sum_{\alpha=1}^{p} A_\alpha \overset{s}{y_\alpha} - f \|.$$

3   Spectral stability analysis of 2D iterative scheme

In this section we consider a two-dimensional iterative scheme (1)

$$\frac{\overset{s+1}{y_\alpha} - \overset{s}{y}}{\tau} + 2A_\alpha\left(\overset{s+1}{y_\alpha} - \overset{s}{y_\alpha}\right) + \sum_{\beta=1}^{2} A_\beta \overset{s}{y_\beta} = f, \quad \alpha = 1, 2. \tag{4}$$

Let denote the error functions

$$\overset{s}{e_\alpha} = \overset{s}{y_\alpha} - y, \quad \alpha = 1, 2, \dots, p, \quad \overset{s}{\tilde{e}} = \overset{s}{\tilde{y}} - y.$$

Then the error functions satisfy the following MAD scheme:

$$\frac{\overset{s+1}{e_\alpha} - \overset{s}{\tilde{e}}}{\tau} + 2A_\alpha\left(\overset{s+1}{e_\alpha} - \overset{s}{e_\alpha}\right) + \sum_{\beta=1}^{2} A_\beta \overset{s}{e_\beta} = 0, \quad \alpha = 1, 2, \tag{5}$$

$$\overset{s}{\tilde{e}} = \frac{1}{2} \sum_{\alpha=1}^{2} \overset{s}{e_\alpha},$$

To apply the discrete von Neumann stability criteria to problem (5), we let

$$\overset{s}{e_\alpha} = \sum_{j=1}^{N-1} \sum_{k=1}^{N-1} \overset{s}{d_{\alpha,jk}} \sin(j\pi x_1)\sin(k\pi x_2), \quad \alpha = 1, 2, \tag{6}$$

where N is the number of grid points in one-dimensional grid. It is well known, that $\sin(j\pi x_\alpha)$ are eigenvectors of the operator A_α, i.e.:

$$A_\alpha \sin(j\pi x_\alpha) = \lambda_j \sin(j\pi x_\alpha),$$

$$8 \le \lambda_1 \le \lambda_2 \le \cdots \le \lambda_{N-1} \le \frac{4}{h^2}. \tag{7}$$

If we replace $\overset{s+1}{e_\alpha}$ and $\overset{s}{e_\alpha}$ in (5) by expressions of the form given by equation (6) and use (7), we get the following matrix equations

$$\overset{s+1}{\mathbf{d}_{jk}} = Q_2 \overset{s}{\mathbf{d}_{jk}}, \tag{8}$$

where $\mathbf{d}_{jk}$ is the column vector of spectral coefficients

$$\mathbf{d}_{jk} = \begin{pmatrix} d_{1,jk} \\ d_{2,jk} \end{pmatrix},$$

and Q_2 is the stability matrix of MAD scheme

$$Q_2 = \begin{pmatrix} \dfrac{0.5 + \tau\lambda_j}{1 + 2\tau\lambda_j} & \dfrac{0.5 - \tau\lambda_k}{1 + 2\tau\lambda_j} \\[3mm] \dfrac{0.5 - \tau\lambda_j}{1 + 2\tau\lambda_k} & \dfrac{0.5 + \tau\lambda_k}{1 + 2\tau\lambda_k} \end{pmatrix}.$$

Now we consider the necessary conditions for the stability of iterative MAD scheme (4). Since Q_2 is not symmetric matrix, the discrete von Neumann stability criteria can not prove that they are also sufficient for stability of MAD scheme.

Theorem 1. *All eigenvalues of stability matrix Q_2 satisfy inequalities*

$$|q_{jk}| < 1, \quad 1 \leq j, k \leq N - 1$$

unconditionally for any values of parameters τ and h.

Proof. Using simple computations we get that eigenvalues q of the amplification matrix Q_2 satisfy the quadratic equation

$$q^2 - q + \frac{\tau(\lambda_j + \lambda_k)}{(1 + 2\tau\lambda_j)(1 + 2\tau\lambda_k)} = 0.$$

Then the eigenvalues of Q_2 are

$$q_{1,2} = \frac{1}{2}\left(1 \pm \sqrt{\frac{(1 - 2\tau\lambda_j)(1 - 2\tau\lambda_k)}{(1 + 2\tau\lambda_j)(1 + 2\tau\lambda_k)}}\right)$$

and they are obviously both less than 1. The theorem is proved.

The convergence order of MDA iterative scheme depends on the parameter τ. We will use quasi - optimal parameter τ_0 which solves the minimization problem

$$w(\tau_0) = \min_{\tau} \max_{\lambda_1 \leq \lambda \leq \lambda_{N-1}} \left|\frac{1 - 2\tau\lambda}{1 + 2\tau\lambda}\right|.$$

Since the function

$$g(a) = \frac{1 - a}{1 + a}$$

is strictly decreasing, we find τ_0 from the equation

$$\frac{1 - 2\tau\lambda_1}{1 + 2\tau\lambda_1} = \frac{2\tau\lambda_{N-1} - 1}{1 + 2\tau\lambda_{N-1}}.$$

After simple computations we get

$$\tau_0 = \frac{1}{2\sqrt{\lambda_1\lambda_{N-1}}} \approx \frac{h}{4\pi}$$

and all eigenvalues of the amplification matrix Q_2 satisfy the inequality

$$|q_{jk}(\tau_0)| \leq \frac{1}{1 + \sqrt{\lambda_1/\lambda_{N-1}}} \approx 1 - \frac{\pi}{2}h. \tag{9}$$

These estimates are similar to convergence estimates obtained for the standard iterative scheme of alternating directions.

4   Spectral stability analysis of 3D iterative scheme

As it was stated in section 3, the von Neumann stability criteria gives only sufficient stability conditions of MAD iterative scheme. In this section we will use the spectral stability analysis for 3D iterative scheme (3). But instead of finding eigenvalues of an amplification matrix Q_3 we use direct numerical experiments and find the optimal value of the parameter τ. Such methodology gives us a possibility to investigate the convergence order of MAD scheme in the usual l_2 norm.

Let consider the model problem

$$\sum_{\alpha=1}^{3} A_\alpha y = (\lambda_j + \lambda_k + \lambda_l) \sin(j\pi x_1) \sin(k\pi x_2) \sin(l\pi x_3), \tag{10}$$

which has the exact solution

$$y = \sin(j\pi x_1) \sin(k\pi x_2) \sin(l\pi x_3).$$

Then the solution of 3D MAD scheme can be represented as

$$\overset{s}{y}_\alpha = \overset{s}{d}_\alpha \sin(j\pi x_1) \sin(k\pi x_2) \sin(l\pi x_3), \quad \alpha = 1, 2, 3,$$

where d_α can be computed explicitly

$$\overset{s+1}{d}_\alpha = \frac{1}{1 + 3\tau\lambda_\alpha} \left(\sum_{\beta=1}^{3} \left(\frac{\overset{s}{d}_\beta}{3} - \tau\lambda_\beta(\overset{s}{d}_\beta - 1) \right) + 3\tau\lambda_\alpha \overset{s}{d}_\alpha \right), \quad \alpha = 1, 2, 3.$$

Then the error of the sth iteration $\overset{s}{d}$ is estimated by the following formula

$$\overset{s}{e} = \left| \frac{\overset{s}{d}_1 + \overset{s}{d}_2 + \overset{s}{d}_3}{3} - 1 \right|.$$

Let $S(\tau, \lambda_j, \lambda_k, \lambda_l)$ be the number of iterations required to achieve the accuracy $\overset{s}{e} \leq \varepsilon$ for given eigenvalues λ_α. Then we investigate the whole interval $[m, M]$, which characterize the stiffness of the problem and compute

$$S(\tau) = \max_{m \leq \lambda_\alpha \leq M} S(\tau, \lambda_j, \lambda_k, \lambda_l).$$

This problem is solved approximately by computing $S(\tau, \lambda_j, \lambda_k, \lambda_l)$ for $(K+1)^3$ combinations of eigenvalues $\lambda_\alpha = m + i(M - m)/K$.

First, we investigated the dependence of the number of iterations $S(\tau)$ on the parameter τ. It was proved that there exists the optimal value τ_0, such that

$$S(\tau_0) \leq S(\tau)$$

Table 1. The number of iterations as a function of τ

ε	τ	$S(\tau)$
0.0001	0.0015	210
0.0001	0.0020	159
0.0001	0.0022	145
0.0001	0.0023	149
0.0001	0.0025	162

and this value satisfies the following condition

$$\max_{m \leq \lambda_\alpha \leq M} S(\tau_0, \lambda_j, \lambda_k, \lambda_l) = S(\tau_0, m, m, m). \tag{11}$$

In Table 1 we present numbers of iterations $S(\tau)$ for different values of τ. These experiments were done with $m = 10, M = 4000$.

The optimal value of the parameter τ depends slightly on ε. In Table 2 we present optimal numbers of iterations $S(\tau_0)$ for different values of ε.

Table 2. The optimal value of τ as a function of ε

ε	τ_0	$S(\tau_0)$
0.001	0.00233	103
0.0001	0.00221	144
0.00001	0.00212	187

Finally, we investigated the dependence of the convergence rate of the MAD iterative scheme on m and M. In Table 3 we present optimal values of τ and numbers of iterations $S(\tau_0)$ for different spectral intervals. We used $\varepsilon = 10^{-4}$ in these experiments.

It follows from results presented in Table 3 that

$$\tau_0 = \frac{c}{\sqrt{mM}}, \quad S(\tau_0) = O\left(\sqrt{\frac{M}{m}}\right).$$

The above conclusion agrees well with results of section 3.

Table 3. The optimal value of τ as a function of m and M

m	M	τ_0	$S(\tau_0)$
10	4000	0.00221	144
10	16000	0.00110	284
10	64000	0.00055	563
10	16000	0.00110	284
40	16000	0.00055	145
90	16000	0.00037	98

5   Error estimates for p-dimensional MAD scheme

In this section we consider p-dimensional iterative scheme (3). Let us introduce the following notation:

$$\overset{s}{y}_{\alpha t} = \frac{\overset{s+1}{y_\alpha} - \overset{s}{y_\alpha}}{\tau}, \quad \overset{s}{\tilde{y}}_t = \frac{\overset{s+1}{\tilde{y}} - \overset{s}{\tilde{y}}}{\tau}, \quad \overset{s}{v}^{(\alpha,\beta)} = \overset{s}{y}_\alpha - \overset{s}{y}_\beta,$$

$$\| \overset{s}{v} \|_3^2 = \sum_{\alpha,\beta=1,\,\alpha>\beta}^{p} \| \overset{s}{v}^{(\alpha,\beta)} \|^2, \quad Q_p(\overset{s}{y}) = \| \overset{s}{r} \|^2 + \frac{1}{p^2\,\tau^2} \| \overset{s}{v} \|_3^2.$$

In the following theorem we estimate the convergence rate of MAD iterative scheme.

Theorem 2. *Iterative scheme (3) produces a sequence converging unconditionally to the solution of problem (2) and the convergence rate is estimated as*

$$Q_p(\overset{s+1}{y}) \le \frac{1}{q} Q_p(\overset{s}{y}), \quad q = \min\left(1 + m\,p\,\tau,\ 1 + \frac{1}{2M\,p\,\tau}\right), \tag{12}$$

where m and M are the spectral estimates of the operator A:

$$m = \min_{1\le\alpha\le p} m_\alpha, \quad M = \max_{1\le\alpha\le p} M_\alpha.$$

Proof. Multiplying both sides of (3) by $\overset{s}{y}_{\alpha t}$ and adding all equalities we get

$$\sum_{\alpha=1}^{p} \left(\frac{\overset{s+1}{y_\alpha} - \overset{s}{\tilde{y}}}{\tau}, \overset{s}{y}_{\alpha t} \right) + p\tau \sum_{\alpha=1}^{p} (A_\alpha \overset{s}{y}_{\alpha t}, \overset{s}{y}_{\alpha t}) + p\left(\sum_{\beta=1}^{p} A_\beta \overset{s}{y}_\beta - f, \overset{s}{\tilde{y}}_t \right) = 0. \tag{13}$$

The first term of (13) can be rewritten as

$$I_1 = \frac{1}{p\tau} \sum_{\alpha=1}^{p} \sum_{\beta=1}^{p} \left(\overset{s+1}{y_\alpha} - \overset{s+1}{y_\beta}, \overset{s}{y_{\alpha t}} \right) + \frac{1}{p\tau} \sum_{\alpha=1}^{p} \sum_{\beta=1}^{p} \left(\overset{s+1}{y_\beta} - \overset{s}{y_\beta}, \overset{s}{y_{\alpha t}} \right)$$

$$= p\|\overset{s}{\tilde{y}_t}\|^2 + \frac{1}{p\tau} \sum_{\alpha,\beta=1,\,\alpha>\beta}^{p} \left(\overset{s+1}{v}(\alpha,\beta), \overset{s}{v_t}(\alpha,\beta) \right)$$

$$= p\|\overset{s}{\tilde{y}_t}\|^2 + \frac{1}{2p}\|\overset{s}{v_t}\|_3^2 + \frac{1}{2p\tau^2} \left(\|\overset{s+1}{v}\|_3^2 - \|\overset{s}{v}\|_3^2 \right). \tag{14}$$

By adding equations (3) we get that

$$\overset{s}{\tilde{y}_t} = -\left(\sum_{\alpha=1}^{p} A_\alpha \overset{s+1}{y_\alpha} - f \right),$$

hence using the third term of (13) we can prove that

$$p\|\overset{s}{\tilde{y}_t}\|^2 + p\left(\sum_{\beta=1}^{p} A_\beta \overset{s}{y_\beta} - f, \overset{s}{\tilde{y}_t} \right) \geq \frac{p}{2} \left(\|\overset{s+1}{r}\|^2 - \|\overset{s}{r}\|^2 \right). \tag{15}$$

The second term of (13) is estimated similarly. Thus the convergence rate estimate (12) follows trivially. The theorem is proved.

As a corollary of Theorem 2 we can find the optimal value of parameter $\tau_0 = 1/p\sqrt{2mM}$, which is obtained from the equation

$$1 + pm\tau = 1 + \frac{1}{2Mp\tau}.$$

6   Seidel-type iterative scheme

In this section we investigate the convergence rate of Seidel-type iterative scheme:

$$\frac{\overset{s+1}{y_\alpha} - \overset{s}{y_\alpha^*}}{\tau} + \sum_{\beta=1}^{\alpha} A_\beta \overset{s+1}{y_\beta} + \sum_{\beta=\alpha+1}^{p} A_\beta \overset{s}{y_\beta} = f, \quad \alpha = 1,2,\ldots,p, \tag{16}$$

$$\overset{s}{y_1^*} = \overset{s}{y_1}, \quad \overset{s}{y_\alpha^*} = 0.5 \left(\overset{s}{y_\alpha} + \overset{s}{y_{\alpha-1}} \right).$$

6.1   Spectral stability analysis of 2D scheme

To apply the discrete von Neumann stability criteria to problem (16), we write the global error as a series:

$$\overset{s}{e_\alpha} = \sum_{j=1}^{N-1} \sum_{k=1}^{N-1} \overset{s}{d_{\alpha,jk}} \sin(j\pi x_1) \sin(k\pi x_2), \quad \alpha = 1,2,$$

Substituting this expansion into (16), we obtain the equation for coefficients

$$\overset{s+1}{\mathbf{d}}_{jk} = Q_2 \overset{s}{\mathbf{d}}_{jk} . \tag{17}$$

where $\mathbf{d}_{jk}$ is the column vector of spectral coefficients and Q_2 is the stability matrix of scheme (16)

$$Q_2 = \begin{pmatrix} \dfrac{1}{1+\tau\lambda_j} & \dfrac{-\tau\lambda_k}{1+\tau\lambda_j} \\[3mm] \dfrac{0.5(1-\tau\lambda_j)}{(1+\tau\lambda_j)(1+\tau\lambda_k)} & \dfrac{0.5+\tau\lambda_j+\tau^2\lambda_j\lambda_k}{(1+\tau\lambda_j)(1+\tau\lambda_k)} \end{pmatrix} .$$

Now we consider the necessary conditions for the stability of scheme (16). The eigenvalues of the amplification matrix Q_2 satisfy the quadratic equation

$$q^2 - \left(1 + \frac{0.5(1-\tau\lambda_j)}{(1+\tau\lambda_j)(1+\tau\lambda_k)}\right)q + \frac{0.5}{1+\tau\lambda_j} = 0 .$$

Theorem 3. *All eigenvalues of stability matrix Q_2 satisfy inequalities*

$$|q_{jk}| < 1, \quad 1 \le j,k \le N-1$$

unconditionally for any values of parameters τ and h.

Proof. Application of the Hurwitz criterion gives that $|q_{jk}| \le 1$ is satisfied if and only if

$$\frac{0.5}{1+\tau\lambda_j} < 1, \quad \left|1 + \frac{0.5(1-\tau\lambda_j)}{(1+\tau\lambda_j)(1+\tau\lambda_k)}\right| < 1 + \frac{0.5}{1+\tau\lambda_j} .$$

Simple computations prove that both inequalities are satisfied unconditionally. The theorem is proved.

6.2   Spectral stability analysis of 3D iterative scheme

Let consider the model problem (10). The solution of 3D scheme (16) can be represented as

$$\overset{s}{y}_\alpha = \overset{s}{d}_\alpha \sin(j\pi x_1)\sin(k\pi x_2)\sin(l\pi x_3), \quad \alpha = 1,2,3,$$

where d_α, $\alpha = 1,2,3$, are computed explicitly

$$\overset{s+1}{d}_\alpha = \frac{1}{1+\tau\lambda_\alpha}\left(\frac{\overset{s}{d}_\alpha + \overset{s}{d}_{\alpha-1}}{2} - \sum_{\beta=1}^{\alpha-1}\tau\lambda_\beta \overset{s+1}{d}_\beta - \sum_{\beta=\alpha+1}^{3}\tau\lambda_\beta \overset{s}{d}_\beta + \sum_{\beta=1}^{3}\tau\lambda_\beta\right) .$$

We estimate the error of the sth iteration $\overset{s}{d}_\alpha$ by the following formula

$$\overset{s}{e} = \max_{1\le\alpha\le3}\left|\overset{s}{d}_\alpha\right| .$$

Table 4. The optimal value of τ as a function of m and M

m	M	τ_0	$S(\tau_0)$
10	4000	0.00173	176
10	16000	0.00089	346
10	64000	0.00045	681
10	16000	0.00089	346
40	16000	0.00042	177
90	16000	0.00028	125

We investigated numerically the dependence of the convergence rate of the iterative scheme (16) on m and M. In Table 4 we present optimal values of τ and numbers of iterations $S(\tau_0)$ for different spectral intervals. We used $\varepsilon = 10^{-4}$ in these experiments.

It follows from results presented in Table 4 that

$$\tau_0 = \frac{c}{\sqrt{mM}}, \quad S(\tau_0) = O\left(\sqrt{\frac{M}{m}}\right).$$

The convergence rate of the Seidel type scheme (16) is the same as of scheme (3).

References

1. Abrashin, V.N.: On the stability of multicomponent additive direction method. Differentsialnyje uravnenyja. **35** (1999) 212–224 (in Russian).
2. Abrashin, V.N., Zhadaeva, N.G.: On the convergence rate of economical iterative methods for stationary problems of mathematical physics. Differentsialnyje uravnenyja. **36** (2000) 1422–1432 (in Russian).
3. Aleinikova, T., Čiegis, R.: Investigation of new iterative methods for solving multidimensional elliptic equations. Differentsialnyje uravnenyja. **29** (1993) 1124–1129 (in Russian).
4. Marchuk, G.I. *Splitting methods.* Nauka, Moscow, 1988. (in Russian).
5. Peaceman, D., Rachford, H.: The numerical solution of parabolic and elliptic differential equations. SIAM, **3** (1955).
6. Samarskii, A.A., Nikolajev, E.S. *Methods for solving difference equations.* Nauka, Moscow, 1978.(in Russian).

Filtration-Convection Problem: Spectral-Difference Method and Preservation of Cosymmetry

Olga Kantur, Vyacheslav Tsybulin

Department of Mathematics and Mechanics,
Rostov State University,
344090 Rostov on Don, Russia
kantur_rsu@mail.ru, tsybulin@math.rsu.ru

Abstract. For the problem of filtration of viscous fluid in porous medium it was observed that a number of one-parameter families of convective states with the spectrum, which varies along the family. It was shown by V. Yudovich that these families cannot be an orbit of an operation of any symmetry group and as a result the theory of cosymmetry was derived. The combined spectral and finite-difference approach to the planar problem of filtration-convection in porous media with Darcy law is described. The special approximation of nonlinear terms is derived to preserve cosymmetry. The computation of stationary regime transformations is carried out when filtration Rayleigh number varies.

1   Introduction

In this work we study the conservation of cosymmetry in finite-dimensional models of filtration-convection problem derived via combined spectral and finite-difference method. Cosymmetry concept was introduced by Yudovich [1, 2] and some interesting phenomena were found for both dynamical systems possessing the cosymmetry property. Particularly, it was shown that cosymmetry may be a reason for the existence of the continuous family of regimes of the same type. If a symmetry group produces a continuous family of identical regimes then it implies the identical spectrum for all points on the family. The stability spectrum for the cosymmetric system depends on the location of a point, and the family may be formed by stable and unstable regimes.

Following [1], a cosymmetry for a differential equation $\dot{u} = F(u)$ in a Hilbert space is the operator $L(u)$ which is orthogonal to F at each point of the phase space i.e $(F(u), L(u)) = 0, u \in R^n$ with an inner product $(\cdot, \cdot)$. If the equilibrium u_0 is noncosymmetric, i.e. $F(u_0) = 0$ and $L(u_0) \neq 0$, then u_0 belongs to a one-parameter family of equilibria. This takes place if there are no additional degeneracies.

A number of interesting effects were found in the planar filtration-convection problem of fluid flow through porous media [1–4]. The investigations in [4] were carried out for finite-dimensional approximations of small size, so it is desirable to

P.M.A. Sloot et al. (Eds.): ICCS 2002, LNCS 2330, pp. 432–441, 2002.

develop appropriate numerical methods for finite-dimensional systems of larger size. It is very important to preserve cosymmetry in finite-dimensional models derived from partial differential equations. It was shown in [5] that improper approximation may lead to the destruction of the family of equilibria. We apply in this work the approach based on spectral expansion on vertical coordinate and finite-difference method in horizontal direction.

2 Darcy convection problem

We will consider the planar filtration-convection problem for incompressible fluid saturated with a porous medium in a rectangular container $\mathcal{D} = [0, a] \times [0, b]$ which is uniformly heated below. The temperature difference δT was held constant between the lower $y = 0$ and upper $y = b$ boundaries of the rectangle and the temperature on the vertical boundaries obeys a linear law, so that a time independent uniform vertical temperature profile is formed. We consider perturbation of the temperature from the basic state of rest with a linear conductive profile.

Because the fluid is incompressible, we introduce a stream function ψ such that horizontal and vertical components of the velocity vector are given as $u = -\psi_y$ and $v = \psi_x$, respectively. The dimensionless equations of the filtration convection problem are:

$$\frac{\partial \theta}{\partial t} = \Delta \theta + \lambda \frac{\partial \psi}{\partial x} + J(\psi, \theta) \equiv F_1 \tag{1}$$

$$0 = \Delta \psi - \frac{\partial \theta}{\partial x} \equiv F_2 \tag{2}$$

where $\Delta = \partial_x^2 + \partial_y^2$ is the Laplacian and $J(\psi, \theta)$ denotes the Jacobian operator over (x, y):

$$J(\psi, \theta) = \frac{\partial \psi}{\partial x} \frac{\partial \theta}{\partial y} - \frac{\partial \psi}{\partial y} \frac{\partial \theta}{\partial x}.$$

The dependent variables $\psi(x, y, t)$ and $\theta(x, y, t)$ denote perturbations of the stream function and temperature, λ is the Rayleigh number given by $\lambda = \beta g \delta T K l / \kappa \mu$, here β is the thermal expansion coefficient, g is the acceleration due to gravity, μ is the kinematic viscosity, κ is the thermal diffusivity of the fluid, K is the permeability coefficient, l is the length parameter. The boundary conditions are:

$$\theta = 0, \qquad \psi = 0 \qquad \text{on} \quad \partial \mathcal{D}, \tag{3}$$

and the initial condition is only defined for the temperature

$$\theta(x, y, 0) = \theta_0(x, y), \tag{4}$$

where θ_0 denotes the initial temperature distribution. For a given θ_0, the stream function ψ can be obtained from (2), (3) as the solution of the Dirichlet problem via Green's operator $\psi = G\theta_x$.

Cosymmetry for underlying system is given by $(\psi, -\theta)$. Really, multiply (1) by ψ and (2) by $-\theta$, sum and integrate over domain $\mathcal{D}$. Then, using integration by parts and Green's formula we derive

$$\int_{\mathcal{D}} (F_1\psi - F_2\theta)dxdy = \int_{\mathcal{D}} (\Delta\theta\psi - \Delta\psi\theta + \lambda\psi_x\psi + \theta_x\theta + J(\theta, \psi)\psi)dxdy = 0. \quad (5)$$

To establish this we also need the following equality

$$\int_{\mathcal{D}} J(\psi, \theta)\psi dxdy = 0. \quad (6)$$

Moreover, the Jacobian J is antisymmetric with respect to its arguments and the equality takes place

$$\int_{\mathcal{D}} J(\psi, \theta)\theta dxdy = 0. \quad (7)$$

For all values of the Rayleigh number there is a trivial equilibrium. The eigenvalues of the spectral problem for the trivial equilibrium are [2]

$$\lambda_{mn} = 4\pi^2 \left(\frac{m^2}{a^2} + \frac{n^2}{b^2} \right), \quad (8)$$

where m, and n are integers. They have multiplicity of two if and only if the diopanthine equation $m^2/a^2 + n^2/b^2 = m_1^2/a^2 + n_1^2/b^2$ has a unique solution with $m_1 = m$ and $n_1 = n$. The lowest eigenvalue corresponds to $m = n = 1$, and when the parameter λ passes λ_{11} a one-parameter family of stationary solutions emerges. This family is a closed curve in the phase space. In [2] it was shown that the spectrum varies along this family and therefore this family can not be an orbit of the action of any symmetry group.

3   Spectral-finite-difference method

The approach based on spectral and finite-difference approximation is applied. We use Galerkin expansion in the direction y and finite-difference method for x. Firstly we take the following

$$\theta(x, y, t) = \sum_{j=1}^{m} \theta_j(x, t) \sin \frac{\pi jy}{b}, \quad \psi(x, y, t) = \sum_{j=1}^{m} \psi_j(x, t) \sin \frac{\pi jy}{b}. \quad (9)$$

After substituting (9) to (1)–(2) and integrating on y we derive:

$$\dot{\theta}_j = \theta_j'' - c_j\theta_j + \lambda\psi_j' - J_j, \quad j = 1 \div m, \quad (10)$$

$$0 = \psi_j'' - c_j\psi_j - \theta_j', \quad j = 1 \div m, \quad (11)$$

here a prime and a dot denote differentiation on x and t respectively, $c_j = j^2\pi^2/b^2$, and for J_j we have

$$J_j = \frac{2\pi}{b} \sum_{i=1}^{m-j} [(i+j)(\theta_{i+j}\psi_i' - \theta_i'\psi_{i+j}) + i(\theta_{i+j}'\psi_i - \theta_i\psi_{i+j}')] \qquad (12)$$

$$+ \frac{2\pi}{b} \sum_{i=1}^{j-1} (j-i)(\theta_i'\psi_{j-i} - \theta_{j-i}\psi_i'), \quad j = 1 \div m.$$

The boundary conditions (3) may be rewritten as

$$\theta_j(t,0) = \theta_j(t,a), \quad \psi_j(t,0) = \psi_j(t,a), \quad j = 1 \div m.$$

We deduce from initial condition (4) the following

$$\theta_j(x,0) = \int_D \theta_0(x,y) \sin\frac{\pi j y}{b} dy, j = 1 \div m.$$

To discretize (10)–(12) on variable x we apply uniform mesh $\omega = \{x_k | x_k = kh, k = 0 \div n, h = a/(n+1)\}$ and the notions $\theta_{j,k} = \theta_j(x_k,t)$, $\psi_{j,k} = \psi_j(x_k,t)$, $J_{j,k} = J_j(x_k,t)$. The centered finite-difference operators are used and we deduce a system of ordinary differential equations

$$\dot\theta_{jk} = \frac{\theta_{j,k+1} - 2\theta_{j,k} + \theta_{j,k-1}}{h^2} - c_j\theta_{jk} + \lambda\frac{\psi_{j,k+1} - \psi_{j,k-1}}{2h} - J_{j,k} \equiv \phi_{1jk}, \quad (13)$$

$$0 = \frac{\psi_{j,k+1} - 2\psi_{j,k} + \psi_{j,k-1}}{h^2} - c_j\psi_{jk} + \frac{\theta_{j,k+1} - \theta_{j,k-1}}{2h} \equiv \phi_{2jk}. \quad (14)$$

The expression for J_{jk}, being discretization of J_j (12) at x_k, will be given below. Finally, the boundary conditions are the following

$$\theta_{j0} = \theta_{jn} = 0, \quad \psi_{j0} = \psi_{jn} = 0. \qquad (15)$$

4 Cosymmetry conservation

One can check that a vector

$$L_h = (\psi_{11}, ..., \psi_{n1}, \psi_{12}, ..., \psi_{nm}, -\theta_{11}, ..., -\theta_{n1}, -\theta_{12}, ..., -\theta_{nm})$$

gives a cosymmetry for (13)–(14). So, a cosymmetric equality must be held

$$\sum_{j=1}^{m} \sum_{k=1}^{n} [\phi_{1jk}\psi_{j,k} - \phi_{2jk}\theta_{j,k}] = 0. \qquad (16)$$

Substitute (13), (14) into (16) and using summation we deduce that linear parts in (13), (14) nullify and the following relation must be preserved

$$\sum_{k=1}^{n} \sum_{j=1}^{m} (J_j\psi_j)_k = 0. \qquad (17)$$

We demand also for $J_{j,k}$ the additional property

$$\sum_{k=1}^{n}\sum_{j=1}^{m}(J_j\theta_j)_k = 0. \tag{18}$$

It should be stressed that usual finite-difference operators do not keep the equalities (17)–(18). To reach correct approximation of nonlinear terms we introduce two operators

$$D_a(\theta,\psi) = \theta'\psi - \theta\psi', \quad D_s(\theta,\psi) = \theta'\psi + \theta\psi'.$$

Then, we rewrite J_j

$$J_j = \frac{2\pi}{b}\left(\sum_{i=1}^{m-j}\chi_{j,i}^1 + \sum_{i=1}^{j-1}\chi_{j,i}^2\right), \tag{19}$$

$$\chi_{j,i}^1 = \frac{2i+j}{2}\left(D_s(\theta_{i+j},\psi_i) - D_s(\theta_i,\psi_{i+j})\right) - \frac{j}{2}\left(D_a(\theta_{i+j},\psi i) + D_a(\theta_i,\psi_{i+j})\right),$$

$$\chi_{j,i}^2 = \frac{j-i}{2}\left(D_s(\theta_i,\psi_{j-i}) + D_a(\theta_{i+j},\psi_i) - D_s(\theta_{j-i},\psi_i) + D_a(\theta_{j-i},\psi_i)\right).$$

Using method of free parameters we derive the special approximation on three-point stencil for D_a D_s

$$d_{a,k}(\theta,\psi) = \frac{\theta_{k+1}-\theta_{k-1}}{2h}\psi_k - \theta_k\frac{\psi_{k+1}-\psi_{k-1}}{2h},$$

$$d_{s,k}(\theta,\psi) = \frac{2\theta_{k+1}\psi_{k+1} + \psi_k(\theta_{k+1}-\theta_{k-1}) + \theta_k(\psi_{k+1}-\psi_{k-1}) - 2\theta_{k-1}\psi_{k-1}}{6h}.$$

5   Numerical Results

We rewrite the (13)–(14) in vector form

$$\frac{d}{dt}\Theta = A\Theta + \lambda B\Psi + L(\Theta,\Psi), \quad 0 = A\Psi - B\Theta, \tag{20}$$

$$\Theta = (\theta_{11},...,\theta_{n1},...,\theta_{1m},...,\theta_{nm}), \quad \Psi = (\psi_{11},...,\psi_{n1},...,\psi_{1m},...,\psi_{nm}).$$

The matrix A consists of m three-diagonal submatrices A_j, nonlinear entries of skew-symmetric matrix $B = \{b_{sr}\}_{s,r=1}^{nm}$ are given by $b_{s,s+1} = -b_{s+1,s} = h/2, s = 1\div nm-1$, and $L(F,G)$ presents the nonlinear terms in (13). The discrete stream function Ψ can be expressed in form of Θ using second equation in (20). It gives the following system of ordinary differential equations:

$$\frac{d\Theta}{dt} = (A + \lambda BA^{-1}B)\Theta + L(\Theta, A^{-1}B\Theta). \tag{21}$$

To carry out computation with (21) we create a code on MATLAB. It allows to analyze convective structures, continue the families of stationary regimes .

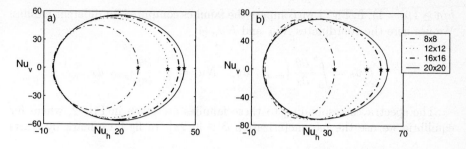

Fig. 1. The families of stationary regimes for different meshes $\lambda = 70$, $\beta = 1.5$ (left), $\lambda = 55$, $\beta = 3$ (right)

As stated in [2], if λ is slightly larger than λ_{11}, then all points of the family are stable. Starting from the vicinity of unstable zero equilibrium we integrate the system (21) up to a point Θ_0 close to a stable equilibrium on the family. We have used here the classical fourth order Runge-Kutta method as time integrator. Then a simplified version of the algorithm for family computation may be formulated in the following steps. Correct the point Θ_0 using the modified Newton method. Determine the kernel of the linearization matrix at the point Θ_0 by singular value decomposition. Predict the next point on the family Θ_0 by using fourth order Runge-Kutta method. Repeat these steps until a closed curve is obtained. This method is based on the cosymmetric version of implicit function theorem [6].

We explored the derived technique to calculate the families consisting both of stable and unstable equilibria. We analyze the case of narrow container ($\beta =$

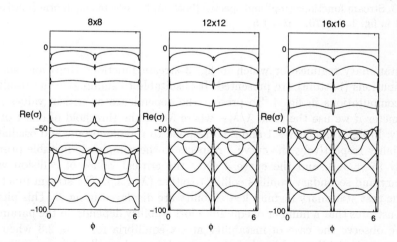

Fig. 2. Spectra for the families computed on different meshes, $\lambda = 55$, $\beta = 3$

$b/a \geq 1, a = 1$). In Fig. 1 we compare the families computed for different meshes. We use here the coordinates Nu_h and Nu_v [4]

$$Nu_h = \int_0^b \frac{\partial \theta}{\partial x} \, |_{x=a/2} \, dy, \quad Nu_v = \int_0^a \frac{\partial \theta}{\partial y} \, |_{y=0} \, dx$$

The spectra corresponding to these families are given in Fig. 2, where for equilibria we use the parameterization $\phi \in [0, 2\pi]$. In fig.1 we mark by starts

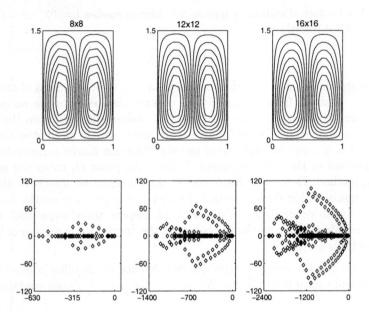

Fig. 3. Stream functions (top) and spectra (bottom) for selected equilibria (marked by stars) in fig. 1, $\lambda = 70$, $\quad \beta = 1.5$

the stationary regimes for which in fig. 3 stream functions (top) and spectra distributions (bottom) are presented. It is suitable to summarize the results of our computations in fig. 4. To present the dependence of critical values from parameter β we use the ratio λ/λ_{11}, where λ_{11} is the threshold of onset of the family. In fig. 4 the curves 1 and 2 correspond to the monotonic and oscillatory instability respectively, the curve 3 respects to the completely instable primary family ($\lambda = \lambda_o$), and the curve 4 gives the critical values of collision when primary and secondary families collide together ($\lambda = \lambda_c$). The stream functions for the first stationary regimes lost stability are displayed in fig. 5. This picture demonstrates that a number of equilibria lost stability depends on the parameter β. We observe the case of instability at six equilibria for $\beta \approx 2.3$ when the monotonic instability takes place in two points and oscillatory one – in four points.

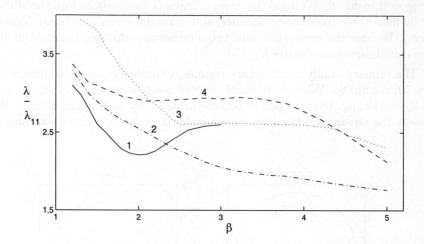

Fig. 4. Critical values for the families of stationary regimes

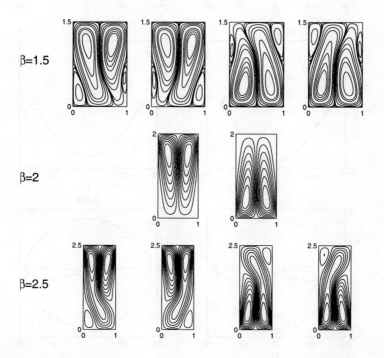

Fig. 5. Stream functions corresponding to first regimes lost stability, $\beta = 1.5, 2, 2.5$

Finally, we demonstrate the scenario of the evolution of the primary family for $\beta = 2$ in fig. 6. We mark by crosses (circles) the regimes lost the stability by monotonic (oscillatory) manner, and draw the secondary family (dotted curve). We use the projection onto two-dimensional unstable manifold of the zero equilibrium (coordinates U_1, U_2).

The primary family of stationary regimes is consists of stable equilibria for $\lambda = 70$, see fig. 6a. When λ increases one can see how the family deforms and at $\lambda_u = 112$ (fig. 6b) two points become unstable by monotonic manner. We present the stream functions for these equilibria in fig. 5. Further two arcs of

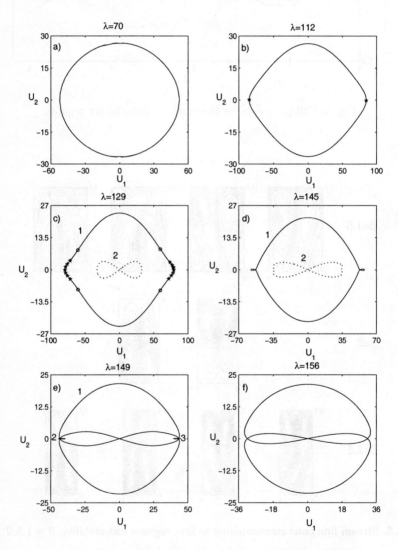

Fig. 6. Primary and secondary families evolution, $\beta = 2$

unstable equilibria are formed, then, for $\lambda = 129$ four regimes lost stability via oscillatory bifurcation (fig. 6c). We mention that for $\lambda > 81.6$ unstable secondary family exists as well (the curve 2 in fig. 6c). In fig. 6d one can see two families for $\lambda = 145$. Collision of primary and secondary families takes place for $\lambda_o = 147.6$. As a result, we see two small families and a combined one, fig. 6. When λ increases the small families collapse and disappear. The combined family for $\lambda = 156$ is completely unstable (fig. 6f).

A quite complete picture of local bifurcations in a cosymmetric dynamical system is presented in [7–9].

Acknowledgements

The authors acknowledge the support by the Russian Foundation for Basic Research (# 01-01-22002 and # 01-01-06340).

References

1. Yudovich, V. I. Cosymmetry, degeneration of solutions of operator equations, and the onset of filtration convection, *Mat. Zametki* **49** (1991) 142–148, 160

2. Yudovich, V. I. Secondary cycle of equilibria in a system with cosymmetry, its creation by bifurcation and impossibility of symmetric treatment of it, *Chaos* **5** (1995) 402–441

3. Lyubimov, D. V. On the convective flows in the porous medium heated from below, *J. Appl. Mech. Techn. Phys.* **16** (1975) 257

4. Govorukhin, V. N. Numerical simulation of the loss of stability for secondary steady regimes in the Darcy plane-convection problem, *Doklady Akademii Nauk* **363** (1998) 806-808

5. Karasözen B., Tsybulin V.G. Finite-difference approximation and cosymmetry conservation in filtration convection problem *Physics Letters A.* **V262** (1999). 321–329.

6. Yudovich, V. I. An implicit function theorem for cosymmetric equations, *Mat. Zametki* **60** (1996) 313–317

7. Yudovich, V. I. Cycle-creating bifurcation from a family of equilibria of dynamical system and its delay, *J. Appl. Math. Mech.* **62**(1998) 19–29, (translated from Russian *Prikl. Mat. Mekh* **62** (1998) 22–34)

8. Kurakin L.G., Yudovich V.I. Bifurcation of a branching of a cycle in n-parametric family of dynamic systems with cosymmetry *Chaos* **7** (1997) 376–386

9. Kurakin L.G., Yudovich V.I. Bifurcations accompanying monotonic instability of an equilibrium of a cosymmetric dynamical system *Chaos* **10** (2000) 311–330

A Comparative Study of
Dirichlet and Neumann Conditions for
Path Planning through Harmonic Functions

Madhuri Karnik, Bhaskar Dasgupta*, and Vinayak Eswaran

Department of Mechanical Engineering,
Indian Institute of Technology, Kanpur — 208 016, India.
{madhuri, dasgupta, eswar}@iitk.ac.in

Abstract. Harmonic functions, by virtue of their extrema appearing only on the domain boundary, are known to have an advantage as a global potential function in the potential field based approach for robot path planning. However, a wide range of possibilities exist for the global harmonic function for a particular situation, depending on the boundary conditions. This paper conducts a comparison of two major classes of boundary conditions, namely Dirichlet and Neumann conditions, and attempts to discover their relative merits and demerits. It is found that the Neumann conditions offer a surer and faster approach to the path planning problem, though suffering from the disadvantage of occasional tendency of the planned path to graze along the domain boundary.

Keywords: Harmonic function, Motion planning, Global potential.

1 Introduction

Harmonic functions are solutions to Laplace's equation. They are known to have a number of properties useful in robotic applications (Connolly and Gruppen [1]). Harmonic functions offer a complete path planning algorithm and paths derived from them are generally smooth. When applied to path planning of robots, they have the advantage over simple potential field based approach, as they exhibit no spurious local minima.

The use of potential functions for robot path planning, as introduced by Khatib [2], views every obstacle to be exerting a repelling force on an end effector, while the goal exerts an attractive force. Koditschek [3], using geometrical arguments, showed that, at least in certain types of domains (like spherical domains with spherical obstacles), there exists potential functions which can guide the effector from almost any point (excepting a set of measure zero) to a given

---

* Author to whom all correspondence should be addressed at the current address: Technische Universität Berlin, Institut für Technische Informatik und Mikroelektronik, Sekretariat EN 10, Einsteinufer 17, D-10587 Berlin, Germany.   Phone: ++49-30-314-25324; Fax: ++49-30-314-21116; Email: dasgupta@cs.tu-berlin.de

P.M.A. Sloot et al. (Eds.): ICCS 2002, LNCS 2330, pp. 442–451, 2002.

point. The usual formulation of potential fields for path planning does not prevent the spontaneous creation of local minima other than the goal. This may cause the robot to terminate its path at such a minimum and achieve a stable configuration short of goal.

Connolly et al [4] and Akishita et al [5] independently developed a global method using solutions to Laplace's equations for path planning to generate a smooth, collision-free path. The potential field is computed in a global manner, i.e. over the entire region, and the harmonic solutions to Laplace's equation are used to find the path lines for a robot to move from the start point to the goal point. Obstacles are considered as current sources and the goal is considered to be the sink, with the lowest assigned potential value. This amounts to using Dirichlet boundary conditions. Then, following the current lines, i.e. performing the steepest descent on the potential field, a succession of points with lower potential values leading to the point with least potential (goal) is found out. It is observed by Connolly et al [4] that this process guarantees a path to the goal without encountering local minima and successfully avoiding any obstacle, as a harmonic function cannot possess an extremum value except at the domain boundary.

In this paper, an attempt has been made to follow the above paradigm for path planning, using the analogy of temperature and heat flux for the potential and path line, respectively, for characterizing the solution of the Laplace equation. The analysis is carried out on two-dimensional domains having square or rectangular outer boundaries with occasional convolutions and obstacles. The finite element program "NASTRAN" is used to solve the Laplace equation and to obtain the temperature values at each node. Considering the robot to be a point, paths based on the potential (i.e. temperature) values are investigated by applying two types of boundary conditions. For Neumann boundary conditions, the start point is taken at a high temperature and the goal point is taken at a lower temperature, while the outer and the inner (obstacle) boundaries are treated as "No flux" (Homogeneous Neumann) boundaries. For Dirichlet boundary conditions, boundaries and the start point are assigned higher temperatures relative to the goal point. Various configurations and dimensions of the obstacles have been tried, with various temperature values assigned to them. The results obtained are analysed for colision-free path of the point robot.

2 Harmonic Function

A harmonic function on a domain $\Omega \subset R^n$ is a function which satisfies Laplace's equation,

$$\nabla^2 \Phi = \sum_{i=1}^{n} \frac{\partial^2 \Phi}{\partial x_i^2} = 0 \qquad (1)$$

where x_i is the i-th Cartesian coordinate and n is the dimension. In the case of robot path construction, the boundary of Ω (denoted by $\partial\Omega$) consists of the outer boundary of the workspace and the boundaries of all the obstacles as well as the start point and the goal point, in a configuration space representation. The spontaneous creation of a false local minimum inside the region Ω is avoided if Laplace's equation is imposed as a constraint on the functions used, as the harmonic functions satisfy the min-max principle [6, 7].

The gradient vector field of a harmonic function has a zero curl, and the function itself obeys the min-max principle. Hence the only types of critical points which can occur are saddle points. For a path-planning algorithm, an escape from such critical points can be found by performing a search in the neighbourhood of that point. Moreover, any perturbation of a path from such points results in a path which is smooth everywhere [1].

3   Configuration Space

In the framework used in the analysis here, the robot, i.e. its configuration, is represented by a point in the configuration space, or C-space. The path planning problem is then posed as an obstacle avoidance problem for the point robot from the start point to the goal point in the C- space.

The C-space examples used in this paper have either square or rectangular outer boundaries, having projections or convolutions inside to act as barriers. Apart from projections of the boundaries, some obstacles inside the boundary are also considered. The C-space is discretized and the coordinates and function values associated with each node are studied with the thermal analysis code of the finite element software NASTRAN (version 9.0). The highest temperature is assigned to the start point whereas the goal point is assigned the lowest for both Neumann and Dirichlet boundary conditions. In some cases with Dirichlet conditions, the start point is not assigned any temperature. The results are processed, for Dirichlet boundary conditions, by assigning different temperature values to the boundaries.

The area occupied by the obstacles, the temperatures or fluxes assigned to the boundaries and the goal points are the variables used to investigate the effectiveness of the technique in the present study.

4   Boundary Conditions

Solutions to the Laplace's equation were examined with two different types of boundary conditions. The following forms of the Dirichlet and Neumann boundary conditions were used.

Dirichlet boundary conditions:

$$\Phi \mid \partial\Omega = c$$

where c is constant. In this case, the boundary is maintained at a particular temperature. The typical obstacle temperature values used are 8°C and 3°C. The values of goal temperature used were 0°C and -150°C[1]. If assigned, the

---

[1] This extremely low temperature is used to examine its effect on the results.

start point is at 10°C. As the boundary temperature is fixed, the heat flow has
to be along the normal to the boundary.
Homogeneous Neumann boundary conditions:

$$\frac{\partial \Phi}{\partial \overline{n}}\bigg| \, \partial \Omega = 0$$

where $\overline{n}$ is the unit vector normal to the boundary. This condition is applied only
to the outer and the obstacle boundaries. The start point (the source) and the
goal point (the sink) are maintained at 10°C and 0°C respectively (in the current
implementation). The Neumann condition constrains the normal component of
the gradient of Φ to be zero at the boundaries. As there is no normal component
of heat flow, the condition forces the flow to be tangential to the boundary. As
the path follows the heat flow, the Neumann condition may lead to a tendency
for a robot to stay close to C-space obstacle/boundary surfaces.

5   Path Planning

Once the harmonic function under the boundary conditions is established, the required path can be traced by the steepest descent method, following the negative
gradient from the start point through successive points with lower temperature
till the goal, which is the point with the lowest temperature. The coordinates
and the nodal gradients of temperature obtained from the finite element analysis
can be used to draw the path.

In the present work, the actual path is not drawn, but the possibility and
efficiency of tracing the path is investigated from the nature of the harmonic
function in the domain. The presence of a significant temperature gradient all
over the C-space indicates that a path can be traced for the point robot by the
use of the gradient information. On the other hand, absence of an appreciable
gradient, even in a portion of the C-space, will prevent the efficient tracing of a
path.

6   Results and Discussion

In this section, we present results for some simple C-space structures and compare the effects of various conditions on the contours of the resulting harmonic
functions. These contours are enough to draw conclusions regarding the possibility and efficiency of developing the actual path from the function[2]. In the
following, results of the Neumann condition are presented first, followed by solutions with Dirichlet conditions. The figures are drawn to show the geometry
of the C-space and the temperature contours. The presence of temperature contours throughout the C-space indicates the presence of temperature gradient in

---

[2] The solution from the software also provides direct information of gradients, which
are to be used for tracing the path. But, that data is not so visually revealing as the
contours, hence have been omitted from presentation here.

the C-space. This shows that the path can be easily and efficiently found using steepest descent method. The contours are labelled from 1 to B; 1 being the contour with the lowest temperature and B with the highest. The values of the minimum temperature (goal point) and the highest temperature (start point or boundary) and the node identities representing these points are also indicated on each figure. The computational time for both cases are the same, namely approximately 0.25 to 0.30 sec of CPU time in a SUN workstation.

6.1 Results with Neumann Boundary Condition

Figures 1–3 show the temperature contours obtained by applying the Neumann boundary condition to the function in three test regions. The start point is maintained at 10°C and the goal point at 0°C in all the cases[3]. The effect of variation in the area and shapes of the obstacle boundaries on the gradient in the C-space is examined. In the first case, the region is obtained by cutting off a convoluted corridor from a unit square. The convolution or inside projection from the boundary itself acts as a barrier and there is no true obstacle. In figure 2, there are two obstacles inside a (unit) square region, and the close proximity of the obstacles induces disturbance in the contour in the neighbourhood. In figure 3, we have an example with both external boundary convolution and an obstacle in a rectangular region of 1.2 × 1.0 units. In all the figures, the origin is at the bottom left corner of the region, and the frame displayed inside shows only the directions.

It is observed that the temperature contours are well-spread in the region and a significant gradient is present throughout the C-space in each of the cases. Even at saddle points with zero gradient, one such possibility occurring between contour lines marked 'A' and '0' in figure 2, a slight perturbation from the point will succeed in resuming a fruitful path tracing. This is an advantage with the Neumann boundary conditions. On the other hand, motion orthogonal to the contour lines, as resulting from gradient-based motion planning, is likely to graze boundaries closely at times and even follow a longer path, rendering the method less than optimal. A large number of other examples tested show similar behaviour.

6.2 Results with Dirichlet Boundary Condition

Figures 4–7 show the temperature contours in the C-space when the Dirichlet boundary condition is applied to the function. All these results show a remarkable difference from the previous results. Here, the contour lines are mostly crowded near the points with assigned temperature values, leaving most of the free-space with very little temperature variation. Figure 4 shows the result with Dirichlet condition (with boundary temperature 3°C) for the same region as in figure 1.

---

[3] Results are completely insensitive to these two temperature values (as long as they are different), as there is only one 'temperature difference' and its unit or scale is arbitrary.

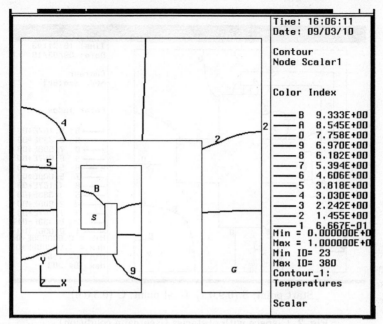

```
Time:  16:06:11
Date:  09/03/10

Contour
Node Scalar1

Color Index

──── B    9.333E+00
──── A    8.545E+00
──── 0    7.758E+00
──── 9    6.970E+00
──── 8    6.182E+00
──── 7    5.394E+00
──── 6    4.606E+00
──── 5    3.818E+00
──── 4    3.030E+00
──── 3    2.242E+00
──── 2    1.455E+00
──── 1    6.667E-01
Min = 0.000000E+0
Max = 1.000000E+0
Min ID= 23
Max ID= 380
Contour_1:
Temperatures

Scalar
```

Start point: S (0.3,0.3), Goal point: G (0.9,0.1)

Fig. 1. C-space with inside boundary projection (Neumann condition)

A comparison of the two figures shows the relative difficulty of path tracing through a gradient-based strategy in the case of the Dirichlet condition solution (figure 4). Changing the boundary temperature has little qualitative effect on the result. As figure 5 shows, the contour curves have got only a little redistributed, when we solve the same problem with boundary temperature maintained at 8°C. Even modifying temperature difference drastically does not spread the contours well enough. For example, figure 6 shows the result of changing the goal point temperature to −150°C. From the result, we see that most of the contour curves are used up in accommodating the temperature difference between the boundary and the goal point, leaving most of the region essentially featureless (less than 1°C temperature difference between curves '0' and 'A'). Different sets of Dirichlet conditions applied to many other example regions, from simple to reasonably complicated, exhibit similar patterns of the contours.

One may notice, that in the case of Dirichlet condition, temperature values are assigned to the boundaries and, for obtaining a harmonic solution, it is not *necessary* to assign temperature values to both start and goal points. In figure 7, we have an example (region the same as that of figure 3) with only two temperature values— boundary temperature 8°C and goal temperature 0°. The result, as expected, shows significant temperature gradient only close to the goal point.

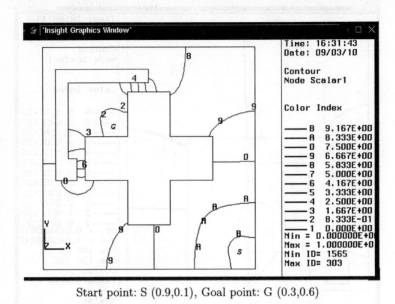

Start point: S (0.9,0.1), Goal point: G (0.3,0.6)

Fig. 2. C-space with obstacles (Neumann condition)

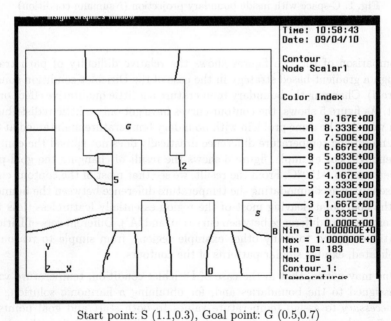

Start point: S (1.1,0.3), Goal point: G (0.5,0.7)

Fig. 3. Rectangular C-space with projection and obstacle (Neumann condition)

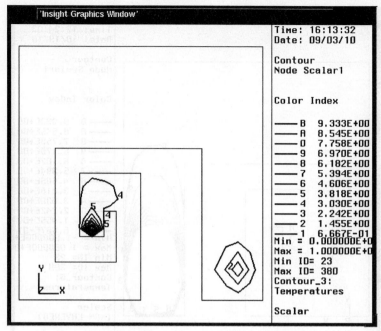

Start point: (0.3,0.3), Goal point: (0.9,0.1)

Fig. 4. C-space with inside boundary projection (Dirichlet condition)

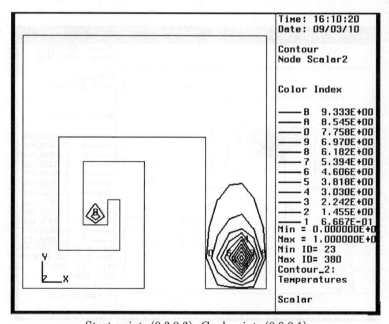

Start point: (0.3,0.3), Goal point: (0.9,0.1)

Fig. 5. Effect of boundary temerature (Dirichlet condition)

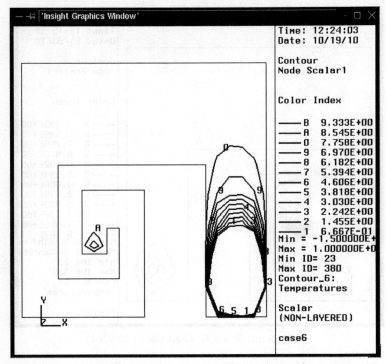

Start point: (0.3,0.3), Goal point: (0.9,0.1)

Fig. 6. Effect of very low temperature at goal point (Dirichlet condition)

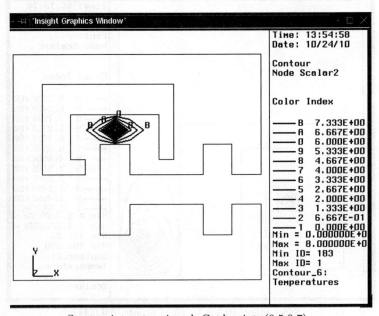

Start point not assigned, Goal point: (0.5,0.7)

Fig. 7. Solution with start point not assigned (Dirichlet condition)

6.3   Discussion

The harmonic solutions presented above show that the two classes of boundary conditions (Neumann and Dirichlet) give quite different potential fields that can be used for path planning in different C-space configurations.

For all the C-spaces analysed with Neumann boundary condition, significant temperature gradient is present all over the C-space making it easier to trace the path of the point robot. But the results obtained with Dirichlet boundary condition are radically different.The remarkable gradient is present only in the region close to the goal point (and also in the vicinity the start point, if it is assigned a temperature value), leaving the rest of the C-space at (almost) constant temperature, with no appreciable gradient. Such a situation makes it difficult to trace a path to the goal point.

7   Conclusions

The comparative studies of Neumann and Dirichlet boundary conditions reveal that one has to judiciously choose the boundary conditions to be applied to Harmonic function used for robot path planning. The Neumann boundary condition gives the significant gradient of temperature (potential) throughout the C-space irrespective of the shape of the obstacles and temperatures assigned to goal point and start point. This makes it an easier and surer method to follow the steepest descent for path planning. But the paths found out by this method may have a tendency to graze the boundary.

As extension of this work, the performance of the method in spaces of higher dimensions may be examined. Investigation of mixed boundary conditions is also an important area of possible further studies.

References

1. Connolly, C. I., Gruppen, R.: On the applications of harmonic functions to robotics. Journal of robotic systems. Journal of Robotic Systems **10** (7) (1993) 931–946
2. Khatib, O.: Real time obstacle avoidance for manipulators and mobile robots. IEEE Transactions on Robotics and Automation **1** (1985) 500–505
3. Koditschek, D. E.: Exact robot navigation by means of potential functions:Some topological considerations. Proceedings of the IEEE International Conference on Robotics and Automation (1987) 1-6
4. Connolly, C. I., Burns, J. B., Weiss, R.: Path planning using Laplace's equation. Proceedings of the IEEE International Conference on Robotics and Automation (1990) 2102–2106.
5. Akishita, S., Kawamura, S., Hayashi, K.: Laplace potential for moving obstacle avoidance and approach of a mobile robot. Japan-USA Symposium on flexible automation, A Pacific rim conference (1990) 139–142.
6. Weinstock, R.: Calculus of variations with applications to physics and engineering. Dover Publications, Inc., New York (1974).
7. Zachmanoglou, E. C., Thoe, D. W.: Introduction to partial differential equations with applications. Dover Publications, Inc., New York (1986).

Adaptation and Assessment of a High Resolution Semi-Discrete Numerical Scheme for Hyperbolic Systems with Source Terms and Stiffness

R. Naidoo[1,2] and S. Baboolal[2]

[1] Department of Mathematics and Physics, M.L. Sultan Technikon, P.O. Box 1334, Durban 4000, South Africa
naidoor@yoda.cs.udw.ac.za

[2] Department of Computer Science, University of Durban-Westville, Private Bag X54001, Durban 4000, South Africa
sbab@pixie.udw.ac.za

Abstract. In this work we outline the details required in adapting the third-order semi-discrete numerical scheme of Kurganov and Levy [SIAM J. Sci. Comput. **22** (2000) 1461-1488.] to handle hyperbolic systems which include source terms. The performance of the scheme is then assessed against a fully discrete scheme, as well as reference solutions, on such problems as shock propagation in a Broadwell gas and shocks in gas dynamics with heat transfer.

1   Introduction

This paper is concerned with the numerical integration of

$$\frac{\partial u(x,t)}{\partial t} + \frac{\partial f(u)}{\partial x} = \frac{1}{\varepsilon}\, g(u), \tag{1}$$

a one-dimensional hyperbolic system of partial differential equations. Here $u(x,t)$ is the unknown m-dimensional vector function, $f(u)$ is the flux vector, $g(u)$ is a continuous source vector function on the right hand side (RHS), with x the single spatial coordinate and t the temporal coordinate and the parameter $\varepsilon > 0$ distinguishes between stiff systems ($\varepsilon << 1$) and standard, non-stiff ones ($\varepsilon = 1$).

Such equationss can be used to model many physical systems, including fluids and gases. In the last decade, particularly following the work of Nessyahu and Tadmor [1], a family of fully-discrete, high-resolution, Riemann-solver-free schemes have been produced in order to numerically solve hyperbolic systems such as the aforementioned. More recently, also based on the same Riemann-solver-free approach, second and third order semi-discrete schemes were devised by Kurganov and Tadmor [2] and Kurganov and Levy [3]. One advantage of the latter is that they can be applied on non-staggered grids and thus ease the implementation of boundary conditions. Here we are particularly interested in the details of adapting the latter so that it be can applied to systems with source

P.M.A. Sloot et al. (Eds.): ICCS 2002, LNCS 2330, pp. 452–460, 2002.

terms including those that are stiff such as (1) above. In order to assess the performance of this scheme we examine its merits against an adaptation [4] following [5] for non-staggered grids, of the fully-discrete scheme of [1] for systems with source terms, as well as against exact or reference solutions for two prototype problems. One such is the problem of shock propagation in a Broadwell gas [6,7] and the other is that of shocks in a model of gas dynamics with heat transfer.

2   The modified numerical scheme

2.1   Derivation with source term

Here the numerical integration of problem (1) is considered on some uniform spatial and temporal grids with the spacings, $\Delta x = x_{j+1} - x_j$; $\Delta t = t^{n+1} - t^n$ (with j and n being suitable integer indices).

For the nonlinear homogeneous case of (1), Kurganov and Levy [3] obtain the third-order semi-discrete scheme

$$\frac{d\bar{u}_j}{dt} = -\frac{1}{2\Delta x}\left[f(u^+_{j+\frac{1}{2}}(t)) + f(u^-_{j+\frac{1}{2}}(t)) - f(u^+_{j-\frac{1}{2}}(t)) - f(u^-_{j-\frac{1}{2}}(t))\right]$$
$$-\frac{a_{j+\frac{1}{2}}(t)}{2\Delta x}\left[u^+_{j+\frac{1}{2}}(t) - u^-_{j+\frac{1}{2}}(t)\right] - \frac{a_{j-\frac{1}{2}}(t)}{2\Delta x}\left[u^+_{j-\frac{1}{2}}(t) - u^-_{j-\frac{1}{2}}(t)\right],$$

$$(2)$$

where

$$a^n_{j\pm\frac{1}{2}} = max\left(\rho\left(\frac{\partial f}{\partial u}(u^-_{j\pm\frac{1}{2}}(t))\right), \rho\left(\frac{\partial f}{\partial u}(u^+_{j\pm\frac{1}{2}}(t))\right)\right), \tag{3}$$

and

$$u^+_{j\pm\frac{1}{2}} := P_{j+1}(x_{j\pm\frac{1}{2}}, t^n); \quad u^-_{j\pm\frac{1}{2}} := P_j(x_{\pm\frac{1}{2}}, t^n). \tag{4}$$

In the above, the forms (4) are respectively the left and right intermediate values at $x_{j+\frac{1}{2}}$ of a piecewise polynomial interpolant $P_j(x, t^n)$ that fit an already computed or known cell average values $\{\bar{u}^n_j\}$ at time level n. Also $\rho(.)$ denotes the spectral radius of the respective Jacobian, defining the maximum local propagation speeds $a^n_{j\pm\frac{1}{2}}$.

They also obtain an extension of the above when the RHS of (1) is of the form $\frac{\partial Q}{\partial x}$ where $Q(u(x,t), u_x(x,t))$ is a dissipation flux satisfying a parabolicity condition [3].

However, to allow for an arbitrary source term, say, $g(u(x,t))$ in (1) (omitting for convenience the stiffness parameter) we must proceed as outlined in [3] and follow through the construction of the scheme with this added detail. Thus, employing the above mentioned uniform spatial and temporal grids and integrating (1) over the cell $I(x) := \{\xi \mid \xi - x \mid \leq \frac{\Delta x}{2}\}$ gives

$$\bar{u}_t + \frac{1}{\Delta x}\left[f(u(x + \frac{\Delta x}{2}, t)) - f(u(x - \frac{\Delta x}{2}, t))\right] = \bar{g} \tag{5}$$

where

$$\bar{u}(x,t) := \frac{1}{\Delta x} \int_{I(x)} u(\xi,t) \, d\xi \tag{6}$$

and

$$\bar{g} := \frac{1}{\Delta x} \int_{I(x)} g(u(\xi,t)) \, d\xi \tag{7}$$

Now assuming the $\{\bar{u}_j^n\}$ are already computed or known cell-averages of the approximate solution at time $t = t^n$ we integrate as in [3] over the control volumes $[x_{j-\frac{1}{2},R}^n, x_{j-\frac{1}{2},L}^n] \times [t^n, t^{n+1}]$, $[x_{j-\frac{1}{2},R}^n, x_{j+\frac{1}{2},L}^n] \times [t^n, t^{n+1}]$ and $[x_{j+\frac{1}{2},L}^n, x_{j+\frac{1}{2},R}^n] \times [t^n, t^{n+1}]$ where

$$x_{j\pm\frac{1}{2},L}^n := x_{j\pm\frac{1}{2}} - a_{j\pm\frac{1}{2}}^n \Delta t; \quad x_{j\pm\frac{1}{2},R}^n := x_{j\pm\frac{1}{2}} + a_{j\pm\frac{1}{2}}^n \Delta t \tag{8}$$

with the piecewise polynomial form in the cell I_j taken as

$$P_j(x,t^n) = A_j + B_j(x - x_j) + \frac{1}{2}C_j(x - x_j)^2. \tag{9}$$

where the constants A_j, B_j, C_j are evaluated as in [3]. These then result respectively, in the weighted averages $\bar{w}_{j-\frac{1}{2}}^{n+1}, \bar{w}_j^{n+1}, \bar{w}_{j+\frac{1}{2}}^{n+1}$ which differ from those in [3] only in the respective additive source terms

$$\frac{1}{2a_{j-\frac{1}{2}}^n \Delta t} \int_{x_{j-\frac{1}{2},L}}^{x_{j-\frac{1}{2},R}} \int_{t^n}^{t^{n+1}} g \, dx \, dt, \tag{10}$$

$$\frac{1}{\Delta x - \Delta t \left(a_{j-\frac{1}{2}}^n + a_{j+\frac{1}{2}}^n \right)} \int_{x_{j-\frac{1}{2},R}}^{x_{j+\frac{1}{2},L}} \int_{t^n}^{t^{n+1}} g \, dx \, dt \tag{11}$$

and

$$\frac{1}{2a_{j+\frac{1}{2}}^n \Delta t} \int_{x_{j+\frac{1}{2},L}}^{x_{j+\frac{1}{2},R}} \int_{t^n}^{t^{n+1}} g \, dx \, dt. \tag{12}$$

Then from the cell averages $\bar{w}_{j+\frac{1}{2}}^{n+1}$ and $\bar{w}_j^{n+1}$ are reconstructed third order piecewise polynomials [3] taken as

$$\tilde{w}_{j+\frac{1}{2}}^{n+1} = \tilde{A}_{j\pm\frac{1}{2}} + \tilde{B}_{j\pm\frac{1}{2}}(x - x_{j\pm\frac{1}{2}}) + \frac{1}{2}\tilde{C}_{j\pm\frac{1}{2}}(x - x_{j\pm\frac{1}{2}})^2, \quad \tilde{w}_j^{n+1}(x) \equiv \bar{w}^{n+1}. \tag{13}$$

where the constants $\tilde{A}_j, \tilde{B}_j$ and $\tilde{C}_j$ are evaluated as in [3]. The new cell averages on the unstaggered grids are obtained from these polynomials by [3]

$$\bar{u}_j^{n+1} = \frac{1}{\Delta x} \left[\int_{x_{j-\frac{1}{2}}}^{x_{j-\frac{1}{2},R}} \tilde{w}_{j-\frac{1}{2}}^{n+1} \, dx + \int_{x_{j-\frac{1}{2},R}}^{x_{j+\frac{1}{2},L}} \tilde{w}_j^{n+1} \, dx + \int_{x_{j+\frac{1}{2},L}}^{x_{j+\frac{1}{2}}} \tilde{w}_{j+\frac{1}{2}}^{n+1} \, dx \right] \tag{14}$$

The semi-discrete form is then defined by the limit

$$\frac{d\bar{u}_j(t)}{dt} = \lim_{\Delta t \to 0} \frac{\bar{u}_j^{n+1} - \bar{u}_j^n}{\Delta t}. \tag{15}$$

Proceeding with (14) and (13) as in [3], the coefficients in the polynomial form simplify resulting in

$$\frac{d\bar{u}_j}{dt} = -\frac{1}{2\Delta x} \left[f(u_{j+\frac{1}{2}}^+(t)) + f(u_{j+\frac{1}{2}}^-(t)) - f(u_{j-\frac{1}{2}}^+(t)) - f(u_{j-\frac{1}{2}}^-(t)) \right]$$

$$-\frac{a_{j+\frac{1}{2}}(t)}{2\Delta x} \left[u_{j+\frac{1}{2}}^+(t) - u_{j+\frac{1}{2}}^-(t) \right] - \frac{a_{j-\frac{1}{2}}(t)}{2\Delta x} \left[u_{j-\frac{1}{2}}^+(t) - u_{j-\frac{1}{2}}^-(t) \right]$$

$$+ \lim_{\Delta t \to 0} \frac{1}{2\Delta x \Delta t} \int_{t^n}^{t^{n+1}} \int_{x_{j-\frac{1}{2},L}}^{j-\frac{1}{2},R} g \, dx \, dt + \lim_{\Delta t \to 0} \frac{1}{2\Delta x \Delta t} \int_{t^n}^{t^{n+1}} \int_{x_{j+\frac{1}{2},L}}^{j+\frac{1}{2},R} g \, dx \, dt$$

$$+ \lim_{\Delta t \to 0} \frac{1}{\Delta t(\Delta x - \Delta t(a_{j+\frac{1}{2}} + a_{j-\frac{1}{2}}))} \int_{t^n}^{t^{n+1}} \int_{x_{j-\frac{1}{2},R}}^{j+\frac{1}{2},L} g \, dx \, dt. \tag{16}$$

We note that the non-smooth parts of the solution are contained over spatial widths of size $2a_{j\pm\frac{1}{2}}^n \Delta t$. Full details with clear sketches are given in [3]. Now, when the limits are taken on the source integrals, the first two vanish as the Riemann fans shrink to zero. At the time, since $\bar{u}^n = \bar{u}(t)$ (and hence g) is a constant over this cell, it can be shown for the other that

$$\lim_{\Delta t \to 0} \frac{1}{\Delta t(\Delta x - \Delta t(a_{j+\frac{1}{2}} + a_{j-\frac{1}{2}}))} \int_{t^n}^{t^{n+1}} \int_{x_{j-\frac{1}{2},R}}^{j+\frac{1}{2},L} g \, dx \, dt = g(u_j^n).$$

Hence the modified semi-discrete scheme with source term $g(u(x,t))$ is

$$\frac{d\bar{u}_j}{dt} = -\frac{1}{2\Delta x} \left[f(u_{j+\frac{1}{2}}^+(t)) + f(u_{j+\frac{1}{2}}^-(t)) - f(u_{j-\frac{1}{2}}^+(t)) - f(u_{j-\frac{1}{2}}^-(t)) \right]$$

$$-\frac{a_{j+\frac{1}{2}}(t)}{2\Delta x} \left[u_{j+\frac{1}{2}}^+(t) - u_{j+\frac{1}{2}}^-(t) \right] - \frac{a_{j-\frac{1}{2}}(t)}{2\Delta x} \left[u_{j-\frac{1}{2}}^+(t) - u_{j-\frac{1}{2}}^-(t) \right] + g(u_j(t)). \tag{17}$$

where the rest of the terms are as in (2)-(3).

To compute with (17), it is convenient to use ODE system solvers, such as Runge-Kutta formulae. For instance, writing (17) in the form,

$$\frac{du_j}{dt} = F_j, \tag{18}$$

where F_j is the vector of the RHS, we can employ the second-order (in time) Runge-Kutta (RK2) scheme [2] for it as:

$$RK2: \begin{cases} U^{(1)} = U^n + \Delta t F(U^n) \\ U^{(2)} = \frac{1}{2}U^n + \frac{1}{2}\left[U^{(1)} + \Delta t F(U^{(1)}) \right] \\ U^{n+1} = U^{(2)} \end{cases} \tag{19}$$

where the U denotes the vector of components u_j, the superscript n and $n + 1$ denote successive time levels, whilst the other $(1,2)$ denote intermediate values.

We shall refer to the scheme (17) with RK2 (19) as the SD3 scheme. We note that such a scheme is generally third order except in regions of steep gradients when it degrades to order two [3]. Since also, RK2 is second order in time, it will make sense when we compare its performance to that of the fully discrete scheme (NNT) for systems with source terms for integration on unstaggered grids [4]:

$$\bar{u}_j^{n+1} = \frac{1}{4}\left(\bar{u}_{j+1}^n + 2\bar{u}_j^n + \bar{u}_{j-1}^n\right) - \frac{1}{16}\left(u_{xj+1}^n - u_{xj-1}^n\right) - \frac{1}{8}\left[u_{xj+\frac{1}{2}}^{n+1} - u_{xj-\frac{1}{2}}^{n+1}\right]$$
$$+\frac{\Delta t}{8}\left[g\left(u_{j+1}^n\right) + 2g\left(u_j^n\right) + g\left(u_{j-1}^n\right) + g\left(u_{j+1}^{n+1}\right) + 2g\left(u_j^{n+1}\right) + g\left(u_{j-1}^{n+1}\right)\right]$$
$$-\frac{\lambda}{4}\left[\left(f_{j+1}^n - f_{j-1}^n\right) + \left(f_{j+1}^{n+1} - f_{j-1}^{n+1}\right)\right] \tag{20}$$

where $\lambda = \Delta t/\Delta x$ and which will be used used with an UNO derivative approximation [1, 4].

2.2 Implementation details

The implementation of the NNT scheme (20) above follows previous reports [1, 4], where in particular we mention that the source term can make the scheme implicit. The latter then requires an iteration at each grid point at every time level. The implementation of the modified semi-discrete scheme SD3 (17) follows closely the prescription given in [3] where in particular we employ the constants given in their equation (2.9) for the non-oscillatory piece-wise polynomial (9). In addition, it is required to compute at every time step the spectral radii (3) of the Jacobians of the flux terms, which we obtained exactly for the small test systems to follow. Finally, with chosen initial and boundary conditions, the solution is advanced with the explicit Runge-Kutta scheme (18)-(19). The codes were written in 64-bit real precision Fortran 77, employing real-time graphics to depict evolving profiles. They were compiled with the Salford FTN95/win32 version 1.28 compiler and run on a PC under MS Windows 2000 and NT.

3 Applications and tests

3.1 Shocks in a Broadwell gas

Here we solve the governing equations for a Broadwell gas [6, 7],

$$\frac{\partial \rho}{\partial t} + \frac{\partial m}{\partial x} = 0,$$

$$\frac{\partial m}{\partial t} + \frac{\partial z}{\partial x} = 0,$$

$$\frac{\partial z}{\partial t} + \frac{\partial m}{\partial x} = \frac{1}{\varepsilon}\left(\rho^2 + m^2 - 2\rho z\right),$$

where ε is the mean free path and $\rho(x,t), m(x,t), z(x,t)$ are the density, momentum and flux respectively. The range $\varepsilon = 1..10^{-8}$ cover the regime from the non-stiff to the highly stiff. In particular, the limit $\varepsilon = 10^{-8}$ requires a renormalization of the variables such as in the form

$$\bar{x} = \frac{1}{\varepsilon} x, \quad \bar{t} = \frac{1}{\varepsilon} t$$

followed by computations on an equivalent finer grid (see for example [8]).

We observe that in the limit $\varepsilon \to 0$ we arrive at

$$z = z_E(\rho, m) = \frac{1}{2\rho}(\rho^2 + m^2)$$

which leads to the equilibrium solution of the governing equations above which then reduce to the Euler equations.

The SD3 (17) and NNT scheme (20) were applied to the above with the two sets ($Rim1$ and $Rim2$) of initial conditions corresponding to several Riemann problems, each distinguished by a specific ε-value:

$$Rim1 : \begin{cases} \rho = 2, m = 1, z = 1; & x < x_J. \\ \rho = 1, m = 0.13962, z = 1; & x > x_J. \end{cases}$$

$$Rim2 : \begin{cases} \rho = 1, m = 0, z = 1; & x < x_J. \\ \rho = 0.2, m = 0, z = 1; & x > x_J. \end{cases}$$

In all calculations absorbing boundary conditions were employed, where in particular, the boundary values were obtained by quadratic extensions of internal point values on a fixed spatial grid, over an integration domain on the X-axis. Results obtained are depicted in Figure 1.

Other parameters used here were $\Delta x = 0.01$, $\Delta t = 0.005$, $x_J = 5$ in (a) and (b) and $\Delta x = 0.02$, $\Delta t = 0.001$, $x_J = 10$ in (c) and (d) for both methods. We observe that in virtually all cases, the semi-discrete scheme give better results than the modified NNT scheme.

3.2   Shocks in an Eulerian gas with heat transfer

Here we solve the Euler equations for the one-dimensional flow of a gas in contact with a constant temperature bath [9]:

$$\frac{\partial \rho}{\partial t} + \frac{\partial (\rho u)}{\partial x} = 0,$$

$$\frac{\partial (\rho u)}{\partial t} + \frac{\partial (\rho u^2 + p)}{\partial x} = 0,$$

$$\frac{\partial (\rho E)}{\partial t} + \frac{\partial (\rho u E + u p)}{\partial x} = -K\rho(T - T_0).$$

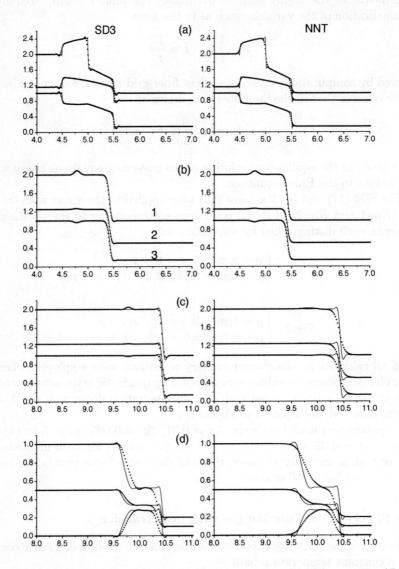

Fig. 1. Broadwell gas shock solutions with (a) $\varepsilon = 1$ ($Rim1$), (b) $\varepsilon = 0.02$ ($Rim1$), (c) $\varepsilon = 10^{-8}$ ($Rim1$) and (d) $\varepsilon = 10^{-8}$ ($Rim2$). Here the curve labelled $1 \sim \rho$, $2 \sim z$, $3 \sim m$. The snap-shot time is $t = 0.5$ in all cases. The dotted lines are computed solutions and the solid ones are 'exact' (refined grid) solutions.

The $\rho, u, T, e, E = e + \frac{1}{2}u^2, p = (\gamma - 1)\rho e$ are the density, flow velocity, temperature in units of e, internal energy, total energy and pressure respectively with $K > 0$ the heat transfer coefficient and T_0 the constant bath temperature.

The initial conditions used were:

$$Rim3 : \begin{cases} \rho = 2.5, u = 1.0, p = 1.0; & x < 50. \\ \rho = 1.0, u = 0.4, p = 0.4; & x > 50. \end{cases}$$

where different $K = 1, 50, 400, 1000$ are employed. Computed results with SD3 and NNT are shown in Figure 2.

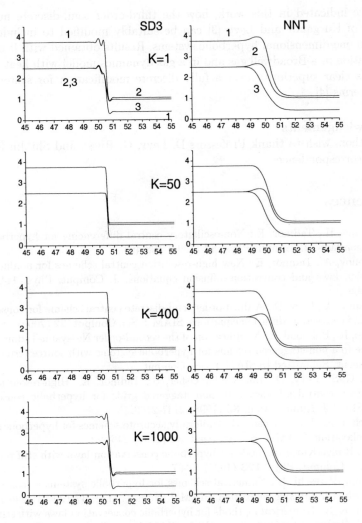

Fig. 2. Shocks in a gas with heat transfer. The curve labelled $1 \sim \rho E$, $2 \sim \rho$, $3 \sim \rho u$. The grid lengths of $\Delta x = 0.1$ and $\Delta t = 0.001$ were used for both methods. The output time is $t = 0.4$ in all cases.

In these we observe SD3 captures the shocks significantly better than does NNT. Although the 'exact' (grid refined ones) are not shown here we mention that they are close to the SD3 ones, except for the oscillations seen in them. We believe the oscillations are due to the derivative calculations from the fitted piecewise polynomials, and can be improved by taking different smoothing coefficients, as suggested in [3]. Further, the NNT curves show poor resolution of the shocks in comparison, and also show far more dissipation, as expected [3].

4   Conclusion

We have indicated in this work, how the third-order semi-discrete numerical scheme of Kurganov and Levy [3] can be suitably modified to include source terms in one-dimensional hyperbolic systems. Results obtained with it on shock propagation in a Broadwell gas and in a gas dynamics model with heat transfer show its clear superiority over a fully discrete modification for systems with source terms [4].

Acknowledgements
The authors wish to thank Professors D. Levy, G. Russo and Shi Jin for their helpful correspondences.

References

1. Nessyahu, H., Tadmor, E.: Non-oscillatory central differencing for hyperbolic conservation laws. J. Comput.Phys. **87** (1990) 408-463.
2. Kurganov, A., Tadmor, E.: New high-resolution central schemes for nonlinear conservation laws and convection-diffusion equations. J. Comput. Phys. **160** (2000) 241-282.
3. Kurganov, A., Levy, D.: A third-order semi-discrete central scheme for conservation laws and convection-diffusion equations. SIAM J. Sci. Comput. **22** (2000) 1461-1488.
4. Naidoo, R., Baboolal, S.: Modification of the second-order Nessyahu-Tadmor central scheme to a non-staggered scheme for hyperbolic systems with source terms and its assessment. Submitted to SIAM J. Sci. Comput.
5. Jiang, G.-S., Levy, D., Lin, C.-T., Osher, S., Tadmor, E.: High-resolution non-oscillatory central schemes with non-staggered grids for hyperbolic conservation laws. SIAM J. Numer. Anal. **35** (1998) 2147-2168.
6. Caflisch, R.E., Jin, S., Russo, G.: Uniformly accurate schemes for hyperbolic systems with relaxation. SIAM J. Numer. Anal. **34** (1997) 246-281.
7. Jin, S.: Runge-Kutta methods for hyperbolic conservation laws with stiff relaxation terms. J. Comput. Phys. **122** (1995) 51-67.
8. Naldi, G., Pareschi, L.: Numerical schemes for hyperbolic systems of conservation laws with stiff diffusive relaxation. SIAM J. Numer. Anal. **37** (2000) 1246-1270.
9. Pember, R. B.: Numerical methods for hyperbolic conservation laws with stiff relaxation II. Higher-order Godunov methods. SIAM J. Sci. Comput. **14** (1993) 824-829.

The Computational Modeling of Crystalline Materials Using a Stochastic Variational Principle [*]

Dennis Cox[1], Petr Klouček[2], and Daniel R. Reynolds[2]

[1] Department of Statistics, Rice University, 6100 Main Street,
Houston, TX 77005, USA
[2] Department of Computational and Applied Mathematics, Rice University,
6100 Main Street, Houston, TX 77005, USA

Abstract. We introduce a variational principle suitable for the computational modeling of crystalline materials. We consider a class of materials that are described by non-quasiconvex variational integrals. We are further focused on equilibria of such materials that have non-attainment structure, i.e., Dirichlet boundary conditions prohibit these variational integrals from attaining their infima. Consequently, the equilibrium is described by probablity distributions. The new variational principle provides the possibility to use standard optimization tools to achieve stochastic equilibrium states starting from given initial deterministic states.

1 INTRODUCTION

Recently, there has been a great deal of research into the modeling and use of so-called Smart Materials. These materials have the ability to undergo internal physical transformations, which may be used to do work in ways and places that traditional engineering materials cannot. Such new materials include composites, ceramics, liquid crystals, biomaterials, ferromagnetics and shape memory alloys. Advances in micro-machines, damping mechanisms, high-resolution displays and superconductors are a few of the applications for which they are being designed.

These materials are singled out by their unique ability to undergo a temperature dependent crystalline phase transformation, known as the Martensitic Transformation. At higher temperatures Shape Memory Alloys exhibit a stiff, cubic crystalline lattice structure known as the Austenitic phase. In lower temperatures, these alloys change to exhibit a more easily deformable tetragonal lattice structure, having many crystallographically equivalent states. The reason for the name comes from the process where, if one begins by establishing a reference configuration in the Austenitic phase, then cools

---

[*] The first author was supported in part by the grant NSF DMS–9971797. The other two authors were supported in part by the grant NSF DMS–0107539, by the Los Alamos National Laboratory Computer Science Institute (LACSI) through LANL contract number 03891–99–23, as part of the prime contract W–7405–ENG-36 between the Department of Energy and the Regents of the University of California, by the grant NASA SECTP NAG5–8136, by the grant from Schlumberger Foundation, and by the grant from TRW Foundation. The computations in this paper were performed on a 16 processor SGI Origin 2000, which was partly funded by the NSF SCREMS grant DMS–9872009.

P.M.A. Sloot et al. (Eds.): ICCS 2002, LNCS 2330, pp. 461–469, 2002.
© Springer-Verlag Berlin Heidelberg 2002

it to the Martensitic phase and deforms the body, the material will resume to the original "remembered" reference shape when the temperature is raised.

At the root of this transformation is the structure Helmholtz free energy for these materials. At high temperatures, this energy has a quadratic structure with respect to the lattice deformation gradient. Thus at the thermodynamically stable minimum, the material will exhibit one stable, undeformed state. At low temperatures, however, the Helmholtz free energy has many minima corresponding to each of the stable Martensitic states (24 in the case of Nickel-Titanium).

Furthermore, various engineering constants such as the specific heat and elastic modulus change depending on which state the material is in. Thus if some of a Shape Memory Alloy is in the Martensitic phase, while the rest is in the Austenitic phase, these constants can vary throughout the body. Hence models using Shape Memory Alloys must encompass the macroscopic scale of the overall system, while also retaining information about the microscopic scale of the crystal lattice.

2   EQUILIBRIUM CONFIGURATIONS OF CRYSTALLINE MATERIALS

Shape memory materials are a notorious example of Crystalline Materials. These alloys develop a striated microcrystalline structure: a mosaic of lamella of coherent compound twins, formed by alternation of phases with symmetry related atomic lattices. Mathematically, the deformation gradients are constrained to attain values from a certain finite set of matrices while subjected to Dirichlet boundary conditions. Their equilibrium properties can thus be described by differential inclusions. Given a set of matrices $\mathcal{A} \subset \mathbb{R}^{m \times n}$, and given a boundary function $g \in W^{1,\infty}(\Omega, \mathbb{R}^m)$, we seek a Lipschitz map u such that

$$\begin{aligned} Du \in \mathcal{A}, &\quad \text{a.e. in } \Omega, \\ u = g, &\quad \text{on } \partial\Omega, \end{aligned} \tag{1}$$

where $\Omega \subset \mathbb{R}^n$, $n = 1, 2, 3$, is an open set, $u : \Omega \mapsto \mathbb{R}^m$, and Du is a matrix of the first derivatives in $\mathbb{R}^{m \times n}$.

The existence of solution(s) to (1) is far from being understood, especially for $m > 2$. We refer to [1] for extensive treatment of this problem. It seems that if $Dg \in$ Interior co-$\mathcal{A}$ then the problem has a dense set of solutions (in the sense of Baire category) while if $Dg \in \partial$ co-$\mathcal{A}$ then the solution can be characterized only in the stochastic sense. If $\text{dist}(Dg, \text{co-}\mathcal{A}) > 0$ then there is no solution. The symbol "co-" stands for various convex hulls of $\mathcal{A}$: quasi-convex, polyconvex, and rank-1 convex hull.

The differential inclusion (1) can be posed as a constrained minimization problem

$$\inf \left\{ \int_\Omega W(Dv(x))\, dx \mid v \in W^{1,\infty}(\Omega, \mathbb{R}^m), v = g, \text{ on } \partial\Omega \right\}, \tag{2}$$

where W is a smooth positive density that vanishes on $SO(n)\mathcal{A}$, $SO(n)$ denotes the set of Simple Orthogonal rotations in $\mathbb{R}^n$.

We address the following problem. Let us assume that $Dg \in \partial$ co-$\mathcal{A}$, and let as assume that this condition guarantees existence of a minimizing sequence $\{v_n\}_{n \in \mathbb{N}}$

of the problem (2) which converges weakly-* to a macroscopic state. The assumption $Dg \in \partial$ co-$\mathcal{A}$ is compatible with lack of lower semicontinuity, i.e. the infimum in (2) is not expected to be attainable. We wish to construct a *stochastic variational principle* which guarantees that standard optmization tools can be used to relax the constrained minimization problem (2).

2.1 LATTICE SYMMETRY PHASE CHANGE

Alloys such as Nickel-Titanium or Indium-Thallium are typical examples of Shape Memory Materials. These alloys exhibit multiscale domain patterns described by gradients of weakly differentiable maps $u : \mathbb{R}^n \to \mathbb{R}^n$ that can come close to the solution of the following *differential inclusion*:

$$
\begin{aligned}
Du &\in \{F_1, F_2\}, \qquad \text{a.e. in } \Omega \subset \mathbb{R}^n, \quad n = 2, 3, \\
F_i &\in M^{n \times n}, \quad i = 1, 2, \\
u(x) &= (\lambda_1 F_1 + \lambda_2 F_2)\, x, \quad x \in \partial\Omega, \\
\lambda_i &> 0, \quad i = 1, 2, \quad \lambda_1 + \lambda_2 = 1, \\
F_2 &= F_1 + a \otimes b, \qquad a, b \in \mathbb{R}^n, \quad (a \otimes b)_{ij} = a_i b_j.
\end{aligned}
\tag{3}
$$

The matrices F_i are assumed to be positive definite and linearly independent. Typical examples of these matrices are associated with crystallographic theories, [3]. We note that there does not exist any functional representation for the solution of (3). This is because it is necessary to create oscillations in the gradients with unlimited frequency to meet the boundary condition between the Austenite and Martensite states.

3   REPRESENTATION OF NON-ATTAINABLE STATES

Typical minimizing sequence of (2) will converge weakly, likely weakly-*, to its expected value, i.e., an average, in an appropriate Sobolev space. Since we do not expect the variational integral in (2) to be weakly lower semicontinous, we assume that

$$
\liminf_{n \to \infty} \int_\Omega W(Du_n(x))\, dx < \int_\Omega W(Du(x))\, dx.
\tag{4}
$$

Here, $\{u_n\}_{n \in \mathbb{N}} \subset W^{1,\infty}(\Omega)$ is a minimizing sequence and $u \in W^{1,\infty}(\Omega)$ is its weak-* limit. Translation of (4) into the framework of material science would imply that the function u representing, e.g., the averaged deformation of a composite or Shape Memory Alloy, does not carry any pointwise information. Yet, a precise description of the spatial composition of deformation gradients is needed to access the elastic properties of these materials, and is also necessary as an input for thermodynamical models [5].

In order to overcome this difficulty we may suppose that for any density $W \in C^0(\Omega)$ which has at least a quadratic growth at infinity there exists a function $\overline{W} \in L^1(\Omega)$ such that

$$
\lim_{n \to \infty} \int_\omega W(Du_n(x))\, dx = \int_\omega \overline{W}(x)\, dx,
\tag{5}
$$

for any ω which is a compact subset of Ω. The function $\overline{W}$ does not remember anything about the equilibrium structure of the material. This is encoded into the density W. Though more importantly, the function $\overline{W}$ does tell us about the pointwise distribution of the elastic energy density W. We can obtain the spatial microstructure of the gradients Du_n for large n by comparing at a given point x the value $\overline{W}(x)$ with $W(s)$ for some $s \in \mathbb{R}^{m \times n}$ belonging to an equlibrium set of W.

The function $\overline{W}$ can be obtained by integral representation using a Radon probablity measure μ_x. Namely,

$$\overline{W}(x) = \int_{\mathbb{R}^{m \times n}} W(y) \, d\mu_x(y). \tag{6}$$

The representation (6) can be derived by a standard Riezs argument [6]. We note that if the minimizing sequence $\{u_n\}_{n \in \mathbb{N}}$ would converge strongly to its weak limit then

$$\mu_x = \delta_{Du(x)}, \qquad \text{for almost all } x \in \Omega, \tag{7}$$

where $\delta_{Du(x)}$ is the Dirac measure assigning 1 to an open subset $\mathcal{D}$ of $\mathbb{R}^{n \times m}$ if $Du(x) \in \mathcal{D}$ and 0 otherwise. This can be seen immediately from the approximation

$$W(Du)(x) \sim \sum_i W(y_i)\delta_{Du(x)}(\Delta y_i) \sim \int_{\mathbb{R}^{m \times n}} W(y) \, d\delta_{Du(x)}(y). \tag{8}$$

The key observation is that the Radon measure μ_x can be approximated. Indeed, we can construct a measure applicable to any Borel subset of the domain Ω by, for almost all x,

$$\mu_{x,R,Du_n}(\mathcal{D}) \stackrel{\text{def}}{=} (\text{meas}(B_R(x)))^{-1} \, \text{meas}\left(\{y \in B_R(x) \mid Du_n(x) \in \mathcal{D}\}\right). \tag{9}$$

It is quite direct to show that [3]

$$\lim_{R \to 0_+} \lim_{n \to \infty} \int_{\mathbb{R}^{n \times m}} f(y)\mu_{x,R,Du_n}(y) = \int_{\mathbb{R}^{n \times m}} f(y) \, d\mu_x(y), \tag{10}$$
$$\text{for any } f \in L^1\left(\mathbb{R}^{n \times m}\right).$$

First since

$$\int_{\mathbb{R}^{n \times m}} f(y)\mu_{x,R,Du_n}(y) \sim \sum_i f(y_i)\mu_{x,R,Du_n}(\Delta y_i), \tag{11}$$

and realizing that

$$\sum_i f(y_i)\mu_{x,R,Du_n}(\Delta y_i) \sim (\text{meas}(B_R(x)))^{-1} \int_{B_R(x)} f(Du_n(y)) \, dy, \tag{12}$$

and defining the averaged measure

$$\mu_{x,R} \stackrel{\text{def}}{=} (\text{meas}(B_R(x)))^{-1} \int_{B_R(x)} \mu_x \, dy, \tag{13}$$

we obtain from (11)-(13), and in view of the weak-$*$ convergence

$$f(Du_n) \rightharpoonup \overline{f} = \int_{\mathbb{R}^{n \times m}} f(y)\mu_x(y),$$ (14)

that

$$\lim_{n \to \infty} \int_{\mathbb{R}^{n \times m}} f(y)\mu_{x,R,Du_n}(y)$$

$$= \lim_{n \to \infty} (\operatorname{meas}(B_R(x)))^{-1} \int_{B_R(x)} f(Du_n(y)) \, dy$$

$$= \lim_{n \to \infty} (\operatorname{meas}(B_R(x)))^{-1} \int_{B_R(x)} \int_{\mathbb{R}^{n \times m}} f(y) \, d\mu_x(y) \, dy$$ (15)

$$= \lim_{n \to \infty} \int_{\mathbb{R}^{n \times m}} f(y)\mu_{x,R}(y) \, dy.$$

It follows from the Lebesque differentiation theorem that

$$\lim_{R \to 0_+} \lim_{n \to \infty} \int_{\mathbb{R}^{n \times m}} f(y)\mu_{x,R}(y) \, dy = \int_{\mathbb{R}^{n \times m}} f(y) \, d\mu_x(y)$$ (16)

which verifies (10).

The Radon measure μ_x has a very natural structure in the case of various alloys undergoing symmetry phase change described in Section 2.1. The structure can be seen quite directly from its approximation (9). It is possible to show that

$$\mu_x = \lambda(x)\delta_{F_1} + (1 - \lambda(x))\delta_{F_2}$$ (17)

in the case (3), [4]. The corresponding smooth and positive energy density W must then vanish on the set

$$\{RF_i^T F_i R^T \mid R \in SO(n)\}.$$ (18)

The function $\lambda = \lambda(x)$ represents the probablity that at a point x the material is deformed with the deformation gradient F_1. Usually, this ratio is called *volume fraction*. It follows directly from the definition (9) that

$$\lambda(x) = \lim_{r \to 0_+} \lim_{R \to 0_+} \lim_{n \to \infty} \frac{\operatorname{meas}(\{y \in B_R(x) \mid \|Du_n(x) - F_1\|\} \le r)}{\operatorname{meas}(B_R(x))}.$$ (19)

For (19) to be valid, the minimizing sequence must become stochastic in its derivative. This represents a major challenge for the optimization, numerical and computational approaches. Desirable methods are those which can generate a stochastic state from the initial deterministic one. It is easy to construct such sequences using self-similar *periodic* construction. It is however impossible to reconstruct such sequences computationally using "off-the-shelf" tools. We address this issue briefly in the next Section 4. We strive to compute the volume fraction using a *stochastic variational principle* that will guarantee that any minimizing sequence becomes asymptotically a *weak white noise*, c.f., Section 5. This seems to be the key selection principle. Such sequences can be obtained computationally despite all the limitations implied by any kind of finite dimensional approximation, which prohibit the convergence otherwise.

4   WEAK WHITE NOISE

We have shown in [2] that the Steepest Descent applied to (2) with $\mathcal{A}$ corresponding to the double-well problem, i. e. $\mathcal{A} = F_1 \cup F_2$, $\text{rank}(F_1 - F_2) = 1$, $n = 2, 3$, does not generate relaxing sequences. In other words, there exists a deterministic limiting state corresponding to a deterministic initial guess and not a limiting stochastic state describing the infimum of the energy. This means in terms of any minimizing sequence $\{u_n\}_{n \in \mathbb{N}}$ that it converges strongly to a function in the norm of some Sobolev space. In this case, the reconstruction procedure (19) for computing the volume fraction cannot provide a reliable approximation.

The structure of the argument used in [2] indicates that the Steepest Descent itself is not at fault. Rather, it is the construction of the minimization problem. The reason is that an energy density that is absolutely stable on $SO(n)\mathcal{A}$ ought to include coupling between the micro- and meso-scales in order to provide a pathway on the microscale local lattice distortion level to connect the absolutely stable states.

In the computational practice, the initial iteration for the Steepest Descent is obtained by a superposition of the averaged state (the weak limit) with superimposed numeric noise. Typically this leads almost immediately to a convergence of the minimizing sequence to the nearest local minimum. The evalutation of the volume fraction based on such sequences using (19) yields results with almost 50% error. We demonstrate this phenomenon in Section 6.

The variational principle we introduce imposes a forcing mechanism prohibiting the minimizing sequences to adhere to any state where possible spatial correlation can occur. This mechanism prohibits the minimizing sequence to converge to any of the many (possibly countably many) local minima of the variational integral (2).

The following Definition establishes how we interpret the notion of weak white noise in the discussed framework.

Definition 41 *We say that a sequence* $\{u_n\}_{n \in \mathbb{N}} \subset W^{1,\infty}(\Omega, \mathbb{R}^m)$*,* $\Omega \subset \mathbb{R}^n$*,* $n = 1, 2, 3$*,* $m = 1, 2, 3$*, that converges weakly-* in $W^{1,\infty}(\Omega, \mathbb{R}^m)$ to a Lipschitz continuous function g, becomes asymptotically weak white noise if the following three conditions hold true. Let*

$$z_n(x) \overset{def}{=} Du_n(x) - Dg(x) \in \mathbb{R}^{m \times n}. \tag{20}$$

The first condition is

$$E_\omega[z_n(x)] = \frac{1}{\text{meas}(\omega)} \int_\omega z_n(x)\, dx \to 0, \qquad \text{for all open } \omega \subset \Omega, \text{ as } n \to \infty.$$

The second and the third conditions are that

for almost all $\tau \in \mathbb{R}$ *the covariance operator* E *has the following two properties*

$$E[z_n(x) \otimes z_n(x + \tau), \omega] = \frac{1}{\text{meas}(\omega)} \int_\omega z_n(x) \otimes z_n(x + \tau)\, dx\sigma^{-1}(\omega)$$

$$\to \begin{cases} I \in \mathbb{R}^{m^2 \times n^2}, & \text{if } \tau = 0, \\ 0 \in \mathbb{R}^{m^2 \times n^2}, & \text{if } \tau \neq 0, \end{cases} \qquad \text{for all open } \omega \subset \Omega, \text{ as } n \to \infty, \tag{21}$$

where σ represents "standard deviation" given by

$$\sigma(\omega) \stackrel{\text{def}}{=} E[z_n \otimes z_n, \omega]^{1/2}. \tag{22}$$

We assume that the functions Du_n are extended periodically onto $\mathbb{R}^{m \times n}$ for $u_n(x + \tau)$ to be defined where $x + \tau \notin \Omega$. We note that E is non-negative.

5 ONE DIMENSIONAL STOCHASTIC VARIATIONAL PRINCIPLE

We define the Principle in one spatial dimension in order to remove unnecessary technical issues. We assume that the density $W = W(u')$ in (2) should be written as

$$W(x, s) \stackrel{\text{def}}{=} W_{\text{meso}}(s) + W_{\text{micro}}(s), \qquad s \in \mathbb{R}^1. \tag{23}$$

The contribution W_{meso} encodes the information about the equlibrium state of a given material and the W_{micro} contribution guarantees that any minimizing sequence becomes *weak white noise* in the sense of Definition 41.

In addition to the coercitivity of the variational integral in (2), we also assume that

$$W_{\text{meso}}(u') \geq \Lambda \operatorname{dist} \{u', \{\pm 1\}\}, \qquad \Lambda > 0. \tag{24}$$

This is enough to guarantee that any minimizing sequence will convenverge macroscopically to a function g, i.e.,

$$u_n \rightharpoonup g \qquad \text{weakly in e.g. } W^{1,2}((0,1)). \tag{25}$$

In order to obtain a computationally feasible problem, we insist that among all (uncountably many) possible minimizing sequences having the propoerty (25), the ones which become asymptotically weak white noise are computationally desirable. Our approach consists in the introduction of a $W_{\text{micro}}(x, s)$ term which can be relaxed only by such sequences. We demonstrate a possible construction on a one dimensional problem in the framework of finite element approximation. Let us assume that u_h is an element of some finite element space defined on a regular mesh with size h. Then we define

$$W_{\text{micro}}(u_h')(x) \stackrel{\text{def}}{=} \frac{1}{N_h} \sum_{k=0}^{N_h/2} \left(z_h^k(x) \bar{z}_h^k(x) - 1 \right)^2, \qquad \text{where}$$

$$z_h^k(x) \stackrel{\text{def}}{=} \frac{1}{2\pi} \int_{-\pi}^{\pi} \frac{u_h'(x) - g_h'(x)}{\sqrt{1 - g_h'^2(x)}} \exp(-ikx) \, dx. \tag{26}$$

Here g_h represents the projection of the weak limit g in the given finite element space. We evaluate the coefficients z_h^k using the fast Fourier Tranform. Let as set $h = 1/n$. We have the following Theorem.

Theorem 51 *There exists asymptotically weak white noises sequences $u_{1/n}'$.*

We give a sketch of the proof. We construct independent random variables Y_{1n}, $Y_{2n}, \ldots Y_{nn}$ taking value ± 1 with $P[Y_{kn} = 1] = (1 + g'(x))/2$ and $P[Y_{kn} = -1] = (1 - g'(x))/2$. Let

$$u'_{1/n}(x) = \sum_{i=1}^{n} Y_i \phi_{[(i-1)/n, i/n)]}(x), \tag{27}$$

where $\phi_A(x)$ denotes the characteristic function of the set A, i.e. $\phi_A(x) = 1$ if $x \in A$ and $\phi_A(x) = 0$ if $x \notin A$. Standard probabilistic arguments are used to show that this is an asymptotically weak white noise with probability one, hence that there exist sequences satisfying the definition. From this discussion we see immediately the following

Theorem 52 *Any asymptotically weak white noise sequence* $u'_{1/n}$ *satisfies*

$$W_{micro}(u'_{1/n}) \to 0 \text{ as } n \to \infty. \tag{28}$$

We conjecture that the converse is also true – i.e. that if $W_{\mathrm{micro}}(u'_{1/n}) \to 0$ then $u'_{1/n}$ is an asymptotically weak white noise sequence in the sense of Definition 41.

6   ONE DIMENSIONAL COMPUTATIONAL EXAMPLE

We present in this section a computational example in which we approximate a solution of the following differential inclusion

$$\begin{aligned}
u'(x) &\in \{-1, 1\}, \qquad \text{a.e. in } (0, 1), \\
u(x) &= 0, \qquad \text{for all } x \in (0, 1).
\end{aligned} \tag{29}$$

We construct the approximate solutions as minimizing sequences $\{u_n\}_{n \in \mathbb{N}} \subset W^{1,4}(0, 1)$ generated by the Steepest Descent minimization of the functional

$$\begin{aligned}
&\mathcal{J}(u_h, h^{3/2}) \\
&\stackrel{\text{def}}{=} \frac{1}{2} \int_0^1 |u_h(x)|^2 \, dx + \frac{1}{4} \int_0^1 \left| u_h'(x)^2 - 1 \right|^2 \, dx + h^{3/2} \int_0^1 W_{\mathrm{micro}}(u_h') \, dx,
\end{aligned} \tag{30}$$

where the density W_{micro} is defined by (26). We note that the differential inclusion (29) is the one dimensional analog of (1) with $\mathcal{A} = \{-1, 1\}$.

The calculations described in the caption to Figure 1 show clear advantage in using the stochastic term in getting the appropriate relaxing sequences.

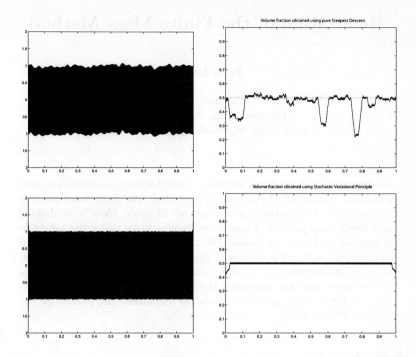

Fig. 1. These calculations are done on the regular mesh with $h = 1/2048$. The upper two pictures show the derivatives (left) and the volume fraction $\lambda = \lambda(x)$ (right) for the sequence obtained by the Steepest descent algorithm applied to (30) without the contribution of the term $h^{3/2} \int_0^1 W_{\text{micro}}(u_h') \, dx$. We note that $\lambda(x) = 1/2$ for any $x \in (0, 1)$ for the Differential inclusion (29). The lower two pictures show the same quantities obtained by the Steepest descent method applied to relaxation of the functional (30).

REFERENCES

1. B. Dacorogna and P. Marcellini, *Implicit partial differential equations*, Birkhäuser, 2000.
2. P. Kloucek, *The steepest descent minimization of double-well stored energies does not yield vectorial microstructures*, Rice University (2001), Technical report 01-04, Department of Computational and Applied Mathematics.
3. Luskin M., *On the computation of crystalline microstructure*, Acta Numerica (1996), 191–257.
4. P. Pedregal, *Variational methods in nonlinear elasticity*, SIAM, Philadelphia, 2000.
5. D. Reynolds, *Vibration damping using martenstic phase transformation*, (2002), Ph.D. Thesis, Rice University.
6. W. P. Ziemer, *Weakly differentiable functions*, Graduate Texts in Math., Springer-Verlag, New York, 1989.

Realization of the Finite Mass Method

Peter Leinen

Mathematisches Institut, Universität Tübingen
leinen@na.uni-tuebingen.de,
http://na.uni-tuebingen.de/~peter

Abstract. The finite mass method, a new Lagrangian method for the numerical simulation of gas flow, is presented. The finite mass method is founded on a discretization of mass, not of space. Mass is subdivided into small mass packets. These mass packets move under the influence of internal and external forces. The right-hand sides of the differential equations governing the motion of the particles are integrals which cannot be evaluated exactly. A Lagrangian discretization of these integrals will be presented that maintains the invariance and conservation properties of the method. An efficient way to implement and parallelize the method is discussed.

1 Introduction

The finite mass method is a new approach to solve the equation of gas dynamics. Contrary to the usual finite volume or finite element methods, the finite mass method is based on a discretization of mass, not of space. Mass is subdivided into small mass packets of finite extension each of which is equipped with finitely many internal degrees of freedom. These mass packets move under the influence of internal and external forces and the laws of thermodynamics and can undergo arbitrary linear deformations.

Numerical methods based on the Lagrangian view of fluid flow like the finite mass method are well-suited for free flows in unbounded space as in astrophysical problems. One of the most popular methods of this type, Monaghan's smoothed particle hydrodynamics (SPH) [4], come from this research area.

The finite mass method is based on the particle model of compressible fluids which has been developed by Yserentant ([5–7]) and is generalized in [1].

Particles do not directly interact by each other but only by global quantities like the mass density. This property is reflected in the data structure. A efficient implementation of the method both on serial and parallel computer architectures is based on a careful analysis of the data flow.

The rest of the paper is organized as follows. In Section 2 the finite mass method is described briefly. It is shown how the particles are built and the mass density is discretized. Derivation of the equation of motion utilizing a Lagrange-function is done in Section 3. In Section 4 we discuss data structures and the data flow in a realization of the method. A parallel implementation for shared and distributed memory architectures is presented.

P.M.A. Sloot et al. (Eds.): ICCS 2002, LNCS 2330, pp. 470–479, 2002.

2   Mass Packets

2.1   Local quantities

The internal mass distribution of a mass packet is described by a continuously differentiable shape function $\psi : \mathbb{R}^d \to \mathbb{R}_+$, d the space dimension, with compact support. We assume that

$$\int \psi(\mathbf{y})\,\mathrm{d}\mathbf{y} = 1\,, \quad \int \psi(\mathbf{y})\mathbf{y}\,\mathrm{d}\mathbf{y} = \mathbf{0}\,, \quad \int \psi(\mathbf{y})\,y_k y_l\,\mathrm{d}\mathbf{y} = J\,\delta_{kl}\,, \quad (1)$$

with the y_k the components of $\mathbf{y}$. The second condition states that the origin of the body coordinates is the center of mass and the third one normalizes the deformation. A suitable choice for the shape function ψ is, for example, the tensor product

$$\psi(\mathbf{y}) = \prod_{k=1}^{d} \widetilde{\psi}(y_k)\,, \quad (2)$$

of the normalized third order B-spline given by

$$\widetilde{\psi}(\xi) = \frac{4}{3} \begin{cases} 1 - 6\xi^2(1 - |\xi|)\,, & 0 \le |\xi| \le 1/2 \\ 2\,(1 - |\xi|)^3 & , \quad 1/2 \le |\xi| \le 1 \\ 0 & , \quad 1 \le |\xi| \end{cases}\,. \quad (3)$$

We allow for linear deformation of the particles. That is the points $\mathbf{y}$ of the particle i move along the trajectories

$$t \;\to\; \mathbf{q}_i(t) + \mathbf{H}_i(t)\mathbf{y}\,, \quad \det \mathbf{H}_i(t) > 0\,. \quad (4)$$

The vector $\mathbf{q}_i(t)$ is the position of the particle and the matrix $\mathbf{H}_i(t)$ determines the linear deformation of the given particle. For a given point $\mathbf{x}$ in space coordinates

$$\mathbf{y} = \mathbf{H}_i(t)^{-1}(\mathbf{x} - \mathbf{q}_i(t)) \quad (5)$$

are the body coordinates at time t.

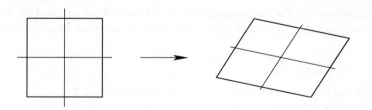

Fig. 1. Transformation of a particle by $\mathbf{q}_i(t) + \mathbf{H}_i(t)\mathbf{y}$

The points $\mathbf{y}$ of the particle i move with the velocity

$$t \;\to\; \mathbf{q}_i'(t) + \mathbf{H}_i'(t)\mathbf{y}\,. \quad (6)$$

With (5) we obtain the velocity field

$$\mathbf{v}_i(\mathbf{x}, t) = \mathbf{q}_i'(t) + \mathbf{H}_i'(t)\mathbf{H}_i(t)^{-1}(\mathbf{x} - \mathbf{q}_i(t)) \tag{7}$$

of the particle i in space coordinates. The masses of the particles can be distinct and are denoted by $m_i > 0$. Of course the m_i are always positive.

2.2   Global quantities

The superposition of the mass density of the single particles

$$\rho(\mathbf{x}, t) = \sum_{i=1}^{N} m_i \left[\det \mathbf{H}_i(t)\right]^{-1} \psi\big(\mathbf{H}_i(t)^{-1}(\mathbf{x} - \mathbf{q}_i(t))\big) \tag{8}$$

is the density of the total mass. With the abbreviation

$$\psi_i(\mathbf{x}, t) = \left[\det \mathbf{H}_i(t)\right]^{-1} \psi\big(\mathbf{H}_i(t)^{-1}(\mathbf{x} - \mathbf{q}_i(t))\big) . \tag{9}$$

this reads as

$$\rho(\mathbf{x}, t) = \sum_{i=1}^{N} m_i \, \psi_i(\mathbf{x}, t) . \tag{10}$$

In a similar way we get the total mass flux density by

$$\mathbf{j}(\mathbf{x}, t) = \sum_{i=1}^{N} m_i \, \psi_i(\mathbf{x}, t) \, \mathbf{v}_i(\mathbf{x}, t) . \tag{11}$$

The total velocity field $\mathbf{v}$ is defined by the relation

$$\mathbf{j}(\mathbf{x}, t) = \rho(\mathbf{x}, t)\mathbf{v}(\mathbf{x}, t) . \tag{12}$$

Beside the mass density we need a second thermodynamic quantity like temperature or entropy to describe the state of a compressible fluid completely. As this is not important for the implementation of the finite mass method we will focus on internal energy ε, which is only a function of the mass density. Ideal gases with

$$\varepsilon(\rho) \sim \Big(\frac{\rho}{\rho_0}\Big)^{\gamma} , \tag{13}$$

represent the most simple example. For air we have $\gamma = 1.4$

2.3   Equation of motion

The total internal energy of the system is

$$V = \int \varepsilon(\rho(\mathbf{x}, t)) \, d\mathbf{x} . \tag{14}$$

The particle i has the kinetic energy

$$E_i(t) = \frac{1}{2} \int m_i \, \psi_i(\mathbf{x}, t) |\mathbf{v}_i(\mathbf{x}, t)|^2 \, \mathrm{d}\mathbf{x} \, ; \tag{15}$$

the total kinetic energy of the system is

$$E(t) = \sum_{i=1}^{N} E_i(t) \, . \tag{16}$$

The equations of motion

$$\frac{\mathrm{d}}{\mathrm{d}t} \frac{\partial \mathcal{L}}{\partial \mathbf{q}_i'} - \frac{\partial \mathcal{L}}{\partial \mathbf{q}_i} = \mathbf{0} \, , \quad \frac{\mathrm{d}}{\mathrm{d}t} \frac{\partial \mathcal{L}}{\partial \mathbf{H}_i'} - \frac{\partial \mathcal{L}}{\partial \mathbf{H}_i} = \mathbf{0} \tag{17}$$

are derived from the Lagrange-function

$$\mathcal{L} = E - V \, , \tag{18}$$

which is a function of $\mathbf{q}_i$, $\mathbf{H}_i$, $\mathbf{q}_i'$ and $\mathbf{H}_i'$. With the normalized forces

$$\mathbf{F}_i = -\frac{1}{m_i} \frac{\partial V}{\partial \mathbf{q}_i} \, , \quad \mathbf{M}_i = -\frac{1}{m_i} \frac{\partial V}{\partial \mathbf{H}_i} \tag{19}$$

the system of ordinary differential equations read

$$\mathbf{q}_i'' = \mathbf{F}_i \, , \quad \mathbf{H}_i'' = \frac{1}{J} \mathbf{M}_i \, . \tag{20}$$

Additional frictional forces are necessary to model shocks correctly. These forces transfer kinetic energy into heat, see [1] for a detailed discussion. These frictional forces can be implemented in the same framework as the pressure forces.

3 Discretization of integrals

The internal energy (14) and therefore the forces are integrals and cannot not be evaluated exactly. Therefore we have to apply a quadrature rule to compute the forces. As the particle model is invariant to translations and rotations, the quadrature rule should preserve these properties. We start from the identity

$$\int f(x)\rho(x) \, \mathrm{d}\mathbf{x} = \sum_{j=1}^{N} m_j \int f(\mathbf{x})\psi_j(\mathbf{x}) \, \mathrm{d}\mathbf{x} \tag{21}$$

$$= \sum_{j=1}^{N} m_j \int f(\mathbf{q}_j + \mathbf{H}_j \mathbf{y})\psi(\mathbf{y}) \, \mathrm{d}\mathbf{y} \, . \tag{22}$$

The integrals on the right-hand side are replaced by a fixed quadrature formula

$$\int g(\mathbf{y})\psi(\mathbf{y})\,d\mathbf{y} \;\rightarrow\; \sum_{\nu=1}^{n}\omega_\nu g(\mathbf{a}_\nu) \tag{23}$$

with weights $\omega_\nu > 0$ and nodes $\mathbf{a}_\nu$ inside the support of the shape function ψ. The overall result is a quadrature rule

$$\int f\rho\,d\mathbf{x} \;\rightarrow\; \sum_{j=1}^{N} m_j\Big[\sum_{\nu=1}^{n}\omega_\nu f(\mathbf{q}_j+\mathbf{H}_j\mathbf{a}_\nu)\Big] =: \int f\,d\mu \tag{24}$$

with weights $m_j\omega_\nu$ and nodes $\mathbf{q}_j+\mathbf{H}_j\mathbf{a}_\nu$. The quadrature rule is based on the given discretization of the mass density and not on a subdivision of space into cells. Figure 2 shows the quadrature rule on the reference configuration and the quadrature points generated by two particles.

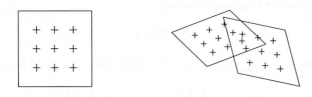

Fig. 2. Quadrature rule

Using the quadrature rule (24) the potential energy (14) is replaced by the fully discrete version

$$V \;=\; \int \tilde{\varepsilon}(\rho)\,d\mu \;, \tag{25}$$

with $\tilde{\varepsilon} = \varepsilon/\rho$. As both $\tilde{\varepsilon}$ and the quadrature points depend on the $\mathbf{q}_i$ and $\mathbf{H}_i$, the forces split into two parts

$$\mathbf{F}_i = \mathbf{F}_i^{(1)}+\mathbf{F}_i^{(2)}, \quad \mathbf{M}_i = \mathbf{M}_i^{(1)}+\mathbf{M}_i^{(2)} \tag{26}$$

of different structure. Into

$$\mathbf{F}_i^{(1)} = -\int \tilde{\varepsilon}'(\rho)\frac{\partial\psi_i}{\partial\mathbf{q}_i}\,d\mu \;, \tag{27}$$

$$\mathbf{M}_i^{(1)} = -\int \tilde{\varepsilon}'(\rho)\frac{\partial\psi_i}{\partial\mathbf{H}_i}\,d\mu \tag{28}$$

all quadrature points $\mathbf{q}_j+\mathbf{H}_j\mathbf{a}_\nu$ contained in the support of the given particle i enter whereas

$$\mathbf{F}_i^{(2)} = -\sum_{\nu=1}^{n}\omega_\nu\,(\nabla\tilde{\varepsilon})(\mathbf{q}_i+\mathbf{H}_i\mathbf{a}_\nu) \tag{29}$$

$$\mathbf{M}_i^{(2)} = -\sum_{\nu=1}^{n}\omega_\nu\,[\,(\nabla\tilde{\varepsilon})(\mathbf{q}_i+\mathbf{H}_i\mathbf{a}_\nu)\,][\,\mathbf{a}_\nu\,]^{\mathrm{T}} \tag{30}$$

depend only on values at the quadrature points $\mathbf{q}_i + \mathbf{H}_i \mathbf{a}_\nu$ assigned to the particle i itself.

These kind of discretization of the internal energy guarantees conservation of energy, momentum, and angular momentum. See [1] for a proof of this property.

4   Realization

The key observation is that the particles do not directly interact by each other but only by global fields like the mass density, for example. This is reflected in our choice of the data structures. We clearly separate the quadrature points in which global fields are stored from the primary particle information.

4.1   Data structures

Our first data structure are the particles. Storage is provided for the masses m_i, the values $\mathbf{q}_i$, $\mathbf{H}_i$, $\mathbf{q}_i'$ and $\mathbf{H}_i'$ and the forces $\mathbf{F}_i$ and $\mathbf{M}_i$. The second data structure are the quadrature points. A quadrature point contains information on its position $\mathbf{x}_Q = \mathbf{q}_i + \mathbf{H}_i \mathbf{a}_\nu$, the associated weights $\omega_Q = m_i \omega_\nu$ and of all relevant field quantities like the mass density $\rho_Q = \rho(\mathbf{x}_Q)$ and $\tilde{\varepsilon}_Q' = \tilde{\varepsilon}'(\rho_Q)$.

4.2   Algorithm

The computation of the forces $\mathbf{F}_i^{(1)}$ and $\mathbf{M}_i^{(1)}$ for a given distribution of the particles splits into four steps.

In an initializing step the position $\mathbf{x}_Q = \mathbf{q}_i + \mathbf{H}_i \mathbf{a}_\nu$ of the quadrature points. and the weights $\omega_Q = m_i \omega_\nu$ are generated. In this step information is transferred from each particle to the quadrature points associated with this particle.

The next step is to compute the values ρ_Q of the mass density at all quadrature points. Information is transferred from a particle i to all quadrature points in the support of ψ_i. This step is realized as a loop over all particles.

In the third step the values $\tilde{\varepsilon}' = \tilde{\varepsilon}'(\rho_Q)$ are computed in a loop over all quadrature points. By this a multiple evaluation of $\tilde{\varepsilon}_Q'$ is avoided.

In the fourth step the forces $\mathbf{F}_i^{(1)}$ and $\mathbf{M}_i^{(1)}$ are finally computed using the results from the steps above. This step is again realized as a loop over all particles. Information is transferred from the quadrature points in the support of ψ_i back to the particle i. Table 1 summarize the four steps.

4.3   Search tree

Steps 2 and 4 crucially depend on an efficient algorithm to find all quadrature points in the support of a given particle. We use adaptive search trees for this purpose.

The root of the tree is a rectangular box containing all quadrature points. A leaf of the tree is further refined if it contains too many quadrature points.

step	operation	data flow
1	$\forall$ particles i $\quad \forall$ quadrature points Q of particle i $\qquad \mathbf{x}_Q \ \leftarrow \ \mathbf{q}_i + \mathbf{H}_i \mathbf{a}_\nu$	$Q(P) \leftarrow P$
2	$\forall$ particles i $\quad \forall$ quadrature points Q at which $\psi_i > 0$ $\qquad \rho_Q \ \leftarrow \ \rho_Q + m_i \psi_i(x_Q)$	$Q \leftarrow P$
3	$\forall$ quadrature points $\qquad \bar{\varepsilon}'_Q \ \leftarrow \ \bar{\varepsilon}'(\rho_Q)$	$Q \leftarrow Q$
4	$\forall$ particles i $\quad \forall$ quadrature points at which $\psi_i > 0$ $\qquad \mathbf{F}_i \ \leftarrow \ \mathbf{F}_i \ - \omega_Q \, \bar{\varepsilon}'_Q \, \dfrac{\partial \psi_i}{\partial \mathbf{q}_i}$ $\qquad \mathbf{M}_i \ \leftarrow \ \mathbf{M}_i \ - \omega_Q \, \bar{\varepsilon}'_Q \, \dfrac{\partial \psi_i}{\partial \mathbf{H}_i}$	$P \leftarrow Q$

Table 1. Computation of the forces

All leaves having a non-empty intersection with a particle establish a superset of the set of quadrature points needed in steps 2 and 4.

To keep the computing time and the amount of memory to store the trees as small as possible, it is important to reorder the quadrature points such that every node of the tree contains only consecutive points. The address of the first and the last of these points is attached to the nodes. This allows to access the quadrature points contained in the domain represented by the node in a very efficient way. Reordering of the quadrature points also improves the data locality in the above algorithm. The nodes of this tree are sorted by means of a Hilbert-like space-filling curve [2, 3, 8], which again helps to improve the data locality. Figure 3 shows the different levels of the adaptive search tree for a given distribution of points and the corresponding space-filling curve. We have allowed at most 5 points in one cell.

5   Parallelization

For ease of presentation here we restrict ourselves to a machine with only two processors. The generalization to the multiprocessor case will be obvious.

The data flow of the algorithm above is the starting point for its parallelization. For the rest of the section we split the set of the particles into two disjoint subsets **P1** and **P2**. By **Q1** and **Q2** we denote the sets of quadrature points associated with the particles in **P1** and **P2**, respectively. The computation of the forces for this situation is shown in Table 2. The interaction between **P1** and **Q2**, for example, is due to the fact that there may be quadrature points from the set **Q2** in the support of particles from **P1**. In a parallel implementation we have to pay attention to this kind of interaction.

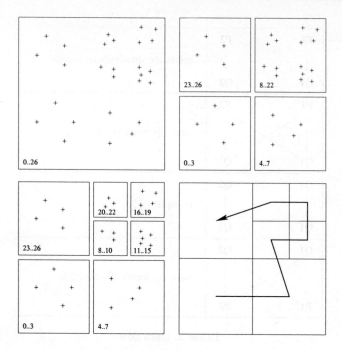

Fig. 3. Adaptive search tree and the space filling curve

5.1   Shared memory implementation

First we will discuss how to get a parallel algorithm for a shared memory architecture. The particle sets **P1** and **P2** and correspondingly the sets **Q1** and **Q2** are now associated with processors 1 and 2. Each processor is allowed to read from all data sets but is only allowed to write to its own data sets. Especially step 2 and step 4 have to be organized in this way. The initialization of the quadrature points as well as the third step are easy to realize in this setting.

The algorithm from Table 2 is organized as shown in Table 3. After the first three steps data have to be synchronized to guarantee that in step 4 the right filed values in the quadrature points are accessed. That is step 4 is not allowed to start before all operations in step 3 are finished.

5.2   Distributed memory implementation

The realization on a distributed memory machine is easy when the particle information is redundantly hold on both processors. The realization of the first three steps then does not differ from the shared memory case because in these three steps every processor accesses only data stored in its own memory. No synchronization is needed after step 3. The computation of the forces (step 4) for the distributed memory case is split into a computation and a communication phase. The computational step ends with information in all particles due to one

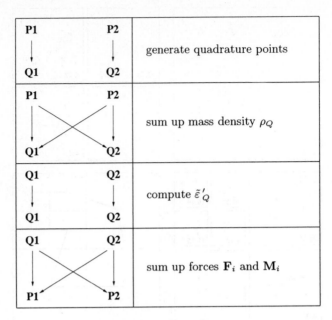

Table 2. Data flow

processor 1			processor 2	
P1 **P2**		generate quadrature points	**P1** **P2**	
Q1 **Q2**			**Q1** **Q2**	
P1 **P2**		compute ρ_Q	**P1** **P2**	
Q1 **Q2**			**Q1** **Q2**	
Q1 **Q2**		compute $\tilde{\varepsilon}'_Q$	**Q1** **Q2**	
Q1 **Q2**			**Q1** **Q2**	
synchronization				
Q1 **Q2**		compute forces	**Q1** **Q2**	
P1 **P2**			**P1** **P2**	

Table 3. Shared memory implementation

set of quadrature points. We obtain the forces if we sum up these values. In Table 4 the implementation of step 4 for the distributed memory case is shown.

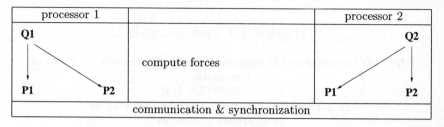

processor 1		processor 2
Q1		Q2
	compute forces	
P1 P2		P1 P2
communication & synchronization		

Table 4. Distributed memory implementation of step 4

For a moderate number of particles this very simple approach is satisfactory since the amount of data stored for the particles is small compared with the data for the quadrature points. In fact we need only copies of those particles which are in the neighborhood of the particles associated with the processor. How to distribute the particles on the processors in a clever and cheap way and how to get information on the copies which are needed, will be discussed in a forthcoming paper.

References

1. Gauger, Chr., Leinen, P., Yserentant, H.: The finite mass method. SIAM J. Numer. Anal., **37** (2000), 1768–1799
2. Hilbert, D.: Über die stetige Abbildung einer Linie auf ein Flächenstück. Math. Annalen, **38** (1891), 459-460.
3. Sagan, H.: Space-filling curves. Springer, 1994
4. Monaghan, J.J.: Smoothed particle hydrodynamics. Ann. Rev. Astron. Astrophys., **30** (1992), 543–574.
5. Yserentant, H.: A particle model of compressible fluids. Numer. Math., **76** (1997), 111–142.
6. Yserentant, H.: Particles of variable size. Numer. Math., **82** (1999), 143–159.
7. Yserentant, H.: Entropy generation and shock resolution in the particle model of compressible fluids. Numer. Math., **82** (1999), 161–177.
8. Zumbusch, G.: On the quality of space-filling curve induced partitions. Z. Angew. Math. Mech., **81**, (2001), 25–28

Domain Decomposition Using a 2-Level Correction Scheme

R.H. Marsden, T.N. Croft, and C.-H. Lai

School of Computing and Mathematical Sciences, University of Greenwich,
Greenwich,
London SE10 9LS, U.K.
{R.H.Marsden, T.N.Croft, C.H.Lai}@gre.ac.uk
http://cms1.gre.ac.uk

Abstract. The PHYSICA software was developed to enable multiphysics modelling allowing for interaction between Computational Fluid Dynamics (CFD) and Computational Solid Mechanics (CSM) and Computational Aeroacoustics (CAA). PHYSICA uses the finite volume method with 3-D unstructured meshes to enable the modelling of complex geometries. Many engineering applications involve significant computational time which needs to be reduced by means of a faster solution method or parallel and high performance algorithms. It is well known that multigrid methods serve as a fast iterative scheme for linear and nonlinear diffusion problems. This papers attempts to address two major issues of this iterative solver, including parallelisation of multigrid methods and their applications to time dependent multiscale problems.

1 Introduction

The PHYSCIA software [6][15] was developed to enable multiphysics modelling allowing for interaction between Computational Fluid Dynamics (CFD) and Computational Solid Mechanics (CSM) and Computational Aeroacoustics (CAA). PHYSICA uses the finite volume method with 3-D unstructured meshes to enable the modelling of complex geometries. Many engineering applications involve significant computational time which needs to be reduced by means of a faster solution method or parallel and high performance algorithms.

It is well known that multigrid methods serve as a fast iterative scheme for linear and nonlinear diffusion problems. There are two major issues in this fast iterative solver. First, multigrid methods are usually very difficult to parallelise and the performance of the resulting algorithms are machine dependent. Early work in parallelisation of multigrid methods include Barkai and Brandt [2], Chan [5], Frederickson [10], Naik [14], etc. Methods developed by these authors concerned the load balancing between processors and the full use of all co-existing coarse level meshes in order to fit into the parallelism requirement. This paper intends to address these issues with particular attention being paid to linear and nonlinear diffusion type of problems in a distributed computing environment. The method is then extended to time-dependent and multiscale problems.

P.M.A. Sloot et al. (Eds.): ICCS 2002, LNCS 2330, pp. 480–489, 2002.

2   Classical Multigrid Methods for Linear Diffusion Problems

The first article which describes a 'true' multigrid technique was probably published by Fedorenko [9]. He formulated a multigrid algorithm for the standard five-point finite difference discretisation of the Poisson equation on a unit square and showed the computational work is of $O(N)$, where N is the number of unknowns. An extension of the concept to a general linear elliptic problem defined in a unit square was given by Bachvalov [1]. Results shown in these articles were not optimal and the method have not been adopted at that time. Practical multigrid algorithms were first published by Brandt [3] in 1973 and, later, in a revised article [4]. There are many excellent review articles and introductory lectures, for examples [17], [16]. The authors do not intend to produce an exhaust list, but related work can be easily found.

2.1   Defect Correction Principle

Suppose one wish to solve the elliptic problem as given below,

$$Lu = f \in \Omega \tag{1}$$

where L is a linear elliptic problem with suitable boundary conditions defined on $\partial\Omega$. The problem is usually discretised by means of a finite volume method, which leads to the discretised system of linear equations,

$$L_h u_h = f_h \in \Omega_h \tag{2}$$

where subscript h denotes a discretised approximation to the corresponding quantity. Here h denotes a typical mesh size being used in the discretisation. Usually smaller h leads to a larger system of linear equations and an iterative method may be used to solve the system. Suppose u_h^* is an approximation obtained by means of an iterative method to the linear system (2), it is possible to compute the defect or residual due to the approximation u_h^* as

$$R_h = f_h - L_h u_h^* \tag{3}$$

For linear operator such as L_h, one obtains the defect equation $L_h(u_h - u_h^*) \equiv L_h u_h - L_h u_h^* = f_h - L_h u_h^* \equiv R_h$, and an iterative scheme may be used to obtain an iterative refinement or a correction v_h, which may be used to correct the approximation u_h^* and compute a better approximation as $u_h^{**} = u_h^* + v_h$. However, such iterative refinement technique does not improve the convergence rate of the iterative method used in the numerical solution process. In essence the convergence rate deteriorates as h decreases.

One purpose of using the concept of multigrid is to avoid the deterioration of the convergence rate. A Fourier smoothing analysis was first introduced by Brandt [4], which explains the role of an iterative method in the context of the above defect correction principle. The smoothing analysis uses the amplification factor, $g(\lambda_\alpha)$, which involves the ratio of the error at the n-th iterative approximation to the error at the $(n-1)$-th iterative approximation, in measuring the growth or decay of a Fourier error mode during an iteration. Here α is the Fourier mode and $\lambda_\alpha = \pi\alpha/m$ is the wave number, where m is the typical number of grid points along a characteristic length of Ω_h and $\alpha = -m+1, -m+2, \ldots, m$. When α increases the wave number, λ_α, increases and the amplication factor, $g(\lambda_\alpha)$, decreases. In other words Fourier error modes with long wavelengths (α close to 1) decay slowly and with short wavelengths decay rapidly. The smoothing factor can now be easily defined as

$$\rho = \max\{|g(\lambda_\alpha)| : \lambda_\alpha = \pi\alpha/m , \qquad \alpha = -m+1, -m+2, \ldots, m\} \qquad (4)$$

and $\rho < 1$ denotes that the iterative method is a smoother, i.e. all Fourier error modes of short wave lengths have been damped out. The remaining part consists of a smoothed part, which is of long wave lengths, should be dealt with at a coarser mesh. This is because long wave lengths on a mesh of size h become relatively shorter wave lengths on a mesh of size $2h$.

Therefore based on Brandt's smoothing analysis, one would like to handle Fourier error modes of long wave lengths on a coarser mesh, say H. The above iterative refinement can then be implemented as the following 2-level algorithm.

Iterate to obtain an approximation u_h^*: $L_h u_h = f_h$;
Compute defect: $R_h := f_h - L_h u_h^*$;
$u_H := I_H^h u_h$; $f_H := I_H^h R_h$;
Solve u_H^*: $L_H u_H = f_H \in \Omega_H$ with suitable boundary conditions;
Apply correction: $u_h := u_h^* + I_h^H u_H^*$;

Usually H is taken as $2h$ and there exists a vast amount of literature in 2-level algorithms. These algorithms are being used recursively to form a multigrid algorithm.

2.2   Parallelisation Issues of Multigrid Methods

It is obvious that as the mesh size is doubled, the number of discrete unknowns decreased by a half. This leads to dummy computation when multigrid methods were implemented on SIMD or vector machines such as the DAP [11], Connection Machines and the CDC Cyber 205 [2]. Such dummy computation cannot be avoided if dynamic data structure is not allowed in the programming language, such as FORTRAN. An attempt to examine a possible parallel network to circumvent the disadvantage was presented by Chan et al [5]. However only theoretical performance analysis was given. The idea of using co-existing coarse grids, such that the union of these coarse grids automatically forms the next finer grid, was also discussed in [10]. Similar concept of using coarse grid correction in conjunction with the finest grid on a unigrid was also examined by McCormick and Ruge [13], and the method was shown to be equivalent to multigrid methods.

Many articles and research reports have been devoted to the implementation of multigrid methods, such as [14] to name just one reference, for a distributed memory machine. The main techniques involve evenly distributing computational load and minimising communication costs. These techniques amount to a data partitioning of the finest level problem being distributed evenly across the processors in the computational system. The finest level problem must be pre-defined making adaptivity extremely difficult in parallel processing.

In the case of using co-existing coarse grids, which may be solved concurrently, may lead to heavy data traffic because these co-existing grids involves mesh points located further away. On the other hand, averaging procedure of all the coarse grids leading to the correction at the fine grid cannot be compared with the fast convergence rate of using the classical multigrid method.

3 Domain Decomposition Methods

The idea of domain decomposition has a long evolving history. Many literatures may be found, and it is not intended to give a full list of these references, but one [7], in this paper. Domain decomposition involves the subdivision of a given problem into a number of subproblems. Each of these subproblems can be solved separately before being combined to give the global solution of the original problem. The subdivision can be done at either the physical problem level or the discretised problem level. At the discretised problem level, the resulting linear system of equations is rearranged as a collection of smaller linear systems which may be solved independently. At the physical problem level, regions governed by different mathematical models or different material properties are identified and decomposed into different subdomains resulting in a number of locally regular subproblems. It should be noted at this stage that the use of a distributed environment is highly suitable for this class of methods [12].

3.1 Block Iterative Methods

Figure 1 shows a rectangular domain which is subdivided into 16 subdomains. Assuming that the subdomains are nonoverlapped and done at the level of the physical problem, then $\bigcup \Omega_i = \Omega$ and $\bigcap \Omega_i = \emptyset$, and the interior boundary or interface, γ, is defined as $\gamma = (\bigcup \partial \Omega_i) \backslash \partial \Omega$. Let $L_i u_i = f_i$ be the subproblem defined in the subdomain Ω_i, $i = 1, \dots, N_s$. A simultaneous update to each of the subdomains may be achieved by means of the block Jacobi algorithm as given below.

> loop
>   for $i := 1 \dots N_s$ do
>   Solve $L_i u_i = f_i$ subject to suitable boundary conditions along $\partial \Omega_i$;
>   end-do
>   Update: Interior boundary conditions along γ;
>   Until solution converged;

This block iterative method may be implemented in parallel but the convergence rate of this method is very slow. There is also a need to update the interior boundary conditions.

A similar technique may be applied to overlapped subdomains. Assuming each of the subdomains in Figure 1 is extended across the interior boundary into its neighbouring subdomains where an overlapped region of thickness of the mesh size h is included into the subdomain Ω_i. Suppose $N(i)$ denotes the numbering of the neighbouring subdomains of Ω_i. It is obvious that $\Omega_i \cap \Omega_{N(i)} \neq \emptyset$ and that $\partial \Omega_i$ lies inside $\Omega_{N(i)}$. Therefore an exchange of information between neighbouring subdomains is sufficient to act as an update to the interior boundaries.

loop
    for $i := 1 \ldots N_s$ do
    Solve $L_i u_i = f_i$ subject to suitable boundary conditions along $\partial \Omega_i$;
    end-do
    Exchange information: Interior boundary conditions along γ;
    Until solution converged;

This block iterative method may be implemented in parallel but the convergence rate of this method is, again, very slow. The update of the interior boundary conditions is achieved by means of exchanging information in neighbouring subdomains.

3.2   A Parallel Multigrid Algorithm

Since the classical multigrid algorithm experiences partitioning issues for parallel or distributed computing environment, the aim here is to seek for an alternative. The concept here is to employ the above block iterative algorithms, either overlapped and nonoverlapped. For the present purpose, only the overlapped version has been studied because the implementation is straight forward. The algorithm sees the subdomains as shown in Figure 1 as a natural set up for the coarsest mesh H. Since each subdomain may be executed autonomously, one natural idea is to use the defect correction principle as discussed above with H not necessarily equal to $2h$. In fact it is purely up to individual subdomain to determine the finest mesh h.

The coarsest level using the classical multigrid method is equivlant to solve the system,

$$L_H u_H = f_H \equiv I_H^{H/2} \ldots I_{2h}^h R_h -$$

$$(I_H^{H/2} \ldots I_{4h}^{2h} L_{2h} u_{2h}^* + I_H^{H/2} \ldots I_{8h}^{4h} L_{4h} u_{4h}^* + \ldots + I_H^{H/2} L_{H/2} u_{H/2}^*) \tag{5}$$

Note that in the present parallel multigrid algorithms it is not possible to evaluate the linear combination of the corrections at all intermediate level. However, it should be clear that the dominant source of error comes from the finest level which is projected onto the coarsest level according to $I_H^{H/2} \ldots I_{2h}^h R_h$. Here the projection can be done by means of a sequence of linear interpolation.

4   Numerical Examples

The numerical example is to find $u(x, y)$ such that $\nabla^2 u(x, y) = 0 \in (0, 20) \times (0, 20)$ subject to $u(x, 0) = 0$, $u(x, 10) = 0$, $u(0, y) = 0$, and $u(20, y) = 100$. A cell centre finite volume technique is used to discretise the problem, which is implemented in PHYSICA. Numerical results are obtained using 2-level correction schemes, one based on classical 2-level algorithm and the other based on the parallel multigrid algorithm. The next coarser level for the classical 2-level algorithm is chosen to be $2h$. The coarsest level of the parallel multigrid algorithm is chosen to be $H = 1.25$ in the new parallel algorithm. For demonstration purposes, a V-cycle multigrid iteration is adopted such that the number of iterations on the coarsest level is chosen to be 6 and that on the finest level is chosen to be 3.

By using a 2-level correction scheme with domain decompsotion method, the present parallel multigrid method, it shows similar computational work to the sequential 2-level multigrid method. When the finest mesh becomes increasingly smaller in its mesh, the computational work seems to be smaller as compare to the sequential 2-level multigrid method. An addition property of the present method is that it is intrinsic parallel, in which each subdomain may be computed simultaneously. Table 1 shows a comparison of the total computational work for the sequential and the parallel multigrid. It should be note that as the parallel multigrid is to be run on distributed computing environment, the projected parallel computational work when all the 16 subdomains are computed simultaneously. The dramatic decrease in the timing can be easily observed.

Two-Level Correction Scheme: Comparison of Computational Work.				
Finest mesh	16×16	32×32	64×64	128×128
	$h = 1.25$	$h = 0.625$	$h = 0.3125$	$h = 0.15625$
Sequential 2-Level	48	151.5	484.5	1533
Multigrid				
Mesh	8×8	16×16	32×32	64×64
Parallel: Sequential run	61.5	120.8	314.5	933.8
Multigrid: Parallel run	10.8	11.2	22	60
Mesh	4×4	4×4	4×4	4×4

Table 1. Comparison of computational work units. 1 Computational work unit is the computational work required to perform 1 iteration on the finest level.

5   An Extension to Multi-Scale Problems

The concept of defect correction is being extended to the problem of sound generation due to fluid motion. This is a multi-scale problem not relating to the size of the subdomains but to the size of the flow variables. The aim here is to solve the non-linear equation

$$\frac{\partial U}{\partial t} + \mathcal{L}\{U\}U \equiv \frac{\partial(\bar{u}+u)}{\partial t} + \mathcal{L}\{\bar{u}+u\}(\bar{u}+u) := 0 , \qquad (6)$$

where $\mathcal{L}\{U\}$ is a non-linear operator depending on U. Here $\mathcal{L}\{U\}$ is the Navier-Stokes operator and u is certain noise signal such that $u \ll \bar{u}$. For a 2-D problem,

$$\bar{u} = \begin{bmatrix} \bar{\rho} \\ \bar{v}_1 \\ \bar{v}_2 \end{bmatrix} \qquad u = \begin{bmatrix} \rho \\ v_1 \\ v_2 \end{bmatrix} ,$$

where ρ is the density of fluid and v_1 and v_2 are the velocity components along the two spatial axes. Using the summation notation of subscripts, the 2-D Navier-Stokes problem $\frac{\partial u}{\partial t} + \mathcal{L}\{u\}u = 0$ may be written as

$$\frac{\partial \rho}{\partial t} + \frac{\partial(\rho v_j)}{\partial x_j} = 0,$$

$$\frac{\partial v_i}{\partial t} + v_j \frac{\partial v_i}{\partial x_j} + \frac{1}{\rho}\frac{\partial P}{\partial x_i} - \frac{\mu}{\rho}\nabla^2 v_i = 0,$$

Expanding $\frac{\partial(\bar{u}+u)}{\partial t} + \mathcal{L}\{\bar{u}+u\}(\bar{u}+u)$ and re-arranging the resulting terms, one obtains

$$\frac{\partial \rho}{\partial t} + \bar{v}_j\frac{\partial \rho}{\partial x_j} + \bar{\rho}\frac{\partial v_j}{\partial x_j} + [v_j\frac{\partial(\bar{\rho}+\rho)}{\partial x_j} + \rho\frac{\partial(\bar{v}_j+v_j)}{\partial x_j}] = -[\frac{\partial \bar{\rho}}{\partial t} + \bar{v}_j\frac{\partial \bar{\rho}}{\partial x_j} + \bar{\rho}\frac{\partial \bar{v}_j}{\partial x_j}],$$

and

$$\frac{\partial v_i}{\partial t} + \bar{v}_j\frac{\partial v_i}{\partial x_j} + \frac{1}{\bar{\rho}}\frac{\partial P}{\partial x_i} - \frac{\mu}{\bar{\rho}}\nabla^2 v_i \qquad (7)$$

$$+[\frac{\rho}{\bar{\rho}}\frac{\partial(\bar{v}_i+v_i)}{\partial t} + (v_j + \frac{\rho}{\bar{\rho}}(\bar{v}_j+v_j))\frac{\partial(\bar{v}_i+v_i)}{\partial x_j}] = -[\frac{\partial \bar{v}_i}{\partial t} + \bar{v}_j\frac{\partial \bar{v}_i}{\partial x_j} + \frac{1}{\bar{\rho}}\frac{\partial \bar{P}}{\partial x_i} - \frac{\mu\nabla^2 \bar{v}_i}{\bar{\rho}}].$$

It can be seen that (6) may be written as

$$\frac{\partial(\bar{u}+u)}{\partial t} + \mathcal{L}\{\bar{u}+u\}(\bar{u}+u) \equiv \frac{\partial \bar{u}}{\partial t} + \mathcal{L}\{\bar{u}\}\bar{u} + \frac{\partial u}{\partial t} + E\{\bar{u}\}u + K[\partial_t, \bar{u}, u], \quad (8)$$

where $\mathcal{L}\{\bar{u}\}$ and $E\{\bar{u}\}$ are operators depending on the knowledge of $\bar{u}$ and $K[\partial_t, \bar{u}, u]$ is a functional depending on the knowledge of both $\bar{u}$ and its derivative and u. Here

$$E\{\bar{u}\}u = \begin{bmatrix} \bar{v}_j\frac{\partial \rho}{\partial x_j} + \bar{\rho}\frac{\partial v_j}{\partial x_j} \\ \bar{v}_j\frac{\partial v_i}{\partial x_j} + \frac{1}{\bar{\rho}}\frac{\partial P}{\partial x_i} - \frac{\mu}{\bar{\rho}}\nabla^2 v_i \end{bmatrix} , \qquad (9)$$

$$K[\partial_t, \bar{u}, u] = \begin{bmatrix} v_j\frac{\partial(\bar{\rho}+\rho)}{\partial x_j} + \rho\frac{\partial(\bar{v}_j+v_j)}{\partial x_j} \\ \frac{\rho}{\bar{\rho}}\frac{\partial(\bar{v}_i+v_i)}{\partial t} + (v_j + \frac{\rho}{\bar{\rho}}(\bar{v}_j+v_j))\frac{\partial(\bar{v}_i+v_i)}{\partial x_j} \end{bmatrix} . \qquad (10)$$

In order to simulate accurately the approximate solution, $\bar{u}$, to the original problem,

$$\frac{\partial(\bar{u}+u)}{\partial t} + \mathcal{L}\{\bar{u}+u\}(\bar{u}+u) = 0 \ ,$$

Let h denote the mesh size and δ_t be a difference approximation to $\frac{\partial}{\partial t}$ being used in the Reynolds averaged Navier-Stokes solver within PHYSICA. Instead of evaluating $\bar{u}^{(n)}$, one would solve the discretised approximation

$$\delta_t \bar{u}_h^{(n)} + \mathcal{L}_h\{\bar{u}_h^{(n)}\}\bar{u}_h^{(n)} = 0$$

to obtain $\bar{u}_h^*$. The residue, R_h, on the fine mesh h may be computed by using a higher order approximation [8] to $-[\delta_t \bar{u}_h^* + \mathcal{L}_h\{\bar{u}_h^*\}\bar{u}_h^*]$. Let H denote the mesh size for the linearised Euler equations solver, where the linearised Euler operator is given by (9), and is chosen to include as many sound signals with a specified long wavelength as possible. Hence $H \neq 2h$. Again instead of evaluating u, one would solve the discretised approximation

$$\delta_t u_H^{(n)} + E_H\{\bar{u}_H^{(n)}\}u_H^{(n)} = R_H^{(n)}$$

to obtain $u_H^{(n)}$. Here $R_H^{(n)}$ is the projection of R onto the mesh H. Let $I_{\{h,H\}}$ be a restriction operator to restrict the residue computed on the fine mesh h to the coarser mesh H. The restricted residue can then be used in the numerical solutions of linearised Euler equations. Therefore the two-level numerical scheme is (for non-resonance problems):

$n := 0;$
Do $n := n + 1$
    Solve $\delta_t \bar{u}_h^{(n)} + \mathcal{L}_h\{\bar{u}_h^{(n)}\}\bar{u}_h^{(n)} = 0$
    $R_H^{(n)} := -I_{\{h,H\}}[\delta_t \bar{u}_h^* + \mathcal{L}\{\bar{u}_h^*\}\bar{u}_h^*$
    $\bar{u}_H^{(n)} := I_{\{h,H\}}\bar{u}_h^{(n)}$
    Solve $\delta_t \bar{u}_h^{(n)} + E_H\{\bar{u}_H^{(n)}\}u_H^{(n)} = R_H^{(n)}$
    $U_H^{(n)} := \bar{u}_H^{(n)} + u_H^{(n)}$(Corrected results do not need to be used in $u_h^{(n+1)}$.)
Until $n = n_{max}$

Here $U_H^{(n)}$ denotes the discretised approximation of the resultant solution on mesh H. Note that $R_H^{(n)}$ cannot be computed as $\delta_t \bar{u}_h^{(n)} + \mathcal{L}\{\bar{u}_h^{(n)}\}I_{\{h,H\}}\bar{u}_h^{(n)}$ because $\mathcal{L}$ is a non-linear operator. Note also that $\delta_t u_H^{(n)}$ involves a number of smaller time steps, each of $\triangle T$, starting from $u_H^{(n-1)}$ such that $u_H^{(n)}$ defines at the same time level as $\bar{u}_h^{(n)}$.

6   Conclusion

This paper provides some earlier experiments of a parallel multigrid algorithm based on the combination of domain decomposition methods and a coarse level correction technique. The convergence of a block Jacobi iterative method in the classical Schwarz iterative scheme is greatly accelerated with the idea of a coarse grid correction. Furthermore, it is found that the coarse grid with mesh size H does not require to be the next coarser level of the finest grid. Due to this property, the finest level in different subdomains may be different from each other. This would enable local refinement to be done in an efficient manner and also enable subproblems to be solved in machines located geographically apart. The technique is also extended to handle a multi-scale problem invovling sound signal propagation.

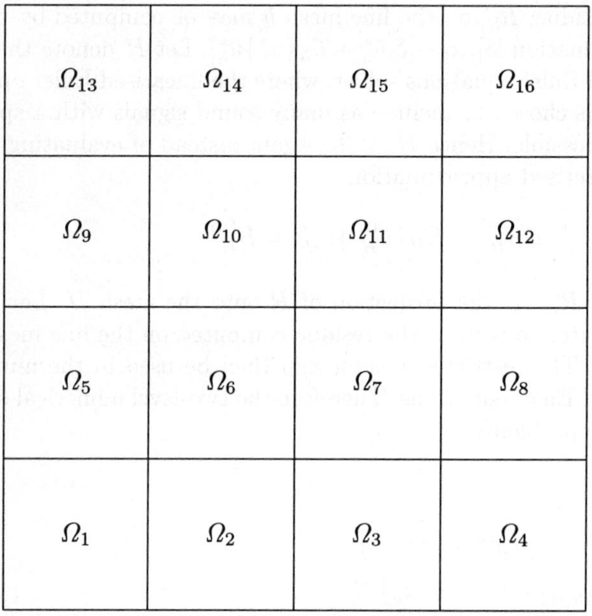

Ω_{13}	Ω_{14}	Ω_{15}	Ω_{16}
Ω_9	Ω_{10}	Ω_{11}	Ω_{12}
Ω_5	Ω_6	Ω_7	Ω_8
Ω_1	Ω_2	Ω_3	Ω_4

Figure 1: Illustration of a nonoverlapped domain decomposition, which consists of 16 subdomains. Each subdomain is denoted as Ω_i where $i = 1, \dots, 16$.

References

1. Bachvalov, N.S.: On the Convergence of a Relaxation Method with Natural Constraints on the Elliptic Operator. USSR Comp. Math. and Math. Phys., **6** (1966) 101–135
2. Barkai, D., Brandt, A.: Vectorised Multigrid Poisson solver for the CDC Cyber 205. Applied Mathematics and Computation, **13** (1983) 215–227
3. Brandt, A.: Multi-level Adaptive Technique (MLAT) for Fast Numerical Solution to Boundary Value Problems. Lecture Notes in Physics, **18** (1973) 82–89
4. Brandt, A.: Multi-level adaptive solutions to boundary value problems. Math. Comp., **31** (1977) 333–390
5. Chan, T.F., Schreiber, R.: Parallel Networks for Multi-Grid Algorithms - Architecture and Complexity. SIAM J. Sci. Stat. Comput., **6** (1985) 698–711

6. Croft, T.N., Pericleous, K.A., Cross, M.: PHYSICA: A Multiphysics environment for complex flow processes. Numerical Methods for Laminar and Turbulent Flow (IX/2), C. Taylor et al. (Eds), Pineridge Press, U.K. (1995)
7. Proceedings of International Conference on Domain Decomposition Methods for Science and Engineering. Vol 9, 11, and 12, DDM.Org.
8. Djambazov, G.S.: Numerical Techniques for Computational Aeroacoustics. Ph.D. Thesis, University of Greenwich (1998)
9. Fedorenko, R.P.: The Speed of Convergence of One Iterative Process. USSR Comp. Math. and Math. Phys., **4** (1964) 227–235
10. Frederickson, P.O., McByran, O.A.: Parallel Superconvergent Multigrid. Cornell Theory Centre Report CTC87TR12 (1987)
11. Lai, C.-H.: Non-linear Multigrid Methods for TSP Equations on the ICL DAP. Annual Research Report, Queen Mary, University of London, (1984)
12. Lai, C.-H.: Domain Decomposition Algorithms for Parallel Computers. High Performance Computing in Engineering - Volume 1: Introduction and Algorithms, H. Power and C.A. Brebbia (Eds), Computational Mechanics Publication, Southampton, (1995) 153–188
13. McCormick, S.F. Ruge, J.W.: Unigrid for Multigrid Simulation. Math. Comp., **41** (1983) 43–62
14. Naik, V.K., Ta'asan, S.: Implementation of Multigrid Methods for Solving Navier-Stokes Equations on a Multiprocessor System. ICASE Report 87-37 (1987)
15. PHYSICA on-line Menu: Three-dimensional Unstructured Mesh Multi-physics Computational Mechanics Computational Modellings. http://physica.gre.ac.uk/physica.html (1999)
16. Trottenberg, U., Oosterlee, C., Schuller, A.: Multigrid. Academic Press, New York (2001)
17. Introduction to Multigrid Methods. ICASE Report 95-11 (1995)

Computational Models for Materials with Shape Memory: Towards a Systematic Description of Coupled Phenomena

Roderick V.N. Melnik[1] and Anthony J. Roberts[2]

[1] University of Southern Denmark,
Mads Clausen Institute, DK-6400, Denmark, rmelnik@mci.sdu.dk
[2] University of Southern Queensland,
Department of Mathematics and Computing, QLD 4350, Australia

Abstract. In this paper we propose a systematic methodology for improving computational efficiency of models describing the dynamics of materials with memory as part of multilayered structures, in particular in thermoelectric shape memory alloys actuators. The approach, based on a combination of the centre manifold technique and computer algebra, is systematic in a sense that it allows us to derive computational models with arbitrary order of accuracy with respect to certain small parameters. Computational results demonstrating the efficiency of the proposed methodology in reproducing the dynamics of austenitic-martensitic phase transformations upon thermoelectric cooling are presented.

1 Introduction

Coupled systems of partial differential equations arise frequently in various contexts and different areas of theory and applications. Their fundamental importance in a better understanding of complex phenomena and processes ranges from such fields as supersymmetric and string theories to fluid-structure interactions, thermoelasticity, electroelasticity, and control, to name just a few.

Models based on coupled systems of PDEs are often inheritably difficult to treat analytically and some such models are amongst the greatest computational challenges in mathematical modelling. The development of efficient computational procedures for such models lags behind the needs in their applicability in sciences and engineering. While for some such models a substantial progress has been achieved in developing efficient numerical methodologies and their rigorous justification (e.g., [10, 12]), efficient computational algorithms and methodologies for many other models, especially those dealing with strongly nonlinear coupled problems, are at the beginning of their development. One problem from this class is at the main focus of the present paper.

In many engineering applications coupled phenomena are in the very essence of the successful design of systems and devices. Classical examples include electromechanical and magnetomechanical systems. As a part of modelling such systems, in some cases we also have to adequately describe the dynamic behaviour

P.M.A. Sloot et al. (Eds.): ICCS 2002, LNCS 2330, pp. 490–499, 2002.
© Springer-Verlag Berlin Heidelberg 2002

of complex materials these systems made of, subject to different loadings. This is known to be one of the most difficult tasks in computational sciences, which is complicated even further as soon as an intrinsic nonlinear interplay between different physical fields, such as mechanical, electric, and/or thermal, is at the heart of the process of interest. Our interest in this paper lies with the materials that under appropriate externally imposed conditions (e.g., through the action of thermal, magnetic, mechanical fields) can recover their shape after being permanently deformed. One class of such materials, known as shape memory alloys (SMA), has become increasingly important on the technological landscape, and the design of new devices and systems requires a better understanding of the dynamics of these materials, including phase transition and hysteresis phenomena [13]. The modelling of SMA dynamics is a challenging field of computational science where interdisciplinary efforts are required. To model adequately SMA-based multilayered structures represents even a greater challenge.

2   Models for SMA-Based Multilayered Thermoelectric Devices and Their Applications

The use of shape memory materials in microtechnological applications, where these materials have a great potential as parts of actuators and sensors, is often impeded by low dynamic responses, especially when an efficient cooling is important [1]. This is the case whenever the control of SMA phase transformations (and a subsequent mechanical power generation) is made by the temperature. Since the time constant for heat transfer is typically large compared to the small time constants required for many high frequency applications, one has to deal with the problem of low rates of cooling. While the heating is easy to implement technologically (typically, by using the Joule effect), the cooling of the SMA sample represents a more difficult task in these applications. The basic problem here is in the fact that by going to smaller scales the efficiency of traditional "cooling" technologies such as thermal convection and conduction, usually applied to SMA samples at larger scales, decreases. That is why experimentalists are looking for other technologies to overcome the problem. One of the promising technologies in this field is connected with the thermoelectric cooling based on the Peltier effect. In electronic industry this technology is used, for example, to cool electronic chips.

Thermoelectric effects have been effectively used for quite a while in power conversion and refrigeration applications, where the efficiency is usually achieved with semiconductor thermoelements. Since the efficiency of thermoelectric devices (heat engines or heat pumps) are limited by the electrical/thermal properties of the semiconductor materials (where the thermal energy is converted into electric power), much progress in the field has been connected with the development in solid-state electronics. For the benefits to the reader we recall that the use of the Peltier effect is based on the fact that whenever we apply an electric current to a system composed of two dissimilar conductors, heat is evolved at one junction and absorbed at the other, that is depending on the current direction, one junction becomes cold and the other becomes hot. This brings difficulties in modelling these devices at a computational level due to a jump in the heat flux at the interface caused by the Peltier effect. The basic idea pertinent to the technicalities of the device we are interested in is shown in Fig.1(a) where $J(t)$ is the current density, and P and N stand for oppositely doped semiconductors (only one thermocouple is shown in this Melcor-type design configuration). The thermoelectric process is thermodynamically reversible, and fundamentals for its modelling were laid by A.F. Ioffe (e.g., [6] and references therein). Based on those early results, semiconductors have been used for localised cooling in various applications. However, it has only recently been noted (e.g., [17, 2]) that the Peltier effect can be used for thermal cycling of the SMA. The potential of this idea in designing large strain SMA actuators have been experimentally exploited, and comparisons with other cooling mechanisms, including natural and forced convection, have been favorable. Therefore, in a way similar to Fig.1(a), the SMA actuator can be made as the cold/hot junction of a thermoelectric couple. As an example, we consider the design configuration shown in Fig. 1(b) (e.g., [2]) with the basic module given in Fig.1(c). As seen in this figure, the SMA layer is typically put in-between two semiconductor layers, usually made of bismuth telluride (TeBi). In this way, SMA can be used for actuating purposes in shape and vibration control problems, as well as in applications ranging from various mechatronics products (e.g., flexible grippers for the assembly of tiny work pieces in the semiconductor industry and the sample collection for microscopic observations in the biochemical labs) to SMA actuators in the area of underwater vehicle design, artificial valves in bioengineering applications, etc.

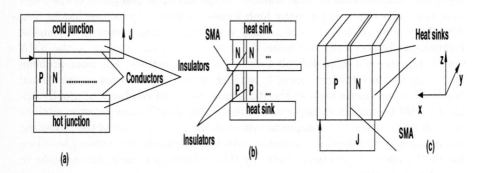

Fig. 1. Design configurations of thermoelectric devices and SMA actuator modules

The modelling of thermoelectrically cooled/heated SMA actuators is typically limited in the literature to thermal analyses only (e.g., [1, 5] and references therein) where the SMA module is described on the basis of the heat equations with different approximations for the internal heat productions. Such models cannot describe the dynamics of phase transitions that would require considering displacements coupled to temperature changes. In addition, until recently such models were based on the assumption of a small ratio between the SMA layer thickness and the thickness of semiconductors (see discussion in [5]). This assumption might be violated in practice. Nevertheless, such models allow us to obtain a first approximation to the temperature at the interface between the SMA and semiconductor layers. This fact will be used in the models discussed below. Following [4], we consider a thermal model for the multilayered SMA actuator based on a system of three heat equations written for each layer

$$k_i \frac{\partial^2 \theta_i}{\partial x^2} + \rho_i J^2(t) - H \frac{P}{A}(\theta_i - \theta_0) = C_v^i \frac{\partial \theta_i}{\partial t}, \tag{1}$$

and coupled together by the flux interface conditions, e.g.,

$$-k_s \frac{\partial \theta_s}{\partial x} = -k_p \frac{\partial \theta_p}{\partial x} + \alpha_p \theta_p J(t). \tag{2}$$

Model (1)–(2) provides a good approximation especially in the case where the temperature in each layer does not vary significantly on the plane perpendicular to the x-axis. In (1), (2) index i stands for N (N-type semiconductor), S (SMA-layer), or P (P-type semiconductor, see Fig.1). We denote the temperature by θ, the thermal conductivity coefficient by k, the perimeter and area of the cross section by P and A, respectively, the heat convection coefficient by H, the electrical resistivity by ρ (so that ρJ^2 is the Joule heat), the heat capacity per unit volume by C_v, and the Seebeck coefficient by α. As in [9], the structure is symmetric, the thickness of the SMA layer is a_0, and the thickness of each semiconductor layer is a. Two other dimensions of the structure, denoted by b and c, are used in the definition of A and P, so that $A = b \cdot c$ and $P = 2(b + c)$ (see details in [9]). By considering a special case where the thermal conductivity of SMA, k_s, is much larger compared to thermal conductivity of semiconductors, k_n, and k_p, (e.g., by assuming a small ratio between the layer thickness of SMA and semiconductor $a_0/a \ll 1$), the strong thermomechanical coupling, intrinsic to the SMA layer, has often been neglected (e.g., [4]). Although the assumption $a_0/a \ll 1$ (dubious in many applications) has been recently removed in [5], most approaches developed in this field so far can account only partly for the thermomechanical coupling in the SMA layer [4, 7, 5]. However, it is this coupling that is responsible for phase transformations, and due to this coupling, the often made assumption that $\theta_s(x, t)$ is independent of x (e.g., [4]) cannot be justified in any situations involving phase transformations. A partial phase transformation (and hence the effect of coupling) has been considered in [7], where the assumption of an almost uniform temperature distribution in the SMA layer has been removed. The authors of [7] used a phenomenological model based on gradual transformations of SMA polycrystals and evolution equations for field variables which were then solved with a Runge-Kutta method. Similarly to their previous works (e.g., [2]), the procedure was limited to stress-free boundaries, and therefore could be applied to a specific form of thermomechanical coupling only. A more general procedure allowing to treat different thermomechanical loadings have been considered recently in [13].

Despite limitations mentioned above, models proposed in [4, 7, 5] are an important step forward, indicating clearly a way of reducing the problem (1), (2) to a relatively simple heat transfer problem with coupling effects implemented at the boundaries, as a result of the dependency of the temperature in semiconductor layers on the heat capacity of the SMA material. Indeed, as shown in [4], this problem can be effectively reduced to an integro-differential equation, which in its turn is reduced further to the solution of a Volterra equation. By considering θ as function of J in the SMA layer, it can be shown that certain conditions, such as bounds on J, should be satisfied in order to achieve a monotonic decay of the SMA temperature. This is an important observation since a major disadvantage in utilising SMA actuators is the low rate of cooling. However, no possible austenite-to-martensite transformations as a result of this cooling has been discussed. It is one of our purposes here to demonstrate that our computational model can reproduced very well such transformations under appropriate cooling conditions. Before proceeding with this task, thermal boundary conditions for the SMA layer should be specified.

3   Describing Coupled Thermomechanical Fields in the SMA Layer

In describing coupled thermomechanical fields in the SMA layer, we follow a two-step procedure. Our first step is equivalent to that described in [5]. This leads us to an approximation of the temperature on the boundary of semiconductor layers, and by using its continuity a fully coupled model for the SMA layer can be formulated. If $J(t)$ is assumed to be constant, the exact solution to the problem at step 1, considered as a "purely" thermal analysis of the multilayered structure, can be found. In this analysis, the heat transfer behaviour along the SMA layer is dominated by the temperature at the interface between the SMA layer and the semiconductor layer. The analysis can be reduced to an integro-differential equation with respect to an auxiliary function, as discussed in [5]. More precisely, we consider equation (1) for $i = s$, that is for the SMA layer, where, due to symmetry, it is sufficient to consider the interval $0 < x < a_0/2$ only. This equation is supplemented by the initial condition $\theta(x,0) = \theta_0$, "symmetry" boundary condition $\dfrac{\partial\theta}{\partial x}(0,t) = 0$, and the "interface" boundary condition at $x = a_0/2$:

$$\frac{\partial\theta}{\partial x} = -\frac{2k}{ak_s}\int_0^t G_1(t-\tau)\left[\frac{d\theta}{dt} + \frac{HP}{C_vA}(\theta-\theta_0)\right]\Bigg|_{t=\tau}d\tau$$

$$+\left(1-\frac{k}{\alpha a}\right)\theta + F(t), \quad t > 0, \tag{3}$$

obtained by using the solution for the semiconductor layer, as explained in [5]. In (3) $G_1(t) = \sum_{n=1}^{\infty} \exp(-\beta_n t)$ and function F has the following form:

$$F(t) = \frac{4\rho k}{C_v k_s a} \int_0^t \sum_{n=1}^{\infty} \exp(-\beta_{2n-1}(t-\tau)) J^2(\tau) d\tau + \frac{k}{ak_s} \theta_0. \tag{4}$$

Indices of the coefficients of the semiconductor adjacent to the SMA layer are omitted in (3) and (4), and $\beta_n = n^2\pi^2 + H$. It is this problem that is reduced to an integro-differential equation. Existence and uniqueness of its solution is established in a standard manner by equivalence with a Volterra equation, and its solution is found numerically by employing a finite difference scheme. Temperature profiles at the interface between the SMA and semiconductor layers can be quite different for different values of the current density J, but once such a profile is found we can proceed to the second step of our procedure.

At the second step, we study the fully coupled thermomechanical dynamic system of the SMA layer by employing a general procedure, similar to that discussed in [13] for a single layer structure. Starting from the 3D coupled model consisting of the equation of motion and the energy balance equation

$$\begin{cases} \rho \dfrac{\partial^2 \mathbf{u}}{\partial t^2} = \nabla \cdot \mathbf{s} + \mathbf{F}, \\[2mm] \rho \dfrac{\partial e}{\partial t} + \rho \tau_0 \dfrac{\partial^2 e}{\partial t^2} - \mathbf{s}^T : (\nabla \mathbf{v}) - \tau_0 \dfrac{\partial}{\partial t}[\mathbf{s}^T : (\nabla \mathbf{v})] - \nabla \cdot (k\nabla \theta) = G, \end{cases} \tag{5}$$

we aim at developing efficient computational models allowing to reproduce austenite-to-martensite phase transformations observed under thermoelectric cooling of SMA actuators. In (5) $\mathbf{u}$ denotes displacements, $\mathbf{s}$ is the stress tensor, $\mathbf{v} = \partial \mathbf{u}/\partial t$ is the velocity vector, e is the internal energy, τ_0 is the relaxation time, $\mathbf{F}$ and G are forcing terms (further details can be found in [8]). Constitutive models used in this paper are based on the general representation of the free energy function in the form

$$\Psi(\epsilon) = \psi^0(\theta) + \sum_{i=1}^{\infty} \sum_{j=1}^{j^i} \psi_j^i \mathcal{I}_j^i \tag{6}$$

with $\mathcal{I}_i^j$ being strain invariants, ψ_i^j being temperature dependent functional coefficients, and ϵ being the strain tensor. Function (6) is made invariant with respect to the symmetry group of austenite, and the upper limits j^i are chosen appropriately to satisfy this condition [8]. Further, we determine the stress component due to mechanical dissipations as $\mathbf{s}^q = \rho \dfrac{\partial \Psi}{\partial \epsilon}$. Thermal dissipations, plasticity, and other effects can be incorporated into the model, but in what follows we describe our methodology for the Landau-Devonshire type constitutive models as an example, where we take $\mathbf{s} = \mathbf{s}^q$. In particular, for the general 3D model we use the Falk-Konopka representation of the free energy function (6) with 10 strain invariant directions, valid for the copper-based SMA materials [8].

4   Reducing Computational Complexity and Preserving Essential Properties of the System with Centre Manifold Models

Model (5), supplemented by appropriate constitutive laws for the SMA materials, boundary and initial conditions, represents a tremendous computational challenge in the general 3D case. Any computational treatment of this model aiming at the description of the dynamics of SMA-based system responses will necessarily require essential simplifications of this model in order to be tractable. Such simplifications should be developed in a systematic way, and below we propose a systematic methodology for improving computational models for the description of SMA dynamics. Our idea is based a combination of the centre manifold technique and computer algebra for developing reduction procedures for the original model (5) on centre manifolds, while retaining essential properties of the system. Such procedures have been used successfully in a number of applications [15, 18, 16], and have recently received an increasing interest in the context of computations of normal forms, an apparently useful methodology in the analysis of nonlinear oscillations.

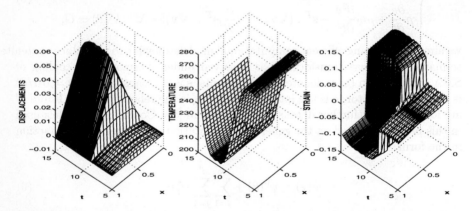

Fig. 2. Austenite-to-martensite phase transformation in the SMA layer of the thermo-electric actuator (from left to right): (a) displacements, (b) temperature, (c) strain.

A quite general theoretical result has been obtained in [3] confirming the existence of an invariant manifold (called a centre manifold) of the invariant set which contains every locally bounded solution. The importance and the need for developing new algorithms in theory and applications of approximation dynamics has been recently re-emphased in a fundamental work [14]. In a series of papers [8, 9, 13] we have initiated applications of the centre manifold technique to studying complex nonlinear coupled systems describing the behaviour of shape memory alloys, and in this paper we apply this technique to modelling multilayered structures such as those depicted in Fig. 1.

Our approach allows considering different patterns of thermomechanical coupling implemented via appropriate choices of the free energy function, constitutive models, initial and boundary conditions. In the dimensions higher than one computational results are virtually absent in this field, and little is know on how to simplify such models as (5), (6) systematically. In the spirit of our previous works [8, 13] we consider a shape memory alloy slab, and associate the qualitative change in system behaviour with the subset of eigenvalues having zero real parts. Then, critical variables are chosen as those that are responsible for the essential behaviour of the system. This leads to the following steps in our computational procedure:

- Analysis of eigenvalues of the cross-slab modes, as required by centre manifold theory;
- Construction of a sub-center manifold based upon the relevant eigenmodes as they evolve slowly.

The low-dimensional invariant manifold should be parametrised by the amplitudes of the critical modes, which constitute a subset of all modes, and determine the essential dynamic behaviour of the system [15]. A computational model is constructed with respect to these amplitudes, and since the leading order structure of the critical eigenmodes are constant across the slab (the longitudinal variations were neglected in the first approximation), we associate these amplitudes with the y-averages of displacements, velocities, and temperature, denoted here as $\mathbf{U} = (U_1, U_2)$, $\mathbf{V} = (V_1, V_2)$, and $\theta' = \theta - \theta_0$ (θ_0 is taken as $300°K$). The connection between these variables, and the variables of the original model is established in an approximate form by employing the low-dimensional invariant manifold and asymptotic sum representations (explicit formal power series representations maybe prohibited even for a computer). Similar to multiple scale techniques, this approach requires a balancing of the order of small effects in the construction of the model, not in its use [16]. The small parameters used in this paper are ∂_x, $E = \|\mathbf{U}_x\| + \|\mathbf{V}_x\|$ and $\vartheta = \|\Theta'\|$, and the computer algebra program has been written to balance their small effects. We treat the strains as small, as measured by E, while permitting asymptotically large displacements and velocities. It is important to emphasise that such approximations can be derived up to arbitrary order of accuracy, as explained in [8]. For the model discussed below, the approximations were derived with errors $\mathcal{O}\left(E^5 + \partial_x^{5/2} + \vartheta^{5/2}\right)$ where the notation $\mathcal{O}(E^p + \partial_x^q + \vartheta^r)$ is used to denote terms involving $\partial_x^{\beta_2} E^{\beta_1} \vartheta^{\beta_3}$ such that $\beta_1/p + \beta_2/q + \beta_3/r \geq 1$. In summary, after the decision on how to parametrise the centre manifold model is taken, and the critical modes are identified in order to project the dynamics onto the "slow" modes of interest, we complete our computational procedure by

- Substituting the asymptotic sums into the governing equations;
- Evaluating residuals by using computer algebra tools to get the result with the required accuracy.

These steps can be performed in a computationally very efficient manner by using an iterative algorithm analogous to that developed in [15]. By applying the above procedure, we have derive the reduced model

$$
\begin{cases}
\rho \dfrac{\partial V_1}{\partial t} = \dfrac{\partial s}{\partial x} + F, \quad \dfrac{\partial U_1}{\partial t} = V_1, \\[2mm]
C_v \dfrac{\partial \Theta'}{\partial t} = k \dfrac{\partial^2 \Theta'}{\partial x^2} + (c_{11} + c_{12}\Theta' - c_{13}(\Theta')^2) \dfrac{\partial U_1}{\partial x} \dfrac{\partial V_1}{\partial x} + \\[2mm]
\quad + (c_{14} + c_{15}\theta') \dfrac{\partial V_1}{\partial x} \left(\dfrac{\partial U_1}{\partial x} \right)^3 + c_{18} \dfrac{\partial V_1}{\partial x} \left(\dfrac{\partial U_1}{\partial x} \right)^5 + g, \\[2mm]
s = (c_1 + c_2\theta' - c_3(\theta')^2) \dfrac{\partial U_1}{\partial x} - (c_4 - c_5\theta') \left(\dfrac{\partial U_1}{\partial x} \right)^3 + c_6 \left(\dfrac{\partial U_1}{\partial x} \right)^5,
\end{cases}
\tag{7}
$$

where coefficients c_k are positive material constants (taken here as in [8]). This model, derived from the general 3D model (5), is exact up to the 4th order with respect to the small parameters, and preserves all essential features of the dynamics of the original system. In the case analysed here, if dissipations are omitted, there is a zero eigenvalue of multiplicity 5 and the rest are purely imaginary. Therefore, the sub-centre manifold has been constructed based on these 5 eigenmodes. Note that for simplicity we consider here the critical eigenvalues that are zeros (see comments on pure imaginary eigenvalues in [15, 16]). The initial conditions for the detailed (original) dynamics also have to be projected onto the low-dimensional manifold, as well as the boundary conditions (see further details on this in [8, 13]). The resulting model (7) is solved by its reduction to a system of differential-algebraic equations, and the displacements and temperature of the slab are recovered by using critical eigenmodes.

The developed computational procedure has been applied to a number of problems dealing with multilayered SMA actuators. Fig. 2 provides one of the results based on this procedure. Profiles for thermal cycling used in these experiments are analogous to those used in thermal analyses of these devices [4, 5]. Starting with high temperature phase (austenite) (represented in Fig. 2(a) and 2(c) by zero displacements and zero strain, respectively) it is shown how the cooling of the SMA layer (Fig. 2(b)) leads to the martensitic phase. Other results of computational experiments demonstrating the efficiency of the centre-manifold-based procedures in capturing all main features of macroscopic phase transitions can be found in [9, 8, 13].

Using our approach it is possible to construct a hierarchy of mathematical models for the description of the dynamics of shape memory alloys. Computational efforts spent for deriving centre manifold models are reduced substantially by employing computer algebra to deal with asymptotic sums in several small parameters chosen to measure the influence of small effects [15]. Since such centre manifold reductions can be derived with arbitrary degree of accuracy, this approach is systematic and deserves more attention in computational science community. With further advances in microtechnologies, for such multilayered structures as discussed in this paper more refined models for semiconductor layers will also be required. Such models by themselves represent a difficult and challenging task in computational sciences (see [11]).

References

1. Abadie, J., Chaillet, N., Lexcellent, C., Bourjault, A.: Thermoelectric Control of SMA Microactuators: a Thermal Model. SPIE Proceedings. **3667** (1999) 326–336
2. Bhattacharyya, A., Lagoudas, D.C., Thermoelectric SMA Actuators and the Issue of Thermomechanical Coupling. J. Phys. IV. **7 C5** (1997) 673–678
3. Chow, S.-N., Liu, W., Yi, Y.: Center Manifolds for Invariant Sets. J. of Diff. Equations. **168** (2000) 355–385
4. Ding, Z., Lagoudas, D.C.: Solution Behavior of the Transient Heat Transfer Problem in Thermoelectric SMA Actuators. SIAM J. of Appl. Math. **57** (1997) 34–52
5. Ding, Z., Lagoudas, D.C.: Transient Heat Transfer Behaviour of 1D Symmetric Thermoelectric SMA Actuators. Math. and Computer Modelling. **29** (1999) 33–55
6. Egli, P.H. (ed.): Thermoelectricity. Wiley, New York & London (1960)
7. Lagoudas, D.C., Ding, Z.: Modeling of Thin Layer Extensional Thermoelectric SMA Actuators. Int. J. Solids Structures. **35** (1998) 331–362
8. Melnik, R.V.N., Roberts, A.J., and Thomas, K.A.: Computing Dynamics of Copper-Based SMA via Centre Manifold Reduction of 3D Models. Computational Materials Science. **18** (2000) 255–268
9. Melnik, R.V.N. and Roberts, A.J.: Modeling dynamics of multilayered SMA actuators. SPIE Proceedings. **4235** (2000) 117–125
10. Melnik, R.V.N.: Generalised Solutions, Discrete Models and Energy Estimates for a 2D Problem of Coupled Field Theory. Applied Mathematics and Computation. **107** (2000) 27–55
11. Melnik, R.V.N., He, H.: Modelling Nonlocal Processes in Semiconductor Devices with Exponential Difference Schemes. J. of Engineering Mathematics. **38** (2000) 233–263
12. Melnik, R.V.N.: Discrete Models of Coupled Dynamic Thermoelasticity for Stress-Temperature Formulations. Applied Mathematics and Computation. **122** (2001) 107–132
13. Melnik, R.V.N., Roberts, A.J., and Thomas, K.A.: Coupled Thermomechanical Dynamics of Phase Transitions in Shape Memory Alloys and Related Hysteresis Phenomena. Mechanics Research Communications. (2001) to appear
14. Pliss, V.A., Sell, G.R.: Approximation Dynamics and the Stability of Invariant Sets. J. of Diff. Equations. **149** (1998) 1–51
15. Roberts, A.J.: Low-Dimensional Modelling of Dynamics via Computer Algebra. Computer Physics Communications **100** (1997) 215–230
16. Roberts, A.J.: Computer Algebra Derives Correct Initial Conditions for Low-Dimensional Dynamical Systems. Computer Physics Communications **126** (200) 187–206
17. Shanin, A.R., Meckl, P.H., Jones, J.D., Thrashner, M.A., Enhanced Cooling of SMA Wires Using Semiconductor "Heat Pump" Modules. J. of Intell. Material Systems and Structures. **5** (1994) 95–104
18. Thomsen J.J.: Vibrations and Stability. Order and Chaos. McGraw-Hill, London (1997)

Calculation of Thermal State of Bodies
with Multilayer Coatings

V. A. Shevchuk

Pidstryhach Institute for Applied Problems of Mechanics and Mathematics,
National Academy of Sciences of Ukraine,
3b Naukova, Lviv, 79053, Ukraine
e-mail: shevchuk@iapmm.lviv.ua

Abstract. The procedure for the calculation of heat conduction process in constructions elements with thin protective multilayer coatings is elaborated. This procedure is based on the essential simplification of solving the initial problem for constructions elements with thin multilayer coatings, and is connected with the modeling of such coatings by thin shells with appropriate geometrical and thermal properties of a coating. In such an approach, the influence of thin coatings on the course of heat processes in a body-coating system is described by special generalized boundary conditions. The efficiency of this approach has been shown by the comparison of results obtained according to this approximate approach with an accurate solution for the test case of an n-layer plate.

1. Introduction

The bodies with thin multilayer coatings represent an important class of non-homogeneous bodies. In practice to protect construction elements from the aggressive influence of environment, special coatings are used. It is essential that such coatings usually have non-homogeneous properties (which is connected with production conditions or caused by functional requirements) – multilayer coatings, in particular.

Several authors have investigated different heat conduction problems for these objects [2, 11, 12]. As a rule, such calculation for bodies with composite coatings is connected with formulating and solving appropriate problems of mathematical physics for multilayer systems, which are cumbersome and ineffective for practical purposes and are usually used as standards in elaborating approximate methods. Furthermore, the specific feature of such compound systems as bodies with composite coatings is the smallness of coating thickness in comparison with that of a substrate. Even numerical analysis of such compound structures, which contain substructures of different dimensionality, encounters significant difficulties [9]. It is natural, therefore, to attempt to build up approximate solutions, which are satisfactorily accurate for practical purposes. There have been suggested various approaches [2, 8, 9, 12] that allow us to take into account the smallness of coating thickness during the calculation of heat processes in the bodies with coatings. One of effective approaches consists in modeling of the influence of thin-walled elements of constructions by special boundary conditions [1, 3, 5, 6, 10]. This approach essentially simplifies solving the problems of finding the thermal state of constructions with thin multilayer coatings.

P.M.A. Sloot et al. (Eds.): ICCS 2002, LNCS 2330, pp. 500–509, 2002.

It is based on the modeling of the coatings by thin shells with appropriate geometrical and thermal properties of a coating.

The elaboration of such generalized boundary conditions can be derived by different techniques. Depending on the type of boundary conditions, the kinds of considered non-linearities and the possibility of preliminary linearization, the following methods can be used:

(i) operator method, which allows us not to accept preliminary hypotheses of the distribution of desired functions over the thickness of coating layers [4, 7];

(ii) approach, based on the application of a priori assumptions of the distribution of desired functions over the thickness of coating layers or on the whole coating [6];

(iii) discrete approach, based on the appropriate difference approximations of normal derivatives in appropriate expansions [3].

In this article, the generalized boundary condition is constructed by the application of operator method, and the efficiency of the approach is illustrated by the example of the solution of a test problem.

2. Problem Statement

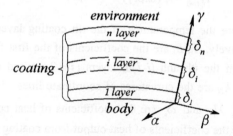

Fig. 1. Scheme of object

The object under investigation is a body with an applied thin multilayer coating with thickness $\delta = \sum_{i=1}^{n} \delta_i$, whose layers are made of different orthotropic materials. Here the n-layer coating is considered as a thin shell, referred to a mixed coordinate system (α, β, γ), which are respectively the lines of principal curvatures of the body-coating interface and the normal to it (Fig. 1). We assume that the edge surface Σ of such a shell is a ruled surface, for which the contour G bounding the body-coating interface S is directrix, and the normal to S in every point of the contour is generatrix.

We write the heat conduction equation for the each thin ith layer as [7]

$$\lambda_i^\gamma \left(\frac{\partial^2 t_i}{\partial \gamma^2} + 2k_i \frac{\partial t_i}{\partial \gamma} \right) + \Delta t_i = \omega_i \frac{\partial t_i}{\partial \tau}, \tag{1}$$

$$\Delta_i = \frac{1}{A_i B_i} \left[\lambda_i^\alpha \frac{\partial}{\partial \alpha} \left(\frac{B_i}{A_i} \frac{\partial}{\partial \alpha} \right) + \lambda_i^\beta \frac{\partial}{\partial \beta} \left(\frac{A_i}{B_i} \frac{\partial}{\partial \beta} \right) \right], \qquad k_i = \frac{k_{1i} + k_{2i}}{2}, \quad i = \overline{1, n}.$$

We assume that the coating is heated by environment according to Newton's law

$$\lambda_n^\gamma \frac{\partial t_n}{\partial \gamma} = \mu(t_e - t_n) \quad \text{at} \quad \gamma = \gamma_n = \sum_{i=1}^{n} \delta_i , \tag{2}$$

$$L_i t_i = \mu_i(t_e - t_i) \text{ on } \Sigma , \ L_i = n_1 \frac{\lambda_i^\alpha}{A_i} \frac{\partial}{\partial \alpha} + n_2 \frac{\lambda_i^\beta}{B_i} \frac{\partial}{\partial \beta} , \quad i = \overline{1, n}; \tag{3}$$

and the ideal thermal contact occurs between the constituents of the coating and the body

$$t_i = t_{i-1} , \quad \lambda_i^\gamma \frac{\partial t_i}{\partial \gamma} = \lambda_{i-1}^\gamma \frac{\partial t_{i-1}}{\partial \gamma} , \quad \text{at} \quad \gamma = \gamma_{i-1} = \sum_{j=1}^{i-1} \delta_j , \quad i = \overline{2, n}; \tag{4}$$

$$t_1 = t_b , \quad \lambda_1^\gamma \frac{\partial t_1}{\partial \gamma} = \lambda_b^\gamma \frac{\partial t_b}{\partial \gamma} \quad \text{at} \quad \gamma = \gamma_0 = 0,$$

and initial conditions are

$$t_i|_{\tau=0} = t_i^0(\alpha, \beta, \gamma) \quad i = \overline{1, n}. \tag{5}$$

Here t_i, t_b, t_e are the temperatures of the ith coating layer, the body and the environment respectively; A_i, B_i are the coefficients of the first fundamental form of the interface between the ith and $(i-1)$th layers ($i = \overline{2, n}$) and of the body-coating interface ($i = 1$); k_{1i}, k_{2i} are the curvatures of coordinate lines; δ_i is the thickness of the ith layer; $\lambda_i^\alpha, \lambda_i^\beta, \lambda_i^\gamma$, and ω_i are the coefficients of heat conductivity and heat capacity; μ, μ_i are the coefficients of heat output from coating surface $\gamma = \gamma_n$ and from ith part of the edge surface Σ; n_1, n_2 are the components of vector of outer normal to the edge surface Σ. The subscripts i, b, and e refer to the ith layer of the coating, the body, and the environment respectively.

3. Derivation of Generalized Boundary Conditions

Instead of the temperature t_i we introduce the variable [4, 7]

$$\theta_i = \exp(k_i \gamma) t_i \quad i = \overline{1, n}. \tag{6}$$

For its determination, we obtain the equation

$$p_i^2 \theta_i + \frac{\partial^2 \theta_i}{\partial \gamma^2} = 0, \quad \gamma \in (\gamma_{i-1}, \gamma_i) \ i = \overline{1, n}; \quad p_i^2 = \frac{1}{\lambda_i^\gamma} \Delta_i - \frac{\omega_i}{\lambda_i^\gamma} \frac{\partial}{\partial \tau} - k_i^2 , \tag{7}$$

and boundary and initial conditions

$$\lambda_n^\gamma\left(\frac{\partial\theta_n}{\partial\gamma} - k_n\theta_n\right) = \mu[\exp(k_n\gamma_n)t_e - \theta_n] \qquad \text{at } \gamma = \gamma_n , \tag{8}$$

$$L_i(\theta_i) = \mu_i[(\exp(k_i\gamma)t_e - \theta_i] \quad \text{on } \Sigma , \qquad i = \overline{1,n}; \tag{9}$$

$$\lambda_i^\gamma\left(\frac{\partial\theta_i}{\partial\gamma} - k_i\theta_i\right) = \exp[(k_i - k_{i-1})\gamma_{i-1}]\lambda_{i-1}^\gamma\left(\frac{\partial\theta_{i-1}}{\partial\gamma} - k_{i-1}\theta_{i-1}\right), \tag{10}$$

$$\theta_i = \exp[(k_i - k_{i-1})\gamma_{i-1}]\theta_{i-1} , \qquad \text{at } \gamma = \gamma_{i-1}, \quad i = \overline{2,n}; \tag{11}$$

$$\lambda_1^\gamma\left(\frac{\partial\theta_1}{\partial\gamma} - k_1\theta_1\right) = \lambda_b^\gamma\left(\frac{\partial t}{\partial\gamma}\right)_b , \qquad \theta_1 = t_b \text{ at } \gamma = \gamma_0, \tag{12}$$

$$\theta_i|_{\tau=0} = \exp(k_i\gamma)t_i^0 \qquad i = \overline{1,n}. \tag{13}$$

Let us introduce notations for the boundary values of the function θ_i and its derivatives:

$$\theta_i^+ = \lim_{\gamma\to\gamma_i-0}\theta_i , \qquad \left(\frac{\partial\theta_i}{\partial\gamma}\right)^+ = \lim_{\gamma\to\gamma_i-0}\left(\frac{\partial\theta_i}{\partial\gamma}\right),$$

$$\theta_i^- = \lim_{\gamma\to\gamma_{i-1}+0}\theta_i , \qquad \left(\frac{\partial\theta_i}{\partial\gamma}\right)^- = \lim_{\gamma\to\gamma_{i-1}+0}\left(\frac{\partial\theta_i}{\partial\gamma}\right).$$

Using the operator method [4, 7], we represent the general solution of Eq. (7) as the solution of an ordinary differential equation

$$\theta_i = \frac{\theta_i^- \sin(p_i(\gamma_i - \gamma)) + \theta_i^+ \sin(p_i(\gamma - \gamma_{i-1}))}{\sin(p_i\delta_i)}, \qquad i = \overline{1,n}. \tag{14}$$

Substituting (14) into (10)-(12), we obtain the following recursive relations for the temperature and its derivative at the interfaces of neighboring layers

$$\tilde{\theta}_1 = D_1\tilde{t}_b , \qquad \tilde{\theta}_i = D_i\tilde{\theta}_{i-1} , \quad i = \overline{2,n}; \tag{15}$$

where $\quad \tilde{\theta}_i = \left[\theta_i^+, \left(\frac{\partial\theta_i}{\partial\gamma}\right)^+\right]^T , \quad i = \overline{1,n}; \qquad \tilde{t}_b = \left[t_b, \left(\frac{\partial t}{\partial\gamma}\right)_b\right]^T ,$

$$D_i = \exp[(k_i - k_{i-1})\gamma_{i-1}]\begin{bmatrix} \cos(p_i\delta_i) + \left(k_i - \dfrac{\lambda_{i-1}^\gamma}{\lambda_i^\gamma}k_{i-1}\right)\dfrac{\sin(p_i\delta_i)}{p_i} & \dfrac{\lambda_{i-1}^\gamma}{\lambda_i^\gamma}\dfrac{\sin(p_i\delta_i)}{p_i} \\[3mm] -p_i\sin(p_i\delta_i) + \left(k_i - \dfrac{\lambda_{i-1}^\gamma}{\lambda_i^\gamma}k_{i-1}\right)\cos(p_i\delta_i) & \dfrac{\lambda_{i-1}^\gamma}{\lambda_i^\gamma}\cos(p_i\delta_i) \end{bmatrix},$$

when using supplement auxiliary notations $k_0 = 0$, $\lambda_0^\gamma = \lambda_b^\gamma$.

Now we write also the representation for the temperature and its derivative through the values of temperature and its derivative at the coating-body interface:

$$\tilde{\theta}_i = F_i \, \tilde{t}_b, \quad i = \overline{1,n}; \tag{16}$$

where

$$F_i = \begin{bmatrix} f_i^{11} & f_i^{12} \\ f_i^{21} & f_i^{22} \end{bmatrix}. \tag{17}$$

From (16), (15) follows:

$$F_1 = D_1, \qquad F_i = D_i F_{i-1}, \quad i = \overline{2,n}. \tag{18}$$

Substitution of (16) at $i = n$ into (8) leads finally after certain transformations to

$$a_{n1} t_b + a_{n2} \left(\frac{\partial t}{\partial \gamma} \right)_b - \mu t_e = 0 \quad \text{on } S, \tag{19}$$

where coefficients $a_n = [a_{n1}, a_{n2}]$ are determined by recursive relations

$$a_{01} = (\mu - \lambda_n^\gamma k_n) \exp(-k_n \gamma_n), \qquad a_{02} = \lambda_n^\gamma \exp(-k_n \gamma_n), \tag{20}$$
$$a_i = a_{i-1} D_{n-i+1} \quad i = \overline{1,n}.$$

Edge (9) and initial (13) conditions are satisfied by integrating

$$\sum_{i=1}^n (L_i + \mu_i) \int_{\gamma_{i-1}}^{\gamma_i} \exp(-k_i \gamma) \theta_i d\gamma - \sum_{i=1}^n \mu_i \int_{\gamma_{i-1}}^{\gamma_i} t_e d\gamma = 0, \tag{21}$$

$$\sum_{i=1}^n \int_{\gamma_{i-1}}^{\gamma_i} \exp(-k_i \gamma) \theta_i |_{\tau=0} d\gamma - \int_{\gamma_0}^{\gamma_n} t_0 d\gamma = 0. \tag{22}$$

Substitution of (14) into (21) and (22) taking account of (11), (12) gives

$$\sum_{i=1}^n \frac{(L_i + \mu_i)(q_i \theta_{i-1}^+ + r_i \theta_i^+)}{(k_i^2 + p_i^2) \sin(p_i \delta_i)} - \delta T_e^\mu = 0, \tag{23}$$

$$\sum_{i=1}^n \frac{(q_i \theta_{i-1}^+ |_{\tau=0} + r_i \theta_i^+ |_{\tau=0})}{(k_i^2 + p_i^2) \sin(p_i \delta_i)} - \delta T_0 = 0, \tag{24}$$

where $\quad T_e^\mu = \dfrac{1}{\delta} \sum_{i=1}^n \mu_i \int_{\gamma_{i-1}}^{\gamma_i} t_e d\gamma, \qquad T_0 = \dfrac{1}{\delta} \int_{\gamma_0}^{\gamma_n} t_0 d\gamma,$

$q_i = \exp(-k_{i-1}\gamma_{i-1})(p_i \exp(-k_i \delta_i) + k_i \sin(p_i \delta_i) - p_i \cos(p_i \delta_i)),$
$r_i = \exp(-k_i \gamma_i)(p_i \exp(k_i \delta_i) - k_i \sin(p_i \delta_i) - p_i \cos(p_i \delta_i)),$

and $\quad \theta_0^+ = \theta_1^- = t_b$ is a supplement auxiliary notation.

Substituting from (16) expression for $\theta_i^+ = f_i^{11} t_b + f_i^{12} \left(\dfrac{\partial t}{\partial \gamma} \right)_b$ into (23) and (24),

we finally obtain

$$a_{s1} t_b + a_{s2} \left(\frac{\partial t}{\partial \gamma} \right)_b - \delta T_e^\mu = 0 \quad \text{on } G, \tag{25}$$

$$a_{\tau 1} t_b|_{\tau=0} + a_{\tau 2} \left(\frac{\partial t}{\partial \gamma} \right)_{b|_{\tau=0}} - \delta T_0 = 0 \quad \text{on } S, \tag{26}$$

where
$$[a_{s1}, a_{s2}]^T = \sum_{i=1}^n \frac{(L_i + \mu_i)}{(k_i^2 + p_i^2) \sin(p_i \delta_i)} \begin{bmatrix} f_{i-1}^{11} & f_i^{11} \\ f_{i-1}^{12} & f_i^{12} \end{bmatrix} \begin{bmatrix} q_i \\ r_i \end{bmatrix},$$

$$[a_{\tau 1}, a_{\tau 2}]^T = \sum_{i=1}^n \frac{1}{(k_i^2 + p_i^2) \sin(p_i \delta_i)} \begin{bmatrix} f_{i-1}^{11} & f_i^{11} \\ f_{i-1}^{12} & f_i^{12} \end{bmatrix} \begin{bmatrix} q_i \\ r_i \end{bmatrix},$$

and f_i^{11}, f_i^{12} are elements of the matrix (17) for $i = \overline{1, n}$; and $f_0^{11} = 1, f_0^{12} = 0$.

Since relations (19), (25), (26) establish the connection between the boundary value of body temperature t_b and its derivative $\left(\dfrac{\partial t}{\partial \gamma} \right)_b$ with given value of environment temperature t_e, they can be interpreted as generalized boundary conditions which take into account the influence of the multilayer coating on the course of heat process in a body.

4. Calculation Variants of Generalized Boundary Conditions

For sufficiently thin coatings, conditions (19), (25), (26) can be simplified by expanding their terms in the series of powers of small δ_i and neglecting the terms containing $\delta_k^m \delta_l^s$ with $m + s \geq 2$. The result is

$$\sum_{i=1}^n \widetilde{\Delta}_i t_b - \lambda_b^\gamma \left(1 - 2K + \frac{\mu}{H} \right) \left(\frac{\partial t}{\partial \gamma} \right)_b + \mu(t_e - t_b) = \Omega \frac{\partial t_b}{\partial \tau}, \quad t_b|_{\tau=0} = T_0 \quad \text{on } S, \tag{27}$$

$$\sum_{i=1}^n (L_i + \mu_i) \delta_i t_b = \delta T_e^\mu \quad \text{on } G.$$

Here $\widetilde{\Delta}_i = \dfrac{1}{A_i B_i} \left[\Lambda_i^\alpha \dfrac{\partial}{\partial \alpha} \left(\dfrac{B_i}{A_i} \dfrac{\partial}{\partial \alpha} \right) + \Lambda_i^\beta \dfrac{\partial}{\partial \beta} \left(\dfrac{A_i}{B_i} \dfrac{\partial}{\partial \beta} \right) \right]$; $\Lambda_i^l = \lambda_i^l \delta_i$ is the reduced thermal conductivity of the ith layer in the direction l $(l = \alpha, \beta, \gamma)$; $\Omega_i = \omega_i \delta_i$; $h_i = \lambda_i^\gamma / \delta_i$ are the reduced thermal capacity and thermal permeability of the ith layer; $\Omega = \sum\limits_{i=1}^{n}\Omega_i$, $K = \sum\limits_{i=1}^{n}k_i\delta_i$, $H^{-1} = \sum\limits_{i=1}^{n}h_i^{-1}$ are the reduced thermal capacity, reduced curvature and reduced thermal resistance of the whole coating.

In the case when the principal curvatures and coefficients of the first fundamental form of the interfaces between layers can be regarded as approximately equal to the corresponding quantities for the body-coating interface ($A = A_i, B = B_i, k = k_i, i = \overline{1,n}$), conditions (27) become

$$\widetilde{\Delta}t_b - \lambda_b^{\gamma}\left(1 - 2K + \frac{\mu}{H}\right)\left(\frac{\partial t}{\partial \gamma}\right)_b + \mu(t_e - t_b) = \Omega\frac{\partial t_b}{\partial \tau}, \quad t_b|_{\tau=0} = T_0 \quad \text{on } S, \tag{28}$$

$$(\mathrm{L}+\mathrm{M})t_b = \delta T_e^{\mu} \quad \text{on } G,$$

where $\quad \widetilde{\Delta} = \dfrac{1}{AB}\left[\Lambda^{\alpha}\dfrac{\partial}{\partial\alpha}\left(\dfrac{B}{A}\dfrac{\partial}{\partial\alpha}\right) + \Lambda^{\beta}\dfrac{\partial}{\partial\beta}\left(\dfrac{A}{B}\dfrac{\partial}{\partial\beta}\right)\right], \quad \Lambda^l = \sum\limits_{i=1}^{n}\Lambda_i^l, \; l = \alpha,\beta;$

$$\mathrm{L} = n_1\frac{\Lambda^{\alpha}}{A}\frac{\partial}{\partial\alpha} + n_2\frac{\Lambda^{\beta}}{B}\frac{\partial}{\partial\beta}, \quad K = k\delta, \quad \mathrm{M} = \sum\limits_{i=1}^{n}\mu_i\delta_i.$$

For the cases when the shell that models the coating is closed, conditions (27)-(28) coincide with the ones, which have been obtained in [10]. Such modeling is also possible when the influence of the edge surface can be neglected due to its small value in the whole heat transfer.

5. Restoration Formulas

After determining the temperature field in a body based on solving the appropriate boundary value problem with generalized boundary conditions, we can find temperature over the thickness of a coating by restoration formulas.

These formulas can be obtained by substitution of change (6) to (14) using (16), (11), and (12). In a general case, they have the form

$$t_i = \exp(-k_i\gamma)\left\{\frac{\exp[(k_i - k_{i-1})\gamma_{i-1}]\sin(p_i(\gamma_i - \gamma))f_{i-1}^{11} + \sin(p_i(\gamma - \gamma_{i-1}))f_i^{11}}{\sin(p_i\delta_i)}t_b + \right. \tag{29}$$

$$\left. + \frac{\exp[(k_i - k_{i-1})\gamma_{i-1}]\sin(p_i(\gamma_i - \gamma))f_{i-1}^{12} + \sin(p_i(\gamma - \gamma_{i-1}))f_i^{12}}{\sin(p_i\delta_i)}\left(\frac{\partial t}{\partial\gamma}\right)_b\right\} \; i = \overline{1,n}.$$

Expanding (29) in power series for small δ_i, and omitting terms containing $\delta_k^m\delta_l^s$ with $m + s \geq 2$, we receive the following calculation variant

$$t_i = t_b + \lambda_b^{\gamma}\left[\sum\limits_{j=1}^{i-1}\frac{1 - 2K_{j-1/2}}{h_j} + \frac{1 - 2K_{i-1/2} + k_i(\gamma_i - \gamma)}{\lambda_i}(\gamma - \gamma_{i-1})\right]\left(\frac{\partial t}{\partial\gamma}\right)_b \quad i = \overline{1,n}; \tag{30}$$

where $\quad K_{i-1/2} = \sum\limits_{j=1}^{i-1}k_j\delta_j + \dfrac{k_i\delta_i}{2}.$

6. Test Problem

The efficiency of suggested approach is illustrated by the comparison of results, obtained according to the above approximate approach with the solution of a test one-dimension problem of non-stationary heat conduction when heating the plate with a threelayer isotropic coating.

Equations of heat conduction and initial conditions for the i-th layer and the body have the form

$$\frac{\partial t_i}{\partial \tau} = a_i \frac{\partial^2 t_i}{\partial x^2}, \tag{31}$$

$$t_i|_{\tau=0} = 0, \qquad i = 1,2,3,b. \tag{32}$$

We assume that the coating-environment boundary surface is heated according to Newton's law and the opposite side of the plate is thermally insulated

$$\lambda_3 \frac{\partial t_3}{\partial x} = \mu(t_3 - t_e) \text{ at } x = -(\delta_1 + \delta_2 + \delta_3) , \tag{33}$$

$$\frac{\partial t_b}{\partial x} = 0 \quad \text{at } x = h . \tag{34}$$

The conditions of ideal thermal contact occur at the interfaces

$$t_3 = t_2, \qquad \lambda_3 \frac{\partial t_3}{\partial x} = \lambda_2 \frac{\partial t_2}{\partial x} \qquad \text{at } x = -(\delta_1 + \delta_2) ; \tag{35}$$

$$t_2 = t_1, \qquad \lambda_2 \frac{\partial t_2}{\partial x} = \lambda_1 \frac{\partial t_1}{\partial x} \qquad \text{at } x = -\delta_1 ;$$

$$t_1 = t_b, \qquad \lambda_1 \frac{\partial t_3}{\partial x} = \lambda_b \frac{\partial t_2}{\partial x} \qquad \text{at } x = 0 .$$

Here $a_i = \lambda_i / \omega_i$ is the coefficient of thermal diffusivity of the ith layer $(i = 1,2,3)$ and of the body $(i = b)$; h is the thickness of the body.

The exact solution of this problem is given in [13]. The approximate solution of the problem on the basis of the above formulated approach is based on the solution of equation (31) in the body domain under the initial condition (32), the boundary condition of insulation (34), and the generalized boundary condition following from (28), which in the considered case has the form

$$\lambda_b \left(1 + \frac{\mu}{H}\right) \frac{\partial t_b}{\partial x} + \mu(t_e - t_b) = \Omega \frac{\partial t_b}{\partial \tau} , \qquad t_b|_{\tau=0} = 0 \quad \text{at } x = 0; \tag{36}$$

$$\frac{1}{H} = \frac{\delta_1}{\lambda_1} + \frac{\delta_2}{\lambda_2} + \frac{\delta_3}{\lambda_3}, \qquad \Omega = \omega_1 \delta_1 + \omega_2 \delta_2 + \omega_3 \delta_3 .$$

Using the Laplace transformation, we represent the solution of this problem in the form

$$\frac{t_b(x,\tau)}{t_e} = 1 - \sum_{j=1}^{\infty} \frac{2\mu h^2 U(x,\kappa_j)\exp(-\kappa_j^2 a_b \tau)}{Z(\kappa_j)}, \quad 0 \le x \le h, \tag{37}$$

$$U(x,\kappa_j) = \cos\kappa_j \cos\frac{x\kappa_j}{h} + \sin\kappa_j \sin\frac{x\kappa_j}{h},$$

$$Z(\kappa_j) = \left[2\Omega a_b + \lambda_b\left(1+\frac{\mu}{H}\right)h\right]\kappa_j \cos\kappa_j + \left[\mu h^2 - \Omega a_b \kappa_j^2 + \lambda_b\left(1+\frac{\mu}{H}\right)h\right]\sin\kappa_j,$$

where κ_j are roots of the equation $\operatorname{tg}\kappa = \dfrac{\left(\mu h^2 - \Omega a_b \kappa\right)}{\lambda_b\left(1+\dfrac{\mu}{H}\right)h\kappa}$.

Substituting (37) in the restoration formulas (30), we obtain the expression for temperature in the coating:

$$\frac{t_i(x,\tau)}{t_e} = 1 - \sum_{j=1}^{\infty} \frac{2\mu h^2 V(x,\kappa_j)\exp(-\kappa_j^2 a_b \tau)}{Z(\kappa_j)}, \quad -(\delta_1 + \delta_2 + \delta_3) \le x \le 0, \tag{38}$$

$$V(x,\kappa_j) = \cos\kappa_j + \lambda_b\left(\sum_{j=1}^{i-1} h_j^{-1} + \frac{x - \gamma_{i-1}}{\lambda_i}\right)\kappa_j \sin\kappa_j.$$

This approximate solution is compared with exact solution according [13] for certain time moments for the four-layer plate characterized by the following values of geometrical and physical parameters: $\delta_1 = \delta_2 = 0.5\delta_3 = 5\cdot10^{-5}m$, $h = 2\cdot10^{-2}m$, $\lambda_1 = 2\,W/m\cdot {}^0C$, $\lambda_2 = 4W/m\cdot {}^0C$, $\lambda_3 = 6W/m\cdot {}^0C$, $\lambda_b = 4W/m\cdot {}^0C$, $a_1 = 5\cdot10^{-6}m^2/S$, $a_2 = 6.4\cdot10^{-6}m^2/S$, $a_3 = 8\cdot10^{-6}m^2/S$, $a_b = 1.2\cdot10^{-5}m^2/S$, $t_e = 100\,{}^0C$, $Bi_1 = \mu(\delta_1 + \delta_2 + \delta_3)/\lambda_1 = 1$.

The analysis of the indicated results obtained for certain time moments (Table 1) shows that the values of temperatures of exact (lower values in the lines) and approximate (upper values in the lines) solutions do not differ essentially.

Table 1. Comparison of exact results and approximate solution

τ sec	$t, {}^0C$							
	$x=-\gamma_3$	$x=-\gamma_2$	$x=-\gamma_1$	$x=0$	$x=0.25h$	$x=0.5h$	$x=0.75h$	$x=h$
2	12.040	11.901	11.796	11.587	3.991	0.955	0.153	0.031
	12.047	11.830	11.722	11.581	3.987	0.953	0.152	0.032
10	24.455	24.333	24.241	24.057	16.015	10.320	6.959	5.852
	24.458	24.271	24.178	24.055	16.013	10.319	6.960	5.856
40	48.287	48.203	48.140	48.014	42.417	38.312	35.806	34.963
	48.294	48.166	48.102	48.018	42.422	38.318	36.813	34.972
100	75.522	75.483	75.453	75.393	72.744	70.801	69.615	69.216
	75.531	75.470	75.440	75.400	72.752	70.810	69.624	69.226
500	99.833	99.833	99.832	99.832	99.814	99.801	99.792	99.790
	99.833	99.833	99.832	99.832	99.814	99.801	99.793	99.790

7. Concluding Remarks

These generalized boundary conditions may be used for an analytical solution (when it is possible) as well as for a numerical one. Since the application of direct numerical methods without the preliminary transformation of initial problems for massive bodies with thin coatings may be impeded (because of the essential difference between geometrical and thermal characteristics of the coating and the base), one may expect the efficiency of application of approaches based on the use of generalized boundary conditions. The suggested approach has the following advantages:

1. It allows us, on the basis of a generalized model, to simplify essentially the calculation of constructions with thin multilayer coatings and to reduce, respectively, the time taken by computation.

2. It gives a possibility of obtaining in certain cases relatively simple analytical solutions of important practical problems for piecewise homogeneous media, which allow us to provide a priori qualitative and quantitative estimation of the thermal state of the constructions without cumbersome calculations.

3. The efficiency of this approach increases with the decrease of the coating thickness in contrast to the application of direct methods without preliminary transformation of initial problems.

References

1. Al Nimr, M.A., Alcam, M.K.: A generalized thermal boundary condition. Int. J. Heat and Mass Transfer. 33 (1997) 157-161
2. Elperin, T., Rudin, G.: Analytical solution of heat conduction problem for a multilayer assembly arising in photothermal reliability testing. Int. Comm. in Heat and Mass Transfer. 21 (1994) 95-104
3. Fleishman, N.P.: Mathematical models of thermal conjugation of media with thin foreign insertions or coatings. Visnyk of L'viv University. Ser. Mech-Math. 39 (1993) 30-34
4. Podstrigach, Ya.S.: On application of operator method to derivation of main relations of theory of heat conduction in thin-walled construction elements. Teplovye Napryazhenia V Elementakh Konstruktsij (Thermal Stresses in Construction Elements). 5 (1965) 24-35
5. Podstrigach, Ya.S., Shevchuk, P.R.: Temperature fields and stresses in bodies with thin coverings. Ibid. 7 (1967) 227-233
6. Podstrigach, Ya.S., Shevchuk, P.R., Onufrik T.M., Povstenko Yu.Z.: Surface effects in solid bodies taking into account coupling physico-mechanical processes. Physico-chemical Mechanics of Materials. 11 (1975) 36-41
7. Podstrygach, Ya.S., Shvets, R.N.: Thermoelasticity of Thin Shells. Naukova Dumka, Kiev (1978)
8. Savula, Ya.H.: Mathematical model of heat conduction through three-dimensional body with thin plate coating. Visnyk of L'viv University. Ser. Mech-Math. 42 (1995) 3-7
9. Savula, Ya.H., Dyyak, I. I., Krevs, V.V.: Heterogeneous mathematical models in numerical analysis of structures. Computers and Mathematics with Applications. 42 (2001) 1201-1216
10. Shevchuk V.A.: Generalized boundary conditions for heat transfer between a body and the surrounding medium through a multilayer thin covering. J. Math. Sci. 81 (1996) 3099-3102
11. Shvets, Yu.I, Prokopov, V.G., Fialko, N. M., et al.: Thermal state of parts with multilayer lacquer coatings. Visnyk of Academy of Sciences of Ukraine. 5 (1987) 39-44
12. Tret'yatchenko, G.N., Barilo, V.G.: Thermal and stressed state of multilayer coatings. Problems of Strength. 1 (1993) 41-43
13. Vendin, S.V.: On calculation of non-stationary heat conduction in multilayer objects with boundary conditions of the third kind. J. Eng. Physics and Thermophysics 65 (1994) 823-825

An Irregular Grid Method for Solving High-Dimensional Problems in Finance

Steffan Berridge* and Hans Schumacher

Department of Econometrics and Operations Research and
Center for Economic Research (CentER),
Tilburg University, Postbus 90153, 5000 LE Tilburg, The Netherlands.
s.j.berridge@kub.nl,j.m.schumacher@kub.nl

Abstract. We propose and test a new method for pricing American options in a high dimensional setting. The method is centred around the approximation of the associated variational inequality on an irregular grid. We approximate the partial differential operator on this grid by appealing to the SDE representation of the stock process and computing the logarithm of the transition probability matrix of an approximating Markov chain. The results of numerical tests in five dimensions are promising.

1 Introduction

The current mathematical approach to option pricing was first introduced by Black and Scholes [1] and Merton [10] in 1973. They showed that pricing a European option on a stock is equivalent to solving a certain parabolic initial boundary value problem, under some simplifying assumptions.

This pricing paradigm has revolutionised the financial world and much work has gone into extending this framework. In particular the pricing of American options, which involves the solution of a parabolic free-boundary problem under the Black-Scholes-Merton assumptions, has drawn much attention since most traded options are of this type. Given the nature of the problem, it is no surprise that numerical methods are nearly always used for pricing American options.

An example of such a product is a put option on a stock, giving the buyer the right to sell a certain stock for a fixed price K at some fixed future date T. In effect this gives the buyer a payoff of $\psi(x) = \max(K - x, 0)$ currency units at the time of exercise where x is the stock price at exercise date T and K is the so-called strike price. If exercise is allowed only at T, the option is called European and if it is also allowed at any time before T it is called American.

In practise we are often confronted with problems involving several state variables such as an option written on several underlying assets or a pricing problem in which we allow some of the model parameters to become stochastic. Pricing an American option in this case amounts to solving a free-boundary problem in a high-dimensional space. Numerical methods give some hope of

---

* Research supported by Netherlands Organisation for Scientific Research (NWO)

P.M.A. Sloot et al. (Eds.): ICCS 2002, LNCS 2330, pp. 510–519, 2002.

finding approximate solutions in this case, but it is well known that the work involved in current grid-based methods grows exponentially with the number of dimensions; this is sometimes called the curse of dimensionality.

An alternative to grid-based methods is Monte Carlo simulation of the corresponding stochastic differential equation (SDE). The European pricing problem can be solved even in a high dimensional setting using this technique since it only involves a numerical integration of the payoff function with respect to the density of the state variables.

American problems in one dimension can often be solved under the Black-Scholes-Merton assumptions by using a method where the time and state variable axes are discretised into a regular grid and a complementarity problem solved at each time-step. Unfortunately this method suffers from the curse of dimensionality.

Apart from finding the option value for a given specification of the state variables, it is also of practical interest to find the sensitivity of the price to the values of the state variables, in particular to find approximations to the first and second derivatives. In specifying numerical schemes we would like to take this into account.

The literature on option pricing has been extensive since 1973. However there is only a limited number of papers dealing with high dimensional problems, and only a few of these concentrate on the American case.

Broadie and Glasserman [2] propose a method based on approximating the stock price dynamics with a stochastic tree in which they obtain two consistent estimators, one biased high and the other low. This leads naturally to a confidence interval method for pricing American options; however the number of nodes in the tree increases exponentially with the number of exercise opportunities. In a later paper [3] they suggest a method based on a stochastic mesh which alleviates this problem; however Fu et al. [6] find that this method has a substantial upward bias in their numerical tests.

Longstaff and Schwartz [9] propose and test a dynamic programming-like method based on estimating the value of continuation at each time step by projecting realised continuation values onto a linear combination of polynomial basis functions. They call this method least squares Monte Carlo (LSM). They provide few theoretical justifications for this procedure, although the results are quite reasonable for the examples they consider. Independently, Tsitsiklis and Van Roy [14] provide theoretical justification for such a method based on the projections onto the set of basis functions (features) being orthogonal, the orthogonality being with respect to a suitably chosen inner product which changes at each time step. The error in the method is bounded by a function of the error inherent in approximating the value function by the features, which is in practice difficult to assess since the value function is exactly what we want to determine.

Stentoft [13] and Moreno and Navas [11] perform further numerical tests on the pricing of American options using the LSM method, with promising results in simple cases. However results for more complex high dimensional examples are mixed (see [11]). Fu et al. [6] perform extensive numerical tests on American option pricing using several alternative Monte Carlo-based methods.

It is clear that a regular grid approach cannot work in high dimensions. Insetad, we approximate the diffusion on an irregular grid. This grid can in principle be arbitrary, but we think of it as having been generated with a pseudo or quasi Monte Carlo method. This has the advantage that the number of points in the grid can be directly controlled, and thus does not have to grow exponentially with the dimension.

We then construct a Markov chain by defining transition probabilities between grid points in a continuous time setting in such a way that the solution to the Markov chain converges to the solution to the PDE as the number of points in the grid increases. This gives us an approximation to the partial differential operator which we use to solve the PDE in the European case and the variational inequality in the American case.

2   Formulation

There are two main paradigms which allow us to formulate and compute values for options.

In the first place we have the SDE paradigm in which the value of the option is obtained as the discounted expected payoff at expiry T. This is the most natural for financial applications since we directly use the financial processes that are specified in our model.

For a European option with payoff function ψ this leads to the pricing formula

$$v_t = e^{-r(T-t)}\mathbb{E}\left(\psi(X_T)\right) \tag{1}$$

at time t where r is the (constant) risk free rate and the expectation is taken under the risk neutral dynamics of the system. This naturally leads to a numerical method using Monte Carlo trials

$$\hat{v}_t = e^{-r(T-t)}\frac{1}{N}\sum_{i=1}^{N}\psi(X_T(\omega))$$

where $X_T(\omega)$ are random variables with the risk neutral densities implied by the SDE. In the American option case we also have to optimise with respect to the exercise date. In this case the pricing problem becomes an optimal stopping problem and

$$v_t = \sup_{\tau \in \mathcal{T}}\mathbb{E}\left(e^{-r(T-\tau)}\psi(X_\tau)\right) \tag{2}$$

where $\mathcal{T}$ is the set of all stopping times with respect to the natural filtration of the underlying process. This American formulation does not lead in such a natural fashion to a pricing method.

In the second place we have the PDE paradigm in which the problem of pricing European options becomes equivalent to solving a linear parabolic initial boundary value problem, backwards in time. Here the initial value is the option payoff at expiry and the coefficients of the PDE are inferred from those of the SDE. If we restrict our region of interest to a subset $\Omega \subset \mathbb{R}^n$ then we want to find $v_0(x)$ such that

$$
\begin{aligned}
v(x,t)\,|_{t=T} &= \psi(x) \quad \text{for } x \in \Omega \\
v(x,t) &= \phi(x,t) \quad \text{for } x \in \partial\Omega,\ t \in [0,T] \\
\frac{\partial v}{\partial t} + \mathcal{L}v &= 0 \quad \text{for } (x,t) \in (\Omega\backslash\partial\Omega) \times [0,T]
\end{aligned}
\tag{3}
$$

where $\mathcal{L}$ is a second degree elliptic operator and $\phi(x,t)$ specifies appropriate boundary conditions. Note that it is not formally necessary to restrict the problem to $\Omega \subset \mathbb{R}^n$ but computationally it is convenient. Such problems possess a unique solution given suitable regularity conditions. This formulation for the price of a European option leads naturally to finite difference and finite element methods in which the most time consuming operation is solving a system of linear equations at successive time points.

For American options the early exercise property manifests itself as a free boundary in the PDE. The problem becomes to find the solution $v(x,t)$ to the variational inequality

$$
\left\{
\begin{aligned}
\frac{\partial v}{\partial t} + \mathcal{L}v &\geqslant 0 \\
v - \psi &\geqslant 0 \quad \text{for } (x,t) \in (\Omega\backslash\partial\Omega) \times [0,T] \\
\left(\frac{\partial v}{\partial t} + \mathcal{L}v\right)(v - \psi) &= 0
\end{aligned}
\right.
\tag{4}
$$

with initial conditions $v(x,t)\,|_{t=T} = \psi(x)$ for $x \in \Omega$ and boundary conditions $v(x,t) = \phi(x,t)$ for $x \in \partial\Omega$, $t \in [0,T]$.

Again some regularity conditions are required for the problem to possess a unique solution (see Jaillet et al. [7]). The most popular numerical method to solve the problem in up to two dimensions is formulated by adapting a finite difference method using projected SOR (PSOR) so that the extra constraint is satisfied at each time-step. The discretised system can be treated as a linear complementarity problem. The existence and uniqueness of solutions at each time-step can be proved when the matrix M_L (see Section 3) multiplying the vector of approximate values $\mathbf{v}(t)$ is of type P (see Cottle, Pang and Stone [4]). The convergence of PSOR for real symmetric positive definite M_L is proved by Cryer [5] and the convergence of the overall computed solution at $t = 0$ (allowing for numerical errors at each time-step) is proved for certain classes of payoff functions ψ by Jaillet et al. [7]. The European and American pricing problems in one dimension are treated in detail in this paradigm by Wilmott, Dewynne and Howison [15] for example.

3   An Irregular Grid Method

The method we propose basically follows the second of the two paradigms mentioned in Section 2, but we also make use of the SDE paradigm and Monte Carlo methodology. We approximate the value function $v(x,t)$ on a grid just as in the PDE method, but, to avoid an exponential growth of the number of grid points, we use an irregular grid. We first approximate the partial differential operator in the space direction via a semidiscrete Markov chain approximation of the SDE. Then we discretise in the second argument, using a time-stepping method which gives rise to a system of linear equations (or a complementarity problem in the American case) at each time step.

It is important to specify our approximation A in a consistent manner, in the sense that the solution obtained via the numerical method converges to the true solution as the number of points in the grid (and the number of time steps) goes to infinity. The construction of such an approximation is the main consideration of Section 4.

In this analysis we assume homogeneous Neumann conditions at the boundary (where the derivative of $v_{.,t}$ in the direction normal to $\partial\Omega$ is zero). We plan to extend the analysis to other types of boundary conditions in future work.

Suppose that $\mathcal{X} = \{x_1, \ldots x_N\}$ is a representative set of points (states) in Ω on which our process can evolve for $t \in [0,T]$. For the moment we can think of $\mathcal{X}$ as being a generic set of low discrepancy points in the sense of Niederreiter [12]. The structure of $\mathcal{X}$ will be specified in more detail later.

Let A be our discrete approximation to the diffusion operator $\mathcal{L}$ on $\mathcal{X}$, let $v_i(t)$ denote the approximated value of $v(x_i,t)$, at points x_i in $\mathcal{X}$, in continuous time and let $\mathbf{v}(t) = (v_1(t), \ldots, v_N(t))'$ and $\psi = (\psi(x_1), \ldots, \psi(x_N))'$. Thus in the European case we now wish to find the solution $\mathbf{v}(0)$ to the system of ODEs

$$\frac{d\mathbf{v}}{dt}(t) + A\mathbf{v}(t) = 0 \quad \text{for } t \in [0,T] \tag{5}$$

with initial condition $\mathbf{v}(T) = \psi$.

We now discretise the time axis so that the problem can be solved numerically at intermediate time steps. For a small time step $\delta t = T/K$ a simple approach is to discretise this system using a θ-method, where $\theta = 0, \frac{1}{2}, 1$ corresponds to explicit, Crank-Nicolson and implicit discretisation respectively. Thus for $k = K-1, \ldots, 0$ we have

$$\frac{\mathbf{v}((k+1)\delta t) - \mathbf{v}(k\delta t)}{\delta t} + \theta A\mathbf{v}(k\delta t) + (1-\theta)A\mathbf{v}((k+1)\delta t) = 0 \tag{6}$$

and the approximate solution is obtained by solving

$$M_L \mathbf{v}^k = M_R \mathbf{v}^{k+1} \qquad k = K-1, \ldots, 0 \tag{7}$$

with initial condition $\mathbf{v}^K = \psi$ where $\mathbf{v}^k$ is the approximation to $\mathbf{v}(k\delta t)$, $K = T/\delta t$ is the number of time steps and the matrices M_L and M_R are equal to $(I - \theta A\delta t)$ and $(I + (1-\theta)A\delta t)$ respectively. Numerically we must solve a system of linear equations at each step, and thus if N is large, we would like the matrix A to be sparse.

In the American case we must include the American constraint $\mathbf{v}^k \geqslant \psi$, so the complementarity problem to solve is

$$
\begin{cases}
M_L \mathbf{v}^k - M_R \mathbf{v}^{(k+1)} \geqslant 0 \\
\mathbf{v}^k - \psi \geqslant 0 \qquad k = K-1, \ldots, 0 \\
\left(M_L \mathbf{v}^k - M_R \mathbf{v}^{(k+1)} \right) \left(v^k - \psi \right) = 0
\end{cases}
\tag{8}
$$

with initial condition $\mathbf{v}^K = \psi$ and where the inequalities are componentwise. Numerically we must solve a complementarity problem at each step and again the sparsity of A is of great interest when it comes to solving such a system for large N. The complementarity problem can be solved using PSOR or linear programming, depending on the nature of the matrices involved.

4   Approximating the Partial Differential Operator

Here we provide a method for approximating the partial differential operator $\mathcal{L}$ on our grid $\mathcal{X}$. To do this we appeal to the SDE representation of the problem.

We assume that the density of the SDE $f_{x_0,t}(x)$ is available for arbitrary initial points x_0 and time horizons t. We first choose a time horizon T_0 at which the density can be approximated suitably well on our grid $\mathcal{X}$ for initial points in $\mathcal{X}$. Before continuing we must specify the nature of the grid $\mathcal{X}$. We suppose that $\mathcal{X}$ consists of points generated from pseudo or quasi Monte Carlo trials with (importance sampling) density $g(x)$ where $g(x) > 0$ whenever $f(x) > 0$ and $x \in \Omega$.

We now let the transition probability between x_i and x_j for time horizon T_0 be approximated by the continuous density $f_{x_i,T_0}(x_j)$ weighted by the importance sampling or empirical density $\tilde{g}(x)$. That is, our transition probability matrix is $P_{T_0} = (p_{ij})$ where $p_{ij} = \mathbb{P}(X_{T_0} = x_j | X_0 = x_i)$ and we set

$$
p_{ij} = \frac{1}{\sum_{k=1}^N \tilde{f}_{x_i,T_0}(x_k)} \cdot \tilde{f}_{x_i,T_0}(x_j)
\tag{9}
$$

and $\tilde{f}_{x_i,T_0} = \frac{f_{x_i,T_0}}{\tilde{g}}$.

As mentioned earlier, we assume homogeneous Neumann boundary conditions; in this case the value function is relatively flat in a region of the boundary, so that the distortion of probabilities in the Markov chain near the boundary will not affect the consistency of the solution.

In the European case, with no time discretisation and choosing $T_0 = T$, we see that solving the option pricing problem using these transition probabilities amounts to Monte Carlo integration with importance sampling,

$$
\mathbf{v}(0) = \int_{\mathbb{R}^n} \psi(x) f_{x_i,t}(x) dx \approx \int_\Omega \psi(x) \left[\frac{f}{g} \right](x) \left[g(x) dx \right]
$$

$$
\approx \sum_{j=1}^N \psi(X_j) \frac{\tilde{f}_{x_i,T}(X_j)}{\sum_{k=1}^N \tilde{f}_{x_i,T}(X_k)} = \sum_{j=1}^N \psi(X_j) p_{ij}
$$

where X_j are iid random variables with density $g(x)$ and the importance sampling function may also incorporate the empirical density of the grid. The approximation $\hat{\mathbf{v}}(0) = P_T \psi$ is good for points x_i in the centre of our grid where the transition probabilities approximate the density well. The American problem is more complicated in that a time discretisation is called for.

The transition probability matrix P now gives us access to an approximation to the partial differential operator on our grid. We note that the evolution of state probabilities in the semidiscrete setting is given by $p(t) = e^{A't} p(0)$ where $p(t)$ is the discrete probability distribution over our grid at time t. Thus we obtain an approximation to $\mathcal{L}$ through finding the following matrix

$$A = \frac{1}{T_0} \log(P_{T_0}) \tag{10}$$

In effect $A = (a_{ij})$ where $a_{ij} = \lim_{\delta t \downarrow 0} \frac{1}{\delta t} [\mathbb{P}(X_{t+\delta t} = x_j | X_t = x_i) - \delta_{ij}]$.

We could also find a transition probability matrix for small time steps δt by taking a root of P_{T_0}, $P_{\delta t} = e^{A \delta t} = (P_{T_0})^{\delta t/T_0}$. This could in principle be useful for directly specifying the matrices M_L and M_R as used in Section 3, and has been found to be a faster operation than computing the logarithm in Matlab.

The more naive approach of calculating the transition probability matrix directly through (9) with $T_0 = \delta t$ has been found to lead to inaccurate results. This suggests that the transition probabilities do not reflect the density of the process over longer horizons when they are calculated in this way.

5   Experimental Results

We conduct experiments to test our proposed method in a Matlab environment, and using no special techniques to accelerate the speed of computation. In particular no attention is paid to the approximately sparse nature of the transition probability matrix when calculating the logarithm. This meant that the maximum feasible grid size was 1500, which is a relatively low number in terms of Monte Carlo integration.

Our experiments are based on results from Stentoft [13], who obtains approximate prices for options written on three and five assets using the LSM method. He considers stock processes driven by correlated Brownian motions.

Specifically our stock prices $\mathbf{S} = (S_1, \ldots, S_n)$ (with Itô correction term included) are given by $\mathbf{S}(t) = \mathbf{S}(0) \exp\{(r\mathbf{1} - \frac{1}{2}\mathrm{diag}(\Sigma))t + R\mathbf{W}(t)\}$ where $\mathbf{W}(t) = (W_1(t), \ldots, W_n(t))'$ is a vector of independent Brownian motions, $\mathbf{1}$ is a vector of ones, $\mathrm{diag}(\Sigma)$ is a vector of the diagonal entries of Σ and R is a Cholesky factor of the covariance matrix $\Sigma = RR'$, the latter giving the covariances of the stock processes in the log domain.

We are given initial stock prices $S_i(0) = 40$ for each i, the correlations between log stock prices are $\rho_{ij} = 0.25$, $i \neq j$, and volatilities[1] are $\sigma_i^2 = 0.04$ for all i, the risk-free interest rate is fixed at $r = 0.06$ and the expiry is $T = 1$.

---

[1] In Stentoft's paper the volatilities are incorrectly specified as $\sigma_i^2 = 0.2$.

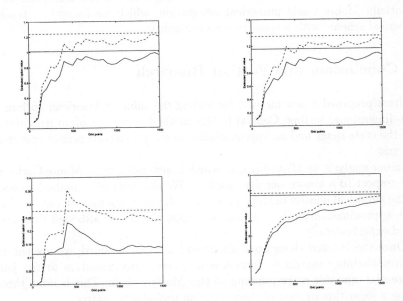

Fig. 1. QMC grid valuation of European (*solid lines*) and American (*dotted lines*) put options on the arithmetic average, geometric average, maximum and minimum respectively over five assets. Stentoft's solutions are drawn as horizontal lines

Furthermore we set $T_0 = T = 1$ in the approximation of the partial differential operator and use implicitness $\theta = \frac{1}{2}$ and $K = 20$ time-steps (in the European case we found that the difference between the ($K = 20$) time-discretised and single time-step calculations was negligible).

Our grid is generated using Sobol points with uniform density $g(\mathbf{x}) \equiv \left(\int_\Omega d\mathbf{x} \right)^{-1}$. The first grid point is set to be the vector of log stock prices at $t = 0$; we only consider the estimated value at this point in our convergence analysis. The region of interest Ω is set to be the rectilinear region (L, U) where $L_i = E(S_i(T)) - 3\sigma_i\sqrt{T}$ and $U_i = E(S_i(T)) + 3\sigma_i\sqrt{T}$.

The payoff functions considered correspond to put options on the maximum, minimum, arithmetic mean and geometric mean which have respective payoff functions $\psi_1(\mathbf{x}) = (K - \max_i(x_i))^+$, $\psi_2(\mathbf{x}) = (K - \min_i(x_i))^+$, $\psi_3(\mathbf{x}) = \left(K - \frac{1}{n} \sum x_i \right)^+$ and $\psi_4(\mathbf{x}) = \left(K - (\prod x_i)^{1/n} \right)^+$.

Figure 1 shows the QMC grid approximation of the option values for increasing numbers of grid points. For most cases we see that the approximation for

1500 grid points is within 10% of Stentoft's estimate; however it appears that more than 1500 points are required to obtain a satisfactory convergence. This is not surprising if we consider the number of trials required to obtain accurate results in numerical integration.

The similarity between American and European convergence patterns is encouraging if we consider that the numerical scheme in the European case is essentially Monte Carlo numerical integration, which we know is a tractable numerical scheme with respect to dimensionality.

6   Conclusions and Further Research

We have proposed a new method for finding the value of American options in a high-dimensional setting. Central to this method is the use of an irregular grid over the state space and an approximation of the partial differential operator on this grid.

In our analysis we allow any grid which is generated using Monte Carlo trials with respect to a known density function. We also suggest a method to correct for the empirical density of these points in order to give a more accurate Markov chain approximation to the SDE on our region of interest, although this was not tested experimentally.

Once the Markov chain approximation has been obtained, we use the transition probability matrix to form a semidiscrete approximation to the partial differential operator corresponding to this Markov chain. This is done through taking a logarithm or root of the transition probability matrix.

Our initial experiments suggest that this method has promising convergence properties. Unfortunately we have not yet been able to use more than 1500 points in our grid since we have not implemented an efficient method for computing the matrix logarithm or root. Such a method could take account of the fact that the logarithm and root of the transition probability matrix would have the same approximately sparse structure as the transition probability matrix itself.

In terms of future research possibilities there are a number of areas which can be explored, in terms of both the numerical and theoretical properties of the algorithm.

To consolidate the numerical results contained in this paper, we would like to increase the number of points in the grid. This relies crucially on being able to determine the matrix logarithm or root in a more efficient manner. Two ways to approach this are to use the approximately sparse nature of the transition probability matrix or to use Krylov subspace methods. Such a procedure may add a further approximation to the algorithm, but the effects of this are expected to be easily outweighed by the efficiency gains.

In the experiments a uniform grid was used, for convenience; however the evolution of the SDE would be better approximated on a grid of radially decreasing density, such as a normally distributed grid. We would also like to explore the possibility of making use of the empirical density of the points to further adjust the transition probabilities. This is expected to improve the accuracy of the transition probability matrix especially when, for example, two points in the grid are very close together.

Other techniques which may prove useful are randomised QMC, which could help us in analysing the standard error of solutions, and various variance reduction techniques such as the use of control variates.

Extra information can also be obtained using our method. In particular we could manipulate the grid in order to be able to easily estimate derivatives of the option value, the so-called Greeks. This would be done by concentrating more points around the initial asset values, either by directly placing extra points or by modifying the importance sampling density.

Further work is also needed to obtain theoretical justification of the proposed method. Such a justification may follow the analysis of Jaillet et al. [7] in the American case and the analysis of Kushner and Dupuis [8] for the more general dynamic programming problem, but would also have to take into account the pseudo or quasi Monte Carlo nature of the solution method.

References

1. Black, F., Scholes, M.: The pricing of options and corporate liabilities. Journal of Political Economy **81** (1973) 637–654
2. Broadie, M., Glasserman, P.: Pricing American-style securities using simulation. Journal of Economic Dynamics and Control **21** (1997) 1323–1352
3. Broadie, M., Glasserman, P.: A stochastic mesh method for pricing high-dimensional American options. Working paper (1997)
4. Cottle, R.W., Pang, J., Stone, R.E.: The Linear Complementarity Problem. Academic Press (1992)
5. Cryer, C.W.: The solution of a quadratic programming problem using systematic overrelaxation. SIAM Journal of Control **9** (1971) 385–392
6. Fu, M.C., Laprise, S.B., Madan, D.B., Su, Y., Wu, R.: Pricing American options: a comparison of Monte Carlo simulation approaches. Journal of Computational Finance **4** (2001) 39–88
7. Jaillet, P., Lamberton, D., Lapeyre, B.: Variational inequalities and the pricing of American options. Acta Applicandae Mathematicae **21** (1990) 263–289
8. Kushner, H.J., Dupuis, P.G.: Numerical Methods for Stochstic Control Problems in Continuous Time. Springer-Verlag (1992)
9. Longstaff, F.A., Schwartz, E.S.: Valuing American options by simulation: a simple least squares approach. Review of Financial Studies **14** (2001) 113–147
10. Merton, R.C.: Theory of rational option pricing. Bell Journal of Economics and Management Science **4** (1973) 141–183
11. Moreno, M., Navas, J.F.: On the robustness of least-squares Monte Carlo (LSM) for pricing American options. Working paper (2001)
12. Niederreiter, H.: Random Number Generation and Quasi-Monte Carlo Methods. SIAM (1992)
13. Stentoft, L.: Assessing the least squares Monte-Carlo approach to American option valuation. CAF Working paper No. 90 (2000)
14. Tsitsiklis, J.N., Roy, B. Van: Regression methods for pricing complex American-style options. Working paper (2000)
15. Wilmott, P., Dewynne, J., Howison, S.: Option Pricing: Mathematical Models and Computation. Oxford Financial Press (1993)

On Polynomial and Polynomial Matrix Interpolation

Petr Hušek and Renata Pytelková

Department of Control Engineering,
Faculty of Electrical Engineering
Czech Technical University in Prague
Technická 2, 166 27, Praha 6, Czech Republic
{husek,dvorako}@fel.cvut.cz

Abstract. The classical algorithms for computations with polynomials and polynomial matrices use elementary operations with their coefficients. The relative accuracy of such algorithms is relatively small and for polynomials of higher order and polynomial matrices of higher dimension the executing time grows very quickly. Another possibility is to use symbolic manipulation package but even this is applicable only for moderate problems. This paper improves a new method based on polynomial interpolation. Its principle is as follows [1]: firstly a sufficient number of interpolation points is chosen, then the interpolated object is evaluated in these points and finally it is recovered from both series of values. The choice of interpolation points is crucial to have a well-conditioned task. Typically, a random choice of real points leads to a badly conditioning for higher order of interpolated polynomial. However, a set of complex points regularly distributed on the unit circle in the complex plane gives a perfectly conditioned task. Moreover very efficient algorithm of fast Fourier transform can be used to recover the resulted polynomial or polynomial matrix. The efficiency is demonstrated on determination of inverse to polynomial matrix.

1 Introduction

Solution of many technical problems leads to computations with polynomials or polynomial matrices. The classical algorithms, which are based on manipulations with coefficients, are distinguished for small relative accuracy and low efficiency. The algorithms based on interpolation-evaluation techniques represent a computationally efficient way to deal with such computations. The interpolation theory for polynomial case is very old. Nevertheless even the problem is very well studied all the algorithms meet the problem of badly conditioned task and could be used only for polynomials of lower degree (up to 30). The generalization to the matrix case appeared only recently. The most general way, which includes all the interpolation schemes, was introduced by Antsaklis [1], generalizing the approach [5]. However, the problem of badly conditioned task remained unsolved even if some ways for its removing were suggested. The different polynomial bases were used in Lagrange and Hermite interpolation ([6]).

P.M.A. Sloot et al. (Eds.): ICCS 2002, LNCS 2330, pp. 520–527, 2002.

The computational aspects of the enhancement to multivariate case are summarized in [3] and [4].

In this paper the conditioned number of corresponding task is improved by choosing a special set of interpolation points in both, polynomial and polynomial matrix case [7,8]. Moreover, the fast Fourier transform (FFT) algorithm is used for evaluation of the interpolated object in the prescribed set of points and its recovery from these values. The proposed algorithm is very efficient and numerically reliable.

2 Polynomial Case

Firstly let us remind, that the term *interpolation* can be used to denote many different mathematical tools. The technique, which will be here understood as interpolation, is described below. The basic idea of all interpolation methods can be summarized in the three following steps:

1. A sufficient number of interpolation points is chosen.
2. The interpolated object is evaluated in this set of points.
3. The interpolated object is recovered from both series of values.

The simplest example of interpolation technique is univariate polynomial interpolation. The corresponding procedure can be stated by the following theorem.

THEOREM 1 (Univariate polynomial interpolation [1])
Given K distinct generally complex scalars s_j, $j = 1, \ldots, K$, and K corresponding complex values b_j, there exists a unique polynomial $r(s)$ of degree $n = K - 1$ for which

$$r(s_j) = b_j, \quad j = 1, \ldots, K. \tag{1}\blacklozenge$$

That is, an nth degree polynomial $r(s)$ can be uniquely represented by the $K = n + 1$ interpolation points (or pairs) (s_j, b_j), $j = 1, \ldots, K$. To see this, the nth degree polynomial $r(s) = r_0 + r_1 s + \cdots + r_n s^n$ can be written as $r(s) = \mathbf{r}\,[1, s, \ldots, s^n]^T$ where $\mathbf{r} = [r_0, r_1, \ldots, r_n]$ is $(1 \times (n+1))$ row vector of the coefficients. The $K = n + 1$ equations can then be written as

$$\mathbf{rV} = \mathbf{r} \begin{bmatrix} 1 & \cdots & 1 \\ s_1 & & s_K \\ \vdots & & \vdots \\ s_1^{K-1} & \cdots & s_K^{K-1} \end{bmatrix} = [b_1, \ldots, b_K] = \mathbf{b}. \tag{2}$$

The matrix $\mathbf{V}$ is called Vandermonde matrix and it is non-singular if and only if the K scalars s_j, $j = 1, \ldots, K$ are distinct. In that case the equation (2) has a unique solution $\mathbf{r}$; that is, there exists a unique polynomial $r(s)$ of degree n which satisfies (1). This proves the theorem 1.

From numerical point of view the solution of equation (2) depends on the condition number of matrix $\mathbf{V}$. This problem is studied in the next section.

3 Condition Number of Vandermonde Matrix

It is well-known fact that the Vandermonde matrix (2) for vector of real numbers is very badly conditioned. Typically, for interpolation points regularly distributed in the interval [-1,1] the critical MATLAB condition number 10^{16} is achieved for $K = 30$. An immense improvement of the condition number of the Vandermonde matrix can be surprisingly achieved using complex interpolation points even if it seems to be contraproductive.

Let us choose the vector of interpolation points as

$$[s_1,...,s_K] = [1,\omega,\omega^2,...,\omega^{K-1}]/\sqrt{K} \tag{3}$$

where ω denotes a primitive K-th root of 1, that is

$$\omega^K = 1, \quad \omega^h \neq 1, \quad h = 1,...,K-1. \tag{4}$$

Such a set of interpolation points is called Fourier points. The corresponding Vandermonde matrix Ω (referred to as Fourier matrix) has the following form:

$$\Omega = \begin{bmatrix} 1 & 1 & 1 & ... & 1 \\ 1 & \omega & \omega^2 & ... & \omega^{K-1} \\ 1 & \omega^2 & \omega^4 & ... & \omega^{2(K-1)} \\ \vdots & \vdots & \vdots & \ddots & \vdots \\ 1 & \omega^{K-1} & \omega^{2(K-1)} & ... & \omega^{(K-1)(K-1)} \end{bmatrix} / \sqrt{K}. \tag{5}$$

Observe that since $\sum_{i=0}^{K-1}(\omega\tilde{\omega})^{ih} = K-1$ for $1 \leq h \leq K$ where $\tilde{\omega} = 1/\omega$ is a primitive root of 1, the Euclidean norm of each column appearing in the matrix Ω is equal to $\sqrt{K}$. Moreover since $\sum_{i=0}^{K}\omega^{hi} = 0$ for $1 \leq h \leq K$, the matrix Ω is unitary and

$$\text{cond}(\Omega) = 1. \tag{6}$$

It means that the Fourier matrix Ω is perfectly conditioned and its inverse can be determined as its complex conjugate

$$\Omega^{-1} = \Omega*. \tag{7}$$

The vector of coefficients of interpolated polynomial can be then obtained as the solution of (2) as

$$r = b\Omega*/\sqrt{K} = b[\tilde{\omega}_{ij}]/K. \tag{8}$$

Another important fact is that in order to evaluate a univariate polynomial in the Fourier points, that corresponds to step 1 of the general interpolation-evaluation algorithm mentioned above, the fast Fourier transform (FFT) algorithm could be applied. The algorithm is well studied, very efficient and numerically reliable. Moreover, as the Fourier matrix is unitary, the inverse FFT algorithm can be used to recover the polynomial coefficients from its values (step 3 of that algorithm or equivalently the product in (8)).

The use of the described algorithm will be shown on an example in section 6 on determination of determinant of polynomial matrix.

In the following section a generalization of the algorithm described above to polynomial matrix case is introduced.

4 Polynomial Matrix Case

The result obtained above can be generalized to polynomial matrix case in many different ways. Nevertheless all of them can be seen as special cases of the basic polynomial matrix interpolation procedure introduced by the following theorem.

Let $\mathbf{S}(s) := \text{block diag}\{[1, s, \ldots, s^{d_i}]^T\}$ where $d_i, i = 1, \ldots, m$ are non-negative integers; let $\mathbf{a}_j \neq 0$ and $\mathbf{b}_j$ denote $(m \times 1)$ and $(p \times 1)$ complex vectors respectively and s_j complex scalars.

THEOREM 2 (Polynomial matrix interpolation [3])
Given interpolation (points) triplets $(s_j, \mathbf{a}_j, \mathbf{b}_j)$, $j = 1, \ldots, K$ and non-negative integers $d_i, i = 1, \ldots, m$ with $K = \sum d_i + m$ such that the $(\sum d_i + m) \times K$ matrix

$$\mathbf{S}_K := [\mathbf{S}(s_1)\mathbf{a}_1, \ldots, \mathbf{S}(s_K)\mathbf{a}_K] \tag{9}$$

has full rank, there exists a unique $(p \times m)$ polynomial matrix $\mathbf{Q}(s)$, with ith column degree equal to $d_i, i = 1, \ldots, m$ for which

$$\mathbf{Q}(s_j)\mathbf{a}_j = \mathbf{b}_j, j = 1, \ldots, K. \tag{10}$$

$\mathbf{Q}(s)$ can be written as

$$\mathbf{Q}(s) = \mathbf{Q}\mathbf{S}(s) \tag{11}$$

where $\mathbf{Q}$ $(p \times (\sum d_i + m))$ contains the coefficients of the polynomial entries. $\mathbf{Q}$ must satisfy

$$\mathbf{Q}\mathbf{S}_K = \mathbf{B}_K \tag{12}$$

where $\mathbf{B}_K = [\mathbf{b}_1, \ldots, \mathbf{b}_K]$. Since $\mathbf{S}_K$ is non-singular, $\mathbf{Q}$ and therefore $\mathbf{Q}(s)$ are uniquely determined.

For $p = m = 1$ $\mathbf{S}_K$ is turned to *Vandermonde matrix*. In the polynomial matrix case, we shall call it *block Vandermonde matrix* because of its structure, that will be shown later.

Both theorems expand step 3 in the general description of interpolation methods introduced before. They show, that the resulted polynomial or polynomial matrix can be found as a solution of the matrix equation (6) or (8) respectively. It means, that the step 3 is the same for every problem solvable by interpolation techniques. The whole difference consists in step 2, which of course can be performed by algorithms working with constant matrices. These algorithms are usually very efficient and therefore guarantee good numerical properties of results computed in a short time.

5   Block Vandermonde matrix

Let us focus on the block Vandermonde matrix S_K appearing in (12) defined by (9). According to theorem 2 the choice of interpolation triplets $(s_j, \mathbf{a}_j, \mathbf{b}_j)$ is arbitrary. The only condition is the full rank of matrix S_K. The postmultiplication of interpolated matrix by the vectors $\mathbf{a}_j$ makes it possible to vary interpolation triplets not only by points s_j but also by the vectors $\mathbf{a}_j$. This is compensated by a higher number of points than they are necessary for element by element interpolation procedure. In section 3 it was shown that Vandermonde matrix has some appropriate properties if evaluated in the Fourier points. Let us try to choose such triplets $(s_j, \mathbf{a}_j, \mathbf{b}_j)$ to maintain this structure in the block Vandermonde matrix.

Let $K = \sum d_i + m$, $j = 1, \ldots, K$, $i = 1, \ldots, m$, $d_0 = 0$, $q = 1, \ldots, r$, $r = 1 + \max d_i$:

$$\text{let } s_j = s_q \, , \ \mathbf{a}_j = \mathbf{e}_i \ \text{for} \ j \in \left[\sum_{k=0}^{i-1} d_k + i, \sum_{k=1}^{i} d_k + i \right];$$

$$q = j - \sum_{k=1}^{i-1} d_k - i + 1, \ \mathbf{e}_i = [0, \ldots, 0, 1, 0, \ldots, 0] \tag{13}$$

where 1 is on the ith position.

This set of interpolation triplets requires only r different interpolation points s_q, which is the lowest possible number and it corresponds to the column-wise interpolation procedure. The block Vandermonde matrix S_K has the following form:

$$S_K = \begin{bmatrix} \mathbf{V}_1 & 0 & \cdots & 0 \\ 0 & \mathbf{V}_2 & \ddots & \vdots \\ \vdots & \ddots & \ddots & 0 \\ 0 & \cdots & 0 & \mathbf{V}_m \end{bmatrix} \tag{14}$$

where each $\mathbf{V}_i, i = 1, \ldots, m$ is the Vandermonde matrix

$$\mathbf{V}_i = \begin{bmatrix} 1 & 1 & \cdots & 1 \\ s_1 & s_2 & \cdots & s_{d_i} \\ \vdots & \vdots & \ddots & \vdots \\ s_1^{d_i} & s_2^{d_i} & \cdots & s_{d_i}^{d_i} \end{bmatrix}. \tag{15}$$

From (14) it immediately follows that the inverse to S_K appearing in solution of (12) can be determined as

$$S_K^{-1} = \begin{bmatrix} V_1^{-1} & 0 & \cdots & 0 \\ 0 & V_2^{-1} & \ddots & \vdots \\ \vdots & \ddots & \ddots & 0 \\ 0 & \cdots & 0 & V_m^{-1} \end{bmatrix}. \tag{16}$$

If the interpolation points s_j are chosen as the Fourier ones then $S_K = \text{diag}\{\Omega_i; i = 1,\ldots,m\}$ and $S_K^{-1} = \text{diag}\{\Omega_i^*; i = 1,\ldots,m\}$ where

$$\Omega_i = \begin{bmatrix} 1 & 1 & 1 & \cdots & 1 \\ 1 & \omega & \omega^2 & \cdots & \omega^{d_i} \\ 1 & \omega^2 & \omega^4 & \cdots & \omega^{2d_i} \\ \vdots & \vdots & \vdots & \ddots & \vdots \\ 1 & \omega^{d_i} & \omega^{2d_i} & \cdots & \omega^{d_i d_i} \end{bmatrix} / \sqrt{d_i + 1}. \tag{17}$$

Again the fast Fourier transform algorithm can be used for evaluation of $Q(s)$ in these points and its recovery from them.

6 Example

Let us use the described procedure to determine inverse to polynomial matrix $A(s)$,

$$A(s) = \begin{bmatrix} s^2 + 1 & s & s + 2 \\ s^2 & 1 & 2s^2 + 3 \\ 0 & s^2 + 2s + 2 & 2s + 1 \end{bmatrix}.$$

The inverse to polynomial matrix $A(s)$ can be determined as

$$A(s)^{-1} = [\det(A(s))]^{-1} \text{adj}(A(s)) = r(s)^{-1} Q(s)$$

where adj stands for adjoint to a matrix.

Firstly let us compute the determinant $r(s)$ using procedure described in section 2.

The first step consists in estimation of degree n of $r(s)$. It is important to note that if the estimated degree is higher than the real one, the coefficients corresponding to high powers will be determined as zero because due to theorem 1 there is only one polynomial of nth degree determined by $n + 1$ points (in fact it should be "of nth or less degree"). Obviously, the estimated degree cannot be less than the real one. The simplest choice is $n = \sum_{i=1}^{m} c_i$, where c_i are the column degrees of $A(s)$ (the highest degrees appearing in the column).

Here $n = 6$ and the Fourier matrix Ω is of the seventh ($K = n + 1 = 7$) order. Now we have to evaluate matrix $A(s)$ in the interpolation points $[1, \omega, \ldots, \omega^6]$ where ω is seventh root of 1. This step is performed by FFT algorithm. Next the vector of values of the determinants $\mathbf{b} = [\det(\mathbf{A}(1)), \det(\mathbf{A}(\omega)), \ldots, \det(\mathbf{A}(\omega^6))]$ is computed using standard procedure for computing determinant of a constant matrix. The recovery of all the coefficients of $r(s)$ (solution of (8)) is performed via inverse FFT applied on the vector $\mathbf{b}$.

The algorithm gives $r(s) = -2s^6 - 3s^5 - 7s^4 - 3s^3 - 8s^2 - 4s - 5$. Now let us determine the polynomial matrix $\mathbf{Q}(s) = \mathrm{adj}(\mathbf{A}(s))$.

As in the previous procedure firstly the column degrees d_i of $\mathbf{Q}(s)$ have to be estimated. The simplest choice is $d_i = (m-1)c, i = 1, \ldots, m$ where $c = \max c_i$. In our case $d_i = 4, i = 1, \ldots, 3$. Let us choose the interpolation points according to (13) as the Fourier points. We need $r = \max d_i + 1 = 5$ scalar complex points $s_j, j = 1, \ldots, 5$, $[s_1, \ldots, s_5] = [1, \omega, \ldots, \omega^4]$ where ω is the fifth root of 1. The total number of interpolation triplets is $K = \sum_{i=1}^{m} d_i + m = 15$. The procedure continues as follows. The polynomial matrix $\mathbf{A}(s)$ is evaluated in the points $s_j, j = 1, \ldots, 5$. Next the values of $\mathbf{Q}(s)$ are computed in those points: $\mathbf{Q}(s_j) = \mathrm{adj}(\mathbf{A}(s_j)), j = 1, \ldots, 5$. Finally, the solution of (12) can be determined by applying inverse FFT algorithm on each entry of $\mathbf{Q}(s_j)$.

The algorithm gives

$$\mathbf{Q}(s) = \begin{bmatrix} -2s^4 - 4s^3 - 7s^2 - 4s - 5 & s^3 + 2s^2 + 5s + 4 & 2s^3 + 2s - 2 \\ -2s^3 - s^2 & 2s^3 + s^2 + 2s + 1 & -2s^4 + s^3 - 3s^2 - 3 \\ s^4 + 2s^3 + 2s^2 & -s^4 - 2s^3 - 3s^2 - 2s - 2 & -s^3 + s^2 + 1 \end{bmatrix}$$

7 Conclusion

In the paper the general concept of algorithms on polynomials and polynomial matrices based on interpolation-evaluation methods was presented. The algorithms are very efficient and numerically reliable because of using the FFT algorithm, very well studied algorithms for constant matrices and perfectly conditioned task. The method was used for computing inverse to polynomial matrix. The experiments reveal that the algorithm is able to deal with polynomial matrices of high dimension with elements of high degrees. For example, determinant of a polynomial matrix (30×30) with elements of degrees 30 is computed in 2 seconds[1]. Adjoint to the same matrix is determined in 10 seconds. The only drawback of interpolation-evaluation methods consists in big storage capacity of a computer needed to store all the values of interpolated object in interpolation points for some algorithms. For example, to determine the determinant above one need to store 810900 complex values (even if the determinant is of degree 401 only). However, to determine the adjoint above the

---

[1] All the computations was done on Pentium II, 64MB, 120MHz.

same number of values has to be stored as the total number of coefficients of the adjoint.

Acknowledgements

This work has been supported by the research program No. J04/98:212300013 "Decision Making and Control for Manufacturing" of the Czech Technical University in Prague (sponsored by the Ministry of Education of the Czech Republic).

References

1. Antsaklis, P.J. and Gao, Z.: Polynomial and Rational Matrix Interpolation: Theory and Control Applications. *International Journal of Control*. Vol. 58, No. 2, pp.349-404 (1993)
2. Bini, D. and Pan, V.: Polynomial and Matrix Computations, Vol. 1, *Fundamental Algorithms*. Birkhäuser, Boston (1994)
3. De Boor, C. and Ron, A.: On Multivariate Polynomial Interpolation. *Constr. Approx.*, **6**, pp.287-302 (1990)
4. De Boor, C. and Ron, A.: Computational Aspects of Polynomial Interpolation in Several Variables. Math. Comp., **58**, pp.705-727 (1992)
5. Davis, P.J.: *Interpolation and Approximation*. Dover Books, 2.edition, (1975)
6. Gasca, M. and Maeztu, J.I. On Lagrange and Hermite interpolation in $\Re^k$. Numer. Math., 39, pp.1-14 (1982)
7. Hušek, P. and Dvořáková, R.: Stability Analysis of Takagi-Sugeno Fuzzy Systems with Linear Input-Output Submodels: Polynomial Approach, in: Proc. of 11th IFAC Workshop Control Applications of Optimization (CAO 2000), St. Petersburg, Russia (2000)
8. Hušek, P. and Štecha, J.: Rational Interpolation. In: *Proceedings of the 1st Europoly Workshop - Polynomial Systems Theory & Applications*. Glasgow. Europoly-the European Network of Excellence no. CP97-7010, pp.121-126 (1999)

Comparing the Performance of Solvers for a Bioelectric Field Problem

Marcus Mohr[1] and Bart Vanrumste[2]*

[1] System Simulation Group of the Computer Science Department,
Friedrich-Alexander-University Erlangen-Nuremberg
Marcus.Mohr@cs.fau.de
[2] Electrical and Computer Engineering, University of Canterbury
b.vanrumste@elec.canterbury.ac.nz

Abstract. The model-based reconstruction of electrical brain activity from electroencephalographic measurements is of constantly growing importance in the fields of Neurology and Neurosurgery. Algorithms for this task involve the solution of a 3D Poisson problem on a complicated geometry and with non-continuous coefficients for a considerable number of different right hand sides. Thus efficient solvers for this subtask are required. We will report on our experiences with different iterative solvers, Successive Overrelaxation, (Preconditioned) Conjugate Gradients, and Algebraic Multigrid, for a discretisation based on cell-centred finite-differences.

1 Introduction

The electroencephalogram (EEG) is a major diagnostical tool to determine the state of the brain. In recent years the model-based analysis of such voltage measurements has substantially gained in importance. Given the measured data, the typical task is to reconstruct the sources inside the brain responsible for the potential field. The results are then used in the planning of brain surgery and even during the surgery itself. Figure 1 shows an application from epilepsy surgery [11]. Here an epileptic focus to be removed by surgery was reconstructed. Another application is to determine important brain areas, which must not be damaged during an operation, and to insert this information into the surgeon's microscope. The reconstruction is denoted the inverse EEG problem, because it is the inverse of the forward problem, where sources are known / given and the potential distribution inside the head and on its surface has to be computed. The forward problem constitutes a classical elliptic boundary value problem to be solved on a 3D representation of the patient's head. All solution approaches to the inverse EEG problem involve the solution of a considerable number of such forward problems.

---

* At the time of this research, the author worked for the Department of Neurology, University of Gent

P.M.A. Sloot et al. (Eds.): ICCS 2002, LNCS 2330, pp. 528–537, 2002.
© Springer-Verlag Berlin Heidelberg 2002

Medical applications naturally require a high level of precision. It is becoming standard nowadays to base the numerical computations on the individual geometry of a patient's head, which can be obtained e.g. from magnetic resonance imaging (MRI). The electric conductivities of the different tissues are usually assumed to be constant and isotropic, but the determination of patient specific anisotropic values from diffusion weighted MRI is ongoing research, [10]. Thus there is growing interest in volume based discretisation techniques which are about to replace the traditional boundary element approach. This transition to volume discretised realistic head models is encumbered by the fact that there is still a lack of fast solvers for the forward problem, see e.g. [5].

In this paper we will report on results with different iterative solvers for a discretisation of the forward problem with a cell-centred finite difference scheme. The list of solvers includes Successive-Overrelaxation as an example of a stationary iteration method, Conjugate Gradients and Preconditioned Conjugate Gradients for Krylov subspace methods and a variant of Algebraic Multigrid. For a description of theses methods see e.g. [6] and [7].

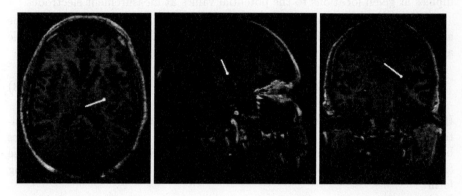

Fig. 1. Axial, saggital, and cordial cuts through an MRI scan of a patient's head. Superimposed is an epileptic focus localised from EEG data.

2   Problem Formulation

In bioelectric field simulation the patient's head is modelled as a volume conductor, i. e. a contiguous, passively conducting medium. Inside the head we have neural current densities which constitute the cause for the electrical and potential field. Due to the temporal behaviour of the sources, which is typically < 1 kHz and the physiological conductivities (e.g. brain 0.2 S/m, skull 0.015 S/m [3]) one can assume a quasi-static behaviour. This leads to the central equation

$$\nabla \cdot (\sigma \nabla \Phi) = \nabla \cdot I_V \ . \tag{1}$$

which relates the potential field Φ to the current densities I_V of the neural sources. The term σ denotes the conductivity tensor. Together with boundary conditions on the current flow through the scalp

$$\sigma \frac{\partial \Phi}{\partial n} = g \qquad (2)$$

this constitutes an elliptic boundary value problem. The classical task of numerical simulation would be to compute the potential field Φ for given source $\nabla \cdot I_V$ and boundary terms g. This is referred to as the *forward problem*. The related *inverse EEG problem* is to reconstruct from g and measurements of Φ at some electrodes either the sources or the potential at some internal interfaces, esp. on the cortex.

There are three standard approaches to the inverse EEG problem: *dipole localisation*, *current density analysis* and *deviation scans*, see e.g. [11, 2]. Important in our context is, that all of these involve the setup of a so called *lead field matrix*. This matrix relates the momentum and orientation of one or more dipoles at given locations to the potential values at measurement electrodes.

Assume that a dipole is given with position r and orientation d and let $\mathcal{E}$ be a set of N electrodes. The vector $M(r, d)$ of potentials generated by the dipole at the electrodes in $\mathcal{E}$ is given by

$$M(r, d) = \hat{L}(r) \cdot d = R \cdot L(r) \cdot d \qquad (3)$$

Here $\hat{L} \in \mathbb{R}^{N \times 3}$ is the lead field matrix. It can be split into two parts. A matrix $L \in \mathbb{R}^{(N-1) \times 3}$ that maps the dipole orientation onto potential differences between the electrodes and a referencing matrix $R \in \mathbb{R}^{N \times (N-1)}$ that turns these $(N-1)$ differences into N scalar potential values. Each row of L is of the following form

$$(\Phi_{AB}^x(r), \Phi_{AB}^y(r), \Phi_{AB}^z(r)) \qquad (4)$$

where $\Phi_{AB}^k(r)$ denotes the potential difference between electrode A and B induced by a unit dipole at position r which is oriented in k-direction. So in principle every entry of the lead field matrix requires the solution of a single forward problem. In all three approaches for the inverse EEG problem the lead field matrix has either to be assembled for a single dipole at a large number of varying positions or for a large number of different dipoles at fixed positions. This constitutes a major part of the computational work. Fortunately this amount of work can (in some cases) drastically be reduced with the help of the reciprocity theorem of Helmholtz, see e. g. [11, 12]. It allows the simple setup of the lead field matrix by interpolation from potential distributions computed in a preparatory step. Selecting $N - 1$ pairs of electrodes from $\mathcal{E}$ this step consist of solving problem (1,2) with one electrode chosen as unit sink and the other one as unit source. This leave us with the solution of $N - 1$ Poisson problems with jumping coefficients on a complicated 3D domain.

3   Discrete Model

In order to solve the boundary value problem (1,2) in terms of the bioelectric potentials we need to pose the problem in a computationally tractable way. Modern medical imaging techniques make it feasible to employ the individual patient's anatomy in the creation of a volume conductor model of the head. Geometry information is typically derived from magnetic resonance imaging (MRI), while the conductivity is assumed to be isotropic and constant for each type of tissue and experimental values from literature are taken.

For our experiments in Sect. 5 we have used MR-images to form a 3D voxel model. In a segmentation process each voxel is assigned to one of the four compartment types scalp, skull, brain and air and constant isotropic conductivities are assumed in each of them. Note however, that the described approach is not limited to a certain number of compartments and in principle can also handle anisotropic conductivities. This option is important, since it is anticipated that in the near future diffusion weighted MRI will allow to approximate the locally varying and anisotropic conductivity tensors on a patient basis, see e.g. [10]. For details on the creation of the employed 3D voxel model see [11].

We discretise the boundary value problem (1,2) by means of Finite Differences. We use the box integration scheme and model the electrodes as point sources within the voxel that is nearest to the position of the centre of the respective electrode. This leads to homogenous Neumann boundary conditions. The resolution of our head models is fine enough that for the given spacing of the two electrodes in a pair will never lie within the same voxel. Let us denote by k_{source} and k_{sink} the index of the voxel containing the source and the sink. Then we get that the discrete approximation Φ^h of (1,2) has to satisfy

$$\left(\sum_{k\in\mathcal{N}_l}\gamma_k\right)\Phi_l^h - \sum_{k\in\mathcal{N}_l}\gamma_k\Phi_k^h = I(\delta_{k_{\text{source}},l} - \delta_{k_{\text{sink}},l}) \tag{5}$$

for every voxel V_l that belongs to the head. Here $\delta_{i,j}$ denotes the Kronecker symbol and $\mathcal{N}_l$ the index set of the six voxels neighbouring V_l. The stencil coefficients γ_j are a mesh-size weighted harmonic average of the conductivities in the central voxel and the corresponding neighbour. For the eastern neighbour γ_j takes the form

$$\gamma_{\text{east}} = \frac{2h_y h_z}{h_x} \cdot \frac{\sigma_l \cdot \sigma_{\text{east}}}{\sigma_l + \sigma_{\text{east}}} \tag{6}$$

and analogously for the other five neighbours. Note that $\gamma_k \geqslant 0$ since σ is non-negative.

Let us denote by A the problem matrix of the linear system resulting from (5). Assuming that we have a contiguous model without isolated head cells it is easy to see from (6) and (5) that A will be a symmetric, positive semidefinite matrix. It is also a singular M-matrix in the classical notation of Berman & Plemmons. We know therefore that $\text{rank}(A) = n - 1$ and the kernel is of dimension one [1]. In fact, it is spanned by the vector $e = (1, \ldots, 1)^T$. Let b be the right hand side we get from (5), then $e^T b = 0$. Thus the problem $Ax = b$ is consistent and possesses infinitely many solutions differing only in an additive constant.

When trying to find the solution of the system, one can either work with the singular problem directly, or introduce an additional constraint, that fixes one element of the set of all solutions and solve the corresponding regular problem. The easiest approach is to fix the value of the potential Φ^h to 0 in one voxel. This leads to a problem with a regular M-matrix and its solution obviously solves the initial problem with $\Phi^h = 0$ in the respective voxel. We will compare these two possibilities.

4   Iterative Solvers

In the literature on bioelectric field problems typically Krylov subspace methods are mentioned for the solution of the related forward problems, see e.g. [2]. We consider representatives of this class and compare them to other methods:

- Successive Over-Relaxation (SOR)
- Conjugate Gradients (CG)
- Conjugate Gradients preconditioned by symmetric SOR (PCG)
- Algebraic Multigrid (AMG)

The SOR method is a representative of the classical stationary methods. It is known to be not the optimal choice as far as convergence is concerned, but it has a very simple structure. Thus it is a good candidate for an optimised implementation.

The CG method is the typical algorithm from the large class of Krylov subspace methods. The convergence of the CG method depends on the condition number, or more precisely on the spectrum, of the problem matrix. It is therefore seldom used without preconditioning. We have chosen a symmetric SOR preconditioner for this purpose.

The last contestant is an algebraic multigrid method. Multigrid methods in general are known to be very efficient solvers for elliptic boundary value problems. They employ a hierarchy of grid levels to treat individual problem components. Unfortunately finding the proper components, i.e. transfer operators, coarsening strategies, etc. can be quite tedious in the case of complex geometries and/or jumping coefficients. Therefore the idea of algebraic multigrid methods is again attracting increased attention. Here a "grid hierarchy" and inter-grid transfer functions are derived automatically from the problem matrix.

While these methods have been developed for regular linear systems, they can also be applied in our semi-definite case. In the case of a consistent right hand side convergence can be guaranteed for SOR and (P)CG, while for AMG theoretical results are more complicated, see [1,9].

5   Numerical Experiments

For our numerical experiments we implemented all algorithms, except AMG, in a grid- and a matrix-based version. As AMG method we took *BoomerAMG*, see [7]. Implementation details can be found in [9].

We compare the performance with respect to two data sets, in the following denoted by data set A and B. They have been created from real patient MRI scans at the Epilepsy Monitoring Unit of the University of Gent, Belgium. The data sets have a different resolution and come from two different subjects. The edge length h of the voxels is the same in all three dimensions. For the conductivities we use the values given in [3]. We scale our linear system by $(\sigma_{skull}h)^{-1}$ and only use the relative values ($\sigma_{scalp} : \sigma_{skull} : \sigma_{brain} = 16 : 1 : 16$). Note that the differences in the conductivities are only one order of magnitude. So we are faced with an interface problem with only moderate jumps. An overview of the properties of the two data sets and of the resulting linear system is given in Tab. 1.

For both patients an EEG was recorded using 27 electrodes, placed according to the international 10–20 standard [8], with three additional electrodes positioned on the temporal lobe areas on both sides. Thus the preparatory step, cf. Sect. 2, that provides the data needed for setup of the lead field matrix consists in the solution of 26 forward problems for as many electrode pairs.

From the four tested methods the CG algorithm is the only one that does not depend on the choice of a special parameter. In case of the SOR and PCG(SSOR) method we have to specify the over-relaxation parameter ω. For AMG a threshold parameter, which allows for the distinction of "weak" and "strong" inter-node connections, must be chosen. This distinction is critical for the construction of the grid hierarchy, see e.g. [7]. We have chosen two of the 26 electrode pairs to test the dependency of the algorithms on these parameters. In these and all following tests we accepted an approximate solution as soon as its residual measured in the Euclidean norm became smaller than 10^{-8}.

Figure 2 shows the dependency of the number of SOR iterations for the case of a singular matrix. We see that this number varies considerable even for the small interval $\omega \in [1.89, 1.99]$. The optimal values seem to be in the vicinity of 1.935 for data set A and 1.970 for data set B. We have employed these values for all subsequent tests. When we apply SOR to the regular matrix, we get basically

Table 1. Details of the two data sets and the corresponding singular linear systems

Data Set	A	B
number of voxels / cube dimensions	65^3	129^3
edge length of voxels	3.92mm	1.96mm
number of head voxels / unknowns	70,830	541,402
number of non-zero matrix entries	482,184	3,738,624
matrix sparsity	0.01%	0.001%

the same picture, with two differences. On the one hand the optimal ω is larger (about 1.9935 for dataset A and about 1.9977 for dataset B). On the other hand also the number of iterations needed to reach the desired accuracy is drastically larger. Even for the nearly optimal ω the numbers are on average 10 resp. 14 times larger, as can be seen in Tab. 2.

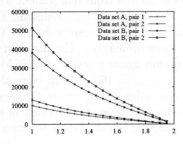 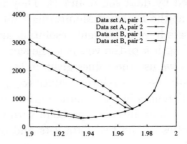

Fig. 2. Number of SOR iterations necessary to satisfy stopping criterion for the singular problem depending on choice of relaxation parameter ω.

The choice of ω appears to be not as important for the PCG method, as can be seen in Fig. 3. The number of iterations varies over smaller intervals and the valleys around the optimal value are more flat, thus choosing a reasonable ω is easier. The optimal values lie in the vicinity of 1.65 and 1.75 for the singular problem and 1.65 and 1.8 for the regular one. Note that again the number of iterations for the singular problem is smaller than for the regular case.

In the AMG algorithm there is quite a number of parameters and algorithmic variants, that have to be fixed, before the method can be applied. These include e. g. cycle type, number of smoothing steps, and so on. We have used a V-cycle with one pre- and one post-smoothing step. The relaxation scheme was a hybrid Gauß-Seidel / Jacobi method. Construction of the grid hierarchy was performed with Ruge-Stüben coarsening. Besides this, we left all other parameters at their default values and investigated the influence of the threshold value α, as shown in Fig. 4 for the singular case. The regular case is not shown, since the results are very similar.

We note two interesting facts. The first is, that the best convergence is achieved for comparatively small values of α. The second fact is that there is a sharp increase in the number of cycles in the interval $\alpha \in [0.50, 0.55]$. This is especially pronounced in the case of data set B. One can show [9] that for $\alpha > 0.53$ we get bad interpolation operators, since skull voxels on the fine grid are only interpolated from scalp / skull voxels on the next coarser grid. Our experiments also indicate that performance will decrease again, when α becomes too small. So, for all further experiments we have settled with $\alpha = 0.05$.

After having determined a set of reasonably optimal parameters for the different methods, we tested the number of iterations needed to satisfy the stopping criterion for all 26 electrode pairs. Table 2 summarises the mean values of all

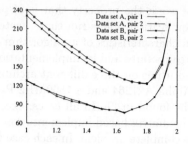
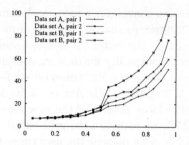

Fig. 3. Number of PCG iterations necessary to satisfy stopping criterion for the singular problem depending on choice of relaxation parameter ω.

Fig. 4. Number of AMG cycles necessary to satisfy stopping criterion depending on choice of threshold parameter α.

pairs. To be able to better compare these values we have determined for each method an estimate of the total amount of arithmetic work involved. This estimate is given in Tab. 2 as a percentage of the amount of work for the SOR method in the semi-definite case.

We see that the CG method performs worst, due to a bad convergence rate. This is considerably improved by preconditioning. The number of iterations drops to roughly 18% of the unpreconditioned case. Due to the higher costs per iteration step this is still about two thirds of the reference case in the most favourable situation. The best performance is achieved by the AMG approach, which also shows the typical feature of a multigrid method, namely that the number of cycles remains constant, independent of the fineness of the discretisation.

Table 2. Iteration counts for different methods, matrix is singular (1) or regular (2)

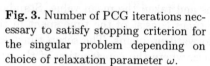

	Number of Iterations					
	Data set					
Method	A			B		
	Mean	% SOR	Std. Dev.	Mean	% SOR	Std. Dev.
SOR (1)	303.5	100	9.6	634.0	100	2.2
SOR (2)	2957.0	974	34.1	8839.0	1400	635.2
CG (1)	407.5	179	9.3	740.7	156	12.5
CG (2)	533.2	234	6.5	986.8	208	9.7
PCG(SSOR) (1)	80.3	88	1.3	126.7	67	1.8
PCG(SSOR) (2)	106.0	116	1.2	165.8	87	1.6
BoomerAMG (1)	7.0	16	0.0	7.0	7	0.0
BoomerAMG (2)	7.0	16	0.0	7.0	7	0.0

Besides convergence rates and amount of work, the property that is of primary interest to the user coming from the application side is run time. Factors determining the latter involve the numerical characteristics of an algorithm as well as its suitability for modern computer architectures and its implementation. We have tested the run times of our four algorithms on three different architectures, a 700MHz AMD Athlon, a 500MHz Alpha A21264 and a 1500MHz Pentium IV. More details of the machines can be found in [9]. In all cases, except for AMG, we compare the grid-based with the matrix-based implementation.

In Fig. 5 we present the user times for the complete problem. In each case we have measured times for 10 program runs and taken the mean value. Standard deviation was always less than 2%. The times consist of the time spent in the setup phase and for the solution of 26 forward problems. The setup times are always negligible except for the AMG case, where the construction of the grid hierarchy adds considerably to the total costs. This varies between 5 and 11% of the total time depending on architecture and problem size.

Concerning the measurements for BoomerAMG we should note that the hypre library, of which it is a part, was developed for solving large, sparse systems of linear equations on massively parallel computers. Thus it is not specially tuned for the sequential environment in which we used it. We also want to point out, that, although we have taken into account performance issues in the implementation of the algorithms, there are still numerous possibilities one could test for a further optimisation of the code. For some ideas on this, see e. g. [4].

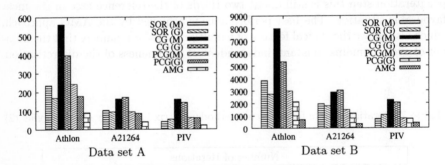

Fig. 5. Run times (user time in seconds) for the complete problem. This includes the setup phase and the solution of 26 singular forward problems.

6   Conclusions

Summing up the contents of this paper, we see three major points. The first one is, that the solution of inverse EEG problems can be sped up considerably by the application of multigrid methods for the solution of the forward problems involved. These appear to offer a much better performance than the Krylov subspace methods typically employed in this context. It remains to be seen however, if this also holds for more sophisticated preconditioners like e.g. ILU, see [6].

The second aspect is that it seems pointless to transform the singular forward problem into an equivalent regular one utilising the approach described in Sect. 3. In our experiments the regularisation of the problem could not improve convergence of the tested methods and, in many cases, even led to a worse behaviour. However, this does not rule out the possibility that other regularisation approaches may lead to better performance.

The question, finally, whether a matrix- or a grid-based implementation of the methods is preferable in view of runtime behaviour, remains open. Our experiments in this context yielded results that vary from architecture to architecture and are ambiguous in themselves even on the same machine.

References

1. Berman, A., Plemmons, R.: Nonnegative Matrices in the Mathematical Sciences, SIAM, Classics in Applied Mathematics **9** (1994)
2. Buchner, H., Knoll, G., Fuchs, M., Rienäcker, A., Beckmann, R., Wagner, M., Silny, J., Pesch, J.: Inverse localization of electric dipole current sources in finite element models of the human head, Electroencephalography and clinical Neurophysiology **102** (1997) 267-278
3. Delbeke, J., Oostendorp, T.: The Conductivity of the Human Skull in vivo and in vitro, Serving Humanity, Advancing Technology, Proceedings of The First Joint BMES/EMBS Conference (1999) 456
4. Douglas, C.C., Hu, J., Kowarschik, M., Rüde, U., Weiß, C.: Cache Optimization for Structured and Unstructured Grid Multigrid, Electronic Transactions on Numerical Analysis **10** (2000) 21-40
5. Ermer, J., Mosher, J., Baillet, S., Leahy R.: Rapidly re-computable EEG forward models for realistic head shapes, Biomag 2000, Proc. of the 12th Internat. Conf. on Biomagnetism, eds.: Nenonen, J., Ilmoniemi, R.J., Katila T. (2001)
6. Hackbusch, W.: Iterative solution of large sparse systems of equations, Springer (1994)
7. Henson, V.E., Meier Yang, U.: BoomerAMG: a Parallel Algebraic Multigrid Solver and Preconditioner, Applied Numerical Mathematics (in press), also available as Technical Report UCRL-JC-141495, Center for Applied Scientific Computing, Lawrence Livermore National Laboratory (2000)
8. Jasper, H.: Report of committee on methods of clinical exam in EEG, Electroencephalography and Clinical Neurophysiology **10** (1958) 370-375
9. Mohr, M.: Comparision of Solvers for a Bioelectric Field Problem, Technical Report 01-2, Lehrstuhl für Informatik 10 (Systemsimulation), Friedrich-Alexander-Universität Erlangen-Nürnberg (2001)
10. Tuch, D., Wedeen, V.J., Dale, A., George, J., Belliveau, W.: Conductivity tensor mapping of the human brain using diffusion tensor MRI, Proc. Natl. Acad. Sci. USA **98(20)** (2001) 11697-11701
11. Vanrumste, B.: EEG dipole source analysis in a realistic head model, PhD thesis, Faculteit Toegepaste Wetenschappen, Universiteit Gent (2001)
12. Weinstein, D., Zhukov, L., Johnson, C.: Lead-field Bases for Electroencephalography Source Imaging, Annals of Biomedical Engineering **28(9)** (2000) 1059-1065

Iteration Revisited
Examples From a General Theory

Poul Wulff Pedersen

Margerittens Kvarter 40, DK-2990 Nivaa, Denmark
p-w-p@get2net.dk

Abstract. For 'any' equation $F(x) = 0$ with an isolated solution x^* which we want to compute, it will be shown that quantities p_n can be defined and <u>computed</u> so that

$$xp_{n+1} := xp_n + p_n \cdot F(xp_n)$$

will 'converge' with 'any' speed needed, i.e. starting with an xp_0 in a neigbourhood of x^*, xp_N will for for some small value of N give x^* within the precision you need. And it will do so even if the 'standard' iteration $x_{n+1} := x_n + F(x_n)$ is divergent.

1 Introduction

Let

$$F(x) = 0, \qquad x, F(x) \in \Re \tag{1}$$

denote an equation with x^* as the solution whose value we want to compute. We assume that x^* is known to exist and that it is the only solution in some interval $I_0 = [L_0; R_0]$. The equation (1) can be brought on an "equivalent iterative form", i.e. an equation $x = f(x)$ with x isolated on one side and with x^* as the only solution in I_0 - for example by adding x on both sides, giving

$$x_{n+1} = x_n + F(x_n), \qquad n = 0, 1, 2, \ldots \tag{2}$$

In the ideal world of the textbooks this iteration will converge; in practice the situation is often different. And it is easy to see what happens:

Comment 1 x_{n+1} *is the x-value where the line from* $(x_n, F(x_n))$ *with slope* $\alpha = -1$ *crosses the x-axis.*

The usual criterium for an iteration to converge is in this case $-2 < F'(x^*) < 0$ instead of $-1 < f'(x^*) < 1$ for $x_{n+1} := f(x_n)$. The center-value for the interval [- 2; 0] is of course the slope $\alpha = -1$ from before.

In the one-dimensional case a suitable function f for an equivalent iteration form $x_{n+1} := f(x_n)$ can of course normally be found without too much trouble, at least as soon as the situation has been analyzed. But, and that is the real problem, you cannot leave this essential part to the average computer. And for N- and infinite-dimensional problems the situation is of course more complicated.

P.M.A. Sloot et al. (Eds.): ICCS 2002, LNCS 2330, pp. 538–547, 2002.

Example 1. The equation

$$x^3 - x - 1 = 0 \quad \text{with} \quad F(x) := x^3 - x - 1$$

has a solution x^* in $I_0 := [1.3 \; ; \; 1.4]$ and there is only this one in I_0. Obvious equivalent iterative forms are

$$x = x^3 - 1, \quad x = \frac{1}{x^2 - 1}, \quad x = \sqrt[3]{x + 1}$$

and it is seen that only the third one gives a convergent iteration. We shall discuss an iteration based on the first one later on.

Instead of trying to change F we shall use 'any' F but define quantities p_n so that

$$x p_{n+1} = x p_n + p_n \cdot F(x p_n), \qquad n = 0, 1, 2, \ldots \tag{3}$$

will give - for some small N - a value $x p_N$ as close to the solution x^* as we need.

For $\sqrt{A}$, $1/\sqrt{A}$ and division $1/A$ (defined by simple algebraic equations) we can compute an ideal p-value $p^* = p^*(x)$ as a series and an iteration of order $m + 1$ is obtained by taking m terms of this series as p. We shall also discuss how to compute interval approximations.

For a function like Arcsin, p^* will be a series in the inverse function sin(x). Or we could compute sin(x) in terms of its inverse Arcsin(x), etc.

In the general case, p_n is found by [inverse] interpolation.

This paper is about DIMension 1, but much can readily be extended to DIM = N (incl. complex numbers) and to dimension ∞.

Example 2. To compute $\sqrt{A}$ we use $F(x) := A - x^2$ and the ideal p-value $p^* = p^*(x)$ - which will take any x into $\sqrt{A}$ - is

$$p^*(x) = \frac{1}{\sqrt{A} + x} = \frac{1}{2x} - \frac{1}{8x}\epsilon + \frac{1}{16x}\epsilon^2 - +\ldots, \quad \epsilon = A - x^2$$

With one term $p(x) := 1/2x$ we get Newton's formula of order 2; with two terms we get a Halley formula of order 3, etc.

Using that for any approximation x, x and A/x will be on each side of $\sqrt{A}$ we get an inclusion by replacing $\sqrt{A}$ in $p^*(x)$ by these.

1.1 p-iteration, deviation functions

Before defining p it may eventually be convinient to modify F into a 'deviation' function.

A deviation function for a solution x^* to $F(x) = 0$ is a function $D = D(x)$ so that

D1 : $D(x)$ is defined and can be computed in an interval $I_0 = [L_0; R_0]$ with x^* as an interior point.

D2 : $D(x) = 0$ if and only $x = x^*$,

D3 : A value $D(x_1)$ or a set of a few values $D(x_1), D(x_2), \ldots$ will give an indication as to where we should look for better approximation to x^*.

D4 : By making $|D(x)|$ sufficiently small, we can make the error $|E(x)| := |x^* - x|$ as small as we need

$$|D(x_n)| < \delta \quad \Rightarrow \quad |E(x_n)| < \epsilon,$$

and a value $\delta = \delta(\epsilon)$ can be computed from ϵ.

D3 is a little vague - examples later on will clarify the situation. Often F from $F(x) = 0$ determining x^* can be used as D. It may eventually be convinient to have two deviation functions, one indicating (mostly) how close an approximation x_n is to x^*, and the other one (mostly) indicating the direction of a continued search.

The error-function E would give an ideal iteration in the sense that

$$x_{n+1} := x_n + E[x_n]$$

gives $x_{n+1} = x^*$ for any start x_n - except that we cannot compute $E(x)$ without x^*.

Change-of-sign : If the solution x^* is not isolated to an interval with a change-of-sign some extra care is needed. A single value $F(x_1)$ will - at most - indicate how close x_1 is to x^* but we cannot tell in which direction we should search for better approximations. And $|F(x_2)| < |F(x_1)|$ does not guarantee that we should continue in the direction from x_1 towards x_2. Think of function values of x^2 type coming from the computer, without the benefit of a drawing to help - and the "computer" has to ensure that you have points x_L, x_R on each side of x^*. Normally the p-method to be described will provide x-values so that $|D(x)|$ and thus also $|E(x)|$ are small, but for the computer it may be a difficult problem to ensure if - or when - we have an inclusion.

1.2   p-iterations

A p-iteration for solving $D(x) = 0$ is an iteration based on a step

$$xp_n := x_n + p_n \cdot D(x_n) \tag{4}$$

where p_n is a number so that the deviation $|D(xp_n)|$ is [much] smaller than $|D(x_n)|$. The ideal p-value p_n^* is a number so that $xp = xp(x_n, p_n^*)$ is the solution x^*, i.e. p_n^* is determined by

$$x^* = x_n + p_n^* \cdot D(x_n) \tag{5}$$

Since x_n is $\neq x^*$, $D(x_n) \neq 0$, and we have

Theorem 1. *For any $x_n \neq x^*$, there exists an ideal p-value $p^* = p^*(x_n)$ determined by*

$$p^*(x_n) := \frac{x^* - x_n}{D(x_n)} \tag{6}$$

Alternatively, we can take x_n to the left side of (4) giving

$$E[x_n] = p^*(x_n) \cdot D(x_n) \tag{7}$$

showing that

Corollary 1. $p^*(x)$ *is the factor that takes deviation $D(x)$ into error $E[x]$.*

1.3 $\mathbf{p^* = p^*(x_n)}$ as a mean-value

From (6) [with $D(x_n) \neq 0$ since $x_n \neq x^*$]

$$p^*(x_n) = \frac{x^* - x_n}{D(x^*) + (x_0 - x^*) \cdot D'(\xi_n)}$$

and, using that $D(x^*) = 0$,

$$p^*(x_n) = \frac{-1}{D'(\xi_n{}^*)} \qquad \xi_n{}^* \in I_n = ITV]x_n; x^*[\tag{8}$$

for some $\xi_n{}^*$ in the open interval $ITV]x_n; x^*[$ between x_n and the solution x^*.

If we have an inclusion for $-1/D'(x)$ over I_n we can compute an inclusion for $p^*(\xi_n)$ and thus for x^*.

If we replace $\xi_n{}^*$ by the approximation x_n we get the Newton iteration

$$NEWT[x_n] = xp_n := x_n - \frac{D(x_n)}{D'(x_n)}$$

The Newton iteration is a for-the-lack-of-better iteration in the sense that we use an endpoint x_n of the interval $ITV]x_n; x^*[$ in which the unknown point ξ_n is known to lie. On the other hand, 'any' point in $]x^*; x_n[$ would not necessarily be better than the end-point x_n.

Concerning ξ, we know [Pof] that in the limit, ξ^* will be midway between x_n and x^*. However, we don't know x^* and we don't know what happens at the beginning of an iteration.

With one more term in Taylor we have

$$p^*(x_n) = \frac{-1}{D'(x^*) + \frac{1}{2} \cdot (x_n - x^*)D''(\xi_2{}^*)}, \qquad \xi_2{}^* \in ITV]x_0; x^*[\tag{9}$$

etc.

Slope α vs p Let x_α denote the x-value where the line from $P_0 = (x_0; D(x_0))$ with slope α intersect the x-axis

$$x_\alpha = x_0 + \frac{-1}{\alpha} \cdot D(x_0)$$

On the other hand, using a p-value p we get

$$xp = x_0 + p \cdot D(x_0)$$

so

$$xp = x_\alpha \quad \text{for} \quad p = \frac{-1}{\alpha}, \quad \alpha = \frac{-1}{p}.$$

Thus, using a p-value p_α is equivalent to using a slope $\alpha = -1/p$. Cf. Newton's formula.

Linear Interpolation. Using the x_α - formula, we can compute the x-value x_0 where the line through $(x_1, D(x_1))$, $(x_2, D(x_2))$ intersects the x-axis and more generally, the x-value x_c where the value is c

$$x_0 = x_1 - \frac{1}{\alpha} D(x_1) \qquad x_c = x_1 - \frac{1}{\alpha} \cdot (D(x_1) - c), \qquad \alpha = \frac{D(x_1) - D(x_2)}{x_1 - x_2}$$

Geometrical interpretation : $x_{n+1} := f(x_n)$ in terms of $F(x_n)$.

The 'standard' iteration

$$x_{n+1} := x_n + F(x_n) \tag{10}$$

obtained by adding x on both sides of $F(x) = 0$, is a p-iteration with $p = 1$, corresponding to $\alpha = -1$. x_{n+1} is therefore the x-value where the line with slope -1 through $(x_n, F(x_n))$ intersects the x-axis (cf. Introduction).

Higher order formulas. For solving $F(x) = 0$ we have [Fr 1]

$$x^* = x_0 - \left[\alpha + \frac{1}{2} a_2 \alpha^2 + \frac{1}{6} \left(3 \cdot a_2{}^2 - a_3 \right) \alpha^3 + \frac{1}{24} \left(15 a_2{}^3 - 10 a_2 a_3 + a_4 \right) \alpha^4 + \ldots \right]$$

where

$$\alpha = \frac{F(x_0)}{F'(x_0)} \qquad a_2 = \frac{F''(x_0)}{F'(x_0)}, \qquad a_3 = \frac{F'''(x_0)}{F'(x_0)}, \qquad a_4 := \frac{F''''(x_0)}{F'(x_0)}, \ldots$$

Taking $F(x_0)$ from α we have an expression for $p^*(x_0)$. With α only from [...] we get Newton's formula

$$xp_1(x_0) := x_0 - \frac{1}{F'(x_0)} \cdot F(x_0)$$

With 2 terms we get a Halley formula of order 3

$$xp_2(x_0) := x_0 - \left[\frac{1}{F'(x_0)} + \frac{F''(x_0) \cdot F(x_0)}{2(F'(x_0))^3} \right] F(x_0)$$

etc.

An error theorem. Let xp_0 denote the value obtained using x_0, p_0

$$xp_0 := x_0 + p_0 \cdot D(x_0)$$

Since

$$x^* = x_0 + p^*(x_0) \cdot D(x_0)$$

we have

$$x^* - xp_0 = (p^*(x_0) - p_0) \cdot D(x_0)$$

or

$$E[xp_0] = EP[p_0] \cdot D(x_0),$$

Here, $E[x]$ is the error $x^* - x$ and $EP[p]$ is the 'error' on p : $EP[p] := p^* - p$. To reduce $E[xp]$ it would f.ex. be enough to compute a better p-value (cf. Ex. 2).

A divergent iteration made convergent. The iteration for $\sqrt{A}$

$$x_{new} := x + x^2 - A$$

is divergent $(f'(x) = 1 + 2 \cdot x > 1)$. But the p-iteration

$$x_{new} := x + p \cdot (x^2 - A), \qquad p = \frac{-1}{2\,x}$$

converges for all A and all starting-values $x_0 > 0$ (Newton).

1.4 p^* for some of the standard functions

Computing $\sqrt{A}$: Define (for $x > 0$)

$$D(x) := A - x^2, \qquad E(x) := \sqrt{A} - x$$

and for

$$xp := x + p \cdot (A - x^2), \qquad p^*(x) = \frac{\sqrt{A} + x}{A - x^2} = \frac{1}{\sqrt{A} + x} \tag{11}$$

For an approximation x to $\sqrt{A}$, $A/x^2 \simeq 1$ and we can define a [small] ϵ by

$$\frac{A}{x^2} = 1 + \epsilon, \qquad \epsilon := \frac{A}{x^2} - 1 = \frac{A - x^2}{x^2}$$

From (11)

$$p^*(x) = \frac{1}{\sqrt{A} + x} = \frac{1}{x} \left[\frac{1}{2} - \frac{1}{8}\epsilon + \frac{1}{16}\epsilon^2 - \frac{5}{128}\epsilon^3 + - \ldots \right] \tag{12}$$

Alternatively (cf Corollary 1)

$$A - x^2 = (\sqrt{A} + x) \cdot (\sqrt{A} - x),$$

$$D(x) = (\sqrt{A} + x) \cdot E(x), \qquad E(x) = \frac{1}{\sqrt{A} + x} \cdot D(x), \tag{13}$$

Intervals. For any approximation x to $\sqrt{A}$, A/x is also an approximation and x, A/x will be on each side of $\sqrt{A}$. Thus, replacing $\sqrt{A}$ in p^* with $ITV[x; A/x]$

$$\sqrt{A} \in ITV\left[x + \frac{1}{[x; A/x] + x} \cdot (A - x^2)\right]$$

giving

$$\sqrt{A} \in ITV\left[x + \frac{1}{2x} \cdot (A - x^2) \; ; \; x + \frac{x}{A + x^2} \cdot (A - x^2)\right] \qquad (14)$$

with length

$$LGT(A, x) = \frac{1}{2x \cdot (A + x^2)}(A - x^2)^2 \simeq \frac{1}{4A\sqrt{A}} \cdot (A - x^2)^2$$

The midpoint is

$$x_{mid} := x + \frac{A + 3x^2}{4x(A + x^2)}(A - x^2) \quad \text{with error} \quad = \frac{-1}{4x(A + x^2)} \cdot (E[x])^4$$

With $A = 2$, $x_0 := 1.414$ we get an interval of length $0.323 \cdot 10^{-7}$ and the midpoint has error $-0.920 \cdot 10^{-16}$. With x_0 as a Newton-value

$$A := 2, \quad x_{00} := NEWT[NEWT[1.414]]$$

we get an interval of length $0.652 \cdot 10^{-65}$ and a midpoint with error $-0.353 \cdot 10^{-131}$.

From a mathematical point of view the two expressions for $p^*(x)$ in (11) are equivalent. But from a numerical point of view, they are completely different

Let $I_0 := [L_0; R_0]$ denote an inclusion for $\sqrt{A}$. Using I_0 in the two p^* expressions, we have

$$p^*(x_0) \in ITV\left[\frac{[L_0; R_0] - x_0}{A - x_0^2}\right], \qquad p^*(x_0) \in ITV\left[\frac{1}{[L_0; R_0] + x_0}\right]$$

and we get two inclusions. The first one

$$\sqrt{A} \in x_0 + ITV\left[\frac{[L_0; R_0] - x_0}{D(x_0)}\right] \cdot D(x_0)$$

where $D(x_0)$ cancels out, giving trivially $[L_0; R_0]$. The other one

$$\sqrt{A} \in x_0 + ITV\left[\frac{1}{[L_0; R_0] + x_0}\right] \cdot (A - x_0^2)$$

with a length

$$LGT_{new} := \frac{R_0 - L_0}{(L_0 + x_0)(R_0 + x_0)}|A - x_0^2| \simeq \frac{|A - x_0^2|}{4A} \cdot LGT_{orig}$$

which is made as small as we want by reducing $A - x_0^2$ - without changing I_0.

Division by A only. We define ϵ

$$\frac{x^2}{A} = 1 - \epsilon, \qquad \epsilon := 1 - \frac{x^2}{A} = \frac{A - x^2}{A}$$

From this

$$\sqrt{A} = x \frac{1}{\sqrt{1 - \epsilon}} = x + \frac{x}{A}\left[\frac{1}{2} + \frac{3}{8}\epsilon + \frac{5}{16}\epsilon^2 + \frac{35}{128}\epsilon^3 + \ldots\right](A - x^2) \qquad (15)$$

Here $D(x) = A - x^2$ so we have the same $p^*(x)$ as before but obviously written in a different form in (15).

Computing $1/\sqrt{A}$. Let x denote an approximation to $1/\sqrt{A}$ so that $1 - A \cdot x^2 \simeq 0$. and we can use

$$D(x) := 1 - A \cdot x^2 = \epsilon$$

Then

$$\frac{1}{A} = x^2 \cdot \frac{1}{1 - \epsilon}, \qquad \frac{1}{\sqrt{A}} = x \cdot \frac{1}{\sqrt{1 - \epsilon}}$$

and

$$p^*(x) = \frac{1}{A}\frac{1}{\frac{1}{\sqrt{A}} + x}, \qquad \frac{1}{\sqrt{A}} = x + x\left[\frac{1}{2} + \frac{3}{8}\epsilon + \frac{5}{16}\epsilon^2 + \frac{35}{128}\epsilon^3 + \ldots\right](1 - A x^2)$$

It is seen that x and $x(2 - A x^2)$ will be on each side of $1/\sqrt{A}$ and using

$$x_{n+1} = x_n + \frac{1}{2} \cdot x_n \epsilon_n \qquad (16)$$

we get an inclusion

$$x_1 := x + 0.5 \, x \, (1 - Ax^2), \qquad x_2 := x + 0.5 \, x \, (2 - Ax^2)(1 - Ax^2)$$

For

$$x_w(x) := \frac{1}{4} \, x_1(x) + \frac{3}{4} \, x_2(x)$$

E^2 error terms will cancel and we get the 3.order formula.

2 Computing Arcsin(t) using its inverse function sin(x)

To compute $\arcsin(t_0)$ for t_0 in $]0 \, ; 1[$ we should 'solve'

$$x = \arcsin(t_0), \qquad t_0 = \sin(x)$$

so we have a deviation function and a p-iteration

$$D(x) := t_0 - \sin(x) \qquad xp = x + p \cdot [t_0 - \sin(x)]$$

with

$$p^*(x) = \frac{\arcsin(t_0) - x}{t_0 - \sin(x)}$$

Let

$$\epsilon := D(x) = t_0 - \sin(x), \qquad t_0 = \sin(x) + \epsilon$$

Then

$$p^*(x) = \frac{\arcsin(sin(x) + \epsilon) - x}{\epsilon} = \sum c_n \cdot \epsilon^n$$

with

$$c_0 = \frac{1}{\cos(x)}, \quad c_1 = \frac{\sin(x)}{2 \cdot \cos(x)^3}, \quad c_2 = \frac{3 - 2 \cdot \cos(x)^2}{6 \cdot \cos(x)^5}$$

$$c_3 = \frac{5 \cdot \sin(x) - 2 \cdot \sin(x) \cdot \cos(x)^2}{8 \cdot \cos(x)^7} \qquad \cdots$$

Here,

$$xp_0(x) := x + \frac{1}{\cos(x)}(t_0 - \sin(x)) \qquad (p = c_0)$$

is the Newton-iteration of order 2. With $p = c_0 + ... + c_n \cdot \epsilon^n$, i.e.

$$xp_n(x) := x + (c_0 + c_1 \cdot \epsilon + ... + c_n \cdot \epsilon^n) \cdot (t_0 - \sin(x)), \quad \epsilon = t_0 - \sin(x)$$

the order is n + 2. Only one value of sin(x) is needed at each iteration step.

Example : $t_0 = 0.7$, $x_0 = 0.75$. Errors are

n	= 0	= 1	= 2
$E_1 =$	0.105(-4)	= -0.640(-7)	= 0.439(-9)
$E_2 =$	0.536(-10)	= -0.169(-21)	= 0.354(-37)
$E_3 =$	0.141(-20)	= -0.314(-65)	= 0.150(-149)

with orders 2, 3, 4.

We can also construct algorithms of any order for π, with only one value of sin(x) needed for each step - essentially by solving $\sin(x) = 0$.

3   Inverse interpolation

To find a solution x^* to $F(x) = 0$ we replace F by a deviation function D and define

$$DP(p) := D(x_0 + p \cdot D(x_0))$$

i.e. deviation as a function of p for a fixed x_0. The goal is to find a p-value p_0^* which will make DP so small that the corresponding x-value $x_0 + p_0^* \cdot D(x_0)$ is so close to the exact solution x^* as we need. For any value of p we can compute $DP(p)$. The inverse problem, to compute the p-value for which DP has a certain value [in this case 0], is solved using [inverse] interpolation : Let $p_1, p_2, \ldots p_N$ denote a set of p-values so that the points

$$x_i := x_0 + p_i \cdot D(x_0), \qquad i = 1, 2, \ldots N$$

are 'close' to x^* [as shown by D being small] and let

$$DP_i := DP(p_i) = D(x_i)$$

be these deviations. Finally, let $POL_N = POL_N(DP)$ denote the interpolation polynomial through (DP_i, p_i). The value

$$p_{00} := POL_N(0)$$

is then used as an approximation to the value p_0^*, and the validity of this choice is controlled by computing $DP(p_{00})$. If $|DP(p_{00})|$ is small enough we have the point $x_{00} := x_0 + p_{00} \cdot D(x_0)$ we wanted; otherwise we modify the set of collocation points, for example by replacing one of the p-values by the new value p_{00}, for which we allready have DP. But we may also use a new set of p-values centered around P_{00} - as used below.

Example 1.1 continued

$$D(x) = x^3 - x - 1, \quad D(1.3) = -0.103, \quad D(1.4) = 0.344$$

$a_1 := POL_1(0; 1.3, 1.4), \quad D(a_1) \simeq -0.007134.$

$a_2 := POL_2(0; a_1 - \Delta, a_1, a_1 + \Delta), \quad \Delta = 10^{-3} \quad D(a_2) \simeq -0.195010^{-7}.$

$a_3 := POL_2(0; a_2 - \Delta, a_2, a_2 + \Delta), \quad \Delta = 10^{-9}, \quad D(a_3) \simeq -0.612210^{-24}.$

a_1 is obtained by linear interpolation (POL_1 of degree 1). a_2 is obtained by interpolation of degree 2 with points [actually p-values corresponding to] $a_1 - \Delta$, a_1, $a_1 + \Delta$; a_1 being the point found with POL_1. a_3 similarly with a polynomial of degree 2 using a_2.

It may be convinient to replace p by $p_1 \cdot APd$ where APd is some approximation to $-1/D'(\xi)$ making $p_1 \simeq 1$. APd can also be used for estimating Error from Deviation and $-1/D'(x)$ could be used for inclusions.

References

1. W. Kahan, An example given during an IBM conference Symbolic Mathematical Computation, Oberlech, Austria 1991.
2. Oliver Abert: Precision Basic in *Precise Numerical Analysis*, 1988. More recent: *Precise Numerical Analysis Using C++*, Academic Press, 1998.
3. Ward Cheney and David Kincaid: Numerical Mathematics and Computing, 1999. Brooks/Cole Publishing Company. (Problem 3.2, 37).
4. Carl-Erik Fröberg: *Numerical Mathematics*, 1985. The 'Higher order formulas' are in Schröder's 1870 paper but I prefer Fröberg's version !.
5. Esteban I. Poffald, *The Remainder in Taylor's Formula*, American Mathematical Monthly, March 1990.

A New Prime Edge Length Crystallographic FFT

Jaime Seguel[1], Dorothy Bollman[2], and Edusmildo Orozco[3]

[1] Electrical and Computer Engineering Department, UPRM
Mayagüez, Puerto Rico
jaime.seguel@ece.uprm.edu
[2] Department of Mathematics, UPRM
Mayagüez, Puerto Rico
bollman@cs.uprm.edu
[3] Doctoral Program in CISE, UPRM
Mayagüez, Puerto Rico
eorozco@cs.uprm.edu

Abstract. A new method for computing the discrete Fourier transform (DFT) of data endowed with linear symmetries is presented. The method minimizes operations and memory space requirements by eliminating redundant data and computations induced by the symmetry on the DFT equations. The arithmetic complexity of the new method is always lower, and in some cases significantly lower than that of its predecesor. A parallel version of the new method is also discussed. Symmetry-aware DFTs are crucial in the computer determination of the atomic structure of crystals from x-ray diffraction intensity patterns.

1 Preliminares

Let N be a positive integer, Z/N the set of integers modulo N, $Z^d/N = Z/N \times \cdots \times Z/N$, the cartesian product of d copies of Z/N, and f a real- or complex-valued mapping defined in Z^d/N. The d-*dimensional discrete Fourier transform* (DFT) of *edge length* N is defined by

$$\hat{f}(\boldsymbol{k}) = \frac{1}{\sqrt{N}} \sum_{\boldsymbol{l} \in Z^d/N} f(\boldsymbol{l}) w_N^{\boldsymbol{k} \cdot \boldsymbol{l}}, \quad \boldsymbol{k} \in Z^d/N; \tag{1}$$

where $w_N = \exp(-2\pi i/N)$, $\cdot$ denotes the dot product, and $i = \sqrt{-1}$. A *fast Fourier transform* [3] (FFT) for $d = 1$ computes (1) in $O(N \log N)$ operations. For $d \geq 2$, the usual FFT method consists of applying N^{d-1} one-dimensional FFTs along each of the d dimensions. This yields $O(N^d \log N)$ operations. Although this complexity bound cannot be improved for general DFT computations, some attempts to reduce the actual operation count and memory space requirements have been made for problems whose data is endowed with redundancies, such as x-ray crystal diffraction intensity data. In this article we review

P.M.A. Sloot et al. (Eds.): ICCS 2002, LNCS 2330, pp. 548–557, 2002.

a method for computing prime edge length DFTs of crystallographic data introduced by Auslander and Shenefelt [1],[2] and propose a new method with lower arithmetic complexity.

We first establish some notation. An expression of the form $[V_k]$ denotes a column vector. A bracketed expression with an ordered pair as a subscript denotes a matrix, and its row and and the column indices, respectively. If A is a set and g a mapping defined in A, its image set is denoted $g(A) = \{g(a) : a \in A\}$. By a *partition* of a set A we understand a collection $\{A_1, ..., A_m\}$ of subsets of A such that for each $j \neq k$, $A_j \cap A_k = \emptyset$, and $A = A_1 \cup \cdots A_m$. An *equivalence relation* on A is a relation $a \approx b$ defined for pairs in $A \times A$ which satisfies reflexivity, symmetry, and transitivity. An equivalence relation *induces* the partition of A consisting of the sets $O(a) = \{b \in A : a \approx b\}$. A subset of A formed by selecting one and only one element from each set in a partition of A is called *fundamental set*. The number of elements in a set A is denoted $|A|$.

A matrix $[W(k,l)]_{(k,l)}$, $0 \leq k < N_1$, $0 \leq l < N_2$ is *Hankel* if $W(k,l) = W(k',l')$ whenever $k + l = k' + l'$. An $N \times N$ matrix $[H(k,l)]_{(k,l)}$ is *Hankel-circulant* if $H(k,l) = H(k',l')$ whenever $k + l = k' + l'$ modulo N. Since a Hankel-circulant matrix is completely determined by its first row, we write $H = hc(\text{first row of } H)$. A Hankel matrix W is said to be *embedded* in a Hankel-circulant H if W is the upper leftmost corner of H. Given an $N_1 \times N_2$ Hankel matrix W there exists at least one $N \times N$ Hankel-circulant into which W can be embedded, for each $N \geq N_1 + N_2 - 1$. For $N = N_1 + N_2 - 1$, this Hankel-circulant is

$$hc(W(0,0), \cdots, W(0, N_2 - 1), W(1, N_2 - 1), \cdots, W(N_1 - 1, N_2 - 1)). \qquad (2)$$

For $N > N_1 + N_2 - 1$, the Hankel-circulant is obtained just by padding the argument of hc with $N - (N_1 + N_2 - 1)$ zeros to the right. An N-point *cyclic convolution* is a product of the form $U = HV$, where H is an $N \times N$ Hankel-circulant. We represent the N-point one-dimensional DFT by its complex matrix

$$F_N = \left[w_N^{kl} \right]_{(k,l)}, \quad 0 \leq k, l \leq N - 1. \qquad (3)$$

It has been shown that for a Hankel-circulant H

$$\Delta(H) = F_N H F_N \qquad (4)$$

is a diagonal matrix whose main diagonal is the DFT of the first row of H. Thus, cyclic convolutions can be computed in $O(N \log N)$ operations through

$$U = F_N^{-1} \Delta(H) F_N^{-1} V. \qquad (5)$$

Equation (5) is often referred as *fast cyclic convolution* algorithm.

2   Symmetries and Crystallographic FFTs

Crystalline structures are determined at atomic level by computing several three-dimensional DFTs of their energy spectrum sampled every few angstroms. Since crystal structures consist of repeating symmetric unit cells, their spectral data is highly redundant. A fast Fourier transform (FFT) method that uses these redundancies to lower the arithmetic count and memory space requirements of a three-dimensional FFT is called a *crystallographic FFT*. In this section we review some basic properties of matrix representations of crystal symmetries, use them to eliminate redundancies from the DFT equations, and outline the Auslander-Shenefelt crystallographic FFT.

We assume throughout that P is prime. A square matrix is a *matrix over* Z/P if all its entries are elements of the set Z/P of integers modulo P. The product of two such matrices modulo P is a matrix over Z/P. A matrix over Z/P is nonsingular if and only if its determinant is not zero modulo P. For a matrix M over Z/P and a nonnegative integer j, M^j denotes the product of M by itself modulo P, j times. In particular, $M^0 = I$, where I is the identity matrix of appropriate size.

A real- or complex-valued mapping f defined on the set Z^d/P of d-dimensional vectors over Z/P is said to be S-symmetric if there exists a $d \times d$ nonsingular matrix S over Z/P such that

$$f(l) = f(Sl) \quad \text{for all } l \in Z^d/P. \tag{6}$$

For example, the mapping f defined in $Z^2/5$ by the two-dimensional data array $[f(k,l)]_{(k,l)}$

$$f(Z^2/5) = \begin{bmatrix} 2.9 & 2.3 & 1.5 & 1.5 & 2.3 \\ 1.2 & 6.0 & 4.3 & 4.6 & 2.8 \\ 1.4 & 3.3 & 5.1 & 4.2 & 1.7 \\ 1.4 & 1.7 & 4.2 & 5.1 & 3.3 \\ 1.2 & 2.8 & 4.6 & 4.3 & 6.0 \end{bmatrix} \tag{7}$$

is S-symmetric where

$$S = \begin{bmatrix} -1 & 0 \\ 0 & -1 \end{bmatrix} = \begin{bmatrix} 4 & 0 \\ 0 & 4 \end{bmatrix} \quad \text{modulo 5.} \tag{8}$$

It is clear from (6) that $f(S^j a) = f(a)$ for all j. Thus, the data redundancies of a S-symmetric mapping constitute subsets of Z^d/P of the form

$$O_S(a) = \{S^j a \text{ modulo } P : j \text{ integer }\}. \tag{9}$$

Such a subset is called an S *orbit* of a. It is readily verified that the relation defined by $a \approx_S b$ if and only if $b = S^j a$ for some integer j is an equivalence relation over Z^d/P. Thus, the set of all S-orbits is a partition of Z^d/P. The number of elements $|O_S(a)|$ is called the *orbit length*. A subset of the form $O_S^{(l_a)}(a) = \{S^j a : 0 \leq j \leq l_a - 1\}$ where $l_a \leq |O_S(a)|$ is said to be an S *segment*. An S-symmetric function is completely determined by its values on a fundamental set $\mathcal{F}_S$ induced by $\approx_S$. For example, the S-orbits and their images under f for (7) are

$$O_S(0,0) = \{(0,0)\}, f(O_S(0,0)) = \{2.9\} \tag{10}$$

$$O_S(0,1) = \{(0,1),\ (0,4)\}\ ,\ f(O_S(0,1)) = \{2.3\} \tag{11}$$
$$O_S(0,2) = \{(0,2),\ (0,3)\}\ ,\ f(O_S(0,2)) = \{1.5\} \tag{12}$$
$$O_S(1,0) = \{(1,0),\ (4,0)\}\ ,\ f(O_S(1,0)) = \{1.2\} \tag{13}$$
$$O_S(1,1) = \{(1,1),\ (4,4)\}\ ,\ f(O_S(1,1)) = \{6.0\} \tag{14}$$
$$O_S(1,2) = \{(1,2),\ (4,3)\}\ ,\ f(O_S(1,2)) = \{4.3\} \tag{15}$$
$$O_S(1,3) = \{(1,3),\ (4,2)\}\ ,\ f(O_S(1,3)) = \{4.6\} \tag{16}$$
$$O_S(1,4) = \{(1,4),\ (4,1)\}\ ,\ f(O_S(1,4)) = \{2.8\} \tag{17}$$
$$O_S(2,0) = \{(2,0),\ (3,0)\}\ ,\ f(O_S(2,0)) = \{1.4\} \tag{18}$$
$$O_S(2,1) = \{(2,1),\ (3,4)\}\ ,\ f(O_S(2,1)) = \{3.3\} \tag{19}$$
$$O_S(2,2) = \{(2,2),\ (3,3)\}\ ,\ f(O_S(2,2)) = \{5.1\} \tag{20}$$
$$O_S(2,3) = \{(2,3),\ (3,2)\}\ ,\ f(O_S(2,3)) = \{4.2\} \tag{21}$$
$$O_S(2,4) = \{(2,4),\ (3,1)\}\ ,\ f(O_S(2,4)) = \{1.7\}. \tag{22}$$

Thus, we may choose $\mathcal{F}_S = \{(0,0),\ (0,1),\ (0,2),\ (1,0),\ (1,1),\ (1,2),\ (1,3),\ (1,4),\ (2,0),\ (2,1),\ (2,2),\ (2,3),\ (2,4)\}$.

It is easy to show that if f is S-symmetric, then $\hat{f}$ is S_*-symmetric, where S_* is the transpose of the inverse of S. Therefore, $\hat{f}$ is also completely determined by its values in a fundamental set $\mathcal{F}_{S_*}$ induced by $\approx_{S_*}$. We call $f(\mathcal{F}_S)$ and $\hat{f}(\mathcal{F}_{S_*})$ *fundamental input data* and *fundamental output data*, respectively. In example (7) the fundamental input data set is

$$f(\mathcal{F}_S) = \{2.9,\ 2.3,\ 1.5,\ 1.2,\ 6.0,\ 4.3,\ 4.6,\ 2.8,\ 1.4,\ 3.3,\ 5.1,\ 4.2,\ 1.7\} \tag{23}$$

Reducing the original input data to a fundamental input data set requires a modification in the DFT equations. Since $f(\mathcal{F}_S(a))$ contains just one $f(a)$ from each $f(O_S(a))$ orbit and each $k \in Z^d/P$, the input datum $f(a)$ is factored out of the terms of $\hat{f}(k)$ indexed by $O_S(a)$ in equation (1). That is,

$$\sum_{l \in O_S(a)} f(l) w_P^{k \cdot l} = f(a) \left(\sum_{l \in O_S(a)} w_P^{k \cdot l} \right) = f(a) K_P(k, a). \tag{24}$$

$K_P(k, a) = \sum_{l \in O_S(a)} w_P^{k \cdot l}$ is called a *symmetrized DFT kernel*. The linear transformation

$$\hat{f}(k) = \sum_{a \in \mathcal{F}_S} K_P(k, a) f(a), \quad k \in \mathcal{F}_{S_*} \tag{25}$$

where the output has also been restricted to a fundamental output set, is called a *symmetrized DFT*. Equation (25) involves

$$\sum_{k \in \mathcal{F}_{S_*}} \sum_{l \in \mathcal{F}_S} |O_S(l)| = \sum_{k \in \mathcal{F}_{S_*}} P^d \le P^{2d} \tag{26}$$

arithmetic operations. Thus, in general, the mere reduction of redundant data does not yield a superior FFT method.

Auslander and Shenefelt [1] proposed computing (25) through *fast cyclic convolutions*. Crucial to this aim is the fact the set of all non-null elements in Z/P, denoted Z/P^*, is a *cyclic group* under multiplication. In particular, there is an element $g \in Z/P^*$ called *generator*, such that for each $a \in Z/P^*$, an integer j can always be found for which $g^j = a$ modulo P. The action of g on Z^d/P produces *g-orbits* of the form $O_g(a) = \{g^j a : 0 \le j \le P - 2\}$. Furthermore, the action of g produces partitions of $\mathcal{F}_S$ and $\mathcal{F}_{S_*}$ formed by g-segments. Let $\mathcal{F}_{gS}$ and $\mathcal{F}_{gS_*}$ be fundamental sets for these partitions. For a pair $(a, b) \in \mathcal{F}_{gS_*} \times \mathcal{F}_{gS}$ let l_a and l_b be the lengths of the g-segments in $\mathcal{F}_{gS_*}$ and $\mathcal{F}_{gS}$, respectively. Then,

$$W_{(a,b)}(k, l) = K_P(g^k a, g^l b), \quad 0 \le k \le l_a - 1. \, 0 \le l_b - 1, \tag{27}$$

is a block in (25). Since clearly for $k + l = k' + l'$, $W_{(a,b)}(k, l) = W_{(a,b)}(k', l')$, each block $W_{(a,b)}$ is Hankel. Each one of these blocks can be computed with the fast cyclic convolution algorithm by embeddings in Hankel-circulants.

Since the length of a g-segment is at most $P - 1$, the Auslander-Shenefelt method involves $O(P^{d-1})$ cyclic convolutions unless the average length of the S-orbits is also a power of P. The latter condition is not satisfied by most of the symmetries of practical importance.

Our strategy towards a more efficient crystallographic FFT is to minimize the amount of cyclic convolutions, even at the cost of increasing their sizes. This is achieved by using M-segments instead of g-segments, where M is a nonsingular matrix over Z/P. This modification and its impact in the computation of (25) are described in the next section.

3   A Framework for Designing Crystallographic FFTs

Let S and M be nonsingular matrices over Z/P. Let k, l, k', and l' be nonnegative integers such that $k + l = k' + l'$. Then for any pair $(a, b) \in \mathcal{F}_{S_*} \times \mathcal{F}_S$,

$$K_P(M^k a, (M^t)^l b) = K_P(M^{k+l} a, b) \tag{28}$$
$$= K_P(M^{k'+l'} a, b) \tag{29}$$
$$= K_P(M^{k'} a, (M^t)^{l'} b). \tag{30}$$

It follows that $W_{(a,b)}(k, l) = K_P(M^k a, (M^t)^l b)$ is Hankel. These matrices will represent all the computations in (25) if and only if $\mathcal{F}_{S_*}$ and $\mathcal{F}_S$ can be partitioned into M- and M^t-segments, respectively. It can easily be shown that this is the case if $MS_* = S_*M$. Let $\mathcal{F}_{MS_*}$ and $\mathcal{F}_{M^t S}$ be fundamental sets for these partitions. Then (25) can be rewritten as

$$\left[\hat{f}(\mathcal{F}_{S_*})\right] = \left[\left[W_{(a,b)}(k, l)\right]_{(k,l)}\right]_{(a,b)} [f(\mathcal{F}_S)], \tag{31}$$

where the nested brackets denote a block matrix, $(a, b) \in \mathcal{F}_{MS_*} \times \mathcal{F}_{M^t S}$, and $0 \le k < l_a$, $0 \le l < l_b$.

Before embedding each $W_{(a,b)}$ in a Hankel-circulant we make two considerations. First, no blocks corresponding to pairs $(0, b)$ and $(a, 0)$ will be embedded. These are $1 \times l_b$ and $l_a \times 1$ matrices and therefore, embedding them in Hankel-circulants is not practical. We call these matrices *border blocks* and compute with them separately. We call the remaining set of blocks the *core symmetric DFT* computations. Second, although other choices are available, we propose a common size $N \times N$ for the Hankel-circulants into which the $W_{(a,b)}$ blocks will be embedded. This N is a positive integer greater than or equal to the sum of the length of the largest input segment plus the length of the largest output segment minus 1. It turns out that the maximum length of the input segments is always equal to the maximum length of the output segments. We denote this common maximum length by L. On the other hand, since the choice of N is guided by the efficiency of the N-point FFT, it is reasonable to assume that $2L - 1 \leq N \leq 2^{\lceil \log_2(2L-1) \rceil}$, where $\lceil x \rceil$ is the smallest integer that is greater than or equal to x.

Let $\mathcal{F}^*_{M^tS}$ and $\mathcal{F}^*_{MS_*}$ be the original fundamental sets without 0. Let $H_{(a,b)}$ be the $N \times N$ Hankel-circulant in which $W_{(a,b)}$ is embedded. Then, the core symmetric DFT computations are represented as $U = HV$ where

$$H = \left[\left[H_{(a,b)}(k,l) \right]_{(k,l)} \right]_{(a,b)}, \tag{32}$$

$(a, b) \in \mathcal{F}^*_{MS_*} \times \mathcal{F}^*_{M^tS}$, and $0 \leq k, l < N$. The input vector V is composed of N-point vector segments V_b, each consisting of the l_b values of $f(O^{(l_b)}_{M^tS}(b))$ followed by $N - l_b$ zeros. Vector U, in turn, is composed by the N-point segments U_a, $a \in \mathcal{F}^*_{MS_*}$, that result from

$$U_a = \sum_{b \in \mathcal{F}^*_{M^tS}} H_{(a,b)} V_b \tag{33}$$

$$= \sum_{b \in \mathcal{F}^*_{M^tS}} F_N^{-1} \Delta \left(H_{(a,b)} \right) F_M^{-1} V_b \tag{34}$$

$$= F_N^{-1} \left(\sum_{b \in \mathcal{F}^*_{M^tS}} \Delta \left(H_{(a,b)} \right) F_M^{-1} V_b \right). \tag{35}$$

Following is a prime edge length symmetric FFT framework based on (35):

Core computations:
Step 1. For each $b \in \mathcal{F}^*_{M^tS}$ compute $Y_b = F_N^{-1} V_b$.
Step 2. For each pair $(a, b) \in \mathcal{F}^*_{MS_*} \times \mathcal{F}^*_{M^tS}$ compute the Hadamard or component-wise product $Z_{(a,b)} = \Delta \left(H_{(a,b)} \right) Y_b$.
Step 3. For each $a \in \mathcal{F}^*_{MS_*}$ compute $X_a = \sum_b Z_{(a,b)}$.
Step 4. For each $a \in \mathcal{F}^*_{MS_*}$ compute $U_a = F_N^{-1} X_a$.

Border computations:
Step 5. $\hat{f}(0) = \frac{1}{\sqrt{P}} \sum_{b \in \mathcal{F}_S} f(b) |O_S(b)|$

Step 6. For each $a \in \mathcal{F}^*_{MS_*}$ compute $\hat{f}(O^{l_a}_M(a)) = \frac{1}{\sqrt{P}} [f(0)]_{l_a} + U^*_a$. Here U^*_a is the column vector formed by the l_a first entries of the vector U_a computed in step 4 .

All parameters required for the actual implementation of the method are determined in a separate precomputation phase. This phase includes the computation of the M^t- and M-orbit segments, and the diagonals $\Delta(H_{(a,b)})$.

Since the border computations involve $O(|\mathcal{F}_S|)$ sums and products, their contribution to the arithmetic complexity of the method is marginal. As for the core computations, let $\Lambda = |\mathcal{F}^*_{M^tS}|$. Then, using N-point FFTs, steps 1 and 4 involve $O(2\Lambda N \log N)$ operations. Step 2 involves $\Lambda^2 N$ complex multiplications and step 3, $\Lambda N(\Lambda - 1)$ complex additions. Therefore, the complexity of the core computation phase is modeled by the two-parameter mapping

$$c(\Lambda, N) = 2\Lambda N \left(\kappa \log N + \Lambda - 1 \right), \qquad (36)$$

where κ is a positive constant. It follows that the order of the arithmetic complexity of a crystallographic FFT derived from the general framework is $O(\Lambda N \log N)$ if

$$\rho_1 = \frac{(\Lambda - 1)}{\log_2 (2L - 1)} \leq 1, \qquad (37)$$

and $O(\Lambda^2 N)$ if

$$\rho_2 = \frac{(\Lambda - 1)}{\lceil \log_2 (2L - 1) \rceil} > 1. \qquad (38)$$

The best case, $\Lambda = 1$, ensures an $O(N \log N)$ symmetric FFT. If ρ_2 is greater than 1, but close to 1 and Λ is relatively small, it is still possible to get a competitive crystallographic FFT. However, the arithmetic count of the crystallographic FFT increases rapidly with Λ, as shown in table 1. The parallel version of the method that is outlined in section 5 is intended to reduce the execution time for cases in which $\Lambda > 1$.

The Auslander-Shenefelt method can be derived from the general framework by setting $M = gI_d$, where I_d is the $d \times d$ identity matrix. From a previous remark we know that the Auslander-Shenefelt algorithm is likely to produce large values for Λ for most symmetries of practical importance. One such example is the symmetry

$$S = \begin{bmatrix} -1 & 0 & 0 \\ 0 & -1 & 0 \\ 0 & 0 & -1 \end{bmatrix}. \qquad (39)$$

In this case, the Auslander-Shenefelt crystallographic FFT gives $\Lambda = P^2 + P$ and $L = (P-1)/2$. Thus, $c(\Lambda, N)$ is always $O(\Lambda^2 N)$. Using the minimum $N = 2L-1$ we see that $\Lambda^2(2L - 1) = (P^2 + P)^2(P - 2)$. Therefore, step 2 of the Auslander-Shenefelt FFT is $O(P^5)$ which is greater than $O(P^3 \log P)$, the complexity order of the usual three-dimensional FFT.

4   Existence of $O(N \log N)$ Crystallographic FFTs

Some theoretical results concerning the existence of $O(N \log N)$ crystallographic FFTs are presented in this section. The lemmas, proofs of which can be found, for instance, in [4], are well-known results that have been used in coding theory, digital signal processing, and the theory of linear autonomous sequential machines, among others.

Lemma 1. For any polynomial $\phi(x)$ over Z/P, there is a positive integer k for which $\phi(x)$ divides $x^k - 1$.

The smallest such k is called the *period* of $\phi(x)$.

Lemma 2. For each irreducible polynomial $\phi(x)$ of degree n over Z/P, $\phi(x)$ divides $x^{P^n - 1} - 1$.

A polynomial is *maximal* if its period is $P^n - 1$.

Lemma 3. The number of maximal polynomials of degree n over Z/P is $\varphi(P^n - 1)/n$, where φ denotes Euler's φ-function.

Similarly, for any nonsingular square matrix M over Z/P, there is a least positive integer k such that $M^k = I$, where I denotes an identity matrix of appropriate size. This k is called the *period* of M. A $d \times d$ M is *maximal* if its period is $P^d - 1$. Thus, the action of a maximal M on Z^d/P produces only two orbits: one consisting of $\{0\}$, and one consisting of all non-null vectors in Z^d/P. The *characteristic polynomial* ϕ_M of a square matrix M is defined by the determinant $M - xI$ modulo P. The Cayley Hamilton Theorem states that every square matrix satisfies its own characteristic equation, i.e., $\phi_M(M) = 0$.

Lemma 4. A nonsingular matrix M over Z/P is maximal if and only if its characteristic polynomial $\phi_M(x)$ is irreducible and maximal.

Corollary 1. All M-orbits are of equal length if the characteristic polynomial of M is irreducible.

These results provide a first example of the existence of an $O(N \log N)$ crystallographic FFT for a symmetry of practical importance. In fact, for any P there will be a maximal matrix M over Z/P. Since the symmetry $S = S_*$ defined in (39) commutes with any 3×3 matrix, it commutes, in particular, with any maximal matrix M over Z^3/P. But since M is maximal, the partitions induced by M^t and M on $\mathcal{F}_S^*$ and $\mathcal{F}_{S_*}^*$, respectively, consist of just one segment. Therefore, since $\Lambda = 1$, the crystallographic FFT derived from the general framework is $O(N \log N)$.

Another interesting case is given by the next theorem.

Theorem 1. Let S be a matrix over Z/P with irreducible characteristic polynomial and let M be a maximal matrix that commutes with S. Then there is exactly one M-segment in $\mathcal{F}_{S_*}$ and its size is $m = (P^d - 1)/k$ where k is the size of the S_*-orbits.

Proof. From the previous corollary, all orbits induced by S_* are of equal length. Let x be any nonzero vector in Z^d/P. Then the sequence $x, Mx, M^2, \cdots, M^{P^d - 2}x$ contains at most k S_*-equivalent elements. Suppose it contains fewer than m S_*-equivalent elements. Let a and b, $a < b$ be the least positive integers for which $M^a x$ and $M^b x$ are S_*-equivalent and so $M^a x = S^c M^b x$ for some c.

Hence $S_*{}^{-c}x = M^{b-a}x$ and so $M^{(b-a)k}M^{b-a}\cdots M^{b-a}x = (S_*{}^{-c})^k x = x$ where $(b-a)k < P^d - 1$, which contradicts that M is maximal.

5   Further Examples and Conclusions

As remarked before, some crystallographic symmetries will not yield $O(N \log N)$ crystallographic FFTs. For such symmetries, the parameter Λ is minimized by finding a matrix M satisfying $MS_* = S_*M$, and whose action produces the largest possible M-segments in $\mathcal{F}_{S_*}$. Such M is called *optimal*. In this section we show a symmetry whose parameter $\Lambda > 1$ and compare the complexity of the crystallographic FFT built with the optimal M with that of the Auslander-Shenefelt algorithm. We also describe a natural parallel version of the method for symmetries whose parameter Λ is greater than 1.

Let us consider the symmetry represented by

$$S = \begin{bmatrix} 0 & 0 & 1 \\ 0 & 1 & 0 \\ -1 & 0 & 0 \end{bmatrix}. \tag{40}$$

Table 1 compares the parameters and the arithmetic complexity of the second step of the crystallographic FFT derived from the general framework using an optimal M with the Auslander-Shenefelt method. We have chosen $N = 2L-1$ for these comparisons. To our knowledge there is no existing method for computing

Table 1. Crystallographic FFT derived from an optimal segmentation of the fundamental sets versus Auslander-Shenefelt crystallographic FFT (ASCFFT) for symmetry (40)

Problem size	Optimal M	Λ	N	$\Lambda^2 N$	ASCFFT Λ	ASCFFT N	ASCFFT $\Lambda^2 E$
$23^3 = 12167$	$\begin{bmatrix} 2 & 0 & 1 \\ 0 & 5 & 0 \\ 22 & 0 & 2 \end{bmatrix}$	24	263	151,488	145	43	904,075
$31^3 = 29791$	$\begin{bmatrix} 4 & 0 & 1 \\ 0 & 3 & 0 \\ 30 & 0 & 4 \end{bmatrix}$	32	479	490,496	257	59	3,896,891
$43^3 = 79507$	$\begin{bmatrix} 2 & 0 & 1 \\ 0 & 3 & 0 \\ 42 & 0 & 2 \end{bmatrix}$	44	923	1,786,928	485	83	19,523,675
$47^3 = 103823$	$\begin{bmatrix} 2 & 0 & 1 \\ 0 & 5 & 0 \\ 46 & 0 & 2 \end{bmatrix}$	48	1103	2,541,312	577	91	30,296,539

the optimal M directly. Our results have been produced computationally.

For $\Lambda > 1$, in a distributed memory system with Λ processors, the symmetric FFT can be performed in parallel, as follows:

Parallel core symmetric FFT

Parallel step 1. For each $b \in \mathcal{F}^*_{M^tS}$ compute in parallel $Y_b = F_N^{-1} V_b$.

Parallel step 2. For each $b) \in \mathcal{F}^*_{MS_*}$ compute in parallel $Z_{(a,b)} = \Delta \left(H_{(a,b)} \right) Y_b$, for all $a \in \mathcal{F}^*_{M^tS}$.

Communication step. For each $a \in \mathcal{F}^*_{MS_*}$, gather all vector segments $Z_{(a,b)}$ in a single processor.

Step 3. For each $a \in \mathcal{F}^*_{MS_*}$ compute in parallel $X_a = \sum_b Z_{(a,b)}$.

Step 4. For each $a \in \mathcal{F}^*_{MS_*}$ compute in parallel $U_a = F_N^{-1} X_a$.

Parallel border computations:

Step 5. $\hat{f}(0) = \frac{1}{\sqrt{P}} \sum_{b \in \mathcal{F}_S} f(b) |O_S(b)|$

Step 6. For each $a \in \mathcal{F}^*_{MS_*}$ compute in parallel $\hat{f}(O_M^{l_a}(a) = \frac{1}{\sqrt{P}} [f(0)]_{l_a} + U_a^*$.
Here U_a^* is the column vector formed by the l_a first entries of the vector U_a computed in step 4 .

The parallel method reduces the time complexity of steps 1 and 4 to $O(N \log N)$ and that of steps 2 and 3 to $O(\Lambda N)$, to the cost of sending $\Lambda N(\Lambda - 1)$ complex number between processors. For large values of Λ, an adquate balance between communications and computations can be achieved by aggregating parallel computations. The symmetric FFT framework can be implemented as a meta-algorithm able to generate crystallographic FFTs that are tailored to a target computer system. The inputs of the meta-algorithm will be the symmetry S and the edge length P of the DFT. The output will be a crystallographic FFT computer program that optimally uses the hardware and the software of the target system, very much in the spirit of the fast Fourier transform method known as the FFTW [5]. Experimental work exploring the potential of this idea is currently underway.

Acknowledgements. This work was partially supported by NIH/NIGMS grant No. S06GM08103 and the NSF PRECISE project of the University of Puerto Rico at Mayagüez

References

1. L. Auslander, and M. Shenefelt, *Fourier transforms that respect crystallogaphic symmetries*, IBM J. Res. and Dev., 31, (1987), pp. 213-223.
2. M. An, J. W. Cooley, and R. Tolimeri, *Factorization methods for crystallographic Fourier transforms*, Advances in Appl. Math., 11 (1990), pp. 358-371.
3. J. Cooley, and J. Tukey, *An algorithm for the machine calculation of complex Fourier series*, Math. Comp., 19 (1965), pp. 297-301.
4. B. Elspas, *The Theory of Autonomous Linear Sequential Networks*, Linear Sequential Switching Circuits, ed. W. Kautz, Holden-Day inc., 1965, 21-61.
5. M. Frigo, S. G. Johnson *An adaptive software architecture for the FFT* ICASSP Conference Proceedings, 3 (1998), pp 1381-1384.

TopoMon: A Monitoring Tool for Grid Network Topology

Mathijs den Burger, Thilo Kielmann, and Henri E. Bal

Division of Mathematics and Computer Science, Vrije Universiteit,
De Boelelaan 1081a, 1081HV Amsterdam, The Netherlands
{mathijs,kielmann,bal}@cs.vu.nl   http://www.cs.vu.nl/albatross/

Abstract. In Grid environments, high-performance applications have to take into account the available network performance between the individual sites. Existing monitoring tools like the Network Weather Service (NWS) measure bandwidth and latency of end-to-end network paths. This information is necessary but not sufficient. With more than two participating sites, simultaneous transmissions may collide with each other on shared links of the wide-area network. If this occurs, applications may obtain lower network performance than predicted by NWS.
In this paper, we describe TopoMon, a monitoring tool for Grid networks that augments NWS with additional sensors for the routes between the sites of a Grid environment. Our tool conforms to the Grid Monitoring Architecture (GMA) defined by the Global Grid Forum. It unites NWS performance and topology discovery in a single monitoring architecture. Our topology consumer process collects route information between the sites of a Grid environment and derives the overall topology for utilization by application programs and communication libraries. The topology can also be visualized for Grid application developers.

1 Introduction

A difficult problem in designing efficient applications for computational Grids is that the wide-area interconnection network is highly complex, dynamic, and heterogeneous. A more traditional computing platform such as a cluster or supercomputer typically has a highly regular network with constant, guaranteed performance. With a Grid, however, the network has an irregular and asymmetric topology, where different links have different speeds, which even change over time. The Network Weather Service (NWS) [23] is a tool that addresses this problem by providing dynamic bandwidth and latency measurements and predictions.

Unfortunately, the information provided by NWS is insufficient for applications with communication patterns where multiple sites compete for the same links. For example, if two pairs of communicating sites are used, NWS will predict performance for each pair separately. If the two communication streams have some links in common, however, these predictions will clearly be too optimistic, as the bandwidth for these links has to be shared by the two streams. The

P.M.A. Sloot et al. (Eds.): ICCS 2002, LNCS 2330, pp. 558–567, 2002.

current NWS system cannot solve this problem, because it lacks topology information and thus cannot know which links are shared. This problem is especially important for applications that use collective communication, where many sites communicate with each other simultaneously. The work on MPICH-G (the MPI implementation of Globus [5]) has shown that our wide-area collective commnication library (MagPIe, [10]) has exactly this problem for shared links [8].

In this paper, we present a new tool, TOPOMON, which augments NWS with topology information about the wide-area network. The added value of integrating both performance and topology information is that applications or communication libraries can not only predict their communication performance, but can also avoid congestion on those Internet links that are shared by multiple, simultaneous data streams.

To foster interoperability with other Grid services, our tool conforms to the Grid Monitoring Architecture (GMA) defined by the Global Grid Forum, recommending a structure of separate processes with sensor, producer, consumer, and directory-service functionality. TOPOMON can analyze which paths between Grid sites overlap with each other, possibly resulting in performance (congestion) problems. The topology is determined by sensors that internally use *traceroute* and is output in the form of XML event descriptions. The NWS measurements and predictions are likewise wrapped using XML. The data from both NWS and topology sensors is sent to a consumer process, which combines the data, for example to compute an optimal spanning tree or to export it for visualization.

The paper is structured as follows. In Section 2, we present TOPOMON's design and implementation. Section 3 illustrates the added value of integrating performance and topology data within TOPOMON's topology consumer process. Section 4 discusses related work, Section 5 concludes.

2 Tool Design and Implementation

In this section, we describe TOPOMON. We present its architecture, the topology sensors, and the performance and topology information transferred to interested consumer processes.

2.1 The Topology Monitoring Architecture

Our goal is to build a Grid network monitoring system that integrates performance data with information about the Grid network topology. The system is designed to run on various different platforms that are used by the sites of a Grid environment. Therefore, TOPOMON's overall architecture (as shown in Figure 1) follows the proposed Grid Monitoring Architecture (GMA) [3]. For portability, we implemented all processes as Java programs that communicate with each other using TCP sockets. (The platform-specific sensors are written in C.)

According to the GMA, different functionality is implemented in specialized processes that interoperate using specific network protocols. We distinguish between *sensors* (performing the actual measurements), *producers* (providing mon-

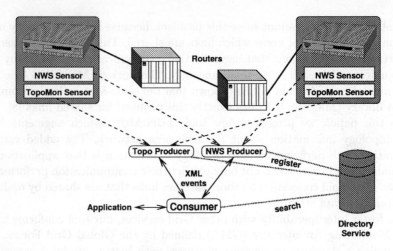

Fig. 1. The TOPOMON architecture

itoring data from one or more sensors to interested *consumers*), and a *directory service* that allows producers and consumers to contact each other.

The Network Weather Service (NWS) [23] already provides excellent performance measurements (and forecasts) for end-to-end network bandwidth and latency. Its architecture also conforms with the ideas of GMA. We therefore decided to augment the existing NWS with network topology information. However, the NWS processes communicate with proprietary protocols using binary representations for the data items exchanged. For interoperability, Grid monitoring data should be exchanged as XML descriptions of monitoring *events* [20].

TOPOMON's implementation builds a wrapper layer around the NWS. The TOPOMON performance ("NWS producer") process presents its performance data as XML descriptions to its consumer processes. It directly wraps the NWS producer processes (called *memory* and *forecaster*), and translates NWS data to XML-event descriptions. As with the original NWS software, the NWS memory processes directly communicate with the NWS network *sensors*, located at the participating Grid sites. Furthermore, TOPOMON also directly uses the NWS directory service, called *name server*.

In addition to wrapping NWS, TOPOMON implements an infrastructure to generate network topology information. At each participating Grid site, a TOPOMON *sensor* determines the network paths to all other sites. The *Topo Producer* process retrieves the topology information from the sensors in their respective data format, and presents it as XML event descriptions. We also implemented a consumer process that retrieves both the topology and NWS performance information, and presents it to application programs. We now describe the topology-related parts of TOPOMON in detail.

2.2 Topology Sensors

TOPOMON's topology sensors explore the structure of the network connecting the sites of a Grid environment. For high-performance applications, we are interested in identifying the end-to-end paths between all sites of a Grid. The performance (bandwidth and latency) of those paths is monitored by the NWS sensors. In addition, we need to identify which parts of the end-to-end paths overlap with each other, as such overlap may cause congestion and performance decrease, if the links are used simultaneously.

As Grid networks are part of the global Internet, identifying end-to-end paths corresponds to observing the respective IP routing information, as used by the *Border Gateway Protocol* (BGP) [16]. However, the routing data available from BGP is restricted to routing between *Autonomous Systems* (AS), which are the organizational units of the Internet, such as ISPs and their customers. In order to maintain global scalability of routing information, BGP data is restricted to entry and exit routers for each AS; routing inside each AS is performed locally. As a consequence, end-to-end paths and their possible intersections, as needed by TOPOMON, can not be derived from BGP data.

Alternatively, path information could be taken from Internet topology repositories, like the ones maintained by CAIDA [19]. Unfortunately, those repositories only provide information about the sites included in their own monitoring efforts, a consequence of the intractable size of the Internet as a whole. As TOPOMON's purpose is to explore the topology exactly between the sites of a Grid environment, general Internet topology repositories do not provide the required information about the specific sites in use.

As a consequence, TOPOMON has to perform its own, active topology discovery. The only feasible approach is to perform probes as done by *traceroute*, or to run the tool directly between the sites of a Grid environment [18]. *traceroute* is directly based on the ubiquitous IP and ICMP protocols. It discovers the routers on a path from one site in the Internet to another one, without relying on any additional network management protocol like SNMP, which is only available within local network installations. We implemented the TOPOMON sensors as wrapper processes around the locally available *traceroute* programs.

2.3 Performance and Topology Event Data

The XML producer-consumer protocol, as proposed by the GGF performance working group [20], describes all performance data as timestamped events of name-value pairs. The respective GGF document prescribes the protocol between producers and consumers, covering both subscription to services and the exchange of event data itself. The messages of this protocol, including the event data, are formatted as XML documents.

The GGF document also provides simple XML schemas for performance data like network roundtrip times. We use similar XML schemas for representing the data generated by NWS. They are shown using example events in Figure 2. For simplicity of presentation, we present example events rather than the formal

```
<NWSBandwidthTcp xmlns="http://www.cs.vu.nl/albatross/TopoMon">
        <SourceHostName>das0fs.cs.vu.nl</SourceHostName>
        <DestinationHostName>das2fs.wins.uva.nl</DestinationHostName>
        <NWSForecast> <Value unit="Mb/s">2.31</Value>
                      <Error>0.010219</Error>
        </NWSForecast>
        <TimeStamp>2001-11-25T16.14.38Z</TimeStamp>
</NWSBandwidthTcp>

<NWSLatencyTcp xmlns="http://www.cs.vu.nl/albatross/TopoMon">
        <SourceHostName>das0fs.cs.vu.nl</SourceHostName>
        <DestinationHostName>huron.cs.unh.edu</DestinationHostName>
        <NWSForecast> <Value unit="ms">12.3</Value>
                      <Error>0.00354</Error>
        </NWSForecast>
        <TimeStamp>2001-11-25T16.23.01Z</TimeStamp>
</NWSLatencyTcp>
```

Fig. 2. XML events for NWS bandwidth and latency data

schema descriptions themselves. The latter are only needed by the producer and consumer processes implementing the protocols.

The figure shows both a bandwidth and a latency data event. Both event schemas are easily human-readable, which is one of the advantages of using XML [20]. The NWS performance events start with a tag indicating their type, followed by information about source and destination host, identifying the pair of NWS sensors that generated this event. The events are completed with the predicted data value, its predicted error, and the event timestamp.

It should be noted that the bandwidth and latency event descriptions are specific to NWS, the network monitoring tool that generates the events. Unfortunately, network performance data is strongly dependent on the measurement method used for generating the data. For example, message roundtrip times measured by NWS and by ping can be very different because of the different protocols involved (TCP vs. ICMP). For later assessment of the data, information about its measurement method has to be added. We are involved in the newly-founded *Network Measurements Working Group* (NMWG) of the GGF. The NMWG aims to develop standards to ensure the compatibility of metrics across measurement systems. Until such standards exist, however, tools like TOPOMON have to define their own metrics.

Figure 3 shows an example event for the traceroute information, as gathered by the topology producers. The traceroute event mainly consists of source and destination host, and of a list of hops, for which the name, IP address, and the measured round-trip times are reported. The XML representation for traceroute data also is easy to understand by humans. Translating the already textual representation of the "raw" data produced by the traceroute sensors into another textual representation has two advantages. First, the very regular XML description is easier to parse by a consumer process. Second, the XML description removes error messages and the (slight) differences in the data formats produced by the traceroute implementations on the various platforms.

```
<Traceroute xmlns="http://www.cs.vu.nl/albatross/TopoMon">
        maxHops="30" probeSize="38">
    <SourceHost>  <Name>das0fs.cs.vu.nl</Name>
                    <IP4Address>130.37.26.4</IP4Address>
    </SourceHost>
    <DestinationHost>  <Name>das2fs.wins.uva.nl</Name>
                    <IP4Address>146.50.13.20</IP4Address>
    </DestinationHost>
    <Hop index="1">  <Host>  <Name>brandaris</Name>
                            <IP4Address>130.37.26.1</IP4Address>
                    </Host>
            <RTT>1.068</RTT>  <RTT>1.032</RTT>  <RTT>1.017</RTT>
    </Hop>
    <Hop index="2">  <Host>  <Name>vnxswitch</Name>
                            <IP4Address>130.37.14.1</IP4Address>
                    </Host>
            <RTT>0.551</RTT>  <RTT>0.406</RTT>  <RTT>0.395</RTT>
    </Hop>
    ...
    <Hop index="8">  <Host>  <Name>das2fs.wins.uva.nl</Name>
                            <IP4Address>146.50.13.20</IP4Address>
                    </Host>
            <RTT>1.584</RTT>  <RTT>2.140</RTT>  <RTT>2.226</RTT>
    </Hop>
    <TimeStamp>2001-11-25T16.15.20Z</TimeStamp>
</Traceroute>
```

Fig. 3. XML event for traceroute information

3   Topology Consumer Functionality

We have implemented a topology consumer process that queries the producer processes and provides the information in preprocessed form to Grid application programs. The consumer thus takes as input the network topology descriptions and the NWS measurement data (see Figure 1), both in XML form. Our consumer process implementation illustrates the capabilities of a Grid monitoring architecture that integrates both performance and topology information. It supports the development of collective communication libraries like MagPIe [10]. Other application domains may develop consumers tailored for their specific needs.

The consumer computes the overall topology of a Grid by combining the topology (traceroute) results between all respective sites, and transitively removing those intermediate nodes with exactly two neighbors – that do not contribute to the determination of overlapping paths. In order to cover the potential path asymmetry between pairs of Grid sites, all links in this topology graph are unidirectional.

All links of the precomputed topology are then attributed with latency and bandwidth values. Latency values are taken from the traceroute data. Bandwidth values are taken from the NWS measurements; for each link l, the maximum of the end-to-end bandwidth values of all connections using l is used as the link's estimated bandwidth. This computation of individual link bandwidth is a conservative estimate, as Internet backbone links may provide higher accumulated bandwidth to multiple, simultaneous transmissions. As NWS currently minimizes the intrusiveness of its measurements, concurrent measurements across shared links are not provided. We are currently developing a scheme to augment NWS by additionally performing such measurements, where the number of additional probes can be limited to those paths where links are actually shared, according to the traceroute results.

The consumer process reports its information to application programs and communication libraries like MagPIe. It can provide end-to-end latency and bandwidth, the path between two endpoints, and the shared parts of two paths between two pairs of endpoints. The consumer can also export the graph connecting the endpoints to the *GraphViz* graph visualization package [7], the latter to support Grid application developers. Figure 4 shows a topology graph between four sites. The endpoints are denoted by shaded boxes. For clarity of presentation, only bandwidth numbers (in Mbit/s) are shown in the figure.

Most importantly, the consumer can also compute (multicast) spanning trees from one endpoint to all other endpoints. As we are interested in application-level communication, only endpoints (Grid sites) can actively forward messages. The consumer process can compute the spanning trees either with minimal latency, or with maximal bandwidth. The former case minimizes the completion time for multicasting a short message to all endpoints, which is mostly independent of the available bandwidth. The latter case optimizes multicast for long messages, which is mostly independent of latency. Computing these spanning trees generalizes our work on MPI-style collective communication [9] from homogeneous to heterogeneous Grid networks.

Spanning trees with minimal latency can be computed based on end-to-end latency values only. For short messages, link bandwidth and thus overlapping paths can be ignored. However, for computing spanning trees with maximal bandwidth, the TOPOMON consumer combines all NWS predictions and traceroute results. All links in the graph (see for example Figure 4) are annotated with their estimated bandwidth. Based on these annotations, the maximum overall bandwidth from one endpoint to the others, can be computed by applying Dijkstra's algorithm to compute a tree of shortest paths, using the reciprocals of the bandwidth values as distance metrics [1].

4   Related Work

Many researchers have found that high-performance Grid applications need to be aware of the interconnection network in use [4, 8, 9, 15, 17]. Performance monitoring systems thus are becoming integral parts of Grid computing platforms. The Global Grid Forum has proposed the GMA architecture [3] that describes the components of monitoring systems and their interactions. In an accompanying document [20], an XML-based producer-consumer protocol has been proposed for the GMA. TOPOMON follows the GMA structure, and also uses the XML-based protocol for portability reasons. For TOPOMON, we had to define additional XML schemas for expressing performance and topology information.

There exist several monitoring systems for the Internet that either measure network performance [2, 13, 14] or explore network topology [6, 19]. However,

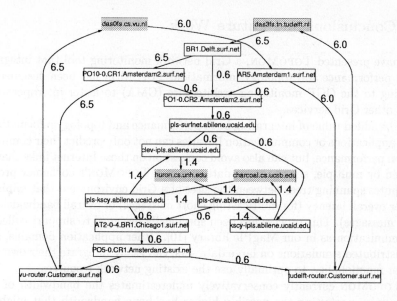

Fig. 4. Topology and link bandwidth (in Mbit/s) derived from TOPOMON data

their usefulness for Grid applications is limited as they only provide data for the nodes involved in their own monitoring efforts. For our purposes, we need information (topology and performance) about exactly those parts of the global Internet that connect the sites of a Grid environment. While Internet-related tools provide useful monitoring techniques, Grid environments have to apply them to their own sites. TOPOMON integrates performance and topology information for given Grid platforms.

Some systems explore the topology of LAN installations in great detail. Most prominently, Remos [4] uses SNMP to collect the necessary information. However, SNMP is not available for general Internet connections. The BGP [16] routing protocol provides global topology information, however only with the granularity of Autonomous Systems, which is insufficient for our purposes.

The most advanced performance monitoring system for Grids is the Network Weather Service (NWS) [23]. However, NWS only provides performance data between pairs of Grid sites. TOPOMON augments this data with information about the topology of the interconnecting networks. Only the combination of both kinds of data allows Grid applications to both predict communication performance and avoid contention between multiple of its own data streams.

Instead of NWS and traceroute, other network sensor tools could be used as well [11, 12, 21]. However, our choice of sensors was determined by low intrusiveness, widespread availability, and for NWS also by application-level measurement using TCP streams. Other sensors might easily be integrated into TOPOMON simply by adapting the respective producer process.

5   Conclusions and Future Work

We have presented TopoMon, a Grid network monitoring tool that integrates both performance and topology information. TopoMon has been designed according to the GGF monitoring architecture (GMA) to foster interoperability with other Grid services.

The added value of integrating both performance and topology information is that applications or communication libraries can not only predict their communication performance, but can also avoid congestion on those Internet links that are shared by multiple, simultaneous data streams. TopoMon's consumer process computes spanning trees between the sites of a Grid environment that minimize either overall latency (for multicasting short messages) or overall bandwidth (for long messages). This process has been specifically designed to support collective communication as in our MagPIe library [10]. Other application domains, such as distributed simulations on large data sets, can easily integrate their own consumer functionality, to optimally use the existing network resources.

TopoMon currently conservatively underestimates the bandwidth of network links, neglecting the possibly higher backbone bandwidth that might be observed by multiple, simultaneous transmissions. This is a direct consequence of using NWS for network measurements. NWS carefully avoids simultaneous measurements both for minimizing intrusiveness and for accuracy [22]. We are currently working on a scheme to augment NWS to additionally perform concurrent measurements that both explores the potential of the Internet backbone and minimizes the necessary, additional measurements. Exploiting the topology of the given network connections is key to this scheme. We are also working on a better graphical representation of the network topology to visualize conflicting data streams. Finally, we are augmenting our MagPIe library with collective communication algorithms that exploit the information derived by TopoMon.

Acknowledgements

We would like to thank Rich Wolski and Martin Swany for the lively exchange of ideas about network monitoring. Jason Maassen, Rob van Nieuwpoort, and Guilleaume Pierre made valuable comments on drafts of this work. The following people kindly granted us access to their systems for performing topology experimentation: Luc Bougé (ENS Cachan), Edgar Gabriel (HLRS), Sergei Gorlatch (TU Berlin), Phil Hatcher (Univ. of New Hampshire), Thomas Kunz (Carleton Univ.), Ludek Matyska (Masaryk Univ.), Jarek Nabrzysky (PNSC), Steven Newhouse (Imperial College), Guillaume Pierre (INRIA Rocquencourt), Satoshi Sekiguchi (ETL), Rich Wolski (UCSB).

References

1. R. K. Ahuja, T. L. Magnanti, and J. B. Orlin. *Network Flows*. Prentice Hall, 1993.
2. Active Measurement Project (AMP). http://amp.nlanr.net.

3. R. Aydt, D. Gunter, W. Smith, M. Swany, V. Taylor, B. Tierney, and R. Wolski. A Grid Monitoring Architecture. Global Grid Forum, Performance Working Group, Grid Working Document GWD-Perf-16-1, 2001.
4. P. Dinda, T. Gross, R. Karrer, B. Lowekamp, N. Miller, P. Steenkiste, and D. Sutherland. The Architecture of the Remos System. In *IEEE Symposium on High Performance Distributed Computing (HPDC10)*, San Francisco, CA, 2001.
5. I. Foster and C. Kesselman. Globus: A Metacomputing Infrastructure Toolkit. *Int. Journal of Supercomputer Applications*, 11(2):115–128, 1997.
6. P. Francis, S. Jamin, C. Jin, Y. Jin, D. Raz, Y. Shavitt, and L. Zhang. IDMaps: A Global Internet Host Distance Estimation Service. *IEEE/ACM Transactions on Networking*, 2001.
7. E. R. Gansner, E. Koutsofios, S. C. North, and K.-P. Vo. A Technique for Drawing Directed Graphs. *IEEE Trans. of Software Engineering*, 19(3):214–230, 1993.
8. N. T. Karonis, B. R. de Supinski, I. Foster, W. Gropp, E. Lusk, and J. Bresnahan. Exploiting Hierarchy in Parallel Computer Networks to Optimize Collective Operation Performance. In *International Parallel and Distributed Processing Symposium (IPDPS 2000)*, pages 377–384, Cancun, Mexico, May 2000. IEEE.
9. T. Kielmann, H. E. Bal, S. Gorlatch, K. Verstoep, and R. F. H. Hofman. Network Performance-aware Collective Communication for Clustered Wide Area Systems. *Parallel Computing*, 27(11):1431–1456, 2001.
10. T. Kielmann, R. F. H. Hofman, H. E. Bal, A. Plaat, and R. A. F. Bhoedjang. MAG-PIE: MPI's Collective Communication Operations for Clustered Wide Area Systems. In *Symposium on Principles and Practice of Parallel Programming (PPoPP)*, pages 131–140, Atlanta, GA, May 1999.
11. B. A. Mah. pchar. http://www.caida.org/tools/utilities/others/pathchar/.
12. Pathrate. http://www.pathrate.org/.
13. V. Paxson, J. Mahdavi, A. Adams, and M. Mathis. An Architecture for Large-scale Internet Measurement. *IEEE Communications*, 1988.
14. PingER. http://www-iepm.slac.stanford.edu/pinger/.
15. A. Plaat, H. E. Bal, R. F. Hofman, and T. Kielmann. Sensitivity of Parallel Applications to Large Differences in Bandwidth and Latency in Two-Layer Inter-connects. *Future Generation Computer Systems*, 17(6):769–782, 2001.
16. Y. Rekhter and T. Li. A Border Gateway Protocol 4 (BGP-4). IETF Network Working Group, RFC 1771, 1995. http://www.faqs.org/rfcs/rfc1771.html.
17. G. Shao, F. Berman, and R. Wolski. Using Effective Network Views to Promote Distributed Application Performance. In *Parallel and Distributed Processing Techniques and Applications (PDPTA)*, 1999.
18. R. Siamwalla, R. Sharma, and S. Keshav. Discovering Internet Topology. In *IEEE INFOCOM*, 1999.
19. Skitter. http://www.caida.org/tools/measurement/skitter.
20. W. Smith, D. Gunter, and D. Quesnel. A Simple XML Producer-Consumer Protocol. Global Grid Forum, Performance Working Group, Grid Working Document GWD-Perf-8-2, 2001.
21. Sprobe. http://sprobe.cs.washington.edu/.
22. M. Swany and R. Wolski. Topology Discovery for the Network Weather Service. Submitted for publication.
23. R. Wolski, N. Spring, and J. Hayes. The Network Weather Service: A Distributed Resource Performance Forecasting Service for Metacomputing. *Future Generation Computing Systems*, 15(5–6):757–768, 1999.

Logistical Storage Resources for the Grid

Alessandro Bassi[1], Micah Beck[1], Erika Fuentes[1], Terry Moore[1], James S. Plank[1]

[1]Logistical Computing and Internetworking Laboratory
Department of Computer Science
University of Tennessee
1122 Volunteer Blvd., Suite 203
Knoxville, TN 37996-3450
{abassi,mbeck,efuentes,tmoore,plank}@cs.utk.edu

1 Introduction

It is commonly observed that the continued exponential growth in the capacity of fundamental computing resources — processing power, communication bandwidth, and storage — is working a revolution in the capabilities and practices of the research community. It has become increasingly evident that the most revolutionary applications of this superabundance use *resource sharing* to enable new possibilities for collaboration and mutual benefit. Over the past 30 years, two basic models of resource sharing with different design goals have emerged. The differences between these two approaches, which we distinguish as the *Computer Center* and the *Internet* models, tend to generate divergent opportunity spaces, and it therefore becomes important to explore the alternative choices they present as we plan for and develop an information infrastructure for the scientific community in the next decade.

Interoperability and scalability are necessary design goals for distributed systems based on resource sharing, but the two models we consider differ in the positions they take along the continuum between total control and complete openness. That difference affects the tradeoffs they tend to make in fulfilling their other design goals. The Computer Center model, which came to maturity with the NSF Supercomputer Centers of the 80s and 90s, was developed in order to allow scarce and extremely valuable resources to be shared by a select community in an environment where security and accountability are major concerns. The form of sharing it implements is necessarily highly controlled – authentication and access control are its characteristic design issues. In the last few years this approach has given rise to a resource sharing paradigm known as information technology "Grids." Grids are designed to flexibly aggregate various types of highly distributed resources into unified platforms on which a wide range of "virtual organizations" can build. [11] By contrast, the Internet paradigm, which was developed over the same 30-year period, seeks to share network bandwidth for the purpose of universal communication among an international community of indefinite size. It uses lightweight allocation of network links via

This work is supported by the National Science Foundation Next Generation Software Program under grant # EIA-9975015 and the Department of Energy Scientific Discovery through Advanced Computing Program DE-FC02-01ER25465.

P.M.A. Sloot et al. (Eds.): ICCS 2002, LNCS 2330, pp. 568–577, 2002.

packet routing in a public infrastructure to create a system that is designed to be open and easy to use, both in the sense of giving easy access to a basic set of network services and of allowing easy addition of privately provisioned resources to the public infrastructure. While admission and accounting policies are difficult to implement in this model, the power of the universality and generality of the resource sharing it implements is undeniable.

Though experience with the Internet suggests that the transformative power of information technology is at its highest when the ease and openness of resource sharing is at its greatest, the Computer Center model is experiencing a rebirth in the Grid while the Internet paradigm has yet to be applied to any resource other than communication bandwidth. But we believe that the Internet model can be applied to other kinds of resources, and that, with the current Internet and the Web as a foundation, such an application can lead to similarly powerful results. The storage technology we have developed called the Internet Backplane Protocol (IBP) is designed to test this hypothesis and explore its implications. IBP is our primary tool in the study of logistical networking, a field motivated by viewing data transmission and storage within a unified framework. In this paper we explain the way in which IBP applies the Internet model to storage, describe the current API and the software that implements it, lay out the design issues which we are working to address, and finally characterize the future directions that this work will take.

2   Background: The Internet Protocol and the Internet Backplane Protocol

The Internet Backplane Protocol is a mechanism developed for the purpose of sharing storage resources across networks ranging from rack-mounted clusters in a single machine room to global networks. [4, 6, 15] To approximate the openness of the Internet paradigm for the case of storage, the design of IBP parallels key aspects of the design of IP, in particular IP datagram delivery. This service is based on packet delivery at the link level, but with more powerful and abstract features that allow it to scale globally. Its leading feature is the independence of IP datagrams from the attributes of the particular link layer, which is established as follows:

3   *Aggregation of link layer packets masks its limits on packet size;*

4   *Fault detection with a single, simple failure model (faulty datagrams are dropped) masks the variety of different failure modes;*

5   *Global addressing masks the difference between local area network addressing schemes and masks the local network's reconfiguration.*

This higher level of abstraction allows a uniform IP model to be applied to network resources globally, and it is crucial to creating the most important difference between link layer packet delivery and IP datagram service. Namely,

> *Any participant in a routed IP network can make use of any link layer connection in the network regardless of who owns it. Routers aggregate individual link layer connections to create a global communication service.*

This IP-based aggregation of locally provisioned, link layer resources for the common purpose of universal connectivity constitutes the form of sharing that has made the Internet the foundation for a global information infrastructure.

IBP is designed to enable the sharing of storage resources within a community in much the same manner. Just as IP is a more abstract service based on link-layer datagram delivery, IBP is a more abstract service based on blocks of data (on disk, tape or other media) that are managed as "byte arrays." The independence of IBP byte arrays from the attributes of the particular *access layer* (which is our term for storage service at the local level) is established as follows:

6   *Aggregation of access layer blocks masks the fixed block size;*

7   *Fault detection with a very simple failure model (faulty byte arrays are discarded) masks the variety of different failure modes;*

8   *Global addressing based on global IP addresses masks the difference between access layer addressing schemes.*

This higher level of abstraction allows a uniform IBP model to be applied to storage resources globally, and this is essential to creating the most important difference between access layer block storage and IBP byte array service:

Any participant in an IBP network can make use of any access layer storage resource in the network regardless of who owns it. The use of IP networking to access IBP storage resources creates a global storage service.

Whatever the strengths of this application of the IP paradigm, however, it leads directly to two problems. First, in the case of storage, the chronic vulnerability of IP networks to Denial of Use (DoU) attacks is greatly amplified. The free sharing of communication within a routed IP network leaves every local network open to being overwhelmed by traffic from the wide area network, and consequently open to the unfortunate possibility of DoU from the network. While DoU attacks in the Internet can be detected and corrected, they cannot be effectively avoided. Yet this problem is not debilitating for two reasons: on the one hand, each datagram sent over a link uses only a tiny portion of the capacity of that link, so that DoU attacks require constant sending from multiple sources; on the other hand, monopolizing remote communication resources cannot profit the attacker in any way - except, of course, economic side-effects of attacking a competitor's resource. Unfortunately neither of these factors hold true for access layer storage resources. Once a data block is written to a storage medium, it occupies that portion of the medium until it is deallocated, so no constant sending is required. Moreover it is clear that monopolizing remote storage resources can be very profitable for an attacker and his applications.

The second problem with sharing storage network-style is that the usual definition of a storage service is based on processor-attached storage, and so it includes strong semantics (near-perfect reliability and availability) that are difficult to implement in the wide area network. Even in "storage area" or local area networks, these strong semantics can be difficult to implement and are a common cause of error conditions. When extended to the wide area, it becomes impossible to support such strong guarantees for storage access.

We have addressed both of these issues through special characteristics of the way IBP allocates storage:

9   *Allocations of storage in IBP can be time limited. When the lease on an allocation expires, the storage resource can be reused and all data structures associated with it can be deleted. An IBP allocation can be refused by a storage resource in response to over-allocation, much as routers can drop packets, and such "admission decisions" can be based on both size and duration. Forcing time limits puts transience into storage allocation, giving it some of the fluidity of datagram delivery.*

10   *The semantics of IBP storage allocation are weaker than the typical storage service.* Chosen to model storage accessed over the network, it is assumed that an IBP storage resource can be transiently unavailable. Since the user of remote storage resources is depending on so many uncontrolled remote variables, it may be necessary to assume that storage can be permanently lost. Thus, *IBP is a "best effort" storage service.* To encourage the sharing of idle resources, IBP even supports "volatile" storage allocation semantics, where allocated storage can be revoked at any time. In all cases such weak semantics mean that the level of service must be characterized statistically.

Because of IBP's limitations on the size and duration of allocation and its weak allocation semantics, IBP does not directly implement reliable storage abstractions such as conventional files. Instead these must be built on top of IBP using techniques such as redundant storage, much as TCP builds on IP's unreliable datagram delivery in order to provide reliable transport.

3   The IBP Service, Client API, and Current Software

IBP storage resources are managed by "depots," or servers, on which clients perform remote storage operations. The IBP client calls fall into three different groups:

Table 1. IBP API Calls

Storage Management	Data Transfer	Depot Management
IBP_allocate IBP_manage	IBP_store, IBP_load IBP_copy, IBP_mcopy	IBP_status

IBP_allocate is used to allocate a byte array at an IBP depot, specifying the size, duration and other attributes. A chief design feature is the use of capabilities (cryptographically secure passwords). A successful IBP_allocate returns a set of three capabilities: one for reading, one for writing, and one for managing the allocated byte array. The **IBP_manage** call allows the client to manipulate the read and the write reference counter for a specific capability, probe the capability itself or change some of its characteristics. The **IBP_store** and **IBP_load** calls are two blocking calls that store and load the data on a particular capability; while with the **IBP_status** call it's possible to query a depot about its status and to change some parameters. The **IBP_copy** and **IBP_mcopy** calls provide data transfer, and will be analyzed in more depth in section 4.

Prototype versions of IBP were available since 1999. Version 1.0 was released in March 2001, and the current release of the code (1.1.1), developed in C, is available for free download at our web site (http://loci.cs.utk.edu/ibp/). It has been successfully tested under Linux, Solaris, AIX, DEC and Apple OS X machines. A Windows version also exists, for both client and depot, and a Java client library is also available.

4   The Data Mover

Since the primary intent of IBP is to provide a common abstraction of storage, it is arguable that third party transfer of data between depots is unnecessary. Indeed, it is logically possible to build an external service for moving data between depots that access IBP allocations using only the `IBP_load` and `IBP_store` calls; however, such a service would have to act as a proxy for clients, and this immediately raises trust and performance concerns. The `IBP_copy` and `IBP_mcopy` data movement calls were provided in order to allow a simple implementation that avoids these concerns, even if software architectures based on external data movement operations are still of great interest to us.

The intent of the basic `IBP_copy` call is to provide access to a simple data transfer over a TCP stream. This call is built in the IBP depot itself, to offer a simple solution for data movement; to achieve the transfer, the depot that receives the call is acting as a client doing an IBP_store on the target depot.

`IBP_mcopy` is a more general facility, designed to provide access to operations that range from simple variants of basic TCP-based data transfer to highly complex protocols using multicast and real-time adaptation of the transmission protocol, depending of the nature of the underlying backbone and of traffic concerns. In all cases, the caller has the capacity to determine the appropriateness of the operation to the depot's network environment, and to select what he believes the best data transfer strategy. A similar control is given over error recovery.

The data mover is a plug-in module to an IBP depot that is activated either by an `IBP_mcopy` call or by an `IBP_datamover` call. The second call is not an API call, but an internal call made b the sending Data Mover. The sender depot is responsible for invoking a Data Mover plug-in on the receiving depot, and it accomplishes this by forking a data mover control process that sends an `IBP_datamover` request, causing the receiving depot to **fork** a symmetric data mover control. Sending and receiving control processes then **exec** the appropriate Data Mover plug-ins for the requested operation and these cooperate to perform the operation, then the plug-in at the sending depot replies to the client and then both plug-ins terminate.

The Data Mover architecture can support a wide variety of operations, including:

2   Point-to-multipoint through simple iterated TCP unicast transfers

3   Point-to-multipoint through simultaneous threaded TCP unicast transfers.

4   Unreliable UDP point-to-mulitpoint utilizing native IP mulitcast

5   Reliable point-to-multipoint utilizing native IP multicast

6   Fast, reliable UDP data transfer over private network links [5]

Experimental results

The design goal of the Data Mover plug-in and the function IBP_mcopy is to provide an optimized point to multipoint transfer tool, as well as a support for different protocols and methods for data transfer. In order to visualize and compare the behavior of the different methods to perform the data movement, two main experiments were completed under a specific environment where each of the nodes involved were interconnected with a stable, fast link within a local area network (LAN).   The subsections 4.1.1 and 4.1.2 describe these experiments and their corresponding results, as well as a comparison of their performance and possible optimizations, using the Data Mover module, with the commonly used methods, such as IBP_copy and TCP respectively.

Point to Multipoint TCP

This subsection concentrates in comparing the transfer of different amounts of data using multiple simultaneous point to point TCP data transfers implemented at user level using threads, and using a single point to multipoint implementation of the Data Mover to transfer the same amounts of data.

As figure 2 reveals the latter approach shows an improvement in the overall transfer time. This experiment consisted of transferring several pieces of data of different sizes from one to various numbers of nodes.

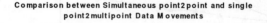

Comparison between Simultaneous point2point and single point2multipoint Data Movements

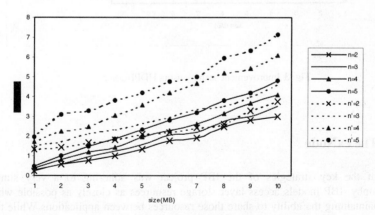

Fig. 2. Dotted lines represent the multiple simultaneous point to point TCP transfers the number of hosts used in different tests is given by n. The continuous lines represent point to multipoint approach, and number of hosts for this cases is given by n .

Point to Point UDP versus TCP

The experiment described in this subsection consists of a comparison between the transfer times using TCP and UDPBlaster point to point Data Movers.

Figure 3 shows the improvement achieved by using UDPBlaster. The Data Mover plug-in could support a variety of protocols and methods, however, for the purpose of this experiment we concentrate on the comparison of TCP versus UDP, to show how the improvement can be achieved within the same data mover using different protocol depending on the characteristics of the backbone being used. It is important to note that since this protocol is still in the experimental phase may behave differently in diverse test environments under different circumstances.

Comparison between different protocols for data transfer

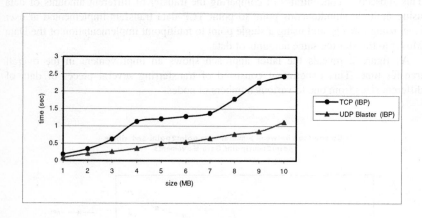

Fig. 3. Improvement seen by using UDPBlaster

5   The exNode

One of the key strategies of the IBP project was adhering to a very simple philosophy: IBP models access layer storage resources as closely as possible while still maintaining the ability to share those resources between applications While this gives rise to a very general service, it results in semantics that might be too weak to be conveniently used directly by many applications. As in the networking field the IP protocol alone does not guarantee many highly desirable characteristics, and needs to be complemented by a transport protocol such as TCP, what is needed is a service at the next layer that, working in synergy with IBP, implements stronger allocation properties such as reliability and arbitrary duration that IBP does not support.

Following the example of the Unix inode, that aggregates disk blocks to implement a file, we have chosen to implement a single generalized data structure, which we call an *external node,* or *exNode,* in order to manage aggregate allocations that can be used in implementing network storage with many different strong semantic properties. Rather than aggregating blocks on a single disk volume, the exNode aggregates storage allocations on the Internet, and the exposed nature of IBP makes IBP byte-arrays exceptionally well adapted to such aggregations. In the present context the key point about the design of the exNode is that it has allowed us to create an abstraction of a network file to layer over IBP-based storage in a way that is completely consistent with the exposed resource approach.

The exNode library has a set of calls that allow an application to create and destroy an exNode, to add or delete a mapping from it, to query it with a set of criteria, and to produce an XML serialization, to allow the use of XML-based tools and interoperability with XML systems and protocols.

A more complete explanation of the exNode and the tools we developed is available online (http://loci.cs.utk.edu/exNode/index.htm). The software library, written in c, is currently in its beta release, while a java version is under development. We plan to make them available together with our IBP distribution in the very near future through our web site.

6   Experience and Applications

Our method in developing the Internet Backplane Protocol is based on implementation and experimentation. A number of simple applications from within our own research group have formed the basis of the experience that has guided our work. In addition, a few external application groups have picked up on early implementations of IBP and contribute by both tracking our releases and giving us needed feedback for future developments. However, it is only with the upcoming release of the exNode library and serialization that we believe a wide application community will find the supporting tools necessary to adopt an IBP-based methodology.

3   *NetSolve is a distributed computation tool created by Casanova and Dongarra [8] to provide remote invocation of numerical libraries for scientific code. One shortcoming of the initial NetSolve architecture is that it is stateless: a series of calls to the same server cannot, for instance, cache arguments or results in order to avoid unnecessary data movement in subsequent calls. One of the approaches taken to optimize NetSolve was to use an IBP depot at the server to implement a cache under the control of the client. Such short-lived use of IBP for value caching is able to make good use of even volatile and time-limited allocations [2]. Experiments conducted by the NetSolve group, using a Client application at the University of California, San Diego requesting a pool of computational*

resources and Data Storage at the University of Tennessee, showed a much-improved efficiency. [1, 2, 9].

4   *TAMANOIR [12] is a project developed by the RESAM lab of the Ecole Normale Superieure of Lyon in the field of Active Networking. It is a framework that allows users to easily deploy and maintain distributed active routers in a wide area network. IBP depots will be among the standard tools available to services implemented within the TAMANOIR framework, along with other basic tools such as the routing manager and stream monitoring tools. It will also be used to implement distribution and caching of services (distributed as Java byte-code modules) that are loaded by TAMANOIR on-demand, freeing TAMANOIR to manage only is own internal cache of services.*

7   Related Work

IBP occupies an architectural niche similar to that of network file systems such as AFS [14] and Network Attached Storage appliances [13], but its model of storage is more primitive, making it similar in some ways to Storage Area Networking (SAN) technologies developed for local networks. In the Grid community, projects such as GASS [7] and the SDSC Storage Resource Broker [3] are file system overlays that implement a uniform file access interface and also impose uniform directory, authentication and access control frameworks on their users.

8   Conclusions

While some ways of engineering for resource sharing, such as the *Computer Center model*, focus on optimizing the use of scarce resources within selected communities, the exponential growth in all areas of computing resources has created the opportunity to explore a different problem, viz. designing new architectures that can take more meaningful advantage of this bounty. The approach presented in this paper is based on the *Internet model of resource sharing* and represents one general way of using the rising flood of storage resources to create a common distributed infrastructure that can share the growing surplus of storage in a way analogous to the way the current network shares communication bandwidth. It uses the Internet Backplane Protocol (IBP), which is designed on the model of IP, to allow storage resources to be shared by users and applications in a way that is as open and as easy to use as possible while maintaining a necessary minimum of security and protection from abuse. IBP lays the foundation for the intermediate resource management components, accessible to every end-system, which must be introduced to govern the way that applications access, drw from, and utilize this common pool in a fully storage-enabled Internet.

Bibliography

1.	D. C. Arnold, D. Bachmannd, and J. Dongarra, "Request Sequencing: Optimizing Communication for the Grid," in *Euro-Par 2000 -- Parallel Processing, 6th International Euro-Par Conference*, vol. 1900, *Lecture Notes in Computer Science*, A. Bode, T. Ludwig, W. Karl, and R. Wismuller, Eds. Munich: Springer Verlag, 2000, pp. 1213-1222.

2.	D. C. Arnold, S. S. Vahdiyar, and J. Dongarra, "On the Convergence of Computational and Data Grids," *Parallel Processing Letters*, vol. 11, no. 2, pp. 187-202, June/September, 2001.

3.	C. Baru, R. Moore, A. Rajasekar, and M. Wan, "The SDSC Storage Resource Broker," presented at CASCON'98, Toronto, Canada, 1998.

4.	A. Bassi, M. Beck, J. Plank, and R. Wolski, "Internet Backplane Protocol: API 1.0," Department of Computer Science, University of Tennessee, Knoxville, TN, CS Technical Report, ut-cs-01-455, March 16, 2001. http://www.cs.utk.edu/~library/2001.html.

5.	M. Beck and E. Fuentes, "A UDP-Based Protocol for Fast File Transfer," Department of Computer Science, University of Tennessee, Knoxville, TN, CS Technical Report, ut-cs-01-456, June, 2001. http://www.cs.utk.edu/~library/2001.html.

6.	M. Beck, T. Moore, J. Plank, and M. Swany, "Logistical Networking: Sharing More Than the Wires," in *Active Middleware Services*, vol. 583, *The Kluwer International Series in Engineering and Computer Science*, S. Hariri, C. Lee, and C. Raghavendra, Eds. Boston: Kluwer Academic Publishers, 2000.

7.	J. Bester, I. Foster, C. Kesselman, J. Tedesco, and S. Tuecke, "GASS: A Data Movement and Access Service for Wide Area Computing Systems," presented at Sixth Workshop on I/O in Parallel and Distributed Systems, May 5, 1999, 1999.

8.	H. Casanova and J. Dongarra, "Applying NetSolve's Network Enabled Server," *IEEE Computational Science & Engineering*, vol. 5, no. 3, pp. 57-661998.

9.	H. Casanova, A. Legrand, D. Zagorodnov, and F. Berman, "Heuristics for Scheduling Parameter Sweep Applications in Grid Environments," presented at 9th Heterogeneous Computing Workshop (HCW'00), May 2000, 2000.

10.	W. Elwasif, J. Plank, M. Beck, and R. Wolski, "IBP-Mail: Controlled Delivery of Large Mail Files," presented at NetStore99: The Network Storage Symposium, Seattle, WA, 1999.

11.	I. Foster, C. Kesselman, and S. Tuecke, "The Anatomy of the Grid: Enabling Scalable Virtual Organizations," *International Journal of SuperComputer Applications*, vol. 15, no. 3, To appear, 2001.

12.	J. Gelas and L. Lefevre, "TAMANOIR : A High Performance Active Network Framework," in *Active Middleware Services*, vol. 583, *The Kluwer International Series in Engineering and Computer Science*, S. Hariri, C. Lee, and C. Raghavendra, Eds. Boston: Kluwer Academic Publishers, 2000.

13.	G. Gibson and R. V. Meter, "Network Attached Storage Architecture," *Communications of the ACM*, vol. 43, no. 11, pp. 37-45, November, 2000.

14.	J. H. Morris, M. Satyanarayan, M. H. Conner, J. H. Howard, D. S. H. Rosenthal, and F. D. Smith, "Andrew: A Distributed Personal Computing Environment," *Communications of the ACM*, vol. 29, no. 3, pp. 184-201, March, 1986.

15.	J. Plank, M. Beck, W. Elwasif, T. Moore, M. Swany, and R. Wolski, "The Internet Backplane Protocol: Storage in the Network," presented at NetStore99: The Network Storage Symposium, Seattle, WA, 1999.

Towards the Design of an Active Grid

Jean-Patrick Gelas and Laurent Lefèvre

RESAM Laboratory UCB - Action INRIA RESO
Ecole Normale Supérieure de Lyon
46, allée d'Italie 69364 LYON Cedex 07 - France
Jean-Patrick.Gelas@ens-lyon.fr, Laurent.Lefevre@inria.fr

Abstract. Grid computing is a promising way to aggregate geographically distant machines and to allow them to work together to solve large problems. After studying Grid network requirements, we observe that the network must take part in the Grid computing session to provide intelligent adaptative transport of Grid data streams.

By proposing new intelligent dynamic services, active network can easily and efficiently deploy and maintain Grid environments and applications. This paper presents the Active Grid Architecture (A-Grid)[1] which focuses on active networks adaptation for supporting Grid middlewares and applications.

We describe the architecture and first experiments of a dedicated execution environment dedicated to the Grid: Tamanoir-G.

1 Introduction

In recent years, there has been a lot of research projects on Grid computing which is a promising way to aggregate geographically distant machines and to allow them to work together to solve large problems ([9, 16, 5, 7, 14, 2, 8, 6]). Most of proposed Grid frameworks are based on Internet connections and do not make any assumption on the network. Grid designers only take into account of a reliable packet transport between Grid nodes and most of them choose TCP/IP protocol.

This leads to one of the more common complaint of Grid designers : networks do not really support Grid applications.

Meantime, the field of active and programmable networks is rapidly expanding [17]. These networks allow users and network designers to easily deploy new services which will be applied to data streams. While most of proposed systems deal with adaptability, flexibility and new protocols applied on multimedia streams (video, audio), no active network efficiently deal with Grid needs.

In this paper we propose solutions to merge both fields by presenting The Active Grid Architecture (A-Grid) which focus on active network adaptation for supporting Grid environments and applications. This active Grid Architecture

---

[1] This research is supported by French Research Ministry and ACI-GRID project JE7 RESAM.

P.M.A. Sloot et al. (Eds.): ICCS 2002, LNCS 2330, pp. 578–587, 2002.
© Springer-Verlag Berlin Heidelberg 2002

proposes solutions to implement the two main kind of Grid configurations : meta-cluster computing and global computing. In this architecture the network will take part in the Grid computing session by providing efficient and intelligent services dedicated to Grid data streams transport.

This paper reports on our experience in designing an Active Network support for Grid Environments. First, we classify the Network Grid requirement depending on environments and applications needs (section 2). In section 3 we propose the Active Grid Architecture. We focus our approach by providing support for the most network requirements from Grid. In section 4, we describe Tamanoir-G, the Active Grid framework and first experiments. Finally, in the last section we present our future works.

2   Network requirements for the Grid

A distributed application running in a Grid environment requires various kinds of data streams: Grid control streams and Grid application streams.

2.1   Grid control streams

We can classify two basic kinds of Grid usage :

– Meta cluster computing :
   A set of parallel machines or clusters are linked together over Wide Area Networks to provide a very large parallel computing resource. Grid environments like Globus[9], MOL[16], Polder[5] or Netsolve[7] are well designed to handle meta-cluster computing session to execute long-distance parallel applications.
   We can classify various network needs for meta-clustering sessions :

   • Grid environment deployment : The Grid infrastructure must be easily deployed and managed : OS heterogeneity support, dynamic topology re-configuration, fault tolerance...
   • Grid application deployment : Two kind of collective communications are needed : multicast and gather. The source code of any applications is multicast to a set of machines in order to be compiled on the target architectures. In case of Java based environments, the *bytecode* can be multicast to a set of machines. In case of an homogeneous architecture, the binaries are directly sent to distant machines. After the running phase, results of distributed tasks must be collected by the environment in a gathering communication operation.
   • Grid support : The Grid environment must collect control data : node synchronization, node workload information... The information exchanged are also needed to provide high-performance communications between nodes inside and outside the clusters.

- Global or Mega-computing : These environments usually rely on thousand of connected machines. Most of them are based on computer *cycles stealing* like Condor[14], Entropia[2], Nimrod-G[8] or XtremWeb[6]. We can classify various network needs for Global-computing sessions :
  - Grid environment deployment : Dynamic enrollment of unused machines must be taken into account by the environment to deploy tasks over the mega-computer architecture.
  - Grid application deployment : The Grid infrastructure must provide a way to easily deploy and manage tasks on distant nodes. To avoid the restarting of distributed tasks when a machine crashes or become unusable, Grid environments propose check-pointing protocols, to dynamically re-deploy tasks on valid machines.
  - Grid support : Various streams are needed to provide informations to Grid environment about workload informations of all subscribed machines. Machine and network sensors are usually provided to optimize the task mapping and to provide load-balancing.

Of course, most of environments work well on both kind of Grid usage like Legion[12], Globus[9], Condor[14], Nimrod-G[8] ...

2.2   Grid application streams

A Grid computing session must deal with various kind of streams :

- Grid application input : during the running phase, distributed tasks of the application must receive parameters eventually coming from various geographically distant equipments (telescopes, biological sequencing machines. . .) or databases (disk arrays, tape silos. . .).
- Wide-area parallel processing : most of Grid applications consist of a sequential program repeatedly executed with slightly different parameters on a set of distributed computers. But with the raise of high performance backbones and networks, new kind of real communicating parallel applications (with message passing libraries) will be possible on a WAN Grid support. Thus, during running phase, distributed tasks can communicate data between each others. Applications may need efficient point to point and global communications (broadcast, multicast, gather. . .) depending on application patterns. These communications must correspond to the QoS needs of the Grid user.
- Coupled (Meta) Application : they are multi-component applications where the components were previously executed as stand-alone applications. Deploying such applications must guarantee heterogeneity management of systems and networks. The components need to exchange heterogeneous streams and to guarantee component dependences in pipeline communication mode. Like WAN parallel applications, QoS and global communications must be available for the components.

Such a great diversity of streams (in terms of messages size, point to point or global communications, data and control messages. . .) requires an intelligence in the network to provide an efficient data transport.

3   Active Grid Architecture

We propose an active network architecture dedicated to Grid environments and Grid applications requirements : the A-Grid architecture.

An active grid architecture is based on a virtual topology of active network nodes spread on programmable routers of the network. Active routers, also called Active Nodes (AN), are deployed on network periphery (edge equipments).

As we are concerned by a wide active routers approach, we do not believe in the deployment of Gigabit active routers in backbones. If we consider that the future of WAN backbones could be based on all-optical networks, no dynamic services will be allowed to process data packets. So, we prefer to consider backbones like high performance well-sized passive networks. We only concentrate active operations on edge routers/nodes mapped at network periphery.

Active nodes manage communications for a subset of Grid nodes. Grid data streams cross various active nodes up to passive backbone and then cross another set of active nodes up to receiver node. The A-Grid architecture is based on Active Node approach : programs, called services, are injected into active nodes independently of data stream. Services are deployed on demand when streams arrive on an active node. Active nodes apply these services to process data streams packets.

3.1   Active Grid architecture

To support most of Grid applications, the Active Grid architecture must deal with the two main Grid configurations :

- Meta cluster computing (Fig. 1) :
  In this highly coupled configuration, an active node is mapped on network head of each cluster or parallel machine. This node manage all data streams coming or leaving a cluster. All active nodes are linked with other AN mapped at backbone periphery. An Active node delivers data streams to each node of a cluster and can aggregate output streams to others clusters of the Grid.
- Global or Mega computing (Fig. 2) :
  In this loosely coupled configuration, an AN can be associated with each Grid node or can manage a set of aggregated Grid nodes. Hierarchies of active nodes can be deployed at each network heterogeneity point.
  Each AN manages all operations and data streams coming to Grid Nodes : subscribing operations of voluntary machines, results gathering, nodes synchronization and check-pointing...

For both configurations, active nodes will manage the Grid environment by deploying dedicated services adapted to Grid requirements : management of nodes mobility, dynamic topology re-configuration, fault tolerance....

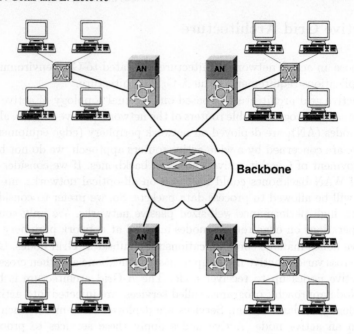

Fig. 1. Meta cluster computing Active Grid Architecture

3.2   Active network benefits for Grid applications

Using an Active Grid architecture can improve the communications needs of Grid applications :

- Application deployment : To efficiently deploy applications, active reliable multicast protocols are needed to optimize the source code or binary deployment and the task mapping on the Grid configuration accordingly to resources managers and load-balancing tools. An active multicast will reduce the transport of applications (source code, binaries, bytecode...) by minimizing the number of messages in the network. Active node will deploy dedicated multicast protocols and guarantee the reliability of deployment by using storage capabilities of active nodes.
- Grid support : the Active architecture can provide informations to Grid framework about network state and task mapping. Active nodes must be open and easily coupled with all Grid environment requirements. Active nodes will implement permanent Grid support services to generate control streams between the active network layer and the Grid environment.
- Wide-area parallel processing : with the raise of grid parallel applications, tasks will need to communicate by sending computing data streams with QoS requests. The A-Grid architecture must also guarantee an efficient data transport to minimize the software latency of communications. Active nodes

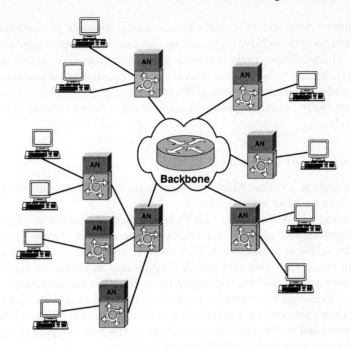

Fig. 2. Global Computing Active Grid Architecture

deploy dynamic services to handle data streams : QoS, data compression, "on the fly" data aggregation...
- Coupled (Meta) Application : the Active architecture must provide heterogeneity of services applied on data streams (data conversion services...). End to end QoS dynamic services will be deployed on active nodes to guarantee an efficient data transport (in terms of delay and bandwidth).

Most services needed by Grid environments : high performance transport, dynamic topology adapting, QoS, on-the-fly data compression, data encryption, data multicast, data conversion and error management must be easily and efficiently deployed on demand on an Active Grid architecture. To allow an efficient and portable service deployment, we will present in next section our approach to propose an active network framework easily mergeable with a Grid environment : The Tamanoir-G Framework.

4   Tamanoir-G : a Framework for Active Grid support

We explore the design of an intelligent network by proposing a new active network framework dedicated to high performance active networking. The Tamanoir-G framework [11] is an high performance prototype active environment based on active edge routers. Active services can be easily deployed in the network and are adapted to architecture, users and service providers requirements.

A set of distributed tools is provided : routing manager, active nodes and stream monitoring, web-based services library... Tamanoir-G is based on a JAVA multi-threaded design to combine performance and portability of services, applications can easily benefit of personalized network services through the injection of Java code.

4.1   Overview of a Tamanoir-G node

An active node is a router which can receive packets of data, process them and forward them to other active nodes.

A Tamanoir-G Active Node (TAN) is a persistent dæmon acting like a dynamic programmable router. Once deployed on a node, it is linked to its neighbors in the active architecture. A TAN receives and sends packets of data after processing them with user services. A TAN is also in charge of deploying and applying services on packets depending on application requirements. When arriving in a Tamanoir-G dæmon, a packet is forwarded to service manager. The packet is then processed by a service in a dedicated thread. The resulting packet is then forwarded to the next active node or to the receiver part of application according to routing tables maintained in TAN.

4.2   Dynamic service deployment

In Tamanoir-G, a service is a JAVA class containing a minimal number of formatted methods. If a TAN does not hold the appropriate service, a downloading operation must be performed.We propose three kind of service deployment. The first TAN crossed by a packet can download the useful service from a service broker. By using an *http address* in service name, TAN contact the web service broker, so applications can download generic Tamanoir-G services to deploy non-personalized generic services. After, next TANs download the service from a previous TAN crossed by packet or from the service broker.

4.3   Multi-protocols support

Most of existing active frameworks propose active services dedicated to UDP streams (like ANTS [18], PAN[15]...). Tamanoir-G environment has been extended to support various transport protocols and specially TCP protocol used by most of Grid middlewares and applications (see figure 3). By this way, Tamanoir-G services can be easily adapted and merged in a Grid Middleware to provide adaptative network services.

4.4   Tamanoir-G Performances

We based our first experiments of Tamanoir-G on Pentium II 350 MHz linked with Fast Ethernet switches. These experiments show that the delay needed to cross a Tamanoir-G active node (latency) remains constant (under 750 μs for simple forwarding operation).

Results presented in figure 3 show bandwidth comparisons of forwarding service of a Tamanoir-G node with UDP and TCP transport. While kernel passive forward operations (*kr*) provide around 50 Mbits/s, active forwarding services running in a JVM obtain good results (around 30 Mbit/s) with Kaffe(*kaffe*) [4], Blackdown(*jvmBkDwn*) [1] and IBM(*jvmIBM*) [3] Java Virtual Machines. We note poor performances obtained with GCJ(*gcj*) [10] compiled version around 17 Mbit/s due to lack of optimizations in GCJ compiler.

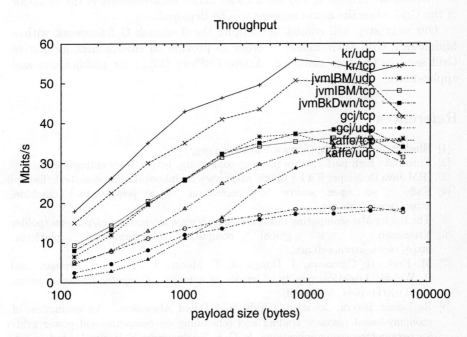

Fig. 3. Bandwidth of a passive (*kr*) or active forwarding operation with Tamanoir-G in UDP or TCP. Comparisons between Java Virtual Machines (*kaffe, jvmBkDwn, jvmIBM*) and compiled approach (*gcj*).

These first experiments show that Tamanoir-G framework can support a Grid environment without adding too much latency to all data streams. So Tamanoir-G can efficiently deploy services on active nodes depending on Grid requirements : QoS, data conversion, multicast, filtering operations,"on the fly" data compression...

5   Conclusion and future works

We have analyzed the Grid computation models in order to determine the main Grid network requirements in terms of efficiency, portability and ease of deployment.

We then studied a solution to answer these problems : the active networking approach, where all network protocols can be deported in the network in a transparent way for the Grid designer and the Grid user.

Most of services needed by Grid environments : high performance transport, dynamic topology adapting, QoS, on-the-fly data compression, data encryption, data multicast, data conversion, errors management must be easily and efficiently deployed on demand by an Active Grid architecture.

We proposed an active network support : the Tamanoir-G Framework. The first results are promising and should lead major improvements in the behavior of the Grid when the active support will be deployed.

Our next step will consist of merging the Tamanoir-G framework with a Middleware Grid environment in order to provide an environment for Active Grid services (reliable multicast, Active DiffServ [13]...) for middlewares and applications.

References

[1] Blackdown JVM. http://www.blackdown.org/.

[2] Entropia : high performance internet computing. http://www.entropia.com.

[3] IBM Java Developer Kit for Linux. http://www.alphaworks.ibm.com/tech/linuxjdk.

[4] Kaffe : an open source implementation of a java virtual machine. http://www.kaffe.org/.

[5] The Polder Metacomputing Initiative. http://www.science.uva.nl/projects/polder.

[6] Xtremweb : a global computing experimental platform. http://www.xtremweb.net.

[7] M. Beck, H. Casanova, J. Dongarra, T. Moore, J. Planck, F. Berman, and R. Wolski. Logistical quality of service in netsolve. *Computer Communication*, 22(11):1034–1044, july 1999.

[8] Rajkumar Buyya, Jonathan Giddy, and David Abramson. An evaluation of economy-based resource trading and scheduling on computational power grids for parameter sweep applications. In C. S. Raghavendra S. Hariri, C. A. Lee, editor, *Active Middleware Services, Ninth IEEE International Symposium on High Performance Distributed Computing*, Pittsburgh, Pennsylvania, USA, aug 2000. Kluwer Academic Publishers. ISBN 0-7923-7973-X.

[9] I. Foster and C. Kesselman. Globus: A metacomputing infrastructure toolkit. *Intl J. Supercomputing Applications*, 11(2):115–128, 1997.

[10] GCJ. The GNU Compiler for the Java Programming Language. http://sourceware.cygnus.com/java/.

[11] Jean-Patrick Gelas and Laurent Lefèvre. Tamanoir: A high performance active network framework. In C. S. Raghavendra S. Hariri, C. A. Lee, editor, *Active Middleware Services, Ninth IEEE International Symposium on High Performance Distributed Computing*, pages 105–114, Pittsburgh, Pennsylvania, USA, aug 2000. Kluwer Academic Publishers. ISBN 0-7923-7973-X.

[12] Andrew Grimshaw, Adam Ferrari, Fritz Knabe, and Marty Humphrey. Legion: An operating system for wide-area computing. *IEEE Computer*, 32(5):29–37, May 1999.

[13] L. Lefèvre, C. Pham, P. Primet, B. Tourancheau, B. Gaidioz, J.P. Gelas, and M. Maimour. Active networking support for the grid. In Noaki Wakamiya Ian W. Marshall, Scott Nettles, editor, *IFIP-TC6 Third International Working Conference on Active Networks, IWAN 2001*, volume 2207 of *Lecture Notes in Computer Science*, pages 16–33, oct 2001. ISBN: 3-540-42678-7.

[14] Miron Livny. Managing your workforce on a computational grid. In Springer Lecture Notes in Computer Science, editor, *Euro PVM MPI 2000*, volume 1908, Sept 2000.

[15] Erik L.Nygren, Stephen J.Garland, and M.Frans Kaashoek. PAN: A High-Performance Active Network Node Supporting Multiple Mobile Code Systems. In *IEEE OPENARCH '99*, March 1999.

[16] A. Reinefeld, R. Baraglia, T. Decker, J. Gehring, D. Laforenza, J. Simon, T. Romke, and F. Ramme. The mol project: An open extensible metacomputer. In *Heterogenous computing workshop HCW'97,IPPS'97*, Geneva, April 1997.

[17] D. L. Tennehouse, J. M. Smith, W. D. Sincoskie, D. J. Wetherall, and G. J. Winden. A survey of active network research. *IEEE Communications Magazine*, pages 80–86, January 1997.

[18] David Wetherall, John Guttag, and David Tennenhouse. ANTS : a toolkit for building and dynamically deploying network protocols. In *IEEE OPENARCH '98*, April 1998.

An Active Reliable Multicast Framework for the Grids*

M. Maimour and C. Pham

Laboratoire RESAM, Université Lyon 1
ENS, Bât. LR5, 46 allée d'Italie
69364 Lyon Cedex 07, France Congduc.Pham@ens-lyon.fr

Abstract. Computational Grids are foreseen to be one of the most critical yet challenging technologies to meet the exponentially growing demands for high-performance computing in a large variety of scientific disciplines. Most of these grid applications imply multiple participants and, in many cases, make an intensive usage of data distribution and collective operations. In this paper, we propose a multicast framework consisting of an active reliable protocol with specialized active services located at the edges of the core network for providing low-latency and low-overhead multicast transfers on computational grid.

1 Introduction

Computational Grids are foreseen to be one of the most critical yet challenging technologies to meet the exponentially growing demands for high-performance computing in a large variety of scientific disciplines. Most of these grid applications imply multiple participants and, in many cases, make an intensive usage of data distribution and collective operations. In the past few years, many grid software environments for gaining access to very large distributed computing resources have been made available (e.g. Condor [1], Globus [2], Legion [3] to name a few) and efficient data distribution is usually a key functionality for these environments to reduce the end-to-end latency.

These multi-point communications could be gracefully and efficiently handled by multicast protocols, that is the process of sending every single packet to multiple destinations, provided that these protocols are well-designed to suit the grid requirements. Motivations behind multicast are to handle one-to-many communications in a wide-area network with the lowest network and end-system overheads. In contrast to best-effort multicast, that typically tolerates some data losses and is more suited for real-time audio or video for instance, reliable multicast requires that all packets are safely delivered to the destinations. Desirable features of reliable multicast include, in addition to reliability, low end-to-end delays, high throughput and scalability. These characteristics fit perfectly the need of the grid computing and distributed computing communities.

* This work is supported in part by the french ACI Grid program and by a grant from ANVAR-EZUS Lyon.

P.M.A. Sloot et al. (Eds.): ICCS 2002, LNCS 2330, pp. 588–597, 2002.

Meeting the objectives of reliable multicast is not an easy task. In the past, there have been a number of propositions for reliable multicast protocols relying on complex exchanges of feedback messages (ACK or NACK) [4–7]. These protocols usually take the end-to-end solution to perform loss recoveries. Most of them fall into one of the following classes: sender-initiated, receiver-initiated and receiver-initiated with local recoveries. In sender-initiated protocols, the sender is responsible for both the loss detection and the recovery (XTP [4]). These protocols usually do not scale well to a large number of receivers due to the ACK implosion problem at the source. Receiver-initiated protocols move the loss detection responsibility to the receivers. They use NACKs instead of ACKs. However they still suffer from the NACK implosion problem when a large number of receivers have subscribed to the multicast session. In receiver-initiated protocols with local recovery, the retransmission of a lost packet can be performed by any receiver (SRM [5]) in the neighborhood or by a designated receiver in a hierarchical structure (RMTP [6], TMTP [7], LMS [8], PGM [9]). All of the above schemes do not provide exact solutions to all the loss recovery problems. This is mainly due to the lack of topology information at the end hosts.

In this paper, we propose a multicast framework consisting of an active reliable protocol with specialized active services located at the edges of the core network (adding complex processing functions inside the core network will certainly slow down the packet forwarding functions) for providing low-latency and low-overhead on computational grid. We assume that the computing resources are distributed across an Internet-based network with a high-speed backbone network in the core (typically the one provided by the telecommunication companies) and several lower-speed (up to 1Gbits/s) access networks at the edge, with respect to the throughput range found in the backbone.

The rest of the paper is organized as follows. Section 2 presents the active reliable multicast concepts. Section 3 describes the generic protocols used in this paper to compare the various active mechanisms involved in reliable multicasting. Then Section 4 presents the performance study and Section 5 concludes.

2 Active networking and the grids

In active networking [10], routers themselves play an active role by executing application-dependent functions on incoming packets. Recently, the use of active network concepts where routers could contribute to enhance the network services by customized (sometimes high-level) functionalities has been proposed in many research areas including multicast protocols [11, 12] and distributed interactive simulations [13]. These results can be very beneficial to the grid community on numerous ways: (i) program and data remote submissions, (ii) information services and naming services, (iii) data replication and caching, (iv) collective operations and clock synchronization for distributed applications, (v) large databases replication...

There are many difficulties, however, for deploying in a large scale an active networking infrastructure. Security and performance are two main difficulties that are usually raised by anti-active networking people. However, active networking has the ability to provide a very general and flexible framework for customizing network functionalities in order to gracefully handle heterogeneity and dynamicity, key points in a computational grid. Therefore we believe it is still worth proposing advanced mechanisms that would globally improve the quality of service on a distributed system such as the grid.

Now, if we look more precisely at multicasting, active services contribute mainly on feedback implosion problems, retransmission scoping and cache of data. Resulting active reliable multicast protocols open new perspectives for achieving high throughput and low latency on wide-area networks:

- the cache of data packets allows for local recoveries of loss packets and reduces the recovery latency.
- the suppression of NACKs reduces the NACK implosion problem.
- the subcast of repair packets to a set of receivers limits both the retransmission scope and the bandwidth usage.

For instance, ARM (Active Reliable Multicast) [11] and AER (Active Error Recovery) [12] are two protocols that use a *best-effort* cache of data packets to permit local recoveries. ARM adopts a *global* suppression strategy: a receiver experiencing a packet loss sends immediately a NACK to the source. Active services in routers then consist in the aggregation of the multiple NACKs. In contrast, AER uses a *local* suppression strategy inspired from the one used by SRM and based on local timers at the receivers. In addition, an active router in ARM would send the repair packet only to the set of receivers that have sent a NACK packet (subcast). In AER, the active router simply multicasts the repair packet to all its associated receivers.

3   The DyRAM Framework

DyRAM is a reliable multicast protocol suite with a recovery strategy based on a tree structure constructed on a per-packet basis with the assistance of routers [14]. The protocol uses a NACK-based scheme with receiver-based local recoveries where receivers are responsible for both the loss detection and the retransmission of repair packets. Routers play an active role in DyRAM which consists in the following active services:

1. the early detection of packet losses and the emission of the NACKs.
2. the NACK suppression of duplicate (from end-hosts) NACKs in order to limit the NACK implosion problem.
3. the subcast of the repair packets only to the relevant set of receivers that have experienced a loss. This helps to limit the scope of the repairs to the affected subtree.
4. the replier election which consists in choosing a link as a replier one to perform local recoveries from the receivers.

Compared to existing active reliable multicast protocols such as ARM and AER, DyRAM has been designed with the following motivations in mind: (i) to minimize active routers load to make them supporting more sessions (mainly in unloading them from the cache of data) and (ii) reduce the recovery latency for enabling distributed applications on the grid. Avoiding cache in routers is performed by a replier-based local recovery strategy. The election of the replier is done dynamically on a per-packet basis by the appropriate active service (number 4) instead of an approximate solution based on timers as in SRM, or a complex Designed Local Receiver discovery as in PGM, both being non-active solutions. Reducing the latency recovery is performed by several mechanisms (actually all 4 services contribute directly or indirectly to reduce the recovery latency; for instance active service number 2 avoids NACK implosion at the source therefore reducing the retransmission time from the source) and especially by the early loss packet detection active service (active service number 1).

3.1 The NACK-based strategy

In DyRAM the receivers are responsible for detecting, requesting and in most cases, retransmitting a lost data packet. A receiver detects a loss by sequence gaps and upon the detection of a loss (or on a receive-timer expiration for delayed packets), a receiver immediately sends a NACK toward the source and sets a timer. Since NACKs and repairs may also be lost, a receiver will re-send a similar NACK when the requested repair has not been received within the timeout interval. Practically, the timeout must be set to at least the estimated round trip time (RTT) to the source. Upon reception of a NACK packet, the source sends the repair packet to the multicast address.

3.2 Active services in DyRAM

At the active routers side, the loss packet detection, the NACK suppression, the subcast and the replier election services can be implemented simply by maintaining information about the received packets and NACKs. This set of information is uniquely identified by the multicast address. For each received NACK, the router creates or simply updates an existing NACK state (NS) structure which has a limited life time.

- seq: the sequence number of the requested packet,
- time: the time when the last valid NACK for this packet has been received,
- rank: the rank of the last received NACK. The last valid NACK has rank 1; the next received one, which is not valid, has rank 2 and so forth ...
- subList: a subcast list that contains the list of links (downstream or upstream) on which NACKs for this packet have arrived.

Packet losses detection In general the repair latency can be reduced if the lost packet could be requested as soon as possible. DyRAM realizes this functionality by enabling routers to detect losses and therefore to generate NACKs to be sent to the source. An active router would detect a loss when a gap occurs in the data packet sequence. On a loss detection, the router would immediately generate a NACK packet toward the source. If the router has already sent a similar NACK for a lost packet then it would not send a NACK for a given amount of time. This "*discard delay*" is set at least to the *RTT* between this router and the source. During this amount of time, all NACK packets received for this data packet from the downstream links are ignored. This loss detection service can be implemented without adding any additional soft state at the routers.

NACK suppression On receipt of a NACK packet, a router would look for a corresponding NS structure. If such a structure exists, the router concludes that at least one similar NACK has already been processed otherwise a new NS structure will be created for this NACK. In the former case the router additionally checks if this NACK is valid, and if so, the router will forward it on the elected replier link (how this election is performed is described later on). Otherwise this NACK will serve to properly update the NS structure (rank and subcast list) and is dropped afterward. We use the rank information in some cases to be sure that all downstream receivers have sent a NACK packet.

Subcast functionality The subcast list in the NS structure is an important component for the subcast functionality. This list contains the set of links (downstream or upstream) from which a NACK has been received. When a data packet arrives at an active router it will simply be forwarded on all the downstream links if it is an original transmission. If the data packet is a repair packet the router searches for a corresponding NS structure and will send the repair on all the links that appear in the subcast list. An NS is created as soon as a valid NACK arrived for which no previous NS was created, and can be replaced only on receipt of the corresponding repair.

Replier Election Local recoveries, when possible, are performed by elected receiver (replier) that have correctly received the loss packet. On reception of a valid NACK, the router initializes a timer noted DTD (Delay To Decide) in order to collect during this time window as much information as possible about the links affected by a loss (updates of the subcast list). On expiration of the DTD timer, the router is able to choose a replier link among those that are not in the subcast list. This link may end up to be the upstream one if all the downstream links appear to be in the subcast list. In an attempt to avoid for the overloading of a particular downstream link, the router always try to choose a different link from the previously elected one (if available) by respecting a ring order among them, thus realizing when possible a load balance.

3.3 Architectural design and deployment

While not relying on active services, LMS [8] and PGM [9]) had previously proposed router-assisted solutions and [17] has investigated the incremental deployment of LMS-based routers. In our case, we push for a fully active solution in the hierarchy of routers of lower-speed access networks. In principle, the nearest router to the backbone is a good location for installing the active services: the cache of packets can serve for several local computing resources for instance.

For the realization of our active approach, the *programmable switch approach* which maintains the existing packet format will be used. Programs are injected separately from the processing of messages. Initially, and using a reliable channel, the source injects the required set of services by sending them to the multicast address. In our case this consists in two services, a data packet service and a NACK service. Afterwards, the source begins the multicast of the data packets. When an active router receives a packet, it first looks for the appropriate service deduced from the packet header.

To dynamically handle the multicast group topology changes, active routers must be able to add or remove active services. An active router that leaves the multicast session has simply to remove the associated services. However when an active router joins a multicast tree, it has to download the required services. This can be achieved by requesting the services from the closest active router which has already received them. If there is no such active router then the services need to be sent from the source.

All the active routers are assumed to perform at least NACK suppressions and the subcast functionality. The active router which is located at the sender's side just before the core network is called the *source router*. A preliminary study, detailed in [15], has shown that the loss detection service is only beneficial if the loss detection capable-router is close enough to the source. Consequently the source router is the best candidate to perform the loss detection service in addition to the two previous services. The other active routers perform the replier election service as seen in the previous section.

4   Simulations

4.1   Network model and metrics

We implemented a simulation model of DyRAM (in the PARSEC language developed at UCLA [16]) and evaluated its performances on a network model. The network model considers one source that multicasts data packets to R receivers through a packet network composed of a fast core network and several slower edge access networks. We will call *source link* the set of point-to-point links and traditional routers that connects the source to the core network. Similarly, a *tail link* is composed of point-to-point links and routers connecting a receiver to the core network (see Fig. 1). Each active router A_i is responsible of B receivers noted $R_{i1}, \cdots, R_{iB}$ forming a local group. We assume that there are l_b backbone links between the source and every active router A_i.

To evaluate our protocol M_S is defined as the number of retransmissions from the source per packet and gives an idea on the load at the source and we use BW which is the average bandwidth consumed per link to represent the load at the network level.

$$M_S = \frac{Number\ of\ retransmissions}{Number\ of\ sent\ packets}, BW = \frac{Bw_{NACK} + BW_{Data}}{N.B + l_b.(N+1)}$$

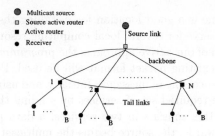

Fig. 1. Network Model.

where BW_{NACK} and BW_{Data} are respectively the bandwidth consumed by the NACK packets and the data packets during the multicast session:

$$BW_{NACK} = \sum_{i=1}^{NackNb} l_{NACK_i}, BW_{Data} = \sum_{i=1}^{DataNb} l_{DP_i}$$

l_{NACK_i} and l_{DP_i} are respectively the number of links crossed by the ith NACK packet and the ith data packet. The third metric is the completion time per packet which is the required time to successfully receive the packet by all the receivers (can be seen as the latency).

4.2   Simulation results

For all the simulations, we set $l_b = 55$. A NACK and a data packet are considered to be of 32 and 1024 bytes respectively (most of data packets on an internet-grid would have size less than 512 bytes and probably 10KBytes at the maximum with GigaEthernet). All simulation model values are normalized to the NACK transmission time (e.g. the time required to send or receive a NACK is set to 1, for a data packet this time is set to 32). For the processing overheads at the routers, we assume that both NACKs and data packets are processed in 32 time units. In our simulation models, we have taken into consideration the fact that the repairs may also be lost.

Local recovery from the receivers Figure 2a and 2b plot for DyRAM the number of retransmissions (M_S) from the source as a function of number the receivers and the completion time. These results have been obtained from simulations of a multicast session with 48 receivers distributed among 12 local groups. The curves show that having local recoveries decreases the load at the source as the number of receivers increases. This is especially true for high loss rates. Putting the recovery process in the receivers requires at least 2 receivers per active router otherwise local recoveries can not be realized. Therefore the local group size (B) is an important parameter. In order to study the impact of B, simulations are performed with the 48 receivers distributed among groups of different sizes and figure 2c shows how much bandwidth (in ratio) can be saved

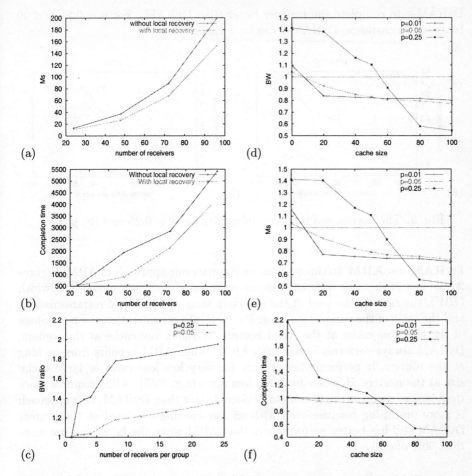

Fig. 2. (a) Load at the source, $p = 0.25$. (b) Completion time, $p = 0.25$. (c) Consumed bandwidth ratio. DyRAM vs. ARM: (d) Consumed bandwidth, (e) Load at the source, (f) Completion time.

with the local recovery mechanism. As the number of receivers per group increases, the consumed bandwidth is reduced by larger ratios for large loss rates. In fact, when the group size is larger it is more likely that the members of the group can recover from each other. For instance, figure 2c shows that with only 6 receivers per group and local recovery we can achieve a gain of 80%.

Early loss packet detection The next experiments show how the early loss detection service could decrease the delay of recovery. To do so, two cases noted DyRAM- and DyRAM+ are simulated. DyRAM- has no loss detection services whereas DyRAM+ benefits from the loss detection service in the source router. Figure 3 plots the recovery delay (normalized to the RTT) for the 2 cases as a function of the number of receivers. In general, the loss detection service allows

DyRAM+ to complete the transfer faster than DyRAM-. For $p = 0.25$ and 96 receivers for instance, DyRAM+ can be 4 times faster.

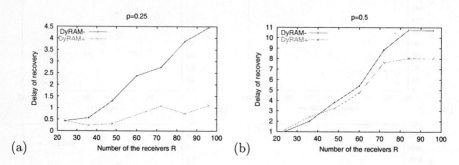

Fig. 3. The normalized recovery delay with (a) $p = 0.25$ and (b) $p = 0.5$

DyRAM vs. ARM In this section, we compare our approach to ARM. Figures 2d, 2e and 2f show for different loss rates the ratio of the consumed bandwidth (BW), the ratio of the load at the source in term of number of retransmissions and the ratio of the completion time. We took into consideration the percentage of the available cache at the ARM routers. Without any cache at the routers, DyRAM always performs better than ARM. When ARM benefits from caching at the routers, it performs better only for very low loss rates or large cache size at the routers. However for large loss rates (e.g. 25%), ARM requires more than 50% of cache at the routers to perform better than DyRAM. Our approach is more promising because even without any caching overhead at the routers DyRAM still has better performances than ARM when the latter can use more than 50% of cache.

5   Conclusions

We presented the DyRAM framework for handling efficiently multicast communications on computational grids. The architecture consists of a core protocol and active services located at the edge of the backbone network. These services are the packet losses detection, the NACK suppression, the subcast and the replier election. Simulation results show that local recovery, associated with early loss detection can provide both low-latency and increased bandwidth compared to a traditional solution.

The network model we considered assumed that there is a single receiver attached to a downstream link of an edge active router. If there are multiple receivers attached through a non active router then more than one receiver may receive a NACK packet (because the replier election is performed on a link basis) and believe to be the elected replier, so duplicate repairs may be sent. There are no simple ways to suppress the duplicate repairs sent by these receivers before they reach the active router. In this kind of topology, we suggest to put active routers as near as possible to the receivers, especially when there are a significant number of receivers. An example of such a situation is the case of several computing centers, each of them with an access router and several attached receivers, located on the same campus in a grid-like environment. In this case, it is preferable to have an active router for each computing center. For a few number of receivers, this problem is not very harmful.

One of our main concerns while designing DyRAM was to propose low-overhead (in terms of processing and memory requirements) and easy to implement solutions for the active services to be put in routers. Based on small data structures we believe the active services introduce only low overheads in routers and that implementations without much performance degradation on the routing and forwarding functions are possible.

References

1. M. Litzkow and M. Livny. Experience With The Condor Distributed Batch System. In *IEEE Workshop on Experimental Distributed Systems, October 1990*.
2. I. Foster and C. Kesselman. Globus: A metacomputing infrastructure toolkit. *Intl J. Supercomputing Applications, 11(2):115-128, 1997*
3. A. Grimshaw, A. Ferrari, F. Knabe and M. Humphrey. Legion: An Operating System for Wide-area computing. *IEEE Computer, 32(5):29-37, May 1999*
4. XTP Forum. *Xpress Transport Protocol Specification*, March 1995.
5. S. Floyd, V. Jacobson, and Liu C. G. A reliable multicast framework for light weight session and application level framing. In *ACM SIGCOMM'95*, pp342–356.
6. S. Paul and K. K. Sabnani. Reliable multicast transport protocol (rmtp). *IEEE JSAC, Special Issue on Network Support for Multipoint Comm.*, 15(3):407–421.
7. R. Yavatkar, J. Griffioen, and M. Sudan. A reliable dissemination protocol for interactive collaborative applications. In *ACM Multimedia'95*.
8. Christos Papadopoulos, Guru M. Parulkar, and George Varghese. An error control scheme for large-scale multicast applications. In *IEEE INFOCOM'98*, pp1188–1996.
9. T. Speakman et al. Pgm reliable transport protocol specification. internet draft.
10. D. L. Tennehouse et al. A survey of active network research. *IEEE Comm. Mag.*, pp80–86, January 1997.
11. L. Wei, H. Lehman, S. J. Garland, and D. L. Tennenhouse. Active reliable multi-cast. In *IEEE INFOCOM'98*.
12. S. K. Kasera et al. Scalable fair reliable multicast using active services. *IEEE Networks, Special Issue on Multicast*, 2000.
13. S. Zabele et al. Improving Distributed Simulation Performance Using Active Networks. In *World Multi Conference, 2000*.
14. M. Maimour and C. Pham. A Throughput Analysis of Reliable Multicast Protocols in an Active Networking Environment. In *6th IEEE Symp. on Comp. and Comm., 2001*.
15. M. Maimour and C. Pham. An analysis of a router-based loss detection service for active reliable multicast protocols. Technical report, RESAM, http://www.ens-lyon.fr/~mmaimour/Paper/TR/TR03-2001.ps.gz, November 2001.
16. R. Bagrodia et al. Parsec: A parallel simulation environment for complex systems. *Computer Magazine, 31(10), October 1998, pp77-85*.
17. C. Papadopoulos and E. Laliotis. Incremental Deployment of a Router-assisted Reliable Multicast Scheme. *Proc. of NGC 2000 Workshop*

A Parallel Quasi-Monte Carlo Method for Solving Systems of Linear Equations

Michael Mascagni[1], and Aneta Karaivanova[1,2]

[1] Department of Computer Science, Florida State University,
203 Love Building, Tallahassee, FL 32306-4530, USA,
mascagni@cs.fsu.edu,
URL: http://www.cs.fsu.edu/~mascagni
[2] Central Laboratory for Parallel Processing, Bulgarian Academy of Sciences,
Acad. G. Bonchev St.,bl. 25 A, 1113, Sofia, Bulgaria,
aneta@csit.fsu.edu,
URL: http://copern.bas.bg/~anet

Abstract. This paper presents a parallel quasi-Monte Carlo method for solving general sparse systems of linear algebraic equations. In our parallel implementation we use disjoint contiguous blocks of quasirandom numbers extracted from a given quasirandom sequence for each processor. In this case, the increased speed does not come at the cost of less thrust-worthy answers. Similar results have been reported in the quasi-Monte Carlo literature for parallel versions of computing extremal eigenvalues [8] and integrals [9]. But the problem considered here is more complicated - our algorithm not only uses an $s-$dimensional quasirandom sequence, but also its $k-$dimensional projections $(k = 1, 2, \ldots, s-1)$ onto the coordinate axes. We also present numerical results. In these test examples of matrix equations, the martrices are sparse, randomly generated with condition numbers less than 100, so that each corresponding Neumann series is rapidly convergent. Thus we use quasirandom sequences with dimension less than 10.

1 Introduction

The need to solve systems of linear algebraic equations arises frequently in scientific and engineering applications, with the solution being useful either by itself or as an intermediate step in solving a larger problem. In practical problems, the order, n, may in many cases be large (100 - 1000) or very large (many tens or hundreds of thousands). The **cost** of a numerical procedure is clearly an important consideration — so too is the **accuracy** of the method.

Let us consider a system of linear algebraic equations

$$Ax = b, \tag{1}$$

* Supported by the U.S. Army Research Office under Contract # DAAD19-01-1-0675

P.M.A. Sloot et al. (Eds.): ICCS 2002, LNCS 2330, pp. 598–608, 2002.

where $A = \{a_{ij}\}_{i,j=1}^{n} \in \mathbb{R}^{n \times n}$ is a given matrix, and $b = (b_1, \ldots, b_n)^t \in \mathbb{R}^n$ is a given vector. It is well known (see, for example, [3,6]) that the solution, x, $x \in R^n$, when it exists, can be found using

- *direct methods*, such as Gaussian elimination, and LU and Cholesky decomposition, taking $O(n^3)$ time;
- *stationary iterative methods*, such as the Jacobi, Gauss- Seidel, and various relaxation techniques, which reduce the system to the form

$$x = Lx + f, \tag{2}$$

and then apply iterations as follows

$$x^{(0)} = f, \ x^{(k)} = Lx^{(k-1)} + f, \ , \ k = 1, 2, \ldots \tag{3}$$

until desired accuracy is achieved; this takes $O(n^2)$ time per iteration.
- *Monte Carlo* methods (MC) use independent random walks to give an approximation to the truncated sum (3)

$$x^{(l)} = \sum_{k=0}^{l} L^k f,$$

taking time $O(n)$ (to find n components of the solution) per random step.

Keeing in mind that the convergence rate of MC is $O(N^{-1/2})$, where N is the number of random walks, millions of random steps are typically needed to achieve acceptible accuracy. The description of the MC method used for linear systems can be found in [1], [5], [10]. Different improvements have been proposed, for example, including sequential MC techniques [6], resolvent-based MC methods [4], etc., and have been successfully implemented to reduce the number of random steps. In this paper we study the quasi-Monte Carlo (QMC) approach to solve linear systems with an emphasis on the parallel implementation of the corresponding algorithm. The use of quasirandom sequences improves the accuracy of the method and preserves its traditionally good parallel efficiency.

The paper is organized as follows: §2 gives the background - MC for linear systems and a brief description of the quasirandom sequences we use. §3 describes parallel strategies, §4 presents some numerical results and §5 presents conclusions and ideas for future work.

2   Background

2.1   Monte Carlo for linear systems - very briefly

We can solve problem (1) in the form (2) with the scheme (3) if the eigenvalues of L lie within the unit circle, [3]. Then the approximate solution is the truncated Neumann series:

$$x^{(k)} = f + Lf + L^2 f + \ldots + L^{(k-1)} f + L^k f, \ k > 0 \tag{4}$$

with a truncation error of $x^{(k)} - x = L^k(f - x)$.

We consider the MC numerical algorithm and its parallel realization for the following two problems:

(a) Evaluating the inner product

$$J(h) = (h, x) = \sum_{i=1}^{n} h_i x_i \tag{5}$$

of the unknown solution $x \in I\!\!R^n$ of the linear algebraic system (2) and a given vector $h = (h_1, \ldots, h_n)^t \in I\!\!R^n$.

(b) Finding one or more components of the solution vector. This is a special case of **(a)** with the vector h chosen to be $h = e(r) = (0, ..., 0, 1, 0, ..., 0)$ where the one is on the r-th place if we want to compute the r-th component.

To solve this problem via a MC method (MCM) (see, for example, [10]) one has to construct a random process with mean equal to the solution of the desired problem. Consider a Markov chain with n states:

$$k_0 \to k_1 \to \ldots \to k_i \to \ldots, \tag{6}$$

with $k_j = 1, 2, \ldots, n$, for $j = 1, 2, \ldots$, and rules for constructing: $P(k_0 = \alpha) = p_\alpha, P(k_j = \beta | k_{j-1} = \alpha) = p_{\alpha\beta}$ where p_α is the probability that the chain starts in state α and $p_{\alpha\beta}$ is the transition probability from state α to state β . Probabilities $p_{\alpha\beta}$ define a transition matrix P. The normalizing conditions are: $\sum_{\alpha=1}^{n} p_\alpha = 1$, $\sum_{\beta=1}^{n} p_{\alpha\beta} = 1$ for any $\alpha = 1, 2, ..., n,$, with $p_\alpha \geq 0$, $p_\alpha > 0$ if $h(\alpha) \neq 0$, $p_{\alpha\beta} \geq 0$, $p_{\alpha\beta} > 0$ if $a_{\alpha\beta} \neq 0$. Define the weights on this Markov chain:

$$W_j = \frac{a_{k_0 k_1} a_{k_1 k_2} \cdots a_{k_{j-1} k_j}}{p_{k_0 k_1} p_{k_1 k_2} \cdots a_{k_{j-1} k_j}} \tag{7}$$

or using the recurrence $W_j = W_{j-1} \frac{a_{k_{j-1} k_j}}{p_{k_{j-1} k_j}}$, $W_0 = 1$.

The following random variable defined on the above described Markov chain

$$\Theta = \frac{h(k_0)}{p_{k_0}} \sum_{j=1}^{\infty} W_j f(k_j) \tag{8}$$

has the property

$$E[\Theta] = (h, f),$$

To compute $E[\Theta]$ we simulate N random walks (6), for each walk we compute the random variable Θ, (8), whose value on the sth walk is $[\Theta]_s$, and take the averaged value:

$$E[\Theta] \approx \frac{1}{N} \sum_{s=1}^{N} [\Theta]_s.$$

Each random walk is finite - we use a suitable stoping criterion to terminate the chain.

2.2   Quasirandom sequences

Consider an s-dimensional quasirandom sequence with elements $X_i = (x_i^{(1)}, x_i^{(2)}, \ldots, x_i^{(s)})$, $i = 1, 2, \ldots$, and a measure of its deviation from uniformity, the *Star discrepancy*:

$$D_N^* = D_N^*(x_1, \ldots, x_N) = \sup_{E \subset U^s} \left| \frac{\#\{x_n \in E\}}{N} - m(E) \right|,$$

where $U^s = [0,1)^s$. We are using the Soboĺ, Halton and Faure sequences, which can be very briefly defined as follows.

Let the representation of n, a natural number, in base b be

$$n = \ldots a_3(n) a_2(n) a_1(n), \qquad n > 0, n \in I\!\!R.$$

Then a one-dimensional quasirandom number sequence (the Van der Corput sequence) is defined as a radical inverse sequence

$$\phi_b(n) = \sum_{i=0}^{\infty} a_{i+1}(n) b^{-(i+1)}, \quad \text{where } n = \sum_{i=0}^{\infty} a_{i+1}(n) b^i,$$

and has star descrepancy $D_N^* = O\left(\frac{logN}{N}\right)$.

In our tests, we use the following multidimensional quasirandom number sequences:

Halton sequence:

$X_n = (\phi_{b_1}(n), \phi_{b_2}(n), \ldots, \phi_{b_s}(n))$, where the bases b_i are pairwise relatively prime.

Faure sequence:

$$x_n^{(k)} = \begin{cases} \sum_{i=0}^{\infty} a_{i+1}(n) q^{-(i+1)}, & k = 1 \\ \sum_{j=0}^{\infty} c_{j+1} q^{-(j+1)}, & k \geq 2 \end{cases},$$

where

$$c_j = \left[\sum_{i \geq j} (k-1)^{i-j} \frac{i!}{(i-j)! j!} a_i(n) \right] \pmod{q}, \quad j \geq 1, \; q \text{ is a prime}(q \geq s \geq 2).$$

Soboĺ sequence: $X_k \in \overline{\sigma}_i^{(k)}, k = 0, 1, 2, \ldots$, where $\overline{\sigma}_i^{(k)}, i \geq 1$ - set of permutations on every $2^k, k = 0, 1, 2, \ldots$ subsequent points of the Van der Corput sequence,

or in binary:

$$x_n^{(k)} = \bigoplus_{i \geq 0} a_{i+1}(n) v_i,$$

where $v_i, \; i = 1, \ldots, s$ is a set of direction numbers.

For the **Halton, Faure, Soboĺ** sequences we have

$$D_N^* = O\left(\frac{log^s N}{N}\right).$$

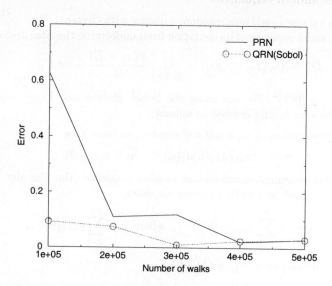

Fig. 1. Accuracy versus number of walks for computing (h, x), where x is the solution of a system with 2000 equations.

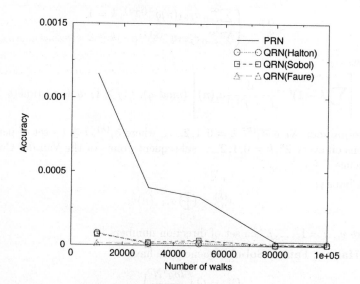

Fig. 2. Accuracy versus number of walks for computing one component, x_{64}, of the solution for a system with 1024 equations.

2.3   Convergence of the quasi-Monte Carlo Method

The MCM for computing the scalar product (h, x) consists of simulating N random walks (6), computing the value of the random variable Θ, (8), for each walk, and then average these values. In quasi-Monte Carlo we generate walks that are in fact not random, but have in some sense better distribution properties in the space of all walks on the matrix elements. The main difference in generating random and quasi-random walks is that we use l single pseudorandom numbers (PRNs) for a random walk of length l, but we use an l-dimensional sequence for the quasirandom walks of length l.

Computing the scalar product, $h^T A^i, f$ is equivalent to computing an $(i + 1)$-dimensional integral, and we analyze it with bounds from numerical integration [8]. We do not know A^i explicitly, but we do know A, and we use quasirandom walks on the elements of the matrix to compute approximately $h^T A^i f$. Using an $(i+1)$-dimensional quasirandom sequence, for N walks, $[k_0, k_1, \ldots, k_i]_s$, $s = 1, \ldots, N$ we obtain the following error bound [8]:

$$\left| h^T A^i f - \frac{1}{N} \sum_{s=1}^{N} \left[\frac{g_{k_0}}{p_{k_0}} W_i f_{k_i} \right]_s \right| \leq C_1(A, h, f) D_N^*,$$

where W_i is defined in (7), and $[z]_s$ is the value of z on the s-th walk. This gives us

$$\left| (h, x) - (h, x^{(k)}) \right| \leq C_2(A, h, f) \, k \, D_N^*.$$

Here D_N^* has order $O((log^k N)/N)$. Remember that the order of the mean square error for the analogous Monte Carlo method is $O(N^{-1/2})$. Figures 1 and 2 illustrate the accuracy versus the number of walks for computing the scalar product (h, x) (h is a given vector with 1 and 0, randomly chosen, and x is the solution of a system with 2000 equations), and for computing one component of the solution of a system with 1024 equations.

3   Parallel strategies

A well known advantage of the MCM is the efficiency by which it can be parallelized. Different processors generate independent random walks, and obtain their own MC estimates. These "individual" MC estimates are then combined to produce the final MC estimate. Such a scheme gives linear speed-up. However, certain variance reduction techniques usually require periodic communication between the processors; this results in more accurate estimates, at the expense of a loss in parallel efficiency. In our implementation this does not happen as we consider preparing the transition probabilities matrix as a preprocessing computation - which makes sense when we plan to solve the same system many times with different rigth-hand side vectors f.

In our parallel implementations we use disjoint contiguous blocks of quasirandom numbers extracted from a given quasirandom sequence for respective processors,

[9]. In this case, the increased speed does not come at the cost of less thrust-worthy answers. We have previously used this parallelization technique for the eigenvalue problem, [8], but the current problem is more complicated. In the eigenvalue problem we need to compute only $h^T A^k h$ using a k-dimensional quasirandom sequence, while here we must compute $\sum_{i=1}^{s} h^T A^i f$ using an s-dimensional sequence and its k-dimensional projections for $k = 1, 2, \ldots, s$.

We solved linear systems with general very sparse matrices stored in "sparse row-wise format". This scheme requires 1 real and 2 integer arrays per matrix and has proved to be very convenient for several important operations such as the addition, multiplication, transposition and permutation of sparse matrices. It is also suitable for deterministic, direct and iterative, solution methods. This scheme permits us to store the entire matrix on each processor, and thus, each processor can generate the random walks independently of the other processors.

4   Numerical examples

We performed parallel computations to empirically examine the parallel effi-ciency of the quasi-MCM. The parallel numerical tests were performed on a Compaq Alpha parallel cluster with 8 DS10 processors each running at 466 megahertz and using MPI to provide the parallel calls.

We have carried out two types of numerical experiments. First, we considered the case when one only component of the solution vector is desired. In this case each processor generates N/p independent walks. At the end, the host processor collects the results of all realizations and computes the desired value. The computational time does not include the time for the initial loading of the matrix because we imagine our problem as a part of larger problem (for example, solving the same matrix equation for diffrent right-hand-side vectors) and assume that every processor independently obtains the matrix. Second, we consider computing the inner product (h, x), where h is a given vector, and x is the unknown solution vector.

During a single random step we use an l-dimensional point of the chosen quasir-andom sequence and its k-dimensional projections ($k = 1, \ldots, l - 1$) onto the coordinate axes. Thus we compute all iterations of $h^T A^k f$ ($1 \le k \le l$) using a single l-dimensional quasirandom sequence. The number of iterations needed can be determined using suitable stoping criterion: for example, checking how "close" two consecuent iterations $h^T A^{k-1} f$ and $h^T A^k f$ are, or by using a fixed number, l, on the basis of $a\ priori$ analysis of the Neumann series convergence. In our numerical tests we use a fixed l - we have a random number of iterations per random step but the first l iterations (which are the most important) are quasirandom, and the rest are pseudorandom.

In all cases the test matrices are sparse and are stored in "sparse row-wise-format". We show the results for two matrix equations at size 1024 and 2000. The average number of non-zero elements per matrix row is $d = 57$ for $n = 1024$

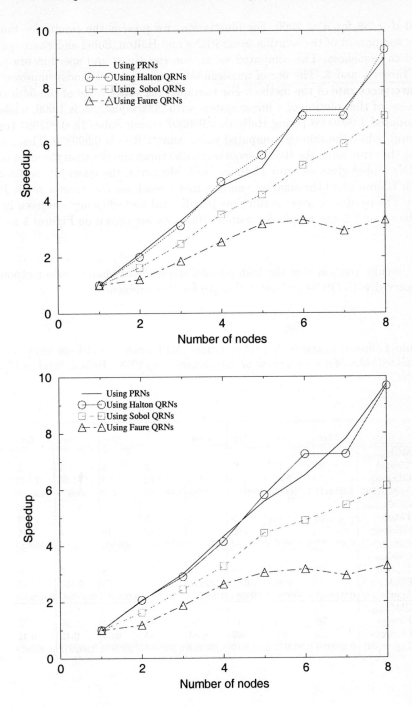

Fig. 3. Speedup when solving linear system with 1024 and 2000 equations using PRNs, Halton, Soboĺ and Faure sequences.

and $d = 56$ for $n = 2000$. For illustration, we present the results for finding one component of the solution using PRNs and Halton, Soboĺ and Faure quasirandom sequences. The computed value, the run times and speedup are given in Tables 1 and 2. The use of quasirandom sequences (all kinds) improves the convergence rate of the method. For example, the exact value of the 54th component of the solution of a linear system with 2000 equations is 1.000, while we computed 1.0000008 (using Halton), 0.999997 (using Sobo/'l), 0.999993 (using Faure), while the average computed value using PRNs is 0.999950. This means that the error using the Halton sequence is 250 times smaller than the error using PRNs (Soboĺ gives an error 17 times less). Moreover, the quasi-MC realization with Halton gives the smallest runing times - much smaller than a single PRN run. The results for average run time (single run) and efficiency are given in the Tables 1 and 2, the graphs for parallel efficiency are shown on Figures 1 and 2.

The results confirm that the high parallel efficiency of Monte Carlo methods is preserved with QRNs and our technique for this problem.

Table 1. Sparse system with 1024 equations: MPI times, parallel efficiency and the estimated value of one component of the solution using PRNs, Halton, Soboĺ and Faure QRNs.

	1pr.	2pr.	3pr.	4pr.	5pr.	6pr.	7pr.	8pr.
MCM$_{\text{pseudo}}$								
Time(s)	36	17	11	8	7	5	5	4
Efficiency		1.05	1.09	1.125	1.02	1.2	1.05	1.125
Appr.x(54)	1.000031	1.000055	.999963	.999966	1.000024	1.000029	.999979	.999990
QMC$_{\text{Halton}}$								
Time(s)	28	14	9	6	5	4	4	3
Efficiency		1	1.03	1.16	1.12	1.16	1	1.16
Appr.x(54)	0.999981	0.999981	0.999981	0.999981	0.999981	0.999981	0.999981	0.999981
QMC$_{\text{Sobol}}$								
Time(s)	42	25	17	12	10	8	7	6
Efficiency		0.84	0.82	0.875	0.84	0.875	0.857	0.875
Appr.x(54)	0.999983	0.999983	0.999983	0.999983	0.999983	0.999983	0.999983	0.999983
QMC$_{\text{Faure}}$								
Time(s)	76	63	41	30	24	23	26	23
Efficiency		0.6	0.62	0.63	0.63	0.55	0.42	0.41
Appr.x(54)	0.999919	0.999919	0.999919	0.999919	0.999919	0.999919	0.999919	0.999919

Table 2. Sparse system with 2000 equations: MPI times, parallel efficiency and the estimated value of one component of the solution using PRNs, Halton, Sobol and Faure QRNs.

	1pr.	2pr.	3pr.	4pr.	5pr.	6pr.	7pr.	8pr.
MCM$_{\text{pseudo}}$								
Time(s)	39	19	13	9	7	6	5	4
Efficiency		1.02	1	1.08	1.11	1.08	1.11	1.21
x(54)	1.000077	1.000008	.999951	.999838	.999999	.999917	1.000044	.999802
QMC$_{\text{Halton}}$								
Time(s)	29	14	10	7	5	4	4	3
Efficiency		1.03	0.97	1.03	1.16	1.21	1.03	1.21
x(54)	1.0000008	1.0000008	1.0000008	1.0000008	1.0000008	1.0000008	1.0000008	1.0000008
QMC$_{\text{Sobol}}$								
Time(s)	49	30	20	15	11	10	9	8
Efficiency		0.82	0.82	0.82	0.89	0.82	0.78	0.76
x(54)	0.999997	0.999997	0.999997	0.999997	0.999997	0.999997	0.999997	
QMC$_{\text{Faure}}$								
Time(s)	79	67	42	30	26	25	27	24
Efficiency		0.59	0.63	0.66	0.61	0.53	0.42	0.41
x(54)	0.999993	0.999993	0.999993	0.999993	0.999993	0.999993	0.999993	0.999993

5   Future work

We plan to study this quasi-MCM for matrix equations with more slowly convergent Neumann series solution, using random walks with longer (and different for different walks) length. In this case, the dimension of of quasirandom sequence is random (depends on the used stoping criterion). Our preliminary numerical tests for solving such problems showed the effectivness of randomized quasirandom sequences, but we have yet to study the parallel behavior of this method in this case.

References

1. J. H. Curtiss, Monte Carlo methods for the iteration of linear operators, *Journal of Mathematical Physics*, **32**, 1954, pp. 209-323.
2. B. Fox, *Strategies for quasi-Monte Carlo*, Kluwer Academic Publishers, Boston/Dordrecht/London, 1999.
3. G. H. Golub, C.F. Van Loon, *Matrix computations*, **The Johns Hopkins Univ. Press**, Baltimore, 1996.
4. Dimov I., V. Alexandrov, A. Karaivanova, Resolvent Monte Carlo Methods for Linear Algebra Problems, *Mathematics and Computers in Simulations*, Vol. . **55**, 2001, pp. 25-36.

5. J.M. Hammersley, D.C. Handscomb, *Monte Carlo methods*, **John Wiley & Sons, inc.**, New York, London, Sydney, Methuen, 1964.
6. J.H. Halton, Sequential Monte Carlo Techniques for the Solution of Linear Systems, *SIAM Journal of Scientific Computing*, Vol.**9**, pp. 213-257, 1994.
7. Mascagni M., A. Karaivanova, Matrix Computations Using Quasirandom Sequences, *Lecture Notes in Computer Science*, (Wulkov, Yalamov, Wasniewsky Eds.), Vol.**1988**, Springer, 2001, pp. 552-559.
8. Mascagni M., A. Karaivanova, A Parallel Quasi-Monte Carlo Method for Computing Extremal Eigenvalues, to appear in: *Lecture Notes in Statistics*, Springer.
9. Schmid W. Ch., A. Uhl, Parallel quasi-Monte Carlo integration using (t, s)-sequences, In: *Proceedings of ACPC'99* (P. Zinterhof et al., eds.), Lecture Notes in Computer Science, **1557**, Springer, 96-106.
10. Soboĺ, I. M., *Monte Carlo numerical methods*, Nauka, Moscow, 1973 (in Russian).

Mixed Monte Carlo Parallel Algorithms for Matrix Computation

Behrouz Fathi, Bo Liu, and Vassil Alexandrov

Department of Computer Science,
University of Reading,
Whiteknights P.O. Box 225
United Kingdom, RG6 6AY
{b.fathivajargah, b.liu, v.n.alexandrov}@rdg.ac.uk

Abstract. In this paper we consider mixed (fast stochastic approximation and deterministic refinement) algorithms for Matrix Inversion (MI)
and Solving Systems of Linear Equations (SLAE). Monte Carlo methods
are used for the stochastic approximation, since it is known that they
are very efficient in finding a quick rough approximation of the element
or a row of the inverse matrix or finding a component of the solution
vector. In this paper we show how the stochastic approximation of the
MI can be combined with a deterministic refinement procedure to obtain MI with the required precision and further solve the SLAE using
MI. We employ a splitting $A = D - C$ of a given non-singular matrix
A, where D is a diagonal dominant matrix and matrix C is a diagonal
matrix. In our algorithm for solving SLAE and MI different choices of D
can be considered in order to control the norm of matrix $T = D^{-1}C$, of
the resulting SLAE and to minimize the number of the Markov Chains
required to reach given precision. Experimental results with dense and
sparse matrices are presented.

Keywords: Monte Carlo Method, Markov Chain, Matrix Inversion, Solution of sytem of Linear Equations, Matrix Decomposition, Diagonal Dominant
Matrices.

1 Introduction

The problem of inverting a real $n \times n$ matrix (MI) and solving system of linear
algebraic equations (SLAE) is of an unquestionable importance in many scientific and engineering applications: e.g. communication, stochastic modelling, and
many physical problems involving partial differential equations. For example,
the direct parallel methods of solution for systems with dense matrices require
$O(n^3/p)$ steps when the usual elimination schemes (e.g. non-pivoting Gaussian
elimination, Gauss-Jordan methods) are employed [4]. Consequently the computation time for very large problems or real time problems can be prohibitive and
prevents the use of many established algorithms.

P.M.A. Sloot et al. (Eds.): ICCS 2002, LNCS 2330, pp. 609–618, 2002.
© Springer-Verlag Berlin Heidelberg 2002

It is known that Monte Carlo methods give statistical estimation of the components of the inverse matrix or elements of the solution vector by performing random sampling of a certain random variable, whose mathematical expectation is the desired solution. We concentrate on Monte Carlo methods for MI and solving SLAEs, since, firstly, only $O(NL)$ steps are required to find an element of the inverse matrix where N is the number of chains and T is an estimate of the chain length in the stochastic process, which are independent of matrix size n and secondly, the process for stochastic methods is inherently parallel.

Several authors have proposed different coarse grained Monte Carlo parallel algorithms for MI and SLAE [7–11]. In this paper, we investigate how Monte Carlo can be used for diagonally dominant and some general matrices via a general splitting and how efficient mixed (stochastic/deterministic) parallel algorithms can be derived for obtaining an accurate inversion of a given non-singular matrix A. We employ either uniform Monte Carlo (UM) or almost optimal Monte Carlo (MAO) methods [7–11]. The relevant experiments with dense and sparse matrices are carried out.

Note that the algorithms are built under the requirement $\|T\| < 1$. Therefore to develop efficient methods we need to be able to solve problems with matrix norms greater than one. Thus we developed a spectrum of algorithms for MI and solving SLAEs ranging from special cases to the general case. Parallel MC methods for SLAEs based on Monte Carlo Jacobi iteration have been presented by Dimov [11]. Parallel Monte Carlo methods using minimum Makrov Chains and minimum communications between master/slave processors are presented in [5, 1]. Most of the above approaches are based on the idea of balancing the stochastic and systematic errors [11]. In this paper we go a step further and have designed mixed algorithms for MI and solving SLAEs by combining two ideas: iterative Monte Carlo methods based on the Jacobi iteration and deterministic procedures for improving the accuracy of the MI or the solution vector of SLAEs.

The generic Monte Carlo ideas are presented in Section 2, the main algorithms are described in Section 3 and the parallel approach and some numerical experiments are presented in Section 4 and 5 respectively.

2   Monte Carlo and Matrix Computation

Assume that the system of linear algebraic equations (SLAE) is presented in the form:

$$Ax = b \tag{1}$$

where A is a real square $n \times n$ matrix, $x = (x_1, x_2, ..., x_n)^t$ is a $1 \times n$ solution vector and $b = (b_1, b_2, ..., b_n)^t$.

Assume the general case $\|A\| > 1$. We consider the splitting $A = D - C$, where off-diagonal elements of D are the same as those of A, and the diagonal elements of D are defined as $d_{ii} = a_{ii} + \gamma_i \|A\|$, choosing in most cases $\gamma_i > 1, i = 1, 2, ..., n$. We further consider $D = B - B_1$ where B is the diagonal matrix of D, e.g. $b_{ii} = d_{ii} i = 1, 2, ..., n$. As shown in [1] we could transform the system (1) to

$$x = Tx + f \tag{2}$$

where $T = D^{-1}C$ and $f = D^{-1}b$. The multipliers γ_i are chosen so that, if it is possible, they reduce the norm of T to be less than 1 and reduce the number of Markov chains required to reach a given precision. We consider two possibilities, first, finding the solution of $x = Tx + f$ using Monte Carlo (MC) method if $\|T\| < 1$ or finding D^{-1} using MC and after that finding A^{-1}. Then, if required, obtaining the solution vector is found by $x = A^{-1}b$.

Consider first the stochastic approach. Assume that $\|T\| < 1$ and that the system is transformed to its iterative form (2). Consider the Markov chain given by:

$$s_0 \rightarrow s_1 \rightarrow \cdots \rightarrow s_k, \tag{3}$$

where the $s_i, i = 1, 2, \cdots, k$, belongs to the state space $S = \{1, 2, \cdots, n\}$. Then for $\alpha, \beta \in S, p_0(\alpha) = p(s_0 = \alpha)$ is the probability that the Markov chain starts at state α and $p(s_{j+1} = \beta | s_j = \alpha) = p_{\alpha\beta}$ is the transition probability from state α to state β. The set of all probabilities $p_{\alpha\beta}$ defines a transition probability matrix $P = \{p_{\alpha\beta}\}_{\alpha,\beta=1}^{n}$ [3, 9, 10]. We say that the distribution $(p_1, \cdots, p_n)^t$ is acceptable for a given vector g, and that the distribution $p_{\alpha\beta}$ is acceptable for matrix T, if $p_\alpha > 0$ when $g_\alpha \neq 0$, and $p_\alpha \geq 0$, when $g_\alpha = 0$, and $p_{\alpha\beta} > 0$ when $T_{\alpha\beta} \neq 0$, and $p_{\alpha\beta} \geq 0$ when $T_{\alpha\beta} = 0$ respectively. We assume $\sum_{\beta=1}^{n} p_{\alpha\beta} = 1$, for all $\alpha = 1, 2, \cdots, n$. Generally, we define

$$W_0 = 1, W_j = W_{j-1} \frac{T_{s_{j-1}s_j}}{p_{s_{j-1}s_j}} \tag{4}$$

for $j = 1, 2, \cdots, n$.

Consider now the random variable $\theta[g] = \frac{g_{s_0}}{p_{s_0}} \sum_{i=1}^{\infty} W_i f_{s_i}$. We use the following notation for the partial sum:

$$\theta_i[g] = \frac{g_{s_0}}{p_{s_0}} \sum_{j=0}^{i} W_j f_{s_j}. \tag{5}$$

Under condition $\|T\| < 1$, the corresponding Neumann series converges for any given f, and $E\theta_i[g]$ tends to (g, x) as $i \rightarrow \infty$. Thus, $\theta_i[g]$ can be considered as an estimate of (g, x) for i sufficiently large. To find an arbitrary component of the solution, for example, the rth component of x, we should choose, $g = e(r) = (0, ..., 1, 0, ..., 0)$ such that
$\underbrace{\qquad}_{r}$

$$e(r)_\alpha = \delta_{r\alpha} = \begin{cases} 1 & \text{if} \quad r = \alpha \\ 0 & otherwise \end{cases} \tag{6}$$

It follows that

$$(g, x) = \sum_{\alpha=1}^{n} e(r)_\alpha x_\alpha = x_r. \tag{7}$$

The corresponding Monte Carlo method is given by:

$$x_r = \hat{\Theta} = \frac{1}{N} \sum_{s=1}^{N} \theta_i [e(r)]_s,$$

where N is the number of chains and $\theta_i [e(r)]_s$ is the approximate value of x_r in the sth chain. It means that using Monte Carlo method, we can estimate only one, few or all elements of the solution vector. We consider Monte Carlo with uniform transition probability (UM) $p_{\alpha\beta} = \frac{1}{n}$ and Almost optimal Monte Carlo method (MAO) with $p_{\alpha\beta} = \frac{|T_{\alpha\beta}|}{\sum_{\beta=1}^{n} |T_{\alpha\beta}|}$, where $\alpha, \beta = 1, 2, \ldots, n$. Monte Carlo MI is obtained in a similar way [3].

To find the inverse $A^{-1} = C = \{c_{rr'}\}_{r,r'=1}^{n}$ of some matrix A, we must first compute the elements of matrix $M = I - A$, where I is the identity matrix. Clearly, the inverse matrix is given by

$$C = \sum_{i=0}^{\infty} M^i, \qquad (8)$$

which converges if $\|M\| < 1$.

To estimate the element $c_{rr'}$ of the inverse matrix C, we let the vector f be the following unit vector

$$f_{r'} = e(r'). \qquad (9)$$

We then can use the following Monte Carlo method for calculating elements of the inverse matrix C:

$$c_{rr'} \approx \frac{1}{N} \sum_{s=1}^{N} \left[\sum_{(j|s_j=r')} W_j \right], \qquad (10)$$

where $(j|s_j = r')$ means that only

$$W_j = \frac{M_{rs_1} M_{s_1 s_2} \cdots M_{s_{j-1} s_j}}{p_{rs_1} p_{s_1 s_2} \cdots p_{s_{j-1} p_j}} \qquad (11)$$

for which $s_j = r'$ are included in the sum (10).

Since W_j is included only into the corresponding sum for $r' = 1, 2, \ldots, n$, then the same set of N chains can be used to compute a single row of the inverse matrix, which is one of the inherent properties of MC making them suitable for parallelization.

The *probable error* of the method, is defined as $r_N = 0.6745\sqrt{D\theta/N}$, where $P\{|\bar{\theta} - E(\theta)| < r_N\} \approx 1/2 \approx P\{|\bar{\theta} - E(\theta)| > r_N\}$, if we have N independent realizations of random variable (r.v.) θ with mathematical expectation $E\theta$ and average $\bar{\theta}$ [6].

3   The Mixed MC Algorithm

The basic idea is to use MC to find the approximate inverse of matrix D, refine the inverse (filter) and find A^{-1}. We can then find the solution vector through A^{-1}. According to the general definition of a regular splitting [2], if A, M and N are three given matrices satisfying $A = M - N$, then the pair of matrices M, N are called regular splitting of A, if M is nonsingular and M^{-1} and N are non-negative.

Therefore, let A be a nonsingular diagonal dominant matrix. If we find a regular splitting of A such as $A = D - C$ than the SLAE $x^{(k+1)} = Tx^{(k)} + f$, where $T = D^{-1}C$, and $f = D^{-1}b$ converge to unique solution x^* if and only if $\|T\| < 1$ [2].

The efficiency of inverting diagonally dominant matrices is an important part of the process enabling MC to be applied to diagonally dominant and some general matrices. The basic algorithms, covering the inversion of diagonally dominant and arbitrary non-singular matrix are given below. First let us consider how we can find Monte Carlo approximate of D^{-1} :

Algorithm1: Finding D^{-1}.

1. **Initial data:** Input matrix A, parameters γ and ϵ.
2. **Preprocessing:**
   2.1 **Split** $A = D - (D - A)$, where D is a diagonally dominant matrix.
   2.2 **Set** $D = B - B_1$ where B is a diagonal matrix $b_{ii} = d_{ii}$ $i = 1, 2, ..., n$.
   2.3 **Compute** the matrix $T = B^{-1}B_1$.
   2.4 **Compute** $\|T\|$, the Number of Markov Chains $N = (\frac{0.6745}{\epsilon} \cdot \frac{1}{(1-\|T\|)})^2$.
3. **For** i=1 to n;
   3.1 **For** j=1 to j=N;
           Markov Chain Monte Carlo Computation:
   3.1.1 **Set** $t_k = 0$(stopping rule), $W_0 = 1$, $SUM[i] = 0$ and $Point = i$.
   3.1.2 **Generate** an uniformly distributed random number *nextpoint*.
   3.1.3 **If** $T[point][netxpoint]! = 0$.
   LOOP
        3.1.3.1 **Compute** $W_j = W_{j-1}\frac{T[point][netxpoint]}{P[point][netxpoint]}$.
        3.1.3.2 **Set** $Point = nextpoint$ and $SUM[i] = SUM[i] + W_j$.
        3.1.3.3 **If** $|W_j| < \gamma$, $t_k = t_k + 1$
        3.1.3.4 **If** $t_k \geq n$, end LOOP.
   3.1.4 **End If**
   3.1.5 **Else** go to step 3.1.2.
   3.2 **End of loop j.**
   3.3 **Compute** the average of results.
4. **End of loop i.**
5. **Obtain** The matrix $V = (I - T)^{-1}$.
6. **Therefore** $D^{-1} = VB^{-1}$.
7. **End** of algorithm.

Consider now the second algorithm which can be used for the inversion of a general non-singular matrix A. Note that in some cases to obtain a very accurate inversion of matrix D some filter procedures can be applied.

Algorithm2: Finding A^{-1}.

1. **Initial data:** Input matrix A, parameters γ and ϵ.
2. **Preprocessing:**
   2.1 **Split** $A = D - (D - A)$, where D is a diagonally dominant matrix.
   2.2 **Set** $D = B - B_1$ where B is a diagonal matrix $b_{ii} = d_{ii}$ $i = 1, 2, ..., n$.
   2.3 **Compute** the matrix $T = B^{-1}B_1$.
   2.4 **Compute** $||T||$, the Number of Markov Chains $N = (\frac{0.6745}{\epsilon} \cdot \frac{1}{(1-||T||)})^2$.
3. **For** i=1 to n;
   3.1 **For** j=1 to j=N;
       Markov Chain Monte Carlo Computation:
       3.1.1 **Set** $t_k = 0$(stopping rule), $W_0 = 1$, $SUM[i] = 0$ and $Point = i$.
       3.1.2 **Generate** an uniformly distributed random number $nextpoint$.
       3.1.3 **If** $T[point][netxpoint]! = 0$.
       LOOP
          3.1.3.1 **Compute** $W_j = W_{j-1}\frac{T[point][netxpoint]}{P[point][netxpoint]}$.
          3.1.3.2 **Set** $Point = nextpoint$ and $SUM[i] = SUM[i] + W_j$.
          3.1.3.3 **If** $|W_j| < \gamma$, $t_k = t_k + 1$
          3.1.3.4 **If** $t_k \geq n$, end LOOP.
       3.1.4 **End If**
       3.1.5 **Else** go to step 3.1.2.
   3.2 **End of loop j**.
   3.3 **Compute** the average of results.
4. **End of loop i.**
5. **Obtain** The matrix $V = (I - T)^{-1}$.
6. **Therefore** $D^{-1} = VB^{-1}$.
7. **Compute** the MC inversion $D^{-1} = B(I - T)^{-1}$.
8. **Set** $D_0 = D^{-1}$ (approximate inversion) and $R_0 = I - DD_0$.
9. **use filter procedure** $R_i = I - DD_i$, $D_i = D_{i-1}(I + R_{i-1})$, $i = 1, 2, ..., m$, where $m \leq k$.
10. **Consider the accurate inversion of D** by step 9 given by $D_0 = D_k$.
11. **Compute** $S = D - A$ where S can be any matrix with all non-zero elements in diagonal and all of its off-diagonal elements are zero.
12. **Main function** for obtaining the inversion of A based on D^{-1} step 9:
    12.1 **Compute** the matrices $S_i, i = 1, 2, ..., k$, where each S_i has just one element of matrix S.
    12.2 **Set** $A_0 = D_0$ and $A_k = A + S$
    12.3 **Apply** $A_k^{-1} = A_{k+1}^{-1} + \frac{A_{k+1}^{-1}S_{i+1}A_{k+1}^{-1}}{1-trace(A_{k+1}^{-1}S_{i+1})}$, $i = k - 1, k - 2, ..., 1, 0$.
13. **Print**the inversion of matrix A.
14. **End** of algorithm.

4   Parallel Implementation

We have implemented the algorithms proposed on a cluster of workstations and a Silicon Graphics ONYX2 machine under PVM. We have applied master/slave approach.

Inherently, Monte Carlo methods for solving SLAE allow us to have minimal communication, i.e. to partition the matrix A, pass the non-zero elements of the dense (sparse) matrix to every processor, to run the algorithm in parallel on each processor computing $\lceil n/p \rceil$ rows (components) of MI or the solution vector and to collect the results from slaves at the end without any communication between sending non-zero elements of A and receiving partitions of A^{-1} or x. The splitting procedure and refinement are also parallelised and integrated in the parallel implementation. Even in the case, when we compute only k components $(1 \leq k \leq n)$ of the MI (solution vector) we can divide evenly the number of chains among the processors, e.g. distributing $\lceil kN/p \rceil$ chains on each processor. The only communication is at the beginning and at the end of the algorithm execution which allows us to obtain very high efficiency of parallel implementation.

In this way we can obtain for the parallel time complexity of the MC procedure of the algorithm $O(nNL/p)$ where N denotes the number of Markov Chains. According to central limit theorem for the given error ϵ we have $N \geq \left(\frac{0.6745}{\epsilon \times (1 - \|T\|)} \right)^2$, L denotes the length of the Markov chains and $L \leq \left(\frac{\log(\gamma)}{\log \|T\|} \right)$, where ϵ, γ show the accuracy of Monte Carlo approximation [3]. Parameters ϵ, γ are used for the stochastic and systematic error. The absolute error of the solution for matrix inversion is $\left\| I - \hat{A}^{-1} A \right\|$, where A is the matrix whose inversion has to be found, and $\hat{A}^{-1}$ is the approximate MI. The computational time is shown in seconds.

5   Experimental Results

The algorithms run on a 12 processor (four 195 MHZ, eight 400MHZ) $ONYX2$ Silicon Graphics machine with 5120 Mbytes main memory. We have carried test with low precision $10^{-1} - 10^{-2}$ and higher precision $10^{-5} - 10^{-6}$ in order to investigate the balance between stochastic and deterministic components of the algorithms based on the principle of balancing of errors (e.g. keeping the stochastic and systematic error of the same order) [7].

Consider now, firstly, a low precision 10^{-1}, approximation to MI for randomly generated dense and sparse matrices and secondly higher precision 10^{-5} approximation to MI for randomly generated dense and sparse matrices.

Matrix Size	Dense Case		
	Time(MC)	Time(total)	Error
100	0.51901	1.135063	0.1
300	2.11397	5.32793	0.1
500	12.7098	69.34915	0.1
1000	59.323708	199.60263	0.1
2000	338.4567	1189.3193	0.1
3000	437.072510	2295.97385	0.1
4000	793.01563	3321.87578	0.1
5000	1033.9867	3979.39237	0.1
6000	1265.5816	5162.56743	0.1
7000	1340.0085	5715.66747	0.1
8000	1645.2306	5991.52939	0.1

Table 1. Dense randomly generated matrices

Matrix Size	Sparse Case		
	Time(MCMC)	Time(total)	Error
100	0.238264	1.127853	0.1
300	0.85422	3.27931	0.1
500	7.95166	29.11630	0.1
1000	47.5126	157.6298	0.1
2000	315.5001	733.28491	0.1
3000	389.2388	1466.9328	0.1
4000	696.9654	1994.95163	0.1
5000	819.3842	2553.76525	0.1
6000	994.8456	3413.86492	0.1
7000	1143.4664	3759.7539	0.1
8000	1389.7674	3955.664901	0.1

Table 2. Sparse randomly generated matrices

Matrix Size	Dense Case	
	Time(total)	Error
100	2.14282	0.002121
300	11.00373	0.000626
500	118.29381	0.000398
1000	799.9064	0.000266
2000	4989.2433	0.000044
3000	5898.4772	0.000038
4000	6723.6863	0.000027
5000	7159.7639	0.000015
6000	7695.0903	0.000006
7000	8023.7381	0.000003
8000	8914.5512	0.00000018

Table 3. Higher precision MI for dense matrices

Matrix Size	Sparse Case	
	Time(total)	Error
100	1.9232	0.000611000
300	8.14498	0.000216000
500	90.25106	0.000163000
1000	763.32165	0.000078000
2000	4758.98103	0.000055000
3000	5678.2463	0.000018000
4000	6318.9001	0.000007200
5000	6855.4118	0.000001400
6000	7057.8491	0.000000510
7000	7528.5516	0.000000220
8000	8018.921	0.000000050

Table 4. Higher precision MI for sparse matrices

6   Conclusion

In this paper we have introduced a mixed Monte Carlo/deterministic algorithms for Matrix Inversion and finding a solution to SLAEs for any non-singular matrix. Further experiments are required to determine the optimal number of chains required for Monte Carlo procedures and how best to tailor together Monte Carlo and deterministic refinement procedures. We have shown also that the algorithms run efficiently in parallel and the results confirmed their efficiency for MI with dense and sparse matrices. Once good enough inverse is found, we can find the solution vector of SLAE. The accuracy of results are comparable with other known computational algorithms.

References

1. Fathi Vajargah B., Liu B. and Alexandrov V., On the preconditioner Monte Carlo methods for solving linear systems. MCM 2001, Salzburg, Austria (presented).
2. Ortega, J., *Numerical Analysis*, SIAM edition, USA, 1990.
3. Alexandrov V.N., *Efficient parallel Monte Carlo Methods for Matrix Computation*, Mathematics and computers in Simulation, Elsevier **47** pp. 113-122, Netherlands, (1998).
4. Golub, G.H., Ch., F., Van Loan, *Matrix Computations,* The Johns Hopkins Univ. Press, Baltimore and London, (1996)
5. Taft K. and Fathi Vajargah B., *Monte Carlo Method for Solving Systems of Linear Algebraic Equations with Minimum Markov Chains.* International Conference PDPTA'2000 Las Vegas, (2000).
6. Sobol I.M. *Monte Carlo Numerical Methods.* Moscow, Nauka, 1973 (in Russian).
7. Dimov I., Alexandrov V.N. and Karaivanova A., *Resolvent Monte Carlo Methods for Linear Algebra Problems,* Mathematics and Computers in Simulation, Vol55, pp. 25-36, 2001.
8. Fathi Vajargah B. and Alexandrov V.N., *Coarse Grained Parallel Monte Carlo Algorithms for Solving Systems of Linear Equations with Minimum Communication,* in Proc. of PDPTA, June 2001, Las Vegas, 2001, pp. 2240-2245.
9. Alexandrov V.N. and Karaivanova A., *Parallel Monte Carlo Algorithms for Sparse SLAE using MPI,* LNCS 1697, Springer 1999, pp. 283-290.
10. Alexandrov V.N., Rau-Chaplin A., Dehne F. and Taft K., *Efficient Coarse Grain Monte Carlo Algorithms for matrix computation using PVM,* LNCS 1497, pp. 323-330, Springer, August 1998.
11. Dimov I.T., Dimov T.T., et all, *A new iterative Monte Carlo Approach for Inverse Matrix Problem,* J. of Computational and Applied Mathematics **92** pp 15-35 (1998).

Numerical Experiments with Monte Carlo Methods and SPAI Preconditioner for Solving System of Linear Equations

Bo Liu, Behrouz Fathi, and Vassil Alexandrov

Computer Science,
The University of Reading,
Whiteknight P.O. Box 225,
Reading, UK
{ b.liu, fathi, v.n.alexandrov}@rdg.ac.uk

Abstract. In this paper we present the results of experiments comparing the performance of the mixed Monte Carlo algorithms and SPAI preconditener with BICGSTAB. The experiments are carried out on a Silicon Graphics ONYX2 machine. Based on our experiments, we conclude that these techniques are comparable from the point of view of robustness and rates of convergence, with the Monte Carlo approach performing better for some general cases and SPAI approach performing better in case of very sparse matrices.

Keywords: Sparse Linear Systems, Preconditioned iterative methods, approximate inverse, SPAI, Monte Carlo methods.

1 Introduction

Consider a system of linear algebraic equations (SLAE) presented in the form:

$$Ax = b \qquad (1)$$

where A is a real square $n \times n$ matrix, $x = (x_1, x_2, ..., x_n)^t$ is a $1 \times n$ solution vector and $b = (b_1, b_2, ..., b_n)^t$.

There are now quite a few deterministic methods based on Krylov subspace methods for solving general linear systems such as GMRES, BI-CGSTAB and QMR. In order to be effective these methods must be combined with a good preconditioner, and it is generally agreed that the choice of the preconditioner is even more crucial than the choice of the Krylov subspace iteration. The search for effective preconditioners is an active research topic in scientific computing. Several potentially successful methods have been proposed in the literature, and the SPAI preconditioner is regard as one of the most promising approaches [4–6]. However Monte Carlo methods are coming to play a role in finding a rough approximate inverse and few efficient parallel Monte Carlo approaches for MI have been presented in the past few years [7, 1]. Nowadays, parallel Monte Carlo

P.M.A. Sloot et al. (Eds.): ICCS 2002, LNCS 2330, pp. 619–627, 2002.

methods can be one of the most promising approaches for solving SLAE and also be an efficient preconditioner. We tested the algorithms solving general systems of linear equations, where the corresponding matrices were generated by simple generator constructing general dense and sparse matrices, which were stored in a row-packed format. Some matrices from the matrix market have also been used. We used Silicon Graphics ONYX2 machine to carry out the experiments.

The ideas of Monte Carlo for Matrix Computation are presented in Section 2, the main algorithm is described in Section 3 and the parallel approach and some numerical experiments are presented in Section 4.

2   Monte Carlo for Matrix Computation

Let us assume that we need to solve a general SLAE in form (1). Consider the general case when $||I - A|| \geq 1$. We consider the splitting $A = B_n - B_1$, where the off-diagonal elements of B_n are the same as A. The diagonal elements of B_n are defined as $b_{ii} = a_{ii} + \gamma_i||A||$, choosing in most cases $\gamma_i > 1, i = 1, 2, ..., n$. We then split $B_n = M - K$, where M is diagonal matrix of B_n. We could transform system (1) to

$$x = Cx + f \qquad (2)$$

where $C = B_n^{-1}B_1$ and $f = B_n^{-1}b$. The multipliers γ_i are chosen so that they reduce the norm of C and reduce the number of Markov chains required to reach a given precision, in a similar way as proposed in [3]. Therefore we consider two possibilities, first, finding the solution of $x = Cx + f$ using Monte Carlo (MC) method and, second, finding A^{-1} and obtaining the solution via $x = A^{-1}f$. In this paper we are more interested in the second idea as a general goal, where we compute the approximate inverse of B_n with MC, then retrieve A^{-1} via a function that will be introduced in the next section.

Consider firstly the stochastic approach. Assume that the system is transformed to its iterative form (2). Consider now the Markov Chain given by:

$$s_0 \to s_1 \to \cdots \to s_k \qquad (3)$$

where the $s_i, i = 1, 2, \cdots, k$, belongs to the state space $S = \{1, 2, \cdots, n\}$. Then for $\alpha, \beta \in S, p_0(\alpha) = p(s_0 = \alpha)$ is the probability that the Markov chain starts at state α and $p(s_{j+1} = \beta | s_j = \alpha) = p_{\alpha\beta}$ is the transition probability from state α to state β. The set of all probabilities $p_{\alpha\beta}$ defines a transition probability matrix $P = \{p_{\alpha\beta}\}_{\alpha,\beta=1}^n$ [7] [8]. We say that the distribution $(p_1, \cdots, p_n)^t$ is acceptable for a given vector g, and that the distribution $p_{\alpha\beta}$ is acceptable for matrix C, if $p_\alpha > 0$ when $g_\alpha \neq 0$, and $p_\alpha \geq 0$, when $g_\alpha = 0$, and $p_{\alpha\beta} > 0$ when $C_{\alpha\beta} \neq 0$, and $p_{\alpha\beta} \geq 0$ when $C_{\alpha\beta} = 0$ respectively. We assume $\sum_{\beta=1}^n p_{\alpha\beta} = 1$, for all $\alpha = 1, 2, \cdots, n$. Generally, we define

$$W_0 = 1, W_j = W_{j-1} \frac{C_{s_{j-1}s_j}}{p_{s_{j-1}s_{j-1}}} \qquad (4)$$

for $j = 1, 2, \cdots, n$.

Consider now the random variable $\theta[g] = \frac{g_{s_0}}{p_{s_0}} \sum_{i=1}^{\infty} W_i f_{s_i}$. We use the following notation for the partial sum:

$$\theta_i[g] = \frac{g_{s_0}}{p_{s_0}} \sum_{j=0}^{i} W_j f_{s_j}. \tag{5}$$

Under the condition $\|C\| < 1$, the corresponding Neumann series converges for any given f, and $E\theta_i[g]$ tends to (g, x) as $i \to \infty$. Thus, $\theta_i[g]$ can be considered as an estimate of (g, x) for i sufficiently large. To find an arbitrary component of the solution, for example, the rth component of x, we should choose,
$g = e(r) = (\underbrace{0, ..., 1}_{r}, 0, ..., 0)$ such that

$$e(r)_\alpha = \delta_{r\alpha} = \begin{cases} 1 & if \quad r = \alpha \\ 0 & otherwise \end{cases} \tag{6}$$

It follows that

$$(g, x) = \sum_{\alpha=1}^{n} e(r)_\alpha x_\alpha = x_r. \tag{7}$$

The corresponding Monte Carlo method is given by

$$x_r = \hat{\Theta} = \frac{1}{N} \sum_{s=1}^{N} \theta_i[e(r)]_s$$

where N is the number of chains and $\theta_i[e(r)]_s$ is the approximate value of x_r in the sth chain. It means that using Monte Carlo method, we can estimate only one, few or all elements of the solution vector. We consider Monte Carlo with uniform transition probability (UM) $p_{\alpha\beta} = \frac{1}{n}$ and Almost optimal Monte Carlo method (MAO) with $p_{\alpha\beta} = \frac{|C_{\alpha\beta}|}{\sum_{\beta=1}^{n} |C_{\alpha\beta}|}$, where $\alpha, \beta = 1, 2, \ldots, n$. Monte Carlo MI is obtained in a similar way [7, 1].

To find the approximate inverse $Q^{-1} = \{q_{rr'}\}_{r,r'=1}^{n}$ of some matrix T, we must first compute matrix

$$F = I - T, \tag{8}$$

where I is the identity matrix. Clearly, the inverse matrix is given by

$$Q = \sum_{i=0}^{\infty} F^i, \tag{9}$$

which converges if $\|F\| < 1$.

To estimate the element $q_{rr'}$ of the inverse matrix Q, we let the vector f be the following unit vector

$$f_{r'} = e(r'). \tag{10}$$

We then can use the following Monte Carlo method for calculating elements of the inverse matrix Q:

$$q_{rr'} \approx \frac{1}{N} \sum_{s=1}^{N} \left[\sum_{(j|s_j=r')} W_j \right], \qquad (11)$$

where $(j|s_j = r')$ means that only

$$W_j = \frac{F_{rs_1} F_{s_1 s_2} \dots F_{s_{j-1} s_j}}{p_{rs_1} p_{s_1 s_2} \dots p_{s_{j-1} s_j}} \qquad (12)$$

for which $s_j = r'$ are included in the sum (11).

Since W_j is included only in the corresponding sum of $r' = 1, 2, \dots, n$, the same set of N chains can be used to compute a single row of the inverse matrix, which is one of the inherent properties of MC making them suitable for parallelization.

The *probable error* of the method, is defined as $r_N = 0.6745\sqrt{D\theta/N}$, where $P\{|\bar{\theta} - E(\theta)| < r_N\} \approx 1/2 \approx P\{|\bar{\theta} - E(\theta)| > r_N\}$, if we have N independent realizations of random variable (r.v.) θ with mathematical expectation $E\theta$ and average $\bar{\theta}$ [8].

3 The parallel approximate methods

It is well-known that Monte Carlo (MC) methods are efficient for solving systems of linear algebraic equations. Initially MC methods have been applied to SLAEs with $||I - A|| < 1$. In the further developments [7, 3, 2] MC methods have been applied to solve SLAE with diagonally dominant and M-matrices. To use Monte Carlo algorithms for solving general systems of linear equations, we developed a new mixed algorithm, where for the system (1), firstly, we construct a diagonal dominant matrix B_n whose off-diagonal elements are the same as those of A, then, secondly we compute the approximate inverse matrix of B_n [3] using the algorithm below.

1. **Initial data:** Input matrix B_n, parameters δ and ϵ.
2. **Preprocessing:**
   2.1 **Split** $B_n = M - K$ where M is a diagonal matrix $m_{ii} = b_{ii}$ $i = 1, 2, ..., n$.
   2.2 **Compute** the matrix $C = M^{-1}K$.
   2.3 **Compute** $||C||$ and the Number of Markov Chains $N = \left(\frac{0.6745}{\epsilon} \cdot \frac{1}{(1-||C||)} \right)^2$.
3. **For** i=1 to n;
   3.1 **For** j=1 to j=N;
       Markov Chain Monte Carlo Computation:
       3.1.1 **Set** $t_k = 0$(stopping rule), $W_0 = 1$, $SUM[i] = 0$ and $Point = i$.
       3.1.2 **Generate** an uniformly distributed random number giving *nextpoint*.
       3.1.3 **If** $C[point][netxpoint] \neq 0$.
       LOOP

3.1.3.1 **Compute** $W_j = W_{j-1} \frac{C[point][nextpoint]}{P[point][nextpoint]}$.

3.1.3.2 **Set** $Point = nextpoint$ and $SUM[i] = SUM[i] + W_j$.

3.1.3.3 **If** $|W_j| < \delta$, $t_k = t_k + 1$

3.1.3.4 **If** $t_k \geq n$, end LOOP.

3.1.4 **End If**

3.1.5 **Else** go to step 3.1.2.

3.2 **End of loop j.**

3.3 **Compute** the average of results.

4. **End of loop i.**

5. **Obtain** The matrix $Z = (I - C)^{-1}$. then the expecting approximate matrix $Y_n = Z \times M^{-1}$.

Finally we retrieve the inversion matrix of A via the following function [2]:

$$Y_{n-1} = Y_n + \frac{Y_n S_n Y_n}{1 - trace(Y_n S_n)}, \tag{13}$$

where Y_n is the computed approximate inverse of B, $S_n = B_n - diag(A)$, n is a sequence of integers ranging from one to the size of the matrix A. It is obvious that the approximate solution vector can be obtained by $A^{-1} * b$. Indeed, before we retrieve A^{-1} we also can apply the following refinement function

$$Y_n^{new} = Y_n^{old}(2 \times I - Y_n^{old} B_n), \tag{14}$$

where I denotes the identity matrix. Clearly the precision of the approximate solution vector can be controlled using the refinement function with just 1 or 2 steps. From the experimental work is evident that the number of steps required to obtain almost exact solution is at most 3 or 4.

On the other hand we studied the SPAI preconditioner, which is based on the idea of Frobenius norm minimization. The approximate inverse matrix of A is computed as matrix M which minimizes $||I - MA||$(or $||I - AM||$ for the right preconditioning) subject to some sparsity constraints, see [4,5]. Once the approximate inverse matrix of A is known, SPAI applies BICGSTAB to solve (1) (see http://www.sam.math.ethz.ch/ grote/spai/)

Now we have introduced both algorithms in brief. We will use both of them to solve general SLAEs.

4   Parallel Implementation and Numerical Results

The Monte Carlo code is written in C using PVM and the SPAI code is written in C using MPI. To test the two approaches we consider matrices from the Matrix Market (http://math.nist.gov/MatrixMarket/) and some general matrices randomly generated by the generator we provide. The generator generates random numbers for the matrices with predefined sparsity. The experiments are carried out on a Silicon Graphics ONYX2 parallel machine with 12 processors (four 195 MHZ and eight - 400 MHZ), with 5120 Mbytes main memory and 32 Kbytes cache in total. We run both algorithms with the corresponding parameters. The computing time of the Monte Carlo calculation and the SPAI preconditioner are shown separately. The full time required to solve the SLAEs for mixed algorithms and SPAI with BICGSTAB is also shown. Some of the test matrices used in this experiment are given below.

- **BWM200.MTX** size n=200, with nz=796 nonzero elements, from Matrix Market web site;
- **CAVITY03.MTX** size n=327, with nz=7327 nonzero elements, from Matrix Market web site;
- **GEN200_DENSE.MC** size n=200, dense, from the generator;
- **GEN200_60SPAR.MC** size n=200, with nz=16000 nonzero elements, from the generator;
- **GEN800_DENSE.MC** size n=800, dense, from the generator;
- **GEN800_70SPAR.MC** size n=800, with nz=192000 nonzero elements, from the generator;
- **GEN2000_99SPAR.MC** size n=2000, with nz=40000 nonzero elements, from the generator;
- **GEN500_90SPAR.MC** size n=500, with nz=25000 nonzero elements, from the generator;
- **GEN1000_95SPAR.MC** size n=1000, with nz=50000 nonzero elements, from the generator;

The default parameters we set up in Monte Carlo methods are

1. $\epsilon = 0.05$ denotes a given stochastic error;
2. $\delta = 0.01$ denotes the accuracy of Monte Carlo approximation;
3. $step = 1$ denotes how many steps are spent on the refinement function, in each single step of using the refinement function two matrix-by-matrix multiplications are computed.

We apply the appropriate parameters in SPAI and show the best performance with the best combination of these parameters. R-MC denotes the residual computing time of Monte Carlo approach which includes the time for the refinement procedure, the retrieval procedure and obtaining the approximate solution vector. MC denotes the time required for Monte Carlo algorithm only. Therefore, TOTAL-MC is the time required for the MC and R-MC. R-SPAI denotes the residual computing time of SPAI, which includes the time of BICGSTAB for obtaining the approximate solution vector (the block procedure and scalar-matrix procedure while using block algorithm). SPAI denotes the time required by the SPAI preconditioner. TOTAL-SPAI is the total time of SPAI and R-SPAI. ERROR denotes the absolute error of the approximate solution vector given below.

$$ERROR = \frac{||x_{exact} - x_{approximate}||_2}{||x_{exact}||_2} \tag{15}$$

Para	BL	SPAI	R-SPAI	TOTAL-SPAI	MC	R-MC	TOTAL-MC
0.5	1×1	0.9303448	85.6012096	86.5315544			
	3×3	0.9759936	75.2054016	76.1813952			
0.6	1×1	0.8856872	220.1436288	221.029316	2.186874	3.164163	5.351037
	3×3	0.9230960	150.3301808	151.2532768			
0.7	1×1	1.3930344	351.6003544	352.9933888			
	3×3	1.1148200	160.762056	161.876876			

Table 1. BWM200.MTX Solution with parameters: ERROR = 10^{-4} and P=10 processors. SPAI is tested with different values of ϵ, and both blocked and non-blocked.

Para	BL	SPAI	R-SPAI	TOTAL-SPAI	MC	R-MC	TOTAL-MC
0.5	1×1	2.9910024	nan	nan			
	3×3	1.1972320	nan	nan			
0.6	1×1	3.1077280	nan	nan	0.9338300	2.346462	3.280292
	3×3	1.2565296	nan	nan			
0.7	1×1	2.6192136	nan	nan			
	3×3	0.9718288	nan	nan			

Table 2. GEN200_DENSE.MC Solution with parameters: ERROR = 10^{-4} and P=10 processors. SPAI is tested with different values of ϵ, and both blocked and non-blocked.

SPAI is run with the default parameter of iterations in BICGSTAB, 500, and nan means we can not obtain the results under the setting, or it is converging slowly.

P#	MC	TOTAL-MC	SPAI	TOTAL-SPAI
2	76.20574	85.813108	0.2917584	nan
5	36.16877	60.095253	0.5471512	nan
8	28.29067	46.830696	0.3889352	nan
12	21.07290	33.248477	1.3224760	nan

Table 3. CAVITY03.MTX ERROR=10^{-4} and P=2, 5, 8, 12 processors respectively.

P#	MC	TOTAL-MC	SPAI	TOTAL-SPAI
2	156.1388	586.202544	130.6839168	nan
5	47.70701	385.626890	67.6120952	nan
8	36.74342	304.847875	24.0445288	nan
12	26.53539	294.639845	23.1435808	nan

Table 4. GEN800_70SPAR.MC ERROR=10^{-4} and P=2, 5, 8, 12 processors respectively.

Matrix	MC	TATOL-MC	SPAI	TOTAL-SPAI
GEN200_60SPAR.MC	1.047589	4.720657	1.1548728	nan
GEN500_90SPAR.MC	7.730741	84.94246	4.1712408	nan
GEN800_DENSE.MC	24.32269	300.3137	29.000558	nan
GEN1000_95SPAR.MC	46.274798	746.9745	9.1504488	nan
GEN2000_99SPAR.MC	302.626325	4649.0219003	6.2901104	nan

Table 5. ERROR=10^{-4} and P=10 processors

The examples show that for some cases MC converges much faster than SPAI. In case of some general matrices it can be seen that SPAI is converging very slowly. There are also cases where SPAI is performing much better than MC. Further experiments are required to carry out detailed analysis.

5   Conclusion

We have presented the results of experiments comparing the efficiency of parallel SPAI preconditioner with BICGSTAB and parallel mixed Monte Carlo algorithms. The experiments are carried out on a Silicon Graphics ONYX2 machine. Based on our experiments, we conclude that these techniques are comparable from the point of view of robustness and rates of convergence, with the Monte Carlo approach being somewhat better for some general cases and SPAI approach performing better in case of very sparse matrices. We can also conclude that both techniques offer excellent potential for use on high performance computers.

References

1. Alexandrov V.N., *Efficient parallel Monte Carlo Methods for Matrix Computation*, Mathematics and computers in Simulation, Elsevier **47** pp. 113-122, Netherlands, (1998).
2. Fathi Vajargah B., Liu B. and Alexandrov V.,*Mixed Monte Carlo Parallel Algorithms for Matrix Computation*, ICCS 2002, Amsterdam, Netherlands.
3. Fathi Vajargah B., Liu B. and Alexandrov V., *On the preconditioner Monte Carlo methods for solving linear systems. MCM 2001, Salzburg, Austria (to appear)*.
4. M.Benze, C.D.Meyer, and M.Tuma, *A sparse approximate inverse preconditioner for the conjugate gradient method*, SIAM, 1996.
5. Marcus J.Grote, Thomas Huckle, *Parallel Preconditioning with Sparse Approximate Inverses*, SIAM J. of Scient. Compute. 18(3), 1997.
6. S.T. Barnard and M. J. Grote, in Proc. *A Block Version of the SPAI Preconditioner*, 9th SIAM Conf. on Parallel. Process. for Sci. Comp., held March 1999.
7. Dimov I., Alexandrov V.N. and Karaivanova A., *Resolvent Monte Carlo Methods for Linear Algebra Problems*, Mathematics and Computers in Simulation, Vol55, pp. 25-36, 2001.
8. Sobol I.M. *Monte Carlo numerical methods*. Moscow, Nauka, 1973 (in Russian).

Measuring the Performance of a Power PC Cluster *

Emanouil I. Atanassov

Central Laboratory for Parallel Processing, Bulgarian Academy of Sciences, Sofia,
Bulgaria
emanouil@copern.bas.bg

Abstract. In this work we present benchmarking results from our Linux
cluster of four dual processor Power Macintosh computers with processors G4/450 MHz. These machines are used mainly for solving large scale
problems in air pollution and Monte Carlo simulations.
We first present some numbers revealing the maximum performance of
an individual machine. The results are from the well known LINPACK
benchmark and an optimized Mandelbrot set computation.
The second set of benchmarking results covers the NAS Parallel Benchmark. These tests are written using MPI and Fortran 77 with some Fortran 90 constructs and are close to the real problems that we solve on
the cluster. We also tested the performance of a free implementation of
Open MP - the Omni Open MP compiler.
The last set of tests demonstrates the efficiency of the platform for parallel quasi-Monte Carlo computations, especially when the vector unit is
used for generating the Sobol' sequences.

1 Description of the Software and Hardware Configuration

Our cluster consists of four dual processor Power Macintosh computers, connected with a BayStack 350 Switch. Each node has 512 MB RAM and 30GB
hard disk space, and has two processors, Power PC G4 at 450 MHz clock frequency, with AltiVec technology. More details about the AltiVec technology can
be found at the website http://www.altivec.org.

We decided to use entirely open source/free source software on these machines. The operating system on the cluster is GNU/Linux, kernel version 2.4.8,
with SMP and AltiVec support enabled. We used the Yellowdog distribution
version 2.0, but we compiled the kernel with the SMP and AltiVec support ourselves. The cluster was upgraded later to version 2.1, but this didn't result in
any significant changes in the benchmarking results, shown here. Since the last
upgrade the cluster had not been restarted.

---

* Supported by a project of the European Commission - BIS 21 under contract ICA1-
CT-2000-70016 and by the Ministry of Education and Science of Bulgaria under
contract NSF I-811/98.

P.M.A. Sloot et al. (Eds.): ICCS 2002, LNCS 2330, pp. 628–634, 2002.

We use in our work the GNU Compiler Collection (GCC), which has compilers for C, C++ and Fortran 77. When one wishes to use the AltiVec unit of the G4 processor, some special compiler support is needed. We use the Motorola patched gcc compiler, which introduces some non-standard extensions to the C language. In our experience use of this compiler and the AltiVec unit pays off only when a small part of the code is critical for the overall performance.

For our parallel programs we mainly use MPI, so we installed two versions of MPI - LAM MPI and MPICH. In the benchmarks and in our daily work we discovered that the LAM version had better results (version 6.5.4 was installed). It had better latency and throughput in almost all cases. That is why only results from this version are shown here. LAM has been compiled with support for OpenSSH, for better security, and with support for System V shared memory, since significant improvement was discovered in the communication between two processor from the same node when shared memory is used instead of TCP/IP.

We tested also the Omni Open MP compiler. More information on it can be found at http://pdplab.trc.rwcp.or.jp/pdperf/Omni/. For the Power PC architecture however, this compiler could be used only to generate parallel programs for the two processors on the same node, while for X86 processors and clusters the programs can operate on the whole cluster.

The cluster is used actively by researchers from the Bulgarian Academy of Sciences, as well as students from Sofia University.

2   Peak Performance Numbers

Power PC G4 is a RISC CPU, which means it has a large set of registers. 32 of them are vector registers, meaning they can store 4 integer or floating point 32 bit values. The CPU can perform one basic operation (like **load, store, xor,** ...) on these registers at once. A significant drawback of the AltiVec unit is that it can perform floating point operations only in single precision, so it can not be used in code that requires double precision floating point operations. In one operation the AltiVec unit deals with 128 bytes of data, so it can perform 4 Flops in 1 processor cycle. It can in some cases perform even more than 4 Flops per cycle, since it has a multiply-and-add operation.

A well known benchmarking test for processor speed is computing a Mandelbrot set. Consider the set coming from the relation $z_{n+1} \rightarrow z_n^4 + c$. Viktor Decyk, Dean Dauger, and Pieter Kokelaar from UCLA developed an AltiVec optimized algorithm, reaching almost 4 Flops per processor cycle. These computations are actively promoted as a benchmark of Power Macintosh computers performance by Apple. Since the source code of the AltiVec version of this code was not freely available, the author developed one such implementation. The code performs approximately 2 additions and 2 multiplications in one clock cycle, leading to an average of about 1533 MFlops.

This result suggests that the Linux OS and the SMP kernel do not impede the performance of the CPU in any way.

The next set of results is connected with the LINPACK benchmark ([6]), which measures essentially maximum speed of the LU algorithm. The author's experience is that the "best" options for compilation are "-O3 -funroll-all-loops", since the G4 CPU has a large set of registers, and all tests shown in the paper were done with these options. When one compiles the reference implementation of the test, he or she gets around 100 MFlops in double precision. However, using the ATLAS library ([1]), one can achieve more than 490MFlops for matrix multiplication and more than 330 MFlops for LU factorization in double precision.

Using the High Performance Linpack (see [5]), and trying various choices for the options at random, we easily found options which allowed us to achieve 300 MFlops rate per processor for the whole cluster, when all 8 processors were used (for matrices of 5000 × 5000 elements).

3   NAS Parallel Benchmark Results

This benchmark suite was developed by NASA with the aim to measure the parallel performance of given parallel architecture. The suite consists of eight tests, coming in four different sizes - W, A, B and C. For a detailed description of these tests see [2].

The test programs implement some important algorithms, and are usually refered by two letter abbreviations. We will write what these abbreviations mean: BT - Block Tridiagonal, CG - Conjugate Gradient, EP - Embarrassingly Parallel, FT - Fourier Transform, IS - Integer Sort, LU - LU Decomposition, MG - Multigrid, SP - Scalar Pentadiagonal.

The benchmark uses MPI and Fortran 77 with some Fortran 90 constructs and is widely accepted as a benchmark in the scientific community and by hardware and software vendors. See for instance [16],[14],[8],[10].

We remind that the rules for benchmarking do not allow changes in the source code, one can only look for the best compiler options. We used the GNU Fortran Compiler - g77, with options "-O3 -funroll-all-loops". We had problems with some tests, since the code uses some Fortran 90 constructs. For this reason the FT test didn't compile. The other tests produced correct results, when they were able to fit in memory.

In some cases the problems in their default sizes were too large to fit into the memory of a single machine, but can be solved if divided in two or four. All the results are given in Table 1,2,3, and 4. For some of the algorithms by Mops the actual MFlops rate achieved by the program is shown, while for the integer algorithms Mops means "millions of basic operations per second". The efficiency shown in the tables is obtained by dividing the Mops rate obtained by the Mops rate of 1 processor and the number of processors. Efficiency is not shown for size C problems, because the respective problems couldn't fit in RAM when only one processor was used.

We also give in Table 5 results from the Open MP version of the NAS Parallel Benchmark, obtained by using the free Omni Open MP compiler. The Mops rates achieved using two processors on the same node are shown, and the efficiency is calculated by dividing the achieved Mops rate by the Mops rate of one processor from Tables 1,2 and 3. These results are not so good, compared with the results from the MPI version. In our institution we mostly use MPI, so we were not very concerned with the Open MP performance of the cluster.

Table 1. Results from the NAS Parallel benchmark, size W

Np	BT		CG		EP		IS		LU		MG		SP	
	Mops	Eff.	Mops	Eff.	Mops	Eff.	Mops	Eff.	Mops	Eff.	Mops	Eff.	Mops	Eff.
1	84.3	1	31.1	1	1.9	1	6.3	1	112.3	1	76.3	1	47.2	1
2			27.7	0.88	1.9	1	4.8	0.77	106.5	0.94	63.8	0.84		
4	84.0	0.99	18.9	0.60	1.9	1	2.3	0.37	100.7	0.89	42.4	0.55		
8			12.9	0.41	1.8	0.97	0.9	0.14	84.6	0.75	28.5	0.37	53.0	1.12

Table 2. Results from the NAS Parallel benchmark, size A

Np	BT		CG		EP		IS		LU		MG		SP	
	Mops	Eff.	Mops	Eff.	Mops	Eff.	Mops	Eff.	Mops	Eff.	Mops	Eff.	Mops	Eff.
1	80.1	1	31.4	1	1.9	1	6.2	1	95.4	1	64.1	1	40.4	1
2			29.0	0.92	1.9	0.99	4.6	0.74	97.6	1.02	58.3	0.91		
4	73.9	0.92	22.0	0.70	1.9	0.99	2.5	0.4	95.9	1.00	44.4	0.69		
8			16.1	0.51	1.9	0.99	1.4	0.22	87.2	0.91	37.6	0.59	42.3	1.05

Table 3. Results from the NAS Parallel benchmark, size B

Np	CG		EP		IS		LU		MG		SP	
	Mops	Eff.	Mops	Eff.	Mops	Eff.	Mops	Eff.	Mops	Eff.	Mops	Eff.
1	20.8	1	1.9	1			88.5	1	68.3	1	45.1	1
2	21.2	1.01	1.9	1			87.7	0.99	62.6	0.92		
4			1.9	1	1.9		87.8	0.99	57.9	0.85	46.2	1.02
8			1.9	0.99					12.5	0.18		

Looking at the numbers, we observe that in most cases the cluster shows acceptable performance. Perhaps only the IS (IS means integer sort) algorithm shows definitely poor performance. The performance drop seen in most algorithms when going from 4 to 8 processors is not unexpected, since the two processors on the same node share the same memory bus and the same network interface. Even when using both processors of the same node instead of two processors on different nodes bears a performance penalty of about 5-20%, when using LAM MPI. Comparing these numbers with the peak performance

Table 4. Results from the NAS Parallel benchmark, size C

Np	CG	EP	IS	LU	MG	SP
	Mops	Mops	Mops	Mops	Mops	Mops
2		1.9		85.4		
4	16.1	1.9		85.9		47.5
8	14.1		1.2	65.3		

Table 5. Results from the Open MP version of the NAS Parallel benchmark

Np	BT		EP		LU		MG		SP		FT	
	Mops	Eff.	Mops	Eff.	Mops	Eff.	Mops	Eff.	Mops	Eff.	Mops	Eff.
W	63.8	0.76	1.7	0.89	91.7	0.27	68.2	0.89	39.1	0.83	67.4	
A	58.0	0.72	1.7	0.89	68.2	0.71	50.8	0.79	35.3	0.83	45.7	
B			1.7	0.89	61.0	0.69	56.4	0.83	35.1	0.78		

numbers from the previous section, one can conclude that some of the algorithms presented here (for instance LU) are not very well optimized with respect to the cache of the particular machine. On the other hand, in real programs one rarely has time to introduce additional parameters and experiment with them, in order to fit more data in cache, so the results shown here can be considered as more "realistic", i.e. close to the performance one would expect for his own programs.

4   Parallel Quasi-Monte Carlo Algorithms

The results shown so far are mostly from standard benchmarks. They can be considered as a prediction for what one could expect from his own code. I was impressed by the huge gain in performance, when the AltiVec unit is used, and decided to develop a vector version of my algorithm for generating the Sobol' sequences.

These sequences were introduced by Sobol in [12] (see also [13]). They are widely used in quasi-Monte Carlo computations, because of the uniformity of their distribution. In Financial Mathematics they are used for computing special kind of multi-dimensional integrals, with the dimension reaching often 360 or more. The reader might look at the paper of Paskov and Traub [15] for more details .

Our algorithm for generation of the Sobol sequences, which will be described elsewhere, satisfies the "guidelines" for using the vector unit - a small part of the code takes most of the CPU time. When one needs less than 2^{22} points, there is no loss of precision if only single precision numbers are generated. Generating more than 2^{22} points, while still gaining from the use of the AltiVec unit, is also possible.

In Table 6 one can see timing results from generating 1 000 000 points in 360 dimensions. These results are compared with the results from a basic version of the algorithm, using ordinary operations. In the programs GB and GV we only made sure the generation of the terms of the sequence actually happens, and is not optimized out by the compiler. In GB we used regular operations and in GV some AltiVec operations were used. In programs SB and SV we show timing results from generating and summing all the terms of the sequence. In SB only regular instructions are used, and in GV the AltiVec instructions are used for both generation and summing the terms, with the summing being done with just the most simple vector algorithm.

Table 6. Timing results for generating and summing 1 million terms of the Sobol' sequences in 360 dimensions

	1			2			4			8		
	Time	Mnps	Eff.	Time	Mnps	Eff.	Time	Mnps	Eff.	Time	Mnps	Eff.
GB	7.3	49.0	1	3.7	96.3	0.98	1.9	189.8	0.97	1.0	363.1	0.92
GV	2.3	159.4	1	1.2	309.9	0.97	0.6	589.9	0.93	0.3	1034.8	0.81
SB	10.8	33.4	1	5.5	65.8	0.99	2.8	130.5	0.98	1.4	252.8	0.95
SV	4.3	83.3	1	2.2	163.6	0.98	1.1	318.2	0.95	0.6	586.6	0.88

By Mnps we denote the number of millions of elements of the sequence generated in one second. We show the total time and the Mnps rate. For each term of the sequence 360 coordinates are needed, so there are 360×10^6 elements to be generated.

Observe the good speed-up obtained by using all the 8 processors of the cluster, and also the fact that SV takes noticeably more time than GV, so for a "real" quasi-Monte Carlo algorithm the generation of the sequence will be performed in a small fraction of the total time. Another observation is that summing is perhaps the simplest operation on the generated sequence that can be performed. Still the difference in time between GB (generation only) and SB (generation and summation), is much more than the time spent for the AltiVec generation program GV.

While good speed-up is usually associated with the Monte Carlo algorithms, the above results show that quasi-Monte Carlo algorithms can also be very efficiently parallelized and vectorized, when the Sobol' sequences are used.

5 Conclusions

The various benchmarking results shown here demonstrate the viability of the Linux cluster built from Power Macintosh machines as a platform for scientific computing. Its vector capabilities can be utilized via standard libraries, specifically optimized for the AltiVec architecture, or by writing hand-tuned subroutines, when the time permits so.

References

1. Automatically   Tuned   Linear   Algebra   Software   (ATLAS)   - http://math-atlas.sourceforge.net/

2. Bailey, D., Barton, J., Lasinski, T., Simon, H., eds.: The NAS Parallel Benchmarks. Technical Report RNR-91-02, NASA Ames Research Center, Moffett Field, CA 94035, January 1991.
3. Bailey, D., Harris, T., Saphir,W., van der Wijngaart, R., Woo, A., Yarrow, M.: The NAS Parallel Benchmarks 2.0 Report NAS-95-020, December, 1995
4. GNU Compiler Collection -http://gcc.gnu.org/
5. High Performance Linpack Benchmark - http://www.netlib.org/benchmark/hpl/
6. Linpack Benchmark - http://www.netlib.org/linpack
7. Linux Kernel Archive - http://www.kernel.org/
8. MPI Performance on Coral - http://www.icase.edu/~josip/MPIonCoral.html
9. NAS Parallel Benchmark - http://www.nas.nasa.gov/NAS/NPB/
10. NAS Parallel Benchmarks on a Scali system - http://www.scali.com/performance/nas.html
11. Omni Open MP Compiler - http://pdplab.trc.rwcp.or.jp/pdperf/Omni/
12. Sobol',I.M.: On the distribution of point in a cube and the approximate evaluation of integrals, USSR Computational Mathematics and Mathematical Physics, 7,86–112, 1967
13. Sobol', I.M.: Quadrature formulae for functions of several variables satisfying general Lipshuz condition, USSR Computational Mathematics and Mathematical Physics, 29,935–941, 1989
14. Origin 2000 on NAS Parallel Benchmark - http://www.nas.nasa.gov/~faulkner/o2k_npb_benchmarks.html
15. Paskov,S. and Traub,J.: Faster Valuation of Financial Derivatives, Journal of Portfolio Management, Vol. 22:1, Fall, 1995, 113–120.
16. Turney, R.D.: Comparison of Origin 2000 and Origin 3000 Using NAS Parallel Benchmarks. NAS Technical Report 01-003
17. Yellowdog Linux - http://www.yellowdog.com/

Monte Carlo Techniques for Estimating the Fiedler Vector in Graph Applications

Ashok Srinivasan and Michael Mascagni

Department of Computer Science, Florida State University,
Tallahassee FL 32306, USA
{asriniva, mascagni}@cs.fsu.edu

Abstract. Determining the Fiedler vector of the Laplacian or adjacency matrices of graphs is the most computationally intensive component of several applications, such as graph partitioning, graph coloring, envelope reduction, and seriation. Often an approximation of the Fiedler vector is sufficient. We discuss issues involved in the use of Monte Carlo techniques for this purpose.

1   Introduction

The Fiedler vector is the eigenvector corresponding to the second smallest non-negative eigenvalue of a matrix. Spectral techniques, based on determining the Fiedler vector of the Laplacian or the adjacency matrix of a graph, are used in graph partitioning [18], matrix envelope reduction [3, 9], seriation [2], and graph coloring algorithms [1]. We introduce terminology and the background for this problem in § 2.

Applications of the above algorithms arise in diverse areas, such as parallel computing, VLSI, molecular dynamics, DNA sequencing, databases, clustering, linear programming, scheduling, and archaeological dating [1, 2, 5, 8, 10, 11, 16, 18]. (A computationally analogous problem, that of determining the second largest eigenvalue, has important applications in statistical and quantum mechanics as well [17].) We outline some applications in § 2.1 and § 2.2.

While spectral techniques for the algorithms mentioned above often give results of high quality, a major computational bottleneck is the determination of the Fiedler vector, which has often limited the application of spectral techniques. We observe that in most of the applications mentioned above, only an approximation to the Fiedler vector is required. Our work is directed at reducing the computational effort through the use of Monte Carlo (MC) techniques. We review MC techniques for computing eigenvalues and eigenvectors in § 3, and in § 4 we discuss ideas for using the properties of graphs to speed up the computations.

We present preliminary experimental results on the use of MC techniques to graph application in § 5. We conclude by suggesting directions for further research, in § 6.

P.M.A. Sloot et al. (Eds.): ICCS 2002, LNCS 2330, pp. 635–645, 2002.
© Springer-Verlag Berlin Heidelberg 2002

2   Background

The eigenvectors of the adjacency matrix and the Laplacian of a graph have important applications. The latter has especially been popular in recent times, and so we shall focus on it. It can be shown that all the eigenvalues of the Laplacian, L are non-negative, and exactly one of them is 0 for a connected graph. The eigenvector e_2 corresponding to the second smallest eigenvalue, λ_2, is called its *Fiedler vector*, and λ_2 is referred to as the *Fiedler value*. In the techniques mentioned in § 1, typically the Fiedler vector is determined and the vertices of the graph are characterized by appealing to the corresponding components of the Fiedler vector, as shown below for graph partitioning and seriation.

2.1   Graph Partitioning

Given a graph $G = (V, E)$ with n vertices, and number of partitions P, the most popular graph partitioning problem partitions the set of vertices V into P disjoint subsets such that each subset has equal size and the number of edges between vertices in different subsets is minimized[1]. Graph partitioning has applications in parallel computing, VLSI, databases, clustering, linear programming, and matrix reordering. This problem is NP-hard, and therefore various heuristics are used to find a good approximation.

The spectral heuristic determines the Fiedler vector of the Laplacian, and components smaller than the median are placed in one subset, say V_1, and the rest in subset V_2. Each of these subsets is recursively partitioned using the spectral method. The spectral method typically gives good quality partitions, but requires enormous computational effort for large graphs. Therefore other methods based on multilevel techniques have become popular, while there has been a simultaneous effort to improve the speed of the spectral method [14, 20].

2.2   Seriation

Consider a set of elements $1, 2, \ldots, n$ and a similarity function f. The seriation problem [2] problem is to determine a permutation π such that $\pi(i) < \pi(j) < \pi(k) \Rightarrow f(i, j) \geq f(i, k)$. Intuitively, a high value of $f(i, j)$ indicates a strong desire for elements i and j to be near each other, and through the permutation π we order the elements so that they satisfy the requirements of the function f. f may be such that it is impossible to find a permutation satisfying it; in that case, minimization of the following penalty function [2] can be attempted: $g(\pi) = \sum_{i,j} f(i, j)(\pi_i - \pi_j)^2$. This minimization is NP-hard, and so approximation techniques are used.

In the spectral method for this problem, a continuous version of the above penalty function minimization is solved. This leads to the following computational technique [2]: The Fiedler vector of the matrix corresponding to the similarity function f is determined. Then the elements are ordered based on the

---

[1] Metrics, other than the number of edges between different partitions, are also used in certain applications.

magnitude of the corresponding component of the Fiedler vector. Applications of this technique arise in DNA sequencing, archaeological dating, and also in a closely related problem of matrix envelope reduction.

3   Monte Carlo for Eigenvector Computations

We note that MC techniques are the ideal algorithms to use when approximate solutions are sufficient. MC computations of eigenvalues and eigenvectors are typically stochastic implementations of the power method [17].

3.1   Power Method

We now summarize the MC algorithm to compute the eigenvector for the eigenvalue of largest magnitude, based on the algorithm given in [6,7]. Consider a matrix $A \in \Re^{n \times n}$ and a vector $h \in \Re^n$. An eigenvector corresponding to the largest eigenvalue can be obtained through the power method as $\lim_{i \to \infty} A^i h$ for a starting vector h that has a non-zero component in the direction of the desired eigenvector. The MC techniques is based on estimating $A^m h$ using Markov chains of length m, for large m. The random walk visits a set of states in $\{1, \ldots, n\}$. Let the state visited in the i th step be denoted by k_i. Then the probability that the start state k_0 is α is given by

$$P_\alpha = \frac{|h_\alpha|}{\sum_{\alpha=1}^n |h_\alpha|}, \tag{1}$$

and the transition probability is given by

$$Pr(k_i = \alpha | k_{i-1} = \beta) = P_{\alpha\beta} = \frac{|a_{\alpha\beta}|}{\sum_{\beta=1}^n |a_{\alpha\beta}|}. \tag{2}$$

Consider random variables W_i defined as follows

$$W_0 = \frac{h_{k_0}}{P_{k_0}}, \quad W_i = W_{i-1} \frac{a_{k_i k_{i-1}}}{p_{k_i k_{i-1}}}. \tag{3}$$

If we let δ denote the Kronecker delta function ($\delta_{ij} = 1$ if $i = j$, and 0 otherwise), then it can been shown [7] that

$$E(W_i \delta_{\alpha k_i}) = (A^i h)_\alpha. \tag{4}$$

Therefore, in order to implement the power method, we evaluate $E(W_i)$ for large i to estimate the k_i th component of the largest eigenvector of A.

3.2   Inverse Power Method

If we need to determine an eigenvector for the eigenvalue of smallest magnitude, then we can apply a stochastic version of the inverse power method, where the desired eigenvector is computed as $\lim_{i\to\infty}(A^{-1})^i h$ for a starting vector h that has a non-zero component in the direction of the desired eigenvector. In the deterministic simulation, we repeatedly solve the following linear system

$$Ax_{k+1} = x_k, \quad x_0 = h. \tag{5}$$

In the MC technique, we replace the deterministic linear solver with a MC solver. There are several techniques for solving a linear system through MC. The basic idea is to write A a as $I - C$, and conceptually use a MC version of the stationary iteration

$$y_k = Cy_{k-1} + h = \sum_{i=0}^{k-1} C^i y_0, \quad y_0 = h, \tag{6}$$

which converges to the the desired solution if the spectral radius of C is less than 1. We can use the MC technique for computing Matrix-vector products, described in Sec 3.1, to determine each $C^i y_0$ on the right hand side of Eqn. (6). This corresponds to the MC estimator introduced by Wasow [19].

4   Improving the Convergence of Monte Carlo Techniques

We discuss ideas to accelerate the convergence of the MC techniques for the graph applications, in this section.

4.1   Deflation

The MC techniques are essentially the power or inverse iteration methods, and therefore, they too retrieve the extreme eigenvalues and eigenvectors. Since we wish to obtain the second smallest one, we deflate the matrix analytically. This can be done for both the adjacency matrix and the Laplacian. We demonstrate it for the Laplacian, since its use is more popular in recent times.

The smallest eigenvalue of the Laplacian is given by $\lambda_1 = 0$ with the corresponding eigenvector $e_1 = (1, 1, \ldots, 1)^T$. We wish to find a non-singular matrix H (See Heath [12], page 127) such that

$$He_1 = \alpha(1, 0, 0, \ldots, 0)^T, \tag{7}$$

for some constant α. Then

$$HLH^{-1} = \begin{bmatrix} \lambda_1 & b^T \\ 0 & B \end{bmatrix}, \tag{8}$$

where B has eigenvalues $\lambda_2, \lambda_3, \ldots, \lambda_n$. Furthermore, if y_2 is the eigenvector corresponding to eigenvalue λ_2 in B, then

$$e_2 = H^{-1} \begin{bmatrix} \alpha \\ y_2 \end{bmatrix}, \quad \text{where } \alpha = \frac{b^T y_2}{\lambda_2 - \lambda_1}. \tag{9}$$

(This requires $\lambda_2 \neq \lambda_1$, which is satisfied due to our assumption that the graph is connected.)

Of course, we wish to find an H such that B and e_2 can be computed fast. Furthermore, since most applications will have sparse graphs, we wish to have B preserve the sparsity. Consider H defined as follows:

$$h_{ij} = \begin{cases} -1, j = 1 \\ 1, i = j, \text{and } i > 1 \\ 0, \text{otherwise} \end{cases}. \tag{10}$$

Then H satisfies Eqn. (7) with $\alpha = -1$. Furthermore, $H^{-1} = H$ and we can show that in Eqn. (8)

$$b_{ij} = l_{i+1,j+1} - l_{1,j+1}, \ 1 \le i, j \le n - 1, \tag{11}$$

$$b_i = -l_{1,i+1}, \ 1 \le i \le n - 1. \tag{12}$$

We can verify that B too is sparse. The number of non-zero elements in L is $\sum_{i=1}^{n} d_i + n$, where d_i is the degree of vertex i. Let V_1 be the set of nodes to which vertex 1 does not have an edge (excluding vertex 1 itself), and V_2 be the set of vertices to which vertex 1 has an edge. The number of non-zero elements in column $i - 1$ of B is $d_i + 1$ if $i \in V_1$, and the number of non-zero entries is $n - 1 - d_i$ if $i \in V_2$ (since the j th row entry of B will be non-zero unless there is an edge between i and j, when $i \in V_2$). Therefore the total number of non-zero entries is given by:

$$\sum_{i \in V_1} (d_i + 1) + \sum_{i \in V_2} (n - 1 - d_i) \tag{13}$$

$$= \sum_{i \in V_1} d_i - \sum_{i \in V_2} d_i + (n - 2)d_1 + n - 1 \tag{14}$$

$$< \sum_{i=2}^{n} d_i + (n - 1)d_1 + n - 1. \tag{15}$$

If we label the vertices so that the vertex with least degree is labeled 1, then from Eqn. (15) we can see that an upper bound on the number of non-zero entries of B is given by

$$2 \sum_{i=2}^{n} d_i + n - 1. \tag{16}$$

Therefore the sparsity of B is decreased by at most a factor of 2 compared with just considering the the submatrix of L that omits the first row and column.

We also need to determine e_2 from y_2. We observe that with our definition of H, we get

$$e_2 = \begin{bmatrix} -\alpha \\ y_2 - \alpha \end{bmatrix}, \text{ where } \alpha = \frac{b^T y_2}{\lambda_2}, \text{ and } \boldsymbol{\alpha} = (\alpha, \dots, \alpha)^T. \tag{17}$$

In fact, if only the relative values of the components is important, as in the applications mentioned in § 1, then we need not compute e_2 explicitly, and can instead use the vector $(0 \ y_2^T)^T$.

4.2 Shifting

We can apply inverse iterations to the matrix B above, both in deterministic and MC techniques, in order to estimate λ_2, and, more importantly for our purpose, y_2. Alternatively, since solving linear systems repeatedly is expensive, we can shift B to enable computation of the desired solution as the eigenvector corresponding to the eigenvalue largest in magnitude. For example, the largest eigenvalue is bounded by $\Delta = 2\sum_{i=1}^{n} d_i$. Therefore we can apply the power method based MC technique to $\Delta I - B$ and obtain the eigenvector corresponding to eigenvalue $\Delta - \lambda_2$.

5 Experimental Results

In this section, we discuss the experimental set up, implementation details, and results of deterministic and MC solutions. We chose the graph partitioning application to test the techniques described above, and used two sample graphs for our tests – *test.graph* from Metis [15] and *hammond.graph* from Chaco [13]. The former is a small graph with 766 vertices and 1314 edges, while the latter, a 2D finite element grid of a complex airfoil with triangular elements, is of moderate size, having 4720 vertices and 13722 edges. All the coding was done on Matlab, and executed on a 1.0GHz Pentium III running Linux. We tested the effectiveness of techniques by partitioning the graphs into two disjoint parts. The ratio of the number of edges cut using our techniques to the number of edges cut using the exact Fiedler vector gives a measure of the effectiveness, with a large number indicating poor performance, and a small one good performance.

Since many of the techniques, such as deflation and shifting, apply to deterministic as well as MC techniques, we first present results demonstrating the effectiveness in deterministic techniques. We first computed the eigenvectors of the two largest eigenvalues of the sparse matrix $\Delta I - L$, using the *eigs* routine in Matlab. The time taken and number of edges cut using the Fiedler vector obtained from this step were used to measure the effectiveness of our techniques. In the rest of the tests, we divided the L matrix by Δ in order for all the eigenvalues of its deflation to be in $(0, 1]$.

We next tested the effectiveness of the deflation technique mentioned in § 4.1 with the deterministic inverse iterations. We could have started out with a random vector for the inverse iterations. However, a random initial vector would have a component in the direction of the the eigenvector for the 0 eigenvalue, which is not conducive to fast convergence. Hence we chose the initial start vector by assigning the first $n/2$ vertices of the graph to the same partition, and the rest to the other one, and then applied the deflation process to get a starting vector for the iterations. We can see from Fig. 1 that this starting vector is more effective than using a random vector, since our scheme leads to identical results

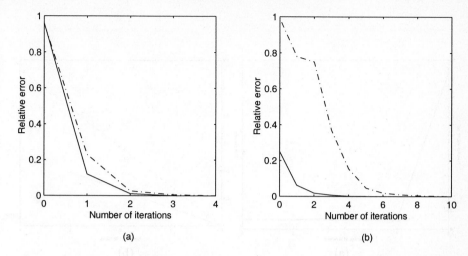

(a)   (b)

Fig. 1. Comparison of the effectiveness of using a random start vector versus our scheme. The solid line shows the relative error for our scheme, while the dash-dotted line shows the relative error for a random start vector with (a) *test.graph* and (b) *hammond.graph*.

with that of the exact Fiedler vector in fewer number of iterations of the linear solver[2].

We can see from Fig. 2 that deterministic inverse iterations give good results with just one linear solve, and agree with the exact Fiedler vector with just a few linear solves. Furthermore, the timing result indicates significant improvement over the direct technique.

We next show in Fig. 3 that deterministic power method applied to $I - B/\Delta$ too is effective. While the number of iterations required is very high, the time taken is low, since matrix-vector multiplication is much faster than solving a linear system. However, note that in the corresponding MC scheme, this would imply that each walk has to have a few hundred steps, and therefore the MC version of this method would not be effective. We therefore restrict our attention to the inverse iteration scheme for MC.

We show results for *test.graph* with MC inverse iterations in Fig. 4. The walk length was fixed at 5 for graphs a, b, c, and plots of edges cut versus number of iterations of the linear solver are presented, for different numbers of simulations. We make the following observations (i) as expected, the accuracy increases with increase in the number of simulations per linear solve, (ii) in terms of time taken (note, 100 iterations with 10 simulations each takes the same time as 1 iteration with 1000 simulations each), it may be useful to use fewer simulations per iteration in return for a larger number of linear solves, and (iii) the results

---

[2] The relative error here is the ratio of number of vertices in a partition inconsistent with that of the exact Fiedler vector, to the maximum that can be in the "wrong" partition.

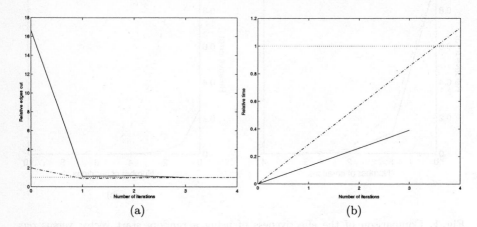

(a)   (b)

Fig. 2. Plot of edges cut and time using the deflated matrix, relative to that of the exact Fiedler vector, for different numbers of inverse iterations. The solid line shows the plot for *test.graph*, while the dash-dotted line plots the curve for *hammond.graph*.

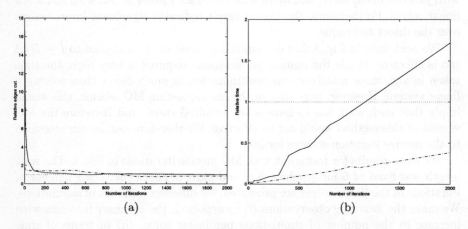

(a)   (b)

Fig. 3. Plot of edges cut and time using the deflation matrix, relative to that of the exact Fiedler vector, for different numbers of power method iterations. The solid line shows the plot for *test.graph*, while the dash-dotted line plots the curve for *hammond.graph*.

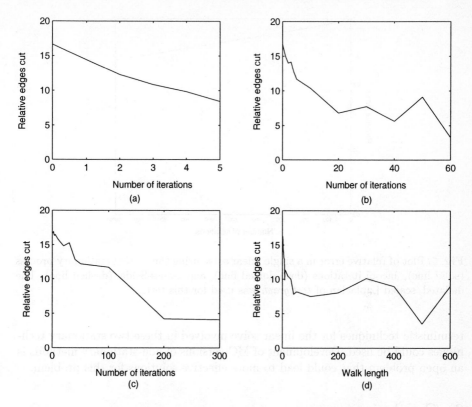

Fig. 4. Plot of edges cut using the MC inverse iterations. Graphs a, b, c use a walk length of 5 and number of iterations of 1000 (a), 100 (b), and 10 (c) respectively. Graph (d) keeps the number of iterations fixed at 3 and number of simulations per iteration at 100.

from the MC technique are not good enough. Graph (d) has varying walk length, holding the number of simulations fixed at 100 and number of iterations of the linear solver at 3.

A reason for the inadequate performance of the MC scheme is that the matrix C of Eqn. 6 has spectral radius close to 1 in our application. This can be improved by changing the stationary iteration scheme. For example, Jacobi iterations can easily be incorporated into the current MC framework. Fig. 5 shows that the relative error[3] in a single linear solve using (deterministic) Jacobi iterations can be much smaller than with the process we have used. However, we need to prove convergence of these methods for the graph under consideration, before using them. Furthermore, the MC methods corresponding to the better stationary schemes, such as Gauss-Seidel and SOR are yet to be developed. Meanwhile, hybrid techniques combining MC for the matrix-vector multiplication and de-

---

[3] The relative error here is $\|x - \hat{x}\|/\|x\|$, where x is the exact solution and $\hat{x}$ is the MC estimate.

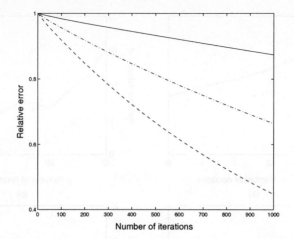

Fig. 5. Plot of relative error in a single linear solve using the current stationary process (solid line), Jacobi iterations (dash-dotted line), and Gauss-Seidel (dashed line). The deflated, scaled Laplacian of *test.graph* was used for this test.

terministic techniques for the linear solve involved in these two stationary techniques could be used. Development of MC versions of non-stationary methods is an open problem that could lead to more effective solutions for this problem.

6   Conclusions

We have outlined ideas for MC to be used in graph partitioning and related applications that require the estimation of the Fiedler vector of the graph Laplacian. We suggested techniques for improvement in convergence, for both deterministic and MC techniques. We demonstrated their effectiveness in deterministic calculations. However, the results with MC are not sufficiently good. We have identified the problem – the type of stationary process used is not suitable for graph applications, and suggested further research into the use of MC linear solvers that use better stationary schemes, or even non-stationary ones. Yet another research issue is to consider variance reduction techniques, for example, based on approximations to the Fiedler vector obtained using multilevel techniques as in [4].

References

1. B. Aspvall and J. R. Gilbert. Graph coloring using eigenvalue decomposition. *SIAM J. Alg. Disc. Meth.*, 5:526–538, 1984.
2. J. E. Atkins, E. G. Boman, and B. Hendrickson. A spectral algorithm for seriation and the consecutive ones problem. *SIAM Journal on Computing*, 28:297–310, 1998.
3. S. T. Barnard, A. Pothen, and H. D. Simon. A spectral algorithm for envelope reduction of sparse matrices. *Numer. Linear Algebra Appl.*, 2:317–334, 1995.
4. S. T. Barnard and H. D. Simon. A fast multilevel implementation of recursive spectral bisection for partitioning unstructured problems. *Concurrency: Practice and Experience*, 6:101–107, 1994.
5. Shashi Shekhar Chang-Tien. Optimizing join index based spatial-join processing: A graph partitioning approach.
6. I. T. Dimov and A. N. Karaivanova. Parallel computations of eigenvalues based on a Monte Carlo approach. *Monte Carlo Methods and Applications*, 4:33–52, 1998.
7. I. T. Dimov and A. N. Karaivanova. A power method with monte carlo iterations. In Iliev, Kaschiev, Margenov, Sendov, and Vassilevski, editors, *Recent Advances in Numerical Methods and Appl. II*, pages 239–247. World Scientific, 1999.
8. B. E. Eichinger. Configuration statistics of Gaussian molecules. *Macromolecules*, 13:1–11, 1980.
9. A. George and A. Pothen. An analysis of spectral envelope reduction via quadratic assignment problems. *SIAM Journal on Matrix Analysis and Applications*, 18:706–732, 1997.
10. D. S. Greenberg and S. Istrail. Physical mapping by STS hybridization: algorithmic stratgies and the challenge of software evaluation. *J. Comp. Biol.*, 2:219–273, 1995.
11. L. Hagen and A. Kahng. Fast spectral methods for ratio cut partitioning and clustering, 1991.
12. M. T. Heath. *Scientific Computing: An introductory survey*. McGraw-Hill, New York, 1997.
13. B. Hendrickson and R. Leland. The Chaco user's guide — version 2.0, 1994.
14. M. Holzrichter and S. Oliveira. A graph based davidson algorithm for the graph partitioning problem. *IJFCS: International Journal of Foundations of Computer Science*, 10:225–246, 1999.
15. G. Karypis and V. Kumar. A fast and high quality multilevel scheme for partitioning irregular graphs. Technical Report 95-035, University of Minnesota, 1995.
16. G. Karypis and V. Kumar. A fast and high quality multilevel scheme for partitioning irregular graphs. *SIAM Journal on Scientific Computing*, 20:359–92, 1998.
17. M. P. Nightingale and C. J. Umrigar. Monte Carlo eigenvalue methods in quantum mechanics and statistical mechanics. In D. M. Ferguson, J. I. Siepmann, and D. G. Truhlar, editors, *Monte Carlo Methods in Chemical Physics*, volume 105 of *Advances in Chemical Physics*, chapter 4. John Wiley and Sons, New York, 1999.
18. A. Pothen, H. D. Simon, and K. Liou. Partitioning sparse matrices with eigenvectors of graphs. *SIAM Journal on Matrix Analysis and Applications*, 11:430–452, 1990.
19. W. Wasow. A note on the inversion of matrices by random walks. *Mathematical Tables and Other Aids to Computation*, 6:78–78, 1952.
20. C. Xu and Y. Nie. Relaxed implementation of spectral methods for graph partitioning. In *Proc. of the 5th Int. Symp. on Solving Irregular Problems in Parallel (Irregular'98), August 1998, Berkeley, CA*, 1998.

Peer-to-Peer Computing Enabled Collaboration

Martin G. Curley

Director, IT People, Intellectual Capital and Solutions,
Intel Corporation,
Leixlip, Co Kildare, Republic of Ireland
Martin.G.Curley@intel.com

Abstract. This paper discusses how peer-to-peer computing is emerging as a disruptive technology for global collaborative solutions. It explains how peer-to-peer computing can enable new collaborative solutions while significantly decreasing IT costs and improving IT asset utilization. An overview of the technology and usage models are discussed whilst the benefits are illustrated through a short case study from Intel. Finally the value proposition for peer-to-peer computing is summarized.

Introduction

Today's workforce faces unprecedented challenges and competitive pressures. Employees need to learn faster and colleagues often need to collaborate cross-functionally and globally, to create new products and achieve operational excellence.

In parallel, information technology is creating new value propositions, which can help businesses solve these problems. For example, Knowledge Management software is helping corporation's to share and leverage knowledge better. The combination of motivated knowledge workers and applied information technology is creating what might be called a bionic organization—one that delivers turbocharged output, with the whole significantly greater than the sum of the parts.

Global Companies like Intel, which have sites distributed across the world rely on global computing and collaboration tools to conduct business. In fact it could be said that IT infrastructure and products and services are the air that modern day global enterprises breathe. There are many factors that influence the effectiveness of global corporations; key modulators include knowledge, skills, desire and collaborative friction. In fact one could postulate that an organization's potential output is somehow related to the product of these variables, i.e.

P.M.A. Sloot et al. (Eds.): ICCS 2002, LNCS 2330, pp. 646–654, 2002.

Maximising Organization Output = Knowledge * Skills * Desire
* (1 – Collaborative friction) * other relevant factors. $\qquad$ (1)

To maximize output and quality from an organization, employees must have the right knowledge and skills to perform their jobs and they must have the ability to share and leverage knowledge easily. They of course must be highly motivated. Geographic and time zone separation are barriers which create collaborative friction. Leo Ellis captures a good example of low collaborative friction in his quote *"Engineers have yet to devise a better inter-office communication system than the water cooler"*. In many instances application of information technology is used in an attempt to reduce organizational collaborative friction. Peer-to-Peer computing is emerging as a very interesting computing platform to help lower collaborative friction.

Defining Peer-to-Peer Computing

In it's narrowest definition Peer-to-Peer Computing is fundamentally about the sharing of resources amongst computers. A broader definition of peer-to-peer is that it is an emerging class of applications, which leverages resources at the edge of a network. These resources include people, knowledge, information, network, computing power, storage. Whatever the definition, Peer-to-Peer computing is emerging as a disruptive technology for both collaboration applications and the underlying IT infrastructure.

It is a disruptive technology for collaboration applications as it is enabling the introduction of new capabilities which were previously cost-prohibitive to implement.

It is a disruptive technology for IT infrastructure as it can significantly increase the utilization of existing IT assets and improves the performance of supported applications.

Intel ® Share and Learn Software

Intel ® Share and Learn Software (INTEL SLS) is a good example of a peer-to-peer computing solution. It is essentially a peer-to-peer file sharing application, which helps solve multiple business problems while leveraging existing IT infrastructure.

INTEL SLS runs on corporate Intranets and using it's advanced content networking solution it can be used for applications such as eLearning and Knowledge Management. It delivers these capabilities while reducing information technology costs by moving network traffic off wide-area networks to much cheaper local area networks and by leveraging inexpensive PC storage. Using INTEL SLS each PC in an enterprise network can be converted to a caching device, significantly increasing performance for certain file transfer operations.

The foundational elements of INTEL SLS are a content index, and content publishing, management and distribution applications. A database server and web server provide a set of services to client users who use the web for retrieving and viewing content while they are connected to a corporate Network. They can use an offline player when they are not connected to the network.

INTEL SLS is content agnostic, meaning that it can manage and distribute a diverse type of files. It is particularly useful for the distribution of rich content files as it allows near seamless propagation of large files throughout a network using a peer-to-peer distribution capability.

The following sections discuss how INTEL SLS and peer-to-peer computing are enabling new collaboration capabilities at Intel Corporation.

Enabling New Collaboration Applications

Knowledge Management

One significant opportunity cost for corporations is that much codified knowledge exists in the form of presentations and text documents stored on individual's PC hard drives. In conventional networks this information and knowledge is not available to others for sharing and reuse.

INTEL SLS introduces a capability where any file can be shared with many others in a corporation, through a personal publishing function. The INTEL SLS Personal Publisher allows the filename and relevant metadata to be easily registered in a index which other users can search based on criteria such as subject, keywords etc. If a user wishes to retrieve the file, the system searches for the closest located copy of that file and initiates a peer transfer from the closest client. The file returned is always the latest version.

This is important, as in effect; an enterprise knowledge management system is being created, leveraging previously unavailable information and knowledge at the edge of a corporation. This creates new opportunities through knowledge sharing and re-use, whilst saving money through using in-expensive client storage. And uniquely to peer-to-peer solutions, the more the system is used the better the performance improves.

Intel itself is using INTEL SLS for its Intel Notebook of Knowledge (INK) program. This program captures codified IT knowledge in the form of white papers, videos and presentations. INTEL SLS using its peer-to-peer services acts as integrated object repository delivering a requested object from the closest client. Using INTEL SLS publisher, any employee can add content from their hard drive to the central index. All files are catalogued using the Dublin Core tagging standard.

e-Learning

The use of rich content, presented in a compelling and immersive fashion can significantly enhance learning effectiveness. Rich content uses technology to blend text, graphics, sound and motion to deliver information. Enabled by new technology, rich content can deliver knowledge and information in a form that enhances understanding and retention by providing emphasis and context. This concept is well supported by the statement of Prof. Fred Hofsteter, Director, Instructional Technology Center, University of Delaware.

"People remember 20% of what they see, 30% of what they hear, 50% of what they see and hear and 80% of what they see, hear and do simultaneously".

Couple rich content with the efficiency of e-Learning, where content is recorded once and used many times (instead of the each-time requirement for facilities, trainer and materials of conventional training) and it creates a win-win scenario. INTEL SLS enables the delivery of rich content and large compressed video files almost seamlessly over an enterprise network.

The Intel IT organization recently converted its new hire integration program to a fully virtual e-Learning program, delivered to new Intel IT employees using Intel SLS. The integration program consists of sixteen modules including technical, business, customer service and interpersonal skills, which new employees take over a six-week period. The net result was that integration satisfaction survey scores improved to 88% from 77% whilst significant travel and training cost savings were achieved. The content is reusable and other Intel IT employees are taking these classes at no extra cost.

Corporate Communications

The use of Oral History is becoming increasingly important in Corporate Communications. Recording a short video clip is often a more effective and efficient method of communication rather than taking the time to write down content and then distribute it via mail or paper memo. Dave Snowden, Director of the Institute of Knowledge Management in EMEA says *"I always know more than I can tell, and I always know more than I can write down".* Hence large corporations are increasingly using short video clips for transfer of key knowledge/messages to employees worldwide.

INTEL SLS introduces the capability to distribute these video clips almost seamlessly throughout the enterprise with no additional bandwidth or infrastructure required. 20-40 MB files download in less than a minute rather than the 20 minutes, which you might typically see across a wide area network. This performance is achieved through the local caching that INTEL SLS provides, thus avoiding multiple wide area network downloads.

Intel Senior Vice-presidents like Mike Splinter and Sean Maloney are using INTEL SLS as another channel to deliver monthly video updates to their teams worldwide. The video quality is close to broadcast quality and conveys key monthly messages and challenges in short clips.

The Corporate IT infrastructure equation

At a macro level IT computing infrastructure costs in a large corporation can be described as the sum of the costs of providing the compute-network-storage platform as well as the costs of managing this environment.

$$\text{Corporate IT Platform (Performance \& Cost)} = \text{Sum (Computing, Storage, Network, Management) Performance \& Costs} \tag{2}$$

A primary function of a Corporate IT organization is to optimize this equation to ensure the best performing corporate computing platform is delivered at the lowest possible cost. For Corporate IT organizations the cost of providing and supporting the enterprise platform continues to escalate which users continually demand better solutions and performance. These user and Corporate IT goals are often conflicting.

Using peer-to-peer computing protocols, utilization of existing corporate IT assets can be increased and equivalent or better performance can be delivered at lower cost. This also allows implementing of new capabilities faster and cheaper than implementing new systems. This is essentially due to the fact that Peer-to-Peer computing allows the substitution of cheaper components for more expensive computing components in the Corporate Infrastructure equation. In-expensive client hard-drives can be used for storing information instead of more expensive network or server attached storage, Computing power on high performance PC's can be used instead of more expensive Server computing power and network traffic can be shifted from expensive Wide Area Networks (WAN) to Local Area Networks (LAN). Peer-to-Peer Computing will likely co-exist as a complimentary computing platform to client-server.

In order to assess the possible impacts of Peer-to-Peer Computing on corporate infrastructure, Intel's IT organization ran a number of internal trials within Intel.

IT trials results

In conjunction with the Intel training organization, Intel ran an internal trial using an eLearning class called American Accent, which consisted of about 60 modules of file sizes ranging from 5 MB to 15 Mbytes. In a trial with 1900 participants in over 50 distributed sites, users of this eLearning saw an average improvement of 4-5X in file delivery times whilst over 80% of the 3000+ files transferred were peer-to-peer transfers which took place over the LAN rather than the WAN. Figure one shows a comparison of transfer times for a large file using a peer-to-peer application compared to a traditional client-server environment.

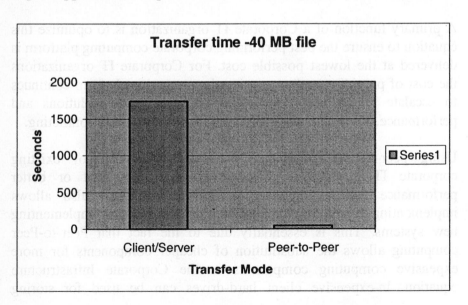

A complimentary driver for the adoption of peer-to-peer computing is the relentless march of Moore's law. Instead of streaming video over a network, an alternative is to compress a video, perform a peer-to-peer transfer, decompress the video and then play it locally on a PC. As computing power increases and CPU architectures become smarter, decompression times are decreasing rapidly – for example a compressed 40 Mbyte video file that took approx 40 seconds to decompress on a 700 MHZ Pentium ® III processor based PC now takes approx 5 seconds to decompress on a new 2.2 GHz Pentium ® 4 processor based PC.

In an another experiment with the distribution of a video speech by Intel's CEO to employees worldwide, multiple thousands of copies of this video were distributed to employees worldwide with greater than 90% of the transfers being peer-to-peer, significantly decreasing file delivery time while lowering networking costs.

Based on the trial results Intel is aggressively adopting Peer-to-Peer computing solutions for collaboration solutions.

Peer-to-Peer Value Proposition

Peer to Peer computing is emerging as a great opportunity for enterprises to take advantage of latent potential in enterprise infrastructure whilst delivering new capabilities and better performance. The adoption of Peer-to-Peer computing drives benefits in the following area.

- Improved Performance
- Lower IT Infrastructure Cost
- Improved IT Asset Utilization
- Smarter Dynamics of Enterprise Systems
- Improved Fault tolerance and scalability
- Ability for more flexible deployment

As the security models of peer-to-peer computing solutions improve, this new computing model is destined to play a significant role in global collaborative solutions in the Twenty First Century.

Conclusion

The new computing model that the peer-to-peer era is ushering in, creates opportunities to enhance collaborative solutions and thus lower collaborative friction to help deliver improved organization performance. In addition the ability of the peer-to-peer computing model to allow better integration of different computing components to deliver improved performance, lower platform costs and better IT asset utilization make it rather unique in terms of computing models. In fact, peer-to-peer computing just may be one of the best-kept computing secrets of the late Twentieth Century.

Working Towards Strong Wireless Group Communications: The Janus Architecture

J. S. Pascoe†‡, V. S. Sunderam‡ and R. J. Loader†‡

†Department of Computer Science
The University of Reading
United Kingdom
RG6 6AY
J.S.Pascoe@reading.ac.uk
Roger.Loader@reading.ac.uk

‡Math & Computer Science
Emory University
Atlanta, Georgia
30322
vss@mathcs.emory.edu

Abstract. Strong wired group communication systems have been employed in numerous applications, often characterised by some degree of *replication*. With the development of wireless technologies, a new series of group applications is emerging and due to the effectively asymmetric characteristics of these networks, existing group communication systems can not be employed directly. Differences between wireless and wired networks in terms of connectivity, latency, bandwidth and other characteristics necessitate alternative algorithms and mechanisms, especially in group communication protocols. In particular, protocols must *adapt* to highly variable network characteristics in the wireless domain, whilst providing standard group communication semantics and delivering traditional qualities of service. In addition, *arbitration* between wired and wireless communication entities is required to reconcile the order-of-magnitude (or more) difference in network parameters between the two.

The Janus project is developing a strong group communications framework for use in wireless environments. It is envisaged that the findings from our research, as well as the software that results, will be of value in a number of emerging domains such as collaborative computing and multiway application sharing that are moving from fully wired to hybrid wireless and land-line networks.

Keywords – Janus, group communication, wireless networking, preemptive adaptation, trouble spots, failure resilience.

1 Introduction

The use of mobile devices such as network-enabled PDA's and notebooks is rapidly growing and becoming a mainstream mode of portable computing and communications. Although some platforms are proprietary (e.g. Palm VIIx), current trends are towards wireless infrastructures supporting IP, facilitating widespread deployment of traditional applications. For example, notebook computers on 802.11 networks are commonplace within buildings, and wireless infrastructures supporting PDA's and hand-helds are becoming increasingly prevalent in metropolitan areas. In these usage scenarios, standard (wired network based) applications execute without the need for modifications to the application or the protocol. Currently, typical applications are mail, web browsing, telnet etc., that are not *substantively* affected by the latency, bandwidth, and connectivity problems that characterize wireless networks. In other words, the pairwise and client-server nature of common applications can tolerate the lowered quality found in wireless

P.M.A. Sloot et al. (Eds.): ICCS 2002, LNCS 2330, pp. 655–664, 2002.
© Springer-Verlag Berlin Heidelberg 2002

networks e.g. a user experiencing lack of coverage while moving in a hallway simply reloads a web page as soon as coverage is re-established, very much like reloading a stalled page on a wired workstation.

However, a number of other current and emerging applications are much more sensitive to the inherently low bandwidth, high latency, and most of all, intermittent connectivity often encountered in wireless networks. This is especially true of applications based on *strong* group protocols, such as collaborative data and document sharing, server-less chat, application multiplexing, and multiway audio. In multi-party protocols, group membership operations are highly sensitive to actual or perceived *joining* and *leaving* by members. Mobile devices encountering areas of disconnection can cause very expensive group operations to be erroneously triggered. Similarly, in several multicast and group communication modes, marked disparities in delivery latencies or bandwidths between different group members can result in severe protocol degradation. For example, a series of reliable multicasts can result in buffer overflows at land-based nodes due to long acknowledgment delays from even a single wireless node. At least in the near future, this large difference in quality between wired and wireless networks is expected to prevail. Therefore, new approaches to providing protocol support for multiway applications in hybrid networks is of crucial importance.

This paper is based on the premise that transport quality in wireless networks is dependent on spatial and environmental factors, i.e. regions within the coverage area that are susceptible to network degradation either due to their location, or due to disturbances (thunderstorms or radiation from a CRT monitor). However, knowledge of these *trouble spots* (or the assumption of their existence) can be used by protocols to preemptively adapt, correct, or otherwise account for the lowered quality when a mobile device encounters such an area. Knowledge of trouble spots is based on both static information gained from prior visits to the area as well as by real-time monitoring of the network from within each mobile device. Adaptation can take many forms, ranging from increased buffering to dynamically changing the QoS of a group communication channel. This project is developing methodologies and protocols, that will support adaptive group communications in hybrid wireless and wired networks.

2   Background

A strong model of group communication differs from weak counterparts by including a formalized notion of *membership*, either through an explicit *group membership service* or as an element of the distributed state. An example of a strong group communication system is ISIS [3] whereas an example of a weak model is IP-multicast [5]. ISIS was a general group communications platform, first proposed by Birman, that introduced the notion of *Virtual Synchrony* [4] and in doing so was successfully employed in over 300 projects. The original virtual synchrony assumed a *primary partition* model and as such, was vulnerable to problems arising through network partitions. This issue inspired several new projects which aimed to address partitionable group communications; two of the more notable being *Totem* [11, 1, 10] and *Transis* [6]. The Totem group developed a new model of virtual synchrony, termed, *Extended Virtual Synchrony* which essentially permitted processes to operate in minority partitions, providing that for a given partition, consensus could be formed on its membership. The Transis system, due to Dolev and Malki, provided a large scale hierarchical multicast service that was tolerant of network partitions and leveraged IP-multicast. Transis contributed extensively to group communications theory, it introduced a number of significant results and later inspired parts of Horus [15], the successor to ISIS. The Horus framework is based around the notion of *protocol composition*, that is, Horus provides a catalog of components that can be quickly composed to provide tailored group communications solutions. This 'building block' approach not only provides genuinely re-usable solutions to group applications, but also supports additional project phases e.g. formal verification. Current projects such as InterGroup [2], Spinglass and Ensemble [14] are exploring the issues of scalability and security in group communications.

3   Architecture

The envisaged model consists of two sub-groups of communicating entities: M entities on wired hosts (that form a group in the traditional sense) and N entities in the wireless domain. Periodically, both wired and wireless hosts may join and leave a *session* or heavyweight group, and within a session, they may join and leave lightweight communication *channels* that support different levels of Quality of Service. Pragmatically, this interaction is facilitated through Janus[1] like *proxies*, that arbitrate between the asymmetric characteristics of the two networks. Thus, a Janus group consists of M wired hosts, N wireless hosts and P proxies.

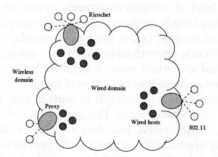

Fig. 1. The envisaged architecture.

The prototype implementation of Janus takes the form of a UDP-based user space protocol. Subsequent projects will then be able to leverage Janus through an API, that is, applications simply link against the Janus libraries and avail the services. In designing Janus, we target three levels of device (PDA, laptop and stationary workstation), just as we target two types of network (802.11 and metropolitan infrastructures). The prototypical work is Linux based and in order to use the system, Janus proxies (configured for the required application) are instantiated on fixed machines in the wired domain. Each wireless device then connects to a proxy and in doing so empowers the proxy to transfer messages, handle synchronization and perform any other function required in the wired domain on its behalf. In sections 4 and 5 we discuss the major challenges that the project is striving to overcome and present an outline of the projects technical direction.

---

[1] Derived from the two-headed Roman god of civil order.

4   Challenges

Preliminary experimentation into the Janus concept has shown that the proposed work is feasible and should contribute a novel framework that will have application in several areas of research. In the process, our experiments have highlighted a number of research issues which are central to the projects foundation. Thus, the fundamental areas of research are described below. It is noteworthy to add, that the maintenance of the *virtual synchrony* semantics is considered paramount in Janus. The virtually synchronous model of group communication has been widely adopted, and as such, it is critical that Janus does not infringe or invalidate this notion both in terms of retrofitting the framework and also leveraging prior research in new projects.

4.1   Wired and Wireless Network Arbitration

One of the key questions being addressed by the project is the notion of arbitration between the wired and wireless domains. Wired and wireless networks roughly exhibit asymmetric characteristics, that is, hosts connected to a wired network can exchange messages through low latency, high bandwidth channels, whereas, wireless networks exhibit a much higher degree of latency and currently provide only a fractional proportion of the bandwidth. Also, it is not possible to make assumptions related to the availability of a wireless host. Experience has shown that wireless networks are riddled with small pockets of non-communication, that is, as a mobile user moves throughout the network, a certain level of intermittent connectivity may be experienced. Preliminary experimentation has largely attributed this effect to geographic phenomenon [8, 12], however, other more dynamic factors (such as the weather) can also cause mobile devices to become arbitrarily uncontactable for *a priori* unknown periods of time. In contrast, hosts connected to a wired network are often highly available with loss of contact being considered an immediate indication of failure.

Given these differences, we postulate that in order to span the two domains, it is necessary to introduce an arbitration entity. This *proxy* will mediate the transfer of messages between the two environments, providing a highly available participant in the wired domain whilst sympathetically adapting to the intermittent connectivity of the wireless device. Note that this differs from the traditional role of a proxy. In other scenarios, proxies are employed as buffers or caches to increase the perceived availability of a resource (e.g. web caching). In Janus, the notion of a proxy is more complex i.e. a Janus proxy will synchronize, co-ordinate and perform required actions in the wired domain for n wireless hosts simultaneously.

	Wired – l	Wired – r	802.11 – s	802.11 – m	Metro. – s	Metro. – m
PL (%)	0	0.2	0	1	0	0.4
L (ms)	0.2	71.4	1.724	2.721	189.3	241.4
BW (KB/s)	7773	48.1	637	524	16.4	13.5

Table 1. Quantitative differences between wired and wireless networks. The packet loss, latency and bandwidth measurements are the average results from sending 1000 ICMP packets across the Internet. The abbreviations *l, r, s* and *m* equate respectively to *local, remote, stationary* and *moving*. Note that the metropolitan area wireless network measurements were made using the Ricochet infrastructure in clear reception areas.

	PDA	Laptop	Desktop
Mobility	very high	less than PDA	not mobile
Network	wireless	wireless & wired	wired
Memory size	$\approx 32Mb$	32 Mb – 256 Mb	≈ 1 Gb
Processor	$\approx 233Mhz$	$\approx 750Mhz$	$\approx 1.5Ghz$
Storage	minimal	abundant	abundant
Battery issues	✓	✓	×

Table 2. Typical characteristics of devices in a Janus group.

Janus is envisioned for use in group applications where a range of mobile devices are required to interface to a network of stationary wired hosts. With the advent of metropolitan area wireless networks and sophisticated mobile computers, the notion of a ubiquitously connected mobile device is becoming more realistic. However, we postulate that mobile computing and communications capabilities will continue to be somewhat different from (lower than) land based services for the near-to-medium term (see tab. 2). This imposes a number of constraints on the Janus client interface. Due to limitations of processing speed, memory and storage, it is not feasible to execute large group communication applications (intended for desktop use) on small portable devices. Indeed, experimentation has shown that a commercial off the shelf Personal Data Assistant can comfortably support a graphical user interface plus minimal amounts of processing. However, such devices are not capable of running large environments or physical simulations. Note, that this is not a general observation; for more powerful mobile computers (e.g. laptops), it is beneficial to utilize the local resources where possible. For this reason, a later project phase will enhance the client component of the Janus framework to support an element of adaptability, migrating processing load where necessary but harnessing local resources where available. This will provide a level of 'future-proofing' as mobile devices develop.

4.2 Fault-tolerance

Traditional fault-tolerance in group communication systems has been based around the reasoned assumption of a *fail-stop* model of failure, that is, once a host has suffered a failure, it does not participate in the group any longer (although it may rejoin as a new member). In a wired environment, this is a justified supposition since the majority of failures are due to host crashes or network partitioning. However, in a wireless environment, mobile hosts are susceptible to *intermittent connectivity* and so a more insightful model of failure is required.

As noted above, our preliminary experimentation with metropolitan area wireless networks revealed that there are pockets of non communication scattered throughout the network. Trouble spots can be categorized as either *static* (which are typically due to geographic phenomenon) or *dynamic* (intermittent / atmospherically related). In either case, the user can experience temporary disconnects (lasting for arbitrary lengths of time) and message garbling (due to problems such as echoing and high path loss [13, 7, 9]). Traditional failure detectors have been based on heartbeat mechanisms viz. each host periodically probes the liveness of the other hosts in its view. Should a heartbeat message remain unacknowledged for a threshold period of time, then this indicates a failure and a membership algorithm is invoked to restore the health of the session. When tested, the use of a heartbeat mechanism in the wireless domain results in a large number

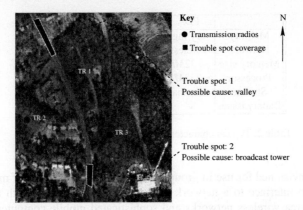

Fig. 2. A geographic view of trouble spots along a road in Atlanta. The trouble spots shown here are typical areas of non communication. When traveling at 35 MPH, they cause a sharp, relatively short drop in communication (that lasts for ≈ 2s). The stretch of road shown here is ≈ 1 mile.

of expensive membership operations i.e. as a host enters a trouble spot, it appears to the remainder of the group as though it has failed. Subsequently, the host emerges from the trouble spot, becomes recontactable only to find that it has been excluded from the group. The only option is for the device to rejoin as a new user, which is both costly and unnecessary. Note, that the situation is exasperated because disconnections can occur for other reasons (e.g. the user deactivates the device to conserve battery power). Again, this will appear to the system as a failure when in fact, it is a calculated action. In summary, traditional failure detectors are unable to differentiate between failed hosts and uncontactable hosts when employed directly in the wireless domain.

It is noteworthy to add, that we do not consider power saving actions or temporary loss of communication due to trouble spots to be anomalous events. Moreover, we postulate that this pattern of interaction should be considered the norm. For the same reasons that Janus opts to preserve virtual synchrony semantics, it is necessary to maintain the illusion of a fail-stop model of behavior in the wired domain. However, Janus will simultaneously exhibit a more sympathetic approach to the wireless pattern of behavior.

4.3 Synchronization at the Wireless Level

As with all asynchronous environments, processes are free to deliver messages at arbitrary rates. In the wired domain, delivery guarantees and virtual synchrony ensure that a session remains synchronized with respect to a logical ordering of events. However, under wireless conditions where latency is orders of magnitude higher, it is quite possible that a disparity can build between users. Consider the example of Jane, Bob and Paul, who all wish to hold a collaborative computing session using wireless hand-held computers on the way to work. Both Jane and Bob are lucky, in that they experience constant uninterrupted wireless connectivity. Paul on the other hand is not so lucky and encounters a number of trouble spots on his route through the network. Although Paul's messages are not necessarily lost, his mobile device delivers messages that were seen previously by Jane and Bob. If left unchecked, this disparity can build to the point where users perceive events at different times. In this example, it could be argued that the application can inherently tolerate a certain level of disparity, however, if Jane, Bob and Paul had been police officers co-ordinating a public order operation, the consequences could be more pronounced. As we envisage frequent wireless disconnections, it is worthwhile to maintain an asynchronous model of communication (i.e. where devices transfer messages when contactable) since a completely synchronous approach would restrict the throughput of all wireless hosts to the rate of the slowest. However, where an explicit end-to-end delivery guarantee is required, the system is forced to wait until the mobile device acknowledges the message.

4.4 Authentication

Secure group communication is required in any group application that involves exchanging sensitive data. Depending on the required level of security, systems may adopt different approaches. One model of security is to screen admission to the group but to relax security once admitted. In terms of Janus, the fundamental authentication issue is in relation to the envisaged model of operation. In the wired domain, failed hosts typically have to re-authenticate when re-joining the session, which, although being expensive is mitigated by the fact that wired hosts are generally more reliable than wireless devices. However, in the wireless network, it is conceivable that an unauthorized device could pose as an authorized device which is in a trouble spot. Thus, in Janus, authentication must not only be performed on entry to the system, but each time a device returns to connectivity. However, full authentication imposes a significant overhead which, in combination with high latency and frequency of trouble spots, can cause excessive delays. Note also, that security requirements vary i.e. a single solution can not suit all applications. Thus, we propose to provide facilities for end users to integrate either standard or custom security schemes in which initial authentication is of a high level, but reconnections are less security intensive.

5   Technical Approach

Through initial experimentation, we believe that the Janus framework is not only feasible, but will provide a practical wireless foundation that can be leveraged into a plethora of emerging projects. Moreover, we consider the goals of the Janus project to be achievable by evolving solutions to the challenges outlined above. In this section, we present a technical overview of the Janus framework, but for brevity, we limit the discussion to the issues of arbitration and synchronization.

5.1   Arbitration: The Janus Proxy

Proxying is a well known distributed systems approach that has been applied in a range of applications, traditionally to increase the perceived availability of some resource. However the projects literature survey provided a level of confidence that the application of proxies in this context is an effective and novel model for addressing the issues discussed above.

The proxy we envisage is diagrammatically expressed in fig. 3. In the wired domain, the proxy executes a group communication process (referred to as the *fixed process*), native to the target application. Through a reliable point-to-point wireless protocol, a mobile device can contact the proxy and request admission to the group. Providing that the device is authenticated appropriately, a fixed process is instantiated in the wired domain

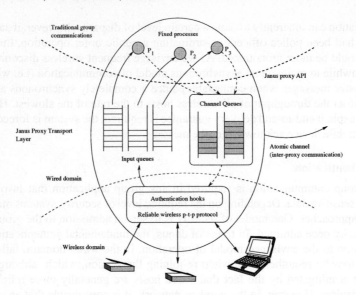

Fig. 3. An architectural view of a Janus proxy.

and the required channels are initialized. Note, that this approach offers two notable advantages, firstly, because the fixed process is viewed as any other wired participant, the semantics of the group communication system are maintained and, secondly, each mobile device becomes individually addressable. Once part of the group, the mobile device can exchange messages with the wired participants through a number of queues. On group admission, each wireless host is provided with an *input queue* viz. a queue of messages transmitted by the mobile device that are in turn presented to the fixed process for wired transmission in the manner native to the target application. Intuitively, one would expect a second queue to be present to pass messages from the wired domain to the mobile device. However, in multichannel group communication architectures, we suggest that it is more effective to store one queue of messages per *channel* and for the proxy to maintain a pointer into the list for each mobile host that is a member of the channel. It is noteworthy to add that another reason this method was adopted was to facilitate a level of synchronization at the wireless host, and as such, this discussion is expanded upon in section 5.2. Finally, when a wireless host wishes to leave the session, the corresponding fixed process is instructed to leave, the applications native leave protocol is executed and the process is terminated along with any redundant queues.

5.2 Synchronization

We postulate that virtual synchrony is prohibitively expensive to apply directly in the wireless domain, unless delivery ordering guarantees are specifically required. However, an element of synchronization is necessary at the wireless level to maintain consistency (at least in terms of user perception). In Janus, two levels of synchrony are supported, but for brevity, only proxy level synchronization is discussed.

Single Proxy Synchronization The Janus proxy facilitates message exchange through queues that are instantiated on a per channel basis and represent the mobile hosts progress with pointers. These are denoted as $current_c^i$ or CU_c^i where i refers to the host and c is the channel. Two additional pointers store the edges of a *synchronization window* for each channel: UE_c for the upper edge and conversely LE_c for the lower edge. Note that we define the window size (WS) in terms of the windows upper and lower edges i.e. $\forall c \in channels \bullet WS_c = UE_c - LE_c$. This is necessary to prevent the synchronization algorithm from deadlocking. Once all hosts have acknowledged a message, both the lower edge and the upper edge pointers are incremented. Thus, messages that occur before LE_c in the message queue are guaranteed to have been received by all mobile hosts connected to a given proxy. Mobile devices are permitted to read messages asynchronously providing that a host i does not reach the upper edge of the window whilst a host j is at the bottom. Thus, the *lag* for a particular channel queue is the difference between the host that has progressed the furthest through that queue and the host that has progressed the least (i.e. $\forall c \in channels \bullet lag_c = max(CU_c) - min(CU_c)$). The event where lag_c becomes equal to WS_c indicates that a *forced resynchronisation* is necessary to maintain consistency amongst the wireless hosts. Resynchronisation is performed by initially identifying those hosts that are at the bottom of the message list and placing them in a set called *slow*. For each host in slow, a query is passed to the proxy in question to see if the host is in a known trouble spot or if it has just emerged from a trouble spot (and so is catching up). Those hosts for which the latter is the case, remain in slow, but all others are moved to a set called *suspect*. At this point, the upper edge of the window is increased to allow operational hosts to continue normal operation i.e. functional hosts are not constrained to the speed of the slowest and do not have to wait for failures to be confirmed. Thus, a fixed window size would not allow the algorithm to iterate and so the proxy would deadlock. On the next resynchronisation, hosts are again classified as slow or suspect. If the number of successive iterations in which a host has been classified as slow exceeds a threshold thr_1, then that host is

---

Algorithm 1: Single Proxy Synchronization
Code for proxy P_i with respect to an arbitrary channel c

Initially $WS_c = WS'$ $(initial window size), thr_1 = slow threshold, thr_2 = suspect threshold$

1: while $(lag_c \neq WS_c)$ // Everything is nominal
   2: w_i reads a message $\Rightarrow CU_c^i = CU_c^i + 1$
   3: $\neg (\exists w_i \in hosts \bullet w_i$ is a member of $c \wedge CU_c^i = LE_c) \Rightarrow$
      4: $(slow \cup suspect = \emptyset \wedge WS > WS') \Rightarrow LE_c = LE_c + 1$ // Shrink window
      5: $(slow \cup suspect = \emptyset \wedge WS = WS') \Rightarrow UE_c = UE_c + 1 \wedge LE_c = LE_c + 1$
6: $(lag_c = WS_c) \Rightarrow$ // Force synchronization
   7: $slow = slow \setminus \{ w_i \mid w_i \in hosts \wedge w_i \in slow \bullet CU_c^i = UE_c \}$
   8: $suspect = suspect \setminus \{ w_i \mid w_i \in hosts \wedge w_i \in suspect \bullet CU_c^i \neq LE_c \}$ // Prune normal hosts
   9: $slow = slow \cup \{ w_i \mid w_i \in hosts \bullet CU_c^i = LE_c^i \}$
     10: $\forall w_i \in slow \bullet count(w_i, slow) = thr_1 \vee \neg (w_i$ is in a known trouble spot$) \Rightarrow$
        $slow = slow \setminus \{ w_i \} \wedge suspect = suspect \cup \{ w_i \}$
   11: $\forall w_i \in suspect \bullet count(w_i, suspect) = thr_2 \Rightarrow kill(fixed process_i)$
   12: $(slow \cup suspect \neq \emptyset) \Rightarrow UE_c = UE_c + K$ // Increase UE_c by some constant K

Fig. 4. The informal Single Proxy Synchronization algorithm.

added to suspect. Similarly, if the number of successive iterations in which a host has been classified as suspect exceeds a threshold thr_2, then the host is deemed to have failed and its fixed process is killed. In the wired domain, this action is viewed as a fail-stop failure and is detected and handled through the applications native mechanisms; thus, the underlying semantics of the group communication system are again upheld. Based on this, we present the single proxy synchronization algorithm in fig. 4.

6 Conclusion

This paper has opened a discussion on the Janus architecture that is developing strong group communications for mobile wireless environments. In this context, strong group communications has been touted as the basis for a number of emerging application domains (e.g. mobile commerce). In section 4 we presented the issues that oppose this goal and subsequently an exemplar solution was proposed to tackle one of these.

Future work will continue the development of these concepts, the envisaged culmination being a usable software framework that can be easily integrated into any traditional strong group communications system.

References

1. D. A. Agarwal. *Totem: A Reliable Ordered Delivery Protocol for Interconnected Local-Area Networks.* PhD thesis, University of California, Santa Barbara, 1994.
2. K. Berket. *The InterGroup Protocols: Scalable Group Communication for the Internet.* PhD thesis, University of California, Santa Barbara, December 2000.
3. K. P. Birman. The Process Group Approach to Reliable Distributed Computing. *Communications of The ACM*, pages 37–53, December 1993.
4. K. P. Birman. *Building Secure and Reliable Network Applications.* Prentice Hall, 1997. Available at: http://www.cs.cornell.edu/ken/.
5. S. Deering. *Host Extensions for IP Multicasting.* Network Working Group, Stanford University, August 1989. Available from: http://www.faqs.org/rfcs/rfc1112.html.
6. D. Dolev and D. Malki. The Transis Approach to High Availability Cluster Communication. In *Communications of the ACM*, April 1996.
7. J. Dunlop and D. G. Smith. *Telecommunications Engineering.* Chapman and Hall, third edition, 1998.
8. P. Gray and J. S. Pascoe. On Group Communication Systems: Inisght, a Primer and a Snapshot. In *Proc. of The 2001 International Conference on Computational Science*, Lecture Notes in Computer Science, May 2001.
9. W. C. Y. Lee. *Mobile Communications Engineering.* McGraw Hill, second edition, 1997.
10. L. E. Moser, Y. Amir, P. M. Melliar-Smith, and D. A. Agarwal. Extended Virtual Synchrony. In *International Conference on Distributed Computing Systems*, pages 56–65, 1994.
11. L. E. Moser, P. M. Melliar-Smith, D. A. Agarwal, R. K. Budhia, and C. A. Lingley-Papadopoulos. Totem: A Fault-Tolerant Multicast Group Communication System. In *Communications of the ACM*, April 1996.
12. J. S. Pascoe, V. S. Sunderam, U. Varshney, and R. J. Loader. Middleware Enhancements for Metropolitan Area Wireless Internet Access. *Future Generation Computer Systems*, Expected: 2002. In press.
13. T. Rappaport. *Wireless Communications Principles and Practice.* Prentice Hall, 1996.
14. O. Rodeh, K. P. Birman, and D. Dolev. The Architecture and Performance of Security Protocols in the Ensemble Group Communication System. Technical Report TR2000-1791, Cornell University, March 2000.
15. R. van Renesse, K. P. Birman, and S. Maffeis. Horus, A Flexible Group Communication System. In *Communications of the ACM*, April 1996.

Towards Mobile Computational Application Steering: Visualizing The Spatial Characteristics of Metropolitan Area Wireless Networks

J. S. Pascoe†‡, V. S. Sunderam‡, R. J. Loader†‡ and G. Sibley‡

†Department of Computer Science
The University of Reading
United Kingdom
RG6 6AY
J.S.Pascoe@reading.ac.uk
Roger.Loader@reading.ac.uk

‡Math & Computer Science
Emory University
Atlanta, Georgia
30322
vss@mathcs.emory.edu
gsibley@emory.edu

Abstract. Computational application steering from mobile devices is an attractive proposition and one that is being actively explored. In wireless networks, particularly metropolitan area ifrastructural networks, a range of unique temporal and spatial performance characteristics may affect the ability to perform computational steering. The network coverage area contains zones, termed *trouble spots*, that are distinct pockets where garbled, poor, or no network coverage exist. They have been experimentally attributed to geographic features (e.g. tunnels and tall buildings). Point-to-point manifestations of trouble spots include stalled web pages, slow ftp connections and data corruption which, although detrimental, do not compromise system usability. However, computational applications such as steering are highly susceptible, and could suffer from serious operational and semantic problems in the presence of trouble spots. Previous experimental work has verified not only the existence of trouble spots in a range of wireless networks, but has also identified some of the issues surrounding their detectability. One of the difficulties encountered during the initial study was the collection of reliable data, primarily due to a lack of tool support. To alleviate this, a visualization package, termed *RadioTool*[1] has been developed; the underlying goal being to investigate the nature of trouble spots more reliably. It is envisaged that the tool will eventually serve as a detection and display mechanism for visualizing network quality as a user moves about a wireless network, and thereby complement and support computational steering applications. This paper describes the features incorporated in RadioTool and describes its use in relation to the exemplar Ricochet radio network.

Keywords – Computational steering, wireless networking, trouble spots

1 Introduction

The Ricochet radio network provides wireless Internet connections of up to 128 Kbps in several cities across the United States. While this paper predominantly focuses on RadioTool, it is noteworthy to explain the motivating factors for its development in the

---

[1] This project was supported in part by NSF grant ACI-9872167 and DoE grant DE-FG02-99ER25379.

P.M.A. Sloot et al. (Eds.): ICCS 2002, LNCS 2330, pp. 665–670, 2002.

overall project. Networks such as Ricochet and GUIDE [10] are facilitating many new directions in mobile computing and demand that legacy applications incorporate additional functionality [12]. In particular, programs that utilize distributed communication must be supplemented with schemes to adapt to connectivity problems or *trouble spots*. To facilitate trouble spot detection, a previous study suggested the following three metrics: *received signal strength indication (RSSI)*, *network response time* and *packet loss*. Furthermore, these metrics facilitate a user process to categorize specific trouble spot instances and in doing so, appropriate resolution strategies can be adopted. This functionality, coupled with a database of trouble spot locations and characteristics facilitates the basis of a preemptive scheme that minimizes the impact of trouble spots.

2   The Ricochet Network

In this section, a brief overview of the Ricochet network is presented. It must, however, be emphasized that the work described herein is architecturally independent of the actual wireless network; in fact, RadioTool and the other associated parts of the system may be used not only in MAN's, but also in wireless local area networks. The prototype implementation of RadioTool used the Ricochet network, but it has been designed to be easily portable to other network types. Ricochet is present in eight major US cities with further expansion into another thirteen being planned in the future. The network offers a fixed bandwidth of 128Kbps and is accessible to the general public through small portable wireless modems. The network itself consists of seven architectural components which are grouped hierarchically to provide the service.

- *Wireless Modems* – Ricochet modems are small devices that can be connected to any portable workstation.
- *Microcell Radios* – Microcell radios communicate with user workstations through wireless modems. In addition, they perform error checking and are responsible for sending acknowledgements.
- *Wired Access Points* – Wired Access Points or WAPs are installed within a 10-20 square mile area. Each WAP collects and converts wireless data into a form suitable for transmission on a wired IP network.
- *Network Interface Facility* – A Network Interface Facility connects a group of WAPs to the Network Operations Centers.
- *Ricochet Gateway* – The Ricochet gateway is part of each Network Operations Center and connects the Ricochet network to other systems.
- *Network Operations Centers* – The Network Operations Centers provide a means for monitoring and control within the network.
- *Name Server* – The name server maintains access control and routing information for every radio and service within the wireless network.

The overall topological structure of the networks wireless component is a mesh, since microcell radios are placed at regular intervals in a 'checkerboard' fashion. The authors argue in [6] that the additional redundancy in a mesh provides a more robust wireless network topology than that of a tree or a star.

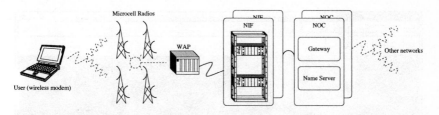

Fig. 1. The Ricochet Communication Architecture

3 Trouble Spot Detection

The project and system described herein is based upon the premise that all wireless networks contain areas of degraded quality, some of which are static (due to topography or structures) while others are dynamic (caused by weather and related factors). While the existence of these "trouble spots" is an accepted fact, one outstanding issue is the pertinent question of how to detect and ascertain the semantics relating to a specific trouble spot. In this regard, it is useful to classify trouble spots into different categories ranging in intensity – from trouble spots that cause occasional garbling and packet loss to those that result in complete disconnection. Detection may be achieved in the first instance by the mobile device continuously monitoring network quality indications, and comparing these values against recent history as well as against a database of stored past values. At this stage, the system must determine (by executing a series of diagnostic tests) which category of trouble spot it has encountered so that it may invoke an appropriate failure handler. In order to achieve this, we propose the following metrics for evaluating the state of a service in a fixed topology wireless network.

- *Received Signal Strength Indication* – determining signal strength is usually accomplished by query the hardware of the network interface and is intuitively central to the detection of trouble spots.
- *Packet loss* – packet loss at the wireless network interface is a function of not only the power of the received signal, but also the path loss and the state of the network. Packet loss is inversely proportional to signal strength and as such, a high packet loss indicates a drop (or imminent drop) in service. Thus, packet loss is an important metric in determining trouble spots of the more severe categories.
- *Response time* – The response time of the network is the duration between a message being transmitted and the reply being received. It can often be useful in confirming trouble spot classification when the packet loss and path loss are both low. In this circumstance, a high response time will indicate that the network problem is related to the wired component of the infrastructure.

The above metrics were suggested by a number of sources, the most notable being [8], [14] and the IETF MANET[2] metrics for the evaluation of ad hoc routing protocol performance [5].

4 RadioTool Technical Operation

RadioTool is an event driven X-Windows (Gtk based) Graphical User Interface and in the prototype version for Ricochet, it utilizes the full range of services offered by

---

[2] Mobile Ad-hoc NETworks working group.

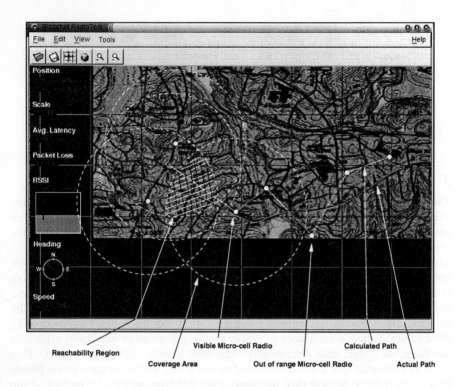

Fig. 2. RadioTool Graphical User Interface

the Ricochet Radio Network API (RRNAPI) [13], that is, an application programming interface developed to gather information specifically from the Ricochet network. By substituting alternate libraries, RadioTool can be utilized in conjunction with all other wireless networks.

Figure 2 depicts the layout of the RadioTool interface. When RadioTool starts, it calls RRN_init, a function within the RRNAPI that dictates how data should be collected. RRNAPI then connects to the wireless modem and launches a thread to regularly query the hardware and update RadioTool's data structures. After invoking RRN_init, RadioTool installs a callback function that is executed once every 250 milliseconds.

Location is measured in latitude / longitude format whereas heading is represented by two-dimensional polar coordinates. Received signal strength indication, packet loss and latency are all measured in standard units. Note that, latency is measured as the average round trip latency to the Ricochet wired access points (i.e. latency measurements do not traverse the Internet).

Furthermore, RadioTool calls RRN_poletops, a function which returns a list of the currently visible micro-cell radios. This list is used to visualize the users position relative to the poletops, as described in section 5.

4.1   Location Determination

Currently, location is determined through one of three methods. The first approximates user location to the position of the nearest microcell radio, that is, the transmitter with the strongest signal. The second method utilizes readings from all reachable radios by taking an average of the contactable radios locations, hence arriving at a 'center of mass' approximation. Method three calculates an offset center of mass based on the relative signal strengths of the MCRs. Thus, in an unbiased system where only two radios are contactable, the hosts location will be approximated to a bisection.

Empirical results (obtained by comparing computed values to those measured via GPS) have shown that method one is the more accurate of the three, typically within 200 meters. Method two performs badly as it is often possible for distant transmitters to be contactable (albeit barely). The hypothesis of the third method is to address this skewing caused by distant radios through a weighting function for the average. In practice, a factor of attraction is empirically set at -125 dB, which is approximately the weakest signal typically observed. Increasing this factor causes method 3 to behave as accurately, and sometimes better, than method 1. Current experimentation is considering which weight results in the best estimate of location.

Additional Methods for Calculating Location Calculating location based on signal strength of known location radios is similar to the problem in mobile robotics of *spatial localization* based on beacon placement [2]. From these studies we can see that there are better ways to calculate location based on RSSI than the averaging scheme used above [2]. In short, there are essentially two general localization methods used in much of the research: proximity based localization [15, 4, 3] and multilateration[3] based localization [1, 7, 9].

Both approaches use a standard method to approximate distance, such as radio frequency localization [15], RSSI [11], or a database of RSSI signatures [1]. Thus, RRNAPI's location determination algorithms are based on the approach that uses signal strength to approximate distance. In either a proximity based approach or a multi-laterizaion based approach, the primary improvement in RRNAPI would be to add the concept of *reachability regions* to the RRN_location function. Reachability regions are areas where different radios coverage areas overlap in the manner of a Venn diagram. For example, if a host can detect two radios, then the reachability region is the intersection of their signal coverage areas. As the number of radios increase, the intersection shrinks thus increasing the accuracy of the approximation.

5 Radio Network Visualization

RadioTool iterates through a list of microcell radios, drawing each radio at its latitude / longitude position as well as indicating an approximate reachability region[4]. RadioTool also stores another list of inactive microcell radios which are shown on the interface but do not correspond to coverage circles. Latitude / longitude grid lines are drawn, along with an indication of scale. The cursor can be used to 'grab' the map and move the users view point; the cursor also displays its position relative to the map it is over.

With the *Map Tool*, geographic views can be loaded as background images. Maps scale and move as the user zooms or moves his view. This feature is useful in determining the position of microcell radios, WAPs and other users. The 'status' overlay (see fig. 2) displays the users location, current scale, packet loss, latency, recent RSSI

---

[3] In multilateration, position is estimated from distances to three or more known points.

[4] Coverage is idealized to a sphere of linearly decreasing signal strength.

measurements, heading, speed, and the modem connectivity statistics. User location is indicated by a small dot that is dynamically updated as the user moves.

6   Conclusions and Future Work

This paper has presented RadioTool, a simple yet robust package for visualizing and collecting data from wireless networks. The insight, information and data that RadioTool provides is being used to develop a solution to the problem of wirelessly steering computational applications. Although RadioTool has some caveats, the data it provides has proved to be largely more reliable than that gathered from ad-hoc experimentation. Future work will focus solely on these goals, the envisaged culmination being a freely available software system.

References

1. P. Bahl and V.N. Padmanabhan. Radar: An in-building user location and tracking system. In *Proc. of the IEEE Infocom 2000*, volume 2, pages 775–785, Tel-Aviv, Israel, March 2000.
2. N. Bulusu, J. Heidemann, , and D. Estrin. Adaptive beacon placement. In *21st International Conference on Distributed Computing Systems*, Phoenix, AZ, April 2001.
3. N. Bulusu, J. Heidermann, , and D. Estrin. Gps-less low cost outdoor localization for very small devices. *IEEE Personal Communications Magazine*, pages 7(5):28–34, October 2000.
4. F.A. Chudak and Shmoys. Improved approximation algorithms for capacitated facility location problems. In *Proceedings of the 10th Anual ACM-SIAM Symposium on Discrete Algorithms*, pages S875–S876, Baltimore, Maryland, USA, January 16-17 1999.
5. S. Corson and J. Macker. *Routing Protocol Performance Issues and Evaluation Considerations*. IETF Mobile Ad Hoc Networking (MANET) Working Group, 1999. RFC 2501.
6. S. Lee et al. Ad hoc Wireless Multicast with Mobility Prediction. In *Proc. IEEE International Conference on Computer Communications and Networks*, pages 4–9, 1999.
7. L. Girod. Development and charcterization of an acoustic rangefinder. Technical Report USC-CS-TR-00-728, University of Southern California, April 2000.
8. F. Halsall. *Data Communications, Computer Networks and Open Systems*. Addison-Wesley, fourth edition, 1995.
9. B. Hofmann-Wellenhoff, H. Lichtenegger, and J. Collins. *Global Positioning System: Theory and Practice*, volume Fourth. Springer Verlag, 1997.
10. Keith Cheverst and Nigel Davies et al. The Role of Connectivity in Supporting Context-Sensitive Applications. In H. W. Gellersen, editor, *Handheld and Ubiqitous Computing*, volume 1707 of *Lecture Notes in Computer Science*, pages 193–207. Springer-Verlag, 1999.
11. F. Koushanfar, A. Savvides, G.Veltri, M. Potkonjak, and M. B. Srivastava. Iterative location discovery in ad hoc sensor networks. Slides from UCLA CS review, April 2000. *http://www.cs.ucla.edu/farinaz/location2_1.ppt.*
12. J. S. Pascoe, G. Sibley, V. S. Sunderam, and R. J. Loader. Mobile Wide Area Wireless Fault Tolerance. Technical report, University of Reading and Emory University, 2001.
13. G. Sibley and V. S. Sunderam. Tools for collaboration in metropolitan wireless networks. In V. Alexandrov, J. Dongarra, B. Juliano, R. Renner, and K. Tan, editors, *Proc. 2001 International Conference on Computational Science (ICCS 2001)*, volume 2073 of *Lecture Notes in Computer Science*, pages 395–403, May 2001.
14. Y. Tu, D. Estrin, and S. Gupta. Worst Case Performance Analysis of Wireless Ad Hoc Routing Protocols: Case Study. Technical report, Univeristy of Southern California, 2000.
15. R. Want, A. Hopper, V. Falcao, , and J. Gibbons. The active badge location system. In *ACM Transactions on Information Systems*, pages 10(1):91–102, January 1992.

Hungarian Supercomputing Grid[1]

Péter Kacsuk
MTA SZTAKI
Victor Hugo u. 18-22, Budapest, HUNGARY
www.lpds.sztaki.hu
E-mail: kacsuk@sztaki.hu

Abstract. The main objective of the paper is to describe the main goals and activities within the newly formed Hungarian Supercomputing Grid (H-SuperGrid) which will be used as a high-performance and high-throughput Grid. In order to achieve these two features Condor will be used as a the main Grid level job manager in the H-SuperGrid and will be combined with P-GRADE, a Hungarian produced high-performance program development environment. The generic Grid middleware will be based on Globus. The paper describes how P-GRADE, Condor and Globus will be integrated in order to form the H-SuperGrid.

1 Introduction

In the March of 2001 a 96-processor Sun Enterprise 10000 supercomputer (462nd position in the top500 list at that time), two 16-processor Compaq Alpha Server supercomputers, and several large PC-clusters were installed in Hungary as major supercomputing resources. All are placed at different institutes and are used by a growing user community from academy. However, even in this early stage of their use it turned out that there exist applications where the capacity of individual supercomputers and clusters are not sufficient to solve the problems at reasonable time. The solution for this problem is to connect these high-performance computing resources by the Hungarian academic network and to use them jointly in a newly formed supercomputing Grid. In order to achieve this goal a new 2-year project, called the Hungarian Supercomputing Grid (H-SuperGrid) project, started in January 2002.

One of the main goals of the project is to establish this Hungarian supercomputing Grid based on the current Hungarian and international results of cluster and Grid computing research. The project is strongly related with two already running Hungarian Grid projects (NI2000 [1], DemoGrid), two international projects (DataGrid [2], LHC Grid) and several national projects from other countries (Condor [3], INFN Grid [4], UK e-science [5]). The SuperGrid project is based on the experience learned on Globus [6] and Condor in the Hungarian NI2000 project and will strongly collaborate with the other running Hungarian Grid project, called DemoGrid (see Figure 1.).

---

[1] The research on the Hungarian Supercomputing Grid is supported by the OTKA T032226 and the OM IKTA-00075/2001 projects.

P.M.A. Sloot et al. (Eds.): ICCS 2002, LNCS 2330, pp. 671–678, 2002.

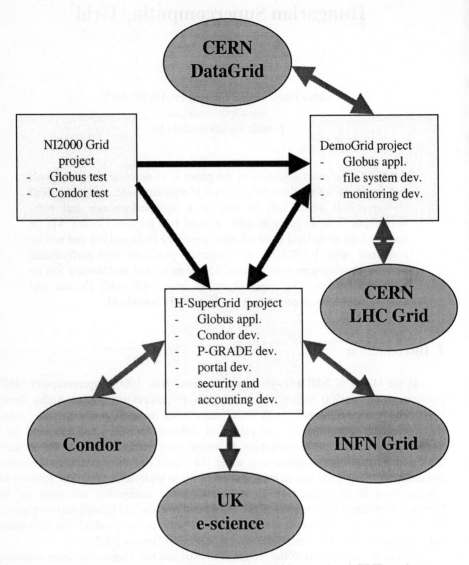

Fig. 1. Links between the Hungarian and International GRID projects

The H-SuperGrid will be used as a high-performance and high-throughput Grid. In order to achieve these two features Condor will be used as a the main Grid level job manager in the H-SuperGrid and will be combined with P-GRADE [7], a Hungarian produced high-performance program development environment.

The heart of the H-SuperGrid is P-GRADE, a graphical parallel programming environment that supports the whole life-cycle of parallel program development and makes parallel programming reasonable easy for scientists and other non-professional programmer end-users. The major goal of P-GRADE is to provide an easy-to-use, integrated set of programming tools for the development of general message-passing applications to be run in heterogeneous computing environments. Such systems currently include supercomputers and clusters. As a result of the H-SuperGrid project P-GRADE will be usable for the Grid and in this way the same programming environment will cover the whole range of parallel and distributed systems.

2 Structure of the H-SuperGrid project

The structure of the H-SuperGrid project is shown in Figure 2. The top layer is the application layer where currently a Monte-Carlo method based application is investigated in the field of simulating nuclear techniques. The user will access the H-SuperGrid by the Grid portal to be developed in the project. The application will be developed in the P-GRADE parallel program development environment which will be connected to Condor in the project. It means that the user can generate directly Condor jobs (containing parallel PVM [8] or MPI [9] program) from the P-GRADE environment. Condor will be used as the Grid level resource manager in the H-SuperGrid. The basic middleware services will come from Globus. The fabric will contain the Hungarian supercomputers and clusters connected by the Hungarian academic network. On the supercomputers and clusters local job schedulers like LSF, PBS, Condor, Sun Grid Engine will be used.

3 P-GRADE

The main advantages of P-GRADE can be summarised as follows. Its visual interface helps to define all parallel activities without knowing the syntax of the underlying message-passing system. P-GRADE generates all message-passing library calls automatically on the basis of graphics. Compilation and distribution of the executables are performed automatically in the heterogeneous environment. Debugging and monitoring information is related directly back to the user's graphical code during on-line debugging and performance visualisation.

P-GRADE currently consists of the following tools as main components: GRAPNEL (GRAphical Parallel NEt Language [7]), GRED (GRaphical EDitor [10]), GRP2C (pre-compiler to produce C/C++ code with PVM/MPI function calls from the graphical program [11]), GRM Monitor (to generate trace file during execution), DIWIDE (distributed debugger [12]), PROVE (performance visualisation tool [14]). P-GRADE is currently used at several universities of three continents for teaching parallel processing. Recently, the SYMBEX COST D23 project accepted P-GRADE to use for parallelising chemistry code and to make it part of a chemistry-oriented metacomputing system to be developed in the SYMBEX project.

P-GRADE has been successfully used by the Hungarian Meteorology Service to parallelise their nowcast program package, called MEANDER. A typical P-GRADE snapshot used in the parallelisation of the CANARI algorithm of MEANDER is illustrated in Figure 3. The Application window of the figure shows how individual processes (master) and the process farm template (Slaves) are graphically represented in GRAPNEL. The overall speed-up achieved with this program was 7.5 using 11 processors on an Origin 2000 supercomputer.

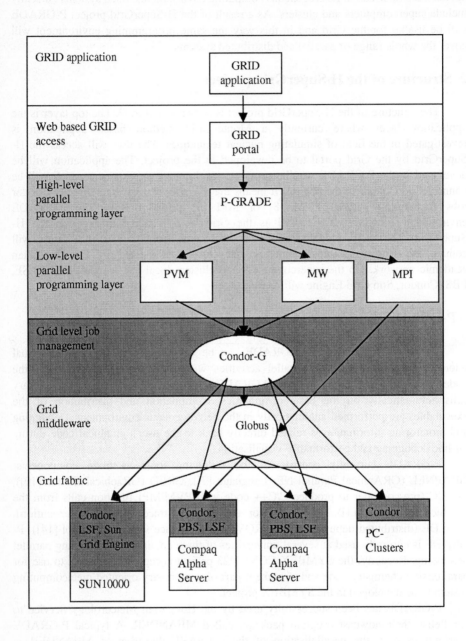

Fig. 2. Components of the Hungarian Supercomputing Grid and their relations

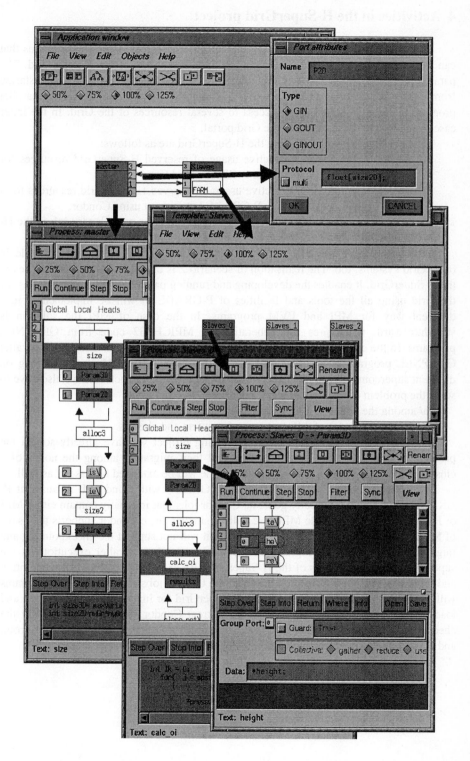

Fig. 3. Graphical parallel code of the CANARI algorithm

4 Activities in the H-SuperGrid project

The main idea of the H-SuperGrid is that the user can start parallel programs that can run in parallel on several supercomputers and/or clusters of the H-SuperGrid. The parallel programs (either PVM or MPI ones) will be developed in P-GRADE and started from P-GRADE either as a Condor job or as an immediately running interactive parallel program if the user gained direct access to several resources of the Grid. In the latter case resources can be reserved by the Grid portal.

The possible scenarios of using the H-SuperGrid are as follows:

1. Development phase: Interactive usage of reserved H-SuperGrid resources for an MPI program without using P-GRADE and Condor.
2. Development phase: Interactive usage of reserved H-SuperGrid resources for a PVM or MPI program using P-GRADE but without using Condor.
3. Production phase: Running a PVM or MPI program as a Condor job in the H-SuperGrid.

Scenario 1 is supported by MPICH-G2 [14] and Globus which are available in other Grid systems, too. The realisation of scenario 2 is one of the novel possibilities in the H-SuperGrid. It enables the developing and running parallel/distributed programs for the Grid using all the tools and facilities of P-GRADE. It will be implemented in a different way for MPI and PVM programs. In the case of MPI the solution is straightforward, it requires the generation of MPICH-G2 code from GRAPNEL programs. In the case of PVM we have to solve the problem how to distribute a parallel GRAPNEL program as a collection of several Parallel Virtual Machines running on different supercomputers and clusters of the H-SuperGrid. For such case we also have to solve the problem of dynamic process migration

a/ among the nodes of a supercomputer or cluster

b/ among the nodes of different supercomputers and clusters.

In a currently running cluster programming project we have already solved the problem of checkpointing PVM programs and their migration among the nodes of a cluster. This checkpointing and migration scheme will be extended for case b/ as well.

Scenario 3 requires the connection of P-GRADE with Condor. It means that P-GRADE will be able to directly generate Condor jobs that internally contain either MPI or PVM programs. Again the MPI case will be simpler since Condor supports the usage of MPI on top of Globus. However, this solution will not support the checkpointing and migration of GRAPNEL programs in the Grid, only their parallel execution on the supercomputers and clusters of the H-SuperGrid. By extending our PVM checkpointing and migration mechanism towards the Grid, the Condor jobs containing PVM programs will be checkpointed and migrated in the H-SuperGrid for improving fault-tolerance and load-balancing capabilities. Notice that the current Condor system cannot support the checkpoint and migration functionalities for general PVM programs. The Condor group and the H-SuperGrid project will strongly collaborate to solve these problems.

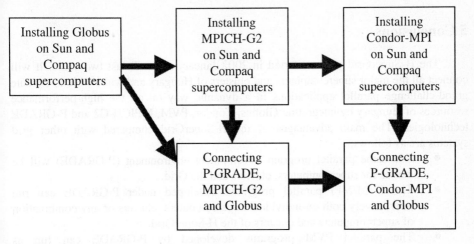

Fig. 4. Workflow diagram to support scenarios 1-3 for MPI programs

All these scenarios will be investigated and implemented in the H-SuperGrid. The workflow diagram of Figure 4 shows the steps of connecting Globus, MPICH-G2, Condor and P-GRADE together to realise scenarios 1-3 for MPI programs generated from GRAPNEL under P-GRADE. The activities to implement scenarios 2 and 3 for PVM programs generated from GRAPNEL are shown in Figure 5.

Besides combining and extending Condor and P-GRADE the other two main tasks of the project are to solve the security problems and to develop an accounting system. These issues will be addressed in a forthcoming paper.

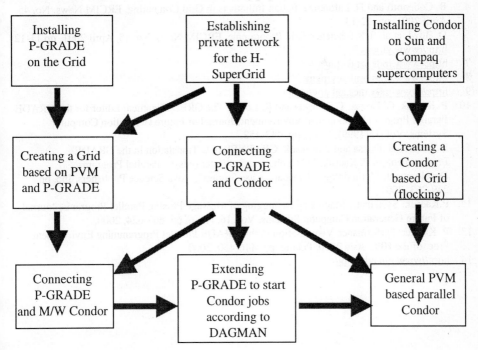

Fig. 5. Workflow diagram to support scenario 2 and 3 for PVM programs

5 Conclusions

The H-SuperGrid project started in 2002 January and will last two years. It will connect all the major supercomputers and clusters of Hungary and will enable to execute grand-challenge parallel applications in a dynamic way on all the high-performance resources of Hungary by integrating Globus, Condor, PVM, MPICH-G2 and P-GRADE technologies. The main advantages of the H-SuperGrid compared with other grid systems are as follows:

- The same parallel program development environment (P-GRADE) will be usable for supercomputers, clusters and the Grid.
- The PVM/MPI parallel programs developed under P-GRADE can run interactively both on individual supercomputers, clusters or any combination of supercomputers and clusters of the H-SuperGrid.
- The parallel PVM programs developed by P-GRADE can run as checkpointable parallel Condor jobs both on individual supercomputers, clusters or any combination of supercomputers and clusters.
- Condor jobs containing parallel PVM programs will be able to migrate among supercomputers and clusters of the H-SuperGrid.

References

1. http://www.lpds.sztaki.hu/
2. http://www.datagrid.cnr.it/
3. http://www.cs.wisc.edu/condor/
4. B. Codenotti and D. Laforenza, Italian Initiatives in Grid Computing, ERCIM News, No. 45, April 2001, pp. 12-13
5. D. Boyd, et al., UK e-Science Grid Programme, ERCIM News, No. 45, April 2001, pp.11-12
6. http://www.globus.org/
7. http://www.lpds.sztaki.hu/
8. http://www.epm.ornl.gov/pvm/
9. http://www-unix.mcs.anl.gov/mpi/
10. P .Kacsuk, G. Dózsa, T. Fadgyas and R. Lovas: The GRED Graphical Editor for the GRADE Parallel Program Development Environment, Journal of Future Generation Computer Systems, Vol. 15(1999), No. 3, pp. 443-452.
11. D. Drótos, G. Dózsa and P .Kacsuk: GRAPNEL to C Translation in the GRADE Environment, in P. Kacsuk, J.C.Cunha and S. Winter (eds): "Parallel Program Development: Methodology, Tools and Case Studies", Nova Science Publishers, Inc. pp. 249-263, 2001.
12. P.Kacsuk: Systematic Macrostep Debugging of Message Passing Parallel Programs, Journal of Future Generation Computer Systems, Vol. 16, No. 6, pp. 609-624, 2000.
13. P. Kacsuk: Performance Visualisation in the GRADE Parallel Programming Environment, Proc. of the HPC'Asia 2000, Peking. pp. 446-450, 2000.
14. http://www.niu.edu/mpi/

The Construction of a Reliable Multipeer Communication Protocol for Distributed Virtual Environments

Gunther Stuer, Frans Arickx, Jan Broeckhove

University of Antwerp
Department of Mathematics and Computer Sciences
Groenenborgerlaan 171, 2020 Antwerp, Belgium.
gunther.stuer@ua.ac.be

Abstract. We present the design and implementation issues of a Reliable MultiPeer Protocol (RMPP). This protocol is suitable for applications in the area of distributed virtual environments and is written in Java. Motivation, protocol classification, design goals and the error recovery algorithm are discussed. This paper concludes by presenting a possible application of the RMPP.

1 Introduction

One of the main bottle–necks in Distributed Virtual Environments (DVE) has always been the availability of sufficient network bandwidth to allow the participating objects to communicate with each other [1]. With the introduction of multicast this problem was partly solved. But because multicast protocols are all based on best effort approaches, message delivery is not guaranteed. In order to achieve this guarantee, reliable multicast protocols were introduced [2,3].

Although there are already many such protocols, none is optimized for DVEs. The problem is that multicast protocols are inherently 1xN protocols, i.e. one sender and many receivers. A typical DVE on the other hand has many participants, each simultaneously sending and receiving each other's information [4]. This is why DVEs rather need an MxN communication protocol, i.e. one with many senders and many receivers. In the literature, these protocols are known as *multipeer protocols* [5].

Multipeer protocols are based on a best effort approach, alike to multicast protocols. To account for guarantees on message reception, a Reliable MultiPeer Protocol (RMPP) is needed.

This paper highlights certain aspects of the development of a RMPP optimized for DVEs. In particular, the design goals, the protocol classification and the error detection algorithm are discussed. In addition a possible application of the RMPP is presented.

P.M.A. Sloot et al. (Eds.): ICCS 2002, LNCS 2330, pp. 679–686, 2002.
© Springer-Verlag Berlin Heidelberg 2002

2 Classification

In the classification of reliable multicast protocols [6], the one presented here is most closely related to the Transport Protocol for Reliable Multicast (TRM) [7]. The RMPP is a message based protocol which means that there is no stream between sender and receiver, but rather that a number of independent messages are transmitted. Every message consists of one or more packets, each one transmitted as a UDP datagram. The most important difference with the TRM is that the RMPP is a member of the multipeer family.

Another way to classify reliable multicast algorithms is on their reliability algorithm. Initialy two groups were identified [8], sender–initiated and receiver–initiated protocols. An extension to this classification was proposed in [9] where tree–based and ring–based protocols were introduced. The most important problem with the latter two protocols is that if one node fails, many other nodes also fail. Because one of the design goals set forth in the next section will be that all nodes need to be independant of each other, tree–based and ring–based approaches are not an option for the RMPP.

In [8] it is shown that receiver–initiated protocols have much better performance than sender–initiated protocols. With receiver–initiated protocols, the receivers are responsible for detecting missing packets. They do this by sending a negative acknowledgement (NACK) to the sender whenever a missing packet is detected. On reception, the sender will re–transmit the missing packets. However, receiver–initiated protocols have two important drawbacks [9]. The first one is the danger of a NACK implosion which can happen when multiple recipients detect that a packet is missing. They will all send a NACK for the same packet with a serious regression in network performance as a result. This problem can be solved by making all receivers wait a random amount of time before actually sending the NACK. If during this delay, they receive a NACK for the same packet send by another receiver, they drop their own request. A consequence if this solution is that all NACK–requests and NACK–responses have to be multicasted to warn all interested recipients.

The second problem is that in theory these protocols need an infinite amount of memory because one can never be sure whether all receivers correctly received a certain datagram. This problem is solved heuristically by assuming that in a virtual environmnent old messages have lost their importance and can be dropped.

With sender–initiated protocols, it is the sender who is responsible for detecting errors or missing packets. To do this, the sender maintains a list of all clients and after each transmission every client has to send an acknowledgement (ACK). If one is missing, an error has occured and unacknowledged datagrams are re–transmitted. This type of protocol does not exhibit the two problems stated above, but it is not very scalable due to the large amount of ACKs.

Taking into account all the advantages and disadvantages of the mentioned protocol families, RMPP is designed to be a receiver–initiated protocol.

3 Design Goals

The primary goal of this research project was the creation of a reliable multipeer system optimized for distributed virtual environments written in Java. As a reference platform we used VEplatform [10], a 100% distributed VR–framework developed in C++ at the University of Antwerp. Note that the demand for a 100% distributed architecture rules out the use of ring–based or tree–based reliable multicast protocols like TRAM developed by Sun Labs [11] which is a part of Sun's JRMS–project [12]. The secondary goal was to emphasise on good design. We considered the architectural aspect to be more important than top notch performance. To achieve this, we extensively used object oriented techniques, UML and design patterns.

The fact that this protocol is tuned for DVEs has three interesting consequences. The first one is that the frequency with which VR–nodes send updates has a maximum of 30 messages a second [13]. This implies that one message is sent for every screen update. When one uses dead–reckoning algorithms it is possible to reduce the update frequency to an average of 1 message per second [14]. This maximum and average value allows for the optimization of the data–structures in the reliability algorithm.

The second one is that the size of a typical VR–message is usually less than 1 KB because only the position, orientation and some state information is transmitted. This makes it possible to optimize buffer sizes.

The third one is that, a message is of no importance anymore after a certain amount of time because the object that sent it probably already altered one or more of the transmitted parameters. This is the reason that *old* messages may be dropped.

4 Error Recovery Algorithm

When one wants to recover from an erroneous or missing packet, it is very important to have a way to uniquely identify this packet. In the RMPP this is done in three steps. The first one is at the level of the VR–participants. Each one is identified by a unique ID, which is basically a random 32–bit number generated during construction. Every datagram transmitted will contain this ID. In this way the receiver can determine where it came from. Every datagram also contains a *message sequence number* (MSN) which uniquely identifies every message sent by a participant. So, the combination (*nodeID, MSN*) uniquely identifies every message in the system. The third level of identification is the *packetNr*. This one uniquely identifies every packet within a message. As such, the 3–tuple (*nodeID, MSN, packetNr*) uniquely identifies every datagram in the system.

Whenever a gap is detected between messages received from the same node, or between two packets from the same message, a *NACK–request* is sent for every message involved. When a whole message is missing, the *NACK–request* contains its *MSN,* and the requested *packetNr* is set to 1, since there is always at least one packet in every message. When one or more packets from a certain message are missing, the

NACK–request contains this message's *MSN* and a list of all missing *packetNrs*. The sender will re–transmit all packets it receives a *NACK–request* for. These are known as *NACK–response* packets. Figure 1 illustrates this algorithm.

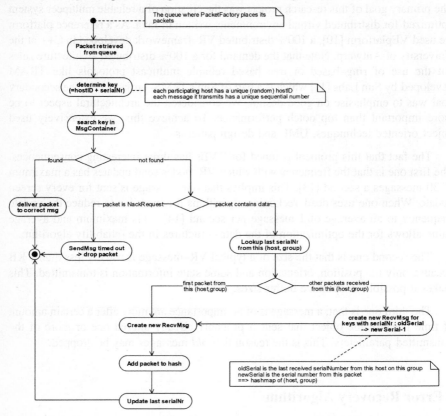

Fig. 1 This figure illustrates how messages are re–assembled at the receiver side. Every packet read from the network is stored in a queue until there are enough resources available to process it. When a packet is the first one for a particular message, the message is created. All other packets are routed towards it based on their key–value which is computed using the packet's unique ID.

When the first packet of a given message arrives, the latter is created at the receiver side and its timer is set to *receiveTimeout*. Whenever another packet for this message arrives, it is added to the message and the timer is reset. When it reaches zero, the RMPP assumes all unaccounted packets lost and sends a *NACK–request* for them. If however all packets were received, the message is considered to be complete.

When a *NACK–request* is sent, the timer of the error producing message is set to *nackTimeout*. If no *NACK–response* is received before the timer ends, another *NACK–request* is sent. This goes on until *maxRequests NACK–requests* have been transmitted after which the message is considered lost and removed from the system. If on the other hand a *NACK–response* is received, the timer is reset to *recvTimeout* and the algorithm starts all over. Figure 2 illustrates this.

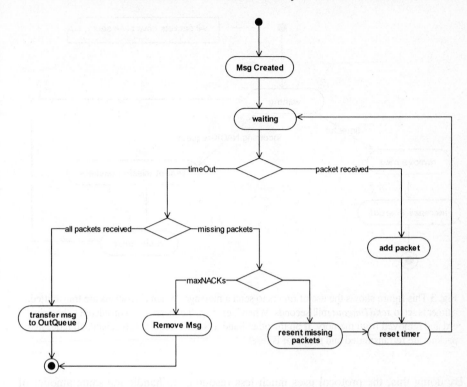

Fig. 2 This figure describes the use of timers when receiving a message. When the message is created and whenever a new packet for it arrives, the timer is reset to *receiveTimeout*. If it times out, the message checks whether it is complete. If not, it checks whether is can send another NACK or whether it should drop this message. If the first option is chosen, a NACK is sent for all missing packets and the timer is reset to *NackTimeout*.

The sender keeps every message in memory for *sendTimeout* seconds. This timer is reset with every incoming *NACK–request*. When it reaches zero, the sender assumes that all receivers did receive the message correctly and removes it from memory. This is illustrated in figure 3.

Java provides no guarantees when it comes to real–time concepts. And indeed, in the first version of the RMPP a major problem was that many timeouts were missed because the Java virtual machine was performing background tasks. To resolve this problem all timeouts mentioned in the previous paragraphs have been made dynamic. Whenever one is missed, its value is increased and whenever one is met with enough spare time, its value is decreased. This way the RMPP assures vivid responses, while still adapting to the characteristics of the network, hardware and underlying operating system.

By taking into account the fact that most VR–messages are small and fit in a single datagram, an important optimization can be made. On reception of such a message, the *recvTimeout* notification can be omitted since we already have the complete message.

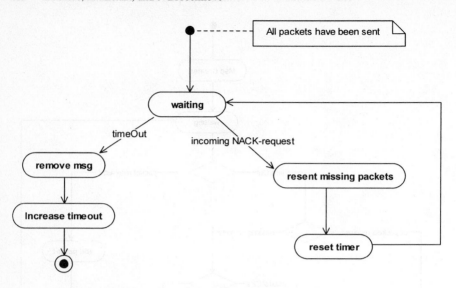

Fig. 3 This figure shows the use of timers to send a message. When all packets are transmitted, a timer is set to *sendTimeout* milliseconds. When it expires, the message is considered to be delivered and removed from memory. If on the other hand a NACK–request is received, all requested packets are retransmitted and the timer is reset.

By doing this, the protocol uses much less resources to handle the same amount of messages.

5 A Possible Application

In this section a possible application of the RMPP is discussed. It is a component of the highly dynamical DVE our research group is considering to construct.

Object mirroring is a possible strategy to minimize communications in a distributed virtual environment. One can use the RMPP to multicast the VR–objects to all interested nodes. With the use of Java classloading it is possible to create a local object from this stream of bytes. This local object serves as a *proxy* for the original one. Figure 4 illustrates this principle where object A gets replicated to objects A' at the receiver sides.

All communication between the original object and its proxies also relies on the RMPP system. When an object wants to update its proxies, an XML message is transmitted which consists of two parts. The first one identifies the target object and the second part contains the actual message.

Using this technique, the virtual world is extremely expandable because each type of object (e.g. table, avatar, room, ...) can define its own message scheme. As such, one does not need to specify the complete protocol from the very beginning.

Fig. 4 VR–Object Mirroring.

6 Conclusions

Our contribution in this paper has been to develop a Reliable MultiPeer Protocol (RMPP) with a weak model of membership similar to the one used in IP Multicast. This protocol is geared towards applications in the area of multi–user DVEs. To achieve our design goals – in particular scaleability and a 100% distributed architecture – a receiver initiated protocol was used.

The presented algorithm is capable of reliably delivering a series of messages conforming to the multipeer communications paradigm. One of the challenging problems in this research was the fact that Java is not tuned towards real–time applications. The current RMPP algorithm account for this problem.

References

1. Zyda, M.J.: Networking Large–Scale Virtual Environments. Proceedings of Computer Animation 1996.

2. Hall K.A.: The implementation and evaluation of reliable IP multicast. Master of Science Theses, University of Tennessee, Knoxville, USA, 1994.

3. Birman K.P.: A Review of experiences with reliable multicast. Software – Practise and Experience, 1999, Vol 29, No. 9, pp 741–774.

4. Sato F., Minamihata K., Fukuoka H., Mizuno T.: A reliable multicast framework for distributed virtual environments. Proceedings of the 1999 International Workshop on Parallel Processing.

5. Wittmann R., Zitterbart M.: Multicast Communications. Academic Press, 2000, chapter 2.

6. Obraczka K.: Multicast Transport Protocols: A survey and taxonomy. IEEE Communications Magazine, 1998, pp. 94–102.

7. Sabata B., Brown M.J., Denny B.A.: Transport Protocol for Reliable Multicast: TRM. Proceedings of IASTED International Conference on Networks, 1998, pp. 143–145.

8. Pingali S.: A comparison of sender–initiated and receiver–initiated reliable multicast protocols. Performance evaluation review, Vol. 22, pp 221–230, 1994.

9. Brain N.L., Garcia–Luna–Aceves J.J.: A Comparison of known classes of reliable multicast protocols. MS thesis, University of California, 1996.

10. Demuynck K.: The VEplatform for distributed virtual reality. Ph.D. thesis, University of Antwerp, Belgium, 2000.

11. Chiu D.M., Hurst S., Kadansky M., Wesley J.: TRAM: A tree–based reliable multicast protocol. Sun Research Labs Technical Report TR–98–66.

12. Hanna S., Kadansky M., Rosenzweig Ph.: The Java Reliable Multicast Service: A Reliable Multicast Library. Sun Research Labs Technical Report TR–98–68.

13. Brutzman D.P., Macedonia M.C., Zyda M.J.: Internetwork Infrastructure Requirements for Virtual Environments. Proceedings of the 1995 symposiom on Virtual Reality Modeling Languages.

14. Demuynck K., Arickx F., Broeckhove J.: The VEplatform system: a system for distributed virtual reality. Future Generation Computer Systems. No 14, 1998, pp. 193–198.

Process Oriented Design for Java: Concurrency for All

Keynote Tutorial

Professor Peter Welch

Computing Laboratory
The University of Kent at Canterbury
Canterbury, Kent
CT2 7NF ,United Kingdom
P.H.Welch@ukc.ac.uk

Abstract

Concurrency is thought to be an advanced topic - much harder than serial computing which, therefore, needs to be mastered first. This seminar contends that this tradition is wrong, which has radical implications for the way we educate people in Computer Science - and on how we apply what we have learnt.

A process-oriented design pattern for concurrency is presented with a specific binding for Java. It is based on the algebra of Communicating Sequential Processes (CSP) as captured by the JCSP library of Java classes. No mathematical sophistication is needed to master it. The user gets the benefit of the sophistication underlying CSP simply by using it.

Those benefits include the simplification wins we always thought concurrency should generate. Although the Java binding is new, fifteen years of working with students at Kent have shown that the ideas within process-oriented design can be quickly absorbed and applied. Getting the ideas across as soon as possible pays dividends - the later it's left, the more difficult it becomes to wean people off serial ways of thought that often fit applications so badly. Concurrency for all (and for everyday use) in the design and implementation of complex, maintainable and scalable computer systems is both achievable and necessary.

P.M.A. Sloot et al. (Eds.): ICCS 2002, LNCS 2330, p. 687, 2002.
© Springer-Verlag Berlin Heidelberg 2002

Collaborative Computing and E-learning

Nia Alexandrov, J. S. Pascoe, and Vassil Alexandrov*

Department of Computer Science,
The University of Reading,
Whiteknights, P.O. Box 225,
Reading, UK
{N.S.Alexandrov, J.S.Pascoe, V.N.Alexandrov}@reading.ac.uk

Abstract. In this paper we present the *Collaborative Computing Frameworks (CCF)* as an integration platform for e-learning. The capabilities of the CCF facilitate mixed modes of delivery and the possibility to integrate already existing e-learning platforms. The CCF features allow to form different groups during the learning process. This coupled with peer-to-peer communication facilities, promotes the efficient implementation of collaborative technologies as an important component of the Dialogue phase of the learning process. The new *Collaborative Computing Transport Layer (CCTL)* will allow wireless devices to be experimented with for the purposes of e-learning. It is envisaged that this direction will dramatically widen the possibilities for content delivery.

Keywords: Collaborative Computing, e-Learning, CCF, CCTL

1 Introduction

In this paper we consider how the *Collaborative Computing Frameworks (CCF)* can be used as an integration platform to deliver e-Learning. The approach is based on the overall functionalities of CCF to enable the construction of efficient and flexible collaboratories [2]. CCF supports application, computation and, data sharing in addition to a wide variety of communication primitives.

There are also systems such as Tango, based on the World Wide Web, which is an open system providing mechanisms for the rapid integration of applications into a multiuser collaborative environment. Tango uses client-server approach and central server to control and distribute information between the clients in contrast with the peer-to-peer approach of CCF.

There are numerous e-Learning platforms focusing on different aspects of the learning process facilitating synchronous and asynchronous communication [10] and ranging from streaming video and audio content to more sophisticated projects based on peer-to-peer approaches such as Intel's *Share and Learn (SLS)* platform [4].

Another category where most of the current e-Learning systems fall are systems supporting only existing applications [10]. These systems do not usually provide a framework for developing collaborative applications.

---

* Supported by European Commission grant 2001-3450/001-001 EDU ELEARN.

P.M.A. Sloot et al. (Eds.): ICCS 2002, LNCS 2330, pp. 688–694, 2002.

The task that we address is to provide a flexible e-Learning platform facilitating mixed content delivery over numerous universities in the European Union [5] and focusing on five strands: Undergraduates, Postgraduates, Multidisciplinary subjects, non-traditional learners, and teachers. Since a variety of e-learning platforms already exist, CCF will be used as an integration platform, serving to provide the underlying transport fabric. Moreover since most of this learning occurs through interaction, it is envisaged that the CCF capabilities will provide an excellent framework. In addition, the CCF supports a peer-to-peer paradigm which is vital for this type of interaction.

In section 2, an overview of the CCF is presented while section 3.1 outlines CCTL and describes its features. Section 4 describes how we are applying the CCF to e-learning.

2   The Collaborative Computing Frameworks

The Collaborative Computing Frameworks (CCF) is a suite of software systems, communications protocols, and methodologies that enable collaborative, computer-based cooperative work [2]. The CCF constructs a virtual work environment on multiple computer systems connected over the Internet, to form a *collaboratory*. In this setting, participants interact with each other, simultaneously access or operate computer applications, refer to global data repositories or archives, collectively create and manipulate documents spreadsheets or artifacts, perform computational transformations and conduct a number of other activities via telepresence. The CCF is an integrated framework for accomplishing most facets of collaborative work, discussion, or other group activity, as opposed to other systems (audio tools, video/document conferencing, display multiplexers, distributed computing, shared file systems, whiteboards) which address only some subset of the required functions or are oriented towards specific applications or situations. The CCF software systems are outcomes of ongoing experimental research in distributed computing and collaboration methodologies.

CCF consists of multiple coordinated infrastructural elements, each of which provides a component of the virtual collaborative environment. However, several of these subsystems are designed to be capable of independent operation. This is to exploit the benefits of software reuse in other multicast frameworks. An additional benefit is that individual components may be updated or replaced as the system evolves. In particular, CCF is built on a novel communications substrate called *The Collaborative Computing Transport Layer (CCTL)*. CCTL is the fabric upon which the entire system is built [8, 9]. A suite of reliable atomic communication protocols, CCTL supports sessions or heavyweight groups and channels (with relaxed virtual synchrony) that are able to exhibit multiple qualities of service semantics. Unique features include a hierarchical group scheme, use of tunnel-augmented IP multicasting and a multithreaded implementation. Other novel inclusions are fast group primitives, comprehensive delivery options and signals.

Although the CCF was initially developed to construct collaboratories for natural sciences research [2] it has an unique features which enable its efficient use in other application areas. In particular the following CCF capabilities are important for e-Learning applications: communication among members of a collaborative work session, shared access to work area and shared access to data and information. Further important features of CCF making it particularly suitable for e-Learning applications are:

- The CCF is not concentrated only on conferencing facilities or application sharing in contrast to most e-Learning systems, it has an integral provision of interfaces and shared file and data object space which enable it to strike the balance between communication, collaborative work and data manipulation.
- The CCF uses a purpose built multicast transport layer (CCTL) designed for distributed and collaborative applications. It provides multiway or many-to-many data exchanges and a variety of qualities of service (unordered, reliable FIFO and atomic) which can be selected to suit the application.
- The CCF has a completely distributed architecture which makes it scalable and increases performance.
- The CCF provides interfaces and a programming API for developing collaborative applications and also facilitates retrofit projects into collaboration unaware applications.
- The CCF includes also multiway mixed audio, synchronous audio/video, totally ordered text chat, annotations to dynamic entities, directory services etc.
- The CCF will be used for collaboration over mobile devices through a CCTL expansion project for wireless networks [7].
- The CCF offers a framework over heterogeneous environment including terrestrial network and wireless devices which other collaborative environments are not offering so far.

3   The Collaborative Computing Transport Layer

CCTL is the communication layer of the CCF and as such it provides *channel* and *session* abstractions to clients. At its lowest level, CCTL utilizes IP multicast whenever possible. Given the current state of the Internet, not every site is capable of IP multicast over WANs. To this extent, CCTL uses a novel tunneling technique similar to the one adopted in the MBone. At each local subnet containing a group member is a multicast relay. This multicast relay (called *mcaster*) receives a UDP feed from different subnets and multicasts it on its own subnet. A sender first multicasts a message on its own subnet, and then sends the tunneled message to remote mcasters at distant networks. The tunneled UDP messages contains a multicast address that identifies the target subnet. TCP-Reno style flow control schemes and positive acknowledgments are used for data transfer, resulting in high bandwidth as well as low latencies. This scheme has proven to be effective in the provision of fast multiway communications both on local networks and on wide area networks. IP multicast (augmented by CCTL flow control and fast acknowledgment schemes) on a single LAN greatly reduces sender load, thus, throughput at each receiver is maintained near the maximum possible limit (approximately 800 kB/s on Ethernet) with the addition of more receivers. For example, with a 20 member group, CCTL can achieve 84% of the throughput of TCP to one destination. If in this situation TCP is used, the replicated transmissions that are required by the sender cause receiver throughput to deteriorate as the number of hosts increases. A similar effect is observed for WANs; table 1 in [9] compares throughput to multiple receivers from one sender using TCP and CCTL.

CCTL offers three types of delivery ordering: *atomic*, *FIFO* and *unordered*. FIFO ordering ensures that messages sent to process q by process p are received in the same order in which they were sent. FIFO guarantees point-to-point ordering but places no constraint on the relative order of messages sent by p and q when received by a third process r.

CCTL offers both reliable and unreliable message delivery. Reliable delivery guarantees that messages sent by a non-faulty process are eventually received (exactly once) by all non faulty processes in the destination set. In a group communication system, this can only be defined in relation to view change operations (membership protocols).

3.1 Architecture

Hierarchical Group Architecture CCTL is logically implemented as a group module, interposed between applications (clients) and the physical network. This module implements the CCTL API and provides session and channel abstractions. Note that channels are light weight groups supporting a variety of QoS semantics. Related channels combine to form a heavy-weight group or session. Recall also that sessions provide an atomic virtually synchronous service called the *default channel*. Sessions and channels support the same fundamental operations (join, leave, send and receive) but many channel operations can be implemented efficiently using the default session channel. Session members may join and leave channels dynamically, but the QoS for a particular channel is fixed at creation. Channels and sessions are destroyed when the last participant leaves.

Fig. 1 shows the CCTL architecture. The group module consists of channel membership, QoS and session sub-modules. The channel membership module enforces view change messages (join and leave). The QoS module also provides an interface to lower-level network protocols such as IP multicast or UDP and handles internetwork routing (IP-multicast to a LAN, UDP tunneling over WANs).

Several researchers have proposed communication systems supporting light-weight groups. These systems support the dynamic mapping of many light-weight groups to a small set of heavy-weight groups. CCTL statically binds light-weight channels to a single heavyweight session, mirroring the semantics of CSCW environments.

As noted above, CCTL implements channel membership using the default session channel. Session participants initiate a channel view change by multicasting a view change request (join or leave) on the default channel. The channel membership sub-module monitors the default channel and maintains a channel table containing name, QoS and membership for all channels in the session. All

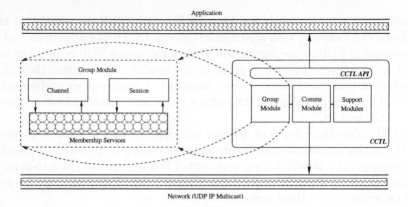

Fig. 1. CCTL Architecture

session participants have consistent channel tables because view change requests are totally ordered by the default channel. This technique simplifies the execution of channel membership operations considerably. For instance, the message ordering imposed by the default channel can be used for ordering view changes. Furthermore, the implementation of channel name services is trivial, requiring a single lookup in the channel table. The architecture of CCTL logically separates channel control transmission (using the default channel) and regular data transmission. This separation increases flexibility by decoupling data service quality from membership semantics.

The hierarchical architecture is also scalable. Glade *et al.* [6] argue that the presence of recovery, name services and failure detectors degrade overall system performance as the number of groups increases. Typically failure detectors periodically poll group members[1]. CCTL performs failure detection on transmission of each multicast message. When a failed process is detected, a unified recovery procedure removes the process from all channels simultaneously, thus restoring the sessions health.

4   CCF as integration platform for e-Learning

Our approach is based on the educational model which assumes that learning is an interactive process of seeking understanding, consisting of three fundamental components: Conceptualization, Construction and Dialogue [3]. We will focus mainly on the construction and dialogue phases since it is known that much significant learning arises through dialogue and debate and so new paradigms of content delivery are needed. There are several approaches, one is to mimic traditional methods using the new technology and the other to develop new paradigms blending the content and technology [12]. Our focus is on the later approach since it is best suited for the dialogue phase. For discussion based modules we will use rich media based collaborative learning approach and for

---

[1] E.g. Horus [11, 1] transmits a heartbeat message every two seconds.

didactic courses (for example, geometry, physics fundamentals, theorem proving) a graphic based more individualized approach will be applied. Also a mixture of the above can be employed in certain cases.

The experience outlined in this paper is from the current European Commission GENIUS project involving four major IT companies (IBM, Intel, BT and Phillips), nine universities across the European Union and other training organizations. Our aim in this project is to address the Information and Communication Technologies (ICT) skills shortage through the following activities: New Curricula Content development (based on the New ICT Curricula guidelines proposed by the Career Space consortium [2]), focusing on five different strands: Undergraduate, Postgraduate, non-traditional learners, multi-disciplinary curricula and training; investigation of different innovative instruction / content delivery mechanisms, corresponding to the pedagogical paradigms based on the new ICT curricula and e-Learning platforms of the partners; development of pilot pan-European collaborative e-learning environment.

Focusing on the collaborative e-learning environment our approach is based on the already existing software, for example, Learning Spaces and Domino of IBM, LearnLink of Mentergy through BT, SLS platform of Intel and CCF developed jointly by Emory and Reading Universities.

To facilitate e-Learning and mixed mode of delivery for the variety of users (traditional and non-traditional learners) outlined above and focus on the Dialogue phase of the learning process we need to rely on efficient collaboration and interaction. Usually we have synchronous or asynchronous mode of delivery. In our case we have to facilitate both. Through programmable APIs, additional modules can be added. Much already exists. We are also faced with the task that different users (partners) use different e-Learning platforms and we have to work together.

Here is where the CCF can provide an efficient integration platform. CCF based on CCTL allows to form different groups which can collaborate. This approach is efficient in defining different groups involved in the e-Learning process such as teachers, tutors, learners etc. The group approach allows, if necessary, different groups to employ to different software or software components. For example, group having streaming video / audio in a synchronous way and group having a discussion using all capabilities of CCF such as white board, chat etc. The first may use, for example, LearnLink and the second one CCF and / or Intel's SLS platform. Further, through the group approach an administrative, management and other groups are defined. Another advantage of the group approach and peer-to-peer to paradigm used by CCF is the possibility to keep the size of the groups manageable and to transfer as shown by Intel's experiments with SLS most of the traffic to the local area network.

With the extra CCF features spanning over wireless devices we can present content and test the limits of heterogeneous collaboration. All the above are widening the possibility for collaborative e-learning.

---

[2] See www.career-space.com.

5 Conclusion

In this paper we presented the generic idea of using the CCF as an integration platform for e-learning. The capabilities of CCF facilitate mixed mode of delivery and the possibility to integrate already existing e-learning platforms. The approach allows to form different groups in the learning process. This coupled with peer-to-peer communication in the CCF allows to efficiently implement collaboration as an important component of the Dialogue phase of the learning process and to reduce the traffic over WANs. In addition the new CCTL features allowing wireless devices to be attached and experimented with for the purposes of e-Learning which widen the possibilities of content delivery. We plan to report some experimental results in the near future.

References

1. K. P. Birman. *Building Secure and Reliable Network Applications*. Prentice Hall, 1997. Out of print, but an online version is available at: http://www.cs.cornell.edu/ken/.
2. S. Chodrow, S. Cheung, P. Hutto, A. Krantz, P. Gray, T. Goddard, I. Rhee, and V. Sunderam. CCF: A Collaborative Computing Frameworks. In *IEEE Internet Computing*, January / February 2000.
3. Lynne Coventry. Videoconferencing in higher education. Technical report, Institute for Computer Based Learning, Herriot Watt University, 1997. Available from: www.man.ac.uk/MVC/SIMA/video3/contents.html.
4. Martin Curley. Peer-to-peer Enabled e-Learning and Collaboration. In *Proc. 2002 International Conference Computational Science*, Lecture Notes in Computer Science. Springer-Verlag, April 2002. In press.
5. Genius - generic e-learning environments and paradigms for the new pan european information and communication technologies curricula. EC project, 2001-2003.
6. B. Glade, K. P. Birman, R. Cooper, and R. Renesse. Light Weight Process Groups In The ISIS system. *Distributed Systems Engineering*, 1:29–36, 1993.
7. J. S. Pascoe, V. S. Sunderam, and R. J. Loader. Working Towards Strong Wireless Group Communications: The Janus Architecture. In *Proc. 2002 International Conference Computational Science*, Lecture Notes in Computer Science. Springer-Verlag, April 2002. In press.
8. I. Rhee, S. Cheung, P. Hutto, A. Krantz, and V. Sunderam. Group Communication Support for Distributed Collaboration Systems. In *Proc. Cluster Computing: Networks, Software Tools and Applications*, December 1998.
9. I. Rhee, S. Cheung, P. Hutto, and V. Sunderam. Group Communication Support for Distributed Multimedia And CSCW Systems. In *Proc. 17th International Conference On Distributed Systems*, May 1997.
10. Trace A. Urdan and Cornelia C. Weggen. Corporate e-Learning; exploring A New Frontier. Equity Research, March 2000.
11. R. van Renesse, K. P. Birman, and S. Maffeis. Horus, A Flexible Group Communication System. In *Communications of the ACM*, April 1996.
12. Various. Teaching at an internet distance: The pedagogy of online teaching and learning. Technical report, University of Illinois, 1999. Available from: www.vpaa.uillinois.edu/tid/report/tid_report.html.

CSP Networking for Java (*JCSP.net*)

Peter H. Welch, Jo R. Aldous and Jon Foster

Computing Laboratory, University of Kent at Canterbury, England, CT2 7NF

P.H.Welch@ukc.ac.uk, jra@dial.pipex.com, jon@jon-foster.co.uk

Abstract. JCSP is a library of Java packages providing an extended version of the CSP/occam model for *Communicating Processes*. The current (1.0) release supports concurrency within a single Java Virtual Machine (which may be multi-processor). This paper presents recent work on extended facilities for the dynamic construction of CSP networks across distributed environments (such as the *Internet*). Details of the underlying network fabric and control (such as machine addresses, socket connections, local multiplexing and de-multiplexing of application channels, acknowledgement packets to preserve synchronisation semantics) are hidden from the JCSP programmer, who works entirely at the application level and CSP primitives. A simple brokerage service – based on channel names – is provided to let distributed JCSP components find and connect to each other. Distributed JCSP networks may securely evolve, as components join and leave at run-time with no centralised or pre-planned control. Higher level brokers, providing user-defined matching services, are easy to bootstrap on top of the basic *Channel Name Server* (CNS) – using standard JCSP processes and networking. These may impose user-defined control over the structure of network being built. JCSP network channels may be configured for the automatic loading of class files for received objects whose classes are not available locally (this uses the network channel from which they were received – the sender must have them). Also provided are *connection* channels (for extended bi-directional transactions between clients and servers) and *anonymous* channels (to set up communications without using any central registry – such as the given CNS). The aims of *JCSP.net* are to simplify the construction and programming of dynamically distributed and parallel systems. It provides high-level support for CSP architectures, unifying concurrency logic within and between processors. Applications cover all areas of concurrent computing – including e-commerce, *agent* technology, home networks, embedded systems, high-performance clusters and *The Grid*.

1 Introduction

JCSP [1, 2, 3, 4, 5] provides direct expression in Java for concurrent systems based on Hoare's algebra of *Communicating Sequential Processes* (CSP [6, 7, 8]). It follows the model pioneered by the **occam** concurrency language [9] in the mid-1980s, which was the first commercial realisation of the theory that was both efficient and secure.

P.M.A. Sloot et al. (Eds.): ICCS 2002, LNCS 2330, pp. 695–708, 2002.

JCSP extends the **occam** model by way of some of the proposed extras (such as shared channels) for the *never implemented* **occam3** language [10] – and by taking advantage of the dynamic features of Java (such as recursion and runtime object construction). These latter, however, throw concurrency security issues (such as race hazards) back on the system designer (that **occam**, with its compile-time memory allocation, could take care of itself). An important and open research issue is how to extend the security mechanisms of **occam** to cover dynamic concurrent systems – but that is not the subject of this paper.

JCSP views the world as *layered networks of communicating processes*, each layer itself being a process. Processes do not interact directly with other processes – only with CSP *sychronisation objects* (such as communication channels, event barriers, shared-memory CREW locks) to which groups of processes subscribe. In this way, *race hazards* are (largely) designed out and *deadlock/livelock* analysis is simplified and can be formalised in CSP for mechanically assisted verification. The strong de-coupling of the logic of each process from any other process means we only have to consider one thing at a time – whether that thing is a process with a serial implementation or a network layer of concurrent sub-processes. A consequence is that CSP concurrency logic scales well with complexity. None of these properties are held by the standard Java concurrency model based on *monitors* and *wait conditions*.

The channel mechanism of CSP lends itself naturally to distributed memory computing. Indeed, this was one of the most exciting attributes of **occam** systems mapped to *transputer* [11] networks. Currently, JCSP only provides support for shared-memory multiprocessors and, of course, for uniprocessor concurrency. This paper reports recent work on extended facilities within JCSP for the dynamic construction, and de-construction, of CSP networks across distributed environments (such as the *Internet* or tightly-coupled clusters).

The CTJ Library [12, 13], which is an alternative to JCSP for enabling CSP design in Java (with an emphasis on real-time applications), already provides *TCP/IP socket* drivers that can be plugged into its channels to give networked communication. Independent work by Nevison [14] gives similar capabilities to JCSP. The work presented here offers a higher level of abstraction that lifts many burdens of dynamic network algorithms from the JCSP programmer, who works only at the application level and CSP primitives.

JCSP.net provides mechanisms similar to the *Virtual Channel Processor* (VCP) of the T9000 *transputer* [11]. Users just *name* networked channels they wish to use. Details of how the connections are established, the type of network used, machine addresses, port numbers, the routing of application channels (e.g. by multiplexing on to a limited supply of socket connections) and the generation and processing of acknowledgement packets (to preserve synchronisation semantics for the application) are hidden. The *same* concurrency model is now used for networked systems as for internal concurrency. Processes may be designed and implemented without caring whether the sychronisation primitives (channels, barriers etc.) on which they will operate are networked or local. This is as it should be.

This paper assumes familiarity with core JCSP mechanisms – channel interfaces, *any-one* concrete channels, *synchronized* communication, buffered *plugins*, processes, parallel process constructors and *alternation* (i.e. passive waiting for one from a set of *events* to happen).

2   Networked JCSP (*JCSP.net*)

We want to use the same concurrency model regardless of the physical distribution of the processes of the system. Figure 1(a) shows a six process system running on a single processor (**M**). Figure 1(b) shows the *same* system mapped to two machines (**P** and **Q**), with some of its channels stretched between them. The semantics of the two configurations should be the same – only performance characteristics may change.

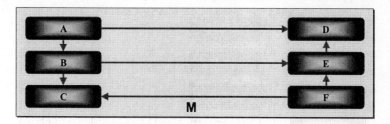

Figure 1(a). Single processor system

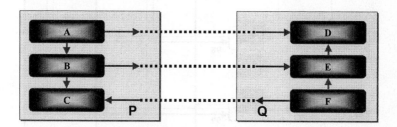

Figure 1(b). Two processor system

2.1 Basic Infrastructure

To implement Figure 1(b), a bi-directional link must be established between the machines and the application channels multiplexed over this. Figure 2 shows broad details of the drivers for such a link. Each application channel is allocated a unique *Virtual Channel Number* (VCN) on each machine using it. For example, the channel connecting processes **A** and **D** has VCN **97** on machine **P** and **42** on machine **Q**.

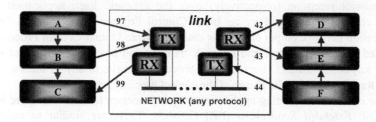

Figure 2. Virtual channel numbers and a link

Application messages sent across the network contain source and target VCNs, as well as the underlying network's requirements (e.g. target network address) and the application data. This information is automatically added and removed as messages fly through the **JCSP.net** infrastructure. Applications processes just read and write normal channel interfaces and may remain ignorant that the implementing channels are networked. The source VCN is needed to route acknowledgement packets back to the source infrastructure to preserve application synchronisation semantics (unless the networked application channel is configured with a jcsp.util *overwriting buffer*).

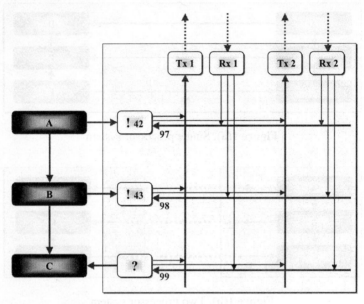

Figure 3. *JCSP.net* crossbar (2 links, 2 writers, 1 reader)

The supporting **JCSP.net** infrastructure on each machine is itself a simple JCSP crossbar, dynamically set up as jcsp.net channels are created. Figure 3 shows this for two *external network links* (to two different machines) and the three *networked application channels* to/from processes **A**, **B** and **C** on machine **P** – figure 1(b). The current state of this crossbar is implemented by just five JCSP *any-one* channels.

Each network link is driven by a pair or processes, **Tx/Rx**, responsible for low-level transmission/reception (respectively) over the network. Each application network input channel is managed by an Input Control Process (ICP), labelled "**?**". Each application network output channel is managed by an Output Control Process (OCP), labelled "**!**". Each control and network transmission process is the server end of a JCSP Any2OneChannel. Each control process can write to any of the **Tx** channels. Each **Rx** process can write to any of the control process channels.

The channels connecting application output processes to their OCPs are instances of JCSP *Extended Rendezvous* (ER) channels. These are similar to normal CSP channels (i.e. unbuffered and fully synchronising). Readers block until a writer writes. Writers block until a reader reads *and* (here is the difference) *subsequently releases it explicitly*. The period between a reader reading and its explicit release of the channel (and blocked writer) is the extended rendezvous. In this period, the reader may do anything except read from that channel again.

As indicated in Figure 3, each OCP has burnt into it the target VCN of the network channel it is supporting. It also locates and holds on to the local **Tx** channel it needs to use – this will not change. Its actions follow a simple cycle. It waits for and reads an object (reference) from an application process, wraps this is a (pooled and recycled) message wrapper containing source and target VCNs, forwards that to the appropriate **Tx** process and waits for an acknowledgement on its server channel. Assuming all is well with that acknowledgment, it releases the application writing process. If there is an error reported by the acknowledgment packet (e.g. network or remote node failure), it releases the writer but causes the writer to throw a suitable exception.

An ICP listens (at the start of its cycle) on its server channel. An arriving packet contains an incoming application object, source VCN and **Tx** link reference (for the link on which the network packet arrived). It outputs the object down the channel to the application reader and, when that has been taken, sends an acknowledgment packet back to the **Tx** link (tagged with the source VCN). The channel between an ICP and its reader is not an ER channel and may be buffered (in accordance to the needs of the application).

Each **Tx** process is set up connected to the correct remote machine. All data from application channels on this machine to the remote one (and acknowledgements of data in the other direction) go via this process. Its behaviour is trivial: wait for packets arriving on its server channel and output them to the network. It abstracts details of the network from the rest of the infrastructure – making *different* types of network easy to manage.

Each **Rx** process cycles by waiting for a network packet to arrive. This will contain a target VCN, which is used to find the relevant ICP server channel and to which the packet is forwarded – end of cycle. Again, details of the type of network is abstracted within this process.

It is crucial to avoid network packets unread by an application process blocking other traffic. Application network channels must not get in each others' way within the infrastructure. A **Tx** server channel needs no buffering for this – we may assume the **Rx** process at the other end is never blocked (see later in this paragraph). An OCP server channel needs no buffering – the OCP will be waiting for acknowledgement packets routed to it. An ICP server channel needs buffering – but this will be limited to the number of network links currently open (and, hence, **Rx** processes that write to it will never block). That limit requires that the same JCSP networked channel cannot be bound to more than one application output channel per machine – which is, in fact, no limitation since any number of application processes on a machine may securely share that channel. This is ensured by the *JCSP.net* channel constructors.

Of course, such reasoning should be formalised and verified to build (even) greater confidence. *JCSP.net* infrastructure, however, is directly modelled by a CSP network so this should not be intractable.

Finally, we re-emphasise that the *JCSP.net programmer* needs no knowledge of the details in this section. JCSP *processes* only see channel *interfaces* and do not care whether the *actual* channels plugged into them are local or networked.

2.2 Establishing Network Channels (the *Channel Name Server*)

JCSP.net does not require networked applications to be pre-planned and mapped on to a pre-arranged configuration of processors. The model is very dynamic – processes find each other simply by knowing (or finding out) the names of the network channels they need to use.

The mechanism is brokered by a *Channel Name Server* (CNS), which maintains a table of network addresses (e.g. IP-address/port-number), channel type (e.g. *streamed, overwriting*) against user-defined channel names (which can have a rich structure akin to URLs). The CNS provides a minimum functionality to make dynamic connections. More sophisticated brokers may be built as user applications, bootstrapped on top of the CNS (see Section 2.5). Once processes know where each other are, network connections can be built without registering at a CNS or higher-level broker (see Section 2.4). If different network types are required, each needs its own CNS.

A CNS must be started first and each application on each machine must be able to find it. By default, this is obtained from an XML file the first time a network channel is created by an application on each machine. Alternatively, there are various ways of setting it directly.

Within a processor, networked application channels are either *to-the-net* or *from-the-net*. They may have *one* or *any* number of application processes attached – i.e. `One2NetChannel`, `Any2NetChannel`, `Net2OneChannel` and `Net2AnyChannel`. The '2Net' channels implement only `write` methods – the 'Net2' channels only `reads`. As usual, only '2One' channels may be used as *guards* in an `Alternative`.

To construct networked channels using the CNS, we only need to know their names. For example, at the receiving end:

```
Net2OneChannel in = new Net2OneChannel ("ukc.foo");
```

allocates an ICP and target VCN for this channel and registers the name `"ukc.foo"` with the CNS, along with that VCN and the network address for a *link-listener* process on this machine (which it constructs and starts if this is the first time). If the CNS cannot be found (or the name is already registered), an exception is thrown.

```
One2NetChannel out = new One2NetChannel ("ukc.foo");
```

The above allocates an OCP and source VCN for this channel and asks the CNS for registration details on `"ukc.foo"`, blocking until that registration has happened (time-outs may be set that throw exceptions). The target VCN from those details is given to the OCP. If a network connection is not already open to the target machine, a connection is opened to the target's *link-listener* process and local **Tx**/**Rx** processes created. Either way, the **Tx** server channel connecting to the target machine is given to the OCP. On the receiving machine, the *link-listener* process creates its local **Tx**/**Rx** processes and passes it the connection information.

Unlucky timing in the creation of opposing direction network channels may result in an attempt to create two links between the same pair of machines. However, this is resolved by the machine with the *'lower'* network address abandoning its attempt.

If network channels are being constructed sequentially by an application, all input channels should be set up first – otherwise, there will be deadlock!

2.3 Networked Connections (Client-Server)

JCSP.net *channels* are uni-directional and support *any-one* communication across the network. This means that any number of remote machines may open a named *output* network channel and use it safely – their messages being interleaved at the *input* end. Establishing *two-way* communication, especially for a prolonged conversation, can be done with *channels*, but is not easy or elegant.

So, *JCSP.net* provides *connections*, which provide logical pairs of channels for use in networked *client-server* applications. For example, at the *server* end:

```
Net2OneConnection in = new Net2OneConnection ("ukc.bar");
```

sets up infrastructure similar to that shown in Figure 3, except that the *Server Control Process* (SCP, instead of ICP) provides *two-way* communication to its attached application process. Similarly:

```
One2NetConnection out = new One2NetConnection ("ukc.bar");
```

sets up a *Client Control Process* (CCP, instead of OCP) for *two-way* communication to its attached application process.

Application processes only see *client* and *server* interfaces (rather than *writer* and *reader* ones) to these connections. Each provides *two-way* communications:

```
interface ConnectionClient {
  public void request (Object o);       // write
  public Object reply ();               // read
  public boolean stillOpen ();          // check (if unsure)
}

interface ConnectionServer {
  public Object request ();             // read
  public void reply (Object o);         // write & close
  public void reply (                   // write & maybe close
    Object o, boolean keepOpen
  );
}
```

JCSP.net *connections* are bi-directional and support *any-one* client-server transactions across the network. This means that any number of remote machines may open a named *client* network connection and use it safely – transactions being dealt with atomically at the *server* end (see Figure 5 in Section 2.4 below).

A transaction consists of a sequence of `request`/`reply` pairs, ending with the server making a *write-and-close* `reply`. If necessary, the client can find out when the transaction is finished – by invoking `stillOpen` on its end of the connection. The server infrastructure locks the server application process on to the remote client for the duration of the transaction. Other clients have to wait. Any application attempt to violate the `request`/`reply` sequence will raise an exception.

No network acknowledgements are generated for any *connection* messages. Clients must commit to read `reply` messages from the server and this is sufficient to keep them synchronised. Note that *connections* may not be buffered.

Note, also, that a connection is not *open* until the first `reply` has been received. Clients may open several servers at a time – but only if all clients opening intersecting sets of servers open them in an agreed sequence. Otherwise, the classic deadlock of *partially acquired resources* will strike.

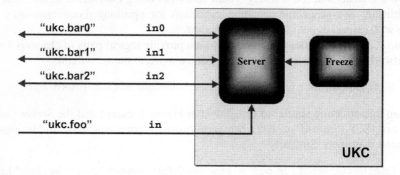

Figure 4. Servicing many events (network connections, network and local channels)

Network connections are *ALTable* at their server ends – i.e. '2One' connections may be used as *guards* in an `Alternative`. For example, `Server` in Figure 4 may wait on events from any of its three networked server *connections* (indicated by the double-arrowed lines), its networked input *channel* and its local input channel.

Finally, we note that JCSP is extended to support *connections* within an application node (i.e. they do not have to be networked). As for channels, processes only see the *interfaces* to their connections and do not need to know whether they are networked.

2.4 Anonymous Channels/Connections

Networked channels and connections do not have to be registered with the CNS. Instead, *input* ends of channels (or *server* sides of connections) may be constructed anonymously. For example:

```
Net2OneChannel in = new Net2OneChannel ();
Net2OneConnection in2 = new Net2OneConnection ();
```

Remote processes cannot find them *by name* using the CNS. But they can be told where they are by communication over channels/connections previously set up with the process(es) that created them. 'Net2' channels and connections contain location information (*network address* and *VCN*). This can be extracted and distributed:

```
NetChannelLocation inLocation = in.getLocation ();
NetConnectionLocation in2Location = in2.getLocation ();

toMyFriend.write (inLocation);
toMyOtherFriend.write (in2Location);
```

Remember that your friends may distribute this further!

A process receiving this location information can construct the *output/client* ends of the networked *channels/connections*. For example:

```
NetChannelLocation outLocation =
    (NetChannelLocation) fromMyFriend.read ();

One2NetChannel out =
    new One2NetChannel (outLocation);
```

and on, perhaps, another machine:

```
NetConnectionLocation out2Location =
    (NetConnectionLocation) fromMyOtherFriend.read ();

One2NetConnection out2 =
    new One2NetConnection (out2Location);
```

Figure 5 shows an example use of anonymous connections. The **UKC** machine holds the *server* end of a CNS-registered connection "ukc.bar" and provides a number of (identical) Server processes for it. This connection has been picked up by applications on machines **P** and **Q**, which have set up *client* ends for "ukc.bar". Now, only one client and one server can be doing business at a time over the shared connection. If that business is a lengthy request/reply sequence, this will not make good use of the parallel resources available.

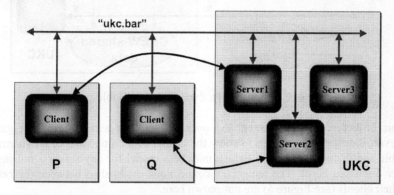

Figure 5. Registered and anonymous connections used in a server farm

Transactions over a *shared resource* should be as brief as possible – in this case: (client) *'gimme a connection'* and (server) *'OK, here's an unregistered one I just made'*. The publicly registered connection is now only used to let a remote client find one of the servers in the **UKC** *farm*. Lengthy *client-server* transactions are conducted over dedicated, and unregistered, connections (shown as curved lines in Figure 5). All such transactions may continue *in parallel*. At the end of a transaction, the server disconnects and discards the unregistered connection – any attempted reuse by the client will fail (and throw exceptions). Note that the unregistered connections will be multiplexed over the link fabric (e.g. *TCP/IP sockets*) that must already be present.

2.5 User Defined Brokers

If you want a matching service more sophisticated than the *simple naming* provided by the CNS, build your own broker as a server for a CNS-registered connection. Anyone knowing (or told) the name of that connection can use your new broker. Applications may then register service details with it, along with the location of their networked server connection (or input channel) supporting it. The broker can be asked for details of registered services or to search for a service matching user requirements. The broker returns the locations of matched services. The networked connections or channels supporting these services – as far as the CNS is concerned – are *anonymous*.

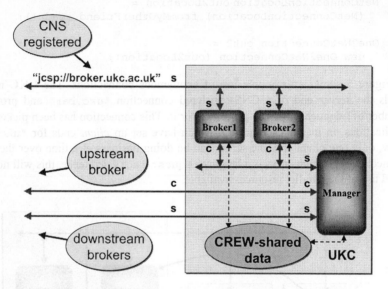

Figure 6. Component of a dynamically scalable server farm

This broker can be structured and operated like the server farm in Figure 5. However, Figure 6 extends this to show the design of a component of a dynamically scalable and distributed service (which may be a model for any server – not just the broker discussed here). The temporary connections established between the brokers and their clients (see Figure 5) are not shown here.

The broker processes are clients on an (internal) connection to their manager. The manager provides service to its local brokers and to the managers of external brokers *downstream*. It also holds a client connection to an external *upstream* manager. The global network of these components forms a *tree topology* of client-servers, which is guaranteed to be *deadlock-free* [15].

Broker processes report their use to their local manager. If the node is in danger of running out, extra ones can be created dynamically by the manager. The manager and its local brokers share a database protected by a *Concurrent-Read-Exclusive-Write* (CREW) lock, which is one of the core capabilities provided by JCSP. The managers in each node regularly keep in touch with each other about the nature of their holdings and can inform their brokers to re-direct clients as necessary.

2.6 Networked Class Loading (Remote Process Launching and Agents)

Objects sent across *JCSP.net* channels or connections currently use Java *serialization*. This means that the receiving JVM has to be able to load – or already have loaded – the classes needed for its received objects. Default *JCSP.net* communications deliver *serialized* data. If an object of an unknown class arrives, a *class-not-found* exception is thrown as *de-serialization* is attempted.

To overcome this, *JCSP.net* channels and connections can be set to download those class files automatically (if the receiving ClassLoader cannot find them locally) – the sending process, after all, must have those files! *JCSP.net* machine nodes cache these class files locally in case they themselves need to forward them.

Consider the simple *server farm* shown in Figure 5. Let the (unregistered) connections dynamically set up between a *client* and *server* process be *class-loading*. Suppose the client has some work it needs to farm out and that that work is in the form of a *JCSP* process. Let out be its networked connection to the remote server:

```
CSProcess work = new MyProcess (...);    // To be done ...
out.request (work);                       // Send to worker.
work = (MyProcess) out.reply ();          // Get results.
```

Results may be safely extracted from the returned work – a non-running process.

At the server end, let in be its networked connection to the remote client:

```
CSProcess work =                 // Class file may
  (CSProcess) in.request ();     // be downloaded.
work.run ();                     // Do the work.
in.reply (work);                 // Send results.
```

If this is the first time the server farm has been given an instance of MyProcess to run, its class will be automatically downloaded from the in connection (and cached).

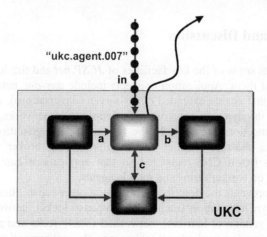

Figure 7. Mobile processes (agents)

The `MyProcess` process launched above was self-contained. It is just as easy to download and launch processes that plug into and engage with local resources – which brings us into *agent* technologies. Figure 7 shows an agent server environment on the **UKC** machine. Agents arrive on the CNS-registered (`"ukc.agent.007"`) input channel, plug into the local network, run and then move on. Code for the actual agent receiving process could take the simple form:

```
while (running) {
  Bond james = (Bond)in.read ();              // Download maybe?
  james.plugin (a, b, c);                     // Connect locally.
  james.run ();                               // Do the business.
  NetChannelLocation escapeRoute =
    james.getNextLocation ();                 // Where next?
  One2NetChannel escape =
    One2NetChannel (escapeRoute);             // Set it up.
  running = james.getBomb ();                 // Leave politely?
  escape.write (james);                       // Goodbye.
  escape.disconnect ();                       // Lose channel.
}
```

In the above loop, `Bond` is just an interface extending `jcsp.lang.CSProcess` with the `plugin`, `getNextLocation` and `getBomb` methods needed later. If the actual process read in is of a class not seen before, its code is downloaded. The agent is then plugged into the local channels and run. As it runs, its behaviour depends on its code, any data it brought with it and what it finds in the local environment. Before it finishes, it must have decided upon where it wants to go next (and whether it wants to *bomb* the loop of this agent handler). After it finishes, the handler gets the next location data, sets up the escape channel, sends the agent down the channel, deletes the escape channel local infrastructure and (maybe) loops around to wait for the next `Bond`.

3   Summary and Discussion

This paper presents some of the key facilities of *JCSP.net* and the JCSP mechanisms used to implement them. Application concepts include *any-one* networked channels and *connections* (that enable extended *client-server* conversations), a *Channel Name Server* (CNS) for the dynamic construction of application networks, and *anonymous* channels and connections (that evade the normal CNS registration). Applications outlined include a scalable parallel and multiprocessing broker for user-defined matching services (itself CNS-registered so that applications can find it), remote process launching on worker farms and mobile agents.

Lack of space prevents the description of several other capabilities: e.g. networked multiple barrier synchronisation (including two-phase locks), networked buckets (a non-deterministic version of barriers), networked *CREW* locks and networked *CALL* channels (which provide a method call interface to channel semantics). The semantics and APIs of their *non-networked* cousins are in the standard JCSP documentation [1]. *JCSP.net* also provides for the migration of *server connection ends* (respectively *input channel ends*) between networked machines in a way that loses no data and is transparent to any of the processes at the *client* (respectively *output*) *ends* – i.e. they are unaware of the migration.

Also only outlined in this paper is the implementing JCSP infrastrucure. The basic software crossbar of concurrent processes supporting standard network channels is explained (Sections 2.1 and 2.2), but not the additional structures needed to deal with *streaming* (buffered) channels, *class-downloading* channels, and *connections* (that, for example, need two *Virtual Channel Numbers* per *Server Control Process*).

Such extra details will be presented in later papers, along with performance results from simple benchmarks and a range of classical parallel and distributed computing demonstrators.

The aims of *JCSP.net* are to provide a *simple* way to build complex, scalable, distributed and dynamically evolving systems. Its primitives are based upon well-established CSP operations, so that its applications may be directly and formally modelled (and *vice-versa*). It allows the expression of concurrency at *all* levels of system – both *between* parallel running devices (whether home, cluster or *Internet* distributed) and *within* a single machine (which may be multi-processor). The same concepts and theory are used regardless of the physical realisation of the concurrency – shared memory and distributed memory systems are programmed in the same way.

This contrasts with the models provided by PVM, MPI and BSP, none of which address concurrency *within* a processing node. We believe that concurrent *processes* are too powerful an idea to leave out of our armoury of system design, analysis, implementation and verification tools. Again, this conflicts with received wisdom that says that concurrent algorithms are *harder* than equivalent serial ones. There is no need to stir up an argument – time and experience will tell. Current experiences with *large* serial applications are not promising. We cite the design illustrated in Figure 3 for an experience of the simplicity wrought, and complex functionality quickly generated, by (CSP based) concurrency.

The status of *JCSP.net* is that it is being alpha-tested within our research group [16] at UKC. Special thanks are accorded to Brian Vinter (of the Southern University of Denmark, currently visiting us on sabatical) for considerable assistance. Details of beta-testing will be posted on the *JCSP* website [1]. We are negotiating to see if commercial support can be provided.

JCSP.net is part of a larger project on language design, tools and infrastructure for scalable, secure and simple concurrency. In particular, we are developing the multi-processing **occam** language with various kinds of dynamic capability (see the *KroC* website [17]) that match the flexibility available to Java systems, but which retain strong semantic checks against major concurrency errors (such as *race hazards*) and ultra-low overheads for concurrency management (some two to three orders of magnitude lighter than those accessible to Java). The Java and **occam** concurrency work feed off each other in many interesting ways – for example, there will be a *KroC.net*.

References

1. Welch, P.H., Austin, P.D.: JCSP Home Page, http://www.cs.ukc.ac.uk/projects/ofa/jcsp/ (2002)
2. Welch, P.H.: Process Oriented Design for Java – Concurrency for All. In: PDPTA 2000 volume 1. CSREA Press (June 2000) 51-57
3. Lea, D.: Concurrent Programming in Java (Second Edition): Design Principles and Patterns. The Java Series, Addison-Wesley (1999) section 4.5
4. Welch, P.H.:, Java Threads in the Light of occam/CSP. In: Architectures, Languages and Patterns for Parallel and Distributed Applications, Proceedings of WoTUG 21. IOS Press (Amsterdam), ISBN 90 5199 391 9 (April 1998) 259-284
5. Welch, P.H.:, Parallel and Distributed Computing in Education. In: J.Palma et al. (eds): VECPAR'98, Lecture Notes in Computer Science, vol. 1573. Springer-Verlag (June 1998)
6. Hoare, C.A.R.:, Communicating Sequential Processes, CACM, 21-8, (1978) 666-677
7. C.A.R.Hoare. Communicating Sequential Processes. Prentice Hall ISBN 0 13 153289 8 (1985)
8. A.W.Roscoe. The Theory and Practice of Concurrency. Prentice Hall, ISBN 0 13 674409 5 (1997)
9. Inmos Limited. occam2.1 Reference Manual, Technical Report. http://wotug.ukc.ac.uk/parallel/occam/documentation/ (1989)
10. Inmos Limited. occam3 Reference Manual, Technical Report. http://wotug.ukc.ac.uk/parallel/occam/documentation/. (1992)
11. May, M.D., Thompson, P.W., Welch, P.H.: Networks, Routers and Transputers. IOS Press, ISBN 90 5199 129 0 (1993)
12. Hilderink, G.H.: CTJ Home Page. http://www.rt.el.utwente.nl/javapp/ (2002)
13. Hilderink, G.H., Broenink, J., Vervoort, W., Bakkers, A.W.P: Communicating Java Threads, In: Bakkers, A.W.P et al. (eds): Parallel Programming in Java, Proceedings of WoTUG 20, Concurrent Systems Engineering Series, vol. 50. IOS Press (Amsterdam), ISBN 90 5199 336 6 (1997) 48-76
14. Nevison, C.: Teaching Distributed and Parallel Computing with Java and CSP. In: Proceedings of the 1st ACM/IEEE International Symposium on Cluster Computing and the Grid, Brisbane, Australia (May 2001)
15. Martin, J.M.R., Welch, P.H.: A Design Strategy for Deadlock-Free Concurrent Systems. Transputer Communications 3(4). John Wiley & Sons, ISSN 1070 454 X (October 1996) 215-232
16. Welch, P.H.: Concurrency Research Group Home Page, Computing Laboratory, University of Kent at Canterbury http://www.cs.ukc.ac.uk/research/groups/crg/ (2002)
17. Welch, P.H., Moores J., Barnes, F.R.M., Wood, D.C.: KRoC Home Page. http://www.cs.ac.ukc.ac/projects/ofa/kroc/ (2002)

The MICROBE Benchmarking Toolkit for Java: a Component-Based Approach

Dawid Kurzyniec and Vaidy Sunderam

Department of Math and Computer Science, Emory University,
1784 North Decatur Road, Suite 100, Atlanta, GA 30322, USA
{dawidk, vss}@mathcs.emory.edu

Abstract. Java technology has recently been receiving increasing attention as a platform for high performance and large scale scientific computing. The MICROBE benchmarking toolkit is being developed to assist in the accurate evaluation of Java platforms. MICROBE is based on the tenet that benchmarking suites, in addition to furnishing benchmark codes, should provide flexibility in customizing algorithms, instruments, and data interpretation to facilitate more thorough evaluation of virtual environments' performance. The MICROBE architecture, projected usage scenarios, and preliminary experiences are presented in this paper.

1 Introduction

The number of areas in which Java technology [9] is adopted has been increasing continuously over last years. In particular, there is strong interest in using Java for so-called *Grande* applications, i.e. large scale scientific codes with substantial memory, network and computational performance requirements [2,20]. The Java Grande Forum [14] has been established to promote and augment Java technology for use in this area. Concerns about performance have traditionally accompanied Java from the very beginning of its evolution and continue to do so, especially in the context of high performance, distributed, and Grid computing. The performance of a Java application depends not only on the program code and the static compiler, as in the case of traditional languages, but is also highly influenced by the dynamic behavior of the Java Virtual Machine (VM). Although the advanced dynamic optimization techniques used in modern virtual machines often lead to application performance levels comparable to those using traditional languages, efficiency in Java is highly dependent on careful coding and following certain implementation strategies. In general, despite the ease and rapid deployment benefits that accrue, obtaining high performance in Java requires significant effort above and beyond the coding process. In this context, the existence and availability of reliable and thorough benchmarking codes, in addition to profiling tools, is of considerable importance to Java community.

Benchmarking (and especially microbenchmarking) of Java codes is much more complicated than benchmarking other languages due to the dynamic nature of Java Virtual Machines. Modern VMs are equipped with so-called Just-In-Time

P.M.A. Sloot et al. (Eds.): ICCS 2002, LNCS 2330, pp. 709–719, 2002.

(JIT) compilers, which translate Java bytecode into optimized native code at run time. Because the optimization process can be time and memory consuming, state-of-the-art Virtual Machines [12,13] often defer compilation (resulting in so-called JIT warm-up time), and adjust the level of optimizations by observing program execution statistics that are collected in real-time. The objective of such VMs is to heuristically self-regulate compilation so that the "hot spots" of the program are strongly optimized while less frequently called methods are compiled without optimization or even not compiled at all. The common optimization techniques adopted include, among others, aggressive code inlining based on the analysis of data flow and the dependency of classes loaded into the VM.

Many benchmark suites try to cope with JIT warm-up issues simply by providing performance averages for increasing time intervals, but in general, the dynamic issues mentioned above can relate the VM performance not only to the time elapsed since the program was started and the number of times a given method was invoked but also to many other factors, like the dependency graph of classes loaded into the Virtual Machine or the amount of available memory. Dynamic garbage collection can also influence benchmark results if it is triggered at unexpected moments; further, background threads collecting run-time statistics can affect program execution in unpredictable ways. Therefore, results from a specific benchmark do not guarantee that equivalent results would be reported by another, similar benchmark – in fact, we observed differences up to an order of magnitude. Without considering dynamic issues and knowing exactly what the benchmark code does, such benchmarking results are of limited value. For benchmarks that do not publish source code, this problem is further exacerbated. Furthermore, different benchmarking approaches are required for server applications, where steady-state performance is in the focus, and client applications, where the VM startup overhead, the dynamic optimization overhead and memory footprint must also be taken into consideration. Also, in some applications (e.g. scientific codes) it is very important if the VM can compile and optimize a method that is executed only once (as the method can include a number of nested loops and have significant impact on overall performance) while in other applications that may be of no importance at all.

All the issues discussed above lead to the conclusion that it is virtually impossible to develop a single, generic and complete benchmarking suite for Java that is appropriate for all kinds of applications even if the area of interest was restricted only to large scale scientific codes. On the other hand, there is a definite interest in understanding Java performance in various, application dependent contexts. Rather than providing yet another standardized benchmarking suite, we propose to the Java community an open, extensible, component-based toolkit based upon separation of benchmarking algorithms, the instruments used for measurement, data transformation routines, and data presentation routines. As these components can be freely assembled and mixed, users can rapidly develop customized benchmarking codes they require. By demonstrating MICROBE version of Java Grande Forum benchmark suite, we show that the toolkit can also be a basis for development of complete, standardized benchmark suites.

2 Related work

A number of Java benchmarks have been developed in recent years. Some of these are microbenchmarks focusing on basic operations like arithmetic, method invocations, or object creation [10,1,11]. Others are computational kernels in both scientific [8,21] and commercial [5,22,19,4,7] domains. Typically, unlike microbenchmarks, these suites are proprietary and do not provide source code. There are also a few benchmarks approximating full-scale applications, and comparative benchmarks that implement the same application in Java and a traditional imperative language.

Perhaps the most comprehensive and important benchmarking framework for scientific applications is the Java Grande Forum Benchmark Suite [3,15], currently at version 2.0. This suite consists of three separate sections. Section 1 contains a set of microbenchmarks testing the overhead of basic arithmetic operations, variable assignment, primitive type casting, object creation, exception handling, loop execution, mathematical operations, method calls, and serialization. Section 2 includes numerical kernels, such as the computation of Fourier coefficients, LU factorization, heap sort, successive over-relaxation, encryption and decryption with the IDEA algorithm, FFT, and sparse matrix multiplication. Finally, Section 3 contains several full applications, including a solver for time-dependent Euler equations, a financial simulation using Monte Carlo methods, and a simple molecule dynamics modeler.

The interesting feature of JGF benchmarks is that they separate instrumentation and data collection from data presentation in a way that enables relatively easy addition of new benchmarks to the suite. Unfortunately, the microbenchmarking algorithms themselves are hard-coded thus very difficult to modify.

A few shortcomings of JGF benchmark suite were pointed out [17,7]. The suite was augmented with new benchmarks [6] including corrected arithmetic and method call microbenchmarks, extended microbenchmarks for object creation and exception handling, new microbenchmarks for thread handling, new kernels including sieve of Eratosthenes, tower of Hanoi, Fibonacci sequence and selected NAS [18] parallel benchmarks.

A common issue of Java benchmarks is that for sake of portability, they usually employ the `System.currentTimeMillis()` method as it is the only time measurement facility currently available in Java. The inaccuracy of this measure forces benchmarks to perform a large numbers of iterations, making them vulnerable to strong run time optimizations [17] and excluding more fine-grained tests, like analysis of JIT compilation process step by step or evaluation of interpreted code.

3 MICROBE Toolkit Architecture

In any benchmarking suite, it is possible to discriminate among several conceptual entities:

- **Operations** that are to be tested, like arithmetic operations, object creation, array sort, or FFT.
- **Algorithms** used to perform the benchmark: for microbenchmarking, there are usually different kind of loops, codes for calibration, etc.
- **Instruments** for quantity measurement: in Java, time is usually measured with `System.currentTimeMillis()` method, whereas memory footprint is measured with `Runtime.totalMemory()` method.
- **Data processors** transforming quantities collected by instruments into some interpretable output, like the number of operations performed per time unit, arithmetic mean and standard error computed for a series of measurements, etc.
- **Data reporters** used to present benchmark results to the user. They can show data in the console window, display graphs, generate tables or HTML output files.

In all the benchmarking suites that we reviewed, these entities are more or less tightly bounded together. Although in a few cases (notably, the JGF Benchmark Suite) it is relatively easy to use custom time measure or data reporting routine, it is still not possible to benchmark the same operation using different or modified algorithms, or using different data processing routines to reinterpret the collected data. The MICROBE toolkit is based on careful separation of all five coefficients leading to independence and orthogonality between them.

3.1 Benchlets and Yokes

In order to separate operations from algorithms, the MICROBE toolkit introduces the notion of *benchlets*. A benchlet is a small piece of code encapsulating only the operation to be tested and implementing the following simple Java interface:

```
interface Benchlet {
    void init(Object[] params);
    void execute();
}
```

The `execute()` method provides the operation to be tested, e.g. binary addition or object creation. Each benchlet has a no-argument constructor; the initialization may be performed in the `init()` method. Additionally, the benchlet can be attributed as *unrolled* or *calibratable* (by implementing appropriate interfaces). The *unrolled* benchlet performs more than one operation inside the `execute()` method. The *calibratable* benchlet designates some other benchlet to rule out benchmarking overheads related to the tested operation, like the empty loop cycles or necessary preliminary steps which would otherwise influence the results.

The algorithms in the MICROBE toolkit are represented by entities called *yokes*. A yoke is able to instantiate a benchlet and control its execution in a certain way, possibly performing some measurements. Instead of encapsulating

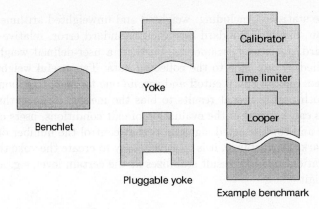

Fig. 1. Benchlets, yokes, and benchmarks

complete, sophisticated algorithms though, yokes focus only on elementary tasks, so the user needs to combine yokes to build more complex ones.

The simplest yoke provided in the toolkit is the *looper*, which repeats the execute() method of an instantiated benchlet within a loop for a specified number of iterations. Most other yokes are not self-sufficient but are rather designed to alter the behavior of other yokes, as denoted by the *pluggable* attribute. A few examples of pluggable yokes include: *calibrator* which adjusts results of measurements for some benchlet with that of its calibrating benchlet, *reseter* which provides benchlets with separate environments by loading them through independent class loaders and only after performing garbage collection, and *time limiter* which restricts the amount of time for which the associated yoke is allowed to run, if only that yoke has the *interruptible* attribute and responds to interrupt request. All standard yokes shipped with the MICROBE toolkit (including the looper yoke) are interruptible. Although this feature requires yokes to continuously monitor the value of a boolean flag, potentially introducing small overheads, it has also the positive side effect of improving the accuracy of calibration, as the calibrator loops (that often tend to be empty) become less vulnerable to dynamic optimization. The relationship between benchlets and yokes is illustrated in Fig. 1.

The modularity of the design enables easy customization of the algorithms formed by collaborating yokes: for instance, it is straightforward to enable or disable calibration, add time constraints, or compare the benchmark behavior as it is loaded through common or separate class loader – features difficult if not impossible to achieve using existing benchmarking suites.

To improve the accuracy and reliability of the results, it is often desirable to repeat a given benchmark several times and collect statistics. To address this need, the MICROBE toolkit provides the *repeater* yoke. This yoke invokes another selected yoke in a loop until the exit condition (specified by the user) is satisfied. During the loop, the repeater yoke can collect arbitrary data series and compute statistics, including: weighted and unweighted arithmetic and geometric mean, median, standard deviation, standard error, relative error, and other standard statistical parameters. Further, a user-defined weight function may be applied at any time to the collected data. Two useful weight functions are predefined: one for sharp cutoff and second one for smooth exponential cutoff, which both permit recent results to bias the metric. Because the results of the statistics can be used in the evaluation of exit conditions, users can develop very specific and sophisticated algorithms that control the number of iterations of a benchmark. For instance, it is relatively easy to create the yoke that repeats a given operation until the result stabilizes at the certain level, e.g. at the third significant digit.

3.2   Data Producers, Consumers, and Filters

Data exchange in the MICROBE toolkit is based upon a variation of event listeners pattern. Yokes do not process or display collected data by themselves, but they may rather have *data outputs* (or *data producers*) to which other entities can attach. Many yokes contain also *data inputs*, or *data consumers*, which can be connected to appropriate data producers. For instance, the calibrator yoke provides inputs accepting the data to be calibrated, and appropriate outputs producing calibrated results. This approach leads to the extreme flexibility – e.g., the calibrator yoke does not have to assume that the measured quantity represents a particular parameter such as computation time or an amount of memory.

The MICROBE toolkit directly supports four kinds of data exchanged between producers and consumers: single *numbers*, *number arrays*, arbitrary *objects*, and *signals*, which do not carry additional information. On their way from one yoke to another, events and data can be processed through *filters* transforming them in numerous ways. The toolkit provides many basic transformation routines, like arithmetic operations on single numbers, *collectors* allowing gathering of several numeric events to be triggered together, array operations computing sum or product of the elements, and many more.

Of particular note are certain kinds of filters which convert signals into numerical values, because various instruments measuring different quantities fit into this category. For example, a *clock* in MICROBE toolkit is nothing but converter of query signal (analogous to pushing the stopper button) into the value expressed in seconds:

```
class Clock implements Transform.Signal2Number {
    public double transform() {
        return System.currentTimeMillis()/1000.0;
    }
}
```

Such a generalization allows the toolkit to be independent of the particular instruments, so users can replace them very easily according to the needs. Also,

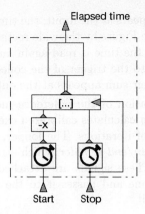

Fig. 2. Difference measurer used to compute time elapsed between two signals

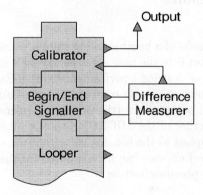

Fig. 3. Example benchmark

because the measurement is reduced just to invocation of simple Java method, new and customized instruments can be developed with little effort.

Apart from the (not very accurate) clock shown above, the MICROBE toolkit provides an additional notion of CPU time measurement based on native code, specifically the `clock()` function from the Standard C Library. Despite the use of native code, portability at the source code level is retained due to the use of Java Native Interface [16] and wide availability of the C library. Measures which are even more accurate but also more platform dependent (like those based upon internal CPU cycle counting) may be developed by users; those can be especially feasible for testing a short term behavior of the Java VM.

Just like yokes, filters can be grouped together to form more sophisticated ones. The Fig. 2 presents a *difference measurer* that computes the time elapsed between two signals using the *clock* described above as an internal instrument (as mentioned, the difference measurer is oblivious to the instrument it uses).

When the "start" signal appears at the input, the time is read and stored by the collector (represented in the Figure by the wide rectangle) after negation. When the "stop" signal is issued, the time is read again and also stored. Because the "stop" signal is connected to the trigger of the collector, these two aggregated values are triggered and their sum appears at the output.

Fig. 3 shows an application of the difference measurer in a example of a benchmark algorithm which calculates calibrated execution time for a benchlet invoked for some number of iterations. The looper, being the innermost yoke used, is wrapped into a *begin/end signaler* which signals the beginning and the end of the computation. These signals are attached to a difference measurer which computes elapsed time and passes it to the calibrator. The calibrated output constitutes the result.

4   Reporting Results

One of the important tasks of a benchmarking suite is to interpret collected data in certain way and report it to the user, e.g., displaying it in the console window, writing it to a text file or drawing a performance graph. Examples of data interpretation include computing a mean over a series of results, transforming time figures into a temporal performance (the number of operations per time unit), etc. The modular structure of the MICROBE toolkit allows virtually any data interpretation to be applied to the benchmark without changing its code. Fig. 4 shows how temporal performance can be computed for the example benchmark algorithm evaluated in previous Section.

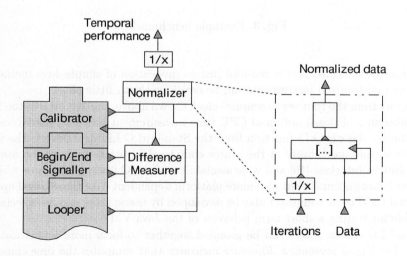

Fig. 4. Data interpreters applied to a benchmark algorithm

The flexible event listener mechanism can also support arbitrary data reporting method as users can decide exactly how (and which) data should be reported by developing proper data consumers and plugging them to appropriate producers. However, although such approach enables unrestricted customization of benchmark behavior, it also requires significant work. In contrast, there is often a need to quickly examine some benchmark code without developing reporting routines intended to be used only once – so, it is desirable that the yokes provide some default notion of data reporting out-of-the-box.

To satisfy this requirement, we added the optional *reporting* attribute to the yoke definition. The *reporting* yoke is the one capable of creating a hierarchically organized report from data collected during benchmark execution. That hierarchical data structure originates from analogous layout of yokes logically nested inside each other. The data report can be then passed to a *Reporter* object in order to exhibit it in some way. At this time we provide only one reporter class which simply displays the collected data in the console window. We are currently working on the improvements to this mechanism, specifically two other reporters: one generating HTML and another generating XML output files.

5   MICROBE and the Java Grande Forum Benchmark Suite

The Java Grande Forum Benchmark Suite is one of the most important benchmarking packages for Java related to large scale scientific computing. To show the appropriateness of MICROBE to become a basis of such benchmark suites, we translated JGF benchmarks into their MICROBE counterparts.

Section 1 of the JGF suite consists of a number of microbenchmarks. We have transformed all of these into benchlets that can be used within the MICROBE toolkit. The algorithm used by the JGF suite for microbenchmarking performs a sequence of successive loops with the number of iterations growing by a factor of two until either the limit of iterations or elapsed time has been reached. We provide an appropriate yoke (called JGFLooper) which implements the same algorithm, but is even more flexible as it fits the generic MICROBE model and can be combined with different yokes to facilitate more sophisticated testing patterns.

Section 2 of the JGF suite consists of various computational kernels. We have developed an appropriate adapter benchlet, which allows MICROBE yokes to execute and measure these kernels. As the adapter benchlet generates signals when approaching subsequent stages of the kernel execution, it is possible to develop some customized yokes especially suited to deal with JGF kernels, although the default ones are sufficient to perform measurements analogous to those of JGF suite.

In Fig. 5, we compare the results obtained by benchmarking the JGF suite (Section 1) and their MICROBE analogs for two different virtual machines. The results reported are geometric mean averages for each microbenchmark. The test platform was a Dell Dimension PC with a 450 MHz Pentium II processor and

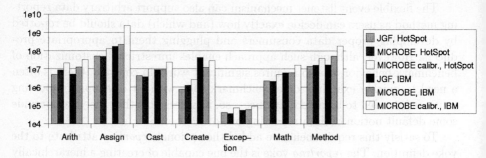

Fig. 5. Performance results [operations per second]

128 MB of memory running Mandrake Linux 7.2. The VMs were SUN HotSpot Client 1.3.0 and IBM 1.3.0 for Linux. As expected, the results for JGF and MI-CROBE are very similar in most of cases, except for the "Create" and "Method" microbenchmarks. That last anomaly is most likely a result of the simplicity of the methods invoked and the objects being created in JGF suite. These aspects, in conjunction with the benchmarking algorithms used, makes the benchmarks very sensitive to run-time optimizations and can introduce performance differences even for very similar codes. It is also worth noting that the calibration – absent in JGF suite but enabled by MICROBE – identified a weakness of the JGF "Assign" test, in which the IBM VM were able to optimize out some variable assignments, resulting in unrealistically high performance figure.

6   Conclusions and Future Work

In this paper, we have described the MICROBE toolkit which facilitates the rapid construction of Java benchmark codes. The toolkit addresses the need observed among the Java community to facilitate testing various aspects of dynamic Java Virtual Machine behavior, especially the issues of dynamic code optimization. We have evaluated the MICROBE version of the Java Grande Forum benchmarks, showing that MICROBE can be used to develop complete benchmark suites.

Currently, we are investigating the possibility of extending the concept of a benchlet to facilitate more sophisticated benchmarks, e.g. emulating the behavior of full blown applications as well as parallel codes. We also recognize the possibility of developing a graphical language which would permit the visual construction of benchlets and yokes in the manner outlined in this paper.

Scientific applications for Java are often integrated with components and libraries written in traditional languages. Such integration can be performed using the Java Native Interface (JNI) [16], which, however, may introduce significant overhead. To address this issue, we intend to establish a full size benchmarking suite for the JNI based on the MICROBE toolkit.

References

1. D. Bell. Make Java fast: Optimize! *JavaWorld*, 2(4), Apr. 1997. http://www. javaworld.com/javaworld/jw-04-1997/jw-04-optimize.html.
2. R. F. Boisvert, J. J. Dongarra, R. Pozo, K. A. Remington, and G. W. Stewart. Developing numerical libraries in Java. In *ACM-1998 Workshop on Java for High-Performance Network Computing*, Stanford University, Palo Alto, California, February 1998. http://www.cs.ucsb.edu/conferences/java98/papers/jnt.ps.
3. J. M. Bull, L. A. Smith, M. D. Westhead, D. S. Henty, and R. A. Davey. A benchmark suite for high performance Java. *Concurrency, Practice and Experience*, 12:375–388, 2000. Available at http://www.epcc.ed.ac.uk/~markb/docs/javabenchcpe.ps.gz.
4. BYTE Magazine. BYTE benchmarks. http://www.byte.com/bmark/bmark.htm.
5. P. S. Corporation. CaffeineMark 3.0. http://www.pendragon-software.com/pendragon/cm3/.
6. Distributed and High-Performance Computing Group, University of Adelaide. Java Grande benchmarks. http://www.dhpc.adelaide.edu.au/projects/javagrande/benchmarks/.
7. O. P. Doederlein. The Java performance report. http://www.javalobby.org/fr/html/frm/javalobby/features/jpr/.
8. J. Dongarra, R. Wade, and P. McMahan. Linpack benchmark – Java version. http://www.netlib.org/benchmark/linpackjava/.
9. J. Gosling, B. Joy, G. Steele, and G. Bracha. *The Java Language Specification*. Addison-Wesley, second edition, 2000. http://java.sun.com/docs/books/jls/index.html.
10. W. Griswold and P. Philips. Excellent UCSD benchmarks for Java. Available at http://www-cse.ucsd.edu/users/wgg/JavaProf/javaprof.html.
11. J. Hardwick. Java microbenchmarks. http://www.cs.cmu.edu/~jch/java/benchmarks.html.
12. Java HotSpot technology. http://java.sun.com/products/hotspot.
13. Jalapeño project home page. http://www.research.ibm.com/jalapeno.
14. Java Grande Forum. http://www.javagrande.org.
15. Java Grande Forum. Java Grande Forum benchmark suite. http://www.epcc.ed.ac.uk/javagrande/.
16. Java Native Interface. http://java.sun.com/j2se/1.3/docs/guide/jni/index.html.
17. J. A. Mathew, P. D. Coddington, and K. A. Hawick. Analysis and development of Java Grande benchmarks. In *ACM 1999 Java Grande Conference*, San Francisco, California, June 12-14 1999. Available at http://www.cs.ucsb.edu/conferences/java99/papers/41-mathew.ps.
18. NASA Numerical Aerospace Simulation. NAS parallel benchmarks. http://www.nas.nasa.gov/Software/NPB/.
19. PC Magazine. JMark 1.01. http://www8.zdnet.com/pcmag/pclabs/bench/benchjm.htm.
20. M. Philippsen. Is Java ready for computational science? In *Proceedings of the 2nd European Parallel and Distributed Systems Conference for Scientific Computing*, Vienna, July 1998. http://math.nist.gov/javanumerics/.
21. R. Pozo. SciMark benchmark for scientific computing. http://math.nist.gov/scimark2/.
22. Standard Performance Evaluation Corporation. SPEC JVM98 benchmarks. http://www.spec.org/osg/jvm98/.

Distributed Peer-to-Peer Control in Harness *

C. Engelmann, S. L. Scott, G. A. Geist

Computer Science and Mathematics Division,
Oak Ridge National Laboratory, Oak Ridge, TN 37831-6367, USA
{engelmannc, scottsl, gst}@ornl.gov
http://www.csm.ornl.gov/dc.html

Abstract. Harness is an adaptable fault-tolerant virtual machine environment for next-generation heterogeneous distributed computing developed as a follow on to PVM. It additionally enables the assembly of applications from plug-ins and provides fault-tolerance. This work describes the distributed control, which manages global state replication to ensure a high-availability of service. Group communication services achieve an agreement on an initial global state and a linear history of global state changes at all members of the distributed virtual machine. This global state is replicated to all members to easily recover from single, multiple and cascaded faults. A peer-to-peer ring network architecture and tunable multi-point failure conditions provide heterogeneity and scalability. Finally, the integration of the distributed control into the multi-threaded kernel architecture of Harness offers a fault-tolerant global state database service for plug-ins and applications.

1 Introduction

Parallel and scientific computing is a field of increasing importance in many different research areas, such as fusion energy, material science, climate modeling and human gnome research. Cost-effective, flexible and efficient simulations of real-world problems performed by parallel applications based on mathematical models are replacing the traditional experimental research.

Software solutions, like Parallel Virtual Machine (PVM) [4] and Message Passing Interface (MPI) [5], offer a wide variety of functions to allow messaging, task control and event handling. However, they have limitations in handling faults and failures, in utilizing heterogeneous and dynamically changing communication structures, and in enabling migrating or cooperative applications.

The current research in heterogeneous adaptable reconfigurable networked systems (Harness) [1]-[3] aims to produce the next generation of software solutions for parallel computing. A high-available and light-weighted distributed

* This research was supported in part by an appointment to the ORNL Postmasters Research Participation Program which is sponsored by Oak Ridge National Laboratory and administered jointly by Oak Ridge National Laboratory and by the Oak Ridge Institute for Science and Education under contract numbers DE-AC05-84OR21400 and DE-AC05-76OR00033, respectively.

P.M.A. Sloot et al. (Eds.): ICCS 2002, LNCS 2330, pp. 720–728, 2002.

virtual machine (DVM) provides an encapsulation of a few hundred to a few thousand physical machines in one virtual heterogeneous machine.

A high availability of a service can be achieved by replication of the service state on multiple server processes at different physical machines [6]-[12]. If one or more server processes fails, the surviving ones continue to provide the service because they know the state. The Harness DVM is conceived as such a high-available service. It will survive until at least one server process is still alive.

Since every member (server process) of a DVM is part of the global state of the DVM service and is able to change it, a distributed control is needed to replicate the global state of the DVM service and to maintain its consistency. This distributed control manages state changes, state replication and detection of and recovery from faults and failures of members.

2   Distributed Peer-to-Peer Control

The distributed control symmetrically replicates the global state of the DVM, which consists of the local states of all members, to all members. A global state change is performed using a transaction, which modifies the local state of one or more members depending on their local and/or the global state during execution. These local state modifications are replicated to all members.

Such a transaction may join or split member groups, add or remove a member, start or shutdown a service at one member or at a member group, change states of services or report state changes of services. The global state is actively replicated (hot-standby) only at a subset of the DVM members, while all other members retrieve a copy of the global state only on demand.

The tradeoff between performance and reliability is adjustable by choosing the number and location of members with active replication. The number of hot-standby members affects the overall DVM survivability and robustness in terms of acceptable fault and recovery rate. The classic client-server model with its single point of failure is realized if there is only one.

The location of hot-standby members additionally addresses network bottlenecks and partitioning. If different research facilities collaborate using a large DVM, every remote site may choose its own hot-standby members. Additionally, any member can become a hot-standby member if a faulty hot-standby member needs to be replaced or the DVM configuration changes.

The hot-standby member group consistently maintains the replicated global state using group communication in an asynchronous connected unidirectional peer-to-peer ring network architecture. The members of this group are connected in a ring of TCP/IP peer-to-peer connections. Messages are transmitted in only one direction and received, processed and sent asynchronously.

3   Group Communication

The distributed control uses group communication services, such as Atomic Broadcast and Distributed Agreement, to simplify the maintenance of consistently replicated state despite random communication delays, failures and recoveries. These services are based on reliably broadcasting messages to all members replicating the state in the hot-standby member group.

3.1   Reliable Broadcast

The Reliable Broadcast is the most basic group communication service. It ensures a message delivery from one member to all members in a group, so that if process p transmits a message m, then m is delivered by all non-faulty processes and all non-faulty processes are notified of any failures.

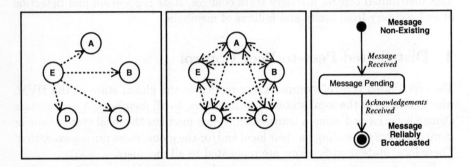

Fig. 1. Reliable Broadcast (Phase I, Phase II, State-Machine)

In order to reliably broadcast a message, one member broadcasts it to all members. Each member who receives it broadcasts an acknowledgement back to all members. A member knows that a message is reliably broadcasted if message and related acknowledgements of all members are received. The Reliable Broadcast performs two phases of broadcast communication.

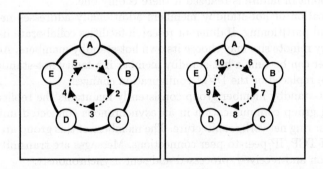

Fig. 2. Reliable Broadcast in a unidirectional Ring (Phase I, Phase II)

A straightforward implementation of this Reliable Broadcast algorithm results in high communication costs (n broadcasts) and is not scalable. What follows is a description of a much more efficient algorithm based on a unidirectional ring network which is used by the distributed control in Harness.

In a unidirectional ring, messages are forwarded from one member to the next to reliably broadcast a message. In phase one, a member sends a message to its neighbor in the ring until a member who has it already received before receives it. In phase two, a member sends the acknowledgement to its neighbor in the ring until a member who has it already received before receives it.

3.2   Atomic Broadcast

The Atomic Broadcast is an extension of the Reliable Broadcast. It additionally ensures a globally unique order of reliably broadcasted messages at all members in a group, so that if process p transmits a message $m1$ followed by a message $m2$, then $m1$ is delivered by all non-faulty processes before $m2$ and all non-faulty processes are notified of any failures.

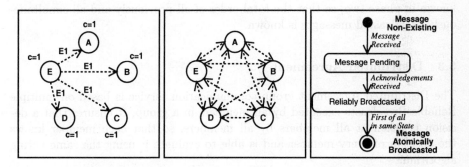

Fig. 3. Atomic Broadcast (Phase I, Phase II, State-Machine)

In order to atomically broadcast a message, the Reliable Broadcast is extended with a message numbering and sorting algorithm.

A global message number is maintained with the message broadcast, so that members cannot reuse it immediately. Every member has a local copy of a global message number counter, which is increased before every Atomic Broadcast and constantly updated with higher received values.

All simultaneously broadcasted messages have the same message number and all sequentially broadcasted messages have increasing message numbers. A static and globally unique member priority, which is assigned by the group during member admission, orders messages with equal numbers.

A message is broadcasted with its globally unique id (number/priority) and sorted into a list on receiving, so that it is atomically broadcasted if all previous messages are reliably broadcasted. One atomically broadcasted message may cause other reliably broadcasted messages to be in the atomically broadcasted state, so that blocks of messages may occur to be atomically broadcasted.

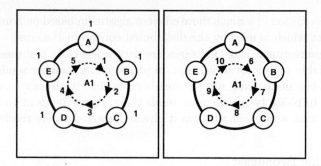

Fig. 4. Atomic Broadcast in a unidirectional Ring (Phase I, Phase II)

In a unidirectional ring, the message id is passed with the message in phase one from one member to the next until received twice. The message id is also passed with the acknowledgement in phase two from one member to the next until received twice. All messages with the same or a lower message number are known in phase two, so that the total order of all previously and all simultaneously broadcasted messages is known.

3.3   Distributed Agreement

The Distributed Agreement group communication service is based on multiple Reliable Broadcasts executed by all members in a group. It ensures that a decision is taken at all members by all members, so that each member knows the decision of every member and is able to evaluate it using the same voting algorithm.

In order to perform a Distributed Agreement on a decision, every member executes a Reliable Broadcast of its decision. A decision is taken at a member by all members if the decisions of all members are reliably broadcasted. Common voting algorithms are: majority, minority (at least one) and unanimous agreement.

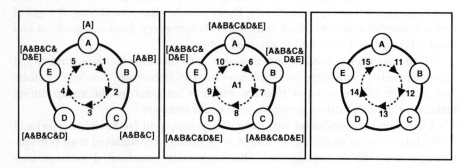

Fig. 5. Distributed Agreement in a unidirectional Ring (Phase I, Phase II, Phase III)

The Distributed Agreement in a unidirectional ring is based on interleaved Reliable Broadcasts. One message is passed around the ring twice to reliably broadcast information appended by all members in the first round. Collective communication is used to avoid n Reliable Broadcasts or n messages.

Passing a message from one member to the next until received twice collects the decisions of all members in phase one. Passing this message around the ring again in phase two broadcasts the final decision and ensures the decision collection of phase one. Passing an acknowledgement around the ring in phase three ensures the final decision broadcast of phase two.

The decisions of all members are collected in phase one using a counter for every class of decision. Their final count is broadcasted in phase two. The voting algorithm compares these values at every member in phase three. This anonymous voting may be replaced by collecting the member names for every class of decision or by a mixture of both.

4 Member Fault Recovery

A single, cascade or multiple member fault does not affect group communication services directly. However, they are based on acknowledgements for received messages from all members and the membership changes when a fault occurs. The underlying communication system notifies all indirectly affected members who are waiting for acknowledgements from faulty members.

Messages may get lost if a faulty member was used as a router. Since all messages are stored until acknowledged, they are sent again if a fault occurs to ensure delivery of messages at least once. Doubled messages are filtered at the receiving member based on the group communication service state.

Single or multiple faults in a ring are recovered by connecting the two correct neighbor members of the faulty member(s) to each other. The neighbor who had sent messages and acknowledgements to a faulty member assumes that all unacknowledged messages and all acknowledgements are lost.

All unacknowledged messages are sent again in the same order. A hop counter contained in every message which is increased when passed to the next member is used to distinguish doubled messages from messages seen twice. A message is ignored if the hop count is less than the stored one.

The last acknowledgement is always stored at a member and sent again. The receiving member regenerates all lost acknowledgements based on his last stored and based on the order the related messages passed previously.

Cascaded faults, which are single or multiple faults occurring during a recovery from previous ones, are handled by restarting the recovery.

5 Membership

The group membership requires an agreement on an initial group state and a linear history of group state changes at all members, so that each member has a complete and consistent replicated copy of the group state.

In order to achieve an agreement on an initial group state, every new member obtains a copy of the group state from a group member. In order to achieve an agreement on a linear history of state changes, every state change is performed as a transaction using an Atomic Broadcast of the state change followed by a Distributed Agreement on the state change execution result.

Transactions consistently replicate the list of members as part of the group state while admitting or removing members. In order to admit a new member, a member of the group connects to the new member and updates it with a copy of the group state. A member removal is based on a controlled member fault. In order to remove a member, the member to be removed disconnects itself from the group. All members expect the member fault and recover easily.

6   Transaction Types

The distributed control extends the management of a symmetrically distributed global state database by enabling transactions to be partially or completely committed or rejected depending on their execution result, which may be different at every member. Transactions may involve local state changes that may fail at one or more members, local state changes at one or more members that cannot fail and global state changes that may fail at all members.

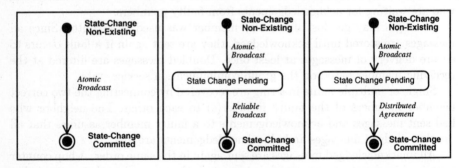

Fig. 6. Transaction Control State-Machines (Type I, II, III)

A linear history of state changes is realized using an Atomic Broadcast of a state change followed by a Distributed Agreement on the execution result, since a state change may involve multiple local state change failures (Type III).

A Reliable Broadcast of the execution result replaces the Distributed Agreement if only one local state change that may fail is involved (Type II). The Atomic Broadcast alone is performed if a transaction involves only local state changes that cannot fail or global state changes that may fail (Type I).

A transaction control algorithm decides the transaction type on a case-by-case basis and guarantees the total order of state change executions and state change commits, so that all state changes are executed and committed/rejected in the same order at all members.

7   Harness Kernel Integration

A first prototype was developed as a distributed stand-alone server application in C on Linux to test and experiment with different variations of the described group communication algorithms. The results were used to refine the algorithms and to define the requirements for the integration of the distributed control into the multi-threaded Harness kernel architecture.

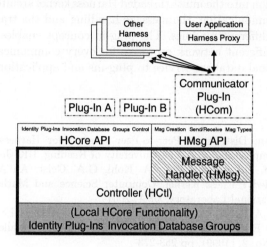

Fig. 7. Scheme of Harness Kernel

The transaction control algorithm manages all transactions in the controller module (HCtl). The central message handler (HMsg) delivers all messages to and from the controller. Communicator plug-ins (HCom) maintain the network connections to communicator plug-ins of other kernels.

Group communication messages are exchanged between a communicator plug-in and the controller using the message handler. State changes are executed by the controller or by a plug-in, while the controller maintains the state database. State change requests are delivered from a plug-in to the controller via the message handler or directly using the kernel API.

8   Conclusions

We have developed a distributed control that automatically detects and recovers single, multiple and cascaded member faults. Its design allows a trade-off between performance and robustness, i.e. tunable multi-point failure conditions. A strictly symmetric global state control and replication ensures fault-tolerance and protects members against denial-of-service and starvation.

The distributed control uses group communication services, such as Reliable Broadcast, Atomic Broadcast and Distributed Agreement, to simplify the maintenance of consistently replicated state despite random communication delays, failures and recoveries. These algorithms were adapted to a peer-to-peer ring network architecture to improve scalability and heterogeneity.

Global state changes are transactions that are executed at all members or at a member group and consistently committed or rejected at the state database of all members depending on the execution results of all members. The transaction control algorithm classifies transactions into different types and chooses the most efficient combination of group communication services.

The integration into the multi-threaded Harness kernel architecture distributes the network communication, the message handling and the transaction control algorithm over different modules. The plug-in concept enables the distributed control to use different network protocols for every communication link, while providing its global database service to plug-ins and applications.

References

1. C. Engelmann: Distributed Peer-to-Peer Control for Harness. Master Thesis, School of Computer Science, The University of Reading, UK, Jan. 2001
2. W.R. Elwasif, D.E. Bernholdt, J.A. Kohl, G.A. Geist: An Architecture for a Multi-Threaded Harness Kernel. Computer Science and Mathematics Division, Oak Ridge National Laboratory, USA, 2001
3. G.A. Geist, J.A. Kohl, S.L. Scott, P.M. Papadopoulos: HARNESS: Adaptable Virtual Machine Environment For Heterogeneous Clusters. Parallel Processing Letters, Vol. 9, No. 2, (1999), pp 253-273
4. G.A. Geist, A. Beguelin, J. Dongarra, W. Jiang, R. Manchek, V. Sunderam: PVM: Parallel Virtual Machine; A User's Guide and Tutorial for Networked Parallel Computing. MIT Press, Cambridge, MA, 1994
5. M. Snir, S. Otto, S. Huss-Lederman, D. Walker, J. Dongarra: MPI: The Complete Reference. MIT Press, Cambridge, MA, 1996
6. F. Cristian: Synchronous and Asynchronous Group Communication. Communications of the ACM, Vol. 39, No. 4, April 1996, pp 88-97
7. D. Dolev, D. Malki: The Transis Approach to High Availability Cluster Communication. Communications of the ACM, Vol. 39, No. 4, April 1996, pp 64-70
8. E. Moser, P. M. Melliar-Smith, D. A. Agarwal, R. K. Budhia and C. A. Lingley-Papadopoulos: Totem: a fault-tolerant multicast group communication system. Communications of the ACM 39, 4 (Apr. 1996), pp 54-63
9. F.C. Gaertner: Fundamentals of Fault-Tolerant Distributed Computing in Asynchronous Environments. ACM Computing Surveys, Vol. 31, No. 1, March 1999
10. S. Mishra, Lei Wu: An Evaluation of Flow Control in Group Communication. IEEE/ACM Transactions on Networking, Vol. 6, No. 5, Oct. 1998
11. T.D. Chandra, S. Toueg: Unreliable Failure Detectors for Reliable Distributed Systems. I.B.M Thomas J. Watson Research Center, Hawthorne, New York and Department of Computer Science, Cornell University, Ithaca, New York 14853, USA, 1991
12. M. Patino-Martinez, R. Jiminez-Peris, B. Kemme, G. Alonso: Scalable Replication in Database Clusters. Technical University of Madrid, Facultad de Informatica, Boadilla del Monte, Madrid, Spain and Swiss Federal Institute of Technology (ETHZ), Department of Computer Science, Zuerich

A Comparison of Conventional Distributed Computing Environments and Computational Grids

Zsolt Németh[1], Vaidy Sunderam[2]

[1] MTA SZTAKI, Computer and Automation Research Institute,
Hungarian Academy of Sciences,
P.O. Box 63., H-1518 Hungary,
E-mail: zsnemeth@sztaki.hu
[2] Dept. of Math and Computer Science,
Emory University,
1784 North Decatur Road, Atlanta, GA 30322, USA
E-mail: vss@mathcs.emory.edu

Abstract. In recent years novel distributed computing environments termed grids have emerged. Superficially, grids are considered successors to, and more sophisticated and powerful versions of, conventional distributed environments. This paper investigates the intrinsic differences between grids and other distributed environments. From this analysis it is concluded that minimally, grids must support *user* and *resource abstraction*, and these features make grids semantically different from other distributed environments. Due to such semantic differences, grids are not simply advanced versions of conventional systems; rather, they are oriented towards supporting a new paradigm of distributed computation.

1 Introduction

In the past decade *de-facto* standard distributed computing environments like PVM [12] and certain implementations of MPI[15], e.g. MPICH, have been popular. These systems were aimed at utilizing the distributed computing resources owned (or accessible) by a user as a single parallel machine. Recently however, it is commonly accepted that high performance and unconventional applications are best served by sharing geographically distributed resources in a well controlled, secure and mutually fair way. Such coordinated worldwide resource sharing requires an infrastructure called a *grid*. Although grids are viewed as the successors of distributed computing environments in many respects, the real differences between the two have not been clearly articulated, partly because there is no widely accepted definition for grids. There are common views about grids: some define it as a high-performance distributed environment; some take into consideration its geographically distributed, multi-domain feature, and others define grids based on the number of resources they unify, and so on. The aim of this paper is to go beyond these obviously true but somewhat superficial views and highlight the fundamental functionalities of grids.

P.M.A. Sloot et al. (Eds.): ICCS 2002, LNCS 2330, pp. 729–738, 2002.

So far, [8] is the only paper that attempts to present an explicit definition for grids. It focuses on *how* a grid system can be constructed, and what components, layers, protocols and interfaces must be provided. Usually the difference between conventional systems and grids are expressed in terms of these technical differences. Our approach is orthogonal to [8] since we try to determine *what* a grid system should provide in comparison with conventional systems without regard to actual components, protocols or any other details of implementation. These details are just consequences of the fundamental differences. We focus on the semantics of these systems rather than their architecture and, in this sense, our approach is novel. The analysis and conclusion presented informally here is a result of formal modeling [18] of grids based on the ASM [2] method. The formal model sets up a simple description for distributed applications running under assumptions made for conventional systems. Through formal reasoning it is shown that the model cannot work under assumptions made for grids; rather, additional functionalities must be present. The formal model in its current form is not aimed at completely describing every details of a grid. Instead, it tries to find the elementary grid functionalities at a high level of abstraction and provides a framework where technical details can be added at lower levels of abstraction, describing the mechanisms used to realize these abstractions.

This paper presents a comparison of "conventional" distributed environments (PVM, MPI) and grids, revealing intrinsic differences between these categories of distributed computing frameworks. Grids are relatively new (as of 1999 - "there are no existing grids yet" [9]) therefore, a detailed analysis and comparison is likely to be of value. In the modeling, an idealistic grid system is taken into consideration, not necessarily as implemented but as envisaged in any papers. Figure 1 serves as a guideline for the comparison. In both cases shown in the figure, cooperating processes forming an application are executed on distributed (loosely coupled) computer systems. The goal of these computing environments is to provide a virtual layer between the application (processes) and the physical platform. In the paper a bottom-up analysis of the virtual layer is presented and it is concluded that the virtual layer, and therefore the way the mapping is established between the application and the physical resources, is fundamentally different in the two framework categories. This fundamental difference is also reflected as technical differences in many aspects, as discussed in this paper. Section 2 introduces the concept of distributed applications and their supporting environments. In Section 3 these environments are compared at the virtual level (see Figure 1.) In Section 4 fundamental differences between grids and conventional systems are introduced, and Section 5 discusses their technical aspects. Finally, Section 6 is an overview of possible application areas.

2   Distributed applications

Distributed applications are comprised of a number of cooperating processes that exploit the resources of loosely coupled computer systems. Although in modern programming environments there are higher level constructs, processes interact with each other essentially through message passing. An application may be distributed

- simply due to its nature.
- for quantitative reasons; i.e. gaining performance by utilizing more resources.
- for qualitative reasons: to utilize specific resources or software that are not locally present.

Distributed computing may be accomplished via traditional environments such as PVM and some implementations of MPI, or with emerging software frameworks termed computational grids. In both cases the goal is to manage processes and resources, provide support for communication or other means of interaction, and in general provide a coherent view to the application by virtually unifying distributed resources into an abstract machine. In both scenarios the structure of an application is very similar. The processes have certain resource needs that must be satisfied at some computational nodes. If all the resource needs of a process are satisfied, it can execute the computation. In this respect, there is not much difference between conventional and grid systems, and any variations are technical ones. The essential difference is in how they acquire the resources – in other words, how they establish a virtual, hypothetical machine from the available resources.

3   The virtual layer

3.1   The pools

A conventional distributed environment assumes a pool of *computational nodes* (see left side in Figure 1). A node is a collection of resources (CPU, memory, storage, I/O devices, etc.) treated as a single unit. The pool of nodes consists of PCs, workstations, and possibly supercomputers, provided that the user has personal access (a valid login name and password) on all of them. From these candidate nodes an actual virtual machine is configured according to the needs of the given application. In general, once access has been obtained to a node on the virtual machine, all resources belonging to or attached to that node may be used without further authorization. Since personal accounts are used, the user is aware of the specifics of the local node (architecture, computational power and capacities, operating system, security policies, etc). Furthermore, the virtual pool of nodes is static since the set of nodes on which the user has login access changes very rarely. Although there are no technical restrictions, the typical number of nodes in the pool is of the order of 10-100, because users realistically do not have login access to thousands of computers.

Some descriptions of grids include : "a flexible, secure, coordinated resource sharing among dynamic collections of individuals, institutions, and resources" [8], "a single seamless computational environment in which cycles, communication, and data are shared, and in which the workstation across the continent is no

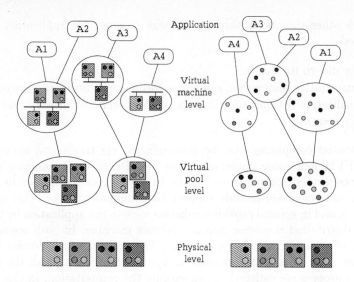

Fig. 1. Concept of conventional distributed environments (left) and grids (right). Nodes are represented by boxes, resources by circles.

less than one down the hall" [14], "widearea environment that transparently consists of workstations, personal computers, graphic rendering engines, supercomputers and non-traditional devices: e.g., TVs, toasters, etc." [13], "a collection of geographically separated resources (people, computers, instruments, databases) connected by a high speed network [...distinguished by...] a software layer, often called middleware, which transforms a collection of independent resources into a single, coherent, virtual machine" [17].

What principles and commonalities can be distilled from these definitions? Grids assume a virtual pool of *resources* rather than computational nodes (right side of Figure 1). Although current systems mostly focus on computational resources (CPU cycles + memory) [11] grid systems are expected to operate on a wider range of resources like storage, network, data, software [14] and unusual ones like graphical and audio input/output devices, manipulators, sensors and so on [13]. All these resources typically exist within nodes that are geographically distributed, and span multiple administrative domains. The virtual machine is constituted of a set of resources taken from the pool.

In grids, the virtual pool of resources is dynamic and diverse. Since computational grids are aimed at large-scale resource sharing, these resources can be added and withdrawn at any time at their owner's discretion, and their performance or load can change frequently over time. The typical number of resources in the pool is of the order of 1000 or more [3]. Due to all these reasons the user (and any agent acting on behalf of the user) has very little or no *a priori* knowledge about the actual type, state and features of the resources constituting the pool.

3.2   The virtual machines

The working cycle of a conventional distributed system is based on the notion of a pool of computational *nodes*. First therefore, all processes must be mapped to nodes chosen from the pool. The mapping can be done manually by the user, by the program itself or by some kind of task dispatcher system. This is enabled and possible because the user and thus, any program on the user's behalf is aware of the capabilities and the features of nodes. There can be numerous criteria for selecting the right nodes: performance, available resources, actual load, predicted load, etc. One condition however, must be met: the user must have valid login access to each node. Access to the virtual machine is realized by login (or equivalent authentication) on all constituent nodes. Once a process has been mapped, requests for resources can be satisfied by available resources on the node. If a user can login to a node, essentially the user is authorized to use all resources belonging to or attached to the node. If all processes are mapped to nodes and all the resource needs are satisfied, processes can start working, i.e. executing the task assigned to the process.

Contrary to conventional systems that try to first find an appropriate node to map the process to, and then satisfy the resource needs locally, grids are based on the assumption of an abundant and common pool of resources. Thus, first the resources are selected and then the mapping is done according to the resource selection. The resource needs of a process are abstract in the sense that they are expressed in terms of resource types and attributes in general, e.g. 64M of memory or a processor of a given architecture or 200M of storage, etc. These needs are satisfied by certain physical resources, e.g. 64M memory on a given machine, an Intel PIII processor and a file system mounted on the machine. Processes are mapped onto a node where these requirements can be satisfied. Since the virtual pool is large, dynamic, diverse, and the user has little or no knowledge about its current state, matching the abstract resources to physical ones cannot be solved by the user or at the application level based on selecting the right nodes, as is possible in the case of conventional environments. The virtual machine is constituted by the selected resources.

Access to the nodes hosting the needed resources cannot be controlled based on login access due to the large number of resources in the pool and the diversity of local security policies. It is unrealistic that a user has login account on thousands of nodes simultaneously. Instead, higher level credentials are introduced at the virtual level that can identify the user to the resource owners, and based on this authentication they can authorize the execution of their processes as if they were local users.

4   Intrinsic differences

The virtual machine of a conventional distributed application is constructed from the nodes available in the pool (Figure 2). Yet, this is just a different view of the physical layer and not really a different level of abstraction. Nodes appear on the virtual level exactly as they are at physical level, with the same names (e.g. $n1$, $n2$ in Figure 2), capabilities, etc. There is an *implicit* mapping from the abstract resources to their physical counterparts because once the process has been mapped, resources local to the node can be allocated to it. Users have the same identity, authentication and authorization procedure at both levels: they login to the virtual machine as they would to any node of it (e.g. *smith* in Figure 2).

On the contrary, in grid systems, both users and resources appear differently at virtual and physical layers. *Resources* appear as entities distinct from the physical node in the virtual pool (right side of Figure 2). A process' resource needs can be satisfied by various nodes in various ways. There must be an *explicit* assignment provided by the system between abstract resource needs and physical resource objects. The actual mapping of processes to nodes is driven by the resource assignment.

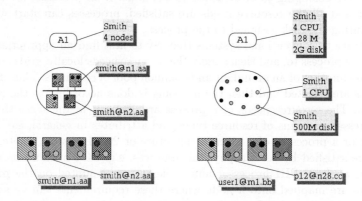

Fig. 2. Users and resources at different levels of abstraction. (Conventional environments on the left, grids on the right.)

Furthermore, in a grid, a user of the virtual machine is different from users (account owners) at the physical levels. Operating systems are based on the notion of processes; therefore, granting a resource involves starting or managing a local process on behalf of the user. Obviously, running a process is possible for local account holders. In a grid a user has a valid access right to a given resource proven by some kind of credential (e.g. user Smith in Figure 2). However, the user is not authorized to log in and start processes on the node to which the resource belongs. A grid system must provide a functionality that finds a proper mapping between a user (a real person) who has the credential to the resources and on whose behalf the processes work, and a local user ID (not necessarily a real person) that has a valid account and login rights on a node (like users *user*1 and *p*12 in Figure 2). The grid-authorized user temporarily has the rights of a local user for placing and running processes on the node.

Thus, in these respects, the physical and virtual levels in a grid are completely distinct, but there is a mapping between resources and users of the two layers.

According to [18] these two fundamental features of grids are termed user- and resource-abstraction, and constitute the intrinsic difference between grids and other distributed systems.

5 Technical differences

The fundamental differences introduced in the previous sections are at too high a level of abstraction to be patently evident in existing grid systems. In practice the two fundamental functionalities of resource and user abstraction are realized on top of several services.

The key to resource abstraction is the selection of available physical resources based on their abstract appearance. First, there must be a notation provided in which the abstract resource needs can be expressed (e.g. Resource Specification Language (RSL) [5] of Globus, Collection Query Language [4] of Legion.) This specification must be matched to available physical resources. Since the user has no knowledge about the currently available resources and their specifics, resource abstraction in a real implementation must be supported at least by the following components that are independent from the application:

1. An *information system* that can provide information about resources upon a query and can support both discovery and lookup. (Examples include the Grid Index Information Service (GIIS) [6] in Globus and Collection [4] in Legion).
2. A local *information provider* that is aware of the features of local resources, their current availability, load, and other parameters or, in general, a module that can update records of the information system either on its own or in response to a request. (Examples are the Grid Resource Information Service (GRIS) [6] in Globus and information reporting methods in Host and Vault objects in Legion [4].)

A user abstraction in a grid is a mapping of valid credential holders to local accounts. A valid (unexpired) credential is accepted through an authentication procedure and the authenticated user is authorized to use the resource. Just as in the case of resource abstraction, this facility assumes other assisting services.

1. A *security mechanism* that accepts global user's certificates and authenticates users. In Globus this is the resource proxy process that is implemented as the gatekeeper as part of the GRAM [7].
2. Local *resource management* that authorizes authentic users to use certain resources. This is realised by the mapfile records in Globus that essentially controls the mapping of users [7]. Authorization is then up to the operating system based on the rights associated with the local account. (In Legion both authentication and authorization are delegated to the objects, i.e. there is no centralized mechanism, but every object is responsible for its own security [16].)

The subsystems listed above, with examples in the two currently popular grid frameworks viz. Globus and Legion, are the services that directly support user and resource abstraction. In a practical grid implementation other services are also necessary, e.g. staging, co-allocation, etc. but these are more technical issues and are answers to the question of *how* grid mechanisms are realized, rather than to the question of *what* the grid model intrinsically contains.

6   Application areas

The differences presented so far reflect an architectural approach, i.e. grids are distinguished from conventional distributed systems on the basis of their architecture. In this section a comparison from the user's point of view is presented, to determine which environment is more appropriate for a given application.

It is a common misconception that grids are used primarily for supporting high performance applications. The aim of grids is to support large-scale resource sharing, whereas conventional distributed environments are based on the resources the user owns. *Sharing*, in this context, means *temporarily* utilizing resources to which the user has no longterm rights, e.g. login access. Similarly, *owning* means having a *permanent* and unrestricted access to the resource.

What benefits follow from resource sharing? Recall, that in general, applications may be distributed for quantitative or qualitative reasons. Resource sharing obviously allows the user to access more resources than are owned, and from a quantitative standpoint, this makes grids superior to conventional environments. The sheer volume of resources made available by coordinated sharing may allow the realization of high performance computation like distributed supercomputing or high-throughput computing [3]. On the other hand starting an application on a grid infrastructure involves several services and layers that obviously introduce considerable overhead in constituting the abstract machine. For example, a resource discovery in the current implementation of GIIS may take several minutes in a grid of 10 sites and approximately 700 processors [1]. If the resource requirements of the application can be satisfied by the resources owned, or the difference between the required and available owned resources is not significant, a conventional distributed environment may perform better than a grid.

Certain devices, software packages, tools that are too specialized, too expensive to buy or simply needed only for a short period of time, can be accessed in a grid. Conventional environments do not provide support for such on-demand applications, whereas the sharing mechanism of grids is an ideal platform for these situations. Even though grids may exhibit some overhead at the resource acquisition phase as discussed before, in this case the potential for obtaining certain non-local resources likely has higher priority than the loss of performance at the startup phase.

Conventional distributed environments usually consider an application as belonging to a single user. This corresponds to the architectural model where users appear identically at different levels of abstraction and processes constituting an application are typically owned by a single user. The shared resource space and the user abstraction of grids allows a more flexible view, where virtual machines may interact and processes of different users may form a single application. Although, grids are not collaborative environments, they provide a good base on which to create one [14][3]. For example the object space of Legion is an ideal platform for objects belonging to different users to interact. Because of the geographical extent of grids, such collaborative environments may be used for virtual laboratories, telepresence, teleimmersion and other novel, emerging applications.

7   Conclusion

With the increasing popularity of so-called grid systems, there are numerous papers dealing with various aspects of those systems like security, resource management, staging, monitoring, scheduling, and other issues. Yet, there is no clear definition of what a grid system should provide; moreover, there are some misleading concepts, too.

The aim of this paper is to identify the fundamental characteristics of grids. We argue that neither the geographically distributed, multi-domain, heterogeneous, large-scale feature of a system nor simply the presence of any of the afore mentioned 'grid services' makes a distributed system grid-like. Rather, grids are semantically different from other, conventional distributed environments in the way in which they build up the virtual layer. The two essential functionalities that a grid must support are: user abstraction and resource abstraction, where the physical world is separated from the virtual one. By user abstraction a user that has a valid credential for using the resources in the virtual pool is associated with local accounts. Resource abstraction means a selection of resources by their specifics in the virtual pool that are subsequently mapped onto nodes of the physical layer. We suggest a semantical definition for grids based on these differences. Technically, these two functionalities are realised by several services like a resource management system, information system, security, staging and so on.

Conventional distributed systems are based on resource *ownership* while grids are aimed at resource *sharing*. Both environments are able to support high performance applications, yet, the sheer volume of resources made available by grids can yield more computational power. Further, the shared virtual environment of grids provides a way to implement novel applications like collaborative systems, teleimmersion, virtual reality, and others that involve explicit cooperation. Conventional systems do not provide direct support for such applications.

Acknowledgments

The work presented in this paper was supported by US Dept. of Energy grant DE-FG02-99ER25379 and by the Hungarian Scientific Research Fund (OTKA) No. T032226.

References

1. G. Allen et al.: Early Experiences with the EGrid Testbed. Proceedings of the IEEE International Symposium on Cluster Computing and the Grid (CCGrid2001), Brisbane, Australia, 2001
2. E. Börger: High Level System Design and Analysis using Abstract State Machines. Current Trends in Applied Formal Methods (FM-Trends 98), ed. D. Hutter et al, Springer LNCS 1641, pp. 1-43 (invited lecture)
3. S. Brunet et al.: Application Experiences with the Globus Toolkit. 7th IEEE Symp. on High Performance Distributed Computing, 1998.
4. S.J. Chapin, D. Karmatos, J. Karpovich, A. Grimshaw: The Legion Resource Management System. Proceedings of the 5th Workshop on Job Scheduling Strategies for Parallel Processing (JSSPP '99), in conjunction with the International Parallel and Distributed Processing Symposium (IPDPS '99), April 1999
5. K. Czajkowski, et. al.: A Resource Management Architecture for Metacomputing Systems. Proc. IPPS/SPDP '98 Workshop on Job Scheduling Strategies for Parallel Processing, 1998
6. K. Czajkowski, S. Fitzgerald, I. Foster, C. Kesselman: Grid Information Services for Distributed Resource Sharing. Proc. 10th IEEE International Symposium on High-Performance Distributed Computing (HPDC-10), IEEE Press, 2001.
7. I. Foster, C. Kesselman, G. Tsudik, S. Tuecke: A Security Architecture for Computational Grids. In 5th ACM Conference on Computer and Communication Security, November 1998.
8. I. Foster, C. Kesselman, S. Tuecke: The Anatomy of the Grid. International Journal of Supercomputer Applications, 15(3), 2001.
9. I. Foster, C. Kesselman: The Grid: Blueprint for a New Computing Infrastructure. Morgan Kaufmann Publishers, 1999.
10. I. Foster, C. Kesselman: Computational grids. In The Grid: Blueprint for a New Computing Infrastructure. Morgan Kaufmann Publishers, 1999. pp. 15-51.
11. I. Foster, C. Kesselman: The Globus Toolkit. In The Grid: Blueprint for a New Computing Infrastructure. Morgan Kaufmann Publishers, 1999. pp. 259-278.
12. A. Geist, A. Beguelin, J. Dongarra, W. Jiang, B. Manchek, V. Sunderam: PVM: Parallel Virtual Machine - A User's Guide and Tutorial for Network Parallel Computing. MIT Press, Cambridge, MA, 1994
13. A. S. Grimshaw, W. A. Wulf: Legion - A View From 50,000 Feet. Proceedings of the Fifth IEEE International Symposium on High Performance Distributed Computing, IEEE Computer Society Press, Los Alamitos, California, August 1996
14. A. S. Grimshaw, W. A. Wulf, J. C. French, A. C. Weaver, P. F. Reynolds: Legion: The Next Logical Step Toward a Nationwide Virtual Computer. Technical report No. CS-94-21. June, 1994.
15. W. Gropp, E. Lusk, A. Skjellum: Using MPI: Portable Parallel Programming with the Message Passing Interface. MIT Press, Cambridge, MA, 1994.
16. M. Humprey, F. Knabbe, A. Ferrari, A. Grimshaw: Accountability and Control of Process Creation in the Legion Metasystem. Proc. 2000 Network and Distributed System Security Symposium NDSS2000, San Diego, California, February 2000.
17. G. Lindahl, A. Grimshaw, A. Ferrari, K. Holcomb: Metacomputing - What's in it for me. White paper. http://www.cs.virginia.edu/ legion/papers.html
18. Zs. Németh, V. Sunderam: A Formal Framework for Defining Grid Systems. Proceedings of 2nd IEEE International Symposium on Cluster Computing and the Grid (CCGrid2002), Berlin, IEEE Computer Society Press, 2002.

Developing Grid Based Infrastructure for Climate Modeling

John Taylor[1, 2], Mike Dvorak[1] and Sheri Mickelson[1]

[1]Mathematics & Computer Science, Argonne National Laboratory, Argonne, Illinois 60439
and Computation Institute, University of Chicago, IL, 60637
[2]Environmental Research Division, Argonne National Laboratory, Argonne, Illinois 60439
{jtaylor dvorak mickelso}@mcs.anl.gov
http://www-climate.mcs.anl.gov/

Abstract. In this paper we discuss the development of a high performance climate modeling system as an example of the application of Grid based technology to climate modeling. The climate simulation system at Argonne currently includes a scientific modeling interface (Espresso) written in Java which incorporates Globus middleware to facilitate climate simulations on the Grid. The climate modeling system also includes a high performance version of MM5v3.4 modified for long climate simulations on our 512 processor Linux cluster (Chiba City), an interactive web based tool to facilitate analysis and collaboration via the web, and an enhanced version of the Cave5D software capable of visualizing large climate data sets. We plan to incorporate other climate modeling systems such as the Fast Ocean Atmosphere Model (FOAM) and the National Center for Atmospheric Research's (NCAR) Community Climate Systems Model (CCSM) within Espresso to facilitate their application on computational grids.

1   Introduction

The Grid is currently considered an emerging technology with the potential to dramatically change the way we use computing resources such as commodity and high performance computers, high speed networks, experimental apparatus and security protocols [1]. In a recent article on the emergence of computational grids, Larry Smarr speculated that computational grids could rapidly change the world, in particular Smarr noted that:

> "Our nation's computer and communication infrastructure is driven by exponentials: it's on steroids. Every aspect – the microprocessor, bandwidth and fiber, disk storage – are all on exponentials, a situation that has never happened before in the history of infrastructure" [1]

The Grid primarily refers to the underlying hardware and software infrastructure needed to undertake computational tasks using distributed resources. Under ideal conditions the Grid would appear to the user as a single system. The details of the Grid infrastructure are ultimately intended to be invisible to the user just as the

P.M.A. Sloot et al. (Eds.): ICCS 2002, LNCS 2330, pp. 739–747, 2002.

infrastructure of the electric grid are largely hidden from the electricity consumer[1]. As advanced computational grids are developed, e.g.: the US National Science Foundation TerraGRID project, they offer the potential to address far more challenging problems in climate science than have been addressed in the past using a single high performance computer system.

2 Grid based infrastructure for climate modeling

In order to develop software to exploit the grid infrastructure we need to be able to:

- access information about the grid components
- locate and schedule resources
- communicate between nodes
- access programs and data sets within data archives
- measure and analyze performance
- authenticate users and resources

All these activities must be undertaken in a secure environment. The Globus project has developed an integrated suite of basic grid services collectively known as the Globus Tooklit [2]. These services and their Application Programming Interface (API) effectively define a Grid architecture on which we can develop Grid applications tools. We adopted the Globus Toolkit API as our underlying architecture for the development of our Grid based tools for climate modeling applications. We selected the Java programming language and the Globus Java Commodity Grid Kit, known as the Java CoG Kit [3], for our software development.

Building the software infrastructure to fully exploit computational grids remains a major challenge. In this paper we discuss current efforts at Argonne National Laboratory (ANL) to build Grid based infrastructure to facilitate climate modeling. We provide an overview of the Grid based software tools that we are currently developing and practical examples of the climate modeling results that we have obtained using this infrastructure.

The emerging Grid technology will allow far more challenging climate simulations to be performed. These simulations may ultimately be used to assess the impacts of global climate change at the regional scale. In a recent IPCC report on *The Regional Impacts of Climate Change* it was concluded that:

> The technological capacity to adapt to climate change is likely to be readily available in North America, but its application will be realized only if the necessary information is available (sufficiently far in advance in relation to the planning horizons and lifetimes of investments) and the institutional and financial capacity to manage change exists [4] (IPCC, 1998).

It was also acknowledged by IPCC that one of the key uncertainties that limit our ability to understand the vulnerability of sub-regions of North America to climate

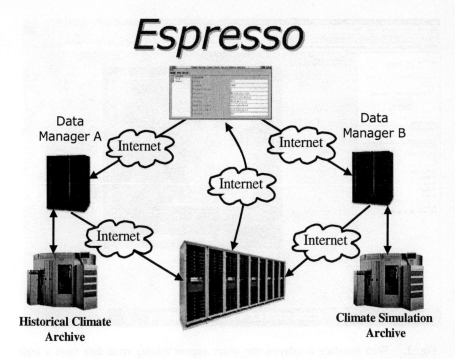

Fig. 1. Espresso allows the user to access remote computational resources and data sets on a climate modeling grid via an easy to use scientific modeling interface. Espresso uses the Globus Java CoG Kit [3] to access Grid services. In the above example we use Espresso to perform a climate model simulation on a remote machine while accessing input data sets from remote data archives.

change, and to develop and implement adaptive strategies to reduce vulnerability, was the need to develop accurate regional projections of climate change, including extreme events [4]. In particular we need to account for the physical-geographic characteristics that play a significant role in the North American climate, e.g. the Great Lakes, coastlines and mountain ranges, and also properly account for the feedbacks between the biosphere and atmosphere [4].

The potential impacts of global climate change have long been investigated based on the results of climate simulations using global climate models with typical model resolutions on the order of hundred's of kilometers [5,6]. However, assessment of impacts from climate change at regional and local scales requires predictions of climate change at the 1-10 kilometer scale. Model predictions from global climate models with such high resolutions are not likely to become widely available in the near future. Accordingly, at ANL we have begun developing a climate simulation capability, at both the global and regional scales, for application on computational grids.

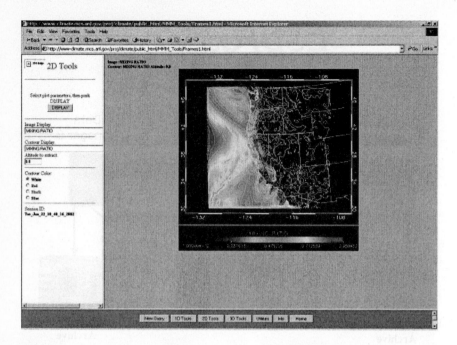

Fig. 2.   Web interface displaying the water vapour mixing ratio data from a high resolution (10km) climate simulation of the western United States at 07Z 3 December 1996. Much detail of the complex topography of the western United States is revealed at this model resolution. The data set from this simulation consists of 10 GB of data, all of which can be viewed via the web analysis tool. The model simulation is for the period 1-10 December 1996 and shows an advancing cold front producing significant precipiation over northern California. All plots are generated and displayed online based on user requests. The next generation of this web tool will allow interactive data analysis in addition to data display.

Our long term goal is to link the predictive global climate modeling capability with the impact assessment and policymaking communities. The regional climate simulation system at Argonne currently includes:

- A Java based Scientific Modeling Interface (Espresso) to facilitate task management on the Grid. See [7,8] for further details on the development of Espresso.
- A high performance version of MM5v3.4 modified to enable long climate simulations on the Argonne 'Chiba City' 512 processor (500 Mhz Pentium III) Linux cluster.
- An interactive web based visualization tool to facilitate analysis and collaboration via the web.
- An enhanced version of the Cave5d software capable of working with large climate data sets.

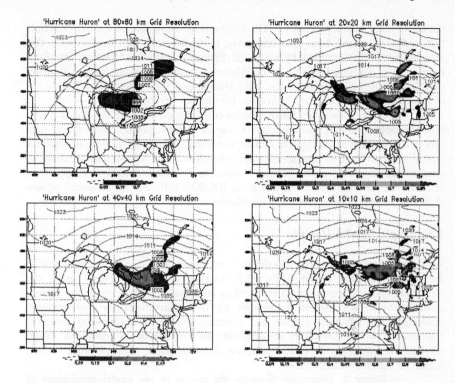

Fig. 3. Precipitation and surface pressure for hour ending 12Z 11 September 1996 at 80, 40, 20 and 10km grid resolution. Precipitation intensifies by nearly an order of magnitude as we go to higher model resolutions and occupies a smaller more sharply defined rainbands. Increasing precipitation intensities will alter the fraction of rainfall allocated to storage in the soil ie soil moisture and runoff which in turn will alter rates of decomposition and photosynthesis particularly under water limited conditions.

The regional climate model used in this study is based on the Pennsylvania State University/National Center for Atmospheric Research (PSU/NCAR) fifth generation mesoscale model (MM5). In brief, MM5 is a three-dimensional, non-hydrostatic, and elastic mesoscale model. It uses finite differences and a time-splitting scheme to solve prognostic equations on an Arakawa type-B staggered grid. Its vertical coordinate, though defined as a function of the reference-state pressure, is similar to a terrain-following coordinate. For case studies, MM5 employs observed wind, temperature, and humidity as the initial and boundary conditions. MM5 incorporates realistic topography and sophisticated physical processes to represent the appropriate forcing for the development of the observed weather system. These physical processes include clouds, long- and shortwave radiation, and the surface fluxes of heat, moisture, and momentum. A more detailed description of MM5 is provided by, Chen and Dudhia [9], Chen et al. [10], Dudhia [11] and Grell et al. [12] .

The NCAR MM5 modeling system consists of six programs: TERRAIN, REGRID, RAWINS, INTERP, MM5, and GRAPH [9]. Each program is executed in the above order interdependently. The programs are composed of a series of scripts that traditionally have been time consuming to modify and execute. Recently, this process has been simplified by using the Espresso Scientific Modeling Interface. Espresso was developed at ANL by the Mathematics and Computer Science Division (MCS) staff, including John Taylor, Mike Dvorak and Sheri Mickelson. The design of Espresso {7,8] has benefited from an earlier climate workbench[13].

In order to facilitate Internet based collaboration we have developed a web browser application that enables access to the output of regional climate model runs generated using the MM5 regional climate modeling system. Fig. 2 above illustrates a typical session. The web browser uses the native MM5 data format, thus avoiding the need to store duplicate copies of model output, and works efficiently with gigabytes of data. The web tool was developed using IDL/ION software. An enhanced version of this web tool is currently awaiting installation at MCS.

3 Example Model Results

We have performed preliminary model runs in climate mode using a recent release of the MM5 modeling system (MM5v3.4) looking at extreme events using the forerunner to the Espresso scientific modeling interface. Mesoscale resolution climate models provide a consistent framework for us to investigate the link between incoming solar radiation, climate and extreme weather. The following series of experiments were undertaken to illustrate the importance of enhanced model resolution to simulating the weather, climate, and atmospheric transport processes that impact on extreme weather events.

In Figure 3, above, we illustrate the results of our model simulations of 'Hurricane Huron'. We have performed four simulations for the period 6-15 September 1996, at a range of model grid resolutions, 80, 40, 20 and 10km over an identical 2000x2000 km region centered over Lake Huron. By performing our model simulation over an identical spatio-temporal domain we can study the effect of grid resolution on model physics.

Figure 3 also illustrates that hourly rainfall intensity increases dramatically, by nearly an order of magnitude, as we go to higher model resolutions. The pattern of rainfall also changes from broad scale low intensity rainfall at 80km grid resolution, to high intensity rainfall with significant spatial structure associated with formation of well defined rain bands. We use NCAR/NCEP Reanalysis Project wind fields to provide boundary and initial conditions.

We have also performed a high resolution (10km) climate simulation using MM5v3.4 over the western United States for the period 1-10 December 1996 using the Climate Workbench [13]. Such high resolution simulations have not previously been undertaken for the western United States. The purpose of this study was to investigate the ability of the MM5v3.4 model to predict intense precipitation over the complex terrain of the Western United States. During this time La Nina conditions persisted in the tropical Pacific. La Nina conditions are usually associated with strong rainfall events over Northern California. Rainfall events associated with La Nina conditions play an important role in increasing water storage levels in Northern California.

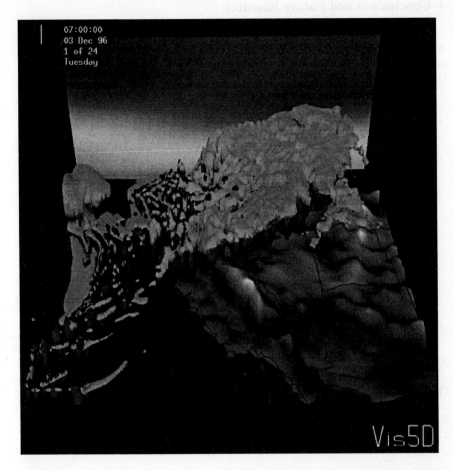

Fig. 4. This figure shows isosurfaces of cloud water and rain water mixing ratios from a high resolution (10km) climate simulation of the western United States at 0700 on 3 Decmber 1996. The figure shows an advancing cold front producing significant precipitation over the Northwest and Northern California. Much of the detail of the complex topography of the Western United States is revealed at this model resolution. The model results above are derived from a MM5V3.4 simulation performed for the period 1-10 December.

Figure 4, above, shows an advancing cold front producing significant precipitation over the Northwest and Northern California for 3 December 1996. The model results look very promising as they are in qualitative agreement with observed precipitation. Further work is needed to fully assess the ability of the MM5 model to correctly predict intense precipation over the complex terrain of the Western United States.

4 Conclusions and Future Research

We will continue to address the key scientific and computational issues in climate modeling [14] and their importance to simulating global climate and the climate of the United States by:

- Developing Grid based tools such as the Espresso Scientific Modeling Interface in order to be able to efficiently manage climate simulations using the full capabilities offered by today's distributed computing environments.
- Contributing to the performance improvement of long term global and regional climate simulations by developing new physical parameterizations suitable for incorporation into climate models.
- Developing computer code capable of running efficiently on the new generation of high performance computers such as commodity clusters.
- Performing challenging climate simulations which test the limits of the available technology.
- Building analysis and visualization tools capable of working with the large data sets (order terrabyte to petabyte) generated by global and regional climate model simulations.
- Defining and delivering quality data products via the web through the development of web based interfaces for analysis and visualization.

Acknowledgments

This work was supported in part by the Laboratory Director Research and Development funding subprogram of the Office of Advanced Scientific Computing Research, U.S. Department of Energy, under Contract W-31-109-Eng-38, and through a NSF Information Technology Research Grant ATM-0121028.

References

1. Foster, I and Kesselman, C. (1999) *The Grid: Blueprint for a New Computing Infrastructure.* Morgan Kaufmann Publishers.
2. The Globus Home Page. http://www.globus.org//
3. von Laszewski, G., Foster, I., Gawor, J., and Lane, P. (2001) A Java commodity grid kit. *Concurrency and Computation: Practice and Experience*, 13:645-662.
4. IPCC (1998) The Regional Impacts of Climate Change, Cambridge Univ. Press, Cambridge.
5. IPCC WGI (1990) Climate Change: The IPCC Scientific Assessment R.A. Houghton *et al.*(eds).,Cambridge Univ. Press, Cambridge, UK.
6. IPCC WGI (1996) Climate Change 1995: The Science of Climate Change R.A. Houghton *et al.*(eds).,Cambridge Univ. Press, Cambridge, UK.
7. Mickelson, S., Taylor, J. and Dvorak, M. (2002) Simplifying the task of generating climate simulations and visualizations. To appear in the *Proceedings of the 2002*

International Conference on Computational Science, Springer-Verlag, Berlin, Germany.

8. Dvorak, M., Taylor, J. and Mickelson, S. (2002) Designing a flexible grid enabled scientific modeling interface. To appear in the *Proceedings of the 2002 International Conference on Computational Science.* Springer-Verlag, Berlin, Germany.

9. Chen, F. and Dudhia, J. 2001: Coupling an Advanced Land-Surface/Hydrology Model with the Penn State/NCAR MM5 Modeling System. Part I: Model implementation and Sensitivity, *Monthly Weather Review,* in press. (See also Pennsylvania State University / National Center for Atmospheric Research, MM5 Home Page http://www.mmm.ucar.edu/mm5/mm5-home.html).

10. Chen, F., K. Mitchell, J. Schaake, Y. Xue, H.L. Pan, V. Koren, Q.Y. Duan, K. Ek, and A. Betts, 1996: Modeling of land-surface evaporation by four schemes and comparison with FIFE observations. *J. Geophys. Res.,* 101, 7251-7268.

11. Dudhia, J., 1993: A nonhydrostatic version of the Penn State / NCAR mesoscale model: Validation tests and simulation of an Atlantic cyclone and cold front. *Mon. Wea. Rev.,* 121, 1493-1513.

12. Grell, G.A., J. Dudhia and D.R. Stauffer, 1994: The Penn State/NCAR Mesoscale Model (MM5). NCAR Technical Note, NCAR/TN-398+STR, 138 pp.

13. Taylor, J. Argonne National Laboratory, Climate Workbench http://wwwclimate.mcs.anl.gov/proj/climate/public_html/climate-workbench.html, 2000.

14. Giorgi, F. and L.O. Mearns (1999) Introduction to special section: Regional climate modeling revisited, *Journal of Geophysical Research,* **104**, 6335-6352.

A Real Application of the Model Coupling Toolkit

Everest T. Ong, J. Walter Larson, and Robert L. Jacob

Argonne National Laboratory, Mathematics and Computer Science Division
9700 S. Cass Ave., Argonne, IL 60439, USA,
eong@mcs.anl.gov

Abstract. The high degree of computational complexity of atmosphere and ocean general circulation models, land-surface models, and dynamical sea-ice models makes coupled climate modeling a grand-challenge problem in high-performance computing. On distributed-memory parallel computers, a coupled model comprises multiple message-passing-parallel models, each of which must exchange data among themselves or through a special component called a *coupler*. *The Model Coupling Toolkit* (MCT) is a set of Fortran90 objects that can be used to easily create low bandwidth parallel data exchange algorithms and other functions of a parallel coupler. In this paper we describe the MCT, how it was employed to implement some of the important functions found in the flux coupler for the Parallel Climate Model(PCM), and compare the performance of MCT-based PCM functions with their PCM counterparts.

1 Introduction

The practice of climate modeling has progressed from the application of atmospheric general circulation models (GCMs) with prescribed boundary condition and surface flux forcing to *coupled earth system models*, comprising an atmospheric GCM, an ocean GCM, a dynamic-thermodynamic sea ice model, a land-surface model, a river runoff-routing model, and potentially other component models—such as atmospheric chemistry or biogeochemistry—in mutual interaction. The high degree of computational complexity in many of these component models (particularly the ocean and atmosphere GCMs) gave impetus to parallel implementations, and the trend in computer hardware towards microprocessor-based, distributed-memory platforms has made message-passing parallelism the dominant model architecture paradigm. Thus, coupled earth-system modeling can entail the creation of a *parallel coupled model*, requiring the transfer of large amounts of data between multiple message-passing parallel component models. Coupled models present a considerable increase in terms of computational and software complexity over their atmospheric climate model counterparts.

A number of architectures for parallel coupled models exist. One strategy is the notion of a *flux coupler* that coordinates data transfer between the other component models and governs the overall execution of the coupled model, which can

P.M.A. Sloot et al. (Eds.): ICCS 2002, LNCS 2330, pp. 748–757, 2002.

be found in the National Center for Atmospheric Research (NCAR) Climate System Model (CSM)[4][5][11] and the Parallel Climate Model (PCM)[1][3]. Another solution is the use of coupling libraries such as CERFACS' Ocean Atmosphere Sea Ice Surface[15], the UCLA Distributed Data Broker[8], and the Mesh-based parallel Code Coupling Interface (MpCCI)[13]. A third solution implements the coupler as an interface between the atmosphere and the surface components such as the Fast Ocean Atmosphere Model[9].

The authors have studied these previous strategies (and others), and decided to create a Fortran90 software toolkit to support the core functions common to coupled models: gridded data domain decomposition and parallel data transfer, interpolation between coordinate grids implemented as sparse matrix-vector multiplication, time averaging and accumulation, representation of coordinate grids and support for global integral and spatial averaging, and merging of output data from multiple component models for input to another component model. The result is the Model Coupling Toolkit (MCT)[12]. Fortran90 was chosen because the vast majority of geophysical models (our target application) are written in Fortran or Fortran90, and we wanted to provide those users with tools that exploit the advanced features of Fortran90, and avoid interlanguage operability issues. The decision to build a toolkit supporting the low-level data movement and translation functions found in flux couplers was driven by our desire to provide maximum flexibility to our users, and to support the widest possible range of coupled model applications.

In Section Two, we describe the PCM flux coupler benchmark, and present performance figures from the original benchmark and the results of the authors' optimization work on this code. In Section Three we describe the Model Coupling Toolkit. In Section Four, we explain how the MCT can be employed to create parallel coupled models, and present MCT-based pseudocode for our implementation of the PCM coupler benchmark. In Section Five, we report performance figures for the MCT implementation of the PCM coupler benchmark and compare them with previous PCM timings.

2   The Parallel Climate Model Coupler Benchmark

The Parallel Climate Model (PCM)[1][3] comprises the NCAR atmospheric model CCM3, the Parallel Ocean Program (POP) ocean model, a dynamic - thermodynamic sea ice model, the LSM land-surface model, and a river-runoff routing model, with a distributed flux coupler coordinating the transfer, inter-grid interpolation, and time- averaging and accumulation of interfacial flux and boundary condition data. Each of these component models is message-passing parallel based on the Message-Passing Interface (MPI) standard. The model is a single-load-image application, and its components are run sequentially in an event-loop as shown in Figure 1.

A unit-tester of the PCM coupler exists, and executes all the coupler's interpolation, time averaging and accumulation, and multi-component data-merging functions for a ten-day period using dummy components in place of the at-

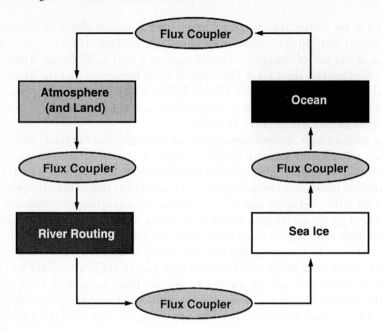

Fig. 1. Schematic of the Parallel Climate Model, showing the component model execution event loop.

mosphere, ocean, *et cetera*. The atmosphere-to-ocean grid operations are: 720 interpolation calls to regrid a bundle of two fields, and two sets of 720 interpolation calls to regrid six fields. The ocean-to-atmosphere interpolation operations are: 240 interpolation calls to interpolate one field (ice fraction), two sets of 240 interpolation calls to regrid six fields of ocean data, and 240 interpolation calls to regrid six fields of sea ice state data (residing on the ocean grid). For our discussion, the PCM coupler benchmark comprises four quantities: (1) total time spent on communications in the atmosphere-to-ocean interpolation, (2) total time spent on floating-point operations in the atmosphere-to-ocean interpolation, (3) total time spent on communications in the ocean-to-atmosphere interpolation, and (4) total time spent on floating point operations in the ocean-to-atmosphere interpolation.

The PCM coupler has been benchmarked for the IBM SP-3[6], assuming an atmosphere with a horizontal grid corresponding to T42 spectral truncation (8192 points) and an ocean with a POPx1 grid (110592 points). This benchmark reflects the coupler design for PCM version one, where the communications employed during the atmosphere-to-ocean and ocean-to-atmosphere interpolation operations sent excess data, and used blocking send and receive operations. Timings on the SP-3 for these operations are presented in Table 1, and are labeled as the "original" values.

The authors optimized the hand-coded PCM communications algorithms by using message-packing to cut bandwidth costs, and employing non-blocking communications[14]. These changes resulted in dramatically reduced timings, as can be seen from the "optimized" values in Table 1.

Table 1. PCM Coupler Benchmark Timings (in seconds)

Number of Processors	Atmosphere-to-Ocean Communications (original)	Atmosphere-to-Ocean Communications (optimized)	Ocean-to-Atmosphere Communications (original)	Ocean-to-Atmosphere Communications (optimized)
16	13.364	2.919	14.542	2.059
32	15.807	2.393	17.138	1.441
64	11.653	2.188	15.090	1.234

3 The Model Coupling Toolkit

The MCT is a set of Fortran90 modules defining derived datatypes and routines that provide core services common to coupled models.

The MCT provides two classes of domain decomposition descriptors: a one-dimensional decomposition (the `GlobalMap`, defined in the module m_`GlobalMap`) and a segmented decomposition for multidimensional and unstructured grids (the `GlobalSegMap`, defined in the module m_`GlobalSegMap`). The MCT provides means for initializing and querying these descriptors, global-to-local indexing services, and the means for exchange of descriptors between component models.

The `MCTWorld` datatype (defined in the module m_`MCTWorld`) provides a registry for component models that describes the processes on which they reside. This allows for automatic translation between local (component model) and global process ID numbers needed for intercomponent model communications. The assignment of communicators and process ID's used by each component model on the communicator `MPI_COMM_WORLD` are determined by invoking the Multi-Process Handshaking (MPH) utility[7].

The `Router` datatype (defined in the module m_`Router`) encapsulates MCT parallel communications scheduling. A `Router` is initialized automatically from a pair of domain decomposition descriptors for the source and destination decompositions. This datatype allows for the automation of the complex parallel intercomponent model data transfers necessary for high performance.

The `AttrVect`, or *attribute vector*, is the MCT's flexible, extensible, and indexible data field storage datatype. This datatype is defined, along with its query and support routines in the MCT module m_`AttrVect`, and has point-to-point and collective communications routines defined by the module m_`AttrVectComms`. This allows the user maximum flexibility in configuring a coupled model. If the component models and coupler are designed properly, the fields passed between component models could be configured at runtime. It is possible to store both real and integer data in this datatype, which allows a user to carry around associated gridpoint indexing information as a means of monitoring processing and an aid to debugging coupled model applications. The storage order in the derived type is field-major, which is compatible with the notion that most communication and computation in a coupler is processing of multiple fields of *point* data. This allows, where possible, for the field storage datatype to be used directly as a buffer for MPI operations and allows for reuse of matrix elements during the interpolation process, thus boosting performance for this process.

The MCT provides support for flux and state data field interpolation via an efficient sparse matrix-vector multiply that works directly with the MCT's `AttrVect` datatype. The MCT `SparseMatrix` datatype, defined in the module `m_SparseMatrix`, encapsulates sparse matrix element storage, along with space for global and local row-and-column indexing information. This module also provides facilities for sorting of matrix elements, which can yield a more regular memory-access pattern and thus boost single-processor performance during matrix-vector multiplication through better cache usage. Communications support for this datatype are provided in the module `m_SparseMatrixComms`. Schemes for automatically computing domain decompositions based on global row or column index information stored in a `SparseMatrix` are provided in the module `m_SparseMatrixToMaps`, and schemes for computing domain decompositions for the `SparseMatrix` are provided in the module `m_SparseMatrixDecomp`. Various parallel matrix-vector multiplication routines are defined in the module `m_MatAttrVectMul`. The MCT currently has no facilities for computing interpolation matrix elements, and it is assumed the user will compute them off-line using a package such as the Spherical Coordinate Regridding and Interpolation Package (SCRIP)[10].

The `GeneralGrid` datatype (defined in the module `m_GeneralGrid`) is the MCT coordinate grid desccriptor, and can support multidimensional and unstructured grids. This datatype allows the user to store geometric information such as grid-cell cross-sectional area and volume weights, and integer data for any number of grid indexing schemes. Grid geometry stored in a `GeneralGrid` is used by the spatial integration and averaging routines defined in the module `m_GlobalIntegrals`.

The `Accumulator` datatype (defined in the module `m_Accumulator`) provides flexible, extensible, and indexible registers for temporal averaging and accumulation of state and flux data. Communications support for this datatype are defined in the module `m_AccumulatorComms`.

The MCT module `m_Merge` provides routines for merging flux and state data from multiple component models for use by another component model, with fractional area weighting and/or masking encapsulated either in the `GeneralGrid` or `AttrVect` datatypes.

The MCT is built upon the NASA DAO Message Passing Environment Utilities (MPEU), which provide support for basic low-level data types upon which MCT classes are built, Fortran90 module-style access to MPI, and tools for error handling, sorting data, and timing. Both MPEU and MCT have their documentation built-in, implemented as extractible prologues compatible with the software package ProTeX, which translates the prologues into LaTeX.

The programming interface to the MCT is described fully in the MCT API Definition Document, which is available via the MCT Web site

```
http://www.mcs.anl.gov/acpi/mct .
```

Currently, the MCT is in beta release. A complete formal release of the MCT will occur in 2002. Work is currently under way to implement the new flux coupler for the Community Climate System Model (CCSM) using MCT[2]

4   Implementation of PCM Flux Coupler Functions using the MCT

Using MCT to couple message-passing parallel models into a coupled parallel model is relatively easy. The user accesses MCT datatypes and modules through Fortran90 use association, creating instantiations of the MCT objects and invoking MCT routines as needed to accomplish the coupling. For the sake of simplicity we consider only the atmosphere-ocean interactions, and present pseudocode where the only coupling activities involved are grid interpolation. In this case, only a subset of the MCT is visible to the user. For example, the `Router` datatype is invoked inside the matrix-vector interpolation routines, and thus is invisible to the user.

The MCT datatypes shared between the ocean, atmosphere, and coupler are defined in the module m_Couplings:

```
module m_Couplings
   use m_AttrVect        ! Field Storage Datatype
   use m_GlobalSegMap    ! Domain Decomposition Descriptors
   use m_SparseMatrix    ! Interpolation Matrices
   ! AttrVect for atmosphere data
   type(AttrVect) :: CouplerInputFromAtmos, CouplerOutputToAtmos
   ! AttrVect for ocean data
   type(AttrVect) :: CouplerInputFromOcean, CouplerInputToOcean
   ! Atmosphere and Ocean GlobalSegMap
   type(GlobalSegMap) :: AtmosGlobalSegMap, OceanGlobalSegMap
   ! Atmosphere-to-Ocean and Ocean-to-Atmosphere SparseMatrix
   type(SparseMatrix) :: AtmosToOceanSMat, OceanToAtmosSMat
end module m_Couplings
```

Pseudocode for the atmosphere, showing how MCT datatypes are used, and routines are invoked is contained in the atmosphere subroutine `ccm()` shown below. The pseudocode for the ocean is analgous, and is not shown for the sake of brevity.

```
subroutine ccm(AtmosInputFromCoupler,AtmosOutputToCoupler)
   use m_MCTWorld        ! MCT Component Model Registry
```

```
  use m_AttrVect            ! Field Storage Datatype
! Initialize MCTWorld registry:
  if(initialization) call MCTWorld_Init()
! Initialize Coupler Atmosphere Domain Decompositions:
  if(initialization) call GlobalSegMap_Init(AtmosGlobalSegMap)
    :
! Unpack atmosphere input from AtmosInputFromCoupler AttrVect
    :
! Atmosphere model integration loop!
  do step = 1,nsteps
    :           :
  enddo
    :
! Pack atmosphere output into AtmosOutputToCoupler AttrVect
end subroutine ccm
```

Pseudocode for the distributed flux coupler, which executes the ocean and atmosphere components, and interpolates data between the atmosphere and ocean grids is shown below in the program cpl.

```
program cpl
    use m_MCTWorld          ! MCT Component Model Registry
    use m_GlobalSegMap      ! Domain Decomposition Descriptors
    use m_AttrVect          ! Field Storage Datatype
    use m_SparseMatrix      ! Interpolation Matrices
    use m_SMatAttrVectMult  ! Parallel Sparse-Matrix-Attribute
                            ! Vector Multiplication Functions
! Initialize MCTWorld registry
    call MCTWorld_Init()
! Initialize input and output Attribute vectors
    call AttrVect_Init(CouplerInputFromAtmos)
    call AttrVect_Init(CouplerOutputToAtmos)
    call AttrVect_Init(CouplerInputFromOcean)
    call AttrVect_Init(CouplerOutputToOcean)
! Initialize Interpolation Matrices:
    call SparseMatrix_Init(AtmosToOceanSMat)
    call SparseMatrix_Init(OceanToAtmosSMat)
    :
! Coupler Time Evolution Loop:
    do icloop=1,ncloops
        if(mod(icloop,1)==0) then
            ! Run the atmosphere model:
            call ccm(CouplerOutputToAtmos, CouplerInputFromAtmos)
            ! Interpolate this data onto the Ocean Grid.  Here the
            ! matrix elements distributed by row (i.e. based on the
            ! ocean grid vector decomposition).
            call SMatAttrVectMul_xdyl(CouplerInputFromAtmos, &
```

```
                           AtmosGlobalSegMap, &
                           AtmosToOceanSMat, &
                           CouplerOutputToOcean )
    endif
    if(mod(icloop,2)==0) then
       ! Run the ocean model:
       call pop(CouplerOutputToOcean, CouplerInputFromOcean)
       ! Interpolate this data onto the Atmosphere Grid.  Here
       ! he matrix elements distributed by column (i.e. based
       ! on the ocean grid vector decomposition).
       call SMatAttrVectMul_xlyd(CouplerInputFromOcean, &
                           OceanGSMap, &
                           OceanToAtmosSMat, &
                           CouplerOutputToAtmos )
    endif
  end do
end program cpl
```

The MCT version of the PCM coupler is very similar to the pseudocode presented above.

5   Performance

Performance results for the MCT implementation of the PCM coupler benchmark are presented in Table 2. The communications routing mechanisms in the MCT are far more flexible than those in the hand-tuned PCM coupler, but comparison of these results with those in Table 1 show the atmosphere-to-ocean and ocean-to-atmosphere communications costs for the MCT implementation are either the same or slightly lower.

Table 2. MCT/PCM Coupler Benchmark Communications Timings (seconds)

Number of Processors	Atmosphere-to-Ocean	Ocean-to-Atmosphere
16	2.909	1.809
32	1.609	1.359
64	1.452	1.156

Floating-point operation timings for the PCM interpolation operations for both the original PCM implementation and the MCT implementation are presented in Table 3. The computation costs are not significantly worse, and usually are significantly better than the original PCM timings. This is very encouraging considering the hand-tuned PCM used f77-style, static arrays, and the MCT implementation employs Fortran90 derived types built on top of allocated arrays. Both the PCM and MCT implementations have been designed so that the loop-order in the interpolation is cache-friendly. The most likely reasons for better computational performance in the MCT implementation are (1) the use of local rather than global arrays for the computations, and (2) the application of sorting of matrix elements in the MCT implementation, both of which improve data-locality and thus improve cache performance.

Table 3. PCM Coupler Benchmark Floating-point Operations Timings (seconds)

Number of Processors	Atmosphere-to-Ocean (PCM)	Atmosphere-to-Ocean (MCT)	Ocean-to-Atmosphere (PCM)	Ocean-to-Atmosphere (MCT)
16	4.636	4.408	2.484	1.616
32	2.470	2.644	1.622	1.180
64	2.188	1.936	1.201	0.984

6   Conclusions

We have described a realistic application of the the Model Coupling Toolkit to implement some of the core functions found in the Parallel Climate Model. The pseudocode presented shows that creating coupled models using the MCT is straightforward. Performance results indicate that the use of advanced features of Fortran90 in this case do not degrade performance with respect to an optimized version of the f77 PCM counterpart. Floating-point performance in the MCT version of the PCM benchmark is slightly better than the original, due to better data locality.

Acknowledgements: We wish to thank many people for the useful discussions and advice regarding this work: Tom Bettge, and Tony Craig of the National Center for Atmospheric Research; and John Taylor and Ian Foster of the Mathematics and Computer Science Division of Argonne National Laboratory, and Jace Mogill and Celeste Corey of Cray Research. This work is supported by the United States Department of Energy Office of Biological and Environmental Research under Field Work Proposal 66204, KP1201020, and the Scientific Discovery through Advanced Computing (SciDAC) Program, Field Work Proposal 66204.

References

1. Bettge, T. , and Craig, A.   (1999).   Parallel Climate Model Web Page. http://www.cgd.ucar.edu/pcm/.

2. Bettge, T. (2000). Community Climate System Model Next-generation Coupler Web Page. http://www.cgd.ucar.edu/csm/models/cpl-ng/.
3. Bettge, T. , Craig, A. , James, R. , Wayland, V. , and Strand, G. (2001). The DOE Parallel Climate Model (PCM): The Computational Highway and Backroads. *Proceedings of the International Conference on Computational Science (ICCS) 2001* , V.N. Alexandrov, J.J. Dongarra, B.A. Juliano, R.S. Renner, and C.J.K. Tan (eds.), Springer-Verlag Lecture Notes in Comupter Science Volume 2073, pp 149-158.
4. Boville, B. A. and Gent, P. R. (1998). The NCAR Climate System Model, Version One. *Journal of Climate*, **11**, 1115-1130.
5. Bryan, F. O. , Kauffman, B. G. , Large, W. G. , and Gent, P. R. (1996) The NCAR CSM Flux Coupler. NCAR Technical Note NCAR/TN-424+STR, National Center for Atmospheric Research, Boulder, Colorado.
6. Craig, A. (2000). Parallel Climate Model IBM SP-3 Benchmarks Web Page. http://www.cgd.ucar.edu/ccr/bettge/ibmperf/timers.txt.
7. Ding, C. H. Q. and He, Y. (2001). MPH: a Library for Distributed Multi-Component Environments, Users' Guide. NERSC Documentation, available on-line at http://www.nersc.gov/research/SCG/acpi/MPH/mph_doc/mph_doc.html.
8. Drummond, L. A. , Demmel, J. , Mechoso, C. R. , Robinson, H. , Sklower, K. , and Spahr, J. A. (2001). A Data Broker for Distributed Computing Environments. *Proceedings of the International Conference on Computational Science (ICCS) 2001* , V.N. Alexandrov, J.J. Dongarra, B.A. Juliano, R.S. Renner, and C.J.K. Tan (eds.), Springer-Verlag Lecture Notes in Comupter Science Volume 2073, pp 31-40.
9. Jacob, R. , Schafer, C. , Foster, I. , Tobis, M. , and Anderson, J. (2001). Computational Design and Performance of the Fast Ocean Atmosphere Model, Version One. *Proceedings of the International Conference on Computational Science (ICCS) 2001* , V.N. Alexandrov, J.J. Dongarra, B.A. Juliano, R.S. Renner, and C.J.K. Tan (eds.), Springer-Verlag Lecture Notes in Comupter Science Volume 2073, pp 175-184.
10. Jones, P. W. (1999). First- and Second-Order Conservative Remapping Schemes for Grids in Spherical Coordinates. *Monthly Weather Reveiw*, **127**, 2204-2210.
11. Kauffman, B. (1999). NCAR Climate System Model (CSM) Coupler Web Page. http://www.ccsm.ucar.edu/models/cpl/.
12. Larson, J. W. , Jacob, R. L. , Foster, I. T. , and Guo, J. (2001). The Model Coupling Toolkit. *Proceedings of the International Conference on Computational Science (ICCS) 2001* , V.N. Alexandrov, J.J. Dongarra, B.A. Juliano, R.S. Renner, and C.J.K. Tan (eds.), Springer-Verlag Lecture Notes in Comupter Science Volume 2073, pp 185-194.
13. Pallas GmbH (2002) MPCCI Web Site. http://www.mpcci.org.
14. Ong, E. T. and Larson, J. .W. (2001). Optimization of the Communications in the PCM Flux Coupler. Technical Memorandum, Mathematics and Computer Science Division, Argonne National Laboratory, Argonne, Illinois, in preparation.
15. Valcke, S. , Terray, L. , Piacentini, A. (2000). OASIS 2.4 Ocean Atmosphere Sea Ice Soil Users' Guide. CNRS/CERFACS Technical Note, available on-line at http://www.cerfacs.fr/globc/.

Simplifying the Task of Generating Climate Simulations and Visualizations

Sheri A. Mickelson,[1] John A. Taylor,[1, 2] and Mike Dvorak[1]

[1]The Computational Institute, University of Chicago, Chicago, IL 60637 and The
Mathematics & Computer Science Division, Argonne National Laboratory, Argonne, Illinois
60439
[2]Environmental Research Division, Argonne National Laboratory, Argonne, Illinois 60439
{mickelso, jtaylor, dvorak}@mcs.anl.gov
http://www-climate.mcs.anl.gov

Abstract. To fully exploit the use of the MM5 modeling system, the scientist must spend several months studying the system, thus losing valuable research time. To solve this problem, we have created a graphical user interface, called Espresso, that allows users to change these values without having to examine the code. This approach dramatically increases the usability of the model and speeds productivity. We have also modified Vis5D to run on tiled displays. Using such displays allows us to view our climate data at much higher resolution.

1 Introduction

Simulating climate data with the MM5 modeling system [1, 2, 3] is a complex task that is highly error prone and time consuming. In order to complete a simulation, many scripts must be edited and run. Often, the same variable and value must be set in different scripts. If a value is changed erroneously in the script, the run may fail or produce false results. This process wastes precious analysis and computational time. Also a great deal of effort must be devoted to studying the computer code in order to use MM5 to its fullest potential. Typically, it takes many months to learn the model before the scientist can do a simple simulation.

These characteristics of MM5 are found in many model and simulation programs. With the "Gestalt of computers" theory becoming more prevalent every day, people are demanding more user-friendly environments [4]. To this end, we have created a graphical user interface, called Espresso, that makes the task of running climate models more straightforward. With Espresso, the scientist no longer has to be proficient in computer science to perform a simulation, and the likelihood that the model will fail because of errors is greatly reduced. Espresso was also designed with flexibility that allows users to add and delete name list variables as well as load in completely different modeling systems.

Planning Espresso, we considered an object-oriented design to be essential. With this in mind, we chose to write the program in Java. In addition, we

P.M.A. Sloot et al. (Eds.): ICCS 2002, LNCS 2330, pp. 758–766, 2002.

incorporated the two packages: Java Swing and the Extensible Markup Language (XML). Swing allows us to add graphical components to the interface quite easily. XML allows users to create new variables easily. An XML configuration file is read in, and Swing components are created to reflect the file.

After a simulation has been created, scientists often wish to visualize the results. To this end, we have enhanced a version of Vis5D that allows scientists to view their data on a high-resolution tiled displays.

2   Requirements of the Espresso Interface

The Espresso interface was based on an existing interface designed by Veronika Nefedova of the Mathematics and Computer Science Division at Argonne National Laboratory. That interface allowed the users to change a set number of variables, select a task and a machine to run on, and then run the selected task. The interface provided the basic functionality to create a climate simulation, but it needed much more flexibility. The new Espresso Interface can be seen in Figures 1 and 2.

2.1   Flexibility

One of the main goals of Espresso was flexibility. Running atmospheric simulations involves many variables, only some of which may be needed. Therefore, the interface must handle variable selection appropriately.

The Espresso interface was designed to reflect the variables that are in an XML file. As a default XML file is read in by the Java code, variables are stored within a tree structure in memory, and then put into a javax.swing.JTree structure to display graphically. Clicking on the tree brings up the variables on the interface where the values can be modified.

Espresso also provides the option of opening different XML files while the program is running. When a new file is opened, the interface is recreated dynamically. The user can override the values from the default XML file with individual test cases.

As well as opening up a new XML file interface, users can also save the interface as an XML file. This feature allows users to set up a case study, save the variable values, and, if desired, disperse the XML file to other scientists. Results can be duplicated and studied by other researchers easily.

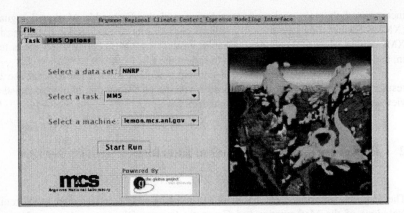

Fig. 1. The first panel of Espresso allows users to select a data set, a task to run, and the machine to run the task on.

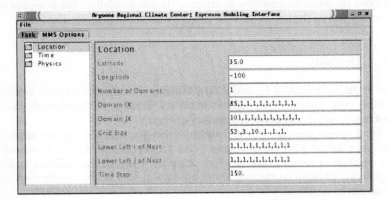

Fig. 2. The second panel displays a tree structure that contains all categories and their variables. When a category is selected, the variables are displayed on the right side where their values can be changed. For compatibility, the format of the data fields are the same as in the original MM5 Fortran namelist.

2.2   Different Modeling Systems

Espresso accommodates different modeling systems. Since the interface parses variables into name lists, it does not matter which model one incorporates into the Espresso modeling system. Some models do require additional information; however, changing the XML file and adding a little Java code can accomplish this task. As we continue to add more modeling systems the need for additional Java code will diminish.

2.3   Object-Oriented Design

An object-oriented design was chosen for easy maintainability. When a program is divided into sections, adding and changing components are simplified. The Espresso interface code was divided into eight classes: a main class, a tabbed pane class, a class for each of the panels, and three classes to implement the parsing and handling of the XML file.

2.4   Error Handling

The previous version of the interface did not check for errors in the input fields. When an incorrect value was entered, the scripts continued to execute, and the user would not be aware of this error until much later. Also, errors were mixed in with a lot of other text output, making them difficult to locate. This situation required significant improvement.

To handle errors in Espresso, we decided to enter parameter ranges and type information in the XML file. For each variable in the XML file, the author must include a minimum and maximum value for the variable and a type (float, integer, array, character, etc). As the XML file is read into the system, these values are checked to make sure they comply with the rules. If a variable does not, a warning dialog box is displayed on the screen and in the terminal output. The same procedure occurs when the user enters invalid input within the GUI, making it much more difficult to enter invalid input into the modeling system.

3   Implementation

While implementing Espresso, we chose the software that was best suited to handle the requirements: Java2 and XML. We used Java2 Swing to create the GUI.

The application first reads in the XML file through the Java tools SAX and DOM. These tools parse the XML file and store the information into a tree data structure. The XML file needed by Espresso provides parameters, values, keywords, minimum and maximum values, and descriptions. The values provided in the XML file are then put through error-handling methods that check whether the values are valid. The graphical interface is then created with Swing from the XML file. The user can interact with this interface as well as change the preset values. After the user selects a data set, task, and machine, the job can be submitted. All the values are then sent to the system's side of Espresso, the scripts are modified, and the jobs are run. The system side of Espresso handles job submission and described in a companion paper, appearing in the volume [5].

3.1.1   Java2

Java was used in the implementation for several reasons. The main reason is its clean object-oriented design and its powerful suite of tools necessary for development of web based applications. To ensure maintainability, we divided Espresso into several different objects. Java also enabled us to use the Swing components to create the interface easily.

Another advantage in using Java is its ability to parse XML files. Specifically, we used Java's SAX and DOM tools. SAX allowed us to read in the file and check for errors that conflicted with the DTD file. DOM allowed us to store the XML information in a tree-linked list in memory. Thus, we were able to traverse the tree quite easily to obtain needed information about the variables.

Espresso is implemented as 100% Java so that we can run across many platforms. This feature increases the flexibility of Espresso as it can be displayed on different types of machine operating systems including Windows and Linux.

3.1.2   Java2 Swing

Swing provides an environment that enables interfaces to be developed quite easily with few commands. Swing includes interface tools such as dialog boxes, text boxes, trees, panels, and tabbed panes that are both consistent and familiar to users [6]. These features increase the ease with which the Espresso user interface can be developed and with which the users can interact.

3.1.3   XML

XML allowed us to create our own format for the configuration file used by the interface. With XML one defines customized tags and how they will be used. Using XML, we defined our own tags to create a readable format. This feature enables the file to be modified easily. It also allows users to add new variables that a user may need to modify on the interface easily. Recompilation and the additional java coding are avoided greatly increasing the flexibility of the Espresso interface. An excerpt of an Espresso XML file is shown in Figure 3.

4   Enhanced Version of Vis5D

Vis5D is a visualization tool created by Bill Hibbard, Johan Kellum, Brian Paul, and others under the Visualization Project at the University of Wisconsin-Madison [7]. This tool visualizes five-dimensional data sets created by climate models. Vis5D provides users with a GUI environment in which to view data objects in a series of different ways [7]. We plan to add the capability to convert model output files from the MM5 and FOAM models to Espresso.

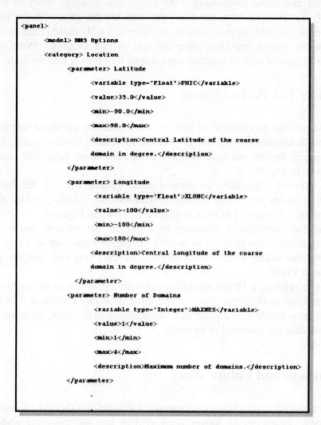

```
<panel>
    <model> MM5 Options
    <category> Location
        <parameter> Latitude
            <variable type="Float">PHIC</variable>
            <value>35.0</value>
            <min>-90.0</min>
            <max>90.0</max>
            <description>Central latitude of the coarse
            domain in degree.</description>
        </parameter>
        <parameter> Longitude
            <variable type="Float">XLONC</variable>
            <value>-100</value>
            <min>-180</min>
            <max>180</max>
            <description>Central longitude of the coarse
            domain in degree.</description>
        </parameter>
        <parameter> Number of Domains
            <variable type="Integer">MAXNES</variable>
            <value>1</value>
            <min>1</min>
            <max>4</max>
            <description>Maximum number of domains.</description>
        </parameter>
```

Fig. 3. The XML file must specify a model. After the model is specified, parameters are defined and are grouped in categories. Each category will appear in its own pane accessible via the tree.

Although Vis5D provided a convenient way to visualize our data, we needed to visualize our data at a larger scale. Tiled displays provide large viewing areas with increased resolution. Researchers in the Futures Lab in the Mathematics and Computer Science Division at Argonne National Laboratory have developed an Active Mural that consists of fifteen XGA projectors producing one image that contains 4,920 x 2,204 resolution [8]. To use this technology, we modified the Vis5D code and compiled it with the WireGL libraries instead of the OpenGL libraries.

WireGL is a version of OpenGL that replaces the OpenGL commands with its own commands. WireGL divides the image into different sections, which are then dispersed to the correct rendering server [9].

In our early experiments, with help from David Jones in the Mathematics and Computer science Division at Argonne National Laboratory, we determined that not all of the OpenGL calls were accepted by WireGL. Problems occurred with destroying contexts and raster positioning. We found that although many of these calls could simply be removed from the code, this was not the case with the raster positioning problems. In order to place text on the screen, Vis5D makes calls to set a raster position on the screen and then place the text in that location. WireGL is unable to handle this type of call, so inserted a pre-drawn font library into the code.

5   Test Case: The Perfect Storm

In this section we provide an example of how we have used the advanced interfaces that we have been developing at Argonne to perform substantial climate simulations. Currently, Espresso is operational and we have completed some basic test runs in order to demonstrate its capability.

With the previous interface, we created three climate runs of the Perfect Storm: the first with 80 km resolution, the second with 40 km resolution, and the third with 20 km resolution. A snapshot of each resolution is shown in Figure 4.

This interface functioned similarly to Espresso but offered much less flexibility. The interface allowed users to modify the preset values set in a text file and then submitted the values to the MM5 scripts. MM5v3 was run, and the data were visualized with Vis5D.

We hope to produce a 10 km resolution run with a grid size of 400x400 with Espresso. This presents a challenge because the input boundary condition files that are generated can grow larger than 2 gigabytes. To accomplish this task, we must edit MM5 to need input data daily instead of monthly.

6   Conclusions and Future Work

We have created a graphical user interface Espresso that simplifies the task of running climate simulations. Scientists no longer need to dive into the modeling code to change simulations; users can now simply edit an XML file or change values within the interface.

We have used Espresso successfully to create high-resolution climate runs with Espresso. The visualization of these runs presents other software challenges, however. To date, we modified the code of Vis5D to run on tiled displays. Thus, we can see climate details at much higher resolution, aiding in the interpretation of the climate modeling results.

Both applications are in their beginning stages of development. We hope to increase Espresso's flexibility by setting it up for other models such as FOAM and CCM. We also intend to add other modules to the interface. We plan to add a progress screen so users can see the status of the current run. We also need to add an interactive topography file for use with the FOAM model so that users can select which area of the world to simulate, can shift continents around to simulate the Earth's climate millions of years ago, or even simulate of other planets such as Mars.

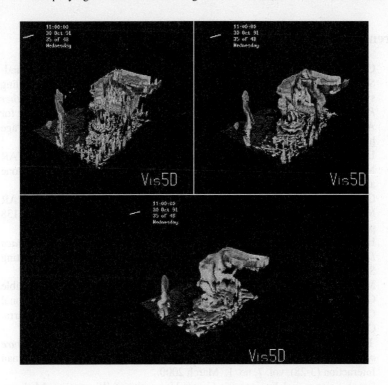

Fig. 4. Clockwise from top left: 20 km resolution climate run with a grid size of 200x250; 40 km resolution climate run with a grid size 100x125; 80 km resolution climate run with a grid size of 50x75. Snapshots are of October 30, 1991, 11:00 AM, of the Perfect Storm climate simulations created with MM5v3. (See http://www-unix.mcs.anl.gov/~mickelso/ for animations of above simulations.)

Vis5D's work involves increasing its functionality. Ultimately we also need to address the problem of the increasing size of the output data sets generated by climate models. New interfaces must be developed to handle this requirement.

Acknowledgments

We thank the staff of the Futures Laboratory at the Mathematics and Computer Science Division at Argonne National Laboratory in Argonne, Illinois. We would also like to thank Veronika Nefedova of the Mathematics and Computer Science Division at Argonne National Laboratory. This work was supported in part by the Laboratory Director Research and Development funding subprogram of the Office of Advanced Scientific Computing Research, U.S. Department of energy, under Contract W-31-109-Eng-38. This work was also supported by the NSF Information Technology Research Grant, ATM-0121028.

References

1. Chen, F. and J. Dudhia. 2001: Coupling an Advanced Land-Surface/Hydrology Model with the Penn State/NCAR MM5 Modeling system: Part I: Model Implementationand Sensitivity, *Monthly Weather Review*, in press. (See also Pennsylvania State University/National Center for Atmospheric Research, MM5 Home Page http://www.mmm.ucar.edu/mm5/mm5-home.html

2. Dudhia, J. 1993. A Nonhydrostatic Version of the Penn State/NCAR Mesoscale Model: Validation Tests and Simulation of an Atlantic Cyclone and Cold Front. Mon. Wea. Rev. 121: 1493-1513

3. Grell, G.A., J Dudhia, and D.R. Stauffer. 1994. The Penn State/NCAR Mesoscale Model (MM5). NCAR technical note, NCAR/TN-398_STR, 138 pp.

4. H. Rex Hartson and Deborah Hix. *Human-Computer Interface Development: Concepts and Systems for Its Management*. ACM Computing Surveys, vol. 21, no. 1. March 1989.

5. Mike Dvorak, John Taylor, Sheri Mickelson (2002). Designing a flexible Grid Enable Scientific modeling Interface. Proc. 2002 International Conference on Computational Science, eds. V.N. Alexandrov, J.J. Dongarra, C. J. K. Tan, Springer-Verlag (in preparation for).

6. Brad Myers, Scott E. Hudson, and Randy Pausch. *Past, Present, and Future of User Interface Software Tools*. ACM Transactions of Computer-Human Interaction (3-28), vol. 7, no. 1. March 2000.

7. Space Science and Engineering Center University of Wisconsin – Madison, Vis5D Home Page http://www.ssec.wisc.edu/~billh/vis5d.html.

8. Futures Lab, Mathematics & Computer Science, Argonne National Laboratory. ActiveMural http://www-fp.mcs.anl.gov/fl/activemural/.

9. Stanford Computer Graphics Lab, WireGL: Software for Tiled Rendering Home Page http://graphics.stanford.edu/software/wiregl/index.html.

On the Computation of Mass Fluxes for Eulerian Transport Models from Spectral Meteorological Fields

Arjo Segers[1], Peter van Velthoven[1], Bram Bregman[2], and Maarten Krol[2]

[1] Royal Netherlands Meteorological Institute (KNMI),
P.O. Box 201, 3730 AE, De Bilt, The Netherlands
{Arjo.Segers,Velthove}@knmi.nl
[2] Institute for Marine and Atmospheric Research (IMAU),
University of Utrecht, The Netherlands
{A.Bregman,M.Krol}@phys.uu.nl

Abstract. A procedure is introduced to compute mass fluxes and mass tendencies for a Eulerian transport model from spectral fields in a consistent way. While meteorological forecast models are formulated in terms of vorticity, divergence, and surface pressure expanded in spherical harmonics, Eulerian transport models, widely used for tracer and climate studies, require wind fields or mass fluxes on a regular grid. Here horizontal and vertical mass fluxes are computed directly from the spectral data to stay as close to the representation in the meteorological model as possible. The presented procedure can be applied to compute fluxes for global as well as limited area models.

1 Introduction

The study of climate and atmospheric composition with three dimensional global models requires accurate simulation of tracer transport. Horizontal winds increasing in magnitude from the surface upwards are the driving force for global transport and mixing of trace gases. The exchange of trace gases between the different atmospheric compartments such as the mid latitude troposphere and stratosphere is a key process for the impact of human activity on climate. For accurate simulation of transport, three dimensional chemistry-transport models (CTMs) are often equipped with sophisticated and expensive advection routines [1]. However, less attention is paid to the numerical procedure to compute mass fluxes required for the advection and to interpolate them to the model grid. A model comparison by [2] showed that performance of CTMs strongly depends on how the meteorological model input is used. Comparison of simulated ozone concentrations with aircraft measurements showed significant differences between chemistry transport models, in spite of a common source of meteorological data (ECMWF) and usage of similar advection and identical chemistry schemes.

The mass fluxes used as input for tracer models are derived from wind fields from a meteorological model. Most tracer models use analyzed or forecasted wind

P.M.A. Sloot et al. (Eds.): ICCS 2002, LNCS 2330, pp. 767–776, 2002.

fields from a weather forecast or climate model. Data from meteorological forecast models have the advantage of being based on assimilation of meteorological measurements. A disadvantage is that the data is only available at discrete time instants 3 or 6 hours apart. Nowadays most operational meteorological forecast models are formulated in terms of vorticity and divergence rather than wind components. Most models represent these fields in terms of spherical harmonics. Tracer transport models however expect input of mass fluxes through boundaries of grid boxes or wind fields at the grid cell centers. Therefore, a method should be developed to compute mass fluxes in a consistent and accurate way from the spectral fields.

A method often applied is interpolation of the spectral wind fields to a high resolution regular grid. Integration of these wind fields over the boundaries of grid boxes then provides the required mass fluxes. If the high resolution grid does not match with the grid of the tracer model, additional interpolations are required. Vertical fluxes are then computed from horizontal divergence and boundary conditions of zero flux through the top and bottom. This method has for instance been used in older versions of CTM2 [3] and TM3 [2]. Although the horizontal fluxes computed in this way are quite accurate, the errors in the vertical fluxes can be quite large. The vertical velocities are very small (a few cm per second) in comparison with the horizontal winds, which are of the order of several to tens of meters per second; this is a consequence of near geostrophic balance [4]. If interpolated wind fields are used to compute the vertical ones, small interpolation errors easily lead to large errors in the vertical wind [4]. In [2] it is stated that for this reason the use of interpolated fields should be avoided.

To avoid unnecessary interpolations, the mass fluxes should be computed from spectral fields directly. The input for the TOMCAT model is computed from vorticity and divergence using integrated spherical harmonic functions [5]. In particular, the vertical fluxes are computed from divergence fields to stay as close to the parent meteorological model as possible. Near the surface, where the surface pressure acts on the hybrid vertical coordinate, the mass fluxes computed for TOMCAT are corrected to preserve the tracer mixing ratio.

This paper proposes a consistent scheme for computation of mass fluxes from spectral fields of vorticity, divergence, and surface pressure, avoiding the use of interpolated wind fields. The next section describes the horizontal and vertical grid of a general Eulerian model, and requirements for the mass fluxes. A short discussion of the integration of spectral fields is given in 3. The algorithm proposed in section 4 computes a consistent set of three dimensional fluxes, in balance with the surface pressure tendency. The method as described here is suitable for computation of mass fluxes for a global tracer transport model with rectangular grid boxes in the lat/lon plane. All equations assume a vertical coordinate system with hybrid σ/pressure levels, for which pressure and σ levels are special cases. Modifications needed for non-regular discrete grids or other vertical coordinate systems are straightforward. The new system of mass fluxes has been evaluated with the TM3 model; a comparison between old and new formulation is provided in section 5.

2   Definition of the tracer model

We consider a general Eulerian transport model, defined on a three dimensional grid of hybrid σ/pressure levels. For the vertical levels it is common practice to copy the set of vertical levels used by the meteorological model or a subset of it. In a system of hybrid σ/pressure coordinates such as used by ECMWF [6], the vertical coordinate is a dimensionless parameter $\eta \in [0, 1]$. All that is required for η is that it is a function $h(p, p_s)$ of the pressure and surface pressure at a certain location, with the boundary $h(0, p_s) = 0$ at the top and $h(p_s, p_s) = 1$ at the surface. In the vertical there are n_{lev} layers with boundaries or half levels at $\eta_{k-1/2}$ and $\eta_{k+1/2}$ and mid levels η_k. At the half levels, the pressure is specified by hybrid coefficients $a_{k+1/2}$ and $b_{k+1/2}$:

$$p(\lambda, \beta, \eta_{k+1/2}) \; = \; a_{k+1/2} \; + \; p_s(\lambda, \beta) \; b_{k+1/2} \qquad \text{[Pa]} \qquad (1)$$

where λ and β denote the longitude and latitude respectively, both in radians. For integration of a field $F(\eta)$ between two half levels we will use the procedure used in the ECMWF model:

$$\int_{\eta_{k-1/2}}^{\eta_{k+1/2}} F(\eta) \, \frac{\partial p}{\partial \eta} \, \mathrm{d}\eta \; \approx \; F_k \, \Delta p_k \; = \; F_k \, (\Delta a_k + p_s \, \Delta b_k) \qquad (2)$$

where $\Delta a_k = a_{k+1/2} - a_{k-1/2}$ and $\Delta b_k = b_{k+1/2} - b_{k-1/2}$. The horizontal grid will be supposed to consist of rectangular cells in the lat/lon plane. A cell i then has a surface area of:

$$A_i \; = \; [\lambda_i^-, \lambda_i^+] \times [\beta_i^-, \beta_i^+] \; = \; \Delta\lambda_i \times \Delta\beta_i \qquad \text{[rad}^2\text{]} \qquad (3)$$

The size and location of the cells is left undefined. With the algorithm introduced in section 4 we will be able to compute mass fluxes through all six boundaries of a box, independent of other boxes. In a simple configuration, the grid cells form a regular pattern with equal spacing in both the latitudinal and longitudinal direction. Irregular grids are however possible too: modern tracer models often use a regular or quasi-regular reduced Gaussian grid which implies irregular spacing in latitudinal direction and for each latitude band a different longitudinal spacing. The tracer model requires three entities which in theory completely define the airmass fluxes between the cells within a time interval $[t_0, t_1]$:

1. the *air mass distribution* at t_0, for example in kg per grid box;
2. the *horizontal mass fluxes* in kg/s through the boundaries of the grid cells, valid for the complete interval $[t_0, t_1]$;
3. the *air mass distribution* at t_1.

Knowledge of these entities implies that the vertical fluxes can be calculated from mass conservation. In here, we will however compute the vertical flux from the meteorological data too, in order to correct for errors in the horizontal fluxes. Mass conservation implies that the net flux into a box should equal the difference in mass between the beginning and end of the interval. The latter is given by the change in surface pressure in the case of time-independent vertical coefficients a_k and b_k.

3   Spectral fields

To compute the variables specified in the previous section, a meteorological model should provide wind fields and surface pressures. Meteorological models such as the ECMWF model solve the continuity equation on the sphere:

$$\frac{\partial}{\partial t}\left(\frac{\partial p}{\partial \eta}\right) + \nabla \cdot \left(v\frac{\partial p}{\partial \eta}\right) + \frac{\partial}{\partial \eta}\left(\dot{\eta}\frac{\partial p}{\partial \eta}\right) = 0 \tag{4}$$

where v is the horizontal wind and ∇ the horizontal gradient operator. The output of the model consists, amongst others, of fields of divergence $D = \nabla \cdot v$, the vertical component of the vorticity $\xi = (\nabla \times v)_\eta$, and the natural logarithm of the surface pressure $\ln(p_s)$. These are available in the form of spectral coefficients. Their value at a certain point in grid space can be computed from the expansion:

$$X(\lambda,\mu) = \sum_{m=-M}^{M} \sum_{n=|m|}^{N} X_n^m\, P_n^m(\mu)\, e^{im\lambda} \tag{5}$$

Here, $\mu = \sin(\beta)$, and P_n^m denote the associated Legendre functions. For computation of mass distributions and fluxes, the spectral fields need to be integrated. An integral in the longitudinal direction can simply be solved analytically. In the latitudinal direction, integrals over the associated Legendre functions can be computed once to be used multiple times for expansion of the spectral sum using integrated rather than ordinary base functions. Since no simple analytic solutions exist for definite integrals of associated Legendre functions, the integrals over P_n^m have to be solved numerically. This approach is for example used for preprocessing meteorological data in the TOMCAT model [5]. Apart from difficulties with oscillating Legendre functions at higher truncations, this method is not suitable for an integral over a product of spectral fields, a computation that is often needed in the hybrid level system when fields are multiplied with the surface pressure before integration (see for example equations (9-11) and (13-14) below). A better method is therefore to evaluate the spectral fields to a high resolution grid, and to use numerical quadrature afterwards.

4   Computation of mass fluxes

The proposed algorithm for computation of the mass fluxes and mass distributions for the tracer model consists of four stages.

4.1   Computation of the mass distribution

The mass distribution in kg per grid box at t_0 and t_1 is computed from the pressure. For the system with hybrid pressure levels, it is sufficient to compute the average surface pressure $(\overline{p_s})_i$ for each cell i:

$$(\overline{p_s})_i = \iint\limits_{A_i} p_s(\lambda,\beta)\, \cos\beta\, d\lambda\, d\beta \, / \, A_i \qquad [\text{Pa}] \tag{6}$$

Together with the hybrid coefficients, this defines the mass in a box i at level k:

$$m_{i,k} = [\; \Delta a_k + \Delta b_k \; (\overline{p_s})_i \;] \; A_i \; R^2 \; / \; g \quad \text{[kg]} \tag{7}$$

where R is the radius of the earth (m), g the acceleration of gravity (m/s^2), and A_i the area of the box defined in (3). If the meteorological model provides a spectral field for the natural logarithm of the surface pressure, it should be evaluated on a fine mesh, followed by transformation into surface pressure and numerical quadrature. Optionally, the mass distributions could be slightly adjusted with a multiplication factor to impose a global mass conservation.

4.2 Computation of the vertical fluxes

In this stage the vertical mass fluxes are computed for the time interval $[t_0, t_1]$. We assume that the temporal resolution of the available spectral data is such that for the center of the time interval the flux could be computed. The computed flux will be used for the entire time interval, without using temporal interpolations.

The vertical flux $\dot{\eta}\partial p/\partial \eta$ is computed by integrating the continuity equation (4) from the top of the model to the desired η:

$$\dot{\eta}\frac{\partial p}{\partial \eta} = -\frac{\partial p}{\partial t} - \int_0^\eta \nabla \cdot \left(v \frac{\partial p}{\partial \eta} \right) \, d\eta \quad \text{[Pa/s]} \tag{8}$$

In the hybrid coordinate system, one can use that the surface pressure tendency is computed from (8) with $\eta = 1$ and boundary condition of zero flux through the surface, so that the flux through the bottom $\eta_{k+1/2}$ of a grid box i at level k is equal to:

$$\Phi_w(A_i, \eta_{k+1/2}) = \frac{R^2}{g} \iint_{A_i} \left(\dot{\eta}\frac{\partial p}{\partial \eta} \right)_{k+1/2} \cos \beta \, d\beta \, d\lambda \tag{9}$$

$$= \frac{R^2}{g} \iint_{A_i} [(\sum_{j=1}^{nlev} \Omega_j) b_{k+1/2} - \sum_{j=1}^{k} \Omega_j] \cos \beta \, d\beta \, d\lambda \quad \text{[kg/s]} \tag{10}$$

where

$$\Omega_j = \nabla \cdot (v_j \, \Delta p_j) = D_j \, (\Delta a_j + \Delta b_j p_s) + \frac{V_j}{\cos \beta} \cdot (\nabla(\ln p_s)) \, p_s \, \Delta b_j \tag{11}$$

In spectral models it is common that the velocity vector $V = v \cos \beta$ is available in spectral form, otherwise it can be derived from the divergence and vorticity. The singularity in $V/\cos \beta$ at the pole is thus circumvented. The asymptotic value of v at the poles can be set to an average of surrounding values. The gradient $\nabla(\ln p_s)$ of the natural logarithm of the surface pressure is hard to obtain from an interpolated pressure field, but easily derived from spherical harmonics.

4.3   Computation of the horizontal fluxes

Similar to the vertical fluxes, the horizontal fluxes are computed from meteorological fields valid at a time centered within $[t_0, t_1]$, and used over the whole
time interval. The flux through a vertically oriented surface S is equal to:

$$\Phi_h = \frac{1}{g} \iint_S n \cdot v \, \frac{\partial p}{\partial \eta} \, d\eta \, dl \qquad [\text{kg/s}] \qquad (12)$$

with v the horizontal velocity vector, n the normal vector on the surface, and
l the horizontal coordinate (m). In the hybrid vertical coordinate system, the
fluxes through a longitudinally or latitudinally oriented surface between two half
levels are given by:

$$\Phi_u(\Delta\beta_i, k) = \frac{R}{g} \int_{\Delta\beta_i} \frac{U(\lambda, \beta, k)}{\cos\beta} \, [\Delta a_k + p_s(\lambda, \beta)\Delta b_k] \, d\beta \qquad [\text{kg/s}] \qquad (13)$$

$$\Phi_v(\Delta\lambda_i, k) = \frac{R}{g} \int_{\Delta\lambda_i} V(\lambda, \beta, k) \, [\Delta a_k + p_s(\lambda, \beta)\Delta b_k] \, d\lambda \qquad [\text{kg/s}] \qquad (14)$$

where U and V are the components of $V = v \cos\beta$. The integrals can be approximated by numerical quadrature after evaluation of V on a fine grid. At the
poles, the limiting values of $U/\cos\beta$ along the meridians need to be evaluated
for computation of Φ_u; the zero length of $\Delta\lambda_i$ ensures vanishing of Φ_v here.

4.4   Conservation of mass

For each grid box in the tracer model, the mass change between the start and
end of the time interval should equal the net flux through the six boundaries:

$$m(t_1) - m(t_0) = [\, \delta_\lambda \Phi_u + \delta_\beta \Phi_v + \delta_\eta \Phi_w \,] \, \Delta t \qquad (15)$$

where the $\delta_\star$ denote the difference operators between opposite boundaries of a
grid box. For two reasons, the balance will not be matched exactly by the variables computed in 4.1-4.3. First, the fluxes through the boundaries are computed
from meteorological fields valid at a discrete time rather than over the complete
interval. Secondly, both the mass distributions and the fluxes are subject to interpolation errors. To produce a consistent set of mass distributions and fluxes,
some of these variables therefore need to be adjusted.

Taking all variables in (15) into account, the largest absolute errors occur
in the net horizontal fluxes. This is because the horizontal velocities are much
stronger than the vertical velocities, and they hardly vary from one box to its
neighbors. The net horizontal flux is therefore computed from differences between
large and almost equal numbers. The adjustment procedure proposed here is to
add small corrections to the horizontal fluxes only, leaving the mass distributions
and vertical fluxes unchanged. The procedure consists of adding a correction flux

$\boldsymbol{F} = (F_u, F_v)$ to the horizontal fluxes $\boldsymbol{\Phi}_h = (\Phi_u, \Phi_v)$ such that for all grid boxes the following expression holds:

$$\boldsymbol{\delta}_h \cdot \boldsymbol{F} = [m(t_1) - m(t_0)]/\Delta t - \boldsymbol{\delta}_h \cdot \boldsymbol{\Phi}_h - \delta_\eta \Phi_w \tag{16}$$

where $\boldsymbol{\delta}_h = (\delta_\lambda, \delta_\beta)$ is the horizontal difference operator. This correction can be applied to every model layer independent of other layers. The system of equations (16) is under-determined, since the number of correction fluxes is about twice the number of grid boxes. Therefore, the correction flux can be obtained from (16) using a least squares solver, able to handle sparse matrices. If the grid is regular (same number of grid cells in each latitudinal or longitudinal band), another procedure could be applied by requiring that the correction flux is divergence free, completely defined by a potential Ψ according to $\boldsymbol{F} = \boldsymbol{\delta}_h \Psi$. With this assumption, system (16) is transformed in a discrete Poisson equation, easily solved with a discrete 2D Fourier transform. The result is a correction field $\boldsymbol{F}$, periodically continued at the boundaries. For a global model, the boundary condition of periodically fluxes in longitudinal direction is valid as such. In the latitudinal direction, the fluxes over the poles should remain zero (periodically with a fixed value), which can be simply achieved by subtraction of the polar flux from each longitudinal column of the grid. Since the corrections are in general small in comparison with the horizontal fluxes (less than a percent), the new horizontal fluxes hardly differ from the original fluxes.

The procedure described here provides balanced mass fluxes for a global tracer model. If mass fluxes are to be derived for a limited area only, the procedure has to be slightly changed. The sub domain is supposed to be a high resolution or zooming area of a global, coarse grid model, for which mass distribution and fluxes are available. Ideally, each cell in the sub domain is covered by one cell of the coarse grid only; in an extreme situation, the sub domain is covered by a single cell of the coarse grid. Mass distribution and fluxes for the high resolution grid are computed following the equations in 4.1–4.3. The first step in balancing the mass is now to bring the computed values in agreement with the values of the coarse grid. That is, if a box of the coarse grid covers n boxes in the high resolution grid, the total mass in the n small boxes should equal the mass in the large box, and similarly for fluxes across the boundaries. A simple scale factor is sufficient here, since the difference between values in the global grid and the sum over high resolution cells is probably small (if not, the global variables are inaccurate and should be recalculated with more accurate quadrature). The second step is to compute the corrections to the horizontal fluxes in the same way as described before, but now for each each cell of the coarse grid, with corrections forced to zero at the boundaries.

5   Model results

The impact of computing mass fluxes directly from spectral fields instead of already interpolated gridded fields has been examined with the TM3 model. The model input for TM3 is derived from ECMWF meteorological data at a temporal resolution of 6 hours and using 31 hybrid σ/pressure levels between the surface and 10 hPa. The horizontal grid of TM3 is regular with a resolution of 2.5° × 2.5°. Mass distributions and mass fluxes are computed for each grid cell, although in the advection routine (second moment scheme [1]) some grid cells around the poles are actually joined to avoid Courant numbers exceeding 1.

The horizontal mass fluxes for TM3 were in the past computed from wind fields interpolated to a grid of 1° × 1° degrees. A small difference between the net mass flux into a column of grid cells and the actual mass change observed from the surface pressure tendency could not be avoided. Therefore, a correction field was added to the horizontal fluxes, distributed over the layers with ratios following the absolute fluxes. Vertical fluxes were only derived afterwards given the corrected horizontal fluxes and the mass tendency in each layer. In the new system, the fluxes are computed from spectral fields (T106), and where necessary interpolated to a grid of 0.25°×0.25° for numerical quadrature. Figure 1 shows an example of the error in vertical fluxes derived from the (uncorrected) horizontal fluxes and mass tendency. The derived vertical fluxes are compared with the vertical fluxes computed using the method described in section 4.2. The error is reduced significantly in almost all layers of the model when using spectral fields instead of interpolated fields. Especially at the top of the model, where the vertical fluxes are small in comparison with the horizontal flow and therefore not so easily derived from horizontal convergence only, the error is reduced by 50% . The errors increase smoothly towards the bottom of the model, and are comparable near the surface where all wind components approach zero. The error in the vertical flux is removed up to machine precision by the mass balance correction described in 4.4.

The difference between mass fluxes computed from interpolated and spectral fields in tracer simulations has been re-examined for experiments described in detail in [2, 7]. In that study, ozone observations from commercial aircraft flights in the mid latitude lower-most stratosphere were compared with simulations with TM3 and other models. It was concluded that the model performance significantly depends on the preprocessing of the fluxes, and that interpolations should be avoided. Figure 2 shows an example of the improvement of the performance if the fluxes are computed from spectral fields. This figure shows the mean plus and minus the standard deviation for the measured and simulated ozone concentrations. The simulations are either based on interpolated winds, with optionally a correction using vertical velocities [2], or on spectral data. The simulated variability of ozone improves significantly in comparison with the simulation based on interpolated fields, especially when the spectral data is used. While in the case of interpolated fields the range of ozone values was almost twice as broad as actually measured, the new simulations roughly cover the same variability range as observed.

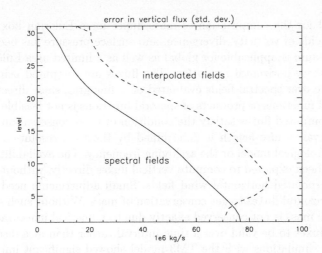

Fig. 1. Error in vertical flux computed from uncorrected horizontal convergence and mass change, if compared with vertical flux computed from spectral fields (section 4.2). Computed for may 26, 1996 over four 6-hour intervals.

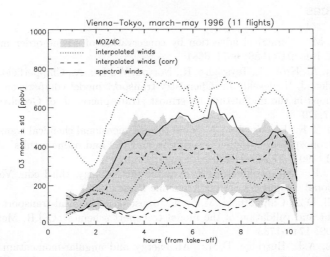

Fig. 2. Example of the impact of the new method for computation of the fluxes on an ozone simulation. Mean plus and minus standard deviation of ozone for flights Vienna-Tokyo during March-May 1996: observations, TM3 with interpolated horizontal winds, corrected with vertical velocities (see [2]), and using spectral fields.

6   Conclusions

A method for the computation of mass fluxes for a Eulerian box model from spectral fields of vorticity, divergence, and surface pressure has been described. The procedure is applicable for global as well as a limited area Eulerian models with arbitrary horizontal resolution. The fluxes are computed using numerical quadrature over spectral fields evaluated on a fine grid, since direct integration of spectral functions or products of spectral functions is not feasible analytically.

The computed fluxes satisfy the condition of mass conservation over a finite time interval, whose length is determined by the time resolution of the driving meteorological model or the archiving frequency. The availability of spectral fields has been explored to compute vertical fluxes directly, without the need for using interpolated horizontal wind fields. Small adjustments need to be made to the horizontal fluxes to get conservation of mass. Without such small adjustments, the mass is not conserved exactly due to numerical inaccuracies, and the need for fluxes to be valid over a time interval rather than at a discrete time.

Ozone simulations with the TM3 model showed significant improvement in the fluxes when derived from spectral data instead of interpolated wind fields. The error in the horizontal mass convergence is reduced by 20-50% in almost all layers of the model. Comparison of TM3 simulations with ozone data observed during aircraft flights showed good agreement of modeled and measured variability in concentrations.

References

1. Prather, M.: Numerical advection by conservation of second-order moments. J. Geophys. Res. **91** (1986) 6671–6681
2. Bregman, A., Krol, M., Teyssèdre, H., Norton, W., Iwi, A., Chipperfield, M., Pitari, G., Sundet, J., Lelieveld, J.: Chemistry-transport model comparision with ozone observations in the midlatitude lowermost stratosphere. J. of Geophys. Res. **106** (2001) 17,479
3. Berntsen, T.K., Isaksen, I.S.: A global three-dimensional chemical transport model for the troposphere: 1. model description and CO and ozone results. J. Geophys. Res. **102** (1997) 21,239–21,280
4. Holton, J.R.: An introduction to dynamic meteorology. third edn. Volume 48 of International Geophysics Series. Academic Press (1992)
5. Stockwell, D.Z., Chipperfield, M.P.: A tropospheric chemical-transport model: Development and validation of the model transport schemes. Q.J.R. Meteorol. Soc. **125** (1999) 1747–1783
6. Simmons, A.J., Burridge, D.M.: An energy and angular-momentum conserving vertical finite-difference scheme and hybride vertical coordinates. Mon. Weath. Rev. **109** (1981) 758–766
7. Bregman, A., Krol, M., Segers, A., van Velthoven, P.: A new mass flux processing method for global models and the effect on ozone distributions in the upper troposphere and lower stratosphere. Submitted to GRL (2002)

Designing a Flexible Grid Enabled Scientific Modeling Interface

Mike Dvorak[1,2], John Taylor[1,2,3], and Sheri Mickelson[1,2]

[1] Mathematics and Computer Science Division,
Argonne National Laboratory, Argonne IL 60439
{dvorak, jtaylor, mickelso}@mcs.anl.gov
http://www-climate.mcs.anl.gov/
[2] Computation Institute,
University of Chicago, Chicago, IL, 60637
[3] Environmental Research Divisions
Argonne National Laboratory, Argonne IL 60439

Abstract. The Espresso Scientific Modeling Interface (Espresso) is a scientific model productivity tool developed for climate modelers. Espresso was designed to be an extensible interface to both scientific models and Grid resources. It also aims to be a contemporary piece of software that relies on Globus.org's Java CoG Kit for a Grid toolkit, Sun's Java 2 API and is configured using XML. This article covers the design and implementation of Espresso's Grid functionality and how it interacts with existing scientific models. We give specific examples of how we have designed Espresso to perform climate simulations using the PSU/NCAR MM5 atmospheric model. Plans to incorporate the CCSM and FOAM climate models are also discussed.

1 Introduction

The Espresso Scientific Modeling Interface (Espresso) is designed to utilize existing Grid computing technology to perform climate simulations [1]. Espresso is also a software tool that gives scientific model users the freedom to eliminate the mundane task of editing shell scripts and configuration files. It empowers the scientist to spend more time performing science and analyzing the output of climate simulations.

Espresso is tailored to the demands of the climate modeler. In the Mathematics and Computer Science (MCS) Division, we make global climate model runs using the Fast Ocean-Atmosphere Model (FOAM) [2]. We also create high resolution meteorological model runs for extended periods of time (e.g. hourly output for years over the United States at 10-52 km resolution) using a regional climate model. Making regional climate simulations requires a robust computing environment that is capable of dealing with resource faults inside complex model codes. Espresso is designed to meet the rigorous demands of a multi-architecture, terra-scale environment.

P.M.A. Sloot et al. (Eds.): ICCS 2002, LNCS 2330, pp. 777–786, 2002.

Moreover, Espresso strives to make the best use of contemporary technology by using: (1) the eXtensible Markup Language (XML) for system and graphical configuration; (2) the Globus.org Java CoG Kit for accessing Grid resources; (3) Sun's Java Web Start for software deployment; (4) a subset of the Apache Software Foundation's Jakarta Project utilities i.e. Regexp. Espresso is a pure Java program that is capable of being run from anywhere on the Internet.

Lastly, Espresso can be utilized by a wide variety of users, not just climate modelers. Espresso is designed for ease of use with different scientific models where model configuration is complex and Grid resources are required. Work is underway to incorporate both the FOAM and the Community Climate System Model [3] into Espresso.

This article focuses on the system side (non-graphical user interface (GUI)) and Grid application design of Espresso. For a detailed look at the design of the GUI, see [4]. It should also be mentioned that Espresso is a second generation tool with the Argonne Regional Climate Workbench [5] as its predecessor. Espresso is different from the Climate Workbench in that: (1) its client side is pure Java; (2) it can be an interface for multiple models; (3) it is fully configurable via text (XML) files; (4) it can run anywhere on the Internet. Lessons learned from the design of the Climate Workbench and its applications have contributed significantly to the design of Espresso.

2   An Example Scenario Using Espresso

While the details of Grid computing for climate science are detailed in papers like [6], the following scenario provides insight on how a Grid enabled interface can be utilized to help climate scientists perform climate simulations. Figure 1 provides an example situation of how a climate scientist might interact with Espresso.

To the user, the most important part of Espresso is the GUI. Espresso's GUI hides all of the implementation details behind buttons and tell-tale output indicators. Users are not forced to learn all of the climate model's idiosyncrasies. Instead, they can concern themselves with the relevant model parameters. The GUI is a secure proxy to the Grid resources. From Espresso, the user can command the supercomputing, storage and analysis resources the user has access to with their authenticated certificate. The user only needs to authenticate once to obtain access to all Grid resources which they are authorized to use. This is the Internet equivalent to logging on to a single computing system.

After the user authenticates, they enter their climate simulation parameters via Espresso's GUI. Figure 1 shows the a scenario in which the climatologist wants to access a historical archive of climate data on a remote server (which could contain terabytes of data) to obtain initial and boundary conditions for a regional climate simulation. Inside the GUI the user simply specifies the historical data set via a combo box. The run dates, geographical grid, and model parameters to run for the simulation is set in text boxes. The user then submits

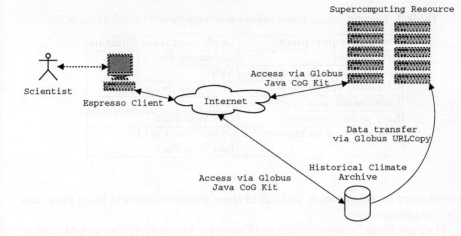

Fig. 1. An Example Scenario Using Espresso to Perform a Climate Run

the job to the Espresso server side component (which could be located anywhere on the Internet) for execution.

Eventually, the Espresso server side component will know how to use a Grid toolkit to obtain the climate data from a remote historical archive. The supercomputing resource will then use third party data transfer to obtain the climate data and proceed with the climate simulation. After the simulation is finished, the Grid toolkit could also be used to analyze and move the data to a different storage server via the Grid. For an overview of Grid computing infrastructure, see [1].

3   Modeling Interface Software Design

Table 1 highlights the most significant design requirements that were required for regional and global climate modeling. These requirements also help to make Espresso sufficiently flexible so that it could be used by other scientific modeling systems.

3.1   Special Considerations for Climate Modeling

Climate modeling places unbounded demands on supercomputing resources. Regional climate modeling intensifies demand on computing resources by increasing spatial resolution and the frequency of atmospheric/oceanographic data output. In MCS, initial and boundary condition files typically run on the gigabyte scale with raw data output running on the terabyte scale. Climate modeling systems also consist of several different data preprocessing programs and executables with many different build options. An effective interface for a climate modeling

Table 1. General design requirements and implementation solutions in Espresso

Design Requirement	Implementation Solution
Grid enabled	Globus.org Java CoG Kit
System-side easily configurable	XML
GUI easily configurable	XML/Java Swing
Distributable case studies	XML
Easy package deployment	Java Web Start
Run anywhere on Internet	Globus Java CoG Kit/ Java Web Start

system must be able to work with all of these preprocessing and build programs in a dynamic manner.

The Penn State University/National Center for Atmospheric Research Mesoscale Model 5 (MM5) is a good example of a scientific model that pushes a computing resource to its limits. We use MM5 for high resolution, regional climate runs. Most of the challenges of running MM5 evolved from porting the atmospheric model to different architectures i.e. MCS's Chiba City, a 512-processor Linux cluster. Other high performance problems are derived from running MM5's complex system of data preprocessing programs. The model is also written in Fortran 77 so variables are not dynamically allocated. Having to run a scientific modeling system with a complex build and preprocessor system placed a high quality design requirement on Espresso. A good review of MM5 is given in [7] [8] [9] and [10].

3.2   Making Espresso Work With Multiple Models

In order to make Espresso usable for a wide group of scientific models, it was necessary to make three broad assumptions:

- Large scientific modeling systems are configured by editing text configuration files.
- By editing these text files and replacing them within the system, the original scientific model can be used in the way that its designers intended.
- No code modifications to the original software.

By stating these underlying assumptions during the design and implementation of Espresso, it has been easy to determine which models will be able to use this interface. An important side effect of these assumptions is that Espresso can easily be used with a scientific model not originally designed to have a GUI. This is the situation for many older Fortran scientific codes.

3.3   General Espresso System Requirements

In order to accommodate the needs of a wide variety of users, Espresso must be extensible. We needed both a Grid functionality component and a GUI that was easily configurable using an input text file. The original Climate Workbench was limited to running only the MM5 model and contained hard-coded fields specific to MM5. It would have been very difficult to extend this interface to new models. Therefore, the interface needed to be customizable via a ASCII text file. Ideally this text configuration file would be written in XML to take advantage of freely available parsing tools.

We required the use of object oriented design and programming techniques in order to incorporate extensibility. The original Climate Workbench modeling interface was written such that it was nearly impossible to extend the functionality of the interface. Along with this design paradigm came the desire to reuse as much code as possible through the inheritance of key scientific modeling components i.e. model variable data structures and general text processing algorithms. Model specific tasks such as the regular expression syntax would be sub-classed.

The most critical design feature to building a "wrapper modeling interface" was embracing the scientific model without structural changes. This approach has substantial advantages with regard to code maintenance. Figure 2 illustrates how Espresso accomplishes this task. Contained within Espresso are only the model's original configuration files that will be used as a template. In step 2, Espresso has modified the configuration files with error checking. Step 3 places the files on the server side supercomputing resource using the Globus URL Copy functionality. These configuration files are "pushed" into place like the final pieces of puzzle, allowing the model to be run as originally intended.

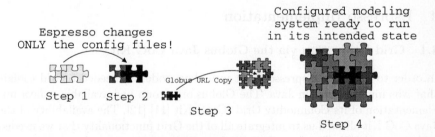

Fig. 2. Espresso runs the remote system by editing the configuration files. These configuration files are then moved back into their original location. The scientific model can be executed in the way intended by the original designers. No structural changes are made to the scientific modeling code when using Espresso.

By making no changes to the scientific modeling system, upgrades can also be performed with minimal effort. This design requirement limited us to only having the configurable text files on the remote system and then copying these files to the server, in the appropriate location. Updating versions of the model code can be achieved with minimal effort using this approach.

The Climate Workbench could be run only on specific machines in MCS. The old version assumed that the Network File System (NFS) was available.

Unfortunately, this limited the interface to run only within the MCS network. We wanted to be able to run Espresso from anywhere via a Java enabled browser. Espresso would have to run as a stand alone application and access all resources via its Grid interface. The Globus Java CoG kit made all of this functionality possible.

Some users may desire to run a non-GUI version of the modeling interface, i.e. a text only version. For testing purposes, this was also a very important feature. Other users may want to perform special Grid computing tasks that would not be feasible within a GUI. This would allow experienced users to take advantage of all of Java and Globus tools described above without the need to enter data via the Interface. Error checking occurs in the GUI so this feature would be lost in the non-GUI version ([4] discusses Espresso's GUI error checking in detail).

3.4   Espresso Server Side Component

The implementation of Espresso's server side component uses several shell scripting languages (TCSH, Python, BASH). For testing and modular purposes, we needed all of these shell scripts to run independently of Espresso. We also wanted to rid the server side of scripts that took long command line arguments. Consequently, we developed additional helper scripts that discovered information about the variables needed to run other scripts e.g. the start and end dates of the model simulation.

4   Espresso Implementation

4.1   Grid Utilization via the Globus Java CoG Kit

In order to make the Espresso client pure Java code, we needed a Grid toolkit that was implemented in Java. The Globus.org group had available a Java implementation of its Commodity Grid (CoG) Kit [11] [12]. The availability of the Java CoG Kit allowed us to integrate all of the Grid functionality that we needed with its Java API. The Java CoG Kit has all the necessary packages in a single Java ARchive (JAR) which can be distributed with the modeling interface. The Java CoG Kit communicates with the C implementation of the Globus Toolkit to perform tasks on the server side.

The two components utilized in the Globus Toolkit are the Globus Resource Allocation Manager (GRAM) and the GridFTP components. GRAM is used to run jobs remotely on different computation resources [13]. In order to execute a job on a resource machine, users use the Resource Specification Language (RSL) to tell the GRAM server what type of job is to be run. Typically, the RSL string sent to a supercomputing resource includes the name of the executable and the location of the standard out and error files (which can be located remotely).

Editing all of the model configuration files on the system side required that we transfer these files to the server side. We used the CoG Kit's Globus URLCopy from the GridFTP [14] component to provide this functionality. The URLCopy class allows both second and third party file transfers in a fashion similar to that of the common FTP client. Authentication on the remote server(s) is handled with the Globus security credentials and there is no need to type in another password once the original Grid Proxy has been established. We plan to use URL Copy's 3rd party copy functionality in future versions of Espresso to move large files from server to server (to which we may add additional error checking).

4.2   Creating a Java Based SED

The Unix "Stream EDitor" (SED) is a commonly used tool to edit streams of text. Since we assumed the scientific models were configured by editing text files, we needed the equivalent functionality of SED operations in Java. The first version of the interface used SED on the Unix server side to edit files.

The Java Foundation Class (JFC) provides the java.util.StringTokenizer class that allows one to parse tokens. Significant additional coding and testing had to be undertaken to mimic SED's functionality. The Apache Software Foundation's Jakarta Project provides us with a regular expressions package written in Java (appropriately named "Regexp") to use with the Java IO package. This allows us to build regular expressions for each model variable. MM5 for example, uses the "Regexp" regular expression "[.*\s—](" + v.getVariableName() + ")\s{1,}=\s{1,}(\d{0,})" to search through the variables in a Fortran name list.

5   Delivering Espresso to the User Community

5.1   Obtaining Access to Grid Resources

In order to use Espresso with Globus Grid resources, you need to establish your Globus credentials. You must first obtain these Globus credentials from the Certificate Authority [16]. Next, you need to have your name added to the "grid-map" on a Globus resource e.g. a mass storage device or a high performance computer. The number of resources that you have access to is limited only by your need to obtain permission to access the resource and Globus being installed on the system.

To use these Grid resources, you are required to validate your Globus credentials via the "grid-proxy-init" utility which asks for you password and validates your certificate for a fixed amount of time (12 hours by default). Once the "grid-proxy-init" is performed, you have complete access to all of your computational resources through the Globus toolkit. Espresso utilizes this functionality by copying files to a file server and running the scientific model on a different machine.

5.2   Accessing Espresso Technology

Delivering updated software to our user community was a concern from the start of the project. We wanted to distribute Espresso with minimal difficulty for both

784   M. Dvorak, J. Taylor, and S. Mickelson

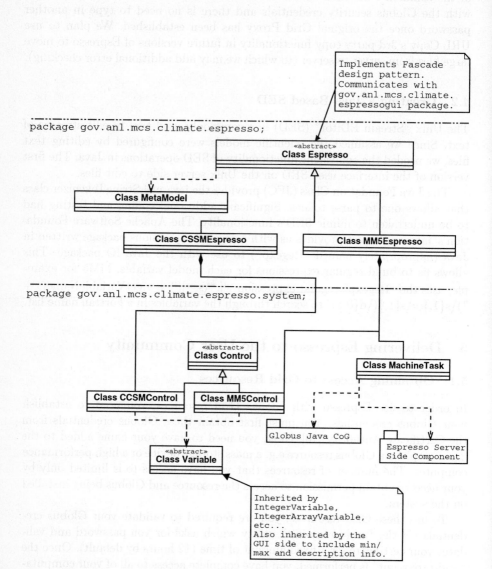

Fig. 3. UML Description of the Espresso System Side Design (Facade design pattern [15])

us (the developers) and the users. Fortunately Sun Microsystems, Inc. has a web-based software solution called Java Web Start [17]. Once users install Web Start (available for Windows, Solaris, Linux and Macintosh OS X), programs are installed, updated and executed via an Internet or desktop link. If Espresso is updated on the web server, Web Start detects a newer version and downloads the update.

Once a historical climate data archive is set up (similar to the diagram in Figure 1), we plan on creating a "case study repository" of XML files. Having a repository of XML files would allow other users to replicate or alter the parameters of other scientists model simulations. Other users could create case studies of interesting weather/climate events and exchange the XML files with other model users. All users could potentially have access to the historical climate data referenced in the "case studies" via the Grid.

6   Conclusion and Future Work

The initial version of the Espresso Scientific Modeling Interface, a scientific modeling productivity tool incorporating Grid technology has been developed. Creating a multipurpose scientific modeling interface that could be applied to many different scientific models was a challenging task. Using existing tools such as the Apache Software Foundation's XML parsers and regular expressions packages, Globus.org's Java CoG Kit and Sun's Web Start technology has allowed us to produce a high quality scientific model interface.

Espresso was intended to be developed iteratively, with the initial focus on the MM5 atmospheric model. Current efforts are being directed toward incorporating FOAM and the CCSM climate models. This will help us to abstract common components of scientific model simulation and analysis. In the coming year, we plan on updating the design of Espresso to simplify the task of adding scientific models to the interface. With increased Grid functionality, Espresso could become an important software tool for performing Grid based climate simulations.

Acknowledgments

We thank that the staff the Mathematics and Computer Science Division and the Computation Institute. We would also like to thank Gregor von Laszewski and Jarek Gawor of the Globus.org group for helping us use the Globus Java CoG Kit. The work was supported in part by the Laboratory Director Research and Development funding subprogram of the Office of Advanced Scientific Computing Research, U.S. Department of Energy, under contract W-31-109-Eng-38. This work was also supported by the NSF Information Technology Research Grant, ATM-0121028.

References

1. Ian Foster and Carl Kesselman. *The Grid: Blueprint For A New Computing Infrastructure.* Morgan Kaufmann Publishers, 1999.
2. Robert Jacob, Chad Schafer, Ian Foster, Michael Tobis, and John Anderson. Computational Design and Performance of the Fast Ocean Atmosphere Model, Version One. In *Computational Science - ICCS 2001*, volume Part I, pages 175–184. International Conference on Computational Science, Springer, May 2001.
3. The CCSM Home Page. http://www.ccsm.ucar.edu/.
4. Sheri Mickelson, John Taylor, and Mike Dvorak. Simplfying the Task of Generating Climate Simulations and Visualizations. Submitted to the *2002 International Conference on Computational Science.*
5. John   Taylor.   Argonne   Regional   Climate   Workbench. http://www-climate.mcs.anl.gov/proj/climate/public_html/.
6. John Taylor, Mike Dvorak, and Sheri Mickelson. Developing GRID based infrastructure for climate modeling. Submitted to the *2002 International Conference on Computational Science.*
7. F. Chen and J. Dudhia. Coupling an Advanced Land-Surface/Hydrology Model with the Penn State/NCAR MM5 Modeling System: Part I: Model Implementation and Sensitivity. *Monthly Weather Review*, 2001. "See also Pennsylvania State University/National Center for Atmospheric Research, MM5 Home Page" http://www.mmm.ucar.edu/mm5/mm5-home.html.
8. F. Chen, K. Mitchell, J. Schaake, Y. Xue, H. L. Pan, V. Koren, Q. Y. Duan, K. Elk, and A. Betts. Modeling Land-Surface Evaporation by Four Schemes and Comparison with FIFE Observations. *Journal of Geophysical Research*, 101:7251–7268, 1996.
9. J. Dudhia. A Nonhydrostatic Version MM5 of the Penn State/NCAR Mesoscale Model: Validation Test and Simulation of an Atlantic Cyclone and Cold Front. *Monthly Weather Review*, 121:1493–1513, 1993.
10. G. A. Grell, J. Dudhia, and D. R. Stauffer. The Penn State/NCAR Mesoscale Model (MM5). Technical Report NCAR/TN-398+STR, National Center for Atmospheric Research, 1994.
11. Gregor von Laszewski, Ian Foster, Jarek Gawor, and Peter Lane. A Java Commodity Grid Kit. *Concurrency and Computation: Practice and Experience*, 13:645–662, 2001.
12. The Globus Toolkit CoG Kit Homepage. http://www.globus.org/cog/java/.
13. Karl Czajkowski, Ian Foster, Nicholas Karonis, Carl Kesselman, Stuart Martin, Warren Smith, and Steve Tueke. A Resource Management Architecture for Metacomputing Systems. Technical report, Proc. IPPS/SPDP '98 Workshop on Job Scheduling Strategies for Parallel Processing, 1998. pp. 62-82.
14. GridFTP:Universal Data Transfer for the Grid. Technical report, Globus Project, September 2000.
15. Erich Gamma, Richard Helm, Ralph Johnson, and John Vlissides. *Design Patterns: Elements of Reusable Object-Oriented Software.* Addison Wesley, 1995.
16. The Globus Homepage. http://www.globus.org//.
17. Java Web Start Home Page. http://java.sun.com/products/javawebstart/.

Parallel Contact Detection Strategies for Cable and Membrane Structures

Jelle Muylle and Barry H.V. Topping

Department of Mechanical and Chemical Engineering
Heriot-Watt University, Edinburgh, United Kingdom

Abstract. The implementation of a parallel simulation procedure for cable and membrane structures is presented. The procedure includes contact detection and enforcement. Two different domain decompositions are simultanously used. For the simulation a static partitioning is computed; for the contact detection a dynamically updated parallel RCB decomposition is used. An overview is also given of the strategy for the detailed geometric contact detection procedures. An example of a falling cloth is included to demonstrate.

1 Introduction

The research presented in this paper must be seen in the framework of computational simulation for analysis and design of structures consisting of cables and tensile membranes in a full three dimensional environment [1, 2].

The particular focus in this paper lies on the contact detection procedures required to deal with collapsing and crumpling membrane parts and with entanglement of cables. Considering the scale of the computational simulation which may arise for large structures it comes as no surprise that parallel and distributed processing have been adopted to keep the execution times within reason.

2 Simulation Components

The simulation process developed by the authors' research group for cable and membrane structures consists of several modular components: the simulation loop, the material model, the partitioning, contact detection, and contact handling. Not included in this list is the pre-processing stage of mesh generation, which for cable and membrane structures does involve a whole range of different complications, as discussed in other publications by the authors [3].

2.1 Simulation Loop

In this research, cable and membrane structures are simulated as a network of non-linear springs and masses whose motion is numerically integrated in an explicit transient dynamic analysis. The forces acting upon the nodes of the

P.M.A. Sloot et al. (Eds.): ICCS 2002, LNCS 2330, pp. 787–796, 2002.
© Springer-Verlag Berlin Heidelberg 2002

network are: gravity; elastic forces due to the elongation of cables or stretching of membrane elements; induced forces over shared cable nodes or shared mebrane element edges for simulation of bending stiffness [2]; contact forces when contacts are detected; and user defined external forces.

The equations governing the system are simply Newton's laws written out for each nodal mass m_i:

$$m_i \cdot \ddot{\mathbf{x}}_i = \sum \mathbf{f}_i \tag{1}$$

where $\ddot{\mathbf{x}}_i$ is the acceleration of node i and $\sum \mathbf{f}_i$ is the total of all force components having an influence on the motion of node i.

A variety of numerical integration schemes are available to derive the unknown position $\mathbf{x}_i^{t+\Delta t}$ of node i at time $t + \Delta t$ from the known positions $\mathbf{x}_i^t$ and forces $\sum \mathbf{f}_i^t$ at time t. An overview of the implemented schemes was previously published [1].

2.2   Material Model

A material model is required to calculate the nodal force components from the elongations at each timestep. Once again a variety of models is available, based on elastic [4] or elasto-plastic [5] assumptions or derived directly from energy functions [6]. The chosen implementation simulates cable elements as non-linear springs and decouples each triangular membrane element into a set of three non-linear springs [7].

The model was enhanced by the authors with a pseudo stiffness implementation that introduces the advantages of bending stiffness for a better simulation of natural fabrics without destroying the efficiency of an explicit time integration scheme [2].

2.3   Contact Detection

Contact detection in this framework involves all the necessary procedures to detect geometrical impossibilities caused by simulation of objects not being restrained in their movement by each other. The procedures that will be suggested must be particularly suited to detect the intersection of any two or more cable or membrane elements at the end of a time step or, in a stricter sense, the proximity of any two or more elements within a distance smaller than the thickness of the elements.

These two types of contact (where the first type is actually a special case of the more general second type) will be termed *intersection contact* and *proximity contact*. To avoid the exhaustive geometric checking of any two elements at each timestep a two stage procedure is used consisting of: a global checking phase and a local checking phase. The global checking phase eliminates the bulk of the combinations and selects a series of element pairs, that are to be examined further for contact. The local checking phase performs the detailed geometric check on the candidate element pairs, selected by the local checking phase. The global checking phase is sometimes called *pre-contact search* [8], or *location phase* [9] or *neighbourhood identification phase* [10]. Several types of solution procedures can be applied to pair up the candidate elements: Solution procedures may be based on hierarchy building, as described in the work of Volino *et al.* [9, 11]. Solution procedures may also be based on spatial subdivision, where the simulation space is divided into subspaces such as buckets, slices or boxes. Finally solution procedures may be based on sorting algorithms where, by efficiently sorting the coordinate values of the nodes, the proximity of elements can be derived. The sorted coordinate lists must however be maintained when the nodes are continuously moving.

The global search procedure adopted in this research is based on the *point-in-box* algorithm [12]. The algorithm, as implemented by the authors [1] consists of the following steps:

– All nodal coordinates are sorted or resorted into three lists corresponding to their three spatial coordinates. Rank vectors, containing the dual information of the sorted lists are also kept.
– Around each element a bounding box is constructed, which is extended to a capturing box by a distance dx, related to the maximum distance any node can travel within one timestep. (See Figure 1.)
– With a binary search the extents of the capturing box in the sorted lists can be found. In each of the three lists a sublist will therefore be marked by the correponding extent of the capturing box.
– Using the rank vectors the three sublists are contracted and all nodes which lie within the capturing box are therefore found.
– The elements that contain any of the captured nodes are said to be in proximity to the element around which the capturing box was constructed. All pairs are then passed on to the local search.

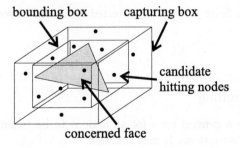

Fig. 1. Construction of the capturing box

The local search or detailed contact check, must then check geometrically if the selected contact pairs did make contact in the last timestep. By relaxing the definition of a contact situation, the detailed contact check can be simplified. For example, in a simulation which involves solid elements, the point-in-box search described above could be used to report the proximity between nodes and the solid elements. The local check could then verify if the node entered into the volume of the solid element. With this kind of *nodal intersection* as basis for contact detection most of the contacts in a simulation with solid elements could be identified. But this is of no good to a simulation including volume-less triangular membrane elements.

If the local search is made to detect the passing of a node through a triangular element, it becomes of more use to membrane simulations. This checking is geometrically more intensive as both triangle and node are moving during the timestep. It involves the solution of a cubic equation to resolve the intersection between the trajectory of the node and the plane through the transforming triangle [12] and a series of geometrical tests to determine if the intersection lies within the triangle.

The great disadvantage of chosing nodal intersection as basis for contact detection is that an entire class of contacts is ignored. Edge contact, such as the one shown in Figure 2 is not detected. Previously published research by the authors [1] presented three methods to allow detection of edge contact. The method that will be explained in Section 4 is a new look on the problem that will not only detect edge contact but will also provide support for contacts involving cable elements and support for the detection of proximity contact situations rather than intersection contact.

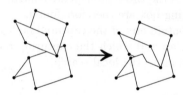

Fig. 2. Edge contact, not detected by nodal intersection

2.4   Contact Handling

Contact handling is a general term for all actions to be taken as a consequence of detected contact situations. It may include:

- correcting geometric impossibilities such as intersections (corrective actions),
- enforcing a constraint placed on the movement of a node,
- or modification of the trajectory of a node that is on a course to intersection by changing its velocity or accelleration components (preventive action).

Two types of contact handling are common: the use of contact forces, or kinematic correction procedures When a contact is detected and contact forces are introduced, it is hoped that within a few timesteps, due to the action of the forces the contact situation will be rectified. With the kinematic correction method a direct change to the position, velocity or acceleration components of a contacting node is made to rectify the situation within the same timestep. Both methods were implemented. However, to enable realistic simulation in the future, the contact handling method selected must be able to guarantee that any geometric intersections of the membrane or cable elements must be avoided at all cost. Geometric intersections could cause serious problems if a volume mesh is present around the cable or membrane model, for simulation of fluid flow. Avoiding geometric intersections does however not exclude the use of contact forces. If a proximity contact check is used whereby the proximity distance is taken as n times the maximum distance any node can travel within one timestep, then the contact handling procedure has n timesteps to alter the course of a node from when it is detected as violating the proximity distance. A carefull choice for the model behind the calculation of the direction and size of the contact force is however required.

A problem faced when implementing the kinematic correction method is how to deal with multiple contacts involving the same nodes or elements. Correcting the position of a node that is involved in e.g. two contacts according to the constraints posed by the first contact situation may cause under or over correction of the second contact situation. Solutions were offered by Volino et al. [11].

3 Implications for Parallel Implementations

The simulation of the cable and membrane structures in motion has been successfully parallellized in a message passing context (MPI). New domain decomposition strategies, based on graph partitioning where developed to take into account the specific nature of cable and membrane structures [13]. A decomposition is therefore used which is optimized for the material simulation steps, in particular the calculation of the nodal forces due the material behaviour. This decomposition only relies on the topology of the mesh and can be determined beforehand. Elements are assigned to each processor in order to achieve load balance of the computational efforts of calculating the forces and integrating the nodal movements. And at the same time the decomposition minimizes the number of subdomain interfaces to reduce the amount of communication due to nodal force components for nodes shared by more than one partition.

The decomposition scheme used for the material simulation steps is however not ideal for parallellizing the contact detection procedure. If the contact detection is carried out by a global-local scheme as detailed above, then in the parallel equivalent of global search procedure, each of the processors should do some of the global searching. If the point-in-box method is used, ideally each subdomain should consist of those elements and nodes which are geometrically close to each other rather than topologically. Such a decomposition exists: *recursive coordinate bisection* (RCB). Moreover RCB has the advantage that all subdomains will be bound by (almost) non-overlapping axis aligned bounding boxes. The parallel version of the contact procedure will therefore be identical to the serial version carried out within each subdomain. The details of obtaining and maintaining the RCB decomposition will be described in Section 5. The scheme on which the authors have based their implementation was first suggested by Sandia National Labs implementation of the PRONTO3D code [10].

The implications of working with two distinct decompositions for a single mesh, will be left to the conclusions in Section 7.1. However it must be mentioned that alternatives are available. Parallel tree searching algorithms do exist [14] but the authors are at present unable to comment on their success or applicability.

Once contacts have been detected by a parallel contact detection algorithm, the contact pairs can be communicated to the processors that own the respective elements for the material simulation. The contact forces or kinematic correction procedure can be carried out in` this decomposition, requiring communication if necessary with other processors should contact be detected between elements belonging to different decompositions.

4   Geometric Contact Detection

The geometric contact detection procedure is based on a library of geometric test-functions which were implemented with adaptive precision arithmetic. Functions were created for measuring distances, and calculating distance between points, lines and planes. Thanks to a consistent system of tolerances and a limited set of internally consistent predicates [15], a stable geometric platform was developed.

Cable elements are encapsulated in a cylinder of radius equal to the proximity distance with two hemispherical caps over the end nodes of the element. Triangular elements are encapsulated between two planes a distance equal to the proximity distance away from the element, in addition to three cylindrical and three spherical caps. The encapsulated forms are compared for intersections and a system of geometric tests classifies the pairs into a number of categories. Figure 3 shows a series of intersection options for the combination of a cable element with a triangular element.

Only six possible configurations are shown in the figure. This excludes all the cases where the cable element is found to be parallel with the triangular element. The classification at the same time allows the contact handling functions to select the correct nodes and interpolation factors for the calculation of the contact forces or kinematic corrections.

5   Parallel Recursive Coordinate Bisection

Recursive coordinate bisection (RCB) (sometimes called orthogonal recursive bisection (URB)) was first introduced as a static partitioning algorithm [16] for unstructured meshes. Since then it has long been superseded by much better static partitioning algorithms, but as a dynamic partitioning/repartitioning algorithm it is still widely used.

RCB is particularly suited for parallelization of contact detection as it is a geometric partitioning method rather than a topological one. The resulting

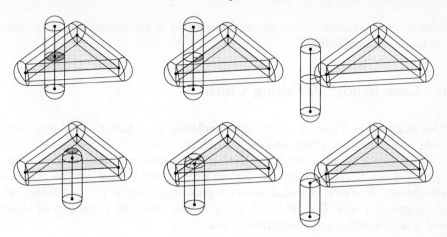

Fig. 3. Geometric Contact Cases for Cable-Membrane Contact

partitions are all bound by disjoint orthogonal bounding boxes. As a repartitioning scheme it shows extremely good performance because small changes in
the node coordinates produce only small changes in the partitioning, and much
of the sorting and searching in the algorithm can be recycled or restarted from
near-optimal values.

The RCB algorithm divides the simulation space up in Cartesian subspaces
attached to each processor with roughly a balanced number of nodes. It does
so by cutting a partition into two sub-partitions with an equal or almost equal
number of nodes. The sub-partitions are bisected recursively until the required
number of partitions is reached. In its most basic form only 2^n partitions can be
created. However by appropriately choosing the ratio of nodes on either side of
the cutting plane, any number of subdomains can be obtained. The basic cutting
step, which is recursively applied, consists of three parts:

1. Selection of the coordinate (x, y or z) upon which the nodes will be judged
   for bisecting.
2. Choosing the cutting point which splits the population of nodes into the
   appropriate ratio ($\frac{1}{2}$ in case of 2^n partitions).
3. Redistribution of the nodes into either one of the subpartitions, depending
   on which side of the cutting point they fall.

A parallel version of the RCB algorithm (pRCB) was implemented by the
authors, based on the ideas of Nakhimovski [17]. As such it becomes a repartitioning algorithm where all processors already own a series of nodes and cooperate in a parallel fashion to determine the RCB of the global set of nodes. The
authors' implementation postpones the actual redistribution of locally owned
nodes to the end and peforms the bisection on destination tags for each node.
Once these have been determined the redistribution can be organized using the

unstructured communication patterns, suggested in references [10, 18] to avoid *all-to-all* communication.

6   Case Study: A Falling Cloth

The example in Figure 4 shows the simulation of a falling cloth suspended from one corner on a four processor machine. The cloth is represented by a unstructured triangular mesh containing 682 nodes and 1262 elements. Image a) of the figure displays the domain decomposition used for the material simulation. A elastic material model representing cotton ($\rho = 409.8$kg/m^3, thickness= 0.47mm, $E = 10^7$N/m^2, $\nu = 0.2$) was used with a timestep of 1msec and a central difference integration scheme.

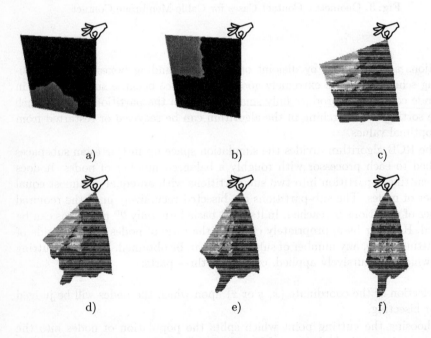

a)   b)   c)

d)   e)   f)

Fig. 4. Case Study: A falling cloth. a) The domain decomposition used for the material simulation. b)-f) The domain decomposition used for contact detection, dynamically updated during the fall of the cloth

It was decided to update the contact detection decomposition only every 50 timesteps. This number was found as an engineering trade-off between the cost of recomputing the RCB decomposition and the extra communication obtained due to overlapping imperfect decompositions. Further efforts are being made to automate the finding of this threshold value.

7   Concluding Notes

7.1   The Implications of Using Two Decompositions

Using a different domain decomposition for the material simulation and the contact detection has quite a number of implications:

- At the end of the calculation of the new position of each node in the material simultation communication has to take place between the processor that owns the node in the material simulation and the processor that owns the node for the contact detection. The communication involves the new coordinates and velocities of the node as both are required for the contact detection.
- As suggested earlier the contact handling is carried out in the material simulation decomposition and not in the contact detection decomposition. All reported contacts therefore will have to be communicated to the material simulation decomposition. The reason for this is that computing the contact response involves far more variables, such as mass and material parameters, which are not known in the contact detection decomposition. Making them all available to the contact detection decomposition throughout the simulation would require far too much extra communication.
- Complications will also occur because of the differences between a decomposition of nodes (with halo elements) and a decomposition of elements (with halo nodes). The halo elements or nodes will always cause unavoidable communication.

It is however suggested that the communication disadvantages of using two decompositions are a reasonable price to pay in order to obtain overall load balance of the entire simulation process [10, 19].

7.2   Looking towards the Future

As a result of to the computational intensity of the geometric procedures in the detailed contact detection phase, the authors' current implementation is still relatively slow compared to other less complicated schemes. The guarantee however that all element intersections can be avoided is of paramount importance when considering realistic simulations.

Research considering the optimization of the communication patterns between the two decompositions is still in progress. The implementation of a robust fully parallel simulation of cable and membrane structures with contact detection and handling is gradually taking shape.

References

[1] Muylle, J., Topping, B.H.V.: Contact detection and enforcement techniques for the simulation of membrane structures in motion. In Topping, B.H.V., ed.: Computational Mechanics: Techniques and Developments, Edinburgh, U.K., Civil-Comp Press (2000) 243–256

[2] Muylle, J., Topping, B.H.V.: A cable and mebrane structure pseudo stiffness implementation. In Topping, B.H.V., ed.: Proceedings of the Eighth International Conference on Civil and Structural Engineering Computing, Stirling, UK, Civil-Comp Press (2001) paper 24.

[3] Topping, B.H.V., Muylle, J., Putanowicz, R., Iványi, P., Cheng, B.: Finite Element Mesh Generation. Saxe-Coburg Publications, Stirling, UK (2002)

[4] Volino, P., Courchesne, M., Magnenat Thalmann, N.: Developing simulation techniques for an interactive clothing system. In: Proceedings VSMM 97, IEEE Computer Society (1997) 109–118

[5] Wu, Y., Thalmann, D., Thalmann, N.M.: Deformable surfaces using physically-based particle systems. In: Proceedings Computer Graphics International 95, Academic Press (1995) 205–216

[6] Baraff, D., Witkin, A.: Large steps in cloth simulation. In: Proceedings Siggraph 1998, New York, ACM Press (1998) 43–54

[7] Topping, B.H.V., Iványi, P.: Computer Aided Design of Cable-Membrane Structures. Saxe-Coburg Publications, Stirling, UK (2002)

[8] Zhong, Z.H.: Finite Element Procedures for Contact-Impact Problems. Oxford University Press, Oxford (1993)

[9] Volino, P., Magnenat Thalmann, N.: Collision and self-collision detection: Efficient and robust solutions for highly deformable surfaces. In Terzopoulos, D., Thalmann, D., eds.: Computer Animation and Simulation '95, Springer-Verlag (1995) 55–65

[10] Attaway, S.W., Hendrickson, B.A.e.a.: A parallel contact detection algorithm for transient solid dynamics simulations using pronto3d. Computational Mechanics 22 (1998) 143–159

[11] Volino, P., Magnenat Thalmann, N.: Accurate collision response on polygonal meshes. In: Proceedings Computer Animation Conference, Philadelphia, IEEE Computer Society (2000)

[12] Heinstein, M.W., Attaway, S.W., Swegle, J.W., Mello, F.J.: A general-purpose contact detection algorithm for nonlinear structural analysis codes. Technical Report SAND92-2141, Sandia National Laboratories, Albuquerque, NM (1993)

[13] Iványi, P., Topping, B., Muylle, J.: Towards a CAD design of cable-membrane structures on parallel platforms. In Bathe, K., ed.: Computational Fluid and Solid Mechanics, Elsevier Science Ltd. (2001) 652–654

[14] Al-furaih, I., Aluru, S., Goil, S., Ranka, S.: Parallel construction of multidimensional binary search trees. IEEE Transactions on Parallel and Distributed Systems 11 (2000) 136–148

[15] Shewchuk, J.R.: Delaunay Refinement Mesh Generation. PhD thesis, Carnegie Mellon University, Pittsburgh, PA (1997)

[16] Berger, M.J., Bokhari, S.H.: A partitioning strategy for nonuniform problems on multiprocessors. IEEE Transactions on Computers 36 (1987) 570–580

[17] Nakhimovski, I.: Bucket-based modification of the parallel recursive coordinate bisection algorithm. Linköping Electronic Articles in Computer and Information Science 2 (1997)

[18] Plimpton, S., Attaway, S., Hendrickson, B., Swegle, J., Vaughan, C., Gardner, D.: Parallel transient dynamics simulations: Algorithms for contact detection and smoothed particle hydrodynamics. Journal of Parallel and Distributed Computing 50 (1998) 104–122

[19] Hendrickson, B., Devine, K.: Dynamic load balancing in computational mechanics. Computer Methods in Applied Mechanics and Engineering 184 (2000) 485–500

A Parallel Domain Decomposition Algorithm for the Adaptive Finite Element Solution of 3-D Convection-Diffusion Problems

Peter K. Jimack and Sarfraz A. Nadeem

Computational PDEs Unit, School of Computing,
University of Leeds, Leeds, LS2 9JT, UK
{pkj, sarfraz}@comp.leeds.ac.uk
http://www.comp.leeds.ac.uk/pkj/

Abstract. In this paper we extend our previous work on the use of domain decomposition (DD) preconditioning for the parallel finite element (FE) solution of three-dimensional elliptic problems [3, 6] and convection-dominated problems [7, 8] to include the use of local mesh refinement. The preconditioner that we use is based upon a hierarchical finite element mesh that is partitioned at the coarsest level. The individual subdomain problems are then solved on meshes that have a single layer of overlap at each level of refinement in the mesh. Results are presented to demonstrate that this preconditioner leads to iteration counts that appear to be independent of the level of refinement in the final mesh, even in the case where this refinement is local in nature: as produced by an adaptive finite element solver for example.

1 Introduction

In [3] we introduce a new two level additive Schwarz (AS) DD preconditioner based upon the existence of a nested sequence of meshes on each subdomain. This preconditioner is proved to be optimal for a particular class of symmetric self-adjoint partial differential equations (PDEs) in both two and three dimensions. In order to construct the preconditioner it is necessary to generate a decomposition of the finite element space, $\mathcal{W}$ say, in the following manner. The description here is given in two dimensions for simplicity but the extension to a tetrahedral mesh in three dimensions is straightforward (and is outlined explicitly in [10]).

Let $\mathcal{T}_0$ be a coarse triangulation of the problem domain, Ω say, consisting of N_0 triangular elements, $\tau_j^{(0)}$, such that $\tau_j^{(0)} = \overline{\tau}_j^{(0)}$,

$$\overline{\Omega} = \bigcup_{j=1}^{N_0} \tau_j^{(0)} \quad \text{and} \quad \mathcal{T}_0 = \{\tau_j^{(0)}\}_{j=1}^{N_0} \, . \tag{1}$$

Also let $\text{diameter}(\tau_j^{(0)}) = O(H)$, and divide Ω into p *non-overlapping* subdomains Ω_i. These subdomains should be such that:

$$\overline{\Omega} = \bigcup_{i=1}^{p} \overline{\Omega}_i \, , \tag{2}$$

P.M.A. Sloot et al. (Eds.): ICCS 2002, LNCS 2330, pp. 797–805, 2002.

$$\Omega_i \cap \Omega_j = \phi \quad (i \neq j), \tag{3}$$

$$\overline{\Omega}_i = \bigcup_{j \in I_i} \tau_j^{(0)} \quad \text{where } I_i \subset \{1, ..., N_0\} \quad (I_i \neq \phi). \tag{4}$$

We now permit $\mathcal{T}_0$ to be refined several times, to produce a family of triangulations, $\mathcal{T}_0, ..., \mathcal{T}_J$, where each triangulation, $\mathcal{T}_k$, consists of N_k triangular elements, $\tau_j^{(k)}$, such that

$$\overline{\Omega} = \bigcup_{j=1}^{N_k} \tau_j^{(k)} \quad \text{and} \quad \mathcal{T}_k = \{\tau_j^{(k)}\}_{j=1}^{N_k}. \tag{5}$$

The successive mesh refinements that define this sequence of triangulations need not be global and may be non-conforming, however we do require that they satisfy a number of conditions:

1. $\tau \in \mathcal{T}_{k+1}$ implies that either
   (a) $\tau \in \mathcal{T}_k$, or
   (b) τ has been generated as a refinement of an element of $\mathcal{T}_k$ into four regular children (eight in 3-d) by bisecting each edge of this parent of τ,
2. the level of any triangles which share a common point can differ by at most one,
3. only triangles at level k may be refined in the transition from $\mathcal{T}_k$ to $\mathcal{T}_{k+1}$.

(Here the level of a triangle is defined to be the least value of k for which that triangle is an element of $\mathcal{T}_k$.) In addition to the above we will also require that:

4. in the final mesh, $\mathcal{T}_J$, all pairs of triangles which share an edge which lies on the interface of any subdomain with any other subdomain have the same level as each other (i.e. the mesh is conforming along subdomain interfaces).

Having defined a decomposition of Ω into subdomains and a nested sequence of triangulations of Ω we next define the restrictions of each of these triangulations onto each subdomain by

$$\Omega_{i,k} = \{\tau_j^{(k)} : \tau_j^{(k)} \subset \overline{\Omega}_i\}. \tag{6}$$

In order to introduce a certain amount of overlap between neighbouring subdomains we also define

$$\tilde{\Omega}_{i,k} = \{\tau_j^{(k)} : \tau_j^{(k)} \text{ has a common point with } \overline{\Omega}_i\}. \tag{7}$$

Following this we introduce the FE spaces associated with these local triangulations. Let G be some triangulation and denote by $\mathcal{S}(G)$ the space of continuous piecewise linear functions on G. Then we can make the following definitions:

$$\mathcal{W} = \mathcal{S}(\mathcal{T}_J) \tag{8}$$

$$\tilde{\mathcal{W}}_0 = \mathcal{S}(\mathcal{T}_0) \tag{9}$$

$$\mathcal{W}_{i,k} = \mathcal{S}(\Omega_{i,k}) \tag{10}$$

$$\tilde{\mathcal{W}}_{i,k} = \mathcal{S}(\tilde{\Omega}_{i,k}) \tag{11}$$

$$\tilde{\mathcal{W}}_i = \tilde{\mathcal{W}}_{i,0} + ... + \tilde{\mathcal{W}}_{i,J}. \tag{12}$$

It is evident that

$$W = \tilde{W}_0 + \tilde{W}_1 + \ldots + \tilde{W}_p \tag{13}$$

and this is the decomposition that forms the basis of the two level additive Schwarz preconditioner (see [13] for example), M say, in [3]. Hence, for a global FE stiffness matrix A, this preconditioner takes the form:

$$M^{-1} = \sum_{i=0}^{p} \tilde{R}_i^T \tilde{A}_i^{-1} \tilde{R}_i \tag{14}$$

where (using the usual local bases for W and $\tilde{W}_i$) $\tilde{R}_i$ is the rectangular matrix representing the L^2 projection from W to $\tilde{W}_i$, and $\tilde{A}_i$ is the FE stiffness matrix corresponding to the subspace $\tilde{W}_i$.

A parallel implementation of the above preconditioner is described in 2-d in [2] and in 3-d in [7], for example. In both of these cases, the coarse grid solve is combined with each of the subdomain solves to yield a preconditioner which is actually of the form

$$M^{-1} = \sum_{i=1}^{p} R_i^T A_i^{-1} R_i \ . \tag{15}$$

Here, R_i and A_i differ from $\tilde{R}_i$ and $\tilde{A}_i$ in (14) since R_i now represents the projection from W to $\tilde{W}_0 \cup \tilde{W}_i$ for $i = 1, \ldots, p$. In addition to this, and following [5], both the 2-d and the 3-d parallel implementations in [2,7] use a restricted version of the AS preconditioner (15). This involves the following simplification:

$$M^{-1} = \sum_{i=0}^{p} D_i A_i^{-1} R_i \ , \tag{16}$$

where D_i is a diagonal matrix with entries of 1 for those rows corresponding to vertices inside subdomain i, 0 those rows corresponding to vertices outside subdomain i, and $\frac{1}{q}$ for those rows corresponding to vertices on the interface of subdomain i (shared with $q - 1$ neighbouring subdomains).

Results presented in [2,7] show that the preconditioner (16) outlined above appears to behave in an optimal manner when applied to a range of problems as the FE mesh is uniformly refined. That is, the number of iterations required to obtain a converged solution appears to be bounded independently of the mesh size h. In particular, in [7] it is demonstrated that when (16) is applied to solve three-dimensional convection-dominated problems using a stabilized (streamline-diffusion) FE discretization, the quality of the preconditioning is excellent.

2   Local Mesh Refinement

In our previous work in three dimensions (e.g. [6–8]) we have focused on applying the finite element method on sequences of uniformly refined meshes. For many practical problems however the solution exhibits local behaviour, such as the existence of shocks or boundary layers, which lead to the requirement for meshes obtained via local, rather than global, refinement. Consider, for example, a convection-diffusion equation of the form

$$-\varepsilon \Delta u + \underline{b} \cdot \nabla u = f \tag{17}$$

on some domain $\Omega \subset \Re^3$. When $|\underline{b}| \gg \varepsilon > 0$ it is common for the solution to involve boundary layers of width $O(\frac{\varepsilon}{|\underline{b}|})$, depending on the precise nature of the boundary conditions. If the solution is smooth away from these layers then it is most efficient, computationally, to concentrate the majority of the degrees of freedom in and around the layers. The following test problem is of the form (17), with a computational domain $\Omega = (0,2) \times (0,1) \times (0,1)$ and exact Dirichlet boundary conditions applied throughout $\partial\Omega$:

$$\left. \begin{array}{l} \underline{b} = (1,0,0)^T \,, \\ f = 2\varepsilon \left(x - \frac{2(1-e^{x/\varepsilon})}{(1-e^{2/\varepsilon})} \right) (y(1-y) + z(1-z)) + y(1-y)z(1-z) \,, \\ u = \left(x - \frac{2(1-e^{x/\varepsilon})}{(1-e^{2/\varepsilon})} \right) y(1-y)z(1-z) \,. \end{array} \right\} \tag{18}$$

The streamline-diffusion discretization of (17) on a mesh, $\mathcal{T}_J$, of tetrahedra seeks a piecewise linear solution of the form

$$u^h = \sum_{i=1}^{N+B} u_i N_i(\underline{x}) \tag{19}$$

where N_i are the usual linear basis functions, N is the number of nodes of $\mathcal{T}_J$ inside Ω and B is the number of nodes of $\mathcal{T}_J$ on $\partial\Omega$. The values of the unknowns u_i are determined so as to satisfy the finite element equations

$$\varepsilon \int_\Omega \nabla u_h \cdot \nabla (N_j + \alpha \underline{b} \cdot \nabla N_j) \, d\underline{x} + \int_\Omega (\underline{b} \cdot \nabla u_h)(N_j + \alpha \underline{b} \cdot \nabla N_j) \, d\underline{x} = $$
$$\int_\Omega f(\underline{x})(N_j + \alpha \underline{b} \cdot \nabla N_j) \, d\underline{x} \tag{20}$$

for $j = 1, ..., N$ (see, for example, [9] for further details). This yields an $N \times N$ non-symmetric linear system of the form

$$K\underline{u} = \underline{f} \,. \tag{21}$$

In [7] it is shown that when the above DD preconditioner, (16), is applied with an iterative method (GMRES, [11,12]) to solve (21) on a sequence of uniformly refined meshes excellent iteration counts are obtained. Table 1, taken from [7], illustrates this for two choices of ε. The precise values for the iteration counts obtained using this preconditioner depend on the particular partition of the coarse mesh into subdomains (see (2), (3) and (4)) that is used. For all of the calculations presented in this paper we partition the domain Ω with cuts that are parallel to the convection direction $\underline{b}$ in (17). For the specific test problem,

Table 1. The performance of the preconditioner (16) on the system (21), the discretization of (17) and (18), using global mesh refinement: figures quoted represent the number of iterations required to reduce the initial residual by a factor of 10^5.

Elements/Procs.	$\varepsilon = 10^{-2}$				$\varepsilon = 10^{-3}$			
	2	4	8	16	2	4	8	16
6144	2	3	3	3	2	3	3	3
49152	3	3	4	4	3	4	4	4
393216	3	4	4	5	3	4	4	4
3145728	4	5	6	6	3	4	4	4

Fig. 1. An illustration of the partitioning strategy, based upon cuts parallel to $\underline{b} = (1, 0, 0)^T$, used to obtain 2, 4, 8 and 16 subdomains, where $\Omega = (0, 2) \times (0, 1) \times (0, 1)$.

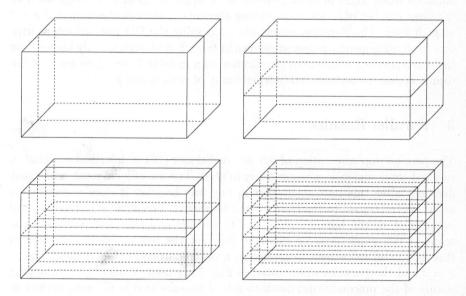

(18), considered here this means that all cuts are parallel to the x-axis, as illustrated in Fig. 1. This strategy is shown in [10] to yield convergence in fewer iterations than is possible with more isotropic partitions when (17) is convection-dominated. Furthermore, in the following section on parallel results, we are able to ensure good load balancing when using this strategy, even in the presence of local refinement.

We now consider the use of a simple *a priori* local mesh refinement strategy for the stabilized finite element solution of (17), (18). The purpose of this is to illustrate the potential for the DD preconditioner (16) to be used successfully within an adaptive FE framework, a discussion of which is included in Section 4. For this particular test problem however we make use of the fact that the only boundary layer in the solution is known to be next to the domain boundary $x = 2$, and that the solution is smooth elsewhere. Hence, beginning with a

Table 2. The performance of the preconditioner (16) on the system (21), the discretization of (17) and (18), using local mesh refinement: figures quoted represent the number of iterations required to reduce the initial residual by a factor of 10^5.

Elements/Procs.	$\varepsilon = 10^{-2}$				$\varepsilon = 10^{-3}$			
	2	4	8	16	2	4	8	16
2560	4	5	5	6	5	6	6	6
9728	5	5	6	6	5	6	6	6
38400	5	6	6	7	5	6	7	7
153088	6	8	8	8	6	7	7	7

mesh of 768 elements, we are able to get results of a similar accuracy to those obtained using uniform mesh refinement by applying local mesh refinement to the same number of levels: only refining elements in the neighbourhood of $x = 2$ at each level. The iteration counts obtained using the DD preconditioner with this mesh refinement strategy are shown in table 2. Although slightly higher than the corresponding numbers of iterations shown in table 1, we again see that the results appear to be bounded independently of both h and p.

3   Parallel Results

The parallel implementation of our preconditioner (16) is described in detail in two and three dimensions respectively in [2, 7]. The use of local mesh refinement does not alter this implementation in any way however it does require extra attention to be paid to the issue of load balancing. This issue is of fundamental importance in the parallel adaptive FE solution of PDEs and is beyond the scope of this paper: see, for example, [14]. For the test problems described in this section all refinement is applied in a neighbourhood of the domain face at $x = 2$ and the partitions illustrated in Fig. 1 are used. Whilst the overlapping nature of the preconditioner leads to a load balance that is far from perfect in the cases $p = 8$ and $p = 16$, it is sufficiently good to yield quite respectable parallel speed-ups, as demonstrated below.

In addition to the potential load imbalance there are at least two more significant factors that affect the performance of our parallel implementation of (16). The first of these is the quality of (16) as a preconditioner, and the second is the parallel overhead associated with inter-processor communications. The numerical tests described in this work were all undertaken on a tightly coupled parallel computer, the SG Origin2000, and so communication overheads are quite moderate in the results presented. The quality of (16) as a preconditioner is considerably more important however. In all of the tables of timings we include not only the parallel solution times but also the sequential solution times for different choices of p. As can be seen, these times vary enormously and are generally significantly slower than the fastest sequential solution time that we are able to obtain (using [11]). For this reason we present two speed-up rows

Table 3. Timings for the parallel DD solver applied to problem (17), (18) with $\varepsilon = 10^{-2}$ using four levels of local mesh refinement.

Processors/Subdomains	1	2	4	8	16
Parallel Time	26.20	17.54	10.54	7.17	5.35
Speed-up	–	1.5	2.9	3.7	4.9
Sequential Time	–	34.64	40.54	50.35	66.96
Parallel Speed-up	–	2.0	3.8	7.0	12.5

Table 4. Timings for the parallel DD solver applied to problem (17), (18) with $\varepsilon = 10^{-3}$ using four levels of local mesh refinement.

Processors/Subdomains	1	2	4	8	16
Parallel Time	20.07	14.02	8.62	6.02	4.87
Speed-up	–	1.4	2.3	3.3	4.1
Sequential Time	–	27.64	33.12	42.05	58.88
Parallel Speed-up	–	2.0	3.8	7.0	12.1

in each table: a regular speed-up which contrasts the parallel solution time with the best sequential solution time, and a parallel speed-up which contrasts the parallel solution time with the sequential time of the p subdomain DD solver.

Tables 3 and 4 present parallel timings for the entire solution algorithm when solving problem (17), (18) on the final mesh obtained using the *a priori* adaptive algorithm (i.e. containing 153088 elements). These timings are for $\varepsilon = 10^{-2}$ and 10^{-3} respectively. It should be noted that equivalent results are presented for the first of these cases in [7] using global mesh refinement to a similar resolution. In that paper parallel solution times of between 505.7 seconds ($p = 2$) and 144.7 seconds ($p = 16$) are reported. The use of local refinement clearly leads to a significant improvement on these times therefore.

An initial assessment of the figures presented in Tables 3 and 4 shows that a speed-up of about 5 has been achieved on 16 processors. Whilst this may be a little disappointing, inspection of the parallel speed-ups achieved gives much greater grounds for optimism. The major cause of loss of efficiency in the overall parallel solution appears to be due to the growth in the solution time of the sequential version of the p subdomain algorithm as p increases. If this can be overcome, through the use of a sequential multigrid solver (e.g. [4]) for each subdomain solve for example, then the parallel speed-ups suggest that there is significant scope for the use of this solution algorithm.

It is also apparent from the results in Tables 3 and 4 that the second most important factor in determining parallel performance is the quality of the load balance across the processors. When $p = 2$ and $p = 4$, as may be observed from Fig. 1, each subdomain has the same boundary area and will therefore have the same number of overlapping elements into neighbouring subdomains. When

$p = 8$ or $p = 16$ this is not the case however: with some subdomains (in the centre) having a larger overlap than others. In these latter cases the individual subdomain problems are of a different size and so the parallel efficiency can be expected to deteriorate. This is clearly observed in Tables 3 and 4 with a parallel speed-up of 3.8 on 4 processors compared to a parallel speed-up of 7.0 on 8 processors and just 12.1/12.5 on 16 processors. The communication overhead associated with the parallel implementation would appear to be the least significant of the three main factors identified above as leading to losses in parallel efficiency.

4   Discussion

The results presented in the previous section demonstrate that the modified additive Schwarz preconditioner, that was successfully applied in parallel to convection-dominated problems in three dimensions using global mesh refinement in [7, 8], may also be successfully applied in parallel to the same class of problem using local mesh refinement. The algorithm is shown to parallelize very efficiently, although there is an additional complexity in ensuring the load balance of the overlapping grids on each subdomain when local refinement is used. Other than this, the major constraint on the efficiency of the algorithm comes from the observed growth in the sequential solution time as p increases. From Table 2 it is clear that the number of iterations taken does not grow with p, so it is apparent that the work per iteration must be growing as the number of subdomains is increased. This is inevitable in the sense that the total size of the overlap region increases with p however there may be scope for improving this aspect of the solution algorithm through the use of a sequential multigrid solver for the subdomain problems.

Whilst the results described in this paper demonstrate that the proposed preconditioner may be applied as part of a parallel adaptive finite element algorithm, we have not explicitly considered such an algorithm here. The examples used in the previous section are based upon a mesh refinement strategy that is known *a priori* and therefore allows a partition of Ω to be made that is guaranteed to lead to reasonably well load-balanced subdomain problems once refinement has taken place. Developing a strategy for incorporating this DD solution method within a robust adaptive mesh refinement procedure, based upon *a posteriori* error estimation for example, is a current area of research. One approach that has been proposed, that involves partitioning the coarse mesh, T_0, based upon an initial *a posteriori* error estimate, is described in [1]. It is our intention to use this technique as a framework for the implementation of a prototype adaptive algorithm that further exploits the DD solution strategy described here.

Acknowledgments

SAN gratefully acknowledges the funding received from the Government of Pakistan in the form of a Quaid-e-Azam scholarship.

References

1. R.E. Bank and M. Holst, *"A New Paradigm for Parallel Adaptive Meshing Algorithms"*, SIAM J. on Scientific Computing, 22, 1411-1443, 2000.
2. R.E. Bank and P.K. Jimack, *"A New Parallel Domain Decomposition Method for the Adaptive Finite Element Solution of Elliptic Partial Differential Equations"*, Concurrency and Computation: Practice and Experience, 13, 327–350, 2001.
3. R.E. Bank, P.K. Jimack, S.A. Nadeem and S.V. Nepomnyaschikh, *"A Weakly Overlapping Domain Decomposition Method for the Finite Element Solution of Elliptic Partial Differential Equations"*, to appear in SIAM J. on Scientific Computing, 2002.
4. R.E. Bank and J. Xu, *"A Hierarchical Basis Multigrid Method for Unstructured Meshes"*, in Tenth GAMM-Seminar Kiel on Fast Solvers for Flow Problems (W. Hackbusch and G. Wittum, eds.), Vieweg-Verlag, Braunschweig, 1995.
5. X.-C. Cai and M. Sarkis, *"An Restricted Additive Schwarz Preconditioner for General Sparse Linear Systems"*, SIAM J. on Scientific Computing, 21, 792–797, 1999.
6. P.K. Jimack and S.A. Nadeem, *"A Weakly Overlapping Parallel Domain Decomposition Preconditioner for the Finite Element Solution of Elliptic Problems in Three Dimensions*, in Proceedings of the 2000 International Conference on Parallel and Distributed Processing Techniques and Applications (PDPTA'2000), Volume III, ed. H.R. Arabnia (CSREA Press, USA), pp.1517–1523, 2000.
7. P.K. Jimack and S.A. Nadeem, *"Parallel Application of a Novel Domain Decomposition Preconditioner for the Stable Finite Element Solution of Three-Dimensional Convection-Dominated PDEs*, in Euro-Par 2001 Parallel Processing: 7th International Euro-Par Conference Manchester, UK, August 2001 Proceedings, ed. R. Sakellariou et al. (Lecture Notes in Computer Science 2150, Springer), pp.592–601, 2001.
8. P.K. Jimack and S.A. Nadeem, *" A Weakly Overlapping Parallel Domain Decomposition Preconditioner for the Finite Element Solution of Convection-Dominated Problems in Three-Dimensions*, to appear in proceedings of Parallel CFD 2001, Egmond aan Zee, NL, 21-23 May, 2001.
9. C. Johnson *"Numerical Solution of Partial Differential Equations by the Finite Element Method"*, Cambridge University Press, 1987.
10. S.A. Nadeem *"Parallel Domain Decomposition Preconditioning for the Adaptive Finite Element Solution of Elliptic Problems in Three Dimensions"*, Ph.D. Thesis, University of Leeds, 2001.
11. Saad, Y.: SPARSKIT: A Basic Tool Kit for Sparse Matrix Computations, Version 2. Technical Report, Center for Supercomputing Research and Development, University of Illinois at Urbana-Champaign, Urbana, IL, USA (1994).
12. Y. Saad, and M. Schultz, *"GMRES: A Generalized Minimal Residual Algorithm for Solving Nonsymmetric Linear Systems"*, SIAM J. on Sci. Comp. 7, 856–869, 1986.
13. B. Smith, P. Bjorstad and W. Gropp, *"Domain Decomposition: Parallel Multilevel Methods for Elliptic Partial Differential Equations"*, Cambridge University Press, 1996.
14. N. Touheed, P. Selwood, P.K. Jimack and M. Berzins, *"A Comparison of Some Dynamic Load-Balancing Algorithms for a Parallel Adaptive Flow Solver"*, Parallel Computing, 26, 1535–1554, 2000.

Parallel Performance in Multi-physics Simulation

Kevin McManus, Mark Cross, Chris Walshaw, Nick Croft, and Alison Williams

Centre for Numerical Modelling and Process Analysis, University of Greenwich,
London SE10 9LS,
k.mcmanus@gre.ac.uk,
WWW home page: http://multi-physics.com

Abstract. A comprehensive simulation of solidification/melting processes requires the simultaneous representation of free surface fluid flow, heat transfer, phase change, non-linear solid mechanics and, possibly, electromagnetics together with their interactions in what is now referred to as 'multi-physics' simulation. A 3D computational procedure and software tool, PHYSICA, embedding the above multi-physics models using finite volume methods on unstructured meshes (FV-UM) has been developed. Multi-physics simulations are extremely compute intensive and a strategy to parallelise such codes has, therefore, been developed. This strategy has been applied to PHYSICA and evaluated on a range of challenging multi-physics problems drawn from actual industrial cases.

1 Introduction

Typically, solidification/melting processes can involve free surface turbulent fluid flow, heat transfer, change of phase, non-linear solid mechanics, electromagnetic fields and their interaction, often in complex three dimensional geometries. The heritage of computational mechanics modelling is such that most Computer Aided Engineering (CAE) software tools have their focus upon one of 'fluids' (CFD)[1][11,5], 'structures' (CSM)[2][3,1] or 'electromagnetics' (CEM)[3][7]. As such, much of the work on phase change processes has addressed either thermofluid or thermomechanical aspects alone. However, as the customer demands ever higher levels of product integrity, then this has its impact upon the need for more comprehensive simulation of all the component phenomena and their interactions. Such 'multi-physics' simulation requires software technology that facilitates the closely coupled interaction that occur amongst the component phenomena. Experience over many years with closely coupled thermomechanical, thermofluid and magnetohydrodynamic applications, has demonstrated that to both capture accurately the loads, volume sources or boundary condition effects from one phenomena in another, in a computationally effective manner,

---

[1] Computational Fluid Dynamics
[2] Computational Structural Mechanics
[3] Computational Electromagnetics

P.M.A. Sloot et al. (Eds.): ICCS 2002, LNCS 2330, pp. 806–815, 2002.

requires all the continuum phenomena to be solved within one software environment.

It is, of course, well recognised that single phenomena simulation based upon CFD, CSM or CEM is computationally intensive, especially for large complex meshes, and may involve many thousands of time-steps in transient problems. The implication here, is that if single phenomenon simulation is a computational challenge, then the challenge is more significant for multi-physics modelling. The authors and their colleagues have been involved in the development of computational procedures and software tools for multi-physics simulation for over a decade. The results of this research have led to the production of the PHYSICA [9, 19] software environment for the computational modelling of multi-physics processes in complex three dimensional geometries. One issue at the core of the development of these multi-physics tools has been the concern to ensure that the software will run effectively on high performance parallel systems. Multi-physics simulation of solidification/melting processes involves a computational load in each control volume of the mesh (representing the geometrical domain) that varies with the cocktail of physics active in the volume, and which may well change with time. As such, from a parallel operation perspective, multi-physics simulation represents a dynamically varying non-homogeneous load over the mesh. A generic approach to this challenge is described in [14].

2   Software Technology Overview

2.1   Generic models and computational procedures

In the context of phase change simulation the following continuum phenomena and their interactions are of key significance:

- free surface transient Navier Stokes fluid flow,
- heat transfer,
- solidification/melting phase change,
- non-linear solid mechanics and possibly
- electromagnetic forces.

It is useful to observe that all the above continuum phenomena can be written in a single form where Table 1 provides a summary of the terms required to represent the equation for each of the above phenomena [9]:

$$\frac{\partial}{\partial t} \int_v \rho A \phi dv = \int_s \Gamma_\phi \nabla \phi \mathbf{n} ds + \int_v Q_v dv - \int_s \mathbf{Q}_s \mathbf{n} ds \qquad (1)$$

The suite of solution procedures chosen for this work is based upon an extension of finite volume (FV) techniques from structured to unstructured meshes. The fluid flow and heat transfer [6], phase change [8] and electromagnetics [18] procedures are based upon cell centred approximations, where the control volume is the element itself. The solid mechanics algorithms used in this work [4,

Table 1. Definition of terms in the generic transport equation

Phenomenon	ϕ	A	Γ_ϕ	Q_v	$\mathbf{Q_s}$
Continuity	1	1	0	S_{mass}	$\rho\mathbf{v}$
Velocity	$\mathbf{v}$	1	μ	$S + \mathbf{J} \times \mathbf{B} - \nabla p$	$\rho\mathbf{v}\mathbf{v}$
Heat Transfer	h	1	k/c	S_h	$\rho\mathbf{v}h$
Electromagnetic Field	$\mathbf{B}$	1	η	$(\mathbf{B}\nabla)\mathbf{v}$	$\mathbf{v}\mathbf{B}$
Solid Mechanics	$\mathbf{u}$	$\frac{\partial}{\partial t}$	μ	$\rho\mathbf{f_b}$	$\mu(\nabla\mathbf{u})^T + \lambda(\nabla \cdot \mathbf{u} - (2\mu+3\lambda)\alpha T)\mathbf{I}$

22] employ a vertex based approximation, so that the control volume is assembled from components of the neighbouring cells/elements to a vertex. The 'cell centred' phenomena are all solved using an extension of the conventional SIMPLE pressure correction procedures originated by Patankar and Spalding [17]. As this is a co-located flow scheme, the Rhie-Chow approximation is used to prevent checker-boarding of the pressure field [21]. The solid mechanics solution procedure involves a formulation as a linear system in displacement and solved in a similar manner to finite element methods [16]. Indeed, this enables the FV procedures to be readily extended to a range of nonlinear behaviours [22]. At this stage, a cautious approach to the solution strategy has been explored. A complete system matrix is constructed regardless of the physical state of each element or node. This approach is key for solidification/melting processes because, as phase change fronts move through the domain, then the local combination of physics changes with time and the transition between solid and liquid is not necessarily well defined. Hence, each phenomena is solved over the entire mesh for each time step.

2.2   The Parallelisation Strategy

Use of a single program multi-data (SPMD) strategy employing mesh partitioning is now standard for CFD and related codes. When the code uses an unstructured mesh, the mesh partitioning task is non-trivial. This work has used the JOSTLE graph partitioning and dynamic load-balancing tool [12, 23]. However, a key additional difficulty with respect to multi-physics simulation tools for phase change problems is that the computational workload per node/mesh element is not constant.

At this exploratory stage of multi-physics algorithm development, a cautious strategy has been followed, building upon established single discipline strategies (for flow, structures, etc) and representing the coupling through source terms, loads, etc [9]. Each discipline may of course involve the solution of one or more variables. A complication here is that separate physics procedures may use differing discretisation schemes. For example, in PHYSICA the flow procedure is cell centred, whereas the structure procedure is vertex centred. It follows that if a parallelisation strategy for closely coupled multi-physics simulation is to be effective, it must be able to achieve a load balance within each single discipline solution procedure.

It is the multi-phase nature of the problem that is a specific challenge here. The parallelisation strategy proposed is then essentially a two-stage process:

a) The multi-physics application code is parallelised on the basis of the single mesh, using primary and secondary partitions for the varying discretisation schemes [14],
b) Determination of a mesh partition that provides the required characteristics of good load balance with low inter-processor communication.

This strategy simplifies both the adaption of the code to run in parallel and the subsequent use of the code in parallel by others. However, it requires a mesh (graph) partitioning tool that has the following capabilities:

− produces load balanced partition for a single (possibly discontinuous) graph with a non-homogeneous workload per node,
− structures the sub-domain partitions so that they minimise inter-processor communication (i.e. the partitions respect the geometry of the problem),
− operates with a graph that is distributed across a plurality of processors (memory).

2.3   Load balancing

A key issue in multi-physics parallelisation is the use of primary and secondary partitions to cope with distinct discretisation techniques (ie. cell centred for flow and vertex centred for solid mechanics). In principle, one could use distinct (i.e. un-related) partitions for each discretisation technique, but this would seriously compromise any parallel scalability, because of the potential communication overheads. The approach is straightforward:

− a mesh entity type (e.g. element) associated with the greatest computational load is selected for a primary partition.
− secondary partitions for the other mesh entity types are derived from the primary partition to satisfy both load balance and minimisation of the overlap depth into neighbouring domains

2.4   Message passing approach

All data access that is not local to a processor (sub-domain) necessitates communication of the required data as a message from the processor that has the data to the processor that requires the data. The overhead of this communication limits parallel performance. The time required for message passing can be characterised in terms of the bandwidth (for long messages) and the communication start-up latency (for short messages). With current technology, processor speed is high in comparison to latency and this significantly restricts parallel performance.

PHYSICA uses a generic 'thin layer' message passing library, CAPLib [13], which provides a highly efficient portability layer that maps onto PVM [20],

MPI [15], shared memory and other native message passing systems. CAPLib is targeted at computational mechanics codes and provides a flexible and compact data model that is very straightforward to apply with no measurable performance overhead [13].

2.5 Portability, Transparency and Scalability

Parallel systems exhibit a diversity of characteristics that are of little or no concern to numerical modellers. The PHYSICA code must port without difficulty or compromise of performance onto a wide range of parallel platforms. Similarly it is essential that no system specific case configuration is required in moving between serial and parallel systems. Serial and parallel geometry and data files are therefore the same, although not necessarily identical. Provided that the user's model stays within the framework prescribed by the 'core' PHYSICA modules, then parallelism is straightforward. The intention is to reduce the difference between running in serial and parallel to be simply the executable instruction. Portability is achieved through the combination of a highly standardised programming language, Fortran77 and the portability layer, CAPlib. Transparency has been achieved by embedding JOSTLE into parallel PHYSICA to provide run-time partitioning.

Scalability may be seen as the extent to which either more processors will reduce run-time or model size may be increased by using more processors. As the number of processors, P, is increased with a fixed problem size the performance will encounter an Amdahl limit [2] because the problem size per processor becomes so small that inherently sequential operations dominate run-time.

2.6 Machine Characterisation

Run time on both the SMP cluster and the NUMA system is dependent upon several factors. Within each system there is competition for access to the memory bus. Execution times are consequently related to the overall load on the machine. In addition, performance of current generation high clock speed, superscalar pipelined processors is closely tied to cache success (hit) rate. Cache hit rates begin to suffer as vector lengths become large in comparison to the cache size and so single processor performance deteriorates. Inter node communication is affected by bottlenecks in the inter-processor communications and so run times vary with activity across the entire machine.

3 High Performance Computing Systems Used

The key objectives of this research programme were to evaluate the parallel scalability performance of the multi-physics simulation code, PHYSICA, for a range of problems on two standard high performance computing systems.

3.1 Compaq Alpha System

The Compaq Alpha System is a symmetric multi-processor (SMP) cluster consisting of three Compaq 4100 quad processor SMP nodes, providing a total of 12 processors with 3Gb of memory. Each Alpha processor is an EV5/6 running at 466Mhz. High speed, low latency inter-node communication is provided by Memory Channel MkI. The inter-processor communication characteristics of this system for MPI calls are approximately 5 nanosecond communication start up latency and 55 Mbytes per second bandwidth.

3.2 SGI Origin 2000 System

These systems are described as cache coherent non-uniform memory access (NUMA) systems. Each computer node has two MIPS R10000 processors with 1Gb of memory and there are up to 32 nodes clustered as a single address space within a cabinet. The overall MPI interprocessor communication characteristics within a box are approximately 5 nanosecond communication start-up latency and a bandwidth of 300 Mbytes per second.

4 Results and Discussion

To provide realistic measures of parallel performance, the test cases discussed here have been developed from actual PHYSICA applications. In order to encourage efficient use and provide meaningful performance figures, PHYSICA reports the run times split into file access time and calculation time. The sum of these two times provides overall run or wallclock time. The test cases aim to provide a realistic view of these times.

4.1 Case 1: Shape casting

In this case the metal-mould arrangement has a complicated three dimensional geometry resulting in a mesh of 82,944 elements and 4 materials illustrated in Figure 1. This case was investigated as a thermo-mechanical problem from which a thermal only parallel test case has been extracted. Run time and parallel speedup results for the Compaq and SGI systems are given in Tables 2 and 3

4.2 Case 2: Aluminium reduction cell

This process involves the interactions between the electric current density, magnetic field, temperature distribution and phase change resulting in a two-phase flow behaviour with thermally induced stresses and deformations. A simplified model of this process considers 'lumped' source terms for the Joule heating, electromagnetic force field and bubble movement. This reduces the modelling to a consideration of the Navier-Stokes fluid flow, heat transfer, phase change, and solid mechanics [10]. These are solved over a mesh of 23,408 elements of mixed

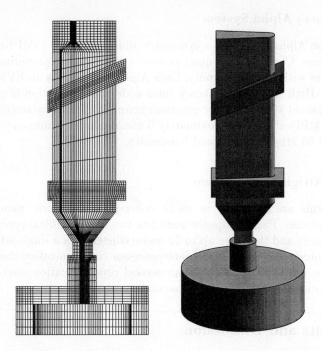

Fig. 1. Mesh and surface of the turbine blade test case

P	t calculation	t file	t overall	Sp calculation	Sp file	Sp overall
1	3644	45	3689	1	1	1
2	1950	72	2022	1.87	0.63	1.82
4	610	73	683	5.97	0.62	5.40
8	323	83	406	11.28	0.54	9.09
12	285	92	377	12.78	0.49	9.79

Table 2. Run times and speedup for the Turbine Blade test case on a Compaq 4100 SMP cluster

P	t overall	Sp overall
1	4177	1
2	1912	2.18
4	1051	3.97
6	761	5.49
8	812	5.14
12	676	6.17
16	657	6.36
32	547	7.64

Table 3. Run times and speedup for the Turbine Blade test case on a SGI Origin system

type with 6 materials illustrated in Figure 2. In this case the linear solvers run to convergence using JCG[4] for heat and flow and BiCG[5] for stress. Run time and parallel speedup results for the Compaq and SGI systems are given in Tables 2 and 3

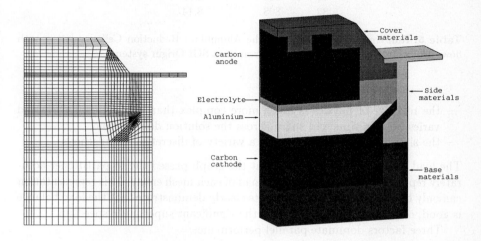

Fig. 2. Mesh and materials for the aluminium reduction cell test case

P	t calculation	t file	t overall	Sp calculation	Sp file	Sp overall
1	3860	52	3912	1	1	1
2	1936	73	2009	1.99	0.71	1.95
4	928	88	1016	4.16	0.59	3.85
8	470	92	562	8.21	0.57	6.96
12	346	120	466	11.16	0.43	8.39

Table 4. Run times and speedup for the Aluminium Reduction Cell with flow, heat transfer, solidification and stress on a Compaq 4100 SMP cluster

5 Conclusion

The work presented in this paper has described the performance results of an investigation into the parallelisation of an unstructured mesh code targeted at multi-physics simulation. Although the code uses finite volume methods, typical of CFD calculations, they are complicated by two factors:

---

[4] Jacobi preconditioned conjugate gradient
[5] Bi-directional conjugate gradient

P	t overall	Sp overall
1	6579	1
2	3707	1.77
4	1524	4.31
8	1310	5.02
16	833	7.90
32	808	8.14

Table 5. Run times and speedup for the Aluminium Reduction Cell simulation with flow, heat transfer, solidification and stress on a SGI Origin system

- the mixture of physics is much more complex than conventional CFD and varies both in time and space across the solution domain, and
- the simultaneous exploitation of a variety of discretisation methods

The load balancing task requires that the graph passed the to partitioner accurately represents the computational load of each mesh entity. This representation can only be an estimate but the results clearly demonstrate that the load balance is good, otherwise we would not see the significant superlinear speed-up.

Three factors dominate parallel performance:

- Load balance
- Communication latency
- File access

Although the test cases employ only static load balancing for dynamically inhomogeneous multi-physics problems the resulting load balance is remarkably good. This is in part due to the cautious solution strategy currently used in PHYSICA and in part due to the reasonably homogeneous mesh element shapes providing reasonable alignment between the primary and secondary partitions.

Latency has a marked effect on parallel performance. This partly why the first example performs so well by removing the norm calculations required to determine convergence in the linear solvers. This is fortunately consistent with SIMPLE type schemes. As computer technology develops the gap between latency and processor performance is increasing and latency effects are expected to become increasingly problematic.

Increasingly the results from modelling take the form of animated images and so a typical PHYSICA run will write significant quantities of data to file during execution. File access remains a problem for parallel performance. Certainly it is possible to stripe the result files across multiple hard drives but striped data has, at some point, to be reconstructed into global data in order to be visualised. Data striping is to a large extent postponing the parallel overhead.

What is clear from this work is that although the current parallelisation strategy is effective for modest numbers of processors, a more scalable strategy will eventually be required. This requires a less conservative sequential solution strategy together with dynamic load balancing parallel solution. Such an effort is now underway.

References

1. Abaqus. URL: http://www.hks.com.
2. G. M. Amdahl. Validity of the single-processor approach to achieving large scale computing capabilities. In *Proc AFIPS*, pages 483–485, 1967.
3. Ansys. URL: http://www.ansys.com.
4. C. Bailey and M. Cross. A finite volume procedure to solve elastic solid mechanics problems in three dimensions on an unstructured mesh. *Int. J Num Meth in Engg*, 38:1757–776, 1995.
5. CFX. URL: http://www.software.aea.com/cfx. AEA.
6. P. Chow, M. Cross, and K. Pericleous. A natural extension of standard control volume CFD procedures to polygonal unstructured meshes. *Appl. Math Modelling*, 20:170–183, 1995.
7. Concerto. URL: http://www.vectorfields.co.uk. Vector Fields.
8. N. Croft, K. Pericleous, and M. Cross. PHYSICA: a multiphysics environment for complex flow processes. *Numerical Methods in Laminar and Turbulent Flows*, Vol IX:1269 – 1280, 1995.
9. M. Cross. Computational issues in the modelling of materials based manufacturing processes. *Journal of Computationally Aided Materials*, 3:100–116, 1996.
10. M .Cross et al. Computational modelling of casting processes - a multi-physics challenge. In G. Irons and A. Cramb, editors, *Brimacombe Memorial Symposium Proceedings*, pages 439–450. Met Soc (Canada), 2000.
11. Fluent. URL: http://www.fluent.com.
12. JOSTLE. URL: http://www.gre.ac.uk/jostle. University of Creenwich.
13. P. Leggett, S. P. Johnson, and M. Cross. CAPLib - a 'thin layer' message processing library to support computational mechanics codes on distributed memory parallel systems. *Advances in Engineering Software*, 32:61–81, 2001.
14. K. McManus, M. Cross, , C. Walshaw, S. Johnson, and P. Leggett. A scalable strategy for the parallelization of multiphysics unstructured-mesh iterative codes on distributed-memory systems. *Int Jnl of High Performance Computing Applications*, 14(2):137–174, 2000.
15. MPI: Message Passing Interface. URL: http://www.mcs.anl.gov/mpi/index.html. Argonne National Laboratory.
16. E. Onate, M. Ceiveia, and O. C. Zienkiewicz. A finite volume format for structural mechanics. *Int J Num Maths in Engg*, 37:181–201, 1994.
17. S. V. Patankar and D. B. Spalding. A calculation procedure for heat, mass and momentum transfer in three dimensional parabolic flows. *Int J Heat Mass Trans*, 15:1787–1806, 1972.
18. K. Pericleous, M. Cross, M. Hughes, and D. Cook. Mathematical modelling of the solidification of liquid tin with electromagnetic stirring. *Jnl of Magnetohydrodynamics*, 32(4):472–478, 1996.
19. Physica. URL: http://www.multi-physics.com. University of Creenwich.
20. PVM Parallel Virtual Machine. URL: http://www.epm.ornl.gov/pvm/pvmhome.html. Oak Ridge National Laboratory.
21. C. Rhie and W. Chow. A numerical study of the flow past an isolated airfoil with trailing edge separation. *JAIAA*, 21:1525–1532, 1982.
22. G. Taylor, C. Bailey, and M. Cross. Solution of elasto-visco-plastic constitutive equalities: a finite volume approach. *Appl Math Modelling*, 19:746–760, 1995.
23. C. Walshaw and M. Cross. Parallel optimisation algorithms for multi-level mesh partitioning. *Parallel Computing*, 26:1635–1660, 2000.

A Parallel Finite Volume Method for Aerodynamic Flows

Nigel Weatherill , Kaare Sørensen , Oubay Hassan , and Kenneth Morgan

Department of Civil Engineering, University of Wales,
Swansea SA2 8PP, Wales, U.K.
{n.p.weatherill, cgsorens, o.hassan, k.morgan}@swansea.ac.uk
http://www.springer.de/comp/lncs/index.html

Abstract. The solution of 3D transient aerodynamic flows of practical interest is obtained by a finite volume approach, implemented on unstructured tetrahedral and unstructured hybrid meshes. The time discretised equation systems are solved by explicit iteration coupled with multigrid acceleration and the procedure is parallelised for improved computational performance. The examples presented involve an inviscid simulation of store release from a complete aircraft configuration and a viscous simulation of flow over an oscillating wing.

1   Introduction

In the aerospace industry, there are many practical applications that require the modelling of compressible time dependent flow with moving boundaries. These include the analysis of flutter, store release, buffeting and the deployment of control surfaces. In this paper, we present an approach for the simulation of such problems. Spatial discretisation of the governing flow equations is achieved by the adoption of an unstructured edge based finite volume method and this provides the geometric flexibility needed for flows involving complex geometries. The algorithmic implementation allows for the use of hybrid meshes of tetrahedra, prisms, pyramids and hexahedra. Hybrid meshes of this type are readily generated by using a modified unstructured tetrahedral mesh generator. For these transient problems, it is attractive to employ an implicit formulation, in which the time step used can be selected on accuracy criteria only. With implicit algorithms, it is important to adopt a formulation that avoids the requirement for working with large matrices and this is accomplished here by the use of a multigrid approach. The coarse meshes are nested and are constructed by agglomeration, with the control volumes merged together through an edge based method. Meshing difficulties encountered in the simulation of problems involving large displacement of the boundaries are overcome by employing a combination of mesh movement and local remeshing. The mesh is first moved, retaining the connectivities, as dictated by the geometry deflection. Any regions of bad mesh quality are then removed, creating holes which are remeshed using the unstructured mesh generator. The practical examples which are included show that the method is robust and applicable to geometries of a complication level experienced in industry.

P.M.A. Sloot et al. (Eds.): ICCS 2002, LNCS 2330, pp. 816–823, 2002.

2 Problem Formulation

For modelling turbulent flows, the governing equations are taken to be the time dependent, Favre averaged [1], compressible Navier–Stokes equations. Here it is assumed that the time averaging suppresses the instantaneous fluctuations in the flow field caused by turbulence, while still being able to capture time dependency in the time scales of interest. The resulting equations can be expressed in integral form, on a three–dimensional Cartesian domain $\Omega(t)$, with closed surface $\partial\Omega(t)$, as

$$\int_{\Omega(t)} \frac{\partial U}{\partial t}\, d\boldsymbol{x} + \int_{\partial\Omega(t)} F^j n_j\, d\boldsymbol{x} = \int_{\partial\Omega(t)} G^j n_j\, d\boldsymbol{x} \tag{1}$$

for $i, j = 1, 2, 3$ and where the summation convention is employed. In this equation, U denotes the averaged vector of the conserved variables and F^j and G^j are the averaged inviscid and viscous flux components in direction x_j respectively. The equation set is closed by assuming the fluid to be calorically perfect and the turbulent viscosity is determined by the Spalart–Allmaras model [2].

3 Discretization Procedure

Hybrid meshes are generated using a merging procedure, constructing complex elements from an initial simplex mesh. Quadrilateral elements are generated on the wall surfaces. This is accomplished by an indirect method of combining triangles, coupled with a split scheme to guarantee meshes consisting of quadrilateral elements only. The advancing layer method [3] is then employed to generate stretched elements adjacent to those boundary surface components which represent solid walls. The height and number of layers is specified by the user in such a way that the expected boundary layer profile should be capable of being adequately represented. The implementation starts from a triangular mesh on the surface of the wall, which means that each quadrilateral element has to be treated as two triangles in the generation process. The layers are constructed by generating points along prescribed lines and connecting the generated points, by using advancing front mesh generation concepts, to form tetrahedral elements. Point generation ceases before the prescribed number of layers is reached if an intersection appears or if the local mesh size is close to that specified in the user–specified mesh distribution function. In these layers, tetrahedra are merged to form prisms, pyramids and hexahedra. The remainder of the domain is meshed using a standard isotropic Delaunay procedure [4]. Tetrahedra are merged to form prisms, pyramids and hexahedra. On the resulting mesh, the governing equations are discretised by using a finite–volume method, in which the unknowns are located at the vertices of the mesh and the numerical integration is performed over dual mesh interfaces [5]. For problems in which the control volumes may change with time, it is convenient to apply the general Leibnitz rule to the time derivative term in equation (1). In this case, the governing equations may be expressed as

$$\frac{\mathrm{d}}{\mathrm{d}t} \int_{\Omega(t)} \boldsymbol{U} \, \mathrm{d}\boldsymbol{x} + \int_{\partial\Omega(t)} \left(\boldsymbol{F}^j - v_j \boldsymbol{U}\right) n_j \, \mathrm{d}\boldsymbol{x} = \int_{\partial\Omega(t)} \boldsymbol{G}^j n_j \, \mathrm{d}\boldsymbol{x} \qquad (2)$$

where $\boldsymbol{v} = (v_1, v_2, v_3)$ is the velocity of the control volume boundary.

3.1   Spatial Discretization

A dual mesh is constructed by connecting edge midpoints, element centroids and face–centroids in such a way that only one node is contained within each control volume [6]. Each edge of the grid is associated with a surface segment of the dual mesh interface between the nodes connected to the edge. This surface is defined using triangular facets, where each facet is connected to the midpoint of the edge, a neighbouring element centroid and the centroid of an element face connected to the edge. With this dual mesh definition, the control volume can be thought of as constructed by a set of tetrahedra with base on the dual mesh. To enable the integration over the surface of the control volumes, a set of coefficients is calculated for each edge using the dual mesh segment associated with the edge. The discrete equations are then formed by looping over the edges in the mesh and sending edge contributions to the nodes. The task of finding edge coefficients that ensure that the resulting scheme is geometrically conservative is not trivial. Here, geometric conservation is assured by modifying the approach of Nkonga and Guillard [7] to handle the case of dual meshes constructed of assemblies of triangular facets. Finally, as the basic procedure is essential central difference in character, a stabilizing dissipation must be added and this is achieved by a method of JST type [8].

3.2   Mesh Movement

For the simulation of a problem involving moving boundaries, the mesh must deform to take account of the movement. This is usually achieved by fixing the mesh on the far field boundary, while moving the mesh nodes on the geometry in accordance with the movement. The interior mesh nodes are moved accordingly to achieve the desired mesh quality. A number of different mesh movement approaches have been investigated in the literature but the approach used here assumes that the edges of the mesh behave like springs connecting the nodes [9]. The nodes on the moving geometry are moved in small increments, with the interior nodes being moved to ensure internal equilibrium for each increment. Typically 50 increments are employed with this procedure. This approach is robust and fast, usually requiring about 10–15% of the total CPU–time required to solve the system time–accurately. With this approach, it is also possible to obtain a certain level of control over the mesh movement, by varying the spring constants of the edges.

3.3   Local Remeshing

Often, it is impossible to avoid remeshing if the necessary mesh quality is to be maintained. The regions in which mesh movement is inappropriate are, usually, relatively small, however, and this is utilised by applying local remeshing only. The regions that need remeshing are found by using a mesh quality indicator which is taken to be the ratio of the volume of an element at the current time level and the element volume in the original mesh. According to the values produced by this indicator, certain elements are removed, creating one or several holes in the mesh. Each of these holes is meshed in the normal manner by using a Delaunay scheme, with the triangulation of the hole surface being taken as the surface triangulation [9]. In certain circumstances, it may not be possible to recover the hole surface triangulation and, in this case, another layer of elements is removed from the original mesh and the process is repeated. The unknown field is transferred from the previous mesh level by linear interpolation.

3.4   Time Discretisation

The spatially discretised equation at node I on the mesh is of the form

$$\frac{d}{dt}(V_I U_I) = R_I \tag{3}$$

where V_I is the control volume associated with node I and R_I denotes the assembled edge contributions to this node. A three level second order accurate time discretisation is employed and this equation is represented in the form

$$\frac{3}{2}[V_I U_I]^{n+1} - 2[V_I U_I]^n + \frac{1}{2}[V_I U_I]^{n-1} = \Delta t R_I^n \tag{4}$$

where the superscript n denotes and evaluation at time $t = t_n$ and $t_n = t_{n-1} + \Delta t$, $t_{n-1} = t_{n-2} + \Delta t$. A first order backward time discretisation scheme is used for simulations which require the use of remeshing.

4   Solution Procedure

4.1   Multigrid Scheme

At every time step, the discretisation procedure results in a numerical solution of an equation system and this solution is obtained by the FAS multigrid scheme [10]. The approach avoids the requirement for linearisation of the discrete system and eliminates the need to store a Jacobian matrix. Assuming nested meshes, the restriction operator that maps the coarse mesh quantities to the fine mesh is constructed as a linear restriction mapping. This implies that the values on the coarse mesh are produced as the weighted average of the fine mesh values. For prolongation, a simple point injection scheme is adopted. The relaxation scheme that is employed in conjunction with the multigrid procedure is a three–stage Runge–Kutta approach with local time stepping, and coefficients 0.6, 0.6 and 1.0.

4.2   Coarse Mesh Generation

The assumption of nested coarser meshes is met by generating the coarser meshes by agglomeration. The agglomeration approach works on the dual mesh, merging control volumes to achieve the local coarseness ratio required [11, 12]. This approach is purely edge–based and can therefore be applied on any type of mesh after the edge–based data structure has been assembled. Since no mesh generation issues are raised in the procedure, the scheme is completely stable. It is also fast and requires little memory overhead.

4.3   Implementation

The computational implementation consists of four main components: (i) mesh movement: a geometry file and the surface movement information is prescribed; the internal nodes of the mesh are positioned employing the mesh movement algorithm; (ii) remeshing: the local quality of the moved mesh is investigated and remeshing is performed if required; (iii) preprocessing: the edge–based data structure is set up from the mesh information, the coarse grids are generated and the inter–grid interpolation arrays for the multigrid procedure are constructed; (iv) equation solving: the output from the preprocessor is accessed and the time dependent discrete equations are solved for one physical time step. These components are loosely coupled, employing a driving script and file communication. Such a loose coupling will, inevitably, adversely affect the solution time compared to implementations which do not require I/O communication. However, since relatively few time steps are required for the implicit schemes, this effect is usually small.

4.4   Parallelisation

Parallelisation of the solution procedure is necessary to make best use of multi–processor computer platforms. For parallelisation, the mesh is created sequentially and then split into the desired number of computational domains using the METIS library [13]. This procedure essentially colours the nodes in the mesh according to the domain to which they belong. A preprocessor renumbers the nodes in the domains and generates the communication data required. The edges of the global mesh are placed in the domain of the nodes attached to the edge, or if the nodes belong to different domains, as occurs with an inter–domain edge, the edge is placed in the domain with the smallest domain number. The edges are renumbered in each domain, placing the inter–domain edges first in the list. On the inter–domain boundaries, the nodes are duplicated so that each domain stores a copy of the nodes appearing in the edge list. These ghost nodes are allocated a place in memory to store the nodal contributions from adjacent domains. If the node belongs to the domain, it is termed a real node. The communication data structure consists of two registers for each domain combination on each grid level. These registers are termed the ghost node register and real node register and simply store the local node indexes of the ghost and real nodes respectively.

The data structure of the boundary edges is created in an analogous fashion. The solution procedure consists, essentially, of loops over edges and loops over nodes and these are readily parallelised. A global parallelisation approach has been followed, in which agglomeration is allowed across domain boundaries. While requiring communication to be performed in the intergrid mappings, this approach yields an implementation equivalent to a sequential code. The approach does not guarantee balanced coarse meshes, but this is of secondary importance as the coarse meshes are significantly smaller than the finest mesh.

5 Examples

Two examples are presented to illustrate the capability of the proposed procedure. The first example is an inviscid simulation involving a complete aircraft configuration while the second example is the simulation of turbulent flow over an wing.

5.1 Store Release

The first example is the simulation of the release of two stores from an aircraft configuration. The stores are assumed to be completely filled with a substance of density 1 250 times that of the free stream air. The free stream Mach number is 0.5, the angle of attack is zero and the Froude number is 1 333. The rigid body movement of the stores is calculated by integrating the pressure field over the stores and using a second order accurate time integration. Local remeshing is employed and each mesh consists of about 2.7 million tetrahedral elements. The resulting surface pressure distribution, at four equally spaced time intervals, is shown in Figure 1.

5.2 Oscillating Wing

The results of a viscous flow simulation over an oscillating ONERA M6 wing are now considered. The free stream Mach number is 0.84 and the initial angle of attack is 3.06 degrees. The Reynolds number, based on the mean geometric chord, is set to 1.2×10^7 and turbulence is triggered at 5% chord. A hybrid mesh, consisting of 1 641 721 prisms, 26 139 pyramids and 1 826 751 tetrahedra is employed. The number of edges in this mesh is 4 961 611. The wing root is held fixed, while the wing tip is subjected to a prescribed sinusoidal movement with 3 degrees amplitude. Between these two sections, the amplitude varies linearly along the wing span and the axis of rotation is along the quarter chord line of the wing. The Strouhal number for this movement is 3.10 and 16 time steps were performed for each cycle. At each time step, 150 multigrid cycles were used. One full movement cycle, employing 16 R14000 cpus for the solver module and one R14000 cpu for the mesh movement and preprocessor modules, requires about 12 hours of wall clock time. A plot of the resulting lift polar is shown in Figure 2.

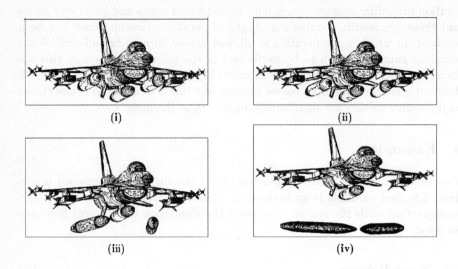

(i)   (ii)

(iii)   (iv)

Fig. 1. Surface pressure distribution, at 0.29 second intervals, following store release from an aircraft

6   Conclusion

A method for solving compressible aerodynamic flow problems involving moving geometries has been described. A parallelised multigrid implicit time stepping approach is adopted which, when coupled with a mesh–movement/local remeshing capability, provides an accurate and robust technique for solving practical aerodynamic problems of this type.

Acknowledgements

Kaare Sørensen acknowledges the sponsorship of The Research Council of Norway, project number 125676/410.

References

1. A. Favre, Equation des gaz turbulents compressibles, *Journal de Mechanique*, **4**, 361–390, 1965
2. P. R. Spalart and S. R. Allmaras, A one equation turbulent model for aerodynamic flows, *AIAA Paper 92–0439*, 1992
3. O. Hassan, K. Morgan, E. J. Probert and J. Peraire, Unstructured tetrahedral mesh generation for three–dimensional viscous flows, *International Journal for Numerical Methods in Engineering*, **39**, 549–567, 1996

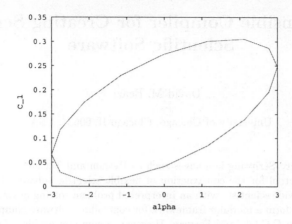

Fig. 2. The lift polar for turbulent flow over an oscillating ONERA M6 wing

4. N. P. Weatherill and O. Hassan, Efficient three–dimensional Delaunay triangulation with automatic boundary point creation and imposed boundary constraints, *International Journal for Numerical Methods in Engineering*, **37**, 2005–2039, 1994

5. K. A. Sørensen, A multigrid accelerated procedure for the solution of compressible fluid flows on unstructured hybrid meshes, *PhD Thesis, C/Ph/251/01, University of Wales Swansea*, 2002

6. M. Vahdati, K. Morgan and J. Peraire, Computation of viscous compressible flows using an upwind algorithm and unstructured meshes, in S. N. Atluri, editor, *Computational Nonlinear Mechanics in Aerospace Engineering*, AIAA, Washington, 479–505, 1992

7. B. Nkonga and H. Guillard, Godunov type method on non–structured meshes for three–dimensional moving boundary problems, *Computational Methods in Applied Mechanical Engineering*, **113**, 183–204, 1994

8. A. Jameson, W. Schmidt and E. Turkel, Numerical simulation of the Euler equations by finite volume methods using Runge–Kutta timestepping schemes *AIAA paper 81-1259*, 1981

9. O. Hassan, E. J. Probert, K. Morgan and N. P. Weatherill, Unsteady flow simulation using unstructured meshes, *Computer Methods for Applied Mechanical Engineering*, **189**, 1247–1275, 2000

10. A. Brandt, Multi–level adaptive solutions to boundary value problems *Mathematics of Computation*, **21**, 333–390, 1977

11. M. H. Lallemand, H. Steve and A. Dervieux, Unstructured multigridding by volume agglomeration: Current status, *Computers & Fluids*, **21**, 397–433, 1992

12. D. J. Mavriplis and V. Venkatakrishnan, A 3D agglomeration multigrid solver for the Reynolds–averaged Navier–Stokes equations on unstructured meshes, *International Journal for Numerical Methods in Fluids*, **23**, 527–544, 1996

13. G. Karypis and V. Kumar, Multilevel k–way partitioning scheme for irregular grids, *Journal of Parallel and Distributed Computing*, **48**, 96–129, 1998

An Extensible Compiler for Creating Scriptable Scientific Software

David M. Beazley

University of Chicago, Chicago IL 60637, USA

Abstract. Scripting languages such as Python and Tcl have become a powerful tool for the construction of flexible scientific software because they provide scientists with an interpreted problem solving environment and they form a modular framework for controlling software components written in C,C++, and Fortran. However, a common problem faced by the developers of a scripted scientific application is that of integrating compiled code with a high-level interpreter. This paper describes SWIG, an extensible compiler that automates the task of integrating compiled code with scripting language interpreters. SWIG requires no modifications to existing code and can create bindings for eight different target languages including Python, Perl, Tcl, Ruby, Guile, and Java. By automating language integration, SWIG enables scientists to use scripting languages at all stages of software development and allows existing software to be more easily integrated into a scripting environment.

1   Introduction

One of the most difficult tasks faced by the developers of scientific software is figuring out how to make high-performance programs that are easy to use, flexible, and extensible. Clearly these goals are desirable if scientists want to focus their efforts on solving scientific problems instead of problems related to software engineering. However, the task of writing such software is usually quite difficult—in fact, much more difficult than most software engineers are willing to admit.

An increasingly popular solution to the scientific software problem is to use scripting languages such as Python or Tcl as a high-level steering language for software written in C, C++, or Fortran [8, 10, 11, 14, 16, 19]. Scripting languages provide a nice programmable user interface that is interactive and which allows more complicated tasks to be described by collections of scripts. In addition, scripting languages provide a framework for building loosely-coupled software components and gluing those components together [13]. This makes it easier to incorporate data analysis, visualization, and other data management facilities into an application without having to create a huge monolithic framework. Scripting languages are also portable and easy to utilize on high performance computing systems including supercomputers and clusters.

In addition to these benefits, the appeal of scripting languages is driven by the piecemeal software development process that characterizes a lot of scientific

P.M.A. Sloot et al. (Eds.): ICCS 2002, LNCS 2330, pp. 824–833, 2002.

software projects. Rarely do scientists set out to create a hyper-generalized program for solving everything. Instead, programs are created to solve a specific problem of interest. Later, if a program proves to be useful, it may be gradually adapted and extended with new features in order to solve closely related problems. Although there might be some notion of software design, a lot of scientific software is developed in an ad-hoc manner where features are added as they are needed as opposed to being part of a formal design specification. Scripting languages are a good match for this style of development because they can be used effectively even when the underlying software is messy, incomplete, lacking in formal design, or under continual development.

Some scientists also view scripting languages as a logical next step in the user interface of their software. For example, when programs are first developed, they are often simple batch jobs that rely upon command line options or files of input parameters. As the program grows, scientists usually want more flexibility to configure the problem so they may modify the program to ask the user a series of questions. They may even write a simple command interpreter for setting up parameters. Scripting languages build upon this by providing an interface in the form of a fully-featured programming language that provides features similar to that found in commercial scientific software such as MATLAB, Mathematica, or IDL. In fact, many scientists add an interpreter to their application just so they can obtain comparable flexibility.

2   The Problem With Scripting

All modern scripting languages allow foreign code to be accessed through a special extension API. However, to provide hooks to existing C, C++, or Fortran code, you usually have to write special wrapper functions. The role of these functions is to convert arguments and return values between the data representation in each language. For example, if a programmer wanted to access the cosine function in the math library from Python, they would write a wrapper like this:

```
PyObject *wrap_cos(PyObject *self, PyObject *args) {
    double x, result;
    if (!PyArg_ParseTuple(args,"d",&x)) return NULL;
    result = cos(x);
    return Py_BuildValue("d",result);
}
```

Even though it's not very difficult to write a few dozen wrappers, the task becomes tedious if an application contains several hundred functions. Moreover, the task becomes considerably more difficult if an application makes use of advanced programming features such as pointers, arrays, classes, inheritance, templates, and overloaded operators because there is often no obvious way to map such features to the scripting language interpreter.

As a result, it is difficult to integrate existing software into a scripting environment without a considerable coding effort. Scientists may also be reluctant to use scripting languages in the early stages of program development since it will be too difficult to keep the wrapper code synchronized with changes in the underlying application. Scientists may also take the drastic step of writing all new software as a hand-written extension of a particular scripting language interpreter. Although this works, it entangles the implementation of the application with the implementation of the interpreter and makes it difficult to reuse the code as a library or to integrate the system into other application frameworks. This arguably defeats the whole point of creating reusable and modular software.

3   SWIG: A Compiler for Extensions

One way to simplify the use of scripting languages is to automatically generate wrapper code using a code generation tool. SWIG (Simplified Wrapper and Interface Generator) is a special purpose compiler that has been developed for this purpose [3]. Originally developed in 1995, SWIG was first used to build scripting language bindings for the SPaSM short-range molecular dynamics code at Los Alamos National Laboratory [4]. This was one of the first large-scale parallel applications to utilize Python as a framework for computational steering and integrated data analysis [5]. Since then, SWIG has been developed as a free software project. It is found in many GNU/Linux distributions and has been used in variety of projects ranging from scientific simulations to commercial video games.

The key feature of SWIG is that it allows existing software to be integrated into a scripting environment with few if any code modifications. Furthermore, the system maintains a clean separation between the underlying application and the interpreter. Because of this, it promotes modularity and allows software to be reused in different settings. The other essential feature is that SWIG generates wrappers from standard C/C++ declarations and is fully automated. This allows the system to be incorporated into a project in a minimally intrusive manner and allows scientists to focus on the problem at hand instead of language integration issues.

3.1   A Simple Example

To illustrate the use of SWIG, suppose that a simulation program defines a few C functions like this:

```
int integrate(int nsteps, double dt);
void set_boundary_periodic();
void init_lj(double epsilon, double sigma, double cutoff);
void set_output_path(char *path);
```

To build a scripting language interface, a user simply creates a special file containing SWIG directives (prefaced by a %) and the C++ declarations they would like to wrap. For example:

```
// file: shock.i
%module shock
%{
#include "headers.h"
%}
int integrate(int nsteps, double dt);
void set_boundary_periodic();
void init_lj(double epsilon, double sigma, double cutoff);
void set_output_path(char *path);
```

This file is then processed by SWIG to create an extension module in one of several target languages. For example, creating and using a Python module often works like this:

```
$ swig -python shock.i
$ cc -c shock_wrap.c
$ cc -shared shock_wrap.o $(OBJS) -o shockmodule.so
$ python
Python 2.1 (#1, Jun 13 2001, 16:09:46)
>>> import shock
>>> shock.init_lj(1.0,1.0,2.5)
>>> shock.set_output_path("./Datafiles")
>>> shock.set_boundary_periodic()
```

In this example, a separate file was used to hold SWIG directives and the declarations to be wrapped. However, SWIG can also process raw header files or header files in which conditional compilation has been used to embed special SWIG directives. This makes it easier for scientists to integrate SWIG into their application since existing source code can be used as SWIG input files. It also makes it easier to keep the scripting language interface synchronized with the application since changes to header files (and function/class prototypes) can be tracked and handled as part of the build process.

3.2 Advanced Features

Although a simple example has been shown, SWIG provides support for a variety of advanced programming features including pointers, structures, arrays, classes, exceptions, and simple templates. For example, if a program specifies a class definition like this,

```
class Complex {
    double rpart, ipart;
public:
    Complex(double r = 0, double i = 0): rpart(r), ipart(i) { };
    double real();
    double imag();
    ...
};
```

the resulting scripting interface mirrors the underlying C++ API. For example, in Python, a user would write the following code:

```
w = Complex(3,4)      # Create a complex
a = w.real()          # Call a method
del w                 # Delete
```

When exported to the interpreter, objects are represented as typed pointers. For example, a `Complex *` in the above example might be encoded as a string containing the pointer value and type such as _100f8ea0_p_Complex. Type information is used to perform run-time type checking in the generated wrapper code. Type checking follows the same rules as the C++ type system including rules for inheritance, scoping, and typedef. Violations result in a run-time scripting exception.

As applications become more complex, SWIG may need additional input in order to generate wrappers. For example, if a programmer wanted to bind C++ overloaded operators to Python operators, they might specify the following:

```
%rename(__add__) Complex::operator+(const Complex &);
%rename(__sub__) Complex::operator-(const Complex &);
%rename(__neg__) Complex::operator-();
...
class Complex {
...
  Complex operator+(const Complex &c) const;
  Complex operator-(const Complex &c) const;
  Complex operator-() const;
};
```

Similarly, if a program uses templates, information about specific template instantiations along with identifier names to use in the target language must be provided. For example:

```
template<typename T> T max(T a, T b) { return  a>b ? a : b; }
...
%template(maxint)     max<int>;
%template(maxdouble) max<double>;
```

3.3   Support for Legacy Software

Unlike wrapper generation tools designed strictly for object-oriented programming, SWIG provides full support for functions, global variables, constants and other features commonly associated with legacy software. In addition, SWIG is able to repackage procedural libraries into an object-based scripting API. For example, if an application used a procedural API like this,

```
typedef struct {
  double re, im;
} Complex;

Complex add_complex(Complex a, Complex b);
double   real_part(Complex a);
```

the following SWIG interface will create a class-like scripting interface:

```
%addmethods Complex {
  Complex(double r, double i) {
    Complex *c = (Complex *) malloc(sizeof(Complex));
    c->re = r;
    c->im = i;
    return c;
  }
  ~Complex() { free(self); }
  double real() { return real_part(*self); }
  Complex add(Complex b) { return add_complex(*self,b); }
};
```

In this case, the resulting scripting interface works like a class even though no changes were made to the underlying C code.

3.4 Customization Features

For advanced users, it is sometimes desirable to modify SWIG's code generator in order to provide customized integration between the scripting environment and the underlying application. For example, a user might want to interface their code with an array package such as Numeric Python [7]. By default, SWIG does not know how to perform this integration. However, a user can customize the code generator using a typemap. A typemap changes the way that SWIG converts data in wrapper functions. For instance, suppose an application had several functions like this:

```
void settemp(double *grid, int nx, int ny, double temp);
double avgtemp(double *grid, int nx, int ny);
void plottemp(double *grid, int nx, int ny, double mn, double mx);
```

Now suppose that a programmer wanted to pass a Numeric Python array as the grid parameter and associated nx and ny parameters. To do this, a typemap rule such as the following can be inserted into the SWIG interface file:

```
%typemap(in) (double *grid, int nx, int ny) {
  PyArrayObject *array;
  if (!PyArray_Check($input)) {
    PyErr_SetString(PyExc_TypeError,"Expected an array");
```

```
        return NULL;
    }
    array = (PyArrayObject *)
            PyArray_ContiguousFromObject(input, PyArray_DOUBLE, 2, 2);
    if (!array) {
        PyErr_SetString(PyExc_ValueError,
                "array must be two-dimensional and of type float");
        return NULL;
    }
    $1 = (double *) array->data;    /* Assign grid */
    $2 = array->dimensions[0];      /* Assign nx   */
    $3 = array->dimensions[1];      /* Assign ny   */
}
```

When defined, all subsequent occurrences of the argument sequence double
*grid, int nx, int ny are processed as a single numeric array object. Even
though the specification of a typemap requires detailed knowledge of the underlying scripting language API, these rules only need to be defined once in order
to be applied to hundreds of different declarations.

To modify the handling of declarations, the %feature directive is used. For
example, if a programmer wanted to catch a C++ exception in a specific class
method and turn it into a Python exception, they might write the following:

```
%feature("except") Object::getitem {
    try { $action } catch (IndexError) {
        PyErr_SetString(PyExc_IndexError, "bad index");
        return NULL;
    }
}
class Object {
    virtual Item *getitem(int index);
};
```

Although %feature is somewhat similar to a traditional compiler pragma,
it is a lot more powerful. For instance, when features are defined for a class
method as shown, that feature is propagated across an entire inheritance hierarchy. Therefore, if Object was a base class, the exception handler defined would
be applied to any occurrence of getitem found in derived classes. Feature declarations can also be parameterized with types. This allows them to be precisely
attached to specific methods even when those methods are overloaded. This behavior is illustrated by the %rename directive example in section 3.2 (which is
really just a special form of %feature in disguise).

4   SWIG Internals

One of SWIG's most powerful features is its highly extensible design. C/C++
parsing is implemented using an extended version of the C preprocessor and

a customized C++ parser. These components differ from a traditional implementation due to issues related to mixing special SWIG directives and C++ declarations in the same file. For instance, certain information from the preprocessor is used during code generation and certain syntactic features are parsed differently in order to make SWIG easier to use. SWIG also does not concern itself with parsing function bodies.

Internally, SWIG builds a complete parse tree and provides a traversal API similar to that in the XML-DOM specification. Nodes are built from hash tables that allow nodes to be annotated with arbitrary attributes at any stage of parsing and code generation [2]. This annotation of nodes is the primary mechanism used to implement most of SWIG's customization features.

To generate code, parse tree nodes are handled by a hierarchical sequence of handler functions that may elect to generate wrapper code directly or forward the node to another handler after applying a transformation. The behavior of each target language is defined by providing implementations of selected handler functions in C++ class. Minimally, a language module only needs to implement handlers for generating low-level function and variable wrappers. For example:

```
class MinimalLanguage: public Language {
public:
    void main(int argc, char *argv[]);
    int  top(Node *n);
    int  functionWrapper(Node *n);
    int  constantWrapper(Node *n);
    int  nativeWrapper(Node *n);
};
```

However, in order to provide more advanced wrapping of classes and structures, a language module will generally implement more handlers.

5   Limitations

SWIG is primarily designed to support software development in C and C++. The system can be used to wrap Fortran as long as the Fortran functions are described by C prototypes. SWIG is also not a full C++ compiler. Certain C++ features such as nested classes and namespaces aren't currently supported. Furthermore, features such as overloaded methods, operators, and templates may require the user to supply extra directives (as illustrated in earlier sections). More complicated wrapping problems arise due to C++ libraries that rely almost entirely on generic programming and templates such as the STL or Blitz++ [17, 18]. Although SWIG can be used to wrap programs that use these libraries, providing wrappers to the libraries themselves would be problematic.

6   Related Work

The problem of creating scripting language extension modules has been explored extensively in the scripting language community. Most scripting languages have tools that can assist in the creation of extension modules. However, few of these tools are designed to target multiple target languages. Scripting language extension building tools can sometimes be found in application frameworks such as Vtk [12]. However, these tend to be tailored to the specific features of the framework and tend to ignore programming features required for them to be more general purpose. A number of tools have been developed specifically for scientific applications. For example, pyfort and f2py provide Python wrappers for Fortran codes and the Boost Python Library provides an interesting alternative to SWIG for creating C++ class wrappers [1, 9, 15].

Work similar to SWIG can also be found in the meta-programming community. For example, the OpenC++ project aims to expose the internals of C++ programs so that tools can use that information for other tasks [6]. Using such information, it might be possible to generate scripting language wrappers in a manner similar to SWIG.

7   Conclusions and Future Work

SWIG is a compiler that simplifies the integration of scripting languages with scientific software. It is particularly well-suited for use with existing software and supports a wide variety of C++ language features. SWIG also promotes modular design by maintaining a clean separation between the scripting language interface and the underlying application code.

By using an automatic code generator such as SWIG, computational scientists will find that it is much easier to utilize scripting languages at all stages of program development. Furthermore, the use of a scripting language environment encourages modular design and allows scientists to more easily construct software that incorporates features such as integrated data analysis, visualization, database management, and networking.

Since its release in 1996, SWIG has been used in hundreds of software projects. Currently, the system supports eight different target languages including Guile, Java, Mzscheme, Perl, PHP, Python, Ruby, and Tcl. Future work is focused on improving the quality of code generation, providing support for more languages, and adding reliability features such as contracts and assertions. More information about SWIG is available at www.swig.org.

8   Acknowledgments

Many people have helped with SWIG development. Major contributors to the current implementation include William Fulton, Matthias Köppe, Lyle Johnson, Richard Palmer, Luigi Ballabio, Jason Stewart, Loic Dachary, Harco de Hilster, Thien-Thi Nguyen, Masaki Fukushima, Oleg Tolmatcev, Kevin Butler, John Buckman, Dominique Dumont, David Fletcher, and Gary Holt. SWIG was originally developed in the Theoretical Physics Division at Los Alamos National

Laboratory in collaboration with Peter Lomdahl, Tim Germann, and Brad Holian. Development is currently supported by the Department of Computer Science at the University of Chicago.

References

1. Abrahams, D.: The Boost Python Library. www.boost.org/libs/python/doc/.
2. Aho, A., Sethi, R., Ullman, J.: Compilers: Principles, Techniques, and Tools. Addison-Wesley, Reading, Massachusetts. (1986)
3. Beazley, D.: SWIG: An Easy to Use Tool for Integrating Scripting Languages with C and C++. In Proceedings of USENIX 4th Tcl/Tk Workshop. (1996) 129-139
4. Beazley, D., Lomdahl, P.: Message-Passing Multi-Cell Molecular Dynamics on the Connection Machine 5. Parallel Computing. **20** (1994) 173-195.
5. Beazley, D., Lomdahl, P.: Lightweight Computational Steering of Very Large Scale Molecular Dynamics Simulations. In Proceeding of Supercomputing'96, IEEE Computer Society. (1996).
6. Chiba, S.: A Metaobject Protocol for C++. In Proceedings of the ACM Conference on Object-Oriented Programming Systems, Languages, and Applications (OOPSLA). (1995) 285-299.
7. Dubois, P., Hinsen, K., Hugunin, J.: Numerical Python. Computers in Physics. **10(3)** (1996) 262-267.
8. Dubois, P.: The Future of Scientific Programming. Computers in Physics. **11(2)** (1997) 168-173.
9. Dubois, P.: Climate Data Analysis Software. In Proceedings of 8th International Python Conference. (2000).
10. Gathmann,F.: Python as a Discrete Event Simulation Environment. In Proceedings of the 7th International Python Conference. (1998).
11. Hinsen, K.: The Molecular Modeling Toolkit: A Case Study of a Large Scientific Application in Python. In Proceedings of the 6th International Python Conference. (1997) 29-35.
12. Martin, K.: Automated Wrapping of a C++ Class Library into Tcl. In Proceedings of USENIX 4th Tcl/Tk Workshop. (1996) 141-148.
13. Ousterhout, J.: Scripting: Higher-Level Programming for the 21st Century. IEEE Computer. **31(3)** (1998) 23-30.
14. Owen, M.: An Open-Source Project for Modeling Hydrodynamics in Astrophysical Systems. IEEE Computing in Science and Engineering. **3(6)** (2001) 54-59.
15. Peterson, P., Martins, J., Alonso, J.: Fortran to Python Interface Generator with an application to Aerospace Engineering. In Proceedings of 9th International Python Conference. (2000).
16. Scherer, D., Dubois, P., Sherwood, B.: VPython: 3D Interactive Scientific Graphics for Students. IEEE Computing in Science and Engineering. **2(5)** (2000) 56-62.
17. Stroustrup, B.: The C++ Programming Language, 3rd Ed. Addison-Wesley, Reading, Massachusetts. (1997).
18. Veldhuizen, T.: Arrays in Blitz++. In Proceedings of the 2nd International Scientific Computing in Object-Oriented Parallel Environments (ISCOPE'98), Springer-Verlag. (1998).
19. White, R., Greenfield, P.: Using Python to Modernize Astronomical Software. In Proceedings of the 8th International Python Conference. (1999).

Guard:
A Tool for Migrating Scientific Applications to the .NET Framework

David Abramson, Greg Watson, Le Phu Dung

School of Computer Science & Software Engineering,
Monash University,
CLAYTON, VIC 3800,
Australia

Abstract. For many years, Unix has been the platform of choice for the development and execution of large scientific programs. The new Microsoft .NET Framework represents a major advance over previous runtime environments available in Windows platforms, and offers a number of architectural features that would be of value in scientific programs. However, there are such major differences between Unix and .NET under Windows, that the effort of migrating software is substantial. Accordingly, unless tools are developed for supporting this process, software migration is unlikely to occur. In this paper we discuss a 'relative debugger' called Guard, which provides powerful support for debugging programs as they are ported from one platform to another. We describe a prototype implementation developed for Microsoft's Visual Studio.NET, a rich interactive environment that supports code development for the .NET Framework. The paper discusses the overall architecture of Guard under VS.NET, and highlights some of the technical challenges that were encountered.

1 Introduction

The new Microsoft .NET Framework is a major initiative that provides a uniform multi-lingual platform for software development. It is based on a Common Language Specification (CLS) that supports a wide range of programming languages and run time environments. Further, it integrates web services in a way that facilitates the development of flexible and powerful distributed applications. Clearly, this has applicability in the commercial domain of e-commerce and P2P networks, which rely on distributed applications.

On the other hand, software development for computational science and engineering has traditionally been performed under the UNIX operating system on high performance workstations. There are good reasons for this. FORTRAN, which is available on all of these systems, is the defacto standard programming language for scientific software. Not only is it efficient and expressive, but also a large amount of software has already developed using the language. Unix has been the operating

P.M.A. Sloot et al. (Eds.): ICCS 2002, LNCS 2330, pp. 834–843, 2002.

system of choice for scientific research because it is available on a wide variety of hardware platforms including most high performance engineering workstations. In recent times high performance PCs have also become a viable option for supporting scientific computation. The availability of efficient FORTRAN compilers and the Linux operating system have meant that the process of porting code to these machines has been fairly easy. Consequently there has been a large increase in the range of scientific software available on PC platforms in recent years. The rise of Beowulf clusters consisting of networks of tightly coupled PCs running Linux has driven this trend even faster.

An analysis of the features available in .NET suggests that the new architecture is equally applicable to scientific computing as commercial applications. In particular .NET provides efficient implementations of a wide range of programming languages, including FORTRAN [11], because it makes use of just-in-time compilation strategies. Further, the Visual Studio development environment is a rich platform for performing software engineering as it supports integrated code development, testing and debugging within one tool.

Some of the more advanced features of .NET, such as Web Services, could also have interesting application in scientific code. For example it would be possible to source libraries dynamically from the Web in the same way that systems like NetSolve [10] and NEOS [9] provide scientific services remotely. This functionality could potentially offer dramatic productivity gains for scientists and engineers, because they can focus on the task at hand without the need to develop all of the support libraries.

Unfortunately the differences between UNIX and Windows are substantial and as a result there is a significant impediment to porting codes from one environment to another. Not only are the operating systems different functionally, but the libraries and machine architectures may differ as well. It is well established that different implementations of a programming language and its libraries can cause the same program to behave erroneously. Because of this the task of moving code from one environment to another can be error prone and expensive. Many of these applications may also be used in mission critical situations like nuclear safety, aircraft design or medicine, so the cost of incorrect software can potentially be enormous. Unless software tools are developed that specifically help users in migrating software from UNIX to the Windows based .NET Framework, it is likely that most scientists will continue to use UNIX systems for their software development.

In this paper we describe a debugging tool called Guard, which specifically supports the process of porting codes from one language, operating system or platform to another. Guard has been available under UNIX for some time now, and we have proven its applicability for assisting the porting of programs many times. We have recently implemented a version of Guard that is integrated into the Microsoft Visual Studio.NET development environment. In addition, we have also demonstrated the ability to support cross-platform debugging between a UNIX platform and a Windows platform from the .NET environment. This has shown that the tool is not only useful for supporting software development on one platform, but also supports the porting of codes between Windows and UNIX. As discussed above, this is a critical issue if software is to be ported from UNIX to the new .NET Framework.

The paper begins with a discussion of the Guard debugger, followed by a description of the .NET Framework. We then describe the architecture of Guard as implemented under Visual Studio.NET, and illustrate its effectiveness in locating programming errors in this environment.

2   Guard – a Relative debugger

Relative debugging was first proposed by Abramson and Sosic in 1994. It is a powerful paradigm that enables a programmer to locate errors in programs by observing the divergence in key data structures as the programs are executing [2], [3], [4], [5], [6], [7], [8]. The relative debugging technique allows the programmer to make comparisons of a *suspect* program against a *reference* code. It is particularly valuable when a program is ported to, or rewritten for, another computer platform. Relative debugging is effective because the user can concentrate on *where* two related codes are producing different results, rather than being concerned with the actual values in the data structures. Various case studies reporting the results of using relative debugging have been published [2] [3], [5], [7], and these have demonstrated the efficiency and effectiveness of the technique. The concept of relative debugging is both language and machine independent. It allows a user to compare data structures without concern for the implementation, and thus attention can be focussed on the *cause* of the errors rather than implementation details.

To the user, a relative debugger appears as a traditional debugger, but also provides additional commands that allow data from different processes to be compared. The debugger is able to control more than one process at a time so that, once the processes are halted at breakpoints, data comparison can be performed. There are a number of methods of comparing data but the most powerful of these is facilitated by a user-supplied *declarative assertion*. Such an assertion consists of a combination of data structure names, process identifiers and breakpoint locations. Assertions are processed by the debugger before program execution commences and an internal graph [8] is built which describes when the two programs must pause, and which data structures are to be compared. In the following example:

```
assert $reference::Var1@1000 = $suspect::Var2@2000
```

the assert statement compares data from Var1 in $reference at line 1000 with Var2 in $suspect at line 2000. A user can formulate as many assertions as necessary and can refine them after the programs have begun execution. This makes it possible to locate an error by placing new assertions iteratively until the suspect region of code is small enough to inspect manually. This process is incredibly efficient. Even if the programs contain millions of lines of code, because the debugging process refines the suspect region in a binary fashion, it only takes a few iterations to reduce the region to a few lines of code.

Our implementation of relative debugging is embodied in a tool called Guard. We have produced implementations of Guard for many varieties of UNIX, in particular Linux, Solaris and AIX. A parallel variant is available for debugging applications on shared memory machines, distributed memory machines and clusters. Currently this is supported with UNIX System V shared memory primitives, the MPICH library, as well as the experimental data parallel language ZPL [12].

The UNIX versions of Guard are controlled by a command line interface that is similar in appearance to debuggers like GDB [16]. In this environment an assert statement such as the one above is typed into the debug interpreter, and must include the actual line numbers in the source as well as the correct spelling of the variables. As discussed later in the paper Guard is now integrated into the Microsoft Visual Studio environment and so is able to use the interactive nature of the user interface to make the process of defining assertions easier.

3   Success Stories

Over the last few years we have used Guard to debug a number of scientific codes that have been migrated from one platform to another or from one language to another (or both). In one case study we used Guard to isolate some discrepancies that occurred when a global climate model was ported from a vector architecture to a parallel machine [7]. This study illustrated that it is possible to locate subtle errors that are introduced when programs are parallelised. In this case both models were written in the same language, but the target architecture was so different that many changes were required in order to produce an efficient solution. Specifically, the mathematical formulation needed to be altered to reduce the amount of message passing in the parallel implementation, and other changes such as the order of the indexes on key array data structures needed to be made to account for a RISC architecture as opposed to a vector one.

In another case study we isolated errors that occurred when a photo-chemical pollution model was ported from one sequential workstation to another [3]. In this case the code was identical but the two machines produced different answers. The errors were finally attributed to the different behaviour of a key library function, which returned slightly divergent results on the two platforms.

In a more recent case study we isolated errors that occurred when a program was rewritten from C into another language, ZPL, for execution on a parallel platform [5]. This case study was interesting because even though the two codes were producing slightly different answers, the divergence was attributed to different floating point precision. However by using Guard it was possible to show that there were actually four independent coding errors – one in the original C code, and three in the new ZPL program.

All of these case studies have highlighted the power of relative debugging in the process of developing scientific code. We believe that many of the same issues will arise when migrating scientific software from UNIX to Windows under the new .NET Framework and that Guard will be able to play an important role in assisting this process.

4   The .NET Framework

The Microsoft .NET Framework represents a significant change to the underlying platform on which Windows applications run [1]. The .NET Framework defines a runtime environment that is common across all languages. This means that it is possible to write applications in a range of languages, from experimental research ones to standard production ones, with the expectation that similar levels of performance and efficiency will be achieved. An individual program can also be composed of modules that are written in different languages, but that interoperate seamlessly. All compilers that target the .NET environment generate code in an Intermediate Language (IL) that conforms to a Common Language Specification (CLS). The IL is in turn compiled into native code using a just-in-time compilation strategy. These features mean that the .NET Framework should provide an efficient platform for developing computational models.

The Web Services features of .NET also offer significant scope for scientific applications. At present most computational models are built as single monolithic codes that call library modules using a local procedure call. More recent developments such as the NetSolve and NEOS application servers have provided an exception to this strategy. These services provide complex functions such as matrix algebra and optimisation algorithms using calls to external servers. When an application uses NetSolve, it calls a local "stub" module that communicates with the NetSolve server to perform some computation. Parameters are sent via messages to the server, and results are returned the same way. The advantage of this approach is that application programmers can benefit by using 'state of the art' algorithms on external high performance computers, without the need to run the codes locally. Further, the load balancing features of the systems are able to allocate the work to servers that are most lightly loaded. The major drawback of external services like this is that the application must be able to access the required server and so network connectivity becomes a central point of failure. Also, building new server libraries is not easy and requires the construction of complex web hosted applications. The .NET Framework has simplified the task of building such servers using its Web Services technology. Application of Web Services to science and engineering programs is an area of interest that requires further examination.

Visual Studio.NET (VS.NET) is the preferred code development environment for the .NET Framework. The VS.NET environment represents a substantial change to previous versions of Visual Studio. Older versions of Visual Studio behaved differently depending on the language being supported – thus Visual Basic used a different set of technologies for building applications to Visual C++. The new VS.NET platform has been substantially re-engineered and as a consequence languages are now supported in a much more consistent manner.

VS.NET also differs from previous versions by exposing many key functions via a set of remote APIs known as 'automation'. This means that it is possible to write a third party package that interacts with VS.NET. For example, an external application can set breakpoints in a program and start the execution without user interaction. A separate Software Development Kit (SDK) called VSIP makes it possible to embed new functions directly into the environment. This allows a programmer to augment VS.NET with new functionality that is consistent with other functions that are already available. This feature has allowed us to integrate a version of Guard with Visual Studio as discussed in the next section.

5   Architecture of Guard

Fig. 1 shows a simplified schematic view of the architecture of Guard under Visual Studio.NET. VS.NET is built around a core 'shell' with functionality being provided by commands that are implemented by a set of 'packages' These packages are conventional COM objects that are activated as a result of user interaction (such as menu selection) within VS.NET, and also when various asynchronous events occur. This component architecture makes it possible to integrate new functionality into the environment by loading additional packages.

Debugging within the VS.NET environment is supported by three main components. The Debugger package provides the traditional user interface commands such as 'Go', 'Step', 'Set Breakpoint', etc. that appear in the user interface. This module communicates with the Session Debug Manager, which in turn provides a multiplexed interface into one or more per-process Debug Engines. The Debug Engines implement low-level debug functions such as starting and stopping a process and setting breakpoints, and providing access to the state of the process. Debug Engines can cause events to occur in response to conditions such as a breakpoint being reached, and these are passed back through the Session Debug Manager to registered event handlers. Each Debug Engine is responsible for controlling the execution of a single process. However the VS.NET architecture supports the concept of remote debugging, so this process may be running on a remote Windows system.

The VS.NET implementation of Guard consists of three main components. A package is loaded into the VS.NET shell that incorporates logic to respond to specific menu selections and handle debugger events. This package executes in the main thread of the shell, and therefore has had to be designed to avoid blocking for any extended time period. The main relative debugging logic is built into a local COM component called the Guard Controller. This is a separate process that provides a user interface for managing assertions and a dataflow interpreter that is necessary to implement the relative debugging process. Because the Guard Controller runs as a separate process it does not affect the response of the main VS.NET thread. The Guard Controller controls the programs being debugged using the VS.NET Automation Interface, a public API that provides functions like set-breakpoint, evaluate expression, etc. We have also built a Debug Engine that is able to control a process running on an external UNIX platform. This works by communicating with the remote debug server developed for the original UNIX version of Guard. The UNIX debug server is based on the GNU GDB debugger and is available for most variants of UNIX. We have modified GDB to provide support for an Architecture Independent Form (AIF) [5] for data structures, which means it is possible to move data between machines with different architectural characteristics, such as word size, endian'ness, etc. AIF also facilitates machine independent comparison of data structures. It is the addition of this Debug Engine that allows us to compare programs executing on Windows and UNIX platforms.

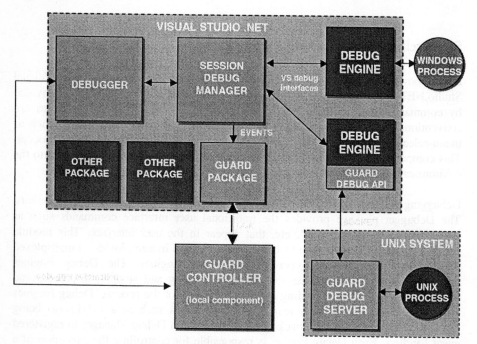

Fig. 1. Architecture of Guard for VS.NET

Fig. 2 shows the Guard control panel when running under VS.NET. When a user wishes to compare two running programs they must first be loaded into a VS.NET "solution" as separate "projects". The solution is then configured to start both programs running at the same time under the control of individual Debug Engines. The source windows of each project can then be tiled to allow both to be displayed at once.

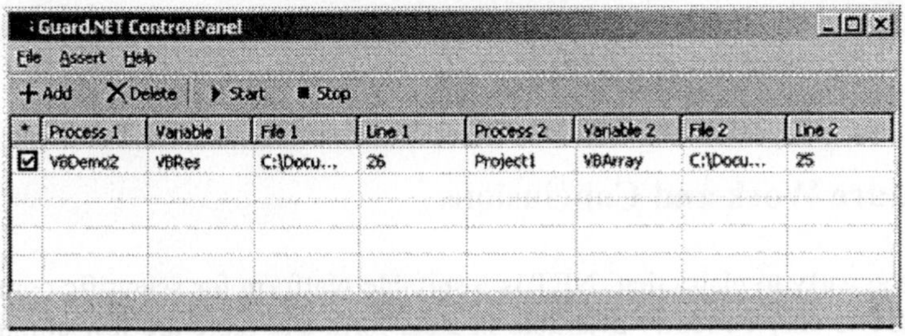

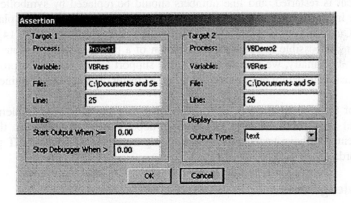

Fig. 2. Guard control panel

A user creates an assertion between the two programs using the Guard Controller, which is started by selecting the "GUARD" item from the "Tools" menu. The Guard Controller has a separate window as shown. An assertion is created in a few simple steps. A new, empty, assertion is created by selecting the "Add" button. Guard displays the dialog box shown in Fig. 3, which allows the user to enter the information necessary to create an assertion. The left hand side of the assertion can be automatically populated with the variable name, line number, source file and program information by selecting the required variable in the appropriate source window and then using a single right-mouse click. The right hand side of the assertion can be filled in using the same technique in the other source window. Finally the user is able to specify properties about the assertion such as the error value at which output is generated, when the debugger should be stopped and the type of output to display. The user can create any number of assertions by repeating this process and then launch the programs using the "Start" button.

Fig. 3. New assertion dialog

Before commencing execution Guard automatically sets breakpoints at the locations in the source files specified by the assertions. During execution Guard will extract the contents of a variable when its corresponding breakpoint is reached and then perform a comparison once data from each half of the assertion has been obtained. Once the appropriate error threshold has been reached (as specified in each assertion), Guard will either display the results in a separate window or stop the debugger to allow interactive examination of the programs' state. Guard currently supports a number of display types include text, bitmaps and the ability to export data into a visualisation package.

6   Future Work and Conclusions

It is far too early to claim that .NET is a suitable platform for scientific computation since .NET is only currently available in Beta form and there are few commercially oriented codes available now, let alone scientific ones. As discussed in the introduction, we believe that .NET offers a number of potential benefits for large numeric models. However, the execution environment is very different from a traditional UNIX platform and so it is critical that as many tools as possible are available to facilitate the transition of existing legacy software. Guard is one such tool because it allows a user to compare two executing programs on two different platforms.

Whilst the implementation of Guard under UNIX is mature and has been used on many case studies, the current version under Visual Studio.NET is still a research prototype. We are planning a number of extensions that will be required if Guard is to be of practical use in supporting migration to .NET. The current user interface is fairly simple and must be made more powerful if it is to be applied to large programs. At present only simple data types and arrays are supported. We need to extend this to encompass the range of types found in scientific codes, such as structures and complex numbers. Assertions need to be able to be saved and restored when the environment is restarted, and line numbers should be replaced by symbolic markers which are independent of the actual numeric line number. We are also planning to integrate Guard into Source Safe [13], Microsoft's equivalent of SCCS [14] or RCS [15] making it possible to compare one version of a program with previous versions automatically. Finally, we plan to enhance the support for multi-process programs to make it feasible to debug programs running on a cluster of Windows machines.

In spite of these shortcomings we have shown that the prototype implementation works and that the technique of relative debugging is feasible in the .NET environment. We have tested this by debugging a number of small .NET programs using Guard.

Acknowledgments

This work has been funded by grants from the Australian Research Council and Microsoft Corporation. We wish to acknowledge the support of a number of individuals at Microsoft for their assistance on various issues related to Visual Studio and .NET. Particular thanks go to Todd Needham, Dan Fay and Frank Gocinski. We also wish to acknowledge our colleagues, A/Professor Christine Mingins, Professor Bertrand Meyer and Dr Damien Watkins for many helpful discussions.

References

1. Meyer, B. ".NET is coming", IEEE Computer, Vol. 34, No. 8; AUGUST 2001, pp. 92-97.
2. Abramson, D.A. and Sosic, R. "A Debugging and Testing Tool for Supporting Software Evolution", Journal of Automated Software Engineering, 3 (1996), pp 369 - 390.
3. Abramson D., Foster, I., Michalakes, J. and Sosic R., "Relative Debugging: A new paradigm for debugging scientific applications", the Communications of the Association for Computing Machinery (CACM), Vol 39, No 11, pp 67 - 77, Nov 1996.
4. Sosic, R. and Abramson, D. A. "Guard: A Relative Debugger", Software Practice and Experience, Vol 27(2), pp 185 – 206 (Feb 1997).
5. Watson, G. and Abramson, D. "Relative Debugging For Data Parallel Programs: A ZPL Case Study", IEEE Concurrency, Vol 8, No 4, October 2000, pp 42 – 52.
6. Abramson, D.A. and Sosic, R. "A Debugging Tool for Software Evolution", CASE-95, 7th International Workshop on Computer-Aided Software Engineering, Toronto, Ontario, Canada, July 1995, pp 206 - 214. Also appeared in proceedings of 2nd Working Conference on Reverse Engineering, Toronto, Ontario, Canada, July 1995.
7. Abramson D., Foster, I., Michalakes, J. and Sosic R., "Relative Debugging and its Application to the Development of Large Numerical Models", Proceedings of IEEE Supercomputing 1995, San Diego, December 95.
8. Abramson, D.A., Sosic, R. and Watson, G. "Implementation Techniques for a Parallel Relative Debugger ", International Conference on Parallel Architectures and Compilation Techniques - PACT '96, October 20-23, 1996, Boston, Massachusetts, USA
9. Czyzyk, J, Owen, J. and Wright, S. "Optimization on the Internet", OR/MS Today, October 1997.
10. Casanova, H. and Dongarra, J. "NetSolve: A Network Server for Solving Computational Science Problems", The International Journal of Supercomputing Applications and High Performance Computing, Vol 11, Number 3, pp 212-223, 1997.
11. http://www.lahey.com/netwtpr1.htm
12. L. Snyder, A Programmer's Guide to ZPL, MIT Press, Cambridge, Mass., 1999.
13. http://msdn.microsoft.com/ssafe/
14. Programming Utilities and Libraries', Sun Release 4.1, Sun Microsystems, 1988.
15. Walter F. Tichy, "RCS-A System for Version Control", Software-Practice & Experience 15, 7 (July 1985), 637-654.
16. Stallman, R. Debugging with GDB – The GNU Source Level Debugger, Edition 4.12, Free Software Foundation, January 1994.

Lithium: A Structured Parallel Programming Environment in Java

M. Danelutto & P. Teti

Dept. Computer Science – University of Pisa – Italy
{Marco.Danelutto@di.unipi.it, tetipaol@libero.it}

Abstract. We describe a new, Java based, structured parallel programming environment. The environment provides the programmer with the ability to structure his parallel applications by using skeletons, and to execute the parallel skeleton code on a workstation network/cluster in a seamless way. The implementation is based on macro data flow and exploits original optimization rules to achieve high performance. The whole environment is available as an Open Source Java library and runs on top of plain JDK.

1 Introduction

The Java programming environment includes features that can be naturally used to address network and distributed computing (JVM and bytecode, multithreading, remote method invocation, socket and security handling, and, more recently, JINI, JavaSpaces, Servlets, etc. [20]). Many efforts have been performed to make Java suitable for parallel computing too. Several projects have been started that aim at providing features that can be used to develop efficient parallel Java applications on a range of different parallel architectures. Such features are either provided as extensions to the base language or as class libraries. In the former case, ad hoc compilers and/or runtime environments have been developed and implemented. In the latter, libraries are supplied that the programmer simply uses within his parallel code. As an example extensions of the JVM have been designed that allow plain Java threads to be run in a seamless way on the different processors of a single SMP machine [2]. On the other side, libraries have been developed that allow classical parallel programming libraries/APIs (such as MPI or PVM) to be used within Java programs [15, 12].

In this work we discuss a new Java parallel programming environment which is different from the environments briefly discussed above, namely a library we developed to support *structured* parallel programming, based on the *algorithmical skeleton* concept. Skeletons have been originally conceived by Cole [5] and then used by different research groups to design high performance structured parallel programming environments [3, 4, 19]. A skeleton is basically an abstraction modeling a common, reusable parallelism exploitation pattern. Skeletons can be provided to the programmer either as language constructs [3, 4] or as libraries [9, 10]. They can be nested to structure complex parallel applications.

P.M.A. Sloot et al. (Eds.): ICCS 2002, LNCS 2330, pp. 844–853, 2002.

```
import lithium.*;
...
public class SkeletonApplication {
  public static void main(String [] args) {
    ...
    Worker w = new Worker();        // encapsulate seq code in a skel
    Farm f = new Farm(w);           // use it as the task farm worker
    Ske evaluator = new Ske();      // declare an exec manager
    evaluator.setProgram(f);        // set the program to be executed
    String [] hosts = {"alpha1","alpha2",
                       "131.119.5.91"};
    evaluator.addHosts(hosts);      // define the machines to be used
    for(int i=0;i<ntasks;i++)
      evaluator.setupTaskPool(task[i]); // prepare input stream
          // this can be done in parallel with parDo() actually
    evaluator.stopStream();             // declare its end
    evaluator.parDo();              // require parallel computation
    while(!evaluator.isResEmpty()) {    // retrieve comput. results
      Object res = evaluator.readTaskPool();
      ...
    }                               // print some statistics
    System.out.println("elapsed time = "+evaluator.getElapsedTime()+
                       "\nstartup = "+evaluator.getStartupTime());
  }
}

public class Worker extends JSkeleton {
  ...
  public Object run(Object task) {  // this method must be implemented
    Object result;                  // it represents the seq skel body
    ... return(result);             // computes an Object res out of
  }                                 // an Object input task
}
```

Fig. 1. Sample Lithium code: parallel application exploiting task farm parallelism

The compiling tools of the skeleton language or the skeleton libraries take care of automatically deriving/executing actual, efficient parallel code out of the skeleton application without any direct programmer intervention [17, 10].

Lithium, the library we discuss in this work, represents a consistent refinement and development of a former work [7]. The library discussed in [7] just provided the Java programmer with the possibility to implement simple parallel applications exploiting task farm parallelism only. Instead, Lithium:

- provides a reasonable set of fully nestable skeletons, including skeletons that model both data and task parallelism;
- implements the skeletons by fully exploiting a macro data flow execution model [8];
- exploits Java RMI to automatically perform parallel skeleton code execution;
- exploits basic Java reflection features to simplify the skeleton API provided to the programmer;
- allows parallel skeleton programs to be executed sequentially on a single machine, to allow functional debugging to be performed in a simple way.

2   Lithium API

Lithium provides the programmer with a set of (parallel) skeletons that include a
Farm skeleton, modeling task farm computations[1], a Pipeline skeleton, modeling computations structured in independent stages, a Loop and a While skeleton,
modeling determinate and indeterminate iterative computations, an If skeleton,
modeling conditional computations, a Map skeleton, modeling data parallel computations with independent subtasks and a DivideConquer skeleton, modeling
divide and conquer computations. All the skeletons are provided as subclasses
of a JSkeleton abstract class.

All skeletons use other skeletons as parameters. As an example, the Farm
skeleton requires as a parameter another skeleton defining the worker computation, and the Pipeline skeleton requires a set of other skeleton parameters, each
one defining the computation performed by one of the pipeline stages. Lithium
user may encapsulate sequential portions of code in a sequential skeleton by creating a JSkeleton subclass[2]. Objects of the subclass can be used as parameters
of other, different skeletons.

All the Lithium parallel skeletons implement parallel computation patterns
that process a stream of input tasks to compute a stream of output results. As
an example, a farm having a worker that computes the function f processes an
input task stream with generic element x_i producing the output stream with the
corresponding generic element equal to $f(x_i)$, whereas a pipeline with two stages
computing function f and g, respectively, processes stream of x_i computing
$g(f(x_i))$.

In order to write parallel applications using Lithium skeletons, the programmer should perform the following, (simple) steps:

1. define the skeleton structure of the application. This is accomplished by
   defining the sequential portions of code used in the skeleton code as JSkeleton
   objects and then using these objects as the parameters of the parallel skeletons (Pipeline, Farm, etc.) actually used to model the parallel behavior of
   the application at hand;
2. declare an *evaluator* object (a **Ske** object) and define the program, i.e. the
   skeleton code defined in the previous step, to be executed by the evaluator
   as well as the list of hosts to be used to run the parallel code;
3. setup a task pool hosting the initial tasks, i.e. a data structure storing the
   data items belonging to the input stream to be processed by the program;
4. start the parallel computation, by just issuing an evaluator parDo() method
   call;
5. retrieve the final results, i.e. the stream of output data computed by the
   program, from a result pool (again, issuing a proper evaluator method call).

Figure 1 outlines the code needed to setup a task farm parallel application
processing a stream of input tasks by computing, on each task, the sequential

---

[1] also known as "embarrassingly parallel" computations

[2] a JSkeleton object is an object having a Object run(Object) method that represents the sequential skeleton body

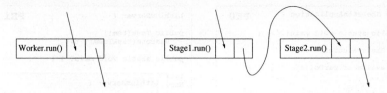

Fig. 2. Macro data flow graphs related to program of Figure 1

code defined in the `Worker run` method. The application runs on three processors (the `hosts` ones). The programmer is not required to write any (remote) process setup code, nor any communication, synchronization and scheduling code. He simply issues an `evaluator.parDo()` call and the library automatically computes the `evaluator` program in parallel by forking suitable remote computations on the remote nodes. In case the user simply wants to execute the application sequentially (i.e. to functionally debug the sequential code), he can avoid to issue all the `Ske` evaluator calls. After the calls needed to build the `JSkeleton` program he can simply issue a `run()` method call on the `JSkeleton` object. In that case, the Lithium support performs a completely sequential computation returning the results that the parallel application would return.

All the skeletons defined in Lithium can be defined and used with API calls similar to the ones shown in the Figure (see [21] or look at the source code available at [22]). We want to point out that a very small effort is needed to change the parallel structure of the application, provided that the suitable sequential portions of code needed to instantiate the skeletons are available. In case we understand that the computation performed by the farm workers of Figure 1 can be better expressed with a pipeline of two sequential stages (as an example), we can simply substitute the lines `Worker w = new Worker();` and `Farm f = new Farm(w);` with the lines:

```
Stage1 s1 = new Stage1();     // first seq stage
Stage2 s2 = new Stage2();     // second seq stage
Pipeline p = new Pipeline(); // create the pipeline
p.addWorker(s1);              // setup first pipeline stage
p.addWorker(s2);              // setup second pipeline stage
Farm f = new Farm(p); // create a farm with pipeline workers
```

and we get a perfectly running parallel program computing the results according to a farm of pipeline parallelism exploitation pattern.

3   Lithium implementation

Lithium exploits a macro data flow (MDF, for short) implementation schema for skeletons. The skeleton program is processed to obtain a MDF graph. MDF instructions (MDFi) in the graph represent sequential `JSkeleton run` methods.

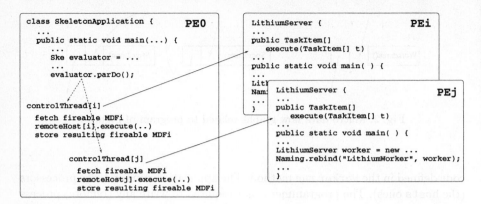

Fig. 3. Lithium architecture

The data flow (i.e. the arcs of MDF graph) is derived by looking at the skeleton nesting structure [6, 8]. The resulting MDF graphs have a single MDFi getting input task (tokens) from the input stream and a single MDFi delivering data items (tokens) to the output stream. As an example, from the application of Figure 1 we derive the MDF graphs of Figure 2: the left one is the one derived from the original application, the right one is the one relative to application with pipelined workers.

The skeleton program is then executed by setting up a server process on each one of the processing elements available and a task pool manager on the local machine. The remote servers are able to compute any one of the fireable MDFi in the graph. A MDF graph can be sent to the servers in such a way that they get specialized to execute only the MDFi in that graph. The local task pool manager takes care of providing a MDFi repository (the *taskpool*) hosting fireable MDFi relative to the MDF graph at hand, and to feed the remote servers with fireable MDFi to be executed.

Logically, any available input task makes a new MDF graph to be instantiated and stored into the taskpool. Then, the input task is transformed into a data flow "token" and dispatched to the proper instruction (the first one) in the new copy of the MDF graph[3]. The instruction becomes fireable and it can be dispatched to one of the remote servers for execution. The remote server computes the MDFi and delivers the result token to one or more MDFi in the taskpool. Such MDFi may (in turn) become fireable and the process is repeated until some fireable MDFi exists in the task pool. Final MDFi (i.e. those dispatching final result tokens/data to the external world) are detected an removed from the taskpool upon `evaluator.readTaskPool()` calls.

Actually, only fireable MDFi are stored in the taskpool. The remote servers know the executing MDF graph and generate fireable complete MDFi to be stored in the taskpool rather than MDF tokens to be stored in already existing, non fireable, MDFi.

---

[3] different instances of MDF graph are distinguished by a progressive task identifier

Remote servers are implemented as Java RMI servers. A remote server implements a `LithiumInterface`. The interface defines three methods: a `String getVersion()` method, used to check compatibility between local task pool manager and remote servers[4], a `TaskItem[] execute(TaskItem[] task)` method, actually computing a fireable MDFi, and a `void setRemoteWorker(Vector SkeletonList)` method, used to specialize the remote server with the MDF graph currently being executed[5]. RMI implementation has been claimed to demonstrate poor efficiency in the past [16] but recent improvements in JDK allowed us to achieve good efficiency and absolute performance in the execution of skeleton programs, as shown in Section 4. Remote RMI servers must be set up either by hand (via some `ssh hostname rmiregistry &` plus a `ssh hostname java Server &`) or by using proper Perl scripts provided by the Lithium environment.

In the local task pool manager a thread is forked for each one of the remote servers displaced on the remote hosts. Such thread obtains a local reference to a remote RMI server, first; then issues a `setRemoteWorker` remote method call to communicate to the server the MDF graph currently being executed and eventually enters a loop. In the loop body the thread fetches a fireable instruction from the taskpool[6], asks the remote server to compute the MDFi by issuing a remote `execute` method call and deposits the result in the task pool (see Figure 3).

The MDF graph obtained from the `JSkeleton` object used in the `evaluator.setProgram()` call can be processed unchanged or a set of optimization rules can be used to transform the MDF graph (using the `setOptimizations()` and `resetOptimizations()` methods of the `Ske` evaluator class). Such optimization rules implement the "normal form" concept for skeleton trees and basically substitute skeleton subtrees by skeleton subtrees showing a better performance and efficiency in the target machine resource usage. Previous results demonstrated that full stream parallel skeleton subtrees can be collapsed to a single farm skeleton with a (possibly huge) sequential worker leading to a service time which is equal or even better that the service time of the uncollapsed skeleton tree [1]. While developing Lithium we also demonstrated that 1) data parallel skeletons with stream parallel only subtrees can be collapsed to dataparallel skeletons with fully sequential workers, and 2) that normal form skeleton trees require a number of processing elements which is not greater that the number of processing elements needed to execute the corresponding non normal form [21].

As the skeleton program is provided by the programmer as a single (possibly nested) `JSkeleton` object, Java reflection features are used to derive the MDF graph out of it. In particular, reflection and `instanceOf` operators are used to understand the type of the skeleton (as well as the type of the nested

---

[4] remote server can be run as daemons, therefore they can survive to changes in the local task pool managers

[5] therefore allowing the server to be run as daemon, serving the execution of different programs at different times

[6] using proper `TaskPool synchronized` methods

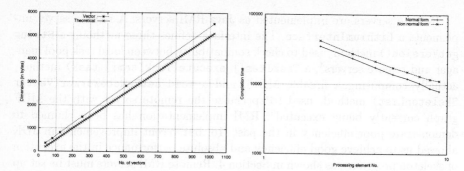

Fig. 4. Serialization overhead (left) and Normal vs. non normal form (right)

skeletons). Furthermore, an `Object[] getSkeletonInfo private` method of the `JSkeleton` abstract class is used to gather the skeleton parameters (e.g. its "body" skeleton). Such method is implemented as a simple `return(null)` statement in the `JSkeleton` abstract class and it is overwritten by each subclass (i.e. by the classes `Farm`, `Pipeline`, etc.) in such a way that it returns in an `Object` vector all the relevant skeleton parameters. These parameters can therefore be inspected by the code building the MDF graph. Without reflection much more info must be supplied by the programmer when defining skeleton nestings in the application code [10].

4   Experiments

We evaluated Lithium performance by performing a full set of experiments on a Beowulf class Linux cluster operated at our Department[7]. The cluster hosts 17 nodes: one node devoted to cluster administration, code development and user interface, and 16 nodes (10 266Mhz Pentium II and 6 400Mhz Celeron nodes) exclusively devoted to parallel program execution. The nodes are interconnected by a (private, dedicated) switched Fast Ethernet network. All the experiments have been performed using Blackdown JDK ports version 1.2.2 and 1.3.

We start considering the overhead introduced by serialization. As data flow tokens happen to be serialized in order to be dispatched to remote executor processes, and as we use Java `Vector` objects to hold tokens, we measured the size overhead of the `Vector` class. Figure 4 (left) reports the results we got, showing that serialization does not add significant amounts of data to the real user data and therefore serialization does not cause significant additional communication overhead.

We measured the differences in the completion time of different applications executed using normal and non normal form. As expected normal form always

---

[7] the Backus cluster has been implemented in the framework of the Italian National Research Council Mosaico Project

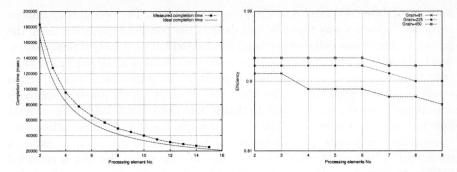

Fig. 5. Ideal vs. measured completion time (left) and Efficiency vs. grain size (right)

performs better that non normal form (see Figure 4 right). The good news are that it performs significantly better and *scales* better (see the behavior when the PE number increases in Fig. 4 right).

Last but not least, we measured the Lithium applications absolute completion time and efficiency related to the computational grain of MDFi. Typical results are drawn in Figure 5. The left plot shows that that Lithium support scales (at least in case of medium to coarse grain computations)[8]. The right plot shows that fairly large grain is required in order to achieve good performance values[9].

All the experiments have been performed using "syntetic" applications. These applications stress different features of the Lithium environment and use skeleton trees (nestings) up to three levels deep, including both task and data parallel skeletons. In addition, we used a couple of simple number crunching applications including Mandelbrot set computation. In all cases the results are similar to the ones presented in the graphs of this Section.

5   Related work

Despite the large number of projects aimed at providing parallel programming environments based on Java, there is no existing project concerning skeletons but the CO_2P_3S one [14, 13]. Actually this project derives from the design pattern experience [11]. The user is provided with a graphic interface where he can combine different, predefined parallel computation patterns in order to design structured parallel applications that can be run on any parallel/distributed Java platform. In addition, the graphic interface can be used to enter the sequential

---

[8] the completion times show an additional decrement from 10 nodes on, as the 11th to 16th nodes are more powerful that the first 10 nodes and therefore take a shorter time to execute sequential portions of Java code.

[9] *grain* represents the average computational grain of MDFi. *grain* = *k* means that the time spent in the computation of MDFi is *k* times the time spent in delivering such instructions to the remote servers plus the time spent in gathering results of MDFi execution from the remote servers

portions of Java code needed to complete the patterns. The overall environment is layered in such a way that the user designs the parallel application using the patterns, then those patterns are implemented exploiting a layered implementation framework. The framework gradually exposes features of the implementation code thus allowing the programmer to perform fine performance tuning of the resulting parallel application. The whole object adopt a quite different approach with respect to our one, especially in that it does not use any kind of macro data flow technique in the implementation framework. Instead, parallel patterns are implemented by process network templates directly coded in the implementation framework. However, the final result is basically the same: the user is provided with an high level parallel programming environment that can be used to derive high performance parallel Java code running on parallel/distributed machines.

Macro data flow implementation techniques, instead, have been used to implement skeleton based parallel programming environments by Serot in the Skipper project [19, 18]. Skipper is an environment supporting skeleton based, parallel image processing application development. The techniques used to implement Skipper are derived from the same results we start with to design Lithium, although used within a different programming enviroment.

6   Conclusions and future work

We described a new Java parallel programming environment providing the user with the possibility to model all the parallel behavior of his applications by using predefined skeletons. This work significantly extends [7] as both data and stream parallel skeletons are implemented, and optimisation rules are provided that improve execution efficiency. Being based on skeletons, the Lithium environment relieves the programmer of all the error prone activities related to process setup, mapping and scheduling, communication and synchronization handling, etc. that must usually be dealt with when programming parallel applications. Lithium is the first full fledged, skeleton based parallel programming environment written in Java and implementing skeleton parallel execution by using macro data flow techniques. We performed experiments with Lithium that demonstrate that good scalability and efficiency values can be achieved. Lithium and is currently available as open source at http://massivejava.sourceforge.net.

References

1. M. Aldinucci and M. Danelutto. Stream parallel skeleton optimisations. In *Proc. of the IASTED International Conference Parallel and Distributed Computing and Systems*, pages 955–962. IASTED/ACTA Press, November 1999. Boston, USA.
2. G. Antoniu, L. Bougé, P. Hatcher, M. MacBeth, K. McGuigan, and R. Namyst. "Compiling Multithreaded Java Bytecode for Distributed Execution". In A. Bode, T. Ludwig, W. Karl, and R. Wismuller, editors, *"EuroPar 2000 - Parallel Processing"*, number 1900 in LNCS, pages 1039–1052. Springer Verlag, 2000.

3. P. Au, J. Darlington, M. Ghanem, Y. Guo, H.W. To, and J. Yang. Co-ordinating heterogeneous parallel computation. In L. Bouge, P. Fraigniaud, A. Mignotte, and Y. Robert, editors, *Europar '96*, pages 601–614. Springer-Verlag, 1996.
4. B. Bacci, M. Danelutto, S. Pelagatti, and M. Vanneschi. SkIE: a heterogeneous environment for HPC applications. *Parallel Computing*, 25:1827–1852, Dec 1999.
5. M. Cole. *Algorithmic Skeletons: Structured Management of Parallel Computations*. Research Monographs in Parallel and Distributed Computing. Pitman, 1989.
6. M. Danelutto. Dynamic Run Time Support for Skeletons. In E. H. D'Hollander, G. R. Joubert, F. J. Peters, and H. J. Sips, editors, *Proceedings of the International Conference ParCo99*, volume Parallel Computing Fundamentals & Applications, pages 460–467. Imperial College Press, 1999.
7. M. Danelutto. Task farm computations in java. In Buback, Afsarmanesh, Williams, and Hertzberger, editors, *High Performance Computing and Networking*, LNCS, No. 1823, pages 385–394. Springer Verlag, May 2000.
8. M. Danelutto. Efficient support for skeletons on workstation clusters. *Parallel Processing Letters*, 11(1):41–56, 2001.
9. M. Danelutto, R. Di Cosmo, X. Leroy, and S. Pelagatti. Parallel Functional Programming with Skeletons: the OCAMLP3L experiment. In *ACM Sigplan Workshop on ML*, pages 31–39, 1998.
10. M. Danelutto and M. Stigliani. SKElib: parallel programming with skeletons in C. In A. Bode, T. Ludwing, W. Karl, and R. Wismüller, editors, *Euro-Par 2000 Parallel Processing*, LNCS, No. 1900, pages 1175–1184. Springer Verlag, August/September 2000.
11. E. Gamma, R. Helm, R. Johnson, and J. Vissides. *Design Patterns: Elements of Reusable Object-Oriented Software*. Addison Wesley, 1994.
12. jPVM. "http://www.chmsr.gatech.edu/jPVM/". The jPVM home page, 2001.
13. S. McDonald, D. Szafron, J. Schaeffer, and S. Bromling. From Patterns to Frameworks to Parallel Programs. submitted to Journal of Parallel and Distributed Computing, December 2000.
14. S. McDonald, D. Szafron, J. Schaeffer, and S. Bromling. Generating Parallel Program Frameworks from Parallel Design Patterns. In A. Bode, T. Ludwing, W. Karl, and R. Wismüller, editors, *Euro-Par 2000 Parallel Processing*, LNCS, No. 1900, pages 95–105. Springer Verlag, August/September 2000.
15. MpiJava. "http://www.npac.syr.edu/projects/pcrc/mpiJava/". The MpiJava home page, 2001.
16. C. Nester, R. Philippsen, and B. Haumacher. "A More Efficient RMI for Java". In *ACM 1999 Java Grande Conference*, pages 152–157, June 1999.
17. S. Pelagatti. *Structured Development of Parallel Programs*. Taylor & Francis, 1998.
18. J. Serot. "Putting skeletons at work. An overview of the SKIPPER project". PARCO'2001 workshop on *Advanced Environments for Parallel and Distributed Computing*, to appear, September 2001.
19. J. Serot, D. Ginhac, R. Chapuis, and J. Derutin. "Fast prototyping of parallel-vision applications using functional skeletons". *Machine Vision and Applications*, 12:217–290, 2001. Springer Verlag.
20. Sun. "The Java home page". http://java.sun.com, 2001.
21. P. Teti. "Lithium: a Java skeleton environment". (*in italian*) Master's thesis, Dept. Computer Science, University of Pisa, October 2001.
22. P. Teti. "http://massivejava.sourceforge.net". home page of the Lithium project at sourceforge.net, 2001.

Using the **TrustME** Tool Suite for Automatic Component Protocol Adaptation

Ralf Reussner[1], Iman Poernomo[1], and Heinz W. Schmidt[2]

[1] Distributed Systems Technology Center (DSTC) Pty Ltd
Monash University, Melbourne, Australia
{reussner|imanp}@dstc.com
[2] Center for Distributed Systems and Software Engineering (DSSE) ,
Monash University, Melbourne, Australia
hws@csse.monash.edu.au

Abstract. The deployment of component oriented software approaches gains increasing importance in the computational sciences. Not only the promised increase of reuse makes components attractive, but also the possibilities of integrating different stand-alone programs into a distributed application. Middleware platforms facilitate the development of distributed applications by providing services and infrastructure. Component developers can thus benefit from a common standard to shape components towards and application designers from using pre-fabricated software components and shared platform services. Although such platforms claim to achieve fast and flexible development of distributed systems, they fall short in key requirements to reliability and interoperability in loosely coupled distributed systems. For example, many interoperability errors remain undetected during development and the adaptation and integration of of third-party components still requires major effort and cost. Partly this problem can be alleviated by the use of formal approaches to automatic interoperability checks and component adaptation. Our Reliable Architecture Description Language (RADL) is aimed at precisely this problem. In this paper we present key aspects of RADL used to specify component-based, compositional views of distributed applications. RADL involves a rich component model, enabling protocol information to be contained in interfaces. We focus on protocol-based notions of interoperability and adaptation, important for the construction of distributed systems with loosely coupled components.
Keywords: component protocol specifications, automatic component adaptation, parameterised contracts, architectural description languages, distributed middleware platforms.

1 Introduction

The deployment of component oriented software approaches gains increasing importance in the computational sciences. Not only the promised increase of reuse makes components attractive, but also the possibilities of integrating different stand-alone programs into a distributed application. Distributed systems are often constructed using middleware platforms. For example, enterprise application

P.M.A. Sloot et al. (Eds.): ICCS 2002, LNCS 2330, pp. 854–863, 2002.

development is increasingly aided by implementations of OMG's CORBA [16] and others, while scientific distributed application development is often based upon PVM or its successor HARNESS [13]. These middleware platforms can facilitate distributed programming, through the provision of services and infrastructure such as distributed name lookup, remote procedure calls, and transactions. In general, middleware supports these features through an underlying component interface model, in which a distributed architecture is decomposed into black-box software units, whose individual functionality is provided by interfaces consisting of method signatures that are understood by means of an informal API [20].

However, errors in design and deployment of distributed middleware-based systems can still arise. This can be partly alleviated by the use of formal, tool-based approaches to configuration of system components. Two feasible aids to configuration are: automatic interoperability checks – checking that components are correctly communicating – and automatic adaptation – enabling a component to correctly communicate with another. System design and implementation can benefit from automatic interoperability checks, because this leads to detection of configuration errors prior to final deployment. Also, automatic component adaptation leads to faster development and less error-prone configurations of interoperating components.

In the TrustME project our approach to these problems is based on a software architectural description language (ADL). We are developing tools that facilitate automatic interoperability checks and adaptation for distributed middleware-based systems. An ADL is used to specify a high-level, compositional view of a distributed software application, specifying how a system is to be composed from components, and how and when components communicate. ADLs usually come equipped with a component model that contains a state-transition style semantics, enabling behavioural analysis [11]. The TrustME ADL is called RADL, short for "Rich" or "Reliable" ADL, because it enriches standard architecture descriptions with interoperability protocol features aiming at reliable interoperability. In RADL a component protocol is considered the set of valid call sequences of component methods. RADL augments a simple component interface model, consisting of a method signature, with additional protocol information and uses architectural composition to associate different fragments of interface specifications to different portions of the overall architecture definition. This leads to a notion of component and system contracts and to parameterised contracts. Our tool, called $CoConut/J$-tool,[1] is able to perform a certain class of component protocol adaptations automatically. This adaptation is based upon parameterised contracts [18], a generalisation of classical software contracts [12].

Through focusing on protocols for checking and adaptation, our work has a strong bearing on concurrent distributed systems architecture and design. There, components communicate through message exchange and it is impor-

---

[1] The $CoConut/J$-tool was originally developed by the first author at the Universität Karlsruhe, Germany as part of the $CoConut/J$-project (Contracts for Components) under DFG-funding.

tant to model the valid sequences of message exchange, i.e., the protocol used for interoperation.

The paper is organised as follows. In section 2 we outline RADL and its component model. In section 3 we provide an overview of parameterised contracts and how they can be concretely applied to our component model defined within the ADL. An example of how our tool implements this work is presented in section 4. Section 5 discusses related work, and section 6 concludes.

2   RADL: An ADL for Component Protocols

The TrustME ADL[2] was originally described in [19], and has been extended further [17]. Like other ADLs, RADL provides a means of defining compositions of component-based systems, but, in contrast to most ADLs, it enables interoperability checking between interfaces of components in an architecture.[3]

The TrustME language decomposes a distributed middleware-based system into hierarchies of components, linked to each other by connections between their interfaces:

- *Components* are self-contained, coarse-grain computational entities, potentially hierarchically composed from other components.
- *Component interfaces* consist of a signature of *services* and protocol definitions. Interfaces are either *provided* or *required*. The former interfaces specify services that are offered by the component, while the latter describe methods that are required by the component.

Our components and interfaces are analogous to their namesakes in Darwin, to components and ports respectively in C2 and ACME, or to processes and ports in MetaH [11]. However, RADL differs from these other languages, in that, by providing a more detailed interface description, it enables automatic interoperability checks between interfaces (most other ADLs treat interfaces as simply a collection of service names).

2.1   Protocol descriptions using finite state machines

The protocol of a provides-interface (i.e., the *provides-protocol*) can be considered as a set of *valid* call sequences to that interface according to the interface signature. A valid call sequence is a call sequence which is actually supported by the component. For example, a file I/O component might provide open, read, close as services, where open-read-close is a valid call sequence, while read-open is

---

[2] TrustME is the product of collaborative research conducted in the DSTC and at Monash University.

[3] The ADL is heterogeneous, possessing a generic and extensible means of defining modelling constructs, and permitting various component interface models. In this paper, we outline the subset of TrustME relevant to describing component protocols within middleware-based architectures.

not. Analogously, the protocol of a requires-interface (i.e., the *requires-protocol*) is a superset of the calls sequences, by which a component calls external services.

In both cases, a protocol is considered as a set of service call sequences. We use finite state machines (FSMs) to specify interface protocols, because FSMs provide a commonly understandable formalism and compositionality and substitutability of components can be checked efficiently due to efficient algorithms to check the inclusion of FSMs. A FSM consists of states and state transitions. State transition diagrams, such as that of Fig. 2, denote FSMs, where state are represented by circles and transitions by arrows. There is one designated state, *start-state*, in which the component is after its creation. Each state is associated with a set of services that are callable in that state. Service calls are given denoted by state transitions, because calls change the state – usually, other services are callable after the call, while others, callable in the old state, are not callable in the new state. A call sequence is only valid if it leads the FSM into a so called *final-state*. These final-states are denoted by solid black bullets within the states. In the left part of Fig. 2 only state 3 is a final-state. So, a call sequence, accepted by this FSM for example is `play-pause-play-stop`, while one cannot use a command sequence like `play-pause-stop`.

2.2   Architectures and interoperability checks

A basic architecture consists of components, with interoperation denoted by connections between provided and required interfaces. For instance, if `VideoStream`, `VideoPlayer` and `SoundPlayer` are components in RADL with provides-interface `provides-` and requires-interface `requires` respectively, then a basic architecture definition of `TrustME` is

```
architecture { VideoStream, VideoPlayer
  bind(VideoStream.requires, VideoPlayer.provides),
  bind(VideoStream.requires, SoundPlayer.provides)
}
```

The second line declares that the two components are part of the system architecture, and the third line declares that their interfaces are connected. More complex architectures can be constructed using hierarchies of components built from other components – see [17, 19] for details. A connection between two interfaces denotes a use relation, this means, call sequences of the requires-interface are sent to the provides-interface. Interoperability checks take the form of ascertaining whether or not the protocol of the provided-interface permits the calls specified by the protocol of the require-interface. Because protocols are given as FSMs, this can be computed easily and efficiently by means of the sublanguage relation between FSMs.

In this way, interoperability checks between components can be considered a form of contracts for components according to B. Meyers design-by-contract principle [12]. A contract for a component specifies, under which precondition a component A has to fulfill the postcondition. Translated to interoperability checks, we check if the environment – say, a component B – offers all functionality A expects. We check the inclusion of the requires-interface of A (precondition) in the provides-interface of B (see point (1) in Fig. 1). Is this check successful, A fits into the environment and will offer all services of its provides-interface (post-condition). Hence, interoperability checks have a boolean outcome: a component fits into system or not. In contrast to other ADLs, based on more complicated behavioural semantics (such as process-calculi), our use of FSMs for protocol description means that such interoperability checks can be efficiently implemented. This is done in the *CoConut/J*-tool, enabling us to perform interoperability checks over entire architectures, by checking each connection between components. Actually, the *CoConut/J*-tool works with counter-constrained FSMs [18], an extension of FSMs which is capable of describing more complicated provides-protocols (such as the one of a stack, where no more **pop** operations must be performed than **push** operation have been performed before).

3   Parameterised Contracts for Protocol Adaptation

Often, a given component is not interoperable with another component. For instance, in the architecture above, VideoStream might generate call sequences that are not accepted by VideoPlayer. A common task for the software architect is to fit a component into its environment by writing adapters to wrap the component with an interface that can interoperate with other given components. In our case, because our interfaces contain protocol information, it is possible to automatically adapt components, using parameterised contracts.

While interoperability tests check the requires-interface of a component against the provides-interface of *another* component, parameterised contracts link the provides-interface of one component to the requires- interface of *the same* component (see points (2) and (3) in Fig. 1). Classical contracts do not reflect this

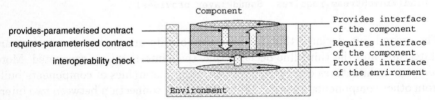

Fig. 1. Interoperability checks (1) and Requires-parameterised Contract (2) and Provides-parameterised Contract (3)

connection between provides- and requires-interfaces. Classical contracts, once formulated statically during component development, cannot change the post- or precondition according to a new reuse context. This motivates the formulation of two kinds of parameterised contracts: (a) provides-parameterised contracts map the provides-interface to a requires-interface. (b) requires-parameterised contracts map the requires-interface to a provides-interface.

Parameterised contracts are a mapping that is bundled with the component and computes the interfaces of the components on demand. The requires-parameterised contract takes as arguments the requires-interface of the component and the provides-interface of the environment. Hence, parameterised contracts are isomorphic mappings between the domain of preconditions and the domain of postconditions. Both domains can be modelled as partially ordered sets (posets). (A mathematical discussion of parameterised contracts can be found in [18].) The intersection of a components requires-interface and the environment provides-interface describes the functionality which is required by the component and provided by the environment. From that information the requires-parameterised contract computes the new provides-interface of the component. Analogously, a provides-interface computes the new requires-interface from the provides interface of the component and the requires-interfaces of its clients. In the $CoConut/J$-tool, we implemented parameterised contracts for component interfaces which contain the provides-/requires-protocol. The reversible mapping between these interfaces are the so-called *service effects finite state machines* (SE-FSMs). Each service s provided by a component is associated with its SE-FSM$_s$. The SE-FSM$_s$ describes all those sequences of calls to other services that can be made by the the service s. A simple example of a SE-FSM is shown in the example in section 4 on the right side in Fig. 2. A transition of an SE-FSM corresponds to a call of an external service. It shows that a call to service play of the video stream can lead to calls to the external services play of a VideoPlayer component and a SoundPlayer component. The latter call is not mandatory as shown by the predicate [ifavail]. (The context of this example is discussed in section 4.) To realise a provides-parameterised contract of a component C we interpret the transition functions of the P-FSM and the SE-FSMs as graphs. Each edge (transition) in the P-FSM-graph corresponds to a service, as mentioned in section 2. We substitute for each transition in the P-FSM-graph the SE-FSM of the corresponding service. The resulting graph contains all the SE-FSMs in exactly those orders that their services can be called by the clients of the component C. This means, that the resulting graph, interpreted as an FSM, describes all sequences of calls to external components. Hence it describes the requires-protocol. If we ensure, that the insertion of SE-FSMs into the P-FSM is reversible (e.g., by marking the "beginning" and "end" of each SE-FSM, we can generate from a given CR-FSM a P-FSM that gives us a requires-parameterised contract. To adapt a component C to an environment E, we build the intersection of CR-FSM$_C$ and P-FSM$_E$. The result is a possibly different CR-FSM$_C'$, which describes the call sequences emitted by C and accepted by E. From that CR-FSM$_C'$ we generate a (possibly) new provides-protocol P-FSM$_C'$ using the requires-parameterised contract as sketched above. A detailed discussion of the implementation of parameterised contracts for protocol adaptation including complexity measures and advanced issues like reentrant code and recursion, can be found in [18].

4   Example Application

As an example we present a distributed multimedia application. From a central server, video streams are sent to various clients over network connections. Such a video stream is not mere data, but also includes functionality such as playing, stopping, manipulation of sound and picture. An important feature is, that the clients are allowed to have different configurations regarding their hardware, operating system and applications software. A user may change these configurations frequently, e.g., by using several platforms, such as desktop computers, mobiles, hand-helds, or systems installed in automatives. These platforms differ significantly in their capabilities of reproduction, e.g., some platforms may not support sound reproduction (e.g., business desktop-systems), others cannot adjust the colours (e.g., some mobiles). Due to that variety of platforms, the server cannot simply send the video stream in thet same way to all clients nor can it assume a unique environment on all client sides for reproduction. Therefore, it is important, that the functionality provided by the video stream is adaptable to the resources available on a concrete client side.

In more technical terms, the functionality of the video stream depends on a video player and a sound player at the client side. As mentioned, the reproduction features of these tools differ significantly. The default protocol of the video stream is presented in Fig. 2 (left). This protocol represents the maximum functionality offered by the video stream. Note that the video stream offers sound and picture

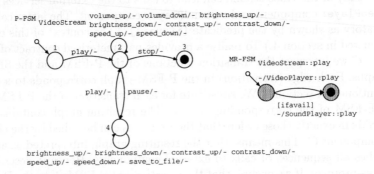

Fig. 2. P-FSM of the video stream (left) and SE-FSM of its service `play` (right)

manipulation while playing. However it offers only picture manipulation while pausing the video.

This protocol has to be adapted to the provides-protocols of the actual video player and actual sound player. Imagine the video stream arrives on a system without any sound support (so no sound player is given) and with a video player satisfying the provides-protocol shown in Fig. 3. Note that this `VideoPlayer` component only offers picture manipulation while playing the video.

The restriction of the video stream provides-protocol according to the concrete client environment (i.e., the `VideoPlayer`) is performed by a requires-parameterised contract. This video stream requires-parameterised contract is computed by the *CoConut/J*-toolsuite.

Fig. 4 shows the *CoConut/J*-tool with the newly computed provides-protocol. This is an adaptation of the video stream to this specific environment. Note that while pausing, it is now only possible to save to a file, but not to manipulate

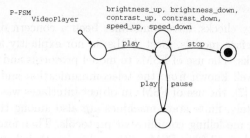

Fig. 3. P-FSM of the `VideoPlayer`

the picture. We should emphasise that this kind of protocol adaptation is not expressible by normal interfaces based on signature lists because the services for picture manipulation still exist in the interface. But they are not available in every state where they have been callable before.

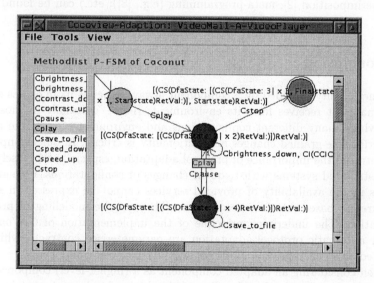

Fig. 4. Adapted P-FSM of the video stream as computed by the *CoConut/J*-tool

5 Related Work

Our focus is more specific than many ADLs, where entire component behaviour is often modelled, using stronger formalisms than finite state machines. For example, Darwin [10] uses the π-calculus [14], Rapide [9] uses partially ordered event sets and Wright [1] can use a form of CSP [6]. However, these approaches have some drawbacks when specifying protocols for architecture interoperability checks. For example, approaches using process calculi to provide more interface information are able to specify real-world protocols, but do not provide efficient algorithms for checking protocol compatibility [21]. While the properties important for architectural system configuration, such as system behaviour analysis and substitutability checks, have sometimes been a concern during the design of component interface models, most ADLs do not explicitly focus on local interoperability checks. The use of FSMs to model protocols and test components systematically is well known from the telecommunication and distributed systems communities [7]. the use of FSMs in object interfaces was firstly described in [15]. Unfortunately, finite state machines are also among the least powerful models concerning modelling complicated protocols. Therefore, more powerful models have been derived from FSMs without loosing their beneficial properties [18]. Non-ADL based interoperability checks, using rich interface specifications, have been realised in some research tools [5, 21]. Besides our approach using interface information for automatic adaptation, [22] use interface information to create adaptors automatically. Overviews and evaluations about other adaptation mechanisms not using interface information (such as delegation, wrappers [4], superimposition [2], meta-programming (e.g., [8]), etc.) can be found in [3, 18].

6   Conclusion

The functionality provided by a component always depends more or less on the functionality it receives from its environment. Hence, especially in those systems, where many different configurations and environments exist, the ability to describe fine-grained changes in functionality is crucial. In our example, we demonstrated the importance of protocol adaptation, especially for loosely coupled distributed systems with frequent changes of configurations. Fine-grained changes in the availability of provided services cannot be expressed in simple signature-list based interface descriptions. They require modelling of protocol information. The underlying principle of the implementation of *CoConut/J*-tool for automatic protocol adaptation are parameterised contracts, which we presented as a generalisation of interoperability between components.

When considering an adapted component as a (higher-level) component, we are able to describe the functionality of this higher-level component in terms of its constituent inner components. Such a compositional interpretation is only possible for layered system architectures. In practice, beside layering, many other composition patterns occur. Hence, future work must be concerned with more general mechanisms to predict properties of the overall software architecture from the properties of the single components and their interaction patterns.

References

1. Robert J. Allen. *A Formal Approach to Software Architecture*. Ph.D. thesis, School of Computer Science, Carnegie Mellon University, Pittsburgh, PE, USA, May 1997.
2. Jan Bosch. Composition through superimposition. In Wolfgang Weck, Jan Bosch, and Clemens Szyperski, editors, *Proceedings of the First International Workshop on Component-Oriented Programming (WCOP'96)*. Turku Centre for Computer Science, September 1996.

3. Jan Bosch. *Design and Use of Software Architectures – Adopting and evolving a product-line approach.* Addison-Wesley, Reading, MA, USA, 2000.
4. Erich Gamma, Richard Helm, Ralph Johnson, and John Vlissides. *Design Patterns: Elements of Reusable Object-Oriented Software.* Addison-Wesley, Reading, MA, USA, 1995.
5. Jun Han. Temporal logic based specification of component interaction protocols. In *Proccedings of the 2nd Workshop of Object Interoperability at ECOOP 2000,* Cannes, France, June 12.–16. 2000.
6. C. A. R. Hoare. *Communicating Sequential Processes.* Prentice/Hall, 1985.
7. Gerald J. Holzmann. *Design and Validation of Computer Protocols.* Prentice Hall, Englewood Cliffs, NJ, USA, 1991.
8. G. Kiczales. Aspect-oriented programming. *ACM Computing Surveys,* 28(4):154–154, December 1996.
9. D.C. Luckham, J.J. Kenney, L.M. Augustin, J. Vera, D. Bryan, and W. Mann. Specification and analysis of system architecture using rapide. *IEEE Transactions on Software Engineering,* 21(4):336–355, Apr 1995.
10. J. Magee, N. Dulay, S. Eisenbach, and J. Kramer. Specifying distributed software architectures. *Lecture Notes in Computer Science,* 989:137–155, 1995.
11. Nenad Medvidovic and Richard N. Taylor. A classification and comparison framework for software architecture description languages. *IEEE Transactions on Software Engineering,* 26(1):70–93, Janurary 2000.
12. Bertrand Meyer. Applying "design by contract". *IEEE Computer,* 25(10):40–51, October 1992.
13. M. Migliardi and V. Sunderam. PVM emulation in the harness metacomputing system: A plug-in based approach. In J. J. Dongarra, E. Luque, and Tomas Margalef, editors, *Proc. of the 6th European PVM/MPI Users' Group Meeting, Barcelona, Spain, September 26–29, 1999,* volume 1697 of *Lecture Notes in Computer Science,* pages 117–124. Springer-Verlag, Berlin, Germany, 1999.
14. R. Milner. The pi calculus and its applications. In Joxan Jaffar, editor, *Proc. of the 1998 Joint International Conference and Symposium on Logic Programming (JICSLP-98),* pages 3–4, Cambridge, June 15–19 1998. MIT Press, Cambridge, MA, USA.
15. Oscar Nierstrasz. Regular types for active objects. In *Proc. of the 8th ACM Conf, on Object-Oriented Programming Systems, Languages and Applications (OOPSLA-93),* volume 28, 10 of *ACM SIGPLAN Notices,* pages 1–15, October 1993.
16. Object Management Group. The CORBA homepage. http://www.corba.org.
17. Iman Poernomo, Ralf Reussner, and Heinz Schmidt. The TrustME language site. Web site, DSTC, 2001. Available at http://www.csse.monash.edu.au/dsse/trustme.
18. Ralf H. Reussner. *Parametrisierte Verträge zur Protokolladaption bei Software-Komponenten.* Logos Verlag, Berlin, 2001.
19. Heinz Schmidt. Compatibility of interoperable objects. In *Information Systems Interoperability,* pages 143–199. Research Studies Press, Taunton, England, 1998.
20. Clemens Szyperski. *Component Software: Beyond Object-Oriented Programming.* ACM Press, Addison-Wesley, Reading, MA, USA, 1998.
21. A. Vallecillo, J. Hernández, and J.M. Troya. Object interoperability. In A. Moreira and S. Demeyer, editors, *ECOOP '99 Reader,* number 1743 in LNCS, pages 1–21. Springer-Verlag, 1999.
22. D. Yellin and R. Strom. Protocol Specifications and Component Adaptors. *ACM Transactions on Programming Languages and Systems,* 19(2):292–333, 1997.

Integrating CUMULVS into AVS/Express

Torsten Wilde, James A. Kohl and Raymond E. Flanery, Jr.

Oak Ridge National Laboratory [1][2]

Keywords: Scientific Visualization, CUMULVS,
AVS/Express, Component-Based Design

Abstract. This paper discusses the development of a CUMULVS interface for runtime data visualization using the AVS/Express commercial visualization environment. The CUMULVS (Collaborative, User Migration, User Library for Visualization and Steering) system, developed at Oak Ridge National Laboratory, is an essential platform for interacting with high-performance scientific simulation programs on-the-fly. It provides run-time visualization of data while they are being computed, as well as coordinated computational steering, application-directed checkpointing and fault recovery mechanisms, and rudimentary model coupling functions. CUMULVS primarily consists of two distinct but cooperative libraries - an application library and a viewer library. The application library allows instrumentation of scientific simulations to describe distributed data fields, and the viewer library interacts with this application side to dynamically attach and then extract and assemble sequences of data snapshots for use in front-end visualization tools. A development strategy will be presented for integrating and using CUMULVS in AVS/Express, including discussion of the various objects, modules, macros and user interfaces.

1. Introduction

Scientific simulation continues to be a field replete with many challenges. Ever-increasing computational power enables researchers to investigate and simulate more and more complex problems on high-performance computers, to obtain results in a fraction of the time or at a higher resolution. The data processed and created by these simulations are huge and require much infrastructure to manipulate and evaluate. Scientific visualization and interactive analysis of complex data during runtime provides a cost-effective means for exploring a wide range of input datasets and physical parameter variations, especially if the simulation runs for days. It can save time and money to discover that a simulation is heading in the wrong direction due to an incorrect parameter value, or because a given model does not behave as expected.

---

[1] Research supported by the Mathematics, Information and Computational Sciences Office, Office of Advanced Scientific Computing research, U. S. Department of Energy, under contract No. DE-AC05-00OR22725 with UT-Battelle, LLC.

[2] This research was supported in part by an appointment to the ORNL Postmasters Research Participation Program which is sponsored by Oak Ridge National Laboratory and administered jointly by Oak Ridge National Laboratory and by the Oak Ridge Institute for Science and Education under contract numbers DE-AC05-84OR21400 and DE-AC05-76OR00033, respectively

P.M.A. Sloot et al. (Eds.): ICCS 2002, LNCS 2330, pp. 864–873, 2002.
© Springer-Verlag Berlin Heidelberg 2002

A proper visualization environment allows scientists to view and explore the essential details of the simulated data set(s) [1]. For 2-dimensional (2D) data sets, a 2D visualization environment is sufficient. But for 3-dimensional (3D) problems, a 3D visualization environment is required to provide access to all the detailed information embedded in the data set.

This paper describes work to integrate CUMULVS[2,3] into the AVS/Express[4] viewer environment, which provides a framework for data visualization, including both 2D and 3D capabilities. AVS/Express is unique in the way that it allows changes to the application structure and functionality during runtime. Applications are constructed by "drag & drop" of modules from the component library. The user can add and/or delete components dynamically, to change the application behavior on-the-fly. This work is important in the sense that the integration of CUMULVS into AVS/Express will enable the user to use runtime scientific data sets for visualization instead of file or static data sets.

2. Background

2.1. CUMULVS

CUMULVS (Collaborative, User Migration, User Library for Visualization and Steering) [2,3] provides an essential platform for interacting with running simulation programs. With CUMULVS, a scientist can observe the internal state of a simulation while it is running via online visualization, and then can "close the loop" and redirect the course of the simulation using computational steering. These interactions are realized using multiple independent front-end "viewer" programs that can dynamically attach to, interact with and detach from a running simulation as needed. Each scientist controls his/her own viewer, and can examine the data field(s) of choice from any desired perspective and at any level of detail. A simulation program need not always be connected to a CUMULVS viewer; this proves especially useful for long-running applications that do not require constant monitoring. Similarly, viewer programs can disconnect and re-attach to any of several running simulation programs. To maintain the execution of long-running simulations on distributed computational resources or clusters, CUMULVS also includes an application-directed checkpointing facility and a run-time service for automatic heterogeneous fault recovery.

CUMULVS fundamentally consists of two distinct libraries that communicate with each other (using PVM[5]) to pass information between application tasks and front-end viewers. Together the two libraries manage all aspects of data movement, including the dynamic attachment and detachment of viewers while the simulation executes. The application or "user" library is invoked from the simulation program to handle the application side of the messaging protocols. A complementary "viewer" library supports the viewer programs, via high-level functions for requesting and receiving application data fields and handling steering parameter updates.

The only requirement for interacting with a simulation using CUMULVS is that the application must describe the nature of its data fields of interest, including their decomposition (if any) across simulation tasks executing in parallel. Using calls to the user library, applications define the name, data type, dimensionality/size, local storage allocation, and logical global decomposition structure of the data fields, so that CUMULVS can automatically extract data as requested by any attached front-end viewers. Given an additional periodic call to the stv_sendReadyData() service routine, CUMULVS can transparently provide external access to the changing state of a computation. This library routine processes any incoming viewer messages or requests, and collects and sends outgoing data frames to viewers.

This manual instrumentation of application data can be alleviated by systems like DynInst [10] which do automatic run-time introspection of codes, however this type of analysis is not sufficient to fully describe distributed data decompositions; some user intervention is needed to specify the implied parallel semantics and the context of the local data in the overall global array. Unlike systems such as DICE [11], where whole copies of each data field are placed in a globally shared file structure using DDD [12] and HDF [13], in CUMULVS the data movement is demand-driven and the viewers dynamically extract only requested subregions of data fields from each application task. This reduces the application overhead in most cases and provides more flexible multi-viewer collaboration scenarios.

When a CUMULVS viewer attaches to a running application, it does so by issuing a "data field request," that includes a selection of desired data or "view" fields (constituting a "view field group"), a specific region of the computational domain to be collected ("view region"), and the frequency with which data "frames" are to be sent back to the viewer. CUMULVS handles the details of collecting the data elements of the view region for each view field. The view region boundaries are specified in global array coordinates, and a "cell size" is set for each axis of the data domain, to determine the stride of elements to be collected for that axis, e.g. a cell size of 2 will obtain every other data element. This feature provides more efficient high-level overviews of larger regions by using only a sampling of the data points, while still allowing every data point to be collected in smaller regions where the details are desired. CUMULVS has been integrated with parallel applications written using PVM [5], MPI [14] and InDEPS [15] and can be applied to applications with other arbitrary communication substrates.

2.2. AVS/Express Visualization Environment

AVS/Express[4] is a commercial environment for visualizing scientific data. It provides the user with a visual programming interface and includes standard modules for the most common visualization functions. Using AVS/Express the user can develop a custom viewer by "drag & drop" of modules (objects) and connecting together specific input and output ports of the objects (see Figure 1). *This concept enables users to create a visualization without the need for programming custom code.* Modules in AVS/Express represent single object instances in an object oriented programming language. They are the basic components of any AVS/Express program. Modules can be grouped into macros in order to create higher-level hierarchical objects. Macros can be grouped with other macros in order to create even higher-order objects. Ultimately, one macro could represent a complete application. Custom module creation can be done for AVS/Express via the following 4 steps:

- Define parameters and values using AVS/Express primitive data types (e.g. integer, real, string) or groups of primitive types and other structures.
- Add methods (functions) for the module processing.
- Define the type of execution for the module methods ~ the user code can be compiled directly into the AVS/Express program or can be compiled as its own distinct program.
- Define execution events for module methods and method behavior.

The developer can specify which parameters are connected to which methods. There are four possible options here:

- notify: The method is called if the parameter value changes.
- read: The method reads the parameter.
- write: The method writes the parameter.
- required (req): The method can only be called if the parameter has a valid value, as checked automatically before event processing.

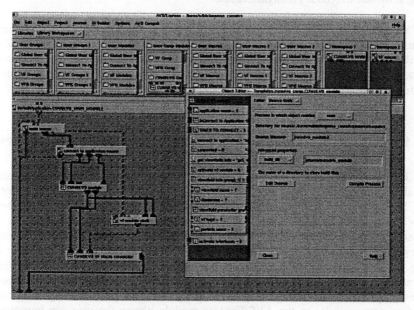

Fig. 1. AVS/Express application creation interface

AVS/Express incorporates a data driven or "Event based" execution paradigm. It responds to events to execute different sets of instructions depending on which event occurs rather than following a pre-defined sequence of instructions. This means a module method can only be executed if a specific parameter has changed. Usually this principle is used where the program states are driven by the graphical user interface (GUI). Figure 2 shows the handling of function return codes in this paradigm. A "hand shake" approach is used, where the caller receives feedback regarding the processing state of the event, such as error-success information. An event change in *Module A* executes method *XY*. Because this event requires some processing in

Module B, the method changes the output port of *A.* This change triggers the input port of *B* (connected to the output port of *A*). *Module B* now reacts to this event by executing method *XYZ.* After finishing the execution *XYZ* writes the status information and/or return values to output port *B.* This changes the value of the corresponding input port of *A,* triggering the execution of the method *check_return_code()* which evaluates the return code and/or values and can inform other modules, or the user, about the status of the event processing.

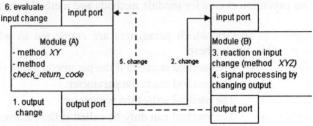

Fig. 2. Module Execution Paradigm Example

Modules can use this principle to verify if another module has been connected. This is important for verifying the program state at all times, especially when dealing with changing connections or new modules at runtime. AVS/Express allows multiple modules to be connected to the same module port. The GUI also can be dynamically extended as the application changes.

3. CUMULVS Interface Design for Integration into AVS/Express

The goal of this work was to integrate CUMULVS into AVS/Express in order to enable runtime data visualization and to create an AVS/Express viewer for the CUMULVS library. CUMULVS already supports several graphical viewers based on AVS5[6], Tcl/Tk[7], VTK[8] and the CAVE[9] environment. The AVS/Express work is especially interesting because of its component-based functionality. It is a complete viewer environment with components for everything needed to view scientific data, e.g. reading, filtering, transforming and visualizing the data. AVS/Express provides a dynamic application structure, e.g. viewers can be customized to an application on-the-fly by adding or deleting components using the visual programming interface. By adding new modules to the AVS/Express module library the user can improve and extend the AVS/Express capabilities and construct custom viewers. The following subsections describe the module structure and GUI for the CUMULVS AVS/Express viewer.

3.1. CUMULVS Module Structure

The CUMULVS functionality is divided into global modules for the AVS/Express viewer by analyzing program functionality paired with the required GUI blocks. Figure 3 shows this global module structure including 4 primary macros.

The *CUMULVS Main Macro* handles all communication with the running application and provides the GUI for specifying the application name and other global parameters. It also provides information about the connected application, like available data fields for viewing, their bounds and data type, and whether the data is particle-based or a mesh decomposition. Because of its central role, this module must communicate with all other modules:

- Sends View Field (VF) information to VF Modules
- Get View Field Group (VFG) information from VFG Module for data collection
- Sends application connection status to VFG Module

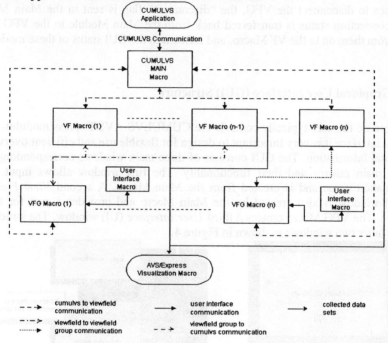

Fig. 3. main object structure

The *VF Macro* gets the VF information from the CUMULVS Main Macro and provides the GUI for selecting one VF from the possible VFs for viewing. Every VF requires one VF Macro, which stores all important information about the selected VF and transfers it to the VFG Macro. One VF Macro can only be connected to one VFG Macro, e.g. if the same VF is required for a second VFG, an additional VF module has to be instantiated for this VF.

The *VFG Macro* combines connected VF Macros into one VFG. The VFG Macro calculates the global bounds and dimension from the connected VF values. For example, the global lower boundary could be the highest VF lower boundary found in the connected VFs and the global upper boundary could be the lowest VF upper boundary found. The VFG Macro provides base values for the *User Interface (UI) Macro*, which sets parameters like boundaries and cell size for each dimension, visualization frequency, etc. The UI Macro checks all user input for errors before sending it to the VFG Macro. When input is forwarded to the Main Module, the data collection is started or appropriate changes to the data collection are made.

In addition to data flow through the system, control parameters are also transferred between macros. Initially, only the Main Macro GUI is activated, but after successful connection to the application the VF Macro is then activated. Activation happens by setting a port connection to "true" (an integer value of 1).

The VFG Macro is activated if at least one connected VF Macro has a valid VF selection. After the user connects the VFG to the application, the VFG information is sent to the Main Module together with the "connect" flag. Likewise if the user chooses to disconnect the VFG, the "disconnect" flag is sent to the Main Module. The connection status is transferred back from the Main Module to the VFG Macro and from there on to the VF Macro, and influences the GUI status of these modules.

3.2. Graphical User Interface (GUI) Structure

Because the user typically controls the CUMULVS-AVS/Express modules via the visual interface it is very important to design for flexible use and efficient overview of the vital information. The GUI consists of three main windows corresponding to the three main macros and their functionality. The first window allows input of the application name and is created from the Main Macro. A second *View Field Info* (VFI) window is also created by the Main Macro and provides a port for the VF Macro. The VFG Macro creates a third *User Interface* (UI) window. The structure of these latter two windows is shown in Figure 4.

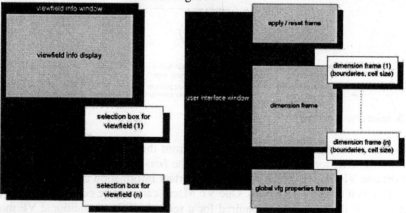

Fig. 4. View Field Info and User Interface Window Structures

The VFI window displays important information about the available VFs, including name and boundaries for each dimension. Each instantiated and connected VF Macro adds a "selection box" for its VF to the VFI window. If the connection from the VF Macro to the Main Macro or the VF Macro itself is deleted, the corresponding "selection box" is also deleted. The VFI window resizes automatically.

The UI window consists of three frames, which are positioned depending on the order of their connection to the window. This portion of the GUI enables the user to change the parameters for the collected VFG data set. The "apply/reset frame" allows the user to apply the changes to the VFG or reset to previous values. The "dimension frame" sets the boundaries and cell size for each dimension. This frame is constructed modularly with one "dimension property frame" per dimension of the VFG, e.g. for a 3 dimensional VFG, 3 such frames are connected. The overall dimension frame is resized automatically. The user can adapt this interface to different problems (applications) on the fly. All global VFG parameters like visualization frequency can be changed using the "global VFG properties frame". Each frame has a well-defined port connection to its parent frame or window. It is therefore possible for users to create their own customized GUIs using the provided frames, or by creating new ones implemented with the given input/output port connection specification.

4. Results

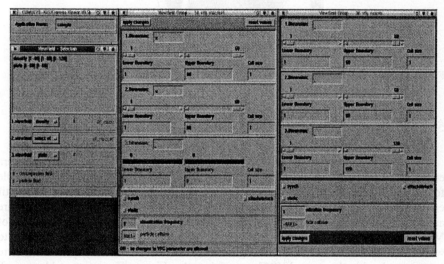

Fig. 5. Screenshot of the CUMULVS-AVS/Express Interface

Figure 5 shows a sample snapshot of the GUI in action. The application name window is in the upper left corner. The next window below that is the "View Field Selection" window. Two data fields from the application are available for viewing; "density" has 3 dimensions with boundaries [1-60][1-60][1-120], and "plate" is 2 dimensional with boundaries [1-60][1-60]. There are three view fields attached. The first two ("VF_macro", "VF_macro#1", see also Figure 6) are connected to one view field group (right interface window, "3d_vfg_macro"). The third view field builds its own VFG (middle interface window, "3d_vfg_macro#1"). The VFG windows are color-coded, with the VF_Macro name in the "View Field Selection" window colored like the VFG to which it belongs. In addition, a special letter code describes the type of the selected view field ("d" for mesh decomposition field and "p" for particle field). The two VFG windows are composed of the three frames described in 3.2. Values are initialized using default VFG parameters. Any user input is validated before changes are submitted to the VFG, after the "apply changes" button is pressed.

Because VF "plate" is two-dimensional in this example, the input for the third dimension is deactivated automatically in the corresponding UI window (left).

Figure 6 shows a 3D visualization of an example application rendering from a simple LaPlace simulation. Surface rendering was used to visualize the data set. The inner surface was solid rendered and the outer surface in 65% transparent.

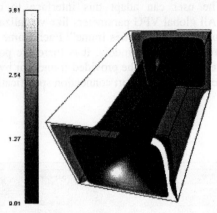

Fig. 6. Visualization of LaPlace Simulation

5. AVS/Express Concerns

During this development several problems with AVS/Express for Linux were encountered. As a result, the CUMULVS viewer plug-in is currently only available for AVS/Express on SGI Irix systems. A crucial bug is related to the motif environment, which causes random crashes of AVS/Express if menus or buttons are accessed. AVS support verifies the problem, and there is an updated version of Motif1.2 available for Linux glibc2.1. Unfortunately this update does not seem to work for other glibc versions like glibc2.2. Also some stability problems were encountered with the new AVS/Express5.1 under Linux. The most stable overall environment seems to be Mandrake7.2, with the Motif update and AVS/Express5.0.

The only know problem with AVS/Express5.1 under SGI Irix is that the user interface input field "UIfield" instantiates with a fixed field width independent of the value set in the object (Figure 5, bottom right window, for visualization frequency).

As a developers note, it is important to point out that there is no clear order of module instantiation and method execution if a complex saved program or macro is instantiated. The order could in fact be opposite to the drag & drop program creation order, and the order of connections is also not guaranteed. It is therefore possible to introduce problems that are only visible during instantiation of the whole program, and not during component testing. Feedback loops can occur involving different high-level macros. Also, setting the "required" flag (see Sect. 2.3) for an array doesn't ensure that all array values will be valid. Missing values inside the array are not detected by AVS/Express. This is especially problematic for pointer arrays where the access of an invalid pointer leads to program termination or critical failure.

6. Summary/Future Work

The integration of the CUMULVS functionality into AVS/Express enables CUMULVS users to take advantage of the powerful component-based AVS/Express viewer environment, and similarly AVS/Express users can collect and visualize data from running parallel/distributed scientific applications using CUMULVS. The event-based execution paradigm and the highly scalable module approach make AVS/Express very flexible, but at a potentially high cost in complexity for internal module communication. The plug-in was tested using simple example applications. The next step will be to use it in real world applications. Future plans include solving the problems with AVS under Linux, integrating a steering interface into the CUMULVS plug-in and improving or rearranging the user interface based on user feedback.

References

[1] K.J. Weiler, "*Topological Structures for Geometric Modeling*", Ph.D. thesis, Rensselaer Polytechnic Institute, Troy, NY, May 1986

[2] G.A. Geist, J.A. Kohl, P.M. Papadopoulos, "*CUMULVS: Providing Fault-Tolerance, Visualization and Steering of Parallel Applications*", INTL Journal of High Performance Computing Applications, Volume II, Number 3, August 1997, pp. 224-236.

[3] J.A. Kohl, P.M. Papadopoulos, "*CUMULVS user guide, computational steering and interactive visualization in distributed applications*", Oak Ridge National Laboratory, USA, Computer Science and Mathematics Division, TM-13299, 02/1999.

[4] "*AVS/Express Developer's Reference*", Advanced Visual System Inc., Release 3.0, June 1996.

[5] G.A. Geist, A. Beguelin, J. Dongarra, W. Jiang, R. Manchek, V. Sunderam, "*PVM: Parallel Virtual Machine*", A User's Guide and Tutorial for Networked Parallel Computing, The MIT Press, 1994.

[6] "*AVS User's Guid*", Advanced Visual Systems, Inc., Waltham, MA, 1992.

[7] J.K. Ousterhout, "*Tcl and the Tk Toolkit*", Addison-Wesley, Reading, MA, 1994.

[8] Will Schroeder, Ken Martin, Bill Lorensen, "*The Visualization Toolkit an object-oriented approach to 3D graphics*", 2nd Edition, Prentice Hall PTR, 1998

[9] CAVERNUS user group, CAVE Research Network Users Society, http://www.ncsa.uiuc.edu/VR/cavernus

[10] DYNINST - An Application Program Interface (API) for Runtime Code Generation http:// www.dyninst.org

[11] J.A. Clarke, J.J. Hare, C.E. Schmitt, "Distributed Interactive Computing Environment (DICE)", Army Research Laboratory, Major Shared Resource Center, http://frontier.arl.mil/clarke/dice.html

[12] J.A. Clarke, J.J. Hare, C.E. Schmitt, "Dice Data Directory (DDD)", Army Research Laboratory, Major Shared Resource Center, see http://frontier.arl.mil/clarke/Dd.html

[13] "Hierarchical Data Format (HDF)", National Center for Supercomputing Applications

[14] M. Snir, S. Otto, S. Huss-Lederman, D. Walker, J. Dongarra, "MPI: The Complete Reference", MIT Press, Cambridge, MA, 1996

[15] R.Armstrong, P.Wyckoff, C.Yam, M.Bui-Pham, N.Brown, "Frame-Based Components for Generalized Particle Methods", High Performance Distributed Computing (HPDC '97), Portland, OR, August 1997, http://glass-slipper.ca.sandia.gov/~rob/poet/

Monitoring System for Distributed Java Applications

Marian Bubak[1,2], Włodzimierz Funika[1], Piotr Mętel[1], Rafał Orłowski[1], and Roland Wismüller[3]

[1] Institute of Computer Science, AGH, al. Mickiewicza 30, 30-059 Kraków, Poland
[2] Academic Computer Centre – CYFRONET, Nawojki 11, 30-950 Kraków, Poland
[3] LRR-TUM – Technische Universität München, D-80290 München, Germany
{bubak,funika}@uci.agh.edu.pl, {metel,witch}@icslab.agh.edu.pl, wismuell@in.tum.de
phone: (+48 12) 617 39 64, fax: (+48 12) 633 80 54, phone: (+49 89) 289 28243

Abstract. The paper presents a concept of an implementation of an extension to the On-Line Monitoring Interface Specification for Java Applications. The extension aims at defining an open interface for providing on-line software development tools. The general-purpose, portable, and extensible approach to handle comprehensive monitoring information from a Java run-time environment is intended to span the existing gap between the needs in Java application development tools and the lack of a uniform environment which provides monitoring support for different kinds of tools, like debuggers or performance analysers. The main goal of the resulting monitoring system is to adapt existing tools and build new ones for monitoring Java distributed applications.

Keywords: Java, monitoring system, monitoring interface, distributed object system.

1 Introduction

As a result of its platform independence, Java has become a wide-spread programming language for distributed applications in heterogeneous environments. As Java applications are getting larger and more and more complex, a lack of programming tools that allow to examine and eventually control the behaviour of these applications is becoming apparent. This is especially true for distributed programming. Although there are some tools for profiling distributed Java applications, like JProf [1] or JaViz [2], tools allowing to observe an application's behaviour in more detail or even to control it, are rare. Java debuggers, for example, usually do not support applications distributed across multiple Java virtual machines (JVMs). A notable exception is JBuilder [3]. Furthermore, the tools are incompatible to each other, i.e. they cannot be used at the same time in order to observe different aspects of a program's execution, like e.g. the high-level communication behaviour on the one hand and the detailed execution behaviour of single threads on the other.

Let us have a closer look at *monitoring tools* in general to examine the reasons for this situation. In this context, the term *monitoring* comprises techniques

P.M.A. Sloot et al. (Eds.): ICCS 2002, LNCS 2330, pp. 874–883, 2002.

and mechanisms to observe and potentially manipulate applications. Tools that monitor distributed applications have to consider all elements of the distributed system and must control the whole application distributed over different machines. Monitoring tools can be classified into two categories *on-line* and *off-line* tools. The on-line tools run concurrently with the application, thus a user can interactively observe and influence the state of the application. In the case of off-line tools, information on application execution is being stored on disk as a trace file. A tool subsequently uses the data gathered in the trace file for analyses activities.

For observing and possibly manipulating a program state, on-line tools need a specialised module which is called *monitoring system*. Usually this module is tightly integrated with the tool but even in the case of Java, it must directly interface with the operating system and in addition depends on the specific implementation of the JVM. As a result, tools are rather complex and are not easily portable to different target platforms. In addition, tools come into conflict with each other when used at the same time to observe the same application. A solution to these problems is to have a clearly separated monitoring system that provides a uniform interface for different kinds of on-line tools. The availability of such a system also greatly facilitates the development of new tools.

However, for distributed Java applications, no suitable monitoring systems exist at the moment. Although there are some approaches (e.g. [4]), they are usually targeted towards resource management or profiling only. Our approach for building a universal monitoring system is based on OMIS [5] and the OCM [6]. The OMIS project defines a standardised interface between the tools and the monitoring system. OMIS is not restricted to a single kind of tools, especially it supports tools oriented towards debugging as well as performance analysis and resource management. Originally OMIS has been designed to support parallel systems, and its referential implementation OCM is applied to PVM.

At the beginning the paper outlines the concepts of the extended version of the OMIS specification that supports Java applications. Next we present the architecture of the monitoring system compliant to the specification, the mechanism of interaction with the Java virtual machine, and a sample scenario of interactions between a tool and the monitoring system. The last part of the paper focuses on implementation concepts.

2   Interface Specification

The specification defines an interface between tool and monitoring system. The interface provides a special language that allows tools to build complex requests by combining primitive ones in order to achieve the needed flexibility. The individual monitoring functions available by requesting them with an expression in this language are called *services*.

The services relate to the defined elements of monitored applications called *system objects*. The set of system objects creates a view of the monitored application that is "visible" by the tool. This view can dynamically change during run-time, according to the application behaviour or on demand from the tool. The specification distinguishes between two kinds system objects: **execution** objects, i.e. *nodes, JVMs, threads* and **application** objects, i.e. *interfaces, classes, objects, methods.*

The interface defines a set of services for each object which provide information about the object (e.g. about a state of a given thread) – *information services*, allow to manipulate the object (e.g. exchange Java class file during program execution) – *manipulation services* and finally services that trigger arbitrary actions whenever a matching event takes place (e.g. when a given method is entered) – *event services.*

Another important feature of the on-line monitoring interface is the definition of a mechanism that reflects relations between objects of the running application, which is called *conversion.* The following example of a request expressed in the language defined by the interface illustrates this feature:

$$:method\_get\_info([class\_id], 1)$$

This request calls a method information service and results in obtaining information on all methods implemented in the specified class. The class identifier *class_id* is expanded by the conversion mechanism into a set of method objects. The last parameter determines what kind of information is returned by the service.

This short description only presents the basic ideas behind the interface specification, which are important to understand the realization concepts of the monitoring system. More detailed information about the interface specification can be found in [7] and [8].

3   Architecture of the Monitoring System

The specification tries to set up a monitoring interface which is as far as possible independent from the Java virtual machine [9], and from the programming library used for applications. However, any instance of the monitoring system must be implemented for the fixed target architecture and use the fixed access interface to the JVM. As the interface specification tries to be abstract, our goal is to identify a generic monitoring system which can easily be ported among different combinations of software and hardware architectures.

Cooperation between tools and the monitoring system is specified to be based on service requests and replies to them. A service request is sent to the monitoring system as a coded string which describes which activities have to be invoked on the occurrence of a specified event. As a matter of fact, service requests program the monitoring system to listen to event occurrences, perform actions, and transfer results back to the tool. Thus, the monitoring system follows an event/action working model. Cooperation with tools is also covered by this model as the receipt of service requests represents an event to which an appropriate action belongs.

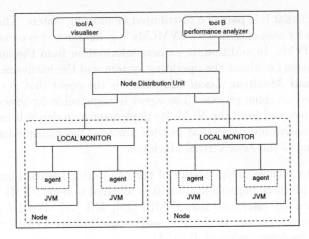

Fig. 1. Architecture of the monitoring environment.

Fig. 1 shows the software module structure of the monitoring system. On the top of the Fig. 1 there are various tools running e.g. on a remote workstation, which communicate with the central component of the monitoring system. This component is called *Node Distribution Unit* (NDU).

Node Distribution Unit is intended to analyse each request issued by a tool and must split it into pieces that can be processed locally by the local monitors on the nodes involved. E.g. a tool may issue the request

: jvm_run_gc([j_1,j_2])

in order to run the garbage collector on the JVMs identified by the tokens j_1 and j_2. In this case, the NDU must determine the node executing JVM process j_1 and the node executing j_2. If the different nodes are different, the request must be split into two separate requests, i.e. : jvm_run_gc([j_1]) and : jvm_run_gc([j_2]), which are sent to the proper nodes. The NDU must also assemble the partial answers which are received from the local monitor processes into a global reply sent to the tool. Addition and removal of nodes is also detected and handled by the NDU.

The distributed part of the monitoring system consists of multiple *Node Local Monitors* (NLM), one per node of the target system and *Java Virtual Machine Local Monitors* (JVMLM), one per process of the JVM (which can be multiple on one node).

Node Local Monitor processes control the JVM processes via agents (JVMLMs) and operating system interfaces. The NLM offers a server interface that is similar to the monitoring interface, with the exception that it only accepts requests that can be handled locally. This approach can simplify the implementation of the local monitor, without paying any attention to the fact that it is part of a distributed monitoring system. The NLM is responsible for cooperation with JVMLMs, e.g. for *start/stop* or *attach/detach* to local JVMs. In addition, it gathers information from the outside of the JVM process i.e. about the operating system and the hardware.

Java Virtual Machine Local Monitor is the *agent* that is embedded in the virtual machine process. The agent is responsible for execution of the requests received from the NLM. Its implementation depends on the virtual machine native interfaces that provide low level mechanisms for interactive monitoring of the Java Virtual Machine.

In order to achieve independence of the monitoring system from a concrete Java Virtual Machine implementation, it is required to build JVMLMs that will use standard JVM interfaces like JVMPI [10], JNI [11], JVMDI [12] or Java bytecode instrumentation. In the next section we consider JVMPI as a possible basis for the implementation of JVMLM.

4   Interaction with JVM

Low level interactions with the Java virtual machine are a critical part of the monitoring system. The component responsible for this activity (JVMLM) has to provide functionality that allows to realize services defined in the interface specification. One of the possible approaches is using the JVMPI interface to access the JVM.

JVMPI defines a mechanism for obtaining profiling data from the Java virtual machine. Profiling data provides information about events that occur during application run time, in the form that enables to track the cost of these events, e.g. it can show what portion of the program allocates the greatest amount of memory. Building the JVMLM on top of JVMPI is possible due to the following features of the interface:

- JVMPI is an interface between JVM and a *profiler agent* that runs in the same process. The agent is implemented as a dynamic library which is loaded by JVM at startup.
- JVMPI defines a set of events that can be sent to the agent by JVM during Java program execution. JVM notifies the agent about such activities like: method *enter/exit*, class *load/unload*. These events are a low level realization of some event services from the monitoring interface specification.
- JVMPI provides a set of callback functions implemented by the Java virtual machine that allow the agent to obtain additional information in response to an event notification and for basic threads and garbage collector manipulations, like: thread *suspend/resume/get backtrace*, garbage collector *enable/disable/run*.

In order to show what kind of information can be obtained via the JVMPI interface we present in Fig. 2 parts of the reports generated by a simple agent during a program execution that simulates a producer - consumer problem.

```
                  Thread start event report
    described by: thread object id, thread id, thread name, thread group
THREAD START (obj=814dd10, id = 4,   name="Finalizer",  group="system")
THREAD START (obj=822de00, id = 6,   name="Producer#0", group="main")
THREAD START (obj=82302a0, id = 7,   name="Consumer#0", group="main")
```

Thread stack trace consists of: trace number, thread id, set of frames.
Frame consists of: class name, method name, source file name, line number.

```
TRACE 15: (thread=4)
    java/lang/Object.wait(Object.java:Native method)
    java/lang/ref/ReferenceQueue.remove(ReferenceQueue.java:112)
    java/lang/ref/ReferenceQueue.remove(ReferenceQueue.java:127)
    java/lang/ref/Finalizer$FinalizerThread.run(Finalizer.java:174)
...
MONITOR DUMP BEGIN                        Thread status:
    THREAD 8,  trace 9,   status: MW    MW - waiting on a monitor
    THREAD 7,  trace 8,   status: MW    CW - waiting on a condition
    THREAD 6,  trace 14,  status: CW    variable
    THREAD 4,  trace 15,  status: CW    R - runnable
    THREAD 2,  trace 17,  status: R     trace number refers to cur-
    ...                                 rent thread stack trace
MONITOR Buffer(8230258)
    owner: thread 6, entry count: 1
    waiting to enter: thread 7, thread 8
MONITOR java/lang/ref/ReferenceQueue$Lock(814f4e8) unowned
    waiting to be notified: thread 4
    ...            Monitor state described by: name, object id which is
                   associated with a monitor, monitor owner thread, entry
                   to monitor count, waiting threads.
MONITOR DUMP END
```

Fig. 2. JVMPI agent reports.

The first report provides information about a thread start event generated by the JVM. The next one shows an example thread stack trace (for the "Finalizer" thread). This is a result of a JVMPI call-back function called by the agent. Very informative is the report of monitors dump. It shows the statuses of threads and states of the Java synchronization monitors.

JVMPI provides only low level information about physical aspects of a running application, but it offers enough data to build higher level services defined in our monitoring interface specification. The whole picture of requests flow in the monitoring system is shown in the next section.

5   Monitoring Scenario

In order to outline the interaction between layers in the monitoring system and to show how the monitoring interface can by used by a tool, below we present a monitoring scenario for an *instrumentation tool*.

The interface specification defines services that allow to instrument Java classes. Instrumentation enables to insert additional bytecode into compiled Java methods that can be executed. This bytecode can realize some monitoring system functionality, for instance, the timestamps of the method call measurements. Fig. 3 shows an architecture of the monitoring system with the code instrumentation tool. The instrumentation tool sends requests to the monitoring system in order to exchange a Java class file before it is loaded by the JVM.

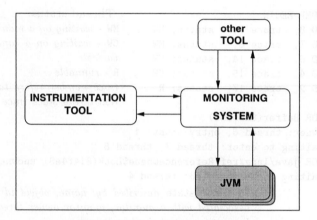

Fig. 3. Instrumentation tool.

Let's assume the tool is interested in having instrumented the *java.lang.Thread* class. Then, the following conditional[1] request can be used:

```
jvm_loaded_hook([j_13],"java.lang.Thread"):
                    class_exchange([$class],newData)
```

The monitoring system executes the *class_exchange* action after the event *jvm_loaded_hook* occurs[2]. The parameters for the event service determine which JVM is monitored (j_13) and which Java class should be considered. Parameter $class in the manipulation service indicates the class hook returned by the event. newData is a reference to the instrumented class file.

An additional request can be sent by a tool to be notified that the class has been correctly loaded by the JVM.

---

[1] A conditional request consists of an event definition and a list of actions that are executed whenever a matching event occurs.

[2] *Class hook* is a pointer to the file of a given class.

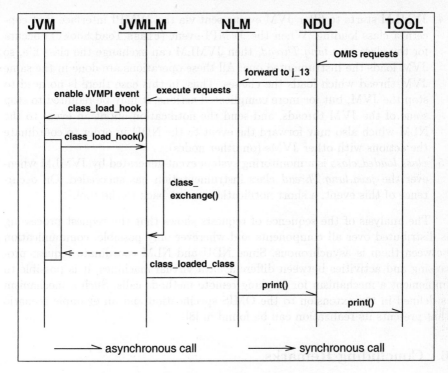

Fig. 4. Sequence diagram of the request execution.

```
class_loaded_class([c_java/lang/Thread_j_13],[],[]):
                    print(["Class",$class,"loaded.OK."])
```

The parameter c_java/lang/Thread_j_13 is a monitoring system representation of class *java.lang.Thread* on JVM with id j_13.

Fig. 4 presents a sequence diagram that shows the stages of execution of these two requests that are input data for the monitoring system. On the output, the message is returned to the tool, but the real effect of those requests is loading the instrumented class file by the JVM. There are only shown the basic components of the monitoring system that are involved in the following scenario:

1. First, the tool sends asynchronously the requests to the monitoring system. In order to minimize communication overhead, these two requests can be sent within a single stream.
2. The central part of the monitoring system (NDU) parses the requests, and – if it is required – splits them and then sends them to a proper NLM. In this example, both requests are sent to the NLM placed on the node where the JVM with id j_13 can be found.
3. NLM instructs the appropriate JVMLM how it should react on the event of *java.lang.Thread* class loading.

4. JVMLM starts to filter JVM events sent via the JVMPI interface that concern a class loading. When the JVMPI event (class_load_hook()) occurs for the class *java.lang.Thread*, then JVMLM can exchange the class file, so JVM loads the instrumented one. All these operations are done in the same JVM thread which loads the classes. Thus, in this case there is no need to stop the JVM, but for more complicated actions, it may be required to stop some of the JVM threads, and send the notification about an event to the NLM which also may forward the event to the NDU in order to coordinate the actions with other JVMs (on other nodes).

5. *class_loaded_class* is a monitoring system event generated by JVMLM whenever the *java.lang.Thread* class instrumentation has succeeded. On occurrence of this event, a short notification is sent back to the tool.

The analysis of the sequence of requests shows that the request processing is distributed over all components and wherever only possible, communication between them is asynchronous. Since NDU and NLM coordinate request processing and activities between different Java virtual machines, it is possible to implement a mechanism for tracking remote methods calls. Such a mechanism is defined in the extension to the OMIS specification and an example scenario that presents its realization can be found in [8].

6   Concluding Remarks

In the paper we presented a model of a monitoring system underlying the functionality of tools used for distributed Java applications. The interface between the tools and the monitoring system is based on an extension to the OMIS specification. The structure of the environment comprises a central component for handling tool requests and replies to them and a set of distributed components for control of JVM processes via profiler agents provided within the JVMPI interface.

The low-level interaction with JVM is intended to supply indispensable information on event occurrences the tools may be interested in during the execution of the program as well as triggering proper actions, based on callback functions provided within JVM.

The monitoring scenario presented had to illustrate a feasible interaction within the monitoring system infrastructure, to reveal the need in establishing a well-defined communication and coordination mechanism between the layers of the monitoring system.

Currently the interface specification is stable, no changes are introduced. At the moment we have implemented a version of performance monitoring tool which realizes tracking the execution of a Java application, based on a subset of the Java-bound extension to OMIS. Now we are working on a design of the reference implementation, and exploring different techniques for realization of low level interaction with JVM, like bytecode instrumentation and JNI. The possible cost of implementation may be assessed on the level of modules designed with UML.

Using standard interfaces like JNI, JVMDI and the possibilities of code instrumentation allows to avoid modifications of JVM when building the monitoring system.

Acknowledgments. The authors are very grateful to Prof. Michael Gerndt for valuable discussions. This work has been carried out within Polish-German collaboration.

References

1. G. Pennington and R. Watson: JProf – a JVMPI Based Profiler
   http://starship.python.net/crew/garyp/jProf.html
2. I.H. Kazi, D.P. Jose, B. Ben-Hamida, C.J. Hescott, C. Kwok, J. Konstan, D.J. Lilja, and P.-C. Yew: JaViz: A Client/Server Java Profiling Tool, *IBM Systems Journal*, **39**(1), (2000) 96-117
   http://www.research.ibm.com/journal/sj/391/kazi.html
3. Borland: JBuilder 3.5 Datasheet
   http://www.borland.com/jbuilder/jb35/datasheet.html
4. P. Bellavista, A. Corradi and C. Stefanelli: Java-based On-line Monitoring of Heterogeneous Resources and Systems. In 7th Workshop HP OpenView University Association, Santorini, Greece, June 2000
   http://lia.deis.unibo.it/Staff/PaoloBellavista/papers/hpovua00.pdf
5. T. Ludwig, R. Wismüller, V. Sunderam, and A. Bode: OMIS – On-line Monitoring Interface Specification (Version 2.0). Shaker Verlag, Aachen, vol. 9, LRR-TUM Research Report Series, (1997)
   http://wwwbode.in.tum.de/~omis/OMIS/Version-2.0/version-2.0.ps.gz
6. R. Wismüller, J. Trinitis and T. Ludwig: A Universal Infrastructure for the Runtime Monitoring of Parallel and Distributed Applications. In Euro-Par'98, Parallel Processing, volume 1470 of Lecture Notes in Computer Science, pages 173-180, Southampton, UK, September 1998. Springer-Verlag
7. M. Bubak, W. Funika, P. Metel, R. Orlowski, and R. Wismüller: Towards a Monitoring Interface Specification for Distributed Java Applications. In Proc. 4th Int. Conf. on Parallel Processing and Applied Mathematics - PPAM 2001, Naleczow, Poland, September 2001. To appear.
8. P. Mętel and R. Orłowski: Extensions to On-line Monitoring Interface Specification for Java Applications. Master Thesis. Institute of Computer Science, Stanislaw Staszic University of Mining and Metallurgy, Kraków, 2001
9. T. Lindholm and F. Yellin, *The Java Virtual Machine Specification*, Addison-Wesley Publishing Co., Reading, MA (1996)
10. Sun Microsystems: Java Virtual Machine Profiler Interface (JVMPI)
    http://java.sun.com/products/jdk/1.2/docs/guide/jvmpi/jvmpi.html
11. Sun Microsystems: The Java Native Interface Specification
    http://java.sun.com/products/jdk/1.1/docs/guide/jni/spec/jniTOC.doc.html
12. Sun Microsystems: Java Virtual Machine Debug Interface (JVMDI)
    http://java.sun.com/products/jdk/1.2/docs/guide/jvmdi/jvmdi.html

A Concept of Portable Monitoring of Multithreaded Programs

Bartosz Baliś[1], Marian Bubak[1,2],
Włodzimierz Funika[1], and Roland Wismüller[3]

[1] Institute of Computer Science, AGH, al. Mickiewicza 30, 30-059 Kraków, Poland
[2] Academic Computer Centre – CYFRONET, Nawojki 11, 30-950 Kraków, Poland
[3] LRR-TUM – Technische Universität München, D-80290 München, Germany
{bubak, funika, balis}@uci.agh.edu.pl,
wismuell@in.tum.de
phone: (+48 12) 617 39 64, fax: (+48 12) 633 80 54 phone: (+49 89) 289-28243

Abstract. Multithreading is potentially a powerful way to achieve high performance in parallel programming. However, there are few tools that support multithreaded programs development. This paper presents an analysis which has led to a concept of building an autonomous monitoring system for multithreaded programs on top of which various tools can be based. Many problems specific to monitoring multithreaded programs are presented, as well as the ideas to solve them, specifically we focus on efficiency, scalability and portability of the monitoring system.
Keywords: Multithreading, monitoring, parallel tools, shared memory

1 Introduction

Multithreaded programming is potentially the most efficient way of parallel programming as threads are faster than processes both in creation, context switching and possibly also communication. Moreover, on many modern architectures which support shared memory, multithreading is a natural way of parallel programming, as threads naturally share address space and usually communicate via shared resources.

On the other hand, multithreading is inherently more difficult and error-prone than multiprocessing. This increases the need for specialized on-line tools supporting development of multithreaded programs. Unfortunately, multithreading support in existing tools is rather poor.

The goal of this paper is to present a concept for a portable, autonomous monitoring system for multithreaded programs focusing on SMP architectures. We present concepts to deal with many problems related to monitoring threads, specifically the monitoring system is designed to be portable, efficient, and scalable. The communication protocol between tools and the monitoring system is based on the OMIS specification [11]. OMIS is a standardized interface for communication between tools and monitoring systems, designed to support a variety of tool classes and extendible to provide new functionality when needed.

P.M.A. Sloot et al. (Eds.): ICCS 2002, LNCS 2330, pp. 884–893, 2002.

We mostly focus on two types of tools: debuggers and performance analyzers, therefore we describe the functionality of the monitoring system to support these kinds of tools. However, the design is going to be generic, so that other types of tools can also be supported.

2 Analysis of Problems in Thread Monitoring

Multithreaded programming poses many problems not present in developing "ordinary" multiprocess parallel applications. Most of the problems derive from the fact that multiple threads share single address space which is not the case with processes. Access to shared data needs synchronization which can be tricky and error-prone. A thread can unexpectedly change a value of some variables which can lead to data corruption. Below we analyze a number of problems connected with monitoring threads.

2.1 Data races

An important class of problems is a *data race* [12]. A race is a situation in which two or more threads access the same piece of data concurrently and at least one access is a write one. Races lead to indeterministic behaviour since data being raced can be accessed in different sequence from execution to execution sometimes (but possibly not every time), probably causing an error. This makes it particularly hard to determine the cause of a program's incorrect behaviour. In addition, due to the *probe effect*, monitoring overhead can easily mask data races which occur in a program when it runs normally. Thus, data races can remain undetected for a long time.

2.2 Performance measurement

Threads communicate between themselves via shared resources, thus this communication is not explicit as in message passing. This makes it harder to measure performance of multithreaded applications (in comparison with a message passing application), since it is hard to measure the amount of communication passed between threads or time needed for this communication, which are essential pieces of information for certain performance measurements.

2.3 Lack of the OS support

Other problems with threads are related to operating system limitations. Some operating systems do not support directly thread monitoring. First of all, standard system interfaces like `ptrace` or the `/proc` file system [16] are supported by the kernel, and - if at all - allow only for dealing with kernel threads. Monitoring user-level threads usually requires a detailed knowledge of the system-specific thread library. There is no standardized interface for monitoring user-level threads. It is up to a particular vendor to provide e.g. a library with interface for monitoring user-level threads (e.g. pthreads). In addition, on some systems the mentioned interfaces are suited only to sequential debugging, since they provide only synchronous communication between debugger and application (i.e., a debugger pauses when the application is running and vice versa - when the application hits a breakpoint, it pauses and transfers the control to the debugger). A debugger for multithreaded programs should be able to independently control each target thread which requires asynchronous communication (neither the whole debugger nor the application process should be stopped, but only the specific thread).

2.4   Shared code image

Another problem stems from the fact that multiple threads often share a single code image. In such a case, if we instrument (e.g. insert a breakpoint) one of these threads, whenever any of the other threads reaches the instrumentation point (e.g. the breakpoint) we need a kind of filtering so as to ensure the proper actions are executed only for the really instrumented thread (e.g. only the proper thread is stopped by the breakpoint). The cost of such a filtering may be very high therefore it must be handled very efficiently.

2.5   Efficiency and scalability

Other issues which must be taken into consideration when implementing a monitoring system for multithreaded applications are efficiency and scalability of the monitoring system. A multithreaded application can potentially create hundreds of threads. For this reason, we must take care that monitoring system's design and implementation are scalable. Specifically, we should ensure that no part of the monitoring system must handle an excessive number of requests at a time. Moreover, as mentioned earlier, threads are more 'lightweight' than processes - their creation time and context switch overhead are much lower than for processes. Therefore, the overhead of monitoring threads should also be proportionally lower than for processes! Otherwise, the monitoring system's influence on the monitored application will be relatively very high.

3   State of the Art

Until today, there were numerous attempts to create tools for multithreaded applications. The bulk of them are debuggers, e.g., gdb, TotalView, kdb, Node-Prism and LPdbx. Wildebeest [1] is an example of a debugger based on gdb, which supports both kernel and user threads. However, it is strictly limited to HPUX platforms and implements only synchronous thread control. TotalView [7] is a commercial debugger which supports a variety of platforms and offers a rich set of debugging capabilities. It is well suited for multithreaded applications and provides even support for applications developed in OpenMP [6]. However, it does not allow for asynchronous thread control unless this feature is supported by the operating system. kdb [4] was designed to address the limitations of other debuggers, specifically it was designed to handle user-level threads and be able to control each target thread independently.

One of the rare efforts to address the performance analysis of multithreaded applications is the Tmon tool [9]. Tmon is a monitoring system combined with visualization module used to present *waiting graphs* for multithreaded applications.

To sum up, none of the mentioned efforts features a separate facility for monitoring multithreaded applications, designed to potentially support multiple classes of tools.

4   Monitoring Concepts

This section provides a more detailed description of our design concepts as well as a deeper insight into the general problems arising in debugging and performance analysis of multithreaded programs. Solutions to most of these issues are proposed.

4.1   Support for debugging

Some of the problems related to debugging multithreaded applications are outlined in Section 2. The two primary problems described there are the issue of the independent control of each target thread and the overhead caused by instrumenting a thread which shares its code image with several other threads. In case of debugging the second problem applies to setting and hitting breakpoints. These two problems can be addressed by placing part of monitoring system directly in the application. With this approach, this *application monitor* can handle events or perform actions on behalf of the *local monitor*, which in turn is the part of the monitoring system residing on each node of the target system. The local monitor thread, instead of establishing a synchronous connection via ptrace or the /proc filesystem, can delegate this task to the application monitor by means of an asynchronous communication (e.g. message passing). Thus, blocking the whole monitor process can be avoided. In a similar way the problem of instrumentation (breakpoints) can also be addressed. The applicability of hitting a breakpoint can be checked locally in the application monitor while the local monitor will only be notified if the breakpoint was hit by a proper thread. The approach described here was successfully used for breakpoints as described in [4] and [10].

It should be noted here that only some tasks can be delegated to the application monitor. Some of them are simply impossible to be handled (e.g. a process cannot stop itself) while some others may be intentionally left to be handled by a separate (local monitor) process (thread). The latter may be dictated by the safety reasons. The monitor code and data inside a user thread may be corrupted by bugs in the application. For this reason, it might be safer to leave some important debugging functions in the local monitor. Only these functions which most affect efficiency should be placed in the user application space.

4.2   Support for performance analysis

In most performance measurements applicable for message-passing applications the key role is played by the time spent on or the volume of communication between processes. In multithreaded programs, it is particularly hard to measure these characteristics, since there is no explicit communication between threads - all communication is done via shared resources. Moreover, this method of communication needs a frequent synchronization to ensure exclusive write access to shared data. This factor also affects performance as due to synchronization some threads may be waiting for others. This *waiting time* can be used as a basis of performance measurements related to multithreaded programs. In [9], an approach is presented to measure waiting time and create *waiting graphs*.

The performance analysis of multithreaded programs has to address the following issues:

- definition of performance measurements for multithreaded applications,
- mechanisms for instrumentation of thread libraries and application code to gather performance data,
- efficiency in gathering performance data.

Definition of performance measurements

First we have to answer the question what we really need to measure, i.e. define the performance measurements for threads. In case of threads we are interested in at least three types of measurements - CPU load related to computations, amount of data passed between threads, and delays related to communication and synchronization, parallelizing operations or remote memory access. The measurements will be defined with regard to constraints, e.g. within code region like loop, function or location, e.g. within particular threads.

Instrumentation

The instrumentation is the main way to gather performance data. It allows for collecting information about interesting events. In case of multithreaded applications the user is interested in the information on accesses to shared objects and operations related to synchronization, i.e., locking/unlocking mutexes or condition waits. For the instrumentation, two primary techniques are considered: *binary wrapping* and *dynamic instrumentation*. The binary wrapping is a technique in which the functions to be instrumented are wrapped with additional code that performs event detection and execution of associated actions [5]. This method has a drawback that it requires relinking the application. Dynamic instrumentation is an advanced technique to patch the code of a running process, directly in the memory. This approach is supported by the DyninstAPI [3]. However, while the DyninstAPI successfully allows to instrument function calls, it does not solve the problem of efficient instrumentation of generic objects, e.g. data shared by multiple threads to detect accesses to this data. While software approaches exist to handle this kind of instrumentation for the purpose of debugging (*watchpoints* of *data breakpoints* [16, 14]) they are not suited for performance analysis, since either their overhead is too high, or they are limited to the observation of a very limited number of memory cells. Performance measurement on shared data can, however, be supported with minimal hardware extensions. A design of a hardware monitor suitable for this task is outlined in [8].

Efficient data gathering

The efficiency problems in gathering performance data derive from the fact that on one hand the data can be generated rapidly, on the other hand, it may be rarely accessed. If we decide to immediately report each event of interest to the monitoring system, the resulting overhead will probably be unacceptable due to frequent context switches and excessive communication. To reduce the number of context switches, we can store the data *locally* and send it to the monitoring system e.g. on an explicit demand. We can further reduce the communication by using *efficient data structures* to store the performance data. Instead of saving a raw trace of events, we can initially process the data and save the summarized information (e.g. only the number of events if this information is sufficient to perform the measurements).

We have a long experience in monitoring message-passing applications with an OMIS-compliant monitoring system OCM and OCM-based tools, among others a performance analyzer PATOP and a debugger DETOP. During the previous work with the OCM and PATOP [2] we used both local storing and specialized data structures to hold performance data, namely *counters* and *integrators*. The support of these features was enabled by placing parts of the monitoring system directly in the context of application processes. The described mechanisms proved to be very efficient. Full performance monitoring overhead was only about 4%.

4.3   Detecting data races

Data races are particularly hard to detect as mentioned in section 2. On the other hand, it is particularly important to detect this kind of bugs, as they can easily cause many other bugs. In [15] a two-phase approach for data race detection is presented. The program being analyzed is executed twice. During the first execution, information about synchronization operations is collected. The second execution is a *replay* of the first one. In this execution, all read and write accesses to global data are instrumented, causing a rather high, but unavoidable intrusion. The information collected during the first execution is used to force the second one to still be *equivalent* to the first one. Thus, the replay will possibly fail (i.e. the internal program flow will differ across the two executions) only if a race condition occurs. This solution avoids the probe-effect otherwise caused by the intrusion of collecting information about memory operations. This approach can be realized using an OMIS-compliant monitoring system. The first execution is already supported by OMIS as it basically consists in instrumenting synchronization operations. The replay operation can also be implemented as an additional service in OMIS.

4.4   Monitoring interface

An autonomous monitoring system must offer a suitable interface that is both flexible and powerful enough to handle very different kinds of tools. In our design, we use OMIS [11] for this purpose, because besides fulfilling these requirements, the existing specification already allows for threads and includes some basic services to inspect and control threads, although those services are currently not implemented in the OMIS-compliant monitoring system OCM. In addition, OMIS specifies a well-defined way to extend it with new objects and services, allowing to smoothly integrate shared data objects or synchronization objects, such as mutexes or condition variables, as well as the associated services. The new services will mainly monitor accesses to these objects, using the instrumentation techniques discussed in section 4.2.

Apart from OMIS, there are few other approaches defining standardized interfaces for monitoring systems. A notable project is DPCL [13], which is entirely based on dynamic instrumentation. DPCL is, however, not well suited for our purpose, because it does not take threads into consideration and defines a more low-level interface as compared to OMIS.

4.5   Design alternatives for a monitoring system

The monitoring system is planned to support multithreaded applications on shared memory multiprocessor architectures. The first question to be answered is what should be the architecture of the monitoring system. In case of the OCM, a monitoring system for clusters of workstations, this architecture is a collection of *local monitors*, one per node in the target system, with one central component, a so called NDU (*Node Distribution Unit*) responsible for accepting a request from a tool, splitting it into sub-requests for local monitors which execute the request and send the (partial) replies back to the NDU which in turn assembles them into a single reply for the tool.

For the SMPs, several architectures of the monitoring system are taken into consideration:[1]

1. Pool of local monitors, each assigned to a specified subset of nodes, and one central component on top of the local monitor to play the role of the NDU, called *service manager*.
2. One local monitor per SMP node and one service manager.
3. One local monitor per SMP node and one service manager per each monitored application.

---

[1] The term 'node' or 'SMP node' refers to a building block of an SMP system, possibly consisting of a couple of processors with local, shared memory, like in cc-NUMA systems.

4. Hybrid of some of the above.

The first two architectures resemble the centralized architecture of the OCM, while the 3rd approach introduces decentralization which should address possible scalability problems. On the other hand, the 2nd and 3rd approaches address the *locality* issue, i.e., each SMP node would have its own dedicated local monitor, while in the 1st approach a single local monitor may handle multiple SMP nodes which in turn reduces the total number of local monitors. Which of these architectures is the most appropriate one depends on the relative importance of the factors scalability, locality, and number of local monitors and cannot be decided a priori.

The second important question is whether the monitoring system should itself be multithreaded or not. At the moment, it seems that it is a natural solution to implement local monitors as threads of a single process (e.g. for the efficiency reasons - fast communication between monitor threads). It may still be necessary to have some parts of the monitoring system as separate processes e.g. to better handle asynchronous control of threads.

Another decision to take is whether there will be a single monitoring system in OS, or there will be one instance of it per application. The OCM follows the latter approach, which is easier to implement as the monitoring system has to deal only with a single application at a time and it runs as a user process instead of root so no concern about access control and permissions is necessary. This approach is also potentially better scalable (only one application to be handled). On the other hand, it introduces more monitoring processes overall in the system so its total overhead could be higher.

In the OCM, some parts of the monitoring system reside directly in the target application process space. These parts are responsible for handling certain events *locally* to reduce the overhead of monitoring. This particularly helps in gathering data for performance measurements as described in Section 4.2. In a monitoring system for SMPs, this solution should also be applied, especially because it can also help in debugging multithreaded applications (Section 4.1).

A generic architecture of the monitoring environment is shown in Fig. 1. In this figure, local monitors and the service manager are threads of a single Global Monitor process. Application consists of a collection of threads with an additional application monitor thread.

5 Conclusion

Monitoring systems for multithreaded programs are faced with a number of severe problems, concerning their efficiency, scalability and portability. In this paper, we have outlined the most important problems that need to be solved and have presented proper solution concepts, as well as design alternatives for the resulting monitoring system. In an ongoing project at Institute of Computer Science, AGH, these concepts will be evaluated in detail by fully implementing this monitoring system. In particular, the project addresses several issues which until now are not adequately solved by existing approaches:

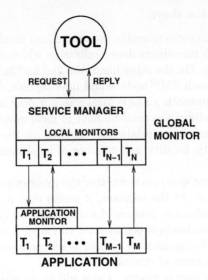

Fig. 1. Possible architecture of the monitoring environment

1. A universal autonomous monitoring facility for multithreaded applications will be implemented. None of current tools for multithreaded programming support features a separate monitoring system.
2. Both kernel and user threads will be supported. Few current products support user-level threads, even fewer do it in a portable way.
3. The problem of asynchronous thread control regardless of platform support will be addressed. Current approaches to solve this problem usually only work when the operating system supports independent thread control.
4. The efficiency and scalability in monitoring threads will be one of the primary concerns. This includes the concept of application monitors which perform certain tasks on behalf of the global monitor. Current projects using this approach do it to a limited extent (e.g. only to handle breakpoints).

References

1. S. S. Adayapalam. In Search of Yeti: Footprint Analysis with Wildebeest. In Mireille Ducassé, editor, *Proceedings of AADEBUG 2000, Fourth International Workshop on Automated Debugging*, Munich, Germany, August 2000. www.irisa.fr/lande/ducasse/aadebug2000/proceedings/04srikanth.ps.gz.
2. M. Bubak, W. Funika, B. Baliś, and R. Wismüller. On-line OCM-based Tool Support for Parallel Applications. In Y. Ch. Kwong, editor, *Annual Review of Scalable Computing*, volume 3, chapter 2, pages 32–62. World Scientific Publishing Co. and Singapore University Press, Singapore, 2001.
3. Bryan Buck and Jeffrey K. Hollingsworth. An API for Runtime Code Patching. *The International Journal of High Performance Computing Applications*, 14(4):317–

329, Winter 2000.
citeseer.nj.nec.com/buck00api.html.

4. P. A. Buhr, M. Karsten, and J. Shih. KDB: A Multi-threaded Debugger for Multi-threaded Applications. In *Proc. of SPDT'96: SIGMETRICS Symposium on Parallel and Distributed Tools*, pages 80–89, Philadelphia, Pennsylvania, USA, May 1996. ACM Press.

5. J. Cargille and B. P. Miller. Binary Wrapping: A Technique for Instrumenting Object Code. *ACM SIGPLAN Notices*, 27(6):17–18, June 1992.

6. J. Cownie and S. Moore. Portable OpenMP debugging with TotalView. In *Proceedings of EWOMP'2000, Second European Workshop on OpenMP*, 2000.

7. TotalView Multiprocess Debugger. WWW-Site of Etnus Inc., Framingham, MA, USA, 1999.
http://www.etnus.com/products/totalview/index.html.

8. Robert Hockauf, Jrgen Jeitner, Wolfgang Karl, Robert Lindhof, Martin Schulz, Vicente Gonzales, Enrique Sanquis, and Gloria Torralba. Design and Implementation Aspects for the SMiLE Hardware Monitor. In *Proceedings of SCI-Europe 2000*, pages 47–56, Munich, August 2000. SINTEF Electronics and Cybernetics.
http://wwwbode.cs.tum.edu/~schulzm/papers/2000-08-gt-scieuro.pdf.

9. M. Ji, E. W. Felten, and K. Li. Performance Measurements for Multithreaded Programs. In *Measurement and Modeling of Computer Systems*, pages 161–170, 1998.
citeseer.nj.nec.com/article/ji98performance.html.

10. P. B. Kessler. Fast Breakpoints. Design and Implementation. *ACM SIGPLAN Notices*, 25(6):78–84, June 1990.
www.acm.org:80/pubs/citations/proceedings/pldi/93542/p78-kessler.

11. T. Ludwig, R. Wismüller, V. Sunderam, and A. Bode. *OMIS — On-line Monitoring Interface Specification (Version 2.0)*, volume 9 of *LRR-TUM Research Report Series*. Shaker-Verlag, Aachen, Germany, 1997. ISBN 3-8265-3035-7.
http://wwwbode.in.tum.de/~omis/OMIS/Version-2.0/version-2.0.ps.gz.

12. R. H. B. Netzer. What are Race Conditions? Some Issues and Formalizations. *ACM Letters on Programming Languages and Systems*, 1:74–88, March 1992.

13. D. Pase. Dynamic Probe Class Library (DPCL): Tutorial and Reference Guide, Version 0.1. Technical report, IBM Corp., Poughkeepsie, NY, July 1998.
http://www.ptools.org/projects/dpcl/tutref.ps.

14. Paul E. Roberts. Implementation and Evaluation of Data Breakpoint Schemes in an Interactive Debugger. Master's thesis, Department of Computer Science, University of Utah, December 1996.
www.cs.utah.edu/flux/papers/perobert_thesis.ps.gz.

15. M. Ronnse and K. De Bosschere. Non-intrusive On-the-fly Data Race Detection Using Execution Replay. In M. Ducasse, editor, *Proceedings of AADEBUG 2000 Fourth International Workshop on Automated Debugging*, Munich, Germany, August 2000.
www.irisa.fr/lande/ducasse/aadebug2000/proceedings/20ronsse.ps.gz.

16. Jonathan B. Rosenberg. *How Debuggers Work: Algorithms, Data Structures, and Architecture*. John Wiley & Sons, 1996.

dproc - Extensible Run-Time Resource Monitoring for Cluster Applications

Jasmina Jancic, Christian Poellabauer, Karsten Schwan, Matthew Wolf, and
Neil Bright

College of Computing
Georgia Institute of Technology
Atlanta, GA 30332
{jasmina, chris, schwan, mwolf, ncb}@cc.gatech.edu

Abstract. In this paper we describe the *dproc* (distributed /proc) kernel-level mechanisms and abstractions, which provide the building blocks for implementation of efficient, cluster-wide, and application-specific performance monitoring. Such monitoring functionality may be constructed at any time, both before and during application invocation, and can include dynamic run-time extensions. This paper (i) presents dproc's implementation in a Linux-based cluster of SMP-machines, and (ii) evaluates its utility by construction of sample monitoring functionality. Full version of this paper can be found at: http://www.cc.gatech.edu/systems/projects/dproc/

1 Introduction

Motivation. Run-time monitoring of large-scale cluster machines is critical to the successful operation of cluster applications. This is because even a single high performance application running on a cluster typically exhibits highly dynamic computational behavior. Moreover, most applications do not run in isolation: they conduct I/O, require real-time data from remote sensors[3], access large-scale remote data contained in digital libraries or share files across the computational grid, support scientific collaboration by remote visualization of their data[18,19], and interact with other computations via the Grid [1][1,2]. Unless run-time monitoring is used to determine the appropriate and dynamic allocation of cluster resources to applications[6,10], high performance is unlikely to be attained.

Run-time monitoring mechanisms are required to dynamically diagnose the performance of cluster programs. The monitoring tools commonly available to cluster programmers, however, are not only used for adjusting programs[22,11] at runtime, but they are also used for diagnosing performance problems at the time of program implementation, for program profiling[20], and even for debugging them[16]. As a result, developers routinely impose a wide range of requirements on such tools, including:

---

[1] http://www.globus.org/

P.M.A. Sloot et al. (Eds.): ICCS 2002, LNCS 2330, pp. 894–903, 2002.

Selective Monitoring of Multiple resources. For cluster machines, at minimum, monitoring must capture usage and availability of both CPUs and network links. Often, additional information is required, as evident from the rich performance data routinely available from current monitoring tools for high performance machines[20, 21]. For large cluster programs, resulting overheads make it infeasible to capture all such data about all nodes at all times. Thus, monitoring must be performed selectively, applied dynamically to precisely the resources and program components under investigation.

Variable granularity. Fine-grain monitoring data is needed for certain optimizations of applications, such as recognizing the precise arrival times of processes at shared barriers, understanding the actual overlap in communication and computation attained by a code[7], or diagnosing the degree of simultaneity in communications and thus, the potential network loads being imposed.Therefore, it should be possible to conduct monitoring at variable frequencies and rates, thereby altering the precision vs. perturbation induced by monitoring.

Flexible and dynamic analysis. It is well-known that monitoring data should be condensed and filtered as 'close' as possible to its points of capture, to reduce monitoring overheads and perturbation[20, 6]. However, the actual analyses to be performed typically depend on what monitoring is currently used for, and such analyses vary in their behavior, some causing little perturbation, others requiring substantial trace data before they may be applied. No single built-in set of analysis routines will satisfy all applications. Furthermore,especially for long running applications, it is not viable to install all monitoring support once, then simply use it. Instead, monitoring should be installed at runtime[20], analyses must be changed as needed[6], and its monitoring overheads should be dynamically controlled.

The dproc approach to performance monitoring. This paper describes kernel-level mechanisms and abstractions that are the building blocks for cluster-wide performance monitoring. Their realizations in a Linux-based cluster of SMP machines are evaluated by construction of dynamically extensible and changeable monitoring functionality. Cluster resources monitored include both node and network attributes, including CPU loads, memory and swap usage, achieved communication bandwidth, loss rate and message round-trip times. The API the tool presents to programmers is an extension of the standard /proc performance interface offered by Linux systems, hence motivating the use of the term dproc for our facilities.

Dproc offers the following functionality:

Selective monitoring via kernel-level publish/subscribe channels. The basic operating system construct offered by dproc is that of *monitoring channel* (monchannel). A single monchannel can capture monitoring information from any number of sources, and it can distribute it to any number of interested parties (sinks). Sources or sinks may reside at user- or at kernel-level. In this fashion, a monchannel can capture monitoring data from multiple resources, and the results of such monitoring can be distributed to whomever requires such data (e.g., performance displays, data storage engines). *Standard API.* Applications need not explicitly handle monchannels. An application accesses dproc entries, which are physically represented by underlying monchannels, through the standard /proc pseudo-file system interface. Kernel modules perform monitoring by publishing data described as *monitoring attributes* (monattributes) on channels and by listening for attribute updates. Applications simply access the dproc entries that correspond to such attributes.

Differential control. Dproc offers simple ways of dynamically varying certain parameters of monitoring actions, such as monitoring rates or frequencies. Specifically, with each monchannel is associated an implicitly defined *control channel* via which control commands are propagated from monchannel sinks to sources. The dproc interface gives applications access to selected control commands via *control attributes* also maintained with dproc entries.

Flexible analysis and filtering. In order to permit monitoring data to be filtered and analyzed when captured at its sources, monchannel creators can define analysis functions, termed *monhandlers*. These handlers are applied to monitoring data at the sources of channels, thus enabling data filtering and condensation. A monhandler is executed every time an information item is submitted to the channel. A simple but nontrivial example of a monhandler is one that provides window-based running averages rather than raw data. To deploy monhandler functions at kernel level, dproc offers a simplified way of linking an appropriate handler function into the local kernel. For deployment across machines, the remote dynamic code generation facility provided by KECho is used.

Runtime configuration. Monchannels, handlers, and control attributes may be created, changed, and deleted at any time during the operation of dproc. In this fashion, new monitoring functionality can be added on the fly, and existing functionality can be altered or removed.

Related work. Cluster monitoring tools typically rely on the use of daemon processes installed on participating cluster nodes[14, 13]. As a result, they cannot capture data with the overheads and granularity offered by dproc. Similarly, when monitoring data is collected and maintained by single (or multiple, hierarchically arranged[6]) monitoring 'master' processes, data may be captured and analyzed efficiently and with high throughput, but such monitoring structures suffer from high latency in data access. This is important when monitoring is used for online program tuning or steering[6, 22].

Dproc could benefit from additional performance information captured at the network or switch levels[5, 23]. At this time, we are implementing network monitoring by inspection of kernel-resident protocol stacks using the kernel-resident libpcap portion of the well-known tcpdump facility.

Higher level services that interpret or analyze monitoring data[15, 21] are not the subject of this research, but would be useful when using monitoring data for runtime program steering[22] or to help programmers tune their cluster applications[21].

Dproc uses the open source nature of the Linux kernel and its ability to dynamically link new modules into the kernel. For other platforms, instrumentation might be performed with runtime binary editing, as described in [20]. Martin et al. [9] demonstrated reduced overheads by embedding monitoring functions into network co-processors rather than operating system kernels. We are experimenting with that approach in a related project [17].

Overview. The remainder of this paper first outlines the software architecture, API, and implementation of dproc. Section 3 evaluates dproc with microbenchmarks and by applying it to improve application performance. Conclusions and future work are outlined in Section 4.

2   Dproc Architecture and Implementation

Overview. Procfs is a standard component of a Linux file system structure, which offers performance characteristics of the local system. For instance, /proc/ meminfo provides statistics about memory and swap usage, buffer and cache sizes and utilization, etc. Unlike standard Unix file systems, viewing files in procfs essentially executes a piece of code that collects this information dynamically, on-demand. Such information is extracted from the kernel data structures, and is updated by the kernel.

Dproc is a distributed extension of /proc that provides hierarchically organized, application-specific views of monitoring information about both local and remote cluster nodes. For instance, viewing /proc/cluster/node1/meminfo will provide information about memory statitistics on node1. Thus, through calls to the local dproc API, an application can view the current values of monitoring attributes about remote nodes. For each such attribute, an application can also specify the attribute's update rate and ranges of values of interest, thereby resulting in fine grain control over the performance vs. overheads of monitoring experienced by applications.

Monitoring attributes are updated via kernel-level monitoring channels that 'push' update events from the sources being monitored (i.e., certain cluster nodes) to their sinks (i.e., other cluster nodes), much like it is done in the object-based model of monitoring for distributed systems described in our earlier research [8]. By providing a separate underlying 'monchannel' for each information item captured and distributed, dproc permits its distribution to be performed at a unique rate and/or its filtering or analysis to be performed in a unique manner.

Prototype. Dproc is not a complete monitoring system. Instead, it provides basic building blocks for constructing customized monitoring functionality for target systems and applications. The dproc prototype used in this paper employs a predefined set of monchannels across cluster nodes. If interested in certain monitoring information, a node subscribes to a monchannel as both a source and a sink, and is thereby able to provide information of this type and also receive it. We have implemented two approaches to monitoring: user- and kernel-level. Both approaches use monchannels to distribute monitoring information. The channels are implemented in user and kernel space, respectively. Dproc reads and writes in user-level are performed through standard read() and write() system calls. In the kernel, dproc entries are accessed directly through a special interface.

Software Architecture. A sample use of dproc monitoring is one in which the current cpu and memory usage is noted for some set of cluster nodes used by an application program. Such information is captured in the kernels of all 'server' nodes and transported to a single dproc subscriber for this information, located on a 'coordinator' node. This is achieved by using a KECho channel created by the 'coordinator' node, which is registered as a 'consumer'. All server nodes are subscribed as 'producers'. Monchannel handler functions comprise the instrumentation resident in the servers' OS kernels. They are executed at rates determined by the monchannel's control attributes. Each time such a function executes, a monitoring event labeled by node id is submitted to the KECho channel. Upon its receipt, 'coordinator' node updates the attribute value for the appropriate node, resident in its local dproc structure.

The dproc API is accessible from any process running on the local machine via a simple system call. A dproc call operates just like a /proc system call, with its resulting overheads corresponding to that of other Linux system calls. Figure 1 shows dproc architecture.

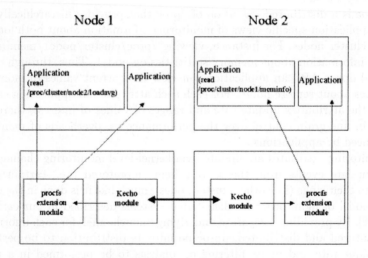

Fig. 1. Dproc Architecture

Runtime monitoring for large-scale cluster machines must provide ways of reducing the potentially large amounts of monitoring data exchanged between cluster nodes. dproc provides multiple ways of reducing monitoring data, including the placement of application-specific filters of monitoring information in data sources. A simple example of this concept is one where such filters dynamically trade off monitoring granularity vs. perturbation by changing the frequency with which certain monitoring attributes are updated. Another example is an extension of remote nodes to compute a specific composite performance measure needed by a parallel graphics application. By computing measures like these remotely rather than locally, reductions in monitoring traffic are realized. This not only reduces traffic, but allows for dynamic load distribution in the system based on resource availability of different nodes.

3   Experimental Evaluation

All experimental results described in this paper are attained on a group of Pentium II quad-processor machines, interconnected via switched Fast Ethernet, and running the Linux version 2.4.0 kernel. We achieve a fine-grained control of the machine loads by running multiple instances of a CPU intensive application. Microbenchmarks are used to evaluate the basic performance of dproc's mechanisms, such as its KECho kernel-kernel communication channels, its runtime API, and its ability to update an API entry in response to the remote update of its value.

3.1   Microbenchmarks

Since information in cluster dproc is collected from remote nodes, the following issues need to be addressed:

- Total monitoring latency, in terms of the minimum delay experienced for updating a remote dproc API in response to a change of some monitored attribute.

- Ability to provide timely access to monitored information, especially during periods of high system load, as monitoring is crucial in such cases. In addition to low access latency, low deviation in access time is also required. These two conditions ensure that monitoring information can be treated as timely data with reasonable confidence.

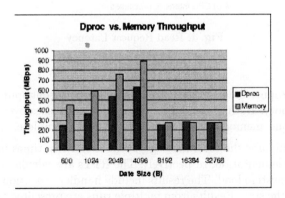

Fig. 2. Dproc Throughput

Figure 2 shows the throughput of dproc, i.e. the time to update a dproc entry as a function of the amount of data. The throughput of dproc is compared to

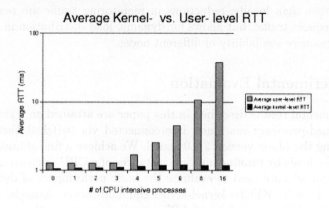

Fig. 3. RTT Variance

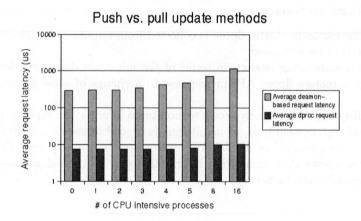

Fig. 4. Read Request Latency

memory throughput. Independent of data size, dproc throughput is comparable to that of memory. Both throughputs decline for data sizes over 4K, as this is the page size, and memory writes now take longer to complete.

Figure 3 compares the total message/event round-trip times in our user and kernel level implementation, respectively. The RTTs are calculated as a function of (symmetric) system load. The results include handler execution times on both the source and the sink. Results from multiple runs are presented to demonstrate the variability of RTT in user and kernel level. Note that we use a logarithmic scale. The results show that a user-level approach to monitoring is especially susceptible to changes in system load, due to the delays experienced on the run queue as user-level threads compete for CPU cycles. In contrast, kernel-level threads are scheduled more frequently, and with much lower variation.

Our experiments also confirm that the RTTs in the user-level approach have a significantly higher standard deviation compared to the kernel-level approach.

Figure 4 shows the delays experienced in accessing monitored information for a daemon-based approach and our user-level implementation as a function of system load. The daemon-based approach is simulated by a server and a client communicating over a socket. This setup approximates a typical monitoring approach which uses a central site to collect, process and distribute monitoring information. We refer to this as a 'pull' based approach, because each interested party has to pull the information from a remote node. In our setup, the client and the server run on the same machine, which is the best-case scenario with the lowest possible network delay. It can be seen that accessing a dproc entry is much more efficient than pulling such information from a remote server via sockets. The access times for dproc are nearly independent on the machine loads, and entirely independent of the network loads.

Preliminary results also show that the perturbation imposed by kernel-level monitoring is negligible for simple monitoring mechanisms used in our experiments (less than 1 percent). We intend to experiment with larger clusters, and quantify the overhead and perturbation imposed by dproc used at both user- and kernel-level.

Microbenchmarks demonstrate that dproc's monitoring overheads and latencies are far better than those experienced by approaches to cluster monitoring that use replicated daemon processes and/or 'pull' monitoring data from remote nodes on demand. In addition to lower latencies, the results demonstrate that monitoring in the kernel provides significantly lower variation in update/access times. This is emphasized when the system is under heavy loads, which implies that user-level daemons experience delays when placed on the run queue. In the kernel approach, those delays are reduced, since monitoring functions are executed by kernel threads.

One advantage of the dproc approach to monitoring is that updates of monitoring attributes are performed asynchronously with application programs' inspections of attribute values. In other words, dproc separates the capture and distribution of monitoring attributes from their inspection by applications. The resulting performance improvements attained by dproc are similar to those attained for parallel programs in which communication is overlapped with computation.

4 Conclusions and Future Work

Our approach to monitoring is based on three principles:(1) monitoring should be reliable during periods of heavy system load, as this is the right time to utilize monitored information and make appropriate changes in the system, (2) monitoring should be dynamically extensible and customizable since no single approach to monitoring will satisfy the needs of all applications, and (3) monitoring overhead and perturbation should be minimized and adjustable. To address the first issue, we built dproc as a kernel-level facility, thereby ensuring that reaction time of the monitoring system, and its variation are minimized. The other two issues are addressed by the extended dproc functionality: filtering and dynamic configuration/extension of the system allow for scaling, reducing and controlling perturbation. Experimental results demonstrate that the dproc approach to monitoring benefits from (1) the fact that information is available locally on the requesting node, essentially implementing a form of caching of remote data, and consequently reducing access time compared to daemon-based (pull) approaches, (2) the richness of information available in the kernel and (3) the immediate thread scheduling in the kernel.

The future development directions of dproc include several extensions to the interface, as well as implementing several guiding examples of how application-specific monitoring can improve performance.

The 'pull'-based model of performance monitoring currently implemented by dproc offers high performance, but does not satisfy uses of monitoring for actions like program steering or adaptation, where an application change may be indicated as soon as some condition has become true. Toward this end, we are currently generalizing the call interface offered to dproc to one that supports both the common 'pull' model and a model in which an application can register its interests in certain values and is signaled when these values meet certain conditions.

We are also interested in developing large-scale parallel codes that exploit customized monitoring tools. One project under way takes a large-scale parallel password cracking algorithm and develops a customized monitoring infrastructure. Utilizing dproc allows the master in this master-worker program to asynchronously determine the progress and performance of the workers (through remote cache miss statistics) and thereby optimize work allocation.

References

1. "Supporting Efficient Execution in Heterogeneous Distributed Computing Environments with Cactus and Globus". G.Allen, T. Dramlitsch, I. Foster, T. Goodale, N. Karonis, M. Ripeanu, E. Seidel and B. Toonen, Proceedings of SC 2001, November 10-16, 2001.
2. "Distance Visualization: Data Exploration on the Grid". I. Foster, J. Insley, G. vonLaszewski, C. Kesselman, M. Thiebaux, (IEEE Computer Magazine, 32 (12):36-43, 1999).
3. Asmara Afework, Michael Benyon, Fabian E. Bustamante, Angelo DeMarzo, Renato Ferreira, Rovert Miller, Mark Silberman, Joel Saltz, Alan Sussman. "Digital Dynamic Telepathology - the Virtual Microscope", In *Proc. of the 1998 AMIA Annual Fall Symposium*, August, 1998.
4. K. Czajkowski, S. Fitzgerald, I. Foster, C. Kesselman, "Grid Information Services for Distributed Resource Sharing." Procedings of the Tenth IEEE International Symposium on High-Performance Distributed Computing (HPDC-10), IEEE Press, August 2001. Networks, Vol. 2, No. 3, 1999
5. A. DeWitt, T. Gross, B. Lowekamp, N.Miller, P.Steenkiste, J. Subhlok, D. Sutherland, "ReMoS: A Resource Monitoring System for Network-Aware Applications". Carnegie Mellon School of Computer Science, CMU-CS-07-194.

6. Greg Eisenhauer, Weiming Gu, Karsten Schwan and Niru Mallavarupu. "Falcon
   – Toward Interactive Parallel Programs: The On-line Steering of a Molecular Dynamics Application", In Proceedings of The Third International Symposium on
   High-Performance Distributed Computing (HPDC-3), San Francisco, August 1994.
   IEEE Computer Society
7. Greg Eisenhauer, Weiming Gu, Thomas Kindler, Karsten Schwan, Dilma Silva
   and Jeffrey Vetter. "Opportunities and Tools for Highly Interactive Distributed
   and Parallel Computing", chapter in Parallel Computer Systems: Performance Instrumentation and Visualization, Rebecca Koskela and Margaret Simmons, editors,
   ACM Press, 1996.
8. Greg Eisenhauer and Karsten Schwan. "An Object-Based Infrastructure for Program Monitoring and Steering", In Proceedings of the 2nd SIGMETRICS Symposium on Parallel and Distributed Tools (SPDT'98), pp. 10-20, August 1998
9. M.E. Fiuczynski, R.P. Martin, T. Owa, and B.N. Bershad. SPINE: An Operating
   System for Intelligent Network Adapters. Proceedings of the Eighth ACM SIGOPS
   European Workshop, pp. 7-12. Sintra, Portugal, September 1998.
10. "A Quality of Service Architecture that Combines Resource Reservation and Application Adaptation". I. Foster, A. Roy, V. Sander, (8th International Workshop
   on Quality of Service, 2000).
11. Ch. Glasner, R. Hügl, B. Reitinger, D. Kranzlmüller, J. Volkert. "The Monitoring
   and Steering Environment" Proc. ICCS 2001, Intl. Conference on Computational
   Science, San Francisco, CA, USA, pp. 781-790 (May 2001).
12. Hart, Delbert; Kraemer, Eileen; Roman, Gruia-Catalin "Interactive Visual Exploration of Distributed Computations," In Proceedings of 11th International Parallel
   Processing Symposium, pp.
13. http://ganglia.mrcluster.org/
14. http://smile.cpe.ku.ac.th/software/scms
15. Jeffrey K. Hollingsworth. Finding Bottlenecks in Large-scale Parallel Programs.
   Ph.D. Dissertation, August 1994. 11-127, Geneva, Switzerland, April 1997
16. D. Kranzlmüller, N. Stankovic, J. Volkert. "Debugging Parallel Programs with
   Visual Patterns" Proc. VL'99, 1999 IEEE Symposium on Visual Languages, Tokyo,
   Japan, pp. 180-181 (Sept. 1999).
17. Rajamar Krishnamurthy, Karsten Schwan, and Marcel Rosu, "A Network Co-
   Processor-Based Approach to Scalable Media Streaming in Servers", International
   Conference on Parallel Processing (ICPP), August 2000.
18. Beth Plale, Volker Elling, Greg Eisenhauer, Karsten Schwan, Davis King, and
   Vernard Martin, Realizing Distributed Computational Laboratories, Int'l Journal
   of Parallel and Distributed Systems and Networks, Vol 2, Num 3, 1999.
19. Randy Ribler, Jeffrey Vetter, Huseyin Simitci, Daniel Reed. "Autopilot: Adaptive
   Control of Distributed Applications",High Performance Distributed Computing,
   August, 1999
20. Ariel Tamches, Barton P. Miller.Fine-Grained Dynamic Instrumentation of Commodity Operating System Kernels Operating Systems Design and Implementation,
   1999
21. TotalView Monitoring software, Etnus LLC. http://www.etnus.com.
22. Jeffrey Vetter and Karsten Schwan, "Techniques for High Performance Computational Steering", IEEE Concurrency, Oct-Dec 1999.
23. Rich Wolski, Neil Spring, and Jim Hayes."The Network Weather Service: A Distributed Resource Performance Forecasting Service for Metacomputing", Journal
   of Future Generation Computing Systems, 1998

A Comparison of Counting and Sampling Modes of Using Performance Monitoring Hardware

Shirley V. Moore

Innovative Computing Laboratory, University of Tennessee
Knoxville, TN 37996-3450 USA
shirley@cs.utk.edu
http://icl.cs.utk.edu/

Abstract. Performance monitoring hardware is available on most modern microprocessors in the form of hardware counters and other registers that record data about processor events. This hardware may be used in counting mode, in which aggregate events counts are accumulated, and/or in sampling mode, in which time-based or event-based sampling is used to collect profiling data. This paper discusses uses of these two modes and considers the issues of efficiency and accuracy raised by each. Implications for the PAPI cross-platform hardware counter interface are also discussed.

1 Introduction

Most modern microprocessors provide hardware support for collecting performance data [2]. Performance monitoring hardware usually consists of a set of registers that record data about the processor's function. These registers range from simple event counters to more sophisticated hardware for recording data such as data and instruction addresses for an event, and pipeline or memory latencies for an instruction. The performance monitoring registers are usually accompanied by a set of control registers that allow the user to configure and control the performance monitoring hardware. Many platforms provide hardware and operating system support for delivering an interrupt to performance monitoring software when a counter overflows a specified threshold.

Hardware performance monitors are used in one of two modes: 1) counting mode to collect aggregate counts of event occurrences, or 2) statistical sampling mode to collect profiling data based on counter overflows. Both modes have their uses in performance modeling, analysis, and tuning, and in feedback-directed compiler optimization. In some cases, one mode is required or preferred over the other. Platforms vary in their hardware and operating system support for the two modes. Some platforms, such as IBM AIX Power3, primarily support counting mode. Some, such as the Compaq Alpha, primarily support profiling mode. Others, such as the IA-64, support both modes about equally well. Either mode may be derived from the other. For example, even on platforms that do not support hardware interrupt on counter overflow, timer interrupts can be

P.M.A. Sloot et al. (Eds.): ICCS 2002, LNCS 2330, pp. 904–912, 2002.
© Springer-Verlag Berlin Heidelberg 2002

used to periodically check for counter overflow and thereby implement statistical sampling in software. Or, if the platform primarily supports statistical profiling, event counts can be estimated by aggregating profiling data. However, the degree of platform support for a particular mode can greatly affect the accuracy of that mode.

Although aggregate event counts are sometimes referred to as "exact counts", and profiling is statistical in nature, sources of error exist for both modes. As in any physical system, the act of measuring perturbs the phenomenon being measured. The counter interfaces necessarily introduce overhead in the form of extra instructions, including system calls, and the interfaces cause cache pollution that can change the cache and memory behavior of the monitored application. The cost of processing counter overflow interrupts can be a significant source of overhead in sampling-based profiling. Furthermore, a lack of hardware support for precisely identifying an event's address may result in incorrect attribution of events to instructions on modern super-scalar, out-of-order processors, thereby making profiling data inaccurate.

Because of the wide range of performance monitoring hardware available on different processors and the different platform-dependent interfaces for accessing this hardware, the PAPI project was started with the goal of providing a standard cross-platform interface for accessing hardware performance counters [1]. For a related project, see [10]. PAPI proposes a standard set of library routines for accessing the counters as well as a standard set of events to be measured. The library interface consists of a high-level and a low-level interface. The high-level interface provides a simple set of routines for starting, reading, and stopping the counters for a specified list of events. The low-level interface allows the user to manage events in *EventSets* and provides the more sophisticated functionality of user callbacks on counter overflow and SVR4-compatible statistical profiling. Reference implementations of PAPI are available for a number of platforms (e.g., Cray T3E, SGI IRIX, IBM AIX Power, Sun Ultrasparc Solaris, Linux/x86, and Linux/IA-64). The implementation for a given platform attempts to map as many of the standard PAPI events as possible to the available platform-specific events. The implementation also attempts to use available hardware and operating system support – e.g., for counter multiplexing, interrupt on counter overflow, and statistical profiling.

Through interaction with the high performance computing community, the PAPI developers have chosen a set of hardware events deemed relevant and useful in tuning application performance. Because modern microprocessors have multiple levels in the memory hierarchy, optimizations that improve memory utilization can have major effects on performance. PAPI provides a large number of events having to do with the memory hierarchy – e.g., cache misses for different levels of the memory hierarchy, and TLB (translation lookaside buffer) misses. PAPI metrics include counts of the various types of instructions completed, including integer, floating point, load, and store instructions. Also included are events for measuring how heavily different functional units are being used, and for detecting when and why pipeline stalls are occurring. The application programmer may be able to use pipeline performance data, together with compiler output files, to restructure application code so as to allow the compiler to do a better job of software pipelining. Another useful measure is the number of mispredicted branches. A high number for this event indicates that something is wrong with the compiler options or that something is unusual about the algorithm. See [1] for a more detailed discussion of uses of PAPI metrics for application performance tuning.

The remainder of the paper is organized as follows: Section 2 discusses usage models of hardware performance monitoring. Section 3 discusses accuracy issues. Section 4 explores implications for the PAPI interface. Section 5 gives conclusions and describes plans for future work.

2 Usage Models

There are basically two models of using performance monitoring hardware:

- the *counting* model, for obtaining aggregate counts of occurrences of specific events, and
- the *sampling* model, for determining the frequencies of event occurrences produced by program locations at the function, basic block, and/or instruction levels.

The first step in performance analysis is to measure the aggregate performance characteristics of the application or system under study [8, 13]. Aggregate event counts are determined by reading hardware event counters before and after the workload is run. Events of interest include cycle and instruction counts, cache and memory access at different levels of the memory hierarchy, branch mispredictions, and cache coherence events. Event rates, such as completed instructions per cycle, cache miss rates, and branch mispredictions rates, can be calculated by dividing counts by the elapsed time.

The profiling model can be used by application developers, optimizing compilers and linkers, and run-time systems to relate performance problems to program locations. With adequate support for symbolic program information, application developers can use profiling data to identify performance bottlenecks in terms of the original source code. Application performance analysis tools can use profiling data to identify performance critical functions and basic blocks. Compilers can use profiling data in a feedback loop to optimize instruction schedules.

For example, on the SGI Origin the `perfex` and `ssrun` utilities are available for analyzing application performance [13]. `perfex` can be used to run a program and report either "exact" counts of any two selected events for the R10000 (or R12000) hardware event counters, or to time-multiplex all 32 countable events and report extrapolated totals. This data is useful for identifying what performance problems exist (e.g., poor cache behavior identified by large number of cache misses). `ssrun` can be used to run the program in sampling mode in order to locate where in the program the performance problems are occurring.

Tools such as vprof [15] and HPCView [7] make use of profiling data provided by sampling mode to analyze application performance. vprof provides routines to collect statistical profiling information, using either time-based or counter-based sampling (using PAPI), as well as both command-line and graphical tools for analyzing execution profiles on Linux/Intel machines. HPCView uses data gathered using ssrun on SGI R10K/R12K systems, or uprofile on Compaq Alpha Tru64 Unix systems, followed by "prof -lines", and correlates this data with program source code in a browsable display.

Aggregate counts are frequently used in performance modeling to parameterize the models. For examples, the methodology described in [14] generates

- a *machine signature* which is a characterization of the rate at which a machine carries out fundamental operations independent of any particular application, and
- an *application profile* which is a detailed summary of the fundamental operations carried out by the application independent of any particular machine.

The method applies an algebraic mapping of an application profile onto a machine signature to arrive at a performance prediction. A benchmark called MAPS (Memory Access Pattern Signature) measures the rate at which a single processor can sustain rates of loads and stores depending on the size of the problem and the access pattern. Hardware performance counters are used to measure cache hit rates of routines and loops in an application which are then mapped onto the MAPS curve. Similarly, the "back-of-the-envelope" performance prediction tool described in [12] makes use of aggregate event counts to construct hardware and software profiles. A given hardware and software profile pair are then combined in algebraic equations to produce performance predictions.

3 Accuracy Issues

Previous work has shown that hardware counter data may not be accurate, especially when the granularity of the measured code is insufficient to ensure that the overhead introduced by counter interfaces does not dominate the event counts [9]. The analysis in [9] made use of three microbenchmarks to study eight MIPS R12000 events. For each of the microbenchmarks, predicted events counts were compared with the measured counts for both the perfex and libperfex interfaces. For the loop benchmark, the counts measured using libperfex were within 5 percent of the predicted counts for four events when the number of loop iterations was at least 250. However, to get the counts generated using perfex within 5 percent of the predicted counts, the number of loop iterations had to be at least 100,000. To relate this work to the PAPI interface on various platforms, we measured the overheads for starting/stopping and for reading the counters in terms of processor cycles. These results, as well as overheads we measured for libperfex, are shown in the table below.

Since the conclusion in [9] is that, given the overhead of the counter interface on a platform, the accuracy of counter data depends heavily on the granularity

	Linux/x86	Linux/IA-64	Cray T3E	IBM Power3	MIPS R12K
PAPI start/stop (cycles/call pr)	3524	22115	3325	14199	24850
PAPI read (cycles/call)	1299	6526	1514	3126	9810
libperfex start/read (cycles/call pr)					5842

of the measured code, we would expect the number of iterations required to get within 5 percent error using PAPI to be close to the 250 required for libperfex on the SGI MIPS R12K, with the exception of the PAPI SGI MIPS R12K interface, which appears to be less efficient that libperfex.

Many profiling tools rely on gathering samples of the program counter value (PC) on a periodic counter overflow interrupt. Ideally, this method should produce a PC sample histogram where the value for each instruction address is proportional to the total number of events caused by that instruction. On modern out-of-order processors, however, it is often difficult or impossible to identify the exact instruction that caused the event.

The Compaq ProfileMe approach addresses the problem of accurately attributing events to instructions by sampling instructions rather than events[5, 6]. An instruction is chosen to be profiled whenever the instruction counter overflows a specified random threshold. As a profiled instruction executes, information is recorded including the instruction's PC, the number of cycles spent in each pipeline stage, whether the instruction caused I-cache or D-cache misses, the effective address of a memory operand or branch target, and whether the instruction completed or if not, why it aborted. By aggregating samples from repeated executions of the same instruction, various metrics can be estimated for each instruction. Information about individual instructions can be aggregated to summarize the behavior of larger units of code. The ProfileMe hardware also supports *paired sampling*, which permits the sampling of multiple instructions that may be in flight concurrently and provides information for analyzing interactions between instructions.

To precisely identify an event's address, the Itanium processor provides a set of *event address registers* (EARs) that record the instruction and data addresses of data cache misses for loads, or the instruction and data addresses of data TLB misses [8]. To use EARs for statistical sampling, one configures a performance counter to count an event such as data cache misses or retired instructions and specifies an overflow threshold. The data cache EAR repeatedly captures the instruction and data address of actual data cache load misses. When the counter overflows, an interrupt is delivered to the monitoring software. The EAR indicates whether or not a qualified event was captured, and if so, the observed event addresses are collected by the software which then rewrites the performance counter with a new overflow threshold. The detection of data cache load misses requires a load instruction to be tracked during multiple clock cycles from instruction issue to cache miss occurrence. Since multiple loads may be in flight simultaneously and the data cache miss EAR can only trace a single load at a time, the mechanism will not always capture all data cache misses. The processor randomizes the choice of which load instructions are tracked to prevent the same data cache load miss in a regular sequence from always being captured, and the accuracy is considered to be sufficient for statistical sampling.

Sampling by definition introduces statistical error. Samples for individual instructions are used to estimate instruction-level event frequencies by multiplying the number of sampled event occurrences by the inverse of the sampling rate. For example, assume an average sampling rate of one sample every S fetched instructions. Let k be the number of samples having a property P. The actual number of fetched instructions with property P may be estimated as kS. Let N be the total number of instructions, and let f be the fraction of those having property P. Then the expected value of kS is fN, and kS will converge to fN as the number of samples increases. However, the rate of convergence may vary depending on the frequency of property P and the coefficient of variation of kS. Infrequent events or long sampling intervals will require longer runs to get enough samples for accurate estimates.

4   Implications for PAPI

The PAPI cross-platform interface to hardware performance counters supports both counting and sampling modes. For counting mode, routines are provided in both the high-level and low-level interfaces for starting, stopping, and reading the counters. For sampling mode, routines are provided in the low-level interface for setting up an interrupt handler for counter overflow and for generating SVR4-compatible profiling data with sampling based on any counter event. Beneath the platform-independent high-level and low-level interfaces lies a platform-dependent substrate that implements platform-dependent access to the counters. To port PAPI to a new platform, only the substrate needs to be reimplemented. Since platform dependencies are isolated in the substrate, changes in the implementation at this level do not affect the platform-independent interfaces, other than making the operations more efficient or providing platform-independent features that had not previously been available on that platform.

The PAPI substrate implementations attempt to use the most efficient and accurate facilities available for native access to the counters. Furthermore, PAPI attempts to use hardware support for counter overflow interrupts and profiling where available. Where hardware and operating system support for counter overflow interrupts and profiling is not available, PAPI implements these features in software on top of hardware support for counting mode. However, the converse has not been attempted – i.e., on platforms such as the Compaq Alpha Tru64 that primarily supports sampling mode, PAPI does not currently implement counting mode in software on top of sampling mode. Although such an implementation is theoretically possible, it raises questions about the accuracy of the resulting event counts since they would be estimated from instruction samples rather than each event being counted by the hardware.

Although the PAPI interface supports profiling based on PC sampling (or, where available, on hardware support for identifying the instruction address for an event), it does not provide access to other information that may be available for the instruction that caused an event, such as data operand addresses or latency information. Nor does PAPI support qualification by opcode or by instruction or data addresses in either counting or sampling modes, although such qualification is available on some platforms such as the IA-64. For example, the Itanium processor provides a way to determine the address associated with a cache miss. It also provides a way to limit cache miss counting to misses associated with a user-determined area of memory. These facilities could enable presentation of data about cache behavior in terms of program data structures at the source code level. Work reported in [3] has shown that such information can be extremely useful in identifying performance bottlenecks caused by bad cache behavior. In [3], the data were obtained through use of a cache simulator which runs considerably slower than the original application (e.g., by a couple of orders of magnitude) and does not model details such as pipelining and multiple instruction issue. Through use of appropriate hardware support (e.g., as on the Itanium), similar data could be obtained more accurately and efficiently.

Although the PAPI library itself does not have any functionality for estimating or compensating for errors, some utility programs have been provided with the PAPI distribution that make some initial attempts. The cost utility measures the overheads in both the number of additional instructions and the number of machine cycles to executing the PAPI_start/PAPI_stop call pair and the PAPI_read call. The calibrate utility runs a benchmark for which the number of floating point operations is known and reports the output of the PAPI_flops call compared with the known number. Error measurement and compensation may be most appropriately implemented at the tool layer rather than at the library layer. However, the PAPI library may be able to provide mechanisms to enable tools to collect the necessary data.

5   Conclusions and Future Work

It is clear that both counting and sampling modes of using hardware performance monitors have their uses and that both should be supported on as many platforms as possible. However, more work is needed to determine which features are most desirable to support in a cross-platform interface and to study accuracy issues related to both models.

Because PAPI presents a portable interface to hardware counters, PAPI is a good vehicle for exploring usability and accuracy issues. PAPI is a project of the Parallel Tools Consortium [11], which provides a forum for discussion and standardization of functionality that may be added in the future. Because of lack of experience with newly available features such as event qualification and data address recording, it seems desirable to experiment with these features before attempting to standardize interfaces to them. The low-level PAPI interface has a routine (PAPI_add_pevent) for implementing programmable events by passing a pointer to a control block to the underlying PAPI substrate for that platform. The routine could be used, for example, to set up event qualification on the Itanium. A corresponding low-level routine (PAPI_read_pevent) has been added to the developmental version of PAPI to allow arbitrary information to be collected. We plan to use programmable events to experiment with new hardware performance monitoring features that are becoming available, with the goal of later proposing standard interfaces to the most useful features. The PAPI_profil call simply generates PC histogram data of where in the program overflows of a specified hardware counter occur. We plan to implement a modified version of this routine that will take a control block as an additional input and allow return of arbitrary information, so as to enable collection of additional information about the sampled instruction (e.g., data addresses, pipeline or memory access latencies). The goal will again be future standardization of the most useful profiling features.

Through the use of microbenchmarks as in [9], we plan to evaluate the accuracy of counter values obtained by the PAPI interface on all supported platforms. Where possible, we will provide calibration utilities that attempt to compensate for measurement errors. We also plan to do statistical studies of the accuracy and convergence rates of profiling data on different platforms, and to investigate the feasibility and accuracy of implementing counting mode in software on top of hardware-supported profiling mode.

For the PAPI software and supporting documentation, as well as pointers to reference materials and mailing lists for discussion of issues described in this paper, see the PAPI web site at http://icl.cs.utk.edu/papi/.

References

1. Browne, S., Dongarra, J., Garner, N., Ho, G., Mucci, P.: A Portable Programming Interface for Performance Evaluation on Modern Processors. International Journal of High Performance Computing Applications **14:3** (Fall 2000) 189–204.
2. Browne, S., Dongarra, J., Garner, N. London, K., Mucci, P.: A Scalable Cross-Platform Infrastructure for Application Performance Optimization Using Hardware Counters. SC'2000. Dallas, Texas. November,2000.
3. Buck, B., Hollingsworth, J.K.: Using Hardware Performance Monitors to Isolate Memory Bottlenecks. SC'2000. Dallas, Texas. November, 2000.
4. Burger, D., Austin, T. M.: The SimpleScalar Tool Set, Version 2.0. University of Wisconsin-Madison Computer Sciences Department Technical Report 1942. June, 1997. http://www.cs.wisc.edu/~mscalar/simplescalar.html
5. Dean, J., Hicks, J., Waldspurger, C. A., Weihl, W. E., Chrysos, G.: *ProfileMe: Hardware Support for Instruction-Level Profiling on Out-of-Order Processors*. 30th Symposium on Microarchitecture (Micro-30). December, 1997.
6. Dean, J., Waldspurger, C. A., Weihl, W. E.: Transparent, Low-Overhead Profiling on Modern Processors. Workshop on Profile and Feedback-Directed Compilation. Paris, France. October, 1998.
7. HPCView: http://www.cs.rice.edu/~dsystem/hpcview/
8. Intel IA-64 Architecture Software Developer's Manual, Volume 4: Itanium Processor Programmer's Guide. Intel, July 2000. http://developer.intel.com/

9. Korn, W., Teller, P., Castillo, G.: Just how accurate are performance counters? 20th IEEE International Performance, Computing, and Communications Conference. Phoenix, Arizona. April, 2001.
10. PCL - the Performance Counter Library: http://www.kfa-juelich.de/zam/PCL/
11. Parallel Tools Consortium: http://www.ptools.org/
12. Pressel, D.: Envelope: A New Approach to Performance Prediction. Department of Defense HPC Users Group Conference. Biloxi, Mississippi. June, 2001.
13. Origin 2000 and Onyx2 Performance Tuning and Optimization Guide. SGI Document number 007-3430-003. July, 2001. http://techpubs.sgi.com/
14. Snavely, A., Wolter, N., Carrington, L.: Modeling Application Performance by Convolving Machine Signatures with Application Profiles. IEEE 4th Annual Workshop on Workload Characterization. Austin, Texas. December, 2001.
15. The Visual Profiler: http://aros.ca.sandia.gov/~cljanss/perf/vprof/

Debugging Large-Scale, Long-Running
Parallel Programs

Dieter Kranzlmüller, Nam Thoai, and Jens Volkert

GUP Linz, Johannes Kepler University Linz,
Altenbergerstr. 69, A-4040 Linz, Austria/Europe,
kranzlmueller@gup.uni-linz.ac.at,
http://www.gup.uni-linz.ac.at/

Abstract. Cyclic debugging depicts error detection techniques, where programs
are iteratively executed to identify the original reason for incorrect runtime behavior. This characteristic is especially problematic for large-scale, long-running
parallel programs concerning the requirements in time and processing resources
and the associated computing costs. A solution to these problems is offered by a
combination of techniques, which use the event graph model as the main representation of parallel program behavior. On the one hand, the number of deployed
processes can be reduced with process isolation, where only a subset of the original processes are executed during debugging. On the other hand, an integrated
checkpointing mechanism allows to extract limited periods of execution time, or
to start subsequent program executions at intermediate points. Additionally, the
event graph offers equivalent program execution in case of nondeterminism, as
well as the possibility to investigate the effects of program perturbation induced
by the observation functionality.

1   Introduction

Debugging is widely accepted as an important part of the software lifecycle, which
tries to improve a program's reliability (and thus quality) by location and correction
of errors. As soon as failures or incorrect results are observed, activities to locate the
original source of the anomaly and to eliminate the reason for the incorrect behavior are
initiated. Unfortunately, the existence of a bug in a program does not imply knowledge
about the error(s) behind it [14]. In fact, the goal of debugging is to deduce the conditions under which the program produced the incorrect output [1]. This characterizes
debugging as a "backward" process, and the corresponding activity is called *flowback*
analysis [2]. Due to the impracticality of reverse execution - some statements, such as
manipulation of file pointers and I/O descriptors, are usually irreversible [7] - flowback
analysis is applied through cyclic debugging.

During cyclic debugging, repeated program executions are utilized to increase the
user's knowledge about the application's behavior. This is achieved by placing breakpoints somewhere in the code, which halt a program at runtime and allow inspection
of the obtained state [4]. Since it is a priori unknown, whether a breakpoint is really
adequate and useful, the placement of breakpoints includes a certain amount of guess-
work. In case a breakpoint is required prior to the reached state - somewhere in the past

P.M.A. Sloot et al. (Eds.): ICCS 2002, LNCS 2330, pp. 913–922, 2002.

of a program's execution - another program run must be initiated from the very beginning. Often, several executions are required to identify the very reason for erroneous behavior.

In order to support these program analysis activities, many useful debugging tools for sequential and parallel software have been developed. Some well-known examples are the command-line debuggers *gdb* and *dbx*, which are included in many operating systems. In addition, advanced debugging tools often incorporate instances of *gdb* and *dbx* for the core debugging activities. Example debuggers in this domain include *xxgdb* or *DDD* [24]. Some representative parallel debugging tools include *P2D2* [9], *Panorama* [16], *PDBG* [5], and the commercial debugger *Totalview* [6], which enrich the capabilities of sequential debuggers with special features for dealing with parallel code.

A particular challenging problem of parallel debuggers is given by the possible scale and the runtime-requirements of parallel software. Parallel and distributed computing is often proposed as the only means to attack computational intensive problems, because sequential computing does not offer sufficient performance. This is especially true for applications of computational science and engineering (CSE), which manifest their need for parallel computing through their requirements in computational speed, amount of memory, and accuracy of results [10]. As a consequence, numerous CSE applications utilize large numbers of processes over a substantial amount of time - days, weeks, or even month - to solve fundamental problems with the available computational resources.

The problem of debugging large-scale, long-running parallel programs is addressed by the approach described in this paper. The principal idea is comparable to program slicing, which reduces a given program by decomposing its execution according to the observed program and data flow [22]. The important idea is to concentrate only on the most important parts of a program - those containing the error - and ignore all other parts. This kind of abstraction is achieved with the event graph model, which represents a particular program execution [11]. Based on this graph, the actually required number of processes during debugging can be reduced to manageable amounts, while the "surroundings" of the selected processes are simulated by the system. For reductions of a program's runtime, the event graph is used as a basis for identifying suitable checkpoint lines. These checkpoints are established to allow the initiation of a program's execution at certain intermediate points instead of the very beginning of the original execution.

This paper is organized as follows: Section 2 defines the event graph and its generation for a parallel program execution with a given set of input data. Some details about reduction of participating processes are given in Section 3, while the combination of checkpointing and debugging is described in Section 4. Some optional extensions to the basic debugging strategy are discussed in 5, before a summary and an outlook on future work in this project concludes the paper.

2   Generating the Event Graph of a Program's Execution

The primary goal of our debugging strategy is reduction of the number of processes and the required execution time without affecting a program's original behavior. Therefore the traditional solution of down-scaling [8] by using smaller problem-sizes with less processes is not feasible, since the behavior of the program may be substantially different for varying problem sizes. In some cases, erroneous behavior may only occur on large scales, and testing with corresponding numbers of processes and problem sizes is therefore inevitable.

A prerequisite for our kind of reduction is exact knowledge of the relations between processes within a parallel program. The relations are captured by the abstract event graph model [11], which is a directed graph $G = (E, \rightarrow)$. The non-empty set of events E is comprised of the events e_p^i of G observed during program execution, with i denoting the sequential order on process p. An event itself is only a virtual item, because it does not require any time for its existence, but instead defines the instant when something happens [20]. It appears to mark a certain activity over the course of time, and therefore emphasizes the activities that are going on [3].

Anything or, to be precise, any state changes of interest occurring during program execution can trigger an event. In particular, the events of interest are established through communication and synchronization statements, which connect events on distinct processes. In parallel programs, process interaction is established either via shared memory or via message passing. The former uses accesses to a common address space to interchange information, while the latter applies dedicated message transfer functions. Although the abstract characteristics of the event graph model allow its application to both kinds of programming paradigms [11], this paper focuses only on the message passing paradigm, which is more scalable due to its loose coupling of processes. The corresponding set of events E consists of communication and synchronization events, as e.g. generated at send and receive statements. In practice, the described approach has been implemented for point-to-point communication functions of the standard message passing interface MPI [17].

The relation " $\rightarrow$ " connecting two events e_p^i and e_q^j is the "happened-before" relation [15]. The expression $e_p^i \rightarrow e_q^j$ describes that there is an edge from event e_p^i to event e_q^j in G with the "tail" at event e_p^i and the "head" at event e_q^j. In concrete, $\rightarrow$ is the transitive, irreflexive closure of the union of the sequential order relation $\xrightarrow{S}$ and the concurrent order relation $\xrightarrow{C}$. The former is implicitly provided as index of each event relative to a particular process, where $e_p^i \xrightarrow{S} e_p^{i+1}$ means, that the ith event e_p^i on any (sequential) process p occurred before the $i+1^{th}$ event e_p^{i+1} on the same process p.

The important relation is the concurrent order relation $\xrightarrow{C}$, which represents interprocess communication and coordination. The concurrent order of events $e_p^i \xrightarrow{C} e_q^j$ defines, that the ith event e_p^i on any process p occurred directly before the jth event e_q^j on another process q, if e_p^i is the sending of a message by process p and e_q^j is the receipt of the same message by another process q, or e_p^i affects the behavior of e_q^j.

Within the event graph, $e_p^i \rightarrow e_q^j$ means, that e_p^i preceded e_q^j and e_q^j occurred after e_p^i. An important characteristic for debugging is that $e_p^i \rightarrow e_q^j$ describes the possibility for event e_p^i to causally affect event e_q^j, or in other words, event e_q^j may be causally affected by event e_p^i. An event e_q^j is causally affected by event e_p^i, if the state of process q at event e_q^j depends on the operation carried out at event e_p^i.

In order to use the event graph for debugging, a parallel program has to be instrumented by adding monitoring code to its executable. Each statement responsible for an event of interest is extended by probes, which deliver necessary information about the occurrence of the event and its relations. During execution of the program with a given input, the probes extract required event graph information and store it in tracefiles for post-mortem processing. This information consists of the following two data structures corresponding to events and relations:

$$\text{event: } e_p^i = (p, i, \text{type}, \text{data})$$
$$\text{happened-before relation: } e_p^i \rightarrow e_q^j = (p, i, q, j)$$

The variables p and i identify the process on which the event occurred and its relative sequential order, respectively. The $type$ identifies the kind of observed event, which is either a send or receive operation. The $data$ field can be used for optional attributes, which are empty in the event graph's most basic form. Please note, that determining the happened-before relation requires some kind of logical clock algorithm as described in [15].

After obtaining the event graph data, a debugging tool may use the data for arbitrary analysis activities. A common practice is the visualization of the event graph as a space-time diagram, with the processes on one axis and the run-time on the second axis. An example space-time diagram for a program's execution is shown in Figure 1. The target program is a finite element solver, which has been executed on 64 processes. The processes are visualized vertically, while the time is displayed along the horizontal axis. Process interaction is visualized with directed edges connecting corresponding sender and receiver operations. The total number of events in this example is 40.000 over approx. 100 seconds of execution time.

As shown in Figure 1, the limits of the display permits only a reduced number of processes and a limited fraction of the program's execution to be visible at a time. Other areas of the event graph can only be investigated through standard graphical manipulations such as zooming and scrolling. In contrast, the techniques described in this paper try to reduce the actual extensions of the event graph in both dimensions. The process isolation approach described in Section 3 reduces the number of processes contained in the event graph. With an integrated checkpointing mechanism, the runtime of the parallel program can be limited as described in Section 4. While these techniques are primarily intended as a reduction of the requirements in terms of computational resources, they also help to simplify the visual representation.

3 Process Isolation Technique

The process isolation approach is initiated, when the reason for erroneous behavior is expected to be restricted to a small subset of all processes. As a starting point, only the actual process exhibiting the error may be a target for process isolation [12]. Whenever backtracking (by repeated executions) detects more and more process interactions,

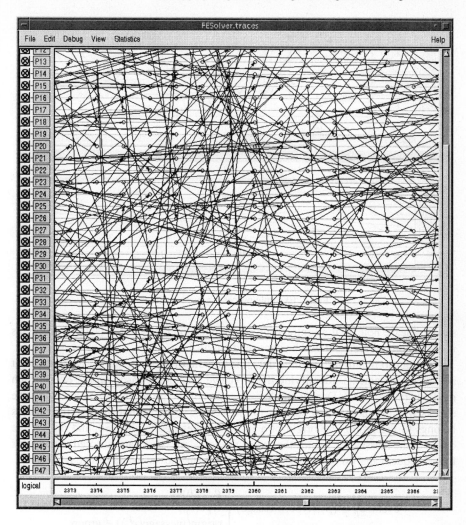

Fig. 1. Example event graph of a parallel program's execution

the number of isolated processes may be increased. In fact, the actual set of processes depends on the suspected distance between the observation point of the erroneous behavior and its original reason, as well as the number of relations to other processes. This characteristic defines the requirements of the process isolation strategy, namely to identify the set of processes and their relations within the given time interval.

The process isolation strategy is initiated by specifying which processes are to be extracted. Based on this specification, the event graph model of the program's execution is analyzed to identify the data required to simulate the surroundings of the selected process group. In case of message passing programs, the interesting data represents all messages, that have been transferred to the process group under consideration. This means that whenever a process of this target group performs a receive operation, the incoming data needs to be available for simulating the process without the original

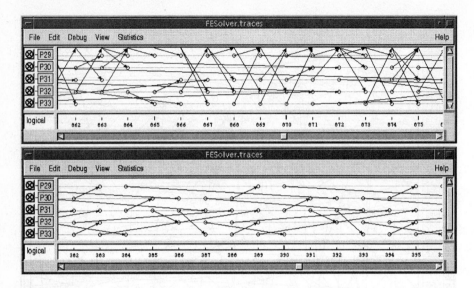

Fig. 2. Example process (group) isolation

sender. Please note, that all intragroup communication between the isolated processes is not required, because it will be generated on the fly.

In order to obtain the required data, the amount of instrumentation is increased at all receive statements. Identifying these receive statements based on the event graph model is rather trivial. During another execution of the program, the monitoring code checks whether the process belongs to the isolated group or not. In case the process belongs to the group, the origin of a incoming message is checked, and only if the origin of the message is outside the group, the actual message data is stored in tracefiles for subsequent analysis phases. In all other cases, the message data can be neglected:

$$\text{event: } e_p^i = (p, i, \text{type}, \text{message}) \begin{cases} p \in \text{group} \\ \text{type}(e_p^i) = \text{``receive''} \\ \text{origin}(\text{message}(e_p^i)) \notin \text{group} \end{cases}$$

With the original event graph and the additional data of the process isolation phase, subsequent executions of only the limited numbers of processes are possible. The correct behavior of these processes will be transparently simulated by the underlying execution environment. This is achieved by intercepting all communication functions and controlling all operations affecting other processes. Process interaction is carried out as specified in the event graph. If data is transferred from outside the isolated process group to a member of the group, the message is obtained from the tracefile and handed to the target process. If data is transferred within the group, regular communication is carried out. All data originating in the group with a destination to a process outside the group can be discarded. Besides, the execution environment must take care of the original process numbering by using virtual process numbers during re-execution.

An example for the process isolation technique is given in Figure 2. The event graph under consideration is the same as in Figure 1, but the number of processes has been limited to a subset of 5 processes (instead of 64 as in the original execution). The top display shows all the communication of the 5 selected processes. The communication operations with messages from processes outside the isolated group are displayed with edges from/to the top of the space-time diagram. These messages are read from the traces instead of transmitting them, when the processes are isolated. The bottom diagram shows the same event graph and the same process group, but only those (intra-group) messages, that are actually transferred between the processes.

4   Combining Checkpointing and Debugging

In contrast to the process isolation approach, the integrated checkpointing approach attempts to reduce the actually required execution time. The term *checkpointing* describes the act of saving the state of a running program, so that it may be reconstructed from that position later in time [19]. In practice, checkpointing is usually applied to fault-tolerant computing; whenever a program's execution fails, the execution of a program may proceed from the last available checkpoint.

The application of checkpointing for debugging large-scale parallel programs has been described in [18, 23]. These approaches permit re-execution of message-passing programs by periodically saving the states as well as the messages in transfer. Additionally, efficient techniques to avoid large amounts of checkpoint storage are included. The drawback of these approaches is that checkpointing is performed at predefined intervals. Thus, it may substantially affect the program's behavior, if the intervals are too narrow. In addition, a complex replay mechanism is needed to avoid inconsistent checkpoints, because program execution may only be initiated at consistent checkpoints.

In contrast, the method implemented in our approach does not suffer from the drawbacks above. By using the event graph model as a representation of parallel program behavior, checkpoints can be placed at virtually any place during the program's execution [21]. The user determines the size of the checkpoint interval by distributing checkpoints arbitrarily, and the system can easily verify checkpoint positions based on the happened-before relation. In addition, the event graph allows to describe arbitrary parallel programs, and is therefore not limited to the message-passing paradigm.

The method operates as follows: Again, the starting point is a given program execution described with an event graph model. The user selects suitable checkpoint positions by selecting arbitrary events of the graph. Upon selection of an event, the remainder of the checkpoint line can be automatically determined with the relations described by the event graph. Therefore, a set of local checkpoints is identified, which collectively serve as a place to obtain the state of the complete program. Besides that, the event graph contains all the required information to identify messages, which are crossing the checkpoint line and are therefore incomplete. In-transmit messages are sent before the checkpoint, but have not been received, while orphan messages are delivered before the sending process has reached the corresponding checkpoint.

An example for placing checkpoints is given in Figure 3. The graph under consideration is again the example finite element solver from Figure 1. As a simplification, we use the isolated group as displayed in Figure 2, although the same approach can be applied for the complete program execution as well. As shown in Figure 3, arbitrary

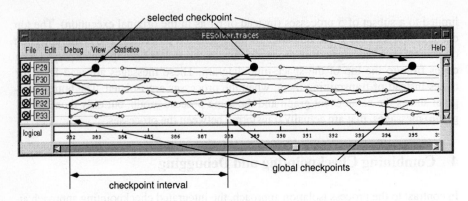

Fig. 3. Placement of checkpoints with the event graph model

local checkpoints have been selected on process P29. The remaining local checkpoints required to construct the global checkpoint are automatically determined by analyzing the event graph. In addition, orphan and in-transit messages are identified.

With the checkpoints defined as above, another execution of the program is initiated at the beginning. During this execution, the integrated checkpointing mechanism extracts the required checkpointing data and stores it to trace files:

$$\text{event: } e_p^i = (p, i, \text{type}, \text{state}) \begin{cases} p \in \text{group} \\ e_p^i = \text{checkpoint} \end{cases}$$

Afterwards, the tracefiles contain the state of each process corresponding to the selected global checkpoint. During each subsequent execution cycle, the user may choose an arbitrary global checkpoint as a starting point for further investigations.

5   Optional Extensions to the Debugging Strategy

The technique described above with its two mechanisms is sufficient to achieve the goals of reducing the number of processes and the time axis during debugging. However, there are some problems existing for any parallel debugging tool, which must also be addressed by our debugging strategy. Two of these issues are the possibility of non-deterministic behavior and the perturbations due to monitor overhead. The event graph model offers some possibilities to decrease the effects of these problems.

The problem of possible nondeterministic behavior is that programs may deliver different results although the same input data is provided. As a consequence, subsequent iterations of the debugging cycle may be meaningless, because previously observed errors may occur only sporadically or vanish completely. This problem can be solved by using the data of the event graph as a prerequisite of the program's re-execution. If the behavior of the program is enforced corresponding to the events and relations contained in the event graph, an equivalent execution is obtained [11].

The perturbation due to the monitor overhead occurs due to the delay induced by the monitoring functionality. Consequently, a minimal monitor overhead reduces the probability of sustained perturbation effects. The most critical problems are changes in event order, which lead to different program behavior and thus different results. With the event graph model, monitor overhead does only occur when generating the event graph data. At this stage, only minimal information (as described in Section 2) is needed, which limits the monitor overhead. Any follow-up executions of the program are controlled by the data of the event graph, which prohibits any changes in event ordering [12]. Additional optimizations in this context are described in [13].

6 Conclusions and Future Work

Cyclic debugging of large-scale parallel programs with long-lasting execution times is hard and tedious, since repeated executions are needed to locate, backtrack, and eliminate erroneous behavior. The process isolation technique and an integrated checkpointing mechanism allows to alleviate this situation by reducing the number of processes and/or the execution time of subsequent debugging cycles. With this improvement, many programs which could not be adequately analyzed in the past due to their requirements of computational resources, are practicable.

The major drawback of the approach are certainly the memory requirements, which are defined by the needs to simulate the surroundings of the extracted process group or to initiate program execution at arbitrary intermediate points. While standard compression mechanisms offer initial support to decrease the amount of event graph data, more optimizations are certainly needed. These consumption of memory space as well as the applicability of the proposed techniques will be investigated in more detail for real-world projects. Fortunately, computational science offers a rich set of complex and large-scale applications, that need to be analyzed by parallel debugging tools.

Acknowledgments: This work was made possible by the support of our colleagues at the GUP Linz. We are most thankful to all of them, especially Christian Schaubschläger, who contributed a large amount to the implementation of the described debugging tools.

References

1. Agrawal, H., DeMillo, R.A., Spafford, E.H., *An Execution Backtracking Approach to Debugging*, IEEE Software, Vol. 8, No. 3, pp. 21–26 (May 1991).
2. Balzer, R.M., *EXDAMS - EXtendable Debugging and Monitoring System*, Proceedings of the AFIPS Spring Joint Computer Conference, pp. 567–580 (1969).
3. Bates, P., Wileden, J.S., *High-Level Debugging of Distributed Systems: The Behavioral Abstraction Approach*, Journal of Systems and Software, Vol. 3, No. 4, pp. 255–264 (Dec. 1983).
4. Choi, J.-D., Miller, B.P., Netzer, R.B., *Techniques for Debugging Parallel Programs with Flowback Analysis*, ACM Transactions on Programming Languages and Systems, Vol. 13, No. 4, pp. 491–530 (Oct. 1991).
5. Cunha, J.C., Loureno, J.M., Anto, T., *An Experiment in Tool Integration: the DDBG Parallel and Distributed Debugger*, EUROMICRO Journal of Systems Architecture, 2nd Special Issue on Tools and Environments for Parallel Processing, Elsevier Science Publisher (1998).
6. Etnus (Dolphin Interconnect Solutions Inc): TotalView 4.1.0, Documentation, Framingham, Massachusetts, USA, (2000).
   http://www.etnus.com/pub/totalview/tv4.1.0/totalview-4.1.0-doc-pdf.tar

7. Feldman, S.I., Brown, Ch.B., *Igor: A System for Program Debugging via Reversible Execution*, Proceedings of the ACM SIGPLAN and SIGOPS Workshop on Parallel and Distributed Debugging (May 1988), University of Wisconsin, Madison, Wisconsin, USA, SIGPLAN Notices, Vol. 24, No. 1, pp. 112–123 (January 1989).
8. Geist, G.A., Beguelin, A., Dongarra, J.J., Jiang, W., Manchek, R., and Sunderam, V.S., *PVM3 - User's Guide and Reference Manual*, Technical Report, Oak Ridge National Laboratory, Tennessee, MIT Press, Cambridge, MA, USA (1994).
9. Hood, R., *The p2d2 Project: Building a Portable Distributed Debugger*, Proc. SPDT'96, ACM SIGMETRICS Symp. on Par. and Distr. Tools, Philadelphia, USA, pp. 127–136 (May 1996).
10. Hossfeld, F., *Teraflops Computing: A Challenge to Parallel Numerics*, in: P. Zinterhof, M. Vajtersic, A. Uhl, (Eds.), "Parallel Computation", Proc. 4th Intl. ACPC Conf., Lecture Notes in Computer Science, Vol. 1557, Springer-Verlag, Salzburg, Austria, pp. 1–12 (Feb. 1999).
11. Kranzlmüller, D., *Event Graph Analysis for Debugging Massively Parallel Programs*, PhD Thesis, GUP Linz, Joh. Kepler Univ. Linz, Austria, (September 2000). http://www.gup.uni-linz.ac.at/~dk/thesis.
12. Kranzlmüller, D., *Incremental Tracing and Process Isolation for Debugging Parallel Programs* Computers and Artificial Intelligence, Vol. 19, No. 6, pp. 569–585 (Nov. 2000).
13. Kranzlmüller, D., Schaubschläger, Ch., Volkert, J., *An Integrated Record&Replay Mechanism for Nondeterministic Message Passing Programs*, Proc. EuroPVM/MPI 2001, 8th European PVM/MPI Users' Group Meeting, Lecture Notes in Computer Science, Vol. 2131, Springer Verlag, Santorini, Greece, pp. 192-200 (September 2001).
14. Krawczyk, H., Wiszniewski, B., *Analysis and Testing of Distributed Software Applications*, in: Wilson, D.R., (Ed.), C3 - Industrial Control, Computers, and Communication Series, Research Studies Press Ltd., Baldock, Hertfordshire, England (1998).
15. Lamport, L., *Time, Clocks, and the Ordering of Events in a Distributed System*, Communications of the ACM, pp. 558 - 565 (July 1978).
16. May, J., Berman, F., *Panorama: A Portable, Extensible Parallel Debugger*, Proc. 3rd ACM/ONR Workshop on Parallel and Distributed Debugging, San Diego, CA, USA (May 1993), reprinted in: ACM SIGPLAN Notices, Vol. 28, No. 12, pp. 96-106 (Dec. 1993).
17. Message Passing Interface Forum, *MPI: A Message-Passing Interface Standard - Version 1.1*,(June 1995). http://www.mcs.anl.gov/mpi/
18. Netzer, R.H.B., Weaver, M.H., *Optimal tracing and incremental reexecution for debugging long-running programs*, Proc. ACM SIGPLAN Conference on Programming Language Design and Implementation, Orlando, FL, pp. 313-325(June 1994).
19. Plank, J.S., *An Overview of Checkpointing in Uniprocessor and Distributed Systems, Focusing on Implementation and Performance*, Technical Report of University of Tennessee, UT-CS-97-372, Jul. 1997.
20. van Rick, M., Tourancheau, B., *The Design of the General Parallel Monitoring System*, Programming Environments for Parallel Computing, IFIP, North Holland, pp. 127-137 (1992).
21. Thoai, N., Kranzlmüller, D., Volkert, J., *Rollback-One-Step Checkpointing and Reduced MessageLogging for Debugging Message-Passing Programs*, Proc. 5th International Meeting on Vector and Parallel Processing VECPAR2002, Porto, Portugal (June 2002).[submitted]
22. Weiser, M., *Program Slicing*, IEEE Transaction on Software Engineering, Vol. 10, No. 4, pp. 352–357 (July 1984).
23. Zambonelli, F., Netzer, R.H.B., *An Efficient Logging Algorithm for Incremental Replay of Message-Passing Applications*, Proc. 13th International Parallel Processing Symposium and 10th Symposium on Parallel and Distributed Processing (1999).
24. Zeller, A., *Visual Debugging with DDD* Dr. Dobb's Journal, No. 332, pp. 21–28 (2001). http://www.ddj.com/ articles/2001/0103/0103a/0103a.htm

Performance Prediction for Parallel Iterative Solvers

V. Blanco[1], P. González[2], J.C. Cabaleiro[3], D.B. Heras[3],
T.F. Pena[3], J.J. Pombo[3], and F.F Rivera[3]

[1] Dept. of Statistics and Computer Science,
La Laguna University, 38071 Tenerife. Spain
Vicente.Blanco@ull.es
[2] Dept. of Electronics and Systems,
A Coruña University, A Coruña. Spain
patricia@dec.usc.es
[3] Dept. of Electronics and Computer Science,
Santiago de Compostela University, 15706 Santiago. Spain
{caba,dora,tomas,juanjo,fran}@dec.usc.es

Abstract. In this paper, an exhaustive parallel library of sparse iterative methods and preconditioners in HPF and MPI was developed, and a model for predicting the performance of these codes is presented. This model can be used both by users and by library developers to optimize the efficiency of the codes, as well as to simplify their use. The information offered by this model combines theoretical features of the methods and preconditioners in addition to certain practical considerations and predictions about aspects of the performance of their execution in distributed memory multiprocessors.

1 Introduction

The complexity of parallel systems makes *a priori* performance prediction difficult. The reasons for the poor performance of codes on distributed memory systems can be varied, and users need to be able to understand and correct performance problems. This fact is especially relevant when high level libraries and programming languages are used to implement parallel codes, as in the case of HPF. A performance data collection, analysis and visualization environment is needed to detect the effects of architectural and system software variations.

Most of the performance tools, both research and commercial, focus on low level message–passing platforms such as MPI or PVM[4], and the most prevalent approach taken by these tools is to collect performance data during program execution and then provide *post–mortem* display and analysis of performance information[12]. Our proposal is different; we present a model that predicts the performance of irregular HPF and MPI codes.

The efficient implementation of irregular codes in HPF is difficult. However, several techniques for handling this problem using intrinsic and library procedures as well as data distribution directives can be applied. An exhaustive HPF

P.M.A. Sloot et al. (Eds.): ICCS 2002, LNCS 2330, pp. 923–932, 2002.

library of iterative methods and preconditioners was developed[2]. A second version of the library was developed using the message–passing programming model for certain kernels of the library to obtain better performance[3]. The model presented in this paper analyses the performance of these codes, and can be used both by users of this library to optimize the efficiency, and by library developers to check inefficiencies.

In the literature, many iterative methods have been presented and it is impossible to cover them all. We chose the methods given below, either because they represent the current state of the art for solving large sparse linear systems [1] or because they present special programming features.

2   A library of iterative methods

2.1   Sparse linear systems

Let us consider applications that can be formulated in terms of the matrix equation $A \cdot x = b$, called linear system, where matrix A and vector b are given, and x must be calculated. The structure of A is highly dependent on the particular application, and some of them give rise to a matrix that is effectively dense and can be efficiently solved using direct factorization–based methods, whereas others generate a matrix that is sparse. For these types of matrices, iterative methods[1] are preferred, especially when A is very large and sparse, due to their efficiency in both memory and work requirements.

We developed $\mathcal{PARAISO}$ (PARAllel Iterative SOlver), that is a lib that includes several iterative methods, such us the Conjugate Gradient (CG), the Biconjugate Gradient (BiCG), the Biconjugate Gradient Stabilized (BiCGSTAB), the Conjugate Gradient Squared (CGS), the Generalized Minimal Residual (GMRES), the Jacobi method, the Quasi–Minimal Residual (QMR) and the Gauss–Seidel Successive Over–Relaxation (SOR). Some preconditioners are also implemented, and can be applied to the target sparse matrix to transform it into one with a more favourable spectrum. These preconditioners are: the Jacobi preconditioner, the Symmetric Successive Over–Relaxation (SSOR), the Incomplete LU factorization (ILU(0)), the Incomplete LU factorization with threshold (ILUT), the Neumann Polynomial preconditioner and the Least Squares Polynomial preconditioner.

We implemented three version of these codes on the AP3000[7]. A F90 version, a HPF version and an enhanced HPF version with those kernels coded in MPI (which we refer to as HPI).

2.2   HPF Implementation

The data–parallel programming model upon which HPF is based requires a well–defined mapping of the data onto local memories in order to achieve an efficient parallel code architecture. Henceforth, we assume that vectors are represented as N–element arrays and the sparse matrix is represented as three one–dimensional arrays, either in CSC or CSR format[9].

The most used operations in paraiso are the dotproduct and daxpy operations. HPF readily supports the inner product operations by an intrinsic function (DOT_PRODUCT), and in addition, the daxpy operation is easily performed using HPF's parallel array assignments. In any parallel implementation that distributes the vectors and the matrix among processor memories, the inner products and sparse matrix–vector multiplications require data communications. The element–wise multiplications in the inner products can be performed locally without any communication overhead, while the merge phase for adding up the partial results from processors involves some communication overhead. However, the data distributions can be arranged so that all of the other computations will be performed on local data only. For each operation we will show a data distribution pattern in order to obtain optimal performance, and how the operation is coded in HPF.

Using N_p processors, daxpy operations can be performed in $\mathcal{O}(N/N_p)$ time on any architecture. On the other hand, the inner products take $\mathcal{O}(N/N_p)$ time for the local phase, but the merge overheads change according to the network architecture.

As an example, we show here the code for a HPF implementation of matrix–vector product. Let's consider the multiplication of an $N \times N$ arbitrarily sparse matrix A, with NNZ non–zero entries, by an $N \times 1$ vector x that gives a $N \times 1$ vector y. Different solutions have been given to solve this problem. One implies the modification of the matrix adding *padding* elements in order to obtain a regular sparse matrix[10]. Other solutions involve the use of HPF extensions to include specific data distributions for sparse matrices[11]. We propose to use HPF intrinsic procedures[2]. Loops are replaced by calls to intrinsic and library procedures, which are inherently parallel. The HPF code for the spmatvec with CSC storage using HPF library procedures is shown below. A detailed description of the code can be found in [3]

```
1   INTEGER, DIMENSION(N+1)     ::  colptr
2   INTEGER, DIMENSION(NNZ)     ::  rowind
3   REAL, DIMENSION(NNZ)        ::  d
4   REAL, DIMENSION(N)          ::  x
5   REAL, DIMENSION(N)          ::  y
6   REAL, DIMENSION(NNZ)        ::  aux
7   LOGICAL, DIMENSION(NNZ)     ::  segment
8
9   !HPF$ ALIGN  (:) WITH x (:) ::  y
10  !HPF$ ALIGN  (:) WITH d (:) ::  rowind, aux, segment
11  !HPF$ DISTRIBUTE (BLOCK)    ::  d, x
12  !HPF$ DISTRIBUTE (*)        ::  colptr
13
14  y = ZERO
15  aux(colptr(:N)) = x
16  aux = COPY_PREFIX(aux, SEGMENT = segment)
17  aux = d * aux
18  y = SUM_SCATTER(aux, y, rowind)
```

2.3   Hybrid Implementation: HPF+MPI

In this section we present the implementation of the main kernels of the iterative methods using MPI (Message Passing Interface). The objective is to take advantage of the flexibility of the message–passing paradigm to optimize irregular computations in these kernels. In this way, the computations that involve vectors (dot products, update of vectors (**daxpy**), etc.) can be efficiently coded in HPF, whereas the irregular kernels are coded in MPI. Henceforth, the version of Paraiso based on this approach will be referred as HPI.

To be able to carry out this approach, the matrix is distributed according to a Block Column Scatter scheme (BCS)[8]. The execution of some functions to redistribute the vectors used in the HPF part of the code (block distribution) to the distributions used in the MPI kernels (cyclic distribution) and vice versa is mandatory. These redistributions require high communication overheads.

The three vectors that represent the sparse matrix are distributed by HPF; it is necessary to transform it into a BCS matrix in order to implement the basic kernels in MPI.

Whereas, the BCS distribution uses a cyclic projection of the matrix onto $P \times Q$ processors. The matrix is partitioned according to a $P \times Q$ template, and each processor takes the non–null entries that fills with its position in the template, as shown in the example in figure 1.

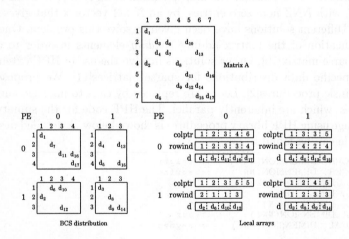

Fig. 1. BCS partition of a sparse matrix in a 2×2 processor mesh.

As the sparse matrix–vector product is the most time consuming operation in each iteration, its implementation should be as efficient as possible. Let us suppose a typical situation in an iterative method:

```
DO iter = 1, MAX_ITER
    HPF operations with vectors
    ...
    CALL HPI_spmatvec( A, x, y )
    ...
    HPF operations with vectors
END DO
```

For an efficient MPI implementation of the sparse matrix–vector product and taking into account the BCS distribution of the matrix, it is necessary to redistribute the input vector, x, from a block to a cyclic by columns distribution. The sparse matrix–vector product is then computed given a cyclic by rows distributed vector. Finally, this vector is redistributed to block in order to obtain the output vector, y. This process is shown in figure 2 and it is summarized as follows:

```
HPI_spmatvec(A, x, y)    /* input vector x, output vector y */
Block2CyclicCols(x, x_cyc_cols);
spmatvec(A, x_cyl_cols, y_cyc_rows);
CyclicRows2Block(y_cyc_rows, y);
```

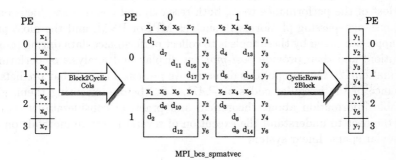

Fig. 2. HPI sparse matrix-vector product on a 2 × 2 processor mesh.

In order to implement the redistribution of the vector from block to cyclic by columns; it is needed a preprocessing stage to obtain the data that will be sent to each processor, their size and their stride. This stage is only executed once in the HPI_init function[3], and given the regularity of the distributions, lacks communications. With this information, the necessary redistributions for each sparse matrix–vector product can be carried out in two stages. In the preprocessing step, the elements that will be sent to the processors in the same row are determined. Given the regularity of the block and cyclic by columns distributions, it can easily be seen that if the first element to be sent is known, the remaining ones are equidistant. In each call to this function a communication step, by rows, is carried out, followed by another one, by columns, to complete the data needed by each processor. Each one of these communication steps is carried out by means of a collective communication. In the case of redistribution from cyclic by rows to block, since there is redundancy of data, only a communication step by columns is needed.

3   Performance Prediction

The complexity of parallel computers makes *a priori* performance prediction difficult. For this reason, performance data collection, analysis and visualization environments are needed to detect the effects of architectural and system software variations.

When programming these systems, the reasons for poor performance of parallel message–passing and data parallel codes can be varied and complex, and users need to be able to understand and correct performance problems. Performance tools can help by monitoring the execution of a program and producing performance data that can be analyzed to locate and understand areas of poor performance. This situation is particularly relevant when high–level libraries and programming languages are used to implement parallel codes, as in the case of HPF. This is true in regular problems, but it is especially important and difficult on irregular codes, such as those included in $\mathcal{PARAISO}$.

Most of the performance tools, both research and commercial, focus on low level message–passing platforms[4], such as MPI or PVM, and the most prevalent approach taken by these tools is to collect performance data during program execution, and then provide *post-mortem* display and analysis of performance information[12]. Our proposal is different, we present a model that predicts performance of the irregular codes of $\mathcal{PARAISO}$ before executing them, giving valuable information about theoretical and practical considerations that can help the user to understand the execution of several iterative methods on their particular sparse linear system.

3.1   Analisis of computations

Execution time is the common measure of computer performance. However, another popular alternative in numerical codes is *million floating point operations per second* (MFLOPS). MFLOPS gauges the capability of a system to deal with floating point math instead of raw instructions. The estimation of the number of FLOPs depends on the target machine. Hence, MFLOPS are not reliable, as the group of floating point operations is not consistent in different systems.

The proposed prediction model counts the number of FLOPs required for every kernel in the library. Based on these kernels it counts the number of FLOPs for one iteration of every method and preconditioner. We can not predict the number of iterations required to achieve convergence, however, the number of FLOPs per iteration gives an idea of the computational cost of any method.

In table 1, the number of FLOPs for different kernels and methods of the library $\mathcal{PARAISO}$ are shown. The number of FLOPs required for the first iteration of each method, which is the most expensive one is also included. In this way, the initial set of residuals and the computation of the required norms for the stopping criterium are considered.

3.2   Analisis of communications

The study of the communication pattern generated by HPF programs is essential for predicting its overhead. The straightforward way to check this pattern is to execute the program with profiling capabilities, taking data from different executions (i.e., different number of processors or different problems).

Table 1. #FLOPs for kernel and methods. n is matrix dimension (N). α is NNZ. r is the restart for the GMRES method

KERNELS		METHODS		
kernel	FLOPs	method	FLOPs 1st iter	FLOPs next iter
spmatvec	2α	CG	$19n + 7\alpha - 3$	$12n + 2\alpha$
spmatvectrans	$2\alpha - 1$	BiCG	$23n + 9\alpha - 4$	$16n + 4\alpha - 1$
dotproduct	$2n - 1$	BiCGstab	$29n + 9\alpha$	$22n + 4\alpha + 3$
jacobi_split	n	CGS	$25n + 9\alpha - 4$	$18n + 5\alpha - 1$
sor_split	$\alpha + n$	GMRES	$(r^2 + 6r + 13)n$	$(r^2 + 6r + 8)n + (2r + 3)\alpha$
stoptest	$2n$		$+(2r + 7)\alpha - ((r^3/3)$	$+(2r + 3)\alpha - ((r^3/3)$
norm_inf	α		$+(9r^2/2) - (5r/6) - 2)$	$+(9r^2/2) - (5r/6) - 1)$
triang	$2\alpha - n$	QMR	$35n + 9\alpha + 16$	$24n + 4\alpha + 18$
		Jacobi	$11n + 4\alpha - 1$	$5n + 2\alpha$
		SOR	$10n + 5\alpha - 1$	$4n + 2\alpha$

We used a profiling tool to trace $\mathcal{PARAISO}$ library. Once the trace information for a given routine is obtained, the communication patterns for the different kernels in $\mathcal{PARAISO}$ library can be extracted. As of an example, the behavior of the matrix–vector kernel is described. next.

The matrix–vector product was implemented in HPF using intrinsic functions in order to achieve high performance as was explained in section 2.2. From the point of view of communications, only lines 15, 16 and 18 produce messages[2].

`aux(colptr(:N))` = x implements the first stage of filling `aux` vector. This HPF line presents an indirection in the left–hand side of the statement, Thus, the HPF compiler cannot detect which elements of vector x will be assigned to the correponding `aux` entry, as they depend on the values of `colptr` which are unknown at compilation time. The compiler that we used solves the situation in two stages: first, vector x is sent to processor 0 and then this processor calculates the corresponding `aux(i)` (since `colptr` is replicated); and then, it sends the result to the processor that owns `aux(i)`. The situation is described in figure 3 for four processors. Note the low efficiency of this approach, as in fact, the HPF line is executed in a sequential way. Similar study was made with routines at lines 16 and 18[3].

The HPI version of $\mathcal{PARAISO}$ integrates MPI coded kernels with HPF coded methods and preconditioners. From the point of view of communications, we only need to predict the MPI kernels, since dense operations with communications, such us the inner product, are performed in HPF and we can use the HPF communication prediction routines in this case.

As is explained in section 2.3, the MPI implementation of the kernels is based on a distributed BCS matrix and a number of redistribution routines to exchange data between the HPF and MPI worlds. The prediction of the performance of these kernels consists of simulating them by using an array of BCS matrices (one for each processor). All the communications in these kernels are based on global MPI communication routines (MPI_Alltoallv, MPI_Allreduce, etc.) which

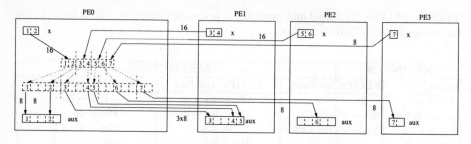

Fig. 3. aux(colptr(:N)) = x using four processors

direct the communication patterns defined in HPLinit. The prediction routines for HPI builds a simulation of these communications patterns and gathers the corresponding communication accounting for every MPI communication routine.

3.3   A model to predict execution times of computations and communications

In order to predict the execution time of parallel irregular codes such as those of $\mathcal{PARAISO}$, it is necessary to consider a great number of events: computations, communications, memory access costs, waiting times, etc.

We construct a model that provides an estimation of the computation time due to FLOPS, the number of cache misses, and the communication times for each message according to its size. The relationship between the number of FLOPs and the actual execution time for every method can be modeled by the following linear expression:

$$t_{comp} = \gamma f + \beta \tag{1}$$

where f is the number of FLOPs and t_{comp} is the execution time in seconds. The values for the parameters γ and β depends on the iterative method. They are shown in table 2. R^2 is the fitting standard deviation.

Table 2. Linear model for computation and communication time. Fitting parameters

Computations				Communications				
Method	γ	β	R^2	Message	Size (doubles)	γ	β	R^2
CG, BiCG,	0.419	0.0013	0.99	Send	0–120	$1.59 \cdot 10^{-7}$	$7.63 \cdot 10^{-5}$	0.39
BiCGStab,				Receive	0–120	$3.21 \cdot 10^{-7}$	$4.59 \cdot 10^{-5}$	0.66
GMRES, CGS				Send	121–1000	$6.25 \cdot 10^{-8}$	$1.15 \cdot 10^{-4}$	0.92
SOR	0.637	0.0014	0.99	Receive	121–1000	$7.44 \cdot 10^{-8}$	$7.41 \cdot 10^{-5}$	0.93
Jacobi	0.726	0.0006	0.99	Send	1001–	$8.02 \cdot 10^{-8}$	$8.91 \cdot 10^{-5}$	1.00
QMR	0.375	0.0012	0.99	Receive	1001–	$1.40 \cdot 10^{-7}$	$9.75 \cdot 10^{-5}$	1.00

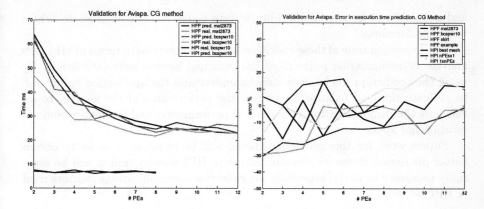

Fig. 4. Validation of the prediction model for the CG method with different matrices

We have carried out a similar study for the cost of the communications, measuring the time for sending and receiving a message of different sizes. Once again, according to a linear model we have:

$$t_{comm} = \gamma m + \beta \qquad (2)$$

where m is the message size in doubles (8 bytes) and t_{comm} is the execution time in seconds. We obtain three different intervals for characterizing this behaviour (shown in table 2). We have used a ping-pong benchmark to establish these measures. Note that, for small messages, the correlation indexes are not high due to the great variance of measuring small runtimes. For larger messages we obtain better results for the fittings.

Finally, we have used a model to predict the number of cache misses found in accessing data for the HPI version of $\mathcal{PARAISO}$. The model[6] is based on a program that simulates the secondary cache of each processor of the AP3000.

4   Results

To validate the prediction model, a number of experiments have been carried out on the AP3000 multiprocessor system. As an example we show the Conjugate Gradient method with a set of matrices of the Harwell–Boeing suite[5]. In figure 4 the predicted and real execution times, and the prediction error for different matrices with the CG method are shown. Note that most of the predictions show errors lower than 15%.

5   Conclusion

A parallel iterative solver library is presented and a performance prediction model for this library is developed. The library has been implement using HPF and a number of kernels of the library were codes in MPI in order to achive better performance.

The execution time of these codes has been characterized in terms of MFLOPs, and the communication patterns of the principal kernels were established. By using the prediction model, it is easy to understand the application behaviour, to evaluate the load balance, to analyze the performance of the kernels, to investigate the communication patterns and performance, and to identify communication hot spots.

Future work for this prediction model will be required in order to obtain better prediction times for communication in HPI version, and it will be especially necessary to model especially the collective communications routines used in HPI kernels.

6   Acknowledgements

The work was supported by CESGA and FECIT under project 1998CP199.

References

1. R. Barret, M. Berry, et al. *Templates for the Solution of Linear Systems: Building Blocks for Iterative Methods*. SIAM, 1994.
2. V. Blanco, J. C. Cabaleiro, P. González, D. B. Heras, T. F. Pena, J. J. Pombo, and F. F. Rivera. A performance analysis tool for irregular codes in HPF. In *Fifth European SGI/Cray MPP Workshop*, Bologna, 1999.
3. V. Blanco, J. C. Cabaleiro, P. González, D. B. Heras, T. F. Pena, J. J. Pombo, and F. F. Rivera. Paraiso project. www.ac.usc.es/~paraiso, jun 2000.
4. S. Browne, J. Dongarra, and K. London. Review of performance analysis tools for mpi parallel programs. www.cs.utk.edu/~browne/perftools-review.
5. I. Duff, R. Grimes, and J. Lewis. Users guide for the harwell-boeing sparse matrix collection. Technical report, CERFACS, 1992.
6. D. Heras, V. Blanco, J. Cabaleiro, and F. Rivera. Modeling and improving locality for the sparse matrix–vector product on cache memories. *High Performance Numerical Methods and Applications*, 2000. special issue in Future Generation Computer Systems.
7. H. Ishihata, M. Takahashi, and H. Sato. Hardware of ap3000 scalar parallel server. *Fujitsu Sci. Tech.*, pages 24–30, 1997.
8. L. F. Romero and E. L. Zapata. Data distributions for sparse matrix vector multiplication. *Parallel Computing*, 21(4):583–605, April 1995.
9. Y. Saad. *Iterative Methods for Sparse Linear Systems*. PWS Publishing Co., 1996.
10. E. Sturler and D. Loher. Parallel solution of irregular, sparse matrix problems using High Performance Fortran. Technical Report TR-96-39, Swiss Center for Scientific Computing, 1996.
11. M. Ujaldon, E. Zapata, B. Chapman, and H. Zima. Vienna Fortran/HPF extensions for sparse and irregular problems and their compilation. *IEEE Transactions on Parallel and Distributed Systems*, 8(10):1068–1083, Oct. 1997.
12. Vampir. Visualization and analysis of mpi programs. www.pallas.de.

Improving Data Locality Using Dynamic Page Migration Based on Memory Access Histograms

Jie Tao, Martin Schulz, and Wolfgang Karl

LRR-TUM, Institut für Informatik,
Technische Universität München, 80290 München, Germany
E-mail: {tao,schulzm,karlw}@in.tum.de

Abstract. This Paper presents an approach which dynamically and transparently improves the data locality of memory references in Non-Uniform Memory Access (NUMA) characterized systems. The approach is based on run-time data redistribution via user-level page migration. It uses memory access histograms gathered by hardware monitors to make correct decisions related to the placement of shared data. First performance experiments on several applications show the potential for a significant gain in speedup. In addition, a graphical user interface has been developed showing the actual data movement thereby helping the user to understand the behavior of the application and to detect performance bottlenecks. This feature complements an already existing Data Layout Visualization tool for the observation of memory locality.

1 Introduction

Due to the excellent price-performance ratio, clusters built from commodity PCs or workstations have established themselves as reasonable alternatives in the area of parallel architectures. In addition, in combination with novel developments in interconnection technologies, they have managed to break into the domain of shared memory multiprocessors, an area which used to be dominated by tightly coupled systems with Uniform Memory Access (UMA) organization, as it is given with Symmetric Multiprocessors (SMPs). Especially, loosely coupled machines with Non-Uniform Memory Access (NUMA) characteristics are becoming increasingly popular because of their scalability and straightforward implementation.

NUMA systems, however, are burdened with an additional performance problem since any memory access to global memory can either be intended for local or remote memory modules with significantly different latency properties. For the programmer, this difference is generally indistinguishable as shared memory programming models work on the assumption of a single uniform global address space. This situation can lead to extensive remote memory accesses, especially with rising numbers of nodes, and thereby to a higher percentage of remote memory access in the overall system. Manual optimizations with respect to data placement can improve data locality, but they can not solve this problem completely since this method is not capable of dealing with applications with dynamically changing access patterns. In this case, also a dynamic approach for the locality optimization needs to be chosen which is capable of significantly reducing remote data accesses via an automatic run-time data redistribution.

P.M.A. Sloot et al. (Eds.): ICCS 2002, LNCS 2330, pp. 933–942, 2002.
© Springer-Verlag Berlin Heidelberg 2002

Such an approach, called the Adaptive Runtime System (ARS), is explored in this paper. ARS is intended to adjust the data distribution at run-time during the execution of an application. It uses memory access histograms, gathered at runtime by a hardware monitor, as the basis for its analysis of access patterns, and dynamically and transparently modifies the location of data. This improves the locality of memory accesses and thus results in better performance.

This work investigates three page migration algorithms which vary in their monitoring information and criteria for making migration decisions. In addition, ARS uses a graphical user interface to provide information about the actual data migrations. This GUI is connected to a Data Layout Visualizer (DLV) [8] which is used to present an application's memory access behavior in a human–readable and easy-to-use way, thus enabling the understanding of an application's access pattern as well as the location of memory access bottlenecks and communication hot spots. These two graphical representation from both the DLV and the ARS therefore complement each other and give the user a good overview of the behavior of the application in either a static or a dynamic scenario.

The remainder of this paper is structured as follows. Section 2 briefly outlines a few previous approaches for improving data locality via data migration. Section 3 discusses the ARS approach, including the framework, the proposed migration algorithms, and the graphical user interface. In Section 4, first experimental results are presented with a comparison of the migration policies. The paper is rounded up with some concluding remarks in Section 5.

2   Related Work

Data locality on NUMA machines has been addressed in many projects over the last years. Among the projects focusing on improving data locality, a few approaches based on page migration have been proposed; most of them, however, only target tightly coupled architectures.

Verghese et.al. [9] study the improvements of performance on CC-NUMA systems, provided by OS supported dynamic migration and replication. This kind of page-migration is based on the information about full-cache misses collected via instrumenting the OS. Hot pages, i.e., pages to which a large number of misses are occurring, are migrated if referenced primarily by one process or replicated if referenced by many processes. Results of their experiments show a performance increase of up to 29% for some workloads.

Nikolopoulos et.al. [5] present two algorithms for moving virtual memory pages to the nodes that reference them more frequently. The purpose of this page movement is the minimization of the worst case latency incurred in remote memory accesses. Their first algorithm works on iterative parallel programs and is based on the assumption that the page reference pattern of one iteration will be repeated throughout the execution of the program. The second proposed algorithm checks periodically for hot memory areas and migrates the pages with excessive remote references. Both algorithms assume compiler support for identifying hot memory areas.

These proposed approaches and systems focus only on the migration algorithms and their implementations. In fact, it is of the same importance to understand the data movement and the behavior of the migration and to report this behavior back to the user in an appropriate way. This requirement, however, has to our knowledge not been followed by any current work.

3   The ARS Approach

ARS is motivated by the fact that shared memory programs running on NUMA machines suffer from excessive remote memory references. While the performance of some applications can be improved by manually optimizing the source code with respect to data placement, others that exhibit dynamically changing access patterns can only be tuned by run-time redistribution of data or computation. ARS implements such a mechanism that migrates shared data during the execution of a program.

The previously proposed approaches make the migration decisions according to the memory access histograms gathered by software with support of the operating system, the compiler, or other memory management mechanisms. This information has to be either inaccurate, incomplete, or associated with a high probe overhead. In order to avoid this problem, the ARS approach establishes its migration decision based on information gathered by hardware monitors with only a minimal probe overhead and without the involvement of compilers, the user, and any system software.

3.1   Framework

Currently, the hardware monitor is designed for our NUMA characterized SMiLE PC clusters. SMiLE stands for *Shared Memory in a Lan–like Environment* and it is a project [3] broadly investigates in SCI–based cluster computing [1]. SCI (Scalable Coherent Interface [1]) is an IEEE–standardized [2] interconnection technology with extremely low latency and very high bandwidth. In order to explore shared memory programming on top of this architecture, a software framework, called HAMSTER [4] (Hybrid-dsm based Adaptive and Modular Shared memory archiTEctuRe), is built within SMiLE enabling the establishment of arbitrary shared memory programming models on top of a single core.

In order to allow an efficient solution for monitoring in this environment, a hardware monitor has been developed. This is necessary because shared memory traffic by default is of implicit nature and performed at runtime through transparently issued loads and stores to remote data locations. In addition, shared memory communication is very fine–grained (normally at word level). This renders code instrumentation recording each global memory operation infeasible since it would slow down the execution significantly and thereby distort the final monitoring to a point where it is unusable for an accurate performance analysis. The only viable alternative is therefore to deploy a hardware monitoring facility.

The SMiLE hardware monitor is designed to be attached to an internal link on current PCI–SCI bridges, the so-called B-Link. This link connects the SCI link chip to the

---

[1] More information at http://smile.in.tum.de/

PCI side of the adapter card and is hence traversed by all SCI transactions intended for or originating from the local node. Due to the bus–like implementation of the B-Link, these transactions can be snooped without influencing or even changing the target system and can then be transparently recorded by the SMiLE hardware monitor. The result of monitoring are the so–called *memory access histograms* which show the number of memory accesses across the complete virtual address space of an application's working set separated with respect to target node IDs. These histograms form the base of a data migration decision by ARS migration mechanisms.

As the hardware monitor is still under development, an event-driven multiprocessor simulator, called SIMT, has been developed within the SMiLE project. SIMT [7] was originally designed to simulate the SMiLE hardware monitor and to provide the exact monitoring information when a hardware monitor is not available. For this purpose, SIMT simulates not only the hardware monitor itself, but also the processor model, the shared memory, the programming interface and models, as well as the parallel execution of applications. Besides that, SIMT comprises functionality enabling a transparent transfer between the simulation platform and a real cluster. Currently, the ARS migration algorithms are implemented on top of SIMT.

3.2   Page Migration Algorithms

Commonly used page migration mechanisms [9, 6] are based on competitive algorithms which migrate a page if the difference between the number of local references and the number of remote references from one node exceeds a predefined threshold. This scheme is easy to implement; a similar one, called *Out-U*, is therefore also applied within ARS. Besides this, we propose two novel page migration algorithms, called *Out-W* and *In-W*, which use a larger decision base and are therefore likely to perform more accurate and timely page migrations. The main difference between them is that they base their migration decisions on different monitoring information: *Out-W* only looks at outgoing memory traffic initiated by the local node, while *In-W* is based on incoming traffic from remote nodes.

The *Out-U* Algorithm   *Out-U* makes decisions whether to move a page from the local node to a remote node. The decision is based on the references performed to the single page from all remote nodes. If the difference between the biggest remote accesses and the average accesses exceeds a threshold, the page is decided to move to the remote node which accesses the page most frequently.

Using this algorithm, however, a correct migration decision can be made only after a large amount of references have been issued, resulting in late migrations and thereby a loss of performance. On the other hand, if a decision is made only based on a small amount of references, many incorrect migrations may be caused. Therefore, we propose another two algorithms which base their migration decisions on references performed on many pages and therefore are able to make a migration decision earlier.

The *Out-W* Algorithm   *Out-W* uses the number of relative references in order to decide the location of a page. The number of relative memory accesses to page P from node

N is calculated as the sum of weighted references from the same node to the pages spatially neighboring page P, using the following formula:

$$R_{PN} = \sum_{i=0}^{n} W_i C_i$$

In this formula, W_i is a weight representing the importance of the i th page to page P and C_i is the number of references to page i, while n is the number of physical pages located on node N. The weight is assigned according to the distance of a page to page P, whereby a closer page is assigned a higher weight due to the spatial locality of memory accesses. Besides that, the neighborhood is restricted to the pages located on the same node of page P. This avoids the overhead of transferring the monitoring information to other nodes, by using only the monitoring information provided by the local hardware monitor.

To determine the location of a page, the numbers of relative references from all remote nodes are compared. If the difference between the highest number and the average accesses exceeds a threshold, the page is decided to move to the remote node owning the biggest number. Here, the same threshold as in *Out-U* is used.

Examine the *Out-U* and the *Out-W* algorithm. Theoretically, spatially neighboring pages have similar access behavior due to the spatial locality of memory accesses. This means that if a node predominately accesses a page, it as well accesses the neighboring pages of this page in the same way. Hence, by using the aggregated monitoring data from a page and its neighboring pages, the information necessary to decide about a page migration can be acquired earlier. This should result in a greater gain in performance if the decision is correct. In order to investigate this, we have analyzed the memory access pattern of all SPLASH2-Benchmark applications [10] and a few other numerical kernels and found that for most applications most pages are frequently accessed only by a single node. This indicates that *Out-W* can make correct decisions because there is only one migration target. In addition, we have observed that for pages accessed by multiple nodes the accesses are not equally distributed. Rather one node normally accesses a page more frequently. Together with an adequate threshold, *Out-W* can therefore also make correct decisions for these pages.

The *In-W* Algorithm While *Out-W* determines whether to move a local page to a remote node, *In-W* decides whether to migrate a remote page to the local node. For this purpose, it uses the monitoring results of memory accesses performed by the local node to other nodes to determine the frequency of a remote page being accessed by the local node.

To make the migration decisions, *In-W* calculates the number of relative references to all remote pages accessed by the local node. It uses the same formula as the *Out-W* algorithm, but involves remote pages into the calculation. If the difference between a relative access and the average exceeds a threshold, the corresponding remote page is decided to be brought to the local node.

In-W can potentially be more accurate than *Out-W* for some data allocation schemes, like *round-robin* which allocates shared data cyclically over all nodes on the system. For these allocation schemes *Out-W* takes pages, which are not directly neighboring in virtual memory, since it only deals with local pages. *In-W*, however, handles all direct neighbors except those located on the local node. It therefore has the potential to better use the spatial locality of the memory accesses. *In-W*, however, is more expensive. All nodes, except the one where a page is located, have the ability to make a decision about the location of this page. Hence, once a migration decision is made by a node, all other nodes must be informed. This increases the communication overhead and the complexity of the management.

3.3   Graphical User Interface

In order to show the run-time page movement, a graphical user interface has been developed and combined with the Data Layout Visualizer (DLV) [8], which has already been implemented within the SMiLE project. It is an on-line tool that provides a set of display windows to show the memory access histograms with different views, allowing programmers to understand the execution behavior of their applications. It also projects the memory addresses back to the data structures within the source code, enabling the optimization of applications resulting in a better data locality at run-time.

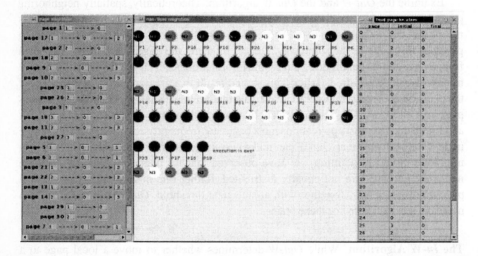

Fig. 1. ARS GUI Display Windows.

The ARS GUI provides several representations to show the actual data migrations, page movements, and data locations. Figure 1 illustrates three sample displays. The *runtime migration* (middle) presents the actual page movements with source node on the top of an item, destination on the bottom, and page number next to the arrow which stands for the direction of moving. Items are dynamically added to the window according to the real time migration. The *page show* (left) illustrates the page movement during the complete execution. The most left rectangle stands for the initial location of an page and the most right stands for the final location of this page, while the rectangle(s) in the middle show intermediate nodes on which the page has resided for some time during the overall runtime of the application. Ping-pong scenarios, such as occurred with page

5, can easily be observed in this view. This enables the evaluation and improvement of the data migration policies. The last window *page location* (right) shows the initial and final location of all shared pages and thereby adds valuable information to the static view given in the DLV.

4 Validation

As mentioned in section 3, the SMiLE hardware monitor is still under development. In order to verify the ARS approach and to evaluate the migration algorithms, we have implemented the first version of ARS on top of SIMT [7] and simulated a number of applications on a 4-node system. These applications are mostly chosen from the SPLASH2-Benchmark suite [10] except the Successive Over Relaxation (SOR) code which is a self-coded numerical kernel used to iteratively solve partial differential equations. The simulated working set size is 2**14 data points for FFT, a 128×128 matrix for LU, 262144 keys for RADIX, 343 molecules for WATER, and a 200×200 grid for SOR.

For all migration schemes we use a constant threshold of two times of the average references performed to a page. This value is chosen to ensure minimal number of Ping-Pongs. As mentioned in section 3.2, we have studied the access patterns of our benchmark applications and found that most pages are predominately accessed by a single node. Also, we have found that across all applications this dominant access is about two times more than the other accesses, and between the other accesses no disparity of two factors can be found. Therfore, using a factor of two in the threshold can guarantee that no remote access exceeds the threshold in the case of dominating local accesses (which are unobservable in the chosen monitoring approach), and on the other hand that migrations are performed in the case of dominating remote accesses. In the next line of this research work, a flexible threshold will be used with the ability of being automatically modified during the run of an application depending on the changing of the application's access pattern.

Since SIMT, in contrast to the final hardware monitor, is capable of providing information about local references, for both *Out-U* and *Out-W* an additional corresponding algorithm is implemented which exploits local access information, in order to examine the relevance of information about local references and to evaluate the ARS migration algorithms. Should the local information be known, the greatest access number will not be compared with the average accesses as it is the case for *Out-U* and *Out-W*, but with the local references.

Figure 2 illustrates the experimental result by performing all migration algorithms on various programs which use *round-robin* as the default allocation policy to initially distribute data. In addition, simulation results of an unoptimized and a manually, but statically optimized run are included for comparison. This kind of optimization is done within the source code by explicitly placing pages on the nodes which most frequently access them.

Examining the migration versions and the transparent default version, it can be seen that all programs run faster after migration, no matter which migration algorithm is deployed. The best performance improvement is gained by the SOR code, where a speedup as high as 1.88 is achieved. This stems from SOR's regular access behavior,

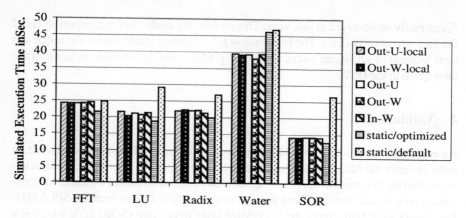

Fig. 2. Simulation time for different programs using round-robin.

where a page is accessed by only one node. When comparing the migration result with the manually optimized code, it can be observed that in most cases the optimized version is better. This stems from the fact that manual optimization introduces an initial correct data placement, timely and without introducing any overheads. However, the WATER code behaves differently, where the migration version performs better than the optimization version. This is caused by the dynamically changing access pattern of WATER, which renders static optimization almost useless.

Application	Out-U-local				Out-W-local				Out-U				Out-W				In-W			
	mig	p-p	mul	err	mig	p-p	mul	err	mig	p-p	mul	err	mig	p-p	mul	err	mig	p-p	mul	err
FFT	41	0	0	0	89	0	0	0	69	0	0	0	27	0	0	0	96	0	0	3
LU	28	0	3	0	30	0	3	0	16	0	1	1	20	0	0	1	30	0	3	0
RADIX	81	0	0	0	191	0	0	0	98	0	1	1	98	0	0	0	106	0	0	0
WATER	26	0	0	0	89	0	29	0	48	0	1	0	119	0	43	0	33	1	1	0
SOR	24	0	0	0	24	0	0	0	25	0	0	1	27	0	0	2	26	0	0	2

Table 1. Migration behavior (mig: total number of migration; p-p: Ping-Pong; mul: multiple migration; err: incorrect migration).

Comparing the individual migration schemes, it can be noted that the distance between the results of migration with or without local access information is insignificant. In some cases, like for RADIX, the migration without local information is even better. The information shown in Table 1 can give an explanation for this behavior. This Table presents the number of total migrations, multiple migrations[2], incorrect migrations[3], and Ping-Pongs. The numbers of incorrect migrations in this table show that the *Out-U* and *Out-W* algorithms scarcely migrate a page mainly accessed by the local node

---

[2] A page is moved to a node and then to another node.

[3] A page is accessed most frequently by the local node but migrated to a remote node.

to a remote node, even though the information about local references is not available. Also, only one Ping-Pong is performed for all applications. Both indicate that the chosen threshold is adequate. In addition, table 1 also explains the abnormal behavior of WATER. It can be seen that many multiple migrations are performed, identifying that pages are alternatively accessed by more nodes. A static optimization placing a page on a fixed node is therefore not suitable and hence the migration result is better.

For the *U*- and the *W*-algorithm, Figure 2 shows that, as expected, *Out-W* outperforms *Out-U* in case of LU and WATER. For FFT, RADIX, and SOR, both algorithms behave similarly. The gain in speedup by *Out-W* for LU and WATER is caused by more migrations which can be seen in Table 1, where we can also observe that these additional migrations are correct. In addition, we have analyzed these migrations using the ARS GUI and found that they are performed in the earlier phase of the program's running. Programs thereby benefit, despite the overhead introduced by the migrations, from the local references that would be remote if no migration was performed. For *In-W*, however, the result is not as expected. In principle, *In-W* should be better than *Out-W* since it should be able to better use the spatial locality of the memory accesses. However, only the SOR code exhibits a gain. This is probably caused by the fact that using the *In-W* algorithm every node can decide whether to move a page to itself. When a page is accessed by multiple nodes, it can happen that the page is migrated and fixed to the first node, but not to the one accessing the page most frequently. For the SOR code, most pages are accessed only by one node, causing the migration not to rely on the node order. The *In-W* scheme behaviors therefore better.

In summary, the results of these first experiments show that a significant improvement has been achieved by ARS's migration approach. It is expected that similar results can be gained when running the applications on actual NUMA machines and with larger and more complex applications.

5   Conclusions

High memory access locality is essential for good performance in NUMA–based environments. This is caused by the sometimes extreme differences in access latencies between local and remote memory modules. In addition to static optimization mechanisms and tools, it is also beneficial to provide dynamic and adaptive mechanisms. These can work without user interaction and require no prior knowledge of the application or even code modifications. In addition, they are also applicable to dynamic or irregular applications in which static optimizations fail.

This work presented a runtime system, called ARS, which is capable of performing such dynamic locality adaptations. It uses memory access histograms, gathered through a hardware monitor with low probe overhead, as input and evaluates this high–level information to perform adequate page migrations. First experiments using three different algorithms implemented within ARS show that ARS is capable of improving the performance significantly in all cases.

As an important feature, ARS also includes a graphical user interface which is capable of reporting the dynamic runtime behavior of the application back to the user in an on–line fashion. This gives the user, together with the DLV, a Data Layout Visualization tool, a deep insight into the memory access patterns of the application and thereby enables further optimizations.

References

1. H. Hellwagner and A. Reinefeld, editors. *SCI: Scalable Coherent Interface: Architecture and Software for High-Performance Computer Clusters*, volume 1734 of Lecture Notes in Computer Science. Springer Verlag, 1999.
2. IEEE Computer Society. *IEEE Std 1596–1992: IEEE Standard for Scalable Coherent Interface*. The Institute of Electrical and Electronics Engineers, Inc., 345 East 47th Street, New York, NY 10017, USA, August 1993.
3. W. Karl, M. Leberecht, and M. Schulz. Supporting Shared Memory and Message Passing on Clusters of PCs with a SMiLE. In A. Sivasubramaniam and M. Lauria, editors, *Proceedings of Workshop on Communication and Architectural Support for Network based Parallel Computing (CANPC) (held in conjunction with HPCA)*, volume 1602 of *LNCS*, pages 196–210, Berlin, 1999. Springer Verlag.
4. M. Schulz. Efficient deployment of shared memory models on clusters of PCs using the SMiLEing HAMSTER approach. In A. Goscinski, H. Ip, W. Jia, and W. Zhou, editors, *Proceedings of the 4th International Conference on Algorithms and Architectures for Parallel Processing (ICA3PP)*, pages 2–14. World Scientific Publishing, December 2000.
5. D. S. Nikolopoulos, T. S. Papatheodorou, C. D. Polychronopoulos, J. Labarta, and E. Ayguade. User-Level Dynamic Page Migration for Multiprogrammed Shared-Memory Multiprocessors. In *Proceedings of the 29th International Conference on Parallel Processing*, pages 95–103, Toronto, Canada, August 2000.
6. V. Soundararajan, M. Heinrich, B. Verghese, K. Gharachorloo, A. Gupta, and J. Hennessy. Flexible Use of Memory for Replication/Migration in Cache-Coherent DSM Multiprocessors. In *Proceedings of the 25th Annual International Symposium on Computer Architecture (ISCA-98)*, pages 342–356, June 1998.
7. J. Tao, W. Karl, and M. Schulz. Using Simulation to Understand the Data Layout of Programs. In *Proceedings of the IASTED International Conference on Applied Simulation and Modelling (ASM 2001)*, pages 349–354, Marbella, Spain, September 2001.
8. J. Tao, W. Karl, and M. Schulz. Visualizing the Memory Access Behavior of Shared Memory Applications on NUMA Architectures. In *Proceedings of the 2001 International Conference on Computational Science (ICCS)*, volume 2074 of LNCS, pages 861–870, San Francisco, CA, USA, May 2001.
9. B. Verghese, S. Devine, A. Gupta, and M. Rosenblum. OS Support for Improving Data Locality on CC-NUMA Compute Servers. Technical Report CSL-TR-96-688, Computer System Laboratory, Stanford University, February 1996.
10. Steven Cameron Woo, Moriyoshi Ohara, Evan Torrie, Jaswinder Pal Singh, and Anoop Gupta. The SPLASH-2 programs: characterization and methodological considerations. In *Proceedings of the 22nd Annual International Symposium on Computer Architecture*, pages 24–36, June 1995.

Multiphase Mesh Partitioning for Parallel Computational Mechanics Codes

C. Walshaw, M. Cross, and K. McManus

School of Computing and Mathematical Sciences, University of Greenwich,
Old Royal Naval College, Greenwich, London, SE10 9LS, UK.
C.Walshaw@gre.ac.uk, URL: http://www.gre.ac.uk/~c.walshaw

Abstract. We consider the load-balancing problems which arise from parallel scientific codes containing multiple computational phases, or loops over subsets of the data, which are separated by global synchronisation points. We motivate, derive and describe the implementation of an approach which we refer to as the multiphase mesh partitioning strategy to address such issues. The technique is tested on example meshes containing multiple computational phases and it is demonstrated that our method can achieve high quality partitions where a standard mesh partitioning approach fails.

Keywords: graph-partitioning, load-balancing, parallel multiphysics.

1 Introduction

The need for mesh partitioning arises naturally in many finite element and finite volume computational mechanics (CM) applications. Meshes composed of elements such as triangles or tetrahedra are often better suited than regularly structured grids for representing completely general geometries and resolving wide variations in behaviour via variable mesh densities. Meanwhile, the modelling of complex behaviour patterns means that the problems are often too large to fit onto serial computers, either because of memory limitations or computational demands, or both. Distributing the mesh across a parallel computer so that the computational load is evenly balanced and the data locality maximised is known as mesh partitioning. It is well known that this problem is NP-hard (i.e. cannot be solved in polynomial time, [3]), so in recent years much attention has been focused on developing heuristic methods, many of which are based on a graph corresponding to the communication requirements of the mesh, e.g. [4].

1.1 Multiphase partitioning – motivation

Typically the load-balance constraint – that the computational load is evenly balanced – is simply satisfied by ensuring that each processor has an approximately equal share of the mesh entities (e.g. the mesh elements, such as triangles

P.M.A. Sloot et al. (Eds.): ICCS 2002, LNCS 2330, pp. 943–952, 2002.
© Springer-Verlag Berlin Heidelberg 2002

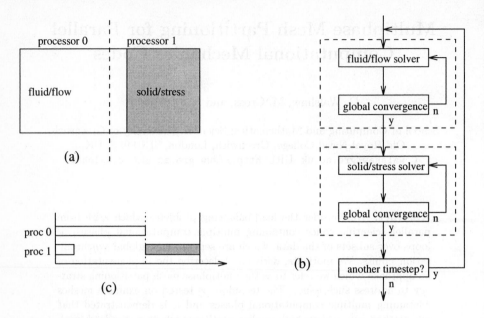

Fig. 1. An example of a multiphysics problem

or tetrahedra, or the mesh nodes). Even in the case where different mesh entities require different computational solution time (e.g. boundary nodes and internal nodes) the balancing problem can still be addressed by weighting the corresponding graph vertices and distributing the graph weight equally. Unfortunately, for some real applications the processor load can also depend on many other factors such as data access patterns but since these are a function of the final partition, it is not possible to estimate such costs *a priori* and we do not address this issue here. We therefore consider only those applications for which a reasonably accurate weighting of the graph, related to computational cost, can be realised. However even for such applications, as increasingly complex solution methods are developed, there is a class of solvers for which such simple models of computational cost break down.

Consider the example shown in Figure 1(a) with a partition for 2 processors indicated by the dotted line. This partition might normally be considered of good quality but for the solution algorithm in Figure 1(b) it is completely unsuitable. As Figure 1(c) shows, during the fluid/flow phase of the calculation, processor 1 has relatively little work to do and indeed during the solid/stress phase processor 0 has no work at all. Furthermore, processor 1 is not able to start the solid/stress calculation until the fluid/flow part has terminated because of the convergence check, a *global synchronisation point* (when all the processors communicate as a group).

In fact it is these **multiple loops** over **subsets** of the mesh entities interspersed by global communications that characterise this modified mesh partitioning problem. If, for example, all the loops in Figure 1(b) were over all the mesh entities (as sometimes happens in codes of this nature when variables are set to zero in regions where a given phenomenon does not occur – e.g. flow in a solid) such balancing problems would not arise. Similarly, if in Figure 1(b) there were no global convergence checks, so that a processor could commence on the stress solution immediately after the flow solution had converged locally, the problem would be removed, although the flow & stress regions might need to be weighted differently. In the simple example in Figure 1 an obvious (and relatively good) load-balancing strategy, therefore, is simply to partition each region (i.e. liquid & solid) of the domain separately so that each processor has an equal number of entities from each region. However, in more complex cases, for example where the regions relating to different computational phases overlap, this may fail to provide a good solution and an advanced strategy is required.

We refer to this modified partitioning problem as the multiphase mesh partitioning problem (MMPP) because the underlying solver has multiple distinct computational subphases, each of which must be balanced separately. Typically MMPPs arise from multiphysics or multiphase modelling (e.g. [7]) where different parts of the computational domain exhibit different physical behaviour and/or material properties. They can also arise in contact-impact modelling, e.g. [6], which usually involves the solution of localised stress-strain finite element calculations over the entire mesh together with a much more complex contact-impact detection phase over areas of possible penetration.

1.2   Overview

In this paper we discuss a strategy for dealing with MMPPs, which uses existing single-phase mesh partitioning algorithms as 'black box' solvers, to partition the problem phase by phase, each partition based on those of the previous phases. The details of this approach are described in Section 2, in particular the necessary vertex classification scheme (§2.1) and an outline of the implementation (§2.2). In Section 3 we present results for the techniques on an illustrative set of example MMPPs. Finally, in Section 4 we summarise the paper and mention some suggestions for further research.

Restrictions on space preclude a full discussion of related work (although some different approaches are reviewed in [10]). However most closely related to the work presented here is the multi-constraint partitioning method of Karypis & Kumar, [5], a different and in some ways more general approach that can be applied to the multiphase partitioning problem. Their idea is to view the problem as a graph partitioning problem with multiple constraints (in this case load-balancing constraints). As here the vertices of the graph have a vector of weights, in this case representing the contribution to each balancing constraint. However, in contrast to the methods presented here, Karypis & Kumar solve the problem in a single computation (rather than on a phase by phase basis).

2 Multiphase partitioning

In this section we describe a strategy which addresses the multiphase partitioning problem, the principle of which is to partition each phase separately, but use the results of previous phases to influence the partition of the current one. The partitioner which we use to carry out the partitioning of each phase is outlined in [10]; however, in principle any partition optimisation algorithm could be used.

2.1 Vertex classification

To talk about multiphase partitioning and more specifically our methods for addressing the problem we need to first classify the graph vertices according to phase. For certain applications the mesh entities (e.g. nodes or elements) will each belong to one phase only, e.g. Figure 2(a), but it is quite possible for a mesh entity, and hence the graph vertex representing it, to belong to more than one phase. For this reason, if F is the number of phases (i.e. the number of distinct computational subphases separated by global synchronisation points – see §1.1), we require for each vertex v that the input graph includes a vector of length F, containing non-negative integer weights that represent the contribution of that vertex to the computational load in each phase. Thus if $|v|_i$ represents the contribution of vertex v to phase i then the weight vector for a vertex v is given by $\mathbf{w} = [|v|_1, |v|_2, \ldots, |v|_F]$ (this is exactly the same as for the multi-constraint paradigm of Karypis & Kumar, [5]). For the example in Figure 2(a) then, the phase 1 mesh nodes would be input with the vector $[1, 0]$ while the phase 2 nodes would be input with the vector $[0, 1]$ (assuming each node contributes a weight of 1 to their respective phases). We then define the vertex *type* to be the lowest value of i for which $|v|_i > 0$, i.e.

$$\text{type}(v) = \begin{cases} \min i \text{ such that } |v|_i > 0 & \text{for } i = 1, \ldots, F \\ 0 \quad \text{if } |v|_i = 0 & \text{for } i = 1, \ldots, F. \end{cases} \tag{1}$$

Thus in the case when the mesh phases are distinct (e.g. Figure 2) the vertex type is simply the phase of the mesh entity that it represents; when the mesh entities belong to more than one phase then the vertex type is the first phase in which its mesh entity is active. Note that it is entirely possible that $|v|_i = 0$ for all $i = 1, \ldots, F$ (although this might appear to be unlikely it did in fact occur in the very first tests of the technique that we tried with a real application, [10]) and we refer to such vertices as type zero vertices. For clarification then, a mesh entity can belong to multiple phases, but the graph vertex which represents it can only be of one type $t = 0, \ldots, F$, where F is the number of phases.

2.2 Multiphase partitioning strategy

To explain the multiphase partitioning strategy, consider the example mesh shown in Figure 2(a) which has two phases and for which we are required to partition the mesh nodes into 4 subdomains. The basis of the strategy is to first

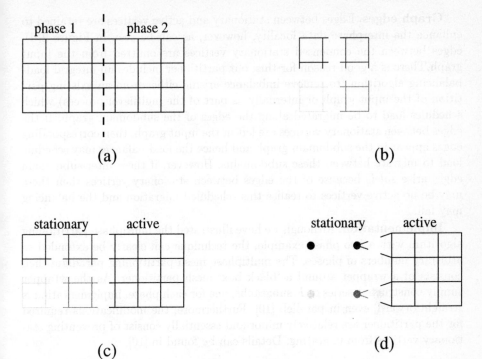

Fig. 2. Multiphase partitioning of a simple two phase mesh: (a) the two phases; (b) the partition of the type 1 vertices; (c) the input graph for the type 2 vertices; (d) the same input graph with stationary vertices condensed

partition the type 1 vertices (each representing a mesh node), shown partitioned in Figure 2(b) and then partition the type 2 vertices. However, we do not simply partition the type 2 vertices independent of the type 1 partition; to enhance data locality it makes sense to include the partitioned type 1 vertices in the calculation and use the graph shown in Figure 2(c) as input for the type 2 partitioning. We retain the type 1 partition by requiring that the partitioner may not change the processor assignment of any type 1 vertex. We thus refer to those vertices which are not allowed to migrate (i.e. those which have already been partitioned in a previous phase) as *stationary* vertices. Non-stationary vertices which belong to the current phase are referred to as *active*.

Vertex condensation. Because a large proportion of the vertices may be 'stationary' (i.e. the partitioner is not allowed to migrate them) it is rather inefficient to include all such vertices in the calculation. For this reason we condense all stationary vertices assigned to a processor p down to a single stationary *super-vertex* as shown in Figure 2(d). This can considerably reduce the size of the input graph.

Graph edges. Edges between stationary and active vertices are retained to enhance the interphase data locality, however, as can be seen in Figure 2(d), edges between the condensed stationary vertices are omitted from the input graph.There is a good reason for this; our partitioner includes an integral load-balancing algorithm (to remove imbalance arising either from an existing partition of the input graph or internally as part of the multilevel process) which schedules load to be migrated along the edges of the subdomain graph. If the edges between stationary vertices are left in the input graph, then corresponding edges appear in the subdomain graph and hence the load-balancer may schedule load to migrate between these subdomains. However, if these inter-subdomain edges arise *solely* because of the edges between stationary vertices then there may be no active vertices to realise this scheduled migration and the balancing may fail.

Implementation. Although we have illustrated the multiphase partitioning algorithm with a two phase example, the technique can clearly be extended to arbitrary numbers of phases. The multiphase mesh partitioning paradigm then consists of a wrapper around a 'black box' mesh partitioner. As the wrapper simply constructs a series of F subgraphs, one for each phase, implementation is straightforward, even in parallel, [10]. Furthermore, the modifications required for the partitioner are relatively minor and essentially consist of preventing stationary vertices from migrating. Details can be found in [10].

3   Experimental results

In this section we give illustrative results by testing the multiphase partitioning strategy on a set of artificial but not unrealistic examples of distinct two-phase problems. By distinct we mean that the computational phase regions do not overlap and are separated by a relatively small interface. Such problems are typical of many multiphysics computational mechanics applications such as solidification, e.g [1]. Further results for the multiphase scheme on other problem types (such as those which arise when different calculations take place on mesh nodes from those taking place on mesh elements, together with some examples from a real-life contact-impact simulation) can be found in [2, 10].

Table 1. Distinct phase meshes

name	V_1	V_2	E	description
512x256	65536	65536	261376	2D regular grid
crack	4195	6045	30380	2D nodal mesh
dime20	114832	110011	336024	2D dual mesh
64x32x32	32768	32768	191488	3D regular grid
brack2	33079	29556	366559	3D nodal mesh
mesh100	51549	51532	200976	3D dual mesh

The example problems here are constructed by taking a set of 2D & 3D meshes, some regular grids and some with irregular (or unstructured) adjacencies and geometrically bisecting them so that one half is assigned to phase 1 and the other half to phase 2. Table 1 gives a summary of the mesh sizes and classification, where V_1 & V_2 represent the number of type 1 & type 2 vertices, respectively, and E is the number of edges. These are possibly the simplest form of two-phase problem and provide a clear demonstration of the need for multiphase mesh partitioning.

The algorithms are all implemented within the partitioning tool JOSTLE[1] and we have tested the meshes with 3 different partitioning variants for 3 different values of P, the number of subdomains/processors. The first of these partitioners is simply JOSTLE's default multilevel partitioning scheme, [8], which takes no account of the different phases and is referred to here as JOSTLE-S. The multiphase version, JOSTLE-M and the parallel multiphase version, PJOSTLE-M, then incorporate the multiphase partitioning paradigm as described here.

The results in Table 2 show for each mesh and value of P the proportion of cut edges, $|E_c|/|E|$, (which gives an indication of the partition quality in terms of communication overhead) and the imbalance for the two phases, λ_1 & λ_2 respectively. These three quality metrics are then averaged for each partitioner and value of P.

As suggested, JOSTLE-S, whilst achieving the best minimisation of cut-weight, completely fails to balance the two phases (since it takes no account of them). On average (and as one might expect from the construction of the problem) the imbalance is approximately 2 – i.e. the largest subdomain is twice the size that it should be and so the application might be expected to run twice as slowly as a well partitioned version (neglecting any communication overhead). This is because the single phase partitioner ignores the different graph regions and (approximately) partitions each phase between half of the processors. Both the multiphase partitioners, however, manage to achieve good balance, although note that all the partitioners have an imbalance tolerance, set at run-time, of 1.03 – i.e. any imbalance below this is considered negligible. This is particularly noticeable for the serial version, JOSTLE-M, which, because of its global nature is able to utilise the imbalance tolerance to achieve higher partition quality (see [8]) and thus results in imbalances close to (but not exceeding) the threshold of 1.03. The parallel partitioner, PJOSTLE-M, on the other hand, produces imbalances much closer to 1.0 (perfect balance).

In terms of the cut-weight, JOSTLE-M produces partitions about 28% worse on average than JOSTLE-S and those of PJOSTLE-M are about 35% worse. These are to be expected as a result of the more complex partitioning problem and are in line with the 20-70% deterioration reported by Karypis & Kumar for their multi-constraint algorithm, [5].

We do not show run time results here and indeed the multiphase algorithm is not particularly time-optimised but, for example, for 'mesh100' and $P = 16$, the run times on a DEC Alpha workstation were 3.30 seconds for JOSTLE-M

[1] available from http://www.gre.ac.uk/jostle

Table 2. Distinct phase results

mesh	$P = 4$			$P = 8$			$P = 16$		
	$|E_c|/|E|$	λ_1	λ_2	$|E_c|/|E|$	λ_1	λ_2	$|E_c|/|E|$	λ_1	λ_2
JOSTLE-S: jostle single-phase									
512x256	0.004	2.000	2.000	0.006	2.000	2.000	0.011	2.000	2.000
crack	0.015	1.906	1.614	0.026	2.434	1.692	0.041	2.445	1.709
dime20	0.001	1.881	1.726	0.003	1.986	2.036	0.004	1.972	2.049
64x32x32	0.023	2.000	2.000	0.038	2.000	2.000	0.052	2.000	2.000
brack2	0.008	1.932	2.096	0.023	1.937	2.138	0.037	1.949	2.145
mesh100	0.008	2.012	1.987	0.016	2.011	2.015	0.025	2.034	2.005
average	0.010	1.955	1.904	0.019	2.061	1.980	0.028	2.067	1.985
JOSTLE-M: jostle multiphase									
512x256	0.004	1.025	1.026	0.009	1.028	1.019	0.013	1.028	1.026
crack	0.016	1.025	1.027	0.030	1.025	1.028	0.055	1.027	1.029
dime20	0.002	1.027	1.015	0.003	1.020	1.025	0.006	1.016	1.018
64x32x32	0.027	1.026	1.029	0.041	1.030	1.029	0.063	1.026	1.030
brack2	0.021	1.010	1.014	0.034	1.030	1.030	0.052	1.029	1.026
mesh100	0.011	1.023	1.021	0.020	1.022	1.029	0.034	1.023	1.029
average	0.013	1.023	1.022	0.023	1.026	1.027	0.037	1.025	1.026
PJOSTLE-M: parallel jostle multiphase									
512x256	0.006	1.000	1.000	0.010	1.000	1.000	0.016	1.000	1.001
crack	0.016	1.000	1.000	0.036	1.000	1.001	0.055	1.000	1.000
dime20	0.002	1.000	1.000	0.004	1.000	1.000	0.007	1.001	1.001
64x32x32	0.029	1.000	1.000	0.046	1.000	1.002	0.066	1.002	1.013
brack2	0.020	1.000	1.001	0.033	1.000	1.002	0.052	1.001	1.005
mesh100	0.011	1.000	1.000	0.021	1.000	1.000	0.033	1.002	1.001
average	0.014	1.000	1.000	0.025	1.000	1.001	0.038	1.001	1.004

and 2.22 seconds for JOSTLE-S. For the same mesh in parallel on a Cray T3E (with slower processors) the run times were 5.65 seconds for PJOSTLE-M and 3.27 for PJOSTLE-S (the standard single-phase parallel version described in [9]). On average the JOSTLE-M results were about 1.5 times slower than those of JOSTLE-S and PJOSTLE-M was about 2 times slower than PJOSTLE-S. This is well in line with the 1.5 to 3 times performance degradation suggested for the multi-constraint algorithm, [5].

4 Summary and future research

We have described a new approach for addressing the load-balancing issues of CM codes containing multiple computational phases. This approach, the multiphase mesh partitioning strategy, consists of a graph manipulation wrapper around an almost unmodified 'black box' mesh partitioner which is used to partition each phase individually. As such the strategy is relatively simple to implement and could, in principle, reuse existing features of the partitioner, such as minimising data migration in dynamic repartitioning context.

We have tested the strategy on examples of MMPPs and demonstrated that it can succeed in producing high quality, *balanced* partitions where a standard mesh partitioner simply fails (as it takes no account of the different phases). The multiphase partitioner does however take somewhat longer than the single phase version, typically 1.5-2 times as long although we do not believe that this relationship can be quantified in any meaningful way. We have not tested the strategy exhaustively and acknowledge that it is not too difficult to derive MMPPs for which it will not succeed. In fact, in this respect it is like many other heuristics (including most mesh partitioners) which work for a broad class of problems but for which counter examples to any conclusions can often be found.

Some examples of the multiphase mesh partitioning strategy in action for contact-impact problems can be found in [2], but with regard to future work in this area, it would be useful to investigate its performance in a variety of other genuine CM codes. In particular, it would be useful to look at examples for which it does not work and either try and address the problems or at least characterise what features it cannot cope with.

References

1. C. Bailey, P. Chow, M. Cross, Y. Fryer, and K. A. Pericleous. Multiphysics Modelling of the Metals Casting Process. *Proc. Roy. Soc. London Ser. A*, 452:459–486, 1995.
2. A. Basermann, J. Fingberg, G. Lonsdale, B. Maerten, and C. Walshaw. Dynamic Multi-Partitioning for Parallel Finite Element Applications. In E. H. D'Hollander *et al.*, editor, *Parallel Computing: Fundamentals & Applications, Proc. Intl. Conf. ParCo'99, Delft, Netherlands*, pages 259–266. Imperial College Press, London, 2000.
3. M. R. Garey, D. S. Johnson, and L. Stockmeyer. Some simplified NP-complete graph problems. *Theoret. Comput. Sci.*, 1:237–267, 1976.
4. B. Hendrickson and R. Leland. A Multilevel Algorithm for Partitioning Graphs. In S. Karin, editor, *Proc. Supercomputing '95, San Diego*. ACM Press, New York, NY 10036, 1995.
5. G. Karypis and V. Kumar. Multilevel Algorithms for Multi-Constraint Graph Partitioning. TR 98-019, Dept. Comp. Sci., Univ. Minnesota, Minneapolis, MN 55455, 1998.
6. G. Lonsdale, B. Elsner, J. Clinckemaillie, S. Vlachoutsis, F. de Bruyne, and M. Holzner. Experiences with Industrial Crashworthiness Simulation using the Portable, Message-Passing PAM-CRASH Code. In *High-Performance Computing and Networking (Proc. HPCN'95)*, volume 919 of *LNCS*, pages 856–862. Springer, Berlin, 1995.
7. K. McManus, C. Walshaw, M. Cross, and S. P. Johnson. Unstructured Mesh Computational Mechanics on DM Parallel Platforms. *Z. Angew. Math. Mech.*, 76(S4):109–112, 1996.
8. C. Walshaw and M. Cross. Mesh Partitioning: a Multilevel Balancing and Refinement Algorithm. *SIAM J. Sci. Comput.*, 22(1):63–80, 2000. (originally published as Univ. Greenwich Tech. Rep. 98/IM/35).

9. C. Walshaw and M. Cross. Parallel Optimisation Algorithms for Multilevel Mesh Partitioning. *Parallel Comput.*, 26(12):1635–1660, 2000. (originally published as Univ. Greenwich Tech. Rep. 99/IM/44).

10. C. Walshaw, M. Cross, and K. McManus. Multiphase Mesh Partitioning. *Appl. Math. Modelling*, 25(2):123–140, 2000. (originally published as Univ. Greenwich Tech. Rep. 99/IM/51).

The Shared Memory Parallelisation of an Ocean Modelling Code Using an Interactive Parallelisation Toolkit

C.S. Ierotheou, S. Johnson, P. Leggett, and M. Cross

Parallel Processing Research Group, University of Greenwich, London SE10 9LS, UK
Email: {c.ierotheou,s.johnson,p.leggett,m.cross}@gre.ac.uk

Abstract. This paper briefly describes an interactive parallelisation toolkit that can be used to generate parallel code suitable for either a distributed memory system (using message passing) or a shared memory system (using OpenMP). This study focuses on how the toolkit is used to parallelise a complex heterogeneous ocean modelling code within a few hours for use on a shared memory parallel system. The generated parallel code is essentially the serial code with OpenMP directives added to express the parallelism. The results show that substantial gains in performance can be achieved over the single thread version with very little effort.

1. Introduction

If oceanographers are to be allowed to continue to do relevant research then they will need to utilise powerful parallel computers to assist them in their efforts. Ideally, they should no be burdened with the task of porting their applications onto these new architectures but instead be allowed to focus their efforts on the quality of ocean models that are required. For example, the development of models that include features with both small spatial scales and large time scales and cover as large a geographic region as possible. The unfortunate reality is that a significant effort is often required to manually parallelise their model codes and this often requires a great deal of expertise. One suggestion is to substantially reduce the effort required for the parallelisation with the introduction of an effective parallelisation toolkit.

Today the shared memory and distributed memory programming paradigms are two of the most popular models used to transform existing serial application codes to a parallel form. For a distributed memory parallelisation of these types of codes it is necessary to consider the whole program when using a Single Program Multiple Data (SPMD) paradigm. The whole parallelisation process can be very time consuming and error-prone. For example, data placement is an essential consideration to efficiently use the available distributed memory, while the placement of explicit communication calls requires a great deal of expertise. The parallelisation on a shared memory system is only relatively easier. The data placement may appear to be less crucial than for a distributed memory parallelisation, but the parallelisation process is still error-prone, time-consuming and still requires a detailed level of expertise. The main goal for developing tools that can assist in the parallelisation of serial application codes is to embed the expertise and the automated algorithms to perform much of the tedious, manual and sometimes error-prone work, and in a small fraction of the time that would otherwise be taken by a parallelisation expert doing the same task manually. In addition to this, the toolkit should be capable of generating generic, portable, parallel source code from the original serial code [1, 2]. The toolkit discussed here was developed at the University of Greenwich and has been supplemented by a directive

P.M.A. Sloot et al. (Eds.): ICCS 2002, LNCS 2330, pp. 953–962, 2002.

generation module [3] from Nasa Ames Research Center. The toolkit can generate either SPMD based parallel code for distributed memory systems or loop distributed directive-based parallel code for shared memory systems.

The aim of this paper is to report on how the toolkit was used and what level of effort was required to parallelise an ocean model code (typified by the Southampton and East Anglia model [4, 5]) for a shared memory based parallel system. A similar manual effort has not been undertaken due to the high cost associated with such a task. Finally, some remarks are made about the quality of the generated code and the parallel performance achieved on a test case.

2. The interactive parallelisation toolkit

The toolkit used in this study has been used to successfully parallelise a number of application codes for distributed memory systems [6, 7] based on distributing arrays across a processor topology. For an SPMD, distributed memory based parallelisation, the mesh over which these equations are solved is used as the basis for the partitioning of the data. The quality of the parallel source code generated benefits from many of the features provided by the toolkit. For example, the dependence analysis is fully interprocedural and value-based (i.e. the analysis detects the flow of data rather than just the memory location accesses) [8] and allows the user to assist with essential knowledge about program variables [9]. There are many reasons why an analysis may fail to accurately determine a dependence graph, this could be due to incorrect serial code, a lack of information on the program input variables, limited time to perform the analysis and limitations in the current state-of-the-art dependence algorithms. For these reasons it is essential to allow user interaction as part of the process, particularly if scalability is required on a large number of processors. For instance, a lack of knowledge about a single variable that is read into an application can lead to a single unresolved equation in the dependence analysis. This can lead to a single assumed data dependence that serialises a single loop, which in turn greatly affects the scalability of the application code. The placement and generation of communication calls also makes extensive use of the interprocedural capability of the toolkit as well as the merging of similar communications [10]. Finally, the generation of readable parallel source code that can be maintained is seen as a major benefit. The toolkit can also be used to generate parallel code for shared memory based systems, by inserting OpenMP [11] directives into the original serial code. This approach also makes use of the very accurate interprocedural analysis and also benefits from a directive browser (section 5) to allow the user to interrogate and refine the directives automatically placed within the code.

3. The Southampton-East Anglia (SEA) model

The SEA model [4, 5] was developed as part of a collaborative project between the Southampton Oceanography Centre and the University of East Anglia. The model is based on an array processor version of the Modular Ocean Model with a reduced set of options (MOMA) [12]. Although MOMA was not itself a parallel code, it could be arranged such that parallelism was exploited and this is what lead to the subsequent development of the SEA model code. The SEA model code can be configured with the aid of a C-preprocessor to execute in parallel on a distributed memory system

[13]. Work has been done relating to the parallelisation of the SEA code onto distributed memory systems [4, 14] but this will not be discussed here. Instead, a serial version of the SEA model code was used as the starting point for this study.

The surface of the model ocean is assumed to be split into a 2-D horizontal grid (i,j). Each (i,j) is used to define a volume of water that extends from the surface of the ocean to the ocean floor (k-dimension). The k-dimension is represented by a series of depths. The main variables solved by the model include the barotropic velocities and associated free-surface height fields as well as the baroclinic velocities and tracers. All other variables can be derived from these. The governing equations are discretised using a finite difference formulation in which the velocities are offset from the tracers and free-surface heights [15].

4.   Parallelising a code using OpenMP

There are a number of different types of parallelism that can be defined using OpenMP, such as task or loop based parallelism. Here we focus on loop based parallelism and the issues in identifying privatisable variables. However, even in utilising these techniques there are essentially two practitioners – those that are CFD specialists (these include authors of the software) and those that are parallelisation experts. The CFD specialists have a deep level understanding of the code, the algorithms and their intended applicability, as well as the data structures used. This aids in identifying parallel loops due to independent calculation for cells in the mesh and also for identifying privatisable variables as they are often used as workspace or temporary arrays in CFD codes. These users tend to make use of implicit assumptions or knowledge about the code during the parallelisation. The parallelisation experts however, have the know how to carry out what is effectively an implicit dependence analysis of the code to examine the data accesses and therefore provide a more rigorous approach to the whole parallelisation process. In doing so, they still typically make assumptions due to the difficulty in manually performing a thorough investigation. Ideally, a user is required that has skills drawn from both disciplines to perform the parallelisation.

Adopting a formal approach to parallelising a code with loop distribution using OpenMP requires a number of considerations. For instance, it is necessary to carry out a data dependence analysis of the code as accurately as it is practical. As well as detecting parallel loops, the analysis should include the ability to identify the scope of variable accesses for privatisation. This means that the analysis must extend beyond the boundaries of a subroutine and therefore be interprocedural. This in itself can be a daunting task when tackled manually by a parallelisation expert, but can be implemented automatically as part of a parallelisation tool. The identification of PARALLEL regions using interprocedural information and the subsequent legal fusion of a series of these regions can be carried out automatically by the toolkit in order to reduce as much as possible the overheads associated with starting up and synchronising PARALLEL regions. In addition, the automatic identification, placement and legal use of NOWAIT clauses for DO loops can help to greatly reduce the overheads for synchronising amongst threads. These tasks require a very detailed analysis of the code and are sometimes viewed as too complex to implement manually, particularly when attempting to consider interprocedural effects.

5. The parallelisation of the SEA model code

The parallelisation of the SEA model code was achieved by executing the toolkit on a supported platform and performing a few simple processes within the toolkit. The first step is to read in the serial source code. The SEA model code is written in FORTRAN and contains nearly 8200 lines of source code. The next step was to perform a dependence analysis of the code using the analyser and this took a few minutes. Inspection of the unresolved questions identified by the analyser was followed by the user addition of some simple knowledge relating to the arrays storing the vertical depths ($0 \leq KMU \leq 32$ and $0 \leq KMT \leq 32$). A subsequent incremental analysis was performed in order to re-evaluate any uncertainties that remained using the additional user knowledge. A database is saved at this stage to allow the user to return at a later date to this point in the parallelisation process. The user can then use the toolkit to automatically generate OpenMP code without examining the quality of the code. Not surprisingly, the performance obtained for this version of the code was comparable to using the SGI auto parallelisation flag $-apo$. The parallel performance of this code was very poor with a speed up of 1.5 for 8 threads.

After reloading the saved database, the user is allowed to investigate in detail any serial loops in the code by using the interactive directives browser (Figure 1). This investigation is essential as it could lead to further parallelism being identified by the toolkit with the additional knowledge provided by the user. Equally important is the failure to look at such loops and allowing them to be executed in serial, as this can significantly affect the scalability of the code (Amdahl's law [16]). The directives browser presents information to allow the user to focus on the precise reasons why, for example, a loop cannot execute in parallel. The browser also provides information about which variables cause recurrences and hence prevent parallelism and which variables could not be legally privatised and therefore inhibit parallelism. This amount of detail in the browser enables both the parallelisation expert and CFD specialist (through their knowledge of independent updates in the algorithms and knowledge of workspace arrays) to address these issues. One possible course of action is to refine the dependence graph. For example, in the SEA code some array accesses define a multi-step update of the solution variables that can be executed in parallel. This may have been too complex to be detected by the dependence analysis alone. Figure 2 shows pseudo code for just such a case in the SEA code. The current state of the analysis has failed to capture this very specific case involving the v array. The serialising array dependence is presented to the CFD expert (through the browser) who is then able to inform the tools (by deleting the assumed dependencies) that this is not the case as the v array updates and uses are independent, thereby enabling the i and j loops to be defined as parallel. This refinement process could be repeated as often as necessary until the user is satisfied with the level of parallelism defined and the toolkit can generate the OpenMP code. The time taken to parallelise the SEA model and generate OpenMP code using the toolkit on an SGI workstation was estimated to be about 2 hours, most of which was spent trying to exploit further parallelism by using the toolkit's interactive browsers. Figure 3 shows an example of the quality of the code generated by the toolkit for the clinic subroutine. In this subroutine alone, it highlights the toolkit's ability to define parallelism at a high level (i.e. at the j loop level which includes 12 nested parallel loops as well as calls to subroutine state), the scope of all scalar and array variables and the automatic identification of THREADPRIVATE common blocks.

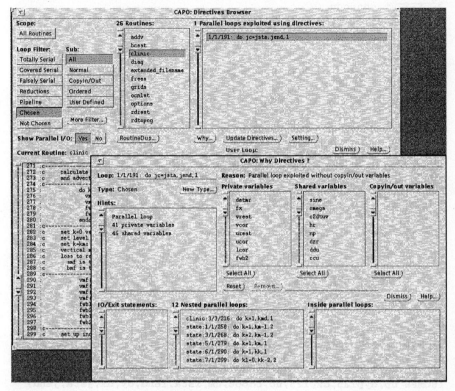

Figure 1 Directive browser displaying loop classification and reasons for loop type

6. Performance of generated parallel code on a test case

An idealised topography for a pseudo global model that extends from 70°S to 70°N was used as the test case for this study. The topography was generated such that the horizontal resolution was modified from the default 182x73 resolution to 722x283, while the number of vertical levels was defined to be 32. It was also indicated that for such a global model cyclic periodic boundary conditions should be applied to the latitudinal boundaries. In the initial parallelisation without any user interaction there were 24 serial and 90 potentially parallel loops (from a total of 114 loops). The toolkit was able to generate code that exhibited a high degree of parallelism as the majority of loops selected for parallel execution were based on the outermost J loop. This ensured that for these loops there was enough computational work to offset the overhead in initialising and synchronising the threads. Following further user interaction (described in section 5) the number of potentially parallel loops had increased to 96.

```
nm=1                                    SUBROUTINE STEP
nc=2                                     nnc=np
np=3                                     nnm=nc
do loop=1,9999999                        nnp=nm
  call STEP                              np=nnp
enddo                                    nc=nnc
                                         nm=nn
SUBROUTINE CLINIC                 200    continue
do j=jsta,jend                           call CLINIC
  do i=ista,iend                         if(mixts.and.eb)THEN
    do k=1,kmc                             nc=nnp
      ... = v(k,i-1:i+1,j,nc),             np=nnm
            v(k,i-1:i+1,j,nm),             mixts=.false.
            v(k,i,j-1:j+1,nc),             goto 200
            v(k,i,j-1:j+1,nm)            endif
    enddo
    do k=1,kmc
      v(i,j,k,np) = ...
    enddo
  enddo
enddo
```

Figure 2 Pseudo code showing the multi-step array accesses for v

```
      common /work/dpdx,dpdy,fue,fuw,fvn,fvs,vmf,vtf,fw,fwb1,
     & fwb2,rhoo,rhpo,rhpp,rhop,maskoo,maskpo,maskmo,...
!$OMP THREADPRIVATE(/work/)
...
!$OMP PARALLEL DO DEFAULT(SHARED) PRIVATE(detmr,fx,vrest,vcor,
!$OMP& urest,ucor,lcor,temp2,temp1,fxb,fxa,fuic,fvjc,k,bmf,
!$OMP& uvmag,smf,kmd,kmc,im1,ip1,i,boxar,boxa,jm1,jp1,j)
!$OMP& SHARED(omega,c2dtuv,acor,fkpm,dy4r,dx4r,dy2r,dx2r,grav,
!$OMP& np,dyp125,dxp125,nc,cdbot,nm,ista,iend,dy,dx,jsta,jend)
      do j=jsta,jend
        do i=ista,iend
          if(kmc.gt.0)then
            call state(t(1,i,j,1,nc),t(1,i,j,2,nc),...)
            call state(t(1,ip1,j,1,nc),t(1,ip1,j,2,nc),...)
            call state(t(1,ip1,jp1,1,nc),t(1,ip1,jp1,2,nc),...)
            call state(t(1,i,jp1,1,nc),t(1,i,jp1,2,nc),...)
```

Figure 3 Illustration of the automatic OpenMP code generated by the toolkit

Table 1 shows a breakdown of the frequency and type of OpenMP directives that were generated by the toolkit in the final generated code (referred to as "static opt0"). In summary, there were a total of 20 PARALLEL regions defined in the code, with just 25 of the 96 potentially parallel loops selected for parallel execution. In addition to the 3 REDUCTION loops that were identified there was also another reduction operation identified that did not conform to the OpenMP specification, but this was still handled by the code generator, choosing instead to use the CRITICAL directive. The parallel platform used was a 64 processor SGI Origin 2000, where each processor was a MIPS R12000 with a 300MHz clock speed. The code was compiled with -mp -O3 flags and Figure 4 below shows the performance of the OpenMP code generated from the toolkit as a result of the parallelisation process described above. An efficiency of 87% on 32 processors and nearly 60% efficiency on 64 processors is quite promising given the fairly small, yet essential, user interaction that was needed.

Table 1 Breakdown of OpenMP directives generated by the toolkit

```
Total number of PARALLEL Regions            : 20
Total number of OMP DO Loops defined        : 25
PARALLEL regions with a single OMP DO loop: 11
REDUCTION Loops                             : 3
ATOMIC/CRITICAL Sections                    : 1
Regions containing FIRSTPRIVATE variables : 2
```

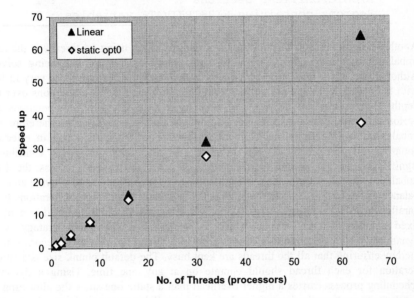

Figure 4 Performance of the OpenMP code generated by the toolkit on an SGI Origin

The possible causes for the degradation in performance could include (i) the execution of serial loops and the effect described by Amdahl's law; (ii) the overhead in starting up and subsequent synchronising of threads; (iii) the granularity of the computation as the number of threads are increased; (iv) the unbalanced workloads assigned to the threads. On examination of the code and with the aid of a profiler (ssrun and prof) a number of optimisations were identified in order to try and improve the performance of the code. Although the toolkit had already completed a fairly comprehensive task in reducing the overhead in starting and synchronising the threads, an additional manual attempt was made to further fuse PARALLEL regions together. This was only possible in a few loops that were contained within IF constructs. As a result of this optimisation the breakdown of the OpenMP directives in the code (Table 2) now show that the number of PARALLEL regions had reduced to 6 but they contained more OMP DO directives. In addition the number of

PARALLEL DO regions were also reduced to 4 as many were fused into larger regions. The performance of the optimised parallel OpenMP code is shown in Figure 5 (static opt1) and shows that there was a small improvement in performance. However, the manual optimisation of fusing the loops did not appear to make a significant contribution to improving the parallel performance.

Table 2 Breakdown of optimised OpenMP directives generated by the toolkit

```
Total number of PARALLEL Regions            : 10
Total number of OMP DO Loops defined        : 25
PARALLEL regions with a single OMP DO loop: 4
REDUCTION Loops                             : 3
ATOMIC/CRITICAL Sections                     : 1
Regions containing FIRSTPRIVATE variables : 2
```

Another possible reason for the limited scalability when using 64 threads is the load imbalance that may exist due to the very nature of the problem being solved. Although there is a maximum of 32 cells in the vertical direction, these vary in line with the ocean depth and are defined using an array such that the iterations over the depth (k-iterations) are different for each i,j cell. In this way, computations are performed only in the ocean regions and only to the necessary depth. Therefore, the unbalanced workloads on each processor together with the reduction in effective computation performed as the number of processors increased contributed significantly to the overall performance. In order to try and redress the load imbalance, the j-iterations need to be distributed in such a way as to provide an even balance of ocean grid cells for each thread. The default scheduling of iterations for a parallel OMP DO is referred to as "static" whereby each thread is assigned a chunk or fixed number of iterations on which to operate. An alternative strategy is to dynamically allocate a number of iterations to a thread as soon as it becomes free, thereby ensuring that all the threads are kept busy. The default chunk size is a single iteration for each thread should operate on at any one time. Using a dynamic scheduling process carries a higher overhead than a static one since the allocation of iterations is continually determined during the parallel execution whereas for a static schedule the allocation of iterations is performed once at the start of the DO loop. Figure 5 shows the effect of using a dynamic scheduling process to execute the 5 parallel loops in the time step loop (dynamic opt1). The performance on 64 threads has now increased to an impressive 75%. In spite of the additional overhead in using a dynamic schedule, it is far outweighed by the superior load balancing achieved. Solving the same problem with a coarser mesh and using the message passing version that has a strategy to deal with the load imbalance [4] shows a similar performance trend.

7. Conclusions

In this paper we have demonstrated that parallelisation tools can be beneficial to both CFD specialists and parallelisation experts by performing most of the tasks needed to parallelise real-world application codes typified by the SEA ocean modelling code. This work also demonstrates the crucial need for an interactive system where the toolkit not only provides valuable information to the user, but also that the user can offer knowledge to supplement and direct the parallelisation process. The interaction is carried out through the use of browsers that filter the vast information acquired by the toolkit and this is presented to the user in a more focused manner. It has also been shown that the generated code is efficient and scalable and that the parallelisation can be done extremely quickly. In comparison, the manual parallelisation of the code using OpenMP has not been undertaken to date because of the prohibitive cost of effort and the expertise that is needed to complete the task.

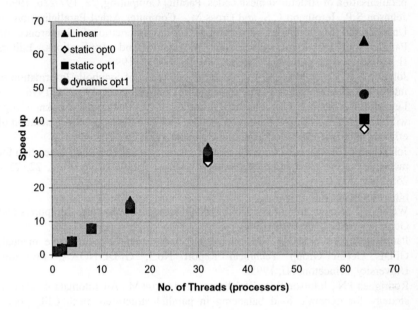

Figure 5 Performance of the optimised OpenMP code generated by the toolkit on an SGI Origin.

8. Acknowledgements

The authors wish to thank the developers of the SEA model code for making the source code available. The authors also wish to thank the many people at the University of Greenwich and NASA Ames who have helped in the development of this toolkit, in particular to Haoqiang Jin (NASA), Jerry Yan (NASA) and also Mallick Arigapudi (SGI) for allowing use of the SGI Origin 2000.

References

1. Evans E.W., Johnson S.P., Leggett P.F., Cross M., Automatic and Effective Multi-Dimensional Parallelisation of Structured Mesh Based Codes. Parallel Computing, 26, 677-703, 2000.
2. Evans E.W., Johnson S.P., Leggett P.F. and Cross M., The automatic code generation of asynchronous communications embedded within a parallelisation tool. Parallel Computing, 23, 1493-1523, 1997.

3. Jin H., Frumkin M., and Yan J. Automatic generation of OpenMP directives and it application to computational fluid dynamics codes. International Symposium on High Performance Computing, Tokyo, Japan, p440, 2000
4. Beare, M.I. and Stevens D.P., Optimisation of a parallel ocean general circulation model, Annales Geophysicae, 15, 1369-1377, 1997.
5. http://www.mth.uea.ac.uk/ocean/SEA/
6. Ierotheou C.S., Johnson S.P., Cross M. and Leggett P.F., Computer aided parallelisation tools (CAPTools) - conceptual overview and performance on the parallelisation of structured mesh codes. Parallel Computing, 22, 197-226, 1996.
7. Johnson S.P., Ierotheou C.S. and Cross M., Computer Aided Parallelisation Of Unstructured Mesh Codes. Proceedings of the International Conference on Parallel and Distributed Processing Techniques and Applications, Editors H.R.Arabnia et al, publisher CSREA, vol. 1, 344-353, 1997.
8. Johnson S.P., Cross M. and Everett M., Exploitation of Symbolic Information In Interprocedural Dependence Analysis. Parallel Computing, 22, 197-226, 1996.
9. Leggett P.F., Marsh A.T.J., Johnson S.P. and Cross M., Integrating user knowledge with information from parallelisation tools to facilitate the automatic generation of efficient parallel Fortran code. Parallel Computing, 22, 259-288, 1996.
10. Johnson S.P., Ierotheou C.S. and Cross M., Automatic parallel code generation for message passing on distributed memory systems. Parallel Computing, 22, 227-258,1996.
11. http://www.openmp.org/
12. Webb, D.J., An ocean model code for array processor computers, Computers and Geosciences, 22 ,569-578, 1996.
13. Pacanowski, R.C., MOM2 documentation, user's guide and reference manual, GFDL Ocean Group Technical Report No.3, GFDL/NOAA, Princeton University, Princeton, NJ, 1995.
14. Rodrigues J.N., Johnson S.P., Walshaw C. and Cross M., An automatable generic strategy for dynamic load balancing in parallel structured mesh CFD code., Parallel Computational Fluid Dynamics, D.Keyes Editor, 345-353, 2000.
15. Mesinger F. and Arakawa A., Numerical methods used in atmospheric models, GARP publications series, No.17, World Meteorological Organisation, 1976.
16. Amdahl G. Validating the single processor approach to achieving large scale computing capabilities. AFIPS conference proceedings,30, 83-485, 1967.

Dynamic Load Equilibration for Cyclic Applications in Distributed Systems

Siegfried Höfinger

Department of Theoretical Chemistry and Molecular Structural Biology,
Molecular Dynamics and Biomolecular Simulation Group,
University of Vienna,
Währingerstr. 17,
A-1090, Vienna, Austria
sh@mdy.univie.ac.at
http://www.mdy.univie.ac.at

Abstract. A Wide Area Network cluster consisting of 5 UNIX-type machines of different architectures and different operating systems is set up via PVM 3.4.3. The 5 nodes are distributed over 4 European research centres located in France, Italy and Austria and only secure shell-type internode communication is allowed. The recently developed concept of Speed Weighted Load Balancing is employed when using GREMLIN - a Quantum Chmistry program - for the electronic structure calculation of alanine. In contrast to the previous static concept, a new dynamic functionality is added, that allows for runtime re-equilibration of all the participating hosts. Even complicated events, such as crash down of contributing hosts may be adressed from this new, dynamic, version of Speed Weighted Load Balancing, without the loss of intermediary data, or a considerable deviation from optimal host performance.

1 Introduction

Distributed Computing has become a major focus in current research and development efforts [1] [2], and it is especially those applications, that efficiently can make use of the steadily growing number of accessible internet computers - nowadays simplified termed "the grid" - that are said to deserve particular attention and need all the concentrated development initiative that is possible. Supercomputer applications in science and technology seem to become feasible in an unforeseen dimension and thus insight into fundamental principles of nature might be gained, if - and only if - these applications really succeed in taking full advantage of all the computational power inherent in the grid of today. Without any doubt Computational Chemistry in general, and Quantum Chemistry in particular, may be considered to be one such example for a scientific discipline, that needs as much computational power as possible, in order to be able to answer the more and more complex growing queries emerging from bio-sciences, chemical industry, etc. Exactly with this future needs in mind *GREMLIN* [3], a Quantum Chemistry program, that solves the *time independent Schrödinger*

P.M.A. Sloot et al. (Eds.): ICCS 2002, LNCS 2330, pp. 963–971, 2002.

equation [4] according to the *Hartree Fock Method* [5] [6], has been developed
with special emphasis on parallel features for shared memory multiprocessor machines [7] as well as distributed cluster architectures of either homogenous type
[3] or even of heterogenous character [8].

In a previous article it was shown [8] that after taking into account the
different node performances of all the individual hosts that formed up a *WAN*-
cluster (Wide Area Network cluster), at the very basic stage of recursive *ERI*-
computation (Electron Repulsion Integrals), which constitutes the major computational task in any Quantum Chemistry program, nearly optimal performance and speed-up could have been established. This previous concept of Speed
Weighted Load Balancing (*SWLB*) could partition the net amount of *ERI* work
into appropriate node-specific fractions based on an input-provided list of relative speed factors, which simply are the proportional constants of all the hosts
performances with respect to the slowest machine. However, this process was
envoked only once in the beginning of the whole iterative procedure and thus
any later change in either network or host conditions would not have been recognized, nor could have been counterbalanced accordingly. Therefore, in this
present work additional dynamic capability of *SWLB* shall be described, which
now offers fast and powerful reactivity to any kind of change in the *WAN*- cluster
environment, where the measures of re-equilibrating become active in between
two subsequent iteration steps of the calculation. It is a frequently observed
characteristic in many large-scale scientific applications, that they repeat a certain major step several times until either an accuracy criterion is met, or an
initially defined overall number of cycles are completed. Therefore it would be of
great advantage, if in between these repeated cycles possible changes in cluster
conditions could be recognized and handled accordingly.

In the remainder of this article a presentation of an extension to static *SWLB*
shall be given, that actually will result in enabling all mentioned dynamic functionality and therefore should contribute considerably to an improvement of
efficiency in distributed computing solutions.

1.1   Problem Description

In the following a brief summary is given that should outline the principal problem arising with electronic structure calculations and their intrinsic need for
computational power. The key process — and also the main time-consuming
one — is the evaluation of *ERIs, the Electron Repulsion Integrals*, which are
6-dimensional, 4-center integrals over the basis functions φ_i

$$ERI = \int_{r_1} \int_{r_2} \varphi_i(r_1)\varphi_j(r_1) \frac{1}{|r_2 - r_1|} \varphi_k(r_2)\varphi_l(r_2) dr_1 dr_2 \tag{1}$$

. The basis functions φ_i are expanded in a series over *Primitive Gaussians* χ_j

$$\varphi_i(r) = \sum_j d_{i,j} \, \chi_j(r) \ , \tag{2}$$

which typically are *Cartesian Gaussian Functions* [10] [11] located at some position (A_x, A_y, A_z) in space.

$$\chi_j(\boldsymbol{r}) = N_j(x - A_x)^l(y - A_y)^m(z - A_z)^n \ e^{-\alpha_j(\boldsymbol{r}-A)^2} \qquad (3)$$

An S-type basis function will consist of primitive gaussians with $l = m = n = 0$, a P-type however of primitives with $l + m + n = 1$, which may be solved at 3 different ways, either $l = 1$ and $m = n = 0$, or $m = 1$ and $l = n = 0$, or $n = 1$ and $l = m = 0$. A D-type specification will likewise mean $l + m + n = 2$ and similarly F-type $l + m + n = 3$.

The principal number of *ERI*s to be considered grows with the 4th power of the number of basis functions, and the latter is proportional to the number of atoms in the molecule. Because modern Quantum Chemistry made clear, how important it is to use extensive, high-quality basis sets, the according number of regarded *ERI*s very soon exceeds conventional RAM and diskspace limits and thus becomes the only limiting factor at all. For example, a simple, small molecule like the amino acid alanine (13 atoms), that has been used in this study, at a basis set description of aug-cc-pVDZ quality [12] [13] (213 basis functions of S, P and D type, 4004 primitive gaussians) leads to a theoretical number of approximately 260×10^6 *ERI*s, which requires about 2.1 GigaByte of permanent or temporary memory and goes far beyond conventional computational resources.

However, an approach called "direct methodology" provides a way to still deal with much larger problem sizes by exploiting the fact that there is partial independence in the mathematical action of all these *ERI*s. In detail this means, that first a certain logical block of related *ERI*s is calculated recursively, where all complicated *ERI*-types $(l + m + n > 0)$ may be deduced from the simpler $(S_i, S_j | S_k, S_l)$ type. In the next step the action of the entire block on all the corresponding Fock-matrix elements — from which there are only a number of $(number\ of\ basis functions)^2$ — is considered, and then the new block of *ERI*s, which overwrites the old one, is formed. Therefore only a small amount of working memory is permanently involved.

Nevertheless one has to respect a hierarchic structure in spawning the space to the final primitive cartesian gaussian functions χ_j, where, following the notation introduced in [3], a certain center $\boxed{i}$ refers to an according block of contracted shells $\rightarrow$ (j)...(k), from which each of them maps onto corresponding intervals of basis functions 1...m and the latter are expanded from primitive cartesian gaussian functions χ_j as seen from equation (2). Therefore, a particular centre quartette $\boxed{i1}$ $\boxed{i2}$ $\boxed{i3}$ $\boxed{i4}$ will entirely define a complete block of integrals, which must be solved together at once. This also becomes the major problem when partitioning the global amount of integrals into equally sized portions, because these blocks may be of considerable different size and one needs to find a well-balanced distribution of centre-quartettes over nodes, that ideally will cause all the various participating hosts to work equally long for the completion of their assigned blocks of integrals.

1.2   Speed Weighted Load Balancing, Static — Dynamic

The goal here is to distribute fractions of the global amount of *ERI*s over all the participating nodes forming up the *WAN*-cluster. If this happens to be done in a way, that all the hosts will actually require exactly the same time to complete their partial tasks, the calculation will be accelerated optimally and one could not do better. Unfortunately a certain block of *ERI*s cannot be considered an independent unit, but is a hierarchic subsection of a set of contracted shells (j)...(k) and these again are a subspace of atomic centres $\boxed{i1}$ $\boxed{i2}$ $\boxed{i3}$ $\boxed{i4}$ and only the last property will be sent forward to the nodes, which will start their work based on received centre-combinations independently and autonomously.

One cannot simply divide the overall numer of theoretical centre quartettes into as many equal sized fractions as there are hosts involved and expect optimal speed up [3]. On the contrary one will have to weigh a certain combination of atomic centres, according to the hierarchic *ERI*-work it will induce. This weightening can be done by estimating the total number of deducable *ERI*s for a certain centre quartette and this is usually done within a neglectable amount of CPU-time. In case of different node performance, which is the general case for *WAN* clusters, the weightening must only be enhanced by the additional factor of per-node-speed, where the relative speed of each node is a multiple of the slowest host speed, but principally the same scheme may be employed [8].

Static Speed Weighted Load Balancing therefore was the implementation of the following steps:

1. Determine the number of participating hosts (NMB) and their according relative speed factors (SPEED[I]). The speed factor is the relative performance of a particular node related to the slowest CPU. So it will either become 1.0 (weakest node), or greater than 1.0 for faster CPUs.
2. Estimate the net amount of computational work (GLOBAL WORK) at the level of quartettes of primitive gaussians to be considered.
3. Form a unit portion (PORTN) of the dimension of

$$\text{PORTN} = \frac{\text{GLOBAL WORK}}{\sum_{I=1}^{\text{NMB}} \text{SPEED[I]}} \qquad (4)$$

4. Loop again over all quartettes of centres and the related contracted shells and basis functions and primitive gaussians either, as if you were calculating GLOBAL WORK, and successively fill the upcoming pair lists for centre quartettes until in the work estimation variable (usually a simple counter, incremented for each new quartette of primitive gaussians) becomes of the size of PORTN*SPEED[I]; then leave the current pair list writing for node I and switch forward to the next node and start with setting up pair lists for this one.

In the *Dynamic* variant of *SWLB* essentially the same steps shall be performed, however, now the whole route of instructions shall be repeated in each of the iterative cycles of the calculation.

Suppose there was a second, completely unrelated process started at some of the nodes, e.g. one, that has got a single CPU only. Then the according SPEED[I] would become just half of SPEED[I], because the rest of the CPU-time would be consumed by the second, unrelated process, that has nothing to do with the cluster calculation. As a consequence, when it comes to summing up all the node data for an intermediary iteration result, all the other nodes J≠I would have to wait for the delayed arrival of the results of node I, and this would go on likewise until the end of the calculation due to the static distribution of the workload.

Only from repeated calls to the partitioning module in each of the iteration cycles, such a serious change in host conditions could be recognized and controlled, and this is what is done in *Dynamic SWLB*. The actual interference with the partitioning module is realized from modifying the speed factors SPEED[I], which means for the example above, that the value of SPEED[I] will be corrected to 0.5*SPEED[I] as long as the 2nd process is active. The changed partitioning scheme would immediately lead to close-to-optimal cluster performance again.

In addition, even drastic changes in the cluster set-up may be covered from *Dynamic SWLB*. For example in the case of crash-down of a certain host I, one only needs to set its corresponding speed factor to SPEED[I]=0, and the calculation will continue with reassigning the current iteration data, but without having to restart the run.

2   WAN Cluster Description

The *WAN*-cluster was set up of five machines located at *GUP* LINZ (A) (2 nodes), *RIST++* SALZBURG (A) (1 node), *ICPS* STRASBOURG (F) (1 node) and *G. Ciamician* BOLOGNA (I) (1 node). Internode communication was realized from *PVM* (Parallel Virtual Machine, rel.3.4.3) [14] based on 1024 bit RSA authenticated ssh-connections. Individual benchmark runs on the small test system glycine/631g (10 centre, 55 basis functions) revealed initial *WAN*-cluster conditions appropriate for *Static SWLB* as shown in table 1, where the network bandwith was measured between the individual nodes and the future master machine (node III). Speed factors were derived from the glycine-benchmark and reflect performance ratios with respect to the slowest machine (node I).

3   Discussion

Dynamic SWLB was used for the *WAN*-cluster calculation of alanine/aug-cc-pVDZ (13 atoms, 213 basis functions) and according speed factors were updated every second iteration. *Static SWLB* (initial two iterations) would lead to a run time behaviour as shown in Fig. 1, where it must be noted that node III not only acts as the master machine, that receives all the data from the remaining hosts and does the serial part of the calculation, but also contributes his CPU to the parallel region. Arrival of parallel thread of node III in Fig. 1 as the first and fastest one is just accidentally. A similar representation on the same scale for *Dynamic SWLB* resulted in the plots shown in Fig. 2 and Fig. 3. Comparison of

Table 1. Speed-Factor and Network-Latency table for the *WAN*-cluster. Speed-Factors represent the relative performance of all the individual hosts in the *WAN*-cluster with respect to the slowest performing CPU (glycine/631g/-271.1538 Hartree in 17 iterations). Network bandwidth was obtained from measuring transfer rates between nodes and the future master-machine (node III).

Physical Location	Architecture/ Clock Speed/ RAM/2L-Cache	Operating System	Relative Speed Factor	Network Bandwidth [kB/s]	Exp.Total Comm. Time [s]
node I G.C. BOLOGNA Italy	INTEL Dual PPro 200 MHz 256 MB/512 KB	LINUX 2.2.14	1.000	166	128
node II ICPS STRASBOURG France	MIPS R10000 200 MHz 20 GB/4 MB	IRIX 64 6.5	1.767	608	35
node III GUP LINZ Austria	INTEL PII 350 MHz 128 MB/512 KB	LINUX 2.2.13	1.943	—	—
node IV GUP LINZ Austria	MIPS R12000 400 MHz 64 GB/8 MB	IRIX 64 6.5	3.609	918	23
node V RIST SALZBURG Austria	ALPHA EV6 21264 500 MHz 512 MB/4 MB	OSF I V 5.0	6.357	592	36

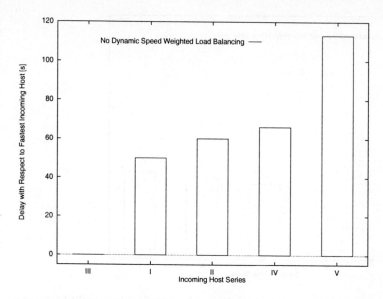

Fig. 1. Representation of the run time behaviour in *Static SWLB* for the iterative data collection phase of the Hartree Fock calculation of alanine/aug-cc-pVDZ with *GREMLIN* in a ssh-connected *WAN*-cluster, made of 5 far-distant machines.

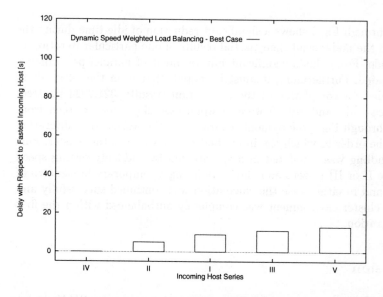

Fig. 2. Representation of the optimal run time behaviour in *Dynamic SWLB* for the iterative data collection phase of the Hartree Fock calculation of alanine/aug-cc-pVDZ with *GREMLIN* in a ssh-connected *WAN*-cluster, made of 5 far-distant machines.

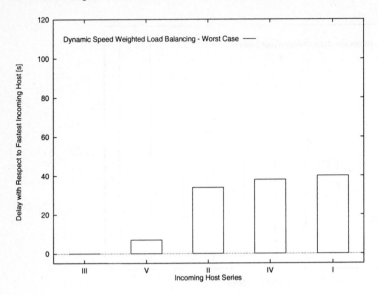

Fig. 3. Representation of the worst case run time behaviour in *Dynamic SWLB* for the iterative data collection phase of the Hartree Fock calculation of alanine/aug-cc-pVDZ with *GREMLIN* in a ssh-connected *WAN*-cluster, made of 5 far-distant machines.

figures Fig.1 through Fig.3 shows a significant reduction of idle-time during the phase when all the nodes send their partial results of one particular iteration to the master node. From this a significant improvement of parallel performance may be concluded. Furthermore, it must be noted, that from the all in all 19 iterations needed for completion of the calculation (result: -321.7924 Hartree) 13 were best-case-like and only 6 were of worst-case character. As seen from figures Fig.1 through Fig.3 the dynamic character of the workload equilibration also changes the order in which the individual nodes arrive at the master. Further event handling was simulated in a sperate run by suddenly setting speed factors of node II or III to zero and thus simulating a temporary breakdown of cluster nodes and in either case the calculation was continued successfully and the change in cluster environment was completely outbalanced within the first subsequent iteration.

4   Conclusion

Taking into account variable node performance in heterogenous *WAN* cluster environments e.g. by the use of *Dynamic Speed Weighted Load Balancing* will substantially enhance overall performance. It also improves the stability and robustness of large grid-applications especially those that perform cyclic iterations and spend almost no time on communication.

Acknowledgement

The author would like to thank Prof. Zinterhof from RIST^{++} Salzburg, Prof. Volkert from GUP Linz Dr. Romaric David from ICPS Strasbourg and Prof. Zerbetto from Ciamician Bologna for providing access to their supercomputer facilities.

References

1. Czajkowski, K., Fitzgerald, S., Foster, I., Kesselman, C.: Grid Information Services for Distributed Resource Sharing. Proc. 10. IEEE Int. Symp. on High-Performance Distributed Computing IEEE Press (2001)
2. Allen, G., Foster, I., Karonis, N., Ripeanu, M., Seidel, E., Toonen, B.: Supporting Efficient Execution in Heterogeneous Distributed Computing Environments with Cactus and Globus. Proc. SC. 2001 (2001)
3. Höfinger, S., Steinhauser, O., Zinterhof, P.: Performance Analysis and Derived Parallelization Strategy for a SCF Program at the Hartree Fock Level. Lect. Nt. Comp. Sc. **1557** (1999) 163–172
4. Schrödinger, E.: Quantisierung als Eigenwertproblem. Ann. d. Phys. **79, 80, 81** (1926)
5. Hartree, D.R.: Proc. Camb. Phil. Soc., **24** (1928) 89
6. Fock, V.: Näherungsmethoden zur Lösung des Quantenmechanischen Mehrkörperproblems. Z. Phys. **61** (1930) 126 **62** (1930) 795
7. Höfinger, S., Steinhauser, O., Zinterhof, P.: Performance Analysis, PVM and MPI Implementation of a DSCF Hartree Fock Program. J. Comp. Inf. Techn. **8** (1) (2000) 19–30
8. Höfinger, S.: Balancing for the Electronic Structure Program GREMLIN in a Very Heterogenous SSH-Connected WAN-Cluster of UNIX-Type Hosts. Lect. Nt. Comp. Sc. **2074** (2001) 801–810
9. Obara, S., Saika, A.: Efficient recursive computation of molecular integrals over Cartesian Gaussian functions. J. Chem. Phys. **84** (7) (1986) 3963–3974
10. Davidson, E.R., Feller, D.: Basis Set Selection for Molecular Calculations. Chem. Rev., **86** (1986) 681–696
11. Shavitt, I.: The Gaussian Function in Calculations of Statistical Mechanics and Quantum Mechanics. Methods in Comp. Phys. ac. New York, **2** (1963) 1–44
12. Dunning Jr., T. H.: J. Chem. Phys. **90** (1989) 1007–1023
13. Woon, D. E., Dunning Jr., T. H.: J. Chem. Phys. **98** (1993) 1358–1371
14. Geist, G., Kohl, J., Manchel, R., Papadopoulos, P.: New Features of PVM 3.4 and Beyond. Hermes Publishing, Paris Sept. (1995) 1–10

3G Medicine – The Integration of Technologies

Andy Marsh

VMW Solutions
9 Northlands Road, Whitenap, Romsey, Hampshire SO51 5RU, UK.
Tel : +44 7703 235 307 Fax : +44 1794 522558
e-mail andy.marsh@vmwsolutions.com

Abstract. Maturing telemedicine technologies, struggling mobile networking revenues and increased personal healthcare awareness have provided the foundations for a new market niche that of '3G Medicine'. During the last 5 years telemedicine (based on internet and web technologies) is becoming a reality both in terms of developing technologies and supportive legislation. Within Europe wireless infrastructures (3G Networking) has received a huge investment and although not well defined in how it will be achieved healthcare has been identified as a major stream of revenue with personal healthcare (e.g. EHCR on the handset) being a key issue especially for the handset manufactures. Combined with an increased awareness not only for outpatients but also for the "well-worried" (healthy and health conscious) 3G Medicine Services will play an important role in personal healthcare management. This paper presents a categorization of 3G Medicine services and the need to integrate technologies.

1 Introduction

During the last 5 years telemedicine has utilized developing technologies and matured into a now usable service acceptable both by patients and medical staff. In essence telemedicine supports the remote application of healthcare services. Isolated medical centers can be connected to hospitals, ambulances can transmit vital sign data to awaiting emergency units, General Practitioners can be keep informed of hospitalized patients and outpatients can be monitored whilst at home. By utilizing the latest mobile technologies (i.e. 3G networking) a new collection of "mobile telemedical" services can be developed referred to hereafter as "3G Medicine". These new services will support not only home care services but also mobile care services for example an outpatient may go about their daily business but still have the confidence they are being continuously monitored.

Traditionally, medical monitors were limited to data acquisition, typically implemented as Holters. Holters are used for 24-48 hour monitoring of ECG, EEG or polysomnography (EEG, EOG, EMG, EKG, heart rate, breathing, body position, snoring, etc.) and recording on cassette tape or flash memory. Recorded signals are then analyzed off line using dedicated diagnostic systems. Increased intelligence coupled with the low power consumption of the new generations of microcontrollers/DSPs combined with wireless technologies makes possible a whole

P.M.A. Sloot et al. (Eds.): ICCS 2002, LNCS 2330, pp. 972–981, 2002.

range of intelligent monitor applications. Combined with an increased awareness not only for outpatients but also for the "well-worried" (healthy and health conscious) 3G Medicine Services will play an important role in personal healthcare management. Subsequently, the development of supportive 3G Medicine products and services will also create a new niche market economy for companies, especially SME's, to develop a range of collaborative technologies. The rationale of this paper is to identify a collaborative framework of integrated technologies and to highlight the requirement of a '3G Medicine Special Interest Group (SIG)' that will ensure conformity for the interoperability of 3G Medical products and services.

The major technologies that need to be integrated are presented in section 2 and a categorization of 3G Medical services is presented in section 3. However, for these new services to be generally accepted and used a certification and conformance body needs to be established as discussed in section 4. Conclusions are drawn in section 5.

2   The need to integrate technologies

2.1   Mobile Wireless Communications

The choice of a wireless technology for a particular application depends on factors such as the range over which the device is expected to operate, the number of other wireless devices that are expected to operate within the same area, and the velocity of the device (that is, whether it is being used by someone sitting in a home or someone in a car traveling down a highway). Wireless technologies tend to be categorized based on the range in which they operate:

1. within a building such as a home or business (Pico cell);
2. within a neighbourhood (micro cell);
3. within a suburban area (macro cell);
4. across the globe.

Technologies And Services Existing Today. Many second-generation mobile technologies exist today each having influence in specific parts of the world. GSM, TDMA (IS 136), and CDMA (IS 95) are the main technologies in the second-generation mobile market. GSM by far has been the most successful standard in terms of it's coverage. All these systems have different features and capabilities. Although both GSM and TDMA based networks use time division multiplexing on the air interfaces, their channel sizes, structures and core networks are different. CDMA has an entirely different air interface.

Technologies And Services Existing Tomorrow. 3rd Generation Wireless, or 3G, is the generic term used for the next generation of mobile communications systems. 3G systems aim to provide enhanced voice, text and data services to user. The main benefit of the 3G technologies will be substantially enhanced capacity, quality and data rates than are currently available. This will enable the provision of advanced services transparently to the end user (irrespective of the underlying network and technology, by means of seamless roaming between different networks) and will bridge the gap between the wireless world and the computing/Internet world, making inter-operation apparently seamless. The third generation networks should be in a position to support real-time video, high-speed multimedia and mobile Internet access. All this should be possible by means of highly evolved air interfaces, packet core networks, and increased availability of spectrum. Although ability to provide high-speed data is one of the key features of third generation networks, the real strength of these networks will be providing enhanced capacity for high quality voice services. The need for landline quality voice capacity is increasing more rapidly than the current 2nd generation networks will be able to support. High data capacities will open new revenue sources for the operators and bring the Internet more closer to the mobile customer. The use of all-ATM or all-IP based communications between the network elements will also bring down the operational costs of handling both voice and data, in addition to adding flexibility. The drive for 3G is the need for higher capacities and higher data rate. Technologies like GPRS (General Packet Radio Service), High Speed Circuit Switched Data (HSCSD) and EDGE fulfill the requirements for packet data service and increased data rates in the existing GSM/TDMA networks. GPRS is actually an overlay over the existing GSM network, providing packet data services using the same air.

2.2 Short-Range Wireless Connectivity

Communication between various devices makes it possible to provide unique and innovative services. Although this inter-device communication is a very powerful mechanism, it is also a complex and clumsy mechanism, leading to a lot of complexity in the present day systems. This makes networking not only difficult but limits its flexibility as well. Many standards exist today for connecting various devices. At the same time, every device has to support more than one standard to make it inter-operable between different devices. Take the example of setting up a network in offices. Right now, entire office buildings have to make provisions for lengths of cable that stretch kilometers through conduits in the walls, floors and ceilings, to workers' desks. In the last few years, many wireless connectivity standards/technologies have emerged. These technologies enable users to connect a wide range of computing and telecommunications devices easily and simply, without the need to buy, carry, or connect cables. These technologies deliver opportunities for rapid ad hoc connections, and the possibility of automatic, unconscious, connections between devices. They will virtually eliminate the need to purchase additional or proprietary cabling to connect individual devices, thus creating the possibility of using mobile data in a variety of applications. Wired LANs have been very successful in the last few years and now with the help of these wireless connectivity technologies, wireless LANs (WLAN) have started emerging as a much more powerful and flexible alternatives to the wired LANs. Until a year ago, the speed of the WLAN was limited to 2 Mbps but with the introduction of these new standards, we are seeing WLANs that can support upto 11 Mbps in the ISM band. There are many such technologies/standards and notable among them are Bluetooth, IrDA, Home RF and IEEE 802.11. These technologies compete in certain fronts and are complementary in other areas. So, given the fact that so many technologies exist, which technology is the best and which solution should one select for a specific application? The premise behind all these standards is to use some kind of underlying radio technology to enable wireless transmission of data, and to provide support for formation of networks and managing various devices by means of high level software.

2.3 Wireless terminals

There is going to be an estimated 60 Million Mobile Internet users in 2004, according to estimates, and this emerging market is now starting to take off. All different parts of the software and Internet industry is moving in to this space with a huge increase in interest over the last 12 months. The recent JavaOne conference in San Francisco also showed this trend, wireless is a major opportunity for Java to make it big. There are lots of debates about Java or NoJava, EPOC vs PalmOS vs PocketPC etc. and maybe these debates are made obsolete by Java? No matter what, there is still one thing that remains to be determined: *What are the devices that we all will use?*

The Wireless Application Protocol. The mass market device for the Mobile Internet applications is likely to be the phone with a WAP browser. Most vendors will include WAP in all their phones. This means that anyone that buys a phone to use for talking is getting the Mobile Internet features built in. WAP might not have the most advanced applications at this point, but it is very easy to use and get into (especially when GPRS takes away the connection setup-time). Even with the advent of more advanced applications, WAP will still be used to access information, much like the web browser on the PC. The main difference is that phones are not going to be used that much for random browsing, but rather information access, straight to the point. With the advent of WAP 2.0, where multimedia and streaming is foreseen, the width of applications increases drastically. The Wireless Application Protocol is a standard developed by the WAP Forum, a group founded by Nokia, Ericsson, Phone.com (formerly Unwired Planet), and Motorola. WAP defines a communications protocol as well as an application environment. In essence, it is a standardized technology for cross-platform, distributed computing. WAP is very similar to the combination of HTML and HTTP except that it adds in one very important feature: optimization for low-bandwidth, low-memory, and low-display capability environments. These types of environments include PDAs, wireless phones, pagers, and virtually any other communications device.

JAVA(TM) wireless toolkit. Sun Microsystems on Nov. 27, 2001 announced that it is offering an enhanced version of the Java (TM) 2 Platform, Micro Edition (J2ME(TM)) Wireless Toolkit(). Live demonstrations of the toolkit, including the early access Japanese version, was presented at the JavaOne(SM) Conference Yokohama, November 28-30, 2001. The updated toolkit securely integrates with major Java technology Integrated Development Environments (IDEs) and is the first toolkit to deliver a set of enhanced new features that enable the design and implementation of Over-The-Air (OTA) provisioning for mobile devices. OTA provisioning is the wireless delivery of applications and services over a secure network to mobile devices, including mobile handsets and PDAs. Named this June by JavaWorld.com as the "Best Java Device Application" by JavaWorld.com, the J2ME Wireless Toolkit will help generate more rich and innovative applications and services for Java technology-enabled devices.

2.4   Information-Standards Organisations in Healthcare Informatics

The necessity of having standards is well understood and highly appreciated. A number of standardisation organisations and committees have been quite active in the development of standards that relate to healthcare informatics:

- American National Standards Institute (ANSI)
- CEN (Comité Europeén de Normalisation) Technical Committee 251
- ISO Technical Committee 215
- American Society for Testing and Materials Committee E31 (ASTM E31)
- Healthcare Informatics Standards Board (HISB)
- Computer-Based Patient Record Institute (CPRI.

2.5   Security

The explosive growth in computer systems and their interconnections via networks, mainly the Internet, has increased the dependence of both organizations and individuals on the information stored and communicated using these systems. This, in turn, has led to heightened awareness of the need to protect data and resources from disclosure, to guarantee the authenticity of data and messages and to protect systems from network-based attacks.

2.6   Wireless medical devices

For millions of Europeans with chronic medical conditions, careful day-to-day health monitoring can help avert catastrophe. Home health gauges abound, but they take readings only at discrete points in time and require a patient's active participation—answering a computer questionnaire, for example. The ideal monitor, however, would record data constantly, and patients wouldn't even notice it's there namely wearable health sensors. "There's no question we're going remote, and we're going wireless," says Credit Suisse First Boston's Robert Hopkins. A non-invasive ambulatory monitoring system can provide clinically relevant parameters with the capability to perform continuous monitoring over hours, days and weeks thus providing a "movie" of health and/or disease versus the "snapshot" resulting from a standard history and physical examination conducted in a physician's office. An ambulatory system could be used to collect and store cardiac, respiratory, blood pressure, posture, activity and emotional measures.

2.7   Wireless healthcare software services

The technology sector of healthcare is entering into a new evolutionary phase. The medical community has an obligation to the public to provide the safest, most effective healthcare delivery system possible. Optimal medical practice will depend on computers. Outcomes analysis that will shape medical treatments in the future will be derived directly from the data entered into Windows Powered mobile devices. If a global database could be obtained using point-of-care devices, existing methods for tracking clinical outcomes would become obsolete. Optimal management of disease prevention and management could be achieved! In addition, these devices, with appropriate software, will allow for foolproof coding and billing. Using Windows Powered mobile solutions, physicians and hospitals can capture additional earned revenue that can amount to millions of lost healthcare dollars per year.

3   A framework for mobile medical services

Based upon the authors experience in identifying a framework for internet/web-based telemedicine [1], a framework for Mobile (M-) medical services has been defined [2] that contains three categories of services namely: **M-Safety, M-Healthcare** and **M-Medicine**. (To support these services there needs to be an investment in hardware R&D to develop the supportive M-Platforms and M-Sensors and also software R&D to incorporate M-Security and Compunetics)

3.1   M-Safety services

The M-Safety (Mobile-Safety) services are a form of the newly developing Mobile Location Services (MLS) but focus on the potential safety implications of location awareness and include child monitoring, location advisory and third party location monitoring services. Three sub-categories of M-Safety services have been identified, namely:

- where the user of the service is informed where they are.
- where another person is informed where the user of the service is.
- where a 3rd party monitors the location of the user of the service.

M-Safety service Category 1:(Pharos). By utilizing the latest mobile phone handsets, with locatable functionalities, a range of services can be developed that verbally informs the user where they are in terms of street names and nearest points of interest. Defined originally for the visually impaired to act as a guidance tool the "Pharos" service is equally applicable in any outdoor Location Based Navigation & Direction environment.

M-Safety service Category 2:(MINDER). By utilizing the latest locatable technologies a range of services can be developed whereby the location of the user is illustrated on a digital map that can be viewed by a web browser. Defined originally for monitoring the movements and whereabouts of children, but equally applicable to any outdoor tracking environment, the " MINDER - Movement INformation DElivered Regularly" service provides regular updates of the outdoor position, displayed on a digital street map, of a personal locatable tag.

M-Safety service Category 3:(Guardian). By utilizing the latest developments in GPS and digital compass technologies a range of services can be developed whereby the indoor location of a user can also be monitored by a 3rd party service provider. Defined originally for the monitoring of the elderly, but equally applicable in any indoor tracking environment, the "Guardian" service [2] provides a reassurance mechanism for family members that for example an elderly people has not fallen over or has become trapped etc.

3.2   M-Healthcare services

The M-Healthcare (Mobile-Healthcare) services are a subset of the newly developing 3G Medical services and focus on the collection and interpretation of personal medical sensor data. a range of M-Healthcare services can be developed which include the recording of personal sensor data (for example ECG) for future comparison, a mechanism to check the data for warning signs (for example high blood pressure) and an automated analysis of the sensor data. Three sub-categories of M-healthcare services have been identified, namely:

- where the results of personal sensors are recorded for later use.
- where an early warning system checks the personal sensor results.
- where the results of personal sensors are automatically diagnosed.
-

M-Healthcare service Category 1:(MEMOIR). By using the memory capabilities of mobile phone handsets a range of services can be developed that record the personal sensor data that can then be used for later comparisons. Defined originally for monitoring ECG over 24 hours, the "MEMOIR services [2] are equally applicable to monitoring any sensor data over long time periods.

M-Healthcare service Category 2:(LOCUM). By using the latest developments in mobile handset technologies that support computational facilities (e.g. Java on the handset) a range of customizable personal early warning services can be developed whereby the data of the users sensors are analyzed against personalized predefined values. If a discrepancy occurs the user is subsequently warned. Defined originally for monitoring the blood pressure of the elderly, but equally applicable in any high risk category, the "LOCUM - LOcally Controlled Ubiquitous Monitoring" service [2] provides a personal healthcare reassurance mechanism.

M-Healthcare service Category 3:(NOMADS). By using the latest developments in wireless communication technologies (such as GPRS) a range of *NOn intrusive Monitoring and Automated Diagnosis Services* (NOMADS) can be developed whereby the data of the users sensors are remotely analyzed. If a problem is diagnosed then a member of the medical community is alerted and the appropriate action initiated. Defined originally for the monitoring of oncology patients the "NOMADS" service [2] is equally applicable for any outpatient analysis.

3.3   M-Medicine services

The M-Medicine (Mobile-Medicine) services are a subset of the newly developing 3G Medical services and focus on 2-way communication between diagnostic medical servers (supported by medical staff) and the users medical sensors. A range of M-Medicine services can be developed which support personalized care (for example informing the user then checking that drugs have been taken with the correct dosage and at the correct time), personalized nursing (for example altering a drug prescription due to updated sensor data) and personalized doctoring (for example modifying a treatment plan). Three sub-categories of M-Medicine services have been identified, namely:

- where the user is informed to do an action that is monitored.
- where the action to be undertaken by the user is modified.
- where the actions to be undertaken by the user are defined.

M-Medicine service Category 1:(3G-Care). By using the latest developments in 3G technologies a range of 3G-Care services can be developed whereby medical staff can remotely instruct and monitor users medical sensors and related supported devices. Defined originally for monitoring drug usage, the "3G-Care " services [2] are equally applicable for instructing the user and monitoring any sensor and/or related supported device usage.

M-Medicine service Category 2:(3G-Nurse). By using the latest developments in 3G technologies a range of 3G-Nurse services [2] can be developed whereby medical staff can remotely instruct, monitor and control users medical sensors and related supported devices enabling modifications of personalized treatment plans and early warning symptoms.

M-Medicine service category 3:(3G-MD). By using the latest developments in 3G technologies a range of 3G-MD services [2] can be developed whereby the Medical Doctor can define a personalized treatment plan and early warning symptoms that can then be downloaded to the users medical sensors and related supported devices.

4   The need for conformance

4.1   A US perspective

The HMO's, that have supporting legislation, have driven the adoption of telemedicine within the US. Internet based applications are being used to improve access to care and the quality of care, reducing the costs of care and the sense of professional isolation for some healthcare practitioners. In this environment the introduction of wireless telemedicine should be introduced via the HMO's expanding their range of services and therefore compatible with the presently installed systems and envisaged wireless LAN systems. In summary, it is envisaged that the introduction of 3G Medical services in the US will again be driven from the HMO's therefore it is essential that some form of standardization and conformance be undertaken in conjunction with existing telemedical services.

4.2   A European perspective

Within Europe telemedicine has not been driven so much by HMO's but more by isolated medical institutions and regional trails. The legislation aspect of telemedicine also within Europe is more complex than in the US especially when National boundaries have to be crossed. However, the telecommunications markets in Europe are more focused and standardized than their US counterparts and it is this that is envisaged to be the driving force behind the introduction of 3G medical services in Europe. In summary, it is essential that some form of standardization and conformance be undertaken in conjunction with developing telecommunication infrastructures and services.

4.3   The 3G Medicine Special Interest Group

By combining the two perspectives above it is clear that for 3G Medicine services to be generally available and accepted worldwide then there needs to a standardization and/or conformance certification group. Similarly to the bluetooth Special Interest Group it is therefore proposed that a 3G Medicine Special Interest Group (SIG) be established with representatives from both HMO's and telecommunications domains both in Europe and the US. Additionally there also needs to be representatives from a number of supporting industrials including platform developers, compunetics (computing and networking) suppliers, security advisors, medical data sensor developers and service developers.

The objective of the **3G Medicine SIG** is to combine 20 areas of expertise:

1. Wireless Medical devices (ERM TG30)
2. Mobile Terminals (PDA, Smart phone)
3. Operating Systems (Linux, Palm OS)
4. Data Storage (M-EHCR)
5. Data Encoding (XML, WML)
6. Programming environments (JAVA)
7. Visualization (MPEG, VRML)
8. Transmission (GPRS, EDGE, UMTS)
9. Collaboration (SMS, WAP, HTTP)
10. Privacy & security (TLS, SSL, PKCS)
11. Compression (JPEG 2000, Wavelet)
12. Archival (HIS, Data warehousing)
13. Knowledge discovery (personalized alarms)
14. Healthcare providers (Doctors, Nurses)
15. Personal healthcare management providers
16. Standardization (R&TTE)
17. Conformance (FDA, EU CE Marking)
18. Legislation (National, EU polices)
19. Service providers (HMO's)
20. User groups (Elderly, Outpatients)

The 3G Medicine SIG will therefore tackle such issues as device availability, possibilities for health with 3G networking, the services that will be required (by health professionals, ambulatory, patients and citizens), the applications that will be developed, the costs (private and public) and compliance with technical issues, legislation and regulatory frameworks.

5   Conclusions

For the successful development of 3G Medical services there needs to be an agreement on what types of categories of services will become available, how different complementary industries can work together to develop these services and how the developed services are certified. The rationale of this paper was to suggest such a framework. In the first instance the author has categorized potential 3G medical services into 3 groups namely M-Safety, M-Healthcare and M-Medicine. Each group is subdivided further relating to how the medical data is utilized. To support the development of these services the author has also identified the necessity of M-Sensors, M-Platforms, M-Compunetics and M-Security. To ensure conformance the author has also suggested that in the first instance a Special Interest Group be established to co-ordinate the development of 3G Medical services and provide certification of developed services.

References

1. Marsh,A, Grandinetti,L, Kauranne, T, "Advanced Infrastructures for Future Healthcare", IOS press, 2000.
2. www.vmwsolutions.com
3. www.benefon.com
4. Marsh, A, May, M., Saarelainen,M., "Pharos: Coupling GSM & GPS-TALK technologies to provide orientation, navigation and location-based services for the blind", IEEE ITAB-ITIS 2000, Washington Nov, 2000.
5. Strouse, K, "Strategies for success in the new telecommunications marketplace", Artech House, USA, 2001.
6. PriceWaterhouseCoopers, "Technology Forecast 2001-2003, 2001
7. Dornan,A, "The essential guide to wireless communications applications", Prentice Hall, USA, 2001.
8. Jupiter Research, "Mobile Commerce Report", 2001.
9. Scheinderman,R, "It's time to get smart", Portable design, March 2001, pp22-24.
10. ARC Group, "3G Industry Survey", 2001

Architecture of Secure Portable and Interoperable Electronic Health Records

Bernd Blobel

Otto-von-Guericke University of Magdeburg, Medical Faculty, Institute of Biometry and
Medical Informatics, Leipziger Str. 44, D-39120 Magdeburg, Germany
bernd.blobel@mrz.uni-magdeburg.de

Abstract. Electronic Health Records (EHR) are moving towards the core application of health information systems. Enabling informational interoperability of
shared care environment including EHR, structure and function of components
used have to follow open standards and publicly available specifications. This
comprises includes also methods and tools applied. After shortly introducing
general aspects of open interoperable component architectures, actual approaches for EHR systems are discussed distinguishing between the one-model
and the dual-model paradigm. The emerging activities for a harmonised multi-
model openEHR as well as its implementation are presented. Special attention
is given to security requirements and solutions. Based on standardised Public
Key Infrastructure (PKI) and security token such as Health Professional Cards
(HPC), policy-defined application security services such as authorisation, access control, accountability, etc., of information recorded, stored and processed
must be guaranteed. In that context, appropriate resource access decision services have to be established. As the European HARP project result, a component-based EHR architecture has been specified and demonstrated for enabling
open, distributed, virtual, and portable EHR implementation with enforcing
fine-grained security services by binding certificates to application components,
by the way enforcing policies.

1   Introduction

For establishing efficient and high quality care of patients, comprehensive and accurate
information about status and processes directly and indirectly related to patient's health
must be provided and managed. Such information concerns medical observations, ward
procedures, laboratory results, medical controlling, account management and billing,
materials, pharmacy, etc. Therefore, health information systems within healthcare establishments (HCE) converge to Electronic Patient Record (EPR) systems as a kernel enabling the management of all the other business processes as specific views on the EPR
and building the informational basis for any communication and co-operation within,
and between, healthcare establishments (HCE). By that way, inter-organisational vir-

P.M.A. Sloot et al. (Eds.): ICCS 2002, LNCS 2330, pp. 982–994, 2002.

tual electronic healthcare records (EHCR)[1] are built. This virtual EHCR has to met shared care requirements of providing any information needed and permitted at the right time to the authorised user at any location in the right format including mobile devices. In that context, it has to fulfil all the needs of the HCE and its principals involved reflecting all the views defined in ISO/IEC 10746-2 "Reference Model – Open Distributed Processing", such as enterprise view, information view, computational view, engineering view, and technological view [9]. These views are different in different HCE with their different scenarios for meeting different requirements under their specific conditions and restrictions. For providing information and functionality needed, EHCR must be structured and operating appropriately.

First, some definitions related to EHCR should be introduced:

- An EHCR is a repository of information about the patient's health available in a computer-readable format.
- An EHCR system is a set of components establishing mechanisms to generate, use, store and retrieve an EPR.
- An EHCR architecture describes a model of generic properties required for any EPR for providing communicable, comprehensive, useful, effective, and legally binding records, which preserve their integrity over the time independent of platforms and systems as well as of national specialities.

Following, different approaches and an optimal way for meeting the requirements and characteristics mentioned is described in more detail by focusing on the architectural and modelling aspects of EHCR.

2   The Generic Component Paradigm for EHCR Architecture

Regarding EHR in general, we have to look for structure and domain knowledge related concepts, but also for the concepts of security, safety and quality. Considering security issues, the concepts of communication security can be distinguished from application security. Quality and safety are related to the latter one. Within one concept, different levels of granularity and abstraction can be defined forming a layered model of services, mechanisms, algorithms, and data [2].

| object = | attributes + operations | (1) |

However, objects require knowledge about object-to-object interactions to be completely useable, reusable, and interoperable on all the different levels of the ISO RM – ODP. Such requirements and conditions of interactions concern pre-conditions, post-conditions, constraints, group behaviour, etc. Within the object-oriented world, the management of object-to-object interactions can, e.g., only be provided at the CORBA COSS level, whereas the healthcare vertical facilities need the specification of this

---

[1] If the record includes also issues beyond patient's care such as, e.g., social aspects and health prevention of citizens, an electronic health record (EHR) is created. In the paper, the EHCR view will be used knowing the validity of the statements also for EHR, however.

knowledge about conditions, relationships, and framework at the business object component architecture level [11].

Figure 1 presents the conceptual schema of the original approach of CORBA. This CORBA conceptual schema enabled already grouping, multi-interfacing, etc.

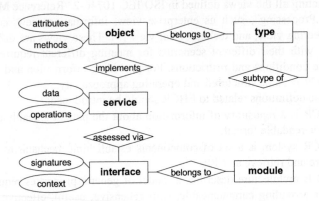

Fig. 1. The CORBA Conceptual Schema

To overcome these problems, a generic component architecture paradigm has been developed by the Magdeburg Medical Informatics Department at 1995. Contrasting to objects, a component is defined as shown in equation (2). The conceptual schema of such generic components is given in figure 2.

component =	attributes + operations + structural constraints + operational constraints + events + multi-interfaces * scenarios + safety + reliability + security + ...	(2)

3   Available EHCR Architecture Models

An EHCR has to meet requirement, that are already investigated e.g., in the context of several EHCR projects. Managing objects, an EHCR arises as dynamic process from clinical practice. It manages a complex workflow connected with medical acts. The EHCR is based on, and supports, electronic communication between all parties involved. It documents any diagnostic and therapeutic measures in a standardised structure. Reducing or avoiding redundancy, an EHCR facilitates an optimised unambiguous presentation of medical concepts, preserving the original context and enabling new ones. It reflects chronology and accommodates future developments and views. For managing an EHCR system, the architecture of such distributed and highly complex component system as well as its behaviour (functionality, set of services) must be designed appropriately.

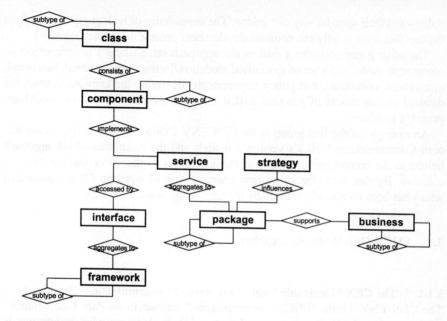

Fig. 2. The Conceptual Schema of the Generic Component Model [2]

Replacing the old relation paradigm of some architecture models for health information systems such as, e.g., the Distributed healthcare Environment (DHE) architecture, the actual EHCR architecture standard models follow the object-oriented or even component-oriented paradigm. However, they are distinguished by a fundamental difference in their approach of establishing the EHCR model. One group intends to develop the complete EHCR architecture within one comprehensive model of structures, functions, and terminology in the classic way covering all the concepts known at the development time. Such an one model approach however reveals some essential weaknesses and problems related to technical, complexity, and management issues which are now shortly resumed [1]:

Considering the technical problems of the one model approach, the mixture of generic and domain-specific knowledge concepts with their own expressions, but also weaknesses in basis class stability must be mentioned.

Regarding the complexity problems, the size of the resulting model leads to difficulties in managing so many concepts in parallel, in completing the model which might be unachievable, in standardising such models and in providing interoperability due to the needed agreement on a huge number of aspects and details.

Related to the management of the one model approach, different developer and user groups dealing with their own concepts expressed in their specific language must be managed, combined and harmonised. The generic part of the EHCR concepts concerns the grammar of the IT-system domain which is specified by computer scientists. The health domain specific concepts representing the domain knowledge are specified and maintained by medical experts. Both groups are characterised by their own terminology and their specific way of thinking. The dependency of both groups results from the fact that there is only one common development process using one formalism.

The other group provides a dual model approach establishing a generic object or component model and a set of specialised models reflecting organisational, functional, operational, contextual, and policy requirements presenting the knowledge about the detailed circumstances of practical EHCR instances overcoming the one model approach's problems.

An example of the first group is the CEN ENV 13606 "Electronic Healthcare Record Communication". HL7's version 3 models and the Australian GEHR approach belong to the second group, despite of the differences explained in detail in the next chapters. By that way, the component characteristic of equation (2) is established which has been completely realised by the GEHR approach.

3.1   EHCR One-Model Approaches

3.1.1   The CEN "Electronic Healthcare Record Communication" Standard

The CEN ENV 13606 "EHCR Communication" defines in its Part 1 an extended component-based EHCR reference architecture [3]. Such an extended architecture is mandated to meet any requirements through the EHCR's complete lifecycle. According to CEN ENV 13606, an EHCR comprises on the one hand a *Root Architectural Component* and on the other hand a *Record Component* established by *Original Component Complexes* (OCC), *Selected Component Complexes*, *Data Items*, and *Link Items*. OCC consist of 4 basic components, such as folders, compositions, headed sections, and clusters. These OCC sub-components can be combined in partially recursive ways. Beside its Part 1 "Extended Architecture", the CEN ENV 13606 offers the Part 2 "Domain Term List", Part 3 "Distribution Rules", and Part 4 "Messages for the Exchange of Information".

3.1.2   The Governmental Computerised Patient Record Project

Launched by a consortium formed by the US Department of Defense, the US Department of Veterans Affairs, and the Indian Health Service, the Governmental Computerised Patient Record (G-CPR) established a model and tools for implementing and managing an proper business as well as technical environment to share patient's information [5]. The main goals concern

- the establishment of a secure technical environment for sharing sensitive personal information,
- the development of a patient focused national information technology architecture,
- the creation of common information model and adequate terminology models to ensure interoperability between disparate systems.

The solution should be based on advanced national and international standards. Using object oriented specifications for interoperability, the approach was rather service oriented than architecture based.

3.2   EHCR Dual-Model Approaches

3.2.1   The GEHR Approach

Based on the European Commission's Third Framework Programme project "Good European Health Record (GEHR)", but also acknowledging the results of other R&D projects and efforts for standards around the globe, the Australian Government launched and funded the Good Electronic Health Record (GEHR) project [6]. The basic challenge towards GEHR is knowledge level interoperability.

The GEHR model consists of two parts: the *GEHR Object Model* (GOM), also called reference model, delivering the EHCR information container needed on the one hand, and the GEHR meta-models for expressing the clinical content on the other hand (figure 3).

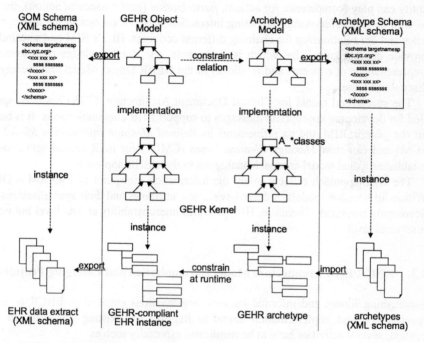

Fig. 3. GEHR Architectural Schema (after T. Beale [1])

Bearing the medical knowledge in the sense of healthcare speciality-specific or the organisation-specific, department-specific or even person-specific views and restrictions, the meta-models are commonly called *Archetypes*. Therefore, the corresponding model is also called archetype model. Because the archetypes are separately developed, they can be instantiated step by step at the technical model level until the complete medical ontology has been specified. In summary, the GEHR approach consists of small flexible pieces like LEGO® bricks which can be combined in a proper, health domain specific way following construction plans defined in archetypes. Summarily, the reference model is the concrete model from which software can be built, and of which EHR data are instances. The archetype model establishes a formalism whose instances are domain concepts which are directly processable by health information systems.

3.2.2   The HL7 Reference Information Model and its Clinical Document Architecture

Within its Version 3 *Message Development Framework*, the well known health industry standard for communication HL7 specified a comprehensive *Reference Information Model* (RIM) covering any information in the healthcare domain in a generic and comprehensive way [8]. The HL7 RIM deals with the associations between the six core classes *entity* (physical information object in the healthcare domain), the *role* the entity can play (competence for action), *participation* (performance of action), the *act* as well as *role relationship* mediating interaction between entities in the appropriate roles and *act relationship* for chaining different activities. HL7's RIM and vocabulary provide domain knowledge which is exploitable, e.g., for knowledge representation (representation of concepts and relations) in the GEHR Object Model and archetypes discussed before.

The specialised model for Clinical Document Architecture (CDA) has been specified for developing appropriate messages to support EHR communications. It is based on the generic RIM and its refinements as *Refined Message Information Model* (R-MIM) and *Common Message Element Types* (CMET) for EHR related scenarios. It establishes a dual model approach analogous to the GEHR approach.

The HL7 approach reflects solely the information viewpoint of ISO RM – ODP. Within information models, it describes classes, attributes and their specialisations for developing messages. Therefore, HL7 provides interoperability at data level but not at functional level.

3.3   EHCR/EHR Architecture Model Harmonisation and Emerging Projects

Establishing formal and informal liaisons, organisations engaged in EHCR or EHR specification and implementation intend to improve the existing standards. In that context, several activities have to be mentioned especially such as

- the recently started improvement of CEN ENV 13606 now called "Electronic Health Record Communication",
- the refinement of G-CPR in the sense of emphasising HL7 communication instead of CORBA service orientation,
- the establishment of the European Commission's EUROREC organisation, and
- the openEHR approach.

Collaborating with HL7, both CEN and openEHR will narrow and harmonise their approaches. Establishing a (initially funded) national EUROREC organisation in all the European Union member states, the EuroRec initiative concerns the improvement of awareness for, and the wider implementation of, EHR in the European practice. In that context, a European Electronic Health Record institute has been founded at November 2001.

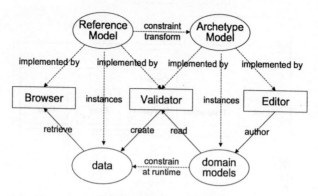

Fig. 4. Meta-Architecture for Implementing and Use of OpenEHR

Regarding implementation and use of the openEHR approach, the model components and system components needed can be presented as shown in figure 4. Based on the models introduced in section 3.2.1, editors are needed to author the domain knowledge in domain models. Read by the openEHR kernel, this information is used to create the object model instances, i.e., the data the principal is interested in to be retrieved and presented by a browser.

4   OpenEHR Package Structure

For implementing openEHR, several system components or packages have to be established. The EHR basic structure is the *Record*. Its sub-packages describe the compositional structure of an EHR. The *Record* package contains the packages *EHR* (incl. EHR extracts), *Transactions* (incl. audit trail), and the related content. The latter contains the *Navigation*, *Entry*, and *Data* packages, whose classes describe the structure and semantics of the contents of transactions in the health record. The *Path* serves for item location. The basic package defines the core classes used in the openEHR approach. *External* refers to external packages providing interoperability with non-EHR systems. The *Party* package addresses the principals involved such as users, systems, components, devices, etc.

Figure 5 presents the package structure of an openEHR system as described.

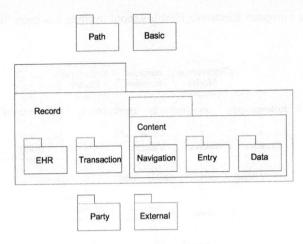

Fig. 5. Package Structure of the openEHR Reference Model

5   The HARP² EHR Implementation

According to the generic component model [2], all views, information content, functionality, implementation environment, and underlying technology but also the proper level of granularity might be modelled in a consistent way. In this way services and the complexity of the running application component can be defined according to the application environment and the user needs. Services concern entry, processing, and presentation of data but also the enforcement of underlying policy for communication and co-operation. The generic component model enables claims change management (viewpoint of the system) and the resolution of the component's complexity by the transition to less complex sub-components as shown in figure 6. Each specific model in the abstraction-granularity space reflects one specific archetype. A theoretical consideration on consistency for state transitions within the generic model have been provided in [2].

The description of the components according to equation (2) is established in archetype schemas using the XML (Extensible Markup Language) standard set. Related to granularity and technology viewpoint, mobile computing has to meet special requirements which are easily enabled by this dynamic selective approach of the proper state of the a complex system.

Within the HARP project, partners from Greece, Germany, Norway, United Kingdom, and the Netherlands specified, developed and implemented the HARP Cross Security Platform (HCSP) for Internet based secure component systems as well as the development methodology and the development tools needed have been specified.

---

² The HARP (Harmonisation for the Security of Web Technologies and Applications) project (Project Number: IST-1999-10923) was funded by the European Commission within the Information Society Technologies (IST) Programme Framework

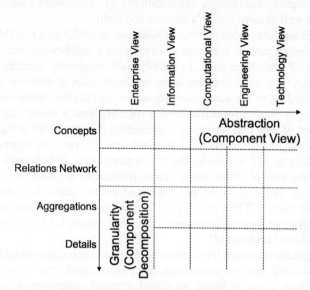

Fig. 6. State Transitions within the Abstraction-Granularity Matrix of Component Systems

5.1 Security Services in the OpenEHR Context

As already mentioned, the archetypes describe conceptual, contextual, organisational, functional, but also legal and ethical framework of the EHCR system and its behaviour. Such framework is also called policy. Archetypes define the domain-specific constraints to be established. Therefore, archetypes enable the description of policies. By refining archetypes, detailed specifications for security services such as authorisation and access control management can be specified and at runtime instantiated. Following the GEHR approach, the overall policy as well as its refinements in special policies and detailed security services sense should be specified in archetype models and expressed in XML schema. The generic meta-models have to be specified using the XML Schema standard.

5.2 The HARP Cross Security Platform

The HARP project's objective is building up entirely secure applications in client-server environments over the Web. Real interoperability leads to a closer connection of both communication and application security services. Communication security services comprise strong mutual authentication and accountability of principals involved, integrity, confidentiality and availability of communicated information as well as some notary's services. As a result of the authentication procedure, authorisation for having access to the other principal has to be decided. Application security services concern accountability, authorisation and access control regarding data and functions, integrity, availability, confidentiality of information recorded, processed and stored as well as some notary's services and audit.

To provide platform independence of solutions in HARP as a real three tiers architecture, the design pattern approach of developing a middleware-like common cross platform (HCSP) has been used. In HCSP, platform-specific security features have been isolated. Using an abstraction layer, communication in different environment is enabled. According to the component paradigm, an interface definition of a component providing a platform-specific service specifies how a client accesses a service without regard of how that service is implemented. So, the HCSP design isolates and encapsulates the implementation of platform-specific services behind a platform-neutral interface as well as reduces the visible complexity. Only a small portion has to be rewritten for each platform The solutions concern secure authentication as well as authorisation of principals even not registered before, deploying proper Enhanced Trusted Third Party (ETTP) services [7]. Especially, it helps to endorse policies by mapping them on processing components. For that reason, HARP components follow the specification of equation (2).

HARP's generic approach implements several basic principles. HARP's solution of embedding security into any application to be instantiated over the web-based environment outlined above is based on object oriented programming principles. It is based on Internet technology and protocols solely. The trustworthiness needed has been provided by applying only certified components which are tailored according to the principal's role. In fine-grained steps, it establishes its complete environment required, avoiding any external services possibly compromised. After strong mutual authentication based on smartcards [4] and TTP services [10], the security infrastructure components are downloaded and installed to be used for implementing the components needed to run the application as well as to transfer data input and output. The SSL (Secure Socket Layer) protocol deployed to initiate secure sessions is provided by the Java Secure Socket Extension API. The applets and servlets for establishing the local client and the open remote database access facilities communicate using the XML standard set including XML Digital Signature. Because messages and not single items are signed, the messages are archived separately for accountability reasons meeting the legislation and regulations for health.

Policies are dynamically interpreted and adhered to the components. All components applied at both server and client site are checked twice against the user's role and the appropriate policy: first in context of their selection and provision and second in context of their use and functionality.

Applet security from the execution point of view is provided through the secure downloading of policy files, which determine all access rights in the client terminal.

This has to be seen on top of the very desirable feature that the local, powerful, and versatile code is strictly transient and subject to predefined and securely controlled download procedures. All rights corresponding to predefined roles are subject to personal card identification with remote mapping of identity to roles and thereby to corresponding security policies with specific access rights.

For realising the services and procedures described, an applet consists of the subcomponents GUI and interface controller, smartcard controller, XML signing and XML processing components, communication component applying the Java SSL extension, and last but not least the data processing and activity controller. Beside equivalent sub-components and an attribute certificate repository at the server side, policy repository, policy solver and authorisation manager have been specified and implemented as a "light weight" Resource Access Decision service (CORBA: RAD).

After exchanging certificates and establishing the authenticated secure session, servlet security is provided from the execution point of view through listing, selecting and finally executing the components to serve the user properly. By establishing an authenticated session that persists for all service selections, a single-sign-on approach can be realised.

HARP enables the implementation of openEHR in a convincing way. Using the open environment of certified Java™ components, portability of the HARP solution to any platform is guaranteed.

In the server-centric approach, a web-accessible middleware has been chosen based on its support of basic security functionality, e.g., MICO/SSL., Apache Web server with mod_ssl, Apache JServ, and Apache Jakata Tomcat.

Combining the server-centric approach of HCSP, its server-centric approach and the network-centric VPN behaviour, the completely distributed HARP Cross Security Platform has been designed.

5.3   Harmonising the HARP Approach and OpenEHR

The HARP approach enables the implementation of any EHR component following constraints defined in archetype models and expressed in XML schema. The HCSP facilitate the instantiation of those components by combining both specifications in the sense of certified components. So, any granularity, any constrain in the sense of domain knowledge, organisational structure, underlying policy, technological requirements for structure and presentation providing portability, etc. are supported properly. The development of HARP rules underlying the XML messages which establish the HARP components (servlets and applets) is based on UML models. Within the HARP project developed independently from openEHR, these models reflect archetypes. For enhancing the current openEHR specification by security archetypes, a harmonisation in concepts and especially terminology used must be performed.

6   Conclusions

EHR architecture and subsequently specified and implemented EHR systems have to meet the shared care paradigm establishing openness, interoperability, scalability, and portability for providing any needed and permitted information to any authorised user at time, location, and format required, including mobile devices. Furthermore, EHR systems have to comply with comprehensive security solutions solely based on available and emerging standards. Actual EHR architecture standards comparably presented in the paper move in the direction requested. Emerging common projects harmonise the different approaches towards an "global" openEHR.

The European HARP project specified and implemented open portable EHR systems enriched with enhanced TTP services and comprehensive development strategies for establishing fine grained application security services. Constraints specified can be bound to components at runtime, enabling different views or supporting specific domain knowledge concepts. By binding attribute certificates to components appropriate policies can be enforced. These constraints such as, e.g., certificates are interpreted at both server and client sides using authorisation services. The HARP Cross Security Platform is solely based on standards including the XML standard set for the establishment of EHR clients and servers as well as their communication.

7   Acknowledgement

The author is in debt to the European Commission for funding and to the EHR community as well as to the HARP project partners for their support and their kind cooperation.

8   References

1.   T. Beale: An Interoperable Knowledge Methodology for Future-Proof Information Systems, 2001
2.   B. Blobel: Application of the Component Paradigm for Analysis and Design of Advanced Health System Architectures. *International Journal of Medical Informatics* **60** (3) (2000) 281-301.
3.   CEN ENV 13606 "Health Informatics – Electronic Healthcare Record Communication", 1999
4.   CEN TC 251 ENV 13729 "Health Informatics - Secure User Identification – Strong Authentication using Microprocessor Cards (SEC-ID/CARDS)", 1999.
5.   G-CPR Project: www.gcpr.gov
6.   GEHR Project: www.gehr.org
7.   The HARP Consortium: http://www.ist-harp.org
8.   Health Level Seven, Inc.: www.hl7.org
9.   ISO/IEC 10746-2 "Information Technology – Open Distributed Processing – Reference Model: Part 2: Foundations".
10. ISO DTS 17090 "Public Key Infrastructure, Part 1 – 3", 2001.
11. Object Management Group, Inc.: CORBA Specifications. http://www.omg.org
12. Object Management Group, Inc.: The CORBA Security Specification. Framingham: Object Management Group, Inc., 1995, 1997.

Designing for Change and Reusability – Using XML, XSL, and MPEG-7 for Developing Professional Health Information Systems

Ad Emmen[1]

[1] Genias Benelux bv, James Stewartstraat 248, NL-1325 JN Almere, The Netherlands, emmen@genias.nl

Abstract. With each new generation of devices, there is a tendency to develop information systems completely from the start, because the old ones were optimised for the previous generation of devices. XML experience has now reached a stage at which it is possible to define guidelines on designing information systems that are device independent, and that have adaptable user interfaces. We show some of these guidelines illustrated by the design of health care provider Internet accessible information systems. We report on a marketplace for medical multimedia objects, called the GridSET Exchange, that is completely based on existing and emerging XML standards. It explores the possibility of using the new MPEG-7 standard for these type of applications. We use a semi-automatic editor generator to construct editors from MPEG-7 fragments.

1. Introduction

With each new generation of devices, there is a tendency to develop information systems completely from the start, because the old ones were optimised for the previous generation of devices. Hence, there is for instance already a large amount of HTML-coded legacy systems even though the Internet is young. Much effort has been put into systems like WAP/WML that did not live up to their expectations.

The original promise of XML way back in 1998 was that everything would become easy, because there was now only one format. It was not quite that simple, of course. But XML experience has now reached a stage at which it is possible to define guidelines on designing information systems that are device independent, and that have adaptable user interfaces.

We show some of these guidelines illustrated by the design of health care provider Internet accessible information systems. In our system, all data, not only the information itself, but also the site lay-out, the configuration, the user management, is stored as XML. To create views for a user, be it an edit screen, or a list for display on a channel, XSLT is used. One can say that XSLT is the technology used to convert the XML information data into knowledge[4].

Because we do no exactly know what will be for instance the parameters of the 3G device on which our information has to be displayed, we store our intended lay-out as

P.M.A. Sloot et al. (Eds.): ICCS 2002, LNCS 2330, pp. 995–1002, 2002.

XSL-FO. XSL-FO turns out to be powerful enough not only to be used as a device independent page description language, but also as a useful format to describe screens.

The first industry standards are now emerging, that were designed after experience with XML was gained, as opposed to the first generation of industry standards that merely consisted of translating some previous existing standards into XML.

In our system we built in parts based on the new MPEG-7 standard. MPEG-7 is an ISO standard for describing meta-data on multi-media objects. MPEG-7 is very elaborate, not only describing the multi-media objects, including for instance video, sound and pictures, themselves, but also has tools for creating collections and describing user information, including browser and search preferences. As such, it provides a possible useful technology for 3G devices too.

Our system is built on top of G-VMP, the Genias Virtual Media Publisher[1]. G-VMP itself is an elaborate content management system focused on publishing magazines, newsletters, for web, print, e-mail and channels, and managing community sites.

We describe the implementation of this system for the GridSET Exchange. This is part of the WebSET project [2]. The WebSET project aims to produce a standardised suite of interactive three-dimensional educational tools, delivered across the WWW. The major focus is put on the use of open technology and standards, as well as the production of learning components which can be used as building blocks for further development in a wide range of application areas, such as surgical training and physiological education. WebSET is supported by the European Commission under contract number IST-1999-10632.

The GridSET Exchange will be used to market reusable learning objects - 3-D models, and course-modules.

People must be able to put these on the web site in a secure way. They should say who can use them under which conditions, and have a way to view the information on the usage of their objects.

On the other hand potential customers will be able to browse the site, get a preview and acquire an object. When payment is necessary, the system will take care of that too.

Although the name is GridSET Exchange, it is in fact a catalogue based market place [3].

2. GridSET Exchange design

The GridSET Exchange's core consists of a repository where the reusable learning objects are stored. At a second level, a meta-data catalogue is build. This catalogue contains information on all the objects, on the creators and the conditions of use. At a third level, the GridSET site engine can be used to access the catalogue, for entering new objects in the repository with simultaneous update to the catalogue; for searching, and for acquiring objects.

Note that this design follows the standard three level-approach of knowledge based systems [4]:

The lowest level is the *data level*, in our case the objects.

At the second level, by adding meta-data, the data is turned into *information*, in our case the catalogue .

At the third level, the information is turned into *knowledge* in our case for instance a search-engine allows for sophisticated queries.

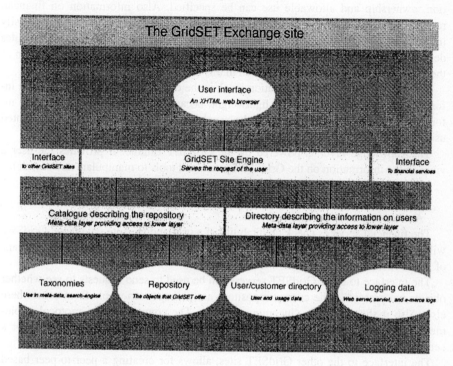

Apart from the repository, the data-level also contains the taxonomies that can be used to augment the information at the meta-data level.

Notice, that, although, strictly spoken, taxonomies are also meta-data, for GridSET, they are data until integrated at the next level. This general also allows taxonomies to have their own format. They can reside on the GridSET server, but can also be maintained elsewhere.

The nexts item at the data level are the directories holding the data on the users, and their usage. This includes data on who they are, access rights, selling and buying data, etc. Again, the separate layer allows to use other types of directory services too.

The last item in the data level consists of logging information. The activities of the server, servlets, and users will be logged.

The information level has two main items. The meta-data catalogue describing the repository is the most important one. It contains all information needed to access the repository. Taxonomies, glossaries and perhaps other information will be integrated to enrich the catalogue. MPEG-7 [5] will be used to implement most, if not all of the catalogue. MPEG-7 features elements to describe the *Content* of an item in the repository. It contains *Collection* elements that will used for grouping. *Controlled vo-*

cabularies will be used to integrate external taxonomies, and *Creation*-elements to describe the information on the person/organisation who did create it.

MPEG-7 is a recently defined ISO standard that describes meta-data about multimedia objects. Apart from detailed descriptions of the properties, information on creation, ownership and allowable use can be specified. Also information on financial results, users and user profiles can be described. Outside taxonomies can be easily included. Although designed by the media researchers and companies, it provides definition schemes that can be useful for Grid information systems too. The design of the GridSET Exchange is an exploration in this field.

The second item in the information level, is the directory containing user/usage information. This can be relatively lightweight, because we do not need that much information for each users, apart from the usual information, that can be implemented using MPEG-7, we need to represent access privileges and usage records.

The knowledge layer is powered by the GridSET Engine. It provides the user a view on the information on the GridSET system and tools to manipulate and search it.

In addition GridSET engine can have two external interfaces:

- An interface to financial services.
- An interface to other GridSET sites.

The financial services check creditability of a user and execute payments. GridSET will use one of the existing services [6] that allow exchange of information by means of XML-based records.

The interface to other GridSET servers can be used to send requests to see whether objects not present in the current repository are perhaps retrievable from somewhere else. Also it should handle incoming requests. The communication protocol implementation under investigation are a DNS model, flocking as used in Condor, or a centralised view of catalogues.

The interface to the other GridSET sites, allows for creating a peer-to-peer based network of GridSET Exchanges. A complete description of this interface is outside the scope of this paper.

The 3-layered approach chosen, allows to create a completely distributed GridSET Exchange system. The repository, for instance, does not have to be on the same computer as the meta-data catalogue. The same holds for the GridSET engine. Also the repository itself can be distributed over many computers, as can the meta-data catalogue.

In general, communication between the components of the system takes place by exchanging XML-messages. Location or access is performed through the URL naming scheme.

3. The GridSET Repository

The GridSET repository contains the reusable learning objects GridSET wants to make available. What can these be? 3-D models, written in VRML or X3D are amongst them. These can be for instance spine models, or torso models. Course modules, or course module fragments are the second type of items. These XML-files can be used as parts of (new) training courses. Third category consists of simulator pro-grammes written in Java, that can be combined with 3-D modules and integrated in training courses.

Additional items can be images, for instance a clinician at work, in any of the usual formats, for instance in JPEG, movies, form instance animations of real operations in for instance MPEG-2, and multi-media presentations in for instance SMIL.

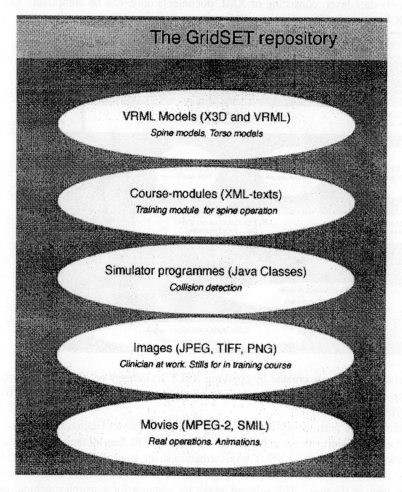

The implementation of the Repository is straight forward. Only a storage mecha-nism with URL type of access with appropriate compression is needed, while the meta-data layer takes care of all the information attached to the objects. A full blown relational database system, for instance, would add too much overhead providing functionality that is already provided by the meta-data layer.

4. The GridSET Site Engine

The GridSET Site Engine serves requests (mainly http requests) from a web browser. Part of these requests are for the static part of the web site that can be handled by a standard web server, the majority of requests will be dynamic in nature and serviced by a Servlet Engine.

The meta-data layer, consisting of XML-documents only, can be completely handled by XSLT-transformation sheets and formatting stylesheets.

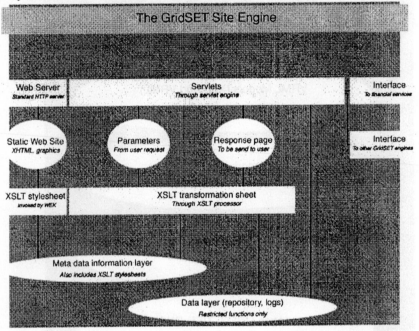

The static web site is created by applying XSLT stylesheets on XML-documents describing the web site content. The lay-out of the dynamic part is also created by XSLT stylesheets.

For the dynamic part, the users sends a request to the Servlet Engine that invokes the appropriate Servlet with the entered user parameters. The Servlet then invokes the XSLT processor to process an XSLT transformation sheet, to process meta-data (read it, update it, etc.) and create a response page for the user. The Servlet Engine then sends this page to the user. This scheme works for instance for a search machine, to get sorted lists of repository items and to update information on organisations and users.

For adding items to the repository, the Servlet and XSLT sheets designed for this purpose have to work closely together. The Servlet.gets the upload request from the user, runs an XSLT stylesheet to check the user's credential and determine where to store the object. Then it stores the object in the repository, runs an XSLT sheet to update the meta-data layer and then sends a response back.

To delete an item from a repository, save-guards ware build in, to prevent illegal deleting.

Presenting a catalogue in different views on the web is rather straight forward. The advantage of using MPEG-7 as our basic format is that we can use XPath together with the XSLT string functions to create a fine grained search engine. This search engine is then used when creating views for the user.

5. The design of the editor

Is getting information out of the system through a web interface not a challenge, getting information in, is. We set out with a client in mind that only had an HTML browser. Reason is that this is the only thing that is guaranteed to work anywhere, anytime. With JavaScript, a much more sophisticated interface can be built, but, unfortunately, the JavaScript implementation by the different browser vendors are too different. Using a Java Applet, is a little better, as long as one restricts to using Java 1.1. But then the interface is not much better than with pure HTML with CSS. X-Forms, will certainly help, but as the standard is not ready, no browser has implemented it to date.

Hence, we use for an editor, that is completely presented in HTML, with additional JavaScript in to ease editing, not as an essential part. The editor definitions and configurations are all stored as pure XML, i.e. when for instance X-Forms catch on, only a small part of the presentation layer will have to be rewritten.

Within GridSET, a lot of customised editor screens are needed. For instance for entering a user, specifying object meta data for different types of objects, specifying allowable usage, etc. The editor screens must also be customisable by user profiles.

The editors are created dynamically, bootstrapped by an initial specification process:

The MPEG-7 schema fragment describing the information that must be editable is put in an XML-configuration file.

The names (labels) to be used for each MPEG-7 item are specified in another part of the XML-configuration file. This naming scheme can be used to create application and user language (English, Dutch, French, etc.) dependent names.

An XSLT stylesheet automatically creates an initial editor for the schema fragment. In most cases, however, there are parts of the schema that although must be specified, should not be editable for the user. With additional special XSLT-templates in the XSLT-stylesheet these are excluded from the user view (but kept as hidden properties with a specified value.

The layout of the editor screen (width of fields, font-type, colours, etc.) is specified in an XSL-FO (Extensible Stylesheet Language - Formatting Objects) file. This file is used by the XSLT stylesheet producing the user view.

When the user creates or edits information, the data that are send back to the user are also handled by an XSLT-stylesheet using the same MPEG-7 schema fragment as basis.

Designing editors with XSLT is extremely fast and can make use of the extensibility build in XML. There is one major drawback, however. XSLT can only handle correct conforming XML documents. Users can easily type things that are not conforming XML. Hence, care has been take to exclude that possibility as much as possible, helping the user to write correct XML. MPEG-7 helps in this way because it is a data oriented XML language rather than a document oriented XML language. This means that an editor can consists of fields to be filled in, rather than that is has to be a WYSIWYG document editor.

6. Conclusion

With the upcoming 3G technologies, that adds to the already available user interfaces, we must try to separate content from presentation. In this paper we described an full XML-based approached that can be used for future generation information systems. A market place implementation for medical multimedia objects, called GridSET has been developed. We have also showed that the MPEG-7 standard can be used as a foundation for the XML descriptions.

7. References

[1] Genias Virtual Media Publisher, http://www.genias.nl/vmp, last accessed January 2002.

[2] The WebSET Consortium project website, http://www.vmwc.org/projects/webset, last accessed January 2002.

[3] Kaplan, S and Sawhney, M. E-Hubs, "The new B2B marketplaces", *Harvard Business Review*, May-June 200, 97-103.

[4] *Long Term Technology Review of the Science & Engineering Base*, Report to the UK Research Councils, April

[5] *MPEG-7, Technology - Multimedia Content Description Interface — Part 5: Multimedia Description Schemes*, ISO/IEC 15938, ISO Organisation, Geneva, 2000.

[6] A. Tokmakoff (ed.), State-of-the-Art in Electronic Accouning, Billing and Payment, Enschede, Telematica Instituut http://www.telin.nl, April 18, 2000.

Personal LocationMessaging

Markku Saarelainen

Oy Arbonaut Ltd, Torikatu 21 c 80100 Joensuu, Finland. www.arbonaut.com,
markku.saarelainen@arbonaut.com

Abstract. When communicating with other people, we are normally aware of our surroundings, as a context of our communication. Geographical representation as a part of messaging has a long history in traditional media. For example, we send postcards with a picture and a postage stamp. The LocationMessaging Server is a messaging platform that provides the possibility to utilize location information to enrich mobile and web based messaging. On one hand, the Web interface of the Server represents a possibility to store, create, organize and deliver messages that include a location reference. A wide range of map material can be used to visualize the location a message created or received contains. On the other hand, mobility gives the freedom to create and receive geographically enriched messages anywhere; LocationMessaging is used to store your own Points of Interest, to inform others about your location, to search and find places and people.

1. Introduction

The LocationMessaging Suite, FleetManager and MINDER form a services family that supports the LocationMessaging needs of different usergroups. The market for location based mobile services is expected to grow to 14 billion euro by 2006 (Applied Business Intelligence). The LocationMessaging Suite is a general solution for mobile location messaging which allows the users to locate their friends or contacts. It is also possible to search the personal surroundings of your current location with keywords, and to find matching Points Of Interest. LocationMessaging is based on SMS messages or WAP, with amendments, which include:

- Geo-coded location documents
- Geo-coded location messages
- GeoSearch

For the user, the functionality is easy to adopt. The users can store and edit LocationMessages in a personal LocationNotebook, or in a public Open LocationNotebook. Incoming messages are stored in an Inbox, and all sent messages, whether sent through the Web or with a mobile terminal, are stored in an Outbox.

Fig. 1: LocationMessaging

P.M.A. Sloot et al. (Eds.): ICCS 2002, LNCS 2330, pp. 1003–1011, 2002.
© Springer-Verlag Berlin Heidelberg 2002

LocationMessaging allows simple answers to, for example, the following tasks:

- **Replying** to "Where R U?"
- **Sharing** "This place is cool!" with friends
- **Agreeing** to "Can we meet ... :-)"
- **Checking** "Did Suzie catch the bus?"
- **Finding out** what other people like you found interesting in this town The

LocationMessaging Services have four common elements:

- LocationNotebook,
- Composer,
- ContactBook and the
- User Profile.

All of these components are available when using the LocationMessaging at the website or with a mobile phone.

LocationNotebook is a service that allows you to store LocationMessages in a database. LocationMessages, your personal Points of Interest, can be created on a website or on a mobile phone using SMS. LocationNotebook allows the user to create folders for the messages and to edit the messages afterwards to suit the needs of the user. This may involve, for example, selecting another icon or adding text to deepen the description of saved data.

Composer is a tool to create LocationMessages. Composing a message takes three steps:

1. selecting send or save functionality,
2. selecting location with referring icon, and
3. writing the text of the message.

At a LocationMessaging website, the Composer can be used with maps and Locationcoding to refer to the location indicated in the LocationMessage. The Locationcoding engine is very useful when an address needs to be converted into coordinates, and further on to a point on a map. Recipients can be selected from a ContactBook, or from a LocationMessaging specific contact database. Alternatively, the phonenumbers or e-mail addresses, respectively, of the recipients can be typed in.

ContactBook is a user defined contact database. By using the ContactBook, users can define the personal contact data, addresses, phone numbers, e-mail accounts and the preferred contact methods of the recipients of their LocationMessages. The ContactBook also manages the highly important setting of permissions for each Contact on if and when they are allowed to apply the friend finder function to the owner of the ContactBook. It is also possible to create Contact groups for sending group messages. Contact groups can include subgroups, and they can overlap partially with other Contact groups.

User Profile is a tool to control and modify each user's own messaging devices. A user can add, edit and priorize the devices by the context the messages are received.

The web interface is used as a default, so that a LocationMessaging website automatically keeps a log of all LocationMessages sent and received, unless the user explicitly disables this function. In addition, the User Profile determines the default values for proximity search radius, emergency messages and for LocationNotebook mailboxes.

All of the messaging environments have common elements, which can be found as generic modules in the Suite or combined as applications, for example, the FleetManager is designed for asset-tracking use so it has additional components used in vehicle tracing whilst MINDER is a solution for personal usage supporting security and safety application such as monitoring the whereabouts of children.

1.1 FleetManager

FleetManager is a generic tool to locate vehicles, workforce and valuable assets – and to communicate with them. Additional parts depend strongly on customers need. Typically the implementation includes proximity search engine, interfacing with the customers point of interest databases and raster and vector map servers.

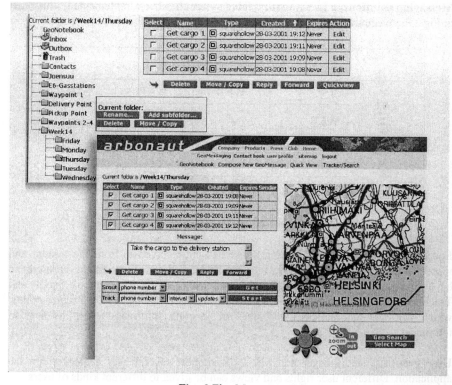

Fig. 2 FleetManager

1.2 MINDER

MINDER (Movement INformation DElivered Regularly) is a version of FleetManager that has been customized for personal location tracking for example determining the location of a child. MINDER can be used to locate personnel outdoors, in urban areas and indoors.

2. General features of application layer

The features of the application layer are modular entities and can be integrated into the existing location information and/or messaging system of any application.

Operating system independency. The LocationMessaging-solutions can be flexibly transferred on to any device, operating system and application environments.

Modules of application layer and their general description

Messaging solutions. a two-way messaging system has been implemented, illustrated in Fig.3, between the Web and mobile terminals :

		Sending		
		WEB	WAP	SMS
Receiving	WEB	✓	✓	✓
	WAP	✓	✓	✓
	SMS	✓	✓	✓
	Email	✓	✓	✓

Fig.3. Cross Media LocationMessaging

Location information-based messages can be sent/received in/from the system and inspected on a map. GPS/GSM-devices can be located and the points in question can be displayed on a map, marked for example with a distinctive symbol. Scout- and Track-features can be used with GPS/GSM-terminals. Also so called Trigger-points, which work as automatic trigger areas of location information, can be sent to terminals

User groups. FleetManager and MINDER support several user groups within the application. Different user rights and views can be given to different kinds of users.

Positioning. The current version of the system supports terminal-specific GPS-positioning. Support for network positioning will be in use as soon as operator(s)

present(s) a functioning and a comprehensive service in Finland. In addition, the system utilises Locationcoding in producing and using location information.

Map management. It is possible to use vector and raster map material in the FleetManager-and MINDER applications in three different ways.

- Customer uses an external map provider (e.g. Genimap or Novo) from whom the maps needed in the service can be obtained for use both in the Internet and in mobile devices
- Customer saves its own digital map material on the image server of Arbonaut, from which the maps needed in the service can be obtained for use for both in the Internet and in mobile devices.
- Arbonaut forms an interface to an existing map server of the user, from which the maps needed in the service can be taken into use for both in the Internet and in mobile devices.

It is possible to use all map materials in the Internet using a browser, which supports Locationcoded navigation, developed by Arbonaut. Maps can be downloaded also to PDA, WAP and Benefon ESC! devices.

User-specific databases. Each user/user group can save LocationMessages into a private database structure, which can be modified user/user group-specifically.

Terminal support. FleetManager and MINDER -applications support the following GPS/GSM devices:

- Benefon ESC! – mobile phone with map display
- Benefon Track – telematics phone for professional use
- Falcom – independent mobile positioning device (Asset tracker)

In addition, normal GSM-terminals and WAP and PDA devices (palmtop) can be connected to the applications.

3. Architectural overview

The LocationMessaging System extends standard Messaging Systems by incorporating a special treatment for Location Based Information. The LocationMessaging System provides:

- Management of personal information
- Access to Community or Public information
- Exchange of information between people
- Map Management to display Location Based Information.

The LocationMessaging System can be divided in the following components:

- LocationMessaging Server
- LocationMessengers
- LocationMessaging Services

The LocationMessaging Server hosts the LocationMessaging Services and the Location Based Information depicted in Fig.4. LocationMessenger is the generic name for a messaging client of a LocationMessaging application. A LocationMessaging System includes other types of clients like Administration Tools. The LocationMessaging System includes interfaces to:

- Map servers through a Map Proxy
- Positioning Systems, especially network based positioning systems
- Traditional messaging systems (e-mail, www, fax, SMS, ...)
- External Content Providers and User Databases

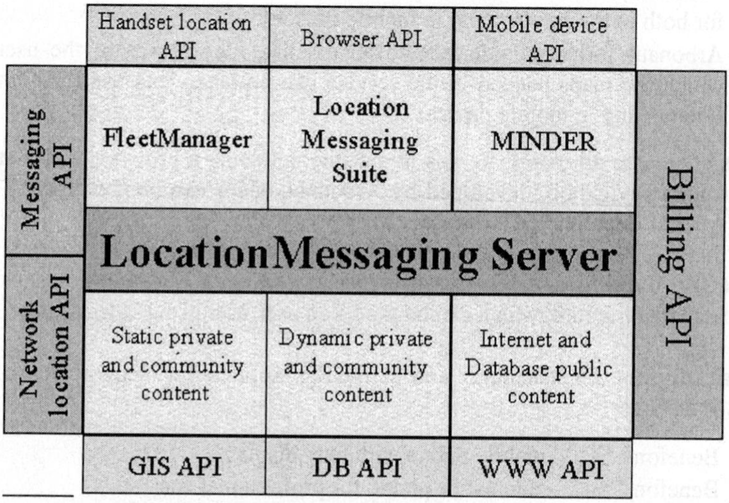

	Handset location API	Browser API	Mobile device API	
Messaging API	FleetManager	Location Messaging Suite	MINDER	Billing API
Network location API	**LocationMessaging Server**			
	Static private and community content	Dynamic private and community content	Internet and Database public content	
	GIS API	DB API	WWW API	

Fig. 4 LocationMessaging Server. Layered structure of the server consists of the core server, service layers (as the Suite, FleetManager and MINDER as well as the content layer) and the communication API and database API layers. In addition there are more customer dependent network location API, messaging API and billing API layers.

4.Application integration interface

The LocationMessaging Suite has been designed to implement and port into various technological environments. Through its adaptable interfaces the LocationMessaging Server can be tailored and harnessed to address the users needs. Typical API's for integration are:

- Existing user database
- Network positioning systems
- Billing interface
- GIS interfaces

5. Friend Finder and Child monitoring

Mobile communication today is free from place, yet our daily actions and decisions are more often defined by our surroundings. Having the possibility to locate ourselves in relation to the services we want to consume and indeed to our social reference group, our friends, families, workgroups and people we share hobbies with gives us more added value and basis to make decisions and to communicate.

Friend finder, an application of MINDER, is a solution designed for every mobile and web user to locate themselves and their friends and to use this information to enrich their messaging.   Based on the fact that the operators network positioning technology is being harnessed to answer each query made by the LocationMessaging Suite, the Friend finder falls into two appearances: the web and the mobile one.

The Friend finder at the Web

> The subscribers can use their private user interface to communicate with their friends. The LocationMessaging system enables the user to make queries about their friend's location. The position information is being derived from the mobile operators network positioning server and the results are brought to the subscriber with indicating icon on a referring maps over the area. The easy usability of the service has been one of the key factors in Friend finder. With the -best practice in semantics- graphical user interface it is easy to select your friend or a group of friends from your contact book and choose suitable friend find option from the service. As a result the list of position query results will be displayed with referring locations.

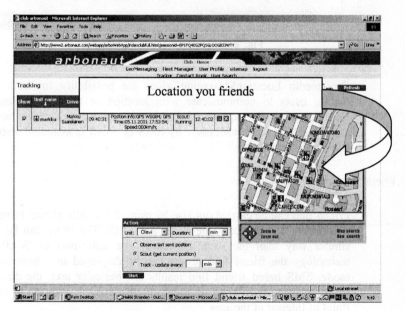

Fig. 5 Friend finder

For each result – the position of the friend – it is possible to center the map according to that position and to see the status (the friend find profile) of the friend. By clicking the referring name of the friend at the result list it is possible to open a certain composer view with your friends contact info already filled in the recipient field and the location of the friend as a starting point to select desired position of your LocationMessage. The message can be sent as a web message, WAP, e-mail or by SMS.

Fig. 6 Web-based LocationMessaging

Crossmedia LocationMessaging gives the possibility for both web and mobile users to communicate with location in two ways. Sending and receiving generally requires subscribing the services in domain, however the receiving the LocationMessage by SMS or email is available also for non-subscribers

Friend find in mobile phones

Friend find can be carried out also by using the mobile phone. Here there are two different scenarios – the WAP and SMS. The WAP can be used in a similar way than the web. Especially the utilization of WAP–mapping technology the friend find results can be displayed in a terminal specific mode. SMS based friend find results are basically text: the direction and distance of the friends location can be displayed in addition to the nearest Point of Interest of the area

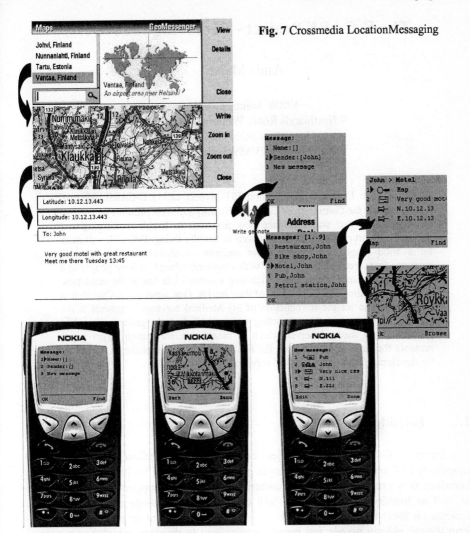

Fig. 7 Crossmedia LocationMessaging

The LocationMessaging system can also be used to provide the foundations to develop a number of Mobile safety types of applications such as child monitoring as depicted in fig .8. MINDER can be used to monitor personnel outdoors (using GPS) in urban canyons (using network positioning) and even indoors.

Fig. 8 Child monitoring using MINDER

The E-CARE Project - Removing the Wires

Andy Marsh

VMW Solutions Ltd,
9 Northlands Road, Whitenap, Romsey,
Hampshire SO51 5RU, UK
Andy.Marsh@VMWSolutions.com

Abstract: The European Commission IST project called E-CARE: *Medical Expert System for Continuity of Care and Healthy Lifestyle* presents innovative health services that will introduce new practices in health monitoring and decision support on health matters as well as healthy living. E-CARE will cater for a wide range of scenarios, from patients on short-term (1-2 months) recovery from treatment to patients with long-term illness, elderly people and people predisposed to diseases, which live a normal life but at the same time need constant attention on the state of their health. One aspect of E-Care that is the major issue for the introduction of 3G Medical services in general is the systematic removal of the connecting wires which link the patients vital data sensors to the medical server which is monitoring the patient. This paper presents the steps adopted within the E-Care project and therefore a blue-print for developing further 3G Medical services.

1.   Introduction

The European Commission IST project called E-CARE: *Medical Expert System for Continuity of Care and Healthy Lifestyle* presents innovative health services that will introduce new practices in health monitoring and decision support on health matters as well as healthy living. E-CARE will cater for a wide range of scenarios, from patients on short-term (1-2 months) recovery from treatment to patients with long-term illness, elderly people and people predisposed to diseases, which live a normal life but at the same time need constant attention on the state of their health.   E-CARE will empower medical doctors to constantly and remotely keep track of their patients' vital parameters (using medical devices with communication abilities), with minimum effort, assisted by an intelligent automated infrastructure. At the same time family and friends of the patients will too have access to the same information, filtered and presented in a comprehensible manner, including latest comments from the doctors. A sophisticated Collaboration Model will manage the whole service and will be aware of each patient's medical record, providing an information channel between the medical staff, the patients and their carers.

The application domains for E-Care have been identified as Elderly, Oncology and Paediatric Oncology. Collectively these domains have identified the follow key ambulatory parameters that are to be monitored:

P.M.A. Sloot et al. (Eds.): ICCS 2002, LNCS 2330, pp. 1012–1018, 2002.

For the Elderly:

Heart rate	Sensitive but non-specific sign of cardiac function.
Activity	Two, dual axes accelerometers, one over the abdomen and the other on lateral aspect of upper thigh indicate activity through measurement of accelerations during walking and running.
Body temperature	Intermittent values of temperature.
.%Vital Capacity expired in one sec (%VC 1 sec)	%VC 1 sec reflects severity of airways obstruction in chronic obstructive pulmonary disease.
Blood Glucouse	Prevent or delay vascular (blood-vessel) or neurological (nerve) complications
EEG	Needed to diagnose the presence and type of seizure disorders, confusion, head injuries, brain tumors, infections, degenerative diseases, and metabolic disturbances that affect the brain

For the Oncology:

ECG waveform	Needed to interpret cardiac arrhythmias; in some instances, requires addition of jugular venous pulse to distinguish atrial contraction to diagnose type of tachyarrhythmia and heart blocks.
Blood pressure	Intermittent values of blood pressure by oscillometric technique.
Medication reminder	Description with audio-visual beeps for time medication reminder prompt.
Arterial oxygen saturation	Assesses amount of oxygen in arterial blood and is a component of sleep study.

For the Paediatric Oncology:

Body temperature	Intermittent values of temperature.
Body weight	Intermittent values of body weight
Pain	Description through Visual Analogue Scale.

One aspect of E-Care that is the major issue for the introduction of 3G Medical services in general is the systematic removal of the connecting wires which link the patients vital data sensors to the medical server which is monitoring the patient. The following steps illustrate the approach adopted within the E-Care project and therefore a blue-print for developing further 3G Medical services:

Step 1: Removing the data logger's RS232 link

The vital data sensors are connected to a data logger that is carried by the patient. The data logger is connected via an RS232 link to a client PC located in the patients home. This home PC is connected via a fixed line connection to the monitoring server that analyses the patient's vital data and generates an alarm, an updated report or an

1014   A. Marsh

emergency call. The first step is to replace the data logger's RS232 link with an Radio
Frequency (RF) link.

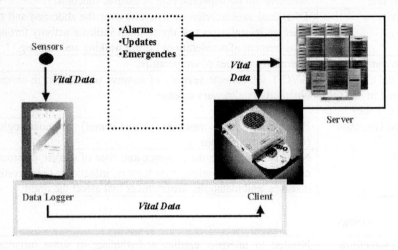

Fig.1. Step 1: Removing the data logger's RS232 link

Step 2: Adding GSM mobility

The RF link will allow the patient, who in this scenario is an elderly person, to move
within the range of the RF transceivers (200 m) and still have the data from their
sensors, which in this case includes movement, temperature and pulse, continuously
monitored. This approach supports home care monitoring but confines the patient to
always be within RF range if they want to be continuously monitored. To overcome
this limitation GSM networking can be used.

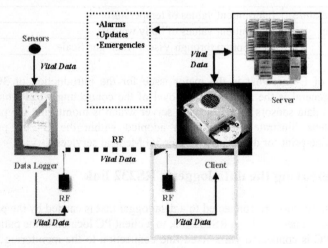

Fig.2. Step 2: Adding mobility

Step 3: Personalizing patient alarms

By using GSM networks, the patient's sensors via the data logger and RF transceivers can be connected to a GSM modem (for example the patients mobile handset) that can periodically transmit using SMS the patient's vital data to a receiving GSM modem connected to a home PC. The home PC can then upload the vital data to the monitoring server. When this service is used for monitoring outpatients, for example oncology patients, the detection of an alarm may be dependent on customized historic data.

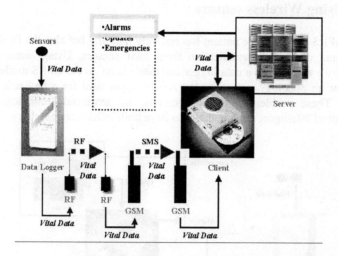

Fig.3. Step 3: Personalizing patient alarms

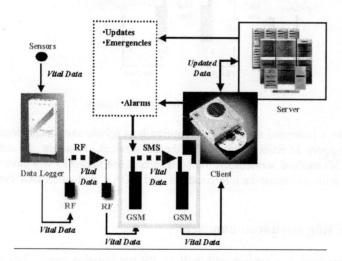

Fig. 4. Step4: Using 2.5G networking

Step 4: Using 2.5G networking

By using the home PC to detect an alarm situation the patient could be warned, on their mobile handset, of an arising situation such as their pulse is to high or even a calendar type of event, such a drug reminder, without the monitoring server being involved. By using SMS communications to transmit the patients data the service is limited. A logical extension would be to use 2.5G networking namely GPRS for the transmission of the vital data.

Step 5: Using Wireless sensors

By using GPRS networks the patient has not only mobility but also the facilities for a larger amount of data communication from their sensors. Using sensors that are connected by a wire link to a data logger and then via RF to the GPRS modem may be tolerable for an oncology outpatient but nearly impractical for a pediatric oncology outpatient. These problems can be overcome by advanced technologies in the development of intelligent sensors that also have built in RF transmission capabilities.

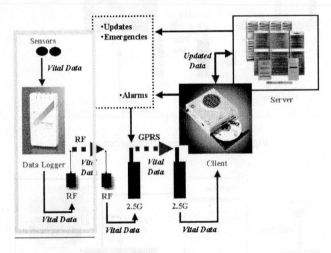

Fig.5. Step 5: Using wireless sensors

The five steps presented above have identified how *off-the-shelf* technologies can be used to support M-Safety and M-Healthcare services. However, to develop the envisaged 3G medical services the 3G communications networks will have to also be combined with advanced mobile handsets and PDA's with increased computation facilities.

Step 6: Using advanced handsets

By using medical data sensors with built-in RF transmission capabilities allows the patient to feel free of the monitoring equipment. As these wireless sensors are being developed supportive handsets will also be developed that will have built-in RF receivers to communicate with the sensors directly. The ideal user scenario would then be to also replace their home PC with an advanced handset that also had computing capabilities to calculate their personalized alarms.

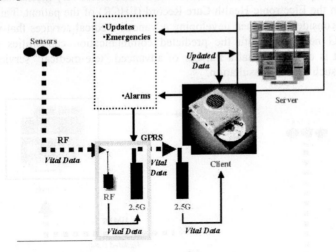

Fig. 6. Step 6: Using advanced handsets

Step 7: Using advanced PDA's

As the computing and networking capabilities of the PDA's are being developed effectively a single device can replace the patient's home PC and their mobile handset. The patients personalized alarms could also be calculated without any networking costs. The patient will then be in greater control of their healthcare whether at work, home or on the move. Additionally, their historic medical record could be kept on their PDA and accessed when required.

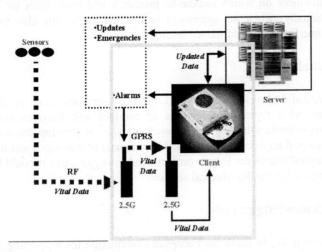

Fig.7. Step 7: Using advanced PDA's

Step 8: 3G Medical services

Recording the patient's data over time on their PDA effectively means that the PDA will contain the Electronic Health Care Record (EHCR) of the patient. This could be taken into consideration when developing advanced medical services that will utilize the full 3G networks. With the predicted communication capabilities of UMTS networks it is envisaged that a number of advanced "tele-medical" services will be developed such as tele-consultation.

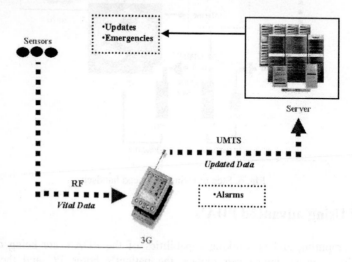

Fig. 8. Step 8: 3G Medical services

The eight steps above have defined a possible roadmap for the development of 3G Medical services. However, for such a range of services to become reality there needs to be an agreement on which standards, protocols and procedures etc that will be used. Additionally, when an agreement is reached there will also be a need to undertake some form of conformance testing.

2.   Conclusions

For the successful development of 3G Medical services there needs to be an agreement on what types of categories of services will become available, how different complementary industries can work together to develop these services and how the developed services are certified. The rationale of this paper was to present the approach adopted within the E-care project. And to suggest that it could be used as a basis for deriving further 3G Medical services.

3.   Acknowledgements

The author would like to thank the European Commission for supporting the E-CARE project (IST-2001-33261).

Automatic Generation of Efficient Adjoint Code for a Parallel Navier-Stokes Solver

Patrick Heimbach[1,*], Chris Hill[1,*], and Ralf Giering[2,*]

[1] Department of Earth, Atmospheric, and Planetary Sciences
Massachusetts Institute of Technology, Cambridge, MA 02139, USA
{heimbach, cnh}@mit.edu
http://mitgcm.org
[2] FastOpt, Martinistr. 21, D–20251 Hamburg, Germany
ralf.giering@fastopt.de
http://www.fastopt.de

Abstract. We describe key computational aspects of automatic differentiation applied to the global ocean state estimation problem. The task of minimizing a cost function measuring the ocean simulation vs. observation misfit is achieved through efficient calculation of the cost gradient w.r.t. a set of controls via the adjoint technique. The adjoint code of the parallel MIT general circulation model is generated using TAMC. To achieve a tractable problem in both CPU and memory requirements, despite the control flow reversal, the adjoint code relies heavily on the balancing of storing vs. recomputation via the checkpointing method. Further savings are achieved by exploiting self-adjointness of part of the computation. To retain scalability of the domain decomposition, handwritten adjoint routines are provided which complement routines of the parallel support package (such as inter-processor communications, global operations, active variable I/O) to perform corresponding operations in reverse mode. The size of the problem is illustrated for the global ocean estimation problem and results are given by way of example.

1 Introduction

In one of the most complex Earth science inverse modeling initiatives ever attempted, the *Estimation of the Circulation and Climate of the Ocean* (ECCO) project is developing greatly improved estimates of the three-dimensional, time-evolving state of the global oceans. To this end, the project is applying advanced, mathematically rigorous, techniques to constrain a state-of-the-art parallel general circulation model (MITgcm) [1–3] with a diverse mix of observations. What emerges is an optimization problem that must be solved to estimate and monitor the state or "climate" of the ocean. Ocean climate is characterized by patterns of planetary scale circulation, the Gulf Stream current for example, and by large scale distributions of temperature and salinity. These quantities can be observed, but only partially, using satellites and oceanographic instruments. Combining,

---

* On behalf of the ECCO Consortium, http://www.ecco-group.org

P.M.A. Sloot et al. (Eds.): ICCS 2002, LNCS 2330, pp. 1019–1028, 2002.

through a formal optimization procedure, the fragmentary observations with a numerical model, which is an a priori expression of the laws of physics and fluid mechanics that govern the ocean behavior, produces a more complete picture of the ocean climate. The ECCO optimization problem proceeds by expressing the difference between a model and observation from the actual ocean in terms of a scalar cost, $\mathcal{J}$, thus,

$$\mathcal{J} = \sum_{i=1}^{n}(M_i - O_i)W_i(M_i - O_i) \ , \tag{1}$$

where M_i refers to a simulated quantity projected onto the ith observational data point O_i, with W_i the associated a priori error estimate. Denoting certain parameters and model state variables as adjustable "controls" C, the simulated state $M(C)$ can be optimized to minimize $\mathcal{J}$ over the n observation points. The optimized controls, C_{opt}, render a numerically simulated ocean state, $M(C_{opt})$, that is spatially and temporally complete and also consistent with observations.

The size of the ECCO optimization problem is formidable. In recent years, with increasing observational and computational capabilities, naturally occurring changes and shifts operating on time-scales of years, for example El Niño–Southern Oscillation (ENSO) [4], and decades, for example the North Atlantic Oscillation (NAO) [5], have become widely appreciated. There is every reason to expect that longer time-scale behaviors are also present in the World oceans. The optimization must, therefore, encompass processes spanning decades to centuries, on global scales; simulating them at spatial and temporal resolutions sufficient to yield state estimates with skill.

Our "smallest" current configuration is characterized by a cost function that spans nine years of planetary-scale ocean simulation and observation. The cost function operates on 10^8 elements and is optimized by corrections to a control vector, C, of size 1.5×10^8. The full Jacobian for this system contains more than 10^{16} elements ($10^8 \cdot 1.5 \times 10^8$) which, even allowing for some sparsity, is fundamentally impractical. Therefore, the reverse mode of automatic differentiation (AD), which allows the computation of the product of the Jacobian and a vector without explicitly representing the Jacobian, plays a central role.

Minimizing $\mathcal{J}$, under the side condition of fulfilling the model equations, leads to a constrained optimization problem for which the gradient

$$\nabla_C \mathcal{J}(C, M(C)) \tag{2}$$

is used to reduce $\mathcal{J}$ iteratively. The constrained problem may be transformed into an unconstrained one by incorporating the model equations into the cost function (1) via the method of Lagrange multipliers. Alternatively and equivalently, the gradient may be obtained through application of the chain rule to (1).

AD [6] exploits this fact in a rigorous manner to produce, from a given model code, its corresponding tangent linear (forward mode) or adjoint (reverse mode) model; see [7]. The adjoint model enables the gradient (2) to be computed in a single integration. The reverse mode approach is extremely efficient for scalar-valued cost functions for which it is matrix free. In practical terms, we are able to develop a system that can numerically evaluate (2), for any scalar $\mathcal{J}$, in roughly four times the compute cost of evaluating $\mathcal{J}$. At this cost, reverse mode AD provides a powerful tool that is being increasingly used for oceanographic and other geophysical fluids applications.

The results that are emerging from the application of reverse mode AD to the ocean circulation problem are of immense scientific value. However, here our focus is on the techniques that we employ to render a computationally viable system, and on providing examples of the calculations that are made possible with a competitive, automatic system for adjoint model development and integration.

The MITgcm algorithm and its software implementation in a parallel computing environment are described in Sect. 2. Section 3 discusses the implications for an AD tool and the attributes of an efficient, scalable reverse mode on a variety of parallel architectures for rendering the calculation computationally tractable. Illustrative applications are presented in Sect. 4 with an emphasis on computational aspects, rather than implications for oceanography or climate. An outlook is given in Sect. 5. For discussion of the scientific aspects of this work refer to reports and data along with animations available at [8].

2 The MIT General Circulation Model

The M.I.T General Circulation Model (MITgcm) is rooted in a general purpose grid-point algorithm that solves the Boussinesq form of the Navier-Stokes equations for an incompressible fluid in a curvilinear framework. The algorithm is described in [2,3]; see [9] for online documentation. The work presented here uses the model's hydrostatic mode, to integrate forward equations for the potential temperature θ, salinity S, velocity vector $\underline{v}$, and pressure p of the ocean using a two phase approach at each time-step.

A skeletal outline of the iterative time-stepping procedure that is used to step forward the simulated fluid state is illustrated in Fig. 1. The two phases **PS** and **DS** are both implemented using a finite volume approach. Discrete forms of the continuous equations are deduced by integrating over the volumes and making use of Gauss' theorem. The terms in **PS** are computed explicitly from information within a local region. **DS** terms are diagnostic and involve an iterative preconditioned conjugate gradient solver.

Finite-volumes provide a natural model for parallelism. Figure 2(a) shows schematically a decomposition into sub-domains that can be computed concurrently. The implementation of the MITgcm code is such that the **PS** phase for a single timestep can be computed entirely by on processor operations. At the end of **PS** communication operations are performed. This communication and **DS** must complete before the next time-step **PS** can start. The implicit step, **DS**, tightly mixes computation and communication. Performance critical communications in MITgcm employ a communication layer in a custom software library called WRAPPER. The performance critical primitives in the WRAPPER layer are illustrated in Fig. 2(b). The operations are all linear combination and permutations of distributed data.

INITIALIZE. Define geometry, initial flow and tracer distributions
FOR each time step n DO
  PS
    Active I/O.
    Step forward state. $\qquad \underline{v}^n = \underline{v}^{n-1} + \Delta t(\underline{G}_v^{n-\frac{1}{2}} - \nabla p^{n-\frac{1}{2}})$
    Get time derivatives. $\qquad \underline{G}_v^{n+\frac{1}{2}} = \mathbf{gv}(\underline{v}, b)$
    Get hydrostatic p. $\qquad p_{hy}^{n+\frac{1}{2}} = \mathbf{hy}(b)$
  DS
    Solve for pressure. $\qquad \nabla_h . H \nabla_h p_s^{n+\frac{1}{2}} = \overline{\nabla_h . \underline{G}_{v_h}^{n+\frac{1}{2}}}^H - \overline{\nabla_h . \nabla_h p_{hy}^{n+\frac{1}{2}}}^H$
END FOR

Fig. 1. The MITgcm algorithm iterates over a loop, with two blocks, **PS** and **DS**. A simulation may entail millions of iterations. In **PS**, time tendencies (G terms) are calculated from the state at previous time levels ($^{n,n-1,}$ $\cdots$). **DS** involves finding a two-dimensional pressure field p_s to ensure the flow $\underline{v}$ at the next time level satisfies continuity. G term calculations for θ (temperature) and S (salinity) have been left out. These have a similar form to the $\mathbf{gv}()$ function and yield the buoyancy, b

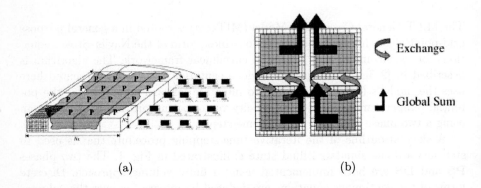

Exchange

Global Sum

(a) $\qquad\qquad\qquad\qquad\qquad$ (b)

Fig. 2. Panel (a) shows a hypothetical domain of total size $N_x N_y N_z$. The domain is decomposed in two-dimensions along the N_x and N_y directions. Whenever a processor wishes to transfer data between tiles or communicate with other processors it calls a special function in a WRAPPER support layer. Two performance critical parallel primitives are provided by the WRAPPER (b). By maintaining transpose forms of these primitives we can efficiently accommodate parallel adjoint computations

3 The Adjoint of MITgcm

MITgcm has been adapted for use with the Tangent linear and Adjoint Model Compiler (TAMC) [7, 10], developed by Ralf Giering. TAMC is a source-to-source transformation tool. It exploits the chain rule for computing the derivative of a function with respect to a set of input variables. Treating a given forward code as a composition of operations—each line representing a compositional element, the chain rule is rigorously applied to the code, line by line. The resulting tangent linear (forward mode) or adjoint code (reverse mode), then, may be thought of as the composition in forward or reverse order, respectively, of the Jacobian matrices of the full forward code's compositional elements. The processed MITgcm code and the adjoint code have about 100k executable lines.

While the reverse mode proves extremely efficient in computing gradients with respect to a scalar cost function, a major challenge is the fact that the control flow of the original code has to be reversed. In the following we discuss some computational implications of the flow reversal, as well as issues regarding the generation of efficient, scalable adjoint code on a variety of parallel architectures.

3.1 Storing vs. Recomputation in Reverse Mode

This is a central issue upon which hinges the overall feasibility of the adjoint approach in the present context of large-scale simulation optimization and sensitivity studies. The combination of four related elements:

- the reverse nature of the adjoint calculation,
- the local character of the gradient evaluations, on which the adjoint operations are performed, for this class of time evolving problem,
- the nonlinear character of the model equations such as the equation of state and the momentum advection terms, and
- the conditional code execution (IF ... ELSE IF ... END IF).

Thus, it is necessary to make available the intermediate model state in reverse sequence. In principle this could be achieved by either storing the intermediate states of the computation or by successive recomputing of the forward trajectory, throughout the reverse sequence computation. Either approach, in its pure form, is prohibitive; storing of the full trajectory is limited by available fast-access, storage media, recomputation is limited by CPU resource requirements which scale as the square of the number of intermediate steps.

TAMC provides two crucial features to balance recomputation and storage. First, TAMC generates recomputations of intermediate values by the Efficient Recomputation Algorithm (ERA) [11]. Secondly, TAMC generates code to store and read intermediate values, if appropriate directives have been inserted into the code. This enables the user to choose between storing and recomputation on every code level in a very flexible way. At the time-stepping level the directives allow for checkpointing that hierarchically splits the time-stepping loop; cf. also [12, 13]. For the MITgcm, a three-level checkpointing scheme, illustrated in Fig. 3, has been adopted as follows:

lev3 The model trajectory is first subdivided into n^{lev3} subsections. The model is then integrated along the full trajectory, and the state stored at every k_i^{lev3}-th timestep.

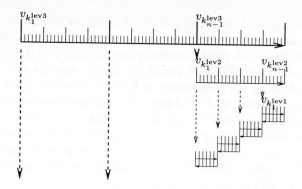

Fig. 3. Schematic view of intermediate dump and restart for 3-level checkpointing

lev2 In a second step, each "lev3"–subsection is divided into n^{lev2} subsections. The model picks up the last "lev3"–saved state $v_{k_{n-1}^{\text{lev3}}}$ and is integrated forward along the last "lev2"–subsection, now storing the state at every k_i^{lev2}-th timestep.

lev1 Finally, the model picks up at the last intermediate saved state $v_{k_{n-1}^{\text{lev2}}}$ and is integrated forward in time along the last "lev1"–subsection. Within this subsection only, the model state is stored at every timestep. Thus, the final state $v_n = v_{k_n^{\text{lev1}}}$ is reached and the model state of all preceding timesteps along the last "lev1"–subsection are available. The adjoint can then be computed back to subsection k_{n-1}^{lev2}.

This procedure is repeated consecutively for each previous subsection carrying the adjoint computation back to initial time k_1^{lev3}.

The 3-level checkpointing requires a total of 3 forward and one adjoint integration, with the latter taking about 2.5 times a forward integration. Thus, a forward/adjoint sweep requires a total of roughly 5.5 times a forward integration. For a given decomposition of the total number of time steps $n_{\text{timeSteps}} = 77,760$ (corresponding to a 9 year integration at an hourly timestep) into a hierarchy of 3 levels of sub-intervals $n_1 = 24$, $n_2 = 30$, $n_3 = 108$ with $n_{\text{timeSteps}} = n_1 \cdot n_2 \cdot n_3$, the storing amount is drastically reduced from $n_{\text{timeSteps}}$ to $n_1 + n_2 + n_3 = 162$ steps. Pure recomputation would incur a computation cost of $77,760^2$.

The two outer loops are stored to disks, while the innermost loop is stored to memory to avoid I/O in this phase of computation. For the outer loops, the storing consists of the model state required to perform the **PS** and **DS** phases of computation. Insertions of store directives at the innermost loop are more intricate and require detailed knowledge of the code. Typical places are near nonlinear expressions. Furthermore, directives may have to be accompanied (or may be avoided), occasionally, by additional measures to break artificial data dependency flows. For instance, a DO-loop which contains multiple consecutive and state-dependent assignments of a variable may be broken into several loops, to enable intermediate results to be stored before further state-dependent calculations are performed.

The character of **DS** has important implications for tangent-linear and adjoint computations. The equation solved in **DS** is self-adjoint. Exploiting this fact in reverse mode derivative calculations provides substantial computing cost savings. By providing a directive, TAMC then uses the original code where needed in the adjoint calculation.

3.2   Parallel Implementation

A number of issues required substantial intervention into the original code to enable correct adjoint code generation which is scalable and both memory and CPU efficient.

- *Exchanges between neighboring tiles* Domain decomposition is at the heart of the MITgcm's parallel implementation. Each compositional unit (tile) representing a virtual processor consists of an interior domain, truly owned by the tile and an overlap region, owned by a neighboring tile, but needed for the computational stencil within a given computational phase. The computational phase within which no communication is required can thus be considerable. Periodically, between computational phases, processors will make calls to WRAPPER functions which can use MPI, OpenMP, or combinations thereof to communicate data between tiles (so-called exchanges), in order to keep the overlap regions up-to-date. The separation into extensive, uninterrupted computational phases and minimum communication phases controlled by the WRAPPER is an important design feature for efficient parallel adjoint code generation. Corresponding adjoint WRAPPER functions were written by hand. TAMC recognizes when and where to include these routines by means of directives.

- *Global arithmetic primitives* Operations within the communication phase, for which a processor requires data outside of the overlap region of neighboring processors, must use communication libraries, such as MPI or OpenMP. So far, all global operations could be decomposed into arithmetic elements involving the global sum as the only global operation (major applications in the context of the MITgcm are the pressure inversion phase and global averages). WRAPPER routines exist, which adapt the specific form of the global sum primitive to a given platform. Corresponding adjoint routines were written by hand and directives to use these are given to TAMC.

- *Active file handling on parallel architectures* Figure 1 also shows an isolated I/O phase that deals with external data inputs that affect the calculation of (1) and (2). This isolation of "active" I/O simplifies AD code transformations. Read and write operations in forward mode are accompanied by corresponding write and read operations, respectively, in adjoint mode required for active variables. MITgcm possesses a sophisticated I/O handling package to enable a suite of global or local (tile- or processor based) I/O operations consistent with its parallel implementation. Adjoint support routines were written to retain compatibility in adjoint mode with both distributed memory and shared memory parallel operation, as implemented in the I/O package of the WRAPPER.

4   Applications

• *Global Ocean State Estimation* The estimation problem currently under way iteratively reduces the model vs. data misfit (1), by successive modification to the controls, C. To infer updates in the controls, the cost gradient (2) is subject to a quasi-Newton variable storage line search algorithm [14]. The updated controls serve as improved initial and boundary conditions in consecutive forward/adjoint calculations. Thus, $\nabla_C \mathcal{J}$, the outcome of the adjoint calculation, is a central ingredient for the optimization problem. By way of example, Fig. 4, taken from [15], depicts the surface heat flux correction estimated from the optimization. The mean changes of the flux relative to the NCEP [16] input fields are large over the area of the Gulf Stream and in the Eastern tropical Pacific. The heat flux corrections inferred here were shown to agree with independent studies of the NCEP heat flux analyses.

• *Sensitivity Analysis* Complementary to the estimation problems, sensitivity studies have been undertaken with the MITgcm and its adjoint which aim at interpreting the adjoint or dual solution of the model state [17]. As an example, Fig. 5 depicts the sensitivity of the North Atlantic heat transport at 24°N to changes in surface temperature.

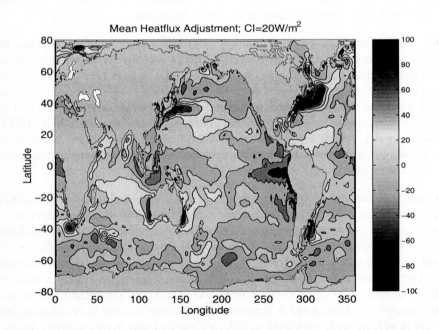

Fig. 4. Mean changes in heat flux relative to the NCEP first guess fields

(a) Sensitivity to temperature (surface) [TW/K]

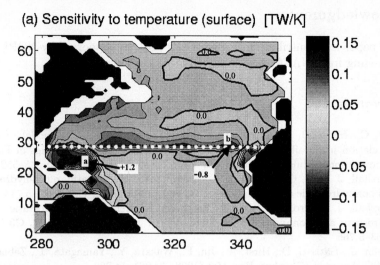

Fig. 5. The sensitivity, $\frac{\partial \mathcal{J}}{\partial \theta}$, of the North Atlantic heat transport at 24°N, $\mathcal{J}$, to changes in temperature, θ, at the ocean surface. At point "a" a persistent, 1 year, unit change in θ will produce a 1.2×10^{12} W increase in annual mean poleward heat transport. The same change at point "b" produces a 0.8×10^{12} W decrease (from [17])

5   Summary and Outlook

Reverse mode AD applications are emerging as powerful tools in oceanographic research. Essential technical features are efficient recomputation algorithms and checkpointing. Scalability of the adjoint code, maintained by hand-written adjoint functions, complements parallel support functions of the forward code. This renders computationally tractable code. AD has been successfully used in solving an unprecedented optimization problem. In addition, a host of physical quantities can be efficiently and rigorously investigated in terms of their sensitivities by means of the dual solution provided by the cost gradient, thus providing novel insight into physical mechanisms. Further reverse mode applications are being pursued using MITgcm's adjoint. They include optimal perturbation/singular vector analyses in the context of investigating atmosphere-ocean coupling. A natural extension of the state estimation problem is the inclusion of estimates of the errors of the optimal controls. The computation of the full error covariance remains prohibitive, but dominant structures may well be extracted from the Hessian matrix. As ambitions grow the ECCO group has recently switched to the TAF tool (Transformation of Algorithms in Fortan), the successor of TAMC, which has enhanced features. ECCO is contributing to the Adjoint Compiler Technology & Standards (ACTS) project, an initiative to increase accessibility to and development of AD algorithms by a larger community through the definition of a common intermediate algorithmic platform, within which AD algorithms can be easily shared among different developers and tools.

Acknowledgments

This is paper is a contribution to the ECCO project, supported by NOPP, and with funding from NASA, NSF and ONR.

References

1. Hill, C., Marshall, J.: Application of a parallel Navier-Stokes model to ocean circulation in parallel computational fluid dynamics. In: Proceedings of Parallel Computational Fluid Dynamics, New York, Elsevier Science (1995) 545–552
2. Marshall, J., Hill, C., Perelman, L., Adcroft, A.: Hydrostatic, quasi-hydrostatic and nonhydrostatic ocean modeling. J. Geophys. Res. **102, C3** (1997) 5,733–5,752
3. Marshall, J., Adcroft, A., Hill, C., Perelman, L., Heisey, C.: Hydrostatic, quasi-hydrostatic and nonhydrostatic ocean modeling. J. Geophys. Res. **102, C3** (1997) 5,753–5,766
4. Neelin, J., Battisti, D., Hirst, A., Jin, F., Wakata, Y., Yamagata, T., Zebiak, S.: ENSO theory. J. Geophys. Res. **103** (1998) 14,261–14,290
5. Marshall, J., Kushnir, Y., Battisti, D., Chang, P., Czaja, A., Hurrell, J., McCartney, M., Saravanan, R., Visbeck, M.: Atlantic climate variability. Int. J. Climatology (2002) (to appear).
6. Griewank, A.: Evaluating Derivatives: Principles and Techniques of Algorithmic Differentiation. SIAM, Philadelphia (2000)
7. Giering, R., Kaminski, T.: Recipes for adjoint code construction. ACM Transactions on Mathematical Software **24** (1998) 437–474
8. ECCO: http://ecco-group.org/. (URL)
9. MITgcm: http://mitgcm.org/sealion/. (URL)
10. Giering, R.: Tangent linear and Adjoint Model Compiler. Users manual 1.4 (TAMC Version 5.2). Technical report, MIT, MIT/EAPS, Cambridge (MA), USA (1999) http://puddle.mit.edu/~ralf/tamc/tamc.html.
11. Giering, R., Kaminski, T.: Generating recomputations in reverse mode AD. In Corliss, G., Faure, C., Griewank, A., Hascoët, L., Naumann, U., eds.: Automatic Differentiation of Algorithms: From Simulation to Optimization. Springer (2002) (to appear).
12. Griewank, A.: Achieving logarithmic growth of temporal and spatial complexity in reverse Automatic Differentiation. Optimization Methods and Software **1** (1992) 35–54
13. Restrepo, J., Leaf, G., Griewank, A.: Circumventing storage limitations in variational data assimilation studies. SIAM J. Sci. Comput. **19** (1998) 1586–1605
14. Gilbert, J., Lemaréchal, C.: Some numerical experiments with variable-storage quasi-Newton algorithms. Math. Programming **45** (1989) 407–435
15. Stammer, D., Wunsch, C., Giering, R., Eckert, C., Heimbach, P., Marotzke, J., Adcroft, A., Hill, C., Marshall, J.: The global ocean circulation and transports during 1992–1997, estimated from ocean observations and a general circulation model. (2002) (to appear).
16. NCEP: http://www.ncep.noaa.gov/. (URL)
17. Marotzke, J., Giering, R., Zhang, K., Stammer, D., Hill, C., Lee, T.: Construction of the adjoint MIT ocean general circulation model and application to atlantic heat transport variability. J. Geophys. Res. **104, C12** (1999) 29,529–29,547

Switchback: Profile-Driven Recomputation for Reverse Mode

Mike Fagan and Alan Carle

Department of Computational and Applied Mathematics
Rice University, 6100 Main Street, Houston, TX 77005–1892, USA
{mfagan, carle}@caam.rice.edu
http://www.caam.rice.edu

Abstract. Many reverse mode codes created by Automatic Differentiation (AD) tools use too much storage. To help mitigate the storage requirement, AD users resort to a technique called *recomputation*. Acceptable performance, however, requires judicious use of the recomputation technique. This work describes our approach to constructing good recomputation for Adifor 3.0-generated reverse mode codes by using a standard execution time profiler.

1 Introduction

Modern Automatic Differentiation (AD) tools have enabled scientists and engineers to easily generate adjoint, or reverse mode programs for computation of gradients. Many reverse mode computations suffer from overuse of both primary and file system memory resources. Consequently, further development of reverse mode codes frequently centers on using *recomputation* to reducing the amount of storage required. Recomputation, however, trades time for space, and must be used judiciously, or else the reverse mode computation will become too slow. AD tools implementing the reverse mode include ADOL-C [1], Odyssée [2], and TAMC/TAF [3].

While working on a shape optimization problem, we discovered that we needed recomputation. We were using Adifor 3.0 [4], the successor of [5], to generate reverse mode code for the computation of geometric derivatives in CFL3D [6], a fluid dynamics code. For the problem instances of interest to us, the reverse mode log exceeded the bounds for the local disk storage. Even after using a standard fixed-point technique to drastically reduce the storage, we still could not fit our computation on the local disk. Ergo, we saw that we must use recomputation to compute the derivatives of interest. The recomputation technique we designed for our problem employed the two main ideas: (a) the construction of local recomputation based on subprograms already present in the user's code and (b) the use of a standard program execution profiling tool to indicate subprograms in which recomputation can be effectively applied.

The remainder of this paper describing our technique is organized as follows. Section 2 gives some general background material on the reverse mode of automatic differentiation, including some discussion of the storage problem. Section 3

P.M.A. Sloot et al. (Eds.): ICCS 2002, LNCS 2330, pp. 1029–1038, 2002.

gives a general overview of the recomputation mechanism, and how it is used to reduce reverse mode storage problems. Section 4 gives the relevant details of the specific Adifor 3.0 reverse mode technique, including the Adifor 3.0 "switchback" recomputation mechanism. Section 5 details how we use a standard execution profiler to select appropriate subroutines for recomputation. Section 6 describes the derivative computation problem under study, including systemic constraints. It also details how we used Adifor 3.0 recomputation to render the problem soluble. Finally, Sect. 7 shows the results of our recomputation transformation, and estimates the performance effect.

2   Background Material

The time cost for a (straightforward) reverse mode computation is proportional to the number of "output", or "dependent" variables, and *does not* depend on the "design" or "independent" variables. This property is what makes reverse mode nearly ideal for optimal control or optimal design applications. For example, our shape optimization problem required the derivatives of each grid point with respect to the lift-over-drag ratio. See [7] for a thorough description of reverse mode resource cost, and a complete treatment of reverse mode mathematics.

The following simple example illustrates the reverse mode. Suppose we are given a simple program

$$y = f(x_1, \ldots, x_n)$$
$$z = g(y, x_1)$$

that computes z given $x_1, \ldots, x_n$. Let $\bar{v} := \partial z / \partial v$ denote the derivative, or adjoint, of the dependent variable z with respect to an intermediate variable v. Then the reverse mode computation for the derivatives of z with respect to the independent variables $x_1, \ldots, x_n$ is given by

Step 1: $\bar{z} = 1, \quad \bar{y} = 0, \quad \bar{x_1} = 0, \quad \ldots, \quad \bar{x_n} = 0,$

Step 2: $\bar{y} = \bar{y} + \dfrac{\partial g}{\partial y} \bar{z}, \quad \bar{x_1} = \bar{x_1} + \dfrac{\partial g}{\partial x_1} \bar{z}, \quad \bar{z} = 0,$

Step 3: $\bar{x_1} = \bar{x_1} + \dfrac{\partial f}{\partial x_1} \bar{y}, \quad \ldots, \quad \bar{x_n} = \bar{x_n} + \dfrac{\partial f}{\partial x_n} \bar{y} \ .$

Note that the function computation order is y, then z, but the reverse mode derivative calculation $\bar{z}$, then $\bar{y}$. This reversal of the function computation ordering is what gives "reverse mode" its name.

An AD tool accepts a user program as input, and outputs an augmented program that computes specified derivatives via reverse mode computation in addition to the original function computation. As illustrated by the preceding example, reverse mode control flow is the reverse of the original function control flow. For simple assignment statements, this control flow reversal is fairly simple. When conditional and unconditional branches, subroutine calls, loops, and other nontrivial program constructs are added to the statement mix, however, the code implementing the reverse control flow can be quite complex. In most real world codes, implementing the control flow reversal requires saving trace information.

In general, a reverse mode computation consists of two phases: a forward, or *logging* phase, and a reverse, or *derivative* phase. During the forward phase, some program state information is saved, but the control flow of the program is unchanged. The forward phase computes the same function outputs as the original unaugmented program. During the reverse phase, however, the control flow of the program is reversed from the normal forward flow in order to compute the derivative accumulations. Derivative accumulations require the partial derivatives of the left hand side of the assignment statement with respect to each variable used on the right hand side of the assignment statement. The responsibility of the forward phase is to log sufficient information to both correctly reverse the control flow and compute the necessary partial derivatives for the assignment statements.

The quasi-mathematical model of assignment statements given in the preceding example oversimplifies a crucial characteristic of computer program assignment statements. The problem can be seen in another simple example where the program fragment (line numbered for convenience)

```
(1) v1 = a ** 2 + b ** 2
(2) a = sin(z)
```

is given. Forward control flow proceeds from statement (1) to statement (2). So, the reverse pass must assure that the derivative accumulations for (2) occur before the derivative accumulations for (1). Also, note that for statement (1), $\partial v1/\partial a = 2a$, where the value of a is the value it had *before* the statement (2) assignment. The forward phase must ensure that the correct value of the partials for statement (1) are computed.

Since assignment statements change the value of a variable, the fundamental problem for reverse mode is saving enough information to compute the partials. There are two primary saving strategies:

1. Compute partials for each assignment in the forward pass, and save them. Use the saved partials during the derivative accumulation of the reverse pass.
2. Save the values of left hand sides of assignments during the forward pass. In the reverse pass, restore the saved values as needed to compute the partials for the given statement.

Clearly, either strategy entails a potentially huge amount of auxiliary storage.

In summary, the *time* cost of reverse mode computation is very favorable for many problems, especially optimization problems. The *space* cost, however, might be problematic.

3   The Recomputation Alternative

To avoid storing values for every single assignment, the reverse mode computation could instead store a *checkpoint*, and when partials are needed, restart the computation using saved checkpoint, recompute the state of the computation [8]. The obvious drawback to such a scheme is the time cost for doing the recomputation.

There are several technical problems associated with employing a recomputation scheme:

- How do we determine what to save as a checkpoint?
- How do we arrange for derivative computation control flow to use the checkpoint?
- How do we determine where to take the checkpoints?

When considering what information to save for checkpoints, we note that the program itself has a natural level of granularity—the subprogram, i.e., subroutines and functions in Fortran. For a given subprogram call site, the natural checkpoint is the input values for that routine. The input values for a routine consist of the read values for routine arguments, as well as global variables that are read by the routine. Given the input values, the subprogram intermediate state can be restored by simply calling the subprogram with the same inputs Consequently, if we focus on checkpointing at the level of a subprogram call, then we need only save the input values to effectively checkpoint the call.

Similarly, the restart mechanism for a subprogram is relatively simple: just call it. The differentiated subprogram components can also be called as needed. An example of this will appear in the following section.

4   Adifor 3.0 Reverse Mode Particulars

In order to ensure reasonable computational *time* performance, Adifor 3.0 reverse mode *by default* stores all left hand sides of relevant assignments, and restores the prior values during reverse mode partial computation. More precisely, given an assignment statement of the form

```
foo = bar * baz
```

Adifor 3.0 generates both forward phase code and reverse phase code. The forward phase code is given by

```
call STORE_r(foo)
foo = bar * baz
```

and the reverse phase code is generated in the form

```
a_bar = a_bar + baz * a_foo
a_baz = a_baz + bar * a_foo
a_foo = 0.0
call LOAD_r(foo)
```

where the prefix a_ is used to denote an adjoint of the corresponding variable.

Adifor 3.0 handles subroutine call control flow by using a switch to select either forward pass or reverse pass computation. To illustrate, suppose we have the following routine with two inputs and two outputs:

```
subroutine SS(in1,in2,out1,out2)
   ! compute stuff
end
```

Then Adifor 3.0 generates an associated reverse mode subroutine that looks like this:

```
subroutine a_SS(dir,in1,a_in1,in2,a_in2,out1,a_out1,
+                out2,a_out2)
   if (dir .eq. FWD) then
      ! compute stuff and log intermediate stuff
   endif
   if (dir .eq. REV) then
      ! restore intermediates, compute partials,
      ! accumulate derivatives.
   endif
end
```

Here, the parameter `dir` controls the forward phase, FWD, or the reverse phase, REV, of the standard reverse mode computation which is driven by

```
call a_SS(FWD,...)
call a_SS(REV,...)
```

The Adifor 3.0 recomputation technique is called the "switchback" transform. A routine that uses switchback uses the forward and reverse components from conventional Adifor processing, but has a different schematic. The switchback version of the preceding SS routine would be:

```
subroutine sb_a_SS(dir,in1,a_in1,in2,a_in2,
+                   out1,a_out1,out2,a_out2)
   if (dir .eq. FWD) then
      call STORE_r(in1) ! save input arg 1
      call STORE_r(in2) ! save input arg 2
      call SS(in1,in2,out1,out2) ! make the (undiff) call
   endif
   if (dir .eq. REV) then
      call LOAD_r(in2) ! restore arguments
      call LOAD_r(in1)
      call a_SS(FWD,...) ! invoke normal fwd pass behavior
      call a_SS(REV,...) ! invoke normal reverse pass behavior
   endif
end
```

Note that constructing a switchback routine uses both the undifferentiated subroutine as well as the differentiated routine.

The user selects routines to which the switchback transform will be applied, and specifies the names in the Adifor control file. In addition, for each selected switchback routine, the *user* must nominate the input values to be saved. This user-based nomination of input values is admittedly crude and tedious, but it will be automated in future Adifor releases.

An actual example of using the switchback mechanism is as follows. Suppose that the following subroutine

```
subroutine flx(a, n, x, flxout)
double precision a, x(n,n), flxout(n), tmp, tmp1
integer n
integer ix(100),iy(100)
double precision z(100),zz(100)
common /flxcomm/ ix,iy,z,zz

tmp = flxx(a,x)
tmp1 = flxz(a,z)

call flxupd(flxout,tmp,tmp1,a,n,x,z)

end
```

is given as part of a larger program where the semantics of the routines flxx, flxz, and flxupd is not known a priori.

After inspecting this routine, and the various calls made by it, we determine that subroutine arguments a, x, and n are read by flx. The variable z is *not* read prior to assignment by flxx or flxz, it is an *output* variable. Similarly, flxout is an output variable, and zz is not used at all. Hence, the switchback mechanism needs only save a, n, and x. This analysis is encoded in the Adifor control script these statements:

```
Switchback
 flx:
  a,n,x
```

Adifor will then ensure that flx uses recomputation rather than logging to compute the necessary partial derivatives for the flx routine.

5   Profiling for Recomputation

Selecting *all* routines for recomputation would result in exceptionally poor performance. In addition, not all routines are good candidates for recomputation. If the checkpoint is nearly as large as the amount of storage used, then nothing has been gained. The ideal candidate for recomputation would be a routine that uses a small number of inputs, but computes a large number of intermediate results, resulting in a large amount of storage used.

To select candidate routines for conversion to switchback, we follow the time-honored computer science paradigm of profiling a code to determine which routines are consuming the majority of the resource of interest. One then focuses intellectual effort on the routines that would benefit the most. For Adifor-generated derivative code, this is especially easy. Since Adifor-generated code stores by calling subroutines, e.g., STORE_r, a standard execution profiler such as Unix gprof will supply information on which routines make the most calls to the storage routine. Sample gprof output for our problem appears in Sect. 6. By focusing the inspection effort on the top "few" routines that use the most storage, some good recomputation candidates can likely be found.

To use profile-driven recomputation for an Adifor-generated code, the following steps are recommended:

1. Run Adifor in reverse mode to obtain default logging-based derivative code
2. Compile the derivative code with profiling support enabled (-pg option on most Unix systems).
3. Run the profile-enabled derivative code on a sample problem.
4. Run the profile post processor to discover which routines call the storage routines the most.
5. Inspect the top 10, say, of these to see if they use a "small" number of input values.
6. Assuming you find some, apply the switchback transform to the candidate routines.

The following section details our use of this recipe for our problem.

6   Reverse Mode Computation for CFL3D

The problem under study was an aerodynamic shape optimization for a wing under steady state flow conditions. The entire application consisted of a custom grid generator, a flow solver, and a single scalar objective function computation, the lift-over-drag ratio. Of these components, only the flow solver presented any difficulty in reverse mode gradient computation. The flow solver was CFL3D version 4. CFL3D [6] is a thin-layer Navier-Stokes solver that supports multiple zones, and uses MPI-based parallelism. CFL3D is approximately 126,000 lines of Fortran 77 code. The code is maintained by NASA Langley.

The grid for this problem was divided into 10 zones of size $65 \times 17 \times 17$. The execution platform was an Origin 2000, with 32 processors, and a relatively small local disk. The I/O system, however, required that all of the processor specific I/O be written on the local disk. In particular, that meant that the logs for the reverse mode gradient computation had to fit onto a small local disk of 500 megabytes. In addition, the disk did not perform well when it was almost full. So, ideally, we would like to keep the logs well under the 500 megabyte limit.

The flow solution ran for 1000 steps, and, using the default reverse mode computation of Adifor 3.0, our logging instrumentation revealed that the log for all 1000 steps would be on the order of 500 Gigabytes *per processor*.

Our first move to reduce the log size was to take advantage of the steady state nature of the computation, and reduce our logging requirements to a single step of the flow solver. This derivative accumulation is iterated over this single step until convergence. This technique is well known to AD users; see [9] for mathematical details. Even when reduced to a single step, our logging instrumentation revealed the following information:

```
blocks = 4129180
integers = 29942967
logicals = 232048
reals =   105
doubles = 46865895
```

This makes the total log size 512, 144, 360 bytes which is still too large to fit into memory.

A sample portion of the profiling run for the CFL3D application shows:

```
-------------------------------------------------
                            20572643/46865895     a_flxt_[15]
                            18484228/46865895     a_fhat_ [22]
                            16314714/46865895     a_diagj_ [59]
                            16314714/46865895     a_diagk_ [60]
                            16314714/46865895     a_diagl_ [61]
                            14901631/46865895     vec_v_ [91]
[6]      78.2    39.15    0.00   46865895        store_d_ [6]
```

The gprof output is dense, but contains the data we want. Line 7, on which store_d_ [6] appears indicates the total number of calls, in this case 46, 865, 895. The lines *above* the store_d_ [6] line indicate routines that call store_d_, and indicate what fraction of the total calls to store_d_ are made by the given routine. For example, line 1 of the sample output shows that routine a_flxt_ called store_d_ 20, 572, 643 times out of the 46, 865, 895 times that store_d_ was called. Furthermore, gprof sorts this output so that the most frequent callers of a given routine appear at the top of the list. In the sample output, the top 6 callers of store_d_ are a_flxt_, a_fhat_, a_diagj_, a_diagk_, a_diagl_, and vec_v_. Note that only the top 6 are listed *in the sample*. The full gprof output lists all the routines, and took several pages of output, as store_d_ is called by 247 routines.

By focusing our attention on the top 10 users of the STORE_d routine as determined by gprof, we found 4 routines that had relatively small number of input values, but consumed a relatively large amount of log. Specifically, a_flxt_,a_diagj_, a_diagk_, and a_diagl_ as shown in the sample output. The a_fhat_routine used a fair amount of common block reads, so it could not be used in our recomputation scheme. We applied the switchback transform to these 4 routines.

7 Results

By applying switchback to 4 routines, we were able to reduce the log size to these values:

```
blocks = 1694684
integers = 15975367
logicals = 223940
reals = 105
doubles = 29520599
```

with total log requirements now $307,741,176$ bytes, which fit comfortably on the local disk. This reduction enabled us to compute our desired gradients using the small local disk for log storage.

Since we could not actually compute derivatives for the given problem without switchback, we cannot say what the performance cost of switchback was. To estimate the performance cost of switchback for this code, we used a much smaller problem, and compared the switchback version to the black box version. NASA researchers generated a $17 \times 9 \times 9$ grid and a $33 \times 9 \times 9$ grid for our experiment. We were able to run both grids on our platform. Table 1 shows that the time penalty for switchback, on this particular problem appears to be about 37%.

Table 1. Timings (sec) of the two versions using Unix system clock

Grid	Black Box	Switchback
$17 \times 9 \times 9$	262.13	359.14
$33 \times 9 \times 9$	556.63	773.57

8 Concluding Remarks

To eliminate excessive storage requirements in a reverse mode application, application developers often resort to recomputation. Overuse of recomputation, however, damages time performance. Consequently, a reverse mode application developer must choose where to apply the recomputation technique. In the work described here, we showed how the time-honored computer science principle of profiling can be used to focus the developer's efforts on the routines with the most potential improvement. The specific AD tool used in the work was Adifor 3.0, but the profiling idea should be generally applicable.

As a final note, our experience with this technique has guided our Adifor development effort to include automatic detection of "read first" variables so that the switchback mechanism can be applied more easily by users.

Acknowledgments

Thanks to Los Alamos Computer Science Institute, and NASA Langley for their support of this research.

References

1. Griewank, A., Juedes, D., Utke, J.: ADOL-C, a package for the automatic differentiation of algorithms written in C/C++. ACM Transactions on Mathematical Software **22** (1996) 131–167
2. Rostaing, N., Dalmas, S., Galligo, A.: Automatic differentiation in Odyssée. Tellus **45A** (1993)
3. Giering, R., Kaminski, T.: Recipes for adjoint code construction. ACM Transactions on Mathematical Software **24** (1998) 437–474
4. Fagan, M., Carle, A.: Adifor 3.0 overview. Technical Report CAAM–TR00–03, Rice University, Department of Computational and Applied Mathematics (2000)
5. Bischof, C., Carle, A., Khademi, P., Mauer, A.: ADIFOR 2.0: Automatic differentiation of Fortran 77 programs. IEEE Computational Science & Engineering **3** (1996) 18–32
6. Rumsey, C.L., Biedron, R.T., Thomas, J.L.: CFL3D: Its history and some recent applications. Technical Report NASA Technical Memorandum 112861, NASA Langley Research Center (1997)
7. Griewank, A.: Evaluating Derivatives: Principles and Techniques of Algorithmic Differentiation. SIAM, Philadelphia (2000)
8. Griewank, A.: Achieving logarithmic growth of temporal and spatial complexity in reverse automatic differentiation. Optimization Methods and Software **1** (1992) 35–54
9. Christianson, B.: Reverse accumulation and attractive fixed points. Optimization Methods and Software **3** (1994) 311–326

Reducing the Memory Requirement in Reverse Mode Automatic Differentiation by Solving TBR Flow Equations

Uwe Naumann

Mathematics and Computer Science Division
Argonne National Laboratory, 9700 S. Cass Avenue, Argonne, IL 60439, USA
naumann@mcs.anl.gov
http://www.mcs.anl.gov

Abstract. The fast computation of gradients in reverse mode Automatic Differentiation (AD) requires the generation of adjoint versions of every statement in the original code. Due to the resulting reversal of the control flow certain intermediate values have to be made available in reverse order to compute the local partial derivatives. This can be achieved by storing these values or by recomputing them when they become required. In any case one is interested in minimizing the size of this set. Following an extensive introduction of the "To-Be-Recorded" (TBR) problem we will present flow equations for propagating the TBR status of variables in the context of reverse mode AD of structured programs.

1 Introduction

The work presented here is a continuation of the results published in [1, 2]. Our aim is to motivate a more formalized view on the problem of generating adjoint code using the reverse mode of AD [3] that requires a minimal amount of memory space when following a "store all" *taping* strategy, which will be explained below.

We consider a single subroutine $F : \mathbb{R}^n \to \mathbb{R}^m$ for computing a vector function $\mathbf{y} = F(\mathbf{x})$. The values of m dependent variables y_j, $j = 1, \ldots, m$, are calculated from the n independent variables x_i, $i = 1, \ldots, n$. The subroutine F represents an implementation of the mathematical model for some underlying real-world application and it will be referred to as the forward code. The forward code is expected to be written in some high-level imperative programming language such as C or Fortran. More generally, it should be possible to decompose F into a sequence of scalar assignments of the form

$$v_j = \varphi_j(v_k)_{k \prec j} , \quad j = 1, \ldots, p + m , \tag{1}$$

such that the result of every intrinsic function and elementary arithmetic operation is assigned to a unique intermediate variable v_j, $j = 1, \ldots, p + m$. m out of these intermediate variables are set to be dependent. Whenever some variable v_j

P.M.A. Sloot et al. (Eds.): ICCS 2002, LNCS 2330, pp. 1039–1048, 2002.

depends directly on another variable v_k we write $k \prec j$. It is assumed that the local partial derivatives

$$c_{ji} = \frac{\partial \varphi_j}{\partial v_i}(v_k)_{k \prec i} \tag{2}$$

of the *elemental* functions φ_j, $j = 1, \ldots, p + m$, exist and that they are jointly continuous in some open neighborhood of the current argument $(v_k)_{k \prec i}$. In this case an augmented version of the forward code can be implemented that computes F itself and the set of all local partial derivatives as defined in (2).

The reverse mode of AD [3, Sect. 3.3] uses these local partial derivatives to compute adjoints

$$\bar{v}_k = \sum_{j:k \prec j} c_{jk} \cdot \bar{v}_j \ , \quad j = p, \ldots, 1 - n \ . \tag{3}$$

"Transposed Jacobian matrix times vector" products $\bar{\mathbf{x}} = F'(\mathbf{x})^T \cdot \bar{\mathbf{y}}$ are computed by initializing the adjoints of the dependent variables $\bar{y}_j \equiv \bar{v}_{p+j}$, $j = 1, \ldots, m$. The Jacobian $F'(\mathbf{x})$ can be accumulated by reverse propagation of the Cartesian basis vectors in $\mathbb{R}^m$ at complexity $O(m)$. In particular, gradients of single dependent variables with respect to all independent variables can be obtained at a computational cost that is a small multiple of the cost of running the forward code; see the *cheap gradient principle* in [3].

Considering (3) we observe that in reverse mode AD the adjoints of all intermediate variables are actually computed in reverse order, i.e. for $j = p, \ldots, 1 - n$. This implies that the local partial derivatives c_{jk} have to be made available in reverse order as well. This can be ensured (a) by storing the arguments of all local partial derivatives on a so-called *tape* before their values get overwritten during the execution of the augmented forward code and by retrieving these values whenever required in the adjoint code [1] or (b) by simply recomputing them "from scratch" [4] when they become required in the adjoint code. Generally speaking, the arguments of all local partial derivatives have to be *recorded*. We will use this term as a place holder for either (a) or (b). Obviously, the former approach may lead to enormous memory requirements for large-scale application programs whereas the latter results in a quadratic computational complexity. Often a mixture of both strategies is employed to achieve reasonable trade-offs between memory requirements and the number of floating-point operations. However, even the efficiency of these *checkpointing* schemes [5] depends on the knowledge about whether some value is actually required or not.

2   TBR Problem

The generation of the adjoint model is done by associating adjoint components $\bar{v}$ with every *active* variable v. In particular, both the independent variables $\mathbf{x}$ and the dependent variables $\mathbf{y}$ are active. Here, the term variable should be understood as a scalar component of some *program variable* that is actually declared in the forward code. An intermediate variable v is active at a given point within the program if $\exists x \in \mathbf{x} : x \prec^* v$ and $\exists y \in \mathbf{y} : v \prec^* y$. Here, $\prec^*$ denotes the transitive closure of the operator $\prec$, i.e. $x \prec^* v$ if there exist $v_0, v_1, \ldots, v_{k-1}, v_k$ such that $x = v_0 \prec v_1 \prec \ldots \prec v_{k-1} \prec v_k = v$. In the following, we assume that the information on the set of active variables is available at every single point in the program. A variable that is not active is called *passive*.

We investigate reverse mode AD of structured programs [6, Sect. 10] by concentrating on the following four different types of statements:

- $s := [v = f(\mathbf{u})]$ – scalar assignments with $f : \mathbb{R}^k \to \mathbb{R}$ as they occur in most imperative programming languages; this restriction helps to keep the notation simple; all results can be generalized for general (vector) assignments $f : \mathbb{R}^{k_1} \to \mathbb{R}^{k_2}$ as they exist for example in Fortran 95 [7];
- $s := [s_1, s_2]$ – cascades of statements;
- $s := [\text{if } (c) \text{ then } s_1 \text{ else } s_2 \text{ fi}]$ – branches where the boolean value c determines whether s_1 or s_2 is executed;
- $s := [\text{while } (c) \text{ do } s_1 \text{ done}]$ – loops where c determines if s_1 is executed followed by another evaluation of c.

$\mathbf{u}$ will be considered as a set of scalar variables, i.e. we will write $w \in \mathbf{u}$ whenever the scalar variable w occurs on the right-hand-side of an assignment $v = f(\mathbf{u})$. c is the value of a scalar boolean function $g : D \to \{\text{true}, \text{false}\}$ over elements of arbitrary data types, i.e. $c = g(\mathbf{v})$ and the values of the arguments of the g determine the value of c and therefore the control flow.

In order to generate a correct adjoint code one has to do the following:

1. The control flow of the forward code has to be reversed.
2. Adjoint versions of every single assignment have to be built.

The former can be achieved in various ways. An exhaustive discussion of these issues is out of the scope of this paper. In the example presented in Sect. 3 we have chosen the following approach:

- Loops $s := [\text{while } (c) \text{ do } s_1 \text{ done}]$ are reversed by counting the number ITER of iterations performed when running the forward code and by executing the adjoint of the loop body s_1 exactly ITER times.
- For branches $s := [\text{if } (c) \text{ then } s_1 \text{ else } s_2 \text{ fi}]$ we push the values of all arguments of c onto the tape whenever they get overwritten during the execution of the forward code. When running the adjoint code these values are popped at the appropriate time to decide whether to execute the adjoint version of s_1 or s_2. If such an argument is overwritten inside s_1 or s_2 then its value has to be retrieved before the execution of the adjoint branch. This is automatically the case if the overwriting takes place after the execution of s in the forward code.

We do not claim this solution to be optimal. However, for structured programs it is a simple method for ensuring a correct reversal of the control flow.

In this paper we will concentrate on the second crucial ingredient of an adjoint code, namely the generation of adjoint versions for all assignments in the forward code. Consider $v = f(\mathbf{u})$ where $\mathbf{u} = \{u_1, u_2, \ldots, u_{n_f}\}$ denotes the set of scalar arguments of f. Reverse mode AD transforms this assignment into the set of adjoint statements

$$\bar{u}_i = \bar{v} \cdot \frac{\partial f}{\partial u_i}(\mathbf{u}) \ , \quad i = 1, \ldots, n_f \ .$$

Three types of values are required for evaluating them correctly:

1. $\text{args}(f'(\mathbf{u}))$ – the arguments of the local partial derivatives $\frac{\partial f}{\partial u_i}(\mathbf{u})$ for $i = 1, \ldots, n_f$;
2. $\text{idxargs}(\mathbf{u})$ – arguments of indices of array-type u_i, $i \in \{1, \ldots, n_f\}$;
3. $\text{idxargs}(v)$ – arguments of indices of v should v be an array element.

A more detailed characterization of $\text{args}(f'(\mathbf{u}))$ has been given in [1] as follows: The TBR status of $w \in \mathbf{u}$ has to be activated if

1. w is a non-linear active argument, e.g. $v = \sin(w)$;
2. w is a passive argument in an active term, e.g. $v = w \cdot a$ where a is an active variable;
3. w is the index of some active element of an array a which occurs non-linearly on the right-hand-side, e.g. $v = a(w+1) \cdot a(w)$;
4. w is the index of some passive element of an array p which occurs in an active term, e.g., $v = p(w) \cdot a$ where a is some active variable;

For a better understanding of this rule it is useful to notice the following comments:

- The decision whether some individual element of an array is active is impossible to make in general. Both static and dynamic *array region analysis* [8] can help to compute some conservative estimate. Without it the activity of one element implies the activity of the whole array.
- 3. actually describes a subset of $\text{idxargs}(\mathbf{u})$. The correct value of the array index is required for the adjoint as well as for restoring the original value that enters the computation of the local partial derivative.
- The indices of passive array elements are only required for restoring the correct arguments of the local partial derivatives. No adjoints are associated with passive variables.

In the following the expression $\mathbf{TBR}(s)$ is used to denote the set of variables whose TBR status is activated as the result of the execution of a statement s. In particular, we define

$$\mathbf{TBR}(s) = \text{args}(f'(\mathbf{u})) \cup \text{idxargs}(\mathbf{u}) \cup \text{idxargs}(v)$$

for scalar assignments $s := [v = f(\mathbf{u})]$. Naturally, this definition can be extended to cover general assignments. Moreover, following the control flow reversal strategy introduced above $\mathbf{TBR}(s)$ is exactly the set of all arguments of the condition c for branches $s := [\text{if } (c) \text{ then } s_1 \text{ else } s_2 \text{ fi}]$. Knowing how to compute $\mathbf{TBR}(s)$ for all statements of F we are able to decide whether the value of some variable has to be recorded. Such variables will be referred to as "tbr-active".

3   Example

Consider the following code fragment (original forward code in lower case on the left-hand-side) which has been augmented by instructions for storing the tape on the right (new statements in upper case). It can be wrapped into a subroutine F computing new values for the elements of a vector x from the corresponding input values.

forward code	augmented forward code
`i=0; j=10`	`i=0; j=10`
	`ITERS=0`
`while (check(j)) do`	`while (check(j)) do`
	`  ITERS=ITERS+1`
`  if (max(i,j)>7) then`	`  if (max(i,j)>7) then`
	`    STORE(x(i))`
`    x(i)=j+sin(x(i))`	`    x(i)=j+sin(x(i))`
`  else`	`  else`
	`    STORE(x(j))`
`    x(j)=j*cos(x(j))`	`    x(j)=j*cos(x(j))`
`  fi`	`  fi`
	`  STORE(i)`
`  i=i+1`	`  i=i+1`
	`  STORE(j)`
`  j=j-1`	`  j=j-1`
`done`	`done`

The variables i and j are assumed to be integers. The control flow is determined by the two boolean values $c_1 := \text{check(j)}$ and $c_2 := \text{max(i,j)} > 7$ where check is some boolean function over the integers and max computes the maximum of two numbers. STORE(w) puts the current value of the variable w onto the top of the stack implementing the tape for variables of the same type as w.

Let us have a closer look at the augmented version of the forward code. The integer variable ITER is introduced to count the number of iterations performed by the while loop. Both the values of i and j are required to generate the adjoint version $\bar{s}$ of the if-statement s and therefore $\mathbf{TBR}(s) = \{i, j\}$. Notice, that neither i nor j is overwritten inside s_1 or s_2. Consequently, their values do not have to be restored before the execution $\bar{s}$ in the adjoint code which is shown below. The fact that they are is a consequence of both i and j being overwritten immediately after the if-statement in the forward code.

For the assignment $s_1 := [v_1 = f_1(\mathbf{u}_1)] \equiv [\text{x(i)} = \text{j} + \sin(\text{x(i)})]$ we observe that

$$\mathbf{TBR}(s_1) = \text{args}(f_1'(\mathbf{u}_1)) \cup \text{idxargs}(\mathbf{u}_1) \cup \text{idxargs}(v_1)$$
$$= \{\text{x(i)}\} \cup \{\text{i}\} \cup \{\text{i}\} = \{\text{x(i)}, \text{i}\} \ .$$

Similarly,

$$\mathbf{TBR}(s_2) = \mathrm{args}(f_2'(\mathbf{u}_2)) \cup \mathrm{idxargs}(\mathbf{u}_2) \cup \mathrm{idxargs}(v_2)$$
$$= \{\mathtt{x(j)}, \mathtt{j}\} \cup \{\mathtt{j}\} \cup \{\mathtt{j}\} = \{\mathtt{x(j)}, \mathtt{j}\}$$

for $s_2 := [v_2 = f_2(\mathbf{u}_2)] \equiv [\mathtt{x(j)} = \mathtt{j} * \cos(\mathtt{x(j)})]$. In both statements the TBR status of the variable written is activated on the right-hand-side which results in the corresponding STORE instructions preceding the statement itself. Notice, that both i and j are stored as a result of being arguments of max(i,j) and as array indices of x. Moreover, j is recorded as an element of $\mathrm{args}(f_2'(\mathbf{u}_2))$.

We assume *joint program reversal mode* [3, Chap. 12] meaning that the adjoint computation is performed immediately after the execution of the augmented forward code. A possible implementation of the adjoint model is given by

```
adjoint code

while (ITERS>0) do
  RESTORE(j)
  RESTORE(i)
  if (max(i,j)>7) then
    RESTORE(x(i))
    adj_x(i)=cos(x(i))*adj_x(i)
  else
    RESTORE(x(j))
    adj_x(j)=-j*sin(x(j))*adj_x(j)
  fi
  ITERS=ITERS-1
done
```

RESTORE(w) puts the value from the top of the stack matching the data type of w into w. The RESTORE statement is always executed before the adjoint version of the statement in front of which the matching STORE statement was performed in the augmented forward code. In certain situations it can be advantageous to store the result of an assignment instead of its arguments; see [1].

Given values for x and adj_x the program consisting of the augmented forward code followed by the adjoint code computes the "transposed Jacobian matrix times adjoint vector" product

$$\mathtt{adj\_x} = F'(\mathtt{x})^T \cdot \mathtt{adj\_x} .$$

Alternatively, we might have recomputed the values of i, j, x(i), and x(j) which would have led to repeated executions of the forward code within the adjoint section while not requiring a tape.

4   TBR Status Flow Equations

In analogy to the approach in [6, Sect. 10] we will consider the following sets:

- $\mathbf{In}_p(s)$ – variables having property p before the execution of a statement s;
- $\mathbf{Out}_p(s)$ – variables having property p after the execution of s;
- $\mathbf{Gen}_p(s)$ – variables gaining property p as the result of executing s;
- $\mathbf{Kill}_p(s)$ – variables loosing property p as the result of executing s;

In particular, we are interested in the case $p = tbr$ (tbr meaning "has to be recorded"), i.e. in $\mathbf{In}_{tbr}(s)$, $\mathbf{Out}_{tbr}(s)$, $\mathbf{Gen}_{tbr}(s)$, and $\mathbf{Kill}_{tbr}(s)$.

The decision to be made is whether the value of a variable v overwritten by some assignment $s := [v = f(\mathbf{u})]$ is required for the evaluation of the adjoint program. If so, it has to be recorded. Below we consider assignments and basic blocks, cascades of statements, branches, and loops – each of them interpreted as a single statement s – under two aspects:

1. Which variables have to be recorded before the execution of s (for assignments only)? Under the restriction to scalar assignments the question is whether the value of the variable on the left-hand-side should be recorded or not.
2. The TBR status of which variables is active after the execution of s (for all statements)?

Assignments. For scalar assignments $s := [v = f(\mathbf{u})]$ the set $\mathbf{Kill}_{tbr}(s)$ is either empty or it contains the single element v. The latter is the case if the TBR status of v is activated before the execution of s or by s itself. Thus,

$$\mathbf{Kill}_{tbr}(s) = (\mathbf{In}_{tbr}(s) \cup \mathbf{TBR}(s)) \cap v \qquad (4)$$

is exactly the set of values that would have to be saved before the execution of s as part of the augmented forward code if we were following a "store all" strategy. For sets containing a single element v only we write v instead of $\{v\}$. Intuitively, we can state that a variable v belongs to $\mathbf{Gen}_{tbr}(s)$ if it is in $\mathbf{TBR}(s)$ but neither in $\mathbf{In}_{tbr}(s)$ nor in $\mathbf{Kill}_{tbr}(s)$, i.e.

$$\mathbf{Gen}_{tbr}(s) = \mathbf{TBR}(s) \setminus \mathbf{In}_{tbr}(s) \setminus \mathbf{Kill}_{tbr}(s) \ .$$

Which can be simplified to get

$$\mathbf{Gen}_{tbr}(s) = \mathbf{TBR}(s) \setminus \mathbf{In}_{tbr}(s) \setminus v \ . \qquad (5)$$

Sequences of set differences are evaluated from left to right. Notice, that this operation is not associative. Both the expressions for $\mathbf{Kill}_{tbr}(s)$ and for $\mathbf{Gen}_{tbr}(s)$ are required for the resolution of

$$\mathbf{Out}_{tbr}(s) = \mathbf{Gen}_{tbr}(s) \cup (\mathbf{In}_{tbr}(s) \setminus \mathbf{Kill}_{tbr}(s)) \ . \qquad (6)$$

This standard data flow equation (see for example [6]) says that a variable is tbr-active after the execution of a statement s if its TBR status became activated by s or if it was tbr-active before s and was not made tbr-passive by s. Substituting (4) and (5) in equation (6) results in

$$\mathbf{Out}_{tbr}(s) = (\mathbf{TBR}(s) \cup \mathbf{In}_{tbr}(s)) \setminus v \ . \qquad (7)$$

Under the restrictions imposed by us assignments s are the only statements for which we are actually interested in $\text{Kill}_{tbr}(s)$. For the remaining types of statements we need to be able to compute $\text{Out}_{tbr}(s)$ from $\text{In}_{tbr}(s)$. Both $\text{Gen}_{tbr}(s)$ and $\text{Kill}_{tbr}(s)$ can be computed recursively from the underlying assignments.

Equation (7) can be generalized to become

$$\text{Out}_{tbr}(s) = \bigcup_{i=1,\ldots,l} \left(\text{TBR}(\mathbf{u}_i) \setminus \left(\bigcup_{j=i,\ldots,l} v_j \right) \right) \cup \text{In}_{tbr}(s) \setminus \left(\bigcup_{i=1,\ldots,l} v_i \right) \quad (8)$$

for cascades of l assignments, i.e. for *basic blocks* $s := [s_i, i = 1, \ldots, l]$ where $s_i := [v_i = f_i(\mathbf{u}_i)]$ and $i = 1, \ldots, l$. Assuming that $\text{In}_{tbr}(s)$ is known equation (8) allows us to compute $\text{Out}_{tbr}(s)$ using structural information on all assignments which is readily available.

Cascades of Statements. The standard data flow equations apply for cascades of statements $s := [s_1, s_2]$, i.e. $\text{In}_{tbr}(s_1) = \text{In}_{tbr}(s)$, $\text{In}_{tbr}(s_2) = \text{Out}_{tbr}(s_1)$, and $\text{Out}_{tbr}(s) = \text{Out}_{tbr}(s_2)$. $\text{Out}_{tbr}(s_i) = \text{Gen}_{tbr}(s_i) \cup (\text{In}_{tbr}(s_i) \setminus \text{Kill}_{tbr}(s_i))$ for $i = 1, 2$ leads to

$$\text{Out}_{tbr}(s) = \text{Gen}_{tbr}(s_2) \cup (\text{Gen}_{tbr}(s_1) \cup (\text{In}_{tbr}(s) \setminus \text{Kill}_{tbr}(s_1))) \setminus \text{Kill}_{tbr}(s_2)$$

which results in the requirement for explicit expressions for $\text{Gen}_{tbr}(s_i)$ and $\text{Kill}_{tbr}(s_i)$ $(i = 1, 2)$. For example, if both s_1 and s_2 are scalar assignments then (4) and (5) can be used to derive more specific expressions. In fact, (8) was derived this way.

Branches. Static TBR analysis is conservative, i.e. for a conditional branch statement $s := [\text{if } (c) \text{ then } s_1 \text{ else } s_2 \text{ fi}]$ we have

$$\text{Gen}_{tbr}(s) = \text{Gen}_{tbr}(s_1) \cup \text{Gen}_{tbr}(s_2)$$
$$\text{Kill}_{tbr}(s) = \text{Kill}_{tbr}(s_1) \cap \text{Kill}_{tbr}(s_2)$$
$$\text{Out}_{tbr}(s) = \text{Out}_{tbr}(s_1) \cup \text{Out}_{tbr}(s_2) \ .$$

Loops. Consider a loop $s := [\text{while } (c) \text{ do } s_1 \text{ done}]$ where s_1 is a cascade of statements as in Sect. 4. We are interested in two types of information:

1. Knowing $\text{In}_{tbr}(s)$ we would like to compute $\text{Out}_{tbr}(s)$.
2. For every assignment $s' \in s_1$ we need to determine $\text{Kill}_{tbr}(s')$.

Both 1. and 2. should be obtained with a minimal computational effort. Below we show that for the TBR status information within structured programs at most two traversals of the code inside loops are required. We will specify the index of the traversal as a superscript, i.e. for example $\text{Out}_{tbr}(s_1)^1$ denotes the set of tbr-active variables after a single analysis of s_1. Furthermore, we assume that the control flow is reversed by counting the number of loop iterations performed by the forward code as in the example in Sect. 1.

Proposition 1 $\mathbf{Out}_{tbr}(s) = \mathbf{Out}_{tbr}(s_1)^1$.

Proof. Under what circumstances is a variable v in $\mathbf{Out}_{tbr}(s)$?

1. If it is in $\mathbf{In}_{tbr}(s)$ and if it is not overwritten inside s_1, i.e. if $v \in \mathbf{In}_{tbr}(s_1)^1 \wedge \nexists s' \in s_1 : v \in \mathbf{Kill}_{tbr}(s')^1$ or
2. if it becomes tbr-active as a result of executing some assignment $s' \in s_1$ and if it is not overwritten by any assignment $s'' \in s_1$ such that $s'' > s'$, i.e. if $\exists s'' \in s_1 : s'' \in \mathbf{Gen}_{tbr}(s'')^1 \wedge \nexists s' > s'' \in s_1 : v \in \mathbf{Kill}_{tbr}(s')^1$.

We use the notation $s'' > s'$ to indicate that the statement s' is executed before s'' when running the forward code. Obviously, both 1. and 2. can be checked by performing a single analysis of the loop body.   □

A major consequence is that in order to determine $\mathbf{In}_{tbr}(s)$ for some statement within the forward code we have to analyze each statement preceding s only once, even if it is part of a (possibly nested) loop.

In general, for $s' := [v = f(\mathbf{u})]$ we will know whether to record v only after the second traversal of s_1.

Proposition 2 $\forall s' \in s_1 : \mathbf{Kill}_{tbr}(s') = \mathbf{Kill}_{tbr}(s')^2$

Proof. Consider $s' \in s_1$ where $s' := [v = f(\mathbf{u})]$.

1. If v is tbr-active during the first traversal, i.e. if

$$(v \in \mathbf{In}_{tbr}(s_1) \vee \exists s'' < s' : v \in \mathbf{Gen}_{tbr}(s')) \wedge$$
$$(\nexists s''' : (s'' < s''' < s' \wedge v \in \mathbf{Kill}_{tbr}(s'''))) ,$$

then the decision can be made during the first traversal.
2. If v is tbr-passive
   (a) because there exists an assignment $s'' < s'$ which overwrites v and $\nexists s''' : s'' < s''' \wedge v \in \mathbf{Gen}_{tbr}(s''')$ then the decision can be made during the first traversal;
   (b) because v is not in $\mathbf{In}_{tbr}(s)$ then due to $\mathbf{In}_{tbr}(s_1) = \mathbf{In}_{tbr}(s) \cup \mathbf{Out}_{tbr}(s_1)^1$ we need a second iteration to cover the case $v \in \mathbf{Out}_{tbr}(s_1)^1$.

   □

5   Summary, Preliminary Results, Outlook

The generation of adjoint code is based on the reversal of the control flow of the forward code. Adjoint versions of every single assignment contained within the latter have to be generated. The reversed control flow leads to the requirement to access the values of certain intermediate variables in reverse order. In general, due to intermediate variables being overwritten in the forward code, this can only be ensured by storing the corresponding values or recomputing them.

In this paper we have presented some flow equations for propagating the information on whether the value of an intermediate variable has to be recorded, i.e. whether the current value is required for the generation of a correct adjoint code. Implementations of these ideas showed promising reductions of the memory requirement when following a pure "store all" strategy. In [1] the ideas presented in this paper were applied to a large industrial thermal-hydraulic code developed at EDF-DER in France (70 000 lines, 500 sub-programs, 1 000 parameters). Using the TBR analysis the tape size could be decreased by a factor of 5. The size of the standard tape generated by *Odyssée* version 1.7 [9] is $213\,920 \cdot 10^6$ scalar values (or $1\,711\,360$ MBytes if every value is a double), whereas the optimized tape contains only $40\,486 \cdot 10^6$ scalar values (or $323\,888$ MBytes).

Our next step will be the generalization of the results presented here to interprocedural TBR analysis and general unstructured programs. In collaboration with colleagues working at INRIA Sophia-Antipolis, France, these ideas are currently being implemented in TAPENADE [10], the successor of *Odyssée*. A general discussion of optimizing the memory requirements in reverse mode AD is in work.

Acknowledgments

This work was supported by the Mathematical, Information, and Computational Sciences Division subprogram of the Office of Advanced Scientific Computing Research, U.S. Department of Energy, under Contract W-31-109-ENG-38.

References

1. Faure, C., Naumann, U.: The taping problem in automatic differentiation. [2] (to appear).
2. Corliss, G., Faure, C., Griewank, A., Hascoët, L., Naumann, U., eds.: Automatic Differentiation of Algorithms: From Simulation to Optimization. Springer (2002) (to appear).
3. Griewank, A.: Evaluating Derivatives: Principles and Techniques of Algorithmic Differentiation. SIAM, Philadelphia (2000)
4. Giering, R., Kaminski, T.: Towards an optimal trade-off between recalculation and taping in reverse mode AD. [2] (to appear).
5. Griewank, A.: Achieving logarithmic growth of temporal and spatial complexity in reverse automatic differentiation. Optimization Methods and Software (1992) 35–54
6. Aho, A., Sethi, R., Ullman, J.: Compilers. Principles, Techniques, and Tools. Addison-Wesley, Reading, MA (1986)
7. International Organization for Standardization: Fortran standard (1997–2000) ISO/IEC 1539, Parts 1–3.
8. Rugina, R., Rinard, M.: Symbolic bounds analysis of pointers, array indices, and accessed memory regions. In: Proceedings of the ACM SIGPLAN'00 Conference on Programming Language Design and Implementation, ACM (2000)
9. Faure, C., Papegay, Y.: *Odyssée* user's guide. Version 1.7. Technical Report 0224, INRIA (1998)
10. Tapenade: `http://www-sop.inria.fr/tropics/`. (URL)

The Implementation and Testing of Time-Minimal and Resource-Optimal Parallel Reversal Schedules

Uwe Lehmann[1] and Andrea Walther[2]

[1] Center for High Performance Computing
Technical University Dresden, D–01062 Dresden, Germany
lehmann@zhr.tu-dresden.de
[2] Institute of Scientific Computing
Technical University Dresden, D–01062 Dresden, Germany
awalther@math.tu-dresden.de

Abstract. For computational purposes such as the computation of adjoint, applying the reverse mode of automatic differentiation, or debugging one may require the values computed during the evaluation of a function in reverse order. The naïve approach is to store all information needed for the reversal and to read this information backwards during the reversal. This technique leads to an enormous memory requirement, which is proportional to the computing time. The paper presents an approach to reducing the memory requirement without increasing the wall clock time by using parallel computers. During the parallel computation, only a fixed and small number of memory pads called checkpoints is stored. The data needed for the reversal is recomputed piecewise by starting the evaluation procedure from the checkpoints. We explain how this technique can be used on a parallel computer with distributed memory. Different implementation strategies will be shown, and some details with respect to resource-optimality are discussed.

1 Introduction

For many applications nonlinear vector functions have to be evaluated by a computer program. For several purposes, e.g. the calculation of adjoints, debugging or interactive control, one may need to reverse the program execution. Throughout this article we assume that the program execution to be reversed can be split into parts, called steps. Furthermore, it is presumed that all steps need the same computation time. Typical examples of such vector functions are iterations or time-dependent evolutions as occur in crash tests and meteorological or oceanographical simulations.

The computing resources needed for the reversal are computing time, memory and computing power, i.e. the number of processors used. The usual way to implement the reversal of a program execution is the following approach. One records a complete execution log, called trace, during the evaluation of the given function [1]. For each arithmetic operation, this execution log contains a code

P.M.A. Sloot et al. (Eds.): ICCS 2002, LNCS 2330, pp. 1049–1058, 2002.

and the addresses of the arguments. Subsequently, the execution log is read backwards to perform the reversal of the program evaluation. Obviously this basic method leads to a memory requirement that is proportional to the runtime of the program execution. Hence, for real-world applications with several thousand steps, the memory requirement can be too large to fit in any computer memory.

To avoid the enormous amount of memory required by the basic approach, checkpoint strategies were proposed and developed for one-processor machines [2, 3]. This checkpointing method naturally increases the computing time because instead of storing all the information required, some of the data is recomputed repeatedly. Using parallel computers, the increased runtime of the reversal can be traded for an increase in computing power [3]. Here, more than one processor is applied to perform the reversal. One important application of such a parallel reversal is given by real time computation, because in this case any increase in temporal complexity has to be avoided.

2   Theoretical Issues

2.1   Definitions and Assumptions

Suppose a given function $F : \mathbb{R}^n \longrightarrow \mathbb{R}^m$ with $x \longmapsto F(x)$ can be decomposed into l parts F_i with $0 \leq i \leq l - 1$, such that $F_i : \mathbb{R}^{n_i} \longrightarrow \mathbb{R}^{n_{i+1}}$ with $x_i \longmapsto x_{i+1} = F_i(x_i)$. These parts F_i are called advancing steps, or steps for short. The argument x_i and the result x_{i+1} define states where i or $i + 1$ is called the state counter. The function F can be viewed as an evolution comprising l steps F_i. Each x_i represents an intermediate state of the evolution process. Usually, these x_i are quite large vectors of constant size.

The goal is to reverse the function F starting at state x_l down to state x_0 as illustrated by Fig. 1. Performing the overall reversal, one would like to achieve an optimal usage of time and computer resources. For every advancing step F_i, a corresponding $\hat{F}_i$ exists. This so-called preparing step not only evaluates F_i but also assembles data needed for the later reversal during the computation of the advancing step F_i. The data is stored in a special data structure called the trace. The trace is needed by the sequential reversal function $\bar{F}_i : \mathbb{R}^{\bar{n}_{i+1}} \longrightarrow \mathbb{R}^{\bar{n}_i}$ with $\bar{x}_{i+1} \longmapsto \bar{x}_i = \bar{F}_i(\bar{x}_{i+1}, x_i)$.

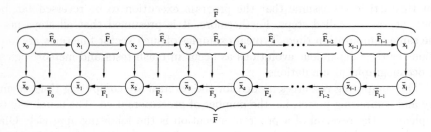

Fig. 1. The Reversal of the Function F

The naïve way to reverse F is to store all information one needs for the reversal onto a global trace. Thus the naïve reversing approach reads: Perform l preparing steps and after that perform l reversing steps as shown in Fig. 1. For real-world problems it may happen that the complete trace of F does not fit into memory. However, it is probably possible to keep a few intermediate states x_i at any time. The states kept in memory can be used as checkpoints in order to perform the desired reversal. In the remainder of this section, some basics of such checkpointing strategies are presented.

As in [4], it is assumed that the time t_i that one advancing step F_i takes is equal to a constant number $t \in \mathbb{R}$ that can be normalised to $t = 1$. Such evolutions F with $t_i = 1$ are said to have uniform step cost. Furthermore, let the value $\hat{t}_i$ be the time needed to perform one preparing step $\hat{F}_i$, and $\bar{t}_i$ the time needed to perform one reversing step $\bar{F}_i$. For simplicity it is assumed in this paper that $t_i = \hat{t}_i$, i.e. $\hat{t}_i = \hat{t} = 1$. If an evolution F has uniform step costs and if there exists a number $\bar{t} \in \mathbb{N}$, such that $\bar{t}_i = \bar{t}$ for all i, then the evolution is called a uniform evolution. Using a parallel schedule, the minimal time to reverse a uniform evolution that comprises l steps is given by

$$t_{\min} = l + l\bar{t} = (\bar{t} + 1)l \ . \tag{1}$$

Reversal schedules achieving this property are called time-minimal. Let $\varrho = c+p$ be the number of resources where c is the maximal number of checkpoints that can be stored at any time and p the maximal number of available processors. Then the maximal number of steps l_ϱ that can be reversed in minimal time with ϱ resources satisfies

$$l_\varrho = \begin{cases} \varrho & \text{if } \varrho \leq 2 \\ l_{\varrho-1} + \bar{t}\, l_{\varrho-2} & \text{else} \end{cases} . \tag{2}$$

The proof for this formula is given in [4]. When $\bar{t} = 1$, the formula (2) recursively yields the ϱ-th Fibonacci numbers. The direct non-recursive approximation

$$l_\varrho \sim \frac{1}{2}\left(1 + \frac{3}{\sqrt{1+4\bar{t}}}\right)\left(\frac{1}{2}(1 + \sqrt{1+4\bar{t}})\right)^{\varrho-1}$$

shows the exponential behaviour of the maximal number of steps l_ϱ that can be reversed for a given ϱ.

Vice versa, the formula (2) can be used to determine the minimal resources needed for the reversal of a given number of steps l. For example: Let the evolution comprise l steps with $l \in (f_\varrho, f_{\varrho+1}]$. Here, f_ϱ denotes the ϱ-th generalised Fibonacci number. Then according to formula (2), the number of required computer resources equals ϱ. For that purpose, it is assumed that each resource can perform an advancing, preparing or reversing step, or can act as a checkpoint. This property is called processor-checkpoint-convertibility in [3].

2.2   Building optimal parallel reversal schedules

For $l = 8$ steps, an optimal reversal schedule is displayed on the left of Fig. 2. The vertical axis in Fig. 2 can be thought of as the time or computational axis.

The horizontal axis denotes the state counter. All vertical solid lines denote checkpoints. The diagonal lines from the top left to the bottom right represent the execution of the step F_i if it is solid or $\hat{F}_i$ if dotted. Both functions compute the state x_{i+1} using the state x_i as an input value (horizontal axis) within one computational time step (vertical axis). The dashed lines represent the reverse steps $\bar{F}_i$ running from state x_{i+1} back to state x_i within one time unit. The three other diagrams show the resources needed. The value c^j denotes the checkpoints, p^j the processors, and s^j the computing resources, i.e. the sum of checkpoints and processors. This function is also called resource profile because it describes the resource requirement at any time of the reversal process.

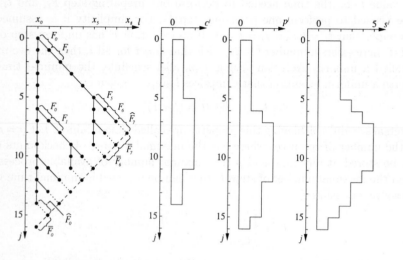

Fig. 2. An optimal schedule for $l = 8$ steps with $\bar{t} = \hat{t} = 1$

The recursive construction instruction of an optimal reversal schedule can be derived from the proof of (2) given in [4]. The first forward integration of the l steps is divided into parts corresponding to the Fibonacci-numbers (left hand diagram in Fig. 3). Hence, a schedule S^0 for l steps with $l \in (f_\varrho, f_{\varrho+1}]$ will be built from the schedule S^1 for $f_{\varrho-1}$ steps and from schedule S^2 for $l - f_{\varrho-1}$ steps as shown in Fig. 3. This is recursively done for all sub-schedules down to trivial single step reversal. By inspection of Fig. 4 we note that, the resources cannot be allocated statically to the various sub-schedule reversals, but must migrate between them during the execution of these subtasks. First the startup computation (dark grey area in the resource profile of Fig. 4) requires two resources. Then as the right and left sub-schedules commence being executed at $j = 5$ and $j = 6$, they both have increasing resource demands until the right sub-schedule begins to wind down at $j = 9$. It then passes resources to the left sub-schedule, which is still winding up. This behaviour is independent of the resource type (checkpoint, processor).

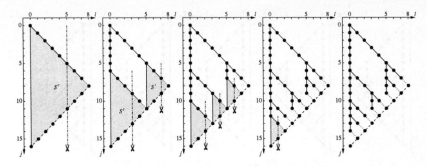

Fig. 3. Recursive construction of an optimal schedule for $l = 8$ steps and $\hat{t} = \bar{t} = 1$

3 Implementation Strategies

This section presents several possible strategies to implement reversal schedules on parallel computers. Furthermore, the strategy used for the numerical example introduced in Sect. 4 is discussed in detail.

3.1 Programming model

The programming model assumes a distributed memory. Here, each part of the memory is assigned to a fixed processor. Hence, one has to worry about how to transfer the data between the processors and how the transfer time influences the algorithm. The most commonly used tools are message passing libraries such as PVM or MPI. By implementing a reversal schedule using the distributed memory model, one can distinguish critical and non-critical communication. If a communication is classified as critical, the data written by one process at the end of one advancing, preparing or reversing step at a time j is needed for the beginning of the next advancing, preparing or reversing step by another process at the same time j. This is independent of the type of data required. If the communication is non-critical, the time between writing and reading of the data set is at least as long as the minimum of the time of an advancing, preparing or reversing step lasts. The data transfer can be carried out asynchronously. Additional temporary memory for the communication might also be needed for the distributed memory programming model since most message passing libraries can only send connected memory regions.

3.2 User provided routines

To use the provided reversal schedule program skeleton, the user has to write three major computing routines [5, 6]. First an advancing routine is needed, which will be referred to as forward(..). This routine has to compute for a given state i the next state $i + 1$. The second user routine does the same, but additionally stores all tracing data required for the reversing while it does the advancing.

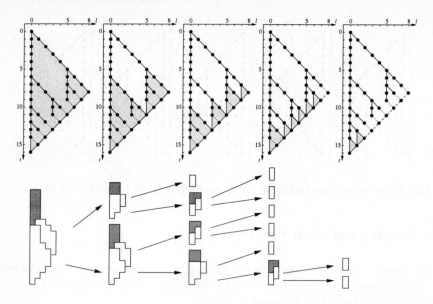

Fig. 4. Recursive construction of the resource profile for an optimal schedule for $l = 8$ steps and $\hat{t} = \bar{t} = 1$

Hence, it computes the state $i+1$ for a given state i and the trace for this computation. This routine is referred to as preparing(..). The third user function is reverse(..), which carries out the reversing step. The input for this routine is the trace for a computation from state i to state $i+1$. If reverse(..) is called for the first time, i.e. it is called at state l, it carries out the required initialisation of so-called adjoint values for the reversal computation. Otherwise, a second input argument is required, namely the values of the reversal computed so far. These values represent the reversal from state l down to state $i+1$.

3.3   Schedules for distributed memory programming models

There are two ways to implement a parallel reversal schedule in a distributed memory style. In both cases it will be assumed that a pool of available processors exists. For simplification $\bar{t}$ is assumed to be 1.

First, one may assign each checkpoint to a fixed processor. This approach, shown in Fig. 5(a), is called checkpoint oriented. A processor receives a checkpoint, stores it and at a predetermined time the processor carries out the advancing up to the state where the next checkpoint has to be written. Once it reaches the state, the processor writes the checkpoint and sends it to the next available processor. Then the processor goes back, i.e. waits at its assigned checkpoint, which it has received, until called upon to advance once more. The number of steps in each advance monotonically decreases until the checkpoint can be vacated after serving as starting point for the preparing step. When using this implementation, all communication is critical. Because of the idle time, this

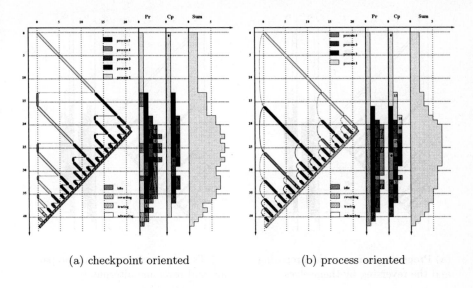

(a) checkpoint oriented   (b) process oriented

Fig. 5. General implementation strategies

scheme can not satisfy the optimal requirement profile [3] if more than three steps have to be reversed.

The second way to assign the processes to the given schedules is shown in Fig. 5(b). It is called process oriented. The computing process receives a checkpoint and carries out the advancing until it reaches its final state. The final state can be the state where the last advancing step ends and the preparing step starts, the state where the preparing step ends and the reversing step starts or the state where the reversing step ends. Along the way, the process writes, stores and sends the needed checkpoints. After the processors have finished the assigned computation they return back to the pool of available processors. Except at the final state, all communication is non-critical. The process oriented implementation fulfils the optimality if the following is true for all schedules and sub-schedules: The checkpoint written at state $i = 0$ is needed before the processor which does the advancing from state $i = 0$ up to state $i = l$ reaches the final state $i = l$. As already mentioned above, there are three ways to carry out the preparing and the reversing step, namely the processor

- stops before the trace has to be written, stores the data in a checkpoint and sends it to a special preparing process (Fig. 6(b));
- carries out the writing of the trace and sends it to a reversing process (Fig. 5(b));
- carries out the writing of the trace and reversing by itself (Fig. 6(a)) and sends the reversing result to the next reversing process.

In the first case, the task preparing and reversing can be assigned to fixed processes or might be combined to one task and must be carried out alternately by

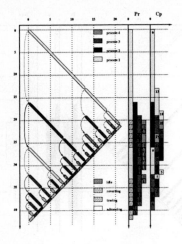

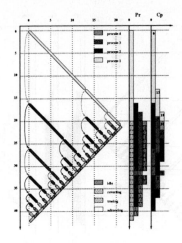

(a) Processes carry out the preparing and the reversing by themselves

(b) Two processes carry out preparing and reversing alternately

Fig. 6. General implementation strategies

two processes (Fig. 6(b)). The size of the data to be sent forms one criteria which implementation strategy should be applied. For large traces, an implementation where no trace segments are sent is preferable (Fig. 6(a) or Fig. 6(b)). If the result of the reversing step is relatively large compared to the size of the result of a preparing step, a fixed reversing processor reduces the communication cost (Fig. 5(b)) This invalidates the previous approach if both trace and adjoint are relatively large.

The decision about which approach is to be used, the checkpoint oriented or the process oriented, depends on the properties of the problem to be reversed. The checkpoint oriented approach might be easier to implement because the writing and sending times of a checkpoint are identical. The disadvantage is that more processors are needed than for an optimal schedule. On the other hand, using a process oriented approach for implementing an optimal reversal schedule, one processor has to store up to three checkpoints temporarily. This is caused by the fact that the sub-schedule S^1 starts before the sub-schedule S^2. The upper bound of three checkpoints is established in the following lemma.

Lemma 1. *Suppose the number of steps to be reversed is l with $l \in (f_\varrho, f_{\varrho+1}]$, where f_ϱ is the ϱ-th Fibonacci number. The schedule is implemented in a process oriented manner. Any process may store as many checkpoints as required, i.e. the checkpoints will be sent as late as possible. Then any processor has to store at most three checkpoints at any time.*

Proof. One only has to show that the process started first satisfies this property. For all other processes, one applies the claim to the sub-schedules. To prove that lemma one defines the number r by $r = l - f_{\varrho-1}$, hence $r \in (f_{\varrho-2}, f_\varrho]$. As shown in [3], the times $t_W(i)$, where the i-th checkpoint is written, are recursively defined by $t_W(1) = 0$, $t_W(2) = r$ and $t_W(i) = t_W(i-1) + f_{\varrho-2i+4}$. The maximal number of checkpoints written is limited above by $\lceil \frac{\varrho}{2} \rceil + 1$. The time when a checkpoint is needed/read is defined by $t_R(i) = t_W(i) + 2f_{\varrho-2i+1}$ [3]. Hence, the lemma is proven if the inequality $t_R(i) < t_W(i+3)$ holds for any i. One has $t_W(i+3) = t_W(i) + f_{\varrho-2i-2} + f_{\varrho-2i} + f_{\varrho-2i+2}$. Furthermore, one obtains

$$
\begin{aligned}
t_R(i) &< t_W(i+3) \\
\Longleftrightarrow \quad t_W(i) + 2f_{\varrho-2i+1} &< t_W(i) + f_{\varrho-2i-2} + f_{\varrho-2i} + f_{\varrho-2i+2} \\
\Longleftrightarrow \quad 0 &< 2f_{\varrho-2i-2}
\end{aligned}
$$

which is true for all $i \leq \lceil \frac{\varrho}{2} \rceil + 1$. $\qquad\square$

4 Example

The implementation of the reversal schedules was tested with the simulation of a simple Formula 1 racing car model. As described in [6], the parallel reversal schedule was used to compute the adjoint of the forward integration of an ODE system with respect to a given road shape. The forward integration was carried out using a four-stage Runge-Kutta scheme and for the adjoint calculation its adjoint Runge-Kutta scheme was used [7]. The computations were measured on a Cray T3E and on a SGI Origin3800. A parallel reversal schedule for $l = 55$ steps was and hence five processors were utilised. A sixth processor (master) was used to organise the program run. In Table 1, the memory requirements and the runtimes are listed for naïve approach and the parallel approach. Compared to the naïve approach, only two percent of the initial memory was needed using the parallel schedules. As the theory has shown, the runtime stays almost the same in all test cases. The slight improvement in computing time may be due to less memory traffic [8].

Table 1. Memory requirement and Runtime

Approach	Values	Memory	Time (T3E)	Time (Origin3800)
Naïve	266010	2128.10 kB $\hat{=}$ 100.0 %	20.27 s $\hat{=}$ 100.0 %	6.71 s $\hat{=}$ 100.0 %
Parallel	5092	40.70 kB $\hat{=}$ 1.9 %	18.91 s $\hat{=}$ 93.3 %	6.04 s $\hat{=}$ 90.0 %

5 Discussions, Conclusions and Further Work

The reversal of evaluation programs may cause problems due to the memory requirement being proportional to the computation time. A theory to reduce the memory requirement and time-minimal parallel reversal schedules were presented. Various possibilities for implementing these parallel reversal schedules were discussed. Thereby, some pitfalls which may cause the loss of resource optimality were shown. Furthermore, we discussed the problem characteristics that determine the choice of the implementation strategy.

As shown by the numerical example, the parallelisation of the reversal process was carried out in time. However, in addition the function may also be parallelised in space. The parallel evaluation of the function has to be autonomous and independent of external influences. In order to use a parallel computer efficiently, the parallelisation in space has to distribute the computational work dynamically between the processors. This is necessary because the number of available processors for the function evaluation will change during the reversal. Thus a repartitioning of the problem may occur, which is generally difficult. Therefore the implementation effort for such two level parallelisation is significant larger.

The advantage of using a distributed memory programming model is that one may apply a shared memory parallelisation within the user functions for the parallelisation in space. Again, the user has to take care while implementing the user functions. The reason for this is that most of the commonly used message passing libraries (MPI, PVM) are not thread-safe by definition. Hence, the parallel execution of shared memory parallelised code can start after the function was called and must end when leaving this function. Possibly improved message passing libraries will remove such a restriction in future.

References

1. van de Snepscheut, J.: What computing is all about. Texts and Monographs in Computer Science. Springer Verlag, Berlin (1993)
2. Griewank, A.: Achieving logarithmic growth of temporal and spatial complexity in reverse automatic differentiation. Optimisation Methods and Software 1 (1992) 35–54
3. Walther, A.: Program Reversal Schedules for Single- and Multi-processor Machines. PhD thesis, TU Dresden, Fakultät für Mathematik und Naturwissenschaften (1999)
4. Walther, A.: An upper bound on the number of time steps to be reversed in parallel. Technical Report IOKOMO–03–2001, Technische Universität Dresden (2001)
5. Benary, J.: DAP – Dresdener Adjugierten Parallelisierungsprojekt. Technical Report IOKOMO–05–1995, Technische Universität Dresden (1995)
6. Walther, A., Lehmann, U.: Adjoint calculation using time-minimal program reversals for multiprocessor machines. Technical Report IOKOMO–09–2001, Technische Universität Dresden (2001)
7. Hager, W.: Runge-Kutta methods in optimal control and the transformed adjoint system. Numer. Math. 87 (2000) 247–282
8. Seidl, S., Nagel, W.E., Brunst, H.: The future of HPC at SGI: Early experiences with SGI SN-1. In Jesson, B.J., ed.: Proceedings of Sixth European SGI/Cray Workshop, Manchester (2000)

Automatic Differentiation for Nonlinear Controller Design

Klaus Röbenack

Institut für Regelungs- und Steuerungstheorie
TU Dresden, Mommsenstr. 13, D–01062 Dresden, Germany
klaus@roebenack.de
http://www.roebenack.de

Abstract. Several new algorithms for nonlinear controller design are based on differential-geometric concepts. Up to now, the feedback of the controller has been computed symbolically. The author proposes a method to compute the feedback using automatic differentiation. With this approach, time-consuming symbolic computation can be avoided.

1 Introduction

During the last decades, several methods for nonlinear controller design based on differential geometry have been developed [1–3]. One import approach is the exact linearization via feedback [2]. This method has been successfully applied to a large number of real-world systems [4–6]. Up to now, the derivatives required by these algorithms have been computed symbolically using computer algebra packages [7–9]. The use of these methods is limited due to a burden of symbolic computations involved [10]. These disadvantages can be circumvented with automatic differentiation. Moreover, the methods presented here can be applied to algorithmic descriptions of the plant to be controlled. This is especially useful since complicated systems are constructed using either traditional programming languages, e.g., C++ in fast circuit simulation [11], or modeling languages such as MODELICA [12, 13] or VHDL-AMS [14, 15].

In Sect. 2 some facts of computing Taylor coefficients by automatic differentiation are given. In Sect. 3, we review controller design for nonlinear state-space systems. The use of automatic differentiation is described in Sect. 4. The design of so-called tracking controllers is addressed in Sect. 5.

2 Univariate Taylor Series

Consider a smooth map $\mathbf{F} : \mathbb{R}^n \to \mathbb{R}^m$ which maps a curve

$$\mathbf{x}(t) = \mathbf{x}_0 + \mathbf{x}_1 t + \cdots + \mathbf{x}_d t^d + O(t^{d+1}) \tag{1}$$

of the vector space $\mathbb{R}^n$ into a curve

$$\mathbf{z}(t) = \mathbf{F}(\mathbf{x}(t)) = \mathbf{z}_0 + \mathbf{z}_1 t + \cdots + \mathbf{z}_d t^d + O(t^{d+1}) \tag{2}$$

P.M.A. Sloot et al. (Eds.): ICCS 2002, LNCS 2330, pp. 1059–1068, 2002.

of $\mathbb{R}^m$ with

$$z_j = \frac{1}{j!} \frac{\partial^j z(t)}{\partial t^j}\bigg|_{t=0} .$$

Each Taylor coefficient $z_j \in \mathbb{R}^m$ is uniquely determined by the coefficients $x_0, \ldots, x_j \in \mathbb{R}^n$. In particular, we have

$$\begin{aligned}
z_0 &= F(x_0) \ , \\
z_1 &= F'(x_0)x_1 \ , \\
z_2 &= F'(x_0)x_2 + \tfrac{1}{2}F''(x_0)x_1x_1 \ , \\
z_3 &= F'(x_0)x_3 + F''(x_0)x_1x_2 + \tfrac{1}{6}F'''(x_0)x_1x_1x_1 \ .
\end{aligned}$$

Using automatic differentiation, the Taylor coefficients $z_0, \ldots, z_d$ can be obtained without symbolic computations of the high-dimensional derivative tensors $F' \in \mathbb{R}^{m \times n^2}$, $F'' \in \mathbb{R}^{m \times n^3}$, $\ldots$, $F^{(d)} \in \mathbb{R}^{m \times n^d}$. The tools ADOL-C [16] and TADIFF [17] use the forward mode to compute this Taylor coefficients. For many functions such as exp, sin, cos, arctan, tanh, the number of operations to be performed is bounded by the square of the degree, d, of the Taylor coefficients, i.e., $\text{OPS}\{z_0, \ldots, z_d\} \lesssim d^2 \cdot \text{OPS}\{F(x_0)\}$, see [18, Sect. 10.2].

3   Exact Linearization via Feedback

In this section we consider the design of a controller for nonlinear input-affine state-space systems

$$\dot{x} = f(x) + g(x)u, \quad y = h(x) \ . \tag{3}$$

The maps $f, g : \mathbb{R}^n \to \mathbb{R}^n$ and $h : \mathbb{R}^n \to \mathbb{R}$ are assumed to be sufficiently smooth. We need *Lie derivatives* $L_f h(x) = \frac{\partial h}{\partial x} f(x)$ of h along f, see [2]. Higher order Lie derivatives $L_f^k h(x)$ are defined by

$$L_f^k h(x) = \frac{\partial L_f^{k-1} h(x)}{\partial x} f(x) \quad \text{with} \quad L_f^0 h(x) = h(x) \ . \tag{4}$$

Similarly, the mixed Lie derivative $L_g L_f^k h(x)$ is given by

$$L_g L_f^k h(x) = \frac{\partial L_f^k h(x)}{\partial x} g(x) \ . \tag{5}$$

First, let us consider the case of an autonomous system

$$\dot{x} = f(x), \quad y = h(x) \ . \tag{6}$$

The time derivatives of the output can be written as

$$\dot{y} = \frac{\partial h}{\partial x} f(x) = L_f h(x), \quad \ddot{y} = \frac{\partial L_f h}{\partial x} f(x) = L_f^2 h(x), \quad \ldots, y^{(k)} = L_f^k h(x) \ .$$

For an initial value $\mathbf{x}_0 = \mathbf{x}(0)$, the output signal y is expressed by the series expansion

$$y(t) = \sum_{i=0}^{\infty} L_{\mathbf{f}}^i h(\mathbf{x}_0) \frac{t^i}{i!} \ ; \tag{7}$$

see [2, p. 140] for details.

Now, we will take the input u into consideration. According to [2], the system (3) is said to have *relative degree* r at $\mathbf{x}_0 \in \mathbb{R}^n$ if $L_{\mathbf{g}} L_{\mathbf{f}}^k h(\mathbf{x}) = 0$ for all $\mathbf{x}$ in a neighborhood of $\mathbf{x}_0$ and all $k \in \{0, \ldots, r-2\}$, and $L_{\mathbf{g}} L_{\mathbf{f}}^{r-1} h(\mathbf{x}_0) \neq 0$. Then, the time derivatives of the output of (3) are

$$y = L_{\mathbf{f}}^0 h(\mathbf{x}), \ \ldots, \ y^{(r-1)} = L_{\mathbf{f}}^{r-1} h(\mathbf{x}), \ y^{(r)} = L_{\mathbf{f}}^r h(\mathbf{x}) + L_{\mathbf{g}} L_{\mathbf{f}}^{r-1} h(\mathbf{x}) \ .$$

A case of particular interest is $r = n$. If the map Φ defined by $\xi_i = \phi_i(\mathbf{x}) = L_{\mathbf{f}}^{i-1} h(\mathbf{x})$ for $i = 1, \ldots, n$ is a diffeomorphism, the change of coordinates $\boldsymbol{\xi} = \Phi(\mathbf{x})$ with $\boldsymbol{\xi} = (\xi_1, \ldots, \xi_n)^T$ transforms (3) into the prime form [19]

$$\dot{\xi}_1 = \xi_2, \ \ldots, \ \dot{\xi}_{n-1} = \xi_n, \ \dot{\xi}_n = L_{\mathbf{f}}^n h(\mathbf{x}) + L_{\mathbf{g}} L_{\mathbf{f}}^{n-1} h(\mathbf{x}) \, u, \ \ y = \xi_1 \ . \tag{8}$$

If we use the state feedback

$$u = \frac{1}{L_{\mathbf{g}} L_{\mathbf{f}}^{n-1} h(\mathbf{x})} \left[-L_{\mathbf{f}}^n h(\mathbf{x}) - \mathbf{k}^\top \boldsymbol{\xi} + v \right] \tag{9}$$

with $\mathbf{k} = (k_0, \ldots, k_{n-1})^T$ and the new input v, one obtains a linear time-invariant system

$$\dot{\xi}_1 = \xi_2, \ \ldots, \ \dot{\xi}_{n-1} = \xi_n, \ \dot{\xi}_n = -\sum_{i=1}^{n} k_{i-1} \xi_i + v, \ \ y = \xi_1 \tag{10}$$

in the new coordinates. The dynamical behavior of (10) is prescribed by its characteristic polynomial $\rho(\lambda) = k_0 + k_1 \lambda + k_2 \lambda^2 + \cdots + k_{n-1} \lambda^{n-1} + \lambda^n$. We can apply methods of linear control theory [20] for the choice of the coefficients k_i. Fig. 1 shows the control scheme.

In general, the relative degree r may be less than n. In terms of the original coordinates, the feedback (9) is given by

$$u = -\frac{1}{L_{\mathbf{g}} L_{\mathbf{f}}^{r-1} h(\mathbf{x})} \left[k_0 L_{\mathbf{f}}^0 h(\mathbf{x}) + \cdots + k_{r-1} L_{\mathbf{f}}^{r-1} h(\mathbf{x}) + L_{\mathbf{f}}^r h(\mathbf{x}) - v \right] \ . \tag{11}$$

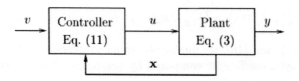

Fig. 1. Control scheme for exact linearization via feedback

4   Computation of Lie Derivatives

A symbolic computation of the Lie derivatives needed in (11) may result in extremely large expressions [10]. In fact, the size of the expressions may increase exponentially in the order of the Lie derivatives. Using automatic differentiation, the computational effort to compute the function values of Lie derivatives increases at most quadratically.

Let r denote the relative degree of (3). First, we will compute a series expansion of the autonomous system (6). We consider the expansion (1) of the solution of the ODE and a map $\mathbf{z} = \mathbf{f}(\mathbf{x})$ with the series expansion of $\mathbf{z}$ such as in (2). Since $\mathbf{x}$ is assumed to be the solution of (6), we have $\mathbf{z} \equiv \dot{\mathbf{x}}$ and $\mathbf{x}_{i+1} = \mathbf{z}_i/(i+1)$ for $i \geq 0$. Starting with an initial value $\mathbf{x}_0 \in \mathbb{R}^n$, we compute the function value $\mathbf{z}_0$ and obtain $\mathbf{x}_1 = \mathbf{z}_0$. Using $\mathbf{x}_0$ and $\mathbf{x}_1$, the Taylor coefficient $\mathbf{z}_1$ can be obtained by a forward sweep of automatic differentiation. We repeat this procedure until we get the Taylor coefficients $\mathbf{x}_0, \ldots, \mathbf{x}_r$ of the solution of (6); see [21] for details. The application of the forward mode to h yields the Taylor coefficients $y_0, \ldots, y_r$ of the associated output signal

$$y(t) = y_0 + y_1 t + y_2 t^2 + \cdots + y_r t^r + O(t^{r+1}) \ . \tag{12}$$

Comparing (7) and (12), we can express the Lie derivatives $L_{\mathbf{f}}^i h(\mathbf{x})$ in terms of Taylor coefficients calculated with automatic differentiation [22]:

$$L_{\mathbf{f}}^i h(\mathbf{x}) = i!\, y_i \quad \text{for} \quad i = 0, \ldots, r \ . \tag{13}$$

Similarly, one can compute the Taylor coefficients of the output of a second autonomous system

$$\dot{\mathbf{x}} = \mathbf{f}(\mathbf{x}) + \mathbf{g}(\mathbf{x}), \quad y = h(\mathbf{x}) \tag{14}$$

for $\mathbf{x}(0) = \mathbf{x}_0$ as well as the associated Lie derivatives $L_{\mathbf{f}+\mathbf{g}}^i h(\mathbf{x})$. Let $\bar{y}_i$ and $\tilde{y}_i$ denote the Taylor coefficients of the output of (6) and (14), respectively. Since the system (3) is assumed to have relative degree r, we have

$$L_{\mathbf{f}}^i h(\mathbf{x}) = i!\, \bar{y}_i = i!\, \tilde{y}_i \quad \text{for} \quad i = 0, \ldots, r-1 \ . \tag{15}$$

The mixed Lie derivative (5) can be obtained from

$$L_{\mathbf{g}} L_{\mathbf{f}}^{r-1} h(\mathbf{x}_0) = L_{\mathbf{f}+\mathbf{g}}^r h(\mathbf{x}_0) - L_{\mathbf{f}}^r h(\mathbf{x}_0) = r!\, (\tilde{y}_r - \bar{y}_r) \ . \tag{16}$$

Using (13), (15), and (16), we finally obtain an expression of the feedback (11) in terms of Taylor coefficients computed by automatic differentiation:

$$u = \frac{1}{r!\,(\bar{y}_r - \tilde{y}_r)} \left[0!\, k_0 \bar{y}_0 + 1!\, k_1 \bar{y}_1 + \cdots + (r-1)!\, k_{r-1} \bar{y}_{r-1} + r!\, \bar{y}_r - v \right] \ . \tag{17}$$

The methods derived here will be applied to two examples. Example 1 demonstrates the basic concepts. Actual performance data is presented in Example 2.

Example 1. Consider the state-space system

$$
\dot{\mathbf{x}} = \begin{pmatrix} x_1 x_2 - x_1^3 \\ x_1 \\ -x_3 \\ x_1^2 + x_2 \end{pmatrix} + \begin{pmatrix} 0 \\ 2 + 2x_3 \\ 1 \\ 0 \end{pmatrix} u, \quad y = x_4 \tag{18}
$$

taken from [2, Ex. 4.1.5, 4.4.1]. Since

$$
y = x_4
$$
$$
\dot{y} = \dot{x}_4 = x_1^2 + x_2
$$
$$
\ddot{y} = 2x_1 \dot{x}_1 + \dot{x}_2
$$
$$
= \underbrace{2x_1(x_1 x_2 - x_1^3) + x_1}_{L_f^2 h(\mathbf{x})} + \underbrace{(2 + 2x_3)}_{L_g L_f h(x)} u
$$

system (18) has relative degree $r = 2$ everywhere except at $x_3 = -1$. Table 1 shows the Lie derivatives and the Taylor coefficients computed at $\mathbf{x} = (1, 2, 3, 4)^T$ with ADOL-C. The coefficients $\mathbf{x}_i$ have been calculated with `forode`. To get the Taylor coefficients of the output we used the C++ function `forward`. From the entry in the last row and last column we obtain the mixed Lie derivative $L_g L_f\, h(\mathbf{x}) = 8$.

Table 1. Lie derivatives associated with Example 1

i	$L_f^i\, h(\mathbf{x})$	$i!\,\bar{y}_i$	$i!\,\tilde{y}_i$	$i!\,(\tilde{y}_i - \bar{y}_i)$
0	x_4	4	4	0
1	$x_1^2 + x_2$	3	3	0
2	$2x_1^2 x_2 - 2x_1^4 + x_1$	3	11	8

To stabilize the example system (18), we used the feedback (17) with $v \equiv 0$. We have chosen the coefficients $k_0 = 1$ and $k_1 = 2$, i.e., the characteristic equation $\rho(\lambda) = k_0 + k_1\lambda + \lambda^2 = 0$ has the roots $\lambda_1 = \lambda_2 = -1$. The trajectories of the closed-loop system, i.e., system (3) with the control (17), for the initial value $\mathbf{x}_0 = \mathbf{x}(0) = (1, 2, 3, 4)^T$ are shown in Fig. 2. We used bold lines for the states $x_1, \ldots, x_4$, and a dashed line for the control u. Note that the trajectories converge to the origin.

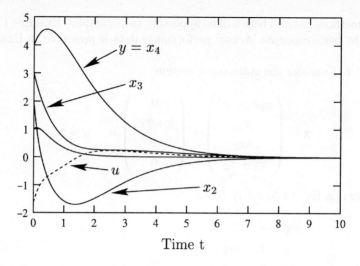

Fig. 2. Trajectories of the closed-loop example system for $0 \leq t \leq 10$

Example 2. The structure of the Translational Oscillator with Rotating Actuator (TORA) benchmark system [23, 24] is sketched in Fig. 3. The system consists of a platform of mass M that can oscillate without damping in the horizontal plan. The platform is connected to a fixed frame by a linear spring with spring constant k. A rotating eccentric mass m on the platform is actuated by a motor. Let x_1 and x_2 denote the (normalized) displacement and the velocity of the platform from the equilibrium position, respectively. Moreover, let x_3 and x_4 be the angle and the angular velocity of the rotor. In these coordinates, the state-space equations (3) are

$$
\begin{aligned}
\dot{x}_1 &= x_1 \\
\dot{x}_2 &= \frac{-x_1 + \varepsilon x_4^2 \sin x_3}{1 - \varepsilon^2 \cos^2 x_3} + \frac{-\varepsilon \cos x_3}{1 - \varepsilon^2 \cos^2 x_3} u \\
\dot{x}_3 &= x_4 \\
\dot{x}_4 &= \frac{\varepsilon \cos x_3 \left(x_1 - \varepsilon x_4^2 \sin x_3 \right) + u}{1 - \varepsilon^2 \cos^2 x_3} \\
y &= x_1 \; ,
\end{aligned}
\tag{19}
$$

where u is the control torque applied to the rotor and y is the output. The parameter ε depends on the physical parameters mentioned above. We used the value $\varepsilon = 0.1$.

We will compare the evaluation time of symbolically computed Lie derivatives $L_{\mathbf{f}}^k h(\mathbf{x}_0)$ for $k = 1, \ldots, 10$ with the time required to obtain these derivatives using automatic differentiation. All tests were performed on a 700 MHz Athlon PC under SuSE Linux 7.2. We used the GNU C++ compiler gcc 2.95.3.

The symbolic computation was carried out with the computer algebra package MuPAD 2.0 [25]. The Lie derivatives have been computed using (4). For

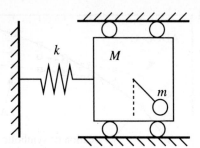

Fig. 3. TORA system configuration

example, the first Lie derivatives are

$$L_\mathbf{f} h(\mathbf{x}) = x_2 \quad \text{and} \quad L_\mathbf{f}^2 h(\mathbf{x}) = \frac{-x_1 + \varepsilon x_4^2 \sin x_3}{1 - \varepsilon^2 \cos^2 x_3} .$$

Subsequently, the results were converted to C code with the MUPAD command generate::C. Depending on the order k of the Lie derivatives, the size of the resulting C source code increases significantly, see Table 2. As shown in Fig. 4, the time to evaluate these expressions after compilation increases drastically with respect to the order of the derivatives involved.

Table 2. Size of the symbolically generated C source code

Order k of $L_\mathbf{f}^k h(\mathbf{x})$	1	2	3	4	5	6	7	8	9	10
Size in bytes	13	73	295	852	2078	4420	8440	15018	25507	40600

Moreover, we computed the function values $L_\mathbf{f}^k h(\mathbf{x}_0)$ by means of automatic differentiation as discussed in Sect. 4, i.e., the coefficients of the series expansion (7) have been computed with Taylor arithmetic. In our example, Fig. 4 shows that ADOL-C is faster than the symbolically generated code for $k \geq 6$, whereas TADIFF is faster for $k \geq 5$. For $k = 10$, ADOL-C needs only 20 percent and TADIFF only 8.6 percent of the CPU time required to evaluate the symbolically generated Lie derivative $L_\mathbf{f}^{10} h(\mathbf{x}_0)$. In this application, TADIFF is approximately twice as fast as ADOL-C; but note that ADOL-C offers more functionality.

Nevertheless, the example system (19) is still very simple. For slightly more complicated example systems [26] the break even point between symbolic and automatic differentiation lies around $k = 2$ or $k = 3$. Full advantages of automatic differentiation can only be exploited for systems with more complicated right hand sides.

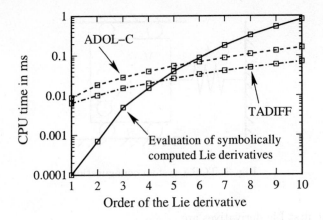

Fig. 4. CPU time needed for evaluation of Lie derivatives

5   Asymptotic Output Tracking

In this section we will discuss the use of automatic differentiation for tracking control. The aim is to design a controller such that the system's output y converges asymptotically to a prescribed reference trajectory y_{ref}. We want to computed the feedback of an output tracking controller by means of automatic differentiation. From a control-theoretic point of view, the problem is addressed in [2, Sect. 4.5].

Imposing the input

$$
\begin{aligned}
u &= \frac{1}{L_{\mathbf{g}} L_{\mathbf{f}}^{r-1} h(\mathbf{x})} \left[-\sum_{i=0}^{r-1} k_i \left(L_{\mathbf{f}}^i h(\mathbf{x}) - y_{\text{ref}}^{(i)} - L_{\mathbf{f}} h(\mathbf{x})^r + y_{\text{ref}}^r \right) \right] \\
&= \frac{1}{L_{\mathbf{g}} L_{\mathbf{f}}^{r-1} h(\mathbf{x})} \left[k_0 (y_{\text{ref}} - y) + \cdots + k_{r-1} (y_{\text{ref}}^{(r-1)} - y^{(r-1)}) + (y_{\text{ref}}^{(r)} - y^{(r)}) \right] ,
\end{aligned}
\tag{20}
$$

the error function $e(t) = y(t) - y_{\text{ref}}(t)$ satisfies the linear differential equation

$$
k_0 e + k_1 e^{(1)} + \cdots + k_{r-1} e^{(r-1)} + e^{(r)} = 0 . \tag{21}
$$

If the coefficients $k_0, \ldots, k_{r-1}$ are chosen in such a way that all roots of the associated characteristic polynomial are in the open complex left half plane, the error converges to zero, i.e., $e(t) \to 0$ for $t \to \infty$. This means that the real output converges to the reference output, i.e., $y(t) \to y_{\text{ref}}(t)$ for $t \to \infty$.

Again, the controller feedback (20) can very efficiently be computed using automatic differentiation. Let $\bar{y}_i$ and $\tilde{y}_i$ denote the ith Taylor coefficient of (6) and (14), respectively, computed as described in Sect. 4. The Taylor coefficients $\breve{y}_0, \ldots, \breve{y}_r$ of $y_{\text{ref}}(\cdot)$ can be obtained directly using Taylor arithmetic (see Sect. 2).

In terms of these Taylor coefficients, the feedback (20) can be written as

$$
\begin{aligned}
u &= \frac{1}{r!\,(\bar{y}_r - \breve{y}_r)} \left[k_0(\bar{y}_0 - \breve{y}_0) + k_1(\bar{y}_1 - \breve{y}_1) + 2\,k_2(\bar{y}_2 - \breve{y}_2) + \cdots \right. \\
&\quad \left. + (r-1)!k_{r-1}(\bar{y}_{r-1} - \breve{y}_{r-1}) + r!\,(\bar{y}_r - \breve{y}_r) \right] \\
&= \frac{1}{r!\,(\bar{y}_r - \breve{y}_r)} \sum_{i=0}^{r} i!\,k_i\,(\bar{y}_i - \breve{y}_i)
\end{aligned}
\tag{22}
$$

with $k_r = 1$.

6 Conclusions

The controller design by exact linearization via nonlinear feedback has been considered. The author presented a method to compute the state feedback of the controller using automatic differentiation. The method is applicable to systems with smooth but complicated nonlinearities. One advantage is the fact, that the system to be controlled must not be given by an explicit symbolic formula but by an algorithm. An other advantage is that higher order derivatives can be computed very fast with automatic differentiation. Moreover, the method presented here is well-suited for rapid prototyping in system design.

References

1. Jakubczyk, B., Respondek, W., Tchoń, K., eds.: Geometric Theory of Nonlinear Control Systems. Wroclaw Technical University Press (1985)
2. Isidori, A.: Nonlinear Control Systems: An Introduction. 3rd edn. Springer (1995)
3. Nijmeijer, H., van der Schaft, A.J.: Nonlinear Dynamical Control Systems. Springer (1990)
4. Joo, S., Seo, J.H.: Design and analysis of the nonlinear feedback linearizing control for an electromagnetic suspension system. IEEE Trans. on Control Systems Technology 5 (1997) 135–144
5. Bolzern, P., DeSantis, R.M., Locatelli, A., Masciocchi, D.: Path-tracking for articulated vehicles with off-axle hitching. IEEE Trans. on Control Systems Technology 6 (1998) 515–523
6. Fujimoto, K., Sugie, T.: Freedom in coordinate transformation for exact linearization and its application to transient behaviour improvement. Automatica 37 (2001) 137–144
7. Rothfuss, R., Zeitz, M.: Einführung in die Analyse nichtlinearer Systeme. In Engell, S., ed.: Entwurf nichtlinearer Regelungen. Oldenbourg-Verlag (1995) 3–22
8. Kugi, A., Schlacher, K., Novaki, R. In: Symbolic Computation for the Analysis and Synthesis of Nonlinear Control Systems. Volume IV of Software for Electrical Engineering, Analysis and Design. WIT-Press (1999) 255–264
9. Kwatny, H.G., Blankenship, G.L.: Nonlinear Control and Analytical Mechanics: A Computational Approach. Birkhäuser (2000)
10. de Jager, B.: The use of symbolic computation in nonlinear control: is it viable? IEEE Trans. on Automatic Control AC-40 (1995) 84–89

11. Grimm, C., Meise, C., Oehler, P., Waldschmidt, K., Fey, F.: AnalogSL: A library for modeling analog power drivers with C++. Forum on Design Languages FDL'01, September 3–7, 2001, Lyon, France (2001)

12. Tummescheit, H.: Object-oriented modeling of physical systems with Modelica. Short Tutorial (2000)

13. Clauß, C., Schneider, A., Schwarz, P.: Objektorientierte Modellierung physikalischer Systeme, Teile 13 und 14. Automatisierungstechnik (2000)

14. Haase, J., Schwarz, P., Trappe, P., Vermeiren, W.: Erfahrungen mit VHDL-AMS bei der Simulation heterogener Systeme. In: 3. ITG/GI/GMM–Workshop "Methoden und Beschreibungssprachen zur Modellierung und Verifikation von Schaltungen und Systemen", February 28–29, 2000, Frankfurt, Germany. (2000) 167–175

15. Klein, W., Griewank, A., Walther, A.: Differentiation methods for industrial strength problems. In Corliss, G., Faure, C., Griewank, A., Hascoët, L., Naumann, U., eds.: Automatic Differentiation of Algorithms: From Simulation to Optimization. Springer (2002) (to appear).

16. Griewank, A., Juedes, D., Utke, J.: ADOL-C, a package for the automatic differentiation of algorithms written in C/C++. ACM Transactions on Mathematical Software 22 (1996) 131–167

17. Bendtsen, C., Stauning, O.: TADIFF, a flexible C++ package for automatic differentiation. Technical Report IMM–REP–1997–07, TU of Denmark, Dept. of Mathematical Modelling, Lungby (1997)

18. Griewank, A.: Evaluating Derivatives: Principles and Techniques of Algorithmic Differentiation. SIAM, Philadelphia (2000)

19. Marino, R., Respondek, W., van der Schaft, A.J.: Equivalence of nonlinear systems to input-output prime forms. SIAM J. Control and Optimization 32 (1994) 387–407

20. Kailath, T.: Linear Systems. Prentice-Hall (1980)

21. Griewank, A.: ODE solving via automatic differentiation and rational prediction. In Griffiths, D.F., Watson, G.A., eds.: Numerical Analysis 1995. Volume 344 of Pitman Research Notes in Mathematics Series. Addison-Wesley (1995)

22. Röbenack, K., Reinschke, K.J.: Reglerentwurf mit Hilfe des Automatischen Differenzierens. Automatisierungstechnik 48 (2000) 60–66

23. Wan, C.J., Bernstein, D.S., Coppola, V.T.: Global stabilization of the oscillating eccentric rotor. In: Proc. 33rd IEEE Conf. Decision and Contr. (1994) 4024–4029

24. Bupp, R.T., Wan, C.J., Cappola, V.T., Bernstein, D.S.: Design of a rotational actuator for global stabilization of translational motion. In: Proc. Symp. Active Contr. Vibration Noise, ASME Winter Mtg. (1994)

25. Oevel, W., Postel, F., Rüscher, G., Wehrmeier, S.: Das MuPAD Tutorium. Springer (1999) Deutsche Ausgabe.

26. Röbenack, K., Vogel, O.: Numerische Systeminversion. Automatisierungstechnik 48 (2000) 487–495

Computation of Sensitivity Information for Aircraft Design by Automatic Differentiation*

H. Martin Bücker, Bruno Lang, Arno Rasch, and Christian H. Bischof

Institute for Scientific Computing
Aachen University of Technology, D–52056 Aachen, Germany
{buecker, lang, rasch, bischof}@sc.rwth-aachen.de
http://www.sc.rwth-aachen.de

Abstract. Given a numerical simulation of the near wake of an airfoil, automatic differentiation is used to accurately compute the sensitivities of the Mach number with respect to the angle of attack. Such sensitivity information is crucial when integrating a pure simulation code into an optimization framework involving a gradient-based optimization technique. In this note, the ADIFOR system implementing the technology of automatic differentiation for functions written in Fortran 77 is used to mechanically transform a given flow solver called TFS into a new program capable of computing the original simulation and the desired derivatives in a simultaneous fashion. Numerical experiments of derivatives obtained from automatic differentiation and finite differences approximations are reported.

1 Introduction

The increasing aircraft traffic has led to a growing interest in maximizing take-off and landing frequencies. A detailed knowledge of the wake flow field is essential to estimate safe-separation distances between aircraft in take-off and landing. However, the complex flow field around an aircraft is still not completely understood and simulations are commonly used to advance the understanding of the underlying physical phenomena. At Aachen University of Technology, a team of engineers, mathematicians, and computer scientists is investigating the fluid-structure interaction at airplane wings to further advance scientific knowledge of the aerodynamics of cruise and high lift configurations. One of the projects aims at optimizing an airfoil with respect to certain design parameters. Traditionally, finding a suitable set of parameters is carried out by running the simulation code over and over again with perturbed inputs. However, this approach may consume enormous computing time and may also require an experienced user to select suitable sets of parameters just to achieve improvement, even without optimality. Here, numerical optimization techniques can help reduce the number

---

* This research is partially supported by the Deutsche Forschungsgemeinschaft (DFG) within SFB 401 "Modulation of flow and fluid–structure interaction at airplane wings," Aachen University of Technology, Germany.

P.M.A. Sloot et al. (Eds.): ICCS 2002, LNCS 2330, pp. 1069–1076, 2002.

of simulation runs, but in particular provide more goal-oriented computational support for a design engineer.

When embedding a pure simulation code in an optimization framework, a crucial ingredient to any gradient-based optimization algorithm is the derivative of the output of the simulation with respect to the set of design parameters. When the simulation is given in the form a high-level programming language such as Fortran, C, or C++, the derivatives can be computed by a technique called algorithmic or automatic differentiation (AD) [1]. Here, a given computer program is automatically transformed into another program capable of evaluating not only the original simulation but also derivatives of selected outputs with respect to selected inputs. In contrast to numerical differentiation, derivatives computed by AD are free of truncation error.

The aim of this note is to demonstrate the feasibility of accurate derivatives of a large-scale simulation in order to apply gradient-based optimization techniques. More precisely, AD is applied to a computational fluid dynamics code called TFS [2, 3] developed at the Aerodynamics Institute, Aachen University of Technology. This simulation code consists of 236 subroutines totaling approximately 24,000 lines of Fortran 77. Other references where automatic differentiation is applied to problems from computational fluid dynamics include [4–6]. The procedure for applying the ADIFOR [7] system implementing the AD technology is outlined in Sect. 3. The problems of a numerical differentiation approach based on finite differences are reported in Sect. 4. The discussion starts with a short review of the functionality of automatic differentiation in Sect. 2.

2   Automatic Differentiation

The term "Automatic Differentiation (AD)" comprises a set of techniques for automatically augmenting a given computer program with statements for the computation of derivatives. That is, given a computer program that implements a function

$$f(\mathbf{x}) = \big(f_1(\mathbf{x}), f_2(\mathbf{x}), \dots, f_m(\mathbf{x})\big)^T \in \mathbb{R}^m \ ,$$

automatic differentiation generates another program that, at any point of interest $\mathbf{x} \in \mathbb{R}^n$, not only evaluates f at a point $\mathbf{x}$ but additionally evaluates its Jacobian

$$J(\mathbf{x}) := \begin{pmatrix} \frac{\partial}{\partial x_1} f_1(\mathbf{x}) & \cdots & \frac{\partial}{\partial x_n} f_1(\mathbf{x}) \\ \vdots & \ddots & \vdots \\ \frac{\partial}{\partial x_1} f_m(\mathbf{x}) & \cdots & \frac{\partial}{\partial x_n} f_m(\mathbf{x}) \end{pmatrix} \in \mathbb{R}^{m \times n}$$

at the same point.

AD technology is applicable whenever derivatives of functions given in the form of a high-level programming language, such as Fortran, C, or C++, are required. The reader is referred to the recent book by Griewank [1] and the proceedings of AD workshops [8–10] for details on this technique. The key idea of automatic differentiation is that any program can be viewed as a—potentially very long—sequence of elementary operations such as addition or multiplication, for which the derivatives are known. Then the chain rule of differential calculus is applied over and over again, combining these step-wise derivatives to yield the derivatives of the whole program. This mechanical process can be automated, and several AD tools are available for transforming a given code into the new code that is called *differentiated code*. In this way, AD requires little human effort and produces derivatives that are accurate up to machine precision.

3   Applying the ADIFOR System to the TFS Code

At the Aerodynamics Institute, Aachen University of Technology, engineers are developing the TFS [2, 3] package for large-scale computational fluid dynamics. The simulation code consists of 236 subroutines totaling approximately 24,000 lines of Fortran 77. The TFS package is capable of simulating incompressible and compressible flows in two and three space dimensions on the basis of finite volume discretization on block-structured grids. For the present study, a version of TFS is taken to compute the two-dimensional flow field around a typical benchmark airfoil known as RAE 2822. A part of the grid of the underlying numerical simulation is depicted in Fig. 1 for the area close to the airfoil.

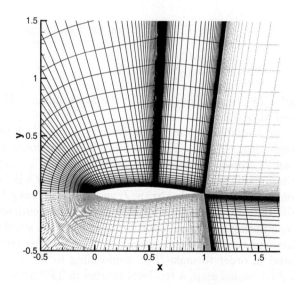

Fig. 1. Detail of a grid for airfoil RAE 2822 employing 4 blocks and 22356 nodes

In this note, we start from a given simulation of the Mach number, M, and the pressure, p, for a given angle of attack, α. That is, there is an implementation using TFS that, for any input α, computes the outputs M and p. From a

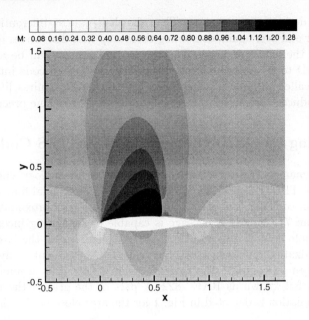

Fig. 2. The Mach number M of the flow field computed by TFS for $\alpha = 2.79°$

mathematical point of view, this TFS simulation evaluates some function

$$\begin{bmatrix} M(\alpha) \\ p(\alpha) \end{bmatrix} = f(\alpha) \ . \tag{1}$$

As an example of a TFS simulation, a plot of the Mach number M is depicted in Fig. 2 evaluated at $\alpha = 2.79°$.

Suppose that we are interested in the derivatives of the Mach number with respect to the angle of attack, i.e., $\partial M/\partial \alpha$. Since TFS is a computer program written in Fortran 77, these derivatives can be computed in a completely mechanical way by using any AD tool for Fortran 77. Notice that a list of available AD tools is currently being compiled at http://www.autodiff.org. In this note, the AD tool ADIFOR is applied to transform TFS into a differentiated version of TFS capable of computing $\partial M/\partial \alpha$ in addition to the original simulation of M.

As a preprocessing step, a few non-standard programming techniques are eliminated by hand in order to make TFS conforming to the Fortran 77 language standard. As a second step, a top-level routine in TFS implementing the function represented by (1) is identified and the program variables corresponding to M and α are indicated to the ADIFOR system. In our specific example, approximately 220 subroutines are fed into the ADIFOR system. ADIFOR recognizes about 100 subroutines as contributing to the derivatives and thus requiring additional code for the derivative computations. The code generated automatically by ADIFOR consists of approximately 20,000 lines of Fortran 77. Finally, a "driver" routine is implemented that initializes the derivative computations and

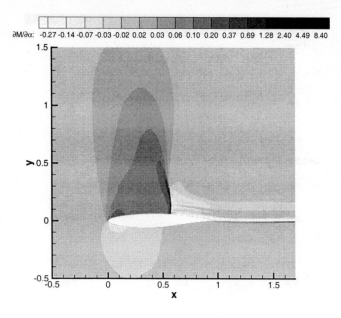

$\partial M/\partial \alpha$:   -0.27 -0.14 -0.07 -0.03 -0.02 0.02 0.03 0.06 0.10 0.20 0.37 0.69 1.28 2.40 4.49 8.40

Fig. 3. AD-generated derivatives $\partial M/\partial \alpha$ evaluated at $\alpha = 2.79^{\circ}$

invokes the differentiated top-level routine. This driver routine is then capable of computing the desired derivatives together with the original simulation. The AD-generated derivatives of the Mach number are depicted in Fig. 3. Here, the dark areas show the largest positive change of M with respect to changes in α. Note that the largest changes occur at the vertical shock and in the wake.

4   Problems with Divided Differences

Automatic differentiation is based on successively applying the chain rule to elementary operations leading to derivative values accurate up to machine precision. In contrast, a traditional alternative for the computation of derivatives is the numerical approximation by divided differences. For the sake of simplicity, we consider a first-order finite difference scheme where the derivative $\partial M/\partial \alpha$ of the Mach number $M(\alpha)$ is approximated by

$$\delta(h) := \frac{M(\alpha + h) - M(\alpha)}{h}$$

involving a step size h. Besides the truncation error, the crucial disadvantage of divided differences (DD) is the need to find a suitable step size.

For the TFS simulation, it turns out that an appropriate step size for the derivatives $\partial M/\partial \alpha$ is extremely hard to determine. When varying the step size h from 10^{-2} to 10^{-8}, the corresponding DD approximations $\delta(h)$ differ so significantly that a quantitative prediction for $\partial M/\partial \alpha$ using DD is not feasible. In

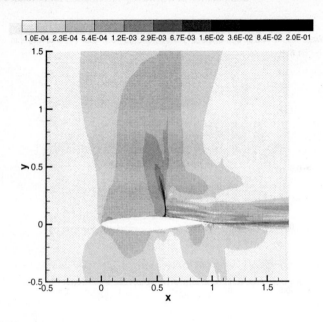

Fig. 4. The difference $|\delta(10^{-5}) - \delta(10^{-6})|/8.446$ at $\alpha = 2.79°$

Fig. 4, the difference of two DD approximations resulting from two different step sizes is given. More precisely, a normalized absolute difference of the DD approximations $\delta(10^{-5})$ and $\delta(10^{-6})$ is shown. The normalization is obtained from dividing the absolute difference by the largest absolute derivative value, 8.446, generated by AD. Recall from Fig. 3 that this scaling factor occurs in the region of the vertical shock.

The difference depicted in Fig. 4 is less than 10^{-4} in the outer region, meaning that, there, the two DD approximations $\delta(10^{-5})$ and $\delta(10^{-6})$ agree rather well. However, in the areas next to the vertical shock and the wake flow field the difference is large reaching a value around 0.2. Thus, the DD approximations of the derivatives are not accurate in these interesting areas.

5   Concluding Remarks

To use a large-scale flow field simulation within a Newton-type optimization framework, derivatives of the flow field with respect to simulation parameters are necessary. Automatic differentiation is a technique for transforming a given simulation code, written in a high-level programming language such as Fortran or C, into another computer program capable of computing not only the given simulation but also some user-specified derivatives. The technique is not only applicable to fairly small programs but scales to large computer codes such as computational fluid dynamics solvers. As opposed to numerical differentiation, automatic differentiation does not involve any truncation error. Therefore, automatic differentiation is currently the only option whenever exact derivatives of functions given in the form of complex computer programs are required.

In this note, the automatic differentiation tool ADIFOR is applied to the TFS package. This simulation package consists of approximately 24,000 lines of Fortran 77. The flow field around the RAE 2822 airfoil is computed using TFS. The derivative of the Mach number with respect to the angle of attack is accurately computed using automatic differentiation whereas an approximation of the same derivatives using a divided difference approach results in reliable derivative values only at regions far from the airfoil. However, in the areas of interest, for instance in the vicinity of the shock, the derivative values produced by divided differences vary significantly with the actual step size being used and, in any case, do not deliver reliable derivative values. At best, divided differences give a qualitative prediction of the derivative field, but they do not permit a detailed understanding of the underlying phenomena.

Work is in progress to further increase the computational efficiency of the code generated by automatic differentiation. Moreover, the integration of the differentiated TFS code into several gradient-based optimization algorithms is currently under investigation.

Acknowledgments

The authors would like to thank Jakob Risch for his notable contribution during the initial phase of this project and Emil Slusanschi for completing the automatic differentiation framework. We would also like to thank the Aerodynamics Institute, Aachen University of Technology, for making available the source code of the flow solver TFS. In particular, Matthias Meinke and Ehab Fares deserve special recognition for their help with the RAE 2822 airfoil. This research is partially supported by the Deutsche Forschungsgemeinschaft (DFG) within SFB 401 "Modulation of flow and fluid–structure interaction at airplane wings," Aachen University of Technology, Germany.

References

1. Griewank, A.: Evaluating Derivatives: Principles and Techniques of Algorithmic Differentiation. SIAM, Philadelphia (2000)
2. Fares, E., Meinke, M., Schröder, W.: Numerical simulation of the interaction of flap side-edge vortices and engine jets. In: Proceedings of the 22nd International Congress of Aeronautical Sciences, Harrogate, UK, August 27–September 1, 2000. ICAS 0212 (2000)
3. Fares, E., Meinke, M., Schröder, W.: Numerical simulation of the interaction of wingtip vortices and engine jets in the near field. In: Proceedings of the 38th Aerospace Sciences Meeting and Exhibit, Reno, NV, USA, January 10–13, 2000. AIAA Paper 2000–2222 (2000)

4. Bischof, C., Corliss, G., Green, L., Griewank, A., Haigler, K., Newman, P.: Automatic differentiation of advanced CFD codes for multidisciplinary design. Journal on Computing Systems in Engineering **3** (1992) 625–638
5. Bischof, C., Green, L., Haigler, K., Knauff, T.: Parallel calculation of sensitivity derivatives for aircraft design using automatic differentiation. In: Proceedings of the 5th AIAA/NASA/USAF/ISSMO Symposium on Multidisciplinary Analysis and Optimization, AIAA 94-4261, American Institute of Aeronautics and Astronautics (1994) 73–84
6. Aubert, P., Di Césaré, N., Pironneau, O.: Automatic differentiation in C++ using expression templates and application to a flow control problem. Computing and Visualization in Science **3** (2001) 197–208
7. Bischof, C., Carle, A., Khademi, P., Mauer, A.: ADIFOR 2.0: Automatic differentiation of Fortran 77 programs. IEEE Computational Science & Engineering **3** (1996) 18–32
8. Griewank, A., Corliss, G.: Automatic Differentiation of Algorithms. SIAM, Philadelphia (1991)
9. Berz, M., Bischof, C., Corliss, G., Griewank, A.: Computational Differentiation: Techniques, Applications, and Tools. SIAM, Philadelphia (1996)
10. Corliss, G., Faure, C., Griewank, A., Hascoët, L., Naumann, U., eds.: Automatic Differentiation of Algorithms: From Simulation to Optimization. Springer (2002) (to appear).

Performance Issues for Vertex Elimination Methods in Computing Jacobians using Automatic Differentiation

Mohamed Tadjouddine[1], Shaun A. Forth[1], John D. Pryce[2], and John K. Reid[3]

[1] Applied Mathematics & Operational Research, ESD
Cranfield University (RMCS Shrivenham), Swindon SN6 8LA, UK
{M.Tadjouddine, S.A.Forth}@rmcs.cranfield.ac.uk
[2] Computer Information Systems Engineering, DOIS
Cranfield University (RMCS Shrivenham), Swindon SN6 8LA, UK
J.D.Pryce@rmcs.cranfield.ac.uk
[3] JKR Associates, 24 Oxford Road, Benson, Oxon OX10 6LX, UK
jkr@rl.ac.uk

Abstract. In this paper, we present first results from EliAD, a new automatic differentiation tool. EliAD uses the **Eli**mination approach for **A**utomatic **D**ifferentation first advocated by Griewank and Reese [*Automatic Differentiation of Algorithms*, SIAM (1991), 126–135]. EliAD implements this technique via source-transformation, writing new Fortran code for the Jacobians of functions defined by existing Fortran code. Our results are from applying EliAD to the Roe flux routine commonly used in computational fluid dynamics. We show that we can produce code that calculates the associated flux Jacobian approaching or in excess of twice the speed of current state-of-the-art automatic differentiation tools. However, in order to do so we must take into account the architecture on which we are running our code. In particular, on processors that do not support *out-of-order execution*, we must reorder our derivative code so that values may be reused while in arithmetic registers in order that the floating point arithmetic pipeline may be kept full.

1 Introduction

In scientific computation, there is a frequent need to compute first derivatives of a function represented by a computer program. One way to achieve this is to use Automatic Differentiation (AD) [1–3], which allows for the computation of derivatives of a function represented by a computer program. The most efficient way to implement AD in terms of run-time speed is usually *source transformation*; here the original code is augmented by statements that calculate the needed derivatives. ADIFOR [4], Odyssée [5], and TAMC [6] are well-established tools for this which make use of the standard forward and reverse modes of AD.

We have developed a new AD tool EliAD [7,8] which also uses source transformation. EliAD is written in Java and uses a parsing front-end generated by the ANTLR tool [9]. In contrast to the AD tools listed above, EliAD uses the

P.M.A. Sloot et al. (Eds.): ICCS 2002, LNCS 2330, pp. 1077–1086, 2002.

vertex elimination approach of Griewank and Reese [2, 10], later generalised to edge and face eliminations by Naumann [11, 12]. Here, we consider only the vertex elimination approach, which typically needs less floating point operations to calculate a Jacobian than the traditional forward and reverse methods implemented by ADIFOR, Odyssée or TAMC. We introduce the vertex elimination approach in Sect. 2.

In Sect. 3 we present results of applying our tool to the Roe flux [13] computation which is a central part of many computational fluid dynamics codes. We used various ways to sequence the elimination. We ran the resulting derivative codes on various machine/compiler combinations and at various levels of compiler optimisation. We found that execution times were not always in proportion to the number of floating point operations and that this effect was very machine-dependent.

Section 4 discusses the performance of certain elimination strategies regarding such machine dependent issues as cache and register utilisation and then points out the importance of statement ordering on certain processors.

2   Automatic Differentiation by Elimination Techniques

AD relies on the use of the chain rule of calculus applied to elementary operations in an automated fashion. The input variables x with respect to which we need to compute derivatives are called *independent* variables. The output variables y whose derivatives are desired are called *dependent* variables. A variable which depends on an independent variable, and on which a dependent variable depends, is called an *intermediate* variable.

To illustrate the elimination approach, consider the code fragment comprising four scalar assignments in the left of

$$
\begin{aligned}
x_3 &= \phi_3(x_1, x_2) & \nabla x_3 &= c_{3,1}\nabla x_1 + c_{3,2}\nabla x_2 \\
x_4 &= \phi_4(x_2, x_3) & \nabla x_4 &= c_{4,2}\nabla x_2 + c_{4,3}\nabla x_3 \\
x_5 &= \phi_5(x_1, x_3) & \nabla x_5 &= c_{5,1}\nabla x_1 + c_{5,3}\nabla x_3 \\
y &= \phi_6(x_4, x_5) & \nabla y &= c_{6,4}\nabla x_4 + c_{6,5}\nabla x_5 \ .
\end{aligned}
\tag{1}
$$

Denoting

$$
\nabla x_i = \left(\frac{\partial x_i}{\partial x_1}, \frac{\partial x_i}{\partial x_2} \right) \text{ and } c_{i,j} = \frac{\partial \phi_i}{\partial x_j} \ ,
$$

we may use standard rules of calculus to write the linearised equations to the right of (1). These linearised equations describe the forward mode of AD and equivalent code would be produced by ADIFOR or TAMC enabling Jacobian calculation on setting $\nabla x_1 = (1, 0)$ and $\nabla x_2 = (0, 1)$. The Computational Graph (CG) for derivative calculation is sketched in Fig. 1. Vertices 1, 2 represent the independents x_1, x_2; vertices 3, 4, 5 the intermediate variables x_3, x_4, x_5; and vertex 6 the dependent y. The edges are labelled with the local derivatives of the equivalent code statements given to the right of (1) as shown. The corresponding

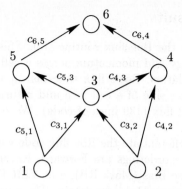

Fig. 1. An example computational graph

matrix representation gives a linear system, to solve for the derivatives, with zero entries in the matrix omitted for clarity:

$$
\begin{bmatrix}
-1 & & & & & \\
 & -1 & & & & \\
c_{3,1} & c_{3,2} & -1 & & & \\
 & c_{4,2} & c_{4,3} & -1 & & \\
c_{5,1} & & c_{5,3} & & -1 & \\
 & & & c_{6,4} & c_{6,5} & -1
\end{bmatrix}
\begin{bmatrix}
\nabla x_1 \\
\nabla x_2 \\
\nabla x_3 \\
\nabla x_4 \\
\nabla x_5 \\
\nabla y
\end{bmatrix}
=
\begin{bmatrix}
-1 & 0 \\
0 & -1 \\
0 & 0 \\
0 & 0 \\
0 & 0 \\
0 & 0
\end{bmatrix}.
$$

We may interpret the elimination approach via either the computational graph or the matrix representation. Via the graph the Jacobian $\frac{\partial y}{\partial(x_1, x_2)}$ is determined by eliminating intermediate vertices from the graph until it becomes bipartite, see [2, 10, 11]. For example vertex 5 may be eliminated by creating new edges $c_{6,1} = c_{6,5} \times c_{5,1}$, $c_{6,3} = c_{6,5} \times c_{5,3}$ and then deleting vertex 5 and all the adjacent edges. We might then eliminate vertex 4 then 3, termed a reverse ordering.

In terms of the matrix representation, vertex elimination is equivalent to choosing a diagonal pivot from rows 3 to 5, in some order. At each step we eliminate all the coefficients under that pivot. For example by choosing row 5 as the pivot row we add the multiple $c_{6,5}$ of row 5 to row 6 and so produce entries $c_{6,1} = c_{6,5} \times c_{5,1}$, $c_{6,3} = c_{6,5} \times c_{5,3}$ in row 6. We see that this is identical to the above computational graph description. By then choosing rows 4 and 3 as pivots we are left with the Jacobian in elements c_{61} and c_{62}.

There are as many vertex elimination sequences as there are permutations of the intermediate vertices. We choose a sequence using heuristics from sparse matrix technology aimed at reducing fill-in, such as the Markowitz criterion studied in [10, 11]. By reducing fill-in such techniques have the desired side-effect of choosing an ordering which minimises the number of floating point operations at each elimination step.

3   Numerical Results

A typical application is the Roe flux routine [13]. This computes the numerical fluxes of mass, energy and momentum across a cell face in a finite-volume compressible flow calculation. Roe's flux takes as input 2 vectors of length 5 describing the flow either side of a cell face and returns as output a length 5 vector for the numerical flux (139 lines of code). We seek the 5×10 Jacobian matrix.

We used EliAD to differentiate the Roe flux code with different vertex elimination orderings. Those orderings are Forward, Reverse, Markowitz (Mark), VLR, Markowitz Reverse Bias (Mark RB), and VLR Reverse Bias (VLR RB) eliminations; see for instance [11, 14] for details. The reverse-bias variants resolve ties when there are several nodes of the same cost by taking the one that appears last in the original statement order [14].

From the input code we generated derivative codes in one of the two following manners.

1. Statement level: local derivatives are computed for each statement, no matter how many variables x_i appear in its right-hand side.
2. Code list: local derivatives are computed for each statement in a rewritten code that has at most two variables x_i in each right-hand side.

Then, we applied the different vertex elimination strategies to the resulting computational graphs (62 intermediate vertices for the first case and 208 intermediates for the second case). We ran the subsequent derivative codes on the following platforms: Silicon Graphics (SGI), Compaq Alpha (ALP), and two SUN machines with different compilers denoted by SUN, NAG and FUJ. The various processor/compiler combinations are described in Table 1.

Table 1. Platforms (processors and compilers)

Label	Processor	CPU	L1-Cache	L2-Cache	Compiler	Options
SGI	R12000	300MHz	64KB	8MB	f90 MIPSPro 7.3	−Ofast
ALP	EV6	667MHz	128KB	8MB	Compaq f95 5.4	−O5
SUN	Ultra10	440MHz	32KB	2MB	Workshop f90 6.0	−fast
NAG	Ultra10	440MHz	32KB	2MB	Nagware f95 4.0a	−O3 −native
FUJ	Ultra1	143MHz	32KB	0.5MB	Fujitsu f90 5.0	−Kfast

Table 2 shows the CPU times required by the calculation of the original function on the different platforms.

Table 3 summarises the ratio between the timings of the Jacobian and original function by using the different methods and platforms. Methods starting with VE are those using a vertex elimination strategy from our AD tool. We write FD for one-sided finite differences, SL for statement level and CL for code list. For TAMC, the postfix ftl denotes the forward mode and ad denotes the reverse mode.

Table 2. Roe flux CPU timings (in μ seconds) of the function, $T(F)$, on different platforms

SGI	ALP	SUN	NAG	FUJ
0.8	0.5	0.9	1.8	6.2

Timings are based on 10,000 evaluations for each method and an average of 10 runs of the executables from the different machines we used. We have checked the accuracy by evaluating the largest difference from the corresponding derivative found by ADIFOR [4]. This was less than 10^{-14} in all cases except finite differences, where it was about 10^{-7}, in line with the truncation error associated with this approximation.

Table 3. Ratios of Jacobian to Function CPU timings, $T(\nabla F)/T(F)$, on various platforms, ratios of floating point operations, $r_{\text{flops}} = \text{FLOPs}(\nabla F)/\text{FLOPs}(F)$, and numbers of lines of code (*#loc*)

No.	Method	SGI	ALP	SUN	NAG	FUJ	r_{flops}	#loc
1	FD (1-sided)	12.1	12.6	12.6	13.3	11.9	11.4	176
2	VE Forward(SL)	7.5	5.5	13.2	12.2	10.2	8.4	1534
3	VE Reverse(SL)	6.5	4.6	8.8	8.8	6.9	6.9	1318
4	VE Mark(SL)	6.4	5.1	9.8	10.4	6.7	7.2	1267
5	VE Mark RB(SL)	6.6	5.0	14.1	10.2	6.6	7.2	1261
6	VE VLR(SL)	6.0	4.5	8.9	9.1	6.1	6.5	1172
7	VE VLR RB(SL)	6.2	4.6	9.0	9.6	6.0	6.5	1170
8	VE Forward(CL)	7.3	5.2	18.7	19.2	14.9	12.1	3175
9	VE Reverse(CL)	6.7	4.7	10.7	9.9	9.5	8.9	2433
10	VE Mark(CL)	6.6	5.0	9.3	11.7	11.2	7.9	2058
11	VE Mark RB(CL)	6.6	5.3	10.2	11.0	10.9	7.7	2012
12	VE VLR(CL)	6.4	5.0	10.1	11.1	10.9	7.8	2116
13	VE VLR RB(CL)	7.2	5.4	10.8	12.4	12.1	8.4	2145
14	ADIFOR	15.9	9.8	31.4	50.5	14.4	15.0	614
15	TAMC-ftl	14.4	10.2	11.9	56.7	14.9	19.9	639
16	TAMC-ad	12.2	8.5	13.2	42.0	10.0	11.9	919

Table 3 shows that on the SGI and ALP platforms the vertex elimination approach for computing Jacobians is about twice as fast as both conventional AD tools and finite differences. On the SUN, NAG and FUJ platforms the vertex elimination approach does not reach the same level of efficiency. We observe that derivative codes for the code list version are often slightly slower than their counterparts differentiated at statement level for ALP and SGI and sometimes much slower on SUN, NAG and FUJ. Denoting by $\text{FLOPs}(F)$ and $\text{FLOPs}(\nabla F)$ the numbers of floating point operations required to compute respectively the function and its Jacobian, we obtained the ratios also shown in Table 3 from the Roe flux test case. We see that on the SGI and ALP platforms the vertex elimination derivative code ran faster than the ratio of floating point operations predicts. On the SUN, NAG and FUJ platforms the reverse is usually true.

4   Performance Issues

The elimination strategies we used are based on criteria aimed at minimising the number of floating point multiplications required to accumulate the Jacobian. Van der Wijngaar and Saphir [15] have shown that, on RISC platforms, neither the number of floating point operations nor even the cache miss rate are sufficient to explain performance of a numerical code; and processor-specific instructions (which can be seen from the generated assembly code) and compiler maturity affect performance.

4.1   Floating Point Performance

If we consider the derivative code speed in terms of floating point operations per clock cycle as shown in Fig. 2, all the elimination strategies are performing over 1 floating point operation per cycle on the SGI and ALP platforms. The Compaq EV6 and R12000 can perform up to 2 floating point operations per clock cycle [16] and a throughput of in excess of one floating point operation per clock cycle may be considered highly satisfactory. On the SUN and NAG platforms a throughput of less than half a floating point operation per clock cycle is achieved. Again the theoretical maximum is 2. It was initially thought that cache misses may explain the poor timing ratios on these platforms.

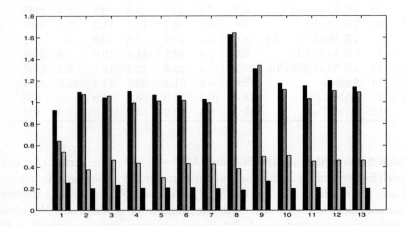

Fig. 2. Number of floating point operations per cycle performed by FD (number 1) and the elimination strategies (numbered 2–13 as they appear in Table 3); the bars represent (from left to right) results obtained from SGI, ALP, SUN, and NAG

4.2   Cache Misses

On the SGI, the outputs from SGI Origin's SpeedShop profiler showed that the derivative codes fit in the instruction cache and they differed from each other only by few cache misses in the primary/secondary data caches. The Compaq Alpha EV6 processor we used has caches comparable to those of the MIPS R12000 (see Table 1). Therefore, we do not expect cache misses to be crucial for the SGI and ALP platforms.

The Sparc Ultra 10 has relatively small caches compared with the R12000 and EV6. On the SUN and NAG platforms we noticed a performance degradation mainly of the code list method. Here, the forward elimination from the code list showed particularly poor performance. Profiling with the Sun Workshop 6.2 tools, we observed a 28% instruction cache miss rate for this forward elimination strategy. Using a SUN Blade 1000 (600 MHz) with double the primary cache size, that percentage came down to 10%. Interestingly the timing ratios remained similar to those from the Ultra 10 indicating that although instruction cache misses were occurring, the root cause of the poor performance lay elsewhere.

4.3   Statement Ordering in the Derivative Code

The SGI and ALP processors may perform 2 floating point operations per clock cycle through their floating point pipeline, provided that the floating point operation being performed uses data currently in a register and does not require any data from the output of an operation ahead of it and still in the pipeline. The SGI platform has a latency of 2 clock cycles before the result of an operation is available whereas for the ALP platform this is 4 clock cycles [16]. This may explain the slightly better results from the SGI compared to the ALP in Fig. 2.

Apart from their larger cache sizes the other main difference between the SGI/ALP platforms and the SUN/NAG/FUJ platforms is that the SGI and ALP processors support *out-of-order execution* [16]. This technique involves maintaining queues of floating point operations ready to be performed and if the one at the head of the current queue requires a value currently being processed in the floating point pipeline or not currently in a register, then the processor switches to another queue. Use of this technique reduces the importance of instruction scheduling by the compiler.

On the Ultra10 and Ultra1 processors of the SUN, NAG and FUJ platforms there is no out-of-order execution. The optimising compiler must perform more intensive optimisation of instruction scheduling in order to maintain good use of the floating point pipeline. The derivative codes we produce are large (see final column of Table 3), contain no loops or branching and hence comprise one large basic block. We therefore expected optimising compilers to schedule floating point instructions effectively since there are no complications regarding the control flow of the program. Since this is obviously not occurring for the SUN, NAG and FUJ platforms, we conjectured that the local optimisation of instruction scheduling performed by the compiler might not be able to maintain a good throughput in the floating point pipeline since statements using the same variables might be separated by several hundred lines of source code. Since an elimination sequence produces a certain set of elimination statements that may be placed in any order that respects their dependencies, we additionally conjectured that we might be able to reorder the statements in the derivative code to make better use of variables currently in registers.

To assess the impact of statement ordering in the derivative code, we perturbed the order of statements without altering the data dependencies within the code. We reordered the assignment statements with the aim of using each assigned value soon after its assignment. We therefore used a modified version of the depth first search algorithm [17]. Namely, we regarded the statements in the derivative code as the vertices of an acyclic graph, with the output statements at the top and an edge from s to t if statement t uses the output from statement s. Then, we arranged the statements in the order produced by a depth-first traversal of this graph. The results are shown in Table 4, which also displays the Table 3 results in brackets. We can see that statement reordering has greatly improved many of the SUN, NAG and FUJ times, has improved most of the SGI times, but has made no significant difference to the ALP times. For instance, looking at the SUN column of Table 4, the statement reordering has improved the generated code using the forward elimination from the code list by 56%. In this case, we observed that the resulting reordered derivative has 34% less loads and 72% less stores than the original derivative code. This indicates that the compiler was not achieving efficient register usage for the original derivative code. We now see that the Jacobian to function CPU time ratios have all significantly improved and we are approaching twice the efficiency of the best of the conventional techniques, AD or finite-differencing, of Table 3.

Table 4. Ratios, $T(\nabla F)/T(F)$, of the reordered derivative codes from the Roe flux test case on different platforms, timings in brackets are those before the reordering (c.f. Table 3)

Method	SGI		ALP		SUN		NAG		FUJ	
VE Forward(SL)	6.7	(7.5)	5.4	(5.5)	9.3	(13.2)	7.8	(12.2)	7.4	(10.2)
VE Reverse(SL)	5.2	(6.5)	4.6	(4.6)	7.7	(8.8)	7.2	(8.8)	6.2	(6.9)
VE Mark(SL)	5.8	(6.4)	5.2	(5.1)	8.6	(9.8)	7.3	(10.4)	6.1	(6.7)
VE Mark RB(SL)	5.8	(6.6)	5.1	(5.0)	7.9	(14.1)	6.8	(10.2)	6.0	(6.6)
VE VLR(SL)	6.4	(6.0)	4.6	(4.5)	8.2	(8.9)	6.3	(9.1)	5.6	(6.1)
VE VLR RB(SL)	6.8	(6.2)	4.5	(4.6)	7.5	(9.0)	6.2	(9.6)	5.6	(6.0)
VE Forward(CL)	6.2	(7.3)	5.2	(5.2)	8.2	(18.7)	13.3	(19.2)	8.5	(14.9)
VE Reverse(CL)	6.4	(6.7)	5.0	(4.7)	8.8	(10.7)	7.5	(9.9)	7.4	(9.5)
VE Mark(CL)	6.6	(6.6)	5.1	(5.0)	7.8	(9.3)	7.3	(11.7)	6.1	(11.2)
VE Mark RB(CL)	6.3	(6.6)	5.4	(5.3)	8.6	(10.2)	7.5	(11.0)	6.1	(10.9)
VE VLR(CL)	7.1	(6.4)	4.9	(5.0)	7.5	(10.1)	6.8	(11.1)	6.2	(10.9)
VE VLR RB(CL)	6.8	(7.2)	5.8	(5.4)	8.1	(10.8)	8.1	(12.4)	6.4	(12.1)

5 Conclusions

Automatic differentiation via vertex elimination allows us to produce Jacobian code which, in conjunction with an optimising compiler, efficiently uses a floating point pipeline on processors that support out-of-order execution. On other platforms tested and that supported a floating point pipeline we found that the compiler was not able to schedule the floating point operations with enough register re-use to allow the floating point processor to perform efficiently. Application of a simple statement reordering strategy based on depth first traversal enabled the compiler to optimise the resulting derivative code more effectively.

We conclude that we have an AD tool that, when assisted by statement reordering on some platforms, produces Jacobian code approaching or exceeding twice the efficiency of the current state-of-the-art for the Roe flux test problem. We are currently testing and evaluating the tool's performance on a number of test problems taken from the MINPACK-2 optimisation test suite [18].

Acknowledgments

We thank EPSRC and UK MOD for funding this project under grant GR/R21882.

References

1. Griewank, A., Corliss, G.: Automatic Differentiation of Algorithms. SIAM, Philadelphia (1991)
2. Griewank, A.: Evaluating Derivatives: Principles and Techniques of Algorithmic Differentiation. SIAM, Philadelphia (2000)
3. Corliss, G., Faure, C., Griewank, A., Hascoët, L., Naumann, U., eds.: Automatic Differentiation of Algorithms: From Simulation to Optimization. Springer (2002) (to appear).
4. Bischof, C., Carle, A., Khademi, P., Mauer, A.: ADIFOR 2.0: Automatic differentiation of Fortran 77 programs. IEEE Computational Science & Engineering **3** (1996) 18–32
5. Faure, C., Papegay, Y.: Odyssée user's guide, version 1.7. Technical Report 0224, INRIA, Unité de Recherche, INRIA, Sophia Antipolis, 2004 Route des Lucioles, B.P. 93, 06902, Sophia Antipolis Cedex, France (1998) See http://www.inria.fr/safir/SAM/Odyssee/odyssee.html.
6. Giering, R., Kaminski, T.: Recipes for adjoint code construction. ACM Transactions on Mathematical Software **24** (1998) 437–474
7. Tadjouddine, M., Forth, S.A., Pryce, J.D., Reid, J.K.: On the Implementation of AD using Elimination Methods via Source Transformation: Derivative code generation. AMOR Report 01/4, Cranfield University (RMCS Shrivenham), Swindon SN6 8LA, England (2001)
8. Tadjouddine, M., Pryce, J.D., Forth, S.A.: On the Implementation of AD using Elimination Methods via Source Transformation. AMOR Report 00/8, Cranfield University (RMCS Shrivenham), Swindon SN6 8LA, England (2000)
9. Parr, T., Lilly, J., Wells, P., Klaren, R., Illouz, M., Mitchell, J., Stanchfield, S., Coker, J., Zukowski, M., Flack, C.: ANTLR Reference Manual. Technical report, MageLang Institute's jGuru.com (2000) See http://www.antlr.org/doc/.

10. Griewank, A., Reese, S.: On the calculation of Jacobian matrices by the Markowitz rule. [1] 126–135
11. Naumann, U.: Efficient Calculation of Jacobian Matrices by Optimized Application of the Chain Rule to Computational Graphs. PhD thesis, Technical University of Dresden (1999)
12. Naumann, U.: Elimination techniques for cheap Jacobians. [3] 241–246
13. Roe, P.L.: Approximate Riemann solvers, parameter vectors, and difference schemes. Journal of Computational Physics **43** (1981) 357–372
14. Tadjouddine, M., Forth, S.A., Pryce, J.D.: AD tools and prospects for optimal AD in CFD flux Jacobian calculations. [3] 247–252
15. der Wijngaart, R.F.V., Saphir, W.C.: On the efficacy of source code optimizations for cache-based processors. NAS Technical Report NAS-00-014, NASA (2000)
16. Goedecker, S., Hoisie, A.: Performance Optimization of Numerically Intensive Codes. SIAM Philadelphia (2001)
17. Knuth, D.E.: The Art of Computer Programming, Volume 1: Fundamental Algorithms. Adison-Wesley (1997)
18. Averick, B.M., Moré, J.J.: User guide for the MINPACK-2 test problem collection. Technical Memorandum ANL/MCS-TM-157, Argonne National Laboratory, Argonne, Ill. (1991) Also issued as Preprint 91-101 of the Army High Performance Computing Research Center at the University of Minnesota.

Making Automatic Differentiation Truly Automatic: Coupling PETSc with ADIC

Paul Hovland, Boyana Norris, and Barry Smith

Mathematics and Computer Science Division
Argonne National Laboratory, 9700 S. Cass Avenue, Argonne, IL 60439, USA
{hovland, norris, bsmith}@mcs.anl.gov
http://www.mcs.anl.gov

Abstract. Despite its name, automatic differentiation (AD) is often far from an automatic process. Often one must specify independent and dependent variables, indicate the derivative quantities to be computed, and perhaps even provide information about the structure of the Jacobians or Hessians being computed. However, when AD is used in conjunction with a toolkit with well-defined interfaces, many of these issues do not arise. We describe recent research into coupling the ADIC automatic differentiation tool with PETSc, a toolkit for the parallel numerical solution of PDEs. This research leverages the interfaces and objects of PETSc to make the AD process very nearly transparent.

1 Introduction

Many varieties of scientific computation, including the numerical solution of nonlinear partial differential equations (PDEs), require derivatives. For complicated functions, it can be a difficult task to implement derivative computations by hand. In contrast, finite difference approximations are simple to implement, but they suffer from both roundoff and truncation error. Furthermore, finding a stepsize that balances these sources of error (thus minimizing the total error) can be difficult. Automatic differentiation (AD) [1, 2] offers an alternative that minimizes human effort and eliminates truncation error. For this reason, automatic differentiation has become a popular tool for scientific computing (see, for example [3, 4]).

One obstacle to widespread adoption of automatic differentiation is that the process is often far from automatic. To achieve acceptable levels of performance, the user may need to specify independent and dependent variables, indicate the derivatives to be computed, and provide information about the structure of the Jacobians or Hessians being computed. Previous work [5–7], however, has demonstrated that when AD is used in conjunction with a toolkit with well-defined interfaces, many of these issues do not arise. This paper describes research into coupling the ADIC [8] automatic differentiation tool with PETSc, a toolkit for the parallel numerical solution of PDEs [9]. This research extends earlier results by directly exploiting the sparsity structure of the Jacobians to be computed. It also provides a strategy for computing Jacobians in parallel,

P.M.A. Sloot et al. (Eds.): ICCS 2002, LNCS 2330, pp. 1087–1096, 2002.

without requiring an AD capability for MPI. Most important, unlike previous work done in coupling AD with numerical toolkits, the use of ADIC within the PETSc environment is fully automated. Thus, application developers can take full advantage of the increased accuracy and potentially better performance of AD-generated derivatives with no extra effort.

The organization of this paper is as follows. Sections 2 and 3 provide brief introductions to ADIC and PETSc, respectively. Section 4 describes how the two tools have been coupled to provide an automatic differentiation process that is nearly transparent to PETSc users. Section 5 illustrates the potential for increased performance and robustness provided by automatic differentiation. Section 6 summarizes our results and describes opportunities for future work.

2   ADIC

ADIC is a tool for the automatic differentiation of ANSI C. Given a set of functions that compute a mathematical function F, ADIC generates a new set of functions that compute F and its Jacobian, $J = F'$. ADIC differentiates statements by using the so-called reverse mode of automatic differentiation and propagates these partial derivatives from independent to dependent variables by using the so-called forward mode. See [8] for more details on ADIC and its implementation.

The behavior of ADIC can be configured with a large number of user-specified options via one or more control files. A control file contains a set of bindings, expressed as key-value pairs, organized in several sections. Some aspects of the coupling of PETSc and ADIC were handled in control scripts, for example, specifying inactive variables and types, and generating different prefixes for multiple versions of the differentiated code that coexist in the final executable.

A number of enhancements to ADIC were inspired by the need to make the AD process fully automated within PETSc. Some of these include the ability to process multiple control scripts, options for renaming generated header files, a facility for deactivating entire structures, and specialized run-time libraries for scalar-valued gradient accumulations (which arise in the matrix-free case).

3   PETSc

PETSc is an object-oriented toolkit for the parallel numerical solution of PDEs. PETSc provides implementations of basic objects, such as matrices and vectors, facilities for managing data associated with both structured and unstructured meshes (distributed arrays and index sets), linear solvers (primarily Krylov methods with a variety of preconditioners), and nonlinear solvers (primarily Newton-type methods).

3.1   Nonlinear Solvers

PETSc provides a collection of Newton-based nonlinear solvers (SNES). These solvers require a nonlinear function (the discretized PDE) whose input and output are a vector. The prototype for this function is

```
int FormFunction(SNES snes,Vec X,Vec F,void *ptr);
```

The solvers also require a Jacobian, or at least the action of the Jacobian on a vector, but PETSc is able to automatically compute an approximation to the Jacobian (or its action) using finite differences. In cases where the inaccuracy of finite differences leads to a degradation in convergence (see, for example, [10]), it is desirable to use analytic derivatives. In such instances, the user can provide a routine for computing the Jacobian or, using the approach described in Sect. 4, PETSc and ADIC can generate code for computing the Jacobian automatically.

3.2   Distributed Arrays and Multigrid Algorithms

PETSc provides several objects to assist in the management of data associated with structured and unstructured meshes. One such object is the distributed array (DA), which provides facilities for managing the field data associated with a single structured grid. The DMMG (Data Management MultiGrid) object manages a hierarchical collection of such objects for use in multigrid algorithms. PETSc provides functions (methods) for exchanging data associated with ghost vertices (generalized gather-scatter operations) and for obtaining a coloring suitable for computing a Jacobian using either finite difference approximations or automatic differentiation.

The coloring provided by PETSc is of the Curtis-Powell-Reed (CPR) variety [11]—two columns of the same color contain no row in which both have a nonzero. Thus, the Jacobian can be approximated by perturbing all columns of the same color simultaneously. Alternatively, the seed matrix for automatic differentiation can be initialized by applying the coloring to an identity matrix; all columns (unit vectors) of the same color are combined into a single column of the seed matrix (see [12] for more details). We note that obtaining an optimal or nearly optimal CPR coloring for Jacobians arising from structured grids is simple [13] while efficient parallel algorithms for coloring Jacobians from unstructured meshes remains an open research topic.

4   Coupling PETSc and ADIC

To produce a routine for computing a Jacobian, one might be inclined to apply ADIC directly to the user's FormFunction routine (see Sect. 3.1). A similar strategy has been effective in other contexts [5, 7]. In the context of PETSc, however, this approach is less appealing. One reason is that the user's FormFunction routine usually contains a number of calls to PETSc utilities, such as the generalized gather-scatter routines for ghost data communication described in Sect. 3.2.

Thus, applying ADIC to `FormFunction` would lead to automatic differentiation of these utility methods. However, because many of these utilities deal with problem setup and data movement and often use MPI, applying ADIC directly would likely lead to unnecessary work plus the added complication of differentiating MPI (and including the appropriate runtime support library). For these reasons, we have followed the example of previous semi-automated approaches to coupling ADIC and PETSc [14, 15] and provide an automated solution based on a domain decomposition approach.

4.1   A Domain Decomposition-Based Strategy

Figure 1 provides an example of a simple nonlinear function.[1] This example illustrates a structure common to most nonlinear functions that use PETSc's DA or DMMG objects. A setup and communication phase precedes a computation phase in which the function is evaluated over the local subdomain. A final phase assigns local values to parallel objects. The first and final phase are essentially problem independent and depend only on the structure of the DA or DMMG grid object being used. Therefore, it is possible to isolate the local computation in a separate routine and have PETSc manage the rest of the nonlinear function evaluation. This approach reduces the amount of user code and simplifies the automatic differentiation process. The user provides a function adhering to the following prototype, an example of which appears in Fig. 2.

```
int FormFunctionLocal(DALocalInfo *info,Field **x,
                      Field **f,void *ptr);
```

4.2   The User's Experience

As intended, the user interface to the AD-enabled PETSc functionality is virtually the same as the interface used for computing derivatives by means of finite differences. In both cases, the user must provide the nonlinear subdomain function evaluation and initialize the nonlinear solver context. If the user's applications uses DMMG objects, the changes are all minor. To use a nonlinear subdomain function, the user replaces a call such as

```
ierr = DMMGSetSNES(dmmg,FormFunction,0);
```

with a call such as

```
ierr = DMMGSetSNESLocal(dmmg,FormFunctionLocal,0,
           ad_FormFunctionLocal, admf_FormFunctionLocal);
```

To indicate that `FormFunctionLocal` should be differentiated using ADIC, the user adds a comment of the form

---

[1] This function comes from PETSc SNES example 5, a solid fuel ignition problem derived from the Bratu problem in the MINPACK-2 test problem collection [16].

```
int FormFunction(SNES snes,Vec X,Vec F,void *ptr)
{
  AppCtx      *user = (AppCtx*)ptr;
  int         err,i,j,Mx,My,xs,ys,xm,ym;
  PetscReal   two = 2.0,lambda,hx,hy,hxdhy,hydhx,sc;
  PetscScalar u,uxx,uyy,**x,**f;
  Vec         localX;

  PetscFunctionBegin;
  err = DAGetLocalVector(user->da,&localX);CHKERRQ(err);
  err = DAGetInfo(user->da,PETSC_IGNORE,&Mx,&My, ... );

  hx     = 1.0/(PetscReal)(Mx-1); hy     = 1.0/(PetscReal)(My-1);
  lambda = user->param;           sc     = hx*hy*lambda;
  hxdhy  = hx/hy;                  hydhx  = hy/hx;

  /* Scatter ghost points to local vector, using a 2-step process */
  err = DAGlobalToLocalBegin(user->da,X,INSERT_VALUES,localX);
  CHKERRQ(err);
  err = DAGlobalToLocalEnd(user->da,X,INSERT_VALUES,localX);
  CHKERRQ(err);
  /* Get pointers to vector data */
  err = DAVecGetArray(user->da,localX,(void**)&x);CHKERRQ(err);
  err = DAVecGetArray(user->da,F,(void**)&f);CHKERRQ(err);
  /* Get local grid boundaries */
  err = DAGetCorners(user->da,&xs,&ys,0,&xm,&ym,0);CHKERRQ(err);

  /* Compute function over the locally owned part of the grid */
  for (j=ys; j<ys+ym; j++) {
    for (i=xs; i<xs+xm; i++) {
      if (i == 0 || j == 0 || i == Mx-1 || j == My-1) {
        f[j][i] = x[j][i];
      } else {
        u      = x[j][i];
        uxx    = (two*u - x[j][i-1] - x[j][i+1])*hydhx;
        uyy    = (two*u - x[j-1][i] - x[j+1][i])*hxdhy;
        f[j][i] = uxx + uyy - sc*PetscExpScalar(u);
      }
    }
  }
  /* Restore vectors */
  err = DAVecRestoreArray(user->da,localX,(void**)&x);CHKERRQ(err);
  err = DAVecRestoreArray(user->da,F,(void**)&f);CHKERRQ(err);
  err = DARestoreLocalVector(user->da,&localX);CHKERRQ(err);
  err = PetscLogFlops(11*ym*xm);CHKERRQ(err);
  PetscFunctionReturn(0);
}
```

Fig. 1. Example of a nonlinear function

```
int FormFunctionLocal(DALocalInfo *info, PetscScalar **x,
                      PetscScalar **f,AppCtx *user)
{
  int          ierr,i,j;
  PetscReal    two = 2.0,lambda,hx,hy,hxdhy,hydhx,sc;
  PetscScalar  u,uxx,uyy;

  PetscFunctionBegin;

  lambda = user->param;
  hx     = 1.0/(PetscReal)(info->mx-1);
  hy     = 1.0/(PetscReal)(info->my-1);
  sc     = hx*hy*lambda;
  hxdhy  = hx/hy;
  hydhx  = hy/hx;

  /* Compute function over the locally owned part of the grid */
  for (j=info->ys; j<info->ys+info->ym; j++) {
    for (i=info->xs; i<info->xs+info->xm; i++) {
      if (i == 0 || j == 0 || i == info->mx-1 || j == info->my-1) {
        f[j][i] = x[j][i];
      } else {
        u       = x[j][i];
        uxx     = (two*u - x[j][i-1] - x[j][i+1])*hydhx;
        uyy     = (two*u - x[j-1][i] - x[j+1][i])*hxdhy;
        f[j][i] = uxx + uyy - sc*PetscExpScalar(u);
      }
    }
  }

  ierr = PetscLogFlops(11*info->ym*info->xm);CHKERRQ(ierr);
  PetscFunctionReturn(0);
}
```

Fig. 2. Example of a nonlinear subdomain function

```
/* Process adiC: FormFunctionLocal */
```

With these changes in place, the user can switch between using automatic differentiation and finite differences using a single runtime option, for example -dmmg_jacobian_mf_ad_operator versus -dmmg_jacobian_mf_fd_operator. When DA objects are used, the interface for initializing the SNES context in the user's driver routine contains a few minor differences for the AD case, mainly in the methods used for obtaining the coloring for the Jacobian. These differences will eventually disappear as the interfaces continue to evolve.

4.3   Behind the Scenes

While the user's experience is largely unchanged with the new subdomain interface and automatic differentiation capability, several additions to PETSc and ADIC were necessary to make nearly total automation possible.

PETSc was augmented to automatically allocate storage for several derivative objects, principally those associated with the input and output vectors (arrays at the subdomain level). PETSc was also modified to initialize the derivatives associated with the input vector, or so-called seed matrix. The initialization uses the CPR coloring discussed in Sect. 3.2. The cost of computing the resulting "compressed" Jacobian is proportional to the number of colors. For a structured grid, this is usually the product of the stencil size and the number of data fields at each grid point.

Once the compressed Jacobian has been computed, the values must be assembled into a PETSc sparse matrix object. Support for extracting a row of the compressed Jacobian was added to ADIC, and PETSc was enhanced to be able to assemble the sparse matrix directly from these compressed rows.

To further simplify the task of using automatic differentiation with PETSc, we extended the PETSc build process to include processing of source files with a python script. This script searches for the phrase "Process adiC," extracts the indicated functions into a separate file, processes them with ADIC using the control scripts described in Sect. 2, and compiles and links the resulting derivative code to provide Jacobian computations in a manner that is virtually transparent to the user.

5   Experimental Results

The potential benefits of using automatic differentiation for derivative computations are illustrated in a simple example from the PETSc distribution, a two-dimensional driven cavity problem using a velocity-vorticity formulation (see ex19.c in the SNES tutorials examples directory for more details). We solved a 100×100 version of this problem with the default nonlinear solver, a GMRES-30 or BiCGStab linear solver, no preconditioner, and derivatives computed using either AD or finite difference approximations (FD) applied in a matrix-free manner (using the -dmmg_jacobian_mf_ad or -dmmg_jacobian_mf_fd option). This approach provides maximal contrast between AD and FD since Jacobian-vector products are approximately twice as expensive to compute using AD, but FD is much more susceptible to roundoff or truncation errors.

Figure 3 compares AD and FD using the default stepsize of $w = 10^{-7}$ by plotting the nonlinear residual norm at each iteration (indicated with a symbol) versus the cumulative elapsed time. Computations were performed on an Athlon 950 workstation with 256MB RAM. The increased accuracy of automatic differentiation results in faster convergence, especially for the matrix-free case. Figure 4 illustrates that the superior performance of AD is largely due to the sensitivity of finite difference approximations to stepsize. If an appropriate noise estimate is selected (using the -snes_mf_err option), FD converges somewhat faster than AD. Thus, a primary advantage of AD over FD in the context of nonlinear solvers is increased robustness, obviating the need to experiment with stepsizes.

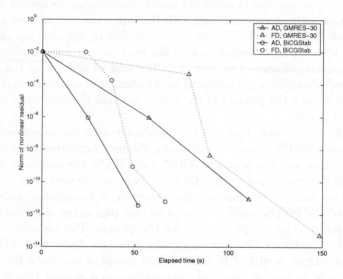

Fig. 3. Performance of matrix-free AD (*solid*) versus matrix-free FD (*dotted*) using GMRES-30 (*triangle*) and BiCGStab (*circle*) linear solvers

6   Conclusions and Future Work

We have presented an integration of automatic differentiation into PETSc, using high-level interfaces to automate fully the use of ADIC to generate the code for computing the Jacobian of a local subdomain function. We described some of the implementation details of this coupling, and we presented a simple application that takes advantage of it. Our experimental results show that frequently the

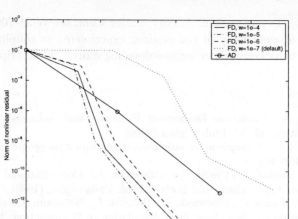

Fig. 4. Convergence of matrix-free FD for various noise estimates (w) using BiCGStab

superior accuracy of AD-generated code leads to convergence in fewer nonlinear and linear iterations than with finite differences.

Future tasks include extending this work to the optimization regime, providing the capability to compute Hessians and gradients of partially separable functions using automatic differentiation. In addition, basic block reverse mode and cross-country preaccumulation strategies could significantly reduce the cost of typical Jacobian computations. The work must also be extended to the case of unstructured meshes. In the case of finite element computations, applying automatic differentiation at the level of element function, and then assembling the full Jacobian from the small, dense element Jacobians may be an effective strategy.

Types, variables, or entire functions that have been specified as inactive in a control file are not augmented with derivative computations. Inactive objects are designated by name, however. In the case of PETSc applications, the user-defined nonlinear function name is not known at the time the ADIC control scripts are created. As shown in Sect. 4, the user specifies the name of the nonlinear function in a special comment. By extending ADIC to allow *active* functions to be designated via a prototype, and generating derivatives only for those functions (and any other objects designated as active), we could eliminate the need for special pre- and postprocessing of the application and differentiated code.

Acknowledgments

This work was supported by the Mathematical, Information, and Computational Sciences Division subprogram of the Office of Advanced Scientific Computing

Research, U.S. Department of Energy, under Contract W-31-109-Eng-38. Po-Ting Wu performed many of the original experiments in coupling ADIC and PETSc. We thank Gail Pieper for proofreading a draft manuscript.

References

1. Griewank, A.: Evaluating Derivatives: Principles and Techniques of Algorithmic Differentiation. SIAM, Philadelphia (2000)
2. Wengert, R.E.: A simple automatic derivative evaluation program. Comm. ACM **7** (1964) 463–464
3. Berz, M., Bischof, C., Corliss, G., Griewank, A.: Computational Differentiation: Techniques, Applications, and Tools. SIAM, Philadelphia (1996)
4. Corliss, G., Faure, C., Griewank, A., Hascoët, L., Naumann, U., eds.: Automatic Differentiation of Algorithms: From Simulation to Optimization. Springer (2002) (to appear).
5. Ferris, M.C., Mesnier, M.P., Moré, J.J.: NEOS and Condor: Solving optimization problems over the internet. ACM Transactions on Mathematical Software **26** (2000) 1–18
6. Gertz, E.M., Gill, P.E., Muetherig, J.: User's guide for SnadiOpt: A package adding automatic differentiation to Snopt. Report NA 01–1, Department of Mathematics, University of California, San Diego (2000)
7. Li, S., Petzold, L.: Design of new DASPK for sensitivity analysis. Technical report, University of California at Santa Barbara (1999)
8. Bischof, C., Roh, L., Mauer, A.: ADIC — An extensible automatic differentiation tool for ANSI-C. Software: Practice and Experience **27** (1997) 1427–1456
9. Balay, S., Gropp, W.D., McInnes, L.C., Smith, B.F.: PETSc users manual. Technical Report ANL–95/11 — Revision 2.1.1, Argonne National Laboratory (2001)
10. Hovland, P.D., McInnes, L.C.: Parallel simulation of compressible flow using automatic differentiation and PETSc. Technical Report ANL/MCS–P796–0200, Mathematics and Computer Science Division, Argonne National Laboratory (2000) (to appear in a special issue of *Parallel Computing* on "Parallel Computing in Aerospace").
11. Curtis, A.R., Powell, M.J.D., Reid, J.K.: On the estimation of sparse Jacobian matrices. J. Inst. Math. Appl. **13** (1974) 117–119
12. Bischof, C., Hovland, P.: Using ADIFOR to compute dense and sparse Jacobians. Technical Report ANL/MCS–TM–158, Argonne National Laboratory (1991)
13. Goldfarb, D., Toint, P.L.: Optimal estimation of Jacobian and Hessian matrices that arise in finite difference calculations. Mathematics of Computation **43** (1984) 69–88
14. Abate, J., Benson, S., Grignon, L., Hovland, P., McInnes, L., Norris, B.: Integrating automatic differentiation with object-oriented toolkits for high-performance scientific computing. Technical Report Preprint ANL/MCS–P818–0500, Mathematics and Computer Science Division, Argonne National Laboratory, Argonne, IL (2000) (to appear in [4]).
15. Wu, P., Bischof, C., Hovland, P.D.: Using ADIFOR and ADIC to provide Jacobians for the SNES component of PETSc. Technical Report ANL/MCS–TM–233, Mathematics and Computer Science Division, Argonne National Laboratory (1997)
16. Averick, B.M., Carter, R.G., Moré, J.J., Xue, G.L.: The MINPACK-2 test problem collection. Preprint MCS–P153–0694, Mathematics and Computer Science Division, Argonne National Laboratory, Argonne, IL (1992)

Improved Interval Constraint Propagation for Constraints on Partial Derivatives [*]

Evgueni Petrov and Frédéric Benhamou

IRIN — Université de Nantes
2 rue de la Houssinière BP 92208
44322 Nantes Cedex 03 France
{evgueni.petrov, frederic.benhamou}@irin.univ-nantes.fr

Abstract. Automatic differentiation (AD) automatically transforms programs which calculate elementary functions into programs which calculate the gradients of these functions. Unlike other differentiation techniques, AD allows one to calculate the gradient of any function at the cost of at most 5 values of the function (in terms of time). Interval constraint programming (ICP) is a part of constraint programming focused on representation and processing of nonlinear constraints. We adapt AD to the context of ICP and obtain an algorithm which transforms elementary functions into constraints specifying their gradient. We describe some experiments with implementation of our algorithm in the logic programming language ECL^{i}PS^e.

1 Introduction

Automatic differentiation (AD) automatically transforms programs which calculate real functions into programs which calculate the gradients of these functions [7,5]. Unlike other differentiation techniques, AD allows one to calculate the gradient at the cost of at most 5 values of the function (in terms of time).

Interval constraint programming (ICP) [1] is a part of constraint programming [6,2] focused on representation and processing of nonlinear constraints. One of classical concepts of ICP is interval constraint propagation which means incremental calculation of multidimensional rectangles bounding solutions to nonlinear constraints.

Applications of AD in ICP are limited to fast calculation of the first coefficients of interval Taylor series of nonlinear constraints [8,4].

We adapt AD to the context of ICP and obtain an algorithm which transforms elementary functions into constraints specifying their gradient. With respect to the techniques of naïve symbolic differentiation, our algorithm generates a smaller number of constraints which are more effectively processed by interval constraint propagation.

---

[*] Financially supported by Centre Franco-Russe Liapunov (Project 06–98), by European project COCONUT IST–2000–26063.

P.M.A. Sloot et al. (Eds.): ICCS 2002, LNCS 2330, pp. 1097–1105, 2002.
© Springer-Verlag Berlin Heidelberg 2002

The paper is structured as follows. Section 2 introduces basic notions of the paper. Section 3 describes our transformation technique. Section 4 states its properties (complexity and accuracy of interval constraint propagation). Section 5 contains the data of experiments with an implementation of our technique in the logic programming language ECL^{i}PS^e [3]. Section 6 concludes the paper.

2   Definitions, notation

We give some definitions first. Real intervals are closed convex subsets of the set $\mathbf{R}$ of real numbers. Real relations are subsets of points of $\mathbf{R}^2$, $\mathbf{R}^3$, etc. Symbol $\mathbf{R}^\infty$ denotes the set of countable sequences of real numbers. A projection function is a function which returns a specific component of these sequences.

Symbols v_0, v_1, v_2, etc. denote the formal variables. Constraints are pairs consisting of a relation and an ordered set of variables. A constraint is primitive, if it relates its variables by the graph of some "basic" function. We assume that (1) every basic function has one or two arguments, (2) the arithmetic operations, the trigonometric functions, the functions exp, log are basic functions. Sequence $p \in \mathbf{R}^\infty$ is a solution to constraint $((v_i, v_j, \ldots), f)$, if it satisfies $(p_i, p_j, \ldots) \in f$.

Real terms are terms constructed of the formal variables, real numbers, and the basic functions. Real terms specify real functions of the sequences from $\mathbf{R}^\infty$. The variables specify the projection functions, real numbers specify the constant functions, compound real terms $f(t)$, $f(t, t')$ specify the composition of basic function f and the functions specified by the real terms t, t'.

Symbol D_ℓ denotes differentiation operator which maps every real function to its partial derivative with respect to argument ℓ.

Templates of real terms are terms which are constructed of the formal variables, variables $\blacksquare_1$, $\blacksquare_2$, real numbers, the basic functions, and which do not contain multiple occurrences of $\blacksquare_1$, $\blacksquare_2$.

Let F, t, t' be a template of real term and two real terms. Expression $F\langle t, t' \rangle$ denotes the result of replacing $\blacksquare_1$, $\blacksquare_2$ with t, t'. If F does not contain $\blacksquare_2$, then t' is omitted. Partial derivatives of basic function f are expressed by templates f'_x, f'_y, i.e., real terms $f'_x\langle v_0, v_1 \rangle$, $f'_y\langle v_0, v_1 \rangle$ specify non-zero partial derivatives of the real function specified by real term $f(v_0, v_1)$. The derivative of unary basic function f is expressed by template f'.

3   AD in terms of constraints

Before going into details, we explain our idea informally in terms of program analysis. Linear blocks are sequences of assignments. Single assignment programs are programs in which every variable is assigned a value only once. As a program transformation technique, AD has an important property: it transforms linear blocks into linear blocks, single assignment programs into single assignment programs. The class of single assignment programs consisting of a signle linear block is equivalent to the class of sets of primitive constraints (because, in such context, every assignment is a valid equation). Thus, AD can be made applicable to functions specified by primitive constraints. In what follows, we give a more detailed description of this idea.

Given a real term which specifies some function h and does not contain other variables than $v_0, v_2, \ldots, v_{2n}$, our algorithm generates such set C of primitive constraints that the projection of the set of solutions to C onto the first $2n + 2$ coordinates is the following subset of $\mathbf{R}^{2n+2}$ (see theorem 1):

$$\{(p_0, \ldots, p_{2n+1}) \mid \forall \ell \in [0, n]\ p_{2\ell+1} = (D_{2\ell}h)(p)\}.$$

3.1 Decomposition algorithm

The following rules define function ad which implements our decomposition algorithm. The rule to be applied should be selected in the "top down" fashion. Because rule 1 does not commute with rules 3, 4, this assumption is essential. Rules 1 through 4 assume that $2i$ is greater than the subscript of any variable occurring in their left hand sides (v_{2i} must be a "free" variable). Function dec maps real terms to sets of primitive constraints. It returns primitive constraints whose solutions annihilate the real function specified by its argument (see implementation in section 4).

Rule 1 (Elimination of multiple occurrences)

$$\mathsf{ad}F\langle v_{2j}, v_{2j}\rangle = \mathsf{ad}F\langle v_{2i}, v_{2i+2}\rangle \cup \{v_{2j} = v_{2i}, v_{2j} = v_{2i+2}, v_{2j+1} = v_{2i+1} + v_{2i+3}\}$$

Rule 2 (Elimination of constants) $\mathsf{ad}F\langle c\rangle = \mathsf{ad}F\langle v_{2i}\rangle \cup \{v_{2i} = c\}$

Rule 3 (Composition-1)

$$\mathsf{ad}F\langle f(v_{2j})\rangle = \mathsf{ad}F\langle v_{2i}\rangle \cup \{v_{2i} = f(v_{2j})\} \cup \mathsf{dec}\,(v_{2j+1} - v_{2i+1}f'\langle v_{2j}\rangle)$$

Rule 4 (Composition-2)

$$\mathsf{ad}F\langle f(v_{2j}, v_{2k})\rangle = \mathsf{ad}F\langle v_{2i}\rangle \cup \{v_{2i} = f(v_{2j}, v_{2k})\} \cup$$
$$\mathsf{dec}\,(v_{2j+1} - v_{2i+1}f'_x\langle v_{2j}, v_{2k}\rangle) \cup \mathsf{dec}\,(v_{2k+1} - v_{2i+1}f'_y\langle v_{2j}, v_{2k}\rangle)$$

Rule 5 (Termination) $\mathsf{ad}v_{2j} = \{v_{2j+1} = 1\}$

In the ECL^{i}PS^e implementation of ad, in order to reduce the number of generated primitive constraints, we simplify the right hand side of rules 3, 4 for the basic functions exp, sqrt, arctan and replace rule 2 with four additional ones which process the cases where one or two arguments of some basic function are real numbers (one instance of rule 3 and three instances of rule 4).

Real terms specify real functions. Our rules reduce calculation of the gradient of one real function (specified by a real term) to calculation of the gradient of some other (specified by some other term). Figures 1, 2, 3 illustrate how our rules change the graphs of these functions in some particular cases.

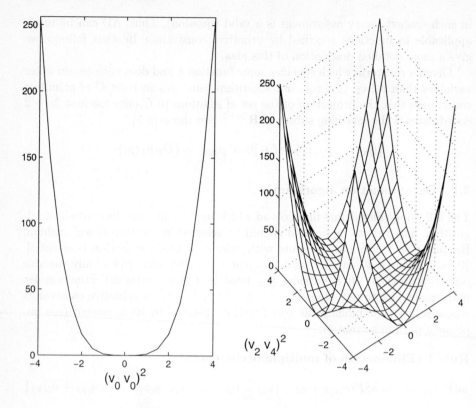

Fig. 1. Application of rule 1 to $(v_0 v_0)^2$.

We conclude this section by calculation of $\text{ad}\left(v_0^2 + (v_0 + v_2)^2\right)$ (Schwefel function 1.2 in 2 dimensions):

$$\text{ad}\left(v_0^2 + (v_0 + v_2)^2\right) \overset{\text{R1}}{=} \{v_0 = v_4, v_0 = v_6, v_1 = v_5 + v_7\} \cup \text{ad}\left(v_4^2 + (v_6 + v_2)^2\right)$$

$$\text{ad}\left(v_4^2 + (v_6 + v_2)^2\right) \overset{\text{R4}}{=} \{v_8 = v_6 + v_2, v_7 = v_9, v_3 = v_9\} \cup \text{ad}\left(v_4^2 + v_8^2\right)$$

$$\text{ad}\left(v_4^2 + v_8^2\right) \overset{\text{R3}}{=} \{v_{12} = v_4^2, v_5 = 2v_{11}, v_{11} = v_4 v_{13}\} \cup \text{ad}\left(v_{12} + v_8^2\right)$$

$$\text{ad}\left(v_{12} + v_8^2\right) \overset{\text{R3}}{=} \{v_{16} = v_8^2, v_9 = 2v_{15}, v_{15} = v_8 v_{17}\} \cup \text{ad}\left(v_{12} + v_{16}\right)$$

$$\text{ad}(v_{12} + v_{16}) \overset{\text{R4}}{=} \{v_{18} = v_{12} + v_{16}, v_{13} = v_{19}, v_{17} = v_{19}\} \cup \text{ad}(v_{18})$$

$$\text{ad}(v_{18}) \overset{\text{R5}}{=} \{v_{19} = 1\}$$

4   Properties of the decomposition algorithm

Let H be a real term which contains variables $v_0, v_2, \ldots, v_{2n}$ only.

Theorem 1 (Correctness). *Let h be the real function specified by real term H. The following statements are true:*

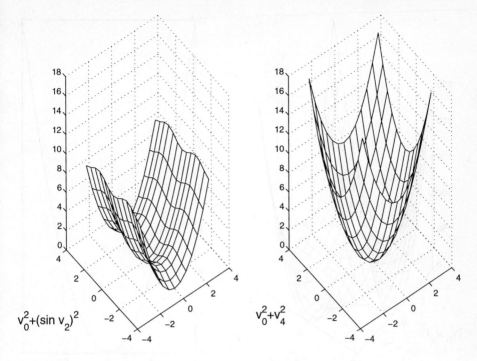

Fig. 2. Application of rule 3 to $v_0^2 + (\sin v_2)^2$.

1. *Every solution p to* $\mathsf{ad}(H)$ *satisfies the equation* $p_{2\ell+1} = (D_{2\ell}h)\,(p)$ *for each variable* $v_{2\ell}$ *from H.*
2. *If all the derivatives* $(D_{2\ell}h)\,(p)$*'s exist for some p, then the constraints* $\mathsf{ad}(H)$, $\{v_{2\ell} = p_{2\ell} \mid 0 \le \ell \le n\}$, $\{v_{2\ell+1} = (D_{2\ell}h)\,(p) \mid 0 \le \ell \le n\}$ *have a common solution.*

Theorem 2 (Number of primitive constraints). *Let function D_i map real terms specifying real functions to real terms specifying their partial derivatives with respect to argument i.*

Let functions dec *(see section 3),* D_i *be defined by the following rules:*

$$\mathsf{dec}F\langle v_j\rangle = \{v_j = 0\} \qquad\qquad \mathsf{D}_\ell\, c = \mathsf{D}_\ell\, v_j = 0$$
$$\mathsf{dec}F\langle c\rangle = \mathsf{dec}F\langle v_i\rangle \cup \{v_i = c\} \qquad \mathsf{D}_\ell\, v_\ell = 1$$
$$\mathsf{dec}F\langle f(v_j)\rangle = \mathsf{dec}F\langle v_i\rangle \cup \{v_i = f(v_j)\} \qquad \mathsf{D}_\ell\, f(t) = f'\langle t\rangle \cdot \mathsf{D}_\ell\, t$$
$$\mathsf{dec}F\langle f(v_j, v_k)\rangle = \mathsf{dec}F\langle v_i\rangle \cup \{v_i = f(v_j, v_k)\} \quad \mathsf{D}_\ell\, f(t, t') = f'_x\langle t, t\rangle \cdot \mathsf{D}_\ell\, t + f'_y\langle t, t\rangle \cdot \mathsf{D}_\ell\, t'$$

Let symbol $||\cdot||$ *denote cardinality. Let real term H contain N basic functions. The following statements are true:*

1. $||\mathsf{dec}(H)|| = N$,
2. $||\mathsf{ad}(H)|| \le \mathrm{const} N$,
3. $||\mathsf{dec}(\mathsf{D}_0(\underbrace{\cos(\cos(\ldots \cos(}_{\cos \text{ occurs } N \text{ times}} v_0)\ldots))))|| = N(N+3)/2$.

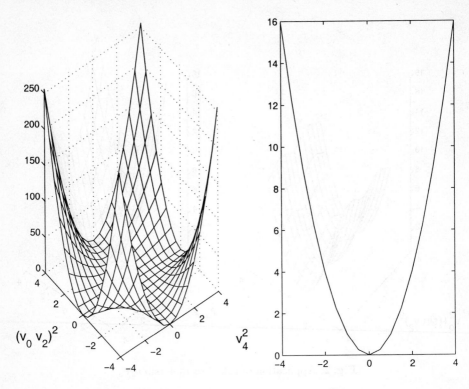

Fig. 3. Application of rule 4 to $(v_0 v_2)^2$.

Theorem 3 (ad and interval constraint propagation). *Let C_ℓ be the set of constraints generated from* $\mathsf{dec}(v_{2\ell+1} - \mathsf{D}_\ell(H))$ *by renaming variables so that different C_ℓ's share variables v_0, v_2, ..., v_{2n} only.*

Let intervals I_0, ..., I_{2n+1} be calculated by the interval constraint propagation algorithm HC3 [1] for the variables v_0, ..., v_{2n+1} and the constraints $\mathsf{ad}(H)$. Let intervals J_0, ..., J_{2n+1} be calculated by the same algorithm for the same variables and the constraints $\bigcup_\ell C_\ell$.

Then $I_0 \times I_1 \times \ldots \times I_{2n} \times I_{2n+1} \subseteq J_0 \times J_1 \times \ldots \times J_{2n} \times J_{2n+1}$.

5   Experiments

We have applied our decomposition algorithm to minimization of generalized Schwefel function 1.2 and 3.2, Rosenbrock function in different dimensions. Each of these functions is specified by a polynomial of low degree and has exactly one local minimum. Consequently, there are no contraindications for minimization of these functions by interval constraint propagation.

We have run the ECL^{i}PS^e implementation [9] of interval constraint propagation algorithm HC3 [1] on the primitive constraints generated by function ad and by naïve symbolic differentiation D_ℓ.

For each objective function, we give a table showing the results of our experiments. Its columns indicate the dimension (column 1), the number of the constraints and the reduction of the initial multidimensional rectangle when HC3 is applied to these constraints for the constraints generated by our transformation technique (columns 2, 3) and by naïve symbolic differentiation (columns 4, 5). The content of columns 3, 5 describes how HC3 changes the initial rectangle. The output rectangles of a diameter comparable with the machine precision are denoted by "exact". The large output rectangles which differ from the initial one at least in one dimension are denoted by "partial". The output rectangle identical to the initial one is denoted by "no reduction".

5.1   Schwefel function 1.2

Generalized Schwefel function 1.2 in n dimensions is specified by real term $H = \sum_{i=1}^{n} \left(\sum_{j=1}^{i} v_j \right)^2$. The minimum is achieved at the point $(0, \ldots, 0) \in \mathbf{R}^n$. The standard initial rectangle is $[-5, 10]^n$.

	AD		Naïve	
n	$\|\|\mathsf{ad}(H)\|\|$	Reduction	$\|\|\mathsf{dec}(H)\|\| + \sum_\ell \|\|\mathsf{dec}(\mathsf{D}_\ell(H))\|\|$	Reduction
1	6	exact	2	exact
2	17	exact	9	exact
3, 4, 5	24, 36, 50	exact	23, 46, 80	partial
6–80	66–6800	exact	127–180280	no reduction

Minimization of Schwefel function 1.2 in the rectangle $[-5, 10]^n$.

The number of primitive constraints generated by AD (column 2) grows quadratically with respect to n (and linearly with respect to $\|\|\mathsf{dec}(H)\|\|$). The number of primitive constraints generated by naïve symbolic differentiation (column 4) grows cubically with respect to n. Besides that, HC3 is more effective on the constraints generated by AD (cf. columns 3, 5).

5.2   Schwefel function 3.2

Generalized Schwefel function 3.2 in n dimensions is specified by real term $H = \sum_{i=2}^{n} \left(v_1 - v_i^2 \right)^2 + (v_i - 1)^2$. The minimum is achieved at the point $(1, \ldots, 1) \in \mathbf{R}^n$. The standard initial rectangle is $[-10, 10]^n$.

	AD		Naïve	
n	$\|\|\mathsf{ad}(H)\|\|$	Reduction	$\|\|\mathsf{dec}(H)\|\| + \sum_\ell \|\|\mathsf{dec}(\mathsf{D}_\ell(H))\|\|$	Reduction
2	12	exact	20	partial
3–10	26–124	partial	40–180	partial
11–80	138–1104	no reduction	200–1580	no reduction

Minimization of Schwefel function 3.2 in the rectangle $[-10, 10]^n$.

The number of primitive constraints grows linearly for AD as for naïve symbolic differentiation (columns 2, 4). However, AD is generates fewer primitive constraints. As far as effectiveness of HC3 is concerned, there is no significant difference between the constraints generated by AD and by naïve symbolic differentiation (cf. columns 3, 5).

5.3   Rosenbrock function

Generalized Rosenbrock function in n dimensions is specified by real term $H = \sum_{i=1}^{n-1} 100 \left(v_i - v_{i+1}^2 \right)^2 + (1 - v_{i+1})^2$. The minimum is achieved at the point $(1, \ldots, 1) \in \mathbf{R}^n$. The standard initial rectangle is $[-2.048, 2.048]^n$. We give also the data for the rectangle $[0.1, 2.048]^n$.

	AD		Naïve	
n	$\|\|\mathsf{ad}(H)\|\|$	Reduction	$\|\|\mathsf{dec}(H)\|\| + \sum_\ell \|\|\mathsf{dec}(\mathsf{D}_\ell(H))\|\|$	Reduction
in rectangle $[-2.048, 2.048]^n$				
2	15	exact	20	partial
3–80	32–1341	partial	40–1580	partial
in rectangle $[0.1, 2.048]^n$				
2–80	15–1341	exact	20–1580	partial

Minimization of Rosenbrock function in different rectangles.

AD generates a smaller number of constraints. With respect to effectiveness of HC3, there is no big difference between the constraints generated by AD and by naïve symbolic differentiation (cf. columns 3, 5). However, this situation changes, if we start interval constraint propagation in the rectangle $[0.1, 2.048]^n$.

6   Conclusion

Automatic differentiation is successfully used in computational mathematics for more than twenty years (as of year 2002). However, applications of AD in interval constraint programming were limited to fast calculation of the first coefficients of Taylor series of nonlinear equations. Up to now, the only means of working with the gradient in optimization problems of ICP was naïve symbolic differentiation in combination with this or that symbolic transformation method. As we have seen in section 4, this approach cannot guarantee an acceptable size of the symbolic expressions for the gradient.

In this paper, we have introduced a technique which allows one to involve the gradient into interval constraint propagation at a low cost in terms of number of constraints. Our experiments indicate that a valuable by-product of our technique is a certain increase in accuracy of interval constraint propagation.

Acknowledgements We thank the referees N° 26 and N° 27 for their comments.

References

1. F. Benhamou. Interval Constraint Logic Programming. In A. Podelski, editor, *Constraint Programming: basics and trends*, volume 910 of *Lecture Notes in Computer Science*, pages 1–21. Springer-Verlag, 1995.
2. R. Dechter. *Principles and Practice of Constraint Programming — CP 2000*. Springer-Verlag Berlin and Heidelberg GmbH & Co. KG, 2000.
3. ECRC. *ECL^{i}PS^e 3.5: ECRC Common Logic Programming System. User's Guide*, 1995.
4. L. Granvilliers and F. Benhamou. Progress in the solving of a circuit design problem. *Journal of Global Optimization*, 2001.
5. A. Griewank. *Evaluating Derivatives: Principles and Techniques of Algorithmic Differentiation*. SIAM Publications, 2000.
6. K. Marriott and P. J. Stuckey. *Programming with Constraints. An Introduction*. The MIT Press, 1998.
7. L. B. Rall. Automatic differentiation: techniques and applications. volume 120 of *Lecture Notes in Computer Science*. Springer, 1981.
8. P. Van Hentenryck, L. Michel, and Y. Deville. *Numerica: a Modelling Language for Global Optimization*. The MIT Press, Cambridge, MA, 1997.
9. T. M. Yakhno, V. Z. Zilberfaine, and E. S. Petrov. Applications of ECL^{i}PS^e: Interval Domain Library. In *Proc. Int. Conf. Practical Application of Constraint Technology*, pages 339–357, Westminster Central Hall, London, UK, 1997.

Author Index

Lecture Notes in Computer Science

For information about Vols. 1–2248
please contact your bookseller or Springer-Verlag